Rangeland Wildlife Ecology and Conservation

Lance B. McNew · David K. Dahlgren ·
Jeffrey L. Beck
Editors

Rangeland Wildlife Ecology and Conservation

Technical Editors: Jill Shaffer and Jim Peek

Editors
Lance B. McNew
Department of Animal and Range Sciences
Montana State University
Bozeman, MT, USA

David K. Dahlgren
Department of Wildland Resources
Utah State University
Logan, UT, USA

Jeffrey L. Beck
Department of Ecosystem Science
and Management
University of Wyoming
Laramie, WY, USA

ISBN 978-3-031-34036-9   ISBN 978-3-031-34037-6 (eBook)
https://doi.org/10.1007/978-3-031-34037-6

This Springer imprint is published by the registered company Springer Nature Switzerland AG
The registered company address is: Gewerbestrasse 11, 6330 Cham, Switzerland

Foreword

Management and conservation are inclusive terms used to convey management, conservation, and ecological understanding of natural resources. Without active management, many of the resources we hold dear, would cease to exist or decline sharply. Effective management enhances conservation and the reestablishment of dwindling populations. Both management and conservation require an understanding of ecology (i.e., the study of interactions between organisms and their environment). Some prefer to discuss management, others conservation, as the primary mechanisms to achieve objectives in the natural world. It really does not matter what human activity is called as long as it does not compromise the ability for humans to live with, maintain, and enhance our natural environment including fish, wildlife, grasslands, shrublands, deserts, and other rangeland resources. That ability rests with an understanding of science.

One of the most contemporary issues in ecology relates to habitat alteration and destruction from anthropogenic factors including climate change. Society is fortunate to have, in most of the West and parts of the eastern United States, rangelands that provide habitat for wildlife, the production of livestock, and management and conservation of other natural resources, biotic and abiotic. Without rangelands, important habitat for numerous flora and fauna would decline and without habitat, biodiversity would certainly decline. Thus it is important to keep abreast of the ecology, use, misuse, and status of rangelands that we are fortunate enough to manage and administer.

The fields of rangeland and wildlife management are brothers in the same fight for the conservation, protection, and management of wildlife and one cannot be completely understood without knowledge of the other and that presents challenges that are often played out within society—the society that dictates what we do with our resources. In the USA, natural resources are governed by society via numerous state and federal laws designed to protect wildlife and the habitats they use. Love of nature is in the bones of North Americans, and few other places occur where wildlife, its habitats, and other natural resources are an essential part of its identity. This identity can be traced back to at least 1842 in the USA when a judge cited the 1215 English Magna Carta to codify that wildlife and fish belong to all the citizens of the USA and that their stewardship was entrusted to the states, thus began a series of decisions that led to the public trust doctrine, which was rooted in Roman law, and the North American Model of Wildlife Conservation, which is based on science.

In 1996, the Society for Range Management published *Rangeland Wildlife*, which provided information about major vertebrates on rangelands in the western USA and provided some insights into their interactions with the millions of head of livestock that share rangeland landscapes. Nearly 30 years have passed since the publication of that work and science has advanced and evolved. In the early years of the wildlife and rangeland professions, there was not as much (or any) emphasis on holistic management and much of the research was related to single species. We are finally in an era of transdisciplinary research that builds on the strengths of other disciplines to gain a better understanding of ecology for enhanced understanding and management—something that Leopold endorsed in the 1930s. Numerous disciplines including administration, agriculture, botany, economics, genetics, human dimensions, policy, population ecology, sociology, and social science all laced with a backdrop of ecology and science are necessary for effective management and conservation. *Rangeland Wildlife Ecology and Conservation* updates the research in the arena over the past 2–3 decades that continues to develop a holistic approach to conservation and management of our rangelands and the biota it supports. Because of the dominant presence of rangelands in the country, and the role of rangelands in the lives of humans and wildlife, this work presents optimism that wildlife and rangeland scientists, managers, and conservationists are advancing the cause our country has been built on and can work together for the benefit of wildlife and society.

Paul R. Krausman
Professor Emeritus
University of Arizona
Tucson, AZ, USA

Acknowledgments

Many people made this book possible. Thanks to the authors and co-authors for sharing their expertise and time with current and future rangeland and wildlife managers. We also thank more than 50 technical reviewers for providing suggestions to improve the manuscript chapters; unfortunately, space precludes us from listing everyone. Lance B. McNew was supported by the Montana Agricultural Experiment Station; David K. Dahlgren was supported by Utah State University Extension and Wildland Resources Department; Jeffrey L. Beck was supported by the Wyoming Agricultural Experiment Station. Cover images provided by professional nature photographer Ken Archer.

We are grateful for technical editors Jill A. Shaffer and James M. Peek whose talents and efforts greatly improved this book.

This book is available thanks to generous sponsorship by the following:

- USDI Bureau of Land Management
- USDA Forest Service
- USDA Natural Resources Conservation Service, Working Lands for Wildlife program
- Pheasants Forever, Incorporated
- Western Association of Fish and Wildlife Agencies
- Brigham Young University
- Montana Agricultural Experiment Station, Montana State University
- Utah State University Cooperative Extension
- Wyoming Agricultural Experiment Station, University of Wyoming

Contents

Part II   Species Accounts

Chapter 1
Introduction to *Rangeland Wildlife Ecology and Conservation*

Lance B. McNew, David K. Dahlgren, and Jeffrey L. Beck

Abstract Rangelands are vast, dynamic, and integral to providing habitat for thousands of vertebrate and invertebrate species, while concurrently serving as the foundation of human food and fiber production in western North America. Reciprocally, wildlife species provide critical services that maintain functional rangeland ecosystems. Therefore, human management of rangelands via fire, grazing, agricultural programs, and policy can enhance, disturb, or inhibit the necessary interactions among natural processes of plants and animals that maintain rangeland ecosystems. As conservation issues involving rangelands have grown in societal awareness and complexity, rangeland managers, wildlife biologists, and others have discovered the need to work more closely together with an increasingly holistic approach, spurring a rapid accumulation of rangeland wildlife information in the early twenty-first century. This book represents a synthesis of contemporary knowledge on rangeland wildlife conservation and ecology. Accordingly, we provide a review of the state of science for new, as well as seasoned, wildlife and rangeland professionals who have stewardship of North America's most undervalued ecosystem.

Keywords Agroecosystems · Ecosystem services · Grasslands · Shrublands · Wildlife

L. B. McNew (✉)
Department of Animal and Range Sciences, Montana State University, Bozeman, MT 59717, USA
e-mail: lance.mcnew@montana.edu

D. K. Dahlgren
Department of Wildland Resources, Utah State University, Logan, UT 84322, USA

J. L. Beck
Department of Ecosystem Science and Management, University of Wyoming, Laramie, WY 82071, USA

© The Author(s) 2023
L. B. McNew et al. (eds.), *Rangeland Wildlife Ecology and Conservation*,
https://doi.org/10.1007/978-3-031-34037-6_1

1

1.1 Introduction

More than half of all lands worldwide, and the majority of lands in the western U.S., are classified as rangelands (Table 1.1). The exact extent of rangelands is difficult to delineate due to variability in the definition of rangelands (Briske 2017), but by any definition rangelands represent collectively the most widespread ecosystem in the western U.S. (Chap. 2). Many picture grasslands when envisioning rangelands. Some classify rangelands as any non-cultivated land grazed by livestock (Menke and Bradford 1992). Others have defined rangelands as 'non-forested lands of low economic activity' (sensu Sayre 2017). In most cases, rangelands in North America represent what was 'left over' after Euro-American settlement and conversion of arable lands in the West during the nineteenth century (Table 1.1). Therefore, rangelands include desert, grassland, and shrubland ecosystems that were unsuitable for cultivation, though they retain economic and social value. Rangelands are held in public or private ownership and provide innumerable goods and services, including significant economic benefit to local communities. For example, nearly 100 million head of cattle spend at least part of their life each year on U.S. rangelands alone. Rangelands also provide habitat for hundreds of vertebrate species and innumerable invertebrates. Thus, rangelands and their management have significant bearing on wildlife in North America and globally.

Wildlife have been a featured player in the history of rangelands (Chap. 3) but are more than that—they are a fundamental piece of the whole that constitute rangeland ecosystems. Wildlife and rangeland management as scientific disciplines share common origins and parallel histories (Chap. 30). Foundations of each were based upon concepts developed in the pioneering field of forestry and focused on sustainable harvest of products—timber, forage, deer, quail. Each field has seen similar progressions in ideas expanding from sustainable harvests of 'valuable' species to adaptive management of functional and resilient ecosystems. This broadening of focus has, no doubt, reflected shifting demographics and stakeholders (van Heezik and Seddon 2005), that have pushed ecologists and managers to think more holistically about rangeland ecosystems as more than the sum of their offtake. Contemporary managers must not only know theories describing population responses of harvest management—either by cow or gun—but should have broader knowledge that includes invasive species ecologies as related to state transitions, policy issues related to threatened and endangered species, functional vs. biological diversity, and so much more. This broadening means that contemporary rangeland and wildlife managers should have training in landscape ecology, community ecology, and rangeland and wildlife policy, in addition to foundational understandings of the biology and ecology of plants and animals. Now layer onto those scientific concepts the fact that rangelands are almost always working lands inextricably linked to a people's sense of place and identity (Chap. 28), and the knowledge required to understand rangeland ecosystems, including rangeland-dependent wildlife, becomes broad and transdisciplinary.

Table 1.1 Key terms used throughout this book

Term	Definition	Source
Great Plains	Area of North America dominated by native grasslands within 12 states (CO, IA, KS, MN, MT, NE, ND, NM, OK, SD, TX, WY) and 3 Canadian provinces (AB, MB, SK)	Chapter 2
Pastureland	Land used primarily for the purpose of producing introduced (nonnative) forage for livestock	Charnley et al. (2014)
Rangeland	Land on which the plant cover is composed principally of native grasses, grass-like plants, forbs, or shrubs suitable for grazing or browsing by native and domestic animals	Briske (2017)
Resilience	The ability of a system (ecological, socio-economical, or social-ecological) or aspects of systems to recover from disturbances and return to its pre-disturbed condition	Walker (2010)
Resistance	The ability of a system (ecological, socio-economical, or social-ecological) or aspects of systems to remain unchanged when subjected to changes or disturbances	Walker (2010)
Social-ecological system	A conceptual framework for describing and studying rangelands cohesively as a combination of social and ecological components, interactions, and processes	Hruska et al. (2017)
Sustainable	A term that describes (1) methods of extraction of renewable natural resources (e.g., grass) that do not diminish the ecological integrity and biodiversity of rangeland ecosystems, and (2) a level of extraction that allows natural resources to recover to similar or higher levels of productivity	Charnley et al. (2014)
West	Area of western North America within 11 western states (AZ, CA, CO, ID, MT, NM, NV, OR, UT, WA, WY), and 3 Canadian Provinces (AB, BC, SK) Rangelands here are often characterized by semi-arid and arid climates	Chapter 2
Working landscapes	Landscapes where people make their living by extracting renewable natural resources, such as grass and trees	Charnley et al. (2014)

That wildlife are integral parts, not just benefactors, of rangeland ecosystems has been understood by native peoples in North America for thousands of years, but not until the late-twentieth century did scientists begin investigating their interactions. In 1996, the Society for Range Management published a volume summarizing information about select vertebrates that inhabited western United States rangelands (Krausman 1996). Although Krausman (1996) still serves as a well-worn reference for rangeland and wildlife managers, a wealth of new information concerning rangeland wildlife has been produced since its publication. For example, a Web of Science search for "rangeland wildlife" produced 790 peer-reviewed publications during 1996–2019 (date of search 10/15/19); by comparison, less than 50 publications were found for the period 1900–1995. As conservation issues have become increasingly more common during this modern Anthropocene, some of the highest profile cases have been with rangeland-dependent wildlife. We are now well past a time when rangeland and wildlife disciplines can remain siloed within their educational and professional pursuits. Our goal for *Rangeland Wildlife Ecology and Conservation* has been to corral the best available science during the last quarter century that addresses rangeland wildlife ecology, conservation, and management into a product that will serve and help integrate professionals of the rangeland and wildlife disciplines.

1.2   What This Book Is

By necessity, if not by design, *Rangeland Wildlife Ecology and Conservation* is a hybrid. Textbooks are traditionally written cover to cover by the same author(s) and attempt to distill major ideas in a discipline to something learnable in a semester; whereas edited volumes in a book series are an assemblage of separate and sometimes disparate articles—often documenting a conference symposium—that synthesize the state of knowledge on a topic. In our hubris to achieve both, we recruited more than 100 subject matter experts to author 30 chapters on topics we identified as needing an updated review—the authors list includes university and federal scientists, state and federal rangeland and wildlife managers, NGO scientists and conservationists, and ranchers. The result of this 3-year effort is both a synthesis of knowledge on major rangeland wildlife topics and a contemporary (2022 c.e.) review of the state of the science that we hope can be used as both a modern textbook in the training of students in rangeland and wildlife science as well as a reference for working professionals.

1.3   What This Book Is Not

Certainly, the *Rangeland Wildlife Ecology and Conservation* is not a full and exhaustive summary of everything rangeland managers and wildlife biologists should know. For example, we acknowledge that soil properties and processes are critical drivers of rangeland ecosystems with important implications for wildlife habitat management; fortunately, a recent excellent review is provided elsewhere (Evans et al. 2017). We have asked our authors to incorporate discussions of management tools (e.g., fire, grazing, conservation programs and policy) into their chapters where appropriate, but this book is not a paint-by-numbers recipe for the management of wildlife on rangelands. That is impossible. Recent work, as demonstrated throughout this book, has highlighted (1) what is unknown and uncertain, and (2) that wildlife interactions and responses to rangeland management are context- and scale-dependent. Proper rangeland management to achieve habitat targets for even a single species in a single rangeland type will vary across space and time due to differences in soils, topology, and precipitation. Instead, we asked authors to synthesize information relative to habitat targets and describe how those may be influenced by managed (e.g., grazing) and unmanaged (e.g., precipitation) conditions so that the content may be principle based and applicable across the distribution of a species. Local expertise is always needed for proper management.

1.4   Organization

Rangeland Wildlife Ecology and Conservation is divided into three parts. In Part I (Chaps. 2–8), rangeland scientists introduce the reader to major concepts in rangeland ecology and management in western North America. Part I is not meant to be an exhaustive review of the ecology and management of rangeland ecosystems; there are excellent texts that do that (e.g., Briske 2017), but we felt that inclusion of this introductory material would be beneficial for wildlife professionals who may not have had previous training in rangeland ecology. Part II (Chaps. 9–26) includes accounts in which subject matter experts present updated reviews and syntheses of representative and well-studied species or guilds thereof. To aid in the use of this book as a text and reference, the chapters in Part II share a common structure and include (1) introductory sections on species life-histories, population dynamics, and habitat requirements, (2) current methods for effective population monitoring, (3) syntheses describing interactions with rangeland management, including livestock grazing and fire, and (4) a summary of current threats to ecosystems. Because rangelands are almost always working landscapes (Table 1.1), we conclude the book in Part III with chapters demonstrating the importance of social-ecological understanding of rangelands, that land, livestock, and wildlife management are intertwined, and how that knowledge can be leveraged into more effective and holistic conservation of rangeland wildlife.

References

Briske DD (2017) Rangeland systems: processes, management, and challenges. Springer Cham. https://doi.org/10.1007/978-3-319-46709-2

Charnley S, Sheridan TE, Sayre N (2014) Status and trends of western working landscapes. In: Charnley S, Sheridan TE, Nabhan G (eds) Stitching the West back together. University of Chicago Press, Chicago, pp 13–32. https://doi.org/10.7208/9780226165851

Evans RD, Gill RA, Eviner VT, Bailey V (2017) Soil and belowground processes. In: Briske DD (eds) Rangeland systems: processes, management, and challenges. Springer Cham. https://doi.org/10.1007/978-3-319-46709-2_4

Hruska T, Huntsinger L, Brunson M, Li W, Marshall N, Oviedo JL, Whitcomb H (2017) Rangelands as social-ecological systems. In: Briske DD (eds) Rangeland systems: processes, management, and challenges. Springer Cham. https://doi.org/10.1007/978-3-319-46709-2_8

Krausman PR (1996) Rangeland wildlife. Society for Range Management, Denver, CO. https://srs.fs.usda.gov/pubs/62529

Menke J, Bradford GE (1992) Rangelands. Agr Ecosyst Environ 42:141–163. https://doi.org/10.1016/0167-8809(92)90024-6

Sayre NF (2017) The politics of scale: a history of rangeland science. The University of Chicago
    Press, Chicago
van Heezik Y, Seddon P (2005) Structure and content of graduate wildlife management and conser-
    vation biology programs: an international perspective. Cons Biol 19:7–14. https://doi.org/10.
    1111/j.1523-1739.2005.01876.x
Walker B (2010) Riding the rangelands piggyback: a resilience approach to conservation. In: du Toit
    JT, Kock R, Deutsch JC (eds) Wild rangelands: conserving wildlife while maintaining livestock
    in semi-arid ecosystems. Wiley-Blackwell, West Sussex, UK, pp 15–29

Part I
Rangeland Ecosystems and Processes

Chapter 2
Rangeland Ecoregions of Western North America

Frank E. "Fee" Busby, Eric T. Thacker, Michel T. Kohl,
and Jeffrey C. Mosley

Abstract The grasslands, deserts, shrublands, savannas, woodlands, open forests, and alpine tundra of western North America where livestock grazed were collectively referred to as 'range' in the nineteenth century. Today these ecosystems are often referred to as rangelands. In the United States, rangelands comprise about 1/3rd of the total land area, mostly in the 17 western states. Large areas of rangeland also occur in Canada and Mexico. Rangelands provide numerous products, values, and ecosystem services including wildlife habitat, clean air, clean water, recreation, open space, scenic beauty, energy and mineral resources, carbon sequestration, and livestock forage. This chapter describes rangeland ecoregions in western North America.

Keywords Rangeland regions · Grasslands · Savannas · Cool deserts · Sagebrush · Hot deserts · Piñon-juniper woodlands · Oak woodlands · Aspen parkland · Ponderosa pine savanna · Mountain rangelands · Alpine tundra

2.1 Introduction

It is unclear when the word range was first used to describe land in the western United States but reports from explorers, ranchers, and scientists in the mid to late 1800s referred to lands where livestock grazed as range. Confusingly, the word has also

F. E. "Fee" Busby (✉) · E. T. Thacker
Department of Wildland Resources, Utah State University, Logan, UT, USA
e-mail: fee.busby@usu.edu

E. T. Thacker
e-mail: eric.thacker@usu.edu

M. T. Kohl
Warnell School of Forestry and Natural Resources, University of Georgia, Athens, GA, USA
e-mail: michel.kohl@uga.edu

J. C. Mosley
Department of Animal and Range Sciences, Montana State University, Bozeman, MT, USA
e-mail: jmosley@montana.edu

© The Author(s) 2023   9
L. B. McNew et al. (eds.), *Rangeland Wildlife Ecology and Conservation*,
https://doi.org/10.1007/978-3-031-34037-6_2

been used to describe the season of use (winter range) and the species of animal using the land (cattle or deer range; Chap. 30). Today, range or rangeland is not thought of as a kind of use but as a kind of land where grasses, forbs, shrubs, sedges, and rushes dominate and the land is valued and managed for wildlife habitat, clean air, clean water, recreation, open space, scenic beauty, energy and mineral resources, carbon sequestration as well as livestock forage (Box 1978; Havstad et al. 2009). Because the word range was used to describe lands throughout the western US, the idea that it referred to various kinds of vegetation developed early. Grasslands, deserts, shrublands, savannas, woodlands, some forests, meadows, and tundra ecosystems are all considered rangeland. Collectively rangelands form the most extensive land type on Earth and make up about 1/3rd (308 million ha) of the land area of the US. Most North American rangeland occurs in the 17 western states and adjacent areas in Canada and Mexico (Havstad et al. 2009).

Rangelands provide important habitat for many birds, herpetofauna, mammals, and insects (e.g., Chaps. 8–26). Sustaining rangeland wildlife requires sustaining rangeland vegetation suitable for wildlife. However, it is important for rangeland habitat managers to reconcile that it is impossible to maximize habitat quality for all wildlife at the same time. Any change in rangeland plant community structure or plant species composition simultaneously favors some wildlife species and disfavors others (Maser and Thomas 1983; Mosley and Brewer 2006). Consequently, habitat management commonly seeks to achieve two goals: (1) provide sufficient variability in vegetation conditions across the landscape to sustain a diverse wildlife community, and (2) make limiting habitat factors for desired wildlife species less limiting (Maser and Thomas 1983; Mosley and Brewer 2006). Changes in rangeland vegetation are dictated by the intensity and frequency of both natural and anthropogenic disturbances, and their interactions. Natural disturbances may include drought, flooding, wildfire, and grazing or browsing by wildlife. Anthropogenic-related disturbances may include chemical or mechanical habitat treatments, prescribed burning, artificial revegetation, and livestock grazing or browsing, which are discussed throughout this book.

Differences in amount, kind, and season of precipitation are the primary factors contributing to the development and distribution of the 25 rangeland ecoregions described in this chapter (Table 2.1; Stephenson 1990). Seven ecoregions occur east of the Rocky Mountains and Sierra Madre Oriental Mountains on the relatively flat landscapes of the Great Plains and Gulf Coastal Plain, while 18 ecoregions occur in the valleys, foothills, and mountains westward from the Rocky Mountains.

The eastern ecoregions, dominated by perennial grasslands and savannas, receive ≥ 70% of their annual precipitation between April and September from storms that originate in the Gulf of Mexico. Warm-season (C4) grasses dominate all but the most northern of these ecoregions. Cool-season (C3) plants dominate the western ecoregions, with most areas receiving ≥ 50% of their annual precipitation between October and April from storms originating in the Pacific Ocean. Many western ecoregions receive 50–70% of their annual precipitation as snow, and plants grow rapidly following snowmelt in spring. Sagebrush shrublands, piñon-juniper woodlands, oak woodlands, and montane ecosystems are the most extensive rangeland types in the

Table 2.1 Major rangeland ecoregions of western North America

East of the rocky mountains	West of the rocky mountains	
	Winter precipitation	Summer precipitation
Great Plains Prairie **Grasslands** Tallgrass Prairie Shortgrass Prairie Northern Mixed-Grass Prairie Southern Mixed-Grass Prairie	**Winter rain** California Annual Grassland California Oak Woodland California Chaparral	Chihuahuan Desert Sonoran Desert Mojave Desert Interior Chaparral Southwestern Oak Woodland
Savannas and Parklands Aspen Parkland Edwards Plateau Tamaulipan Thornscrub	**Winter snow** Salt Desert Shrub Sagebrush Steppe Great Basin Sagebrush Piñon-Juniper Woodland Mountain Brush Montane Grassland Montane Sagebrush Steppe Ponderosa Pine Savanna Montane and Subalpine Meadow Alpine Tundra	

Ecoregions are listed in the order they are presented in text

western ecoregions. West of the Sierra Nevada Mountains, annual grasses and forbs are major components of the vegetation and grow throughout the winter. These areas receive ≥ 80% of their annual precipitation as rain between October and April.

Ecoregions in the southwestern US and northern Mexico receive most of their precipitation from July to October during the North American monsoon. These ecoregions also have hotter air temperatures than areas located farther north, and C4 grasses, shrubs, and succulents (plants with CAM photosynthesis) are common.

Classification of rangelands into ecoregions helps us to understand ecological relationships at a large-scale level but is not sufficient for rangeland management. For management purposes, an ecological site system, including state-and-transition models (described in Chap. 5), has been developed for most rangelands in the US (Caudle et al. 2013). Within a rangeland ecoregion, an ecological site is a distinctive kind of land with specific soil and physical (primarily climate and topography) characteristics that differ from other kinds of land in its ability to produce a distinctive kind and amount of vegetation and its ability to respond similarly to management actions and natural disturbances. The state-and-transition model for a site identifies: (1) multiple stable vegetation states, (2) plant communities that can exist within a state, (3) pathways that indicate changes such as a fire and recovery from fire that can occur between plant communities, (4) reversible transitions between states, (5) thresholds or ecological constraints such as soil erosion that change soil water holding capacity, and (6) irreversible transitions that occur when thresholds are crossed (Fig. 2.1; see Chap. 5). Discussion included in an ecological site description and its state-and-transition model provide guidance to rangeland managers on which interventions

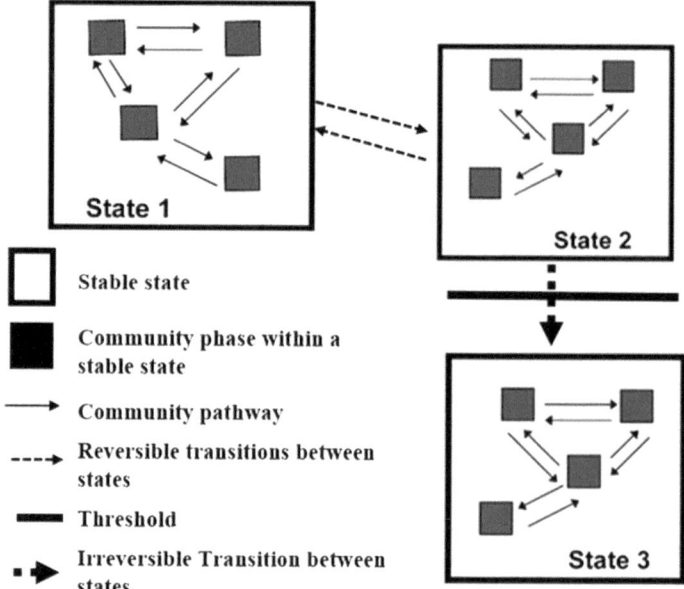

Fig. 2.1 Conceptual state and transition model incorporating the concepts of multiple stable state, communities within states, community pathways between communities within states, reversible transitions, thresholds, and irreversible transitions (modified from Stringham et al. 2003)

(e.g., grazing, prescribed fire, mechanical treatments, herbicides, etc.) are likely to be successful. Opportunities for management intervention are severely limited or lost when a threshold has been crossed and a rangeland plant community transitions into a different state (Stringham et al. 2003; Briske et al. 2005).

Climate, soils, and topographic position mediate vegetation dynamics and define the ecological potential of rangeland vegetation at a broader ecoregion-scale. Rangeland ecoregions provide ecological sideboards that constrain rangeland habitat management options. In this chapter we highlight the physiognomy and ecology of 25 major rangeland ecoregions in western North America (Table 2.1). Each ecoregion provides critical habitat for wildlife. Common plant names are presented here, matched with their scientific names in Table 2.2.

2.2 Rangelands East of the Rocky Mountains

The Great Plains and the Gulf Coastal Plain extend from the Rocky Mountains and Sierra Madre Oriental Mountains toward the Mississippi River and from the Gulf of Mexico and northeastern Mexico to southern Canada. Dominant vegetation in the Great Plains and Gulf Coast Prairie was grassland before much of it was converted to cropland agriculture. North and south of the grasslands lies savannas where grasses and woody plants coexist. Aspen parkland savanna lies north while the Edwards Plateau and the Tamaulipan thornscrub savannas are located south of the grasslands (Fig. 2.2). Warm-season (C4) grasses dominate the rangeland ecoregions east of the Rocky Mountains except the northern mixed-grass prairie where cool-season (C3) grasses become codominant. Cool-season grasses dominate the aspen parkland (Sims and Risser 2000).

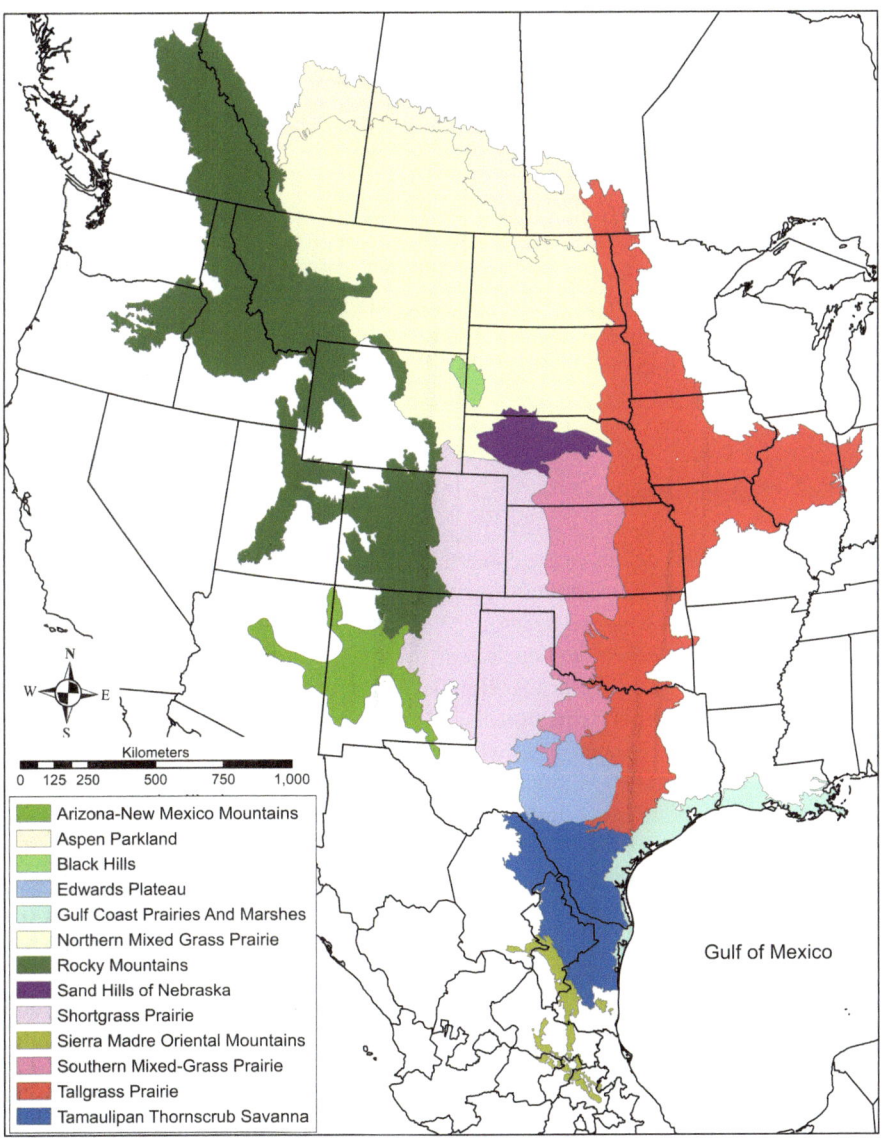

Fig. 2.2 Rangeland ecoregions in the eastern half of western North America

2.2.1 Great Plains Prairie Grasslands

Across the Great Plains elevations range from 1200–1500 m along the Rocky Mountain foothills to 200–400 m along a line running from central Texas to southcentral Manitoba. The Gulf Coast Prairie occurs as a narrow strip of land along the southern coast of Texas (Fig. 2.2). Topography of the Great Plains is described as rolling plains although scattered mountains such as the Black Hills occur in the region. The Gulf Coast Prairie with elevations ranging from sea level to 50 m has little topographic relief. Temperature, precipitation, and evapotranspiration decrease from south to north while precipitation effectiveness increases. Precipitation also decreases from east to west. Based on these differing climatic conditions the prairie is divided into tallgrass, northern-mixed grass, southern mixed-grass, and shortgrass prairies (Sims and Risser 2000; Anderson 2006).

Today ~65% of the prairie grasslands is used for cropland or another non-range use (Comer et al. 2018). The greatest loss of rangeland is in the tallgrass and northern mixed-grass prairies. Despite the amount of land that has been converted, the Great Plains prairie grasslands support ≈50% of the US beef cow herd (Klemm and Briske 2021) and 75% of the beef cattle in Canada (Wang et al. 2017). Most of the livestock operations in the Great Plains graze less than 100 animals and produce both livestock and crops (Mitchell 2000).

2.2.1.1 Tallgrass Prairie

The tallgrass prairie occurs in the eastern Great Plains, extending from central Texas into Manitoba (Fig. 2.2). Annual precipitation averages 500–1000 mm with most falling during the summer (Anderson 2006). Approximately 90% of annual herbaceous production is completed each year by 1 September (Stephenson et al. 2019; Smart et al. 2021). Tallgrass prairie flora is dominated by four warm-season grasses (Samson and Knopf 1994; Griffith et al. 2004). Big bluestem, Indiangrass, and switchgrass grow in the more moist areas while little bluestem dominates drier sites. Green needlegrass and porcupine grass are important cool-season grasses that grow in the northern third of the tallgrass prairie. Grasses produce the larger amount of biomass, but the number of forb species exceeds the number of grass species. Kentucky bluegrass and smooth bromegrass are non-native cool-season grasses that have invaded the northern part of the tallgrass prairie. Caucasian bluestem and yellow bluestem (collectively "Old World bluestems") are non-native warm-season grasses that have invaded in the south. *Sericea lespedeza*, a forb originally introduced from Asia for erosion control, is invading the tallgrass prairie from Texas to Nebraska. Woody plants such as blackjack oak, post oak, and especially eastern red cedar are expanding throughout the tallgrass prairie (Lauenroth et al. 1999; DeKeyser et al. 2013). About 86% of the tallgrass prairie has been converted to cropland (Comer et al. 2018). Large areas where tallgrass prairie remain are where soils are not suitable for cultivation (e.g., the Flint Hills in Kansas; Anderson 2006).

2.2.1.2 Shortgrass Prairie

The shortgrass prairie occurs on the flat-to-rolling dry plains of western Kansas and Oklahoma, eastern Colorado and New Mexico, and the High Plains of Texas (Fig. 2.2). Precipitation averages 300–500 mm (Lauenroth 2008). About 62% of the shortgrass prairie has been converted to cropland or other uses (Comer et al. 2018). The largest areas of land conversion are in the High Plains of Texas, eastern Colorado, and western Oklahoma and Kansas where irrigation water from the Ogallala aquifer is available (Lauenroth 2008).

Blue grama and buffalograss are the dominant grasses in the shortgrass prairie. Sideoats grama, galleta, threeawns, tobosagrass, and sand dropseed are other prominent warm-season grasses. Needle and thread, New Mexico feathergrass, prairie junegrass, western wheatgrass, and sun sedge are conspicuous cool-season plants (Lauenroth 2008). Semi- or subshrubs (herbaceous stems but woody at the base) such as broom snakeweed and prairie sagewort grow throughout the shortgrass prairie. Forbs can be abundant during wet years but seldom comprise large proportions of shortgrass prairie plant communities. Prickly pear cactus is common on dry sites. Honey mesquite and various species of juniper have increased on shortgrass prairie rangelands in Texas and New Mexico to such an extent that many areas are now savannas (Sims and Risser 2000; Lauenroth 2008). Sand shinnery oak, a native, low-growing deciduous shrub forms dense thickets on sandy soil sites in the southern High Plains of Texas, New Mexico, and Oklahoma (Peterson and Boyd 1998; Haukos 2011).

2.2.1.3 Northern Mixed-Grass Prairie

The northern mixed-grass prairie is found in Alberta, Montana, North Dakota, Saskatchewan, South Dakota, and Wyoming. The boundaries are the Rocky Mountains on the west, shortgrass prairie and the Nebraska sandhills on the south, tallgrass prairie on the east, and aspen parkland on the north (Fig. 2.2). Northern mixed-grass prairie has the most diverse flora among the Great Plains grasslands (Barker and Whitman 1988; Lavin and Siebert 2011), with plant species that also exist in the tallgrass and shortgrass prairies and in the cool deserts located farther west. Large expanses of northern mixed-grass prairie remain, except in Manitoba where most rangeland has been converted to cropland (Coupland 1992). Annual precipitation averages 350–500 mm and peaks in April–June with 90% of annual herbaceous production completed each year by 1 July (Vermeire et al. 2009; Smart et al. 2021).

The most abundant cool-season grasses include bluebunch wheatgrass, bottlebrush squirreltail, green needlegrass, Idaho fescue, needle and thread, porcupine grass, prairie junegrass, Sandberg bluegrass, slender wheatgrass, and western wheatgrass. Big bluestem, blue grama, little bluestem, and sideoats grama are prevalent warm-season grasses in the eastern portion of the northern mixed-grass prairie. Shrubs such as shrubby prairie rose, silver buffaloberry, and snowberry grow in low-lying areas where snow accumulates. Other notable shrubs include plains silver sagebrush, Wyoming big sagebrush, and yellow rabbitbrush. Numerous native forbs occur in northern mixed-grass prairie but rarely comprise large proportions of the plant communities. Cheatgrass, crested wheatgrass, field brome, Kentucky bluegrass, medusahead, smooth bromegrass, and ventenata are non-native grasses that have invaded much of this ecoregion (DeKeyser et al. 2013). Non-native invasive forbs such as knapweeds, leafy spurge, and yellow toadflax also are widely distributed within the northern mixed-grass prairie.

2.2.1.4   Southern Mixed-Grass Prairie

Southern mixed-grass prairie receives 530–870 mm annual precipitation, and 90% of annual herbaceous production is completed each year by 1 September (Vermeire et al. 2009; Smart et al. 2021). Southern mixed-grass prairie is bordered by northern mixed-grass prairie to the north, the Edwards Plateau to the south, tallgrass prairie to the east, and shortgrass prairie to the west (Fig. 2.2). Warmer temperatures, greater mid- to late summer precipitation, a longer growing season, and dominance by warm-season grasses distinguish southern mixed-grass prairie from northern mixed-grass prairie. Approximately 70% of the southern mixed-grass prairie in Texas has been converted to cropland (Comer et al. 2018).

Important warm-season grasses include sideoats grama, little bluestem, bristlegrass, dropseeds, silver bluestem, threeawns, and white tridens. Important cool-season grasses are Texas wintergrass in the southern part of this ecoregion, with needle and thread and western wheatgrass prevalent in the northern part. Big bluestem, Indiangrass, and switchgrass grow on moist sites throughout this ecoregion (Lauenroth et al. 1999; Sims and Risser 2000).

The southern mixed-grass prairie was dominated by grasses prior to European settlement, although escarpments and canyons were dominated by woody species including honey mesquite, eastern red cedar, lotebush, and redberry juniper. Post-settlement fire suppression enabled woody plants to expand into the grassland with honey mesquite and redberry juniper invading in New Mexico and Texas and eastern red cedar in the northern part of this ecoregion in Oklahoma and Kansas. Blue grama occurs throughout this ecoregion and becomes more abundant with increased grazing pressure and severe drought, as do buffalograss, red grama, and threeawns (Wright and Bailey 1980; Griffith et al. 2004).

2.2.2   Savannas and Parklands

Savannas and parklands occur where woody and herbaceous plants are co-dominant and woody canopy is sufficiently open to allow growth of grasses and other herbaceous species. Low-intensity ground fires maintain the codominance and openness of the savanna by reducing but not eliminating woody plant cover. Woody plants that are capable of regrowing from root buds located below ground are favored (Fowler and Beckage 2020). With fire suppression, woody plants increase in density and open savannas become closed woodlands or forests (Archer et al. 1988; Staver et al. 2011). Three savannas occur east of the Rocky Mountains and Sierra Madre Oriental Mountains: aspen parkland north of the northern mixed-grass prairie, Edwards Plateau south of the southern mixed-grass prairie, and Tamaulipan thornscrub in southern Texas and northeastern Mexico (Fig. 2.2).

2.2.2.1 Aspen Parkland

Aspen parkland occurs as a mosaic of aspen groves and interspersed grassland from North Dakota, across Saskatchewan, to south-central Alberta (Strong and Leggat 1992; Padbury et al. 1998; Fig. 2.2). Aspen parkland is a transition zone, bounded to the north and east by boreal forest, to the south by northern mixed-grass prairie, and to the west by Rocky Mountain foothills. Annual precipitation averages 400–500 mm with 80+% occurring from late spring to early summer. The topography is mostly level to undulating, with aspen groves growing on moist north-facing slopes and depressions, and grassland occupying the drier hilltops and south-facing slopes. Balsam poplar often co-dominates with aspen in the wettest areas. Moderated air temperatures and longer frost-free periods beneath aspen grove canopies generate abundant and diverse understory vegetation (Powell and Bork 2007). Understory shrubs include chokecherry, serviceberry, snowberry, and wild rose. Noteworthy understory herbaceous species include bluegrasses, sedge, and western meadowrue. In the interspersed grassland, plains rough fescue dominates. Subdominant grasses include porcupine grass, prairie junegrass, and slender wheatgrass, and grassland forbs include geranium, goldenrod, and western yarrow. Before European settlement, fire prevented aspen from encroaching into the grassland. Post-settlement fire suppression has enabled aspen to expand (Bailey and Wroe 1974; Anderson and Bailey 1980). Heavy livestock grazing pressure has reduced or eliminated plains rough fescue in many locations, and non-native orchardgrass and smooth bromegrass have become widespread. Much of the aspen parkland ecoregion (> 80%) has been converted to highly productive cropland (Comer et al. 2018).

2.2.2.2 Edwards Plateau

The Edwards Plateau ecoregion is in central Texas, south of the shortgrass prairie and southern mixed-grass prairie, west of the tallgrass prairie, and east of the Chihuahuan Desert (Fig. 2.2). The eastern portion of the Edwards Plateau has weathered into low, rounded hills and valleys known locally as "The Hill Country" (Jordan 1978). Topography in the western portion of the Edwards Plateau is flat to gently rolling, dissected by steep-sloped canyons. Mean annual precipitation varies west to east, from 480 to 790 mm. Most soils are shallow and rocky. Woody plant density increases and understory plants decrease in the absence of fire or other disturbance (Fuhlendorf and Smeins 1997; Griffith et al. 2004).

Vegetation in the Edwards Plateau ecoregion is a mix of woodland and savanna. Ashe juniper co-dominates with Texas live oak in eastern portions of the Edwards Plateau. Prominent savanna grasses in the eastern Edwards Plateau include big bluestem, blue grama, Indiangrass, little bluestem, sideoats grama, silver bluestem, and switchgrass. These grasses decrease in abundance as tree canopy increases, enabling hairy grama, curlymesquite, Texas wintergrass, and threeawns to gain dominance. In the western part of the Edwards Plateau, Ashe juniper co-dominates with honey mesquite. Savanna grasses in the western Edwards Plateau include black grama, blue grama, dropseeds, little bluestem, lovegrasses, and sideoats grama. With increased herbivory and increased woody plant cover, buffalograss, curlymesquite, Texas grama, Texas wintergrass, and threeawns increase (Fuhlendorf and Smeins 1997; Griffith et al. 2004).

2.2.2.3 Tamaulipan Thornscrub

Tamaulipan thornscrub occurs south of the Edwards Plateau and includes the South Texas Plain in southern Texas and adjacent areas in northeastern Mexico (Fig. 2.2). Plant species diversity is high in Tamaulipan thornscrub due to its location at the confluence of subtropical, desert, and coastal ecoregions. Elevations range from near sea level to 800 m, and annual precipitation averages 600–750 mm. Small trees and shrubs, many with thorns or spines, dominate the vegetation including algerita, Berlandier's wolfberry, blackbrush acacia, catclaw acacia, guajillo acacia, lotebush, prickly pear cactus, spiny hackberry, and Texas persimmon. Associated grasses include bristlegrass, cane bluestem, lovegrasses, multiflowered false rhodesgrass, pink pappusgrass, sideoats grama, silver bluestem, and thin paspalum. Tobosagrass grows on heavy clay soils. Grasses on drier sites include buffalograss, curlymesquite, hooded windmillgrass, red grama, Texas grama, and threeawns (Archer et al. 1988; Griffith et al. 2004). Several introduced perennial grasses have become invasive, including bermudagrass, buffelgrass, Lehmann lovegrass, and yellow bluestem (Wied et al. 2020). Tanglehead is a native perennial grass that also has become invasive (Bielfelt and Litt 2016; Wester et al. 2018).

2.3 Rangelands West of the Rocky Mountains

This region extends from southern British Columbia and southwestern Alberta to northern Mexico. The eastern boundary is the eastern slopes of the Rocky Mountains, Arizona-New Mexico Mountains, and the Sierra Madre Oriental Mountains. The western boundary is the Pacific Ocean (Fig. 2.3). Large areas in California, Washington, and Idaho have been converted to cropland, but when compared with the Great Plains the western rangelands are relatively intact.

Unlike the Great Plains where there is little topographic relief, the region westward from the Rocky Mountains is dominated by valleys and mountains.

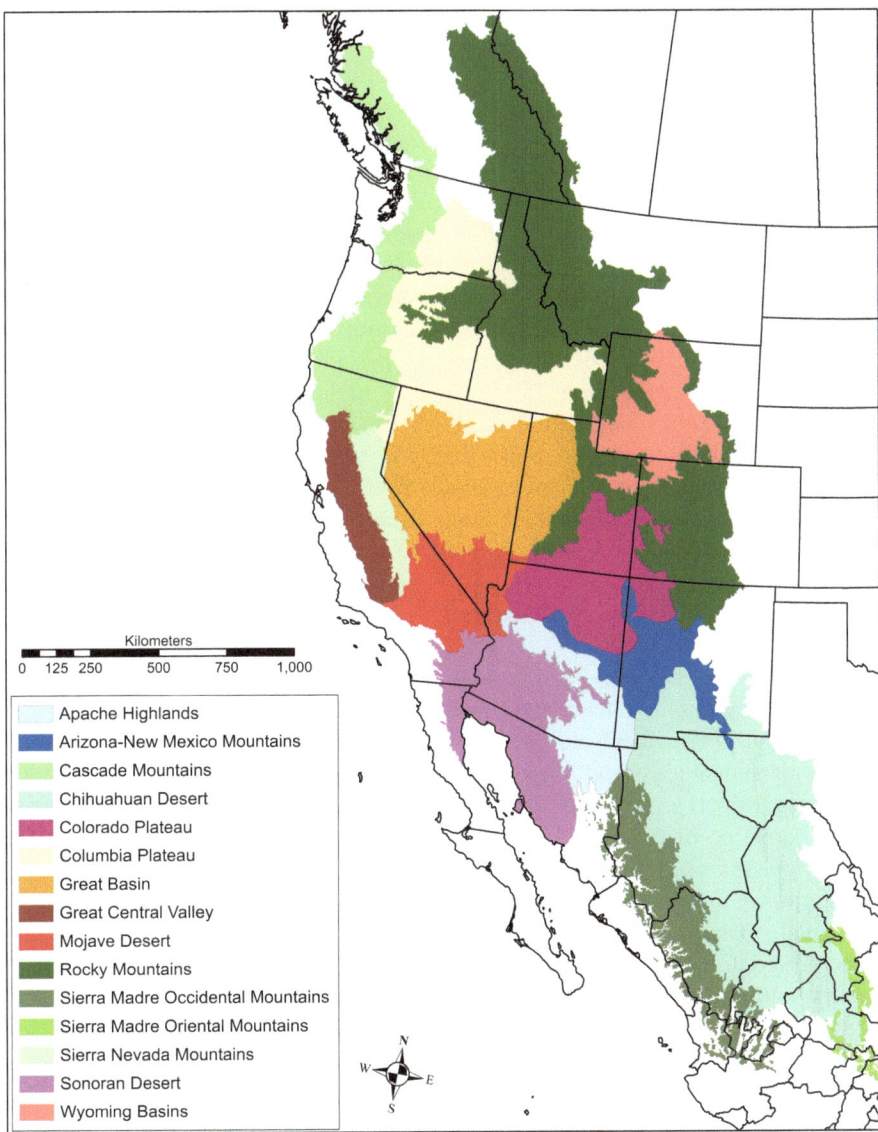

Fig. 2.3 Rangeland ecoregions in the western half of western North America

Precipitation increases, temperatures decrease, and vegetation changes as elevation increases. Deserts, shrublands, and grasslands dominate the valleys. Sagebrush, chaparral, piñon-juniper, and oak woodlands dominate the mid-elevations, and montane meadows, shrublands, and savannas dominate the upper elevations. Alpine tundra is found on the highest mountains.

2.3.1 Rangeland Regions Receiving Precipitation as Winter Rain

Mediterranean climates are characterized as having warm wet winters and hot dry summers. In North America, Mediterranean rangelands are found in the Great Central Valley of California and surrounding foothills (Fig. 2.3). Ecoregions considered here are California annual grasslands, oak woodlands, and chaparral. Annual grasses and forbs dominate the annual grasslands, where topography is flat to rolling with elevation ranging from 15 to 150 m. Oak savannas-woodlands occur on rolling to steeply sloping hills with annual grasses and forbs dominating the understory and open meadows. Chaparral occurs on rocky slopes. Precipitation in the Central Valley annual grasslands ranges from 150 to 400 mm. In the oak woodlands and chaparral, annual precipitation varies from 200 to 815 mm, increasing with elevation and from south to north. Annual precipitation is highly variable but approximately 80+% of the precipitation in all three ecoregions occurs from October to April. Severe droughts are common (Barbour and Minnich 2000; Rundel et al. 2016).

2.3.1.1 California Annual Grassland

The California annual grassland ecoregion is centered near the Great Central Valley (Fig. 2.3). Topography is flat to rolling with elevation ranging from 15 to 150 m. Non-native annual grasses and forbs from Spain and other Mediterranean regions dominate California annual grassland. Prior to Spanish settlement in 1769, this grassland was dominated by native perennial grasses including California oatgrass, nodding needle-grass, purple needlegrass, wheatgrasses, and wildryes (Burcham 1957). Non-native annual grasses were widespread by the early 1800s (Wagner 1989). Heavy live-stock grazing pressure and severe drought contributed to the conversion from native perennials to non-native annuals (Burcham 1957). A change in precipitation pattern exacerbated drought and grazing effects. Peak precipitation shifted from summer to winter, which favored the non-native annual grasses and disfavored native bunch-grasses (Axelrod 1973; Raven and Axelrod 1978). Hundreds of non-native annual grass and forb species grow in this ecoregion today, but only a few are widespread (Baker 1989). Ripgut brome, soft chess, slender wild oat, and wild oats are the domi-nant annual grasses. Associated annual forbs include burclover, filaree, and longbeak stork's bill. Medusahead, a non-native annual grass, is a significant problem (Nafus and Davies 2014).

2.3.1.2 California Oak Woodland

California oak woodland occupies a transition zone between California annual grass-land and montane forest at the upper margins. Blue oak, coast live oak, and valley oak are the most widespread oak species. In blue oak woodland and savanna, blue oak associates with California foothill pine, coast live oak, and valley oak. In valley oak woodland and savanna, valley oak associates with coast live oak, blue oak, California black oak, walnut, and California sycamore. In coast live oak woodland and savanna, coast live oak associates with blue oak. The same annual grasses and forbs that grow in the lower elevation annual grasslands grow in open stands of the oak woodlands. Conspicuous understory shrubs include black sage, ceanothus, manzanita, and narrowleaf goldenbush (Bartolome 1987; Allen-Diaz et al. 2007). A major threat to California oak woodland is the plant disease known as sudden oak death which is caused by the non-native, fungus-like, soil-borne pathogen *Phytophthora ramorum* (Rizzo et al. 2002).

2.3.1.3 California Chaparral

California chaparral is widespread throughout the state of California. More than 1000 plant species inhabit California chaparral. Herbaceous species comprise about 75% of the species, but plant communities are dominated by 1.5–4-m tall closely spaced shrubs that have thick, leathery, evergreen leaves (Rundel 2018). Ceanothus, chamise, and manzanita commonly inhabit drier sites. Buckthorn, scrub oak, and sumac occupy wetter sites (Keeley 2000). The thick, leathery leaves of chaparral plants increases their drought resistance, but their leaves also contain flammable resins and oils. Pre-European settlement fire return interval is estimated to have been 30–90 years depending on stand density and location. Due to human interference the return interval has decreased to 5–10 years. Both historical and current fires are high-severity, stand-replacing fires. Annual grasses and forbs proliferate soon after fire, but shrubs recover within 10 years because chaparral shrubs resprout following top-kill by fire or other disturbance, and seed dormancy is broken by heat or chemicals in smoke (Mooney and Miller 1985; Keeley et al. 2012).

2.3.2 Rangeland Regions Receiving Precipitation as Winter Snow

Most precipitation falls as snow during the winter in the Colorado Plateau, Columbia Plateau, Great Basin, Wyoming Basins, and at higher elevations in all western mountain ranges from Canada to Mexico. Temperatures are cool and snow accumulates throughout the winter. Soil water is generally at its maximum following spring snowmelt, allowing rapid growth of plants (West and Young 2000). In the valleys and lower foothills of these rangelands, salt desert shrublands and sagebrush shrublands have been described as cold deserts, but they are only cold in the winter (West 1983b; West and Young 2000). Summers in these regions are hot although temperatures decrease as elevation increases. Invasive annual grasses have increased the fire frequency and severity in salt desert shrub, sagebrush steppe, and Great Basin sagebrush ecoregions (West and Young 2000). Rangeland ecoregions in the following sections are arranged from lower to higher elevations.

2.3.2.1   Salt Desert Shrub

Salt desert shrub occurs in the Colorado Plateau, Great Basin, and Wyoming Basins (Fig. 2.3) on sites where soils are alkaline, saline or both (West 1983a; Blaisdell and Holmgren 1984). This vegetation type also occurs on similar sites in the northern mixed-grass prairie. Annual precipitation averages 130–330 mm. Shadscale and winterfat are important shrubs throughout this ecoregion. Gardner's saltbush co-occurs with shadscale and winterfat in the Wyoming Basins, whereas shadscale and winterfat associate with mat saltbush and valley saltbush in the Colorado Plateau. Greasewood occurs in areas with a seasonally high-water table (West and Young 2000; Duniway et al. 2018). Bottlebrush squirreltail, Indian ricegrass, needle and thread, and Sandberg bluegrass are the dominant native cool-season grasses. Warm-season grasses are more prevalent in the southern part of this ecoregion and include blue grama, galleta, purple threeawn, and sand dropseed. Buckwheat, desert prince's-plume, and scarlet globemallow are important native forbs. The salt desert shrub ecoregion has been invaded by cheatgrass, bur buttercup, halogeton, and Russian thistle. Wildfire is infrequent but areas dominated by cheatgrass can burn during years of above-average precipitation (Blaisdell and Holmgren 1984; Duniway et al. 2018).

2.3.2.2   Sagebrush Steppe

The sagebrush steppe occurs in the Columbia Plateau and Wyoming Basins (Fig. 2.3; West and Young 2000). Wyoming big sagebrush is typically dominant, with basin big sagebrush dominant in deeper, well-drained soils. Associated shrubs include horsebrush, rubber rabbitbrush, winterfat, and yellow rabbitbrush. Herbaceous under-stories are dominated by bluebunch wheatgrass, bottlebrush squirreltail, Columbia needlegrass, Idaho fescue, Indian ricegrass, needle and thread, Sandberg bluegrass, slender wheatgrass, and Thurber needlegrass (Miller and Eddleman 2000; West and Young 2000). Because of fire suppression, western juniper has invaded areas of sagebrush steppe (Miller et al. 2005, 2019). This ecoregion also has been invaded by non-native annual grasses including cheatgrass, medusahead, and ventenata (Davies et al. 2011; Jones et al. 2020).

2.3.2.3   Great Basin Sagebrush

Sagebrush-dominated rangeland at low-to-mid elevations in the Great Basin and Colorado Plateau comprises the Great Basin sagebrush ecoregion (Fig. 2.3; West and Young 2000). Wyoming big sagebrush and basin big sagebrush are the dominant shrubs. Rubber rabbitbrush and yellow rabbitbrush often co-occur and can dominate after fire. Broom snakeweed, horsebrush, and winterfat also are important shrubs. On sites with sandy loam or sandy soils, blackbrush, fourwing saltbush, Mormon tea, and sand sagebrush co-occur with Wyoming big sagebrush. Antelope bitterbrush grows with Wyoming big sagebrush on rocky soils adjacent to higher elevation piñon-juniper woodland (West 1983c; West and Young 2000). Basin wildrye, bluebunch wheatgrass, bottlebrush squirreltail, muttongrass, prairie junegrass, and Sandberg bluegrass are notable grasses. In the southern third of this ecoregion, associated warm-season grasses include blue grama, galleta, purple threeawn, and sand dropseed (West 1983c; West and Young 2000). Important forbs are blue flax, globemallow, and western yarrow (West and Young 2000; Leger and Baughman 2015). Utah juniper has invaded areas of Great Basin sagebrush (Miller et al. 2008, 2019). Numerous non-native invasive plant species have invaded the Great Basin sagebrush ecoregion, including cheatgrass, medusahead, halogeton, bur buttercup, Russian thistle, mustards, and pepperweed (Pyke et al. 2016; Boyd et al. 2021).

2.3.2.4 Piñon-Juniper Woodland

Piñon-juniper woodlands occupy mid-elevation foothills in the Apache Highlands, Arizona-New Mexico Mountains, Colorado Plateau, Columbia Plateau, Great Basin, and Wyoming Basins (Fig. 2.3; Adams 2018). Annual precipitation averages 300–600 mm across this expansive and diverse ecoregion. Approximately 50–60% of the precipitation falls during the winter in the Columbia Plateau, Great Basin, and Wyoming Basins while 60+% falls during the summer in the southern regions. Varied species of juniper and piñon dominate together or alone, depending upon climate, soils, and topography (West 1999). Western juniper dominates alone in the Columbia Plateau. Utah juniper dominates alone in Wyoming, but co-occurs with singleleaf piñon in the Great Basin, and with two-needle piñon in the Colorado Plateau. Alligator juniper and oneseed juniper associate with two-needle piñon in the Arizona and New Mexico Mountains. Redberry juniper grows with Mexican piñon in the Arizona Highlands and mountain foothills in northern Mexico. In Northern Rocky Mountain foothills, Rocky Mountain juniper dominates alone or co-dominates with limber pine (Adams 2018).

Where juniper and piñon co-occur, juniper is more drought resistant, the first to expand into new areas, and usually more abundant at lower elevations. Piñons dominate or co-dominate more mesic and higher elevation sites (Romme et al. 2009; Miller et al. 2019). Piñon-juniper physiognomy in the Colorado Plateau, Arizona-New Mexico Mountains, and Apache Highlands is usually a persistent woodland with high tree densities, whereas the woodland is typically more savanna-like in the Columbia Plateau, Great Basin, Rocky Mountain foothills, and Wyoming Basins.

Juniper and piñon woodland has increased in area and tree density during the past 150 years, while understory grass, forb, and shrub cover has decreased (Burkhardt and Tisdale 1969; Romme et al. 2009). Within the Columbia Plateau, Great Basin, and Wyoming Basins, prominent understory shrubs include antelope bitterbrush, black sagebrush, curl-leaf mountain mahogany, low sagebrush, Mormon tea, most subspecies of big sagebrush, rubber rabbitbrush, and yellow rabbitbrush. Grasses include bluebunch wheatgrass, Idaho fescue, Indian ricegrass, prairie junegrass, needle and thread, Sandberg bluegrass, and bottlebrush squirreltail. In the Colorado Plateau, blue grama and galleta also characterize piñon-juniper woodlands (West 1983c; Miller et al. 2019). Further south in Arizona, New Mexico, and Mexico, understory grasses include Arizona cottontop, black grama, blue grama, hairy grama, muhly, Rothrock grama, sideoats grama, slender grama, and threeawns. Understory shrubs include desert ceanothus, manzanita, Mexican cliffrose, shrub live oak, and true mountain-mahogany (Gottfried and Severson 1993; Floyd et al. 2004).

2.3.2.5 Mountain Brush

Mountain brush occurs above piñon-juniper woodlands and below coniferous forests at elevations between 1500 and 2500 m. The largest areas of mountain brush occur in Colorado and Utah on the western slopes of the Rocky Mountains (Fig. 2.3; Vankat 2013a). Annual precipitation ranges from 380 to 560 mm. Gambel oak, the most widespread plant in this ecoregion, reproduces from rhizomes or resprouts from its root crown following top-kill by fire or other disturbance (Harper et al. 1985; Tiede-mann et al. 1987). Growth of this species varies from dense shrub stands with sparse understory, to open plant communities with Gambel oak growing in clumps (mottes) and shrubs and herbaceous plants growing in the interspaces. Sometimes individual mottes cover several hectares. Other shrubs include antelope bitterbrush, bigtooth maple, chokecherry, mountain big sagebrush, mountain snowberry, and serviceberry. Grasses include bluebunch wheatgrass, Idaho fescue, muhly, muttongrass, needle-grass, slender wheatgrass, Thurber fescue, and western wheatgrass. Important forbs are balsamroot, beardtongue, blue flax, geranium, lupine, mule's ear wyethia, and western yarrow. Bulbous bluegrass, cheatgrass, knapweeds, thistle, and yellow toad-flax are notable invasive plants in mountain brush (Vankat 2013a; Kaufmann et al. 2016).

2.3.2.6 Montane Grassland

Montane grassland occurs on mountain and foothill slopes of the Rocky Mountains, eastern Cascade Mountains, and Sierra Nevada Mountains (Fig. 2.3). This ecoregion extends westward from the Cypress Hills in southwestern Saskatchewan to southern Alberta and British Columbia; south to include northeastern Oregon, southeastern Washington, and northern Nevada; eastward to include Idaho, Colorado, and Wyoming; and north to include central and western Montana (Mueggler and Stewart 1980). Annual precipitation averages 400–650 mm. Vegetation is dominated by cool-season perennial bunchgrasses including bluebunch wheatgrass, Idaho fescue, and slender wheatgrass. These grasses co-occur with foothills rough fescue in western Montana and southern British Columbia. Native forbs such as geranium, lupine, and western yarrow are common (Brewer et al. 2007; Thrift et al. 2013). In southern Alberta, Parry oatgrass replaces bluebunch wheatgrass in association with Idaho fescue and foothills rough fescue (Looman 1969). Heavy grazing pressure weakens bluebunch wheatgrass, foothills rough fescue, and Idaho fescue. In turn, Parry oatgrass, purple threeawn, and timber oatgrass increase along with invasive grasses and forbs. Non-native invasive grasses include cheatgrass, Kentucky bluegrass, medusahead, smooth bromegrass, timothy, and ventenata. Non-native invasive forbs include leafy spurge, spotted knapweed, and sulphur cinquefoil.

2.3.2.7 Montane Sagebrush Steppe

Montane sagebrush steppe occurs on mountain and foothill slopes of the Rocky Mountains (Fig. 2.3) at elevations of 1050–3050 m. This ecoregion occurs in Alberta, British Columbia, California, Colorado, Idaho, Montana, Nevada, Oregon, Utah, Washington, and Wyoming. Annual precipitation averages 250–700 mm. Montane sagebrush steppe generally has a longer summer growing season and warmer winters than montane grassland. The shrub layer in montane sagebrush steppe is usually dominated by mountain big sagebrush, mountain silver sagebrush, or snowfield sagebrush. Threetip sagebrush dominates or co-dominates some sites, and horsebrush and antelope bitterbrush are frequent associates. Cool-season perennial bunchgrasses such as bluebunch wheatgrass, foothills rough fescue, or Idaho fescue usually dominate the herbaceous layer. Foothills rough fescue and Idaho fescue are better adapted to mesic sites, and bluebunch wheatgrass better-suited to drier sites. Other common grasses include Columbia needlegrass, mountain brome, prairie junegrass, Sandberg bluegrass, and slender wheatgrass. Forbs are more abundant in montane sagebrush steppe than in lower elevation sagebrush steppe. Agoseris, buckwheat, balsamroot, cinquefoil, fleabane, hawksbeard, Indian paintbrush, lupine, and western yarrow are conspicuous forbs (Mueggler and Stewart 1980).

Historical records indicate that fires set by Native Americans and lightning occurred in the montane sagebrush steppe. However, evidence varies on how often fires occurred, with fire interval estimates ranging from 15 to 50 years or more (Welch and Criddle 2003; Moffet et al. 2015). Post-settlement fire suppression has increased sagebrush density and canopy cover, decreased herbaceous productivity, and decreased plant species diversity, especially forbs. Threetip sagebrush, snowfield sagebrush, and mountain silver sagebrush resprout after fire but mountain big sagebrush is easily killed by fire. However, after fire, mountain big sagebrush can reestablish readily from seed, which enables sagebrush to recover from fire much more quickly in montane sagebrush steppe than in lower elevation sagebrush areas in the Columbia Plateau, Colorado Plateau, Great Basin, and Wyoming Basins (Innes and Zouhar 2018).

2.3.2.8 Ponderosa Pine Savanna

Ponderosa pine savanna occurs throughout the mountains of the western US and in northern mixed-grass prairie (Figs. 2.2 and 2.3; Franklin and Dyrness 1973; Peet 2000). In the Sierra Nevada Mountains, Jeffrey pine forms savannas similar to those of ponderosa pine in the Rocky Mountains. Ponderosa pine savanna generally occurs in drier environments and is the climax tree species. Annual precipitation in the ponderosa pine savanna ecoregion averages 280–760 mm. From north to south, winter precipitation decreases and summer precipitation increases. Older ponderosa pine trees have thick bark that helps them tolerate surface fires that historically occurred every 5–30 years. Without fire or other disturbance, tree density increases rapidly and park-like savannas can become "dog-hair thickets" (Covington et al. 1997). Increased tree canopy cover, coupled with increased duff on the soil surface, reduces understory productivity. In the southern portion of ponderosa pine savanna, Arizona fescue, pine dropseed, muhly, muttongrass, and New Mexico feathergrass are the most common understory perennial grasses. Black dropseed, blue grama, Kentucky bluegrass, and threeawns increase with increased grazing pressure (Milchunas 2006; Strahan et al. 2015). In northern ponderosa pine savanna, important grasses include bluebunch wheatgrass, Idaho fescue, little bluestem, pinegrass, sedge, sideoats grama, and Thurber fescue. Understory shrub associates in northern ponderosa pine savanna include antelope bitterbrush, Bolander silver sagebrush, chokecherry, low sagebrush, mountain big sagebrush, mountain silver sagebrush, serviceberry, and Wyoming big sagebrush (Skovlin et al. 1975; Graham and Jain 2005).

2.3.2.9 Montane and Subalpine Meadow

Montane and subalpine meadows (hereafter mountain meadows) occur on nearly level, high-elevation, low-lying terrain in the Rocky Mountains, Cascade Mountains, and Sierra Nevada Mountains (Fig. 2.3; Franklin and Dyrness 1973; Barbour and Minnich 2000; Peet 2000). Soil substrates are alluvium, colluvium, or glacial outwash. Mountain meadows vary in size from one to several hundred hectares, adjoined in the montane zone by lodgepole pine or Douglas-fir forest. Subalpine meadows are commonly associated with subalpine fir or spruce forest.

Mountain meadow vegetation is a diverse mixture of grasses, sedges, rushes, and forbs. Shrubs are usually absent, although adjacent willow communities may occur along streams. Mountain big sagebrush, shrubby cinquefoil, or snowfield sagebrush may be present along drier meadow margins. Mean annual precipitation varies from 500 to 1250 mm. Mountain meadows are often categorized into three types based on soil moisture regime: wet meadows, moist meadows, and dry meadows (Hall 1973). Wet meadows remain wet at or near the soil surface throughout the growing season. Moist meadows have freely available water within the rooting zone throughout the growing season, but the soil surface is dry by late summer. Dry meadows are moist to wet in spring and early summer, but the soil surface is moderately to severely dry by mid-summer. Elephanthead lousewort, rushes, sedges, and tall mountain shootingstar characterize wet meadows. Alpine timothy, cinquefoil, groundsel, sedge, and tufted hairgrass occur in moist meadows. Dry meadows are often dominated by bluejoint reedgrass, Idaho fescue, sedge, timber oatgrass, thickstem aster, and western yarrow (Ratliff 1985; Mosley et al. 1989).

Conifers have encroached into many mountain meadows during the past 150 years. No single causal agent is responsible. Jakubos and Romme (1993) suggest that warmer and wetter conditions from the late 1800s to mid-twentieth Century aided conifer expansion, particularly into dry meadows. Pocket gopher (Family Geomyoidae) activity and grazing by wild and domestic ungulates can also promote conifer seedling establishment by reducing vegetative cover and exposing bare mineral soil. Heavy cattle and sheep grazing in the late 1800s and early 1900s prevented many seedlings from surviving, but reduced stocking rates and fire suppression beginning in the 1920s allowed surviving seedlings to grow into trees (Ratliff 1985; Taylor 1990). Conifer expansion also is affected by weather. In mesic and wet meadows where soil moisture inhibits conifers in most years, drought decreases soil moisture and herbaceous competition, thereby benefiting tree seedling establishment (Butler 1986). In dry meadows, conifer recruitment benefits when wet years immediately follow years with favorable conifer seed production (Dyer and Moffett 1999).

2.3.2.10 Alpine Tundra

Alpine tundra occurs from Canada to Mexico on snowcapped peaks, cliffs, rocky slopes, plateaus, and in glaciated valleys. Annual precipitation ranges from 750 mm at 55 degrees north latitude in west-central Alberta (elevation 500 m; Janz and Storr 1977) to 2250 mm at 25 degrees north latitude in central Mexico (elevation 3800 m; Beaman and Andersen 1966; Ramírez-Amezcua et al. 2016). Plant growth is limited because of the cold temperatures, short growing seasons, and winds associated with high elevations. Annual plants are rare. Trees can only survive where they are sheltered by rock formations or snow cover. Where trees do occur they have a stunted, twisted growth form referred to as Krumholtz. Forbs, grasses, and shrubs avoid desiccation and damage from cold and wind by growing close to the ground and maintaining little living plant material aboveground during winter (Billings 2000; Litaor et al. 2008).

Alpine bluegrass, alpine fescue, Cusick's bluegrass, and purple reedgrass are often abundant. Other common plant species include cinquefoil, Lewis' flax, mountain sorrel, rush, sedge, Townsend daisy, whitlowgrass, and woodrush. Alpine avens is common in alpine tundra of the Rocky Mountains but does not occur in the Sierra Nevada Mountains (Fowler et al. 2014; Rundel 2011).

2.3.3 Rangeland Regions Receiving Precipitation as Summer Rain

Low-elevation rangelands occurring from southern Nevada and eastern California to western Texas and northern Mexico are termed warm deserts because they are warmer and drier than the cold deserts to the north (i.e., salt desert shrub, Great Basin sagebrush, and lower elevation sagebrush steppe). Three warm deserts are recognized—Chihuahuan, Mojave, and Sonoran—due to differences in topography, climate, and vegetation. The harsh environment of the warm deserts is largely due to their location near the 30th parallel north where descending air creates hot and dry conditions. The warm deserts are also in rain shadows of surrounding mountains (MacMahon 2000).

Mean annual precipitation varies from 80 mm at low elevations of the Mojave Desert and Sonoran Desert to 380 mm in the eastern portion of the Chihuahuan Desert where elevation is greater. Summer monsoon precipitation is important in all three deserts but the proportion of summer to winter precipitation varies among the three warm deserts. The Mojave Desert receives 65–75% of its annual precipitation during winter. The Arizona and California portions of the Sonoran Desert have a bimodal precipitation pattern, receiving 50–60% of annual precipitation during winter. A greater proportion of total precipitation is received during summer in the Mexican portion of the Sonoran Desert. Precipitation is greater in the eastern Sonoran Desert due to increased elevation and orographic lifting which results in more summer thunderstorms. In the Chihuahuan Desert, peak precipitation occurs during summer (Adams and Comrie 1997; Sheppard et al. 2002).

Soils throughout the three warm deserts are shallow and alkaline, with subsoil petrocalcic layers often present (Duniway et al. 2007; Stefanov and Green 2013). Creosotebush is one of the few plant species that is widely distributed in all three warm deserts, growing on soils that are coarse, well-drained, alkaline, non-saline and often underlain by a petrocalcic layer. Creosotebush most commonly associates with tarbush in the Chihuahuan Desert, with triangle bursage in the eastern Sonoran Desert, and with burrobush in the Mojave Desert and western part of the Sonoran Desert (MacMahon 2000; Schafer et al. 2012).

2.3.3.1   Chihuahuan Desert

The Chihuahuan Desert occupies southern New Mexico, southwestern Texas, and northern Mexico between the Sierra Madre Oriental and Occidental Mountains (Fig. 2.3). Both desert scrub and desert grassland occur in the Chihuahuan Desert. Desert scrub varies from low-diversity, creosotebush-dominated plant communities on gently sloping plains, to more diverse shrub communities on upland rocky bajadas (i.e., alluvial fans that extend from mountain foothills). Important desert scrub plants include honey mesquite, tarbush, ocotillo, crown of thorns, agave, Spanish-bayonet, and many cacti of varied genera (Brown 1982a; Alvarez et al. 2011).

Desert grasslands are widespread throughout the Chihuahuan Desert, occurring on flat lowlands where soils are more developed and annual precipitation is greater than desert scrub areas. Grama grasses are prevalent, with black grama dominant on sandy loam uplands. Alkali sacaton and tobosagrass dominate areas with clay soils. Dropseeds and threeawns are widespread (Brown 1982b). Creosotebush, honey mesquite, tarbush, and yuccas increased in density in Chihuahuan Desert grassland after large numbers of domestic livestock were introduced into the ecoregion in the late 1800s (Gibbens et al. 1992). Fire suppression also enabled woody plants to increase (Drewa and Havstad 2001), resulting in less herbaceous cover and more soil erosion (Dinerstein et al. 2000).

2.3.3.2   Sonoran Desert

The Sonoran Desert occurs in southwestern Arizona, southeastern California, and in northern Mexico (Fig. 2.3). Vegetation on upper bajadas includes saguaro cactus, paloverde, ocotillo, desert ironwood, barrel-shaped cactus, prickly pear cactus, cholla cactus, creosotebush, and triangle bursage. Western honey mesquite occupies drier sites whereas honey mesquite grows on the most favorable upland sites. Velvet mesquite occupies riparian areas. Arizona cottontop and grama grasses (e.g., black grama, blue grama, Rothrock grama, sideoats grama, and slender grama) are abundant on lower slopes (MacMahon 2000; Medeiros and Drezner 2012). Broad, flat valleys are dominated by creosotebush, brittlebush, and burrobush. In Mexico, saguaro cactus is replaced with Mexican giant cardón. Invasive annual cool-season grasses such as Arabian schismus, Mediterranean grass, and red brome have invaded the Sonoran Desert (Evens et al. 2007; Steers and Allen 2012). Buffelgrass and Lehmann love-grass, both non-native perennial grasses, also have become invasive (Van Devender et al. 1997; Brenner 2010).

2.3.3.3   Mojave Desert

The Mojave Desert is located in southern Nevada, southeastern California, north-eastern Arizona, and southwestern Utah (Fig. 2.3). In addition to creosotebush, important plant communities in the Mojave Desert are characterized by black-brush, burrobush, or Joshua tree. Blackbrush communities in the northern portion of the Mojave Desert are the transition zone to the Great Basin, occupying upland terraces, ridges, open plains, and alluvial slopes (Bowns 1973; Brooks and Matchett 2003). Joshua trees are most abundant in the southern portion of the Mojave Desert. Burrobush-dominated communities, in association with creosotebush, Mojave buck-wheat, and Mormon tea, are prevalent in eastern portions of the Mojave Desert. Big galleta, bush muhly, desert needlegrass, galleta, and Indian ricegrass are noteworthy native perennial grasses in the Mojave Desert (Rasmuson et al. 1994; Sirchia et al. 2018). Arabian schismus, cheatgrass, Mediterranean grass, and red brome are non-native annual grasses that have invaded the Mojave Desert. Filaree, a non-native cool-season forb, also has invaded much of the Mojave Desert (Brooks and Matchett 2003; Underwood et al. 2019).

2.3.3.4 Interior Chaparral

Interior chaparral occurs almost entirely in central Arizona in the foothills bordering the Sonoran Desert on the north and the Mojave Desert on the east (Fig. 2.3). Average annual precipitation ranges from 380 to 640 mm (Cable 1975; Carmichael et al. 1978) with about half occurring during the summer. Shrub live oak comprises 25–80% of the total shrub cover on most sites. Associated shrubs include buckthorn, desert ceanothus, fragrant sumac, and manzanita (Pase and Brown 1982; Vankat 2013b). Important grasses, now largely confined to rocky, protected sites because of historical livestock grazing practices, include blue grama, black grama, cane bluestem, deergrass, and threeawns. Forbs are not abundant except for brief periods after fire. Nonnative grasses, including buffelgrass, Lehmann lovegrass, red brome, and weeping lovegrass have invaded interior chaparral (Carmichael et al. 1978; Vankat 2013b).

2.3.3.5 Southwestern Oak Woodland

The northern portion of southwestern oak woodland is located in the Apache Highlands of southeastern Arizona, southwestern New Mexico and northern Mexico, bounded by the Sonoran Desert to the west and the Chihuahuan Desert to the east (Fig. 2.3; McPherson 1992). The southern portion of the ecoregion is located in the Sierra Madre Occidental Mountains of Mexico (Fig. 2.3), where southwestern oak woodland forms expansive savannas. Precipitation is distributed evenly between early spring and mid-to-late summer, averaging 350–600 mm annually. Emory oak is widespread in oak woodland in Arizona and New Mexico. In northern Mexico, Emory oak associates with Arizona white oak, Mexican blue oak, and several species of juniper. Further south in Mexico, Emory oak associates with Chihuahuan oak. Herbaceous understories are dominated by warm-season perennial grasses, including blue grama, hairy grama, sideoats grama, slender grama, bullgrass, common wolfstail, green sprangletop, and threeawns. Annual forbs emerge briefly each year coincident with early spring rains and again with summer rains. Non-native plants are rarely present in this ecoregion (McClaran and McPherson 1999; Ffolliott et al. 2008).

Appendix

See Table 2.2.

Table 2.2 Common and scientific names of plants referenced in this chapter, rangeland ecoregions of western North America

Common name	Scientific name	Growth form	Longevity	Origin	Comment
Agave	*Agave* spp.	Succulent	Perennial	Native	
Agoseris or false dandelion	*Agoseris* spp.	Forb	Perennial	Native	
Algerita	*Mahonia trifoliolata*	Shrub	Perennial	Native	
Alligator juniper	*Juniperus deppeana*	Shrub/tree	Perennial	Native	
Alkali sacaton	*Sporobolus airoides*	Grass/C4	Perennial	Native	
Alpine avens	*Geum rossii*	Forb	Perennial	Native	
Subalpine big sagebrush	*Artemisia spiciformis*	Shrub	Perennial	Native	
Alpine bluegrass	*Poa alpina*	Grass/C3	Perennial	Native	
Alpine fescue	*Festuca brachyphylla*	Grass/C3	Perennial	Native	
Alpine timothy	*Phleum alpinum*	Grass/C3	Perennial	Native	
Antelope bitterbrush	*Purshia tridentata*	Shrub	Perennial	Native	
Arabian schismus	*Schismus arabicus*	Grass/C3	Annual	Introduced	Invasive
Arizona cottontop	*Digitaria californica*	Grass/C4	Perennial	Native	
Arizona fescue	*Festuca arizonica*	Grass/C3	Perennial	Native	
Arizona white oak	*Quercus arizonica*	Tree	Perennial	Native	
Ashe juniper	*Juniperus ashei*	Shrub	Perennial	Native	Invasive
Aspen	*Populus tremuloides*	Tree	Perennial	Native	
Balsam poplar	*Populus balsamifera*	Tree	Perennial	Native	
Balsamroot	*Balsamorhiza* spp.	Forb	Perennial	Native	
Barrel-shaped cactus	*Ferocactus* spp.	Succulent	Perennial	Native	
Basin big sagebrush	*Artemisia tridentata* ssp. *tridentata*	Shrub	Perennial	Native	
Basin wildrye	*Leymus cinereus*	Grass/C3	Perennial	Native	
Beardtongue or penstemon	*Penstemon* spp.	Forb	Perennial	Native	
Berlandier's wolfberry	*Lycium berlandieri*	Shrub	Perennial	Native	
Bermudagrass	*Cynodon dactylon*	Grass/C4	Perennial	Introduced	Seeded/Invasive
Big bluestem	*Andropogon gerardii*	Grass/C4	Perennial	Native	
Big galleta	*Hilaria rigida*	Grass/C4	Perennial	Native	
Big sagebrush	*Artemisia tridentata*	Shrub	Perennial	Native	
Bigtooth maple	*Acer grandidentatum*	Shrub/tree	Perennial	Native	

(continued)

Table 2.2 (continued)

Common name	Scientific name	Growth form	Longevity	Origin	Comment
Blackbrush (Mojave)	*Coleogyne ramosissima*	Shrub	Perennial	Native	
Blackbrush acacia (Tamaulipan Thornscrub)	*Vachellia rigidula* = *Acacia rigidula*	Shrub	Perennial	Native	
Black dropseed	*Sporobolus interruptus*	Grass/C4	Perennial	Native	
Black grama	*Bouteloua eriopoda*	Grass/C4	Perennial	Native	
Blackjack oak	*Quercus marilandica*	Shrub/tree	Perennial	Native	Invader
Black sage	*Salvia mellifera*	Shrub	Perennial	Native	
Black sagebrush	*Artemisia nova*	Shrub	Perennial	Native	
Bluebunch wheatgrass	*Pseudoroegneria spicata*	Grass/C3	Perennial	Native	
Blue flax	*Linum perenne*	Forb	Perennial	Native	
Blue grama	*Bouteloua gracilis*	Grass/C4	Perennial	Native	
Bluegrasses	*Poa* spp.	Grass/C3	Perennial	Native	
Bluejoint reedgrass	*Calamagrostis canadensis*	Grass/C3	Perennial	Native	
Blue oak	*Quercus douglasii*	Tree	Perennial	Native	
Bolander silver sagebrush	*Artemisia cana* ssp. *bolanderi*	Shrub	Perennial	Native	
Bottlebrush squirreltail	*Elymus elymoides*	Grass/C3	Perennial	Native	
Bristlegrass	*Seteria* spp.	Grass/C4	Perennial	Native	
Brittlebush	*Encelia farinose*	Shrub	Perennial	Native	
Broom snakeweed	*Gutierrezia sarothrae*	Semi-shrub	Perennial	Native	Invader
Buckthorn	*Rhamnus* spp.	Shrub	Perennial	Native	
Buckwheat	*Eriogonum* spp.	Forb	Perennial	Native	
Buffalograss	*Bouteloua dactyloides* = *Buchloe dactyloides*	Grass/C4	Perennial	Native	
Buffelgrass	*Cenchrus ciliaris* = *Pennisetum ciliare*	Grass/C4	Perennial	Introduced	Seeded/Invasive
Bulbous bluegrass	*Poa bulbosa*	Grass/C3	Perennial	Introduced	Seeded/Invasive
Bullgrass	*Muhlenbergia emersleyi*	Grass/C4	Perennial	Native	

(continued)

Table 2.2 (continued)

Common name	Scientific name	Growth form	Longevity	Origin	Comment
Burclover	*Medicago polymorpha*	Forb	Annual	Introduced	Invasive
Bur buttercup	*Ceratocephala testiculata*	Forb	Annual	Introduced	Invasive
Burrobush or white bursage	*Ambrosia dumosa*	Shrub	Perennial	Native	
Bush muhly	*Muhlenbergia porteri*	Grass/C4	Perennial	Native	
California black oak	*Quercus kelloggii*	Tree	Perennial	Native	
California foothill pine	*Pinus sabiniana*	Tree	Perennial	Native	
California oatgrass	*Danthonia californica*	Grass/C3	Perennial	Native	
California sycamore	*Platanus racemosa*	Tree	Perennial	Native	
Cane bluestem	*Bothriochloa barbinodis*	Grass/C4	Perennial	Native	
Catclaw acacia	*Senegalia greggii =* *Acacia greggii*	Shrub	Perennial	Native	
Caucasian bluestem	*Bothriochloa bladhii*	Grass/C4	Perennial	Introduced	Seeded/ Invasive
Ceanothus	*Ceanothus* spp.	Shrub	Perennial	Native	
Chamise	*Adenostoma fasciculatum*	Shrub	Perennial	Native	
Cheatgrass	*Bromus tectorum*	Grass/C3	Annual	Introduced	Invasive
Chihuahuan oak	*Quercus chihuahuensis*	Tree	Perennial	Native	
Chokecherry	*Prunus virginiana*	Shrub	Perennial	Native	
Cholla cactus	*Cylindropuntia* spp.	Succulent	Perennial	Native	
Cinquefoil	*Potentilla* spp.	Forb	Perennial	Native	
Coast live oak	*Quercus agrifolia*	Tree	Perennial	Native	
Columbia needlegrass	*Achnatherum nelsonii*	Grass/C3	Perennial	Native	
Common wolfstail	*Lycurus phleoides*	Grass/C4	Perennial	Native	
Creosotebush	*Larrea tridentata*	Shrub	Perennial	Native	
Crested wheatgrass	*Agropyron cristatum*	Grass/C3	Perennial	Introduced	Seeded/ Invasive
Crown of thorns	*Koeberlinia spinosa*	Shrub	Perennial	Native	
Curl-leaf mountain-mahogany	*Cercocarpus ledifolius*	Shrub	Perennial	Native	

(continued)

Table 2.2 (continued)

Common name	Scientific name	Growth form	Longevity	Origin	Comment
Curlymesquite	*Hilaria belangeri*	Grass/C4	Perennial	Native	
Cusick's bluegrass	*Poa cusickii* ssp. *epilis*	Grass/C3	Perennial	Native	
Deergrass	*Muhlenbergia rigens*	Grass/C4	Perennial	Native	
Desert ceanothus	*Ceanothus greggii*	Shrub	Perennial	Native	
Desert ironwood	*Olneya tesota*	Shrub/tree	Perennial	Native	
Desert needlegrass	*Pappostipa speciosa* = *Achnatherum speciosum*	Grass/C3	Perennial	Native	
Desert prince's-plume	*Stanleya pinnata*	Forb	Perennial	Native	
Douglas-fir	*Pseudotsuga menziesii*	Tree	Perennial	Native	
Dropseeds	*Sporobolus* spp.	Grass/C4	Perennial	Native	
Eastern red cedar	*Juniperus virginiana*	Shrub/tree	Perennial	Native	Invasive
Elephanthead lousewort	*Pedicularis groenlandica*	Forb	Perennial	Native	
Emory oak	*Quercus emoryi*	Tree	Perennial	Native	
Field brome	*Bromus arvensis*	Grass/C3	Annual	Introduced	Invasive
Filaree or redstem stork's bill	*Erodium cicutarium*	Forb	Annual	Introduced	Invasive
Fleabane	*Erigeron* spp.	Forb	Perennial	Native	
Foothills rough fescue	*Festuca campestris*	Grass/C3	Perennial	Native	
Fourwing saltbush	*Atriplex canescens*	Shrub	Perennial	Native	
Fragrant sumac or skunkbush sumac	*Rhus aromatica*	Shrub	Perennial	Native	
Galleta	*Hilaria jamesii*	Grass/C4	Perennial	Native	
Gambel oak	*Quercus gambelii*	Shrub/tree	Perennial	Native	
Gardner's saltbush	*Atriplex gardneri* var. *gardneri*	Shrub	Perennial	Native	
Geranium	*Geranium* spp.	Forb	Perennial	Native	
Globemallow	*Sphaeralcea* spp.	Forb	Perennial	Native	
Goldenrod	*Solidago* spp.	Forb	Perennial	Native	
Greasewood	*Sarcobatus vermiculatus*	Shrub	Perennial	Native	
Green needlegrass	*Nassella viridula*	Grass/C3	Perennial	Native	
Green sprangletop	*Leptochloa dubia*	Grass/C4	Perennial	Native	

(continued)

Table 2.2 (continued)

Common name	Scientific name	Growth form	Longevity	Origin	Comment
Groundsel or butterweed	*Senecio* spp.	Forb	Perennial	Native	
Guajillo acacia	*Senegalia berlandieri = Acacia berlandieri*	Shrub	Perennial	Native	
Hairy grama	*Bouteloua hirsuta*	Grass/C4	Perennial	Native	
Halogeton	*Halogeton glomeratus*	Forb	Annual	Introduced	Invasive
Hawksbeard	*Crepis* spp.	Forb			
Honey mesquite	*Prosopis glandulosa* var. *glandulosa*	Shrub	Perennial	Native	Invasive
Hooded windmillgrass	*Chloris cucullata*	Grass/C4	Perennial	Native	
Horsebrush	*Tetradymia* spp.	Shrub	Perennial	Native	
Idaho fescue	*Festuca idahoensis*	Grass/C3	Perennial	Native	
Indiangrass	*Sorghastrum nutans*	Grass/C4	Perennial	Native	
Indian paintbrush	*Castilleja* spp.	Forb	Perennial	Native	
Indian ricegrass	*Achnatherum hymenoides*	Grass/C3	Perennial	Native	
Jeffrey pine	*Pinus jeffreyi*	Tree	Perennial	Native	
Joshua tree	*Yucca brevifolia*	Succulent	Perennial	Native	
Juniper	*Juniperus* spp.	Shrub/tree	Perennial	Native	
Kentucky bluegrass	*Poa pratensis*	Grass/C3	Perennial	Introduced	Invasive
Knapweeds	*Centaurea* spp.	Forb	Perennial	Introduced	Invasive
Leafy spurge	*Euphorbia esula*	Forb	Perennial	Introduced	Invasive
Lehmann lovegrass	*Eragrostis lehmanniana*	Grass/C4	Perennial	Introduced	Seeded/ Invasive
Lewis' flax	*Linum lewisii*	Forb	Perennial	Native	
Limber pine	*Pinus flexilis*	Tree	Perennial	Native	
Little bluestem	*Schizachyrium scoparium*	Grass/C4	Perennial	Native	
Lodgepole pine	*Pinus contorta*	Tree	Perennial	Native	
Longbeak stork's bill	*Erodium botrys*	Forb	Annual	Introduced	Invasive
Lotebush	*Ziziphus obtusifolia*	Shrub	Perennial	Native	
Lovegrasses	*Eragrostis* spp.	Grass/C4	Perennial/ Annual	Native/ Introduced	

(continued)

Table 2.2 (continued)

Common name	Scientific name	Growth form	Longevity	Origin	Comment
Low sagebrush	*Artemisia arbuscula* ssp. *arbuscula*	Shrub	Perennial	Native	
Lupine	*Lupinus* spp.	Forb	Perennial	Native	
Manzanita	*Arctostaphylos* spp.	Shrub	Perennial	Native	
Mat saltbush	*Atriplex corrugata*	Shrub	Perennial	Native	
Mediterranean grass	*Schismus barbatus*	Grass/C3	Annual	Introduced	Invasive
Medusahead	*Taeniatherum caput-medusae*	Grass/C3	Annual	Introduced	Invasive
Mexican blue oak	*Quercus oblongifolia*	Tree	Perennial	Native	
Mexican giant cardón	*Pachycereus pringlei*	Succulent	Perennial	Native	
Mexican cliffrose	*Purshia mexicana*	Shrub	Perennial	Native	
Mexican piñon	*Pinus cembroides*	Shrub/tree	Perennial	Native	
Mojave buckwheat	*Eriogonum fasciculatum*	Shrub	Perennial	Native	
Mormon tea	*Ephedra* spp.	Shrub	Perennial	Native	
Mountain big sagebrush	*Artemisia tridentata* ssp. *vaseyana*	Shrub	Perennial	Native	
Mountain brome	*Bromus carinatus*	Grass/C3	Perennial	Native	
Mountain silver sagebrush	*Artemisia cana* ssp. *viscidula*	Shrub	Perennial	Native	
Mountain snowberry	*Symphoricarpos oreophilus*	Shrub	Perennial	Native	
Mountain sorrel	*Oxyria digyna*	Forb	Perennial	Native	
Muhly	*Muhlenbergia* spp.	Grass/C4	Perennial	Native	
Mule's ear wyethia	*Wyethia amplexicaulis*	Forb	Perennial	Native	
Multiflowered false rhodesgrass	*Trichloris pluriflora*	Grass/C4	Perennial	Native	
Mustards	*Brassica* spp.	Forb	Perennial/ Annual	Native/ Introduced	May be invasive
Muttongrass	*Poa fendleriana*	Grass/C3	Perennial	Native	
Narrowleaf goldenbush	*Ericameria linearifolia*	Forb	Perennial	Native	
Needle and thread	*Hesperostipa comata*	Grass/C3	Perennial	Native	
Needlegrass	*Achnatherum* spp.	Grass/C3	Perennial	Native	
New Mexico feathergrass	*Hesperostipa neomexicana*	Grass/C3	Perennial	Native	

(continued)

Table 2.2 (continued)

Common name	Scientific name	Growth form	Longevity	Origin	Comment
Nodding needlegrass	*Nassella cernua*	Grass/C3	Perennial	Native	
Oak	*Quercus* spp.	Shrub/tree	Perennial	Native	
Ocotillo	*Fouquieria splendens*	Semi-succulent	Perennial	Native	
Oneseed juniper	*Juniperus monosperma*	Shrub/tree	Perennial	Native	
Orchardgrass	*Dactylis glomerata*	Grass/C3	Perennial	Introduced	Seeded/invasive
Paloverde	*Parkinsonia* spp.	Shrub/tree	Perennial	Native	
Parry oatgrass	*Danthonia parryi*	Grass/C3	Perennial	Native	
Pepperweed	*Lepidium* spp.	Forb	Annual	Introduced	Invasive
Pine dropseed	*Blepharoneuron tricholepis*	Grass/C4	Perennial	Native	
Pinegrass	*Calamagrostis rubescens*	Grass/C3	Perennial	Native	
Pink pappusgrass	*Pappophorum bicolor*	Grass/C4	Perennial	Native	
Piñon pine	*Pinus* spp.	Shrub/tree	Perennial	Native	
Plains rough fescue	*Festuca hallii*	Grass/C3	Perennial	Native	
Plains silver sagebrush	*Artemisia cana* ssp. *cana*	Shrub	Perennial	Native	
Ponderosa pine	*Pinus ponderosa*	Tree	Perennial	Native	
Porcupine grass	*Hesperostipa spartea*	Grass/C3	Perennial	Native	
Post oak	*Quercus stellata*	Tree	Perennial	Native	Invasive
Prairie junegrass	*Koeleria macrantha*	Grass/C3	Perennial	Native	
Prairie sagewort	*Artemisia frigida*	Semi-shrub	Perennial	Native	
Prickly pear cactus	*Opuntia* spp.	Succulent	Perennial	Native	
Purple needlegrass	*Nassella pulchra*	Grass/C3	Perennial	Native	
Purple reedgrass	*Calamagrostis purpurascens*	Grass/C3	Perennial	Native	
Purple threeawn	*Aristida purpurea*	Grass/C4	Perennial	Native	
Redberry juniper (Arizona)	*Juniperus coahuilensis*	Shrub/tree	Perennial	Native	
Redberry juniper (Texas)	*Juniperus pinchotii*	Shrub/tree	Perennial	Native	Invasive
Red brome	*Bromus rubens*	Grass/C3	Annual	Introduced	Invasive
Red grama	*Bouteloua trifida*	Grass/C4	Perennial	Native	

(continued)

Table 2.2 (continued)

Common name	Scientific name	Growth form	Longevity	Origin	Comment
Ripgut brome	*Bromus diandrus*	Grass/C3	Annual	Introduced	Invasive
Rocky Mountain juniper	*Juniperus scopulorum*	Shrub/tree	Perennial	Native	
Rothrock grama	*Bouteloua rothrockii*	Grass/C4	Perennial	Native	
Rubber rabbitbrush	*Ericameria nauseosa*	Shrub	Perennial	Native	Invasive
Rush	*Juncus* spp.	Grass-like	Perennial	Native	
Russian thistle	*Salsola* spp.	Forb	Annual	Introduced	Invasive
Saguaro cactus	*Carnegiea gigantea*	Succulent	Perennial	Native	
Sandberg bluegrass	*Poa secunda*	Grass/C3	Perennial	Native	
Sand dropseed	*Sporobolus cryptandrus*	Grass/C4	Perennial	Native	
Sand sagebrush	*Artemisia filifolia*	Shrub	Perennial	Native	
Sand shinnery oak	*Quercus havardii*	Shrub	Perennial	Native	
Scarlet globemallow	*Sphaeralcea coccinea*	Forb	Perennial	Native	
Scrub oak	*Quercus berberidifolia*	Shrub	Perennial	Native	
Sedge	*Carex* spp.	Grass-like	Perennial	Native	
Sericea lespedeza	*Lespedeza cuneata*	Forb	Perennial	Introduced	Seeded/ Invasive
Serviceberry	*Amelanchier* spp.	Shrub	Perennial	Native	
Shadscale	*Atriplex confertifolia*	Shrub	Perennial	Native	
Shrub live oak	*Quercus turbinella*	Shrub	Perennial	Native	
Shrubby cinquefoil	*Dasiphora fruticosa*	Shrub	Perennial	Native	
Shrubby prairie rose	*Rosa arkansana*	Shrub	Perennial	Native	
Sideoats grama	*Bouteloua curtipendula*	Grass/C4	Perennial	Native	
Silver bluestem	*Bothriochloa laguroides*	Grass/C4	Perennial	Native	
Silver buffaloberry	*Shepherdia argentea*	Shrub	Perennial	Native	
Singleleaf piñon	*Pinus monophylla*	Shrub/tree	Perennial	Native	
Slender grama	*Bouteloua repens*	Grass/C4	Perennial	Native	
Slender wheatgrass	*Elymus trachycaulus*	Grass/C3	Perennial	Native	
Slender wild oats	*Avena barbata*	Grass/C3	Annual	Introduced	
Smooth bromegrass	*Bromus inermis*	Grass/C3	Perennial	Introduced	Seeded/ Invasive
Snowberry	*Symphoricarpos* spp.	Shrub	Perennial	Native	

(continued)

Table 2.2 (continued)

Common name	Scientific name	Growth form	Longevity	Origin	Comment
Snowfield sagebrush	*Artemisia spiciformis*	Shrub	Perennial	Native	
Soft chess	*Bromus hordeaceus*	Grass/C3	Annual	Introduced	Invasive
Spanish-bayonet	*Yucca harrimaniae*	Succulent	Perennial	Native	
Spiny hackberry or granjeno	*Celtis ehrenbergiana = C. pallida*	Shrub/tree	Perennial	Native	
Spotted knapweed	*Centaurea stoebe*	Forb	Perennial	Introduced	Invasive
Spruce	*Picea* spp.	Tree	Perennial	Native	
Subalpine fir	*Abies lasiocarpa*	Tree	Perennial	Native	
Sulphur cinquefoil	*Potentilla recta*	Forb	Perennial	Introduced	Invasive
Sumac	*Rhus* spp.	Shrub	Perennial	Native	
Sun sedge	*Carex inops* ssp. *heliophila*	Grass-like	Perennial	Native	
Switchgrass	*Panicum virgatum*	Grass/C4	Perennial	Native	
Tall mountain shootingstar	*Primula jeffreyi*	Forb	Perennial	Native	
Tanglehead	*Heteropogon contortus*	Grass/C4	Perennial	Native	Invasive
Tarbush	*Flourensia cernua*	Shrub	Perennial	Native	
Texas grama	*Bouteloua rigidiseta*	Grass/C4	Perennial	Native	
Texas live oak	*Quercus fusiformis*	Tree	Perennial	Native	
Texas persimmon	*Diospyros texana*	Shrub/tree	Perennial	Native	
Texas wintergrass	*Nassella leucotricha*	Grass/C3	Perennial	Native	
Thickstem aster	*Eurybia integrifolia*	Forb	Perennial	Native	
Thin paspalum	*Paspalum setaceum*	Grass/C4	Perennial	Native	
Thistle	*Carduus* spp., *Centaurea* spp., and *Cirsium* spp.	Forb	Perennial	Native/ Introduced	Invasive
Threeawns	*Aristida* spp.	Grass/C4	Perennial	Native	
Threetip sagebrush	*Artemisia tripartita*	Shrub	Perennial	Native	
Thurber fescue	*Festuca thurberi*	Grass/C3	Perennial	Native	
Thurber needlegrass	*Achnatherum thurberianum*	Grass/C3	Perennial	Native	
Timber oatgrass	*Danthonia intermedia*	Grass/C3	Perennial	Native	
Timothy	*Phleum pratense*	Grass/C3	Perennial	Introduced	Seeded/ Invasive
Tobosagrass	*Hilaria mutica*	Grass/C4	Perennial	Native	

(continued)

Table 2.2 (continued)

Common name	Scientific name	Growth form	Longevity	Origin	Comment
Townsend daisy	*Townsendia leptotes*	Forb	Perennial	Native	
Triangle bursage	*Ambrosia deltoidea*	Shrub	Perennial	Native	
True mountain-mahogany	*Cercocarpus montanus*	Shrub	Perennial	Native	
Tufted hairgrass	*Deschampsia cespitosa*	Grass/C3	Perennial	Native	
Two-needle piñon	*Pinus edulis*	Shrub/tree	Perennial	Native	
Utah juniper	*Juniperus osteosperma*	Shrub/tree	Perennial	Native	Invasive
Valley oak	*Quercus lobata*	Tree	Perennial	Native	
Velvet mesquite	*Prosopis velutina*	Shrub/tree	Perennial	Native	
Ventenata	*Ventenata dubia*	Grass/C3	Annual	Introduced	Invasive
Walnut	*Juglans* spp.	Tree	Perennial	Native	
Weeping lovegrass	*Eragrostis curvula*	Grass/C4	Perennial	Introduced	Seeded/Invasive
Western honey mesquite	*Prosopis glandulosa* var. *torreyana*	Shrub/tree	Perennial	Native	Invasive
Western meadowrue	*Thalictrum occidentale*	Forb	Perennial	Native	
Western juniper	*Juniperus occidentalis*	Shrub/tree	Perennial	Native	Invasive
Western yarrow	*Achillea millefolium*	Forb	Perennial	Native	
Western wheatgrass	*Pascopyrum smithii*	Grass/C3	Perennial	Native	
White tridens	*Tridens albescens*	Grass/C4	Perennial	Native	
Whitlowgrass	*Draba* spp.	Forb	Perennial	Native	
Wild oat	*Avena fatua*	Grass/C3	Annual	Introduced	
Willow	*Salix* spp.	Shrub/tree	Perennial	Native	
Winterfat	*Krascheninnikovia lanata*	Semi-shrub	Perennial	Native	
Wild rose	*Rosa* spp.	Shrub	Perennial	Native	
Wildryes and wheatgrasses	*Elymus* spp. and *Leymus* spp.	Grass/C3	Perennial	Native	
Woodrush	*Luzula* spp.	Grass-like	Perennial	Native	
Wyoming big sagebrush	*Artemisia tridentata* ssp. *wyomingensis*	Shrub	Perennial	Native	
Yellow bluestem (King Ranch bluestem)	*Bothriochloa ischaemum*	Grass/C4	Perennial	Introduced	Invasive

(continued)

Table 2.2 (continued)

Common name	Scientific name	Growth form	Longevity	Origin	Comment
Yellow rabbitbrush	*Chrysothamnus viscidiflorus*	Shrub	Perennial	Native	Invasive
Yellow toadflax	*Linaria vulgaris*	Forb	Perennial	Introduced	Invasive
Yucca	*Yucca* spp.	Succulent	Perennial	Native	

Source for common and scientific names is the Integrated Taxonomic Information System (ITIS; https://www.itis.gov, National Museum of Natural History 2023). Warm-season (C4) and cool-season (C3) photosynthetic pathways of grasses and CAM pathways for succulents are indicated under growth form. Longevity and origin information is from a variety of sources. The comment column is reserved to indicate plants that are generally considered invasive, although in some locations they may not be invasive. Seeded/invasive refers to plants that were purposely introduced to North America but have become invasive on some sites

References

Adams RP (2018) *Juniperus* of Canada and the United States: taxonomy, key and distribution. Lundellia 21:1–34. https://doi.org/10.25224/1097-993X-21.1

Adams DK, Comrie AC (1997) The North American monsoon. Bull Amer Meteor Soc 78:2197–2213. https://doi.org/10.1175/1520-0477(1997)078%3c2197:TNAM%3e2.0.CO;2

Allen-Diaz B, Standiford R, Jackson RD (2007) Oak woodlands and forests. In: Barbour M et al (eds) Terrestrial vegetation of California, 3rd edn. University of California Press, Berkeley, pp 313–338

Alvarez LJ, Epstein HF, Li J et al (2011) Spatial patterns of grasses and shrubs in an arid grassland environment. Ecosphere 2(9):30. https://doi.org/10.1890/ES11-00104.1

Anderson RC (2006) Evolution and origin of central grasslands of North America: climate, fire, and mammalian grazers. Torrey Bot Soc 133:626–627. https://doi.org/10.3159/1095-5674(2006)133[626:EAOOTC]2.0.CO;2

Anderson HG, Bailey AW (1980) Effects of annual burning on grassland in the aspen parkland of east-central Alberta. Can J Bot 58:985–996. https://doi.org/10.1139/b80-121

Archer S, Scifres C, Bassham CR et al (1988) Autogenic succession in a subtropical savanna: conversion of grassland to thorn woodland. Ecol Monogr 58:111–127. https://doi.org/10.2307/1942463

Axelrod DI (1973) History of the Mediterranean ecosystem in California. In: di Castri F, Mooney HA (eds) Mediterranean type ecosystems. Springer, Berlin, pp 225–277. https://doi.org/10.1007/978-3-642-65520-3_15

Bailey AW, Wroe RA (1974) Aspen invasion in a portion of the Alberta Parklands. J Range Manage 28:263–266

Baker HG (1989) Sources of the naturalized grasses and herbs in California grasslands. In: Huenneke LF, Mooney HA (eds) Grassland structure and function. Springer, Dordrecht, pp 29–38. https://doi.org/10.1007/978-94-009-3113-8_3

Barbour MG, Minnich RA (2000) California upland forests and woodlands. In: Barbour MG, Billings WD (eds) North American terrestrial vegetation. Cambridge University Press, Cambridge, UK, pp 161–202

Barker WT, Whitman WC (1988) Vegetation of the northern Great Plains. Rangelands 6:266–272

Bartolome JW (1987) California annual grassland and oak savannah. Rangelands 9:122–125

Beaman JH, Andersen JW (1966) The vegetation, floristics and phytogeography of the summit of Cerro Potosi, Mexico. Amer Midl Nat 75:1–33. https://doi.org/10.2307/2423480

Bielfelt BJ, Litt AR (2016) Effects of increased *Heteropogon contortus* (tanglehead) on rangelands: the tangled issue of native invasive species. Rangel Ecol Manage 69:508–512. https://doi.org/10.1016/j.rama.2016.06.006

Billings WD (2000) Alpine vegetation. In: Barbour MG, Billings WD (eds) North American terrestrial vegetation. Cambridge University Press, Cambridge, pp 537–572

Blaisdell JP, Holmgren RC (1984) Managing Intermountain rangelands: salt-desert shrub ranges. USDA Forest Service General Technical Report INT-GTR-163. Intermountain Forest and Range Experiment Station, Ogden, UT. https://doi.org/10.2737/INT-GTR-163

Bowns JE (1973) An autecological study of blackbrush (*Coleogyne ramosissima* Torr.) in southeastern Utah. PhD Dissertation, Utah State University

Box TW (1978) Food, fiber, fuel, and fun from rangelands. J Range Manage 31:84–86

Boyd CS, Davis DM, Germino MJ et al (2021) Invasive plant species. In: Remington TE et al (eds) Sagebrush conservation strategy—challenges to sagebrush conservation. US Geological Survey Open-File Report 2020-1125, pp 99–115. https://doi.org/10.3133/ofr20201125

Brenner JC (2010) What drives the conversion of native rangeland to buffelgrass (*Pennisetum ciliare*) pasture in Mexico's Sonoran Desert?: the social dimensions of a biological invasion. Human Ecol 38:495–505. https://doi.org/10.1007/s10745-010-9331-5

Brewer TK, Mosley JC, Lucas DE et al (2007) Bluebunch wheatgrass response to spring defoliation on foothill rangeland. Rangel Ecol Manage 60:498–507. https://doi.org/10.2111/1551-5028(2007)60[498:BWRTSD]2.0.CO;2

Briske DD, Fuhlendorf SD, Smeins FE (2005) State-and-transition models, thresholds, and rangeland health: a synthesis of ecological concepts and perspectives. Rangel Ecol Manage 58:1–10. https://doi.org/10.2111/1551-5028(2005)58%3C1:SMTARH%3E2.0.CO;2

Brooks ML, Matchett JR (2003) Plant community patterns in unburned and burned blackbrush (*Coleogyne ramosissima* Torr.) shrublands in the Mojave Desert. West North Amer Nat 63:283–298

Brown DE (1982a) Chihuahuan Desert scrub. Desert Plants 4:169–179

Brown DE (1982b) Semi desert grassland. Desert Plants 4:123–131

Burcham LT (1957) California range land. California Department of Natural Resource Division of Forestry, Sacramento

Burkhardt JW, Tisdale EW (1969) Nature and successional status of western juniper vegetation in Idaho. J Range Manage 22:264–270. https://doi.org/10.2307/3895930

Butler DR (1986) Conifer invasion of subalpine meadows, central Lemhi Mountains, Idaho. Northwest Sci 60:166–173

Cable DR (1975) Range management in the chaparral type and its ecological basis: the status of our knowledge. USDA Forest Service Research Paper RM-RP-155. Rocky Mountain Forest and Range Experiment Station, Fort Collins, CO

Carmichael RS, Knipe OD, Pase CP et al (1978) Arizona chaparral: plant associations and ecology. USDA Forest Service Research Paper RM-RP-202, Rocky Mountain Research Station, Fort Collins, CO. https://doi.org/10.5962/bhl.title.98412

Caudle D, DiBenedetto J, Karl MS et al (2013) Interagency ecological site handbook for rangelands. USDI Bureau of Land Management, USDA Forest Service, USDA Natural Resources Conservation Service, Washington, DC

Comer PJ, Hak JC, Kindscher K et al (2018) Continent-scale landscape conservation design for temperate grasslands of the Great Plains and Chihuahuan Desert. Nat Areas J 38:196–211. https://doi.org/10.3375/043.038.0209

Coupland RT (1992) Mixed prairie. In: Coupland RT (ed) Ecosystems of the world, natural grassland: introduction and western hemisphere. Elsevier, Amsterdam, pp 151–182

Covington WW, Fule PZ, Moore MM et al (1997) Restoring ecosystem health in ponderosa pine forests of the Southwest. J for 95:23–29. https://doi.org/10.1093/jof/95.4.23

Davies KW, Boyd CS, Beck JL et al (2011) Saving the sagebrush sea: an ecosystem conservation plan for big sagebrush plant communities. Biol Conserv 144:2573–2584. https://doi.org/10.1016/j.biocon.2011.07.016

DeKeyser ES, Meehan M, Clambey G et al (2013) Cool season invasive grasses in northern Great Plains natural areas. Nat Areas J 33:81–90. https://doi.org/10.3375/043.033.0110

Dinerstein E, Olson D, Atchley J et al (2000) Ecoregion-based conservation in the Chihuahuan Desert: a biological assessment. World Wildlife Fund, Washington, DC

Drewa PB, Havstad KM (2001) Effects of fire, grazing, and the presence of shrubs on Chihuahuan Desert grasslands. J Arid Environ 48:429–443. https://doi.org/10.1006/jare.2000.0769

Duniway MC, Herrick JE, Monger HC (2007) The high water-holding capacity of petrocalcic horizons. Soil Sci Soc Amer J 71:812–819. https://doi.org/10.2136/sssaj2006.0267

Duniway MC, Geiger EL, Minnick TJ et al (2018) Insights from long-term ungrazed and grazed watersheds in a salt desert Colorado Plateau ecosystem. Rangel Ecol Manage 71:492–505. https://doi.org/10.1016/j.rama.2018.02.007

Dyer JM, Moffett KM (1999) Meadow invasion from high-elevation spruce-fir forest in south-central New Mexico. Southwest Nat 44:444–456. https://doi.org/10.2307/3672342

Evens JM, Hartman S, Kobaly R et al (2007) Vegetation survey and classification for the Northern and Eastern Colorado Desert coordinated management plan (NECO). California Native Plant Society Vegetation Program, Sacramento

Ffolliott PF, Gottfried GJ, Stropki CL (2008) Vegetative characteristics and relationships in the oak savannas of the Southwestern Borderlands. USDA Forest Service Res Paper RMRS-RP-74. Rocky Mountain Research Station, Fort Collins, CO. https://doi.org/10.2737/RMRS-RP-74

Floyd ML, Hanna DD, Romme WH (2004) Historical and recent fire regimes in piñon-juniper woodlands on Mesa Verde, Colorado, USA. For Ecol Manage 198:269–289. https://doi.org/10.1016/j.foreco.2004.04.006

Fowler NL, Beckage B (2020) Savannas of North America. In: Scogings PF, Sankaran M (eds) Woody plants and large herbivores. John Wiley, New York, pp 123–149

Fowler JF, Nelson BE, Hartman RL (2014) Vascular plant flora of the alpine zone in the southern Rocky Mountains, U.S.A. J Bot Res Inst Texas 8:611–636

Franklin JF, Dyrness CT (1973) Natural vegetation of Oregon and Washington. USDA Forest Service General Technical Report PNW-GTR-8. Pacific Northwest Forest Range Experiment Station, Portland, OR

Fuhlendorf SD, Smeins FE (1997) Long-term vegetation dynamics mediated by herbivores, weather, and fire in a *Juniperus-Quercus* savanna. J Veg Sci 8:819–826. https://doi.org/10.2307/3237026

Gibbens RP, Beck RF, McNeely RP et al (1992) Recent rates of mesquite establishment in the northern Chihuahuan Desert. J Range Manage 45:585–588. https://doi.org/10.2307/4002576

Gottfried GJ, Severson KE (1993) Distribution and multi-resource management of piñon-juniper woodlands in the southwestern United States. In: Aldon EF, Shaw DW (eds) Managing piñon-juniper ecosystems for sustainability and social needs. USDA Forest Service General Technical Report RM-GTR-236. Rocky Mountain Forest and Range Experiment Station, Fort Collins, CO, pp 108–116. https://doi.org/10.2737/RM-GTR-236

Graham RT, Jain TB (2005) Ponderosa pine ecosystems. In: Ritchie MW et al (tech coords) Ponderosa pine: issues, trends, and management. USDA Forest Service General Technical Report PSW-GTR-198. Pacific Southwest Research Station, Albany, CA, pp 1–31

Griffith G, Bryce S, Omernik J et al (2004) Ecoregions of Texas. Environmental Protection Agency, Corvallis, OR

Hall FC (1973) Plant communities of the Blue Mountains in eastern Oregon and southeastern Washington. US Forest Service R6 Area Guide 3-1. Pacific Northwest Region, Portland, OR

Harper KT, Wagstaff FJ, Kunzler FM (1985) Biology and management of the Gambel oak vegetation type: a literature review. USDA Forest Service General Technical Report INT-GTR-179. Intermountain Forest and Range Experiment Station, Ogden, UT

Haukos DA (2011) Use of tebuthiuron to restore sand shinnery oak grasslands of the southern high plains. In: Hasaneen MNAE (ed) Herbicide mechanisms and mode of action. Intech, Rijeka, pp 103–124

Havstad K, Peters D, Allen-Diaz B et al (2009) The western United States rangelands: a major resource. Wedin WF, Fales SL (eds) Grassland quietness and strength for a new American agriculture. ASA, CSSA, SSSA Madison ,WI, pp 75–85. https://doi.org/10.2134/2009.grassl and.c5

Innes RJ, Zouhar K (2018) Fire regimes of mountain big sagebrush communities. In: Fire effects information system [Online]. USDA, Forest Service, Rocky Mountain Research Station, Missoula Fire Sciences Laboratory

Integrated Taxonomic Information System (2023) National Museum of Natural History, http://www.itis.gov

Jakubos B, Romme WH (1993) Invasion of subalpine meadows by lodgepole pine in Yellowstone National Park, Wyoming, U.S.A. Arctic Alpine Res 25:382–390. https://doi.org/10.2307/1551921

Janz B, Storr D (1977) The climate of the contiguous mountain parks: Banff, Jasper, Yoho, Kootenay. Canada Department of the Environment, Toronto

Jones LC, Davis C, Prather TS (2020) Consequences of *Ventenata dubia* 30 years post invasion to bunchgrass communities in the Pacific Northwest. Invasive Plant Sci Manage 13:226–238. https://doi.org/10.1017/inp.2020.29

Jordan TG (1978) Perceptual regions in Texas. Geog Rev 68:293–307. https://doi.org/10.2307/21504

Kaufmann MR, Huisjen DW, Kitchen S et al (2016) Gambel oak ecology and management in the Southern Rockies: the status of our knowledge. Southern Rockies Fire Science Network Publication 2016-1, Fort Collins, CO

Keeley JE (2000) Chaparral. In: Barbour MG, Billings WD (eds) North American terrestrial vegetation. Cambridge University Press, Cambridge, pp 204–253

Keeley JC, Fotheringham FC, Rundel PW (2012) Postfire chaparral regeneration under Mediterranean and non–Mediterranean climates. Madroño 59:109–127. https://doi.org/10.3120/0024-9637-59.3.109

Klemm T, Briske DD (2021) Retrospective assessment of beef cow numbers to climate variability throughout the U.S. Great Plains. Range Ecol Manage 78:273–280. https://doi.org/10.1016/j.rama.2019.07.004

Lauenroth WK (2008) Vegetation on the shortgrass prairie. In: Lauenroth WK, Burke IC (eds) Ecology of the shortgrass steppe. Oxford University Press, New York, pp 70–83

Lauenroth WK, Burke IC, Gutmann MC (1999) The structure and function of ecosystems in the central North American grassland region. Great Plains Res 9:223–259

Lavin M, Siebert C (2011) Great Plains flora? Plant geography of eastern Montana's lower elevation shrub-grass dominated vegetation. Nat Res Env Issues 16:2

Leger EA, Baughman OW (2015) What seeds to plant in the Great Basin? Comparing traits prioritized in native plant cultivars and releases with those that promote survival in the field. Nat Areas J 35:54–68. https://doi.org/10.3375/043.035.0108

Litaor MI, Williams M, Seastedt TR (2008) Topographic controls on snow distribution, soil moisture, and species diversity of herbaceous alpine vegetation, Niwot Ridge, Colorado. J Geophys Res 113:10. https://doi.org/10.1029/2007JG000419

Looman J (1969) The fescue grasslands of western Canada. Vegetatio 19:128–145. https://doi.org/10.1007/BF00259007

MacMahon JA (2000) Warm deserts. In: Barbour MG, Billings WD (eds) North American terrestrial vegetation. Cambridge University Press, Cambridge, pp 285–322

Maser C, Thomas JW (1983) Wildlife habitats in managed rangelands—the Great Basin of southeastern Oregon: introduction. USDA Forest Service General Technical Report PNW-160. Pacific Northwest Forest and Range Experiment Station, Portland, OR. https://doi.org/10.2737/PNW-GTR-160

McClaran MP, McPherson GR (1999) Oak savanna in the American Southwest. In: Anderson RC et al (eds) Savannas, barrens and rock outcrop plant communities of North America. Cambridge University Press, Cambridge, pp 275–287

McPherson GR (1992) Ecology of oak woodlands in Arizona. In: P.F. Pfolliott et al (tech coords) Proceedings of the symposium on ecology and management of oak and associated woodlands:

perspectives in the southwestern United States and northern Mexico. USDA Forest Service General Technical Report RM-218. Rocky Mountain Research Station, Fort Collins, CO, pp 24–33

Medeiros AS, Drezner TD (2012) Vegetation, climate, and soil relationships across the Sonoran Desert. Ecosci 19:148–160. https://doi.org/10.2980/19-2-3485

Milchunas DG (2006) Responses of plant communities to grazing in the southwestern United States. USDA Forest Service General Technical Report RMRS-GTR-169. Rocky Mountain Research Station, Fort Collins, CO. https://doi.org/10.2737/RMRS-GTR-169

Miller RF, Eddleman LL (2000) Spatial and temporal changes of sage grouse habitat in the sagebrush biome. Oregon State Univ Agr Exp Sta Tech Bull 151

Miller RF, Bates JD, Svejcar TJ et al (2005) Biology, ecology, and management of western juniper. Oregon State Univ Agr Exp Sta Corvallis Tech Bull 152

Miller RF, Tausch RJ, McArthur ED et al (2008) Age structure and expansion of piñon-juniper woodlands: a regional perspective in the Intermountain West. USDA Forest Service Res Paper RMRS-RP-69. Rocky Mountain Research Station, Fort Collins, CO. https://doi.org/10.2737/RMRS-RP-69

Miller RF, Chambers JC, Evers E et al (2019) The ecology, history, ecohydrology, and management of piñon and juniper woodlands in the Great Basin and Northern Colorado Plateau of the western United States. USDA Forest Service General Technical Report RMRS-GTR-403. Rocky Mountain Research Station, Fort Collins, CO. https://doi.org/10.2737/RMRS-GTR-403

Mitchell JE (2000) Rangeland resource trends in the United States: a technical document supporting the 2000 USDA Forest Service General Technical Report RMRS-GTR-68. Rocky Mountain Research Station, Fort Collins, CO. https://doi.org/10.2737/RMRS-GTR-68

Moffet CA, Taylor JB, Booth DT (2015) Postfire shrub cover dynamics: a 70-year fire chronosequence in mountain big sagebrush communities. J Arid Environ 114:116–123. https://doi.org/10.1016/j.jaridenv.2014.12.005

Mooney HA, Miller PC (1985) Chaparral. In: Chabot BF, Mooney HA (eds) Physiological ecology of North American plant communities. Chapman Hall, New York, pp 213–231. https://doi.org/10.1007/978-94-009-4830-3_2

Mosley JC, Brewer TK (2006) Targeted livestock grazing for wildlife habitat improvement. In: Launchbaugh KL, Walker JW (eds) Targeted grazing: a natural approach to vegetation management and landscape enhancement. American Sheep Industry Association, Centennial, CO, pp 115–128

Mosley JC, Bunting SC, Hironaka M (1989) Quadrat and sample sizes for frequency sampling mountain meadow vegetation. Great Basin Nat 49:241–248

Mueggler WF, Stewart WL (1980) Grassland and shrubland habitat types of western Montana. USDA Forest Service General Technical Report INT-66. Intermountain Forest and Range Experiment Station, Ogden, UT

Nafus AM, Davies KW (2014) Medusahead ecology and management: California annual grasslands to the Intermountain West. Invasive Plant Sci Manage 7:210–221. https://doi.org/10.1614/IPSM-D-13-00077.1

Padbury GA, Acton DF, Stushnoff CT (1998) The ecoregions of Saskatchewan. University Regina Canadian Plains Research Center, Regina

Pase CP, Brown DE (1982) Interior Chaparral. Desert Plants 4:95–99

Peet RK (2000) Forest and meadows of the rocky mountains. In: Barbour WB, Billings WD (eds) North American terrestrial vegetation. Cambridge University Press, New York, pp 75–121

Peterson R, Boyd C (1998) Ecology and management of sand shinnery communities: a literature review. USDA Forest Service General Technical Report RMRS-GTR-16. Rocky Mountain Research Station, Fort Collins, CO. https://doi.org/10.2737/RMRS-GTR-16

Powell GW, Bork EW (2007) Effects of aspen canopy removal and root trenching on understory microenvironment and soil moisture. Agrofor Syst 70:113–124. https://doi.org/10.1007/s10457-007-9051-z

Pyke DA, Chambers JC, Beck JL et al (2016) Land uses, fire, and invasion: exotic annual *Bromus* and human dimensions. In: Germino MJ et al (eds) Exotic brome-grasses in arid and semiarid ecosystems of the western US. Springer, Switzerland, pp 307–337

Ramírez-Amezcua Y, Steinmann VW, Ruiz-Sanchez E et al (2016) Mexican alpine plants in the face of global warming: potential extinction within a specialized assemblage of narrow endemics. Biodivers Conserv 25:865–885. https://doi.org/10.1007/s10531-016-1094-x

Rasmuson KE, Anderson JE, Huntly N (1994) Coordination of branch orientation and photosynthetic physiology in the Joshua tree (*Yucca brevifolia*). Great Basin Nat 54:204–211

Ratliff RD (1985) Meadows in the Sierra Nevada of California: state of knowledge. USDA Forest Service General Technical Report PSW-84. Pacific Southwest Forest and Range Experiment Station, Berkeley, CA. https://doi.org/10.2737/PSW-GTR-84

Raven PH, Axelrod DI (1978) Origin and relationships of the California flora. University of California Press, Berkeley

Rizzo DM, Garbelotto M, Davidson JM et al (2002) *Phytophthora ramorum* and sudden oak death in California: I. Host relationships. In: Standiford R, McCreary D (eds) USDA Forest Service General Technical Report PSW-GTR-184, pp 733–740

Romme WH, Allen CD, Bailey JD et al (2009) Historical and modern disturbance regimes, stand structures, and landscape dynamics of piñon-juniper vegetation of the western United States. Range Ecol Manage 62:203–222. https://doi.org/10.2111/08-188R1.1

Rundel PW (2011) The diversity and biogeography of the alpine flora of the Sierra Nevada, California. Madroño 58:153–184. https://doi.org/10.3120/0024-9637-58.3.153

Rundel PW (2018) California chaparral and its global significance. In: Underwood E et al (eds) Valuing chaparral. Springer, New York, pp 1–27. https://doi.org/10.1007/978-3-319-68303-4_1

Rundel PW, Arroyo MTK, Cowling RM et al (2016) Mediterranean biomes: evolution of their vegetation, floras, and climate. Ann Rev Ecol Evol Syst 47:383–407. https://doi.org/10.1146/annurev-ecolsys-121415-032330

Samson F, Knopf F (1994) Prairie Conservation in North America. Biosci 44:418–421. https://doi.org/10.2307/1312365

Schafer JL, Mudrak EL, Haines CE et al (2012) The association of native and non-native annual plants with *Larrea tridentata* (creosote bush) in the Mojave and Sonoran Deserts. J Arid Environ 87:129–135. https://doi.org/10.1016/j.jaridenv.2012.07.013

Sheppard PR, Hughes MK, Comrie AC et al (2002) The climate of the US Southwest. Clim Res 21:219–238. https://doi.org/10.3354/cr021219

Sims PL, Risser PG (2000) Grasslands. In: Barbour WB, Billings WD (eds) North American terrestrial vegetation. Cambridge University Press, New York, pp 323–356

Sirchia F, Hoffmann S, Wilkening J (2018) Joshua tree status assessment. US Fish and Wildlife Service

Skovlin JM, Harris RW, Strickler GS et al (1975) Effects of cattle grazing methods on ponderosa pine-bunchgrass range in the Pacific Northwest. USDA Forest Service Tech Bull 1531

Smart AJ, Harmoney K, Scasta JD et al (2021) Critical decision dates for drought management in central and northern Great Plains rangelands. Rangel Ecol Manage 78:191–200. https://doi.org/10.1016/j.rama.2019.09.005

Staver AC, Archibald S, Levin SA (2011) The global extent and determinants of savanna and forest as alternative biome states. Sci 334:230–232. https://doi.org/10.1126/science.1210465

Steers RJ, Allen EB (2012) Impact of recurrent fire on annual plants: a case study from the western edge of the Colorado Desert. Madroño 59:14–24. https://doi.org/10.3120/0024-9637-59.1.14

Stefanov WL, Green D (2013) Geology and soils in deserts of the southwestern United States. In: Malloy R et al (eds) Design with the desert: conservation and sustainable development. Taylor & Francis, Boca Raton, FL, pp 37–57

Stephenson NL (1990) Climatic control of vegetation distribution: the role of the water balance. Amer Nat 135:649–670. https://doi.org/10.1086/285067

Stephenson MB, Volesky JD, Schacht WH et al (2019) Influence of precipitation on plant production at different topographic positions in the Nebraska Sandhills. Rangel Ecol Manage 72:103–111. https://doi.org/10.1016/j.rama.2018.09.001

Strahan RT, Laughlin DC, Bakker JD et al (2015) Long-term protection from heavy livestock grazing affects ponderosa pine understory composition and functional traits. Rangel Ecol Manage 68:257–265. https://doi.org/10.1016/j.rama.2015.03.008

Stringham TK, Krueger WC, Shaver PL (2003) State and transition modeling: an ecological process approach. J Range Manage 56:106–113

Strong WL, Leggat KR (1992) Ecoregions of Alberta. Alberta Forestry Lands and Wildlife, Edmonton, AB

Taylor AH (1990) Tree invasion in meadows of Lassen Volcanic National Park, California. Prof Geogr 42:457–470. https://doi.org/10.1111/j.0033-0124.1990.00457.x

Thrift TM, Mosley TK, Mosley JC (2013) Impacts from winter-early spring elk grazing in foothills rough fescue grassland. West N Amer Nat 73:497–504. https://doi.org/10.3398/064.073.0402

Tiedemann AR, Clary WP, Barbour RJ (1987) Underground systems of Gambel oak (*Quercus gambelii*) in central Utah. Amer J Bot 74:1065–1071. https://doi.org/10.2307/2443947

Underwood EC, Klinger RC, Brooks ML (2019) Effects of invasive plants on fire regimes and postfire vegetation diversity in an arid ecosystem. Ecol Evol 9:12421–12435. https://doi.org/10.1002/ece3.5650

Van Devender TR, Felger RS, Burquez A (1997) Exotic plants in the Sonoran Desert region, Arizona and Sonora. In: 1997 Symposium proceedings, California Exotic Pest Plant Council

Vankat JL (2013a) Gambel oak shrublands. In: Vankat JL (ed) Vegetation dynamics on the mountains and plateaus of the American Southwest. Springer, London, pp 372–402. https://doi.org/10.1007/978-94-007-6149-0_6

Vankat JL (2013b) Interior chaparral shrublands. In: Vankat JL (ed) Vegetation dynamics on the mountains and plateaus of the American Southwest. Springer, London, pp 404–441. https://doi.org/10.1007/978-94-007-6149-0_6

Vermeire LT, Heitschmidt RK, Rinella MJ (2009) Primary productivity and precipitation-use efficiency in mixed-grass prairie: a comparison of northern and southern U.S. sites. Rangel Ecol Manage 62:230–239. https://doi.org/10.2111/07-140R2.1

Wagner FH (1989) Grazers, past and present. In: Huenneke LF, Mooney HA (eds) Grassland structure and function. Springer, Dordrecht, pp 151–162. https://doi.org/10.1007/978-94-009-3113-8_13

Wang T, Luri M, Janssen L et al (2017) Determinants of motives for land use decisions at the margins of the Corn Belt. Ecol Econ 134:227–237. https://doi.org/10.1016/j.ecolecon.2016.12.006

Welch BL, Criddle C (2003) Countering misinformation concerning big sagebrush. USDA Forest Service Research Paper RMRS-RP-40. Rocky Mountain Research Station, Ogden, UT. https://doi.org/10.2737/RMRS-RP-40

West NE (1983a) Intermountain salt-desert shrubland. In: West NE (ed) Temperate deserts and semi-deserts. Elsevier, Amsterdam, pp 375–397

West NE (1983b) Overview of North American temperate deserts and semi-deserts. In: West NE (ed) Temperate deserts and semi-deserts. Elsevier, Amsterdam, pp 321–330

West NE (1983c) Western intermountain sagebrush steppe. In: West NE (ed) Temperate deserts and semi-deserts. Elsevier, Amsterdam, pp 351–374

West NE (1999) Distribution, composition, and classification of current juniper-piñon woodlands and savannas across western North America. In: Monsen SB, Stevens R (eds) Proceedings: ecology and management of piñon-juniper communities within the Interior West. USDA Forest Service RMRS-Proc-9. Rocky Mountain Research Station, Fort Collins, CO, pp 20–23

West NE, Young JA (2000) Intermountain valleys and lower mountain slopes. In: Barbour MG, Billings WD (eds) North American terrestrial vegetation. Cambridge University Press, Cambridge, pp 255–284

Wester DB, Bryant FC, Tjelmeland AD et al (2018) Tanglehead in southern Texas: a native grass with an invasive behavior. Rangelands 40:37–44. https://doi.org/10.1016/j.rala.2018.03.002

Wied JP, Perotto-Baldivieso HL, Conkey AAT et al (2020) Invasive grasses in South Texas range-lands: historical perspectives and future directions. Invasive Plant Sci Manage 13:41–58. https://doi.org/10.1017/inp.2020.11

Wright HA, Bailey AW (1980) Fire ecology and prescribed burning in the Great Plains: a research review. USDA Forest Service General Technical Report INT-GTR-77. Intermountain Forest and Range Experiment Station, Ogden, UT

Chapter 3
A History of North American Rangelands

Nathan F. Sayre

Abstract North America's diverse grassland, savanna, steppe and desert ecosystems evolved in the absence of domesticated livestock. The arrival of cattle, sheep, goats, pigs and horses after 1492 transformed many ecosystems while enabling European soldiers, missionaries and settlers to conquer the continent. The decimation of indigenous populations by warfare, disease and economic dependency further transformed rangelands by removing Native management practices, especially the use of fire. The history of rangelands since then has been one of recursive efforts to commodify and territorialize rangeland resources—including wildlife, grass, soil fertility and the land itself—for market production and exchange. Many former rangelands have been lost altogether, by conversion to forest cover (due to fire suppression) or to agricultural uses (especially in the Great Plains), and invasive exotic plant species have radically altered large areas of rangelands in California, the Great Basin, and other regions. Nonetheless, North American rangelands remain both vast and invaluable for wildlife. The Western Range system of public land grazing leases, which emerged from the devastating overgrazing of the late nineteenth century, succeeded in stabilizing range conditions and linking land use and management across large landscapes of mixed ownerships. With accelerating urbanization, the rise of environmentalism, and structural shifts in the livestock industry since World War II, however, the Western Range has begun to unravel, exposing rangelands to development and fragmentation. Climatic variability in the form of droughts, floods and extreme fire conditions, more so than aridity per se, has frustrated efforts to extract value from rangelands from the outset, and climate change promises to amplify these phenomena going forward.

Keywords Ecological imperialism · Fire · Fur trading · Livestock industry · Environmentalism · Urbanization · Western Range

N. F. Sayre (✉)
Department of Geography, University of California, Berkeley, CA 94720, USA
e-mail: nsayre@berkeley.edu

3.1 Introduction: Rangelands and History

A comprehensive history of North America's rangelands has yet to be written. The volumes that come closest are probably Sherow's (2007) *Grasslands of the United States* (although it omits California and the Southeast) and *The Western Range,* also known as Senate Document No. 199, which was a 620-page "letter" from the Secretary of Agriculture published in 1936. It was replete with facts, including historical facts for the period since about 1800, but it was motivated by a pitched bureaucratic rivalry between the Agriculture and Interior Departments (see Sect. 3.6), and it is by now quite dated. Historians generally organize their research by place or region rather than land type, and they may omit environmental issues altogether, while the vibrant sub-field of environmental history has rarely made rangelands a particular focus. Sociologists and political scientists have studied the political-bureaucratic dimensions of federal rangeland administration, and more humanistic or interdisciplinary scholars have explored rangeland conservation in relation to cultural identity and community values, but history is not prominent in these works. Textbooks in range science often include one or two historical chapters, but these usually focus on disciplinary or industry matters rather than the lands themselves. Finally, geographers have written historical accounts of range livestock production, and there are scores of monographs on the history of ranches and range livestock production in specific regions.

A proper history of rangelands involves more than assembling facts from this corpus of existing scholarship, however. The concept of rangelands itself must be examined and elaborated for analytically coherent historiographic use. Although *rangeland* is now typically defined trans-historically as a set of land types based primarily on vegetation and cover (see Chap. 2), *range* has a history that is conceptually, ecologically and politically significant (Sayre 2017). Etymologically, *range* dates to the late fifteenth century (immediately prior to European expansion) and derives from the Old French verb *renger,* which referred to the movement of herders and livestock across large, open areas. Some scholars still define rangelands this way, for example as "land where people have intervened to manage the vegetation with livestock for economic gain" (Menke and Bradford 1992). Insofar as pre-Columbian North America lacked domesticated livestock, application of the term *range* before the early 1500s could be considered anachronistic (Bowling 1942; Crosby 1986). This is not simply of academic or terminological importance, moreover, because the arrival of cattle, sheep, goats, horses and other livestock was transformational. Their activities triggered widespread changes in ecosystems, as we will see, but the full effects went much further. Richard White (1994, p. 238) is not alone in his view that, "Without domesticated animals, Europeans would have neither survived nor conquered" in the New World. Livestock performed work on several levels, enabling activities as diverse as cultivation, transport and warfare as well as representing ideals of civilization, property and land use (Seed 1995; Anderson 2004). This breadth of roles and capacities made range livestock production "the principal means whereby Europeans colonized and exploited the natural resources of sub-Saharan Africa,

Australia, North and South America" (Grice and Hodgkinson 2002, p. 2). In short, by virtue of their intrinsic relation to livestock, rangelands are not simply sites of historical events, or places with histories; they are inescapably implicated in the conquest and settlement of North America by European- and African-descended peoples.

Put another way, North American rangelands are not static biophysical or evolutionary givens, but rather the product of intertwined social and ecological processes. These processes continue to operate, moreover, both on rangeland ecosystems and in how they are understood. Vegetation and land cover can change significantly over time, and parts of North America that are not classified as rangelands today, would once have met the current definition. In the Great Plains, for example, more than 96% of the tallgrass prairies and three-fifths of the mixed-grass prairies have been plowed and replaced by croplands (Samson et al. 1998), removing them from rangeland status. Large areas of the northeastern and southeastern United States were savannas at the time of European contact, but they gradually transformed into closed canopy forests due to the removal of Native American fire management practices (Mann 2005; Noss 2013). In sum, range and rangelands have become "a residual category, comprising everything (other than ice-covered lands) that doesn't fit into more specific types such as forest, urban, or cropland… they might best be understood as nonforested places where intensive economic activities have not (yet) taken root" (Sayre 2017, pp. 2–3). A history of rangelands must encompass and account for these losses.

This chapter presents a necessarily abbreviated history of North American rangelands from the immediate pre-Columbian period to the present. The focus is on how different groups of people have viewed, valued, used and altered these diverse lands, and the factors that have driven and shaped these changes. I hope to shed light on how and why North America's rangelands are both vast and diminished, mythologized and marginalized, contentious and misunderstood. On the one hand, the history of rangelands has been a story of manifold losses—the conquest and dispossession of Native Americans, the wholesale destruction of beaver, bison, wolves, grizzly bears, pronghorn, elk, prairie dogs and other wildlife, widespread conversion to non-native vegetation, and the disappearance of millions of hectares for agriculture, industry and urban development. On the other hand, the rangelands that remain are nonetheless among the continent's most ecologically intact landscapes: neither cultivated, irrigated, paved over nor built up, they are put to human use and transformed thereby, yet also relatively natural—working wilderness, so to speak (Sayre 2005). It should be no surprise, then, that wildlife has been central throughout this history, whether as subsistence resources, commercial products, agricultural pests or conservation causes.

In a brilliant essay, Richard White (1994) approached the history of the American West as a transformation from "animals as people" (as many Native Americans understood them) to "animals as enterprise." Expanding on many of White's points, I interpret this transformation as a series of efforts to *commodify* and *territorialize* rangelands. Beginning with the first European expeditions, myriad public and private entities have worked to identify, locate, map, exploit, control and regulate rangelands' diverse resources. Compared to other parts of the continent, however, rangelands have often proved recalcitrant to these efforts, even down to the present day. Initially this was due to Native American resistance, but a more lasting obstacle has been spatial scale: the extent of rangelands is vast, and the costs of control and extraction are high relative to most of the commodity values they yield (the exceptions being mining, oil and gas). Many historians have emphasized aridity as the defining feature of the American West (Webb 1931; Stegner 1954; Worster 1985). Equally important, however, has been the variability of rangelands over space and time: the unpredictable rainfall and droughts, fires and floods that attend rangelands from Mexico to Canada. This variability, exacerbated by climate change, is likely to be a hallmark of North American rangelands in the decades ahead.

3.2   The Late Indigenous Period

Beginning with mid-nineteenth century writers and artists such as Henry David Thoreau, James Fenimore Cooper, Henry Wadsworth Longfellow and George Catlin, the conventional wisdom for generations of scholars was that North America was 'pristine,' 'wild,' and 'natural' at the time of European contact (Denevan 1992). In stark contrast to the humanly transformed landscapes of the Old World, the Americas were thought to have been thinly populated by Native Americans, whose societies had made little or no impact on the continents' landscapes and ecosystems. This view has by now been thoroughly debunked and replaced by a three-part thesis: (1) Native peoples made widespread, significant and intentional impacts on American ecosystems (Dobyns 1981, 1983; Cronon 1983); (2) infectious diseases introduced by Europeans devastated Native populations, reducing their pre-contact numbers by as much as 90% and curtailing their ecological impacts proportionally, often well in advance of European peoples themselves (Crosby 1986); (3) Euro-Americans failed to recognize these facts, preferring to imagine an empty continent free for the taking and mistaking conditions circa 1750—when the total hemispheric population was still only about 30% of what it had been in 1492—as original, normative and time-less (Wolf 1982; Denevan 1992). The idea of America as untrammeled wilderness, then, is not only empirically false but theoretically flawed and ethically bankrupt—a self-serving delusion that legitimates settler colonialism and erases Native agency. All three parts of the thesis are directly relevant to the history of today's rangelands.

According to present scholarly understanding, Native Americans sustained many rangelands by conscious and willful actions, especially involving the use of fire (Stewart et al. 2002). Motivated primarily by subsistence needs, Native practices included sophisticated habitat management strategies for both plants and wildlife, as Anderson (2005, 2007) and Lightfoot and Parrish (2009) have shown in detail for California's diverse ecosystems. Writing about early colonial New England, Cronon (1983, p. 52) ventured the idea that "the Indians were practicing a more distant kind of husbandry of their own," one that did not involve keeping livestock. "Rather than domesticate animals for meat, Indians retooled ecosystems to encourage elk, deer, and bear. Constant burning of undergrowth increased the numbers of herbivores, the predators that fed on them, and the people who ate them both" (Mann 2005, p. 282). At a landscape scale, burning maintained a heterogeneous mosaic of habitat conditions while reducing the risks of dangerous wildfires (Fuhlendorf et al. 2008; Chap. 6, this volume). In specific locales, fire could favor desired plants for food or medicinal purposes, eliminate or reduce insect pests, or enhance conditions for hunting or self-defense; it could also serve as a means of hunting or warfare. In many settings, repeated burning shifted the structure and composition of vegetation communities away from trees and other woody plants and towards grasses and herbs. In this way, Native burning opened up forests and expanded bison habitat eastward from present-day Iowa and Illinois to New York and Georgia (Mann 2005). Noss (2013) argues that most of the longleaf pine forest of the southeastern coastal plains, from east Texas and Louisiana to Florida and northward through the Carolinas, was likewise maintained in savanna condition by repeated fires. Empirically and ecologically, it can be difficult or impossible to disentangle people from lightning as ignition sources, intentional from unintentional ignitions, or resource management from other motives for burning; some scholars dispute the ubiquity of Native fire impacts in specific sub-regions of the western U.S. (Vale 2002). But allocating causality between the two poles of a nature/human binary may be beside the point. What matters is that both human and biotic communities were adapted to frequent, widespread burning. This may be especially true for rangelands, but it was not limited to them (Pyne 1982).

Stretching from Mexico to Canada, the Great Plains merit specific mention as North America's largest and most archetypical rangelands. Mann (2005, p. 282) contends that "Native Americans burned the Great Plains and Midwest prairies so much and so often that they increased their extent; in all probability, a substantial portion of the giant grassland celebrated by cowboys was established and maintained by the people who arrived there first." Adapted to fire and grazing, native prairie grasses sustained an estimated 20–30 million bison at the time of initial European contact (Flores 2016a). Plains Indians developed religious and cultural systems as well as livelihood skills and social practices that orbited around the enormous bison herds. "Many Plains tribes, it seemed, thought of bison in human terms—they had families and societies, opinions and memories" (Flores 2016a, p. 38). As the staple food of northern Plains tribes, pemmican figured prominently in myths, origin stories and rituals. Made from a complex mix of different bison fats, melted and poured into sacks of pulverized dried bison meat, pemmican would count today as a kind of miracle food: succulent, high in both fat and protein, and virtually non-perishable. A mature bison yielded about ninety pounds of pemmican, or one large, brick-like bag (itself made of bison hide); when consumed, pemmican provided some 3500 cal per pound. Its invention in the northern Great Plains roughly 5–6000 years ago was "a key moment in the cultural history of the region, as pemmican's massive energy stores and durability…encouraged longer-distance travel, warfare, the elaboration of plains trade patterns and greater food security" (Colpitts 2015, p. 10).

Linda Black Elk (2016, p. 3) writes that rangelands "are central to the lives of Indigenous peoples, and they have been so for millennia." Native Americans, she explains, approach the land in terms of ecological interrelatedness, or the belief that

"we, as human beings, are related to everything and everyone—from huge cottonwood trees to the cool wind, and from barking prairie dogs to the fertile soil." Native
peoples understood animals as "other-than-human persons with whom relationships
were social and religious instead of purely instrumental... Indian religions made
hunting holy and gave human-animal relations a depth and complexity largely lacking
among Europeans. In hunting, some persons died so that others might live" (White
1994, p. 237). This worldview stands in stark contrast to the market and profit orientation that would infiltrate the Plains tribes and ultimately dispossess them over the
course of the nineteenth century (see Sect. 3.5 below). Notably, however, and unlike
much of the rest of North America, the dominant plants of the Great Plains were not
displaced by Old World species, even after the Native Americans who lived there
had been conquered and their management practices discontinued. Crosby (1986,
p. 290) observes that bison and perennial grasses "formed a tight partnership... each
sustaining and perpetuating the other and fending off the entry of any great number
of exotic plants and animals." Cattle occupied the niche vacated by bison, and as
Hart and Hart (1997, p. 10) point out, "much of the Great Plains before European
settlement looked about like it looks now," dominated by native perennial grasses
such as blue grama, buffalograss and galleta grass.

3.3   Fur Trading

The earliest sustained forays of Europeans into North America's interior rangelands
were motivated by "frontier capitalism's insatiable appetite for killing wild animals"
(Flores 2016a, p. 35)—that is, the commercial gains to be had from animals whose
populations had in some cases erupted with the decline of Native American hunting
pressure. In what might be termed "accumulation by extermination," hunters and
trappers pursued wildlife not for subsistence but for faraway markets, extracting just
those parts that could be economically transported and sold, often leaving much
of the carcass behind. Thus, did large portions of North America first encounter
market forces, amplified by stark differentials of power, trade and geography. In
many cases, fur trading incorporated Native Americans for their knowledge, skills
and labor (Dolin 2010). Russian, British and American traders pushed sea otters on
the Pacific coast to the brink of extinction between the 1780s and 1850, conscripting
Aleut and Kodiak men to do the work and shipping the pelts primarily to China.
Starting from the St. Lawrence Seaway and Great Lakes region, the northern beaver
trade spread west of the Missouri River under French control in the mid-eighteenth
century before passing into the hands of the British Hudson's Bay Company after
the Seven Years' War. Meanwhile, a mix of Anglo-, Franco- and Mexican–American
trappers worked the southern Rockies—often without the sanction of Spanish or
Mexican authorities—sending furs eastward along the Santa Fe Trail. As competition
between the Hudson's Bay Company and the American Fur Company intensified in
the early 1800s, beavers disappeared entirely from large parts of their former range,
with untold effects on watersheds. The slaughter stopped more or less by accident

in the 1830s, after European hat makers secured advantageous terms for Chinese silk (a side effect of the opium trade) and beaver felt passed out of fashion, thereby collapsing prices (White 1994).

The literature on the demise of the bison is too large to review here, but a brief summary is warranted (White 1994; Isenberg 2000; Flores 2003, 2016a; Colpitts 2015; Cunfer and Waiser 2016). Market demand for bison hides—initially as robes and subsequently as leather for industrial belts—drove the trade. A period of wetter than normal conditions in the first two decades of the nineteenth century may have helped expand the bison population, while the forced relocation of some 87,000 Native Americans from the Southeast to the southern Plains increased the regional subsistence demand and the number of potential hunters. Before the railroad reached the Great Plains, most of the hunting labor was provided by Native American men, and Native women did virtually all of the work to process the hides into robes. Robes soon became a primary source of cash income, the mechanism by which "nineteenth century Native peoples all over the continent were snared into dependency by the global economy" (Flores 2016a, p. 40). Roughly 100,000 robes were exported annually through New Orleans in the 1820s, and nearly that many again through Saint Louis in the 1840s. "By 1840, commercial production had reached about ninety thousand robes a year on the northern plains, and trade robes represented about 25% of the total buffalo kill of the plains" (White 1994, p. 246). Drought conditions ensued, peaking in the decade after 1855 and culling bison numbers by perhaps as much as 40–60%; bovine diseases introduced by cattle may have added to the mortality. With the railroad came professional Anglo-American hunters, who took more than four million bison from the southern Great Plains between 1872 and 1874, effectively eliminating the herd there. In the northern plains, where pemmican was the fuel for the Hudson's Bay Company's human-powered, waterborne transcontinental trade, the company used its monopoly to drive down the price it paid for pemmican, diminishing the real income of northern plains tribes and thereby impelling them to kill ever more bison, even as the herds dwindled (Colpitts 2015). By 1884, the northern herd, too, had been all but exterminated.

In summary, the destruction of Native American peoples by disease, warfare, dispossession and dependency had significant ecosystem effects across North America, including its current and erstwhile rangelands. In Crosby's (1986) famous formulation, European conquest of the Americas was ecological imperialism, empowered by Old World crops, weeds, livestock, rodents, insects and pathogens to which neither Native Americans nor native American ecosystems were adapted. Forests filled in as fires became less common, and some prey species of wildlife grew more abundant, at least in the short term. As Mann (2005, p. 362) notes, "ecologists and archaeologists increasingly agree that the destruction of Native Americans also destroyed the ecosystems they managed... By 1800 the hemisphere was thick with artificial wilderness." The resulting bounty, perceived by many Anglo-Americans as limitless, served as a windfall for colonists, market hunters and merchants, as wildlife were converted *en masse* into commodities and shipped to urban centers around the world. "The nineteenth-century Great Plains was a slaughterhouse. In the years from the 1820s to the 1920s, this single American region experienced the largest wholesale destruction of animal life discoverable in modern history" (Flores 2016b, p. 6).

3.4 Livestock

Columbus brought horses, cows, goats, sheep, pigs and chickens on his second voyage to the Americas in 1493, and by 1512 a cattle industry had been established in the West Indies, whence animals were later shipped to Florida and the Mississippi valley (Bowling 1942). Gregorio de Villalobos brought cattle to mainland North America in 1521, at what is now Veracruz, Mexico, where he founded the first of 233 *estancias* granted by the Spanish Crown over the ensuing century (Sluyter 2012). In 1540, the Coronado expedition set out from Compostela in what is now Nayarit, Mexico, with several hundred horses, 5000 sheep and 150 cattle; the cattle may have been the first to enter the present-day United States, but it is doubtful that any were still alive when the expedition reached present-day Kansas two years later (Wagoner 1952; Wildeman and Brock 2000). Another Spaniard, Juan de Oñate, brought sheep, goats and cattle when he founded *Nuevo México* in 1598; by 1700, the Navajo had become expert livestock raisers (Weisiger 2009), and Spaniards in New Mexico were exporting surplus sheep to Old Mexico annually by the late eighteenth century (White 1994). Elsewhere on the continent, the French introduced livestock into the St. Lawrence valley in 1541, and the first English cattle arrived in 1611 at Jamestown, Virginia. The Carolinas would emerge as the source area for the development and expansion of Anglo cattle ranching in the Southeast, which spread through the coastal plains to Texas and scattered locations in the Ohio and lower Mississippi River valleys between 1650 and 1850 (Jordan 1981). In California, Spanish missionaries introduced livestock from Mexico in the late eighteenth century, and *ranchos* multiplied rapidly there following the secularization of mission lands by the Mexican government in 1833 (Cleland 1941). Sheep were particularly important in the Pacific Northwest, where a range livestock industry developed after 1850, initially with animals from California and supplemented soon thereafter with breeds imported from eastern states via the Oregon Trail, although some Merino sheep are reported to have arrived by ship via Australia (Carman et al. 1892).

Some of the people who arrived in North America after 1492 came from places with significant rangelands, such as the Iberian Peninsula and North Africa, and they brought with them knowledge about how to raise livestock on the grasslands they found in the New World. Terry Jordan (1993) examined the development of range cattle production in North America on the basis of material culture and techniques of animal husbandry. He identified livestock systems that descended from the Old World and evolved in various ways as they diffused: a suite of overlapping Mexican systems that spread north and west from the Veracruz area; an Anglo-Texan system that blended traits from the American South and northeastern Mexico, spreading north and west from the Gulf Coast plains; and a Californian system, rooted in Spanish and Mexican practices, which expanded inland from the belt of missions, presidios and rancherias along the California coast. Such typologies are heuristic, and Jordan (1993, p. 308) cautioned that "Each cattle frontier was unique and far more accidental than predictable, the result of chance juxtapositions of peoples and places." A more lasting contribution may be his demonstration of the pluralistic, not to say multi-cultural, makeup of early cattle ranching. "The first Texas cowboys," as White (1994, p. 243) notes, "were Indians," and African-Americans, Native Americans, and Mexican-Americans were far more numerous among the cowboy work force of the late nineteenth century than depicted in Hollywood Westerns.

Extending Jordan's efforts, Andrew Sluyter (2012) has documented the key roles of Africans and their descendants, including slaves and former slaves, in adapting techniques of animal husbandry, horseback riding, and the management of land and water to enable range livestock production in New Spain, Louisiana, the Caribbean, and parts of South America. Old World plants also played supporting roles in many regions, colonizing areas disturbed by livestock grazing and displacing native vegetation in places where large grazing animals had previously been absent. As Sluyter (2012, p. 5) explains:

> Along with the cattle came grasses. Many millennia of association between livestock and grasses in Africa, Asia, and Europe ensured a greater symbiosis than that between the cattle and the grasses of the Americas. The non-American grasses were not only more palatable and nutritious, but the cattle preferentially propagated them, favoring them when grazing, carrying their seeds inland from the coast, and fertilizing them with manure.

Several African grasses spread through the tropics in Mexico, while Bermuda grass (also originally from Africa) colonized a subtropical belt from South Carolina to Texas. California's native grasses were widely displaced by Eurasian annual species by the nineteenth century (d'Antonio et al. 2007).

The Great Plains were more resistant to Old World plant invasions, and the interior of the continent was not so quickly overtaken by Europeans or their livestock, with one exception (Haines 1938). Beginning in Santa Fe around 1630,

> Indians spread horses rapidly and widely across North America. West of the Rockies, they transported the animal to the Snake River valley by 1700 and the Columbia Plateau by 1730. East of the Rockies, the horse reached the central Great Plains by the 1720s and western Canada by the 1730s… Indians used horses for transport, war, hunting, and more rarely, food. For most groups, a life without horses became unimaginable. (White 1994, pp. 238–239)

Empowered by horses, Native Americans stymied Spanish, Mexican and U.S. settlement of interior North America for centuries. "Rangelands were where native tribes succeeded the longest in resisting US conquest: the Comanche and others in Texas until 1875, the Sioux in the northern Great Plains until 1881, and the Apache in Arizona and New Mexico until 1886" (Sayre 2018, p. 342). From northern Mexico to Canada, and from the Great Plains through the Great Basin, tribes maintained complex and shifting relations of raiding, warfare, alliance and trade both among themselves and with European and Euro-American traders and settlers (Isenberg 2000; Blackhawk 2006; DeLay 2008; Colpitts 2015). Broadly speaking, tribes impeded state territorialization of rangelands, even while participating in the commodification of selected wildlife, livestock and animal products.

3.5   U.S. Expansion, Conquest and Settlement

Some 2,144,000 km^2 of territory, including much of the Great Plains, came into nominal possession of the United States with the Louisiana Purchase of 1803. The U.S. annexed Texas in 1845, and Mexico ceded another 1,370,000 km^2, also largely rangelands, under the Treaty of Guadalupe Hidalgo in 1848, which ended the Mexican–American War. As just mentioned, however, effective conquest and settlement of most of this area did not take place until the closing third of the nineteenth century. Expeditions into the Great Plains led by Zebulon Pike (1806–07) and Stephen Long (1818–19) reinforced a widespread perception of the region as a wasteland or "Great American Desert" unfit for agricultural settlement, as the limited surface waters and near-total absence of trees failed to conform to European notions of a civilizable landscape. The Gold Rush drew migrants from around the world to California after 1848, and more limited commerce and migration took place along the Santa Fe, Oregon, and other stagecoach trails throughout midcentury. But Native American resistance and political gridlock over slavery stymied policies in support of interior western settlement up to the Civil War.

The post-war period, by contrast, witnessed dramatic transformations of rangelands in demographic, political-economic and biophysical terms. In 1862, with Southern representatives absent, Congress passed the first of the Homestead Acts; the same year, President Lincoln created the US Department of Agriculture (USDA). These would become the institutional foundations for settlement beyond the Mississippi River. Inspired by Jeffersonian agrarianism, the policy goal was settlement by as many independent, landowning families as possible, in contrast to both the plantation South and aristocratic Europe; tacitly but effectively, the model settler was a white, male, Christian, English-speaking, American citizen (Carman et al. 1892; Sayre 2018). The Homestead Acts eventually transferred some 650,000 km^2 of public land into private hands, nearly all of it for commercial agriculture, in parcel sizes that were generally too small for economical use as rangelands. Meanwhile, the USDA provided scientific know-how and support, not only for farmers but also for loggers and ranchers operating on those parts of the public domain that were never successfully privatized. The economic basis followed shortly after the war in waves of migration, mining, ranching, timber-cutting, farming and railroad building, all fueled by investment capital from the east coast and Europe. With the partial exception of California (Walker 2001), the West became a colonial hinterland of the East, serving both as a source of natural resources for industrial development and as a destination for surplus capital produced by that development. "Across the telegraph wires came the instructions and information that coordinated eastern financial markets and western production sites. Along the railroads traveled the raw materials of the West and the finished products of the East" (White 1991, p. 236). As both cause and effect of late-century boom-and-bust capitalism, the western frontier was prone to crises at all scales, from farm foreclosures and corporate bankruptcies to the international depressions of 1873 and 1893. But it nonetheless resulted in the territorialization of the region into a system of property, investments, and land use oriented to national and global market production.

3.5.1 The Open Range and the Cattle Boom

The most legendary face of frontier expansion on North American rangelands, and the force behind its breakneck speed, was the Cattle Boom, which swept across the Great Plains in near lockstep with the decimation of the bison. It was actually two, overlapping and intersecting booms. One commenced immediately after the Civil War and was essentially bovine mercantilism: over the ensuing two decades, some 5.2 million ownerless, semi-feral Longhorn cattle that had built up in Texas during the war were rounded up and trailed north to urban markets, military forts, Indian reservations, and railhead towns, where they fetched prices as much as ten times what they cost (Webb 1931; Paul 1988). This was the boom of mythic cowboys, cattle trails and stampedes (McCoy 1951). The second boom picked up steam in the mid-1870s and effectively swallowed the first boom by the early 1880s. After smaller western banks failed in the 1873 panic, larger eastern firms and investors from as far away as Scotland jumped in to capitalize on high regional interest rates, free grass on unfenced rangelands, and surging national and international demand for beef (Dale 1930; Atherton 1961). This boom was the capitalist, financialized 'Beef Bonanza' (Brisbin 1881) of cattle barons, overnight fortunes and aristocratic pretensions. "[T]he Western range cattle industry during the last two decades of the nineteenth century was operated basically on borrowed capital" (Gressley 1966, p. 145), including some $45 million from Great Britain by the 1880s and another $284 million from the eastern US by the end of the century (Frink 1956; Graham 1960). Ahead of the homesteaders, with millions of hectares open to the first taker,

> Every man was seized with the desire to make the most that was possible out of his oppor-
> tunities while they lasted. He reasoned that there was more grass than his own cows could
> possibly eat. There was plenty of stock water for five times as many cows as were now on the
> range. There was no rent to pay, and not much in the way of taxes, and while these conditions
> lasted every stockman thought it well to avail himself of them. Therefore all bought cows to
> the full extent of their credit on a rising market and at high rates of interest. (Bentley 1898,
> p. 8)

Bank loans, mortgages and stock issues compelled ranchers to produce for the market, both to secure credit and to repay debts. In the 1870s, responding to the demands of the nascent packing industry as well as the admonitions of their faraway investors, cattle producers began to cross their Texas Longhorns with "improved" British breeds such as Herefords and Shorthorns, which yielded higher quality cuts of meat, especially when finished on corn. The perfection of cheap barbed wire in 1874 facilitated controlled breeding, but it also increased ranchers' costs and was illegal to install on the public domain, creating much uncertainty and sometimes violent conflict over informal 'range rights.' Meanwhile, the Union Stock Yards of Chicago and its Big Four meat processors (Armour, Swift, Morris, and Schwartzschild and Sulzberger (S&S)) pioneered advances in slaughtering and refrigerated transport that drove processing costs down, democratizing beef consumption and boosting demand. But the processors also used their monopoly position and outright collusion to exert downward pressure on prices paid to farmers and ranchers (Virtue 1920; Pacyga 2015; Specht 2019). This prompted further herd expansion, along the lines described by Bentley above. "Economy, culture, and ecology all combined to create conditions that led to an explosion in the numbers of cattle" (White 1991, p. 220).

The boom collapsed from the combined effects of over-expansion and bad weather. Drought in the southern Great Plains killed large numbers of livestock in 1883–84; many owners shipped their herds north and west in search of pasture, only to see them wiped out by severe winter storms in 1886–87. As of 1888, "[m]any thousands of animals were lying dead all over the range, starved and frozen; the survivors were riding in boxcars to the stockyards for rapid liquidation by their owners" (Worster 1992, p. 41). The last ripples of the boom washed across New Mexico and Arizona, where cattle numbers exploded between 1885 and 1891 and collapsed in the drought of 1891–93 (Sayre 1999). Coupled with the 1893 depression, it was an ecological-economic crisis. Scores of cattle companies went bankrupt. Vast areas of rangeland were reduced to dirt, triggering acute surface and gully erosion, altering fire regimes, and initiating widespread, long-term vegetation changes across the Southwest (Cooke and Reeves 1976; Bahre and Shelton 1996). Comparably severe vegetation changes would unfold across large parts of the northern shrub/steppe over the ensuing century (Sayre 2017). The fact that cattle grazing is routinely included among the official causes of decline for wildlife listed as threatened or endangered in the West is often due to impacts inflicted long ago.

3.5.2   Landownership

The mosaic of public, private and other landownership types[1] that characterizes North American rangelands today, dating from this period, can be loosely arranged by the availability of water and fertile soil, interacting with government policies and market forces.[2] The driest, highest, and/or least fertile areas defied settlement altogether and remained in the public domain, eventually passing into the administration of the

---

[1] Other landownership types include provincial and first nations lands in Canada; tribal and state lands in the US; and communal and indigenous lands in Mexico. The details of the three countries' landownership systems exceed the space available here, so I focus on the US case for simplicity. It is worth noting that enormous areas of rangelands in northern Mexico were privatized and sold to American capitalists in the late nineteenth century, facilitated by the Porfirio Diaz regime; revulsion at the land-grabbing helped to motivate the Mexican Revolution (Hart 2002).

[2] Texas is a partial exception in terms of landownership, because it entered the union in possession of its unsettled lands and disposed them to private owners on terms other than the Homestead Acts,

USDA's Forest Service or the Department of Interior's Bureau of Land Management (see Sect. 3.6). Important exceptions occurred where desert or semi-desert lands could be put under large-scale irrigation following passage of the 1902 Newlands Reclamation Act, such as in the Gila and Salt River valleys of Arizona, the Imperial and Sacramento-San Joaquin valleys of California, and the Palouse prairies of eastern Oregon and Washington. Here rangelands were lost to cultivation, often attended by speculation or fraud and ending up in the hands of large private landowners (Reisner 1987).

At the opposite end of the spectrum, in the wettest parts of the Great Plains, the soil was among the most fertile on Earth but there was generally too much water, or it was distributed in space and time such as to limit cultivation. The installation of drainage tiles—permeable pipes buried below plow depths to accelerate spring drying—spread rapidly across Illinois and Iowa in the 1870s and '80 s, often underwritten by banks or speculators who then sold the lands to prospective farmers (Prince 1997). Extending a model first developed in the Ohio Valley in the 1830s, the resulting farms used livestock to consume their copious corn harvests and convert them into moveable, saleable commodities (Hudson 1994). The aggregate result was a self-reinforcing cycle: farmers bought drained land on credit, and abundant yields pushed corn prices down, prompting farmers to cultivate ever more land to cover their debts. As the tall-grass prairie disappeared under the plow, calf production was displaced westward into the drier, mixed- and short-grass prairies of the western Great Plains (Dale 1930).

Intermediate on the spectrum were higher elevation valleys with mountain streams subject to diversion onto fertile floodplains. In these settings—scattered throughout the Great Basin, Rocky Mountains and Southwest—homesteaders successfully settled the flattest, most fertile fraction of the landscape and left the surrounding mountains and uplands in public ownership (Scott et al. 2001). Over time, the private lands became increasingly devoted to pasture or hay crops for winter feeding to herds of livestock that grazed on the surrounding public lands in the warmer months (Starrs 2000).

Finally, the most nettlesome cases were those where dry farming was possible in some years but not others, especially the Southern Great Plains and the belt of lands lying between the 100th and the 102nd meridians (Stegner 1954; Worster 1979). With about 50 cms of average annual precipitation, these lands appeared arable enough to induce land rushes among immigrant homesteaders hungry for farms of their own; by 1890, six million people inhabited the Great Plains. But these areas also periodically experienced multi-year droughts that devastated crops, bankrupted settlers and exposed the plowed fields to severe wind erosion. By the 1930s, one-third of the southern Great Plains—some 13.4 million hectares of former short-grass prairie—were sod-busted, setting the stage for the infamous Dust Bowl. Many failed homesteads reverted to public ownership either by tax default or through the New Deal's Rural Resettlement Administration.

---

resulting in a near-total absence of federal lands today. In terms of farming and ranching as land uses, however, it broadly resembles neighboring states.

With minor adjustments, the aggregate outcome for US rangelands was the pattern
of landownership and land use still visible today: near-total conversion to private
ownership and crop agriculture east of the 100th meridian, and a complex mosaic
to the west. As of 1940, some 16.1% of the seventeen Western states was farmland,
ranging from three percent or less in Nevada, Utah, New Mexico and Arizona to just
over half of Kansas and North Dakota (Stoddart 1945). Nearly all of the other 83.9%
remained rangeland, roughly half private and half in public ownership, but unevenly
distributed, with the public rangelands skewing towards higher, drier, and generally
less productive areas (Secretary of Agriculture 1936).

3.6 The Western Range

By 1890, North American rangelands were enfolded into a market-oriented,
continent-spanning "cattle-beef complex" that encompassed Midwestern corn farms,
cattle ranches across the West, Chicago packing plants, and refrigerated railroad meat
distribution to cities throughout the East (Specht 2019). The first stirrings of the
conservation movement were beginning to be felt in Washington, D.C.: Inspired by
the near-extinction of the bison, widespread clear-cutting of forests, and the destruction wrought by the Cattle Boom, prominent eastern scientists such as William
Hornaday (1889) were openly condemning market forces for the annihilation of
wildlife and their habitats. (In 1900, Congress would pass the Lacey Act, the first
federal law regulating interstate traffic in wildlife.) Out West, most of the land suitable for dry farming had been claimed under the Homestead Acts and plowed—it
was no longer rangeland at all. Hopeful settlers would continue to file entries into
the 1930s, but it was already evident that large areas would remain in (or revert to)
the public domain for lack of reliable water and/or arable soil, and that in many
cases their chief value was in fact a public one, as timber sources and watersheds for
downstream settlements.

   How should these lands be administered and managed? Congress answered this
question in a series of loosely coordinated steps for different subsets of the federal
domain. Rangelands fell principally into two of these: areas withdrawn under the
Forest Reserves Act of 1891, and the residual public domain (Calef 1960; Voigt 1976;
Rowley 1985).[3] Both were administered by the Department of Interior's General
Land Office (GLO), and both were already being grazed by livestock. But they
would follow quite different paths after 1894, when the GLO, facing pressure from
conservationists, banned all grazing on the Forest Reserves (Rowley 1985). The move
set off a political skirmish that ricocheted across the continent for the next half-
century and ultimately reterritorialized the open range, replacing it with a system
of exclusive leasehold tenure for private livestock producers to utilize the forage

---

[3] That is, lands not withdrawn for other purposes such as the military, Indian Reservations, national
parks, and lands granted to states. All of these categories included rangelands, but this fact was
generally incidental to their administration.

in fenced allotments of public rangelands. This system can be termed the Western Range, after the landmark 1936 USDA report of the same name (see below).

It was the dawn of the Progressive Era, and the debate over the Forest Reserves was waged in the language of science and the public good. The Senate asked the National Academy of Sciences to appoint a committee, which borrowed the words of John Muir ("hoofed locusts") to condemn livestock—especially sheep—for damaging the forests (NAS 1897). Cattle and sheep producers complained, and the USDA enlisted its premier botanist, Frederick Coville, to study the matter. Coville (1898) conducted a detailed survey in the Cascade Mountains of Oregon and systematically refuted the Academy's claims, and five years later he sat on the second Public Lands Commission,[4] convened by President Teddy Roosevelt. "The great bulk of the vacant public lands throughout the West," the commission wrote, "are, and probably always must be, of chief value for grazing" (Coville et al. 1905, p. xx). Some 120 million hectares were "theoretically open commons, free to all citizens," but in practice were subject to "tacit agreements" that were routinely violated. "Violence and homicide frequently follow," often between cattle and sheep producers. The commission's conclusion was an early articulation of the Tragedy of the Commons:

> The general lack of control in the use of public grazing lands has resulted, naturally and inevitably, in overgrazing and the ruin of millions of acres of otherwise valuable grazing territory. Lands useful for grazing are losing their only capacity for productiveness, as, of course, they must when no legal control is exercised. (Coville et al. 1905, p. xxi)

The commission's report led directly to passage of legislation that transferred the Forest Reserves to the USDA and created the US Forest Service to manage them. The law further authorized the Secretary of Agriculture to lease these lands to livestock producers and to charge them a fee for that use, as well as to stipulate terms and conditions for management. With a stroke of Roosevelt's pen, his close friend Gifford Pinchot, head of the USDA's Division of Forestry, was suddenly in charge of some 38 million hectares of land.

There is a large literature on the history of the Forest Service, but relatively little of it focuses on rangelands (but see Rowley 1985). Grazing wasn't the new agency's primary concern, after all: forests and timber, fire protection, and watersheds were all higher priorities. Western settlement had been attended and abetted by a proliferation of federal government entities tasked with developing scientific knowledge and information about the nation's land and natural resources. The goal in virtually every instance was to increase the output and efficiency of commercial agriculture for the benefit of settlers. At a time when European scholars and universities dominated the sciences, however, rangelands were an afterthought. Unlike forests, mines and farmlands, there was no established science for "unimproved" pastures and ranges. Basic taxonomic investigations of western U.S. range grasses only began in the 1880s, and the first formal program dedicated to "grass and forage plant investigations," the

---

[4] The first such commission was convened in 1879; its members included Clarence King and John Wesley Powell. Their report (Williamson et al. 1880) used the term "pasturage lands" to refer to rangelands, and noted that they were the least valuable lands, per acre, in the public domain, but also for that reason the most accessible to ordinary citizens with minimal capital (Sayre 2017).

USDA's Division of Agrostology, wasn't founded until 1895. American plant ecology was born in large measure into this vacuum. Charles Bessey, Frederic Clements, and their students and successors at the University of Nebraska dominated the field well into the twentieth century (Tobey 1981), producing an applied "science of empire" (Robin 1997) to address the needs of western rangelands.

That rangeland science unfolded under the administration of the Forest Service was more or less accidental, but also consequential. The goals of the Western Range were those of Progressive Era conservation, distilled in Pinchot's words as "the greatest good of the greatest number for the longest time" (Pinchot 1947). But in practice, this elegant utilitarian motto was rather contradictory. "The greatest good" effectively meant the greatest economic output, measured in profits and embodied in livestock, but no one knew how to calculate such an optimum. Even seemingly simple tasks such as mapping and measuring forage resources posed staggering logistical challenges, and highly variable interannual rainfall, on top of widely divergent range conditions, made relations with lessees perennially contentious, especially regarding stocking rates. "The greatest number" meant as many lessees as possible, but this too depended on forage production, and having too many risked repeating the errors of the open range period. To cull the pool, the Forest Service required permittees to own nearby private land sufficient to support their herds through the winter ("commensurate property"), effectively disqualifying poorer, non-landowning producers—many of whom were from minority groups (Sayre 2018). Finally, "the longest time" was an imponderable criterion. No one knew if rangelands could recover from acute overgrazing, or how long it might take, although the Public Lands Commission had confidently asserted that "Lands apparently denuded of vegetation have improved in condition and productiveness upon coming under any system of control which affords a means of preventing overstocking and of applying intelligent management to the land" (Coville et al. 1905, p. xxi).

In theory, exclusive access and security of tenure gave lessees a rational self-interest in conserving range resources on their allotments. But realizing exclusive access ran counter to maximizing profits. It required either the employment of full-time herders or the construction of fences, and both were prohibitively expensive. To study the matter, Pinchot and Coville sponsored an experiment in 1907–09, with an outcome that was predetermined: the high cost of fencing could be justified economically provided that it rendered herders unnecessary and thereby reduced producers' labor costs. But herders also protected livestock from wolves, grizzly bears, and the like, so eliminating herders would also require the West-wide elimination of predators. The Forest Service was already actively engaged in predator control on its lands, and in 1914 Congress authorized and funded the USDA's Bureau of Biological Survey (BBS) to do so throughout the West (Cameron 1929). Between 1915 and 1920, the BBS reported killing 128,513 predatory animals by hunting and trapping, and an unknown but probably larger number by poisoning. Wolves and grizzly bears were extirpated from large parts of their former ranges. Similar campaigns were launched against prairie dogs and a long list of other "pests," numbering in the hundreds of millions. Meanwhile, most of the fences needed to demarcate grazing allotments would not be built until the Civilian Conservation Corps subsidized the effort with a massive supply of cheap labor during the Depression (Sayre 2017).

The effects of the Western Range on rangeland ecosystems were mixed. The number of cattle and sheep grazing on the National Forests spiked during World War I, and the agency faced continuous resistance from lessees and livestock associations about stocking reductions. But over time, control of numbers and seasons of use were gradually achieved, and some indications of range recovery could be found, at least relative to the still unregulated, open range of the remaining public domain. Probably the greatest impacts of the new system, though, would not become evident till decades later. As early as 1920, the Forest Service had evidence that grazing reduced the incidence, intensity and spread of wildfires—and fire protection had become the agency's foremost concern since the politically embarrassing "Big Blow-up" of 1910 (Pyne 1982). Grazing for fuels management became de facto policy within the agency by the end of the 1920s. New stock roads, bridges, and water systems served both to open up access to additional forage for lessees and to expand the footprint of fire protection, and static stocking rates ensured heavy grazing (relative to forage production) during drier years, when fire risks were high (Sayre 2017). In the long-term, however, the effects of fire suppression included much denser forest stands, compositional shifts, and greater susceptibility to catastrophic crown fires.

Another product of the Western Range was Aldo Leopold, who joined the Forest Service fresh out of the Yale School of Forestry in 1909. For a brief period in 1914–15, he worked in the Office of Grazing for the Southwestern Region, where he encountered the concept of carrying capacity. "The discovery would reverberate through his work for the rest of his life" (Meine 1988, p. 136), shaping his interpretation of predator–prey interactions on the Kaibab plateau and informing his landmark textbook, *Game Management* (Leopold 1933). He was deeply involved with state hunting regulations, and he came to see hunters and private land owners as important allies in advancing conservation. Finally, he was among the first to question the wisdom of unrestrained fire suppression. Based on his observations in the Southwest, he wrote: "Until very recently we have administered the southern Arizona Forests on the assumption that while overgrazing was bad for erosion, fire was worse, and that therefore we must keep the brush hazard grazed down to the extent necessary to prevent serious fires. In making this assumption we have accepted the traditional theory as to the place of fire and forests in erosion, and rejected the plain story written on the face of Nature" (Leopold 1924, p. 6).

The Forest Service had come into being in 1905 with political support from sheep and cattle producers and their well-connected livestock associations, who had been persuaded that they had more to gain than to lose in paying fees to secure exclusive access to forage on the forests (Steen 1977; Rowley 1985). But for the lower, drier, and generally less productive lands that remained in the public domain, it would take another generation before such a coalition could be forged (Merrill 2002). Inspired by the success of the Mizpah-Pumpkin Creek Grazing District in Montana (Muhn 1987), livestock producers agreed to support the Taylor Grazing Act of 1934, which applied fencing and leases to the GLO's 63 million hectares of grazing lands. But it did not transfer those lands to the USDA, and in the years surrounding its passage an extraordinary bureaucratic struggle took place, largely behind the scenes. Secretary of Interior Harold Ickes lobbied President Franklin Roosevelt to reverse the earlier transfer and restore the National Forests to Interior, which he proposed to rename the Department of Conservation (Merrill 2002). Secretary of Agriculture Henry Wallace parried Ickes's efforts, however, arguing that the new Taylor Grazing Districts belonged in the care of the Forest Service (notwithstanding the near-absence of forests on those lands): *The Western Range* (Secretary of Agriculture 1936) was a 620-page briefing paper-cum-lobbying effort, mustering every piece of available evidence to support the contention that National Forest rangelands had improved since 1905, while the other 243 million hectares of the nation's grazing lands had remained degraded or worse. Roosevelt was reported to side with Ickes at first, but in the end, he did nothing, leaving the Western Range divided between two agencies with distinct land bases, institutional cultures and legislative mandates. The Bureau of Land Management did not receive an organic act to guide its management authority until 1976, when the Federal Land Policy and Management Act directed the agency to practice sustained multiple use.

3.7   Environmentalism and (Ex)urbanization

The post-World War II period saw the politics of rangelands fracture along new fault lines even as the Western Range consolidated. Range conditions on the Taylor Act lands generally improved by the late 1950s, then remained unchanged for the next quarter-century (Hadley et al. 1977). The new grazing districts were administered initially by the Division of Grazing, then the Grazing Service, and finally by the Bureau of Land Management, which absorbed and extinguished the GLO in 1946. The bureaucratic reorganizations reflected more than internal adjustments, though, as the new lessees and their livestock associations mounted a bid to devolve the new grazing districts into state, county or private ownership. Thus, was the modern Rangeland Conflict born: The cattle and wool growers provoked the ire of Bernard DeVoto, editor of *The New Republic*, who penned a series of articles denouncing their effort as a "land grab" and recasting the American cowboy from hero into despoiler of the nation's patrimony. DeVoto struck a chord with conservationists and everyday citizens in the East and also out West—he himself was a Utahn and prolific Western historian—and the episode signaled a lasting shift in the politics of public lands grazing. As the environmental movement grew out of the 1960s, helping to motivate passage of the Clean Air Act (1963), the National Environmental Policy Act (1970), the Clean Water Act (1972) and the Endangered Species Act (1973), ranchers and environmentalists increasingly saw each other as diametrically opposed. More recently, the demands of the Sagebrush Rebellion of the 1970s and '80s and the Malheur National Wildlife Refuge occupation in 2016 were remarkably similar to those of the livestock associations in DeVoto's day.

   Progressive faith in science to resolve political problems lingered, but it began to falter on the rangelands themselves. The discipline of range science, which had grown up as a step-child within the Forest Service, found greater professional autonomy after the Depression as employment opportunities multiplied in the BLM, the Soil Conservation Service, and the academy; in 1948, a new Society for Range Management came into being, cleaving away from forestry and agronomy. According to the scientists, controlling and reducing stocking rates was supposed to lead to range restoration, based on Frederic Clements's (1916, 1920) theory of plant succession and Arthur Sampson's (1919) influential adaptation of Clementsianism to range management. And indeed, stocking rates have declined on both Forest Service and BLM lands. But shrub encroachment persisted in large areas: juniper throughout much of the region, mesquite in the Southwest, and sagebrush in the Great Basin. Severe drought in the 1950s exacerbated fears that conditions were worsening. Facing pressure from lessees not to cut stocking rates, the USDA launched large-scale projects to restore grasses by mechanically or chemically removing shrubs, treating hundreds of thousands of hectares with little or no long-term success; indeed, the grasses that were seeded included a number of non-native species that later became problems in their own right (Sayre 2017). The role of fire suppression in ongoing vegetation change, meanwhile, was scrupulously avoided for decades, with the Forest Service sometimes actively preventing publication of fire research in prominent journals (Pyne 1982).

Demographic and technological changes have strongly affected rangelands and livestock production since the mid-twentieth century. Air conditioning, interstate highways and cheap energy enabled rapid suburban growth nationwide, especially in California and the Southwest. Population stagnated or decreased throughout the Great Plains, except in and around larger urban areas, as the labor demands on farms and ranches declined and young people migrated to cities for work. The average household grew smaller in terms of people, but larger in terms of house and parcel size; nationwide, the area of exurban development (4–16 ha/household) increased five-fold, from 5 to 25% of the conterminous US between 1950 and 2000 (Brown et al. 2005). Residential development sidesteps the ecological dependence of agriculture on fickle rainfall, capitalizing instead on warm climate, expansive views and low market prices for agricultural land. In the eight interior Western states, total farm and ranch land peaked in 1964 at 108 million hectares, then declined by an average of roughly 400,000 ha per year through 1997. Some 650,000 ha of grazing land went out of production (including public lands) every year in the 1990s; over the period 1982–1997, about 45% of lost grazing land was converted to urban uses (Knight et al. 2002).

Livestock production has also changed dramatically, albeit mostly on former rangelands converted to agriculture. Post-war surpluses of ammonia from decommissioned munitions factories flooded the market with cheap fertilizer in the late 1940s and '50s, and new chemical pesticides also came online. When applied to new hybrid varieties of corn and sorghum, the chemical inputs sent yields skyrocketing throughout the Plains states; cheap grain, in turn, opened up profit opportunities in concentrated livestock feeding (Nall 1982; Corah 2008; Ogle 2013). As feedlots concentrated in the southern Great Plains, processing plants gravitated towards them, taking advantage of non-union workforces and technological advances in slaughter to reduce costs and increase scale (Skaggs 1986; Stanley 1994). Declining margins have driven consolidation in farms and ranches through the US, with mid-sized operations decreasing dramatically (MacDonald 2018).

3.8 Conclusion

The aggregate effect of all these trends has been to marginalize rangelands *as rangelands* still further than before, ecologically, economically, socially and politically (Sayre et al. 2013). The Western Range system of leases, for all of its other weaknesses, did succeed in linking the management and use of private and public lands together in large, relatively contiguous parcels; as the Public Lands Commission reasoned, security of tenure would incentivize conservation as long as the "chief value" of the land was for grazing. As of 2000, some 45 million hectares of private lands were dependent on federal grazing permits for at least some of their forage (Gentner and Tanaka 2002). Now, however, private land values exceed what livestock production can justify nearly everywhere and often by wide margins, and nearly all ranches depend on off-farm income or wealth to remain solvent (Torell et al. 2004). The greatest threat to rangelands and their biodiversity is no longer livestock grazing but weed invasions, fragmentation and development (Hansen et al. 2005).

Historians have long emphasized the aridity of rangelands in the western United States as a key factor in the nation's settlement, as it defied the Jeffersonian, yeoman farmer model embedded in the Homestead Acts. This thesis requires modification in light of more recent scholarship, however. Biophysically, many North American rangelands (such as the Great Plains) were more resilient to Old World plants and livestock than other biomes, and it was their climatic variability, rather than aridity per se, that resulted in the greatest obstacles to Euro-American settler colonialism. Climate change is now increasing variability throughout the West, magnifying the challenges of drought, floods, fire, and water provision for urban, exurban, and rural areas alike. The lessons to be learned from the history of North American rangelands will only grow more salient, then, as more and more places come to experience comparable degrees of variability.

References

Anderson VD (2004) Creatures of empire: how domestic animals transformed early America. Oxford University Press, New York

Anderson MK (2005) Tending the wild: Native American knowledge and the management of California's natural resources. University of California Press, Berkeley CA

Anderson MK (2007) Native American uses and management of California's grasslands. In: Stromberg M, Corbin JD, D'Antonio CM (eds) California grasslands: ecology and management. University of California Press, Berkeley CA

Atherton LE (1961) The cattle kings. Indiana University Press, Bloomington

Bahre CJ, Shelton ML (1996) Rangeland destruction, cattle and drought in southeastern Arizona at the turn-of-the-century. J Southwest 38:1–22. https://www.jstor.org/stable/40169964

Bentley HL (H (1898) Cattle ranges of the Southwest: a history of the exhaustion of the pasturage and suggestions for its restoration. USDA Farmer's Bulletin No. 72. Government Printing Office, Washington

Black Elk L (2016) Native science: understanding and respecting other ways of thinking. Rangelands 38:3–4. https://doi.org/10.1016/j.rala.2015.11.003

Blackhawk N (2006) Violence over the land: Indians and empires in the early American West. Harvard University Press, Cambridge MA

Bowling GA (1942) The introduction of cattle into colonial North America. J Dairy Sci 25:129–154. https://doi.org/10.3168/jds.S0022-0302(42)95275-5

Brisbin JS (1881) The beef bonanza; or, how to get rich on the plains. J. B. Lippincott & Co., Philadelphia

Brown DG, Johnson KM, Loveland TR, Theobald DM (2005) rural land-use trends in the conterminous United States, 1950–2000. Ecol Appl 15:1851–1863. https://doi.org/10.1890/03-5220

Calef W (1960) Private grazing and public lands. University of Chicago Press, Chicago

Cameron J (1929) The Bureau of Biological Survey: its history, activities and organization. The Johns Hopkins Press, Baltimore, MD

Carman EA, Heath HA, Minto J (1892) Special report on the history and present condition of the sheep industry of the United States. USDA-Bureau of Animal Industry, Washington, DC

Cleland RG (1941) The cattle on a thousand hills: Southern California, 1850–1870. The Huntington Library, Los Angeles

Clements FE (1916) Plant succession: an analysis of the development of vegetation. Carnegie Institution of Washington, Washington, DC

Clements FE (1920) Plant indicators: the relation of plant communities to process and practice. Carnegie Institution of Washington, Washington DC

Colpitts G (2015) Pemmican empire: food, trade, and the last bison hunts in the North American Plains, 1780–1882. Cambridge University Press, New York

Cooke RU, Reeves RW (1976) Arroyos and environmental change in the American South-West. Clarendon Press, Oxford UK

Corah LR (2008) ASAS centennial paper: development of a corn-based beef industry. J Anim Sci 86:3635–3639. https://doi.org/10.2527/jas.2008-0935

Coville FV (1898) Forest growth and sheep grazing in the Cascade mountains of Oregon. USDA-Division of Forestry Bulletin No. 15. Government Printing Office, Washington DC

Coville FV, Hatton JH, Potter AF, Roosevelt T (1905) Report of the Public Lands Commission with appendix. Senate Document No. 189. Government Printing Office, Washington, DC

Cronon W (1983) Changes in the land: Indians, colonists, and the ecology of New England. Hill and Wang, New York

Crosby AW (1986) Ecological imperialism: the biological expansion of Europe, 900–1900. Cambridge University Press, New York

Cunfer G, Waiser B (2016) Bison and people on the North American Great Plains: a deep environmental history. Texas A&M University Press, College Station TX

Dale EE (1930) The range cattle industry. University of Oklahoma Press, Norman, OK

D'Antonio CM, Malmstrom C, Reynolds SA, Gerlach J (2007) Ecology of invasive non-native species in California grassland. In: Stromberg M, Corbin JD, D'Antonio CM (eds) California grasslands: ecology and management. University of California Press, Berkeley CA

DeLay B (2008) War of a thousand deserts: Indian raids and the U.S.-Mexican War. Yale University Press, New Haven CT

Denevan WM (1992) The pristine myth: the landscape of the Americas in 1492. Ann Assoc Am Geogr 82:369–385. https://doi.org/10.1111/j.1467-8306.1992.tb01965.x

Dobyns HF (1981) From fire to flood: historic human destruction of Sonoran Desert riverine oases. Ballena Press, Socorro NM

Dobyns HF (1983) Their number become thinned: native American population dynamics in eastern North America. University of Tennessee Press, Knoxville, TN

Dolin EJ (2010) Fur, fortune, and empire: the epic history of the fur trade in America. W.W. Norton & Co, New York

Flores D (2003) The natural West: environmental history in the Great Plains and Rocky Mountains. University of Oklahoma Press, Norman OK

Flores D (2016a) Reviewing an iconic story: environmental history and the demise of the bison. In: Cunfer G, Waiser B (eds) Bison and people on the North American Great Plains: a deep environmental history. Texas A&M University Press, College Station TX

Flores DL (2016b) American Serengeti: the last big animals of the great plains. University Press of Kansas, Lawrence, KS

Frink M (1956) When grass was king; contributions to the western range cattle industry study. University of Colorado Press, Boulder CO

Fuhlendorf SD, Engle DM, Kerby J, Hamilton R (2008) Pyric herbivory: rewilding landscapes through the recoupling of fire and grazing. Conserv Biol 23:588–598. https://doi.org/10.1111/j.1523-1739.2008.01139.x

Gentner BJ, Tanaka JA (2002) Classifying federal public land grazing permittees. J Range Manag 55:2–11. https://doi.org/10.2307/4003256

Graham R (1960) The investment boom in British-Texan cattle companies 1880–1885. Business History Review 34:421–445. https://doi.org/10.2307/3111428

Gressley GM (1966) Bankers and cattlemen. Knopf, New York

Grice AC, Hodgkinson KC (2002) Global rangelands: progress and prospects. CABI Publishers, Wallingford UK

Hadley RF, Dasgupta B, Reynolds NE et al (1977) Evaluation of land-use and land-treatment practices in semi-arid western United States. Philos Trans R Soc Lond B 278:543–554. https://doi.org/10.1098/rstb.1977.0061

Haines F (1938) The northward spread of horses among the Plains Indians. American Anthropol 40:429–437. https://www.jstor.org/stable/662040

Hansen AJ, Knight RL, Marzluff JM et al (2005) Effects of exurban development on biodiversity: patterns, mechanisms, and research needs. Ecol Appl 15:1893–1905. https://doi.org/10.1890/05-52213

Hart JM (2002) Empire and revolution: the Americans in Mexico since the Civil War. University of California Press, Berkeley CA

Hart RH, Hart JA (1997) Rangelands of the Great Plains before European settlement. Rangelands 19:4–11

Hornaday WT (1889) The extermination of the American bison, with a sketch of its discovery and life history. Government Printing Office, Washington DC

Hudson JC (1994) Making the corn belt: a geographical history of middle-western agriculture. Indiana University Press, Bloomington IN

Isenberg AC (2000) The destruction of the bison: an environmental history, 1750–1920. Cambridge University Press, New York

Jordon TG (1981) Trails to Texas: southern roots of western cattle ranching. University of Nebraska Press, Lincoln NE

Jordan TG (1993) North American cattle-ranching frontiers: origins, diffusion, and differentiation. University of New Mexico Press, Albuquerque NM

Knight RL, Gilgert WC, Marston E (2002) Ranching west of the 100th meridian: culture, ecology, and economics. Island Press, Washington, DC

Leopold A (1924) Grass, brush, timber, and fire in southern Arizona. J Forest 22:1–10

Leopold A (1933) Game management. C. Scribner's Sons, New York

Lightfoot KG, Parrish O (2009) California Indians and their environment: an introduction. University of California Press, Berkeley CA

MacDonald JM (2018) CAFOs: farm animals and industrialized livestock production. In: Oxford research encyclopedias of environmental science. https://doi.org/10.1093/acrefore/978019938 9414.013.240

Mann CC (2005) 1491: New revelations of the Americas before Columbus. Alfred A. Knopf, New York

McCoy JG (Joseph G, Worrall H (1951)) Historic sketches of the cattle trade of the West and Southwest. Long's College Book Co, Columbus, OH

Meine C (1988) Aldo Leopold: his life and work. University of Wisconsin Press, Madison, WI

Menke J, Bradford GE (1992) Rangelands. Agr Ecosyst Environ 42:141–163. https://doi.org/10.1016/0167-8809(92)90024-6

Merrill KR (2002) Public lands and political meaning: ranchers, the government, and the property between them. University of California Press, Berkeley CA

Muhn JA (1987) The Mizpah-Pumpkin Creek Grazing District: its history and influence of the enactment of a public lands grazing policy, 1926–1934. Montana State University, Bozeman MT

Nall GL (1982) The cattle-feeding industry on the Texas high plains. In: Dethloff HC, May IM Jr (eds) Southwestern agriculture: pre-Columbian to modern. Texas A&M University Press, College Station TX

NAS (1897) Report of the committee appointed by the National Academy of Sciences upon the inauguration of a forest policy for the forested lands of the United States to the Secretary of the Interior. Government Printing Office, Washington DC

Noss RF (2013) Forgotten grasslands of the South: natural history and conservation. Island Press, Washington DC

Ogle M (2013) In meat we trust: an unexpected history of carnivore America. Houghton Mifflin Harcourt, New York

Pacyga DA (2015) Slaughterhouse: Chicago's union stock yard and the world it made. University of Chicago Press, Chicago

Paul RW (1988) The far west and the great plains in transition, 1859–1900. Harper & Row, New York

Pinchot G (1947) Breaking new ground. Harcourt, Brace, New York

Prince HC (1997) Wetlands of the American midwest: a historical geography of changing attitudes. University of Chicago Press, Chicago

Pyne SJ (1982) Fire in America: a cultural history of wildland and rural fire. Princeton University Press, Princeton NJ

Reisner M (1987) Cadillac desert: the American West and its disappearing water. Penguin Books, New York

Robin L (1997) Ecology: a science of empire? University of Washington Press, Seattle

Rowley WD (1985) US forest service grazing and rangelands. Texas A&M University Press, College Station TX

Sampson AW (1919) Plant succession in relation to range management. USDA Bulletin No. 791. Government Printing Office, Washington, D.C

Samson FB, Knopf FL, Ostlie WR (1998) Grasslands. In: Status and trends of the nation's biological resources, vol 2. US Geological Survey, Reston, VA

Sayre N (1999) The cattle boom in southern Arizona: towards a critical political ecology. J Southwest 41:239–271. https://www.jstor.org/stable/40170135

Sayre NF (2005) Working wilderness: the Malpai Borderlands Group and the future of the western range. Rio Nuevo Publishers, Tucson AZ

Sayre NF (2017) The politics of scale: a history of rangeland science. The University of Chicago Press, Chicago

Sayre NF (2018) Race, nature, nation, and property in the origins of range science. In: Lave R, Biermann C, Lane SN (eds) The Palgrave handbook of critical physical geography. Palgrave Macmillan, Cham Switzerland. https://doi.org/10.1007/978-3-319-71461-5_16

Sayre NF, McAllister RRJ, Bestelmeyer BT et al (2013) Earth stewardship of rangelands: coping with ecological, economic, and political marginality. Front Ecol Environ 11:348–354. https://doi.org/10.1890/120333

Scott JM, Davis FW, McGhie RG et al (2001) Nature reserves: do they capture the full range of America's biological diversity? Ecol Appl 11:999–1007. https://doi.org/10.1890/1051-0761

Secretary of Agriculture (1936) The western range. Government Printing Office, Washington DC

Seed P (1995) Ceremonies of possession in Europe's conquest of the New World, 1492–1640. Cambridge University Press, Cambridge UK

Sherow JE (2007) Grasslands of the United States: an environmental history. ABC-CLIO, Santa Barbara CA

Skaggs JM (1986) Prime cut: livestock raising and meatpacking in the United States, 1607–1983. Texas A&M University Press, College Station TX

Sluyter A (2012) Black ranching frontiers: African cattle herders of the Atlantic world, 1500–1900. Yale University Press, New Haven CT

Specht J (2019) Red meat republic: a hoof-to-table history of how beef changed America. Princeton University Press, Princeton, NJ

Stanley K (1994) Industrial and labor market transformation in the U.S. meatpacking industry. In: McMichael P (ed) The global restructuring of agro-food systems. Cornell University Press, Ithaca NY

Starrs PF (2000) Let the cowboy ride: cattle ranching in the American West. Johns Hopkins University Press, Baltimore MD

Steen HK (1977) The U.S. Forest Service: a history. University of Washington Press, Seattle WA

Stegner W (1954) Beyond the hundredth meridian: John Wesley Powell and the second opening of the West. Houghton Mifflin, Boston

Stewart OC, Lewis HT, Anderson K (2002) Forgotten fires: Native Americans and the transient wilderness. University of Oklahoma Press, Norman OK

Stoddart LA (1945) Range land of America and some research on its management. Utah State Agricultural College, Logan UT

Tobey RC (1981) Saving the prairies: the life cycle of the founding school of American plant ecology, 1895–1955. University of California Press, Berkeley CA

Torell LA, Rimbey NR, Harris L (2004) Current issues in rangeland resource economics. Utah Agricultural Experiment Station Research Report 190. Utah State University, Logan, UT

Vale TR (2002) Fire, native peoples, and the natural landscape. Island Press, Washington, DC

Virtue GO (1920) The meat-packing investigation. Quart J Econ 34:626–685. https://doi.org/10.2307/1885160

Voigt W (1976) Public grazing lands: use and misuse by industry and government. Rutgers University Press, New Brunswick, NJ

Wagoner JJ (1952) History of the cattle industry in southern Arizona, 1540–1940. University of Arizona, Tucson AZ

Walker RA (2001) California's golden road to riches: natural resources and regional capitalism, 1848–1940. Ann Assoc Am Geogr 91:167–199. https://doi.org/10.1111/0004-5608.00238

Webb WP (1931) The Great Plains. Grosset & Dunlap, New York

Weisiger ML (2009) Dreaming of sheep in Navajo country. University of Washington Press, Seattle WA

White R (1991) "It's your misfortune and none of my own": a history of the American West. University of Oklahoma Press, Norman OK

White R (1994) Animals and enterprise. In: Milner CA, O'Connor CA, Sandweiss MA (eds) The Oxford history of the American West. Oxford University Press, New York

Wildeman G, Brock JH (2000) Grazing in the southwest: history of land use and grazing since 1540. In: Jemison R, Raish C (eds) Livestock management in the American Southwest: ecology, society, and economics. Elsevier Science, Amsterdam and New York

Williamson JA, King C, Britton AT, Donaldson T, Powell JW (1880) Report of the public lands commission. US Government Printing Office, Washington DC

Wolf ER (1982) Europe and the people without history. University of California Press, Berkeley CA

Worster D (1979) Dust Bowl: the southern plains in the 1930s. Oxford University Press, New York

Worster D (1985) Rivers of empire: water, aridity, and the growth of the American West. Pantheon Books, New York

Worster D (1992) Under western skies: nature and history in the American West. Oxford University Press, New York

Chapter 4
Western Rangeland Livestock Production Systems and Grazing Management

Timothy DelCurto, Samuel A. Wyffels, Martin Vavra, Michael J. Wisdom, and Christian J. Posbergh

Abstract Rangeland wildlife ecology and conservation is strongly influenced by domestic livestock systems. Domestic livestock production on rangelands in North America is dominated by ruminant livestock, with beef cattle being the largest industry. Rangeland ruminant livestock production systems are unique in that land/animal managers develop production systems that attempt to optimize the use of limited-nutrition forage bases. This involves the strategic selection of calving/lambing dates to coincide with forage resources and labor limitations. Likewise, the species, breed, and age of animal is selected to be productive in sometimes suboptimal nutrition and environmental conditions. In addition, the role of this industry in the conservation and enhancement of wildlife diversity and ecosystem services is important now and paramount in future management goals. Grazing systems that are unique to the needs of ecosystems are designed to enhance soils, vegetation, and wildlife diversity. In addition, understanding how wild and domestic animals utilize landscapes of varying topography is an ongoing area of research. Continued investigations into how animals use landscapes, grazing distribution/behavior, botanical composition of diets, and dietary strategies will be important in designing management approaches for all animals that are dependent on rangeland resources. The paradigm of sustainable management of livestock systems needs to view herbivory as a tool to manage vegetation for optimal biological integrity and resiliency. Only by the optimization of biological processes within plant communities on rangelands, will managers create systems that benefit both livestock and wildlife.

Keywords Beef production · Domestic sheep production · Grazing behavior · Grazing systems · Livestock-wildlife interactions

T. DelCurto (✉) · S. A. Wyffels · C. J. Posbergh
Department of Animal and Range Science, Montana State University, Bozeman, MT 59717-2900, USA
e-mail: timothy.delcurto@montana.edu

M. Vavra · M. J. Wisdom
Pacific Northwest Research Station Starkey Experimental Forest, USDA Forest Service, La Grande, OR 59850, USA

© The Author(s) 2023
L. B. McNew et al. (eds.), *Rangeland Wildlife Ecology and Conservation*,
https://doi.org/10.1007/978-3-031-34037-6_4

4.1 Western Forage-Based Livestock Production Systems

Livestock grazing is the dominant land use in areas of the western U.S. that are not suited for farming. With the exception of the Great Plains, a significant portion of the western U.S. is characterized by high elevation rangelands exceeding 1000 m (Fig. 4.1a). The Rocky Mountains are a key feature of many western states, and the associated mountain plateaus provide important summer forage resources for the domestic sheep and beef cattle industries. The region's geological features are often characterized by shallow/rocky soils, rugged terrain, and steep slopes. Because of the dominance of high elevation regions throughout the western U.S., many areas have limited growing seasons, with the relative length of growing periods being dependent on adjacent topography, climatic patterns, and elevation.

In addition, most of the western U.S. is characterized by arid and semi-arid environments with precipitation zones of less than 50 cm (20 inches; Fig. 4.1b). Therefore, western livestock producers typically operate in areas with growing seasons of less than 120 days and precipitation patterns that limit native and introduced forage production. The limited precipitation and highly variable patterns of rain/snow events often lead to seasonal shortages of forage and hay for livestock production. Furthermore, short growing seasons and irregular precipitation patterns lead to forage resources that are often limited in nutritional quality and quantity. Therefore, many livestock producers need to consider supplemental inputs to meet their animals' nutritional needs, although the need for supplemental inputs may vary from year to year (DelCurto et al. 2000).

The western beef industry is very extensive in its land use, with optimal production being a function of the resources on each ranching unit and management's success in matching the type of cow and production expectations to the available resources (Putman and DelCurto 2020). Successful beef producers are not necessarily the ones who wean the heaviest calves, obtain 95% conception, or provide the most optimal winter nutrition. Instead, successful producers demonstrate economic viability despite the multiple economic, environmental, and social pressures on the industry. Management practices that promote the ecological, economic and social sustainability of livestock production are paramount for the survival of the western livestock industry and rural communities that are dependent on their success.

4.1.1 Kinds and Classes of Livestock

The U.S. beef cattle industry. The United States is the world's largest and most efficient beef cattle producer (pounds of beef per year). Over the past decade, the U.S. consistently produced between 24 and 27 billion pounds of beef per year, despite severe droughts that have plagued the southwest and western regions of the country during this time. The closest world competitor is Brazil, but the type of cattle differ dramatically in respect to age, breed composition, and, as a result, quality. Simply

a Topography

b Annual Precipitation

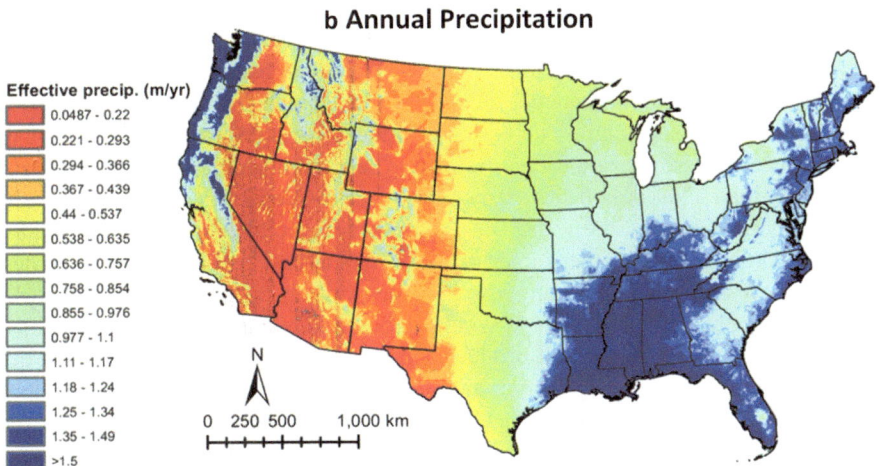

Effective precip. (m/yr)

- 0.0487 - 0.22
- 0.221 - 0.293
- 0.294 - 0.366
- 0.367 - 0.439
- 0.44 - 0.537
- 0.538 - 0.635
- 0.636 - 0.757
- 0.758 - 0.854
- 0.855 - 0.976
- 0.977 - 1.1
- 1.11 - 1.17
- 1.18 - 1.24
- 1.25 - 1.34
- 1.35 - 1.49
- >1.5

N

0   250 500   1,000 km

Fig. 4.1 Topography (**a**) and rainfall (**b**) of the United States (U.S.). Color bands denote the regional elevation ranges and annual precipitation amounts. The western U.S. is dominated by areas that exceed 1000 m in elevation and less than 50 cm of rainfall

put, the United States does not have significant competition from other countries with respect to production capabilities or product quality.

The Western Beef Industry. For the purpose of this chapter, the authors discuss regions of the United States with substantial rangeland areas and ecotypes, which include the 11 western states and the Great Plains states of Texas, Oklahoma, Kansas,

Nebraska, South Dakota and North Dakota (see Chap. 2). These 17 western-most states represent approximately 61% of the U.S. beef cow inventory, with approximately 19.2 million beef cows (Table 4.1). Texas is the clear leader in beef cow/calf production with approximately 4.57 million beef cows (14% of the U.S. Herd) with cash receipts for cow/calf sales generating 4.6 billion dollars annually (USDA-NASS 2019). In addition, the six Great Plains states and Montana represent almost 46% of the U.S. cow herd with all seven states in the top 10 of cow/calf production. Beef cattle productivity of this region is due in large part to the productivity of the rangeland forages reflected in the tallgrass, mixedgrass and shortgrass rangeland ecosystems. These rangelands are relatively low in elevation with continental weather patterns that yield high amounts of precipitation with most coming from April to October. In turn, this type of precipitation amount and distribution allow for greater forage production and forage quality. These rangelands often have both warm season and cool season forages, which can be strategically used to expand the window of adequate nutrition for cow/calf production particularly during lactation.

The rangeland ecotypes of the Great Plains vary in species composition and forage production as a function of precipitation (Fig. 4.1b; Table 4.1). Generally, moving from east to west, there is a transition from tallgrass prairie to mixed grass to shortgrass prairie over a range of precipitation that ranges from 100 to 25 cm (see Chap. 2). In addition, most regions see a gradient of predominantly warm season grasses to cool season grasses with corresponding changes in overall production from east to west. The differences in vegetation and productivity result in differing beef cattle management/production strategies across the Great Plains Region. In the Tallgrass Prairie Region, a substantial stocker cattle industry exists to capture value of the predominantly warm season forages that yield substantial gains with yearlings from May thru July. This region has tremendous forage production but forage quality limits production when the warm season grasses reach advanced stages of phenological maturity. The northern and southern mixedgrass prairie is more balanced with respect to warm and cool season grasses, which results in lower rangeland productivity but greater nutritional windows of adequate nutrition. As a result, this region is dominated by cow/calf production because of the opportunity to meet beef cattle requirements during lactation and, perhaps, run yearling animals on the late spring and summer forage base. The Great Plains, which include the tallgrass, mixedgrass, and shortgrass prairies represent rangeland vegetation that has coevolved with greater herbivory (specifically American bison [*Bison bison*] and Rocky Mountain elk [*Cervus canadensis*]) than other regions of the West. As a result, these areas are generally more resilient in respect to impacts of herbivory on rangeland vegetation. Cow/calf production systems throughout the 17 western-most states often reflect the forage resources of the region that often focuses on optimizing the use of forage resources, minimizing the needs for supplemental inputs, while optimizing beef cattle production and sustainability of the forage resources.

The 10 western states (excluding Montana) are home to 15% of the U.S. beef cattle herd (4.75 million cows) yet over 40% of the U.S. land area. These states have the greatest amount of federal lands (tribal lands, USDI National Parks and Bureau of Land Management, and USDA Forest Service) and the ranching areas

Table 4.1 Beef and sheep ranches, animal numbers, and rangeland ecotypes for the 17 western states (excluding Alaska and Hawaii)

State	Beef cow ranches	Beef cows	Sheep ranches	Sheep and lambs	Wool prod. (lbs)	Rangeland ecotypes
Arizona	5560	199,325	7509	177,392	723,394	Intermountain Lower Slopes (Juniper), Intermountain Mixed Conifer, Sonoran Desert, Apache Highlands, Colorado Plateau
California	10,254	682,372	3807	475,291	2,308,728	Sagebrush Steppe, Intermountain Lower Slopes (Juniper), Intermountain Mixed Conifer, Mojave Desert, Sonoran Desert
Colorado	12,407	806,216	1731	414,672	2,156,447	Shortgrass Prairie, Intermountain Lower Slopes (Juniper), Intermountain Mixed Conifer
Idaho	8149	497,984	1447	248,289	1,642,897	Sagebrush Steppe, Intermountain Lower Slopes (Juniper), Intermountain Mixed Conifer, Columbia Plateau
Kansas	23,682	1,499,843	1234	73,526	231,604	Southern Mixedgrass Prairie, Tallgrass Prairie, Shortgrass Prairie
Montana	10,290	1,487,789	1383	218,544	1,755,718	Northern Mixedgrass Prairie, Sagebrush Steppe, Intermountain Lower Slopes (Juniper), Intermountain Mixed Conifer
Nebraska	17,707	1,896,454	1153	63,043	257,861	Southern Mixedgrass Prairie, Tallgrass Prairie, Shortgrass Prairie, Sand Hills
Nevada	1356	2,48,515	328	76,074	626,919	Great Basin Sagebrush, Sagebrush Steppe, Mojave Desert, Columbia Plateau
New Mexico	8991	482,320	4047	105,896	722,079	Shortgrass Prairie, Great Basin Sagebrush, Intermountain Lower Slopes (Juniper), Intermountain Mixed Conifer, Colorado Plateu

(continued)

Table 4.1 (continued)

State	Beef cow ranches	Beef cows	Sheep ranches	Sheep and lambs	Wool prod. (lbs)	Rangeland ecotypes
North Dakota	8245	984,687	573	70,182	356,975	Northern Mixedgrass Prairie, Tallgrass Prairie, Aspen Parkland
Oklahoma	46,080	2,129,403	2216	69,094	76,542	Southern Mixedgrass Prairie, Tallgrass Prairie, Shortgrass Prairie
Oregon	11,584	538,702	3263	177,646	866,184	Sagebrush Steppe, Great Basin Sagebrush, Intermountain Lower Slopes (Juniper), Intermountain Mixed Conifer, Salt Desert Shrub, Columbia Plateau
South Dakota	12,613	1,927,126	1337	233,006	1,568,559	Northern Mixedgrass Prairie, Tallgrass Prairie, Sand Hills, Black Hills
Texas	134,250	4,572,742	14,672	729,438	2,224,455	Southern Mixedgrass Prairie, Shortgrass Prairie, Tallgrass Prairie, Edwards Plateau and Tamaulipas Thornscrup Savannas
Utah	6508	338,572	1898	300,749	2,415,7675	Great Basin Sagebrush, Sagebrush Steppe, Colorado Plateau, Intermountain Lower Slopes (Juniper), Intermountain Mixed Conifer
Washington	9295	239,154	2281	52,329	287,441	Sagebrush Steppe, Intermountain Lower Slopes (Juniper), Intermountain Mixed Conifer, Columbia Plateau
Wyoming	5035	715,568	859	367,702	2,796,792	Sagebrush Steppe, Northern Mixedgrass Prairie, Intermountain Lower Slopes (Juniper), Intermountain Mixed Conifer
United States	729,046	31,722,039	101,387	5,391,252	25,095,459	

Beef cow numbers reflect producing females and do not include dairy cattle. Data based on the 2017 Census of Agriculture (USDA NASS 2019) are primarily confined to the arid and semi-arid regions. Federal ownership of lands range from 30% in Washington to approximately 87% in Nevada. Perhaps compared to the Midwest, western beef production does not seem that important. However, beef cattle production and hay production for ruminant livestock are cornerstones of the rural economies in the western U.S. as well as Great Plains states (Tanaka et al. 2007).

Beef cattle producers in the western region are faced with many challenges. First, their ranch resources are often limited in forage quality and quantity, both of which are dynamic and dependent on climatic conditions. Thus, western ranch managers often select cattle based on their ability to thrive in environments with limited nutritional resources. Often, western beef producers select cattle with smaller frames, low to moderate milk production, and the ability to be reproductively efficient in a limited nutrition environment. These producers also tend to select calving dates that optimize beef cattle production with available forage resources. As a result, greater than 80% of western beef producers calve in the spring. Ranches that market calves at weaning tend to calve a month or two before the onset of green forage. In contrast, a growing number of producers who retain ownership or keep calves as yearlings are moving calving dates to match the onset of green forage more closely. By calving in April/May, producers try to match the cow's nutritional requirements as closely to the forage resources as possible and minimize supplemental inputs.

Despite efforts to match cow type and production to rangeland environments, most western livestock producers are dependent on supplemental and harvested forage during portions of the year. High elevation rangelands/ranches often have extended periods of snow cover. During these periods, harvested forages are necessary. Ranches that provide feed during the December through March winter period often require a minimum of 2 tons of harvested forage per cow. While a great deal of effort is made to reduce the reliance on harvested forage, most of the alternatives such as stockpiled forage, straw, and other crop residues, are limited by nutritional quality and need substantial inputs to meet the nutritional demands of the cow/calf. Strategic supplementation is essential for these producers and often critical to their success (DelCurto et al. 2000; Kunkle et al. 2000).

The U.S. Sheep Industry. The first permanent U.S. domestic sheep flock was established in Virginia in the early 1600s (Bell 1970). From there, the American sheep industry continued to grow and eventually peaked at an estimated 56 million head in 1945 (USDA 2004). Over the last several decades, the U.S. sheep industry has contracted drastically in size, and in 2021 it was reported that there were approximately 5.17 million total sheep (USDA 2021). However, there appears to be a diminished rate of decline in recent years, suggesting that the observed exponential decay of the U.S. sheep inventory may be close to its lower asymptote.

Most recently, the U.S. was ranked 50th in the world in total sheep inventory, substantially smaller than China (1st; 163 million head), India (2nd; 74.2 million head), and Australia (3rd; 65.7 million head), to name a few of the world leaders (FAO 2017). Not surprisingly, the current number of sheep in the U.S. is small compared to swine (71.7 million head; USDA 2017b) and beef and dairy cattle (103 million head; USDA 2017c). However, the number of sheep operations across the nation (101,387) ranks only behind beef cattle (729,046), as more Americans raise sheep than dairy cattle (54,599) and hogs (66,439; USDA 2019).

Traditionally, the bulk of the U.S. sheep population has been located in the 24 states west of the Mississippi River. Today, an estimated 80% of the country's sheep are found in the West, with Texas (1st; 730,000 head), California (2nd; 555,000 head), and Colorado (3rd; 445,000 head) being among the leaders (USDA 2021). Furthermore, the plurality of the total U.S. sheep inventory (43%) is found on large operations (> 1000 head; USDA 2012), which are more typical of the western sheep industry.

Many eastern states (MI, NY, OH, PA, VA, WI, and "Other States") exhibited positive growth in their sheep inventory from 2001 to 2007, whereas the inventory in all but two western states (OK and MO) continued to decline during this period (NRC 2008). Despite eastern states being home to only 19% of the total U.S. inventory in 2012, they contained 39% of the nation's sheep producers (USDA 2012). Therefore, recent trends suggest that the makeup of the U.S. sheep industry is shifting toward smaller flocks. For example, the proportion of operations with < 100 head, 100 to 999 head, and > 1000 head in 1974 was 77%, 20%, and 3%, respectively (NRC 2008), contrasted with 93%, 6%, and 1%, respectively, in 2017 (USDA 2019).

Production characteristics. Sheep have been bred to produce one or more of three products: wool, meat, and milk. The majority of the world's dairy sheep are located in the Mediterranean countries of southern Europe and northern Africa (FAO 2017). Sheep milk is typically processed into high-quality cheeses, and Roquefort, Pecorino Romano, and Manchego styles can often be found in U.S. urban and suburban super-markets. While the U.S. dairy sheep industry has been growing over the last several decades, it is still relatively small (Thomas et al. 2014). Therefore, the two major U.S. sheep commodities are wool and lamb.

Advances in textile technologies allow today's wool products to range in appli-cation from next-to-skin to protective outerwear suitable in all temperatures. Wool's durability and odor resilience are ideal for both the working class and outdoor enthusi-asts. Additionally, its fire-retardant properties are capable of protecting U.S. military men and women where synthetic fibers (e.g., nylon, polyester, etc.) fail. Throughout the country and world, the western states are known for producing a high-quality wool clip (i.e., total quantity of wool shorn in an area for one year). Colorado marketed the most wool in 2020 (1.14 million kg [2.5 million lbs]), followed by Utah (1.09 million kg [2.1 million lbs]) and California (0.90 million kg [2.0 million lbs]). The heaviest average individual fleece weights came from sheep in Nevada (4.2 kg [9.2 lbs]), Montana (4.1 kg [8.9 lbs]), and Utah (4.1 kg [8.9 lbs]; USDA NASS 2021).

Sheep are shorn once per year, generally in winter or spring before pregnant ewes give birth. The highest average returns from wool on a unit basis in 2020 were garnered in Washington ($2.50/lb), Wyoming ($2.35/lb), Nevada ($2.30/lb), and Montana ($2.20 lb; USDA NASS 2017a). Therefore, the average revenue from a Montana fleece was over $19 per head in 2020. However, receipts from the sale of wool represented just 5 to 13% of the total revenue for the average U.S. sheep producer from 2010 to 2015 (LMIC 2016). Though the sale of wool is a timely income source for the extensively managed operations prevalent in the western states, the success of most sheep operations in the U.S. hinges on the value of their lamb crop.

An estimated 50% of lambs are born in April and May on operations with 500 or more breeding ewes (USDA APHIS 2014). Sheep producers benefit from the ewe's ability to give birth to and raise multiple lambs at a time. The states with the highest lambing percentage in 2020 were Iowa (141%), Minnesota (138%), and South Dakota (132%; USDA NASS 2021). As with most commodities in agriculture, if the sheep producer wants more output (e.g., a greater lambing percentage), they need to supply more input (e.g., better genetics and increased nutrition). Therefore, the highest lambing percentages in the U.S. tend to come from Midwestern states where harvested feeds are more abundant and less expensive.

The average age and weight of lambs at weaning were 4.5 months and 33.8 kg (74.5 lbs), respectively, on western and central U.S. sheep operations in 2010. Additionally, these operations marketed their non-replacement lambs shortly after weaning at an average age and weight of 5.7 months and 42.8 kg (94.3 lbs), respectively (USDA APHIS 2012). From there, most lambs are placed in a dry lot and fed a high concentrate diet until they are finished, which was at an average live weight of 61.2 kg (135 lbs) in recent years (NRC 2008). California and Colorado have traditionally been the largest lamb feeding states, with an estimated 250,000 and 235,000 lambs on feed, respectively, in 2020 (USDA NASS 2021). Like the U.S. sheep inventory, the average per capita consumption of lamb in the U.S. has continued to decline and was below 1 pound per person in 2015 but has increased to 1 pound as of 2020. This is especially concerning considering Americans consumed an average of 34.1 kg (75.1 lbs) of poultry, 23.3 kg (51.4 lbs) of beef, and 21 kg (46.3 lbs) of pork available per person in 2015 (USDA ERS 2017). Efforts to promote American lamb, especially within the younger, more diverse U.S. population, have increased in recent years.

There are many reasons, both anecdotal and substantiated, for the contraction of the U.S. sheep industry. Throughout most of the history of domestic and international sheep production, wool was the major product, and sheep meat was, more or less, a byproduct (USDA ERS 2004). With technological advances in the 1960s, less expensive manmade fibers began to outcompete wool in the textiles market. Since then, sheep-producing nations have mostly switched their emphasis to improving lamb production while maintaining a quality wool clip. Although the U.S. is the largest meat and poultry consuming nation globally, attempts to promote lamb and increase its consumption have largely been unsuccessful. Despite these realities, sheep production in the U.S. can still be quite profitable. For example, it was estimated that the typical Wyoming region sheep operation had an average profitability of $28.11 per ewe per year from 2010 to 2015 (LMIC 2016), the equivalent of a per cow profitability of $140.56 per year.

4.1.2 Public Land Ownership in the Western U.S.

Western livestock industries are also dependent on the continued use of public lands for livestock grazing. Most ranches have a mosaic of pastures and rangelands (both private and public) that provide the resources for a 12-month forage resource base. Approximately 20% of the animal unit months for western livestock production are derived from public lands. While that may not seem like a large amount, when one considers that 60% of beef production is derived from ranches of 100 head of producing cows or more, approximately 1/3 of the forages for these ranches, on average, come from public lands (four months of grazing). For many areas of the West, such as the Southwest and lower elevation rangelands in the Great Basin, many ranches graze public lands for the majority of the calendar year. The greatest challenge related to public land management is managing these lands for multiple values and uses. Other values include recreation (hunting, camping, hiking, and fishing), conservation for wildlife, and the overriding desire to preserve lands for future generations.

Due to the arid to semi-arid nature of rangelands in the western U.S., these lands are often more sensitive to disturbance or overuse and, as a result, are more likely to be damaged by improper livestock use. Currently, livestock producers must be vigilant regarding public land stewardship and respect other public land values or services. Current concerns often relate to threatened and endangered species, riparian area structure and function, and differences of opinion with the public with respect to other values and ecosystem services. Other significant challenges include the fate of ranches with significant esthetic and wildlife recreational value. Numerous ranches that have changed ownership in the recent past have been purchased by investment groups for their investment and/or recreation value rather than income from beef cattle production. For these ranches, recreation and esthetic values often are prioritized over beef production goals. In addition, ranches located in desirable vacation locations (ski areas, near national parks, close to urban areas) often have property tax increases that challenge the profitability of the ranch. Many producers in these types of locations take advantage of conservation easements because of shared values and the lowering of property taxes.

Many western land grant universities and associated USDA–ARS research locations are devoting substantial resources to evaluate grazing as a tool to improve public and private land vegetation diversity and structure (Bailey et al. 2019). Specifically, studies with various species of livestock have suggested that targeted grazing could be a tool to reduce noxious weeds and, in turn, encourage more desirable vegetation. Likewise, livestock grazing is being used to control fuels to reduce the occurrence and severity of wildfires on public lands (Davies et al. 2010; Bailey et al. 2019). Perhaps, the future of public land grazing will focus more on the use of domestic herbivores to manage vegetation for more desirable outcomes.

4.2 Great Plains and Western Rangeland Livestock Management Techniques and Systems

Livestock managers need to provide forage resources for their animals over the 12-month production cycle (Raleigh 1970; Vavra and Raleigh 1976). Most livestock operations in the Great Plains and western U.S. utilize spring-time calving and lambing as the basis of their production cycle. For many beef cattle producers, calving one to two months before the onset of green forage is preferred because it matches cow nutrient requirements with forage resources and optimizes the weaning weights of beef calves. Some producers, however, have moved calving dates to coincide with the onset of green forage to minimize the need for nutritional inputs and closely match cow requirements with forage resources. These ranches will often retain weaned heifers/steers to capture body weight gain during the backgrounding or yearling stage of production.

Great Plains production systems are designed to optimize the use of forage resources. Most ranches implement spring calving with the greatest period of nutrient demand during lactation, coinciding with the onset of spring forage. Spring calving dates will vary from February/March to May, depending in large part on if the livestock/ranch manager plans to market calves as weaned calves or retain ownership and market as yearlings. These production systems also vary with precipitation amounts and distribution patterns, as well as vegetation characteristics. Because of the lower elevations and continental weather patterns, these regions can usually be grazed for a greater proportion of the year and, as a result, these regions have less reliance on harvested forages and hays. However, they are often challenged during the winter period with Arctic storm systems that can dramatically influence production systems in the region. In fact, weather system extremes often cause substantial problems for producers in this region and represents significant economic losses.

Great Basin, Intermountain, and Northern Mixed-grass native rangelands are often grazed in late-spring, summer, and early fall, then livestock are typically brought back to their base units before the onset of winter conditions. This allows managers to market calves during the fall/early winter period, provide supplemental feed for the winter, and manage calving/lambing at or near the ranch's headquarters. Predators are an increasing problem for western livestock producers (see Chap. 24: Large Carnivores). Predation by wolves (*Canis lupus*) and grizzly bears (*Ursus arctos horribilis*) on cattle has expanded in the Intermountain Region, which has increased the need to manage calving and the early post-partum period to minimize risks related to predation. Likewise, the rangeland sheep industry struggles with predation due to large raptors (primarily golden eagles [*Aquila chrysaetos*]), black bears (*U. americanus*) coyotes (*C. latrans*), mountain lions (*Puma concolor*), and wolves. The timing and location of lambing are a challenge for these producers, with an increasing need for security for the animals during the most vulnerable part of their production cycle. The use of guard dogs and other security measures has increased dramatically in recent years (Mosley et al. 2020).

One of the significant management challenges for Great Plains and Western livestock producers is selecting animals that optimize production in limited nutrition environments. The ideal ewe/cow can convert forage resources to pounds of lamb/calf weaned with minimal supplemental nutrients and be reproductively successful. In addition, animals that fit the physical requirements of extensive remote rangeland sites are often preferred. Low-to-moderate milk production and moderate body sizes are often preferred with beef cattle because of the rugged terrain and limited nutrition. In the Great Plains, producers often have less challenging terrain and increased forage production, however, the quality of forage with significant warm season forage component is of lower quality compared to cool season forages, which necessitates the need for supplemental inputs. Strategic supplementation is critical in utilizing low-quality high-fiber forages during the fall and winter grazing period (DelCurto et al. 2000). Likewise, sheep breeds with greater flocking instincts and wool traits are preferred over larger carcass-based breed types. Crossbreeding to create heterosis and, as a result, increased vigor is beneficial to both livestock industries. For both sheep and cattle industries, selection for good feet and legs as well as other physical attributes for an animal that can traverse rugged terrain over several years without physical breakdown are important selection criteria.

Finally, most of the Great Plains and western livestock industry is moving to 12-month management systems that reduce the need and reliance on harvested forages (hays; Putman and DelCurto 2020). The cost of equipment and labor associated with haying are major challenges for beef producers in these regions. Moving to management systems that extend the grazing season into the late fall and winter period reduces the need for additional labor, equipment, and reliance on fossil fuels in the production system, which align with current and future trends in animal agriculture.

4.3   Wild and Domestic Ruminant Ecology

Ruminant animals, both wild and domestic, have co-evolved with grasslands for millions of years (Van Soest 1994). As a result, ruminants have an important function in grassland ecosystems. Most wild populations of ruminants occupy unique ecosystem niches that co-exist with other ruminants, allowing for sustainable maintenance of wildlife populations and the grassland ecosystems that are essential for their survival. Likewise, understanding how domestic ruminants co-exist with wild ruminants on rangelands is important for long-term management of both wild and domestic animal populations (McNaughton 1985).

Ruminant animals vary in respect to their ecological niche, and, as a result, so does their dietary selection strategy and associated digestive physiology (Fig. 4.2; Cheeke and Dierenfeld 2010). Generalist grazers are larger in body size (specifically ruminal-reticular size) relative to intermediate and selective grazers. In addition, grazers such as bison (*Bison bison*) and cattle have large muzzles, dentition adapted to optimize bite size, and large ruminal-reticular size or volume to accommodate larger fill and greater ruminal fermentation retention time (Cheeke and Dierenfeld

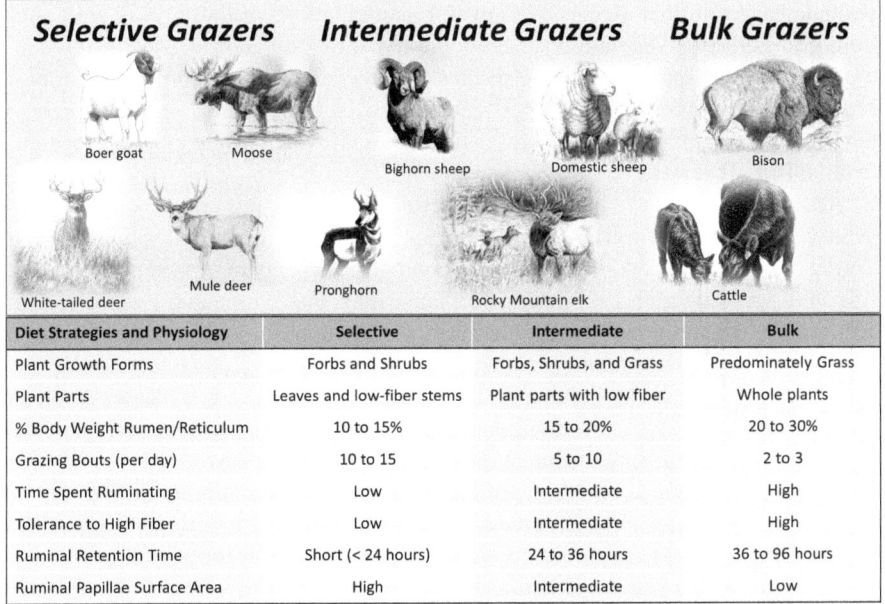

Fig. 4.2 Ruminant dietary preference and physiologic attributes. Ruminants have co-evolved with grasslands to occupy unique ecological niches (adapted from Cheeke and Dierenfeld 2010)

2010). Furthermore, larger grazers primarily exhibit diurnal grazing patterns with two to three grazing bouts per day and have a greater reliance on rumination to assist with the breakdown of high-fiber, low-quality forage resources (Van Soest 1994). These animals have a general preference for grass and "grasslike" species such as sedges even when available forbs and shrubs are of higher nutrient composition (Clark et al. 2013; Damiran et al. 2019). Simply put, these animals have evolved to consume high-fiber, low-quality forages and their ecological niche relates to the optimal use of vegetation with moderate to high levels of cellulose (Van Soest 1994).

In contrasts, selective and intermediate ruminants require higher quality diets and, as a result, have feeding strategies that only use grasses when young and succulent (Fig. 4.2). Rumen size relative to overall body size is smaller and their digestive physiology is adapted to more specialized diets demonstrated by smaller muzzles, increased frequency of grazing bouts, reduced reliance on rumination, and the ability to optimize the use of moderate to high quality vegetation (Van Soest 1994; Cheeke and Dierenfeld 2010). In a study evaluating mule deer (*Odocoileus hemionus*), Rocky Mountain elk, and cattle diets; Damiran and co-workers (2019) found that cattle primarily selected for grasses whereas mule deer and elk diets focused on forbs and shrubs when late summer grazing in mixed-conifer forest understories. In addition, estimates of dietary quality suggested that cattle selected the lowest quality diet but with greater intake rates (grams per minute) than Rocky Mountain elk (intermediate) and mule deer (selective) diets. The difference in dietary strategies suggest that these animals have minimal dietary overlap and manipulations of these ruminant populations can influence vegetation successional dynamics. In addition, the difference in dietary strategies also relate to the species tolerance or ability of ruminal microbes to break down fiber (cellulose). Understanding the distribution and use of all grazers on the landscape is important in designing management systems that optimize vegetation and wildlife diversity.

All ungulates graze selectively, so across a given landscape, there are plants grazed and those ungrazed. A large number of studies show persistent heavy grazing during the growing season decreases the competitive ability of grazed plants (Augustine and McNaughton 1998). Grazed plants may lose vigor and even die, either scenario giving ungrazed plants the competitive opportunity to increase. Consequences include a decline in palatable plant species production (Hobbs 1996) and a shift in plant community composition to unpalatable or invasive plants (Augustine and McNaughton 1998). A simultaneous decline in animal production may occur due to the lower nutritional value of the forage crop. Thus, season of use plays a vital role in the maintenance or decline of forage plants and plant communities.

Timing, duration, and intensity of grazing all interact to impact the health of forage plants. In the Intermountain West, continuous season-long grazing will negatively affect forage plants (Milchunas and Lauenroth 1993), given summer drought limits the opportunity for regrowth. However, when grazing duration is limited, there is often the opportunity for regrowth (Ganskopp et al. 2007). The amount of regrowth is probably dependent on soil moisture and plant species. However, Ganskopp et al. (2007) found no relationship between soil moisture and regrowth but did notice some species regrew better than others did. Grazing northern Great Basin grasses during vegetative, boot stage and flowering caused respective declines in fall standing crop of 34%, 42%, 58% in one year and 34%, 54%, and 100% reductions the next (Ganskopp et al. 2007). Ganskopp et al. (2004) reported similar results with boot stage-grazed Idaho fescue (*Festuca idahoensis*), bluebunch wheatgrass (*Psuedorogeneria spicatum*), and bottlebrush squirreltail (*Elymus elymoides*) plants. The detrimental effects of repeated boot-stage grazing of cool-season grasses are well documented. From vegetative through flowering stages, a decline in forage quality occurs, but the grazing animal's nutritional requirements are met in most cases.

Introduced species with a tolerance for defoliation may be grazed during the growing season on native ranges deferred until after seed ripening. Crested wheatgrass (*Agropyron desertorum*), common in the Intermountain West, provides such an option. During the growing season, nutritional quality and palatability are adequate (Cruz and Ganskopp 1998) for livestock production. Care must be taken not to underutilize crested wheatgrass as the development of "wolfy plants" and decreased grazing efficiency of the affected pasture could occur (Ganskopp et al. 1992; Romo 1994). In a study conducted on the Zumwalt Prairie, a remnant of the Palouse Prairie, Wyffels and DelCurto (2020) reported the bunchgrass prairie contained 13% non-native species (Kentucky bluegrass [*Poa pratensis*], intermediate wheatgrass [*Thinopyrum intermedium*], and brome species (*Bromus* spp.)) and these species accounted for 20–50% of the botanical composition of cattle diets during the late spring early summer grazing period. The results of these studies suggest that non-native species can be used to reduce herbivory of native bunchgrasses, which would be particularly beneficial during the growing season when native bunchgrasses are most vulnerable to the negative effects of defoliation.

Once the forage plant has completed its life cycle for the year and seed has been produced, grazing has little effect on the plant's physiological well-being unless it is overly excessive (Holechek et al. 1998). However, plant residue during dormancy plays a critical role in protecting the plant. Hyder (1953) found maintaining 200 kg/ha of residual forage maintained or improved range conditions on most sites in southeastern Oregon. Late summer, fall, and winter grazing may be practiced on ranges requiring an improvement in vigor. Unfortunately, the forage's nutritional quality by late summer is marginal and may decline further as fall progresses. This, in turn, will result in the livestock/range manager providing supplemental inputs for optimal livestock production.

Grazing impacts may also influence seed production, establishment, and survival of young plants and longevity of older plants. Miller et al. (1994) summarized these impacts. Heavy grazing generally decreases seed production. Young plants, one to two years old, are the most susceptible to mortality caused by grazing. The longevity of plants is variable and dependent on species and grazing history. Heavy grazing effects have generally been compared to no grazing. The impacts of light to moderate grazing have not been adequately described. Given proper stocking rate control and a grazing system that provides growing season rest, these impacts can be mitigated.

Grazing bunchgrass communities during dormancy creates additional challenges for land managers and ranch managers. Grazing distribution can become problematic when riparian areas are green and upland communities are dormant (Parsons et al. 2003; DelCurto et al. 2005). Use of pastures without sensitive riparian areas is encouraged (also deferment or rest), as well as the use of management tools to move cattle away from riparian communities. These tools may include "off-stream" water (Porath et al. 2002), herding, and strategic use of supplements (Bailey et al. 2001; Tanaka et al. 2007). More recently, research efforts have focused on new technologies utilizing electronic animal identification (EID), global positioning systems and activity monitors to observe and manage livestock distribution on extensive rangeland ecosystems (DelCurto and Olson 2010; Bailey et al. 2021). Fenceless livestock systems using GPS technology have been demonstrated to be effective in managing cattle grazing in extensive environments (Ranches et al. 2021; Boyd et al. 2022). In addition, these systems have also been effective in protecting areas recently burned from grazing without the need for temporary fencing (Boyd et al. 2022).

Annual grasses, with cheatgrass (*Bromus tectorum*) being the most notable, are usually managed to disrupt physiological processes and prevent seed production or even kill the plants. Cheatgrass provides adequate nutrition after germination in the fall and early spring (Cook and Harris 1952), which coincides with the species sensitivity to grazing effects.

4.4 Grazing Systems and Season of Use

When evaluating the impact of domestic livestock on pastures and grasslands, one of the most important aspects is the management of grazing which is commonly referred to as "grazing systems." The actual grazing management on a given ranch often incorporates multiple types of grazing system approaches. In addition, grazing systems are generally specific to a geographic area with unique vegetation communities that, in turn, have unique needs with respect to the maintenance or improvement of that plant community (Table 4.2). For more detailed information on grazing systems, outstanding reviews have been provided by numerous authors (Holechek et al. 1998; Fuhlendorf and Engle 2001; Heitschmidt and Taylor 2003; Kothmann 2009; Briske et al. 2011; Holechek et al. 2020).

Grazing systems have been initiated with the explicit purpose of manipulating the season of use so that periodic rest or deferment occurs during the growing season. The main goal is to allow forage plants to periodically complete their annual growth cycle and replenish nutrient reserves without being defoliated. It is generally unreasonable to expect stocking rates to be increased when moving from a season-long grazing pattern to a grazing system (Holechek et al. 1998). Typical systems used in the Intermountain West are deferred rotation and rest rotation. In some cases, other specialized systems may be used or the aforementioned modified for a specific goal such as riparian zone restoration. Specialized systems, such as intensive early stocking, have been developed with a focus of optimizing stocker cattle weight gains per acre (ha) in more productive rangelands such as the Tallgrass Prairie (Smith and Owensby 1978; Owensby and Auen 2018).

For the Great Plains Region, grazing system recommendations often differ from the more arid regions in the western U.S. Specifically, continuous grazing is often recommended for the Tallgrass Prairie and regions of the Mixed-grass Prairie (Fuhlendorf and Engle 2001; Briske et al. 2011). In general, these rangeland ecotypes usually are more homogeneous with less vegetation and topographic diversity on a landscape basis as compared to the more westerly regions of the U.S. Grazing encourages greater heterogeneity of the vegetation and, as a result, may provide more diverse habitat opportunities for wildlife. In addition, the greater precipitation and fire frequency potential for this region result in a more resilient and productive rangeland system.

For many plant species in the Intermountain West, the most critical period for detrimental grazing effects is floral initiation through the development of seed (Holechek et al. 1998). This period is critical because the plant's demand for photosynthetic products is high, and the opportunity for regrowth is low due to declining soil moisture conditions in arid and semi-arid rangeland communities. As a result of repeated grazing at this time, the capacity of forage plants to produce both root and shoot growth the next year may be diminished, especially if the plants are heavily grazed. The development of modern grazing systems incorporates this knowledge of plant physiology, and animal behavior, so physiological damage to forage plants is minimized. Unfortunately, the best time to graze to maximize animal production is when

Table 4.2 Summary of grazing systems utilized on rangelands with common acronyms, defining features, regions of use, unique attributes, as well as references for each system

Grazing system	Defining features	Region of use	Unique attributes	References
Pastoral grazing	Herding, no fences	Common in Africa and Asia. Modified use with western U.S. range sheep industry	Requires a herder and involves daily herding of the animals. Often used when protection from predation is needed	Koocheki and Gliessman (2005) Török et al. (2016)
Continuous season long grazing (CSL)	Single pasture or paddock with long-term grazing by an animal group	Common worldwide and potentially beneficial on homogeneous rangeland areas such as Tallgrass and Mixed-Grass Prairie regions	Simple for animal manager and can create heterogeneous rangeland vegetation and vegetation structure. Allows for animal selection and repeat defoliations	Briske et al. (2011) Fuhlendorf and Engle (2001) Heitschmidt and Taylor (2003) Holechek et al. (1998) Kothmann (2009)
Rotational grazing (R)	Two or more pastures or paddocks. Animals are rotated from pasture to pasture	Common worldwide	Allows for concentrating use (increasing stock density), and the deferment of use when in other pastures. Low to moderate decrease in vegetation selection	Briske et al. (2011) Heitschmidt and Taylor (2003) Holechek et al. (1998) Kothmann (2009)
Deferred rotation grazing (DR)	Two or more pastures or paddocks. Order of pasture grazing is changed each year	Common in western U.S. particularly with USDA Forest Service and Bureau of Land Management (BLM) federal lands	Form of Rotational Grazing. Reduces the likelihood of use at the same time each year. Good for areas with vegetation less tolerant of defoliation	Briske et al. (2011) Heitschmidt and Taylor (2003) Holechek et al. (1998) Kothmann (2009)
Rest rotation grazing (RR)	Two or more pastures or paddocks. One of the pastures is rested each year	Used on arid and semi-arid rangelands of the western US. Used in areas with poor condition rangelands due to annual grass/forb invasions and post-fire management	Form of Rotational Grazing. Usually used on poor condition rangelands that need rest for the reestablishment of perennial grasses	Briske et al. (2011) Heitschmidt and Taylor (2003) Holechek et al. (1998) Kothmann (2009)

(continued)

Table 4.2 (continued)

Grazing system	Defining features	Region of use	Unique attributes	References
Deferred rest rotation grazing (DRR)	Two or more pastures or paddocks. Order of grazing and rest is changed each year	Used on arid and semi-arid rangelands of the western US. Used in areas with poor condition rangelands due to annual grass/forb invasions and post-fire management	Benefits of Deferred and Rest Rotation Systems. Used for poor condition rangelands. Can be used with restoration efforts to decrease annual grasses/forbs or post-fire management	Briske et al. (2011) Heitschmidt and Taylor (2003) Holechek et al. (1998) Kothmann (2009)
High intensity low frequency grazing (HILF)	Multiple pastures/paddocks with high numbers of animals on a pasture for a short duration with long rest between grazing events	Specialized system of grazing that works best in moderate to high precipitation zones or pastures with irrigation. Limited application in arid and semi-arid rangelands in extensive production systems	Allows for pasture/paddock vegetation defoliation and regrowth between cycles provided the plants have adequate moisture. More uniform use of all plant species by grazing animal. Systems allows for enhancing plant succession	Briske et al. (2011) Heitschmidt and Taylor (2003) Holechek et al. (1998) Kothmann (2009)
Short duration (SD)	Similar to HILF but has shorter rest period between grazing bouts on the same pasture	Specialized system of grazing that works best in moderate to high precipitation zones or pastures with irrigation. May also work with long growing seasons and precipitation during the growing season	High stock densities with several repeat defoliations. Herd effect reported to improve range conditions. More uniform use of all plant species within the pasture	Heitschmidt and Taylor (2003) Heitschmidt and Walker (1983) Holechek et al. (1998)
Intensive early stocking (IES)	Refers to the intensive (higher-density, shorter-duration without changing stocking rate) use of rangelands early in the growing season	Used in the tallgrass prairie and mixed-grass prairie of the south-central Great Plains. Often used in conjunction with yearly or seasonal burning events	System designed to optimize yearling cattle gains when the vegetation quality and quantity is optimized for steer/heifer body weight gains	Smith and Owensby (1978) Owensby and Auen (2018)

forage plants are green and growing. Modern grazing systems incorporate use during the growing season in some years to foster animal performance and annual rest or growing season deferment during other years to allow forage plants to maintain vigor and reproduce.

Most of the western U.S. is characterized as having a short grazing history and suffers from a lack of seasonal rainfall, so forage plants are more susceptible to physiological damage from grazing (Milchunas and Lauenroth 1993). Additionally, with season-long grazing in large landscape pastures, animals have preferred grazing areas, and these patches may be heavily impacted by grazing animals while others are underutilized (Teague and Dowhower 2003). These areas typically occur near water and where forage is plentiful. Even under light stocking, these areas will receive excessive use (Holechek et al. 1998; Teague and Dowhower 2003). In contrast, the Great Plains represent vegetation communities that co-evolved with grazing ruminants (bison, elk, deer, pronghorn [*Antilocapra americana*], etc.). As a result, the level of use and impact of herbivory on the plants and plant communities differs from the more arid rangelands of the Intermountain West.

Deferred rotation grazing involves not grazing at least one pasture during the growing season. The simplest form is a two-pasture system where each pasture is deferred during the first half of the grazing season every other year (Holechek et al. 1998). Vegetation response under this system has been slightly better than season-long grazing on bunchgrass ranges (Skovlin et al. 1976). Under rest rotation grazing, one pasture receives a year of nonuse while grazing is distributed among the other pastures in the system (Hormay 1970). For much of the western U.S., a typical rotation system is made up of three or four pastures used during the late spring to early fall period. The rested pasture receives use after the growing season the year following rest. Rest rotation grazing resulted in Idaho fescue's improved vigor compared to that grazed season-long (Ratliff and Reppert 1974). The development of grazing systems must include the critical economic component. Grazing systems generally involve either substantial initial investment in fences and water developments or significant annual expenditures for increased herding of livestock. The benefits of developing grazing systems must be compared to the costs of instituting these systems.

Critical to the success of a rotation grazing system is stocking rate control (Holechek et al. 1998). Depending on the number of pastures in the system, more animals are concentrated in one pasture than if the entire range was used season-long or continuously. However, most rotational grazing systems do not change stocking rate when expressed on a season long basis with the stock density increase being a function of the number of pastures. Increases in stocking rate over season-long levels may not be practical. Holechek et al. (1998) reviewed the literature and reported that stocking rates for livestock in the Intermountain West should be established, resulting in 25–40% utilization of preferred forage species. Failures of rotational grazing systems are usually related to heavy stocking rates (Holechek et al. 1998).

Rangelands dominated by annual grasses like cheatgrass require entirely different grazing systems to ensure maintenance or restoration of the perennial plant community. Mosley and Roselle (2006) provide insight into the design of a targeted grazing system for cheatgrass:

- Targeted grazing can be used to disrupt fine fuel continuity and reduce fuel loads.
- Annual invasive grasses can be suppressed when livestock grazing reduces the production of viable seeds.
- Seedheads of invasive grasses must be removed while the grasses are still green.
- It may be necessary to graze annual grasses two or three times in the spring.
- In mixed stands of annual grasses and perennial plants, livestock should be observed closely to avoid heavy grazing of any desirable perennial plants.
- Livestock perform well on annual grasses in the spring, producing weight gains similar to those from uninfested ranges.
- Targeted grazing can be integrated with prescribed fire, herbicides, and mechanical treatments to improve efficacy of control.
- Applying targeted grazing before artificial seeding can help in restoration efforts.

Targeted grazing systems designed to suppress invasive annual species should be an area of focused research because the threat to ecosystem integrity is great. Targeted grazing differs from traditional grazing management in that the goal of targeted grazing is to apply defoliation and/or trampling to achieve specific vegetation management objectives such as reduction of a noxious/invasive plant species (Bailey et al. 2019). By using specific dietary strategies of grazing ruminants, rangeland managers can exert pressure on individual plants by herbivore defoliation and, as a result, move the vegetation community towards a more desirable plant community composition (Lehnhoff et al. 2019). In addition to targeted grazing of cheatgrass, research has shown promise with targeted grazing on invasive annuals such as medusahead rye (*Taeniatherum caput-medusae*; DiTomaso et al. 2008; Brownsey et al. 2017). In contrast, species such as ventenata (*Ventenata dubia*) have been more challenging because of extremely low palatability regardless of growth stage (McCurdy et al. 2017). Other notable species would include potentially toxic rangeland plants such as larkspur (*Delphinium* spp.) where early sheep use has been shown to decrease the risks with subsequent cattle grazing (Pfister et al. 2010).

As mentioned previously, mixed species grazing may have management applications where multiple herbivore species with divergent dietary strategies may more uniformly use diverse vegetation communities (Walker 1997). Understanding distribution patterns of grazing ruminants as a function of season, weather extremes, and dietary strategies will be important in accounting for the impact of wild and domestic ruminants on rangeland landscapes and vegetation communities. Both domestic livestock and wildlife have been demonstrated to modify riparian vegetation which, in turn, may alter riparian hydrologic function (DelCurto et al. 2005; Averett et al. 2017). Additionally, one of the most important considerations is how ruminants modify plant communities and, in turn, how that influences fire ecology. Riggs and co-workers (2015) suggested that historical herbivory modifies future biomass and fire behavior over time. Specifically, multi-species herbivory lengthens the landscape fire-return interval for most vegetation communities. However, the effects are site-specific, and contingent on future climatic conditions and fire-suppression efforts.

4.5   Ruminant Animal Grazing Behavior

Most of the arid to semi-arid rangeland in the western U.S. is used as extensive pastures with ample opportunity for livestock to freely disperse over areas of diverse topography. Generally, animal use is first influenced by abiotic factors such as distance to water and slope (Coughenour 1991). Other factors are more subtle but important to predict animal distribution on the landscape. Early season use often leads beef cattle to use south-facing aspects or areas with early-maturing annual grasses (DelCurto et al. 2005). These sites are often the first areas to initiate growing and provide areas of the highest nutrient density per bite early in the grazing period. Deeper-rooted perennials such as bluebunch wheatgrass and Idaho fescue may be preferentially selected over Sandberg bluegrass (*Poa secunda*) due to the ability to remain green longer than the more shallow-rooted grass.

Grazing distribution patterns on diverse landscapes also indicate that cattle often prefer to do most grazing away from cool air sinks such as riparian meadows when phenology of the upland forage is vegetative (Parsons et al. 2003; DelCurto et al. 2005). In recent studies evaluating the botanical composition of diets among diverse plant communities, beef cattle showed strong preferences for grass species even though forbs and shrubs may have had a higher nutrient density (Walburger et al. 2007; Clark et al. 2013; Wyffels and DelCurto 2020). Generally, as stocking rate (use) increases and upland forages become dormant, foraging efficiency decreases (Damiran et al. 2013). The overall decrease in foraging efficiency may be due to the inability to find preferred species, resulting in increasing search time and smaller amounts consumed per bite of the preferred species. Monitoring daily grazing behavior without measuring forage intake will not provide the meaningful insight needed to understand the complex interrelationships that exist in the grazing ruminant (Krysl and Hess 1993). Krysl and Hess (1993) also state that harvesting efficiency allows further evaluation of supplementation regimens and the energetic cost of grazing, which is an essential element in understanding the effects of grazing behavior on ecosystem function.

The understanding of climate change and climatic extremes is also an important consideration for both wild domestic ruminants in rangeland ecosystems. Understanding how animals respond to drought (Roever et al. 2015) and heat stress is important, particularly in the management of cattle use near streams (DelCurto et al. 2005). Roever and coworkers (2015) indicated that cattle during drought will consolidate distribution patterns with increased reliance on riparian areas. In addition, research relative to ecological fit of domestic ruminants is important to the optimal production and use of native rangelands (Sprinkle et al. 2020). Likewise, in the interior Pacific Northwest, Intermountain West, and upper Great Plains, understanding how ruminants respond to cold stress is important for optimal management of landscape use and nutrient needs (Wyffels et al. 2019, 2020a, b; Parsons et al. 2021). When providing supplemental inputs, managers need to focus on optimizing the use of forage resources as well as encouraging optimal grazing distribution on extensive rangeland pastures or paddocks. Research in the Northern Mixed-grass Prairie has demonstrated that supplement intake patterns vary as a function of environmental extremes and are also impacted by cow age (Wyffels et al. 2020a, b; Parsons et al. 2021).

4.6   Other Disturbance Factors

It is difficult to evaluate wild and domestic animal interactions and impacts on vegetation diversity without discussing other disturbance factors such as fire and/or logging. Generally, fire will cause significant declines in forbs, shrubs, and trees while promoting a grass understory. Similarly, logging will open up the canopy, which encourages grasses and early successional shrubs and forbs on western Intermountain forests of North America. Combinations of logging (thinning) and understory controlled-burns have been shown to improve diets of elk and cattle early in the grazing season whereas diets in late summer and early fall were lower in quality with the treated areas (Long et al. 2008; Clark et al. 2013).

In a study evaluating overstory tree type and stand age on understory vegetation composition and quality by forage classes, Davis and coworkers (2019; Fig. 4.3) reported only limited differences due to overstory tree type and no differences due to stand age with respect to vegetation crude protein (CP) and plant fiber composition (neutral detergent fiber and acid detergent fiber). However, fibrous fractions of the vegetation were substantially lower in the understories of ponderosa pine (*Pinus ponderosa*) and Douglas fir (*Pseudotsuga menziesii*) overstories. In contrasts, graminoids and non-forested sites (meadows and grasslands) had dramatically higher fiber and lower crude protein in the late summer sampling periods. Similarly, Walburger and coworkers (2007) reported that timber harvest and previous herbivory had no effects on the quality of diets selected by cattle. In addition, cattle grazing forested rangelands in northeastern Oregon preferred a diet that was dominated by graminoids despite the fact that forbs and shrubs had higher CP and lower fiber content. However, as graminoid production and/or availability decreased, such as in heavily timbered areas, cattle increased consumption of forbs and shrubs.

4.7   Interactive Effects with Wildlife

Domestic livestock grazing (sheep and cattle) can have both positive and negative impacts on wildlife habitat. The intensity of use plays a confounding role in analyzing effects as residual vegetation left after grazing may be a key consideration

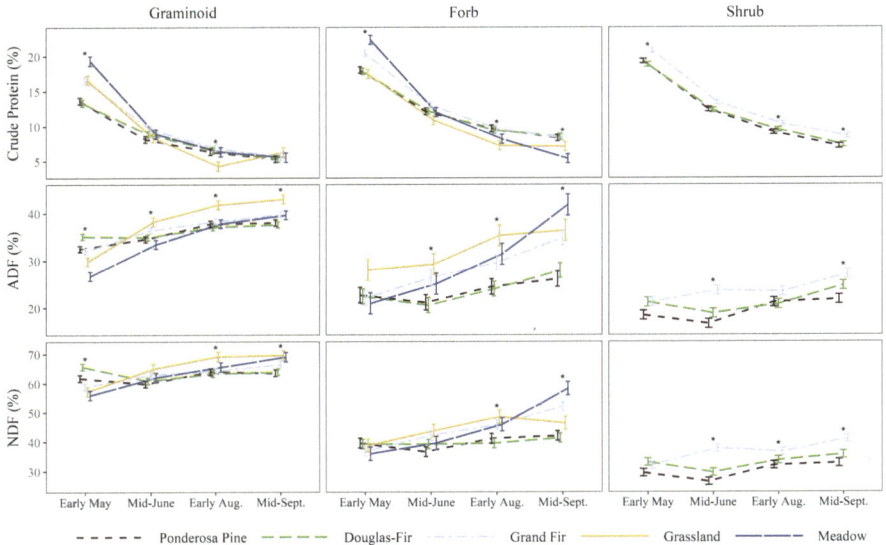

Fig. 4.3 Crude protein and fiber composition of forage growth forms (graminoids, forbs, and shrubs) from May to September in diverse mixed conifer overstories, meadows, and grasslands (adapted from Davis et al. 2019)

for wildlife habitat. Limiting grazing to vegetative, boot, or flowering stages may provide residual forage through regrowth (see previous discussion, Ganskopp et al. 2007). The regrowth can then provide forage and cover for various species of wildlife. During the boot stage of growth, grazing has been shown to improve the nutritional quality of forage (regrowth) available to ungulates in the fall or winter (Ganskopp et al. 2004). However, forage biomass available is less than similar ungrazed forage (Ganskopp et al. 2007).

Season of livestock grazing can also play a significant role in altering plant community composition (Severson and Urness 1994). Bitterbrush (*Purshia tridentata*) is a shrub species palatable to wild ungulates including mule deer, elk, and pronghorn. Ganskopp et al. (1999) reported grazing during the boot stage of bunchgrasses improved both the diameter and volume of bitterbrush plants, but grazing after the flowering of grasses resulted in an extensive use of bitterbrush. Previously, Lesperance et al. (1970) suggested that to prevent overconsumption of bitterbrush on mule deer wintering range, cattle grazing should be limited to early in the season when grasses are green and palatable. Early season grazing of meadows allows regrowth of forbs, potentially improving foraging conditions for sage-grouse (*Centrocercus urophasianus*; Evans 1986).

Season of use considerations can be adjusted to critical life events like bird ground-nesting and pronghorn fawning. This could include deferment to preserve habitat or grazing prior to the event to create habitat. Moderate and low stocking rates of cattle grazing during the nesting season on bunchgrass communities in northeastern Oregon caused no adverse impacts to ground-nesting songbirds (Johnson et al. 2011). These stocking rates generally provided suitable habitat for all species studied compared to the no grazing treatment. However, high stocking rates did not provide suitable habitat for ground-nesting birds. Stocking rates utilized 9.5 (elk use), 20, 32, and 46% of the available forage, respectively, for zero, low, medium, and high beef cattle stocking rates.

Domestic livestock and wildlife disease transmission is also a challenge for livestock, wildlife, and rangeland management in the future. The most commonly cited concern is pneumonia transmission from domestic sheep to bighorn sheep (*Ovis canadensis*; Wehausen et al. 2011; Carpenter et al. 2014). These authors provide evidence that contact between domestic and bighorn sheep may be factors in disease transmission and, in turn, a primary factor in the limited success to re-establish bighorn sheep populations throughout the West. Others argue that the direct causes of respiratory disease in wild sheep are not clearly elucidated and, despite the dramatic decline in rangeland domestic sheep numbers over the past three decades, bighorn sheep have not recovered. In addition, mortality due to pneumonia is greatest with early post-partum lambs between 1 and 3 months of age (Cassirer et al. 2013) with adult bighorn sheep demonstrated to be long-term carriers of pathogens that might cause pneumonia.

One major area of concern to wildlife in close proximity to domestic sheep is the transmission of *Mycoplasma ovipneumoniae*, which is thought to be the agent that predisposes bighorn sheep to pneumonia (Besser et al. 2008). *Mycoplasma ovipneumoniae* is a respiratory pathogen that infects animals in the Caprinae subfamily and can lead to secondary infections. While the disease has a global distribution, the prevalence in the U.S. domestic sheep population has been estimated at 88.5% of operations with at least one individual testing positive via PCR for *M. ovipneumoniae* (Manlove et al. 2019). While not presented as a major concern for western domestic sheep production it has been estimated that *M. ovipneumoniae* at current prevalence levels is associated with a 4.3% reduction in annual lamb production with lower average daily gain in lambs exposed (Manlove et al. 2019; Besser et al. 2019).

Exposure to *M. ovipneumoniae* is primarily the result of interactions between infected domestic sheep and wild bighorn sheep. Based on experiments that co-mingled bighorn sheep with domestic sheep free of *M. ovipneumoniae* and those infected, the *M. ovipneumoniae* negative co-mingled bighorn sheep presented a significantly higher survival rate than those that co-mingled with *M. ovipneumoniae* positive domestic sheep (Besser et al. 2012; Foreyt and Jessup 1982; Foreyt 1989, 1990; Lawrence et al. 2010). This led to significant restrictions on sheep grazing in bighorn sheep habitats to limit the potential for mass die-offs in bighorn sheep populations. However, *M. ovipneumoniae* has been reported in populations of wild Rocky Mountain goats (*Oreamnos americanus*) and other species outside the Caprinae subfamily such as moose (*Alces alces*), caribou (*Rangifer tarandus*), and mule deer which may also serve as transmission pools to bighorn sheep populations (Wolff et al. 2019; Highland et al. 2018).

Another concern that will certainly increase in the future will be the passive transfer of brucellosis from bison to elk to cattle relative to the Greater Yellowstone Ecosystem region (Mosley and Mundinger 2018). Brucellosis infections seem to have limited long-term impacts on bison and elk wildlife populations yet could cause considerable impacts on humans including big game hunters, ranchers, and veterinarians. Brucellosis, primarily transferred via placental and mammary fluid/tissue, was largely eradicated with the mandatory pasteurization of milk and milk products in the 1930s, as well as current "bangs" vaccination programs in the beef cattle industry. The spread of brucellosis via elk populations, however, has created considerable concern for the beef cattle industry and wildlife managers. Efforts to reduce the interaction between elk populations and cattle during the late gestational stages of elk (March through May) may be key management considerations to reduce the transmission of brucellosis. In addition, confined winter feeding of elk should be reconsidered due to the higher incidence of brucellosis in winter fed Rocky Mountain elk (Brennen et al. 2017).

4.8   Sustainable Livestock Systems of the Future

The Great Plains and Western U.S. rangelands have historically been managed to accommodate livestock production. However, Congress has altered the framework that governs land management with the passage of the Multiple Use Act (1968), National Environmental Policy Act (1969), Clean Water Act (1972), and the Threatened and Endangered Species Act (1973). The continued use of public and private rangelands across the western region depends on our ability to develop sustainable systems that maintain or enhance biological diversity of forages, riparian function, and wildlife. Grazing livestock nutrition and management must develop systems for economic viability that also maintain biological diversity (vegetation and wildlife) and the industry's traditions and integrity (DelCurto and Olson 2010; Fig. 4.4). Research that is grounded in economic and ecologic sustainability should be encouraged and supported. Recent reviews evaluating the management of livestock distribution and applied management strategies for optimal distribution on arid rangelands provide a relevant background for this discussion (Bailey 2005; DelCurto et al. 2005; DelCurto and Olson 2010; Bailey et al. 2019, 2021; Holechek et al. 2020).

Future sustainable livestock production systems will need to incorporate significant management paradigm shifts to be successful. Specifically, optimal use will be a function of landscape use patterns of livestock and wildlife, where we manage the vegetation for optimization of biological processes (Vavra 2005). Specifically, there is a need to focus on the amount of vegetation needed on a landscape (post grazing) to optimize the success of that plant and plant community with a focus on photosynthetic processes, and, particularly in arid and semi-arid environments, the capture, storage and release of water. Optimal use by livestock will necessarily be related to leaving behind sufficient foliage for the plant to regenerate in the future, provide sufficient vegetation biomass to maintain or enhance soil organic matter value, and provide for enhanced soil microbial populations. Perhaps, optimal use for biological processes will hinge on vegetation remaining rather than vegetation removed by the herbivore. In addition, future management systems need to account

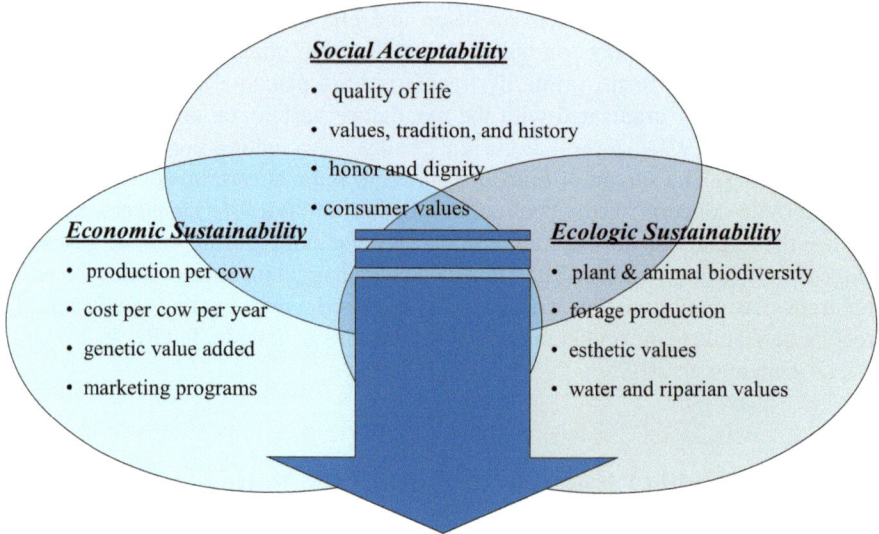

Sustainable Range Beef Production Systems

Fig. 4.4 Sustainable western rangeland livestock production systems will have to embrace economic viability, ecological integrity, and social values to be successful in the future (adapted from DelCurto and Olson 2010)

for all herbivores, which will encompass insects to large ruminant generalist grazers, as well as livestock in respect to forage system management.

Managers need to strive for systems that promote deep-rooted perennial species that optimize nutritional opportunities for all ruminants, as well as promote healthy and stable soils for optimal production and water holding capacity. Utilizing grazing systems/principles that promote desired vegetation succession while still capturing economic value will be paramount to these efforts (Bailey et al. 2021). In turn, promoting vegetation that optimizes photosynthetic processes and water use/conservation, will be more productive and diverse (grass/forb/shrub/tree) providing greater production and nutritional opportunities for domestic livestock and wildlife. These systems, in turn, will be more resilient and adaptive to climate change processes that are especially challenging in western rangeland environments (Holechek et al. 2020). In addition, providing structure for improved habitat cover as well as improved nutrition opportunities (food) over a greater portion of the year, will benefit both wild and domestic species with respect to protection from predation, increased reproductive success and production.

Although we are faced with obvious challenges with respect to sustainable rangeland management, the integration of knowledge relative to plant responses to herbivory, herbivore ecology and grazing strategies, and dynamic grazing behavior patterns will aid in developing better and more sustainable grazing management strategies (Raynor et al. 2021). The use and application of precision technologies which will include GPS applications, virtual fencing (Boyd et al. 2022), and activity monitors that are downloaded to a cell phone or computer in "real-time" (Bailey et al. 2021) will be critical technologies of the future. Using domestic livestock to manage fuels, selectively graze undesirable plant species while minimizing impacts on desirable plants will be paramount to our success. Future systems will continue to expand the grazing season, reducing the reliance on fossil fuel and labor, while promoting less growing season use when deferred and/or rotation grazing systems are used or fewer animals are grazed in continuous grazing. Simply put, grazing to promote biological processes with respect to soils and vegetation communities will provide the most benefit to both wildlife and domestic livestock production.

References

Augustine DJ, McNaughton SJ (1998) Ungulate effects on the functional species composition of plant communities: herbivore selectivity and plant tolerance. J Wildl Manage 62:1165–1183. https://doi.org/10.2307/3801981

Averett JP, Endress BA, Rowland MM, Naylor BJ, Wisdom MJ (2017) Wild ungulate herbivory suppresses deciduous woody plant establishment following salmonid stream restoration. For Ecol Manage 391:135–144. https://doi.org/10.1016/j.foreco.2017.02.017

Bailey DW (2005) Identification and creation of optimum habitat conditions for livestock. Rangelnd Ecol Manag 58(2):109–118. https://doi.org/10.2111/03-147.1

Bailey DW, Kress DD, Anderson DC, Boss DL, Miller ET (2001) Relationship between terrain use and performance of beef cows grazing foothill rangeland. J Anim Sci 79:1883–1891. https://doi.org/10.2527/2001.7971883x

Bailey DW, Mosley JC, Estell RE, Cibils AF, Horney M, Hendrickson JR, Walker JW, Launchbaugh KL, Burritt EA (2019) Synthesis paper: targeted livestock grazing: prescription for healthy rangelands. Rangel Ecol Manag 72(6):865–877. https://doi.org/10.1016/j.rala.2019.11.001

Bailey DW, Trotter MG, Tobin C, Thomas MG (2021) Opportunities to apply precision livestock management on rangelands. Front Sustain Food Syst 5:6111915. https://doi.org/10.3389/fsufs.2021.611915

Bell DS (1970) Trends in the industry of the United States: effects of breed type and economic circumstances. Res Cir 179

Besser TE, Cassirer EF, Potter KA, VanderSchalie J, Fischer A, Knowles DP, Herndon DR, Rurangirwa FR, Weiser GC, Srikumaran S (2008) Association of *Mycoplasma ovipneumoniae* infection with population-limiting respiratory disease in free-ranging Rocky Mountain bighorn sheep (*Ovis canadensis canadensis*). J Clin Microbiol 46(2):423–430. https://doi.org/10.1128/jcm.01931-07

Besser TE, Highland MA, Baker K, Cassirer E, Anderson NJ, Ramsey JM, Mansfield K, Bruning DL, Wolff P, Smith JB, Jenk JJA (2012) Causes of Pneumonia epizootics among bighorn sheep, Western United States, 2008–2010. J Emerg Infect Dis 18(3):406–414. https://doi.org/10.3201/eid1803.111554

Besser TE, Levy J, Ackerman M, Nelson D, Manlove K, Potter KA, Busboom J, Benson M (2019) A pilot study of the effects of *Mycoplasma ovipneumoniae* exposure on domestic lamb growth and performance. PLoS ONE 14(2):423–430. https://doi.org/10.1371/journal.pone.0207420

Boyd CS, O'Connor R, Ranches J, Bohnert DW, Bates JD, Johnson DD, Davies KW, Parker T, Doherty KE (2022) Virtual fencing effectively excludes cattle from burned sagebrush steppe. Rangel Ecol Manag 81:55–62. https://doi.org/10.1016/j.rama.2022.01.001

Brennan A, Cross PC, Portacci K, Scurlock BM, Edwards WH (2017) Shifting brucellosis risk in livestock coincides with spreading seroprevalence in elk. PLoS ONE 12:e0178780. https://doi.org/10.1371/journal.pone.0178780

Briske DD, Derner JD, Milchunas DG, Tate KW (2011) An evidence-based assessment of prescribed grazing practices. In: Conservation benefits of rangeland practices: assessment, recommendations, and knowledge gaps. USDA NRCS, pp 21–74

Brownsey P, James JJ, Barry SJ, Becchetti TA, Davy JS, Doran MP, Forero LC, Harper JM, Larsen RE, Larson-Praplan SR, Zhang J (2017) Using phenology to optimize timing of mowing and grazing treatments for medusahead (*Taeniatherum caput-medusae*). Rangel Ecol Manag 70:210–218. https://doi.org/10.1016/j.rama.2016.08.011

Carpenter TE, Coggins VL, McCarthy C, O'Brien CS, O'Brien JM, Schommer TJ (2014) A spatial risk assessment of bighorn sheep extirpation by grazing domestic sheep on public lands. Prev Vet Med 114:3–10. https://doi.org/10.1016/j.prevetmed.2014.01.008

Cassirer EF, Plowright RK, Manlove KR, Cross PC, Dobson AP, Potter KA, Hudson PJ (2013) Spatio-temporal dynamics of pneumonia in bighorn sheep. J Anim Ecol 82:518–528. https://doi.org/10.1111/1365-2656.12031

Clark A, DelCurto T, Vavra M, Dick BL (2013) Botanical composition and diet quality of beef cattle grazing at three stocking rates following fuels reduction in mixed conifer forests. Rangel Ecol Manag 66:714–720. https://doi.org/10.2111/rem-d-12-00122.1

Cheeke PR, Dierenfeld ES (2010) Comparative animal nutrition and metabolism. CABI, pp 22–44. https://doi.org/10.1079/9781845936310.0277

Cook CW, Harris LE (1952) Nutritive value of cheatgrass and crested wheatgrass on spring ranges in Utah. J Range Manage 5:331–337. https://doi.org/10.2307/3894038

Coughenour MB (1991) Spatial components of plant-herbivore interactions in pastoral, ranching, and native ungulate systems. J Range Manage 44:530–542. https://doi.org/10.2307/4003033

Cruz R, Ganskopp D (1998) Seasonal preference of steers for prominent northern Great Basin grasses. J Range Manage 51:557–565. https://doi.org/10.2307/4003376

Damiran D, DelCurto T, Findholt SL, Johnson BK, Vavra M (2013) Comparison of bite-count and rumen evacuation techniques to estimate cattle diet quality. Rangel Ecol Manag 66:106–109. https://doi.org/10.2111/rem-d-12-00046.1

Damiran D, DelCurto T, Findholt SL, Johnson BK, Vavra M (2019) The effects of previous grazing on the subsequent nutrient supply of ungulates grazing late-summer mixed-conifer rangelands. Sustain Agri Res 8(526–2020–553):13–27. https://doi.org/10.5539/sar.v8n4p13

Davies KW, Bates JD, Svejcar TJ, Boyd CS (2010) Effects of long-term livestock grazing on fuel characteristics in rangelands: an example from the sagebrush steppe. Rangeland Ecol Manag 63:662–669. https://doi.org/10.2111/rem-d-10-00006.1

Davis NG, Wyffels SA, Damiran D, Darambazar E, Vavra M, Riggs RA, DelCurto T (2019) Influence of forest series and stand age on the growing season nutritional dynamics of plant growth forms in a mixed-conifer forest. Transl Anim Sci 3(S1):1724–1727. https://doi.org/10.1093/tas/txz067

DelCurto T, Olson KC (2010) Issues in Grazing Livestock Nutrition. In: Hess BW, DelCurto T, Bowman JGP, Waterman RC (eds) Proceedings 4th grazing livestock nutrition conference. Western Section American. Society of Animal Science, Champaign, Il, pp 1–10

DelCurto T, Olson KC, Hess B, Huston E (2000) Optimal supplementation strategies for beef cattle consuming low-quality forages in the western United States. J Anim Sci Symp Proc 77:1–16. https://doi.org/10.2527/jas2000.77e-suppl1v

DelCurto T, Porath M, Parsons CT, Morrison JA (2005) Management strategies for sustainable beef cattle grazing on forested rangelands in the Pacific Northwest. Invited Synthesis Paper Rangel Ecol Manag 58:119–127. https://doi.org/10.2111/1551-5028(2005)58%3c119:MSFSBC%3e2.0.CO;2

DiTomaso JM, Kyser GB, George MR, Doran MP, Laca EA (2008) Control of medusahead (*Taeniatherum caput-medusae*) using timely sheep grazing. Invasive Plant Sci Manag 1(3):241–247. https://doi.org/10.1614/ipsm-07-031.1

Evans CC (1986) The relationship to cattle grazing to sage-grouse use of meadow habitat on the Sheldon National Wildlife Refuge (thesis), University of Nevada, Reno, NV, 92 p

FAO (2017) http://www.fao.org/faostat/en/#data/QA. Accessed October 2017

Foreyt WJ (1989) Fatal *Pasteurella haemolytica* pneumonia in bighorn sheep after direct contact with clinically normal domestic sheep. Am J Vet Res 50(3):341–344

Foreyt WJ (1990) Pneumonia in bighorn sheep: effects of *Pasteurella haemolytica* from domestic sheep and effects on survival and long-term reproduction. In: Proceedings of the biennial symposium of the northern wild sheep and goat council, pp 92–101

Foreyt WJ, Jessup DA (1982) Fatal pneumonia of bighorn sheep following association with domestic sheep. J Wildl Dis 18(2):163–168. https://doi.org/10.7589/0090-3558-18.2.163

Fuhlendorf SD, Engle DM (2001) Restoring heterogeneity on rangelands: ecosystem management based on evolutionary grazing patterns: we propose a paradigm that enhances heterogeneity instead of homogeneity to promote biological diversity and wildlife habitat on rangelands grazed by livestock. Bioscience 51(8):625–632. https://doi.org/10.1641/0006-3568(2001)051[0625:rhorem]2.0.co;2

Ganskopp D, Angell RA, Rose J (1992) Response of cattle to cured reproductive stems in a caespitose grass. J Range Manage 45:401–404. https://doi.org/10.2307/4003091

Ganskopp D, Svejcar T, Taylor F, Farstvedt J, Paintner K (1999) Seasonal cattle management in 3 to 5 year old bitterbrush stands. J Range Manage 52:166–173. https://doi.org/10.2307/4003512

Ganskopp D, Svejcar TJ, Vavra M (2004) Livestock forage conditioning: bluebunch wheatgrass, Idaho fescue, and bottlebrush squirreltail. J Range Manage 57:384–392. https://doi.org/10.2458/azu_jrm_v57i4_ganskopp

Ganskopp D, Aguilera L, Vavra M (2007) Livestock forage conditioning among six northern Great Basin grasses. Rangel Ecol Manag 60:71–78. https://doi.org/10.2111/05-230r1.1

Heitschmidt R, Walker J (1983) Short duration grazing and the Savory grazing method in perspective. Rangelands 5(4):147–150

Heitschmidt RK, Taylor CA (2003) Livestock Production. In: Heitschmidt and Stuth (eds) Grazing management: an ecological perspective. CBLS, 119 Brentwood Street, Marietta, OH, pp 161–178

Highland MA, Herndon DR, Bender SC, Hansen L, Gerlach RF, Beckmen KB (2018) *Mycoplasma ovipneumoniae* in wildlife species beyond Subfamily *Caprinae*. J Emerg Infect Dis 24(12):2384–2386. https://doi.org/10.3201/eid2412.180632

Hobbs NT (1996) Modification of ecosystems by ungulates. J Wildl Manage 60:695–713. https://doi.org/10.2307/3802368

Holechek JL, Pieper RD, Herbal CH (1998) Range management principles and practices. Prentice Hall, Upper Saddle River, New Jersey, p 542

Holechek JL, Geli HM, Cibils AF, Sawalhah MN (2020) Climate change, rangelands, and sustainability of ranching in the western United States. Sustainability 12:4942. https://doi.org/10.3390/su12124942

Hormay AL (1970) Principles of rest-rotation grazing and multiple use land management. USDA Forest Service Training Text 4 (2200) Washington, D.C.

Hyder DN (1953) Grazing capacity as related to range condition. J for 51:206

Johnson TL, Kennedy PL, DelCurto T, Taylor RV (2011) Bird community responses to cattle stocking rates in a Pacific Northwest bunchgrass prairie. Agric Ecosys Env 144:338–346. https://doi.org/10.1016/j.agee.2011.10.003

Koocheki A, Gliessman SR (2005) Pastoral nomadism, a sustainable system for grazing land management in arid areas. J Sust Agri 25(4):113–131. https://doi.org/10.1300/j064v25n04_09

Kothmann M (2009) Grazing methods: a viewpoint. Rangelands 31(5):5–10. https://doi.org/10.2111/1551-501x-31.5.5

Krysl LJ, Hess BW (1993) Influence of supplementation on behavior of grazing cattle. J Anim Sci 71:2546–2555. https://doi.org/10.2527/1993.7192546x

Kunkle WE, Johns JT, Poore MH, Herd DB (2000) Designing supplementation programs for beef cattle fed forage-based diets. J Animal Sci Symp Proc. https://doi.org/10.2527/jas2000.002188 12007700es0012x

Lawrence PK, Shanthalingam S, Dassanayake RP, Subramaniam R, Herndon CN, Knowles DP, Srikumaran S (2010) Transmission of *Mannheimia haemolytica* from domestic sheep (*Ovis aries*) to bighorn sheep (*Ovis canadensis*): unequivocal demonstration with green fluorescent protein-tagged organisms. J Wildl Dis 46:706–717. https://doi.org/10.7589/0090-3558-46.3.706

Lehnhoff EA, Rew LJ, Mangold JM, Seipel T, Ragen D (2019) Integrated management of cheatgrass (*Bromus tectorum*) with sheep grazing and herbicide. Agronomy 9:315. https://doi.org/10.3390/agronomy9060315

Lesperance AL, Tueller PT, Bohman VR (1970) Competitive use of the range forage resource. J Anim Sci 30:115–120. https://doi.org/10.2527/jas1970.301115x

LMIC (2016) U.S. baseline lamb cost of production model. http://lmic.info/sites/default/files/gen eral_files/Project%20_Summary.pdf. Accessed October 2017

Long RA, Rachlow JL, Kie JG, Vavra M (2008) Fuels reduction in a western coniferous forest: effects on quantity and quality of forage for elk. Rangel Ecol Manag 61(3):302–313. https://doi.org/10.2111/07-046.1

Manlove K, Branan M, Baker K, Bradway D, Cassirer EF, Marshall KL, Miller RS, Sweeney S, Cross PC, Besser TE (2019) Risk factors and productivity losses associated with *Mycoplasma ovipneumoniae* infection in United States domestic sheep operations. Prev Vet Med 168:30–38. https://doi.org/10.1016/j.prevetmed.2019.04.006

McCurdy DE, Watts CJ, Chibisa GE, Prather TS, Laarman AH (2017) Feed processing affects palatability of ventenata infested grass hay. J Anim Sci 95:295. https://doi.org/10.2527/asasann.2017.295

McNaughton SJ (1985) Ecology of a grazing ecosystem: the Serengeti. Ecol Monogr 55:259–294. https://doi.org/10.2307/1942578

Milchunas DG, Lauenroth WK (1993) Quantitative effects of grazing on vegetation and soils over a global range of environments. Ecol Monogr 63:327–366. https://doi.org/10.2307/2937150

Miller RF, Svejcar TJ, West NE (1994) Implications of livestock grazing in the intermountain sagebrush region: plant composition. In: Vavra M, Laycock WA, Pieper RD (1994) Ecological implications of livestock herbivory in the west. Society for Range Management, Denver, CO, 297 pp

Mosley JC, Roselle L (2006) Chapter 8: Targeted grazing to suppress invasive annual grasses. In: Launchbaugh K, Walker J (eds) Targeted grazing: a natural approach to vegetation management. University of Idaho, Moscow ID, 199 pp

Mosley JC, Mundinger JG (2018) History and status of wild ungulate populations on the Northern Yellowstone Range. Rangelands 40:189–201. https://doi.org/10.1016/j.rala.2018.10.006

Mosley JC, Roeder BL, Frost RA, Wells SL, McNew LP, Clark PE (2020) Mitigating human conflicts with livestock guardian dogs in extensive sheep grazing systems. Rangel Ecol Manag 73(5):724–732. https://doi.org/10.1016/j.rama.2020.04.009

NRC (2008) Changes in the sheep industry in the United States: Making the transition from tradition. National Academies Press, Washington, D.C. https://doi.org/10.17226/12245

Owensby CE, Auen LM (2018) Steer and pasture productivity influenced by intensive early stocking plus late season grazing. Crop, for and Turfgrass Manage 4(1):1–7. https://doi.org/10.2134/cftm2017.02.0011

Parsons CT, Momont PA, DelCurto T, McInnis M, Porath ML (2003) Cattle distribution patterns and vegetation use in mountain riparian areas. J Range Manage 56:334–341. https://doi.org/10.2458/azu_jrm_v56i4_parsons

Parsons CT, Dafoe JM, Wyffels SA, DelCurto T, Boss DL (2021) The influence of residual feed intake and cow age on beef cattle performance, supplement intake, resource use, and grazing behavior on winter mixed-grass rangelands. Animals 11(6):1518. https://doi.org/10.3390/ani11061518

Pfister JA, Gardner DR, Panter KE (2010) Consumption of low larkspur (*Delphinium nuttallianum*) by grazing sheep. Rangel Ecol Manag 63:263–266. https://doi.org/10.2111/rem-d-09-00084.1

Porath ML, Momont PA, DelCurto T, Rimbey NR, Tanaka JA, McInnis M (2002) Offstream water and trace mineral salt as management strategies for improved cattle distribution. J Anim Sci 80:346–356. https://doi.org/10.2527/2002.802346x

Putman DH, DelCurto T (2020) Forage systems for arid areas. In: Moore KJ, Collins M, Nelson CJ, Redfearn DD (eds) Forages: the science of grassland agriculture II, 7th ed. https://doi.org/10.1002/9781119436669.ch24

Raleigh RJ (1970) Symposium on pasture methods for maximum production in beef cattle: manipulation of both livestock and forage management to give optimum production. J Anim Sci 30:108–114. https://doi.org/10.2527/jas1970.301108x

Ranches J, O'Connor R, Johnson D, Davies K, Bates J, Boyd C, Bohnert DW, Parker T (2021) Effects of virtual fence monitored by global positioning system on beef cattle behavior. Transl Anim Sci 5(S1):S144–S148. https://doi.org/10.1093/tas/txab161

Ratliff RD, Reppert JN (1974) Vigor of Idaho fescue grazed under rest rotation and continuous grazing. J Range Manage 27:447–449. https://doi.org/10.2307/3896719

Raynor EJ, Gersie SP, Stephenson MB, Clark PE, Spiegal SA, Boughton RK, Bailey DW, Cibils A, Smith BW, Derner JD, Estell RE (2021) Cattle grazing distribution patterns related to topography across diverse rangeland ecosystems of North America. Rangel Ecol Manag 75:91–103. https://doi.org/10.1016/j.rama.2020.12.002

Riggs RA, Keane RE, Cimon N, Cook R, Holsinger L, Cook J, DelCurto T, Baggett LS, Justice D, Powell D, Vavra M (2015) Biomass and fire dynamics in a temperate forest-grassland mosaic: integrating multi-species herbivory, climate, and fire with the FireBGCv2/GrazeBGC system. Ecol Modelling 296:57–78. https://doi.org/10.1016/j.ecolmodel.2014.10.013

Roever CL, DelCurto T, Rowland M, Vavra M, Wisdom MJ (2015) Cattle grazing in semiarid forestlands: habitat selection during periods of drought. J Anim Sci 93:3212–3225. https://doi.org/10.2527/jas.2014-8794

Romo JT (1994) Wolf plant effects on water relations, growth and productivity in crested wheatgrass. Can J Plant Sci 74:767–771. https://doi.org/10.4141/cjps94-137

Severson KE, Urness PJ (1994) Livestock grazing: a tool to improve wildlife habitat. In: Vavra M, Laycock WA, Pieper RD (eds) Ecological implications of livestock herbivory in the west. Society for Range Management, Denver, CO, pp 232–249

Skovlin JM, Harris RW, Strickler GS, Garrison GA (1976) Effects of cattle grazing methods on ponderosa pine-bunchgrass range in the Pacific Northwest. U.S. Dep Agr Tech Bull No 1531:40

Smith EF, Owensby CE (1978) Intensive early stocking and season long stocking of Kansas Flint Hills range. J Range Manage 31:14–17. https://doi.org/10.2307/3897624

Sprinkle JE, Taylor JB, Clark PE, Hall JB, Strong NK, Roberts-Lew MC (2020) Grazing behavior and production characteristics among cows differing in residual feed intake while grazing late season Idaho rangelands. J Anim Sci 2020:1–9. https://doi.org/10.1093/jas/skz371

Tanaka JA, Rimbey NR, Torell LA, Taylor DT, Bailey D, DelCurto T, Walburger K, Welling B (2007) Grazing distribution: the quest for the silver bullet. Rangelands 29:38–46. https://doi.org/10.2458/azu_rangelands_v29i4_tanaka

Teague WR, Dowhower SL (2003) Patch dynamics under rotational and continuous grazing management in large, heterogeneous paddocks. J Arid Environ 53:211–229. https://doi.org/10.1006/jare.2002.1036

Thomas DL, Berger YM, McKusick BC, Mikolayunas CM (2014) Dairy sheep production research at the University of Wisconsin-Madison, USA—a review. J Anim Sci Biotechnol 5:22–23. https://doi.org/10.1186/2049-1891-5-22

Török P, Valkó O, Deák B, Kelemen A, Tóth E, Tóthmérész B (2016) Managing for species composition or diversity? Pastoral and free grazing systems in alkali steppes. Agric Ecosyst Environ 234:23–30. https://doi.org/10.1016/j.agee.2016.01.010

USDA, Animal and Plant Inspection Service (2012) Sheep 2011 Part I: references of sheep management practices in the United States, 2011

USDA, Animal and Plant Inspection Service (2014) Lambing management practices on U.S. sheep operations, 2011

USDA, Economic Research Service (2004) Trends in U.S. sheep industry. AIB-787
USDA, Economic Research Service (2017) https://www.ers.usda.gov/data-products/food-availabil
  ity-per-capita-data-system/. Accessed October 2017
USDA, National Agricultural Statistics Service (2019) Census of agriculture. AC-17-A-51
USDA, National Agricultural Statistics Service (2017a) Sheep and goats (January 2017a). ISSN:
  1949-1611
USDA, National Agricultural Statistics Service (2017b) Quarterly hogs and pigs (September 2017b).
  ISSN: 1949-1921
USDA, National Agricultural Statistics Service (2017c) Cattle (July 2017c). ISSN: 1948-9099
USDA, National Agricultural Statistics Service (2021) Quick stats. https://quickstats.nass.usda.gov
Van Soest PJ (1994) Nutritional ecology of the ruminant. Cornell University Press
Vavra M (2005) Livestock grazing and wildlife: developing compatibilities. Rangel Ecol Manag
  58:128–134. https://doi.org/10.2458/azu_rangelands_v58i2_vavra
Vavra M, Raleigh RJ (1976) Coordinating beef cattle management with the range forage resource.
  J Range Manage 29:449–452
Walburger KJ, DelCurto T, Vavra M (2007) Influence of forest management and previous herbivory
  on cattle diets. Rangel Ecol Manag 60:172–178. https://doi.org/10.2111/05-223r3.1
Walker JW (1997) Multispecies grazing: the ecological advantage. West Sect Proc Amer Soc Anim
  Sci 48:7–10
Wehausen JD, Kelley ST, Ramey RR (2011) Domestic sheep, bighorn sheep, and respiratory disease:
  a review of the experimental evidence. Calif Fish Game 97(1):7–24
Wolff PL, Blanchong JA, Nelson DD, Plummer PJ, McAdoo C, Cox M, Besser TE, Muñoz-Gutiérrez
  J, Anderson CA (2019) Detection of *Mycoplasma ovipneumoniae* in pneumonic mountain goat
  (*Oreamnos americanus*) kids. J Wildl Dis 55(1):206–212. https://doi.org/10.7589/2018-02-052
Wyffels SA, DelCurto T (2020) Influence of beef cattle stocking density on utilization of vegetative
  communities in a late-spring, early-summer native bunchgrass prairie. J Agric Studies 8:400–
  410. https://doi.org/10.5296/jas.v8i4.17462
Wyffels SA, Petersen MK, Boss DL, Sowell BF, Bowman JG, McNew LB (2019) Dormant season
  grazing: effect of supplementation strategies on heifer resource utilization and vegetation use.
  Rangelnd Ecol Manag 72:878–887. https://doi.org/10.1016/j.rama.2019.06.006
Wyffels SA, Boss DL, Sowell BF, DelCurto T, Bowman JG, McNew LB (2020a) Dormant season
  grazing on northern mixed grass prairie agroecosystems: does protein supplement intake, cow
  age, weight and body condition impact beef cattle resource use and residual vegetation cover?
  PLoS ONE 15:e0240629. https://doi.org/10.1371/journal.pone.0240629
Wyffels SA, Dafoe JM, Parsons CT, Boss DL, DelCurto T, Bowman JG (2020b) The influence of
  age and environmental conditions on supplement intake by beef cattle winter grazing northern
  mixed-grass rangelands. J Anim Sci 98(7):skaa217. https://doi.org/10.1093/jas/skaa217

Chapter 5
Manipulation of Rangeland Wildlife Habitats

David A. Pyke and Chad S. Boyd

Abstract Rangeland manipulations have occurred for centuries. Those manipulations may have positive or negative effects on multiple wildlife species and their habitats. Some of these manipulations may result in landscape changes that fragment wildlife habitat and isolate populations. Habitat degradation and subsequent restoration may range from simple problems that are easy to restore to complex problems that require multiple interventions at multiple scales to solve. In all cases, knowledge of the wildlife species' habitat needs throughout their life history, of their population dynamics and habitat-related sensitivities, and of their temporal and spatial scale for home ranges and genetic exchange will assist in determining appropriate restoration options. Habitat restoration will begin with an understanding of the vegetation's successional recovery options and their time scales relative to wildlife population declines. We discuss passive and active manipulations and their application options. Passive manipulations focus on changes to current management. Active manipulations may include removal of undesirable vegetation using manual harvesting, mechanical, chemical, or biological methods while desirable vegetation is enhanced through the reintroduction of desirable wildlife habitat structure and function. These techniques will require monitoring of wildlife and their habitat at both the landscape and site level in an adaptive management framework to learn from our past and improve our future management.

Keywords Adaptive management · Climate change · Landscapes · Monitoring · Passive versus active management · State and transition models · Wildlife habitat management · Vegetation manipulations

D. A. Pyke (✉)
U.S. Geological Survey, Forest and Ecosystem Science Center, Corvallis, OR 97331, USA
e-mail: david_a_pyke@usgs.gov

C. S. Boyd
USDA Agricultural Research Service, Eastern Oregon Agricultural Research Center, Burns, OR 97720, USA

L. B. McNew et al. (eds.), *Rangeland Wildlife Ecology and Conservation*,
https://doi.org/10.1007/978-3-031-34037-6_5

5.1 Introduction

Early hominins likely began manipulating their environment soon after they learned to control fire between about 1.5 and 0.4 million ybp (Gowlett 2016). They may have noticed benefits of improved hunting and gathering after wildfires thus leading to intentional fires to gain those benefits. One of the earliest documented cases of manipulating habitats for the benefit of wildlife was during the thirteenth century reign of Kublai Khan (Valdez 2013). Native Americans commonly used fires to clear lands for wildlife use and hunting (Lewis 1985). The classic example of fire to control woody plant encroachment onto the tall-grass prairie, benefitted bison among other ungulate wildlife (Lewis 1985). Europeans as they colonized the Americas applied their previous experiences generally relying on only conservation on game reserves and limited hunting controls while generally lacking knowledge on how to manipulate habitat to benefit wildlife (Leopold 1933).

The early 1900s began an awakening for information on how to actively manage wildlife, as populations of some wildlife species were declining, and public lands were being overused. Land improvement began with soil conservation, forest and grazing management. Leopold (1933) argued these were tools for managing and improving wildlife habitat. He advocated concepts of plant successional theory of the day and recognized land manipulations via planting, livestock grazing use and non-use, fire use and prevention, and mechanical tools (e.g., plowing, mowing, etc.) for manipulating vegetation in the context of improving or sustaining wildlife habitats. Recent additions to this toolbox include chemical and microbiological treatments (Pyke et al. 2017). More recently, animal monitoring technology has been useful in detailing information on what plant communities wildlife species use seasonally. When managers couple wildlife use locations with functional and structural formations of plants into communities within landscapes, managers begin to understand how specific manipulations may improve or decrease a wildlife species' population. However, manipulations geared to benefit one species in the ecosystem, may be detrimental to others with differing habitat requirements (Fulbright et al. 2018).

Understanding animal movements, life history, and habitat use has been greatly improved by the use of remote sensing and geographic information systems (GIS) that allows managers to depict animal spatial movements over time. These assist managers in understanding spatial and temporal elements of wildlife population dynamics and in understanding the scale at which manipulations to landscapes, whether intended for the benefit of wildlife or not, may ultimately impact how wildlife use or avoid certain habitats over time. Depending on the wildlife species even small human influences, such as power poles, may create roosts for predators and result in potential prey avoiding surrounding lands, even if the vegetation community provides the necessary plant species composition to become sufficient habitat (Leu and Hanser 2011). Therefore, it is important for managers to understand landscape scale impacts of habitat manipulations.

In this chapter, we address important concepts relating to wildlife habitat management in rangeland settings through manipulations of plant communities within spatial and temporal contexts that align with wildlife habitat requirements. Initially, we define wildlife habitat in a spatial and temporal context that impacts habitat quantity and quality and discuss the applied ecology of rangeland plant communities. Lastly, we address the various types of manipulations typically used in rangelands. Because livestock grazing systems and fire are presented elsewhere in this book (Chaps. 4 and 6, respectively), we will limit our discussion of these tools to their uses in manipulating habitat.

5.2   Concepts

Across the world, ecosystems have been fundamentally altered due to current and historical anthropogenic activities, and the rate of change is increasing (Millennium Ecosystem Assessment 2005). Over the last 100 years, policy making for and management of wildlife habitat in the United States has seen dramatic change with respect to both specific issues and the general nature of natural resource management challenges. Historically, such challenges have related strongly to easily identifiable disruptions of ecosystem pattern and process that were amenable to policy-based solutions (Grier 1982; Boyd et al. 2014). While many such policies continue to play a defining role in topical management of wildlife habitat, new factors such as climate change and its indirect effects have been associated with broader disruptions of ecosystem processes, creating strong impetus for a more expanded notion of conserving not just habitats of individual species, but the ecosystems in which those habitats exist (Benson 2012; Evans et al. 2013). In this section, we synthesize traditional concepts in conservation of wildlife habitat and explore how these concepts are developing and changing to meet a new generation of challenges facing stewards of rangeland wildlife habitat.

5.2.1   What is Rangeland Wildlife Habitat?

In its most basic form, the term "habitat" represents where an animal lives, and resources it uses while there. Those basic resources fall under the categories of food, water, and cover, which are collectively used by animals to meet basic needs including survival in the face of predation, amelioration of thermal stress, and meeting nutritional demands of metabolic maintenance, growth, and reproduction.

Habitat needs of wildlife species play out within spatially and temporally variable rangeland environments. Because of this variability, wildlife species must not only occupy a home range that is large enough to contain the habitat needs described above, but the size of that home range may vary in accordance with yearly conditions (Anderson et al. 2005). Within an animals' home range, different habitats may be better suited to specific life history needs (e.g., breeding, summer, or winter habitat). The spatial dispersion of these seasonal habitats can create seasonal move-ment patterns within the larger home range (Connelly et al. 2011). The existence of seasonal habitats, and movement between these habitats may be related to weather and climate extremes (e.g., summer vs winter habitat) but is often associated with spatio-temporal variability in plant phenology and production, in association with temperature gradients (e.g., elevation) and rainfall distribution patterns (Holdo et al. 2009; Le Corre et al. 2017; Pratt et al. 2017). Anthropogenic factors such as infrastruc-tural development and hunting activities can have strong influence on the geography of movements between seasonal habitats (Gates et al. 2012; Amor et al. 2019).

Wildlife habitat can be thought of as occurring across a range of conceptual scales, from the geographical range of a species to the within-site habitat characteristics important to that species. These scales collectively represent a hierarchy of needs wherein the importance of smaller scale habitat characteristics is predicated on the existence of sufficient habitat elements at larger scales (Johnson 1980). At the largest practical management scale for most rangeland managers, landscape cover refers to the dominant overhead cover components expressed as a fractional percentage of landscape area. These data are useful both in large scale management planning and for assessing links between habitat properties and populations for species with large home ranges (Aldridge et al. 2008). Generally, landscape cover is measured through remote sensing where the reflectance of vegetation functional groups (e.g., shrubs, perennial grasses, etc.) or prominent species dominate the wavelengths of pixels in images and are used as cover attributes in landscape analyses (e.g., Jones et al. 2018). These data can also be collated to more broadly determine cover of higher order biotic and plant associations (e.g., Brown et al. 2007). In addition, contemporaneous tech-nology surrounding remotely sensed landscape cover is developing rapidly, allowing for higher resolution data to detect individual species and biological soil crusts (Karl et al. 2017). Moreover, data storage and retrieval technology has advanced to the point that retrospective fractional cover estimates are now available going back to the late 1980s using historical Landsat imagery (Allred et al. 2021) providing the ability to track temporal variation over larger scales. These data also create a broad spectrum of opportunities for both managers and researchers to retrospectively assess the effectiveness of habitat treatment practices and relationships between landscape cover attributes and population dynamics of wildlife species.

At local scales, a key attribute of habitat is to provide cover associated with a diversity of needs including nesting, brood-rearing, fawning/calving, breeding, roosting, and thermal regulation. Cover, generally in the form of vegetation, must occur in sufficient amounts to allow for species' survival and reproduction. Cover may act to decrease visibility of animals and nests (Conover et al. 2010), but can also act to disrupt air circulation patterns and reduce the ability of predators to find prey using olfactory cues (Fogarty et al. 2017). Cover also acts as a barrier to thermal extremes that could otherwise result in decreased fitness or death of wildlife species. For example, Guthery et al. (2001, 2010) reported that heat stress can result in decreased breeding activity and even death of northern bobwhite (*Colinus virginianus*), and that these consequences can be abated by habitat that serves as thermal refugia. At the other end of the spectrum, cover can also act to mitigate physiological stresses of winter thermal extremes for ungulate species such as mule deer (*Odocoileus hemionus*; Webb et al. 2013).

Cover for wildlife comes in two basic structural forms: horizontal and vertical. Horizontal cover (also known as "horizontal foliar density") refers to the degree of interception created by vegetation when habitat is viewed in a horizontal plane. The degree of interception will vary by height from ground level and the cumulative horizontal cover profile at a site is often referred to as "vertical structure" (Nudds 1977). Vertical structure can be a good predictor of habitat use by prey species (e.g., Holbrook et al. 2016) and is also an important determinant of habitat selection and reproductive success of many avian species (Hagen et al. 2007; Kennedy et al. 2009). Measuring vertical structure is accomplished via the use of a photoboard or pole painted in contrasting bands; vegetation obstruction of the board or pole (Griffith and Youtie 1988) is determined at a fixed distance using either digital photography or field estimates (Nudds 1977; Limb et al. 2007). Vertical, canopy, or foliar cover refers to the amount of land surface area obscured by vegetation when viewed from above. Canopy cover shapes wildlife habitat suitability through its influence on shading, which effects thermal properties of the habitat (Guthery et al. 2010), understory plant dynamics (Boyd and Bidwell 2002) and microenvironments (Royer et al. 2012), and is also the primary attribute impacting the ability of a habitat to protect prey species from overhead predators (Matthews et al. 2011). Canopy cover is also applied to both vegetation and non-vegetational components of habitat such as rock and bare ground, which can be important in describing both the ecological context of a habitat, as well as habitat suitability for some species (Conway et al. 2012; Pyke et al. 2014). In practice, the thermal and hiding cover afforded by a habitat will be a function of species requirements and the interactive effect of both horizontal and vertical cover attributes (Culbert et al. 2013).

A major function of an animal's habitat is to provide energy and nutrients necessary for survival, growth, and reproduction. Energy and nutrient sufficiency is a function of both the nutrient requirements of a species, which are subject to temporal variation in association with life history stage, as well as the dynamics of plant species composition, nutrient quality, and production in space and time (see discussion of the latter below). Links between animal performance at a given life history stage and the nutrients/energy provided by the habitat can be both direct and indirect. Nutritional limitations may directly induce weight loss, result in impaired growth and development, and decrease reproductive success (Boyd et al. 1996), particularly during periods of thermal extremes (DelGiudice et al. 1990, 1991). Insufficiency of nutrients/energy may indirectly affect individual animals and perhaps populations by negatively impacting physiological status of affected individuals and increasing the likelihood of mortality from disease or predation (Lochmiller 1996). Abiotic factors, such as thermal extremes or drought conditions can interactively exacerbate effects of nutritional limitations of habitats by reducing nutrient/energy availability and inducing physiological stress that increases an animal's nutrient/energy demand (Lochmiller 1996; Dabbert et al. 1997).

5.2.2 Climate, Weather, and Soil Influences on Rangeland Communities

Climate and weather factors are critically important in determining plant community responses to disturbance factors, as well as a plant community's potential for restoration success. In fact, weather, and to some extent climate variability, are the most frequent "it depends" caveats associated with generalizations of rangeland treatment effects or recovery trajectories of associated plant communities. Re-establishment of desired vegetation following disturbance often fails in rangeland ecosystems (Pyke et al. 2013) and the likelihood of success has been strongly tied to precipitation amount (Hardegree et al. 2011), timing, and frequency (Pyle et al. 2021) relative to the needs of seeded or recovering species, and all of the preceding factors interact with soil temperature (James et al. 2019) to determine recovery outcome.

Climate and weather have strong effects on rangeland productivity and composition, and by extension, the manipulation of rangeland wildlife habitats. The term "climate" refers to the long-term (e.g., averaged across years) patterns of precipitation, temperature, and other atmospheric properties for a given location. Climate differs from "weather" in that the latter refers to short-term variation (e.g., within year or shorter) in these same properties. At the continental scale, inter-annual to multi-decadal oscillations in temperature and precipitation are strongly influenced by recognizable ocean temperature patterns and circulation (Wang 2021). These ocean–atmosphere phenomena include the Pacific Decadal Oscillation, the El Niño Southern Oscillation, and the Atlantic Multidecadal Oscillation (McCabe et al. 2004; Guilyardi et al. 2009). While mechanics of how ocean temperature patterns influence terrestrial climate and weather are beyond the scope this chapter, both the effects and occurrence of these patterns are somewhat predictable and have been incorporated into management decision making on rangelands (e.g., Raynor et al. 2020). Climate is also changing in association with greenhouse gas emissions; predicted changes in climate, including more frequent droughts and severe weather, suggest that the influence of climate and weather on rangeland plant community dynamics will increase over time (Polley et al. 2017) and portend future challenges for management of rangeland plant communities and wildlife habitats. The extent to which ongoing climate change via greenhouse gas emission is influencing the occurrence of ocean–atmosphere phenomena is not well understood at present. That said, it is likely that some of the effects of climate change on rangelands (e.g., increased air temperatures) could interact with ocean-atmospheric associated events such as drought to decrease rangeland plant productivity (Schlaepfer et al. 2017). Alternatively, the ongoing increase in atmospheric CO_2 may be differentially increasing the production potential for some plant species, leading to altered successional dynamics and the potential for increasing rangeland fuel loads (Ziska et al. 2005). The bottom line is that substantial uncertainty exists regarding interrelationships between future climate and rangeland plant communities, reinforcing the need for active and adaptive management of rangeland wildlife habitats.

While climate factors associated with ocean-atmospheric events have some degree of predictability, the predictability of shorter-term weather conditions relevant to restoration projects or recovery from disturbance has proven more difficult and the useful accuracy of most forecasting techniques does not extend beyond 7–10 days (Hardegree et al. 2018). That said, current seasonal climate forecasts provide some level of generalization of weather conditions for periods up to several months (Doblas-Reyes et al. 2013) and new, more restoration-oriented products are emerging (e.g., Bradford and Andrews 2021).

While short-term forecasting of weather can be difficult, qualitative generalizations of site-associated temperature and moisture potential can be assessed using abiotic characteristics such as soils, elevation, slope, and aspect. For some rangelands, soil temperature and moisture regimes have been used by managers to assess the capacity for plant communities to both recover from disturbances such as fire or grazing (i.e., resilience), as well as their capacity to resist biotic change due to stressors such as invasive plant species (i.e., resistance; Chambers et al. 2014, 2016a, b). While these classifications can be useful from a management planning standpoint, site specific management should take into account current variability in climate and weather factors as well as biotic conditions of a site (Miller et al. 2014).

Soils quite literally form the biogeochemical foundation upon which rangeland wildlife habitats and other ecosystem services are built, and specific soil properties have strong influence on both plant community composition, and the resulting habitat structure (Evans et al. 2017). Soil texture is a fundamental property of the soil environment and has a strong role in influencing water availability for plants. Infiltration of water into the soil profile decreases as soil particle size goes from coarse to fine (i.e., in order of decreasing particle size: sand, silt, clay; Lowery et al. 1996). Water infiltration into the soil not only provides a supply of water to plants but also helps to prevent overland flow and surface soil erosion (Evans et al. 2017). The relationship between water holding capacity, or the ability of soil to trap and hold water, is inverse to that of water infiltration, with finer textured soils being more capable of retaining water. The impact of trading water infiltration potential for water holding capacity is moderated by annual precipitation. In arid regions, coarse soils can decrease evaporative loss, which off-sets reduced water holding capacity and increases water available to plants. In more mesic areas with less evaporative loss, the increased water holding capacity of finer textured soils results in increased soil water available for plants (Austin et al. 2004; Evans et al. 2017). Soil organic matter content is correlated positively with water holding capacity and can, to some extent, moderate the effects of particle size on soil water storage.

Plant species distributions within rangeland habitats are also influenced by soil chemistry. For example, saline soils support halophytic plants to the exclusion of non-salt tolerant species, while shinnery oak (*Quercus havardii*) mottes can create acidic soil conditions that approximate the pH of forest soils (Wiedeman and Pendound 1960). Soil pH, along with particle size and organic matter, can also modulate the persistence and efficacy of herbicides; although the specific effects are dependent on herbicide type (Duncan and Scifres 1983). Lastly, soil depth can influence water storage capacity of a site as well as competitive interactions between plants. In general, soil water storage decreases, and competition for belowground resources increases as depth to restrictive layer (e.g., bedrock) decreases; this accentuates the importance of understanding soil characteristics in predicting management outcomes. For example, Miller et al. (2005) reported that with sufficient rooting depth, perennial bunchgrasses were maintained during juniper (*Juniperus* spp.) woodland expansion in sagebrush (*Artemisia* spp.) steppe habitat, but in shallower soils bunchgrasses declined dramatically or were entirely absent with juniper expansion.

5.2.3 Rangeland Vegetation Dynamics

Understanding how and why rangeland plant communities and the associated wildlife habitats change over time allows managers to infer impacts on constituent wildlife, anticipate and act on opportunities for habitat improvement, and mitigate undesired changes. Change in rangeland plant communities can be broadly classified in terms of equilibrium and non-equilibrium succession. Under the non-equilibrium succession paradigm, vegetation dynamics are driven by stochastic, abiotic factors (e.g., precipitation) and herbivore density rarely reaches the level necessary to have strong impact on successional change in habitat conditions (Vetter 2005). In contrast, equilibrium succession refers to the idea that changes in plant community composition are mediated via density-dependent biotic feedbacks between herbivores (i.e., wildlife or livestock) and plant communities they utilize as habitat.

These paradigms have strong implications to policies relating to land use and management, and recognizing these differences is more than just an academic exercise. For example, biotic control of successional processes suggests that policies that control herbivore density (e.g., grazing regulations or wildlife harvest regulations) will stimulate desired changes in habitat conditions. Alternatively, abiotic control of succession would argue for policies that promote preemptive management to increase rangeland plant community resilience to episodically-stressful environmental conditions. While equilibrium dynamics undoubtedly play a role in successional change in some rangeland systems (particularly at small spatio-temporal scales), non-equilibrium dynamics are now recognized as the driving force behind plant succession in most rangeland ecosystems (Briske 2017).

Management toward or maintenance of desired habitat conditions involves using specific tools or processes to manipulate vegetation composition and structure. Equilibrium and non-equilibrium dynamics can have strong influences on the types of problems or challenges managers must overcome and implications these problems or challenges create for management planning and actions. *Simple* habitat management problems are those problems with solutions that are relatively invariant in space and time. From a habitat management standpoint, these problems are typically associated with plant communities undergoing equilibrium succession (Boyd and Svejcar 2009). For these problems, generalized solutions have broad management utility. An example of a simple problem might be reducing shrub fuels in an equilibrium system using a brush-beating technique. Results of brush beating are likely to be both successful and predictable in space and time (due to the equilibrial nature of the system) to the extent that treating 4 ha is synonymous with reducing the size of the problem by 4 ha for the effective life of the treatment.

Complex habitat management problems are those where the nature of the problem, and by extension appropriate management actions, will vary depending on the location and when the action will occur (i.e., space and time; Boyd and Svejcar 2009). Complex problems are usually associated with non-equilibrium succession. For example, restoration of perennial plants in arid or semi-arid rangeland systems is typically a complex problem. Choice of management techniques (or whether to even attempt restoration) in such systems is driven strongly by abiotic factors such as precipitation and temperature patterns that vary strongly in space and time. In this case, generalized solutions do not have broad management utility. Instead, habitat manipulations involving complex problems in non-equilibrium systems require a diversity of management techniques and tools to cope with a diversity of abiotically-driven habitat management challenges.

The process of setting habitat management objectives and selecting appropriate management actions to achieve or maintain those conditions in non-equilibrium systems can be guided by using state-and-transition models. State and transition models (Stringham et al. 2003) describe a range of potential plant community phases that dynamically shift in plant dominance or habitat structure within a relatively stable state (Fig. 5.1). Shifts, also known as pathways, among community phases within a state are generally viewed as reversable and influenced by both management and non-management factors. Movements between states are known as transitions and are relatively irreversible. Additionally, some states are sufficiently persistent that their existence represents what could be considered a new "novel ecosystem" (DiTomaso et al. 2017). For example, the invasion of exotic annual grasses in the Great Basin region of the United States has created vast areas of rangeland with near-monoculture abundance of these species. Because these species promote, and can persist in the presence of increased wildfire, these annual grass-dominated areas are extremely stable; some consider such areas to be novel ecosystems and suggest a management focus that recognizes the ecology (and management implications) of this alternative state as a new reference state (Davies et al. 2021).

Putting it all together, state and transition models represent an organized framework for managing plant communities and their associated wildlife habitats in an ecologically based manner. In reality, a seemingly infinite number of states could be present for a plant community assemblage because community phases are represented as static plant composition, but are merely a gradation of shifts in plant dominance that occur annually. Thus, the goal of constructing state-and-transition models for management is to assign this variability into as few states and phases as necessary so the model is sufficiently practical for management use, while retaining sufficient complexity to represent ecologically important plant community dynamics. The utility of these models for managing rangeland plant communities and their associated wildlife habitats can be increased by assigning values to states that are consistent with either measured population densities of target wildlife species (Holmes and

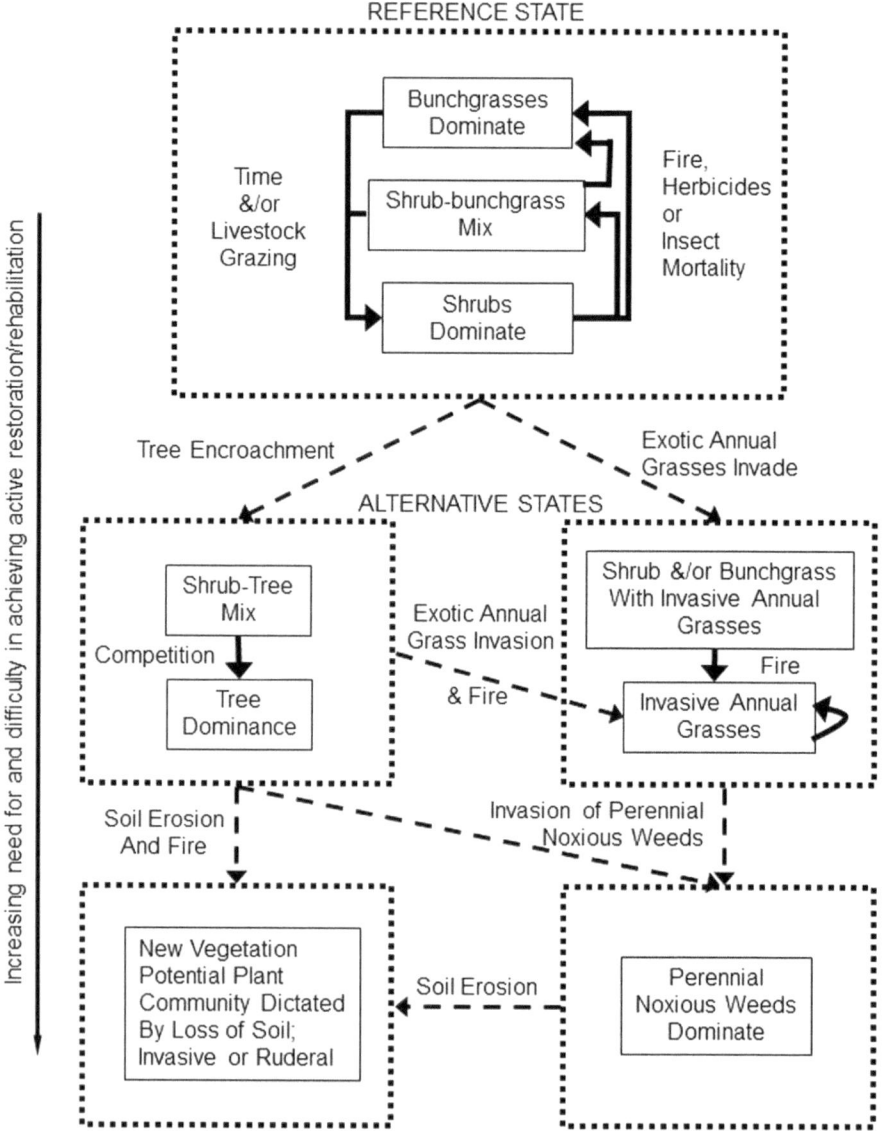

Fig. 5.1 Generalized shrub-grassland community with five *vegetation states (dotted rectangles)* with *community phases (solid rectangles)* within each state. *Pathways (solid arrows)* depict shifts in habitat dominance or structure within a state driven by biotic and abiotic influences. *Transitions (dashed arrows)* depict relatively irreversible changes in habitat dominance or structure

Miller 2010) or assigning qualitative values that represent the likelihood that habitat structure and plant composition of a state will service year-long or seasonal habitat needs of target wildlife species (Boyd et al. 2014).

For non-equilibrium rangeland wildlife habitats, knowledge of the plant community's resilience and resistance to disturbance will help define and guide management options. In this case *resilience* is defined as the capacity of ecosystems to reorganize and regain their fundamental structure, processes, and functioning (i.e., to recover) when altered by stressors like drought and disturbances such as fire or inappropriate livestock grazing (Holling 1973; Chambers et al. 2016a). *Resistance*, in turn can be defined as the capacity of ecosystems to retain their fundamental structure, processes, and functioning when exposed to stress (e.g., invasive species) or disturbance (e.g., fire; Folke et al. 2004; Chambers et al. 2016a). Characterizing the resilience and resistance of rangeland wildlife habitats involves examining both the abiotic and biotic environments. There are a host of abiotic factors that influence resilience and resistance of plant communities including temperature, precipitation, and a wide variety of soil factors. In practice a useful index to abiotic resilience and resistance can be created by characterizing soil temperature and moisture regimes across the area of interest into descriptive categories. For example, Chambers et al. (2014) characterized resilience and resistance of plant communities within the sagebrush ecosystem along a gradient from warm and dry to cold and moist; resilience and resistance increase along this gradient in accordance with increasing elevation and plant community productivity. These categories can be combined with habitat needs of a species or groups of species and geospatially depicted to help guide habitat management at broad spatial scales. For example, Chambers et al. (2016a, b) created a matrix that included all combinations of low, medium, and high resilience and resistance, combined with low, moderate, and high landscape cover of sagebrush. The resulting cells of the matrix create categories that can be geospatially depicted to guide management planning for the greater sage-grouse at large spatial scales (Fig. 5.2).

Utility of using resilience and resistance to inform habitat management will be increased by supplementing knowledge of contributing abiotic factors with current assessments of biotic properties, particularly at the project implementation scale. These biotic properties relate to the abundance of plant species within a community that have disproportionately strong influence on resilience and resistance. A good example is the influence that native perennial bunchgrasses have on resilience and resistance of sagebrush plant communities. These species effectively occupy space and utilize resources within the soil profile such that their abundance is highly and inversely correlated with probability of invasion by exotic annual grass species that are prevalent throughout the sagebrush biome (Chambers et al. 2007; Davies 2008). Thus, the abundance of perennial bunchgrasses can be used as a metric to identify and prioritize for management those areas within a landscape that are most likely to experience undesired change following disturbance. Additionally, the pre-treatment abundance of these species can be used to gauge the potential for unintended and undesired effects of active management treatments such as prescribed fire (Bates et al. 2000). At larger scales, assessment of biotic properties important to resilience

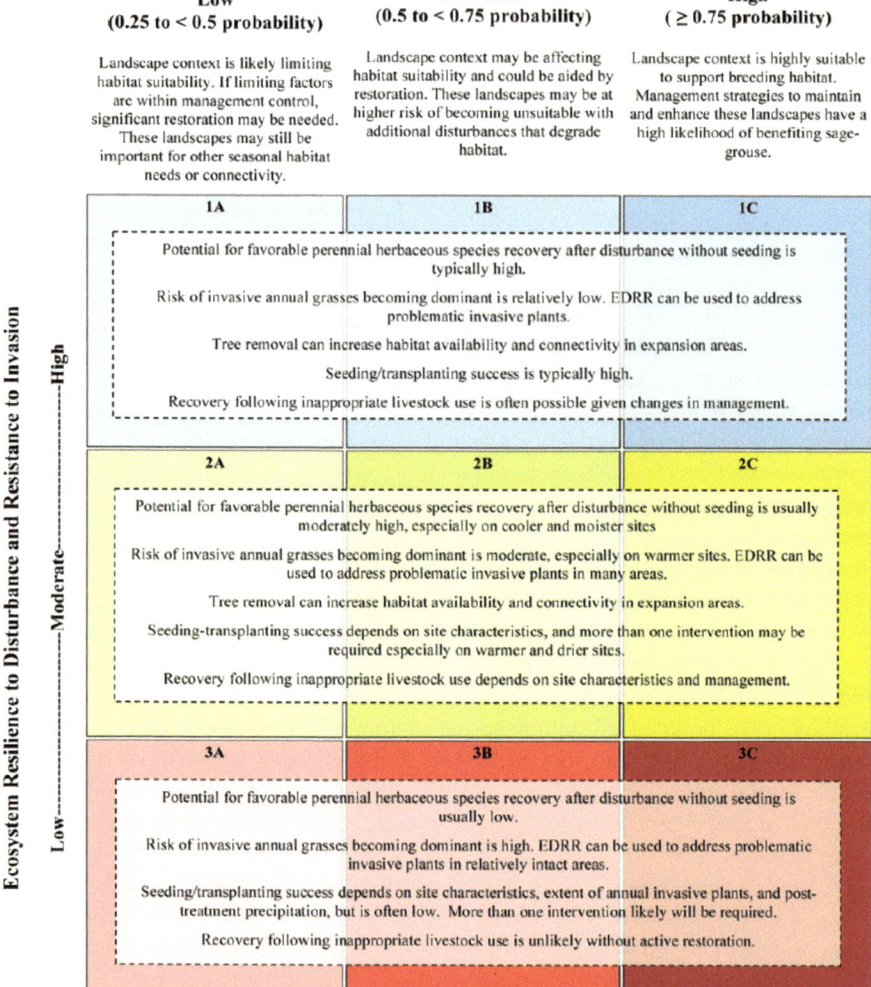

Fig. 5.2 Matrix depicting plant community resistance and resilience combined with landscape suitability for greater sage-grouse habitat. *Rows* indicate generalized recovery potential (resilience) and resistance to change during stress (e.g. exotic annual grass invasion). Increasing dominance of the landscape by sagebrush (depicted in *columns*) broadly suggests increasing suitability for greater sage-grouse. *Cells* within the matrix can be geospatially depicted to broadly inform decisions regarding management of greater sage-grouse habitat. Taken from Chambers et al. (2017)

and resistance will benefit greatly from emerging geospatial technologies such as the Rangeland Analysis Platform (Allred et al. 2021). These technologies not only allow managers and researchers to assess the abundance of vegetation functional groups across broad geographies, but can also be used to retroactively explore how plant communities responded to disturbances and management treatments.

5.2.4 Point-Based Versus Process-Based Habitat Management

One of the most basic challenges for contemporary rangeland wildlife habitat managers is to determine the relative priorities associated with management of ecosystem dysfunction versus the needs of individual species or groups of species of concern, and determining where those priorities do and do not intersect. As discussed earlier, the term "wildlife habitat" and by extension, wildlife habitat management, is an inherently species-specific, and often a life-history phase-specific premise; for example, we might use prescribed fire as a tool to create plant community structure suitable for nesting needs of black-capped vireo (*Vireo atricapilla*). Such management has generally been tied to specific micro-habitat requirements representing point-in-time vegetation conditions. We refer to this as *point-based management* (Table 5.1). In the case of vireos, fire can be beneficial to nesting habitats because this shrub-nesting species is picky about the height of shrubs in which it nests. When shrubs become higher than a desired height, the habitat is no longer suitable for nesting (Grzybowski 1995) and fire can be used as a tool to reduce shrub height. We can therefore think of point-based management as *practices applied to specific geographies that are intended to result in the floristic composition, structure, or spatial arrangement of plant communities needed to meet specific habitat needs of a species at a particular moment.*

Point-based activities define much of our history with wildlife habitat management on rangelands and the attraction to this type of management is multi-fold. For example, point-based management is easy to administer where land ownership boundaries define project areas (e.g., on private lands), and knowledge of species habitat requirements provides a clear picture of desired changes to habitats, which in turn suggests appropriate tools for the job. That said, if prescribed fire is needed to maintain proper nesting habitat for black-capped vireos, then how did this species successfully evolve (i.e., it successfully nested and reproduced) within these habitats for millennia? The answer probably relates to the fact that fire frequency in black-capped vireo habitat has decreased in modern times, creating conditions that favor sustained growth of woody plant species (Grzybowski 1995). While point-based management using prescribed fire may indeed create benefit to geographically specific vireo nesting habitats, managers should consider whether point-based habitat deficiencies are merely symptomatic of higher order issues such as declining fire frequency. This is an important distinction because if local habitat deficiencies

Table 5.1 Contrasts between form-based and process-based approaches to management of rangeland wildlife habitats

Characteristic	Management type	
	Point-based	Process-based
Goal	Modify habitat conditions to align with species habitat requirements	Modify ecosystem processes to create enabling conditions that influence desired future outcomes
Success metrics	Direct management effects on habitat composition and structure	Indirect management effects on ecosystem processes
Spatial focus	Plant community	Landscape
Temporal focus	Short term change	Long term change
Diversity of impact	Individual species or small groups of species	Groups of species to species guilds
Frequency of management inputs	Opportunistic	Persistent

are associated with disruption of ecosystem processes like fire, then point-based treatments may be creating islands of source habitat within landscapes that can act as habitat sinks and may also serve to obfuscate or even disincentivize management of ongoing system-level dysfunction, ultimately leading to reduced ecosystem resilience (Hiers et al. 2016). Evaluating the importance of local vs. landscape factors can be guided by frameworks (e.g., Pyke et al. 2015, 2017) that consider the spatial ecology of primary threats to plant communities and associated habitats, home range of the target species, types and locations of seasonal habitats, and the likely response of target habitats based on abiotic characteristics.

When fundamental ecosystem issues such as disruptions in fire frequency are driving undesired changes to habitat of desired wildlife species, a different management paradigm is required. In contrast to point-based management, the goal of *process-based* management is to *modify ecosystem processes to create enabling conditions that influence desired future habitat attributes* (Table 5.1). Effects of process-based management will differ from point-management in that they are indirect, often play out at relatively larger temporal and spatial scales, are more likely to impact a larger number of species, and are likely to require persistent management inputs over time. The need for a process-based approach to management of wildlife habitats is becoming increasingly wide-spread due to both direct effects of anthropogenic disturbance on ecosystem processes, and through the indirect effects of climate change (Walker and Salt 2006).

A good example of process-based management would be the conservation of low to mid elevation sagebrush habitats in the western US. The range of sagebrush plant communities has decreased dramatically since European arrival due to a variety of factors including agricultural conversions, oil and gas development, and housing development. Within sagebrush habitats the spatial footprint of wildfire has increased dramatically in recent decades, in part due to the dramatic expansion of exotic annual grass species such as cheatgrass (*Bromus tectorum*), which can create near continuous coverage of fine fuels that desiccate earlier in the growing season than native grasses; effectively lengthening the fire season. Sagebrush (*Artemisia* sp.) species within the region are easily killed by fire (Young and Evans 1978) and are difficult to restore following fire (Mietier et al. 2018), creating a conservation crisis for a host of sagebrush dependent wildlife species including the greater sage-grouse (Boyd et al. 2014). Improving habitat conditions for sagebrush dependent wildlife benefits from a process-based approach to create enabling conditions, namely treatments aimed at reducing fire occurrence and size, which allow for both active and passive restoration of degraded habitats, and maintenance of intact habits. As noted above, in the absence of enabling conditions, point-based restorative treatments run the risk of creating sink habitats within dysfunctional landscapes, and the benefits of successful point-based restoration attempts are time limited in accordance with fire dynamics (Boyd et al. 2017). Once enabling conditions have been achieved via process-based management, point-based treatments can then be used to impact habitats within the landscape to benefit sagebrush dependent wildlife species (Pyke et al. 2015). This sequencing of management emphases is consistent with hierarchical habitat selection by wildlife species (Johnson 1980) and can help bring clarity to management planning and increase effectiveness of conservation efforts in a growing diversity of complex and dysfunctional ecosystems.

5.3 Landscape Context for Wildlife Habitat Manipulations

5.3.1 Rangeland Loss and Fragmentation

Rangeland by definition is "land supporting indigenous vegetation that either is grazed or that has the potential to be grazed, and is managed as a natural ecosystem" (SRM 1998). Changes in land uses may modify vegetation to maintain a desired plant community that will benefit the new land use or they can completely replace the natural ecosystem with a simplified community of plants based on human desires (e.g., crops). Exurban, suburban, and urban development provide decreasing levels of natural plant communities with increasing levels of buildings and human infrastructure. In northeastern Colorado, ground- and shrub-nesting bird species diversity declined in density with movement from rangeland to exurban developments and while domesticated cats and dogs increased along the same gradient (Maestas et al. 2003). In 2019, half of the top ten states in percent population growth were states with non-federal rural lands dominated by rangelands (Table 5.2). Current rangeland watersheds with the greatest projected housing development through 2030 are around southern California cities, Las Vegas, Nevada and Phoenix, Arizona and will result from exurban development (Reeves et al. 2018). Some of this exurban development will lead to conversion of farmland to ranchettes, whereas rangelands are then converted nearly simultaneously to farmlands (Emili and Greene 2014).

Table 5.2 Top ten states in percent growth of population between 2018 and 2019 (US Census Bureau 2019; USDA 2020)

Rank	State	2018	2019	Percent growth (%)	Percent of rural land that is rangeland (%)
1	Idaho	1,750,536	1,787,065	2.1	36.7
2	Nevada	3,027,341	3,080,156	1.7	85.1
3	Arizona	7,158,024	7,278,717	1.7	82.9
4	Utah	3,153,550	3,205,958	1.7	64.7
5	Texas	28,628,666	28,995,881	1.3	59.2
6	South Carolina	5,084,156	5,148,714	1.3	0.0
7	Washington	7,523,869	7,614,893	1.2	21.9
8	Colorado	5,691,287	5,758,736	1.2	61.1
9	Florida	21,244,317	21,477,737	1.1	9.9
10	North Carolina	10,381,615	10,488,084	1.0	0.0

There is a flux between rangeland and farmland area in some locations of the US due to economic fluctuations of crop prices, disaster payment, and conservation incentive policies (e.g., Conservation Reserve Program) to convert farmland to rangeland and the reverse with consequences to wildlife habitat and populations (Rashford et al. 2011; Drummond et al. 2012; Smith et al. 2016; Lark et al. 2020). Coupled with human development comes the need for roads, irrigation, power and water lines, fences and often changes in the vegetation. These manipulations create the potential for wildlife habitat fragmentation even when they do not directly impact the majority of rangeland plant communities (Reeves et al. 2018). This human footprint can have substantial impacts on some wildlife species (Leu and Hanser 2011). For wildlife with large landscape patches of habitat, synanthropic predators of these wildlife species may increase with greater human activities or infrastructures on the landscape and threaten population survival of prey species. An example is increased Corvid predation on greater sage-grouse nests with increased human activity or structures (Coates et al. 2016). Alternatively, direct losses of habitat for these wildlife prey species may reduce land available for critical life history stages or may isolate their populations through removals of corridors between habitat patches.

Past, current, and future social and economic needs have and will continue to shape land uses, while new technologies may allow spatial placement of land manipulations in habitat-friendly locations minimizing habitat losses while allowing resource extraction or land uses. For example, horizonal drilling for oil and gas with multi-bore well pads located on or near existing roads or human infrastructure corridors (Thompson et al. 2015; Germaine et al. 2020) may minimize wildlife habitat impacts.

Livestock grazing occurs throughout rangelands and infrastructures to manage and promote livestock production may also impact wildlife. In southern Alberta, Canada, there are 77% more km of fence than all roads combined including unimproved roads. For example, fences intended to impede movement of sheep will also impede movement of pronghorns (Gates et al. 2012) and modelling indicates that fences restrict habitat area available to pronghorns (Reinking et al. 2019). Mineral licks, both natural and human-placed licks are common attractants for wildlife (Kreulen 1985; Robbins 1993). Seasonal gestational benefits of mineral licks are suspected for some wild ungulates (Ayotte et al. 2006), however recent information indicates they are potential locations for transmission of wildlife diseases (Payne et al. 2016; Plummer et al. 2018). Seeps and spring development is another livestock-related development that has potential beneficial and detrimental impacts for wildlife. Water developments of springs or seeps that capture and pipe water to troughs may result in dewatering of these areas and in reducing the wetland vegetation associated with these sites impacting wetland-dependent wildlife and insects especially in arid rangelands (Parker et al. 2021). Well-designed water developments that spread water across landscapes and are available to wildlife may have benefits to some wildlife (Bleich et al. 2005; Gurrieri 2020).

5.3.2 Broad-Scale Decisions

A review of the literature indicated that only about 10% of terrestrial restoration projects considered landscape characteristics in locating projects (Gilby et al. 2018). Considering landscape requirements and threats for wildlife species at a broad scale, usually greater than a typical size of a restoration project (tens to hundreds of hectares) can increase effectiveness of vegetation manipulations for creating habitat that benefits one or more populations of the species.

It is important to recognize that all wildlife species have broad and site scale habitat needs while simultaneously recognizing that multiple species may overlap in landscapes and coexist during certain times while other species may use the same geographic locations, but at different times or seasons. For example, Garcia and Armbruster (1997) evaluated the U.S. Bureau of Reclamation, Lonetree Wildlife Management Unit in North Dakota for four proposed habitat manipulations to improve gadwall (*Mareca strepera*) habitat while maintaining sharp-tailed grouse (*Tympanuchus phasianellus*) habitat. They modelled four scenarios and incorporated economic costs of manipulations into their results on gadwall and sharp-tail grouse. Their model was limited to a few populations found on the refuge. Other models use a regional approach with multiple populations and varying habitats, but these are rare (Doherty et al. 2016). Rarer still are models that consider optimal locations for restoration across broad scales (Ricca et al. 2018; Ricca and Coates 2020).

Creating vegetation goals that meet the animal's vegetation community and structural needs alone may not create wildlife habitat without considering other landscape factors that may restrict the animal's use or movement. For example, the habitat manipulation goal for a shrub-obligate animal might be to clear trees that are roosting habitat for predators and to create shrub habitat through releasing understory shrubs and herbaceous vegetation from competition with trees. But if this cleared patch is not connected to an adjacent shrub habitat without trees, the animal may never use the treated area because there is no connection to safe habitat. The vegetation objective of clearing trees and releasing shrubs and herbaceous vegetation could be achieved, but the wildlife objective would not because the manager failed to consider the connecting landscape of treeless area necessary to provide the animal access to the cleared patch. A decision framework for landscape-level habitat manipulations may assist in providing managers with queries to consider for optimizing animal benefits from habitat manipulations.

5.3.2.1 Does the Animal's Population Cover a Broad Scale of Land Types?

Affirmative answers to one or more of the following questions will indicate the potential that an animal's range covers a broad scale of land types that some people refer to as a landscape species.

1. Does the animal depend seasonally on multiple vegetation communities for population survival?
2. Does the animal migrate seasonally?
3. Is the animal's seasonal or annual home range larger than the typical manipulation project?
4. Will habitat use of a manipulated area depend on current use of adjacent areas?
5. Will spatial gradients of environmental variables impact the achievement of manipulation goals?

5.3.2.2 Define Regional or Broad Scale Landscape Objectives for Habitat Manipulations

Landscape objectives should be defined with the knowledge of how to monitor to determine movement toward or away from the objective over time. These objectives will likely deal with metrics obtained over large spatial or temporal scales. For vegetation components, remotely-sensed data is often used to determine changes in vegetation dominance over time and vegetation patch inter-relationships with surrounding habitat patches. Coupled with vegetation metrics, it would be optimal to determine any animal population or use objectives related to vegetation manipulations within treated regions or landscapes (Pilliod et al. 2022). Examples of objectives include:

1. Increase connectivity among populations or seasonal habitat by 5% within the region in 10 years.
2. Develop a system of fire breaks to protect priority habitat and to maintain no net loss of habitat and population levels in the region for the next 10 years.

5.3.2.3 Identify Components Necessary to Meet Landscape Objectives

Generally, managers do not have the capacity (physical or financial) to restore all landscapes and sites that require restoration within region. A priority structure will aid planning and hopefully target manipulations to locations within the region where the likelihood of achieving objectives will be the greatest. This identification process is a triage of the entire landscape. The first step in this process is to identify data layers that define landscape or regional objectives for the habitat and the animal. These objectives and data layers may include, but are not limited to:

1. Increasing connectivity among seasonal habitats or among separated populations might require maps of existing habitats and barriers for movement between habitat or populations.
2. Conserving high quality habitat from future threats through risk maps of known threats (e.g., fire, invasive species, development, climate change).
3. Mapping potential habitat locations for beneficial manipulations.

5.3.2.4 Identify Existing Habitat, Potential Habitat, and Wildlife Population Trends Associated with Those Habitats

This stage provides data on the current state of populations across the landscape and the habitat quality of those populations. This information is useful in determining population strongholds where habitat connections might create new avenues for genetic exchange among separated populations. Maps of current vegetation relative to potential vegetation can aid decisions on where manipulations may produce habitat and create corridors for population interchange. Knowledge of state and transition successional models and maps of current and potential vegetation and of soils and their associated descriptions of ecological dynamics may be useful at this stage.

5.3.2.5 Identify Landscapes with Locations that Best Meet Habitat Criteria

This step is accomplished either through a series of map overlays to examine unions of spatial criteria or through a series of models using these criteria. Typically, results are a series of gradations illustrating locations where manipulations are likely to benefit populations on one end to those likely to negatively impact populations on the other (Figs. 5.2 and 5.3). Information similar to Fig. 5.2 provides managers with management options and potential outcomes, while Fig. 5.3 incorporates the potential outcomes to the animal's population given other factors that might regulate the animal's use of the landscape.

5.4 Site-Scale Habitat Manipulations

Rangeland manipulations conducted at specific sites may be intended for improving habitat for wildlife or they may have other intended goals (e.g., livestock forage production, fuel reduction, watershed health) that have wildlife consequences.

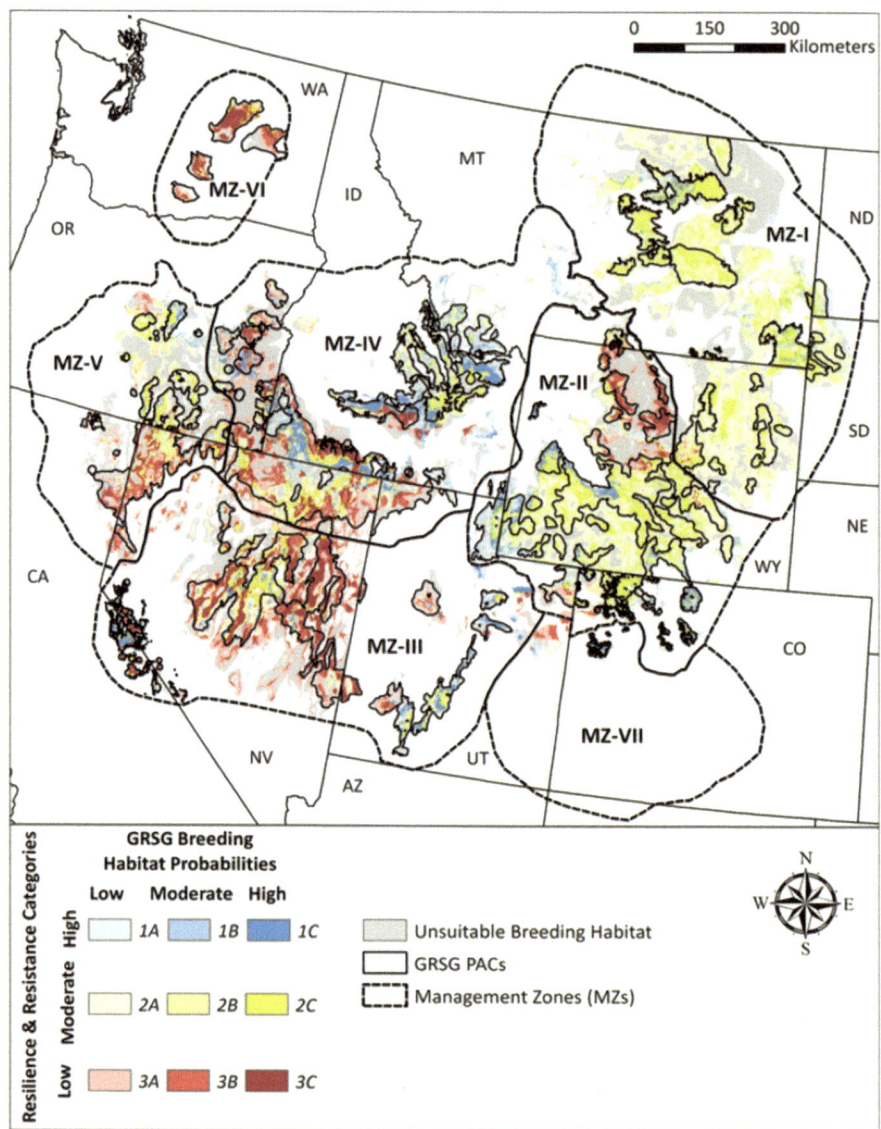

Fig. 5.3 The union of Greater Sage-grouse (*GRSG*) breeding habitat probabilities (*A, B, C*) with sage-grouse habitat resilience and resistance (*1, 2, 3*) within each of the seven management zones (*MZ—dashed polygons*) and Priority Conservation Areas (*GRSG PAC*) across the current range for the GRSG in the USA. Taken from Chambers et al. (2017)

These consequences can range from beneficial to detrimental depending on the type of manipulation, its extent and intensity, its location relative to other habitat requirements, and the wildlife species.

Outcomes of rangeland manipulations will depend upon a variety of factors, such as, treatment objectives, methods and configuration, weather, climate, and post-treatment management. Ideally, site-level habitat manipulations are formulated with the idea of providing information useful for adaptive habitat management. We provide six important considerations for a manipulation to be effective (Pyke et al. 2017).

5.4.1 Develop Site-Specific Management and Sampling Objectives

Properly written objectives will provide the spatial and temporal elements of the proposed effective habitat manipulation and will provide guidance for data collection and level of change necessary to determine manipulation success (effectiveness monitoring). A properly written habitat manipulation objective typically includes the following (Elzinga et al. 1998):

1. The target plant species, groups of species, or ecological conditions (e.g., a plant species, all shrubs, or bare soil) that will be measured to determine success,
2. Location of the manipulation,
3. The measurement attribute (e.g., cover, density, height),
4. The action of change (e.g., increase, decrease, limit, or maintain),
5. The quantity or qualitative state of the anticipated change,
6. The time frame for success.

5.4.2 Consider Ecological Site Characteristics

Ecological sites comprise "a land classification system that describes vegetation, ecological potential, and ecosystem dynamics of land areas" (https://www.nrcs.usda.gov/wps/portal/nrcs/main/national/landuse/rangepasture/ Accessed 04/18/2021). This system of land classification was developed by the U.S. Natural Resources Conservation Service and has become standardized for use across multiple Federal land management agencies (Caudle et al. 2013). An individual ecological site is "a distinctive kind of land with specific soil and physical characteristics that differ from other kinds of land in its ability to produce a distinctive kind and amount of vegetation and its ability to respond similarly to management actions and natural disturbances" (https://www.nrcs.usda.gov/wps/portal/nrcs/detail/national/landuse/rangepasture/?cid=stelprdb1068392 Accessed 04/17/2021).

Ecological site data for a location are identified through the Web Soil Survey (https://websoilsurvey.nrcs.usda.gov/ Accessed 04/17/2021) where an interactive map allows the user to outline an area of interest for the habitat manipulation.

Specific ecological site descriptions (ESD) of individual ecological sites are found at the Ecosystem Dynamics Interpretive Tool (EDIT; https://edit.jornada.nmsu.edu/catalogs/esd Accessed 04/17/2021). Each ESD includes a state and transition model for the ecological site that describes stressors that may shift vegetation dynamics to alternative stable states.

State and transition models provide information on the vegetation community dominance in plant community phases in the reference state and in alternative states. The current vegetation at the manipulation site is compared to these ranges of plant communities in the array of states in the state and transition models to determine the potential for a habitat manipulation to achieve the habitat objective. Manipulations that may drive the community to one of the phases in the references state, as opposed to those in an alternative state, are most likely to provide the greatest resilience to further disturbances and resistance to invasive plants (Chambers et al. 2017).

5.4.3   Determine Land Use and Disturbance History

Past, present, and future land uses and the previous disturbance history may provide managers with information regarding the time period necessary for successfully achieving habitat objectives. In some cases, previous disturbances or land uses may have led to the current vegetation at the site and may require changes in these uses to achieve the objective. Before implementing a manipulation to a site, managers might consider if previous manipulations have been done to the site and if those were successful. On Bureau of Land Management property, the Land Treatment Digital Library provides available spatial information on historic land manipulations and reports on their success in meeting objectives (https://ltdl.wr.usgs.gov/ Accessed 04/17/2021). Some disturbances may have led to a loss of ecological potential through the loss of soil erosion as an example. This loss of potential may determine whether the proposed manipulation can create the proposed habitat. Another GIS-based tool to assist managers in making decisions to move forward with manipulations at a proposed site is the Land Treatment Exploration Tool (https://www.usgs.gov/centers/forest-and-rangeland-ecosystem-science-center/science/land-treatment-exploration-tool Accessed 04/17/2021).

In addition, Interpreting Indicators of Rangeland Health (Pellant et al. 2020) is a fast on-site assessment of the soil, hydrology, and biotic potential that can assist managers in determining if site potential has been lost. Ratings of departures from reference conditions (the potential for the site) that are more severe than moderate, especially for soil and site stability and hydrologic function, may provide an indication that attaining the ecological potential for this location may not be possible; even with revegetation, the soil or water on the site may no longer function at a level that can support the potential vegetation and managers may be left with alternative states and with questions if desirable habitat can be created with ecological processes in which the site contains.

5.4.4 Consider the Role of Pre- and Post-treatment Weather

Weather is a critical element in regulating plant responses, but it is outside the control of the manager attempting to modify or create wildlife habitat. The weather before a habitat manipulation may dictate existing plant's vigor which relates to nutrient status of the plant and the storage of nutrients in stems and roots immediately before a manipulation that may partially cut or damage plants requiring regrowth after the disturbance. If the manipulation is intended to reduce the damaged plant for as long as possible, then weather before the disturbance that reduces the plant's vigor may delay regrowth and extend the habitat objective, such as reducing woody plants. However, if the objective is to increase a group of plants through growth or seed production and establishment, but plants are in poor vigor, then the disturbance may not achieve its objective (Hardegree et al. 2012). In the future, models may incorporate past weather and future weather predictions to assist in projecting plant responses to habitat manipulation (Hardegree et al. 2016).

5.4.5 Evaluate Plant Removal Methods and Associated Effects

5.4.5.1 Passive Manipulations

Passive forms of manipulations generally involve changes in current land management with an expectation that plant community dynamics will respond with changes in plant dominance to create the desired wildlife habitat. For example, changes in livestock management may include changes in stocking rates including elimination of use, changes in livestock periods of use, distribution, or changes in the type of livestock grazing the area.

Targeted grazing is a passive form of manipulation where a class of animal grazes for a set season and duration at a set stocking rate to shift plant species or lifeform dominance in an area (Frost and Launchbaugh 2003; Bailey et al. 2019). Targeted grazing for fuel reductions typically require fencing or herding animals to graze live and standing dead plants that may become fuels for wildfires (Fig. 5.4a). In addition, animals can learn to feed on plants they may not prefer normally or to avoid plants they may normally prefer through conscious and subconscious learning. This type of targeted grazing requires diet conditioning (i.e., training). Conditioning is a natural process that young animals learn from their mothers in utero or from milk and then is reinforced by following their mothers and eating the same foods while experiencing similar flavors and nutritional responses (Nolte and Provenza 1992; Nolte et al. 1992). Diet conditioning can also be used to teach animals to avoid certain plants (Lane et al. 1990) or novel plants that are previously unknown (Walker et al. 1992; Dietz et al. 2010). Supplements with polyethylene glycol, protein, and

energy may increase the use of some plants by animals, but these are species- and animal-specific (Bailey et al. 2019).

Some evidence suggests that livestock grazing before wildlife arrive to an area may increase the wildlife forage use of the area. Bailey et al. (2019) document several studies indicating that livestock grazing improves forage for wildlife. However, most studies only documented the improved nutrient levels of the forage, not increased wildlife use of these locations. Crane et al. (2016) provides an exception by demonstrating increased elk use in areas previously grazed vs. ungrazed by cattle. There are many hypotheses for creating habitats through restoration and manipulation of the current environment, but the full set of ecosystem complexities are rarely tested (Hilderbrand et al. 2005). When manipulating a community to create habitat for

◄Fig. 5.4 a Targeted Grazing—Cattle being used to graze cheatgrass in Nebraska to reduce cheatgrass seed production and population and help recovery of mid-grass prairie. **b** Prescribed Fire—Ruby Lake National Wildlife Refuge uses prescribed fire to reduce undesirable plants and release desirable vegetation for waterfowl. Mechanical Removal—**c** Bull Hog masticating juniper tree in Utah and **d** Cut, drop and leave is one form of woody plant removal practiced in Oregon's Bureau of Land Management lands. **e** Pelleted herbicide tebuthiuron being used to thin shrubs in Washington. This same method is used to aerially broadcast seeds for restoration of desirable plants. **f** Biocontrol—Salt Cedar (*Tamarix chinensis*) has infested many riparian areas of the Southwestern US, but introductions of tamarisk leaf beetle (*Diorhabda elongata*) (inset) often control this invasive plant. Photo Credits. 4A. Julie Kray, USDA ARS, Fort Collins Colorado—Photo is in Scottsbluff, Nebraska. 4B. US Fish and Wildlife Service—Ruby Lake National Wildlife Refuge, Ruby Valley, Nevada https://usfws.medium.com/using-prescribed-fire-to-improve-habitat-and-save-wildlife-c836453d51b0. 4C. Onaqui, Utah SageSTEP Project site—Photo by Brad Jessop, Bureau of Land Management Utah 2006. 4D. Middle of Nevada—Photo taken on June, 2011, Natural Resources Conservation Service media folder—https://www.nrcs.usda.gov/Internet/FSE_MEDIA/nrcs144p2_036837.jpg. 4E. Moses Coulee SageSTEP project site, Washington—Photo taken by Scott Shaff, U.S. Geological Survey—November 24, 2008. 4F. Photo and inset photo from Glen Canyon National Recreation Area, Utah—Photo National Park Service Photo—Date unknown for both photos. Main Photo—https://www.nps.gov/glca/learn/nature/images/Tamarisk-Minimally-Impacted-by-TLB-web.jpg; inset—https://www.nps.gov/glca/learn/nature/images/saltcedar-leaf-beetle.JPG

wildlife, monitoring for wildlife use and ultimately population trends would be helpful for adaptive management (Pilliod et al. 2022).

5.4.5.2 Active Manipulations

Active manipulations are necessary when passive management changes and successional processes are inadequate to meet objectives, whether for wildlife or for other reasons. Active manipulations include fire-, mechanical-, and chemical/biological/microbial-induced modifications to physical or biological components of the ecosystem (Fig. 5.4b–f).

Prescribed fires can be useful tools when they remove or reduce undesirable vegetation and encourage growth and dominance of desirable plants while not making the community vulnerable to undesirable physical or biological components of the ecosystem (e.g., soil erosion, hydrophobic soils, invasive plants). Tolerance of and susceptibility to fire depends on whether the entire plant is consumed by fire and can regrow after a fire (Pyke et al. 2010). The Fire Effects Information System (FEIS; https://www.feis-crs.org/feis/ Accessed 21 April 2021) provides information on the susceptibility of individual plant species to fire; useful information for a manager deciding whether to use fire for creating habitat.

Prescribed fires are modulated through adjustments in fire: (1) intensity by manipulating fuel amount and packing, (2) duration by the size and cellular density of fuel or by the fire type (e.g., surface vs. crown fire or backing vs. head fires), (3) extent and patchiness of burned areas (Pyke et al. 2017). The heat created (intensity) by fire and duration of that heat will determine its effect on plants and seeds (Whelan

1995). Larger fires that kill plants with limited seed banks or regrowth mechanisms will increase the time required for those plants to disperse to the site and recover, especially for plants with limited dispersal mechanisms. Consult with trained fire manager in developing fire objectives to meet habitat objectives.

Mechanical and Chemical manipulations (e.g., Fig. 5.4c–e) use several potential pieces of equipment to modify vegetation on rangelands (https://greatbasinfirescie nce.org/revegetation-equipment-catalog-draft/ Accessed 21 April 2021). Methods of habitat manipulations can range from those that remove all plants to those that are more selective for removing or thinning species or lifeforms. Mechanical equipment that operates entirely above the soil surface is intended to remove or reduce height and cover of vegetation. Shrubs with limited resprouting ability or without adventitious or perennating buds on remaining live, woody tissue will be reduced in dominance more than those with these resilience mechanisms; similar to the effect of fire (Pyke et al. 2010). The FEIS provides information on resprouting ability of plants. Mechanical equipment that digs into the soil kills or reduces the dominance of all plant life forms impacted with the exception of plants with strong adventitious buds on roots or rhizomes. Some equipment, such as tractor-pulled anchor chains, not only removes large trees and shrubs, but also remove some herbaceous plants (grasses, grasslike and forbs) when they dig into the ground. Plows and harrows cause similar effects. These areas of soil exposure may result in soil erosion and invasive plant establishment and spread, especially in years immediately after treatment. Before treatment, consider if invasive plants already exist on the site and might increase and spread with soil disturbing treatments. Seeding with desirable plants and using herbicides focused on invasive plants may be necessary to limit invasive species and encourage desirable plant establishment and growth.

Miller et al. (2014) suggests considering a series of questions to weigh the monetary and ecological costs and benefits of using mechanical treatments to manipulate plant communities. These include: (1) will equipment create unacceptable soil compaction? Wet, fine-textured soils are more susceptible to compaction than dry, course-textured soils. Mechanical manipulations in the dry season or when soils are frozen may reduce the severity of soil compaction. (2) Will the mechanical manipulation create unacceptable amounts of mineral soil exposed to raindrop impact and will these patches be on steep slopes? Bare soil is vulnerable to invasions of undesirable plants and to soil erosion. Larger patches of bare soil are susceptible to wind- or water-induced erosion, whereas the steeper the land's slope, the greater the potential for water-induced soil erosion. (3) Will the manipulation disturb biological soil crusts (biocrusts)? Biocrusts are soil surface lichens, mosses, algae, and cyanobacteria that adhere to soil particles and protect soil from wind- and water-induced erosion. In some arid and semi-arid environments, biocrusts can also fix nitrogen for use by other organisms in the ecosystem (Belnap and Lange 2003). (4) Will the mechanical treatment damage existing perennial grasses and forbs? If the intention of the mechanical treatment is to reduce woody plants, then the resilience of the remaining plant community constituents and their resistance to invasive plants is important. If the mechanical treatment impacts community components that are necessary for community resilience and resistance, then the resulting community after the treatment may achieve its objective of reducing woody plants, but may ultimately degrade the site through loss of soil or reduced hydrologic capacity. (5) Will the treatment provide a seedbed for seedling establishment? If a mechanical treatment is accompanied by a reseeding treatment, then a seedbed for seedling establishment is important, but recognize that if the community already has invasive plants, the mechanical treatment may enhance invasive plant establishment and create a competitive environment for the reseeded desirable plants. (6) Will changing the timing of treatments influence plant response positively or negatively? Consider what is the optimum manipulation time to reduce potential negative and maximize positive outcomes.

Herbicides can be selective, affecting only certain plant life forms, or non-selective (broad-spectrum) potentially affecting all plant life forms. Some broad-spectrum herbicides can become selective for certain plant groups by manipulating the timing or application rate. In addition, each herbicide is registered for uses on different types of lands. Be certain when selecting an herbicide that it is registered for use on rangelands and follow all label instructions. New herbicides are being tested and released annually. Work closely with a licensed herbicide applicator in selecting, planning, and applying an herbicide.

Herbicides rarely eradicate a target plant species or group, but they often reduce targeted species for a period of time. The removal of a target plant will often leave a void for other plants to fill. If desirable plants do not fill those vegetation gaps, undesirable plants, even the original target plant, may re-establish and dominate the site. Seeding chemically treated areas with desirable vegetation may be necessary in environments where residual vegetation is not sufficient to fill voids left by removed vegetation.

Biocontrols are sometimes used to reduce undesirable plants (McFadyen 1998). Targeted grazing is a form of biological control, but insects are the most common form of biocontrol of weedy plants. In addition, biocontrols can include microbial pathogens (e.g., fungi, bacteria, and viruses; Harding and Raizada 2015). Insects are generally released by hand at a site, while microbes are often applied using methods similar to herbicide applications since they can be mixed with water, pelletized, or coated on seeds or degradable inert biological forms such as rice hulls.

Effectiveness of biocontrols has been variable. Effective biocontrols generally do not eradicate the target plant. Complete elimination of the target would likely eradicate the biocontrol agent too. Therefore, biocontrols may reduce undesirable plant species to low levels and should the target plant increase, the biocontrol's population would ideally increase as their food source increases. Provided the biocontrol agent reduces the target plant, a concomitant objective should be for desirable vegetation to increase to fill the void left through the death of the undesirable plant.

Revegetation (Figs. 5.4e and 5.5a–d) is used when desirable vegetation populations are insufficient to provide propagules to fill the void in an adequate timeframe after undesirable plants are removed. The timeframe will vary depending on the site's resilience and resistance; sites with low values often need propagules to establish and dominate in less than ten years and those with high values having larger timeframes. Managers may consider whether to seed or plant juvenile plants. Plant species selected for creating wildlife habitat through revegetation is a union of the group of plants defined as habitat species and plants that have the potential for existing and successfully reproducing on the site. The best source of information for selecting native species is the ecological site description for the site. Examine the plant community phases found in the state and transition model and select the plant community phase that matches the ideal life-forms to provide habitat composition and structure for the target wildlife species (e.g., trees, shrubs, grasses and grass-like and forbs). Include in the revegetation mixture plant species that would dominate the site and are currently in insufficient numbers for the site.

The geographic source of the propagule used in a revegetation project is important for establishment and for sustaining future generations of plants on the site. Foresters have known for decades that seed source is important for matching a tree's genetics to the environment where it will be grown (Johnson et al. 2004). They use seed zones for collecting and planting reforestation projects. Rangeland provisional seed zones are proposed for some regions (Bower et al. 2014; https://www.fs.fed.us/wwe tac/threat-map/TRMSeedZoneData.php Accessed 04/23/2021) and when used may improve revegetation success. Climate change has sparked considerations for using assisted migration techniques to move species or ecotypes within species from lower to higher elevations or latitudes (Loss et al. 2011). Although these approaches have been considered hypothetically, they are mostly in the testing phases (Wang et al. 2019).

Plantings and Seedings After selecting the species and propagule source, the type of revegetation method is determined. Seedings are either broadcasted (Fig. 5.4e; aerial or ground-based) or drilled (Fig. 5.5a, b). Plantings can come in several forms (Shaw 2004) and are most often conducted with woody species. Small container-grown plants are started in greenhouses, hardened to the environment, and transplanted at the site with their roots contained within a potting soil. Bare-root plantings are initially grown in gardens in a loose compost soils, then the plant and roots are extracted from the soil immediately before planting at the revegetation site. Cuttings of shrub branches are taken from live plants and the cut branch is planted in the soil and allowed to root. This is a common technique for shrubs in riparian areas because branches can produce adventitious roots in moist soil. Wildings are small plants extracted, with their soil, from an existing site and planted at a new location. This is a good approach for salvaging plants that might be destroyed where human development would require plant removal before development. Planting techniques are often labor intensive, but may provide greater establishment than plants germinating and establishing from seeds.

Seeding projects are the most common form of revegetation (Hardegree et al. 2011; Pilliod et al. 2017). Drill seeding is generally considered the most successful seeding method because the seed drill places seeds at the appropriate depth in the soil for germination and emergence of the seedling. Broadcasting seeds, when used alone without other soil disturbing techniques (e.g., anchor chains, or harrows), places seeds on or slightly above (if litter exists) the soil surface where they are vulnerable to predation or displacement by wind and water (Stevens and Monsen 2004). Drill seeding often requires some site preparation (e.g., fire) to remove any larger woody vegetation that would limit the use of a tractor or would bind in the

Fig. 5.5 Rangeland drills are designed to seed multiple species at different depths of soil. Traditional rangeland drill (**a**) that places seeds in furrows (**c**). In contrast, minimum-till rangeland drill (**b**) leaves the soil flat (**d**) after placing seeds. Photo credits. **a, b, c** and **d**. Location likely Mountain Home, Idaho in 2006. Photo by US Forest Service. Image currently on Great Basin Fire Science Exchange, Revegetation Equipment Catalog, but originally in Joint Fire Science Final Report, Project #07-1-3-12 by Dr. Nancy Shaw, USFS, https://www.fs.usda.gov/rm/pubs_other/rmrs_2011_shaw_n003.pdf. Photo now found on https://revegetation.greatbasinfirescience.org/wp-content/uploads/2021/01/LRangelandDrillRightMinTillDrill_SoilDisturbance_USFS-294x300.jpg

seed drill. If tractors and drills are limited by obstacles or terrain, aerial seeding is the best seeding method.

Emerging seeding technologies are being tested and may prove helpful in increasing seedling emergence, establishment, and competition with invasive species and decreasing seed predation. Coating seeds with hormones to hasten or delay germination may insure that germination occurs at the ideal time of the season or may allow a bet-hedging strategy with seeds germinating over a longer timeframe than normal (Madsen et al. 2016, 2018; Davies et al. 2018). Seeds encompassed in pellets with activated carbon may allow simultaneous herbicide applications of preemergent herbicides to reduce invasive plants while the pellet absorbs and retains the herbicide allowing safe germination of desired species (Brown et al. 2019). Coating seeds with materials that prevent animals from eating seeds may alleviate seed predation common with broadcast seeds (Pearson et al. 2019).

Restoration of biocrusts is another emerging field that may become common for arid and semiarid environments where biocrusts are an important ecosystem component for rangeland health. Biocrust production and application are most common for cyanobacteria that can be commercially increased for applications, whereas research for moss and lichen restoration is in its infancy (Antoninka et al. 2020).

5.4.6 *Effectiveness Monitoring for Adaptive Resource Management*

Adaptive resource management (ARM) is an evolutionary process where the best management decisions are enacted to achieve desired outcomes (i.e. objectives) and the outcomes are tested (i.e., monitored) along with environmental variables that may influence outcomes to determine their effectiveness at one or more timeframes and across numerous similar sites. If objectives were not met and an alternative management action is suspected to improve achieving objectives, then the alternative is enacted and the process is repeated (Fig. 5.6; Reever-Morghan et al. 2006; Williams et al. 2009; Pilliod et al. 2021). Rangeland manipulations applied to lands with a goal of improving wildlife habitat should incorporate monitoring the habitat and the associated wildlife populations to determine if the predicted habitat was achieved and if wildlife populations are responding in the predicted manner (Pilliod et al. 2021, 2022). This is not a trivial component of manipulations and often is an expensive, time-consuming component that requires adequate planning and funds to accomplish. When done correctly, ARM will incorporate data from multiple sites using compatible methods and producing adjustments to the previous manipulation model or to formulate alternative models to improve effectiveness of manipulations to produce wildlife habitat.

5.5 Conclusions

Manipulations of rangeland ecosystems in the twenty-first century should not be viewed as impacting singular resources, but rather consider the complexity of the resource being manipulated and multiple physical and biological components that may respond to manipulations. Wildlife species may use multiple types of habitats across large landscapes or they may limit use to a narrow range of conditions in

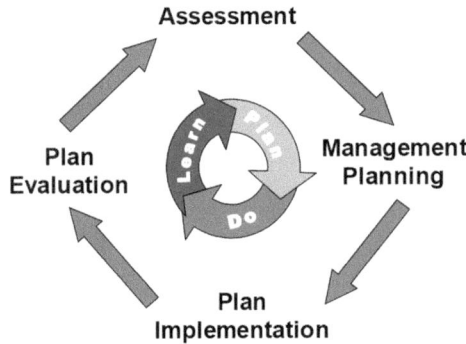

Fig. 5.6 Adaptive management begins with *assessing* the ecological status of current plant communities as well as factors that limit succession to a more desirable state. In *management planning*, objectives are formulated, and important ecological processes determined. These processes suggest specific management tactics. This information is translated into a spatially explicit plan (*Plan Implementation*) that indicates what will be done and where it will occur on the landscape. Following plan implementation, research and monitoring are used to *evaluate* management impacts and assess the validity of assumptions made in the planning process. This adaptive management process links different conservation elements into an iterative cycle of *planning, doing, and learning* that allows for management in the face of uncertainty and is necessary when managing complex problems in non-equilibrium rangeland ecosystems

small isolated environments. Regardless, rangeland manipulations at a local scale may influence wildlife at larger spatial extents, therefore, it is important to consider broad-scale responses even when manipulations are focused at local levels to prevent unintended consequences to wildlife species from the applied manipulation. Rangeland managers have many tools for planning and implement manipulations with more tools arriving in the future.

References

Aldridge CL, Nielsen SE, Beyer HL et al (2008) Range-wide patterns of greater sage-grouse persistence. Divers Distrib 14:983–994. https://doi.org/10.1111/j.1472-4642.2008.00502.x
Allred BW, Bestelmeyer BT, Boyd CS et al (2021) Improving Landsat predictions of rangeland fractional cover with multitask learning and uncertainty. Methods Ecol Evol 12:841–849. https://doi.org/10.1111/2041-210X.13564
Amor JM, Newman R, Jensen WF et al (2019) Seasonal home ranges and habitat selection of three elk (*Cervus elaphus*) herds in North Dakota. PLoS ONE. https://doi.org/10.1371/journal.pone.0211650
Anderson DP, Forester JD, Turner MG et al (2005) Factors influencing female home range sizes in elk (*Cervus elaphus*) in North American landscapes. Landsc Ecol 20:257–271. https://doi.org/10.1007/s10980-005-0062-8
Antoninka A, Faist A, Rodriguez-Caballero E et al (2020) Biological soil crusts in ecological restoration: emerging research and perspectives. Rest Ecol 28(S2):S3–S8. https://doi.org/10.1111/rec.13201

Austin A, Yahdijan L, Stark JM et al (2004) Water pulses and biogeochemical cycles in arid and semiarid ecosystems. Oecologia 141:221–235. https://doi.org/10.1007/s00442-004-1519-1

Ayotte JB, Parker KL, Arocena JM, Gillingham MP (2006) Chemical composition of lick soils: functions of soil ingestion by four ungulate species. J Mamm 87:878–888. https://doi.org/10.1644/06-MAMM-A-055R1.1

Bailey DW, Mosley JC, Estell RE et al (2019) Targeted livestock grazing: prescription for healthy rangelands. Rangel Ecol Manage 72:865–877. https://doi.org/10.1016/j.rama.2019.06.003

Bates J, Miller RF, Svejcar TJ (2000) Understory dynamics in cut and uncut western juniper woodlands. J Range Manag 53:119–126. https://doi.org/10.2458/azu_jrm_v53i1_bates

Belnap J, Lange OL (eds) (2003) Biological soil crusts: structure, function, and management. Springer-Verlag, Berlin. https://doi.org/10.1007/978-3-642-56475-8

Benson MH (2012) Intelligent tinkering: the Endangered Species Act and resilience. Ecol Soc 17(4):28. https://doi.org/10.5751/ES-05116-170428

Bleich VC, Kie JG, Loft ER et al (2005) Managing rangelands for wildlife. In: Braun CE (ed) Techniques for wildlife investigations and management. Wildlife Society, Bethesda, pp 873–897

Bower AD, St. Clair JB, Erickson V (2014) Generalized provisional seed zones for native plants. Ecol Appl 24:913–919. https://doi.org/10.1890/13-0285.1

Boyd CS, Bidwell TG (2002) Effects of prescribed fire on shinnery oak (*Quercus havardii*) plant communities in western Oklahoma. Restor Ecol 10:324–333. https://doi.org/10.1046/j.1526-100X.2002.01021.x

Boyd CS, Svejcar TJ (2009) Managing complex problems in rangeland ecosystems. Rangel Ecol Manag 62:491–499. https://doi.org/10.2111/08-194.1

Boyd CS, Collins WB, Urness PJ (1996) Relationship of dietary browse to intake in captive muskoxen. J Range Manag 49:2–7. https://doi.org/10.2307/4002717

Boyd CS, Johnson DD, Kerby JD et al (2014) Of grouse and golden eggs: can ecosystems be managed within a species-based regulatory framework? Rangel Ecol Manag 67:358–368. https://doi.org/10.2111/REM-D-13-00096.1

Boyd CS, Kerby JD, Svejcar TJ et al (2017) The sage-grouse habitat mortgage: effective conifer management in space and time. Rangel Ecol Manage 70:141–148. https://doi.org/10.1016/j.rama.2016.08.012

Bradford JB, Andrews CM (2021) Ecological drought forecast tool for drylands. https://www.usgs.gov/news/ecological-drought-forecast-tool-drylands. Accessed 10 Feb 2022

Briske DD (2017) Rangeland systems: foundation for a conceptual framework. In: Briske DD (ed) Rangeland systems. Springer Series on Environmental Management. Springer, Cham. pp 1–21. https://doi.org/10.1007/978-3-319-46709-2_1

Brown DE, Unmack PJ, Brennan TC (2007) Digitized map of biotic communities for plotting and comparing distributions of North American animals. Southwest Nat 52:610–616. https://doi.org/10.1894/0038-4909(2007)52[610:DMOBCF]2.0.CO;2

Brown VS, Ritchie AL, Stevens JC et al (2019) Protecting direct seeded grasses from herbicide applications: can new extruded pellet formulations be used in restoring natural plant communities. Rest Ecol 27:488–494. https://doi.org/10.1111/rec.12903

Caudle D, DiBenedetto J, Karl MS et al (2013) Interagency ecological site handbook for rangelands. U.S. Department of the Interior, Bureau of Land Management, U.S. Department of Agriculture, Natural Resources Conservation Service, and U.S. Department of Agriculture, Forest Service, Washington

Chambers JC, Roundy BA, Blank RR et al (2007) What makes Great Basin sagebrush ecosystems invasible by *Bromus tectorum*? Ecol Monogr 77:117–145. https://doi.org/10.1890/05-1991

Chambers JC, Miller RF, Board DI et al (2014) Resilience and resistance of sagebrush ecosystems: implications for state and transition models and management treatments. Rangel Ecol Manag 67:440–454. https://doi.org/10.2111/REM-D-13-00074.1

Chambers JC, Maestas JD, Pyke DA et al (2016a) Using resistance and resilience concepts to manage persistent threats to sagebrush ecosystems and greater sage-grouse. Rangel Ecol Manag 70:149–164. https://doi.org/10.1016/j.rama.2016.08.005

Chambers JC, Beck JL, Campbell S et al (2016b) Using resilience and resistance concepts to manage threats to sagebrush ecosystems, Gunnison sage-grouse, and greater sage-grouse in their eastern range: a strategic multi-scale approach. U.S. Department of Agriculture, Forest Service, Rocky Mountain Research Station, General Technical Report RMRS-GTR-356, Fort Collins. https:// doi.org/10.2737/RMRS-GTR-356

Chambers JC, Beck JL, Bradford JB et al (2017) Science framework for conservation and restoration of the sagebrush biome: linking the Department of the Interior's integrated rangeland fire management strategy to long-term strategic conservation actions, Part 1. Science basis and applications. U.S. Department of Agriculture, Forest Service, Rocky Mountain Research Station, General Technical Report RMRS-GTR-360, Fort Collins. https://doi.org/10.2737/ RMRS-GTR-360

Coates PM, Ricca MA, Prochazka BG et al (2016) Wildfire, climate, and invasive grass interactions negatively impact an indicator species by reshaping sagebrush ecosystems. Proc Natl Acad Sci USA 113:12745–12750. https://doi.org/10.1073/pnas.1606898113

Connelly JW, Hagen CA, Schroeder MA (2011) Characteristics and dynamics of greater sage-grouse populations. In: Knick ST, Connelly JW (eds) Greater sage-grouse: ecology and conservation of a landscape species and its habitats. Studies in Avian Biology 38:53–67. https://doi.org/10. 1525/california/9780520267114.003.0004

Conover MR, Borgo JS, Dritz RD et al (2010) Greater sage-grouse select nest sites to avoid visual predators but not olfactory predators. The Condor 112:331–336. https://doi.org/10.1525/cond. 2010.090172

Conway WC, Smith LM, Ray JD (2012) Shorebird habitat use and nest-site selection in the playa lakes region. J Wildl Manage 69:174–184. https://doi.org/10.2193/0022-541X(2005)069%3c0 174:SHUANS%3e2.0.CO;2

Crane KK, Mosley JC, Mosley TK et al (2016) Elk foraging site selection on foothill and mountain rangeland in spring. Rangel Ecol Manage 319–325. https://doi.org/10.1016/j.rama.2016.04.001

Culbert PD, Radeloff VC, Flather CH et al (2013) The influence of vertical and horizontal habitat structure on nationwide patterns of avian biodiversity. Auk 130:656–665. https://doi.org/10. 1525/auk.2013.13007

Dabbert CB, Lochmiller RL, Teeter RG (1997) Effects of acute thermal stress on the immune system of the northern bobwhite (*Colingus virginianus*). Auk 114:103–109. https://doi.org/10. 2307/4089069

Davies KW (2008) Medusahead dispersal and establishment in sagebrush steppe plant communities. Rangel Ecol Manage 61:110–115. https://doi.org/10.2111/07-041R2.1

Davies KW, Boyd CS, Madsen MD et al (2018) Evaluating a seed technology for sagebrush restoration across an elevation gradient: support for bet hedging. Rangel Ecol Manage 71:19–24. https:/ /doi.org/10.1016/j.rama.2017.07.006

Davies KW, Leger EA, Boyd CS, Hallett LM (2021) Living with exotic annual grasses in the sagebrush ecosystem. J Environ Manage 288:112417. https://doi.org/10.1016/j.jenvman.2021. 112417

DelGiudice GD, Mech LD, Seal US (1990) Effects of winter undernutrition on body composition and physiological profiles of white-tailed deer. J Wildl Manage 54:539–550. https://doi.org/10. 2307/3809347

DelGiudice GD, Seal US, Mech LD (1991) Indicators of severe undernutrition in urine of free-ranging elk during winter. Wildl Soc Bull 19:106–110

Dietz TH, Scott CB, Campbell EJ et al (2010) Feeding redberry juniper (*Juniperus pinchotii*) at weaning increases juniper consumption by goats on pasture. Rangel Ecol Manage 63:366–372. https://doi.org/10.2111/08-247.1

DiTomaso JM, Monaco TA, James JJ, Firn J (2017) Invasive plant species and novel rangeland systems. In: Briske DD (ed) Rangeland systems. Springer Series on Environmental Management. Springer, Cham. pp 429–465. https://doi.org/10.1007/978-3-319-46709-2_13

Doblas-Reyes FJ, Garcia-Serrano J, Lienert F et al (2013) Seasonal climate predictability and forecasting: status and prospects. Wiley Interdiscip Rev Climate Change 4:245–268. https://doi.org/10.1002/wcc.217

Doherty KE, Evans JS, Coates PS et al (2016) Importance of regional variation in conservation planning: a rangewide example of the greater sage-grouse. Ecosphere 7(10):e01462. https://doi.org/10.1002/ecs2.1462

Drummond MA, Auch RF, Karstensen KA et al (2012) Land change variability and human-environment dynamics in the United States Great Plains. Land Use Policy 29:710–723. https://doi.org/10.1016/j.landusepol.2011.11.007

Duncan KW, Scifres CJ (1983) Influence of clay and organic matter of rangeland soils on tebuthiuron effectiveness. J Range Manage 36:295–297. https://doi.org/10.2307/3898472

Elzinga CL, Salzer DW, Willoughby JW (1998) Measuring and monitoring plant populations. Bureau of Land Management, Technical Reference 1730-1, Denver

Emili LA, Greene RP (2014) New cropland on former rangeland and lost cropland from urban development: the "replacement land" debate. Land 3:658–674. https://doi.org/10.3390/land3030658

Evans DM, Goble DD, Scott JM (2013) New priorities as the Endangered Species Act turns 40. Front Ecol Environ 11:519–519. https://doi.org/10.1890/1540-9295-11.10.519

Evans R.D, Gill RA, Eviner VT, Bailey V (2017) Soil and belowground processes. In: Briske DD (ed) Rangeland systems. Springer Series on Environmental Management. Springer, Cham. pp 130–168. https://doi.org/10.1007/978-3-319-46709-2_4

Fogarty DT, Elmore RD, Fuhlendorf SD, Loss SR (2017) Influence of olfactory and visual cover on nest site selection and nest success for grassland birds. Ecol Evol 7:6247–6258. https://doi.org/10.1002/ece3.3195

Folke C, Carpenter S, Walker B et al (2004) Regime shifts, resilience, and biodiversity in ecosystem management. Annu Rev Ecol Syst 35:557–581. https://doi.org/10.1146/annurev.ecolsys.35.021103.105711

Frost RA, Launchbaugh KL (2003) Prescription grazing for rangeland weed management. Rangelands 23:43–45

Fulbright TE, Davies KW, Archer SR (2018) Wildlife responses to brush management: a contemporary evaluation. Rangel Ecol Manag 71:35–44. https://doi.org/10.1016/j.rama.2017.07.001

Garcia LA, Armbruster M (1997) A decision support system for evaluation of wildlife habitat. Ecol Modelling 102:287–300. https://doi.org/10.1016/S0304-3800(97)00064-1

Gates CC, Jones P, Suitor M et al (2012) The influence of land use and fences on habitat effectiveness and distribution of pronghorn in the grasslands of North America. In: Somers M, Hayward M (eds) Fencing for conservation—restriction of evolutionary potential or a riposte to threatening processes? Springer, New York. pp 277–294. https://doi.org/10.1007/978-1-4614-0902-1_15

Germaine SS, Assal T, Freeman A, Carter SK (2020) Distance effects of gas field infrastructure on pygmy rabbits in southwestern Wyoming. Ecosphere 11:e03230. https://doi.org/10.1002/ecs2.3230

Gilby BL, Olds AD, Connolly RM et al (2018) Spatial restoration ecology: placing restoration in a landscape context. Bioscience 68:1007–1019. https://doi.org/10.1093/biosci/biy126

Gowlett JAJ (2016) The discovery of fire by humans: a long and convoluted process. Phil Trans R Soc B 371:20150164. https://doi.org/10.1098/rstb.2015.0164

Grier JW (1982) Ban of DDT and subsequent recovery of reproduction in bald eagles. Science 218:1232–1235. https://doi.org/10.1126/science.7146905

Griffith B, Youtie BA (1988) Two devices for estimating foliage density and deer hiding cover. Wildl Soc Bull 16:206–210

Grzybowski JA (1995) Black-capped vireo (*Vireo atricapillus*). In: Poole A, Gill F (eds) The birds of North America, No. 181. The Academy of Natural Sciences, Philadelphia, The American Ornithologists' Union, Washington. https://doi.org/10.2173/tbna.181.p

Guilyardi E, Wittenberg A, Fedorov A et al (2009) Understanding El Niño in ocean-atmosphere general circulation models: progress and challenges. Bull Am Meteorol Soc 90:325–340. https://doi.org/10.1175/2008BAMS2387.1

Gurrieri JT (2020) Rangeland water developments at springs: best practices for design, rehabilitation, and restoration. U.S. Department of Agriculture, Forest Service, Rocky Mountain Research Station, General Technical Report RMRS-GTR-405, Fort Collins. https://doi.org/10.2737/RMRS-GTR-405

Guthery FS, Land CL, Hall BW (2001) Heat loads on reproducing bobwhites in the semiarid subtropics. J Wildl Manage 65:111–117. https://doi.org/10.2307/3803282

Guthery FS, Rybak AR, Fuhlendorf SD et al (2010) Aspects of the thermal ecology of bobwhites in north Texas. Wildl Monogr 159:1–36. https://doi.org/10.2193/0084-0173(2004)159[1:AOTTEO]2.0.CO;2

Hagen CA, Pitman JC, Sandercock BK et al (2007) Age-specific survival and probable causes of mortality in female lesser prairie-chickens. J Wildl Manage 71:518–525. https://doi.org/10.2193/2005-778

Hardegree SP, Jones TA, Roundy BA et al (2011) Assessment of range planting as a conservation practice. In: Briske DD (ed) Conservation benefits of rangeland practices: assessment, recommendations, and knowledge gaps. Allen Press, Lawrence, pp 171–212

Hardegree SP, Cho J, Schneider JM (2012) Weather variability, ecological processes, and optimization of soil micro-environment for rangeland restoration. In: Monaco TA, Sheley RL (eds) Invasive plant ecology and management: linking processes to practice. CABI, Cambridge, pp 107–121. https://doi.org/10.1079/9781845938116.0107

Hardegree SP, Sheley RL, Duke SE et al (2016) Temporal variability in microclimatic conditions for grass germination and emergence in the sagebrush steppe. Rangel Ecol Manage 69:123–128. https://doi.org/10.1016/j.rama.2015.12.002

Hardegree SP, Abatzoglou JT, Brunson MW et al (2018) Weather-centric rangeland revegetation planning. Rangel Ecol Manage 71:1–11. https://doi.org/10.1016/j.rama.2017.07.003

Harding DP, Raizada MN (2015) Controlling weeds with fungi, bacteria and viruses: a review. Front Plant Sci 6:659. https://doi.org/10.3389/fpls.2015.00659

Hiers JK, Jackson ST, Hobbs RJ (2016) The precision problem in conservation and restoration. Trends Ecol Evol 31:820–830. https://doi.org/10.1016/j.tree.2016.08.001

Hilderbrand RH, Watts AC, Randle AM (2005) The myths of restoration ecology. Ecol Soc 10:19. www.ecologyandsociety.org/vol10/iss1/art19/. Accessed 14 July 2022

Holbrook JD, Squires JR, Loson LE et al (2016) Multiscale habitat relationships of snowshoe hares (*Lepus americanus*) in the mixed conifer landscape of the Northern Rockies, USA: cross-scale effects of horizontal cover with implications for forest management. Ecol Evol 7:125–144. https://doi.org/10.1002/ece3.2651

Holdo RM, Holt RD, Fryxell JM (2009) Opposing rainfall and plant nutritional gradients best explain the wildebeest migration in the Serengeti. Am Nat 173:431–445. https://doi.org/10.1086/597229

Holling CS (1973) Resilience and stability in ecological systems. Annu Rev Ecol Syst 4:1–23

Holmes AL, Miller RF (2010) State-and-transition models for assessing grasshopper sparrow habitat use. J Wildl Manage 74:1834–1840. https://doi.org/10.2193/2009-417

James JJ, Sheley RL, Leger EA et al (2019) Increased soil temperature and decreased precipitation during early life stages constrain grass seedling recruitment in cold desert restoration. J Appl Ecol 56:2609–2619. https://doi.org/10.1111/1365-2664.13508

Johnson DH (1980) The comparison of usage and availability measurements for evaluating resource preference. Ecology 61:65–71. https://doi.org/10.2307/1937156

Johnson GR, Sorenson FC, St. Clair JB (2004) Pacific Northwest forest seed zones. Native Plants J 5:131–140. https://doi.org/10.2979/NPJ.2004.5.2.131

Jones MO, Allred BW, Naugle DE et al (2018) Innovation in rangeland monitoring: annual, 30 m plant functional type percent cover maps for U.S. rangelands, 1984–2017. Ecosphere https://doi.org/10.1002/ecs2.2430

Karl JW, Herreck JE, Pyke DA (2017) Monitoring protocols: options, approaches, implementation, benefits. In: Briske DD (ed) Conservation benefits of rangeland practices: assessment, recommendations, and knowledge gaps. Allen Press, Lawrence, pp 527–567. https://doi.org/10.2979/NPJ.2004.5.2.131

Kennedy PL, DeBano SJ, Bartuszevige AM, Lueders AS (2009) Effects of native and non-native grassland plant communities on breeding passerine birds: implications for restoration of northwest bunchgrass prairie. Restor Ecol 17:515–525. https://doi.org/10.1111/j.1526-100X.2008.00402.x

Kreulen DA (1985) Lick use by large herbivores: a review of benefits and banes of soil consumption. Mamm Rev 15:107–123. https://doi.org/10.1111/j.1365-2907.1985.tb00391.x

Lane MA, Ralphs MH, Olsen JD et al (1990) Conditioned taste aversion: potential for reducing cattle loss to larkspur. J Range Manage 43:127–131. https://doi.org/10.2307/3899029

Lark TJ, Spawn SA, Bougie M, Gibbs HK (2020) Cropland expansion in the United States produces marginal yields at high costs to wildlife. Nat Commun 11:4295. https://doi.org/10.1038/s41467-020-18045-z

Le Corre M, Dussault C, Cote SD (2017) Weather conditions and variation in timing of spring and fall migrations of migratory caribou. J Mammal 98:260–271. https://doi.org/10.1093/jmammal/gyw177

Leopold A (1933) Game management. Scribner's Sons, New York

Leu M, Hanser SE (2011) Influences of the human footprint on sagebrush landscape patterns. In: Knick ST Connelly JW (eds) Greater sage-grouse: ecology and conservation of a landscape species and its habitats. Studies in Avian Biology 38:253–271. https://doi.org/10.1525/california/9780520267114.003.0014

Lewis HT (1985) Why Indians burned: specific versus general reasons. In: Lotan JE, Kilgore RM, Fischer WC et al (compilers) (eds) Proceedings—symposium and workshop on wilderness fire, Missoula, November 1983. U.S. Department of Agriculture, Intermountain Forest and Range Experiment Station INT-GTR-182, Ogden, Utah, pp 75–80

Limb RF, Hickman KR, Engle DM et al (2007) Digital photography: reduced investigator variation in visual obstruction measurements for southern tallgrass prairie. Rangel Ecol Manage 60:548–552. https://doi.org/10.2111/1551-5028(2007)60[548:DPRIVI]2.0.CO;2

Lochmiller RL (1996) Immunocompetence and animal population regulation. Oikos 76:594–602. https://doi.org/10.2307/3546356

Loss SR, Terwilliger LA, Peterson AC (2011) Assisted colonization: integrating conservation strategies in the face of climate change. Biol Conserv 144:92–100. https://doi.org/10.1016/j.biocon.2010.11.016

Lowery B, Hickey WJ, Arshad MA, Lal R (1996) Soil water parameters and soil quality. In: Doran JW, Jones AJ (eds) Methods for assessing soil quality, vol 49. Soil Science Society of America, Madison. pp 143–155. https://doi.org/10.2136/sssaspecpub49.c8

Madsen MD, Davies KW, Boyd CS et al (2016) Emerging seed enhancement technologies for overcoming barriers to restoration. Rest Ecol 24(S2):S77–S84. https://doi.org/10.1111/rec.12332

Madsen MD, Svejcar L, Radke J, Hulet A (2018) Inducing rapid seed germination of native cool season grasses with solid matrix priming and seed extrusion technology. PLoS ONE 13(10):e0204380. https://doi.org/10.1371/journal.pone.0204380

Maestas JD, Knight RL, Gilgert WC (2003) Biodiversity across a rural land-use gradient. Conserv Biol 17:1425–1434. https://doi.org/10.1046/j.1523-1739.2003.02371.x

Matthews TW, Tyre AJ, Taylor JS et al (2011) Habitat selection and brood survival of Greater Prairie-Chickens. In: Sandercock BK, Martin K, Segelbacher G (eds) Ecology, conservation, and management of grouse. Studies in Avian Biology 39:179–191

McCabe GJ, Palecki MA, Betancourt JL (2004) Pacific and Atlantic Ocean influences on multi-decadal drought frequency in the United States. Proc Natl Acad Sci 101:4136–4141. https://doi.org/10.1073/pnas.03067381

McFadyen REC (1998) Biological control of weeds. Annu Rev Entomol 43:369–393

Mietier EP, Rew LJ, Rinella MJ (2018) Establishing Wyoming big sagebrush in annual brome-invaded landscapes with seeding and herbicides. Rangel Ecol Manage 71:705–713. https://doi.org/10.1016/j.rama.2018.06.001

Millennium Ecosystem Assessment (2005) Millennium ecosystem assessment. Ecosystems and Human Well-Being: Biodiversity Synthesis, World Resources Institute, Washington

Miller RF, Bates JD, Svejcar TJ et al (2005) Biology, ecology, and management of western juniper. Tech. Bull. 152. Oregon State University, Agricultural Experiment Station, Corvallis. https://catalog.extension.oregonstate.edu/sites/catalog/files/project/pdf/tb152.pdf. Accessed 15 July 2022

Miller RF, Chambers JC, Pellant M (2014) A field guide for selecting the most appropriate treatment in sagebrush and piñon-juniper ecosystems in the Great Basin: evaluating resilience to disturbance and resistance to invasive annual grasses, and predicting vegetation response. U.S. Department of Agriculture, Forest Service, Rocky Mountain Research Station, General Technical Report RMRS-GTR-322, Fort Collins. https://doi.org/10.2737/RMRS-GTR-322

Nolte DL, Provenza FD (1992) Food preferences in lambs after exposure to flavors in milk. Appl Anim Behav Sci 32:381–389. https://doi.org/10.1016/S0168-1591(05)80030-9

Nolte DL, Provenza FD, Callan R, Panter KE (1992) Garlic in the ovine fetal environment. Physiol Behav 52:1091–1093. https://doi.org/10.1016/0031-9384(92)90464-D

Nudds TD (1977) Quantifying the vegetative structure of wildlife cover. Wildl Soc Bull 5:113–117

Parker SS, Zdon A, Christian WT et al (2021) Conservation of Mojave Desert springs and associated biota: status, threats, and policy opportunities. Biodivers Conserv 30:311–327. https://doi.org/10.1007/s10531-020-02090-7

Payne A, Chappa S, Hars J et al (2016) Wildlife visits to farm facilities assessed by camera traps in a bovine tuberculosis-infected area in France. Eur J Wildl Res 62:33–42. https://doi.org/10.1007/s10344-015-0970-0

Pearson DE, Valliant M, Carlson C et al (2019) Spicing up restoration: can chili peppers improve restoration seeding by reducing seed predation. Rest Ecol 27:254–260. https://doi.org/10.1111/rec.12862

Pellant M, Shaver PL, Pyke DA et al (2020) Interpreting indicators of rangeland health, version 5. U.S. Department of the Interior, Bureau of Land Management, Technical Referance 1734-6, Denver, Colorado. pp 1-186. https://www.blm.gov/sites/blm.gov/files/documents/files/Interpreting%20Indicators%20of%20Rangeland%20Health%20Technical%20Reference%201734-6%20version%205_0.pdf. Accessed 15 July 2021

Pilliod DS, Welty JL, Toevs GR (2017) Seventy-five years of vegetation treatments on public rangelands of the Great Basin of North America. Rangelands 39:1–12. https://doi.org/10.1016/j.rala.2016.12.001

Pilliod DS, Pavlacky DC, Manning ME et al (2021) Adaptive management and monitoring. In: Remington TE, Deibert PA, Hanser SE et al (eds) Sagebrush conservation strategy—challenges to sagebrush conservation. US Geological Survey, Open-File Report 2020–1125, Reston p 223–239. https://doi.org/10.3133/ofr20201125

Pilliod DS, Beck JL, Duchardt CJ, Rachlow JL, Veblen KE (2022) Leveraging rangeland monitoring data for wildlife: from concept to practice. Rangelands 44:87–98. https://doi.org/10.1016/j.rala.2021.09.005

Plummer IH, Johnson CJ, Chesney AR, Pedersen JA, Samuel MD (2018) Mineral licks as environmental reservoirs of chronic wasting disease prions. PLoS ONE 13(5):e0196745. https://doi.org/10.1371/journal.pone.0196745.DOI:10.1371/journal.pone.0196745

Polley HW, Bailey DW, Nowak RS, Stafford-Smith M (2017) Ecological consequences of climate change on rangelands. In: Briske DD (ed) Rangeland systems. Springer Series on Environmental Management. Springer, Cham, pp 228–260. https://doi.org/10.1007/978-3-319-46709-2_7

Pratt AC, Smith KT, Beck JL (2017) Environmental cues used by greater sage-grouse to initiate altitudinal migration. Auk 134:628–643. https://doi.org/10.1642/AUK-16-192.1

Pyke DA, Brooks ML, D'Antonio C (2010) Fire as a restoration tool: a decision framework for predicting the control or enhancement of plants using fire. Rest Ecol 18:274–284. https://doi.org/10.1111/j.1526-100X.2010.00658.x

Pyke DA, Wirth TA, Beyers JL (2013) Does seeding after wildfires in rangelands reduce erosion or invasive species? Restor Ecol 21:415–421. https://doi.org/10.1111/rec.12021

Pyke DA, Shaff SE, Lindgren AI et al (2014) Region-wide ecological responses of arid Wyoming big sagebrush communities to fuel treatments. Rangel Ecol Manage 67:455–467. https://doi.org/10.2111/REM-D-13-00090.1

Pyke DA, Knick ST, Chambers JC et al (2015) Restoration handbook for sagebrush steppe ecosystems with emphasis on greater sage-grouse habitat—Part 2. Landscape level restoration decisions. US Geological Survey Circular 1418 Reston. https://doi.org/10.3133/cir1418

Pyke DA, Chambers JC, Pellant M et al (2017) Restoration handbook for sagebrush steppe ecosystems with emphasis on greater sage-grouse habitat—Part 3. Site level restoration decisions: U.S. Geological Survey Circular 1426, Reston. https://doi.org/10.3133/cir1426

Pyle LA, Sheley RL, James JJ (2021) Timing and duration of precipitation pulses and interpulses influence seedling recruitment in the Great Basin. Rangel Ecol Manage 75:112–118. https://doi.org/10.1016/j.rama.2020.12.004

Rashford BS, Walker JA, Bastian CT (2011) Economics of grassland conversion to cropland in the prairie pothole region. Conserv Biol 25:276–284. https://doi.org/10.1111/j.1523-1739.2010.01618.x

Raynor EJ, Derner JD, Hoover DL et al (2020) Large-scale and local climatic controls on large herbivore productivity: implications for adaptive rangeland management. Ecol Appl 30(3):e02053. https://doi.org/10.1002/eap.2053

Reever Morghan KJ, Sheley RL, Svejcar TJ (2006) Successful adaptive management: the integration of research and management. Rangel Ecol Manage 59:216–219. https://doi.org/10.2111/05-079R1.1

Reeves MC, Krebs M, Leinwand I et al (2018) Rangelands on the edge: quantifying the modification, fragmentation, and future residential development of U.S. rangelands. U.S. Department of Agriculture, Forest Service, Rocky Mountain Research Station, General Technical Report RMRS-GTR-382, Fort Collins. https://doi.org/10.2737/RMRS-GTR-382

Reinking AK, Smith KT, Mong TW et al (2019) Across scales, pronghorn select sagebrush, avoid fences, and show negative responses to anthropogenic features in winter. Ecosphere 10(5):e02722. https://doi.org/10.1002/ecs2.2722.10.1002/ecs2.2722

Ricca MA, Coates PS, Guftafson KB et al (2018) A conservation planning tool for Greater Sage-grouse using indices of species distribution, resilience, and resistance. Ecol Appl 28:878–896

Ricca MA, Coates PS (2020) Integrating ecosystem resilience and resistance into decision support tools for multi-scale population management of a sagebrush indicator species. Front Ecol Evol 7:493. https://doi.org/10.3389/fevo.2019.00493

Robbins CT (1993) Wildlife feeding and nutrition, 2nd edn. Academic Press, San Diego

Royer PD, Breshears DD, Zou CB et al (2012) Density-dependent ecohydrological effects of pinon-juniper woodland canopy cover on soil microclimate and potential soil evaporation. Rangel Ecol Manage 65:11–20. https://doi.org/10.2111/REM-D-11-00007.1

Schlaepfer DR, Bradford JB, Lauenroth WK et al (2017) Climate change reduces extent of temperate drylands and intensifies drought in deep soils. Nature Comm 8:14196. https://doi.org/10.1038/ncomms14196.DOI:10.1038/ncomms14196

Shaw N (2004) Production and use of planting stock. In: Monsen SB, Stevens R, Shaw NL, (compilers) (eds) Restoring western ranges and wildlands, Volume 3. U.S. Department of Agriculture, Forest Service, Rocky Mountain Research Station, General Technical Report RMRS-GTR-136-vol-3, Fort Collins, Colorado, pp 745–768. https://doi.org/10.2737/RMRS-GTR-136-V3

Smith JT, Evans JS, Martin BH et al (2016) Reducing cultivation risk for at-risk species: predicting outcomes of conservation easements for sage grouse. Biol Conserv 201:10–19. https://doi.org/10.1016/j.biocon.2016.06.006

SRM (Society for Range Management) (1998) Glossary of terms used in range management, 4th edn. Society for Range Management, Denver

Stevens, R, Monsen SB (2004) Mechanical plant control. In: Monsen SB, Stevens R, Shaw NL, (compilers) (eds) Restoring western ranges and wildlands, vol 1. U.S. Department of Agriculture, Forest Service, Rocky Mountain Research Station, Gen Tech Rep RMRS-GTR-136-vol-1, Fort Collins, Colorado, pp 65–87. https://doi.org/10.2737/RMRS-GTR-136-V1

Stringham TK, Krueger WC, Shaver PL (2003) State and transition modeling: an ecological process approach. J Range Manage 56:105–113. https://doi.org/10.2307/4003893

Thompson SJ, Johnson DH, Niemuth ND, Ribic CA (2015) Avoidance of unconventional oil wells and roads exacerbates habitat loss for grassland birds in the North American great plains. Biol Conserv 192:82–90. https://doi.org/10.1016/j.biocon.2015.08.040

US Census Bureau (2019) U.S. population estimates continue to show the nation's growth is slowing. Census Bureau Press Release December 30, 2019, CB19-198, Washington. https://www.census.gov/newsroom/press-releases/2019/popest-nation.html. Accessed 06 May 2021

USDA (U.S. Department of Agriculture) (2020) Summary Report: 2017 National Resources Inventory, Natural Resources Conservation Service, Washington, and Center for Survey Statistics and Methodology, Iowa State University, Ames https://www.nrcs.usda.gov/wps/portal/nrcs/main/national/technical/nra/nri/results/. Accessed 06 May 2021

Valdez R (2013) Exploring our ancient roots—Genghis Khan to Aldo Leopold: the origins of wildlife management. Wildl Prof Summer Issue 50–53

Vetter S (2005) Rangelands at equilibrium and non-equilibrium: recent developments in the debate. J Arid Environ 62:321–341. https://doi.org/10.1016/j.jaridenv.2004.11.015

Walker B, Salt D (2006) Resilience thinking—sustaining ecosystems and people in a changing world Island Press, Washington

Walker JW, Hemenway K, Hatfield PG, Glimp HA (1992) Training lambs to be weedeaters: studies with leafy spurge. J Range Manage 45:245–249. https://doi.org/10.2307/4002971

Wang C (2021) Three-ocean interactions and climate variability: a review and perspective. Clim Dyn 53:5119–5136. https://doi.org/10.1007/s00382-019-04930-x

Wang Y, Pedersen JLM, Macdonald SE et al (2019) Experimental test of assisted migration for conservation of locally range-restricted plants in Alberta. Canada. Glob Ecol Conserv 17:e00572. https://doi.org/10.1016/j.gecco.2019.e00572

Webb SL, Dzialak MR, Kosciuch KL, Winstead JB (2013) Winter resource selection by mule deer on the Wyoming-Colorado Boarder prior to wind energy development. Rangel Ecol Manage 66:419–427. https://doi.org/10.2111/REM-D-12-00065.1

Whelan RJ (1995) The ecology of fire. Cambridge University Press, New York

Wiedeman VE, Penfound WT (1960) A preliminary study of the shinnery in Oklahoma. Southw Nat 5:117–122

Williams BK, Szaro RC, Shapiro CD (2009) Adaptive management: The U.S. Department of the Interior technical guide. US Department of the Interior, Washington.

Young JA, Evans RA (1978) Population dynamics after wildfires in sagebrush grasslands. J Range Manage 31:283–289. https://doi.org/10.2307/3897603

Ziska LH, Reeves JB III, Blank RR (2005) The impact of recent increases in atmospheric CO_2 on biomass production and vegetative retention of cheatgrass (*Bromus tectorum*): implications for fire disturbance. Glob Change Biol 11:1325–1332. https://doi.org/10.1111/j.1365-2486.2005.00992.x

Chapter 6
Role and Management of Fire in Rangelands

J. Derek Scasta, Dirac Twidwell, Victoria Donovan, Caleb Roberts, Eric Thacker, Ryan Wilbur, and Samuel Fuhlendorf

Abstract Fire is a fundamental ecological process in rangeland ecosystems. Fire drives patterns in both abiotic and biotic ecosystem functions that maintain healthy rangelands, making it an essential tool for both rangeland and wildlife management. In North America, humanity's relationship with fire has rapidly changed and shifted from an era of coexistence to one that attempts to minimize or eliminate its occurrence. Prior to Euro-American settlement, Indigenous people's coexistence with fire led to regionally distinct fire regimes that differed in terms of their fire frequency, intensity, severity, seasonality, and spatial complexity. As the relative occurrence of prescribed fire and wildfire continue to change in North American rangelands, it is necessary for wildlife managers to understand the complex social-ecological interactions that shape modern fire regimes and their conservation outcomes. In this chapter, we discuss the fire eras of North American rangelands, introduce foundational relationships between fire and wildlife habitat, and discuss potential futures for fire in wildlife management.

Keywords Anthropocene · Burning · Disturbance · Mitigation · Prescribed fire · Pyric-herbivory · Suppression · Wildland · Wildfire

J. D. Scasta (✉) · R. Wilbur
Department of Ecosystem Science and Management, University of Wyoming, Laramie, WY, USA
e-mail: jscasta@uwyo.edu

D. Twidwell · V. Donovan · C. Roberts
Department of Agronomy and Horticulture, University of Nebraska, Lincoln, NE, USA

E. Thacker
Department of Wildland Resources, Utah State University, Logan, UT, USA

S. Fuhlendorf
Department of Natural Resource Ecology and Management, Oklahoma State University, Stillwater, OK, USA

© The Author(s) 2023   147
L. B. McNew et al. (eds.), *Rangeland Wildlife Ecology and Conservation*,
https://doi.org/10.1007/978-3-031-34037-6_6

6.1 Introduction

Aldo Leopold famously laid out the tools for wildlife management including the axe, plow, cow, fire, and gun (Leopold 1949). Contemporary wildlife managers still use a variety of these basic tools for directly manipulating habitats, including fire. Fire plays a foundational role in shaping rangeland ecosystem structure and function and thus, wildlife habitat. For instance, fire can drive the stimulation of aspen (*Populus tremuloides*) resprouts that provide browse and fawning cover (Pojar and Bowden 2004; Margolis and Farris 2014; Krasnow and Stephens 2015; Walker et al. 2015), as well as maintain open grasslands for gallinaceous and other ground-nesting birds (Briggs et al. 2002; Hagen et al. 2004; Hovick et al. 2014; Lautenbach et al. 2017). While fire is used for both livestock and wildlife management in rangelands, objectives often differ. For example, a rancher in the eastern Great Plains ecoregion might burn the same pasture every year in the early spring in order to stimulate perennial grass dominance and production (Anderson et al. 1970). In contrast, a wildlife manager in the same region might burn different portions of a pasture at different times of the year for a variety of objectives, such as burning some patches in the fall to stimulate forbs for quail, burning some patches in the spring for forage and browse production for herbivores, and leaving some unburned patches as refugia for other species (Weir and Scasta 2017). Manipulation of fire at different times and at different spatial scales alters the structure and function of an ecosystem by, for instance, variably depressing or enhancing certain plant species or manipulating the balance and availability of plant species (Towne and Craine 2016). An understanding of the complex interactions between different components of fire regimes (the pattern of fires over space and time, including fire frequency, intensity, severity, and seasonality; see Table 6.1) and rangeland abiotic and biotic components is needed to inform decisions by wildlife managers in how to integrate fire into a management plan (Limb et al. 2016). For a historical review of the fire regime concept see Krebs et al. (2010).

Perspectives on fire management in North America is ever evolving, having shifted from a tool historically used for survival and land-stewardship activities, to a force that must be suppressed, to a contemporary tool for ecosystem manipulation. These shifting fire management perspectives have had lasting legacies on rangeland systems. Indigenous use of fire by Native American and First Nation tribes historically included applications for survival, including hunting, warfare, and agriculture, and survival (Roos et al. 2018; Nikolakis et al. 2020) and today include applications for land stewardship. Fire as a survival tool was supplanted by the Euro-American settlers' approach of fire suppression, an idea reinforced by federal suppression policies (Busenberg 2004; Roos et al. 2018). For example, educational campaigns such as Smokey Bear's iconic "only you can prevent forest fires" slogan set the precedent of United States fire management with implications for wildlife for over half a century (Donovan and Brown 2007). This "pyropolitical" campaign started in 1944 and has influenced the notion that fire needed to be eliminated from the landscape (Minor and Boyce 2018), creating a paradigm ingrained in post-European North American culture that there was no "good" fire. In contemporary times, however,

Table 6.1 Key terms and definitions for metrics of a fire regime

Term	Definition
Frequency	Indication of how frequently an area burns and may also be indicated as the fire return interval and elapsed time since a fire occurred. May be calculated several ways including fire rotation (or fire cycle; defined as time to burn an area equal to the area of interest), mean fire interval (or fire return interval; defined as the average period between fires under a presumed historical regime), annual probability of fire (defined as the average fraction of the landscape expected to burn annually), and fire frequency (defined as the number of fires in a given time period)
Intensity	Rate at which fire produces thermal energy; for example fire line intensity is calculated as $I = HWR$ following Byram's equation where I is fire line intensity, H is a specific heat yield constant, W is the amount of fuel consumed, and R is the rate of spread. Fire temperature may be considered as a proxy for intensity
Seasonality	Time of year when fire is most common to occur naturally (i.e. wildfire) or be applied prescriptively (i.e. prescribed fire); typically winter, spring, summer, or fall
Severity	Relative amount of alteration, disruption, or damage a site experiences due to a fire. May include mortality of plants, structural and compositional changes to the plant community, soil burn severity, etc

wildlife managers have come to recognize the facilitating nature of fire and have begun to integrate fire disturbances into wildlife management plans. Ultimately, the shifting social perception of fire to allowances for useful applications has begun to transform fuel loads, wildlife habitats, and ecosystems. Future management of wildfires and prescribed fires must embrace the social perspectives of fire in order to optimize impacts on wildlife.

Both wildfires and prescribed fires shape the context of wildlife management today. Wildfires are unplanned and often burn outside of human control. As such, they may burn under drier and windier weather conditions than prescribed fires, generating fires that are hotter (often > 1000 °C) and thus burn with greater intensity, and with higher levels of fuel consumption. In contrast, prescribed fires are planned fires conducted to achieve targeted management objectives. Due to safety concerns, prescribed fires are generally conducted during more mild weather conditions that are moister and less windy. Thus, prescribed fires tend to burn cooler (often between 400 and 700 °C), with lower intensity, and with a lower level of fuel consumption. However, there are some instances where prescribed fires have been safely implemented during drier conditions or with added fuel loads to generate higher fire intensities (> 1000 °C) in order to achieve rangeland and wildlife management objectives, such as to restore rangelands that are experiencing woody encroachment (e.g., Twidwell et al. 2016). In both cases, prescribed fires require substantial preparation, including developing a written plan, acquiring necessary approvals, installing sufficient fire guards, maintaining functioning equipment, and having a trained crew (Weir 2009). The spatial and temporal patterns of wildfire and prescribed fire events combined make up the fire regime that shapes rangeland habitat.

In this chapter, we use Twidwell et al. (2021) *Cultural Fire Eras* in rangelands to provide a historical overview of how shifts in people's relationship with fire are associated with major changes in rangeland fire regimes. We then describe how fire regime components can shape wildlife habitats. Finally, we discuss the potential for future fire management ideologies to emerge, the implications of competing ideologies for wildlife, and the challenges of creating critical ranges of complexity and spatiotemporal heterogeneity (defined as variation of landscape features and in the context of wildlife it is habitat features that vary relative to fire variation) necessary for rangeland wildlife persistence.

6.2   Cultural Fire Eras on North American Rangelands

6.2.1   The 'Coexistence Era' of Fire Management

Historically, people coexisted and thrived with fire across many of the Earth's flammable biomes (Bond and Keeley 2005). Prior to Euro-American settlement of North American rangelands (~ 20,000–200 years ago), Indigenous people's land stewardship using fire helped to promote dynamic rangeland systems that spanned much of North America. Human fire ignitions promoted frequent fire in areas like the Great Plains ecoregion, making it one of the most pyrogenic systems on Earth. Here, there was a high level of heterogeneity in applications of fire characteristics of occurrence, intensity, spatial arrangement, severity, and seasonality. The shifting spatial arrangement of burned and unburned areas followed by spatially heterogeneous grazing created by roaming herds of American bison (*Bison bison*) and other mammals, created a diversity of rangeland habitats that, in turn, harbored a diversity of wildlife populations and plant communities.

The occurrence of fire is often described in terms of the fire regime and specifically the frequency of fire (Table 6.1). Fire frequency, or fire return intervals, varied greatly depending on location. In the Great Plains ecoregion, the average fire return interval ranged from 1 to 3 years in eastern tallgrass prairies to 3–8 years in shortgrass and mixed-grass prairies (Guyette et al. 2012). In contrast, rangelands of the Great Basin (i.e., the Western Deserts, Grasslands, Shrublands, and Woodlands ecoregion) had average fire return intervals that, in some places, could exceed 100 years (Guyette et al. 2012; Mensing et al. 2006). Fire intensities ranged from low to extreme, both within and among fire events. High-intensity fires were used by indigenous people for hunting and warfare (Stewart 2002, 1951), whereas lower-intensity fires were used for clearing vegetation, attracting game species, and for agricultural purposes (Higgins 1986). Fires ranged from very small for clearing around camp to very large (> 500,000 ha) with great variation in size. Given the variation in fire size and intensity there was also variation in severity of the effects of the fire in terms of plant mortality, vegetation structure, botanical composition. Thus it is important to understand the variation of fire effects and ecological processes generated by these practices that have been shown to be important for plant and wildlife populations because certain species prefer different scales and relative time since the fire disturbance (Hutto et al.

2016; Roberts et al. 2020). Moreover, the concept of pyrodiversity (considered the variation or heterogeneity in fire frequency, seasons, spatial arrangement, fire type, severity, and intensity) has been suggested to be important for biodiversity and may serve as a framework for conservation (Fuhlendorf et al. 2006; Kelly et al. 2017).

6.2.2   The 'Suppression and Wildfire Eras' of Fire Management

Following Euro-American settlement and the displacement of indigenous Indigenous people from their lands, fire patterns in North American rangelands underwent a drastic shift (Twidwell et al. 2021). Sharp declines in human ignitions due to the differing land management practices of invading European settlers, along with the later introduction of extensive fire suppression efforts, led to a massive decrease in the number of fires across North American rangelands (Fuhlendorf et al. 2018). This change was not necessarily immediate nor uniform as some native fire cultures persisted longer than others and some Euro-American fire cultures established such as the Celtic descendants in the southeastern U.S. (Doolittle and Lightsey 1979; Putz 2003; Stambaugh et al. 2013); yet such changes were detectable in the landscape through charcoal analysis and tree scars (Brown and Sieg 1999; Scasta et al. 2016b). Grazing became static and constrained to fenced pastures while fire was de-coupled from human land management, eliminating the shifting mosaic of fire and grazing interactions. The Great Plains ecoregion, once one of the most burned biomes in the world, became an area with a low probability of fire occurrence (Donovan et al. 2017).

The era of fire suppression is recognized as one of unrealistic expectations underpinning a failure in ecosystem management. Briefly, the generally applied policy of suppressing all fires resulted a ubiquitous fuel load increase and subsequent wildfire risk escalation across numerous ecosystems (Calkin et al. 2015; Pyne 2007; Twidwell et al. 2013b). As a result, wildfires, rather than the purposeful human ignitions used in land stewardship activities to manipulate ecosystems during the *Coexistence Era*, now dominate North American rangelands. As human populations have expanded and the climate has changed in North America, there has been a surge in total area burned by wildfire in rangeland and forested ecosystems, despite increases in suppression costs over time (Fig. 6.1; Donovan et al. 2017; Hanes et al. 2019). The Great Basin has seen a shift, with some areas changing from 100-year fire return intervals to fires occurring every 5–10 years (D'Antonio and Vitousek 1992). This increase in fire frequency outside of historical fire return intervals can degrade wildlife habitat, for example through the loss of sagebrush shrubs, and drive local wildlife extirpation (e.g., Pedersen et al. 2003). Wildfires have surged in the Great Plains ecoregion, associated in part with woody species that have been able to proliferate through grasslands due to a lack of fire activity (Donovan et al. 2017, 2020). While contemporary fire frequency is still lower than historical frequency in the region, the change in frequency represents stochastic disturbances that can pose a greater risk to human populations than the fires stewarded by humans in the past (Twidwell et al. 2013b). Of particular concern is the significant increases in wildfire size and the greater prevalence of very large wildfires (> 20,234 ha) in western regions (Stavros et al. 2014). Exacerbating this trend is the warming global climate and enhanced aridity (Williams and Abatzoglou 2016) and increasing wildfire season length due to earlier warming in the spring (Westerling 2016; Westerling et al. 2006). For example, wildfire seasons and the duration of burns have increased in length—from 2003 to 2012 seasons were 84 days longer than those from 1973 to 1982 (Westerling et al. 2006), with an increase of average wildfire burn time from 6 days (1973–1982), to 20 days (1983–1992), to 37 days (1993–2002), to > 50 days (2003–2012). In 2020, many very large wildfires occurred, such as the Mullen Fire along the border of Colorado and Wyoming that engulfed more than 70,000 ha (Fig. 6.2). As climate changes and novel shifts in drought conditions occur, these wildfires are likely to further increase in intensity, severity and extent (e.g., Scasta et al. 2016b).

6.2.3 The 'Contemporary Era' of Fire Management

Today, fire regimes in North American rangeland systems are overwhelmingly driven by wildfire occurrences rather than fires used for land stewardship (Fig. 6.3). Only one rangeland-dominated ecoregion in the central and western U.S.—the Flint Hills of Kansas—has a fire regime dominated by prescribed fire rather than wildfire (Fig. 6.3). Prescribed fires throughout the rest of the U.S. Great Plains typically occur on < 1% of the land area. In cases where contemporary prescribed fires are applied in rangelands, they represent a greatly dampened range of variability in the size, frequency, intensity, severity, and seasonality of the fire regimes stewarded by Indigenous people before Euro-American settlement. Prescribed fires are small, typically ranging from 10 to 160 ha (Weir et al. 2016, 2015). Restrictions tied to 'safe fire conditions' limit the weather conditions under which prescribed fires can burn, leading to low and homogenous fire intensities and generating highly restricted prescribed fire seasonality. Thus, while prescribed fires are discussed commonly as the backbone of fire management, prescribed fires impact minimal land area, and their functional variability has been reduced to a shadow of its role during the Coexistence Era.

At the turn of the twenty-first century, diverse stakeholders have created Prescribed Burn Associations (PBAs) and Prescribed Fire Councils (PFCs) to create cultural and attitudinal change towards the willingness to utilize prescribed fire as a management tool and apply prescribed fires on their landholdings. For example, over 60 PBAs have now been created throughout the Great Plains ecoregion—from southern Texas to South Dakota, pooling equipment, funds, experience, and training on how to safely apply prescribed fire (Weir et al. 2016). PFC organizations also have spread nationally to provide educational outlets to private landowners to better understand the benefits of prescribed fire (Wilbur and Scasta 2021). Whereas PBAs and PFCs strive to

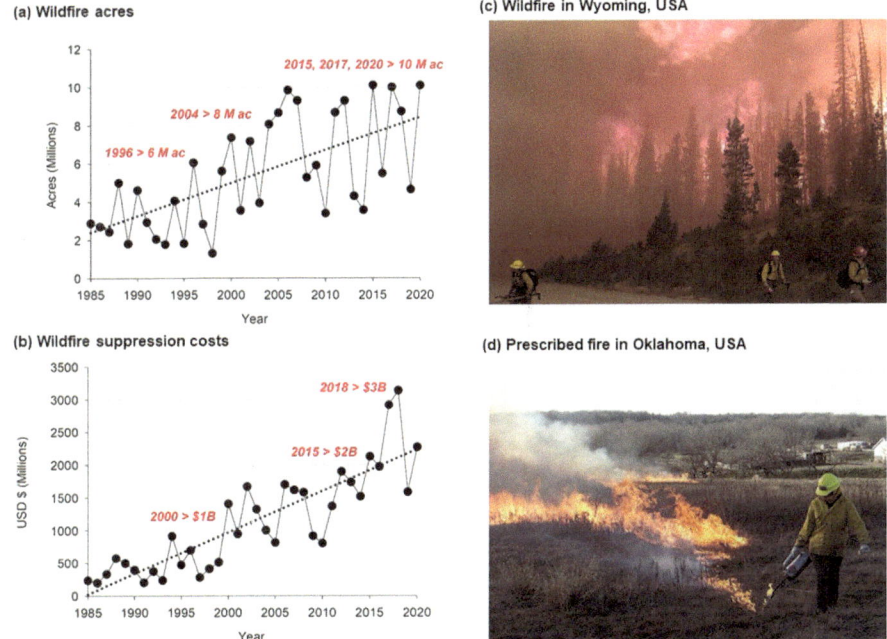

(a) Wildfire acres

(c) Wildfire in Wyoming, USA

(b) Wildfire suppression costs

(d) Prescribed fire in Oklahoma, USA

Fig. 6.1 Annual wildland fire statistics from the National Interagency Fire Center covering all 50 U.S. states for **a** total wildland fire acres and **b** federal firefighting costs for suppression only. Data are publicly available at https://www.nifc.gov/fire-information/statistics. **c** In 2020, total wildland fire acres in the U.S. exceeded 4 million hectares and included large conflagrations such as the Mullen Fire along the border of Colorado and Wyoming. In this photo taken at 6:15 pm on September 24, 2020, crews were attempting to slow the fire and prevent its jumping the road by lighting a smaller and controlled fire, but the fire was moving too quickly and crews were pulled to safety. Photo credit to Josh Shroyer (Operations Section Chief Type 2, Flaming Tree Solutions LLC, Riverton, WY). **d** Lighting a prescribed fire in the southern Great Plains. Photo credit to John Derek Scasta

instill confidence and share resources in prescribed burning, Rangeland Fire Protection Associations (RFPAs) and Good Neighbor Authority (GNAs) projects foster collaborative efforts between federal, state, local, and tribal agencies to suppress the spread of wildfire and to facilitate restoration projects (i.e., hazardous fuel reduction) (Taylor 2005; Twidwell et al. 2013b; Abrams et al. 2017; Stasiewicz and Paveglio 2017; Bertone-Riggs et al. 2018). This grass-roots rise in privatized fire management organizations represents the largest restoration of prescribed fire within North America's rangelands, but still mostly operate within a landscape where purposeful human ignitions for fire management have been extirpated (Twidwell et al. 2013b).

(a) Aspen regeneration after fire

(c) Prescribed fire to reduce Juniper trees

(b) Elk foraging in recent burned area

(d) Training future fire managers

Fig. 6.2 a Example of the diversity of plant species and regeneration strategies 15 years after a wildfire on the Bridger-Teton National Forest of Wyoming. Aspen (*Populus tremuloides*) has a resprouting strategy (note the recruitment of new resprouts around adult trees), whereas sagebrush (*Artemisia tridentata*) has a reseeding strategy. Photo credit to John Derek Scasta. **b** Elk (*Cervus canadensis*) foraging in a recent prescribed fire in southern Wyoming. The burned area was dominated by aspen and serviceberry (*Amelanchier* species). Photo credit to John Derek Scasta. **c** Example of prescribed fire employed to reduce juniper (*Juniperus* species) encroachment in the Great Plains. Photo credit to Caleb Roberts. **d** Training future fire managers requires hands-on opportunities and collaborations to develop the social license (i.e., the approval within a local community and group of stakeholders such that an activity can start or continue) to burn. Photo credit to John Derek Scasta

6.3   Influence of Fire on Wildlife Habitat

Human-driven shifts in fire regimes have imposed novel pressures on wildlife habitat. In North American rangelands, wildlife and biodiversity are dependent on the extent to which fire shaped ecological complexity historically (Fuhlendorf and Engle 2001; Pyne 2001). Many rangeland songbirds, for example, rely on prescribed fire to curtail the encroachment of woody species and to maintain grasslands of varying plant heights and litter denseness to provide breeding habitat for a diverse avian assemblage

Fig. 6.3 Rangeland ecosystems driven by wildfire versus prescribed fire in the western and central U.S. (figure developed by the authors)

(reviewed in Chaps. 8–26). Biodiversity in grassland birds, for instance, is highly dependent on heterogeneity created by patterns in fire regime characteristics like fire frequency and fire seasonality (Hovick et al. 2015a). Species often select for fire outcomes tied to fire severity, like the black-backed woodpecker (*Picoides arcticus*), which is dependent on severely burned areas (Hutto 2008). It is the fire regime that is most important to understand for wildlife management, not the impacts of a single fire event. Ecological complexity owed to fire, both historically and today, is the aggregate result of multiple fire events that vary in terms of their intensity, severity, spatial pattern and extent, thereby leading to a temporal signature of fire's role in a given region. Even the rare occurrence of fire can leave legacies that impart unique niche space that can last for decades after the event (Roberts et al. 2019). This is what drove cross-scale variation that occurs within and among patches, landscapes, and biomes that created sufficient diversity in habitats to host the entire suite of rangeland biota (Fuhlendorf et al. 2012; Fuhlendorf and Engle 2001; Roberts et al. 2020). Changes in these fire regime components have therefore been linked to losses in grassland biodiversity and abundances of grassland biota, which have declined at a greater rate than in any other ecosystem type (Brennan and Kuvlesky 2005; Newbold et al. 2016; Rosenberg et al. 2019). Similar effects have been noted in open forested ecosystems such as ponderosa pine. To advance our understanding of fire relative to wildlife management, we briefly overview the direct effects of a single fire event, which are described in terms of first-order and second-order fire effects and shape the immediate response of vegetation to fire occurrence, and then contrast perspectives that will impact future fire regimes and future wildlife management efforts.

6.3.1   Understanding First-Order and Second-Order Fire Effects

The effects of fire on ecosystem properties, and thus wildlife habitat, are characterized often as *first-order* and *second-order* fire effects. **First-order fire effects** are the direct or immediate consequences of fire, and include biomass consumption, tree crown scorch, soil heating, and smoke production. These occur during the fire. **Second-order fire effects** include interactions with many other non-fire factors that influence post-fire ecosystem responses. Whereas first-order fire effects are owed to immediate consequences of fire, second-order fire effects characterize responses over longer time periods, from days to months to years or longer. Examples of second-order fire effects include vegetation succession, changes in vegetation productivity, and forest regeneration. The concept of ordered effects on plants has also been applied to animals (Whelan et al. 2002; Engstrom 2010). The direct impacts of heat transfer or first-order effects can cause injury or mortality heat exposure, toxic effects of smoke, and oxygen depletion. The processes that occur after the fire, particularly soil, water, and plant responses, can influence starvation, predation, or immigration ultimately influencing population viability.

   Second-order fire effects are complex. For rangeland soils, the amount of heating is generally less than 5% which is radiated into the soil profile (i.e., the vertical section of the soil from the surface of the ground downward to where soil meets the underlying rock layer) and this radiating heat interacts with non-fire factors to influence soil organic matter, soil microbiology, nutrient cycling, and erosion potential. For plants, second-order effects can be categorized relative to life history strategies associated with post-fire recovery (i.e., recovery from fire damage or mortality). Some plants persist and regenerate on-site from established plants; these use asexual 'resprouting' life strategies relying on below-ground and/or above-ground plant parts (Bond and Midgley 2003). Other plants are *colonizers* or *reseeders* and rely on recruitment to regenerate from seed. *Residual colonizers* regenerate on-site from seedbank present at the time of the burn, whereas *off-site colonizers* regenerate from seed brought in by animals or other sources from unburned areas in the surrounding landscape. See Table 6.2 for important rangeland plant species and the associated life history strategy (Midgley 1996).

Table 6.2 Common plants in western North America and associated strategy for responding to fire disturbance

Resprouters	Colonizers/reseeders
Aspen (*Populus tremuloides*)	Budsage (*Artemisia spinescens*)
Greasewood (*Sarcobatus vermiculatus*)	Eastern red cedar (*Juniperus virginiana*)
Honey mesquite (*Prosopis glandulosa*)	Lodgepole pine (*Pinus contorta*)
Nuttall's Saltbush (*Atriplex nuttallii*)	Mountain big sagebrush (*Artemisia tridentata* susp. *vaseyana*)
Sand sagebrush (*Artemisia filifolia*)	Ponderosa pine (*Pinus ponderosa*)
Winterfat (*Krascheninnikovia lanata*)	Shadscale (*Atriplex confertifolia*)
Yucca (*Yucca glauca*)	Wyoming big sagebrush (*Artemisia tridentata*)

Highly specialized regeneration strategies for both life history strategies have developed, including conifers with serotinous cones that open and release seeds following fire (*Pinus contorta* in the Rocky Mountains ecoregion; Knapp and Anderson 1980) and layering, which occurs via asexual reproduction when vertical stems droop and root upon contact with the soil (*Symphoricarpos orbiculatus*; Scasta et al. 2014). All of these affect forage availability and habitat characteristics that in turn affect wildlife. For instance, species such as aspen (*Populus tremuloides*) can be stimulated by fire to resprout, often quite aggressively (Krasnow and Stephens 2015; Fig. 6.2), and this provides high quality browse and fawning habitat for ungulates like mule deer (*Odocoileus hemionus*) and nesting cover for ruffed grouse (*Bonasa umbellus*) (Pojar and Bowden 2004; Rusch and Keith 1971).

Powerful models of first-order fire effects rely on complex, physics-based models of fire behavior coupled with knowledge of key factors that affect first-order fire effects. To model first-order effects, these models most commonly assume homogeneous or spatially constant conditions relevant to scales of individual plants or homogeneous landscape patches (Reinhardt and Dickinson 2010). Second-order effects are dependent on first-order effects, so knowledge of second-order effects is dependent on understanding first-order effects and interactions with additional non-fire drivers. Existing assumptions of homogeneous conditions limit applicability for many wildlife studies, so it is rare for investigations to link second-order effects to first-order impacts. Even the leading fire models do not sufficiently capture the complexity owed to describing their interdependencies and instead model one or the other (e.g., First Order Fire Effects Model; FOFEM). This is one reason that the fire regime concept has been useful in simplifying key aspects of these complex relationships to characterize major departures in key landscape-scale fire metrics and corresponding fire effects important to wildlife such as escalation of fire frequency in the Great Basin.

6.3.2   Control of Invasive Plants with Prescribed Burning

The first- and second-order effects of prescribed fire are the primary points of consideration used to strategically manage invasive species and structure habitat features for wildlife (DiTomaso et al. 2006). For the control of annual plants, the strategic application of fire to kill plants before seeds become viable or before they disperse may be effective. Because the majority of fire-associated heat radiates upward, seeds in the soil bank are not usually exposed to lethal temperatures and thus seeds still in the canopy would be most susceptible. For example, early summer burning has been shown to reduce barb goatgrass (*Aegilops triuncialis*) in California when flames kill exposed seeds (Marty et al. 2015). Fire also may alter the soil-litter environment; for example, fall burning may lead to reductions of litter and subsequently of annual bromes (*Bromus* species) in the northern mixed-grass prairie and along such ecotones transitioning to sagebrush steppe (Whisenant 1990; Estep 2020; Symstad et al. 2021). Similarly, annual forbs such as yellow starthistle (*Centaurea solstitialis*), which is widespread throughout California, Oregon, Washington, and Idaho, can be controlled with repeated early summer burns (DiTomaso et al. 1999). Perennial forbs such as leafy spurge (*Euphorbia esula*) present a more difficult challenge and may require the integration with fire of other methods, such as herbicide applications (DiTomaso et al. 2006). Cool-season perennial grasses such as Kentucky bluegrass (*Poa pratensis*) and smooth brome (*Bromus inermis*) may be controlled with fire if applied when tillers are elongating on the target species but concurrently when the desirable native warm-season grasses are still dormant (DiTomaso et al. 2006). Obligate-seeding woody species that rely solely on recruitment of seeds for regeneration (such as eastern red cedar [*Juniperus virginiana*] and Wyoming big sagebrush [*Artemisia tridentata*]) are very susceptible to fire (Beck et al. 2009, 2012; Twidwell et al. 2013a), whereas woody species that use asexual regeneration strategies such as resprouting may be difficult to control or may be stimulated by fire (such as mesquite (*Prosopis glandulosa*; in the Tamaulipan Vegetation ecoregion, southern Great Plains ecoregion, and the Western Deserts, Grasslands, Shrublands, and Woodlands ecoregion) and roughleaf dogwood (*Cornus drummondii*; in the Great Plains ecoregion) (Drewa 2003; Heisler et al. 2004; Starns et al. 2021) unless fires are conducted in more extreme conditions that overcome plant species persistence (Twidwell et al. 2016). The timing of fire, fuel load, topographic position, and post-fire moisture conditions of fire applications also must be considered because such features can influence the flammability and response of a species (Keeley and McGinnis 2007; Weir and Scasta 2014).

First order and second-order fire effects always result in both positive and negative responses for wildlife (Table 6.3). This is critical to understand to avoid oversimplifications of complex fire-plant-wildlife interactions. Some species will respond positively to the immediate post-fire environment and preferentially select recently burned areas, whereas others avoid it and move to later successional habitat stages. For instance, greater prairie chickens (*Tympanuchus cupido*) in the Great Plains ecoregion have been shown to move lek locations in response to dynamic fire patterns so that leks fall near the edges of recently burned patches providing recently disturbed habitat and nearby unburned refugia habitat (Hovick et al. 2015b). In a heterogeneous landscape, an array of patches facilitate a shifting mosaic of vegetation structure that varies spatially and temporally with the temporal pattern of fire in the region. In a homogeneous landscape, there may only be short vegetation across the landscape (if fire is applied homogeneously) or only tall vegetation (if fire is suppressed homogeneously).

Positive and negative effects of fire often are associated with human values or conservation goals rather than natural biological outcomes. Some examples of positive effects include enhanced forage quality, reduced parasite and disease exposure, altered animal distribution, reduced dominance by invasive plants, and greater diversity of habitat and food resources (Fig. 6.2). In tallgrass prairie, forage crude protein may be 4 times higher in recently burned areas, resulting in a strong attraction for large herbivores such as American bison (Allred et al. 2011). This type of focal grazing and intense disturbance of plants may alter plant primary nutrients as well as secondary compounds and rate of plant defoliation compared to unburned areas (Scasta et al. 2021). The reduction by fire of ecto-parasites such as ticks, which reduces all life stages, may consequently reduce risk of Lyme disease transmission (Scifres et al. 1988; Mather et al. 1993; Stafford et al. 1998; Cully 1999). Similar positive effects are noteworthy for endo-parasites. For example, native sheep (*Ovis dalli stonei*) in western Canada that were able to access recent burns had up to 10 × lower lungworm (*Protostrongylus* spp.) loads (Seip and Bunnell 1985); in the southeastern US, fire is thought to disrupt microhabitat of gastropod hosts of the meningeal brain worm (*Parelaphostrongylus tenuis*) which has negatively affected elk (*Cervus canadensis*) and deer (Weir 2009); and for amphibians, internal nematodes with free-living terrestrial stages are reduced (Hossack et al. 2013b).

Some examples of negative effects include direct mortality, loss of habitat structure, and/or loss of food resources. These vary with fire intensity, spatial extent, and post-fire moisture availability. Direct mortalities are thought to be relatively rare for both prescribed fires and wildfires (Lyon et al. 1978; Means and Campbell 1981; Russell et al. 1999; Smith 2000) but do occur, with slow-moving species at

Table 6.3 First-order and second-order fire effects lead to plant-wildlife-fire interactions and directional positive (+) and/or negative (−) ecological outcomes for different taxa

Examples of plant-wildlife-fire interactions	Directional ecological outcome(s)
Fire kills a snake due to heat exposure and later gets eaten by scavengers	+ *scavengers*; − *snake*
Small mammals move to escape fire, resulting in immediate food resource availability for raptors	+ *raptor*; − *small mammals*
Fire kills eastern redcedar resulting in increased greater prairie chicken habitat but reduced cedar waxwing habitat	+ *prairie chicken*, + *grasslands*; − *eastern redcedar*, − *cedar waxwing*

greatest risk. However, even slow-moving species, such as amphibians and reptiles, have strategies for escaping fire morality by finding refugia, such as going underground in tunnels or new burrows; these species may even respond positively to fire due to changes in the thermal environment (Hossack and Corn 2007; Hossack et al. 2009, 2013a). The post-fire moisture and subsequent recovery of vegetation for food and cover is also critical. For example, Clapp and Beck (2016) reported that translocated sheep in Wyoming had a 30% reduction in survival after wildfires and droughts, suggesting that interactions with other abiotic features can influence outcomes. However, these same sheep were shown to select low- and high-severity fire areas after the fire and drought events (Donovan et al. 2021a), with female use of low- and high-severity burned areas increasing with greater time since fire, while males tended to decrease use of areas that burned at high severity with greater time since fire. The loss of habitat structure and food resources also can impose long-term effects such as the case for Wyoming big sagebrush and sage-grouse (*Centrocercus urophasianus*; a sagebrush obligate species) that can take decades to centuries to recover to pre-fire levels (Beck et al. 2012).

It is also important to note how animals immediately respond to fires. Some animals flee while others may hide below ground. Pausas (2019) suggests four mechanisms of animal responses to fire including resistance (by using thick cover or having physiological tolerance), refugia (by detecting fire and being able to quickly react and hide), avoidance (by selecting habitat with reduced flammability generally), and crypsis (by altering feeding and or hiding capacity after fire such as a darker cryptic color in neonates or adults or entering torpor). Such responses are initiated by olfactory, auditory, and visual cues of fire (Nimmo et al. 2021). For more information on refugia see Meddens et al. (2018).

Ultimately, the placement of human values onto fire effects for flora and fauna—whether positive or negative—are often short-term, sometimes biased, and usually oversimplify fire to a binary (positive or negative) outcome. In reality, fire's role in nature is more complex. A comprehensive understanding of fire is difficult, if not impossible, without looking beyond individual fire events and considering how fire regime characteristics have changed in terms of their spatial pattern and extent, frequency of occurrence over time, critical ranges of variation in fire function, and the relative contribution of these factors to the degree of vegetation heterogeneity that occurs on landscapes (Twidwell et al. 2021).

6.3.3 Spatial Scales of Fire–Akin to Wildlife Home Range Size

Fire effects cannot be deeply understood, or applied in wildlife management, without careful consideration of scale. Scientists and managers are becoming increasingly aware of the importance of scale in solving academic and managerial debates about how to best apply and manage fire for a variety of outcomes for plants and animals following fire. Scale refers to the spatial and temporal dimensions used to study phenomena in ecology such as migration at large landscape scales or habitat attributes for nesting. Scale also refers to the analytical and quantitative dimensions that might differ between observers or independent studies. Spatial scale is described analytically in terms of both grain (the resolution of the data) and extent (the size of the landscape encompassed for observation). A single fire event is often described in terms of its extent (the perimeter of the burned area) but vegetation is never 100% consumed in a fire event. Plant parts, whole plants, or entire patches in the landscape escape fire and remain unburned and the scale at which these are measured represents the grain. Both the minimum resolution associated with variability in burned/ unburned patterns at fine scales and the total size, or spatial extent, of the fire are important to wildlife. Fire regimes, like other disturbance regimes, operate across a sufficient range of scales (through both space and time), and changes in its scale of function represent some of the greatest challenges to wildlife managers.

One way to better understand the context of animal responses to the scale of fire is to think of it relative to a species' home range size. Home range is the spatial extent at which an animal operates; it can fluctuate with an animal's size, seasonal needs, age, and sex as well as the spatial distribution of hiding and nesting cover, brood-rearing habitat, thermal regulation resources, food, and water (Burt 1943; Hayne 1949; Powell and Mitchell 2012). The scale and distribution of fire can affect all these drivers of home range size and core use areas differentially in a wildlife community. For example, ornate box turtles (*Terrapene carolina*) have a home range of < 10 ha, indicating that its resource needs are fulfilled within that spatial extent but also suggesting that larger fires could alter habitat resources at a spatial scale much larger than a single individual. In contrast, the more expansive seasonal home ranges of mule deer (*Odocoileus hemionus*; > 2878 ha (Kie et al. 2002)) and seasonal migrations of more than 110 km (Sawyer et al. 2019) suggest that mule deer are capable of shifting use areas in relation to fire. While identifying important seasonal use is important, defining home ranges solely by seasonality fails to recognize the spatial extent over which species may operate during the annual cycle and how the timing of a fire may affect a species. Moreover, the patchiness of burn severity in a burned area can also influence animal home range because the distribution of fire-altered resources can significantly influence home range sizes for wildlife in nearly every North American ecosystem (e.g., Gullion 1984; Kie et al. 2002; Patten et al. 2011; Fuhlendorf et al. 2017b). Because home range considerations differ among wildlife species, every ecosystem type exhibits unique and complex spatial signatures of fire on wildlife communities. The same complexities needed to understand the spatial context of a species' home range are also necessary to understand fire, the occurrence of variable fire effects within and across ecosystem types, and then how wildlife are likely to respond.

6.4 Competing Ideologies for Future Fire Management

Future fire management has the potential to alter outcomes for wildlife and wildlife management in rangelands depending on the ideologies that are embraced (Twidwell et al. 2021; Fuhlendorf et al. 2012). Perceptions tied to fire as a management tool, the range of variability in fire regimes utilized for management, the spatial and temporal scales of fire, and the spatial scales of fire management planning all have impacts on how we manage rangeland wildlife in the future. Following Twidwell et al.'s (2021) framework we present five ideologies that currently exist or that could further emerge, and that may shape or constrain wildlife management. These ideologies reflect competing sociopolitical values, potentially further changing the fire regimes that wildlife evolved with during the *Coexistence Era* from the changes imposed during the *Wildfire and Suppression Era* that dominates today. These five ideologies acknowledge that some form of disturbance is necessary to maintain rangeland wildlife diversity, but they differ by (1) how/if they utilize fire as a key disturbance, (2) how much they seek to control fire's range of variability, (3) the spatial and temporal scales at which they allow fire to function, and (4) the spatial scales of management planning (Fig. 6.4).

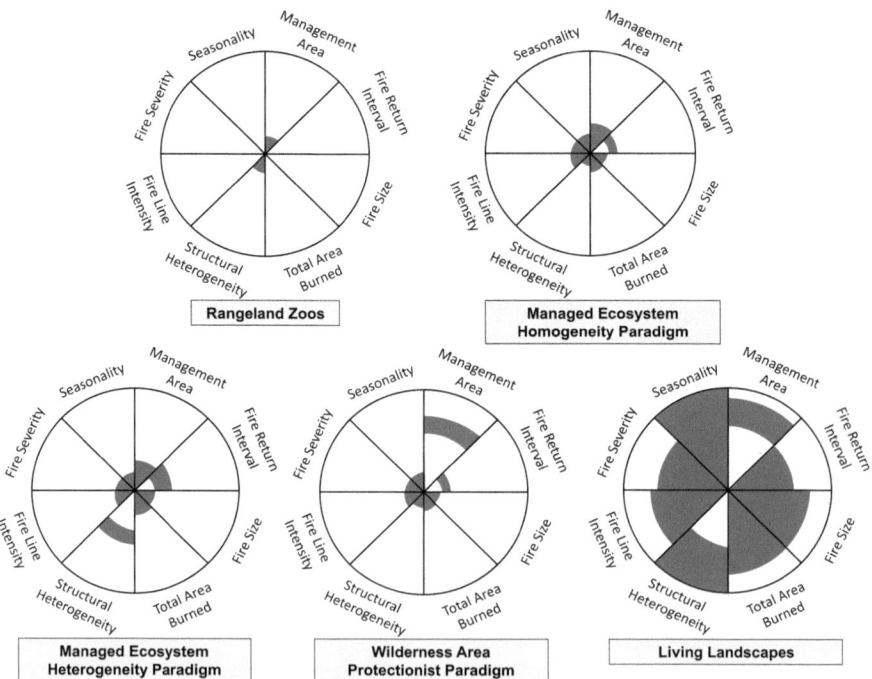

Fig. 6.4 Five fire management ideologies that compare the range of variation in fire regime characteristics, outcomes, and management area, applicable to North American rangelands (figure developed by the authors)

6.4.1 Ideology 1: 'Rangeland Zoos'

Given anthropogenic pressures this century, wildlife are being more intensively managed to safeguard iconic species in ecosystems with extensive land use conversion through such measures such as translocations to support viable populations of species vulnerable to extinctions or extirpations (IUCN 2013) and vaccination programs for endangered species (Haydon et al. 2006). As these types of intensive management efforts continue, a network of small, zoo-like "ghosts of rangelands past" are expected. Unlike typical classical zoos, these areas may be very small remnants of habitat that will still mirror habitat requirements through intensive management that tailors vegetation structure and composition to rangeland wildlife species' needs. Maintaining these zoo-like systems to support a small portion of their endemic biodiversity will require intensive and expensive management to coerce the system into an idealized state while reinforcing fire and fuel feedbacks are absent. Fire, if used at all, will occur only under the tightest range of variation and likely only as a demonstration of its occurrence as a historically relevant process. Thus, unlike other future fire ideologies, rangeland zoos will largely avoid the use of fire, focusing on very small spatial scales for management.

Species that require large expanses of heterogeneous rangelands may be locally extirpated or require intensive management. Even for species that require smaller-scale rangeland habitat features, the limited size of rangeland zoos will be unlikely to support even a minimum viable population without human intervention (With et al. 2008). Species will require assisted dispersal, colonization, and migration. Pressure will mount from threats in the surrounding habitat matrix. Outside of the boundaries of rangeland zoos, ecosystems will continue to undergo conversion and collapse to alternative, undesirable vegetative states (e.g., woodlands, nonnative annual grasslands) or to be developed for alternative land uses. Examples of the latter occur today in the last remaining (small) remnants of tallgrass prairie, where intensive management with a suite of regularly applied chemical and mechanical treatments are needed to preserve a semblance of the region's past biodiversity.

6.4.2 Ideology 2: 'Managed Ecosystem—Homogeneity Paradigm'

The Managed Ecosystem—Homogeneity Paradigm shares many similarities to Rangeland Zoos but operates at larger scales and has livestock production as a secondary concern. The Homogeneity Paradigm is the classical rangeland approach used to create uniform vegetation utilization and is based on the utilitarian perspective of benefiting livestock production while limiting tradeoffs to other ecosystem services (e.g., wildlife; Fuhlendorf et al. 2012). As a result, fire management is used to create specific vegetation conditions that are relatively homogenous in terms of vegetation composition and structure. Fire regime characteristics are constrained to

low intensities, strict spatial scales, strict seasonality, and single property or jurisdictional boundaries. Under the Homogeneity Paradigm, wildlife abundance and
diversity may be reduced due to homogeneity of the vegetation, creating a narrower
range of habitat conditions that benefit only that select suite of species that prefers
conditions associated with a heavily grazed and burned land use (Fuhlendorf et al.
2012). Species that require expansive, more heterogeneous vegetation characteristics
may experience range contractions and therefore require translocations or assisted
dispersal. Species that require vegetation structure from high-intensity fires may be
locally extirpated (Hutto 2008; Hutto et al. 2016). Without the full range of variability,
positive feedbacks that lead to undesirable vegetation state transitions can emerge
(i.e., transition from a grassland or shrubland to a woodland; Twidwell et al. 2016,
2019). By constraining variation, rangeland management production predictability
in the short-term and local scales for unpredictability in the long-term and at broad
scales. This can lead to the need for the Rangeland Zoos ideology if fire management
is delayed or slowed during, for example, drought years or if surrounding rangelands
are experiencing state transitions.

The homogeneity paradigm has been the predominant approach used to manage
rangelands and its wildlife over the past several decades. The hallmark of the Homogeneity Paradigm for ecosystem management is the assumption that rangelands can
be maintained without cross-scale heterogeneity. The idea implicit in this assumption is that rangelands can be held in a static state with a precise amount of human-
directed disturbance per the succession-retrogression theory of rangeland management. Importantly, fire may or may not be part of management strategies; when used,
it is a tool to maintain stasis. Much of North America's current rangelands have been
managed as static, fire-excluded systems for decades while maintaining their identities and many wildlife species. For example, the Nebraska Sandhills excluded fire for
decades and remain one of the most intact grasslands in North America—although
there are early warnings of vegetation state transitions to woodlands (Donovan et al.
2018; Fogarty et al. 2020). Once external factors like climate change or woody plant
encroachment appear, managing for static systems becomes infeasible except at small
scales at specific locations. This is due to the increasingly intensive management
required to counteract the positive feedbacks of the surrounding collapsed rangeland
state. Such practices can also have inadvertent negative outcomes relative to fire.
For instance, throughout the Great Plains ecoregion, this static view of management
that utilizes fire suppression is a major reason that woody encroachment is a primary
management concern across the region (Briggs et al. 2005; Twidwell et al. 2013b),
which in turn has been linked to increasing wildfire (Donovan et al. 2020).

6.4.3 Ideology 3: 'Managed Ecosystems—Heterogeneity Paradigm'

Counter to the utilitarian perspective of classical rangeland management that stresses uniform forage production for livestock use, the Heterogeneity Paradigm of ecosystem management recognizes that critical ranges of vegetation variability (or heterogeneity) are the foundation for biological diversity (Fuhlendorf et al. 2012, 2017a). The Heterogeneity Paradigm acknowledges that fire played a role, whether frequent or infrequent, in shaping the vegetation heterogeneity upon which wildlife diversity depends. To date, however, the application of the Heterogeneity Paradigm has been applied primarily in ecosystems that burned more frequently in the past and where fire was the primary driver of spatial heterogeneity (e.g., grasslands of the Great Plains). This has been accomplished within managed areas by embracing 'pyric herbivory,' which refers to herbivory driven by fire and that creates a shifting mosaic of fire, grazing, and vegetation structure through space and time (Fuhlendorf et al. 2009).

The Heterogeneity Paradigm has increased coexistence of livestock and wildlife in many managed rangelands and removed the perspective that they are competing interests (Fuhlendorf et al. 2017a). Fire results in high-quality forage for livestock while also maintaining a greater range of vegetation structures that promote wildlife diversity (Fuhlendorf and Engle 2001; Hovick et al. 2014; Scasta et al. 2016a). Elapsed time-since-fire is an integral consideration of management; however, other characteristics associated with fire regimes are tightly restricted or not considered as part of fire management objectives, such as fire intensity, seasonality, and the spatial scales at which fires are applied, nor is the vegetation heterogeneity that might derive from these characteristics (Donovan et al. 2021b; Hutto et al. 2016; Roberts et al. 2020). For example, fire is generally avoided in sage-brush dominated systems due to negative effects on sage obligate wildlife species.

The application of the Heterogeneity Paradigm is much lower than application of the Homogeneity Paradigm, so its outcomes are limited to the extent of heterogeneity created within geographically expansive homogeneous rangelands or converted ecosystems. As a result, wildlife species that range over expansive areas and require large-scale heterogeneity will still likely experience range contractions and may require assisted dispersal. The landscape heterogeneity promoted by pyric herbivory will better enhance rangeland stability (Holling and Meffe 1996), however, it may be unable to promote high levels of rangeland ecosystem resilience to large-scale global change drivers without greater adoption and geographic influence.

6.4.4 Ideology 4: 'Wilderness Area—Protectionist Paradigm'

Like the Homogeneity Paradigm of rangeland management, the Protectionist Paradigm used in wilderness areas has been the prevailing focus of large-scale wildlife management efforts on public lands. Wilderness areas often focus on a protectionist viewpoint with a fixed endpoint at large scales. Disturbances are generally minimized in order to maintain an idealized habitat configuration based on a historical reference condition (or an idealized climax community). Major disturbances, like wildfire, are usually suppressed before they can 'damage' too much of the idealized community configuration. Charismatic species are often the main focus of conservation (Brambilla et al. 2013; Colléony et al. 2017) and may require more intensive habitat and population management to create the conditions necessary for their persistence. While the larger size of wilderness areas enhances the conservation of far-ranging wildlife species, management efforts generally favor habitat generalists and the few species that are adapted to persist in the climax community. Less charismatic species reliant on fire and its interaction with other disturbance processes are less abundant or extirpated. In an effort to offset tradeoffs to fire-dependent species, management tactics like prescribed fire are utilized under a limited range of conditions and at small scales.

6.4.5   Ideology 5: 'Living Landscape'

Rangeland management under the Living Landscapes Paradigm builds off of the Heterogeneity Paradigm and, of all the ideologies, places the fewest restrictions on fire. Fire and its full range of variability are treated as an integral system process in rangelands, rather than something that degrades or destroys habitat under more 'extreme' conditions. This management style shifts away from applying wildlife management principles to the advantage of a few charismatic species, to applying principles on maintaining the integrity of ecosystems and ecosystem processes. Large expanses of rangelands are maintained by allowing fire variability to play out across their full range of variability (temperatures, seasons, humidity, wind speeds, sizes, frequencies, intensities) and interact with dynamic patterns in variables like grazing, topography, and climate, with fire only be restricted in the wildland-urban interface. This creates heterogeneous conditions across multiple scales, allowing for landscapes that are heterogeneous at local, regional, and biome levels and that can support a wide range of rangeland plant and wildlife species.

The Living Landscapes and Heterogeneity Paradigm are the two ideologies that embrace variability in fire as a critical driver of wildlife habitat and they are closest to the ideologies embraced during the *Coexistence Era*. Expanding on these ideologies in fire management will require large-scale landowner collaborations and land planning to allow for fires to burn across multiple property boundaries to avoid small, fragmented burn patterns of limited utility in rangeland management objectives. Corridors for fire spread, along with other ecological processes like wildlife migration and dispersal, could help facilitate large-scale interactions and connectivity unhindered by roads and other human development. In terms of fire, these corridors could help to reduce the impacts of fragmentation on fire size and intensity (Brudvig et al. 2012). Regional planning could emphasize fire districts to assist in planning of prescribed fires to develop alternative fuel break types, like existing crops or agriculture, to protect human developments from wildfires (Donovan et al. 2020). Further expansion of prescribed fires within the wildland-urban interface would need to consider risks to anthropogenic developments (Radeloff et al. 2018; Theobald and Romme 2007), and land-use planning would instead need to operate under the assumption that fires, whether planned or unplanned, will inevitably occur. The focus here is to identify areas with the potential to foster coexistence of rangelands, wildlife, fire, and people once again. While examples of re-creating critical ranges of variability in fire regime functioning for wildlife management is exceedingly rare, particularly given novel global change threats mounting today, recent applications have led to increased biodiversity and enhanced conservation outcomes (e.g., Twidwell et al. 2021).

6.4.6 Practical Applications of Competing Ideologies

To maintain the full suite of the habitat needs of rangeland wildlife and also to meet society's needs in the face of global climate change, critical ranges of variability in fire function need to be understood—and managed accordingly—across North America (Bowman and Legge 2016; Hutto et al. 2016; Roberts et al. 2020). Only two existing ideologies focus on that perspective and attempt to manage for complexity by understanding fire's contribution to it (the Managed Ecosystems—Heterogeneity Paradigm and Living Landscapes ideologies). However, all five ideologies will likely be necessary given the rate and prevalence of state transitions and rapid global change impacting North American rangelands (Garmestani et al. 2020; McWethy et al. 2019). For example, the Rangeland Zoos ideology may be necessary where rangelands have already succumbed or are completely surrounded by alternative regimes (e.g., Attwater's prairie chicken conservation in coastal prairie, TX). Nevertheless, investments in those areas should not be replicated at large-scales where greater complexity in wildlife habitat management is possible. Rangelands were forged by fire functioning across spatial and temporal scales, and many species evolved to be highly dependent on specific fire dynamics (Pausas and Parr 2018; Twidwell et al. 2021). Although we may never return to these historical fire dynamics in all North American rangelands, restoring large-scale fire dynamics in as many locations as possible is the best and most economical approach to maintaining rangeland wildlife diversity and resilience (Fuhlendorf et al. 2012).

6.5 Conclusions

In order to acquire a modern understanding of fire and wildlife, it is necessary to understand how fire historically functioned and how that function is changing across spatial and temporal scales. Cultural perspectives about wildfire and prescribed fire have dictated the policies, regulations, management, and resources dedicated to fire management in the past and will continue to do so into the future (Calkin et al. 2005; Canton-Thompson et al. 2008; Ryan et al. 2013). These decisions affect how wildlife function within rapidly changing ecosystems and the potential to mitigate impacts from invasive species or a changing climate. Public views of fire take on multiple lenses that are influenced by news media, liability, knowledge, personal experience, collective attitudes, and cultural norms (Paveglio et al. 2011; Toledo et al. 2013; Joshi et al. 2019; Bendel et al. 2020). Prescribed fire application requires cultural and attitudinal acceptability, along with resources and planning assistance to foster prescribed fire as a management tool to reduce hazardous fuels and invasive species (Kreuter et al. 2008; Weir 2010; Toledo et al. 2013). Considering the effects fire can have on a landscape, there are also concerns about the impacts prescribed fire may have on wildlife species because of a lack of literacy on the complexity of how fire interacts with non-fire factors (Bowker et al. 2008; Elmore et al. 2009; Coon et al. 2018). Legacies of fire can generate unique and more diverse wildlife communities that can persist for decades (Roberts et al. 2019; Donovan et al. 2021b), and there is great opportunity for wildlife managers to incorporate knowledge of how fire regimes have been altered in recent decades to set unique conservation priorities and improve wildlife outcomes.

References

Abrams J, Davis EJ, Wollstein K (2017) Rangeland fire protection associations in great basin rangelands: a model for adaptive community relationships with wildfire? Hum Ecol 45:773–785. https://doi.org/10.1007/s10745-017-9945-y

Allred BW, Fuhlendorf SD, Engle DM et al (2011) Ungulate preference for burned patches reveals strength of fire–grazing interaction. Ecol Evol 1:132–144. https://doi.org/10.1002/ece3.12

Anderson KL, Smith EF, Owensby CE (1970) Burning bluestem range. J Range Manage 23:81–92. https://doi.org/10.2307/3896105

Beck JL, Connelly JW, Reese KP (2009) Recovery of greater sage-grouse habitat features in Wyoming big sagebrush following prescribed fire. Restor Ecol 17:393–403. https://doi.org/10.1111/j.1526-100X.2008.00380.x

Beck JL, Connelly JW, Wambolt CL (2012) Consequences of treating Wyoming big sagebrush to enhance wildlife habitats. Rangeland Ecol Manag 65:444–455. https://doi.org/10.2111/REM-D-10-00123.1

Bendel C, Toledo D, Hovick T et al (2020) Using behavioral change models to understand private landowner perceptions of prescribed fire in North Dakota. Rangeland Ecol Manag 73:194–200. https://doi.org/10.1016/j.rama.2019.08.014

Bertone-Riggs T, Cyphers L, Davis EJ et al (2018) Understanding good neighbor authority: case study from across the west. Rural Voices Cons Coal 1–32. https://doi.org/10.1016/j.resconrec.2003.11.001

Bond WJ, Keeley JE (2005) Fire as a global 'herbivore': the ecology and evolution of flammable ecosystems. Trends Ecol Evol 20:387–394. https://doi.org/10.1016/j.tree.2005.04.025

Bond WJ, Midgley JJ (2003) The evolutionary ecology of sprouting in woody plants. Int J Plant Sci 164(S3):S103–S114. https://doi.org/10.1086/374191

Bowker JM, Lim SH, Cordell HK et al (2008) Wildland fire, risk, and recovery: results of a national survey with regional and racial perspectives. J Forest 106:268–276. https://doi.org/10.1093/jof/106.5.268

Bowman DM, Legge S (2016) Pyrodiversity—why managing fire in food webs is relevant to restoration ecology. Restor Ecol 24:848–853. https://doi.org/10.1111/rec.12401

Brambilla M, Gustin M, Celada C (2013) Species appeal predicts conservation status. Biol Cons 160:209–213. https://doi.org/10.1016/j.biocon.2013.02.006

Brennan LA, Kuvlesky WP Jr (2005) North American grassland birds: an unfolding conservation crisis? J Wildl Manag 69:1–13. https://doi.org/10.2193/0022-541X(2005)069%3c0001:NAGBAU%3e2.0.CO;2

Brown PM, Sieg CH (1999) Historical variability in fire at the ponderosa pine-Northern Great Plains prairie ecotone, southeastern Black Hills, South Dakota. Ecoscience 6(4):539–547. https://doi.org/10.1080/11956860.1999.11682563

Briggs JM, Hoch GA, Johnson LC (2002) Assessing the rate, mechanisms, and consequences of the conversion of tallgrass prairie to *Juniperus virginiana* forest. Ecosystems 5:578–586. https://doi.org/10.1007/s10021-002-0187-4

Briggs JM, Knapp AK, Blair JM et al (2005) An ecosystem in transition: causes and consequences of the conversion of mesic grassland to Shrubland. Bioscience 55:243–254. https://doi.org/10.1641/0006-3568(2005)055[0243:AEITCA]2.0.CO;2

Brudvig LA, Wagner SA, Damschen EI (2012) Corridors promote fire via connectivity and edge effects. Ecol Appl 22:937–946. https://doi.org/10.1890/11-1026.1

Burt WH (1943) Territoriality and home range concepts as applied to mammals. J Mammal 24:346–352. https://doi.org/10.2307/1374834

Busenberg G (2004) Wildfire management in the United States: the evolution of a policy failure. Rev Policy Res 21:145–156. https://doi.org/10.1111/j.1541-1338.2004.00066.x

Calkin DE, Gebert KM, Jones JG et al (2005) Forest Service large fire area burned and suppression expenditure trends, 1970–2002. J Forest 103:179–183. https://doi.org/10.1093/jof/103.4.179

Calkin DE, Thompson MP, Finney MA (2015) Negative consequences of positive feedbacks in US wildfire management. Forest Ecosyst 2:1–10. https://doi.org/10.1186/s40663-015-0033-8

Canton-Thompson J, Gebert KM, Thompson B et al (2008) External human factors in incident management team decision making and their effect on large fire suppression expenditures. J Forest 106:416–424. https://doi.org/10.1093/jof/106.8.416

Clapp JG, Beck JL (2016) Short-term impacts of fire-mediated habitat alterations on an isolated bighorn sheep population. Fire Ecol 12:80–98. https://doi.org/10.4996/fireecology.12.03.999

Colléony A, Clayton S, Couvet D et al (2017) Human preferences for species conservation: animal charisma trumps endangered status. Biol Cons 206:263–269. https://doi.org/10.1016/j.biocon.2016.11.035

Coon JJ, Morton LW, Miller JR (2018) A survey of landowners in the Grand River Grasslands: managing wildlife, cattle, and non-native plants. University of Illinois Department of Natural Resources and Environmental Sciences, Thesis

Cully JF (1999) Lone star tick abundance, fire, and bison grazing in tallgrass prairie. J Range Manage 52:139–144. https://doi.org/10.2307/4003507

D'Antonio CM, Vitousek PM (1992) Biological invasions by exotic grasses, the grass/fire cycle, and global change. Annu Rev Ecol Syst 23:63–87. https://doi.org/10.1146/annurev.es.23.110192.000431

DiTomaso JM, Kyser GB, Hastings MS (1999) Prescribed burning for control of yellow starthistle (*Centaurea solstitialis*) and enhanced native plant diversity. Weed Sci 47:233–242. https://doi.org/10.1017/S0043174500091669

DiTomaso JM, Brooks ML, Allen EB et al (2006) Control of invasive weeds with prescribed burning. Weed Tech 20:535–548. https://doi.org/10.1614/WT-05-086R1.1

Donovan GH, Brown TC (2007) Be careful what you wish for: the legacy of Smokey Bear. Front Ecol Environ 5:73–79. https://doi.org/10.1890/1540-9295(2007)5[73:BCWYWF]2.0.CO;2

Donovan VM, Burnett JL, Bielski CH et al (2018) Social–ecological landscape patterns predict woody encroachment from native tree plantings in a temperate grassland. Ecol Evol 8:9624–9632. https://doi.org/10.1002/ece3.4340

Donovan VM, Dwinnell SPH, Beck JL et al (2021a) Fire-driven landscape heterogeneity shapes habitat selection of bighorn sheep. J Mammal 102:757–771. https://doi.org/10.1093/jmammal/gyab035

Donovan VM, Roberts CP, Wonkka CL et al (2021b) Collapse, reorganization, and regime identity: breaking down past management paradigms in a forest-grassland ecotone. Ecol Soc 26:27. https://doi.org/10.5751/ES-12340-260227

Donovan VM, Wonkka CL, Twidwell D (2017) Surging wildfire activity in a grassland biome. Geophys Res Lett 44:2017GL072901. https://doi.org/10.1002/2017GL072901

Donovan VM, Wonkka CL, Wedin DA et al (2020) Land-Use type as a driver of large wildfire occurrence in the U.S. Great Plains. Remote Sens 12:1869. https://doi.org/10.3390/rs12111869

Doolittle ML, Lightsey ML (1979) Southern woods-burners: a descriptive analysis (No. 151). US Department of Agriculture, Forest Service, Southern Forest Experiment Station

Drewa PB (2003) Effects of fire season and intensity on Prosopis glandulosa Torr. var. glandulosa. Int J Wildland Fire 12(2):147–157. https://doi.org/10.1071/WF02021

Elmore RD, Bidwell TG, Weir JR (2009) Perceptions of Oklahoma residents to prescribed fire. In: Proceedings of the 24th tall timbers fire ecology conference: the future of prescribed fire: public awareness, health, and safety. Tall Timbers Research Station, Tallahassee, Florida, USA

Engstrom RT (2010) First-order fire effects on animals: review and recommendations. Fire Ecol 6(1):115–130. https://doi.org/10.4996/fireecology.0601115

Estep CE (2020) Wyoming big sagebrush survival and herbaceous community response to prescribed burns across an invasion gradient of annual brome. University of Wyoming

Fogarty DT, Roberts CP, Uden DR et al (2020) Woody plant encroachment and the sustainability of priority conservation areas. Sustainability 12:8321. https://doi.org/10.3390/su12208321

Fuhlendorf SD, Engle DM (2001) Restoring heterogeneity on rangelands: ecosystem management based on evolutionary grazing patterns. Bioscience 51:625–632. https://doi.org/10.1641/0006-3568(2001)051[0625:RHOREM]2.0.CO;2

Fuhlendorf SD, Harrell WC, Engle DM et al (2006) Should heterogeneity be the basis for conservation? Grassland bird response to fire and grazing. Ecol Appl 16(5):1706–1716. https://doi.org/10.1890/1051-0761(2006)016[1706:SHBTBF]2.0.CO;2

Fuhlendorf SD, Engle DM, Kerby J et al (2009) Pyric herbivory: rewilding landscapes through the recoupling of fire and grazing. Cons Biol 23:88–598. https://doi.org/10.1111/j.1523-1739.2008.01139.x

Fuhlendorf SD, Engle DM, Elmore RD et al (2012) Conservation of pattern and process: developing an alternative paradigm of rangeland management. Rangeland Ecol Manag 65:579–589. https://doi.org/10.2111/REM-D-11-00109.1

Fuhlendorf SD, Fynn RW, McGranahan DA et al (2017a) Heterogeneity as the basis for rangeland management. In: Rangeland systems. Springer, Cham, pp 169–196. https://doi.org/10.1007/978-3-319-46709-2_5

Fuhlendorf SD, Hovick TJ, Elmore RD et al (2017b) A hierarchical perspective to woody plant encroachment for conservation of prairie-chickens. Rangeland Ecol Manag 70(1):9–14. https://doi.org/10.1016/j.rama.2016.08.010

Fuhlendorf SD, Davis CA, Elmore RD et al (2018) Perspectives on grassland conservation efforts: should we rewild to the past or conserve for the future? Philos T R Soc B 373(1761):20170438. https://doi.org/10.1098/rstb.2017.0438

Garmestani A, Twidwell D, Angeler DG et al (2020) Panarchy: opportunities and challenges for ecosystem management. Front Ecol Environ 18:576–583. https://doi.org/10.1002/fee.2264

Gullion GW (1984) Grouse of the North Shore. Willow Creek Press, Oshkosh WI

Guyette RP, Stambaugh MC, Dey DC et al (2012) Predicting fire frequency with chemistry and climate. Ecosystems 15:322–335. https://doi.org/10.1007/s10021-011-9512-0

Hagen CA, Jamison BE, Giesen KM et al (2004) Guidelines for managing lesser prairie-chicken populations and their habitats. Wildlife Soc B 32(1):69–82. https://doi.org/10.2193/0091-764 8(2004)32[69:GFMLPP]2.0.CO;2

Hanes CC, Wang X, Jain P et al (2019) Fire-regime changes in Canada over the last half century. Can J Forest Res 49(3):256–269. https://doi.org/10.1139/cjfr-2018-0293

Haydon DT, Randall DA, Matthews L et al (2006) Low-coverage vaccination strategies for the conservation of endangered species. Nature 443(7112):692–695. https://doi.org/10.1038/nature 05177

Hayne DW (1949) Calculation of size of home range. J Mammal 30(1):1–18. https://doi.org/10. 2307/1375189

Heisler JL, Briggs JM, Knapp AK et al (2004) Direct and indirect effects of fire on shrub density and aboveground productivity in a mesic grassland. Ecology 85(8):2245–2257. https://doi.org/ 10.1890/03-0574

Higgins KF (1986) Interpretation and compendium of historical fire accounts in the northern Great Plains. US Department of the Interior, Fish and Wildlife Service

Holling CS, Meffe GK (1996) Command and control and the pathology of natural resource management. Cons Biol 10:328–337. https://doi.org/10.1046/j.1523-1739.1996.10020328.x

Hossack BR, Corn PS (2007) Responses of pond-breeding amphibians to wildfire: short-term patterns in occupancy and colonization. Ecol Appl 17(5):1403–1410. https://doi.org/10.1890/ 06-2037.1

Hossack BR, Eby LA, Guscio CG et al (2009) Thermal characteristics of amphibian microhabitats in a fire-disturbed landscape. Forest Ecol Manag 258(7):1414–1421. https://doi.org/10.1016/j. foreco.2009.06.043

Hossack BR, Lowe WH, Corn PS (2013a) Rapid increases and time-lagged declines in amphibian occupancy after wildfire. Cons Biol 27(1):219–228. https://doi.org/10.1111/j.1523-1739.2012. 01921.x

Hossack BR, Lowe WH, Honeycutt RK et al (2013b) Interactive effects of wildfire, forest management, and isolation on amphibian and parasite abundance. Ecol Appl 23(2):479–492. https:// doi.org/10.1890/12-0316.1

Hovick TJ, Elmore RD, Fuhlendorf SD (2014) Structural heterogeneity increases diversity of non-breeding grassland birds. Ecosphere 5(5):1–13. https://doi.org/10.1890/ES14-00062.1

Hovick TJ, Elmore RD, Fuhlendorf SD et al (2015a) Spatial heterogeneity increases diversity and stability in grassland bird communities. Ecol Appl 25(3):662–672. https://doi.org/10.1890/14-1067.1

Hovick TJ, Allred BW, Elmore RD et al (2015b) Dynamic disturbance processes create dynamic lek site selection in a prairie grouse. PLoS ONE 10(9):e0137882. https://doi.org/10.1371/jou rnal.pone.0137882

Hutto RL (2008) The ecological importance of severe wildfires: some like it hot. Ecol Appl 18:1827–1834. https://doi.org/10.1890/08-0895.1

Hutto RL, Keane RE, Sherriff RL et al (2016) Toward a more ecologically informed view of severe forest fires. Ecosphere 7:e01255. https://doi.org/10.1002/ecs2.1255

IUCN (2013) Guidelines for reintroductions and other conservation translocations. https://www. iucn.org/content/guidelines-reintroductions-and-other-conservation-translocations

Joshi O, Poudyal NC, Weir JR et al (2019) Determinants of perceived risk and liability concerns associated with prescribed burning in the United States. J Environ Manage 230:379–385. https:/ /doi.org/10.1016/j.jenvman.2018.09.089

Keeley JE, McGinnis TW (2007) Impact of prescribed fire and other factors on cheatgrass persistence in a Sierra Nevada ponderosa pine forest. Int J Wildland Fire 16(1):96–106. https://doi.org/10. 1071/WF06052

Kelly LT, Brotons L, McCarthy MA (2017) Putting pyrodiversity to work for animal conservation. Cons Biol 31(4):952–955. https://doi.org/10.1111/cobi.12861

Kie JG, Bowyer RT, Nicholson MC et al (2002) Landscape heterogeneity at differing scales: effects on spatial distribution of mule deer. Ecology 83(2):530–544. https://doi.org/10.1890/0012-965 8(2002)083[0530:LHADSE]2.0.CO;2

Knapp AK, Anderson JE (1980) Effect of heat on germination of seeds from serotinous lodgepole pine cones. Am Midl Nat 104:370–372. https://doi.org/10.2307/2424879

Krasnow KD, Stephens SL (2015) Evolving paradigms of aspen ecology and management: impacts of stand condition and fire severity on vegetation dynamics. Ecosphere 6(1):1–16. https://doi.org/10.1890/ES14-00354.1

Krebs P, Pezzatti GB, Mazzoleni S et al (2010) Fire regime: history and definition of a key concept in disturbance ecology. Theor Biosci 129(1):53–69. https://doi.org/10.1007/s12064-010-0082-z

Kreuter UP, Woodard JB, Taylor CA et al (2008) Perceptions of Texas landowners regarding fire and its use. Rangeland Ecol Manag 61(4):456–464. https://doi.org/10.2111/07-144.1

Lautenbach JM, Plumb RT, Robinson SG et al (2017) Lesser prairie-chicken avoidance of trees in a grassland landscape. Rangeland Ecol Manag 70(1):78–86. https://doi.org/10.1016/j.rama.2016.07.008

Leopold A (1949) A sand county almanac. Oxford University Press, New York, NY

Limb RF, Fuhlendorf SD, Engle DM et al (2016) Synthesis paper: assessment of research on rangeland fire as a management practice. Rangeland Ecol Manag 69(6):415–422. https://doi.org/10.1016/j.rama.2016.07.013

Margolis EQ, Farris CA (2014) Quaking aspen regeneration following prescribed fire in Lassen Volcanic National Park, California, USA. Fire Ecol 10(3):14–26. https://doi.org/10.4996/fireecology.1003014

Marty JT, Sweet SB, Buck-Diaz JJ (2015) Burning controls barb goatgrass (*Aegilops triuncialis*) in California grasslands for at least 7 years. Invas Plant Sci Mana 8(3):317–322. https://doi.org/10.1614/IPSM-D-14-00043.1

Mather TN, Duffy DC, Campbell SR (1993) An unexpected result from burning vegetation to reduce Lyme disease transmission risks. J Med Entomol 30(3):642–645. https://doi.org/10.1093/jmedent/30.3.642

McWethy DB, Schoennagel T, Higuera PE et al (2019) Rethinking resilience to wildfire. Nat Sustain 2(9):797–804. https://doi.org/10.1038/s41893-019-0353-8

Means DB, Campbell HW (1981) Effects of prescribed burning on amphibians and reptiles. Proceedings of the Prescribed Fire and Wildlife in Southern Forests Symposium, Belle Baruch Forest Science Institute, Clemson University, Georgetown, South Carolina. (pp. 89–96).

Meddens AJ, Kolden CA, Lutz JA et al (2018) Fire refugia: what are they, and why do they matter for global change? Bioscience 68(12):944–954. https://doi.org/10.1093/biosci/biy103

Mensing S, Livingston S, Barker P (2006) Long-term fire history in great basin sagebrush reconstructed from macroscopic charcoal in spring sediments, Newark Valley, Nevada. West N Am Naturalist 66:64–77. https://doi.org/10.3398/1527-0904(2006)66[64:LFHIGB]2.0.CO;2

Midgley JJ (1996) Why the world's vegetation is not totally dominated by resprouting plants; because resprouters are shorter than reseeders. Ecography 19:92–95. https://doi.org/10.1111/j.1600-0587.1996.tb00159.x

Minor J, Boyce GA (2018) Smokey Bear and the pyropolitics of United States forest governance. Polit Geogr 62:79–93. https://doi.org/10.1016/j.polgeo.2017.10.005

Newbold T, Hudson LN, Arnell AP et al (2016) Has land use pushed terrestrial biodiversity beyond the planetary boundary? A global assessment. Science 353:288–291. https://doi.org/10.1126/science.aaf2201

Nikolakis W, Roberts E, Hotte N et al (2020) Goal setting and Indigenous fire management: a holistic perspective. Int J Wildland Fire 29(11):974–982. https://doi.org/10.1071/WF20007

Nimmo DG, Carthey AJ, Jolly CJ et al (2021) Welcome to the Pyrocene: animal survival in the age of megafire. Glob Change Biol 27(22):5684–5693. https://doi.org/10.1111/gcb.15834

Patten MA, Pruett CL, Wolfe DH (2011) Home range size and movements of Greater Prairie-Chickens. Stud Avian Biol 39:51–62. https://doi.org/10.1525/9780520950573-006

Pausas JG, Parr CL (2018) Towards an understanding of the evolutionary role of fire in animals. Evol Ecol 32:113–125. https://doi.org/10.1007/s10682-018-9927-6

Pausas JG (2019) Generalized fire response strategies in plants and animals. Oikos 128(2):147–153. https://doi.org/10.1111/oik.05907

Paveglio T, Norton T, Carroll MS (2011) Fanning the flames? Media coverage during wildfire events and its relation to broader societal understandings of the hazard. Hum Ecol Review 18:41–52

Pedersen EK, Connelly JW, Hendrickson JR et al (2003) Effect of sheep grazing and fire on sage grouse populations in southeastern Idaho. Ecol Model 165:23–47. https://doi.org/10.1016/S0304-3800(02)00382-4

Pojar TM, Bowden DC (2004) Neonatal mule deer fawn survival in west-central Colorado. J Wildl Manag 68(3):550–560. https://doi.org/10.2193/0022-541X(2004)068[0550:NMDFSI]2.0.CO;2

Powell RA, Mitchell MS (2012) What is a home range? J Mammal 93(4):948–958. https://doi.org/10.1644/11-MAMM-S-177.1

Putz FE (2003) Are rednecks the unsung heroes of ecosystem management? Wild Earth 13(2/3):10–15

Pyne SJ (2001) Fire: a brief history. University of Washington Press, Seattle, WA, USA

Pyne SJ (2007) Problems, paradoxes, paradigms: triangulating fire research. Int J Wildland Fire 16:271–276. https://doi.org/10.1071/WF06041

Radeloff VC, Helmers DP, Kramer HA et al (2018) Rapid growth of the US wildland-urban interface raises wildfire risk. P Natl A Sci 115(13):3314–3319. https://doi.org/10.1073/pnas.1718850115

Reinhardt ED, Dickinson MB (2010) First-order fire effects models for land management: overview and issues. Fire Ecol 6(1):131–142. https://doi.org/10.4996/fireecology.0601131

Roberts CP, Donovan VM, Nodskov SM et al (2020) Fire legacies, heterogeneity, and the importance of mixed-severity fire in ponderosa pine savannas. Forest Ecol Manag 459:117853. https://doi.org/10.1016/j.foreco.2019.117853

Roberts CP, Donovan VM, Wonkka CL et al (2019) Fire legacies in eastern ponderosa pine forests. Ecol Evol 9(4):1869–1879. https://doi.org/10.1002/ece3.4879

Roos CI, Zedeño MN, Hollenback KL et al (2018) Indigenous impacts on North American Great Plains fire regimes of the past millennium. P Natl A Sci 115(32):8143–8148. https://doi.org/10.1073/pnas.1805259115

Rosenberg KV, Dokter AM, Blancher PJ et al (2019) Decline of the North American avifauna. Science 366:120–124. https://doi.org/10.1126/science.aaw1313

Rusch DH, Keith LB (1971) Ruffed grouse-vegetation relationships in central Alberta. J Wildl Manag 35:417–429. https://doi.org/10.2307/3799692

Russell KR, Van Lear DH, Guynn DC (1999) Prescribed fire effects on herpetofauna: review and management implications. Wildlife Society Bulletin 27(2):374–384. https://www.jstor.org/stable/3783904

Ryan KC, Knapp EE, Varner JM (2013) Prescribed fire in North American forests and woodlands: history, current practice, and challenges. Front Ecol Environ 11(s1):e15–e24. https://doi.org/10.1890/120329

Sawyer H, Merkle JA, Middleton AD et al (2019) Migratory plasticity is not ubiquitous among large herbivores. J Anim Ecol 88(3):450–460. https://doi.org/10.1111/1365-2656.12926

Scasta JD, Engle DM, Harr RN et al (2014) Fire induced reproductive mechanisms of a *Symphoricarpos* (Caprifoliaceae) shrub after dormant season burning. Bot Stud 55(1):1–10. https://doi.org/10.1186/s40529-014-0080-4

Scasta JD, Duchardt C, Engle DM et al (2016a) Constraints to restoring fire and grazing ecological processes to optimize grassland vegetation structural diversity. Ecol Eng 95:865–875. https://doi.org/10.1016/j.ecoleng.2016.06.096

Scasta JD, Weir JR, Stambaugh MC (2016b) Droughts and wildfires in Western U.S. rangelands. Rangelands 38:197–203. https://doi.org/10.1016/j.rala.2016.06.003

Scasta JD, McCulley RL, Engle DM et al (2021) Patch burning tall fescue invaded grasslands alters alkaloids and tiller defoliation with implications for cattle toxicosis. Rangeland Ecol Manag 75:130–140. https://doi.org/10.1016/j.rama.2020.12.009

Scifres CJ, Oldham TW, Teel PD et al (1988) Gulf coast tick (*Amblyomma maculatum*) populations and responses to burning of coastal prairie habitats. Southwest Nat 33:55–64. https://doi.org/10.2307/3672088

Seip DR, Bunnell FL (1985) Nutrition of Stone's sheep on burned and unburned ranges. J Wildl Manag 49(2):397–405. https://doi.org/10.2307/3801541

Smith JK (2000) Wildland fire in ecosystems: effects of fire on fauna. Gen Tech Rep RMRS-GTR-42-vol. 1. Ogden, UT: U.S. Department of Agriculture, Forest Service, Rocky Mountain Research Station, 83 p

Stafford KC III, Ward JS, Magnarelli LA (1998) Impact of controlled burns on the abundance of *Ixodes scapularis* (Acari: Ixodidae). J Med Entomol 35(4):510–513. https://doi.org/10.1093/jmedent/35.4.510

Stambaugh MC, Guyette RP, Marschall J (2013) Fire history in the Cherokee Nation of Oklahoma. Hum Ecol 41(5):749–758. https://doi.org/10.1007/s10745-013-9571-2

Starns HD, Wonkka CL, Dickinson MB et al (2021) *Prosopis glandulosa* persistence is facilitated by differential protection of buds during low-and high-energy fires. J Environ Manage 303:114141. https://doi.org/10.1016/j.jenvman.2021.114141

Stasiewicz AM, Paveglio TB (2017) Factors influencing the development of Rangeland Fire Protection Associations: exploring fire mitigation programs for rural, resource-based communities. Soc Natur Resour 30(5):627–641. https://doi.org/10.1080/08941920.2016.1239296

Stavros EN, Abatzoglou J, Larkin NK et al (2014) Climate and very large wildland fires in the contiguous western USA. Int J Wildland Fire 23(7):899–914. https://doi.org/10.1071/WF13169

Stewart OC (2002) Forgotten fires: Native Americans and the Transient Wilderness. University of Oklahoma Press

Stewart OC (1951) Burning and natural vegetation in the United States. Geogr Rev 41:317–320. https://doi.org/10.2307/211026

Symstad AJ, Buhl DA, Swanson DJ (2021) Fire controls annual bromes in northern great plains grasslands—up to a point. Rangeland Ecol Manag 75:17–28. https://doi.org/10.1016/j.rama.2020.11.003

Taylor CA Jr (2005) Prescribed burning cooperatives: empowering and equipping ranchers to manage rangelands. Rangelands 27:18–23. https://doi.org/10.2111/1551-501X(2005)27%3c18:PBCEAE%3e2.0.CO;2

Theobald DM, Romme WH (2007) Expansion of the US wildland–urban interface. Landscape Urban Plan 83:340–354. https://doi.org/10.1016/j.landurbplan.2007.06.002

Toledo D, Sorice MG, Kreuter UP (2013) Social and ecological factors influencing attitudes toward the application of high-intensity prescribed burns to restore fire adapted grassland ecosystems. Ecol Soc 18(4):1–9. https://doi.org/10.5751/ES-05820-180409

Towne EG, Craine JM (2016) A critical examination of timing of burning in the Kansas Flint Hills. Rangeland Ecol Manag 69(1):28–34. https://doi.org/10.1016/j.rama.2015.10.008

Twidwell D, Fuhlendorf SD, Taylor CA Jr et al (2013a) Refining thresholds in coupled fire–vegetation models to improve management of encroaching woody plants in grasslands. J Appl Ecol 50(3):603–613. https://doi.org/10.1111/1365-2664.12063

Twidwell D, Rogers WE, Fuhlendorf SD et al (2013b) The rising Great Plains fire campaign: citizens' response to woody plant encroachment. Front Ecol Environ 11(s1):e64–e71. https://doi.org/10.1890/130015

Twidwell D, Rogers WE, Wonkka CL et al (2016) Extreme prescribed fire during drought reduces survival and density of woody resprouters. J Appl Ecol 53(5):1585–1596. https://doi.org/10.1111/1365-2664.12674

Twidwell D, Wonkka CL, Wang H-H et al (2019) Coerced resilience in fire management. J Environ Manage 240:368–373. https://doi.org/10.1016/j.jenvman.2019.02.073

Twidwell D, Bielski CH, Scholtz R et al (2021) Advancing fire ecology in 21st century rangelands. Rangeland Ecol Manag 78:201–212. https://doi.org/10.1016/j.rama.2020.01.008

Walker SC, Anderson VJ, Fugal RA (2015) Big game and cattle influence on aspen community regeneration following prescribed fire. Rangeland Ecol Manag 68(4):354–358. https://doi.org/10.1016/j.rama.2015.05.005

Weir JR (2009) Conducting prescribed fires: a comprehensive manual. Texas A&M University Press

Weir JR (2010) Prescribed burning associations: Landowners effectively applying fire to the land. In: Proceedings of the 24th Tall Timbers fire ecology conference. The future of prescribed fire: Public awareness, health, and safety, pp 44–46

Weir JR, Scasta JD (2014) Ignition and fire behaviour of *Juniperus virginiana* in response to live fuel moisture and fire temperature in the southern Great Plains. Int J Wildland Fire 23(6):839–844. https://doi.org/10.1071/WF13147

Weir JR, Scasta JD (2017) Vegetation responses to season of fire in tallgrass prairie: a 13-year case study. Fire Ecol 13(2):137–142. https://doi.org/10.4996/fireecology.130290241

Weir JR, Twidwell D, Wonkka CL (2015) Prescribed burn association activity, needs, and safety record: a survey of the Great Plains. Great Plains Fire Science Exchange 6:19

Weir JR, Twidwell D, Wonkka CL (2016) From grassroots to national alliance: the emerging trajectory for landowner prescribed burn associations. Rangelands 38(3):113–119. https://doi.org/10.1016/j.rala.2016.02.005

Westerling AL (2016) Increasing western US forest wildfire activity: sensitivity to changes in the timing of spring. Philos T R Soc B 371(1696):20150178. https://doi.org/10.1098/rstb.2015.0178

Westerling AL, Hidalgo HG, Cayan DR et al (2006) Warming and earlier spring increase western US forest wildfire activity. Science 313(5789):940–943. https://doi.org/10.1126/science.1128834

Whelan RJ, Rodgerson L, Dickman CR et al (2002) Critical life processes of plants and animals: developing a process-based understanding of population changes in fire-prone landscapes. In: Bradstock RA, Williams JE, Gill AM (eds) Flammable Australia: the fire regimes and biodiversity of a continent. Cambridge University Press, United Kingdom, pp 94–124

Whisenant SG (1990) Postfire population dynamics of *Bromus japonicus*. Am Midl Nat 123:301–308. https://doi.org/10.2307/2426558

Wilbur R, Scasta JD (2021) Participant motivations for the Wyoming Prescribed Fire Council (PFC): emergence from a regional void. Rangelands 43:93–99. https://doi.org/10.1016/j.rala.2020.12.006

Williams AP, Abatzoglou JT (2016) Recent advances and remaining uncertainties in resolving past and future climate effects on global fire activity. Curr Clim Chang Rep 2(1):1–14. https://doi.org/10.1007/s40641-016-0031-0

With KA, King AW, Jensen WE (2008) Remaining large grasslands may not be sufficient to prevent grassland bird declines. Biol Cons 141:3152–3167. https://doi.org/10.1016/j.biocon.2008.09.025

Chapter 7
Water Is Life: Importance and Management of Riparian Areas for Rangeland Wildlife

Jeremy D. Maestas, Joseph M. Wheaton, Nicolaas Bouwes, Sherman R. Swanson, and Melissa Dickard

Abstract Water scarcity and climatic variability shape human settlement patterns and wildlife distribution and abundance on arid and semi-arid rangelands. Riparian areas–the transition between water and land–are rare but disproportionately important habitats covering just a fraction of the land surface (commonly < 2% in the western U.S.). Riparian areas provide critical habitat for fish and other aquatic species, while also supporting the vast majority (70–80%) of terrestrial wildlife during some portion of their life cycle. Diverse riparian types serve as vital sources of water and late summer productivity as surrounding uplands dry during seasonal drought. The health and function of rangeland riparian systems are closely tied to hydrology, geomorphology, and ecology. Riparian areas have attracted intense human use resulting in their widespread degradation. Conservation actions, including improved livestock grazing management and restoration, can help maintain and enhance riparian resilience to drought, wildfire, and flooding. This chapter provides readers with an introduction to the importance of riparian areas in rangelands, their nature and ecology, functions for wildlife, and prevailing management and restoration approaches.

Keywords Riparian · Water · Resilience · Wildlife · Restoration · Grazing · Beaver · Hydrology

J. D. Maestas (✉)
USDA Natural Resources Conservation Service, West National Technology Support Center, Portland, OR 97232, USA
e-mail: jeremy.maestas@usda.gov

J. M. Wheaton · N. Bouwes
Department of Watershed Sciences, Utah State University, Logan, UT 84322, USA

S. R. Swanson
Department of Agriculture, Veterinary and Rangeland Sciences, University of Nevada-Reno (Emeritus), Reno, NV 89557, USA

M. Dickard
Bureau of Land Management, National Operations Center, Denver, CO 80225-0047, USA

© The Author(s) 2023
L. B. McNew et al. (eds.), *Rangeland Wildlife Ecology and Conservation*,
https://doi.org/10.1007/978-3-031-34037-6_7

7.1 Introduction

On the range, water is life. Water scarcity and climatic variability are defining features of arid and semi-arid rangelands that shape human settlement patterns and wildlife distribution and abundance (Donnelly et al. 2018, Chap. 3). Riparian areas are the transition zones between water and land and are a critically important anomaly on rangelands. Relative to the broader landscape context, riparian areas are reservoirs of moisture and productivity lying in stark contrast with drier uplands. Commonly occupying only < 2% of the land base today in the western U.S., riparian areas provide disproportionately important resources to wildlife and people (Thomas et al. 1979; Patten 1998; McKinstry et al. 2004). Due to their outsized value and myriad threats, many riparian systems have been degraded or reduced in size over the last two hundred years. Yet, due to the persistence of water and associated plant communities that drive recovery after disturbance, riparian areas are inherently resilient. With active restoration, the potential land that could support riparian conditions may be several times larger than the current footprint (Wheaton et al. 2019a; Macfarlane et al. 2018).

Knowledge of riparian ecology, management, and restoration is increasingly vital for rangeland managers challenged with sustaining productive grazing lands, wildlife populations, clean water, and recreation in the face of growing water demands and climate change (Seavy et al. 2009; Perry et al. 2011). In this chapter, we introduce the importance of managing riparian areas in rangeland ecosystems to motivate and inform conservation and stewardship of these critical habitats. This is not a how-to guide for management but rather an introduction to the subject matter to increase awareness and summarize current knowledge of key riparian concepts, properties, risks, opportunities, and related science. Specifically, we describe what rangeland riparian areas are and why they matter, their nature and ecology, functions for wildlife, and prevailing management and restoration approaches.

7.2 What Are Riparian Areas and Why Are They Important?

Riparian is an adjective typically defined as "relating to or living or located on the bank of a natural watercourse" (Merriam-Webster 2021), although in an ecological sense, riparian more broadly encompasses diverse ecosystems characterized by the preponderance of water. Riparian areas are habitats that occur along river and stream corridors, meadows and bogs, seeps and springs, wetlands and lakes (Fig. 7.1, NRC 2002). All riparian areas share unique soil and vegetation characteristics that are strongly influenced by free or unbound water in the soil which make them distinct from upland areas. Water availability and source, disturbance regimes, and site conditions, such as geology, soils, and topography, are among the key factors influencing riparian types and plant communities. Biology—especially in the form of vegetation

Meadows, Bogs

Springs, Seeps

Streams, Rivers

Wetlands

Fig. 7.1 Types of riparian areas embedded within rangeland watersheds. Riparian areas are rare but diverse ecosystems that occur along river and stream corridors, meadows and bogs, springs and seeps, wetlands, and lakes

and North American beaver (*Castor canadensis*)—also interacts with physical conditions to shape riparian form and function (Castro and Thorne 2019; Wheaton et al. 2019a). From a hydrologic standpoint, riparian systems are the interface between open water bodies (channels and ponds) and land connected through surface or subsurface flow (Wilcox et al. 2017). Hydrology can be characterized by flow type: standing water (lentic) or flowing water (lotic), and flow duration: year-round (perennial), seasonal (intermittent), or precipitation-dependent (ephemeral). Perhaps more than any other ecosystem, the structure and function of riparian areas are influenced by interactions between watercourses and the surrounding lands requiring a holistic watershed perspective for management (Gregory et al. 1991).

Riparian areas perform many functions with on- and off-site effects that yield important ecosystem services to society (NRC 2002). Broad categories of functions include: (1) water and sediment transport and storage, (2) carbon and nutrient cycling, and (3) habitat and food web maintenance. For example, healthy riparian areas play a crucial role in proper watershed function helping capture, store, and slowly release water thereby attenuating floods and supporting base flows during dry seasons (Elmore 1992). Riparian areas also serve as 'kidneys' of the landscape, supporting vegetation buffers that improve water quality by trapping sediment, cycling nutrients and chemicals, and filtering pollutants from the watershed (Hook 2003; Mayer et al. 2006; Swanson et al. 2017). Finally, riparian areas function as habitat, providing food, cover, water, refugia, and movement corridors for rangeland wildlife.

Particularly in arid ecosystems, riparian areas are hotspots of biodiversity supporting 70–80% of vertebrate species during some stage of their life cycle (Thomas et al. 1979; Brinson et al. 1981; Naiman et al. 1993; Knopf 1985). Wildlife use of riparian systems is disproportionate to its availability (Thomas et al. 1979; McKinstry et al. 2004). This is best documented for migratory birds (Knopf et al. 1988) where species richness in riparian habitats can be 10–14 times higher than adjacent uplands (Stevens et al. 1977; Hehnke and Stone 1979). Over half of all bird species are completely dependent upon riparian areas in the desert southwest (Johnson and Jones 1977). Of course, riparian habitats are essential to fish and other aquatic and semi-aquatic species in water-limited rangelands (Saunders and Fausch 2007). Perhaps less appreciated is how important seasonal floodplain habitats are for many species of fish (Opperman et al. 2017). Numerous species of mammals (small and large), amphibians and reptiles, and invertebrates, including pollinators, depend on healthy riparian habitats at some point during the year. For example, on rangelands of southeastern Oregon, 288 of 363 terrestrial species (80%) are directly dependent upon riparian areas or use them more than other habitats (Thomas et al. 1979).

Euro-American fur trapping and homesteading patterns also highlight the historical importance of riparian areas to people and working lands across western North America. In the early 1800's, abundant beaver in western waterways attracted fur trappers and fueled Euro-American expansion (Dolin 2011; Goldfarb 2018). Later in the century, homesteading pioneers sought out water and associated riparian areas to allow agriculture in an environment where soils and climate make land less arable. While most rangelands in the Intermountain West are publicly owned, 50–90% of riparian areas are privately-owned (Donnelly et al. 2018). In grasslands of the Great Plains, the proportion of privately-owned riparian areas is even higher. For over two centuries, riparian systems have helped supply reliable water that is the lifeblood of commerce and rural communities, supporting irrigation of crops, livestock grazing, development, mining, recreation, and other activities.

Riparian areas contribute to overall rangeland resilience and resistance to sudden change during drought, wildfire, and flooding, providing a buffer against increasing climate variability and extreme weather events (Seavy et al. 2009). Management and restoration of degraded riparian areas improves drought resilience by boosting soil moisture storage and floodplain vegetation productivity (Silverman et al. 2018; Fesenmeyer et al. 2018). Fully functioning riparian areas are better able to resist wildfire damage (Fairfax and Whittle 2020), providing refugia for wildlife and livestock in burned landscapes in an era of increasing fire size and frequency, and better preparing watersheds for extreme floods (Perry et al. 2011). Restoring riparian areas is key to increasing resilience and maintaining the capacity of these systems to provide ecosystem services in the face of climate change and long-term drought (Seavy et al. 2009; Capon et al. 2013; Williams et al. 2022).

7.3   Reading the Riparian Landscape

Riparian areas are an expression of watershed-scale processes because they are located in relatively low spots on the landscape where water and sediment collect (Wilcox et al. 2017). Geology, hydrology, and biology interact as primary drivers of riparian form and function (Castro and Thorne 2019). While some riparian types are primarily groundwater-driven (e.g., springs), surface water flow and sediment movement within watersheds are particularly important in shaping the condition and extent of most riparian areas (NRC 2002). Diverse vegetation communities arise from variation in water availability, flow and disturbance regimes, soils, and climate. A mosaic of habitats can exist ranging from persistently saturated wetlands, to ephemerally inundated floodplains, to meadows supported mainly by high water tables and wetted only during periods of runoff. Given the high degree of connectivity (vertically and laterally) of riparian systems to both adjacent water bodies and surrounding uplands that feed and define them, a holistic geomorphic perspective is helpful when assessing function, condition, and potential.

On rangelands, the bulk of riparian areas are associated with fluvial systems, such as streams and rivers. Best viewed as *riverscapes*, stream and riverine landscapes are composed of floodplains and channels that together comprise the valley bottom (Fig. 7.2, Ward 1998; Wheaton et al. 2019a). The valley bottom reflects the maximum possible extent that could be occupied by riparian vegetation. Valley bottoms may consist of one or more of these building blocks: active channels and active floodplains (i.e., areas experiencing flooding thereby capable of supporting riparian vegetation), and inactive floodplains (i.e., disconnected areas that could plausibly flood and support riparian vegetation if conditions improved). Not all riverscapes have all building blocks and the valley setting helps determine what is natural and expected. For example, a steep confined gorge may only have a single active channel. In contrast, some riverscapes in lower gradient settings (e.g., wet meadows) have only active floodplains and no natural channels. Sometimes riverscapes are separated from hillslopes by other geomorphic features like terraces (i.e., valley bottoms from a relic or historic flow regime), moraines (i.e., accumulated debris from past glacial activity), or fans from tributaries (e.g., alluvial fans). Assessing riverscapes to identify the valley bottom building blocks and other related geomorphic features helps set appropriate expectations for the intrinsic potential of riparian areas in fluvial systems (Fyrirs and Brierley 2013; Wheaton et al. 2015).

Rangeland managers are often tasked with assessing riparian condition, a challenging responsibility given the diverse and dynamic nature of these systems. However, hydrologic, soil/geomorphic, and biological attributes and processes provide clues to riparian area condition and function (Dickard et al. 2015; Wheaton et al. 2019a; Gonzalez and Smith 2020). In general, properly functioning riparian areas are dynamic environments that have adequate space and other characteristics to accommodate water runoff, dissipate energy, and adjust to change (Fig. 7.2a). They can expand and contract within the confines of their natural valley bottom in response to disturbances like floods and droughts (Whited et al. 2007). They also support the kinds and amounts of vegetation needed to stabilize soils, capture sediment, and slow and attenuate surface and subsurface flows. Properly functioning ecological processes in riparian areas, like beaver dam activity, vegetation recruitment, and wood accumulation, provide structural elements that increase habitat complexity and build resilience (Silverman et al. 2018; Wheaton et al. 2019a).

A variety of impairments occurring at watershed to site-specific scales can reduce riparian condition. Watershed land cover changes due to catastrophic wildfire, improper grazing, woodland expansion, development, or other factors impact

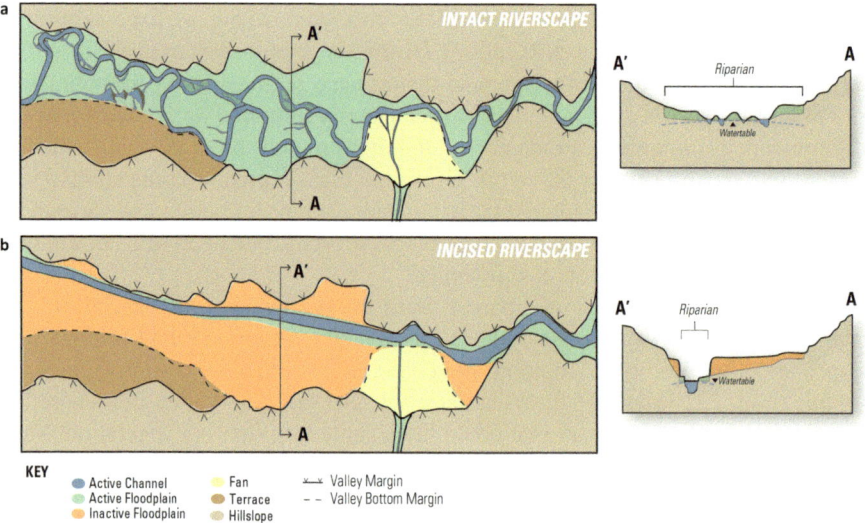

Fig. 7.2 Planform and cross-sectional views showing geomorphology of a hypothetical riverscape in an intact (**a**) and degraded scenario (**b**). Reading the riverscape valley bottom is important as it reflects the maximum potential extent of riparian vegetation. Intact riverscapes (**a**) support a higher water table and more riparian vegetation, while incised or channelized riverscapes (**b**) result in a lower water table with less riparian potential. The valley bottom includes the area that could plausibly flood in the contemporary flow regime and is made up of the active channel(s), active floodplain, and/or inactive floodplain. Restoration to improve connectivity of inactive floodplains may be required in degraded systems to fully realize riparian potential. Figure adapted by Adrea Wheaton from Wheaton et al. (2019b) and licensed under a Creative Commons Attribution 3.0 United States License

watershed hydrologic function and can lead to riparian degradation. Intensive land uses located within valley bottoms like roads and associated infrastructure, water diversion, and cultivated agriculture can directly limit riparian functions and potential extent (i.e., the area capable of supporting riparian vegetation). Channel incision—downcutting of channel bed elevation through erosion—is a widespread symptom of degradation in rangeland watersheds resulting in predictable changes in riparian condition (Cluer and Thorne 2013; Dickard et al. 2015; Gonzalez and Smith 2020). Incision reduces the potential for riparian vegetation within valley bottoms by lowering the water table and disconnecting channels from active floodplains (Fig. 7.2b). Learning to recognize impairments and the degree of incision are critical to informing management actions and appropriate expectations for riparian areas (Skidmore and Wheaton 2022).

7.4 Ecology of Riparian Areas

7.4.1 Vegetation

A defining attribute of riparian ecology is the distinct vegetation arising from increased water availability. In rangelands, this riparian vegetation often forms a green strip adjacent to drier uplands during annual summer droughts. Vegetation is indicative of hydrologic function in riparian areas (Table 7.1, Lichvar et al. 2012, 2016; USACE 2018). Hydrophytic, or water-loving, vegetation includes plants specifically adapted to grow in low oxygen (i.e., anaerobic) environments. Specific plant species occur along a gradient of wetted conditions, such as, obligate species in water or saturated soils, facultative wetland species in frequently wet soils, or facultative species in soils that fluctuate between wet and dry (USACE 2018). Other vegetation less capable of withstanding anaerobic conditions are facultative upland and upland species that typically occur in drier environments and can be used to identify non-wetland areas (USACE 2018). Common herbaceous riparian plants include graminoids such as sedges (*Carex* spp.), rushes (*Juncaceae* spp.), and wet-adapted grasses (*Poaceae* spp.). Typical woody riparian species include a diversity of willows (*Salix* spp.), cottonwoods (*Populus* spp.), and alders and birches (*Betulaceae* spp.). Riparian plant composition is highly variable and a product of climate, hydrologic regime, soils and geomorphology, land use, species distribution, and other factors (NRC 2002; Hough-Snee et al. 2014). Healthy riparian plant communities exist in a wide variety of forms ranging from relatively simple and stable herbaceous mats consisting of just a few species, to dynamic and diverse woody-dominated riparian shrublands and forests.

Vegetation plays a critical role in the structure and function of riparian systems as well as the broader rangeland water cycle (Wilcox et al. 2017). Above ground, plants provide surface roughness to redistribute flow patterns and facilitate deposition and soil building (Manners et al. 2013). Many riparian plants have adaptations to withstand stream or overland flows, such as cordlike rhizomes, fibrous root masses, coarse

Table 7.1 Vegetation provides clues to hydrologic function in riparian areas

Wetland indicator status	Designation	Qualitative description
Obligate (OBL)	Hydrophyte	Almost always occurs in wetland
Facultative Wetland (FACW)	Hydrophyte	Usually occurs in wetlands, but may occur in non-wetland areas
Facultative (FAC)	Hydrophyte	Occurs in wetland and non-wetland areas
Facultative Upland (FACU)	Non-Hydrophyte	Usually occurs in non-wetland areas, but may occur in wetlands
Upland (UPL)	Non-Hydrophyte	Almost never occurs in wetlands

The National Wetland Plant List (Lichvar et al. 2012, 2016; USACE 2018) characterizes plant species by the frequency in which they are found in wetlands

leaves, and strong flexible crowns (Winward 2000). Structurally forced changes in flow patterns produce physically diverse and complex habitats and enhance resilience to disturbance (Corenblit et al. 2007; Wheaton et al. 2019a). For example, in-stream wood accumulation affects water velocities and depths, sediment erosion and deposition, and provides organic material essential to support diverse aquatic species (Wheaton et al. 2015). Below ground, riparian plants tend to have very dense root systems (Manning et al. 1989) that bind soil particles, provide stability, and slow runoff. Healthy plant roots create macropores and soil organic matter that improve infiltration of surface runoff and water-holding capacity of soils, increasing residence time of water. Riparian vegetation creates unique microclimates that affect water temperatures and humidity. Forested riparian areas support large wood accumulation important to stream processes and aquatic habitat.

Riparian vegetation can be prone to damage owing to disproportionate use by livestock, free-roaming horses, wildlife, and people. While erosion and deposition are natural processes in riparian systems, degradation of vegetation can result in predictable changes affecting form and function (Cluer and Thorne 2013). If hydrophytic vegetation is damaged, the resulting loss of structure often leads to reductions in riparian stability. Improper grazing and trailing by livestock, free-roaming horses, and big game can reduce plant health and vigor, leaving riparian areas more vulnerable to accelerated erosion. Stream systems may incise, and wet meadows may develop channels which cause the water table to drop, drying out the soil and converting the riparian area to more upland vegetation with less robust root structure (Wyman et al. 2006). Careful management of grazing animals in riparian areas is required to support a variety of functions and values.

7.4.2 Beavers: Ecosystem Engineers

The North American beaver is a keystone riparian species with a unique role in shaping the form, function, and ecology of many riverscapes. Often referred to as an ecosystem engineer, beaver have a profound influence on riparian environments, producing diverse habitats required by many other wildlife species. Modifications occur through foraging primarily on woody plants and activities associated with dam building including construction of ponds, lodges, canals, tunnels, and burrows. These activities trigger a cascade of effects altering riparian function and diversity, aquatic food webs, geomorphology, hydrology, and biogeochemical cycling (see reviews in: Brazier et al. 2020; Wohl 2021). The functional role of beaver in riparian areas has been increasingly recognized as an integral part of stream evolution, riparian ecology, and conservation (Stoffyn-Egli and Willison 2011; Pollock et al. 2014; Castro and Thorne 2019; Jordan and Fairfax 2022).

Prior to Euro-American arrival, an estimated 60–400 million beaver occupied North America (Seton 1929). Beaver dam activity created some 25–250 million ponds (Pollock et al. 2003) with a total surface area of approximately 230,000 square miles, equivalent to the land area of Arizona and Nevada combined (Butler and Malanson

2005; Goldfarb 2018). Highly prized for their fur, beaver trapping began in the late 1500s, and by the early 1800s they were harvested to near extinction (Dolin 2011). With the removal of these industrious rodents, unmaintained dams often breached leading to the draining of hundreds of millions of ponds and wetlands (Ott 2003; Goldfarb 2018). While the arrival in the late 1800s of irrigated agriculture, timber harvest, overgrazing, and mining impacted western watersheds, legacy effects of the loss of beaver and their dam-building activities are often underestimated and forgotten (Wohl 2021). In the early 1900s, the important role of beaver in shaping riverscapes started to be recognized with conservation and relocation efforts occurring over the next century. Populations are now estimated between 6 and 12 million and, although still far less than their historic abundance, beavers have re-colonized much of their former range across North America (Naiman et al. 1988).

Beaver basically require water and vegetation to make a living, occupying diverse riparian ecosystems from boreal forest to deserts (Brazier et al. 2020; Larsen et al. 2021). Beaver are herbivores, consuming herbaceous plants and cambium (i.e., the inner bark of woody plants). Their diet varies greatly across different environments and seasons. In colder climates, for example, herbaceous plants may comprise a majority of their diet, especially in the summer. In lotic systems, during winter they often switch to more woody species that are harvested and stored in deep water caches that can be accessed below ice from underwater entrances to their lodges (Milligan and Humphries 2010). As central place foragers, beaver select vegetation based on size and palatability depending on availability and distance from their lodge or pond (Mahoney and Stella 2020). Highly preferred plant species, such as aspen and willow, have coevolved with beaver and regenerate, often with increased vigor, after they have been cut (Runyon et al. 2014). Beaver may greatly reduce preferred riparian vegetation within their home range forcing them to move to new areas to access forage while the depleted stand regenerates, acting much like coppice foresters seeking to stimulate plant growth by cutting back plants (Hall 1970). As a result, beaver create a shifting mosaic of riparian habitat types within riverscapes that support varying structural diversity, plant composition and richness, and seed or sprout production (Mahoney and Stella 2020).

While beaver can be awkward on land, they are powerful swimmers that prefer water deep enough to evade predators. In some systems, water is sufficiently deep to provide this protection, rendering dam building unnecessary. In shallower streams common across rangelands, however, beaver typically build dams to slow down and deepen water to escape predation and allow for more extensive harvest and transport of wood and vegetation within the refuge of water. Beaver often incorporate larger woody material into dams and lodges, along with smaller branches that are consumed or stored in the bottom of the ponded area. The influx of woody material and the creation of dams and canals enhance hydraulic and geomorphic complexity (e.g., sediment sorting) producing physically diverse habitats. Slower ponded areas with accumulated fine sediment can create anaerobic conditions that alter biogeo-chemical cycles (Naiman et al. 1988; Murray et al. 2021). Diversity in hydraulics, physical habitats, substrate, and riparian structure, creates more complex habitat often supporting a higher diversity of stream fauna (Burchsted et al. 2010; Bouwes et al. 2016).

Beaver dam activity is an important ecological process influencing stream evolution creating multithreaded, complex reach types (i.e., "stage 0" in Cluer and Thorne 2013). Furthermore, structural changes caused by beaver dam activity can accelerate recovery of incised riverscapes by facilitating widening, aggradation (i.e., deposition of material by current), and floodplain reconnection (Fig. 7.3, Pollock et al. 2014). By increasing both aggradation and water depth, beaver dams enhance frequency and duration of floodplain inundation even during baseflows. Higher water surfaces increase water table elevations and create greater hydraulic gradients resulting in elevated exchange of surface and groundwater (Majerova et al. 2015). Thus, dams increase vertical and lateral hydrologic connectivity allowing water to be stored during high flow events (Westbrook et al. 2020). Stored water is more slowly released over the descending limb of the annual hydrograph resulting in improved drought resilience (Fesenmeyer et al. 2018; Silverman et al. 2018). In a similar but compressed time frame, beaver dams and dam complexes help attenuate high flows (Westbrook et al. 2020) and reduce unit stream power, dissipating energy by spreading flow onto adjacent floodplains reducing the likelihood of channel incision (Pollock et al. 2014). Greater inundation and soil moisture not only increase vegetation recruitment and vigor but also improve riparian resistance to fire and drought (Fairfax and Whittle 2020). Thus, beaver confer resiliency to riverscapes that can help them withstand multiple disturbances that are likely to become more intense with climate change.

7.4.3 Riparian Functions for Wildlife

Biotic and abiotic factors combine in riparian areas to create habitat for aquatic and terrestrial wildlife found nowhere else on rangelands. The disproportionate diversity and abundance of life in riparian systems stems from their ability to provide water, food, cover, refuge, and migration corridors needed to fulfill annual life cycle requirements (Fig. 7.4). Greater availability of water in riparian areas than uplands gives rise to higher productivity of vegetation and insects, and a more dependable source for water consumption, which becomes increasingly important during seasonal summer drought. Diverse vegetation communities provide breeding, nesting, rearing, loafing, feeding, and escape cover needed by most terrestrial rangeland wildlife at some point in the year (Thomas et al. 1979). Habitat connectivity is also provided by riparian corridors and wetlands interspersed across rangeland watersheds, facilitating movement, dispersal, and migration.

Reliance on riparian areas by migratory birds is well documented (Johnson et al. 1977; Knopf et al. 1988). Each year in the spring, millions of neotropical songbirds return to breed in western riparian areas after wintering in Mexico, and Central and South America. Over 80% of migrant birds breed in riparian areas in some locations (Knopf 1985). Riparian vegetation and insects supply necessary food and cover for nesting, rearing, and fledging. Landbird abundance and richness are closely tied to

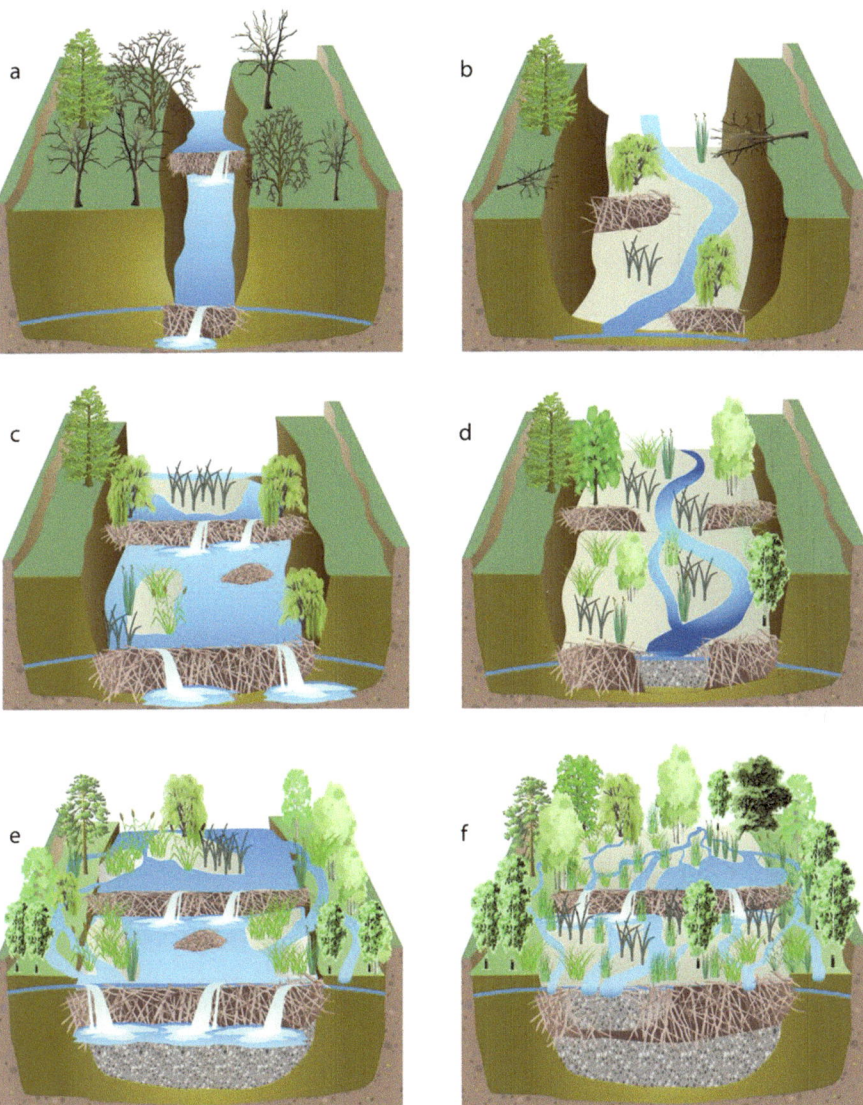

Fig. 7.3 Beaver dam activity accelerates stream evolution from an incised (**a**) to anastomosing riverscape (**f**) (from Pollock et al. 2014). This series illustrates the typical progression: **a** beaver build dams in an incised channel that has been disconnected from the active floodplain, **b** high stream power results in beaver dams failing by end cutting, forcing water to erode the bank leading to channel widening, **c** inset floodplains begin to form in the widened trench and a widened channel facilitates sediment capture, **d** sediment captured behind the dams also aggrades the channel and facilitates riparian vegetation establishment, **e** dams raise the surface water reconnecting the stream to the formerly active floodplain, and **f** beaver dam activity creates well-connected and vegetated riparian area across the valley bottom

Fig. 7.4 Examples of riparian functions for wildlife. Riparian areas provide critical habitat needed by most rangeland wildlife to fulfill annual life cycle requirements, such as **a** breeding cover for amphibians, **b** nesting and brood-rearing cover for birds, **c** water, forage, and calving cover for big game, and **d** refugia from wildfire. Photo credits **a–d**: Jeremy Maestas, Richard Van Vleck, Nathan Seward, Joseph Wheaton

structural heterogeneity of riparian vegetation (Anderson and Ohmart 1977; Rich 2002; Brand et al. 2008; Cubley et al. 2020). Given this connection, assessment of breeding bird presence and abundance can serve as an indicator of riparian health and plant structural complexity (Rich 2002). Following the breeding season, riparian areas provide valuable stopover habitats for neotropical birds during the fall migration south. Waterfowl, shorebirds, and waterbirds all rely on wetland, meadow, and playa riparian habitats in rangelands for breeding and stopovers supporting spring and fall migrations (Donnelly et al. 2019; Haukos and Smith 1994).

Riparian areas are important in meeting seasonal habitat needs for imperiled resident birds and can influence landscape carrying capacity (Donnelly et al. 2018). Sage-grouse (*Centrocercus urophasianus*), known for their dependency on sagebrush steppe, "follow the green line" as uplands dry out in the summer to reach riparian areas and other mesic habitats (i.e., areas with adequate moisture for plant growth), such as high elevation habitats and irrigated fields, that provide abundant forbs and insects to feed growing chicks (Chap. 10). Mesic resource abundance and drought resilience influences the distribution and abundance of sage-grouse where other life requisites have been met (Donnelly et al. 2016, 2018). Columbian sharp-tailed grouse (*Tympanuchus phasianellus columbianus*) utilize riparian zones as well for winter food and cover, foraging on the buds of woody vegetation (Giesen and Connelly 1993). Numerous other more common landbirds including non-migratory songbirds, quail, and raptors also use riparian habitats in various seasons.

A host of aquatic and semi-aquatic species would not exist in rangelands without riparian ecosystems. Aquatic and riparian zones are intricately linked through exchange of invertebrate prey, plant material, and water affecting food webs and habitat quantity and quality. Riparian vegetation supports terrestrial invertebrate inputs to streams that can constitute half of the food resources for fish like salmonids (Baxter et al. 2005; Saunders and Fausch 2007). Riparian leaf fall and woody debris also retain and support aquatic food webs and provide for added aquatic habitat complexity for a diversity of aquatic species and various life stages (Gregory et al. 1991). Conversely, emergence of aquatic insects also feed terrestrial wildlife including bats, birds, reptiles, and amphibians. Wetlands, beaver ponds, and floodplains support unique amphibians year-round (Munger et al. 1998; Arkle and Pilliod 2015). Healthy and hydrologically functioning riparian areas ensure habitat connectivity that allows water-reliant species to disperse and migrate. Although much less well known, rangeland springs are home to many species of springsnails (*Pyrgulopsis* spp.), one of the most abundant and diverse groups of endemic organisms in the region (Hershler et al. 2014), and a diversity of other aquatic biota (Sada et al. 2001).

Small and large mammals rely on riparian areas for water, foraging, cover, movement, and migration corridors. Riparian areas provide productive foraging habitat for bats (Holloway and Barclay 2000), and riparian woodlands supply roost sites in otherwise treeless rangelands (Williams et al. 2006; Trubitt et al. 2018). Mule deer (*Odocoileus hemionus*) and elk (*Cervus canadensis*) frequent riparian zones, especially during summer months after fawning and calving (McCorquodale 1986; Morano et al. 2019). In the Great Basin, where mule deer habitat selection is largely driven by forage availability and water, deer select habitats closer to riparian areas especially in areas that are hotter and drier (Morano et al. 2019). In late summer, riparian areas provide productive and nutritious forage, shade, and water for all grazing animals, wild and domestic.

Riparian ecosystems play a crucial role in reducing wildlife vulnerability to climate change, providing refugia and adaptation opportunities (Seavy et al. 2009; Capon et al. 2013). Riparian vegetation provides shade and microclimates that give thermal refuge to animals adjusting to warmer air or water temperatures. Beaver dam activity produces fire-resistant riparian corridors that provide important refugia during wildfire, especially for species unable to physically escape the spread of flames (Fairfax and Whittle 2020). Well-connected riverscapes provide opportunities for water-dependent species to move to more hospitable parts of the watershed as climate conditions dictate.

7.5 Management and Restoration

Proper stewardship of riparian resources through a holistic watershed approach helps ensure maintenance of vital ecosystem services and goods for society. While there are many conservation actions land managers can take, promotion of ecological, hydrologic, and geomorphic processes underpins all successful approaches (NRC 2002; Goodwin et al. 1997). At the most fundamental level, water availability sets the stage for riparian ecology, so protection and restoration of hydrology is of paramount importance. Livestock grazing is a nearly ubiquitous land use on rangelands that has direct and indirect effects on riparian function, from influencing watershed hydrology to direct changes to riparian vegetation (Kauffman and Krueger 1984; Elmore 1992; Belsky et al. 1999; Wilcox et al. 2017). Widespread channel incision in riparian areas reflects a legacy of degradation that may be unrelated to current management (Chambers and Miller 2004) but often requires active intervention to reverse (Zeedyk and Clothier 2009). In this section, we highlight a few of the prevailing concepts and strategies for riparian management, protection, and restoration.

7.5.1 Grazing Management

Negative effects of overgrazing in riparian areas are well documented (Kauffman and Krueger 1984; Belsky et al. 1999). Less well understood are the complex relationships between contemporary grazing and legacy effects of historical riparian degradation due to other factors, such as the fur trapping era of the 1800s (Ott 2003), unregulated grazing during the late 1800s to early 1900s prior to reserving National Forests or the Taylor Grazing Act of 1934, water diversion and development during most of the 1900s, and more recently invasive species, climate change, and wild and free-roaming horses. "Passive" restoration through establishment of riparian exclosures that remove livestock grazing can result in riparian improvement but may not be sufficient to fully restore valley bottoms that have sustained drastic geomorphic alteration. In part, this is because the protection is usually limited to an already diminished remnant riparian area. Furthermore, livestock exclusion is often not feasible or desirable on western working lands where most riparian areas are in private ownership and support grazing operations. Fortunately, much has been learned in recent decades about compatible strategies for managing grazing to maintain and promote riparian functions and values (Text box 7.1; Platts 1991; Elmore 1992; Wyman et al. 2006; Swanson et al. 2015).

Text Box 7.1: Riparian recovery through improved grazing

On Susie Creek, Nevada, riparian conditions improved dramatically following changes in grazing strategies by the Maggie Creek Ranch and Bureau of Land Management (BLM) (Swanson et al. 2015; Charnley 2019). This photo series chronicles recovery following a switch from decades of growing-season-long cattle grazing (a) to a combination of spring and/or fall grazing, hot season grazing and periods of rest from grazing over 28 years (b–f). In 1992, the riparian area in this wide gully (formed by incision in about 1910) was fenced into a riparian pasture and no longer grazed throughout the growing season every year. Subsequently, riparian plants expanded, slowed water forces, captured sediment, and stabilized streambanks resulting in a vegetated flood-plain and elevated water table (b). With willows available, beavers accelerated the process of rehydrating the gully, creating expansive areas of ponded water and wetland vegetation (c). Beaver activity led to conversion of the willow community to other types of riparian plant communities, including a short-lived cattail marsh (d). A functional and well vegetated floodplain within the old gully continues to evolve (e–f). Riparian recovery improved resilience to flooding (including a rain on snow event in 2017), wildfire, and drought (Fesenmeyer et al. 2018). In the severe drought of 2020, abundant green riparian vegetation thrived with below ground summer water storage and, in the fall, water came to the surface in more locations. In this high-sediment watershed, aggradation in the widened gully is raising the water table across the valley bottom and providing critical green forage and water for wildlife and livestock, especially during summer and fall. The BLM and ranchers continue to fine tune and adjust management based on observed changes in weather/flows, vegetation, sediment deposition, and management effects in this dynamic riparian area. The Susie Creek story is just one of many examples of the outcomes possible following collaborative, private-public efforts to improve grazing management (also see: https://www.youtube.com/watch?v=kSctr0aQOso and https://iwjv.org/new-video-changing-a-landscape-to-a-lifescape/) Photos by: Carol Evans, retired fisheries biologist, Elko District, Bureau of Land Management (1988–2016).

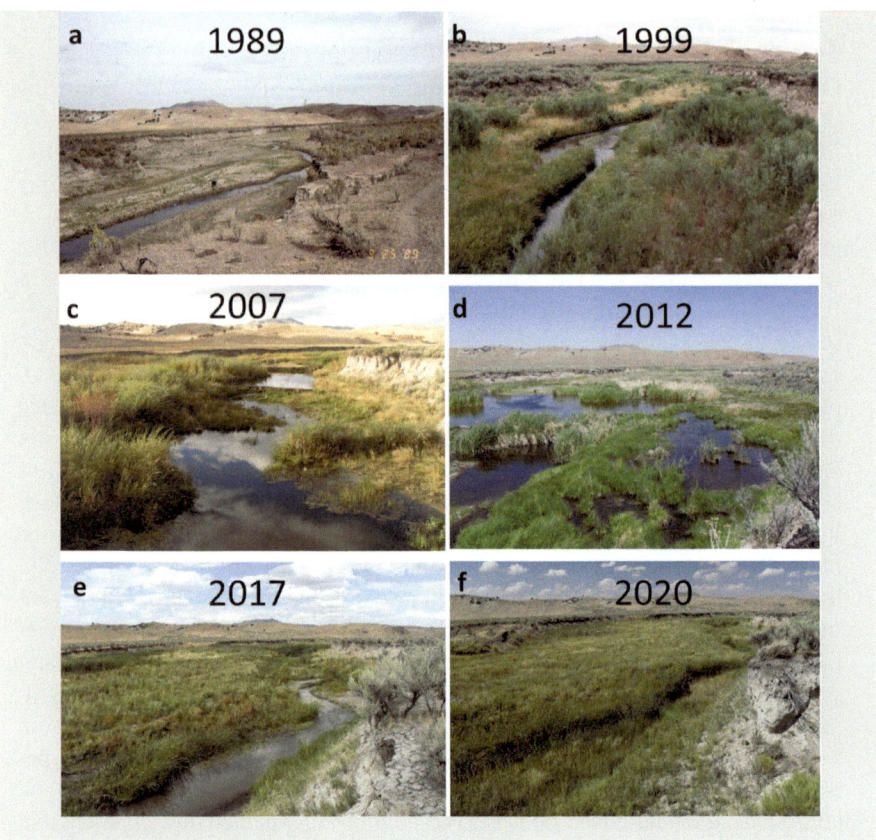

Similar to wildlife reliance on riparian resources, livestock on the range often depend on these systems to complete their life cycle. Livestock need water daily and riparian areas can be a primary source to meet those requirements. When upland plants dry out in hot summers or dry seasons, riparian areas remain green providing more nutritious forage which further attracts livestock to valley bottoms. Overgrazing can occur when plants are stressed by repeated defoliation without adequate time for recovery, or when trampling damages soils, which weakens riparian plant roots, destabilizes streambanks, and lowers the water table. Addressing overgrazing often involves changes in stocking rate (or the number of grazing animals in a given area for a specific time period) and/or livestock distribution (i.e., where animals are allowed to graze) and timing (i.e., when animals are allowed to graze). Stocking rate reductions often have to be severe to effect change in riparian conditions, which may be impractical in working landscapes. This is because riparian valley bottoms are typically only a small fraction of larger grazing pastures where stocking rates are set, and livestock tend to stay concentrated near water regardless of how many animals are present. Therefore, manipulation of livestock distribution and timing are often important strategies in minimizing overgrazing in rangeland riparian areas.

Riparian area grazing management is more likely to be successful if it enables control of, and variation in, periods of grazing and recovery, livestock distribution, and intensity of use (Swanson et al. 2015). Effective grazing strategies prevent repeated or excess damage to valley bottom soils and plants when they are most susceptible to grazing-related stresses (Wyman et al. 2006). Grazing management should be designed to balance grazing periods with opportunities for plant growth and recovery, and/or providing retention of adequate leaf area on individual plants post-grazing (Swanson et al. 2015). Rotation or variation in timing of grazing prevents stress in the same season year after year so plants can successfully complete all phases of their annual life cycle. By actively managing livestock, grazing impacts can be controlled to ensure plant growth or regrowth before, during or after grazing. Grazing management actions and monitoring are most effective when they embrace the interdependence of public and private lands.

Grazing managers have access to a wide variety of strategies for riparian-focused management to accomplish objectives and allow recovery (Table 7.2). A fundamental choice driving management actions, grazing criteria, and methods for short-term or implementation monitoring is whether to build management around: (a) schedules of grazing and recovery, or (b) limiting utilization levels within the growing season (Boyd and Svejcar 2004, 2012). To ensure appropriate management and enable sufficient flexibility to adapt management as riparian areas change, a plan should be written around a set of core principles that inform selection of locally targeted grazing use indicators. Such principles allow for flexibility and success in each pasture. Grazing use indicators and criteria should fit the chosen treatments and strategies to achieve resource objectives (University of Idaho Stubble Height Study Team 2004). Table 7.2 couples a suite of effective strategies with relevant implementation monitoring indicators. Rationale for how and why strategies work is described in Swanson et al. (2015), and recommendations for monitoring long-term effectiveness and short-term implementation is provided in Burton et al. (2011), Dickard et al. (2015), Swanson et al. (2018), and Gonzalez and Smith (2020).

Fencing is an important tool for facilitating many of the strategies described in Table 7.2, but establishing physical fences creates additional costs and liabilities for land managers (Knight et al. 2011) and impacts wildlife behavior and movements (Jakes et al. 2018) so new fences should be carefully planned. Alternatives to physical fencing, such as herding and stockmanship techniques (Cote 2019) or manipulating water/salt/supplement placement, may also be appropriate to effectively implement riparian grazing strategies (Table 7.2). Virtual fencing is an increasingly viable technology for achieving riparian improvements while minimizing risks associated with traditional fencing (Campbell et al. 2018; Ranches et al. 2021; Boyd et al. 2022). For example, virtual fencing could be used to allow only part of a stream, part of a large wetland, or each one of a series of springs in a spring complex to be accessed by livestock at any one time, thereby decreasing the duration of grazing and increasing recovery time. While a variety of techniques can be used to manage livestock, riparian functions may continue to be impaired in some instances by wild and free-roaming horses or grazing wildlife unless animal populations are maintained at an appropriate management level (USDI 2010).

Table 7.2. Riparian grazing management strategies that often support riparian functions and recovery

Grazing management strategies	Rationale	Implementation monitoring focus
Use short grazing periods	Grazed plants are not re-grazed	Dates of use/nonuse Refine by watching for animals selecting regrowth, nonuse is during the growing season, and observing plant development
Provide long recovery periods	All plants recover before subsequent grazing event	
Allow for regrowth before winter	Vegetation growth provides protects streambanks in spring high water	
Vary season of use from year to year	Graze in different seasons or phenology stages every year	
Provide occasional growing season rest	Year-long opportunity for plants to grow leaves and roots	
Graze at moderate to light intensity	Plants maintain leaf area to sustain growth and growing points	Utilization, woody species use or stubble height
Allow woody plants to grow	Woody plants grow above grazing height	Woody species use
Establish riparian pasture	Fenced pasture with specific riparian objectives	Varies with grazing strategy used in pasture
Employ herding, stockmanship techniques	Place animals away from riparian areas or in use areas each with planned use periods	
Ensure pastures are cleared of stock	All livestock moved during pasture moves to ensure riparian recovery periods	Check for stragglers
Graze large pasture when upland plants are green or microclimate preferred	Green uplands or favorable temperature or breezes attract livestock relieving pressure on riparian plants	Grazing utilization mapping, animal tracking
Provide off-stream water. Scatter salt & supplement. Select for hill climbers	Improve grazing distribution across pasture	

Strategies near the top work better with a high degree of animal control, while strategies near the bottom focus more on distribution of grazing away from limited riparian areas within a large pasture. See Swanson et al. (2015) for more details

7.5.2 Protection and Restoration

7.5.2.1 Riparian Planting

Where riparian plant communities have been lost or severely reduced, changes in management alone may be insufficient to recover them so revegetation is required. Re-establishment of riparian vegetation, often referred to as *buffers*, can improve water quality and shading, habitat for fish and wildlife, and overall riparian function. Matching the appropriate plant species to site conditions is key to success. Hydrology is a primary determinant of vegetation composition and unique species are adapted to varying degrees of inundation and soil moisture (Hoag et al. 2008). All too often, riparian planting is done in degraded riparian areas without first addressing the hydrology, geomorphology, or grazing management needed to sustain riparian vegetation, leading to chronically low success rates. Realistic planting zones, based on the elevational and lateral relationships of vegetation to surface and subsurface water, should be used to guide planting plans (Hoag et al. 2001). Consideration of species composition and structural diversity is also important in meeting life history needs of wildlife, especially birds (Gardner et al. 1999). A common misperception is that all riparian areas support woody vegetation, but many systems naturally do not because of water source, gradient, soils, climate, or other factors. Site-specific conditions, along with local reference areas, should be used to guide species selection and placement (Hoag et al. 2001). Techniques for revegetation include use of live cuttings from certain woody species (e.g., willows, cottonwoods), transplanting of bare root or container stock plants, and seeding (Gardner et al. 1999). Degraded riparian areas can be havens for invasive species (Stohlgren et al. 1998), such as reed canarygrass (*Phalaris arundinacea*), salt cedar (*Tamarix* spp.), and many others, so planting efforts typically require measures to control competition and follow-up management of weed infestations. Long-term success of revegetation hinges on compatible grazing management, invasive species control, and promotion of essential riparian processes (e.g., hydrology, disturbance regime) needed to support and recruit new native vegetation through time (Stromberg 2001).

7.5.2.2 Floodplain Reconnection

Levees, dams, roads, constructed channels, channel incision, and other impairments have greatly reduced the proportion of active floodplain within valley bottoms of many riverscapes (Tockner and Stanford 2002; Skidmore and Wheaton 2022). Thus, a common restoration strategy is to reconnect former floodplains. Multiple restoration approaches are used to achieve this goal, such as channel reconstruction and remeandering (e.g., Natural Channel Design, Rosgen 2011), levee and riprap breaching or removal, floodplain lowering, dam and barrier removal, road decommissioning, and increased instream flows/flood regimes from dam releases. These approaches to floodplain reconnection, typically applied in larger streams and rivers, require engineering design and expertise, heavy equipment operators, and special permits especially where infrastructure vulnerability is high. While important, these approaches are expensive (Bernhardt et al. 2007) and may not be practical to address the scale of degradation of riverscapes, particularly across vast range-lands. Alternative approaches that rely on grading, but are still cost effective and process-based, include the "geomorphic grade line" approach to achieving Stage 0 (Powers et al. 2018). Approaches and techniques selected to enlarge or reconnect floodplains depend on stream size and type, hydrology, accessibility, risks, budgets, and timelines for recovery.

Smaller headwater streams of rangeland watersheds are often overlooked for restoration opportunities but represent the vast majority of riverscape miles across much of the West. Here, low-tech processed-based restoration approaches (Wheaton et al. 2019a, b; Zeedyk and Clothier 2009) can be effective, cost-efficient, and scalable solutions to achieve floodplain reconnection (Text Box 7.2). For example, simple hand-built structures (e.g., post-assisted log structures, beaver dam analogues) can be used to mimic and promote important processes of wood accumulation and beaver dam activity that increase connectivity with active floodplains (Pollock et al. 2014; Wheaton et al. 2019a, b).

Text Box 7.2 Low-tech process-based restoration

Bridge Creek, Oregon, was historically subject to intensive grazing, timber harvest in the upper watershed, and beaver trapping. Large storm events in the early 1900s led to massive valley fills, followed quickly by channel incision resulting in a lowered water table and loss of riparian vegetation. Much of the creek became a highly simplified channel disconnected from its floodplain, representative of incised streams commonly found in rangelands. Early management to reverse degradation included removal of livestock grazing, allowing some willow recovery along the channel margins. Beaver reoccupied the creek with increasing vegetation. Sediment aggraded quickly behind beaver dams, but because of the force of high flows in the incised trench and the small woody material available, the dams were short-lived.

In a watershed-scale experiment, Bouwes et al. (2016) piloted low-tech process-based restoration using beaver dam analogues (BDAs) to improve aggradation, beaver dam longevity, and floodplain connectivity. Researchers installed 121 BDAs in Bridge Creek and compared hydrologic, geomorphic, and ecological responses to a control watershed [see sample treatment reach (a) and reference reach (b)]. Within 4 years, the BDAs accelerated beaver dam activity nearly eightfold. The increase in dam density and stability captured sediment, raised the stream bed, and reconnected the channel with its floodplain leading to higher groundwater storage, surface water temperature diversity (Weber et al. 2017), and increased fish habitat quantity and complexity. Importantly, these changes produced population-level benefits for threatened steelhead (*Oncorhynchus mykiss*), representing a rare example in the scientific literature documenting a positive fish population response following restoration.

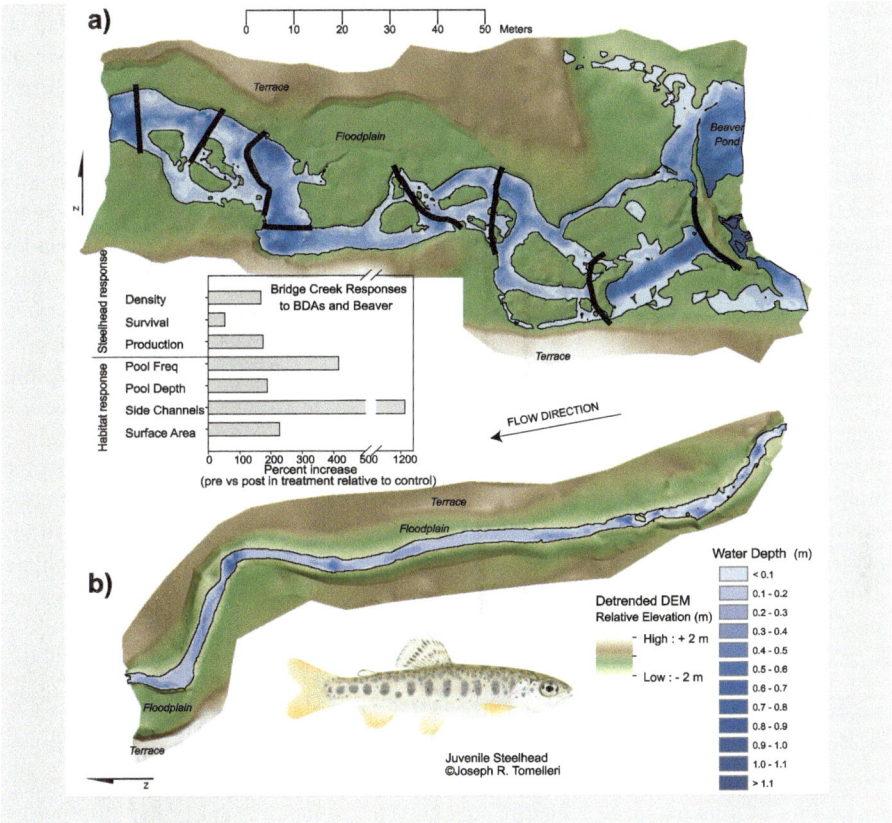

7.5.2.3 Wet Meadow Protection and Restoration

Wet meadows can be particularly vulnerable to loss and degradation due to channel incision, improper grazing and animal trailing, roads, intentional drainage, land use conversion, and altered watershed hydrology. A variety of strategies are used to protect functional meadows that remain and restore those that have been impacted. Given the preponderance of wet meadows in private ownership (Donnelly et al. 2018), conservation easements can be an important tool used to keep wet meadows intact, preventing loss of wetland and meadow habitats to other land uses (e.g., tillage agriculture, development, water diversion). Incentive-based easement programs, such as those administered by the U.S. Department of Agriculture Natural Resources Conservation Service, offer compensation to landowners who voluntarily agree to protect wet meadows and forego certain activities. Easements have been used to protect critical rangeland wildlife habitats across the West, such as playas and prairie potholes for waterfowl, meadows for sage-grouse, and mule deer migration corridors (Doherty et al. 2013; Copeland et al. 2014; NRCS 2015).

Channel incision is a widespread impairment affecting meadow function that has been the focus of restoration in the western U.S. for over a century (Kraebel and Pillsbury 1934; Ramstead et al. 2012). Early in the gully erosion process, simple low-tech restoration methods can be effectively used to stop headcut advancement and reconnect meadows to floodplain surfaces (Text Box 7.3, Zeedyk and Clothier 2009; Maestas et al. 2018). As the incision trench deepens, more intensive restoration involving heavy equipment and earthwork is often conducted, such as, pond-and-plug techniques where gullies are partially filled and water returned to the historic meadow elevation (Rosgen 1997; Zeedyk and Vrooman 2017; Rodriguez et al. 2017). Alternatively, process-based restoration approaches (e.g., geomorphic grade line or low-tech) have been used in systems where it is possible to leverage natural processes of erosion and deposition to help aid in rebuilding healthy, connected floodplains (Powers et al. 2018; Wheaton et al. 2019a). Roads in valley bottoms can cause or exacerbate incision so relocating them to uplands outside or along the margin of the valley bottom or creating hardened crossings may be part of restoration planning (Zeedyk 2006). Restoration success is linked to grazing strategies that promote wetland and meadow vegetation, and practices like riparian pasture fencing, offsite water, and drift fencing may be needed to discourage livestock and wild ungulates from congregating and trailing in meadow bottoms (Wyman et al. 2006; Maestas et al. 2018).

Text Box 7.3 Treating channel incision in springs and meadows

Channel incision in headwater meadows and springs lowers the water table, drying out riparian and wetland vegetation. Many gullies begin at knickpoints, or headcuts, where enough flow concentrates to erode a pour-over hole and creating an abrupt change in elevation (a). Surface water runoff shifts from sheetflow above the headcut to concentrated flow below which accelerates runoff and erosion resulting in gully formation. Once started, headcuts migrate up valley until a hard point is reached. If headcuts are caught early, simple low-tech treatments can be implemented to protect and restore meadows and springs (Zeedyk and Clothier 2009). Hand-built structures made of natural materials installed at headcuts protect plant roots from further erosion while structures placed in downstream gullies slow flow, trap sediment, and raise water tables (b). These techniques have been used in a variety of rangeland settings, such as the desert southwest and Colorado's Gunnison Basin where diverse partners are working to improve riparian resiliency and brood habitat for imperiled Gunnison sage-grouse (*Centrocercus minimus*, TNC 2017; Maestas et al. 2018; Silverman et al. 2018). Photo credits a–b: Shawn Conner, Jeremy Maestas.

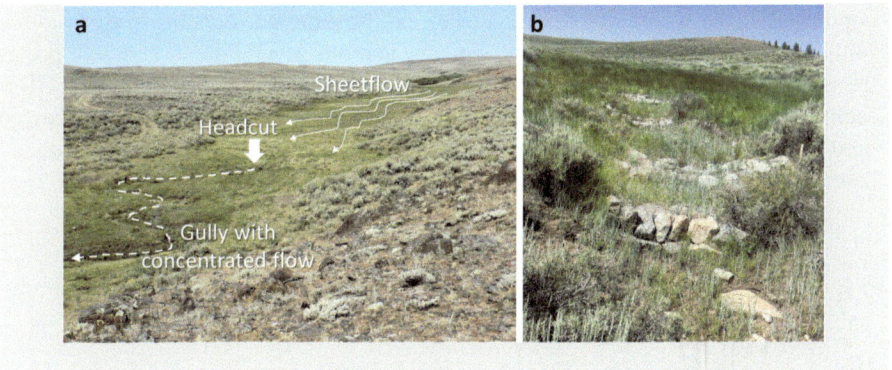

7.5.2.4   Spring Development and Protection

Springs are interspersed throughout rangelands and play an important role in providing reliable water for livestock and habitat for wildlife, but they are prone also to degradation due to animal congregation, trampling, and improper grazing. Spring development to capture flow for stock water can also contribute to loss of riparian vegetation around springs. However, spring developments remain an essential practice to facilitate grazing management strategies to improve livestock distribution and overall rangeland health, including riparian goals. Thoughtful planning of water developments can protect ecological values while providing sufficient drinking water for livestock to enable effective grazing strategies such as grazing within a shorter duration and moving for recovery (Swanson et al. 2015). Gurrieri (2020) provides a review of considerations and sample designs for modern spring developments.

An overarching goal of spring developments should be to provide the needed livestock drinking water while returning as much as possible to the system and protecting riparian soils and vegetation. Removal of water from a spring impacts riparian hydrology and vegetation, so ideally livestock water should be sourced from the most productive and resilient sites possible. Where available, streams may be better alternatives for sourcing water than more fragile spring systems. Consulting a hydrologist about source flow rates and the amount needed to sustain the system is recommended. There are a variety of techniques that can be used in project design to minimize negative impacts on springs during development. For example, installing and maintaining a float valve on a trough allows the water to be automatically turned off when the trough is full and a shut-off valve allows trough water to be withdrawn from that area so all livestock move to the next trough and use area, leaving unneeded water in the spring system. Splitters can also be used to ensure that only a portion of water is diverted from the system. Adding pipelines and locating the trough away from the source concentrates trampling away from riparian soils and vegetation. Outdated, unused, or poorly maintained livestock water developments are abundant across western rangelands, but rehabilitation efforts can improve site conditions if the water source is intact or can be restored. Fencing spring sources (either using exclosures or pastures) on new and old developments can help protect them from damage by grazing animals. Wildlife-friendly fence designs (Paige 2015), including virtual fencing where feasible (Campbell et al. 2018), should be considered to allow continued spring use by dependent wildlife. Long term maintenance is essential and should be included in project planning.

7.6 Summary

Historically, the importance of riparian areas and their connection to uplands has been overlooked or minimized in rangeland ecology and management. Wildlife biologists were among the first to raise awareness of the disproportionate value of riparian areas in sustaining biodiversity (Johnson and Jones 1977; Thomas et al. 1979). Habitat supplied by riparian areas, in the form of food, cover, water, refugia, and corridors, helps support unique aquatic and semi-aquatic species, as well as most terrestrial species found in rangelands. In recent decades, a growing body of science has shown how vital these ecosystems are for supplying a wide variety of goods and services, ranging from water quality and flood control to wildfire and drought resilience (NRC 2002; McKinstry et al. 2004; Silverman et al. 2018; Fairfax and Whittle 2020). Today, riparian ecology is rightfully recognized as an integral component of holistic rangeland management and is among the top resource issues being addressed by agencies and landowners across private and public rangelands (BLM 2021; NRCS 2021).

Enhancing the resilience of riparian areas will only become more essential for climate change mitigation and adaptation (Seavy et al. 2009; Capon et al. 2013; Reed et al. 2020). Increasing climate variability and extreme weather events make inhabiting already harsh rangeland environments more difficult for wildlife and people. Healthy riparian systems provide natural infrastructure to buffer against climate change effects and meet growing water demands. Because many western riparian areas have been degraded for so long (e.g., more than two centuries in some cases), a shifting baseline of riparian expectations under-represents what is possible (Wheaton et al. 2019a). With a current knowledge of ecology, hydrology, and geomorphology, rangeland conservationists are better equipped to implement management and restoration strategies necessary to realize the full riparian potential. Building riparian resiliency serves as a unifying goal to reduce vulnerability of rangeland wildlife and human communities in a changing world.

Acknowledgements We wish to thank Carol Evans, Chad Boyd, and Dave Dahlgren for their contributions, insights, and feedback on this chapter. This work was supported in part by the U.S. Department of Agriculture, Natural Resources Conservation Service. Special thanks to the NRCS Working Lands for Wildlife (https://www.wlfw.org/) team and partnership for intellectual contributions to this chapter, funding support for open access, and efforts to scale up riparian conservation on rangelands. The findings and conclusions in the publication are those of the authors and should not be construed to represent any official USDA or U.S. Government determination or policy.

References

Anderson BW, Ohmart RD (1977) Vegetation structure and bird use in the lower Colorado River valley. In: Johnson RR, Jones Jr DA (eds) Importance, preservation and management of riparian habitat: a symposium. U.S. Department of Agricultural Forest Services General Technical Report RM-43, pp 23–34

Arkle RS, Pilliod DS (2015) Persistence at distributional edges: Columbia spotted frog habitat in the arid Great Basin, USA. Ecol Evol 5:3979–3994

Baxter CV, Fausch KD, Saunders WC (2005) Tangled webs: reciprocal flows of invertebrate prey link streams and riparian zones. Fresh Biol 50:201–220

Belsky AJ, Matzke A, Uselman S (1999) Survey of livestock influences on stream and riparian ecosystems in the western United States. J Soil Water Cons 54:419–431

Bernhardt ES, Sudduth EB, Palmer MA et al (2007) Restoring rivers one reach at a time: results from a survey of US river restoration practitioners. Rest Ecol 15:482–493

Bouwes N, Weber N, Jordan CE et al (2016) Ecosystem experiment reveals benefits of natural and simulated beaver dams to a threatened population of steelhead (*Oncorhynchus mykiss*). Sci Rep 6:28581

Boyd CS, Svejcar TJ (2004) Regrowth and production of herbaceous riparian vegetation following defoliation. J of Range Manag 57(5):448–454. https://doi.org/10.2111/1551-5028(2004)057[0448:RAPOHR]2.0.CO,2

Boyd CS, Svejcar TJ (2012) Biomass production and net ecosystem exchange following defoliation in a wet sedge community. Range Ecol Manag 65(4):394–400. https://doi.org/10.2111/REM-D-11-00159.1

Boyd CS, O'Connor R, Ranches J, Bohnert DW, Bates JD, Johnson DD, Davies KW, Parker T, Doherty KE (2022) Virtual fencing effectively excludes cattle from burned sagebrush steppe. Rangeland Ecol Manag 81:55–62

Brand LA, White GC, Noon BR (2008) Factors influencing species richness and community composition of breeding birds in a desert riparian corridor. The Condor 110(2):199–210. https://doi.org/10.1525/cond.2008.8421

Brazier RE, Puttock A, Graham HA et al (2020) Beaver: nature's ecosystem engineers. WIREs Water 8(1):e1494. https://doi.org/10.1002/wat2.1494

Brinson MM, Swift BL, Plantico RC et al (1981) Riparian ecosystems: their ecology and status. FWS/OBS-81/17. U.S. Fish and Wildlife Service Biological Services Program, Washington, DC

Burchsted D, Daniels M, Thorson R et al (2010) The river discontinuum: applying beaver modifications to baseline conditions for restoration of forested headwaters. Bioscience 60(11):908–922

Bureau of Land Management (BLM) (2021) Wetland and riparian areas. https://www.blm.gov/programs/natural-resources/wetlands-and-riparian-health. Accessed 3 July 2021

Burton TA, Smith SJ, Cowley ER (2011) Multiple indicator monitoring (MIM) of stream channels and streamside vegetation (Technical Reference No. 1737-23. BLM/OC/ST-10/003+1737+REV). Denver, CO, USA: US Department of the Interior, Bureau of Land Management, National Operations Center, p 155. https://www.blm.gov/learn/blm-library/agency-publications/technical-references

Butler D, Malanson G (2005) The geomorphic influences of beaver dams and failures of beaver dams. Geomorphology 71(1–2):48–60

Campbell DLM, Haynes SJ, Lea JM, Farrer WJ, Lett C (2018) Temporary exclusion of cattle from a riparian zone using virtual fencing technology. Animals 9(1):5. https://doi.org/10.3390/ani901 0005

Capon SJ, Chambers LE, Mac Nally R et al (2013) Riparian ecosystems in the 21st century: hotspots for climate change adaptation? Ecosystems 16:359–381. https://doi.org/10.1007/s10021-013-9656-1

Castro JM, Thorne CR (2019) The stream evolution triangle: integrating geology, hydrology, and biology. River Res Appl 35(4):315–326. https://doi.org/10.1002/rra.3421

Chambers JC, Miller JR (eds) (2004) Great Basin riparian ecosystems: ecology, management and restoration. Island Press, Covelo, CA

Charnley S (2019) If you build it, they will come: ranching, riparian revegetation, and beaver colonization in Elko County, Nevada. Res. Pap. PNW-RP-614. Portland, OR: U.S. Department of Agriculture, Forest Service, Pacific Northwest Research Station, p 39. https://www.fs.usda.gov/treesearch/pubs/57954

Cluer B, Thorne C (2013) A stream evolution model integrating habitat and ecosystem benefits. River Res Appl 30:135–154. https://doi.org/10.1002/rra.2631

Copeland HE, Sawyer H, Monteith KL et al (2014) Conserving migratory mule deer through the umbrella of sage-grouse. Ecosphere 5:117. https://doi.org/10.1890/ES14-00186.1

Corenblit D, Tabacchi E, Steiger J et al (2007) Reciprocal interactions and adjustments between fluvial landforms and vegetation dynamics in river corridors: a review of complementary approaches. Earth Sci Rev 84(1):56–86. https://doi.org/10.1016/j.earscirev.2007.05.004

Cote S (2019) Manual of stockmanship: a complete livestock handling guide for the range, feedlot, dairy and farm operation. Hudson, Salt Lake City, Utah, USA

Cubley ES, Bateman HL, Merritt DM et al (2020) Using vegetation guilds to predict bird habitat characteristics in riparian areas. Wetlands. https://doi.org/10.1007/s13157-020-01372-8

Dickard M, Gonzales M, Elmore W et al (2015) Riparian area management: proper functioning condition assessment for lotic areas (Technical Report No. 1737-15 v.2). Denver, CO, USA: US Department of the Interior, Bureau of Land Management. https://www.blm.gov/learn/blm-library/agency-publications/technical-references

Doherty KE, Ryba AJ, Stemler CL et al (2013) Conservation planning in an era of change: state of the U.S. Prairie Pothole Region. Wild Soc Bull 37:546–563

Dolin EJ (2011) Fur, fortune and empire: the epic history of the fur trade in America. WW Norton & Company

Donnelly JP, Naugle DE, Hagen CA et al (2016) Public lands and private waters: scarce mesic resources structure land tenure and sage-grouse distributions. Ecosphere. https://doi.org/10.1002/ecs2.1208

Donnelly JP, Allred BW, Perret D et al (2018) Seasonal drought in North America's sagebrush biome structures dynamic mesic resources for sage-grouse. Ecol Evol 00:1–14. https://doi.org/10.1002/ece3.4614

Donnelly JP, Naugle DE, Collins DP et al (2019) Synchronizing conservation to seasonal wetland hydrology and waterbird migration in semi-arid landscapes. Ecosphere. https://doi.org/10.1002/ecs2.2758

Elmore W (1992) Riparian responses to grazing practices. In: Naiman RB (ed) Watershed management: balancing sustainability and environmental change. Springer-Verlag, New York, pp 442–457

Fairfax E, Whittle A (2020) Smokey the beaver: beaver-dammed riparian corridors stay green during wildfire throughout the western USA. Ecol Appl. https://doi.org/10.1002/eap.2225

Fesenmyer KA, Dauwalter DC, Evans C et al (2018) Livestock management, beaver, and climate influences on riparian vegetation in a semi-arid landscape. PLoS ONE. https://doi.org/10.1371/journal.pone.0208928

Fryirs KA, Brierley GA (2013) Geomorphic analysis of river systems: an approach to reading the landscape, 1st edn. Blackwell Publishing Ltd., Chichester, U.K

Gardner PA, Stevens R, Howe FP (1999) A handbook of riparian restoration and revegetation for the conservation of land birds in Utah with emphasis on habitat types in middle and lower elevations. Utah Division of Wildlife Resources Publication Number 99–38

Giesen KM, Connelly JW (1993) Guidelines for management of sharp-tailed grouse habitats. Wild Soc Bull 21:325–333

Goldfarb B (2018) Eager: the surprising, secret life of beavers and why they matter. Chelsea Green Publishing, White River Junction, VT

Gonzalez MA, Smith SJ (2020) Riparian area management: Proper functioning condition assessment for lentic areas. 3rd ed. Technical Reference 1737-16. U.S. Department of the Interior, Bureau of Land Management, National Operations Center, Denver, Colorado. http://www.blm.gov/wo/st/en/info/blm-library/publications/blm_publications/tech_refs.html

Goodwin CN, Hawkins CP, Kershner JL (1997) Riparian restoration in the western United States: overview and perspective. Rest Ecol 5:4–14

Gregory SV, Swanson FJ, McKee WA et al (1991) An ecosystem perspective of riparian zones. Bioscience 41:540–551. https://doi.org/10.2307/1311607

Gurrieri JT (2020) Rangeland water developments at springs: best practices for design, rehabilitation, and restoration. Gen. Tech. Rep. RMRS-GTR-405. Fort Collins, CO: U.S. Department of Agriculture, Forest Service, Rocky Mountain Research Station, p 21. https://www.fs.usda.gov/rmrs/publications/rangeland-water-developments-springs-best-practices-design-rehabilitation-and

Hall JG (1970) Willow and aspen in the ecology of beaver in Sagehen Creek, California. Ecology 41(3):484–494

Haukos DA, Smith LM (1994) The importance of playa wetlands to biodiversity of the Southern High Plains. Land Urban Plan 28:83–98

Hehnke M, Stone CP (1979) Value of riparian vegetation to avian populations along the Sacramento River system. In: Johnson RR, McCormick JF (eds) Strategies for protection and management of floodplain wetlands and other riparian ecosystems, General Technical Report WO-12, US For Ser, Washington, DC, pp 228–235

Hershler R, Liu H-P, Howard J (2014) Spring snails: a new conservation focus in western North America. Bioscience 64:693–700

Hoag JC, Berg FE, Wyman SK et al (2001) Riparian planting zones in the Intermountain West. USDA-NRCS Aberdeen Plant Materials Center. Aberdeen, Idaho

Hoag C, Tilley D, Darris D et al (2008) Field guide for the identification and use of common riparian woody plants of the Intermountain West and Pacific Northwest Regions. USDA-NRCS Aberdeen Plant Materials Center, Boise, Idaho

Holloway GL, Barclay RMR (2000) Importance of prairie riparian zones to bats in southeastern Alberta. Ecoscience 7:115–122

Hook PB (2003) Sediment retention in rangeland riparian buffers. J Env Qual 32:1130. https://doi.org/10.2134/jeq2003.1130

Hough-Snee N, Roper B, Wheaton JM et al (2014) Riparian vegetation communities of the American Pacific Northwest are tied to multi-scale environmental filters. River Res Appl. https://doi.org/10.1002/rra.2815

Jakes AF, Jones PF, Paige LC, Seidler RG, Huijser MP (2018) A fence runs through it: a call for greater attention to the influence of fences on wildlife and ecosystems. Bio Conserv 227:310–318

Johnson RR, Jones Jr DA (eds) (1977) Importance, preservation and management of riparian habitat: a symposium. U.S. Department of Agricultural Forest Services General Technical Report RM-43

Johnson RR, Haight LT, Simpson JM (1977) Endangered species vs. endangered habitats: a concept. In: Johnson RR, Jones Jr DA (eds) Importance, preservation and management of riparian habitat: a symposium. U.S. Department of Agricultural Forest Services General Technical Report, RM-43, pp 68–74

Jordan CE, Fairfax E (2022) Beaver: the North American freshwater climate action plan. Wiley Interdisciplinary Rev Water e1592

Kauffman JB, Krueger WC (1984) Livestock impacts on riparian ecosystems and streamside management implications—a review. J of Range Manag 37:430–438

Knight KB, Toombs TP, Derner JD (2011) Cross-fencing on private US rangelands: financial costs and producer risks. Rangelands 33(2):41–44

Knopf FL (1985) Significance of riparian vegetation to breeding birds across an altitudinal cline. In: Johnson RR, Ziebell CD, Patten DR et al (eds) Riparian ecosystems and their management: reconciling conflicting uses. U.S. Department of Agricultural Forest Services General Technical Report RM-120, pp 105–111

Knopf FL, Johnson RR, Rich T et al (1988) Conservation of riparian ecosystems in the United States. Wils Bull 100:272–284

Kraebel C, Pillsbury AF (1934) Guideline for watershed improvement measures: Handbook of erosion control in Mountain Meadows. https://www.nrcs.usda.gov/Internet/FSE_DOCUME NTS/nrcs144p2_053967.pdf. Accessed 21 Dec 2020

Larsen A, Larsen JR, Lane SN (2021) Dam builders and their works: Beaver influences on the structure and function of river corridor hydrology, geomorphology, biogeochemistry and ecosystems. Earth Sci Rev 218:103623. https://doi.org/10.1016/j.earscirev.2021.103623

Lichvar RW, Melvin NC, Butterwick M et al (2012) National wetland plant list indicator rating definitions. ERDC/CRREL TN-12-1. Army Corps of Engineers, Engineer Research and Development Center, Hanover, NH. http://wetland-plants.usace.army.mil/nwpl_static/data/ DOC/NWPL/pubs/2012b_Lichvar_et_al.pdf. Accessed 3 Aug 21

Lichvar RW, Banks DL, Kirchner WN et al (2016) The National Wetland Plant List: 2016 wetland ratings. Phytoneuron 2016-30: 1-17. Published 28 April 2016. ISSN 2153 733X http://wetland-plants.usace.army.mil/nwpl_static/data/DOC/lists_2016/National/ National_2016v2.pdf. Accessed 3 Aug 21

Macfarlane WW, Gilbert JT, Gilbert JD, Saunders WC, Hough-Snee N, Hafen C, Wheaton JM, Bennett SN (2018) What are the conditions of riparian ecosystems? Identifying impaired floodplain ecosystems across the western U.S. Using the Riparian Condition Assessment (RCA) Tool. Environmental Management. https://doi.org/10.1007/s00267-018-1061-2

Maestas JD, Conner S, Zeedyk B et al (2018) Hand-built structures for restoring degraded meadows in sagebrush rangelands: Examples and lessons learned from the Upper Gunnison River Basin, Colorado. Range Technical Note No. 40. USDA-NRCS, Denver, CO. https://doi.org/10.13140/ RG.2.2.33628.08321

Mahoney MJ, Stella JC (2020) Stem size selectivity is stronger than species preferences for beaver, a central place forager. Ecol Manag 475:118331

Majerova M, Neilson BT, Schmadel NM et al (2015) Impacts of beaver dams on hydrologic and temperature regimes in a mountain stream. Hydrol Earth Syst Sci 19(8):3541–3556

Manners R, Schmidt J, Wheaton JM (2013) Multiscalar model for the determination of spatially explicit riparian vegetation roughness. J Geophys Res Earth Surf 118:65–83. https://doi.org/10. 1029/2011jf002188

Manning ME, Swanson SR, Svejcar T et al (1989) Rooting characteristics of four intermountain meadow community types. J Range Manag 42(4):309–312

Merriam-Webster (2021) Riparian. https://www.merriam-webster.com

Mayer PM, Reynolds SK, McCutchen MD et al (2006) Riparian buffer width, vegetative cover, and nitrogen removal effectiveness: A review of current science and regulations. EPA/600/R-05/ 118. U.S. Environmental Protection Agency, Cincinnati, OH

McKinstry MC, Hubert WA, Anderson SH (eds) (2004) Wetland and riparian areas in the Intermountain West: ecology and management. University of Texas Press, Austin

McCorquodale SM (1986) Elk habitat use patterns in the shrub steppe of Washington. J Wild Man 50:664–669

Milligan HE, Humphries MM (2010) The importance of aquatic vegetation in beaver diets and the seasonal and habitat specificity of aquatic-terrestrial ecosystem linkages in a subarctic environment. Oikos 119(12):1877–1886

Morano S, Stewart KM, Dilts T et al (2019) Resource selection of mule deer in a shrub-steppe ecosystem: influence of woodland distribution and animal behavior. Ecosphere. https://doi.org/10.1002/ecs2.2811

Munger JC, Gerber M, Madrid K et al (1998) US National Wetland Inventory classification of the occurrence of Columbian spotted frogs (*Rana luteiventris*) and Pacific treefrogs (*Hyla regilla*). Cons Biol 12:320–330

Murray D, Neilson BT, Brahney J (2021) Source or sink? Quantifying Beaver Pond Influence on Non-Point Source Pollutant Transport in the Intermountain West. J Env Manag 285:112127

Naiman RJ, Johnston CA, Kelley JC (1988) Alteration of North American streams by beaver. Bioscience 38(11):753–762

Naiman RJ, Decamps H, Pollock M (1993) The role of riparian corridors in maintaining regional biodiversity. Ecol Appl 3:209–212. https://doi.org/10.2307/1941822

National Research Council (NRC) (2002) Riparian areas: functions and strategies for management. The National Academies Press, Washington, DC. https://doi.org/10.17226/10327

Natural Resources Conservation Service (NRCS) (2015) Outcomes in conservation: Sage Grouse Initiative. US Department of Agriculture

Natural Resources Conservation Service (NRCS) (2021) A framework for conservation action in the Sagebrush Biome. Working Lands for Wildlife, USDA-NRCS. Washington, D.C. https://wlfw.rangelands.app. Accessed 3 July 2021

Opperman JJ, Moyle PB, Larsen EW, Florsheim JL, Manfree AD (2017) Floodplains: processes and management for ecosystem services. University of California Press, Oakland, CA

Ott J (2003) "Ruining" the rivers in the Snake Country: the Hudson's Bay Company's Fur Desert Policy. Oreg Hist Quar 104:166–195

Paige C (2015) A wyoming landowner's handbook to fences and wildlife: practical tips for fencing with wildlife in mind. Wyoming community foundation, Laramie, WY. pp 56. https://extension.colostate.edu/wp-content/uploads/2022/01/A-Wyoming-Landowners-Handbook-to-Fences-and-Wildlife_2nd-Edition_-lo-res.pdf

Patten DT (1998) Riparian ecosystems of semi-arid North America: Diversity and human impacts. Wetlands 18:498–512. https://doi.org/10.1007/BF03161668

Perry LG, Andersen DC, Reynolds LV et al (2011) Vulnerability of riparian ecosystems to elevated CO_2 and climate change in arid and semiarid western North America. Glob Change Biol 18:821–842. https://doi.org/10.1111/j.1365-2486.2011.02588.x

Platts WS (1991) Livestock grazing. In: Meecham WR (ed) Influences of forest and rangeland management on salmonid fisheries and their habitats. Bethesda, MD, USA: American Fisheries Society Special Publication 19, pp 389–423

Pollock MM, Beechie TJ, Wheaton JM et al (2014) Using beaver dams to restore incised stream ecosystems. Bioscience 64(4):279–290

Pollock MM, Heim M, Werner D (2003) Hydrologic and geomorphic effects of beaver dams and their influence on fishes. Amer Fish Soc Sym 37:213–233

Powers PD, Helstab M, Niezgoda SL (2018) A process-based approach to restoring depositional river valleys to Stage 0, an anastomosing channel network. River Res Appl 35:3–13. https://doi.org/10.1002/rra.3378

Ramstead KM, Allen JA, Springer AE (2012) Have wet meadow restoration projects in the Southwestern U.S. been effective in restoring geomorphology, hydrology, soils, and plant species composition? Env Evid 1:11. https://doi.org/10.1186/2047-2382-1-11

Ranches J, O'Connor R, Johnson D, Davies K, Bates J, Boyd C, Bohnert DW, Parker T (2021) Effects of virtual fence monitored by global positioning system on beef cattle behaviour. Transl Anim Sci 5:S144–S148. https://doi.org/10.1093/tas/txab161

Reed CC, Merrill AG, Drew WM et al (2020) Montane meadows: a soil carbon sink or source? Ecosystems. https://doi.org/10.1007/s10021-020-00572-x10.1007/s10021-020-00572-x

Rich TA (2002) Using breeding land birds in the assessment of western riparian systems. Wild Soc Bull 30:1128–1139

Rodriguez K, Swanson S, McMahon A (2017) Conceptual models for surface water and groundwater interactions at pond and plug restored meadows. J Soil Water Cons 72(4):382–394. https://doi.org/10.2489/jswc.72.4.382

Rosgen DL (1997) A geomorphological approach to restoration of incised rivers. In: Wang SSY, Langendoen EJ, Shields JFD (eds) Proceedings of the Conference on management of landscapes disturbed by channel incision. ISBN 0-937099-05-8

Rosgen DL (2011) Natural channel design: Fundamental concepts, assumptions, and methods. In: Simon A, Bennett SJ, Castro JM (eds) Stream restoration in dynamic fluvial systems: scientific approaches, analyses, and tools. Geophysical Monograph Series 194:69–93. Washington, D.C.: American Geophysical Union

Runyon MJ, Tyers DB, Sowell BF et al (2014) Aspen restoration using beaver on the Northern Yellowstone Winter Range under reduced ungulate herbivory. Rest Ecol 22(4):555–561

Sada DW, Williams JE, Silvey JC, et al (2001) A Guide to managing, restoring, and conserving springs in the western United States. Riparian Area Management TR 1737-17 Denver, CO, USA: US Department of the Interior, Bureau of Land Management. https://www.blm.gov/learn/blm-library/agency-publications/technical-references

Saunders WC, Fausch KD (2007) Improved grazing management increases terrestrial invertebrate inputs that feed trout in Wyoming rangeland streams. Trans Am Fish Soc 136:1216–1230

Seavy NE, Gardali T, Golet GH et al (2009) Why climate change makes riparian restoration more important than ever: recommendations for practice and research. Ecol Restor 27:330–338. https://doi.org/10.3368/er.27.3.330

Seton E (1929) Lives of game animals, vol 4, Part 2, Rodents, etc, Doubleday, Doran, and Company, Garden City, New York, USA

Silverman NL, Allred BW, Donnelly JP et al (2018) Low-tech riparian and wet meadow restoration increases vegetation productivity and resilience across semi-arid rangelands. Rest Ecol 27(2):269–278. https://doi.org/10.1111/rec.12869

Skidmore P, Wheaton JM (2022) Riverscapes as natural infrastructure: meeting challenges of climate adaptation and ecosystem restoration. Anthropocene 38:100334

Stevens LE, Brown BT, Simpson JM et al (1977) The importance of riparian habitat to migrating birds. In: Johnson RR, Jones Jr DA (eds) Importance, preservation and management of riparian habitat: a symposium. U.S. Department of Agricultural Forest Services General Technical Report RM-43, pp 156–164

Stoffyn-Egli P, Willison JHM (2011) Including wildlife habitat in the definition of riparian areas: the beaver (*Castor canadensis*) as an umbrella species for riparian obligate animals. Env Rev 19:479–494

Stohlgren TJ, Bull KA, Otsuki Y et al (1998) Riparian zones as havens for exotic plant species in the central grasslands. Plant Ecol 138:113–125. https://doi.org/10.1023/A:1009764909413

Stromberg JC (2001) Restoration of riparian vegetation in the south-western United States: importance of flow regimes and fluvial dynamism. J Arid Env 49:17–34

Swanson S, Wyman S, Evans C (2015) Practical grazing management to maintain or restore riparian functions and values. J Range Appl 2:1–28. ISSN: 2331-5512 http://journals.lib.uidaho.edu/index.php/jra/article/view/16

Swanson S, Kozlowski D, Hall R et al (2017) Riparian proper functioning condition (PFC) assessment to improve watershed management for water quality. J Soil Water Cons 72(2):190–204. https://doi.org/10.2489/jswc.72.2.168

Swanson S, Schultz B, Novak-Echenique P et al (2018) Nevada rangeland monitoring handbook, 3rd edn. University of Nevada Cooperative Extension Special Publication SP-18-03, p 122. https://extension.unr.edu/publication.aspx?PubID=2944

The Nature Conservancy (TNC) (2017) Gunnison Basin wet meadow and riparian restoration and resilience-building project. Executive Summary. https://www.conservationgateway.org/ConservationByGeography/NorthAmerica/UnitedStates/Colorado/Documents/2017.06.08_ExecSummary_GunnisonWetMeadows-sm%20Final.pdf. Accessed 13 Dec 2020

Thomas JW, Maser C, Rodiek JE (1979) Wildlife habitats in managed rangelands-the Great Basin of southeastern Oregon. Riparian Zones. USDA Forest Services Genral Technical Report PNW-80

Tockner K, Stanford JA (2002) Riverine flood plains: present state and future trends. Env Cons 29:308–330. https://doi.org/10.1017/S037689290200022X

Trubitt RT, Hovick TJ, Gillam EH et al (2018) Habitat associations of bats in a working rangeland landscape. Ecol Evol 9:598–608. https://doi.org/10.1002/ece3.4782

University of Idaho Stubble Height Study Team (2004) University of Idaho stubble height study report. Moscow, ID, USA: University of Idaho Forest, Wildlife and Range Experiment Station, p 26. https://digital.lib.uidaho.edu/digital/collection/rangecoll/id/41/

U.S. Department of Interior (USDI) (2010) Wild horse and burros management handbook. Bureau of land management handbook H-4700-1, pp 80. https://www.blm.gov/policy/handbooks

U.S. Army Corps of Engineers (USACE) (2018) National Wetland Plant List, version 3.4. http://wetland-plants.usace.army.mil/

Ward JV (1998) Riverine landscapes: biodiversity patterns, disturbance regimes and aquatic conservation. Biol Cons 83:269–278. https://doi.org/10.1016/S0006-3207(97)00083-9

Weber N, Bouwes N, Pollock MM et al (2017) Alteration of stream temperature by natural and artificial beaver dams. PLoS ONE 12(5):e0176313

Westbrook CJ, Ronnquist A, Bedard-Haughn A (2020) Hydrological functioning of a beaver dam sequence and regional dam persistence during an extreme rainstorm. Hyd Proc 34(18):3726–3737. https://doi.org/10.1002/hyp.13828

Wheaton J, Fryirs K, Brierley GJ, Bangen SG, Bouwes N, O'Brien G (2015) Geomorphic mapping and taxonomy of fluvial landforms. Geomorphology 248:273–295. https://doi.org/10.1016/j.geomorph.2015.07.010

Wheaton JM, Bennett SN, Bouwes N et al (eds) (2019a) Low-tech process-based restoration of riverscapes: design manual. Version 1.0. Utah State University Restoration Consortium. Logan, UT. https://doi.org/10.13140/RG.2.2.19590.63049/2

Wheaton JM, Wheaton A, Maestas J et al (2019b) Low-tech process-based restoration of riverscapes: pocket field guide. Utah State University Restoration Consortium. https://doi.org/10.13140/RG.2.2.28222.13123/1

Whited DC, Lorang MS, Harner MJ, Hauer FR, Kimball JS, Stanford JA (2007) Climate, hydrologic disturbance, and succession: drivers of floodplain pattern. Ecology 88:940–953

Wilcox BP, Le Maitre D, Jobbagy E et al (2017) Ecohydrology: processes and implications for rangelands. In: Briske D (eds) Rangeland systems. Springer Series on Environmental Management. Springer, Cham. https://doi.org/10.1007/978-3-319-46709-2_3

Williams JA, O'Farrell MJ, Riddle BR (2006) Community structure and habitat use by bats in a riparian corridor of the Mojave Desert of southern Nevada. J Mammal 87:1145–1153

Williams AP, Cook BI, Smerdon JE (2022) Rapid intensification of the emerging southwestern North American megadrought in 2020–2021. Nature Climate Change, pp 1–3

Winward AH (2000) Monitoring the vegetation resources in riparian areas. General Technical Report RMRSGTR-47. Ogden, UT: U.S. Department of Agriculture, Forest Service, Rocky Mountain Research Station, p 49. https://www.fs.usda.gov/treesearch/pubs/5452

Wohl E (2021) Legacy effects of loss of beavers in the continental United States. Env Res Lett 16(2):025010

Wyman S, Bailey D, Borman M et al (2006) Riparian area management: grazing management processes and strategies for riparian-wetland areas. Technical Reference 1737-20. BLM/ST/ST-06/002+1737. U.S. Department of the Interior, Bureau of Land Management, National Science and Technology Center, Denver, CO, p 105

Zeedyk B (2006) Harvesting water from low-standard rural roads. A Joint Publication of the Quivira Coalition, Zeedyk Ecological Consulting, LLC, The Rio Puerco Management Committee—Watershed Initiative, and the New Mexico Environment Department – Surface Water Quality Bureau

Zeedyk WD, Vrooman S (2017) The plug and pond treatment: restoring sheetflow to high elevation slope wetlands in New Mexico. New Mexico Environment Department, Surface Water Quality Bureau Wetlands Program

Zeedyk B, Clothier V (2009) Let the water do the work: induced meandering, an evolving method for restoring. Chelsea Green Publishing, White River Junction, VT

Chapter 8
Rangeland Biodiversity

Torre J. Hovick, Courtney J. Duchardt, and Cameron A. Duquette

Abstract In its simplest form, biodiversity is defined as species richness (the number of species in a given area). More complex definitions include the variety of life on Earth, from genes to ecosystems, and include the ecological and evolutionary processes that sustain that life. As in other ecosystems, biological communities in rangelands are influenced by a number of different abiotic and biotic drivers or "filters" at both broad and fine scales, and an understanding of these processes is critical for maintaining ecosystem services as well as addressing widespread biodiversity declines. In rangeland ecosystems specifically, the primary threats to biodiversity are habitat loss, fragmentation, and degradation through mismanagement, which includes suppression or mis-application of historical disturbance regimes. Restoring heterogeneity to rangelands by mimicking historical disturbance regimes has been shown to benefit biodiversity, but the exact role of disturbance varies widely throughout North American rangelands. As such, careful consideration of the type, duration/periodicity, intensity, and spatial and temporal extent and configuration of these disturbances is necessary when managing for site-specific biodiversity outcomes. It is important to consider the effects of both inherent (i.e., either natural or historical) and human-caused variability on rangeland plant and wildlife communities. In the future, practitioners should promote management practices that maintain and enhance biodiversity to maximize ecosystem functions and services that improve the quality and quantity of economic (e.g., livestock production, carbon banking) and ecological (e.g., biodiversity, sustainability) outcomes in North American rangelands.

T. J. Hovick (✉) · C. A. Duquette
School of Natural Resource Sciences, North Dakota State University, Fargo, ND 58108, USA
e-mail: Torre.hovick@ndsu.edu

C. A. Duquette
e-mail: Cameron.Duquette@ndsu.edu

C. J. Duchardt
Department of Natural Resource Ecology and Management, Oklahoma State University, Stillwater, OK 74078, USA
e-mail: Courtney.Duchardt@okstate.edu

Keywords Climate · Diversity · Fire · Grazing · North America · Rangeland · Soil · Threats

8.1 Overview

In the broadest sense, biodiversity refers to the variety of life at each ecological scale from the genome to the biome (Gaggiotti et al. 2018). We use the term biodiversity to refer to the genetics of a population or the species within an ecological community; in both cases, greater biodiversity would indicate greater variation among the units of interest (genes or species, respectively). While biodiversity includes all taxa (e.g., plants, fungi, animals), in this chapter we focus primarily on animals, giving the bulk of our attention to vertebrate diversity but also discussing arthropod diversity and plant diversity as it relates to supporting animal diversity (see Chap. 26 for further treatment of rangeland insects). We also focus largely on species diversity; although genetic diversity is becoming increasingly important to consider, we do not expand on that here and we refer readers to other discussions of this topic (Allendorf and Luikart 2007; Costa and Delotelle 2008).

Animal biodiversity influences many different aspects of ecosystems. For example, the diversity of animal species within biological communities can affect ecological stability (Ives and Carpenter 2007), and communities can shift with the removal of one or a few species (Paine 1966). Genetic diversity is also important, and tracking this diversity is critical in endangered or reintroduced populations to ensure successful recruitment and to avoid inbreeding (Allendorf and Luikart 2007). From a management perspective, *biodiversity hotspots* provide high return on investment opportunities for conservation and management because focused efforts in these areas can protect a large suite of species (Marchese 2015).

Across spatial and temporal scales, biodiversity conservation has become a major focus because species diversity is declining worldwide (WWF 2018; Fig. 8.1). As of 2021, current estimated extinction rates are 10–100 times higher than the average rate over the past 10 million years (IPBES 2019). It seems all but certain that these declines will become more ubiquitous under likely scenarios of global climate change and human population growth (WWF 2018). Understanding the drivers and consequences of these declines, and ameliorating them, will require an understanding of the natural processes that support biodiversity in rangelands, the success and failures of historical and current management strategies, and the best ways to evaluate conservation and measure success.

Rangeland biodiversity is largely determined by climate and *disturbance regimes*—the spatial and temporal characteristics of events that shape a system over time such as fire, grazing, and extreme weather events—and earlier chapters in this book highlight the central role of climatic variation and disturbance in shaping North American rangeland ecosystems (e.g., see Chaps. 2 and 6). Rangeland wildlife in North America co-evolved within the context of dynamic climate and disturbance regimes, leading to species adaptations that facilitate coping with

Fig. 8.1 Rangeland biodiversity is a key conservation target worldwide. Clockwise from top left, examples of taxa native to North American rangelands that are of conservation concern include American bison (*Bison bison*), western meadowlark (*Sturnella neglecta*), regal fritillary (*Speyeria idalia*), Texas horned lizard (*Phrynosoma cornutum*), swift fox (*Vulpes velox*), and Great Plains toad (*Anaxyrus cognatus*). All taxa examples have undergone population declines and most are considered species of greatest conservation need in states where they occur

extreme changes in resource availability and vegetation structure over time (Knopf and Samson 1997). For example, as an adaptation to uncertain resource availability, many wildlife populations exhibit annual migrations across states and countries [e.g., mule deer (*Odocoileus hemionus*), see Chap. 17; pronghorn (*Antilocapra americana*), see Chap. 19], seasonally or in altitude [e.g., greater sage-grouse (*Centrocercus urophasianus*), see Chap. 10], or even across continents (e.g., rangeland songbirds, see Chap. 12; waterfowl, see Chap. 13). Other species evolved to be more nomadic, continuously following resources as they became available across the plains [e.g., American bison (*Bison bison*); see Chap. 23]. As such, rangeland wildlife species often exhibit lower levels of ***site fidelity*** relative to taxa in forested ecosystems in order to take advantage of an ever-shifting landscape (Jones et al. 2007; Jonzén et al. 2011). While climatic variability plays a central role in defining North American rangelands, other forms of disturbance also have important roles in maintaining the complexity of rangelands (Knopf and Samson 1997; Fig. 8.2). For example, fire, roaming bison herds, and burrowing mammals historically served to alter plant communities and landscape structure in multiple ways (e.g., increased bare ground, reduced vegetation structure, altered forage quality and soil nutrient content, greater structural heterogeneity) throughout North American rangelands. But, some of these same disturbances (fire, bison herbivory) were much less influential farther west within the sagebrush steppe because reduced herbaceous biomass in these systems (driven by different timing and frequency of precipitation) did not historically facilitate frequent fire or dense grazer populations (Innes and Zouhar 2018).

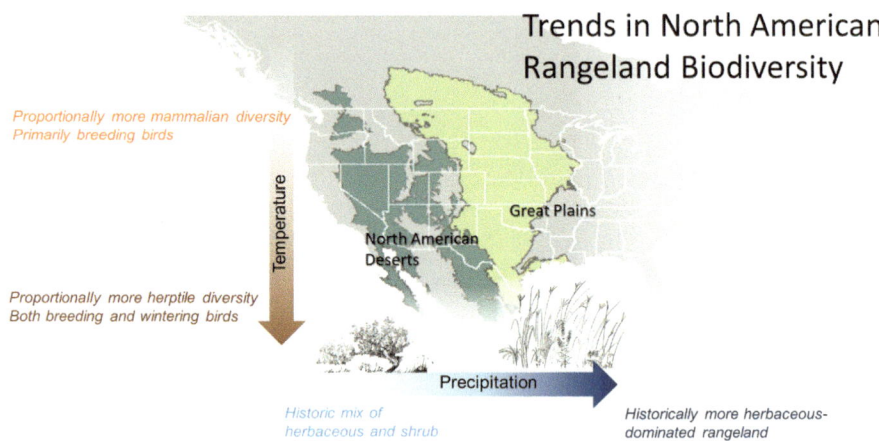

Fig. 8.2 North American rangeland biodiversity is driven in large part by gradients in increasing moisture from west to east and increasing temperature from north to south as well as the seasonal distribution of these weather gradients. As a consequence, rangelands vary greatly in predominant herbaceous taxa and vegetation structure, which in turn influences community structure of flora and fauna and overall biodiversity

Although many North American wildlife species co-evolved in the context of frequent disturbance, the unpredictability of resources and heterogeneity inherent in these disturbance regimes has long been considered incompatible with livestock production goals (Fuhlendorf et al. 2017). As a result, the suppression of natural disturbances, especially fire, has occurred in disturbance-dependent rangelands (particularly, grasslands) of North America from European settlement to the recent past. Furthermore, suppression of disturbances for the perceived benefit of livestock has been linked to detrimental changes in rangeland wildlife communities, threatening biodiversity (Fuhlendorf et al. 2012; see Chap. 6). Conversely, climate change and invasive plants have increased disturbance frequency and intensity in western rangelands (particularly, shrubsteppe ecoregions), killing fire-intolerant big sagebrush (*Artemisia tridentata*) and reducing habitat availability for sagebrush *obligates* like sage-grouse (*Centrocercus spp.*, Chap. 10; Bagne et al. 2012; DiTomaso et al. 2017).

If we view biodiversity simply as species richness (i.e., the number of species present), rangelands often have lower richness compared to other biomes with most biodiversity hotspots found in parts of the tropics, and especially tropical forests. The broad-scale drivers of these relationships are discussed elsewhere (MacArthur 1958; Brown 1995), but include the relationship of increasing niche space with greater vertical structure provided within forested ecosystems (MacArthur and MacArthur 1961) as well as links between greater solar energy, increased vegetation productivity and resource availability (Clarke and Gaston 2006), all of which facilitate greater species richness. However, species richness is just one dimension of biodiversity. This metric does not capture the *evenness* (i.e., all species having similar abundance versus a few highly abundant species and many rare) of species within the community, the conservation status of those species, the role they play in the community [e.g., *keystone species* (Paine 1969), *ecosystem engineers* (e.g., Chap. 15), whether they are components of a *mutualism*], or how genetically and/or functionally unique they are among regional and global taxa. Understanding rangeland biodiversity in North America requires considering all these aspects while keeping in mind the other services we utilize and expect from our rangeland systems, including livestock production.

Within this chapter, we take a broad view of rangeland biodiversity in North America, examining the mechanisms that shape and limit biodiversity (Sect. 8.2), how to measure and manage for biodiversity in rangelands (Sects. 8.3 and 8.4), and forces that threaten rangeland biodiversity (Sect. 8.5).

8.2 Processes that Influence Rangeland Biodiversity

One way to conceptualize the formation of ecological communities, and the diversity therein, is through the lens of abiotic and biotic filters (Götzenberger et al. 2012; Kraft et al. 2015). Put simply, this recognizes that the presence of a species in a given location is a function both of landscape and regional-scale drivers of species range [e.g., temperature gradients or drought (Choat et al. 2012; DeBello et al. 2013; Keddy and Laughlin 2022)], but also finer-scale drivers including species interactions like intraspecific or interspecific competition (Connell 1983; Chen et al. 2010), herbivory (Moolman and Cowling 1994), or predation (e.g., keystone species, Paine 1966). In rangelands, climatic gradients are major abiotic filters of species ranges. For example, we see much higher abundance of reptiles in southern rangelands as compared to northern (or high elevation) rangelands because of thermal limits of ectotherms (Fig. 8.2). The same north–south temperature gradient drives a transition in dominant photosynthetic pathway of grasses from C3 to C4 in hotter southern grasslands (Teeri and Stow 1976). Precipitation gradients east to west across North American rangelands also drive major shifts in vegetation, which in turn influence wildlife communities. For example, tall grasses transition to short grasses as average precipitation decreases and precipitation variability increases moving east to west (Anderson 2006). Moving further west, as precipitation regimes shift from spring/ summer-dominated to winter-dominated, we observe a shift towards shrublands; sagebrush (*Artemisia* spp.) becomes more common where snowmelt is a dominant source of soil moisture (Schlaepfer et al. 2012), with desert shrublands occurring in the warmer parts of the southwest that experience extreme drought in late spring and summer (Gao and Reynolds 2003). Other abiotic filters include nutrient and hydrological cycling processes. Some important filters that influence species occurrence, like soils, are more difficult to categorize as biotic or abiotic because they are a combination of the two, but soil type plays a major role in vegetation communities, and thus can help in determining wildlife occupancy and community composition

(Evans et al. 2017). The natural and anthropogenic disturbances that shape range-lands are another major filter (Fuhlendorf et al. 2017), as some species benefit from disturbances while others are less disturbance tolerant. Finally, species interactions like predation, competition, and mutualisms all serve as fine-scale biotic filters that determine where certain species can occur. As an example in rangelands, researchers have noted that coyotes (*Canis latrans*) and swift fox (*Vulpes velox*; Fig. 8.1) rarely co-occur because the larger coyote behaviorally excludes, and sometimes even depre-dates, swift fox (Kitchen et al. 1999). Below we discuss some of the major drivers of biodiversity in rangelands, and how they impact rangeland wildlife.

8.2.1 *Climate*

Climate, or long-term weather patterns, can be characterized as average temperature and precipitation over time. However, in complex landscapes like the western United States, this simplistic description may be inadequate to describe the many factors influencing the climate. More appropriately, climate also includes factors like timing of precipitation, amount of sunshine, average wind speed and direction, number of days above freezing, weather extremes, and ocean currents. Climate is a major factor in determining biomes and critical for shaping overall species diversity (Begon et al. 2006) and macroecological theory suggests that patterns of diversity (i.e., species richness) are limited by ambient energy at high latitudes and moisture at low latitudes (Hawkins et al. 2003). Across North American rangelands, we see climate driving taxonomic composition of wildlife communities, with proportionally more mammals and migratory breeding birds in more northern rangelands, and more herpetofauna and resident or wintering birds in southern rangelands (Valentine-Darby 2010; Fig. 8.2).

Over broad spatial scales, regional climate is a determinant of biodiversity and plant-biomass productivity, and it is foundational in determining the ***fundamental niche*** of animal species (Hutchinson 1957). The fundamental niche is determined by the potential tolerances and requirements of individuals. How those interact with the conditions, resources, and individuals around them to shape actual occurrence determines an organism's ***realized niche*** (Hutchinson 1957). Niches have multiple dimensions that represent species tolerances of various biotic and abiotic factors, and the overall availability of niches or niche space plays a major role in determining the biodiversity in a system.

Temperature is one of the most important components of climate that influences biodiversity. In particular, the importance of extreme high-temperature events in influencing species distribution and fitness has long been acknowledged (Begon et al. 2006). Therefore, temperature regulation or amelioration of thermal extremes can be an important landscape function (Hovick et al. 2014a; Melin et al. 2014). For example, variation in vegetation composition can alter the variability in thermal environments (e.g., by providing shade) thereby allowing animals greater oppor-tunity for selecting suitable thermal conditions (Carroll et al. 2015; Londe et al.

2020). This is increasingly important to understand in the face of climate change as temperature extremes become more frequent and organisms increasingly experience warmer temperatures (IPCC 2013a, b). Despite knowledge of how climate determines a species' distribution, survival, and reproduction, there are still relatively few studies focused on how temperature affects wildlife habitat selection and survival (Elmore et al. 2017). However, this field of research has been growing in recent years with more studies examining the influence of management on vegetation, and in turn, the influence on thermal environments and how that affects habitat selection of wildlife (Hovick et al. 2014a; Carroll et al. 2015; Raynor et al 2018).

Precipitation is another major component determining a region's climate with the interannual variability, seasonal distribution, and annual total all impacting animal populations (Pearce-Higgins et al. 2015). In general, temperate species of the US have evolved in highly variable environments and are therefore highly adaptable to variation in precipitation and temperature (Bonebrake and Mastrandrea 2010). However, it is predicted that precipitation regimes are going to change in many regions of the world (IPCC 2013a, b), and overall there is limited understanding of how such alterations will affect biodiversity. Both extended drought and large rain events are expected to become more common throughout much of the US (IPCC 2013a, b), and these events will have varying impacts on biodiversity but are most likely to have a negative influence on shorter temporal scales (Albright et al. 2009).

8.2.2 Soils

Soils can be viewed as one of the underlying templates upon which rangeland biodiversity is structured. Soil properties (e.g., particle size, pH) serve as a filter for vegetation composition and structure, both of which have direct impacts on the species that can occupy an area. Soil type is also important for many species including burrowing organisms like prairie dogs (*Cynomys* spp.) (Reading and Matchett 1997; Chap. 15), and these burrowing organisms may in turn alter soil structure via soil mixing, or "bioturbation" (Barth et al. 2014). Not only are soils the foundation for above-ground biodiversity, they are alive with a wealth of below-ground biodiversity. Decaëns et al. (2006) predict that 25% of all species live in soil for some part of their lifecycle, including protists, nematodes, earthworms, and arthropods (Yarwood et al. 2020; Chap. 26). Some of these species, including arbuscular mycorrhiza (i.e., fungal associations in plant roots that form plant-fungal mutualisms), are critical for fixing nitrogen in rangeland plants, and are especially important to consider when trying to restore rangeland flora (Miller et al. 2012; Duell et al. 2022). In many ways, our rangeland soils remain a vast frontier on a microscopic scale. For example, we know very little about the capacity of rangeland soils to store carbon belowground (Fynn et al. 2010), but the USDA-Agricultural Research Service has estimated that rangelands in the United States have the capacity to store 19 million metric tonnes of carbon per year (Schuman and Derner 2004). Carbon storage is a major focus in efforts to reduce greenhouse gasses to mitigate climate change, and thus, carbon storage potential in rangelands, which is driven both by abiotic and biotic factors in soil (Hungate et al. 2017), is important as we consider economic valuations in these systems and the bottom up influence of soils on rangeland biodiversity (Ritten et al. 2012).

8.2.3 Herbivory

Grazing is a dynamic process that interacts with other disturbances as well as topoedaphic (i.e., soils and topography) and vegetation features across landscapes to form patterns that impact ecosystem functions and biodiversity (Collins et al. 1998; Tews et al. 2004). Historically, rangelands of North America were shaped by fire, herbivores, and their predators for nearly 10,000 years (Knapp et al. 1999; Anderson 2006). American bison, in particular, are considered keystone species that were critical in shaping the flora and fauna of North America's Great Plains (Knapp et al. 1999). Estimates on the number of bison inhabiting the Great Plains before the 1800s range from 30 to 60 million that roamed in herds large enough to span from horizon to horizon (Flores 1991). These nomadic herds followed fires created by lightning and Native Americans, feeding primarily on grasses and often leaving forbs ungrazed (Fahnestock and Knapp 1993; Damhoureyeh and Hartnett 1997). In addition to grazing patterns that can influence biodiversity, bison herds can alter nutrient cycling through fecal and urine inputs that can change plant species composition (Blair 1997). This is the result of the effects on nitrogen cycling which can be critical in grasslands because nitrogen availability often limits plant productivity in these landscapes (Seastedt et al. 1991; Turner et al. 1997). Another aspect of bison behavior that contributed to the diversity of grasslands is wallowing (discussed further below). Despite their abundance and influence on the landscape, bison numbers dwindled from tens of millions to just a few thousand near the end of the nineteenth century due to overexploitation by European settlers who were expanding westward (Flores 1991). Since their near extinction, the complex landscapes that contained roaming herds of bison have been replaced by highly parcelized and fragmented landscapes that resulted from early settlement and legislation such as the Homestead Act of 1862. This fragmentation has also had negative impacts on other extant native ungulate grazers or browsers that occur in North American rangelands including pronghorn (*Antilocapra americana*), elk (*Cervus canadensis*), mule deer (*Odocoileus hemionus*), and white-tailed (*O. virginianus*) deer (Chaps. 17–20). Non-native grazers on North American rangelands include feral equids (*Equus* spp.; Chap. 22) as well as domestic livestock (Chap. 4).

Globally, livestock grazing is the most widespread and pervasive anthropogenic land use on rangelands (Alkemade et al. 2013), occurring on approximately 60% of the world's agricultural lands (Alexandratos and Bruinsma 2012). Despite these large numbers, livestock consumption by humans has more than doubled over the last half century and is projected to increase by another 70% by 2050 (Thornton 2010; Alexandratos and Bruinsma 2012). Given the large amount of land that is used for livestock production, it is not surprising that livestock herbivory has profound impacts on rangeland biodiversity. Grazing by domestic livestock can affect an ecosystem in many different ways, including altering plant community composition and diversity (Augustine and McNaughton 1998; Allred et al. 2012). Because grazed rangelands provide habitat for many wildlife species, livestock management decisions in these areas can have profound impacts on wildlife and biodiversity (Fuhlendorf et al. 2012; Alkemade et al. 2013). Some have suggested that livestock in rangeland ecosystems act as ecosystem engineers due to their direct and indirect influences on vegetation structure and the availability of resources to other organisms (Jones et al. 1997; Derner et al. 2009). Previous research investigating the influence of livestock grazing on wildlife has suggested a negative influence on some species (e.g., Tetraonidae spp.; Dettenmaier et al. 2017), whereas other studies have shown how restoring disturbance patterns, particularly grazing and burning, can have a positive influence on biodiversity (Fuhlendorf et al. 2006; Hovick et al. 2015; Duchardt et al. 2016; He et al. 2019). The influence of livestock grazing on wildlife is largely dependent upon the spatial and temporal distribution of the grazer and may also be influenced by livestock type, timing and frequency of grazing, grazing duration, livestock distribution across the landscape, seasonality, stocking rate, and the evolutionary history of grazing at a given site (Dettenemaier et al. 2017; see Chap. 4).

In addition to the influence of historical and contemporary grazing patterns by large herbivores, many native, smaller herbivores also play an important role in shaping rangeland ecosystems in North America. For example, prairie dogs and other rodents, rabbits and hares (*Leporidae*), and grasshoppers, locusts, and crickets (*Orthoptera*) have the ability to manipulate vegetation structure and composition in grasslands that influences biodiversity. These organisms are often thought of as pests in rangelands, but their importance as prey, ecosystem engineers, and nutrient cyclers should not be overlooked (Belovsky and Slade 2000; Augustine and Baker 2013; see Chaps. 15 and 26).

8.2.4 Fire

Fire as a disturbance process is critical in shaping world biomes and biodiversity patterns. Fire plays a large role in maintaining the structure and function of fire-prone ecosystems, which includes many rangelands (Bond and Keeley 2005). Moreover, fire influences global ecosystem patterns and processes, including vegetation distribution and structure, the carbon cycle, and climate (Bowman et al. 2009). The consequences of fire suppression can be significant for biological systems, and may result in a loss of biodiversity, alteration of ecosystem function, and changes in community structure and composition (Swetnam et al. 1999; Bond and Keeley 2005; Nowacki and Abrams 2008). In general, it is important to think of fire similarly to soils and climate in the sense that the biota in every region have evolved and are adapted to a particular regime (e.g., ranging from no fire to frequent fire) and alterations to those regimes can be detrimental. Fire regimes include important factors such as fire-return intervals, seasonality, intensity, and severity. Variation in these factors, known as *pyrodiversity*, can have a strong influence on biodiversity (Beale et al. 2018; Fig. 8.3). Management of rangelands focused on maintaining or enhancing biodiversity may have limited success without restoring historical fire patterns, including variable fire season and fire intensity and combining these with other disturbances such as grazing across broad landscapes (Fuhlendorf et al. 2012).

Envisioning fire as an ecological process is important for its application in conservation and land managers should try to maintain historical fire regimes in native ecosystems that are fire-adapted. For example, this means targeting three to five-year fire-return-intervals in tallgrass prairies throughout the central United States (Allen and Palmer 2011; Ratajczak et al. 2016), which are made possible because of sufficient precipitation and the amount of biomass that creates adequate fuel loads to sustain fires at this interval. In these systems, cessation of fire for as little as ten years can lead to state shifts from tallgrass prairie to eastern red cedar (*Juniperus virginiana*)-dominated woodland (Briggs et al. 2002; Ratajczak et al. 2016). On the other end of the spectrum, too-frequent burning, as applied in much of the Flint Hills of the Great Plains in the form of annual burning, can lead to reduced litter accumulation and loss of native forb species as well as favoring a less diverse grassland breeding bird community (Hovick et al. 2015; McGranahan et al. 2018). Moving west from tallgrass prairie systems, precipitation declines and fire-return-intervals generally increase to 5–20 years in mixed and shortgrass prairie due to a reduction in annual biomass production that can act as fuels for fires (Zouhar 2021).

Fig. 8.3 Fire is a major driver of rangeland biodiversity. Both wildfires and prescribed fires (top left) influence forage quality as well as vegetation structure. Recent burns provide foraging habitat for a number of bird species including Swainson's hawk (*Buteo swainsoni,* top right), American golden plover (*Pluvialis dominica,* bottom left), and upland sandpiper (*Bartramia longicauda,* bottom right)

Moving even further west into the sagebrush steppe, fire return intervals were extremely long historically (decades to centuries), as evidenced by extremely slow recovery of current sagebrush systems post-fire (Baker 2006). In these landscapes, invasions of nonnative annual grasses [e.g., cheatgrass (*Bromus tectorum*) and medusahead (*Taeniatherum caput-medusae*)] have created a positive feedback loop that has led to higher fire frequency than historically observed, through accumulation of high *fine fuel* loads (Balch et al. 2017). Collectively, this combination of annual grass invasion and alteration to fire regimes along with habitat loss and degradation associated with other factors such as agricultural, industrial, and urban development have reduced the extent of the sagebrush ecosystem by nearly 50% (Schroeder et al. 2004; Davies et al. 2011). These changes have led to decreased native plants and wildlife populations and reduced diversity in sagebrush systems (Crawford et al. 2004; Shipley et al. 2006; Davies et al. 2018; Mahood and Balch 2019). Many sagebrush obligate species including sage-grouse, pygmy rabbit (*Brachylagus idahoensis*), Brewer's sparrow (*Spizella breweri*), sagebrush sparrow (*Artemisiospiza nevadensis*), and sage thrasher (*Oreoscoptes montanus*) have declined because of changes in these historical disturbance patterns (Mutter et al. 2015; Oh et al. 2019; Smith et al. 2019). Sagebrush restoration is needed because these systems provide numerous *ecosystem services* and *functions* (Prevéy et al. 2010), including favorable microclimates for seed germination and establishment and habitat for wildlife of conservation concern. However, the success of sagebrush restoration is closely tied to the reinstatement of fire regimes that mimic historical intensities and return intervals to maximize the future conservation of biodiversity. Although we have highlighted threats of annual grass invasion and altered fire regimes to Wyoming big sagebrush (*A. t. wyomingensis*) habitat, which typically occurs in lower and drier elevations, we note that higher-elevation sagebrush species such as mountain big sagebrush (*A. t. vaseyana*) suffer from invasion by upslope coniferous species such as Juniper (*Juniperus* spp.) and pinyon pine (*Pinus monophylla* and *edulis*) due to long-term fire suppression (Davies et al. 2011). This change in woody species composition reduces herbaceous vegetation (Davies et al. 2011), which subsequently lowers biodiversity of higher trophic levels. As such, this is yet another example where managers must try to reinstate prescribed fire that replicates historical regimes to reduce juniper expansion.

8.2.5 Other Disturbances

In addition to the disturbances of herbivory and fire, many herbivorous species have secondary impacts on habitat structure and rangeland biodiversity through behaviors like wallowing, burrowing, and vegetation clipping. As mentioned above, bison wallowing, which involves individuals rolling on the ground, creates depressions with compacted soils that often collect water and provide habitat for amphibians (Gerlanc and Kauffman 2003), and facilitate increased arthropod diversity at a landscape scale (Nickell et al. 2018). With the vast numbers of bison that once occupied the Great Plains, these soil depressions were probably abundant and widespread features of the landscape prior to European settlement (England and DeVos 1969).

Another suite of ecosystem engineers in North American rangelands are burrowing rodents that disturb soil and alter vegetation communities, providing unique habitats for a number of taxa [Fig. 8.4; see Chap. 15]. Beyond herbivory and burrowing, two species of prairie dog (the black-tailed (*C. ludovicianus*) and Mexican (*C. mexicanus*)] live at very high densities and are colonial, actively clipping vegetation on colonies to maintain visibility of predators (Hoogland 1995). This additional disturbance makes these colonies especially unique structurally, increasing avian diversity at a landscape scale where they occur (e.g., Duchardt et al. 2018), and leading to extreme community shifts in both birds and mammals when prairie dogs are removed from the landscape (Duchardt et al. 2023a, b).

Other potential disturbances in rangelands can include flooding or drought, which may occur over relatively short (e.g., a few days or weeks of flooding) to long (e.g., multiple decades of drought) intervals (Vose et al. 2015) and may be considered as a component either of climate or disturbance. Other events of discrete weather

Fig. 8.4 Biodiversity associated with ecosystem engineering by the presence of black-tailed prairie dogs (*Cynomys ludovicianus*) within a biome. Clockwise from top: prairie rattlesnake (*Crotalus viridis*) and black widow (*Latrodectus hesperus*) occupying a prairie dog burrow, mountain plover (*Charadrius montanus*) reliant on short-structure grasslands created by prairie dogs, black-tailed prairie dog on burrow entrance, and pronghorn (*Antilocapra americana*) foraging on prairie dog colony

like hailstorms and tornadoes (e.g., Carver et al. 2017), as well as disease, can also lead to individual mortality or habitat alteration, shifting species interactions and influencing biodiversity in the impacted system.

8.2.6 Interactions Among Drivers

Climate, soils, herbivory, and fire all interact to shape rangeland wildlife biodiversity. Collectively, these factors have been acting over millennia to shape ecosystems that are now largely influenced by anthropogenic forces. Human action or inaction has a major influence on where and when disturbances occur. For example, fire is a pattern-driving process on rangelands that interacts with other disturbances to contribute to vegetation heterogeneity (Bond and keeley 2005; Fuhlendorf et al. 2009). In particular, the interaction of fire and grazing, or *pyric-herbivory*, is a critical process in rangelands that can affect patterns of wildlife colonization and influence site selection for many species throughout their life history, ultimately shaping biodiversity in many rangeland systems (Fuhlendorf et al. 2012). It has been argued that interacting grazing and fire may best be viewed as a single disturbance process in ecosystems that evolved with it, and that the resulting heterogeneity from this interaction is the foundation of biodiversity in grassland systems (Fuhlendorf and Engle 2001; Fuhlendorf et al. 2009). Wildlife grazing, movement, and defecation also alter hydrologic and nutrient cycling, while nutrient and water availability inevitably shape wildlife habitat use. Some of these interactions, such as the fire-grazing interaction, have received attention in recent years, while others of these interactions are not yet understood, and require further study.

8.3 Methods for Evaluation and Monitoring Biodiversity

Whether we want to further evaluate the roles of abiotic and biotic filters in influencing diversity, or quantify the response of focal taxa to management, we must decide among many different methods of evaluating rangeland biodiversity. Biodiversity can be measured in a myriad of ways, depending on one's goals, and each method categorizes the value of communities differently. The simplest metric, species richness, is generated by counting the number of species within a site (Magurran 1988). This measure of diversity treats all species equally, regardless of their abundance, conservation needs, functional traits, relative abundance, or evenness, with respect to other species (Krebs 1999). Various diversity indices, such as the Simpson (1949) and Shannon (1948) indices, factor richness and evenness into a composite measure of site-level diversity (Buckland et al. 2005). In addition to types of diversity metrics, diversity can be viewed at multiple spatiotemporal scales. *Alpha diversity* quantifies species richness at a particular site and is likely the most familiar diversity metric for ecologists (Sepkoski 1988). **Beta diversity** represents the differences in community composition between sites (also called "species turnover", Sepkoski 1988), and **gamma diversity** summarizes diversity in a region and encompasses aspects of both alpha and beta diversity (Angeler and Drakare 2013).

Though the alpha diversity metric is commonly used to study biodiversity declines, doing so may neglect important components of biodiversity. For example, despite the low species richness in an advanced closed-canopy state, woody encroachment in rangelands often raises the site-level species richness of birds at low- to- moderate levels of tree cover due to the addition of generalists and non-grassland species (Sirami et al. 2009; Andersen and Steidl 2019). From the perspective of alpha diversity, increasing woody cover in grasslands may enhance local biodiversity. However, if woody encroachment displaces some grassland species and replaces them with species more tolerant of woody vegetation over large scales, grassland specialists may become rare or absent and communities may become homogenized, which would reduce beta and gamma diversity. This highlights the importance of accounting for the level of specialization that a given species has on rangeland habitat: obligate grassland species require grasslands for most or all of their life history, whereas *facultative* species may use grasslands but are more generalist in their habitat preferences (Vickery et al. 1999). The importance of obligate versus facultative species becomes apparent when considering beta and gamma diversity in this example: the displacement of grassland specialists reduces regional diversity even while enhancing site-level species richness (Andersen and Steidl 2019). As climatic and land-use changes place new and varied extinction pressures on rangeland biota, a focus on regional and landscape-level biodiversity metrics (beta or gamma diversity) becomes more important so that the homogenization of biotic communities across large scales through the replacement of specialist species by generalists can be avoided.

Collecting data to measure biodiversity of just one taxa (e.g., birds, insects, mammals) can be time consuming, with multi-taxa surveys requiring even more time and effort. When possible, multi-species and multi-taxa surveys are desirable, but in some cases surveying for one or a few species can provide a surprising amount of information. Information about **umbrella species** is often a good indicator of the presence of other associated taxa. Species like sage-grouse and northern bobwhite (*Colinus virginianus*) require relatively large contiguous tracts of habitat, and often the presence of these species is correlated with that of many other grassland birds (Crosby et al. 2015; Carlisle et al. 2018). As such, protecting habitat of these umbrella species may also benefit a number of other species (often termed "background species"). Despite the fact that protections for umbrella species may benefit other species using the same habitat, researchers and managers should be cognizant that very specific habitat requirements of other target species may be overlooked, as sometimes these umbrellas can have unexpected "holes". For example, umbrella reserves focusing on greater sage-grouse did not outperform randomly generated protected areas at providing habitat for 40 of 52 species considered (Carlisle et al. 2018). Broad area protections for sage-grouse failed to preserve habitat for species with specific requirements and species that are not associated with the larger habitat type (sagebrush) at fine spatial scales, such as the Wyoming pocket gopher (*Thonomys* *clusius*; Carlisle et al. 2018). Consideration not only of spatial overlap but also niche overlap (e.g., similar nesting substrate; Duchardt et al. 2023b) will help managers in determining appropriate management for umbrella and background species.

Many evaluations of biodiversity use taxonomic groupings as the units of assessment; however, a simple tabulation of species in an environment ignores unique species identities and traits, and may not accurately quantify the level of ecosystem services and functions provided. *Functional traits*, which are physical characteristics of an organism with links to ecosystem processes and services (de Bello et al. 2010), can be used by scientists to quantify the effects of management on biodiversity and the ability of a landscape to provide ecosystem services (Garnier and Navas 2012; Keddy and Laughlin 2022). For example, bee (*Anthophila* spp.) researchers have used functional traits such as diet specificity and tongue length, sociality, body size, and nesting preferences to investigate the effects of cheatgrass invasion and livestock grazing on bee diversity (Thapa-Magar et al. 2020). Functional dispersion (i.e., variability in functional traits in a suite of biota) may be used as a biodiversity index instead of species richness or evenness due to asymmetrical taxonomic representation in the bee community; functional traits thus capture biodiversity in a way that is more directly relatable to provisioning ecosystem services and mechanistic drivers of diversity loss (Thapa-Magar et al. 2020). Researchers in forested systems have recently shown that structural diversity better predicts primary productivity than species diversity, making it a useful tool for inventorying ecological services and functioning (LaRue et al. 2019). Whether this relationship holds in rangelands is an open and promising research question. Despite the push towards using functional traits to measure diversity, these methods should be seen as companions to species identities. Above all, it is important to consider the purpose of categorizing diversity, and choosing the best classification method for the job.

8.4 Managing Rangelands for Biodiversity

8.4.1 Brief History

Conservation and management of rangelands began in the late 1800s and early 1900s in response to overexploitation of these resources. Initially, conserving soil and plant communities (especially in areas affected by the Dust Bowl) was the primary goal of rangeland management with more of a "habitat" focus than conservation of biodiversity. The Society for Range Management was founded in 1948 with the guiding principle of proper distribution of grazing animals to prevent negative impacts of overstocking and to determine the proper *carrying capacity* (Holecheck et al. 2004). The focus on even animal distribution and moderate use was successful at minimizing soil loss and degradation, but it largely ignored wildlife in rangeland systems and some even viewed managing for wildlife as antithetical to livestock production (Stoddart and Smith 1943; Sampson 1952; Fuhlendorf and Brown 2016). It was not until the latter half of the twentieth century that grazing and conservation began to be viewed as compatible (Bakker and Londo 1998), but the intricacies of grazing management have still led to broad assessments that often label livestock grazing as bad for wildlife and biodiversity more generally (Fuhlendorf et al. 2012; Dettenmaier et al. 2017). Even as perspectives shift, much of grazing management still focuses on uniform grazing practices, which limits biodiversity and favors generalist species that can utilize areas that are moderately disturbed and have vegetation structure that reflects these practices (Fuhlendorf et al. 2006; Hovick et al. 2015; Duchardt et al. 2016).

Simultaneous to efforts being made to change grazing practices in the US, game management and consideration for an ecosystems approach to conservation was beginning to gain momentum (Leopold 1933). Eventually, this sportsman-guided movement led to the Pittman-Robertson Federal Aid in Wildlife Restoration Act, which has become one of the most instrumental pieces of legislation for the conservation of biodiversity. This act utilizes an 11% excise tax on all hunting weapons and ammunition. The resulting conservation funds are collected by the federal government and distributed to the states based on the number of hunting licenses sold, human population, and land area (Burger et al. 2006). While initially created with game species in mind, this act has also benefitted non-game species and biodiversity broadly across North American ecosystems, largely via the umbrella species concept as discussed above.

Most recently, rangeland management has been undergoing a paradigm shift, moving away from the early ideas of uniform distribution and moderate disturbance to a more nuanced approach focused on the conservation of disturbance processes. By restoring disturbance processes such as fire, grazing, and their interaction, complex patterns of vegetation structure are generated that can provide greater resources for the conservation of biodiversity (Fuhlendorf et al. 2012). However, many challenges still exist when attempting to unify profitable livestock production with biodiversity conservation (Samson et al. 2004). To describe these multi-objective management scenarios, Polasky et al. (2005) introduced the concept of '*working landscapes*'— rangelands simultaneously managed for livestock production and conservation— with the goal of achieving multiple stakeholder objectives on rangelands. This view of working landscapes has become central to biodiversity conservation on both public and private rangelands.

Many of the Great Plains states have ≥ 90% private ownership (NRCS 2021) with a focus on agricultural production, and while western rangeland occurs largely on public lands (managed by the U.S. Forest Service or Bureau of Land Management in the U.S., or by province-specific governments in Canada), even these are managed for "multiple uses" (including livestock grazing by permittees). In Mexico, though rangelands are largely composed of private parcels and communal lands, government efforts and collaborations with non-profits are also seeking to manage these rangelands to simultaneously benefit livestock and wildlife (PACP-Ch 2011; Villareal et al. 2019). While land ownership may vary, the paradigm that these landscapes serve the dual purposes of supporting livestock and wildlife is now shaping rangeland management across North America (NRCS 2021).

8.4.2 Shifting Paradigms

Modern approaches of landscape ecology and adaptive management suggest that embracing variability is important for promoting biodiversity and multi-functionality in rangeland working landscapes (Fuhlendorf and Brown 2016). However, mismanagement and overstocking of rangelands during the early twentieth century resulted in a focus on moderate and homogenous disturbance in rangelands that was largely detrimental to biodiversity conservation (Holechek et al 2004; Fuhlendorf et al. 2012). Such uniformity was achieved through techniques such as cross fencing to reduce pasture size, increasing livestock density, adding watering facilities to improve uniformity of use, and implementing supplemental feeding (Vallentine 1990; Bailey et al. 2008). Moreover, many rangeland managers adopted rotational grazing (see Chap. 4), with the goals of (1) improving plant species composition or productivity by allowing a rest period during the growing season, (2) reducing animal selectivity by increasing stock density, and (3) ensuring uniform animal distribution through water location and fencing (Savory 1978 but see Briske et al. 2008). Rotational systems have been regularly modified in attempts to attain livestock production and forage goals (Vallentine 1990) but all emphasize uniformity of livestock utilization with minimal thought given to biodiversity (Savory 1978; di Virgilio 2019). There are many factors (e.g,. livestock density, duration of grazing period, precipitation) that can influence the impacts that rotational grazing has on rangeland biodiversity, but in general, the consequences have been negative for wildlife and livestock productivity alike (Briske et al. 2008; di Virgilio et al. 2019). Although management that achieves uniform grazing distribution and moderate forage utilization can benefit soil from erosion, protect water quality, and provide habitat for some generalist wildlife species, rotational management may not meet the objectives of an ever-diversifying pool of stakeholders, such as providing habitat requirements for organisms that rely on vegetation characteristics that result from highly disturbed (e.g., heavily grazed or burned) or undisturbed rangelands (Fuhlendorf et al. 2012).

Scientific research on rangelands has followed similar trends as agricultural research, in which the simplification and reduction of complex systems into homogenous units for the benefit of simplified analyses and understanding has been a goal (Fuhlendorf and Brown 2016). The focus has been on controlling variability rather than embracing or promoting inherent and imposed heterogeneity in rangelands that can benefit biodiversity (Fuhlendorf et al. 2017). However, a growing body of research suggests that vegetation structural heterogeneity enhances biodiversity in working landscapes (Benton et al. 2003; Fuhlendorf et al. 2012; Hovick et al. 2014b, 2015), and these findings support the earlier theoretical underpinnings of the habitat heterogeneity hypothesis (MacArthur and MacArthur 1961). As further evidence, a meta-analysis of the relationship between animal species diversity and vegetation heterogeneity found that over 80% of all studies surveyed found a positive relationship between heterogeneity in vegetation and faunal diversity (Tews et al. 2004). Therefore, management focused on conserving natural disturbance processes such as grazing and fire can create patterns of complex vegetation structure and composition that promote biodiversity (Tews et al. 2004; Fuhlendorf et al. 2012).

Benefits for biodiversity from managing for heterogeneity can be exemplified by the responses of grassland birds (see Chap. 12). Grassland birds evolved with dynamic disturbances, which created spatially and temporally distinct patterns in vegetation structure, sometimes referred to as a shifting grassland mosaic (Fuhlendorf and Engle 2001; Askins et al. 2007). Because of this, grassland bird species have very specific preferences in terms of breeding habitat structures (Cody 1985a, b). Efforts to conserve grassland bird populations have begun to focus on the maintenance or restoration of these spatiotemporal patterns to create heterogeneous vegetation structure that is beneficial to the suite of grassland bird species (Walk and Warner 2000; Fuhlendorf et al. 2006). Spatial heterogeneity of vegetation structure at appropriate scales (i.e., patches at the territory scale) provides greater breadth of niches and increases the variety of grassland bird communities that can occur across the landscape compared to traditional approaches that create minimal structural diversity (i.e., homogeneity; Fuhlendorf et al. 2006; Hovick et al. 2015). Moreover, interacting fire and grazing that promotes vegetation heterogeneity may also be beneficial for over-wintering, non-breeding birds (Hovick et al. 2014b; Fig. 8.5), and migrating grasslands birds (Hovick et al 2017a, b). While the importance of disturbance regimes has received the greatest support in tallgrass prairie, evidence that mosaics of vegetation structure with differing disturbance histories and sources generates greater gamma diversity in birds has also been supported in northern mixed grass prairie (Duquette et al. 2023), southern sand-shinnery rangelands (Londe et al. 2021), shortgrass prairie (Skagen et al. 2018), and at the ecotone between the Great Plains and sagebrush steppe (Duchardt et al. 2018). Different sources of disturbance may be at play (native versus domestic herbivores, fire, burrowing rodents) and the proportion of disturbed and undisturbed landscapes may vary across North American rangelands, but the role of disturbance in creating a shifting structural mosaic of vegetation at appropriate scales seems nearly universal in supporting the conservation of biodiversity, especially rangeland birds. Collectively, this body of evidence is one of the most compelling cases for why the new paradigm of management in grasslands should focus on restoring disturbance processes to promote patterns of vegetation heterogeneity that can help conserve biodiversity (Fuhlendorf et al. 2012).

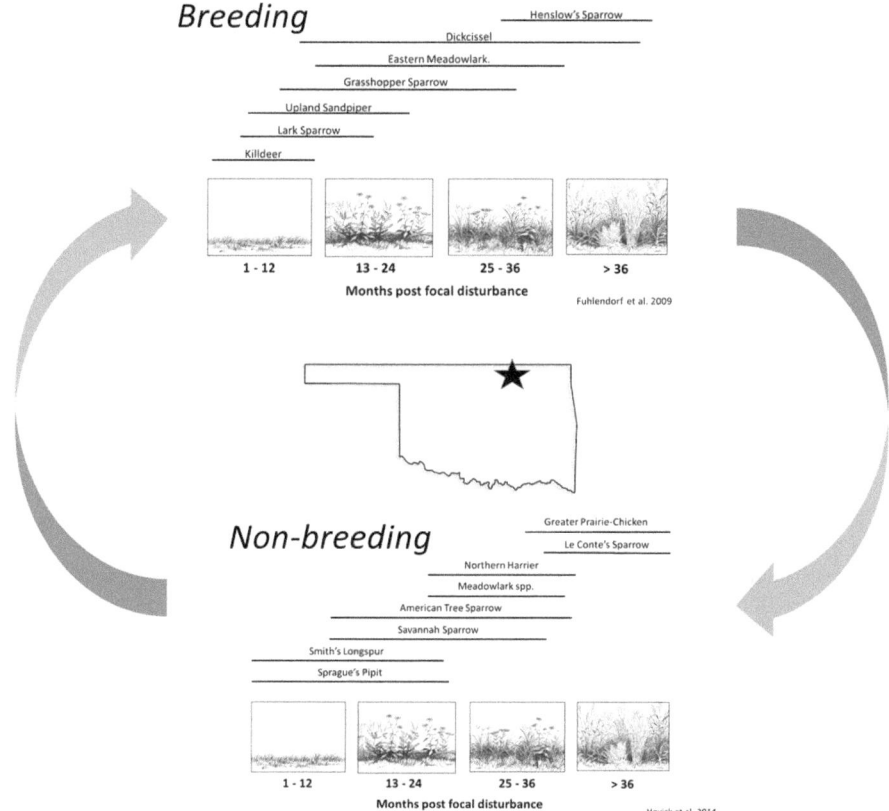

Fig. 8.5 The interaction between fire and grazing creates a structural gradient of vegetation over time post-disturbance that influences the diversity of the breeding and non-breeding grassland bird community. Figure adapted from Fuhlendorf et al. (2009) and Hovick et al. (2014a, b)

8.5 Threats

Rangeland biodiversity faces a number of threats from emerging sources (Fig. 8.6). Many of these constitute an interaction between anthropogenic and natural drivers, and result in simplified or fragmented landscapes. Climate change, habitat loss/ overexploitation, and invasive plants/woody encroachment are three of the main drivers of biodiversity declines in western rangelands, and each affect rangeland communities uniquely (Allred et al. 2015; Kreuter et al. 2016; Stephens et al. 2018).

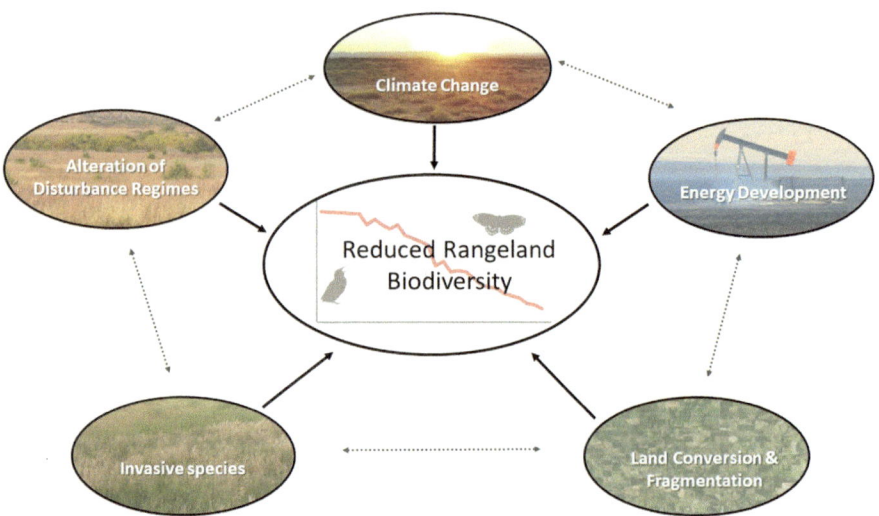

Fig. 8.6 Major threats to North American rangeland biodiversity include climate change, energy extraction and development, land fragmentation and conversion to row crop agriculture, invasive species including cheatgrass (*Bromus tectorum*) and eastern redcedar (*Juniperus virgiana*), and alteration of disturbance regimes such as fire suppression in the east or increasing fire severity and frequency in the west. Solid lines indicate direct effects on biodiversity, while dashed lines represent the potential for interactions among threats (e.g., fragmentation facilitates invasive species spread, while some invasive species alter disturbance regimes)

8.5.1 Climate Change

Climate change is defined as significant and persistent alterations to the mean or variability of climate regime components such as temperature, precipitation, and wind (Allen et al. 2019). Though climate change can refer to natural variation in trends, recent sharp deviations from long-term patterns are the result of anthropogenic greenhouse gas emissions (Allen et al. 2019). Increases in greenhouse gas concentrations in recent years have a broad range of impacts globally, including alterations to temperature and precipitation regimes, carbon sequestration rates, and photosynthetic capacity (Monzón et al. 2011; Staudinger et al. 2013). These sweeping changes have had broad impacts to species niches. If species cannot adapt, they must either shift their range along climatic gradients or risk extinction (Pecl et al. 2017; Roman-Palacios and Wiens 2020). Though often predictable at broad spatial and temporal scales, the effects of climate change are heterogeneous, even within discrete physiographic regions such as the North American Plains (Motha and Baier 2005).

Climate change has the potential to impact biodiversity on several levels. The widespread and variable shifting of species ranges has the potential to create novel species interactions (Gilman et al. 2010). Range shift theory states that biotic drivers (e.g., competition, predation, etc.) govern species ranges at the trailing edge of their range, whereas abiotic factors (e.g., precipitation, frost-free days) restrict expansion at the leading edge of a species range (Anderegg and HilleRisLambers 2019). However, more interactions between biotic and abiotic drivers complicate this relationship, and abiotic stressors can predispose species to competitive disadvantage and vice versa (Sirén and Morelli 2020).

Though specific predicted effects of climate change on rangeland biota are lacking, climate change will likely negatively affect sagebrush, a keystone species in western US rangelands. Models of sagebrush cover under climate change scenarios generally predict declines at southern latitudes, with neutral or positive effects of warming temperatures at mid- to high latitudes (Rigge et al. 2021; Zimmer et al. 2021). However, increasing temperatures in sagebrush systems interact with invasive species to increase the risk of wildfire and associated mortality of sagebrush plants (Bishop et al. 2020; Schlaepfer et al. 2021). As such, sagebrush cover may increase in some areas while decreasing in others, with unknown impacts on associated wildlife species. More broadly, climate change may lead to other vegetation compositional shifts, such as between forbs and graminioids (Teyssonneyre et al. 2002) or C3 and C4 grasses (Morgan et al. 2011), which will influence habitat structure for wildlife.

Above we discussed climate as an abiotic filter not only in terms of averages but also as a function of timing and variation in temperature and precipitation. While increases in the mean value of climate variables like temperature and rainfall will be impactful to biodiversity, changes to the variability and intensity of climate weather events will also be impactful. For example, sagebrush sparrow nest survival has been shown to suffer under extreme wet and extreme dry conditions, indicating that future climate regimes with more variability would be detrimental to this species, even if average conditions were to stay the same (Schroeder et al. 2022). Similarly, predicted increases in the frequency and intensity of extreme weather events such as hail, flooding, and tornadoes, are projected to cause more wildlife mortality events (Carver et al. 2017). Inherent in this increased variability of climate is decreased predictability of seasonal events by people and wildlife. When timing of events like last frost, bud burst, and first significant rainfall of the year become more difficult to anticipate or track, this can create trophic mismatches (Post et al. 2008). Demonstrated examples include temporal gaps in flower blooming and pollinator emergence, caterpillar availability and breeding bird behavior, and ungulate migration and forage greenup (Post and Forchhammer 2008; Hindle et al. 2015; Burgess et al. 2018). Due to adaptation to inherently variable conditions, rangeland wildlife may be buffered from phenological mismatch somewhat compared to taxa from other biomes, but the extent to which rangeland taxa exhibit fitness consequences of mismatches is largely an open inquiry. In any case, landscape diversity and heterogeneity has been demonstrated to buffer the negative effects of phenological mismatch by increasing the spatial variability in event timing, further underscoring the consideration of heterogeneity in rangeland wildlife management (Hindle et al. 2015; Ohler et al. 2020).

8.5.2 *Habitat Loss and Overexploitation*

Habitat loss, here defined as the removal of lands that previously provided food, water, and cover for wildlife species (National Wildlife Federation) is occurring rapidly in North American rangelands. For example, rates of grassland conversion to crops in the Northern Great Plains are currently analogous to deforestation rates in tropical rainforests in Southeast Asia, Oceania, and South America (Wright and Wimberly 2013). Oil and gas development in North America directly removed an additional 3 million hectares of habitat between 2000 and 2012 (Allred et al. 2015). Past periods of intense row crop conversion in the southern and eastern Great Plains are currently being mirrored in new areas such as the prairie pothole region due to crop improvements, biofuel demand, and increases in commodity prices (Johnson 2013; Hendrickson et al. 2019). In addition to current and previous direct habitat loss, remaining rangelands are fragmented and experiencing declines in biodiversity as a result (Wimberly et al. 2018).

Contemporary cropland conversion often occurs on marginal land by necessity, as highly productive land has largely already been plowed (Lark et al. 2020). However, rangelands with low agricultural potential often serve as habitat for a high diversity of species, resulting in an uneven cost–benefit ratio of conversion (Lark et al. 2020). For example, in a study looking at cropland conversion in the Midwest, the initial stem densities of milkweed (*Asclepias* spp.) were 3.4-times greater on rangeland that was later chosen to be converted to agriculture compared to the average of unconverted land in the area, while high wetland densities in converted lands allowed potential access by twice as many breeding waterfowl pairs as the Midwest natural lands average (Lark et al. 2020). Once converted, these areas produced below average crop yields, suggesting that marginal croplands were often not marginal wildlife habitat before conversion, and that the value of remaining unconverted rangelands as wildlife habitat often exceeds its value as potential cropland.

The effects of grassland conversion to agriculture on biodiversity have been documented in numerous taxa. Steep declines in global avian diversity have been linked to agricultural intensification, including increased use of pesticides and land conversion (Rosenberg et al. 2019). This has also been linked to mortality of insects that grassland birds depend on as a food source (Rosenberg et al. 2019). Conversion to cropland and intensification of agricultural practices have resulted in documented declines in bees, butterflies and moths (*Lepidoptera*), and grasshoppers (*Orthoptera*; Raven and Wagner 2021).

In contrast to agricultural conversion in the central Great Plains, western rangelands are most vulnerable to habitat loss due to energy infrastructure. Western rangelands are key areas for the extraction of oil and natural gas and the development of "green" energy sources such as wind, solar, and biofuels (Kreuter et al. 2016). The infrastructure supporting these energy sources, including roads, pump jacks, pipelines, compressor stations, turbines, solar panels, power lines, and tanker trucks, have the potential to fragment, disturb, and deteriorate rangeland wildlife habitat.

Many rangeland species, such as mule deer, sage grouse, and rangeland songbirds are area-sensitive and experience declines or local extirpations in energy extraction landscapes (Hess and Beck 2012; Northrup et al. 2015; Shaffer and Buhl 2016).

Dissimilar to agricultural production, energy development impacts can partially be mitigated through flexibility in siting criteria. By factoring biodiversity metrics into energy infrastructure placement, developers can optimize tradeoffs between energy production goals and biodiversity conservation (Thomas et al. 2018). Mitigation tools are available that estimate the amount of grasslands and wetlands needed to support breeding pairs of grassland birds and waterfowl displaced by wind, oil, gas, or transportation infrastructure (Shaffer et al. 2019). New technologies allow site planning operations to minimize the surface footprint and fragmentary effect of energy capture activities. For example, horizontal well drilling allows for multidirectional oil and gas exploration from a single well pad, allowing for a smaller surface footprint on the landscape (Thompson et al. 2015). In addition, conservation plans for imperiled rangeland species such as sage- grouse reduce or prohibit the construction of new oil and gas wells during sensitive life stages such as lekking (Patricelli et al. 2013).

Regardless of cause, habitat loss requires a diversity of approaches to mitigate negative impacts. One method to protect rangeland diversity and preserve habitat is by restricting undesirable use. This can be accomplished in various ways, including enrolling in private conservation easements or purchase by private conservation organizations (Cameron et al. 2014). It is important to note that rangeland biodiversity conservation often does not benefit from "full protection", as complete grazing cessation can have negative biodiversity outcomes (Toombs et al. 2010). That is, protection of rangelands should not necessarily be thought of as the removal of all disturbances. Rather, the preservation of rangeland biodiversity depends on informed, monitored, and responsible use. Whether at the programmatic level or via actions by private landowner, preventing rangeland habitat loss can be achieved through diverse stakeholder input and responsible working lands management.

8.5.3 Invasive Plants

By altering climate conditions, manipulating natural disturbance regimes, changing land use, and transporting propagules, humans have greatly increased the incidence of invasive plants in rangelands. In many cases, humans have also removed native vegetation from pastureland, replacing it with forage species that are perceived as higher in quality, and associated with greater weight gains in livestock (e.g. Svejcar and Vavra 1985). These homogenous "improved" pastures are typically poorer habitat for wildlife [e.g., Tall fescue (*Festuca arundinacea*), Washburn et al. 2000; Nelson et al.; crested wheatgrass (*Agropyron cristatum*), Rockwell et al. 2021], which is unsurprising given the importance of heterogeneity for rangeland biodiversity discussed above. In many cases these species do not readily spread without direct human assistance or outcompete native flora, and are thus not considered invasive, but some may become invasive in some locations.

Invasive plants are typically characterized as being both non-native and likely to cause environmental, economic, or medical harm (Barney et al. 2013). Invasive plants often proliferate in new systems due to high seeding rates, escape from competitors and pests that regulate them in native systems (Callaway and Aschehoug 2000; Mitchell and Power 2003; Gaskin et al. 2021). These factors often combine to allow invasive plants to achieve high densities in invaded areas, decreasing plant diversity as a result (Rout and Callaway 2009). Though hundreds of invasive species have been described in North American rangelands, here we focus on a few common and impactful examples.

Cheatgrass (*Bromus tectorum*) is a cool-season annual grass that was mostly accidentally introduced to western rangelands in the late 1800s, causing sweeping structural changes and becoming the dominant vegetation in large parts of its introduced range (Knapp 1996). Cheatgrass is uniquely impactful as an invasive species because it has a very different fuel structure from native perennial bunchgrass systems, increasing the size and frequency of fire beyond historic levels (D'antonio and Vitousek 1992). Though certain species seem to benefit from cheatgrass introduction, overall rodent abundance and diversity in cheatgrass-invaded systems decreases at high levels of invasion, likely due to reduced structural heterogeneity and loss of sagebrush cover (Freeman et al. 2014; Holbrook et al. 2016; Kleuver et al. 2019). In a similar fashion, grassland-associated bird species tended to decline in abundance as native perennial bunchgrasses were replaced by cheatgrass (Earnst and Holmes 2012). In that same study, shrubland-associated birds were less sensitive as long as adequate shrub cover remained (Earnst and Holmes 2012), but as discussed above, because cheatgrass facilitates reduced fire return intervals there is evidence that in the long-term cheatgrass also negatively impacts sagebrush birds (Knick et al. 2005). Similar declines have been reported in several sagebrush keystone species, including badger (*Taxidea taxus*, Holbrook et al. 2016) and greater sage grouse (Lockyer et al. 2015).

Many western rangelands are water limited, making riparian corridors both essential and sensitive landscape features. For this reason, the invasion of saltcedar (*Tamarisk* spp.) into western watercourses is viewed as particularly serious. Saltcedar is a salt-tolerant, deep-rooted deciduous shrub capable of forming dense stands along watercourses (DiTomaso et al. 2017). Replacement of native cottonwood/willow (*Populus/Salix*) riparian habitat reduces regional (gamma) avian diversity by displacing unique species (Brand et al. 2008), while effects on lizard and small mammal communities appear to be mixed (Bateman and Ostoja 2012). However, tamarisk is used extensively by the endangered southwestern willow flycatcher (*Empodinax traillii extimus*), possibly complicating control efforts (Owen et al. 2005).

Impacts of invasive species on rangelands including those above, as well as others (e.g., *Lespedeza cuneata, Festuca arundinacea, Pyrus calleryana, Taeniatherum caput-medusae*) could fill a whole book, but these examples highlight the multifaceted issue. Invasive status on its own does not equate to uniform deleterious impacts to wildlife; there will usually be some 'winners' following invasive plant introduction. However, it is important to consider the traits or the invasive plant and changes to native systems. When invasive plants displace natives, biodiversity often suffers as a result (Powell et al. 2013).

8.5.4 Woody Encroachment in Rangelands

The potential for non-native plant introductions to have negative effects on diversity is easily understood. However, the proliferation of native plants beyond historic levels can be equally deleterious. Though many grasslands have a native and historic shrub or tree component, the extent of woody plant cover in southern and western rangelands has increased due to a variety of factors (Bestelmeyer et al. 2018). Periods of overgrazing, climate change, alterations to fire regimes, and soil erosion have increased woody cover in many rangelands beyond their historical levels (Staver et al. 2011; Bestelmeyer et al. 2018; Archer et al. 2017). Encroachment of woody species into rangelands has a variety of effects from both a livestock-production and ecological standpoint.

Eastern red-cedar and honey mesquite (*Prosopis glandulosa*) are two prominent woody encroaching species with profound effects on the central Great Plants and southwestern rangelands, respectively. Eastern red-cedar spread is thought to be primarily the results of alterations to historic fire return intervals (Ratajczak et al, 2016), and in the absence of fire, grasslands can be converted to closed-canopy woodlands in as little as 40 years (Zhou et al. 2015). Increases in honey mesquite dominance are thought to be due to increased dispersal by cattle and freedom of seedlings from fire mortality (Buffington and Herbel 1965). Once established, woody encroachers have varied and significant impacts on rangeland biodiversity; aside from the direct displacement of grasses by less palatable trees or shrubs, the proliferation of woody cover in rangelands can alter the spatial patterning of nutrients, causing positive feedback mechanisms that promote a state change to low diversity woodland ecosystems, particularly in arid environments (Zhou et al. 2018). In many rangeland systems, the number of encroaching woody species is low, which can have the effect of reducing plant species richness and associated niche diversity (Archer et al. 2017). Though in theory woody encroachment should add to structural complexity and increase diversity, in practice woody encroachment has caused widespread declines in the diversity of herbaceous vegetation through competition for water, light, and nutrients (Van Auken 2009; Ratajczak et al. 2016). Predictably, woody encroachment is broadly detrimental to grassland bird species, while benefitting some shrubland species, especially at low levels of encroachment (Coppedge et al. 2004). Similarly, most sagebrush small mammal species responded negatively to encroachment from upslope woody plants (pinyon-juniper encroachment), with negative effects outweighing those of simultaneous cheatgrass invasion (Hamilton et al. 2019).

Although woody encroachment can have extremely deleterious effects on rangeland biodiversity, it is important that we avoid a black-and-white approach that conducts woody plant removal without regard for historical woody distribution. Indeed, in some landscapes we are seeing the dangers of such a strategy—pinyon juniper removal to improve greater sage-grouse habitat in the Great Basin has been linked to even steeper declines in the pinyon jay (*Gymnorhinus cyanocephalus*), an obligate species in pinyon-juniper systems currently (as of 2022) being petitioned for federal listing (Boone et al. 2018). In response to this emerging challenge, researchers are now rushing to identify target areas for conifer removal that do not negatively affect the pinyon jay.

In many of these systems, emerging evidence indicates that the reinstatement of historical abiotic and biotic filters may help slow woody spread or allow for a shift back to grasslands; as discussed above, Ratajczak et al. (2016) identified fire intervals that will likely prevent state-shift to woodlands, and other researchers have noted that prairie dogs reduce and sometimes eliminate mesquite from their colonies (Ponce-Guevara et al. 2016; Hale et al. 2020). Reinstatement of historical filters will also help to avoid removal of woody vegetation where it historically occurred, helping managers to avoid pitfalls such as those previously mentioned declines in pinyon jays.

8.6   Looking Ahead

Worldwide biodiversity is in decline and rangelands make up a large proportion of landcover (30–50%; Olson et al. 2001a, b; Briske 2012) across the globe (WWF 2018). As such, rangelands merit efforts towards biodiversity conservation, ensuring future provisioning of ecosystem services and maintenance of ecosystem functions. In rangelands, much of the variation in biodiversity is driven by climatic gradients and interacting disturbance processes. Managing for historic regimes that promote heterogeneity is paramount. North American rangelands are complex systems with large amounts of variation in species composition and richness, and this variation should be included in methods to evaluate diversity to ensure we can track changes in biodiversity trends. Despite the complexities and variation in North American rangelands, they all face a suite of threats including land conversion for energy development and agriculture, woody plant encroachment and plant invasions resulting from alterations in disturbance regimes, and climate change. To address these broad issues, research is shifting away from small, site-scale questions to larger gradient landscape-scale questions, but these findings need to be made available to managers in a way that is useful and actionable. Finally, if biodiversity conservation in rangelands is to be successful, land managers need to be adaptive and focus on the temporal and spatial scales of disturbance processes that resemble historic disturbance regimes to benefit native wildlife. These issues and a more in-depth discussion of the ecology and management of rangeland wildlife taxa are presented in the chapters that follow.

Glossary of Terms BOX 8.7

Abiotic filters[1,2] Non-living components of community assembly, including precipitation, temperature, topography, disturbances (e.g., fire or severe weather), and soil structure.

Biodiversity[3] The variety of life at each ecological scale from the genome to the biome.

Biodiversity hotspots[4] Areas of diversity with particularly high species richness. In terms of conservation, these areas should be evaluated and ranked based on some level of uniqueness and endemism of the species or land features present.

Biotic filters[5,6] Living components of an ecosystem that drives community assembly, including herbivory, bioturbation, and species interactions like competition or predation.

Carrying capacity[7] Average number of livestock and/or wildlife that may be sustained on a management unit compatible with management objectives for the unit.

Disturbance[8,9,10] A temporary change to an ecological system that alters the ecosystem. Examples of disturbances in rangelands include fire, herbivory, bioturbation, and severe weather events. Disturbances may be natural/historical (e.g., fire and some herbivory) or novel and human-caused (e.g., energy extraction).

Disturbance regime[11,12,13] The spatial and temporal characteristics of events that shape a system over time. For example, the characteristics such as frequency, intensity, and seasonality that describe fire or grazing over longer time intervals such as decades or millennia.

Driver[14] Any natural or human-induced factor that directly or indirectly causes a change in an ecosystem.

Ecosystem engineer[15] A species that mechanically changes an ecological system in a way that influences other species. Examples of ecosystem engineers in rangelands include prairie dogs and bison.

---

[1] Bedell (1998)

[2] Keddy and Laughlin (2022

[3] Morin (2011)

[4] Myers (1988)

[5] Bedell (1998)

[6] Keddy and Laughlin (2022

[7] Bedell (1998)

[8] Morin (2011)

[9] Briske (2012)

[10] Caro (2010)

[11] Morin (2011)

[12] Briske (2012)

[13] Caro (2010)

[14] Morin (2011)

[15] Morin (2011)

Ecosystem function[16] These are biological, geochemical, and physical processes that take place within an ecosystem. Examples may include soil retention, climate regulation, or nutrient cycling.

Ecosystem services[17,18] The components of nature that are directly consumed or enjoyed by humans and increase well-being. These are categorized into four groups: (1) provisioning (2) regulating (3) supporting and (4) cultural.

Evenness[19] This is generally reported as part of a diversity index (i.e., Shannon or Simpson, or separately as Pielou's evenness), and it quantifies the numerical value of each species represented in a system. Systems with higher evenness have similar representation of individuals, whereas systems with lower evenness have individuals with greatly varying abundances.

Facultative[20] Able to persist in multiple environmental conditions or habitats, but often with preference for one type.

Fine fuels[21] The herbaceous plants available for combustion in a system.

Functional trait[22,23] Component of an organism's phenotype that determines its effect on ecosystem processes and its response to environmental factors (de Bello et al. 2010). Examples of types of functional traits in rangeland wildlife include body size or sociality of pollinators, structural habitat preferences or diet requirements (e.g., granivores, insectivores, carnivores) of birds, or gut morphology of ungulates.

Fundamental niche[24] The full range of environmental conditions and resources an organism can possibly occupy and use in the absence of competition and geographic barriers.

Heterogeneity[25,26] Variability in a given trait at a specified scale. For rangeland wildlife, heterogeneity in structure of vegetation is a critical driver of species diversity.

Keystone species[27,28,29] A species with an outsized role in its community relative to biomass or population size. The removal of a keystone species typically results in major cascading changes in species abundance and community structure. Examples of keystone species in rangelands include bison, prairie dogs, top predators,

---

[16] Morin (2011)
[17] Morin (2011)
[18] Briske (2012)
[19] Morin (2011)
[20] Morin (2011)
[21] Briske (2012)
[22] Keddy and Laughlin (2022
[23] Morin (2011)
[24] Morin (2011)
[25] Morin (2011)
[26] Briske (2012)
[27] Morin (2011)
[28] Paine (1969)
[29] Caro (2010)

and some pollinators. Note this is different than the "Key species" described by Bedell (1998).

Levels of diversity (Alpha, Beta, and Gamma)[30] Alpha diversity, or site-level diversity, refers to any diversity metric calculated at the scale of a single site, whereas Gamma diversity refers to total diversity among all sites in a given study system and is often used to describe landscape-scale diversity. Beta diversity essentially refers to species turnover among sites and can be used as a metric of the community uniqueness of sites or strata within a landscape.

Mutualism[31] A type of species interaction wherein both species benefit from one another. Plant-pollinator relationships are an example of a mutualism.

Obligate[32] restricted to a specific environmental condition or habitat.

Pyrodiversity[33] The variation in the timing, spatial extent, and intensity of fire regimes.

Pyric herbivory[34] The interaction of fire and grazing where fire determines where grazing occurs and grazing patterns determine the extent and intensity of future fires.

Realized niche[35] The environment that a species occupies and lives in. This is the result of barriers and competition that constrain an organism into an area where they have access to resources to live and reproduce.

Scale[36] The dimensions used to study a phenomenon, often referring to temporal (days, weeks, years, centuries) or spatial (nest site, home range, landscape, region).

Site Fidelity (also called "philopatry" or "site tenacity")[37] The tendency for an animal to return to the same place, generally in reference to breeding locations. For example, the tendency for some migrating bird species to return to the same location each year for nesting activities.

Umbrella species[38] A species that typically has specific and large-scale habitat requirements, where management for that species has the potential to benefit a suite of other species utilizing the same habitat resources. Examples of umbrella species in rangeland include *Galliformes* such as sage-grouse, and prairie chickens and sharp-tailed grouse (*Tympanuchus* spp.).

Working landscapes[39] These are rangelands simultaneously managed for multiple stakeholder objectives, including livestock production and conservation.

Resources: 8. Turner and Gardner (2001)

---

[30] Morin (2011)

[31] Morin (2011)

[32] Morin (2011)

[33] Briske (2012)

[34] Briske (2012)

[35] Morin (2011)

[36] Morin (2011)

[37] Cody (1985a, b)

[38] Caro (2010)

[39] Polasky et al. (2005)

References

Albright TP, Pidgeon AM, Rittenhouse CD et al (2009) Effects of drought on avian community structure. Glob Chang Biol 16:2158–2170. https://doi.org/10.1111/j.1365-2486.2009.02120.x

Alexandratos N, Bruinsma J (2012) World agriculture towards 2030/2050: The 2012 revision. Rome: UN Food and Agricultural Organization (FAO). https://doi.org/10.22004/ag.econ.288998

Alkemade R, Reid RS, van den Berg M et al (2013) Assessing the impacts of livestock production on biodiversity in rangeland ecosystems. Proc Natl Acad of Sci 110:20900–20905. https://doi.org/10.1073/pnas.101101310

Allen MS, Palmer MW (2011) Fire history of a prairie/forest boundary: more than 250 years of frequent fire in a North American tallgrass prairie. J Veg Sci 22:436–444. https://doi.org/10.1111/j.1654-1103.2011.01278.x

Allen M, Antwi-Agyei P, Aragon-Durand F et al (2019) Technical summary: global warming of 1.5°C. An IPCC Special Report on the Impacts of Global Warming of 1.5°C Above Pre-Industrial Levels and Related Global Greenhouse Gas Emission Pathways, in the Context of Strengthening the Global Response to the Threat of Climate Change, Sustainable Development, and Efforts to Eradicate Poverty. https://www.ipcc.ch/site/assets/uploads/sites/2/2018/12/SR15_TS_High_Res.pdf

Allendorf FW, Luikart GH (2007) Conservation and the genetics of populations. Blackwell Publishing, Oxford, UK

Allred BW, Fuhlendorf SD, Smeins F et al (2012) Herbivore species and grazing intensity regulate community composition and an encroaching woody plant in semi-arid rangeland. Basic Appl Ecol 13:149–158. https://doi.org/10.1016/j.baae.2012.02.007

Allred BW, Smith WK, Twidwell D et al (2015) Ecosystem services lost to oil and gas in North America. Science 348:401–402. https://doi.org/10.1126/science.aaa4785

Anderegg LDL, HilleRisLambers J (2019) Local range boundaries vs. large-scale trade-offs: climatic and competitive constraints on tree growth. Ecol Lett 22:787–796. https://doi.org/10.1111/ele.13236

Anderson RC (2006) Evolution and origin of the Central Grassland of North America: climate, fire, and mammalian grazers. J Torrey Bot Soc 133:626–647. https://doi.org/10.3159/1095-5674(2006)133[626:EAOOTC]2.0.CO;2

Andersen EM, Steidl RJ (2019) Woody plant encroachment restructures bird communities in semiarid grasslands. Biol Conserv 240:108276. https://doi.org/10.1016/j.biocon.2019.108276

Angeler DG, Drakare S (2013) Tracing alpha, beta, and gamma diversity responses to environmental change in boreal lakes. Oecologia 172:1191–1202. https://doi.org/10.1007/s00442-012-2554-y

Archer SR, Andersen EM, Predick KI et al (2017) Woody plant encroachment: causes and consequences. In: Briske DD (ed) Rangeland systems: processes, management and challenges. Springer, New York, NY, USA, pp 25–84

Askins RA, Chávez-Ramírez F, Dale BC et al (2007) Conservation of grassland birds in North America: understanding ecological processes in different regions. Ornithol Monogr 64. American Ornithologists' Union, Washington, D.C., USA

Augustine DJ, Baker BW (2013) Associations of grassland bird communities with black-tailed prairie dogs in the North American Great Plains. Conserv Biol 27:324–334. https://doi.org/10.1111/cobi.12013

Augustine DJ, McNaughton SJ (1998) Ungulate effects on the functional species composition of plant communities: Herbivore selectivity and plant tolerance. J Wild Manage 62:1165–1183. https://doi.org/10.2307/3801981

Bagne K, Ford P, Reeves M (2012) Grasslands. U.S. Department of Agriculture, Forest Service, Climate Change Resource Center. www.fs.usda.gov/ccrc/topics/grasslands/

Baker WL (2006) Fire and restoration of sagebrush ecosystems. Wild Soc Bull 34:177–185. https://doi.org/10.2193/0091-7648(2006)34[177:FAROSE]2.0.CO;2

Bakker JP, Londo G (1998) Grazing for conservation management in historical perspective. In: WallisDeVries MF, Van Wieren SE, Bakker JP (eds) Grazing and conservation management. Conservation biology series, vol 11. Springer, Dordrecht

Bailey DW, Vanwagoner HC, Weinmeister R et al (2008) Evaluation of low-stress herding and supplement placement for managing cattle grazing in riparian and upland areas. Rangel Ecol Manag 61:26–37. https://doi.org/10.2111/06-130.1

Balch JK, Bradley BA, Abatzoglou JT et al (2017) Human-started wildfires expand the fire niche across the United States. Proc Natl Acad Sci 114:2946–2951. https://doi.org/10.1073/pnas.161 7394114

Barth CJ, Liebig MA, Hendrickson JR et al (2014) Soil change induced by prairie dogs across three ecological sites. Soil Sci Soc Am J 78:2054–2060. https://doi.org/10.2136/sssaj2014.06.0263

Barney JN, Tekiela DR, Dollette ESJ et al (2013) What is the "real" impact of invasive species? Front Ecol Environ 11:322–329. https://doi.org/10.1890/120120

Bateman HL, Ostoja SM (2012) Invasive woody plants affect the composition of native lizard and small mammal communities in riparian woodlands. Anim Conserv 15:294–304. https://doi.org/10.1111/j.1469-1795.2011.00517.x

Beale CM, Mustaphi CJ, Morrison TA et al (2018) Pyrodiversity interacts with rainfall to increase bird and mammal richness in African savannahs. Ecol Lett 21:557–567. https://doi.org/10.1111/ele.12921

Bedell TE (1998) Glossary of terms used in range management. Soc Range Manage

Begon M, Townsend CR, Harper JL (2006) Ecology: from individuals to ecosystems, 4th edn. Blackwell, Malden, Massachusetts, USA

Belovsky GE, Slade JB (2000) Insect herbivory accelerates nutrient cycling and increases plant production. Proc Natl Acad Sci 97:14412–14417. https://doi.org/10.1073/pnas.250483797

Benton TG, Vickery JA, Wilson JD (2003) Farmland biodiversity: is habitat heterogeneity the key? Trends Ecol Evol 18:182–188. https://doi.org/10.1016/S0169-5347(03)00011-9

Bestelmeyer BT, Peters DPC, Archer SR et al (2018) The grassland-shrubland regime shift in the southwestern United States: misconceptions and their implications for management. Biosci 68:678–690. https://doi.org/10.1093/biosci/biy065

Blair JM (1997) Fire, N availability and plant response in grasslands: A test of the transient maxima hypothesis. Ecol 78:2359–2368. https://doi.org/10.1890/0012-9658(1997)078[2359:FNAAPR]2.0.CO;2

Bishop TBB, Nusink BC, Molinari RL et al (2020) Earlier fall precipitation and low severity fire impacts on cheatgrass and sagebrush establishment. Ecosphere 11:e03019. https://doi.org/10.1002/ecs2.3019

Bond WJ, Keeley JE (2005) Fire as a global "herbivore": the ecology and evolution of flammable ecosystems. Trends Ecol Evol 20:387–394. https://doi.org/10.1016/j.tree.2005.04.025

Bonebrake TC, Mastrandrea MD (2010) Tolerance adaptation and precipitation changes complicate latitudinal patterns of climate change impacts. Proc Natl Acad Sci 107:12581–12586. https://doi.org/10.1073/pnas.0911841107

Boone JD, Ammon E, Johnson K (2018) Long-term declines in the Pinyon Jay and management implications for piñon–juniper woodlands. In: Shuford WD, Gill RE Jr., Handel CM (eds) Trends and traditions: Avifaunal change in western North America. Studies of Western Birds 3. Western Field Ornithologists, Camarillo, CA, pp 190–197. https://doi.org/10.21199/SWB3.10

Bowman DMJS, Balch JK, Artaxo P et al (2009) Fire in the earth system. Science 324:481–484. https://doi.org/10.1126/science.1163886

Brand LA, White GC, Noon BR (2008) Factors influencing species richness and community composition of breeding birds in a desert riparian corridor. The Condor 110:199–210. https://doi.org/10.1525/cond.2008.8421

Briggs JM, Hoch GA, Johnson LC (2002) Assessing the rate, mechanisms, and consequences of the conversion of tallgrass prairie to Juniperus virginiana forest. Ecosystems 5:578–586

Briske D (ed) (2012) Rangeland systems: processes, management, and challenges. Springer, Cham, Switzerland. https://doi.org/10.1007/978-3-319-46709-2_5

Briske DD, Derner JD, Brown JR et al (2008) Rotational grazing on rangelands: reconciliation of perception and experimental evidence. Rangel Ecol Manag 61:3–17. https://doi.org/10.2111/06-159R.1

Brown JH (1995) Macroecology. University of Chicago Press, Chicago, USA

Buckland S, Magurran A, Green R et al (2005) Monitoring change in biodiversity through composite indices. Philos Trans R Soc Lond B Biol Sci 360:243–254. https://doi.org/10.1098/rstb.2004.1589

Buffington LC, Herbel CH (1965) Vegetational changes on a semidesert grassland range from 1858 to 1963. Ecol Monogr 35:139–164

Burger LW Jr, McKenzie D, Thackston R et al (2006) The role of farm policy in large-scale conservation: Bobwhite and buffers. Wild Soc Bull 34:986–993. https://doi.org/10.2193/0091-7648(2006)34[986:TROFPI]2.0.CO;2

Burgess MD, Smith KW, Evans KL et al (2018) Tritrophic phenological match-mismatch in space and time. Nat Ecol Evol 2:970–975. https://doi.org/10.1038/s41559-018-0543-1

Callaway RM, Aschehoug ET (2000) Invasive plants versus their new and old neighbors: a mechanism for exotic invasion. Science 290:521–523. https://doi.org/10.1126/science.290.5491.521

Cameron DR, Marty J, Holland RF (2014) Whither the Rangeland?: protection and conversion in California's rangeland ecosystems. PLoS One 9. https://doi.org/10.1371/journal.pone.0103468

Carlisle JD, Keinath DA, Albeke SE et al (2018) Identifying holes in the greater sage-grouse conservation umbrella. J Wild Manage 82:948–957. https://doi.org/10.1002/jwmg.21460

Caro T (2010) Conservation by proxy: indicator, umbrella, keystone, flagship, and other surrogate species. Island Press. Washington D.C, USA

Carroll JM, Davis CA, Elmore RD et al (2015) Thermal patterns constrain diurnal behavior of a ground-dwelling bird. Ecosphere 6, art222. https://doi.org/10.1890/ES15-00163.1

Carver AR, Ross JD, Augustine DJ et al (2017) Weather radar data correlate to hail-induced mortality in grassland birds. Remote Sens Ecol Conserv 3:90–101. https://doi.org/10.1002/rse2.41

Chen L, Mi X, Comita LS et al (2010) Community-level consequences of density dependence and habitat association in a subtropical broad-leaved forest. Ecol Lett 13:695–704. https://doi.org/10.1111/j.1461-0248.2010.01468.x

Choat B, Jansen S, Brodribb TJ et al (2012) Global convergence in the vulnerability of forests to drought. Nature 491:752–755. https://doi.org/10.1038/nature11688

Clarke A, Gaston KJ (2006) Climate, energy and diversity. Proc R Soc B Biol Sci 273:2257–2266. https://doi.org/10.1098/rspb.2006.3545

Cody ML (1985a) Habitat selection in grassland and open country birds. In: Cody ML (ed) Habitat selection in birds. Academic Press, Orlando, Florida, USA, pp 191–226

Collins SL, Knapp AK, Briggs JM et al (1998) Modulation of diversity by grazing and mowing in native tallgrass prairie. Science 280:745–747. https://doi.org/10.1126/science.280.5364.745

Connell JH (1983) On the prevalence and relative importance of interspecific competition: evidence from field experiments. Am Nat 122:661–696

Coppedge BR, Engle DM, Masters RE et al (2004) Predicting juniper encroachment and CRP effects on avian community dynamics in southern mixed-grass prairie, USA. Biol Conserv 115:431–441. https://doi.org/10.1016/S0006-3207(03)00160-5

Costa R, Delotelle R (2008) Genetic diversity—understanding conservation at genetic levels. In: Van Dyke F (ed) Conservation biology. Springer, Netherlands, pp 153–184

Crawford JA, Olson RA, West NE et al (2004) Ecology and management of sage-grouse and sage-grouse habitat. J Range Manag 57:2–19. https://doi.org/10.2111/1551-5028(2004)057[0002:EAMOSA]2.0.CO;2

Crosby AD, Elmore RD, Leslie DM Jr et al (2015) Looking beyond rare species as umbrella species: Northern Bobwhites (*Colinus virginianus*) and conservation of grassland and shrubland birds. Biol Conserv 186:233–240. https://doi.org/10.1016/j.biocon.2015.03.018

Cody M (1985b) Habitat selection in birds. Academic Press, Cambridge, MS, USA

D'antonio CM, Vitousek PM (1992) Biological invasions by exotic grasses, the grass/fire cycle, and global change. Annu Rev Ecol Evol Syst 23:63–87. https://www.jstor.org/stable/2097282

Damhoureyeh SA, Hartnett DC (1997) Effects of bison and cattle on growth, reproduction, and abundances of five tallgrass prairie forbs. Am J Bot 84:1719–1728. https://doi.org/10.2307/2446471

Davies KW, Boyd CS, Beck JL et al (2011) Saving the sagebrush sea: an ecosystem conservation plan for big sagebrush plant communities. Biol Conserv 144:2573–2584. https://doi.org/10.1016/j.biocon.2011.07.016

Davies KW, Boyd CS, Madsen MD et al (2018) Evaluating a seed technology for sagebrush restoration efforts across an elevation gradient: support for bet hedging. Rangel Ecol Manag 71:19–24. https://doi.org/10.1016/j.rama.2017.07.006

de Bello F, Lavorel S, Díaz S et al (2010) Towards an assessment of multiple ecosystem processes and services via functional traits. Biodivers Conserv 19:2873–2893. https://doi.org/10.1007/s10531-010-9850-9

de Bello F, Lavorel S, Lavergne S et al (2013) Hierarchical effects of environmental filters on the functional structure of plant communities: a case study in the French Alps. Ecography 36:393–402. https://doi.org/10.1111/j.1600-0587.2012.07438.x

Decaëns T, Jiménez JJ, Gioia C et al (2006) The values of soil animals for conservation biology. Eur J Soil Biol 42:23-S38. https://doi.org/10.1016/j.ejsobi.2006.07.001

Derner JD, Lauenroth WK, Stapp P et al (2009) Livestock as ecosystem engineers for grassland bird habitat in the western Great Plains of North America. Rangel Ecol Manag 62:111–118. https://doi.org/10.2111/08-008.1

Dettenmaier SJ, Messmer TA, Hovick TJ et al (2017) Effects of livestock grazing on rangeland biodiversity: a meta-analysis of grouse populations. Ecol Evol 7:7620–7627. https://doi.org/10.1002/ece3.3287

di Virgilio A, Lambertucci SA, Morales JM (2019) Sustainable grazing management in rangelands: over a century searching for a silver bullet. Agric Ecosyst Environ 283:106561. https://doi.org/10.1016/j.agee.2019.05.020

DiTomaso JM, Monaco TA, James JJ et al (2017) Invasive plant species and novel rangeland systems. Briske DD (ed) Rangeland systems: processes, management and challenges. Springer, New York, NY, USA, pp 429–466

Duchardt CJ, Miller JR, Debinski DM et al (2016) Adapting the fire-grazing interaction to small pastures in a fragmented landscape for grassland bird conservation. Rangel Ecol Manag 69:300–309. https://doi.org/10.1016/j.rama.2016.03.005

Duchardt CJ, Porensky LM, Augustine DJ et al (2018) Disturbance shapes avian communities on a grassland-sagebrush ecotone. Ecosphere 9:e02483. https://doi.org/10.1002/ecs2.2483

Duchardt CJ, Augustine DJ, Beck JL et al (2023a) Disease and weather induce rapid shifts in a rangeland ecosystem mediated by a keystone species (Cynomys ludovicianus). Ecological Applications 33:1–19. https://doi.org/10.1002/eap.2712

Duchardt CJ, Monroe AP, Edmunds DR et al (2023b) Assessing the efficacy of the sagebrush umbrella at eastern range edge. Landscape Ecology. 38:1447–1462. https://doi.org/10.1007/s10980-022-01586-7

Duquette CA, Hovick TJ, Geaumont BA et al (2023) Embracing inherent and imposed sources of heterogeneity in rangeland bird management. Ecosphere 3:e304

Duell EB, O'Hare A, Wilson GWT (2022) Inoculation with native soil improves seedling survival and reduces non-native reinvasion in a grassland restoration. Restor Ecol 1–9. https://doi.org/10.1111/rec.13685

Earnst SL, Holmes AL (2012) Bird-habitat relationships in interior Columbia Basin shrub steppe. Condor 114:15–29. https://doi.org/10.1525/cond.2012.100176

Elmore RD, Carroll JM, Tanner EP et al (2017) Implications of the thermal environment for terrestrial wildlife management. Wildl Soc Bull 41:183–193. https://doi.org/10.1002/wsb.772

England RE, DeVos A (1969) Influence of animals on pristine conditions on the Canadian grasslands. J Range Manag 22:87–94

Evans RD, Gill RA, Eviner VT et al. (2017) Soil and belowground processes. Pages 131–168. In Briske DD (ed) Rangeland systems: processes, management, and challenges. Springer Series on Environmental Management. https://doi.org/10.1007/978-3-319-46709-2_5

Fahnestock JT, Knapp AK (1993) Water relations and growth of tallgrass prairie forbs in response to selective herbivory by bison. Int J Plant Soil Sci 154:432–440. https://doi.org/10.1086/297126

Flores D (1991) Bison ecology and bison diplomacy: the southern plains from 1800 to 1850. J Am Hist 78:465–485

Freeman ED, Sharp TR, Larsen RT et al (2014) Negative effects of an exotic grass invasion on small-mammal communities. PLoS ONE 9:e108843. https://doi.org/10.1371/journal.pone.0108843

Fuhlendorf SD, Engle DM (2001) Restoring heterogeneity on rangelands: ecosystem management based on evolutionary grazing patterns. Biosci 51:625–632. https://doi.org/10.1641/0006-3568(2001)051[0625:RHOREM]2.0.CO;2

Fuhlendorf SD, Brown JR (2016) Future directions for usable rangeland science: from plant communities to landscapes. Rangel 38:75–78. https://doi.org/10.1016/j.rala.2016.01.005

Fuhlendorf SD, Harrell WC, Engle DM et al (2006) Should heterogeneity be the basis for conservation? Grassland bird response to fire and grazing. Ecol Appl 16:1706–1716. https://doi.org/10.1890/1051-0761(2006)016[1706:SHBTBF]2.0.CO;2

Fuhlendorf SD, Engle DM, Kerby J et al (2009) Pyric herbivory: Rewilding landscapes through the recoupling of fire and grazing. Conserv Biol 23:588–598. https://doi.org/10.1111/j.1523-1739.2008.01139.x

Fuhlendorf SD, Engle DM, Elmore RD et al (2012) Conservation of pattern and process: developing an alternative paradigm of rangeland management. Rangel Ecol Manag 579–589. https://doi.org/10.2111/REM-D-11-00109.1

Fuhlendorf SD, Fynn RWS, McGranahan DA et al (2017) Heterogeneity as the Basis for Rangeland Management. In: Briske DD (ed) Rangeland systems: processes, management and challenges. Springer, New York, NY, USA, pp 169–176

Fynn AJ, Alvarez AP, Brown JR et al (2010) Soil carbon sequestration in United States rangelands. Grassland Carbon Sequestration: Manage Policy Econ 11:57–104

Gaggiotti OE, Chao A, Peres-Neto P et al (2018) Diversity from genes to ecosystems: a unifying framework to study variation across biological metrics and scales. Evol Appl 11:1176–1193. https://doi.org/10.1111/eva.12593

Gao Q, Reynolds JF (2003) Historical shrub-grass transitions in the northern Chihuahuan Desert: modeling the effects of shifting rainfall seasonality and event size over a landscape gradient. Glob Chang Biol 9:1475–1493. https://doi.org/10.1046/j.1365-2486.2003.00676.x

Garnier E, Navas ML (2012) A trait-based approach to comparative functional plant ecology: concepts, methods and applications for agroecology. A review. Agron Sustain Dev 32:365–399. https://doi.org/10.1007/s13593-011-0036-y

Gaskin JF, Espeland E, Johnson CD et al (2021) Managing invasive plants on Great Plains grasslands: a discussion of current challenges. Rangel Ecol Manag 78:235–249. https://doi.org/10.1016/j.rama.2020.04.003

Gerlanc NM, Kaufman GF (2003) Use of bison wallows by anurans on Konza Prairie. Am Midl Nat 150:158–168. https://doi.org/10.1674/0003-0031(2003)150[0158:UOBWBA]2.0.CO;2

Gilman SE, Urban MC, Tewksbury J et al (2010) A framework for community interactions under climate change. Trends Ecol Evol 25:325–331. https://doi.org/10.1016/j.tree.2010.03.002

Götzenberger L, de Bello F, Bråthen KA et al (2012) Ecological assembly rules in plant communities-approaches, patterns and prospects. Biol Rev 87:111–127. https://doi.org/10.1111/j.1469-185X.2011.00187.x

Hale SL, Koprowski JL, Archer SR (2020) Black-tailed prairie dog (Cynomys ludovicianus) reintroduction can limit woody plant proliferation in grasslands. Front Ecol Evol 8:1–11. https://doi.org/10.3389/fevo.2020.00233

Hamilton BT, Roeder BL, Horner MA (2019) Effects of sagebrush restoration and conifer encroachment on small mammal diversity in sagebrush ecosystem. Rangel Ecol Manag 72:13–22. https://doi.org/10.1016/j.rama.2018.08.004

Hawkins BA, Field R, Cornell HV et al (2003) Energy, water, and broad-scale geographic patterns of species richness. Ecol 84:3105–3117. https://doi.org/10.1016/j.rama.2018.08.004

He T, Lamont BB, Pausas JG (2019) Fire as a key driver of Earth's biodiversity. Biol Rev 94:1983–2010. https://doi.org/10.1111/brv.12544

Hendrickson JR, Sedivec KK, Toledo D et al (2019) Challenges facing grasslands in the Northern Great Plains and North Central Region. Rangel 41:23–29. https://doi.org/10.1016/j.rala.2018.11.002

Hess JE, Beck JL (2012) Disturbance factors influencing greater sage-grouse lek abandonment in north-central Wyoming. J Wildl Manage 76:1625–1634. https://doi.org/10.1002/jwmg.417

Hindle BJ, Kerr CL, Richards SA et al (2015) Topographical variation reduces phenological mismatch between a butterfly and its nectar source. J Insect Conserv 19:227–236. https://doi.org/10.1007/s10841-014-9713-x

Holbrook JD, Arkle RS, Rachlow JL et al (2016) Occupancy and abundance of predator and prey: implications of the fire-cheatgrass cycle in sagebrush ecosystems. Ecosphere 7:e01307. https://doi.org/10.1002/ecs2.1307

Holechek J, Pieper RD, Herbel CH (2004) Range management: principles and practices, 5th edn. Prentice Hall, Upper Saddle River, NJ, USA, p 587

Hoogland JL (1995) The black-tailed prairie dog: social life of a burrowing mammal. University of Chicago Press, Chicago

Hovick TJ, Elmore RD, Allred BW et al (2014a) Landscapes as a moderator of thermal extremes: a case study from an imperiled grouse. Ecosphere 5:1–12. https://doi.org/10.1890/ES13-00340.1

Hovick TJ, Elmore RD, Fuhlendorf SD (2014b) Structural heterogeneity increases diversity of non-breeding grassland birds. Ecosphere 5:62. https://doi.org/10.1890/ES14-00062.1

Hovick TJ, Elmore RD, Fuhlendorf SD et al (2015) Spatial heterogeneity increases diversity and stability in grassland bird communities. Ecol Appl 25:662–672. https://doi.org/10.1890/14-1067.1

Hovick TJ, Carroll JM, Elmore RD et al (2017a) Restoring fire to grasslands is critical for migrating shorebird populations. Ecol Appl 27:1805–1814. https://doi.org/10.1002/eap.1567

Hovick TJ, McGranahan DA, Elmore RD et al (2017b) Pyric-carnivory: raptor use of prescribed fires. Ecol Evol 1–7. https://doi.org/10.1002/ece3.3401

Hungate BA, Barbier EB, Ando AW et al (2017) The economic value of grassland species for carbon storage. Sci Adv 3:1–9. https://doi.org/10.1126/sciadv.160188

Hutchinson GE (1957) Concluding remarks. population studies: animal ecology and demography. Cold Spring Harb Symp Quant Biol 22:415–427. https://doi.org/10.1007/BF02464429

Innes RJ, Zouhar CAK (2018) Fire regimes of mountain big sagebrush communities. Fire Effects Information System, U.S. Department of Agriculture, Forest Service, Rocky Mountain Research Station, Missoula Fire Sciences Laboratory (Producer). https://www.fs.fed.us/database/feis/fire_regimes/mountain_big_sagebrush/all.html

IPCC (Intergovernmental Panel on Climate Change) (2013a) Climate change 2013a: the physical science basis. Summary for policymakers. Fifth assessment report, p 36. https://doi.org/10.1017/CBO9781107415324.004

IPCC (Intergovernmental Panel on Climate Change) (2013b) Global Warming of 1.5°C. An IPCC Special Report on the impacts of global warming of 1.5°C above pre-industrial levels and related global greenhouse gas emission pathways, in the context of strengthening the global response to the threat of climate change, sustainable development, and efforts to eradicate poverty

IPBES (2019) Global assessment report on biodiversity and ecosystem services of the intergovernmental science-policy platform on biodiversity and ecosystem services. In: Brondizio ES, Settele J, Díaz S, Ngo HT (eds) IPBES secretariat, Bonn, Germany

Ives AR, Carpenter SR (2007) Stability and diversity of ecosystems. Science 317:58–62. https://doi.org/10.1126/science.1133258

Johnson CA (2013) Wetland loss due to row crop expansion in the Dakota Prairie Pothole Region. Wetlands 33:175–182. https://doi.org/10.1007/s13157-012-0365-x

Jones CG, Lawton JH, Shachak M (1997) Positive and negative effects of organisms as ecosystem engineers. Ecology 78:1946–1957. https://doi.org/10.1890/0012-9658(1997)078 [1946:PANEOO]2.0.CO;2

Jones SL, Dieni JS, Green MT et al (2007) Annual return rates of breeding grassland songbirds. Wilson J Ornithol 119:89–94. https://doi.org/10.1676/05-158.1

Jonzén N, Knudsen E, Holt RD et al (2011) Uncertainty and predictability: the niches of migrants and nomads. Animal Migration. Oxford University Press, pp 90–109

Keddy P, Laughlin D (2022) A framework for community ecology. Cambridge University Press, Cambridge, UK

Kitchen AM, Gese EM, Schauster ER (1999) Resource partitioning between coyotes and swift foxes: space, time, and diet. Can J Zool 77:1645–1656. https://doi.org/10.1139/z99-143

Kleuver BM, Smith TN, Gese EM (2019) Group effects of a non-native plant invasion on rodent abundance. Ecosphere 10:e02544. https://doi.org/10.1002/ecs2.2544

Knapp PA (1996) Cheatgrass (*Bromus tectorum L*) dominance in the Great Basin Desert: history, persistence, and influences to human activities. Glob Environ Change 6:37–52. https://doi.org/10.1016/0959-3780(95)00112-3

Knapp AK, Blair JM, Briggs JM et al.(1999) The keystone role of bison in north American tallgrass prairie bison increase habitat heterogeneity processes. Bioscience 49:39–50. http://www.jstor.org/stable/https://doi.org/10.1525/bisi.1999.49.1.39

Knick ST, Holmes AL, Miller RF (2005) The role of fire in structuring sagebrush habitats and bird communities. Stud Avian Biol 30:1–13

Knopf FL, Samson F (1997) Ecology and conservation of great plains vertebrates. Springer Verlag, New York

Kraft NJB, Adler PB, Godoy O et al (2015) Community assembly, coexistence and environmental filtering metaphor. Funct Ecol 29:592–599. https://doi.org/10.1111/1365-2435.12345

Krebs CJ (1999) Ecological methodology. Benjamin/Cummings

Kreuter UP, Iwaasa ADAW, Theodori GL et al (2016) State of knowledge about energy development impacts on North American rangelands: an integrative approach. J of Environ Manage 180:1–9. https://doi.org/10.1016/j.jenvman.2016.05.007

LaRue EA, Hardiman BS, Elliott JM et al (2019) Structural diversity as a predictor of ecosystem function. Environ Res Lett 14:114011. https://doi.org/10.1088/1748-9326/ab49bb

Lark TJ, Spawn SA, Bougie M et al (2020) Cropland expansion in the United States produces marginal yields at high costs to wildlife. Nat Commun 11:1–11. https://doi.org/10.1038/s41467-020-18045-z

Leopold A (1933) Game management. Charles Scribner's Sons, USA, p 481

Lockyer ZB, Coates PS, Casazza ML et al (2015) Nest-site selection and reproductive success of greater sage-grouse in a fire-affected habitat of northwestern Nevada. J Wildl Manage 79:785–797. https://doi.org/10.1002/jwmg.899

Londe DW, Carroll JM, Elmore RD et al (2021) Avifauna assemblages in sand shinnery oak shrublands managed with prescribed fire. Rangel Ecol Manag 79:164–174. https://doi.org/10.1016/j.rama.2021.08.009

Londe, DW, Elmore RD,Davis CA et al (2020) Structural and compositional heterogeneity influences the thermal environment across multiple scales. Ecosphere 11. https://doi.org/10.1002/ecs2.3290

MacArthur RH (1958) Population ecology of some warblers in northeastern coniferous forests. Ecology 39:599–619. https://doi.org/10.2307/1931600

MacArthur RH, MacArthur JW (1961) On bird species diversity. Ecology 42:594–598. https://doi.org/10.2307/1932254

Magurran AE (1988) Ecological diversity and its measurement. Princeton University Press, Princeton, USA

Mahood AL, Balch JK (2019) Repeated fires reduce plant diversity in low-elevation Wyoming big sagebrush ecosystems (1984–2014). Ecosphere 10. DOI: https://doi.org/10.1002/ecs2.2591

Marchese C (2015) Biodiversity hotspots: a shortcut for a more complicated concept. Glob Ecol Conserv 3:297–309. https://doi.org/10.1016/j.gecco.2014.12.008

Melin M, Matala MM, Mehtatalo L et al (2014) Moose (*Alces alces*) reacts to high summer temperatures by utilizing thermal shelter in boreal forests—an analysis based on airborne laser scanning of the canopy structure at moose locations. Glob Chang Biol 20:1115–1125. https://doi.org/10.1111/gcb.12405

McGranahan DA, Hovick TJ, Elmore RD et al (2018) Moderate patchiness optimizes heterogeneity, stability, and beta diversity in mesic grassland. Ecol Evol 8:5008–5015. https://doi.org/10.1002/ece3.4081

Maresh Nelson SB, Coon JJ, Schacht SH, Miller JR (2019) Cattle select against the invasive grass tall fescue in heterogeneous pastures managed with prescribed fire. Grass and Forage Science 74:486–495. https://doi.org/10.1111/gfs.12411

Miller RM, Wilson GWT, Johnson NC (2012) Arbuscular mycorrhizae and grassland ecosystems. Chapter 3. In: Biocomplexity of plant-fungal interactions. Ed. D. Southworth. Wiley-Blackwell, pp 59–84

Mitchell CE, Power AG (2003) Release of invasive plants from fungal and viral pathogens. Nature 421:625–627

Monzón J, Moyer-Horner L, Palamar MB (2011) Climate change and species range dynamics in protected areas. Bioscience 61:752–761. https://doi.org/10.1525/bio.2011.61.10.5

Moolman HJ, Cowling RM (1994) The impact of elephant and goat grazing on the endemic flora of South African succulent thicket. Biol Conserv 68:53–61. https://doi.org/10.1016/0006-3207(94)90546-0

Morgan JA, Lecain DR, Pendall E et al (2011) C4 grasses prosper as carbon dioxide eliminates desiccation in warmed semi-arid grassland. Nature 476:202–205. https://doi.org/10.1038/nature10274

Morin P (2011) Community ecology. Blackwell Science, West Sussex, UK

Motha RP, Baier W (2005) Climate variability on agriculture in the temperate regions: North America. Clim Change 70:137–164

Mutter M, Pavlacky DC, Van Lanen NJ et al (2015) Evaluating the impact of gas extraction infrastructure on the occupancy of sagebrush-obligate songbirds. Ecol Appl 25:1175–1186. https://doi.org/10.1890/14-1498.1

Myers N (1988) Threatened biotas: "hotspots" in tropical forests. Environmentalist 8:187–208

Nickell Z, Varriano S, Plemmons E et al (2018) Ecosystem engineering by bison (*Bison bison*) wallowing increases arthropod community heterogeneity in space and time. Ecosphere 9. https://doi.org/10.1002/ecs2.2436

Northrup JM, Anderson CR, Wittemyer G (2015) Quantifying spatial habitat loss from hydrocarbon development through assessing habitat selection patterns of mule deer. Glob Chang Biol 21:3961–3970. https://doi.org/10.1111/gcb.13037

Nowacki GJ, Abrams MD (2008) The demise of fire and "mesophication" of forests in the eastern United States. Bioscience 58:123–138. https://doi.org/10.1641/b580207

Natural Resources Conservation Service (NRCS) (2021) A framework for conservation action in the Great Plains Grasslands Biome. Working Lands for Wildlife, USDA-NRCS. Washington, D.C. Available at: https://wlfw.rangelands.app

Oh KP, Aldridge CL, Forbey JS et al (2019) Conservation genomics in the sagebrush sea: population divergence, demographic history, and local adaptation in sage-grouse (*Centrocercus spp.*). Genome Biol Evol 11:2023–2034. https://doi.org/10.1093/gbe/evz112

Ohler L, Lechleitner M, Junker RR (2020) Microclimatic effects on alpine plant communities and flower-visitor interactions. Sci Rep 10:1366. https://doi.org/10.1038/s41598-020-58388-7

Olson DM, Dinerstein E, Wikramanayake ED et al (2001a) Terrestrial ecoregions of the world: a new map of life on Earth. Bioscience 51:933–938. https://doi.org/10.1641/0006-3568(2001)051[0933:TEOTWA]2.0.CO;2

Olson DM, Dinerstein E, Wikramanayake ED et al (2001b) Terrestrial ecoregions of the world: a new map of life on Earth. Bioscience 51(11):933–938

Owen JC, Sogge MK, Kern MD (2005) Habitat and sex differences in physiological condition of breeding southwestern willow flycatchers (*Empidonax trailliiextimus*). Auk 122:1261–1270. https://doi.org/10.1093/auk/122.4.1261

Paine RT (1966) Food web complexity and species diversity. Am Nat 100:65–75. https://www.jstor.org/stable/2459379

Paine RT (1969) The Pisaster-Tegula interaction: prey patches, predator food preference, and intertidal community structure. Ecology 50:950–961. https://doi.org/10.2307/1936888

PACP-Ch (2011) Plan de Acción para la Conservación y Uso Sustentable de los Pastizales del Desierto Chihuahuense en el Estado de Chihuahua 2011–2016. In: Guzmán-Aranda JC, Hoth J, Blanco YE (eds) Gobierno del Estado de Chihuahua, México

Patricelli GL, Blickley JL, Hooper SL (2013) Recommended management strategies to limit anthropogenic noise impacts on greater sage-grouse in Wyoming. Hum Wildl Interact 7:230–249. https://www.jstor.org/stable/24874869

Pearce-Higgins JW, Ockendon N, Baker DJ et al (2015) Geographical variation in species' population responses to changes in temperature and precipitation. Proc Royal Soc B 282:20151561. https://doi.org/10.1098/rspb.2015.1561

Pecl GT, Araújo MB, Bell JD et al (2017) Biodiversity redistribution under climate change: impacts on ecosystems and human well-being. Science 355. https://doi.org/10.1126/science.aai9214

Polasky S, Nelson E, Lonsdorf E et al (2005) Conserving species in a working landscape: land use with biological and economic objectives. Ecol Appl 15:1387–1401. https://doi.org/10.1890/03-5423

Ponce-Guevara E, Davidson A, Sierra-Corona R et al (2016) Interactive effects of black-tailed prairie dogs and cattle on shrub encroachment in a desert grassland ecosystem. PLoS One 11. https://doi.org/10.1371/journal.pone.0154748

Post E, Forchhammer MC (2008) Climate change reduces reproductive success of an Arctic herbivore through trophic mismatch. Philos Trans R Soc Lond B Biol Sci 363:2369–2375. https://doi.org/10.1098/rstb.2007.2207

Post E, Pedersen C, Wilmers CC et al (2008) Warming, plant phenology and the spatial dimension of trophic mismatch for large herbivores. Proc Royal Soc B 275:2005–2013. https://doi.org/10.1098/rspb.2008.0463

Powell KI, Chase JM, Knight TM (2013) Invasive plants have scale-dependent effects on diversity by altering species-area relationships. Science 339:316–318. https://doi.org/10.1126/science.1226817

Prevéy JS, Germino MJ, Huntly NJ (2010) Loss of foundation species increases population growth of exotic forbs in sagebrush steppe. Ecol Appl 20:1890–1902. https://doi.org/10.1890/09-0750

Ratajczak Z, Briggs JM, Goodin DG et al (2016) Assessing the potential for transitions from tallgrass prairie to woodlands: are we operating beyond critical fire thresholds? Rangel Ecol Manag 69:280–287. https://doi.org/10.1016/j.rama.2016.03.004

Raven PH, Wagner DL (2021) Agricultural intensification and climate change are rapidly decreasing insect biodiversity. Proc Natl Acad Sci 118. https://doi.org/10.1073/pnas.2002548117

Raynor EJ, Powell LA, Schacht WH (2018) Present and future thermal environments available to Sharp-tailed Grouse in an intact grassland. PLoS ONE 13:1–20. https://doi.org/10.1371/journal.pone.0191233

Reading RP, Matchett R (1997) Attributes of black-tailed prairie dog colonies in north central Montana. J Wildl Manage 61:664–673. https://doi.org/10.2307/3802174

Rigge M, Shi H, Postma K (2021) Projected change in rangeland fractional component cover across the sagebrush biome under climate change through 2085. Ecosphere 12:e03538. https://doi.org/10.1002/ecs2.3538

Ritten JP, Bastian CT, Rashford BS (2012) Profitability of carbon sequestration in Western Rangelands of the United States. Rangel Ecol Manag 65:340–350. https://doi.org/10.2111/REM-D-10-00191.1

Rockwell SM, Wehausen B, Johnson (2021) Sagebrush Bird Communities Differ with Varying Levels of Crested Wheatgrass Invasion. J Fish Wildl Mgmt 12: 27–39. https://doi.org/10.3996/JFWM-20-035

Roman-Palacios C, Wiens JK (2020) Recent responses to climate change reveal the drivers of species extinction and survival. Proc Natl Acad Sci 117:4211–4217. https://doi.org/10.1073/pnas.1913007117

Rosenberg KV, Dokter AM, Blancher PJ et al (2019) Decline of the North American avifauna. Science 366:120–124. https://doi.org/10.1126/science.aaw1313

Rout ME, Callaway RM (2009) An invasive plant paradox. Science 324:734–735. https://doi.org/10.1126/science.1173651

Sampson AW (1952) Range management: principles and practices. New York, NY, USA: John Wiley & Sons, p 570

Samson FB, Knopf FL, Ostlie WR (2004) Great plains ecosystems: past, present and future. Wildl Soc B 32:6–15. https://doi.org/10.2193/0091-7648(2004)32[6:GPEPPA]2.0.CO;2

Savory A (1978) A holistic approach to ranch management using short duration grazing. In: Proceedings of the first international rangeland congress, pp 555–557

Schlaepfer DR, Lauenroth WK, Bradford JB (2012) Effects of ecohydrological variables on current and future ranges, local suitability patterns, and model accuracy in big sagebrush. Ecography 35:374–384. https://doi.org/10.1111/j.1600-0587.2011.06928.x

Schlaepfer DR, Bradford JB, Lauenroth WK et al (2021) Understanding the future of big sagebrush regeneration: challenges of projecting complex ecological processes. Ecosphere 12:e03695. https://doi.org/10.1002/ecs2.3695

Schroeder MA, Aldridge CL, Apa AD et al (2004) Distribution of sage-grouse in North America. Condor 106:363–376. https://doi.org/10.1650/7425

Schroeder VM, Robinson WD, Johnson DD et al (2022) Weather explains differences in sagebrush-obligate songbird nest success under various grazing regimes. Glob Ecol Conserv 34. https://doi.org/10.1016/j.gecco.2022.e02010

Schuman GE, Demer JD (2004) Carbon sequestration by rangelands: management effects and potential, in proceedings of the western regional cooperative soil survey conference. USDA Natural Resources Conservation Service, Casper, Wy., Jackson, Wy

Seastedt TR, Briggs JM, Gibson DJ (1991) Controls of nitrogen limitation in tallgrass prairie. Oecologia 87:72–79. https://doi.org/10.1007/BF00323782

Sepkoski JJ (1988) Alpha, beta, or gamma: Where does all the diversity go? Paleobiology 14:221–234. https://www.jstor.org/stable/2400884

Shaffer JA, Buhl DA (2016) Effects of wind-energy facilities on breeding grassland bird distributions. Conserv Biol 30:59–71. https://doi.org/10.1111/cobi.12569

Shaffer JA, Loesch CR, Buhl DA (2019) Estimating offsets for avian displacement effects of anthropogenic impacts. Ecological Applications, 29, e01983

Shannon CE (1948) A mathematical theory of communication. Bell Labs Tech J 27:379–423. https://doi.org/10.1002/j.1538-7305.1948.tb01338.x

Shipley LA, Davila TB, Thines NJ et al (2006) Nutritional requirements and diet choices of the pygmy rabbit (Bachylagus idahoensis): a sagebrush specialist. J Chem Ecol 32:2455–2474. https://doi.org/10.1007/s10886-006-9156-2

Simpson EH (1949) Measurement of diversity. Nature 163:688. https://doi.org/10.1038/163688a0

Sirami C, Seymour C, Midgley G et al (2009) The impact of shrub encroachment on savanna bird diversity from local to regional scale. Divers Distrib 15:948–957. https://doi.org/10.1111/j.1472-4642.2009.00612.x

Sirén APK, Morelli TL (2020) Interactive range-limit theory (iRLT): an extension for predicting range shifts. J Anim Ecol 89:940–954. https://doi.org/10.1111/1365-2656.13150

Skagen SK, Augustine DJ, Derner JD (2018) Semi-arid grassland bird responses to patch-burn grazing and drought. J Wildl Manage 82:445–456. https://doi.org/10.1002/jwmg.21379

Smith IT, Rachlow JL, Svancara LK et al (2019) Habitat specialists as conservation umbrellas: do areas managed for greater sage-grouse also protect pygmy rabbits? Ecosphere 10. https://doi.org/10.1002/ecs2.2827

Staudinger MD, Carter SL, Cross MS et al (2013) Biodiversity in a changing climate: a synthesis of current and projected trends in the US. Front Ecol Environ 11:465–473. https://doi.org/10.1890/120272

Staver AC, Archibald S, Levin SA (2011) The global extent and determinants of savanna and forest as alternative biome states. Science 80:230–232. https://doi.org/10.1126/science.1210465

Stephens T, Wilson SC, Cassidy F et al (2018) Climate change impacts on the conservation outlook of populations on the poleward periphery of species ranges: a case study of Canadian black-tailed prairie dogs (Cynomys ludovicianus). Glob Chang Biol 24:836–847. https://doi.org/10.1111/gcb.13922

Stoddart LA, Smith AD (1943) Range management. McGraw-Hill Book Company, New York, NY, USA, p 547

Swetnam TW, Allen CD, Betancourt JL (1999) Applied historical ecology: using the past to manage for the future. Ecol Appl 9:1189–1206. https://doi.org/10.1890/1051-0761(1999)009[1189:AHEUTP]2.0.CO;2

Teeri JA, Stowe LG (1976) Climatic patterns and the distribution of C4 grasses in North America. Oecologia 23:1–12. https://doi.org/10.1007/BF00351210

Tews J, Brose U, Grimm V et al (2004) Animal species diversity driven by habitat heterogeneity/diversity: the importance of keystone structures. J Biogeogr 31:79–92. https://doi.org/10.1046/j.0305-0270.2003.00994.x

Teyssonneyre F, Picon-Cochard C, Falcimagne R et al (2002) Effects of elevated CO_2 and cutting frequency on plant community structure in a temperate grassland. Glob Chang Biol 8:1034–1046. https://doi.org/10.1046/j.1365-2486.2002.00543.x

Thapa-Magar KB, Davis TS, Kondratieff B (2020) Livestock grazing is associated with seasonal reduction in pollinator biodiversity and functional dispersion but cheatgrass invasion is not: variation in bee assemblages in a multiuse shortgrass prairie. PLoS ONE 15:1–18. https://doi.org/10.1371/journal.pone.0237484

Thomas KA, Jarchow CJ, Arundel TR et al (2018) Landscape-scale wildlife species richness metrics to inform wind and solar energy facility siting: an Arizona case study. Energy Policy 116:145–152. https://doi.org/10.1016/j.enpol.2018.01.052

Thompson SJ, Johnson DH, Niemuth ND et al (2015) Avoidance of unconventional oil wells and roads exacerbates habitat loss for grassland birds in the North American great plains. Biol Conserv 192:82–90. https://doi.org/10.1016/j.biocon.2015.08.040

Thornton PK (2010) Livestock production: recent trends, future prospects. Philos Trans R Soc Lond B Biol Sci 365:2853–2867. https://doi.org/10.1098/rstb.2010.0134

Toombs TP, Derner JD, Augustine DJ et al (2010) Managing for biodiversity and livestock. Rangelands 32:10–15. https://doi.org/10.2111/RANGELANDS-D-10-00006.1

Turner CL, Blair JM, Schartz RJ et al (1997) Soil N availability and plant response in tallgrass prairie: effects of fire, topography and supplemental N. Ecology 78:1832–1843. https://doi.org/10.2307/2266105

Turner M, Gardner R (2001) Landscape ecology in theory and practice. Springer, New York, NY, USA

Vallentine JF (1990) Grazing management. Academic Press, San Diego, California USA

Valentine-Darby P (2010) Southern plains network inventory and monitoring program

Van Auken O (2009) Causes and consequences of woody plant encroachment into western North American grasslands. J Environ Manage 90:2931–2942. https://doi.org/10.1016/j.jenvman.2009.04.023

Vickery PD, Herkert JR, Knopf FL et al (1999) Grassland birds: an overview of threats and recommended management strategies. In: Bonner RE Jr, Pashley DN, Cooper R Ithaca (eds) Strategies for bird conservation: creating the partners in flight planning process. Cornell Laboratory of Ornithology, NY

Villareal ML, Haire SL, Bravo JC et al (2019) A mosaic of land tenure and ownership creates challenges and opportunities for transboundary conservation in the US-Mexico Borderlands. Case Stud Environ. https://doi.org/10.1525/cse.2019.002113

Vose JM, Clark JS, Luce CH et al (2015) Effects of drought on forests and rangelands in the United States: a comprehensive science synthesis. In: General Technical Report WO-93b. Gen Tech Rep WO-93b Washington, DC US Department of Agriculture, Forest Service Washington Office

Walk JW, Warner RE (2000) Grassland management of the conservation of songbirds in the Midwestern USA. Biol Conserv 94:165–172. https://doi.org/10.1016/S0006-3207(99)00182-2

Washburn BE, Barnes TG, Sole JD (2000) Improving northern bobwhite habitat by converting tall fescue fields to native warm-season grasses. Wildl Soc Bull 28:97–104. http://www.jstor.org/stable/4617289

Wimberly MC, Narem DM, Bauman PJ et al (2018) Grassland connectivity in fragmented agricultural landscapes of the north-central United States. Biol Conserv 217:121–130. https://doi.org/10.1016/j.biocon.2017.10.031

World Wildlife Federation (2018) Living planet report—2018: aiming Higher. In: Grooten M, Almond REA (eds) WWF, Gland, Switzerland. ISBN 978-2-940529-90-2

Wright CK, Wimberly MC (2013) Recent land use change in the Western Corn Belt threatens grasslands and wetlands. Proc Natl Acad Sci 110:4134–4139. https://doi.org/10.1073/pnas.1215404110

Yarwood SA, Bach EM, Busse M et al (2020) Forest and rangeland soil biodiversity. In: Pouyat RV, Page-Dumroese DS, Patel-Weynand T, Geiser LH (eds) Forest and rangeland soils of the united states under changing conditions. Springer International Publishing

Zhou CB, Caterina GL, Will RE et al (2015) Canopy interception for a tallgrass prairie under juniper encroachment. PLoS ONE 10:e0141422. https://doi.org/10.1371/journal.pone.0141422

Zhou Y, Boutton TW, Wu XB (2018) Woody plant encroachment amplifies spatial heterogeneity of soil phosphorus to considerable depth. Ecology 99:136–147. https://doi.org/10.1002/ecy.2051

Zimmer SN, Grosklos GJ, Belmont P et al (2021) Agreement and uncertainty among climate change impact models: a synthesis of sagebrush steppe vegetation projections. Rangel Ecol Manag 75:119–129. https://doi.org/10.1016/j.rama.2020.12.006

Zouhar K (2021) Fire regimes of plains grassland and prairie ecosystems. In: Fire Effects Information System. U.S. Department of Agriculture, Forest Service, Rocky Mountain Research Station, Missoula Fire Sciences Laboratory (Producer). Available: www.fs.fed.us/database/feis/fire_regimes/PlainsGrassland_Prairie/all.html

Part II
Species Accounts

Chapter 9
Prairie Grouse

Lance B. McNew, R. Dwayne Elmore, and Christian A. Hagen

Abstract Prairie grouse, which include greater prairie-chicken (*Tympanuchus cupido*), lesser prairie-chicken (*T. pallidicinctus*), and sharp-tailed grouse (*T. phasianellus*), are species of high conservation concern and have been identified as potential indicator species for various rangeland ecosystems. Greater prairie-chickens are found in scattered populations in isolated tallgrass prairie throughout the Midwest, but primarily occur in the more expansive tallgrass and mixed-grass prairies in the Great Plains. Lesser prairie-chickens occur in mixed-grass, short-grass, and arid shrublands of the southern Great Plains. Sharp-tailed grouse occur in mixed-grass, shortgrass, shrub steppe, and prairie parkland vegetation types and are broadly distributed across the northern Great Plains, portions of the Great Basin, and boreal parkland areas of Alaska and Canada. Due to reliance on a variety of rangeland types, consideration of management and anthropogenic activities on rangelands are critical for prairie grouse conservation. Grazing is one of the more prominent activities that has the potential to affect prairie grouse by altering plant structure and composition, and recent research has attempted to identify the mechanisms of grazing effects on prairie grouse. Fire is another important disturbance affecting grouse habitat, especially considering how the current distribution and intensity of fire differs from what occurred historically. Additionally, human infrastructure in the form of roads and energy development, as well as land conversion and degradation such as tillage and tree encroachment can fragment and reduce habitat for prairie grouse. Finally, weather including drought, extended rain, and temperature extremes are common across the distribution of prairie grouse. Although not directly under management control, the effects of weather are an overarching factor that need to be considered in conservation planning. This chapter will summarize the life-histories

L. B. McNew (✉)
Department of Animal and Range Sciences, Montana State University, Bozeman, MT 59717, USA
e-mail: lance.mcnew@montana.edu

R. Dwayne Elmore
Department of Natural Resource Ecology and Management, Oklahoma State University, Stillwater, OK 74078, USA

C. A. Hagen
Department of Fisheries, Wildlife and Conservation Sciences, Oregon State University, Corvallis, OR 97331, USA

© The Author(s) 2023   253
L. B. McNew et al. (eds.), *Rangeland Wildlife Ecology and Conservation*,
https://doi.org/10.1007/978-3-031-34037-6_9

and habitat requirements of prairie grouse, discuss how rangeland management and other human activities affect them, highlight major threats to prairie grouse and provide recommendations for future management and research.

Keywords Greater prairie-chicken · Lesser prairie-chicken · Rangeland wildlife · Sharp-tailed grouse

9.1 General Life History and Population Dynamics

Prairie grouse collectively refer to three species of grouse in the genus *Tympanuchus*: greater prairie-chicken (*T. cupido*), lesser prairie-chicken (*T. pallidicinctus*), and sharp-tailed grouse (*T. phasianellus*; Fig. 9.1), classified within the order Galliformes, family Phasianidae, and sub-family Tetraoninae. Where generalities exist across all three species, we will refer to them collectively as prairie grouse, whereas the specific species is referenced when reviewing information appropriate only for that species. Generally, prairie grouse have relatively fast life-histories with high reproductive effort and short lifespans (typically < 3 years). Home range sizes are variable but can be large (> 2500 ha) relative to other galliforms (Patten et al. 2011; Robinson et al. 2018). While prairie grouse often stay relatively close to established leks (see Sect. 9.1.1), they can disperse great distances to find habitat (Earl et al. 2016). Their populations can fluctuate dramatically between years due to weather, but overall trends are influenced by longer term changes in vegetation conditions. The life cycle of prairie grouse is typically partitioned into broad seasonal delineations of lekking, nesting, brood-rearing, and non-breeding seasons.

9.1.1 Lekking

Prairie grouse are polygynous and have a lek-mating system in which courtship and mating is generally limited to lek sites known as 'booming' (greater prairie-chicken), 'gobbling' (lesser prairie-chicken), and 'dancing' (sharp-tailed grouse) grounds during the spring breeding period (March–May), although males often display at leks in the fall as well (Emlen and Oring 1977). This mating system has multiple potential benefits for females. Lekking is associated with strong female mate choice allowing females to efficiently choose from multiple potential mates. Secondary sexual characteristics, including brightly colored air sacs and eye combs (all species) and elongated pinnae (in both prairie-chicken species) are used by females to identify desirable mates (Robel 1966; Bergerud et al. 1988). Additionally, vocalizations and vigor of display are used as cues of general fitness by females with males in the center of the lek typically doing most of the breeding (Behney et al. 2012). Male prairie grouse provide no parental investment after mating and

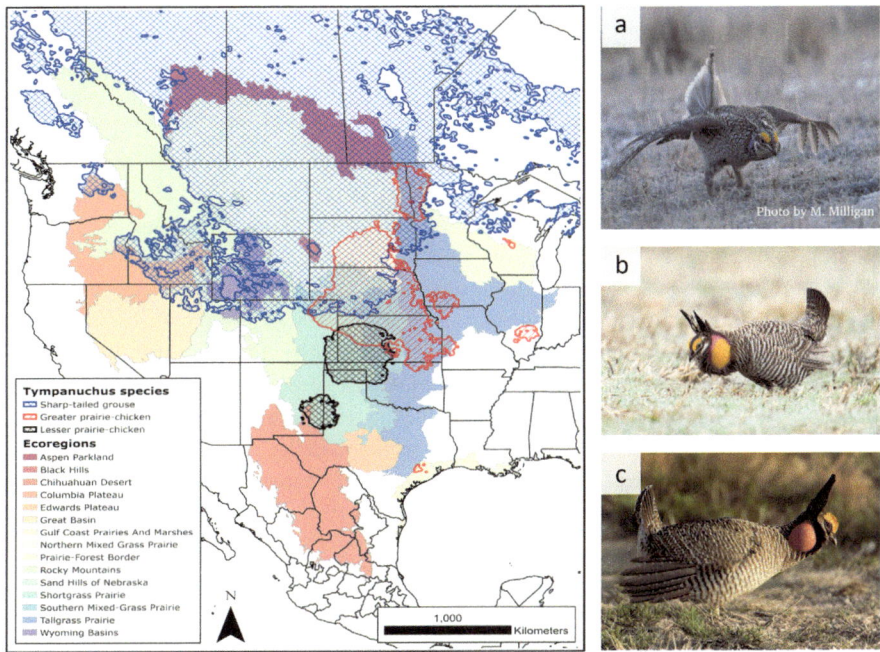

Fig. 9.1 Distribution of three species of prairie grouse in North America. Sharp-tailed grouse (**a**), greater prairie-chickens (**b**), and lesser prairie-chickens (**c**) occupy grasslands and shrublands from Texas to Alaska. Most populations occur on rangelands managed for livestock production. *Map credit* M. Solomon; *Photo credits* M. Milligan and N. Richter

are closely associated with leks throughout much of the year (Schroeder and White 1993).

Lek locations are generally considered to be stable from year to year, and there is evidence that females prefer to visit established leks over newly formed leks (Schroeder and Braun 1992; Haukos and Smith 1999). However, lek locations can move in response to vegetation conditions when disturbance patterns are dynamic (Hovick et al. 2015a). Females may visit more than one lek, and clusters of leks are important for sustaining populations (Schroeder 1991; Hagen et al. 2017). Often lek locations are used for prioritizing conservation actions because they are conspicuous on the landscape (easily mapped) and most annual prairie grouse activity (e.g., nesting and brood-rearing) occurs within 5 km of a lek (Schroeder 1991; Boisvert et al. 2005; Winder et al. 2015).

9.1.2 Nesting

Prairie grouse have high reproductive potential with high nest initiation rates, large clutch sizes and high egg viability (Connelly et al. 1998; McNew et al. 2011a). As a result, nest success has consistently been identified as one of the most important vital rates affecting prairie grouse populations (Wisdom and Mills 1997; Hagen et al. 2009; Gillette 2014). Most females, regardless of age, will initiate at least one nest per year. Clutch sizes average 10–12 eggs but can be highly variable across climatic gradients (McNew et al. 2017). Additionally, clutch size can be smaller for greater prairie-chickens during early and warm springs suggesting that external environmental cues may be related to nest initiation (Londe et al. 2021b). If the initial clutch is lost during laying or early in incubation, females will often renest, and clutch sizes of renests are typically smaller than initial nests (Hagen and Giesen 2005; Johnson et al. 2020). The probability of renesting declines as the nesting season progresses (Pitman et al. 2006a; McNew et al. 2011a). The incubation period for prairie grouse averages 25–28 d (Hagen and Giesen 2005; Johnson et al. 2020). Generally, the peak of hatching occurs during May–June; local environmental conditions can impact average incubation and hatch dates of nests (McNew et al. 2011b; Londe et al. 2021b).

Nest success, the proportion of nests that hatch $\geq$ 1 egg, varies across years and sites for a variety of reasons, including differences in weather, age structure of the population, predator populations, and differences in local and landscape habitat conditions. Weather has been found to strongly influence nest fate for greater prairie-chickens in the southern Great Plains (Hovick et al. 2015b; Londe et al. 2021b) and sharp-tailed grouse in the northern Great Plains (Milligan et al. 2020a). Nest success can also vary between first nests and renest attempts (Hovick et al. 2014a, b; Williamson 2009) in relation to seasonal variation in nesting habitat conditions (McNew et al. 2011a). Additionally, higher nest success for adults ($\geq$ 2nd breeding season) than yearlings (1st breeding season) has been reported (Bergerud et al. 1988), although others have observed no difference in nest success between adult and yearling prairie grouse (Apa 1998; Collins 2004; Milligan et al. 2020a; Londe et al. 2021b). Egg viability, or the proportion of eggs that hatch within successful nests, is typically high with at least 90% of eggs hatching (Meints 1991; Pitman et al. 2006b; McNew et al. 2011b).

9.1.3 Brood-Rearing

Female prairie grouse rear one brood (a group of chicks) per year. Chicks are precocial and follow the female away from the nest shortly after hatching. Chicks cannot thermoregulate for up to two weeks after hatching (Bergerud et al. 1988), making them vulnerable to environmental conditions and dependent on the female to provide temperature regulation from both hot and cold conditions. Broods remain with the female and often stay relatively close to the nesting area throughout the summer (Marks and Marks 1988; Gratson 1988; Meints 1991). Daily summer movements have ranged from 45–276 m (Hart et al. 1950; Pitman et al. 2006a, b).

9.1.4 Chick Survival

Chick survival is a key determinant of population dynamics and may be an even more limiting factor than nest success for some populations (Hagen et al. 2009). As in most galliforms, the highest period of chick mortality is during the first 2 weeks after hatch, largely due to the inability of chicks to fly or thermoregulate (Bergerud et al. 1988). Survival probability of chicks increases rapidly after this period as they become less dependent on the female for thermoregulation and can escape predators more effectively. As a result, chicks are vulnerable to three main sources of mortality: predation, starvation, and exposure. Reported survival rates of chicks to 35 days of age range from 0.13 to 0.67 (reviewed in Hagen and Giesen 2005; Milligan et al. 2018; Johnson et al. 2020) and likely vary due to local habitat conditions and weather. Estimates of juvenile survival from 35-d to 1-y of age are generally lacking due to the difficulty in capturing and monitoring juvenile prairie grouse and is a research need.

9.1.5 Non-breeding

The non-breeding season, delineated generally as the period August–February, is the least understood portion of prairie grouse life history. Like other galliforms, research has been focused on the reproductive period, due to both perceived importance of reproductive success to annual recruitment and availability of field researchers during summer months. Survival is generally higher for female prairie grouse during the non-breeding season as compared to the breeding season (Winder et al. 2014a; Robinson et al. 2018; Milligan et al. 2020b). Habitat requirements can be dramatically different during the non-breeding season with grouse selecting different vegetation types (Pirius et al. 2013; Hiller et al. 2019; Londe et al. 2019) and exhibiting contrasting behavioral avoidance of some anthropogenic structures as compared to the breeding season (Londe et al. 2019; Sect. 9.4). During the non-breeding season, prairie grouse may flock to feed on crops, sometimes traveling great distances between grassland cover and crop fields, but the effects of movement, concentration of birds, and use of crops on vital rates is poorly understood (Robinson et al. 2018).

9.1.6 Survival

Typically, annual survival rates reported for adult prairie grouse have ranged from 0.17–0.43 (McNew et al. 2017) but was observed to be as high as 0.71 in South Dakota for greater prairie-chickens (Robel et al. 1972). Reported differences in survival between adults and yearlings or between sexes are variable with some studies showing no differences (Boisvert 2002; Winder et al. 2018; Milligan et al. 2020b) and others reporting significant differences (Hagen et al. 2005, lesser prairie-chicken; Matthews et al. 2016, translocated sharp-tailed grouse). Increased female mortality is more likely to occur during the nesting season and brood-rearing season (Hagen et al. 2007a; Winder et al. 2014a, 2018; Milligan et al. 2020b), and male mortality typically increases during the lekking period (Collins 2004). Winter mortality depends on the severity of the winter. In Idaho, sharp-tailed grouse survival rates ranged from 0.29 in a severe winter to 0.86 in a mild winter (Ulliman 1995). In southern populations, prairie-chickens generally have high overwinter survival (McNew et al. 2012b; Pirius et al. 2013). Causes of adult mortality include predation (Hagen et al. 2007a; Winder et al. 2018; Milligan et al. 2020b) and collisions with powerlines, fences, and vehicles (Wolfe et al. 2007; Robinson et al. 2018). Maximum reported lifespan of sharp-tailed grouse was 7.5 years, although life-expectancy is < 3 years on average (Connelly et al. 1998). Although estimates for juvenile (i.e., stage from fledging to first breeding season) overwinter mortality for greater prairie-chickens or sharp-tailed grouse are lacking, juvenile survival of lesser prairie-chickens has been reported (0.70; Pitman et al. 2006a). Further work is needed to understand survival of juvenile prairie grouse.

9.1.7 Seasonal Movements and Dispersal

Prairie grouse are generally considered non-migratory, and seasonal shifts between summer and fall home ranges are often small (< 10 km; Johnson et al. 2020; Stinson and Schroeder 2012). However, longer seasonal shifts up to 50 km have been reported for lesser prairie-chickens (Earl et al. 2016). On average, female prairie grouse move < 2 km from their lek of capture to nesting sites (Schroeder 1991). Movements from nesting to brood-rearing areas are generally short (< 2 km) as well (Collins 2004; Hoffman et al. 2015). However, some females moved more than 3.5 km to brood-rearing sites, potentially due to drought conditions and lack of resources (Collins 2004). Little is known about natal dispersal in prairie grouse because studies of radio-marked juvenile grouse are lacking. Females appear to be the primary dispersers and males remain more localized and perhaps recruit to leks near natal areas (Pitman et al. 2006a; Earl et al. 2016).

9.1.8   Population Dynamics

Like other game birds, prairie grouse have population cycles that are linked to habitat quality and weather. Interannual variability in habitat quality can be driven by weather (Londe et al. 2021b), especially precipitation (Ross et al. 2016; Fritts et al. 2018) as well as rangeland management. However, these relationships are constrained by landscape-level factors such as patch size, habitat fragmentation, or habitat composition (Hagen et al. 2020). Historically, prairie grouse exhibited cyclical "boom or bust" patterns that were largely dependent upon precipitation or drought conditions. Given the short life span and high reproductive output of prairie grouse, populations can fluctuate from year to year. This, combined with lack of precision in population size estimates, makes management decisions based on short term (year to year) changes in population indices less reliable and emphasizes the need to evaluate longer term trajectories. Although prairie grouse have likely always fluctuated dramatically as environmental conditions varied, these cyclical patterns have been exacerbated as available habitat was diminished and carrying capacity reduced. Recent evidence indicates that extreme drought coupled with less available habitat leads to slower population recovery and perhaps an inability to rebound to previous abundance levels (Ross et al. 2016; Fritts et al. 2018).

Due to high reproductive effort, reproductive success has a disproportionately large influence on overall population dynamics. Sensitivity analyses of stable populations of lesser prairie-chickens and sharp-tailed grouse show that changes in nest and brood survival have the largest contributions to population dynamics (Hagen et al. 2009; Gillette 2014). However, the relative importance of adult survival and fecundity varied among populations of greater prairie-chickens, suggesting that human land use patterns can affect the comparative influence among vital rates on population dynamics (McNew et al. 2012b; Sullins et al. 2018). Nevertheless, management prescriptions that improve reproductive success and recruitment are more likely to effectively recover prairie grouse populations than those directed at adult survival (McNew et al. 2012b; Milligan et al. 2018).

9.2   Current Species and Population Status

9.2.1   Greater Prairie-Chickens

Three subspecies of greater prairie-chickens historically occurred in North America (Johnson et al. 2020). The extinct heath hen (*T. c. cupido*) once occupied areas of New England in grasslands and shrublands maintained by frequent fire. The Attwater's prairie-chicken (*T. c. attwateri*) is currently on the brink of extinction and is maintained in two isolated locations in Texas via captive breeding (Silvy et al. 1999). Only about 1% of the coastal grasslands in Texas that once supported Attwater's prairie-chicken remains (Smeins et al. 1991), providing limited carrying capacity. The greater prairie-chicken (*T. c. pinnatus*) has also shown significant population declines in the last several decades across its continually shrinking distribution (Braun et al.1994; Johnson et al. 2020). Greater prairie-chicken populations, which were once known to occur in 20 states and 4 provinces, are listed as threatened or extirpated in at least 15 states and provinces (Braun et al. 1994; Svedarsky et al. 2000). The earliest documentation by Euro-American settlers indicates that greater prairie-chickens were primarily found in the Midwestern portions of the United States. Some anecdotal notes suggest that grain crops initially caused distribution expansions west and north of historical distribution within the Great Plains (Johnson and Joseph 1989) and Johnsgard and Wood (1968) document that, except for the Flint Hills of Kansas and southeastern Nebraska, most large contemporary populations of greater prairie-chickens occur in areas that were not known to be occupied by this species until after Euro-American settlement.

9.2.2 Lesser Prairie-Chickens

Lesser prairie-chickens have experienced significant declines in distribution and population size since Euro-American settlement. Historically, the estimated distribution of lesser prairie-chickens extended over 180,000 km² across western Kansas and Oklahoma, eastern Colorado and New Mexico, and north-central Texas. Lesser prairie-chicken populations now occupy 17% of their historical distribution (Garton et al. 2016; Fig. 9.1). Sympatric overlap and hybridization with greater prairie-chickens north of the Arkansas River in northwestern Kansas (Bain and Farley 2002) is likely due to the conversion of former cropland to mixed grass prairie through programs such as the Conservation Reserve Program (Dahlgren et al. 2016); although genetic evidence suggests multiple periods of sympatry during the evolutionary history of these species (DeYoung and Williford 2016). Recent comprehensive population analyses have demonstrated long-term declines during the last century until apparent population stabilization in the mid-1990s (Garton et al. 2016). Regional populations exhibited signs of recovery during the early 2000s, but a range-wide drought during 2011–2013 reduced populations by 50% (McDonald et al. 2014). As a result of this rapid decline and ongoing threats, in 2014, the U.S. Fish and Wildlife Service (USFWS) listed the lesser prairie-chicken as threatened under the Endangered Species Act (1973; ESA). However, the listing was vacated by the U.S. District Court for the Western District of Texas in 2015 due to "substantial efforts already being made by state wildlife agencies, industries, and private landowners to restore and conserve lesser prairie-chicken habitat" (USFWS 2016). Annual population surveys up to 2020 indicated that the species has nearly recovered to 2011 pre-drought abundance levels (Nasman et al. 2020). Nevertheless, in November 2022, the USFWS relisted the lesser prairie-chicken under the ESA, this time into two Distinct Population Segments (DPS); the Southern DPS of lesser prairie-chickens occurring in New Mexico and Texas was listed as endangered and the Northern DPS occuring in Colorado, Kansas, and Oklahoma was listed as threatened.

9.2.3 Sharp-Tailed Grouse

Sharp-tailed grouse are the most widespread of the prairie grouse (Schroeder et al. 2004; Fig. 9.1), historically distributed across 21 states and 8 Canadian provinces (Aldrich 1963; Johnsgard 1973). There are six subspecies of sharp-tailed grouse, two of which are native to rangelands of western North America; Columbian sharp-tailed grouse (*T. p. columbianus*) and plains sharp-tailed grouse (*T. p. jamesi*; Connelly et al. 1998). Originally, the two subspecies were thought to be separated by the Continental Divide, with *T. p. jamesii* distribution limited to grasslands east of the Rocky Mountains. However, recent genetic evidence suggests that plains sharp-tailed grouse occupied intermountain valleys west of the Continental Divide (Warheit and Dean 2009). A third subspecies, prairie sharp-tailed grouse (*T. p. campestris*), occurs in rangeland and parkland of the upper Midwestern U.S. and Canada (Johnsgard 2016). The three northern races of sharp-tailed grouse (*T. p. caurus*, *T. p. kennicotti*, and *T. p. phasianellus*) occur in forest-dominated landscapes where information on rangeland management is lacking and are not covered here.

Distribution-wide, sharp-tailed grouse are currently considered stable (BirdLife International 2012; Panjabi et al. 2012). However, populations are extirpated from Kansas, Illinois, California, Oklahoma, Iowa, Nevada, New Mexico and Oregon (Johnsgard 1973). Declines are mainly attributed to habitat loss and fragmentation associated with conversions of rangeland to cultivation and other human development (Connelly et al. 1998; Schroeder et al. 2004). Columbian sharp-tailed grouse occur in remnant populations in British Columbia, Colorado, Idaho, Utah, Washington, and Wyoming (Johnsgard 2016). Columbian sharp-tailed grouse have been petitioned twice for threatened or endangered species listing, however, both instances resulted in a 'not warranted for listing' determination (U.S. Fish and Wildlife Service 2000, 2006). The species is listed as threatened by the State of Washington (Stinson and Schroeder 2012), a species of concern by the Province of British Columbia (Leupin and Chutter 2007) and by USFWS, and as a sensitive species by the Bureau of Land Management (BLM) and United States Forest Service (USFS).

9.3 Population Monitoring

Monitoring efforts for prairie grouse vary considerably and have generally been the responsibility of state and provincial wildlife agencies. In general, Great Plains and Midwestern states allocate twice as much effort monitoring game birds, including prairie grouse, as western states (Sands and Pope 2010). Similarly, there is variation among states relative to data collection on spatial–temporal population fluctuations.

9.3.1 Lek Surveys

Prairie grouse populations are typically monitored using lek surveys either at established leks or along survey routes (e.g., Utah Department of Natural Resources 2002; South Dakota Game, Fish, and Parks 2010). Spring lek count surveys provide estimates of relative abundance (Cannon and Knopf 1981; Reese and Bowyer 2007; Garton et al. 2016; Hoffman et al. 2015). Long term monitoring must have established protocols and consistent survey effort each year, otherwise comparisons across years are inappropriate (Luukkonen et al. 2009; Hoffman et al. 2015). Generally, population estimates are calculated by doubling the maximum count of males on leks in spring; this method assumes that all males attend leks and the sex ratio is at parity (Schroeder et al. 2008).

Recent work has highlighted several potential biases associated with using unadjusted counts of birds (e.g., lek counts) that are observed imperfectly (Royle and Dorazio 2008; Walsh et al. 2004). Certainly, raw lek counts should not be used to describe true population sizes or evaluate short-term population dynamics. Evaluations of scale-associated biases of lek counts as population indices are lacking for prairie grouse. However, recent work evaluating biases associated with lek survey protocols for sage-grouse indicate that lek count data generally correlate with annual abundance of males (Fedy and Doherty 2011), especially when (1) leks are surveyed multiple times each spring and the maximum or peak count is used as the index, and (2) inferences about trends are evaluated at large spatial scales (e.g., $\geq$ 50 leks; Fedy and Aldridge 2011). However, single counts timed to match the peak of attendance at more leks appear to have greater use than multiple counts at a smaller number of leks when the objective is monitoring at the scale of a population (e.g., hunting district; Fedy and Aldridge 2011). Research aimed at understanding covariate effects (e.g., habitat management) on population trends at local spatial scales, however, should strive to separate observation error (e.g., imperfect detection) from process variance (e.g., population size or growth rates) (Dail and Madsen 2011; Blomberg and Hagen 2020).

Recently, distribution-wide population monitoring for lesser prairie-chickens began aerial surveys to estimate annual abundance (McDonald et al. 2014). A probabilistic spatially balanced sampling frame composed of 15 $\times$ 15 km grid cells is laid over the regional distributions of the species. Grid cells are randomly selected for aerial surveys within each season. Two helicopter transects are flown per grid cell per year and all birds observed along each transect are counted by two observers, one located in the front and one in the rear of the helicopter. This double-observer method enables rigorous estimates of detection probability and abundance. Additionally, this grid has been adapted to estimate spatio-temporal changes in prairie-chicken occupancy as a function of landscape, habitat, and climatic covariates (Hagen et al. 2020).

9.3.2 Harvest Surveys

Surveys of grouse hunters by mail, phone, or at check stations are used by several states to index or estimate harvest (Sands and Pope 2010). Hunter reported harvests are used by management agencies to estimate harvest rates, estimate hunter effort, and index annual population sizes (e.g., South Dakota Fish, Game, and Parks 2010). Hunter-reported harvests should be considered an index of harvest, rather than true harvest, as respondents may inflate the number of birds they harvest (Atwood 1956; Martinson and Whitesell 1964). Generally, the total number of birds harvested, even if indexed accurately, is related more to hunter effort (number of hunters and days afield) than it is to population size. Thus, reported harvest rates are sometimes divided by reported hunter effort (e.g., harvest per hunter days) and this ratio used to index annual population sizes and temporal population trend (Beaman et al. 2005). This approach assumes that biases in reported hunter effort (rounding of effort, typically up) are consistent over time and across management units. Another major source of potential error associated with harvest surveys of hunters is non-response bias. Unsuccessful hunters who tend to not respond to surveys can have major influences on estimated harvest rates; however intensive resampling of non-respondents by mail and phone can minimize bias or allow surveyors to calculate correction factors (Aubry and Guillemain 2019). Harvest information provided by hunters at mandatory check stations are considered more reliable (Dahlgren et al. 2021).

9.3.3 Wing and Feather Collections

Several wildlife agencies use wing, tail, and scalp feather collections from harvested birds to provide information on sex and age-composition of the population, either through volunteer collection containers ("wing barrels") at common bird-hunting areas (Hoffman 1981) or from targeted mail-in programs (e.g., Alaska Department of Fish and Game; Idaho Department of Fish and Game). The collection of wings and tails from hunter-harvested birds is inexpensive relative to other monitoring programs that require active brood surveys or intensive telemetry-based studies. However, harvest metrics garnered from age- or sex-ratios estimated from hunter-harvested wings and tails likely do not accurately reflect population processes of interest (e.g., productivity, recruitment) because of differences in harvest vulnerability among sex and age-classes (Pollock et al. 1989). Empirical assessment of these biases for prairie grouse are lacking but have been demonstrated for other species of upland game birds [ruffed grouse (*Bonasa umbellus*): Fischer and Keith 1974; northern bobwhite (*Colinus virginianus*): Roseberry and Klimstra 1992; sage-grouse: Hagen et al. 2018]. Harvest rates of juveniles are generally higher than adults; therefore, age-ratios (juvenile:adult females) from harvested birds will yield upward biased estimates of true productivity and result in incorrect inferences regarding population dynamics, habitat quality, and other ecological processes of interest. Age-ratios from hunter-harvested wings may provide relative estimates of productivity across management units and years if harvest effort and harvest vulnerability of both juveniles and adults are consistent across space and time. This seems unlikely given annual variability in hunter effort (e.g., 5–30% annually; Oregon Department of Fish and Wildlife 2020), variability in harvest vulnerability (Caudill et al. 2017), and habitat/location effects on harvest rates (e.g., Breisjøberget et al. 2018; Davis et al. 2018). Further convoluting the use of hunter-collected materials is that biases due to vulnerability may or may not change during a single season (Flanders-Wanner et al. 2004). Overall, the use of uncorrected age- and sex-ratios from hunter-harvested wings and tails is tenuous. At a minimum, assessments of bias due to systematic changes in harvest-age ratios is a prerequisite of population-level analyses (Flanders-Wanner et al. 2010; Hagen et al. 2018). Information provided by hunter-harvested wings and tails are better used to monitor the distribution and timing of harvest, rather than the sex- or age-structure of the population.

9.3.4 Combining Multiple Datasets

Wildlife agencies often collect multiple independent datasets (e.g., lek counts, hunter surveys, wing/tail collections, habitat indices) (Broms et al. 2010). In some cases, a formal decision-making system like adaptive harvest management is used (Dahlgren et al. 2021). Integrated population models (IPMs; Schaub and Abadi 2011) were developed specifically to (1) more fully identify and account for the uncertainties in population parameters and (2) account for inherent biases in each data set when estimating population processes of interest (e.g., rates of population changes, Broms et al. 2010). These models are highly adaptable to a variety of data types, including traditional lek counts, productivity indices (e.g., brood counts, wing collections), harvest numbers that are collected at scales of a management unit, as well as localized data from intensive demographic study (e.g., nest survival, annual survival). Recent work has highlighted the use of IPMs to address a variety of scientific questions for grouse (Coates et al. 2014, 2018; McCafferty and Lukacs 2016; Ross et al. 2018; Milligan and McNew 2022). To date, however, IPMs have not been formally applied to state or region-wide population monitoring programs.

9.4 Habitat Associations

As their name implies, prairie grouse are obligate grassland/shrubland birds. The size, composition, and arrangement of seasonal habitat requirements is critical to maintain viable populations of prairie grouse (Temple 1992; Hoffman et al. 2015). Sensitivity to isolation becomes more pronounced as habitat patches become smaller, especially if barriers prevent movement of individuals among semi-isolated subpopulations (Temple 1992). Specific habitat requirements are discussed below, but there are some general patterns that are consistent across species. Prairie grouse are associated with large expanses of prairie (i.e., grasslands and shrublands) that are relatively unfragmented. For example, large scale crop cultivation and anthropogenic development is associated with population declines for all species of prairie grouse (McNew et al. 2012b; Garton et al. 2016; Runia et al. 2021) and tree encroachment reduces the availability and quality of prairie-chicken habitat (Fuhlendorf et al. 2017).

9.4.1 Greater Prairie-Chickens

Greater prairie-chickens are primarily found in tallgrass and mixed-grass prairies of the eastern Great Plains (Fig. 9.1). Habitat selection is similar throughout the species' distribution with less variation than observed with either lesser prairie-chickens or sharp-tailed grouse (Winder et al. 2015). However, greater prairie-chickens show seasonal variation in habitat selection that is associated with their diverse life-history (Londe et al. 2019; Svedarsky et al. 2022). Lek sites are areas characterized by low vegetation (< 15 cm) proximal to nesting cover and this can include crop fields, areas with intensive livestock use, mowed areas, recent burns, or areas with shallow depth to bedrock (Svedarsky et al. 2022). Lek sites are often at higher elevations in landscapes that have variable elevation (Hovick et al. 2015c) and occur in areas with relatively low proportions of cropland and forests (Niemuth 2000, 2003). In the southern Great Plains, greater prairie-chickens show strong avoidance of tree cover during all seasons (Merrill et al. 1999; Lautenbach et al. 2017; Londe et al. 2019).

Females tend to nest within 2 km of active leks (Hovick et al. 2015b), although this may be due to males choosing to lek near nesting cover due to the presence of females (Beehler and Foster 1988). A nest consists of a shallow depression generally with overhead grass cover (Hovick et al. 2015a; Matthews et al. 2013; Fig. 9.2) and intermediate litter depth (Svedarsky 1979). At local scales (i.e., within breeding season home ranges) nest site selection and success of prairie-chickens are strongly associated with the height and density of herbaceous vegetation. Visual obstruction reading (VOR), an index of herbaceous biomass and nest concealment (Robel et al. 1970), is a predominant measure of nesting habitat quality and is commonly associated with both female preference and nesting success (McNew et al. 2014, 2015; Powell et al. 2020). Nest success is often maximized at intermediate measures of VOR. For example, an optimum VOR of 27 cm was reported in Minnesota (Svedarsky 1979), and nest success decreased when VOR exceeded 40 cm (Buhnerkempe et al. 1984). In Kansas, nest site selection and success were maximized when VOR was 30–60 cm (McNew et al. 2014, 2015). An intermediate optimal of VOR indicate some degree of disturbance by fire or grazing is beneficial for nesting prairie-chickens, and this optimal can be realized through moderate livestock stocking rates (Kraft et al. 2021), and specialized rangeland management regimes (McNew et al. 2015; see Sect. 9.5). As nest site selection occurs prior to the current year's growth of most grasses, residual nesting cover from the previous season is critical. This has large implications for the distribution, area, and timing of grazing and prescribed fire.

Fig. 9.2 Nest sites and brood habitat of prairie grouse. Top: **a** hatched sharp-tailed grouse nest in grass cover, **b** sharp-tailed grouse nest in snowberry, and **c** greater prairie-chicken nest in tallgrass prairie. Bottom: brood habitat of **d** greater prairie-chickens, **e** lesser prairie-chickens, and **f** sharp-tailed grouse are rich in forbs and insects. *Photo credits* D. Elmore and M. Milligan

Brood-rearing habitat is characterized by areas with abundant forbs and insects that are open at ground level to accommodate the movements of chicks while providing overhead screening cover from predators (Svedarsky 1988; Matthews et al. 2011; Fig. 9.2). Females often move broods from nesting cover to suitable brood-rearing areas and the proximity of nesting and brood-rearing cover likely plays a role in brood success and therefore productivity. The plant communities association with brood-rearing vary regionally (reviewed in Svedarsky et al. 2022). For example, brood habitat has been described as recently-disturbed lowland areas with abundance sedges (*Carex* spp.), mixed upland vegetation dominated by forbs, and cool-season CRP fields in Minnesota, South Dakota, and Nebraska, respectively (Svedarsky 1979; Norton et al. 2010; Matthews et al. 2013). The plant community type is less important than micro-habitat conditions that provide a mix of cover and food resources (forbs and insects). Disturbances, including fire and livestock grazing, are known to create conditions favorable for broods across the species' distribution (Svedarsky 1979; Londe et al. 2021a). Heterogeneity of brood cover may also be important; brooding females select patches with higher overhead cover during the heat of the day and limit movement although areas that are more open were used during early morning when temperatures were lower (Londe et al. 2021a).

In general, information on habitat selection during the non-breeding season is lacking for greater prairie-chickens. In southern portions of their distribution, greater prairie-chickens use tallgrass prairie, including areas recently burned during the non-breeding period (Londe et al. 2019). Although they may use available grain crops, greater prairie-chickens in the southern Great Plains do not require them (Horak 1985). In northern portions of their distribution, use of agronomic crops appears more prevalent and has been suggested as leading to expansion of historic distribution (Kobriger 1965). Crops are used during fall and winter if available and can make up a substantial portion of the diet of greater prairie-chickens (Korschgen 1962; Rosenquist and Toepfer 1995). Nevertheless, land uses that reduce large areas of grassland are detrimental to greater prairie-chickens (Runia et al. 2021).

9.4.2 Lesser Prairie-Chickens

Habitat associations of lesser prairie-chicken are generally more xeric plant communities relative to those of other prairie grouse and vary latitudinally. Lesser prairie-chickens occur in four ecoregions. The mixed-grass prairie ecoregion extends from southwest Kansas through western Oklahoma into the northeast panhandle of Texas. Vegetation in this region consists of mid and tall grasses and often co-dominated by sand sagebrush (*Artemesia filifolia*) with patchy distributions of deciduous shrubs. The short-grass prairie ecoregion is contained within western Kansas and north of the Arkansas River and estimated to host most of the species' current abundance (McDonald et al. 2014). Vegetation here is largely sod-forming short-grasses interspersed with considerable acreage of Conservation Reserve Program (CRP) fields. Structurally and compositionally, these fields are similar to mixed grass prairie. The sand shinnery oak prairie ecoregion occupies portions of northwest Texas and eastern New Mexico. Dominated by shinnery oak (*Quercus havardii*) and mid-tall grass species, this region is perhaps the most susceptible to frequent and severe drought. The sand sagebrush prairie ecoregion spans from southeast Colorado along the Arkansas River into Kansas, and Oklahoma. Vegetation is similar to that of the shinnery oak in terms of soil types and herbaceous plant species, but sand sagebrush is the dominant or co-dominant vegetation.

Despite subtle differences in regional life-histories across ecoregions, lesser prairie-chicken females select nest sites based on similar vegetation structure regardless of plant species present (Hagen et al. 2013). Typically, nests have overhead cover which may be grass or shrub cover depending on the site (Hagen et al. 2013). Greater vegetation density as measured by VOR (25–40 cm) have been linked to nest site selection and nest success across the species' distribution, and in some cases optimum values have been identified (Hagen et al. 2013, Grisham et al. 2014, Lautenbach et al. 2021). Alternatively, brood habitat tends to have less dense vegetation, more bare ground, and generally has abundant forbs (Bell et al. 2010; Hagen et al. 2006, 2013; Fig. 9.2). Although lesser prairie-chickens have been observed using crop fields (e.g., alfalfa) during the reproductive stages, relatively large tracts of prairie, including shrublands, are required for nesting and rearing broods (Hagen et al. 2004).

Non-reproductive stages and winter habitat use is remarkably similar across the species distribution. Harvested croplands that contain waste grain are often selected for foraging sites when adjacent to prairie (Jones 1963; Hagen et al. 2007b). Shinnery oak plant communities provide acorns and insect galls for food (Jones 1963; Riley et al. 1993). Prairies, including shrubland, are used by lesser prairie-chickens for roosting and loafing when not feeding (Hagen et al. 2007b). In extreme winter conditions, lesser prairie-chickens have been observed feeding on the buds of deciduous shrubs along riparian corridors (Schwilling 1955). Home range size tends to double during the non-breeding season as resource availability declines (Hagen et al. 2007b; Robinson et al. 2018).

9.4.3 Sharp-Tailed Grouse

Sharp-tailed grouse occur throughout relatively large but variable subclimax brush or shrub-grassland rangeland communities in North America (Aldrich 1963; Johnsgard 2002; Fig. 9.1). As such, specific habitat needs are quite variable throughout the distribution making inference for management difficult to extend beyond a certain geographic area. Nevertheless, some general habitat conditions exist. Sharp-tailed grouse habitat consists of large tracts of native prairie (grasslands and shrublands depending on location and subspecies), wooded draws, and sometimes interspersed with conifers or cropland (Swenson 1985). Like prairie-chickens, habitat of sharp-tailed grouse is generally dominated by herbaceous vegetation (grasses and forbs; Connelly et al. 1998), especially during nesting and brood-rearing periods. Unlike prairie-chickens, during all seasons sharp-tailed grouse use deciduous shrubs which produce high energy food from berries, buds, and leaves (Evans and Dietz 1974), or cover for nests, broods, and adults (Northrup 1991).

Although sharp-tailed grouse have been observed in wheat and alfalfa fields, sharp-tailed grouse prefer native grassland/shrublands (Niemuth and Boyce 2004; Burr et al. 2017; Milligan et al. 2020a). Like prairie-chickens, sharp-tailed grouse nest sites are often characterized by relatively dense cover of grasses but will also nest within stands of shrubs (e.g., snowberry [*Symphoricarpos* spp.]; Pepper 1972; Marks and Marks 1988; Fig. 9.2); the availability of shrub thickets may offset effects of heavy grazing or drought which can limit herbaceous nest cover (Prose 1987; Kirby and Grosz 1995). Milligan et al. (2020a) found that both nest site selection and nest survival increased asymptotically with VOR with nest survival maximized when visual obstruction was 20–30 cm. Notably, positive effects of VOR extended only 6 m from nests suggesting cover can be relatively patchy for successful nesting. Similarly, brood-rearing habitat is characterized as grasslands having high heterogeneity in herbaceous biomass (e.g., VOR) and composition (e.g., % cover of forbs, grass, bare ground) that provide a combination of concealment from predators, food, and thermal cover for precocial chicks (Manzer and Hannon 2008; Goddard et al. 2009; Geaumont and Graham 2020).

Deciduous and coniferous uplands and riparian areas become increasingly important during the non-breeding season (Nielsen 1982; Northrup 1991; Deeble 1996). Shrubby draws and riparian areas are thought to provide food resources as well as thermal cover (Swenson 1985). Boisvert et al. (2005) observed that home range sizes increased during the fall and winter and included more diverse vegetation types, including crop fields where grouse will feed on waste grain. Nevertheless, research evaluating overwinter habitat use and its effect on survival are lacking.

9.5   Rangeland Management

9.5.1   Livestock Grazing

Grazing by livestock is the predominant land use of rangelands occupied by prairie grouse in the U.S. and southern Canada. Livestock do not directly affect prairie grouse demography; that is, livestock have not been demonstrated to kill or displace adults or young prairie grouse. Further, trampling of nests by cattle is infrequent (Pitman et al. 2006b; McNew et al. 2015; Milligan et al. 2020a). However, grazing by livestock can have indirect effects on prairie grouse through manipulation of vegetation and grazing infrastructure (e.g., fences and water tanks). As prairie grouse habitat is principally influenced by the amount, distribution, and types of vegetation on rangelands, management practices and uses that alter plant composition and structure can influence populations.

Not surprisingly, the efficacy of livestock grazing systems for influencing vegetation structure and composition (Briske et al. 2008) and wildlife habitat quality varies widely in the literature (Krausman et al. 2009; Schieltz and Rubenstein 2016). A wide range of stocking rates, season of use, and species of grazer can directly influence the structure and composition of vegetation and indirectly affect prairie grouse. Further, variability in average annual rangeland productivity and yearly variation due to climate lead to dramatic differences in grazing effects. For example, a livestock grazing system that improves habitat quality for nesting prairie grouse in a tallgrass prairie ecosystem (Fuhlendorf et al. 2006; Coppedge et al. 2008) may not have similar effects in the semi-arid mixed-grass prairie (Augustine and Derner 2015). Even within a single grassland ecosystem such as mixed-grass prairie, researchers have found variable responses to livestock grazing management that are apparently influenced by site-specific productivity and precipitation (Kraft et al. 2021).

The quality of prairie grouse habitat is determined by the spatial–temporal composition, structure, and productivity of vegetation that is largely driven by interactions between weather, disturbance (e.g., grazing, fire), and topo-edaphic features. Until recently, evaluations of the effects of grazing were based on correlations and perceptions of managers (e.g., Kessler and Bosch 1982; Klott and Lindzey 1990) or study designs with simplified 'grazed' or 'ungrazed' treatments (e.g., Kirby and Grosz 1995). More recent research (c.a. 2000–present) has focused on evaluating specific livestock grazing attributes to isolate the effects of weather and livestock grazing in the context of complex working landscapes (Table 9.1). Precipitation within seasons and across multiple years can affect prairie grouse demography directly, but also indirectly through the effects of grazing by livestock on the structure and composition of vegetation (Grisham et al. 2013). For example, season long (i.e., the entirety of the growing season) grazing aimed at 50% forage utilization had positive effects on population growth of lesser prairie-chickens prior to a drought but no measurable effects after a drought in the Sandhills of New Mexico (Fritts et al. 2018).

Stocking rate (i.e., the number of animals on a given amount of land over a certain period; Chap. 4) is probably the single most influential livestock management decision affecting habitat quality for prairie grouse because it is the primary driver of rangeland vegetation biomass, composition, and structure (Briske et al. 2008). Most prairie grouse co-evolved with large herbivores, many of which were nomadic and created pulses of heavy grazing followed by low to no grazing, often for multi-year periods. Low to moderate levels of grazing by livestock have limited effects on prairie grouse. For example, short-duration grazing with low–moderate forage utilization had no measurable effect on nest site selection or nest survival for lesser prairie-chickens (< 25% utilization; Fritts et al. 2016) or sharp-tailed grouse ($\leq 50\%$ utilization; Milligan et al. 2020a). High stocking rates at the pasture scale, however, can result in a lack of quality nesting and brood-rearing cover, especially under stocking regimes designed to homogenize livestock utilization across a management unit (e.g., intensive early stocking and annual burning; McNew et al. 2012b, 2015). Importantly, the effects of stocking rate on prairie grouse habitat will vary spatially and temporally due to differences in soil conditions and precipitation. The effects of an animal unit month (AUM) at a site that produces 2,000 kg ha^{-1} of herbaceous vegetation will differ significantly from a site with 800 kg ha^{-1}. Overall, livestock grazing systems that facilitate, rather than reduce, variation in the composition and structure of vegetation at multiple spatial scales should be the focus of management (Fuhlendorf et al. 2017; Kraft et al. 2021; Sect. 9.5.3) as this not only is similar to historic disturbance patterns with which prairie grouse co-evolved, but also provides prairie grouse with options to meet their various life history requirements.

9.5.2   Fire

Like grazing, fire can have both positive and negative effects on prairie grouse depending on frequency, size, pattern of burning and grazing, weather conditions, and interactions with other disturbance. If fire occurs in patchy distributions in time and space such that grassland heterogeneity matches prairie grouse habitat requirements, then it will benefit prairie grouse. However, large and frequent fires that remove most nesting cover have been shown to have negative effects on greater prairie-chickens because females select nesting areas characterized by moderate levels of herbaceous

Table 9.1 Responses of lesser (LEPC), greater (GPCH) prairie-chickens, and sharp-tailed grouse (STGR) to grazing management as reported in the primary peer-reviewed literature, 1970–2021. +, −, and 0 indicated a positive, negative, or no effect of a particular grazing treatment

Species	Study	Rangeland type, location	Management/grazing system	Stocking rate (AUM ha⁻¹)	Treatment effect	Measured response			
						Productivity	Adult survival	Space use/habitat selection	Population trend/viability
LEPC	Lautenbach et al. in press	Mixed grass prairie	Patch-burn grazing	0.8–1.0	System			+	
	Kraft et al. (2021)	Mixed and short grass prairies	Rotational	0.06–0.54	Stocking density (AU ha⁻¹)	0	0	+	
	Fritts et al. (2016)	Sand shinnery oak-teb treat	Rotational	0.02–0.07	Grazed vs. Ungrazed	0		0	
	Fritts et al. (2018)	Sand shinnery oak-pre drought	Season long	0.06	Pre drought	+	+		+
				0.02	Drought correction	0	0		0
GPCH	McNew et al. (2015), Winder et al. (2017, 2018)	Tallgrass, Flint Hills, KS	Patch-burn grazing	2–6	System	+	+	+	
	McNew et al. (2015), Winder et al. (2018)	Tallgrass, Flint Hills, KS	Variable	2–6	Stocking rate (AUM ha⁻¹)	−	−		
STGR	Kirby and Grosz (1995)	Northern mixed-grass, southcentral ND	Rotational	2.4–2.7	Grazed vs. Ungrazed	+			
	Milligan et al. (2020a, b, c); Milligan and McNew (2022)	Northern mixed-grass, eastern MT	Rest rotation w/ deferment	< 2.0	System	0	0	0	0

(continued)

Table 9.1 (continued)

Species	Study	Rangeland type, location	Management/ grazing system	Stocking rate (AUM ha^{-1})	Treatment effect	Measured response			
						Productivity	Adult survival	Space use/ habitat selection	Population trend/viability
	Milligan et al. (2020a, b, c)	Northern mixed-grass, eastern MT	Variable	< 2.0	Stocking rate (AUM ha^{-1})		0		

Note Only studies which directly assessed specific grazing effects on prairie grouse responses are included

biomass (see Sect. 9.4) available in patches that were burned 2–4 years ago (Hovick et al. 2014a; McNew et al. 2015; Lautenbach et al. 2021). Conversely, a lack of fire can often lead to tree encroachment, degrading habitat for greater prairie-chickens (Londe et al. 2019). Similarly, lesser prairie-chickens are negatively affected by lack of fire as tree encroachment is a primary threat (Lautenbach et al. 2017). Much of the research focused on fire and prairie grouse has occurred in the southern Great Plains, however as most prairies in the Great Plains are fire dependent systems that are capable of growing trees in the absence of periodic disturbance, some level of fire is likely needed to maintain prairie-chicken habitat regardless of the location. Sharp-tailed grouse are the most tolerant of trees and northern populations are adapted to parkland prairie vegetation in forested mosaics. The effect of prescribed fire on sharp-tailed grouse is poorly understood. However, the winter use of both deciduous and coniferous trees by sharp-tailed grouse across their distribution suggests a more restrictive use of prescribed fire in wintering areas may be warranted.

9.5.3 Managing for Heterogeneity

Vegetation heterogeneity, the variability in the composition and structure of plant communities over space and time, is correlated with diversity of both plants and animals (Wiens 1976; Tews et al. 2004). Heterogeneity may be variation in seral stages but also includes within seral stage variability. In fact, heterogeneity occurs and can be measured over multiple spatial scales that can be generalized into four categories relevant to land managers: landscape (> 100 km^2), ranch (10–100 km^2), among pasture (1–10 km^2), and within pasture (< 1 km^2; Toombs et al. 2010) and this hierarchy of scales in heterogeneity is important for prairie grouse. Historically, a combination of fire and herbivory by large herds of nomadic grazers maintained a shifting mosaic of heterogeneity at the landscape scale within rangelands (Fuhlendorf et al. 2006). Today, landscape-scale heterogeneity is determined by human land use patterns that are driven by a variety of economic and social-cultural traditions (see Chaps. 3 and 28). At ranch and pasture scales, herbivore grazing interacts with physical conditions (e.g., soils, topography, weather) to determine habitat heterogeneity within remaining grassland/shrubland habitats (Toombs et al. 2010; McNew et al. 2015).

Research has highlighted the importance of patch-level (i.e., within or among pasture) heterogeneity in grassland vegetation for prairie grouse (McNew et al. 2015; Winder et al. 2018; Sullins et al. 2018; Londe et al. 2019; Lautenbach et al. 2021). Rangeland management designed to create or restore patch-level structural heterogeneity to rangelands, such as patch-burn-grazing, has been successfully applied to grasslands in the southern Great Plains and have had positive effects on prairie-chickens relative to grazing systems designed to homogenize forage utilization by livestock (Fig. 9.3). For example, nest survival, adult survival, and habitat use by greater prairie-chickens were increased on rangelands managed with patch-burn grazing relative to intensive early stocking and annual spring burning in the Flint

Hills of Kansas (McNew et al. 2015; Winder et al. 2017, 2018). Similar results have been reported for lesser prairie-chickens in mixed-grass prairie of southcentral Kansas (Lautenbach et al. 2021).

Prescribed fire may not be a socially acceptable management tool in some places (Sliwinsky et al. 2018) and research has evaluated whether specialized livestock grazing management can improve habitat heterogeneity for prairie grouse without the use of prescribed fire. Rest-rotation systems that include season-long deferment (e.g., Hormay and Evanko 1958; Chap. 4) in the short-grass prairies of western Kansas provided vegetation heterogeneity that was selected by lesser prairie-chickens (Kraft et al. 2021). In contrast, rest-rotation grazing had no apparent effect on sharp-tailed grouse demography or space use relative to traditional season-long grazing in mixed-grass prairie of eastern Montana (Milligan et al. 2020a, b, c). In these studies, habitat heterogeneity was not influenced by grazing system because stocking rates were low to moderate and study areas were variable in topography and soil conditions (Fig. 9.3). For example, mean and variation of VOR, a key vegetation metric associated with

Fig. 9.3 Heterogeneous habitat of prairie grouse. Heterogeneity results from both intrinsic constraints on vegetation (e.g., soil, topography, precipitation) and from extrinsic forces (e.g. grazing, fire, herbicide) resulting from rangeland management. Habitat of greater prairie-chickens in Oklahoma (**a**), lesser prairie-chickens in Kansas (**b**), Columbian sharp-tailed grouse in Oregon (**c**) managed with combinations of prescribed fire and grazing. Bottom-right: heterogeneous habitat of plains sharp-tailed grouse in northern mixed-grass prairies grazed moderately by livestock in eastern Montana (**d**). Heterogeneity in the composition and structure of vegetation at multiple spatial scales benefits prairie grouse. *Photo credits* D. Elmore, M. Milligan, and N. Richter

nest survival of prairie grouse (Sect. 9.4), did not differ among pastures managed with season-long, summer rotational, or rest-rotation grazing management (Milligan et al. 2020a; Smith et al. 2020).

The disparity in results of heterogeneity-focused management suggest that relative effects of grazing on habitat selection and demography of prairie grouse are likely spatially variable and dependent on habitat conditions considered at broader spatial scales (McNew et al. 2013). Therefore, best management practices for even a single species within a single ecosystem may vary. Further, although one grazing system or stocking rate may favor certain life history aspects (e.g., nesting), it may not create conditions favorable for another (e.g., brood rearing). The important point is that no one grazing system should be broadly prescribed. As prairie grouse require habitat heterogeneity, rangeland management that promotes vegetation heterogeneity should be the goal (Fuhlendorf et al. 2017), particularly in grasslands that lack inherent heterogeneity due to topo-edaphic variation.

Although vegetation heterogeneity is important to prairie grouse, a hierarchy of habitat requirements constrains the effectiveness of habitat management prescriptions (Johnson 1980; Fuhlendorf et al. 2017). Despite local conditions that may be favorable, landscape factors such as tree encroachment, anthropogenic development, and conversion to crop may make local conditions irrelevant and doom prairie grouse populations (Hagen and Elmore 2016). So, although local management does matter, it does so only in the context of broader landscape characteristics (Toombs et al. 2010; Sect. 9.8).

9.6 Effects of Disease

Infectious disease is not thought to be a limiting factor to prairie grouse, but parasites are widespread with some populations having consistently high parasite loads (Peterson 2004). Although little evidence exists that parasites regulate prairie grouse populations, it is possible they could negatively affect populations that are already stressed (Peterson 2004). Population cycles of European red grouse (*Lagopus lagopus*) have been linked to parasitic nematodes (Hudson 1986). Currently there is no evidence of direct or indirect linkages between rangeland management and infectious agents of prairie grouse (Peterson 2004). Unlike greater sage-grouse and ruffed grouse, outbreaks of West Nile virus in prairie grouse have not been reported, or at least no population-level impacts have been observed. In western Europe, tick-borne flavivirus can cause significant economic impacts to livestock and grouse (Burrell et al. 2016); however, prairie grouse are not known to share diseases with livestock in North America.

9.7 Ecosystem Threats

9.7.1 Habitat Conversion

Historically, loss of habitat due to conversion to other land uses and land cover has been the primary threat to prairie grouse. Most of the tallgrass prairie in the Midwest has been converted to other uses and large portions of tallgrass and mixed grass prairies in the Great Plains have likewise been converted or fragmented. Much of this conversion has been to crop production and introduced grasses. Although some conversion is still occurring, the vast majority of arable land was altered decades ago which has dramatically lowered carrying capacity for prairie grouse. There is some evidence that introduction of crops up to some threshold may have allowed for greater prairie-chicken distribution expansion in the Great Plains (Johnson and Joseph 1989), yet overall, the conversion to crops has been negative for prairie grouse due to losses of large areas of rangelands. Additionally, tree encroachment due to fire suppression is a significant cause of land conversion that has negatively affected prairie grouse (Fig. 9.4). This has been particularly problematic in the southern Great Plains for both greater and lesser prairie-chickens (Falkowski et al. 2017; Lautenbach et al. 2017; Londe et al. 2019; Hagen et al. 2020). However, tree encroachment is occurring in the northern Great Plains as well where it threatens greater prairie-chickens and to a lesser extent sharp-tailed grouse (Berger and Baydack 1992). Urbanization has reduced significant amounts of prairie grouse habitat (Runia et al. 2021). In most areas occupied by prairie grouse, human density is low and not at high risk of urban development. However, even low-density housing and associated road and powerline network is problematic as prairie grouse have been shown to be sensitive to human development and avoid anthropogenic structures (Pitman et al. 2005). Habitat conversion can range from complete habitat loss such that prairie grouse populations are extirpated, to varying degrees of habitat loss and fragmentation that reduces carrying capacity.

9.7.2 Energy Development

Research results have been mixed in how prairie grouse respond to energy development (Hovick et al. 2014b; Lloyd et al. 2022), which is not surprising given the range of scales, vegetation types, seasons, and structure types evaluated. Research suggests that prairie-chickens avoid roads, powerlines, and oil/gas wells (Hagen 2010; Hovick et al. 2014b; Plumb et al. 2019) with degrees of avoidance varying among structure types and season. For example, greater prairie-chickens in Oklahoma avoided powerlines, roads, and high-density oil wells during the non-breeding period, with few effects noted during the breeding season (Londe et al. 2019). Nesting greater prairie-chickens in Nebraska avoided roads, but habitat selection was not affected by proximity to wind turbines (Harrison et al. 2017; Raynor et al. 2019). Lesser

Fig. 9.4 Threats to prairie grouse include loss and fragmentation of grasslands/shrublands, energy development, and rangeland degradation resulting from improper management. *Photo credits* D. Elmore and L. McNew

prairie-chickens in Kansas were not affected by a wind facility, although the authors caution that potential effects may have happened prior to data collection as the site was already impacted (LeBeau et al. 2020). Female greater prairie-chickens shifted core use areas away from wind turbines after construction (Winder et al. 2014b) but there was no effect on adult survival (Winder et al. 2014a) or nest site selection and survival within three years of development (McNew et al. 2014). Lek density of lesser prairie-chickens was negatively related to roads and active oil/gas wells in Texas (Timmer et al. 2014). In contrast, sharp-tailed grouse nest success in high density oil and gas areas within the Bakken Oil Field was nearly twice as high as in low-density areas, presumably due to reduced predator occupancy in high-density areas (Burr et al. 2017).

Some of the disparity in results among studies may be due to variability in experimental design, duration of study, and other mediating environmental conditions (e.g., habitat conditions, predator communities) that likely constrain the observable impacts of energy development on grouse (Lloyd et al. 2022). Overall, it appears that prairie grouse can tolerate some level of energy development, but as much of the Great Plains is at risk of becoming industrialized, fragmentation may exceed prairie grouse tolerances. Additional research with standardized designs that occur across gradients of mediating factors are needed. Inferences offered by studies that incorporate pre-construction data, include some form of control, and that are based on longer time-series should be prioritized. Long study durations (> 5 years) are especially important for prairie grouse because high site fidelity to historic leks may result in a delayed response to energy development (Lloyd et al. 2022). Further, little research has been conducted during the non-breeding season, yet limited data suggest this may be the period when prairie grouse are most sensitive to anthropogenic development (Londe et al. 2019).

9.7.3 Invasive Species

Human land use including livestock grazing, conversion to introduced forages, road construction, and vehicle travel have all contributed to invasive species becoming established across rangelands of North America. Some of these invasive species exist at low to moderate density and are used by prairie grouse and therefore not generally considered problematic for grouse conservation. Such species include dandelion (*Taraxacum* spp.), salsify (*Tragopogon dubius*), alfalfa (*Medicago sativa*), and kochia (*Kochia scoparia*). However, several species are highly problematic due to their aggressive nature and ability to shift plant communities and suppress more desirable vegetation. These include sericea lespedeza (*Lespedeza cuneata*), Old World bluestem (*Bothriochloa bladhii* and *B. ischaemum*), and exotic bromes (*Bromus* spp.). These plants often form large monotypic stands which are incapable of providing all the habitat requirements for prairie grouse. For example, although the exotic sericea lespedeza is sometimes used as brood cover during the heat of the day, its seed passes through the gut of galliforms undigested and it can displace more desirable forbs (Baldwin-Blocksome 2006). Control methods of various invasive species vary with some being vulnerable to grazing and or fire. Herbicide (e.g., 2, 4-D, triclopyr) can be effective at killing invasive plants, however collateral damage to nontarget plants is often a substantial problem. Spot application vs pasture level spraying can reduce collateral damage. Additionally, the use of selective herbicides may lessen collateral damage to desired plants. Biological controls have proven effective for some invasive species such as saltcedar (*Tamarix spp.*), leafy spurge (*Euphorbia esula*), and musk thistle (*Carduus nutans*). Regardless of the control used, the goal should be to target problem plants if they are reducing habitat quality for prairie grouse while minimizing loss of desirable plant species.

9.7.4 Climate Change

Areas occupied by prairie grouse are expected to undergo dramatic climatic shifts by the end of the century. Below we have summarized model output for the distribution of prairie grouse as obtained from Climate Wizard accessed on 30 March 2021

(Girvetz et al. 2009). Rangelands of southern populations of greater and lesser prairie-chickens are expected to become drier by 2100, particularly during summer months. In contrast, northern distributions of greater prairie-chickens and sharp-tailed grouse are projected to become wetter by 2100. These changes can have both positive and negative effects on prairie grouse survival and reproduction depending on exact timing and distribution of precipitation (Londe et al. 2021b). Temperature is also expected to depart from current conditions; the entire distribution of prairie grouse is expected to be warmer during every month of the year. Portions of the northern Great Plains are expected to depart the most from current temperature during the winter months. Extreme temperature departures are expected throughout much of the Great Plains, across the mountain west, and into the Pacific Northwest during the summer. These predictions suggest increased frequency of flash droughts, extended drought, and reduced snow retention, all of which have implications for prairie grouse conservation. Changes in atmospheric CO_2 will also affect plant composition and dominance in the future. While plant composition is affected directly by management, land use, soils and other factors, CO_2 can facilitate some plants such as C3 pathway woody plants (Archer et al. 1995) and may exacerbate tree encroachment in some areas where prairie grouse occur.

The resulting regional effects of changing climates on prairie grouse are unknown but cast uncertainty on whether current species' distributions will be within the range of environmental tolerances (i.e., niche) for prairie grouse. Increasing temperatures are especially relevant for grouse that evolved and primarily occur in northern climes. Southern populations of prairie grouse may be particularly at risk given evidence that temperature and precipitation directly affect vital rates within and across years (Bell et al. 2010; Grisham et al. 2013; Hovick et al. 2014a, b; Londe et al. 2019). Increased productivity of northern rangelands due to greater precipitation and CO_2 levels may alter habitat management recommendations including the timing, intensity, and duration of livestock grazing and application of prescribed fire to prevent forestation and maintain prairies (Symstad and Leis 2017; Brookshire et al. 2020). Although there is uncertainty with any climate model, it is important to note that change is predicted for most areas where prairie grouse occur. This change should be considered in conservation planning to allow for flexibility in management as well as mitigation for climate change through increased habitat quality, quantity, and spatial distribution.

9.8 Conservation and Management Actions

9.8.1 Reversing the Loss and Fragmentation of Grassland

Despite the tremendous variation in vegetation and climatic regimes both among and within the distributions of prairie grouse, one trait is shared among the species—they require large and relatively intact rangeland (i.e., shrubland and/or grassland) landscapes, of which we have few remaining. Although the exact size of landscapes necessary for population persistence is unknown, it is likely tens of thousands of acres based on characteristics of stable populations. Except for sharp-tailed grouse habitat in the far north, the majority of these landscapes are privately owned, and require broad coalitions and partnerships to implement conservation at meaningful scales (Elmore and Dahlgren 2016). The threats facing prairie grouse are as large and diverse as the landscapes on which they depend, devising relevant conservation actions for these species requires a strategic approach (Gerber 2016). First, there must be a recognition that we may not be able to conserve it all and pragmatism is needed to identify the most important areas for conservation. Several efforts have been initiated to prioritize prairie grouse conservation at local, state, and regional levels (Fandel and Hull 2011; Van Pelt et al. 2013). Typically, such efforts first identify population core areas based on breeding bird density or species distribution modeling (Niemuth 2011). Once populations have been mapped and prioritized, then landscapes can be targeted for conservation actions to maintain or increase prairie (Hagen and Elmore 2016; Sullins et al. 2019).

First order goals, such as mapping exercises that demonstrate the extent of the threats (e.g., woodland conversion, energy development) to each core landscape enable managers to strategically manage appropriate resources to maintain that landscape (Sullins et al. 2019; Schindler et al. 2020). Then, second order goals, like land management to improve vegetation communities for prairie grouse and promote heterogeneity would be prudent (Hagen et al. 2013; Hagen and Elmore 2016). Local scale conservation must be implemented in the context of broader landscapes; if surrounded by larger threats, even the best local scale management will be in vain. Landscapes that will be largely converted to anthropogenic development, crops, or tree cover are doomed for prairie grouse regardless of local management.

Nearly ubiquitously, prairie grouse occur in working landscapes and additional resources associated with conservation must be mitigated through cost-share and technical assistance programs to help incentivize landowner participation (Santo et al. 2020; Schindler et al. 2020). Here again, spatial targeting tools can assist in identifying landscapes in which specific outreach (e.g., direct mailings) and extension (e.g., town hall meetings) efforts can be focused to maximize landowner participation in conservation efforts (Sullins et al. 2019; Schindler et al. 2020). Finally, monitoring programs to assess ecological and socio-economic outcomes from conservation actions are vital to ensure effectiveness overtime and to adapt implementation as necessary.

9.8.2 Habitat Management

Practices and principles within rangeland management are fundamentally dependent upon geographic location (Holechek et al. 2011). The broad spatial extent of North America's prairie ecosystems accentuates the importance of recognizing innate variability when managing rangelands. Variable productivity among prairie ecosystems, driven largely by regional climate, influences the vegetative characteristics within a specific landscape (Holechek et al. 2011). Even within a single ecosystem, annual variability in precipitation from one growing season to the next significantly affects vegetation structure and composition (Lwiwski et al. 2015). Without accounting for this variation, management actions may not meet wildlife habitat goals.

Despite the temporal and spatial variation within rangelands, prairie grouse have basic life history requirements that must be fulfilled for sustainable populations. All prairie grouse require some level of vegetation heterogeneity to meet nesting, brood rearing, and non-breeding needs. Grazing/rest, prescribed fire, mechanical disturbance, and herbicide application can all be used to meet these habitat requirements but there is no uniform prescription (Sect. 9.5). Managers should seek to understand habitat requirements of target species and apply appropriate disturbances and rest as needed depending on landscape context and environmental variability. This necessitates active and adaptive management across landscapes and years. Management that seeks stability or uniformity is likely to fail to meet prairie grouse objectives in rangelands that are inherently dynamic. Optimal management would be flexible and nimble to ensure that all parts of prairie-grouse habitat requirements are met at sufficiently large scales. Finally, as habitat selection is a hierarchical process, landscape features that render areas unusable, such as those impacted by human development or tree encroachment, will make smaller scale management within those landscapes irrelevant for prairie grouse (Hagen and Elmore 2016; Fuhlendorf et al. 2017).

9.8.3 Standardizing Population Monitoring

Managers have historically monitored populations of prairie grouse with ground-based lek counts (Bibby et al. 2000) and road-based lek surveys (Best et al. 2003). However, specific protocols vary among states. For example, Hagen et al. (2017) reported that spring monitoring of lesser prairie-chicken populations varied considerably among each of the 5 states where the species occurs, with some states surveying a set of annually monitored leks (e.g., Colorado) and others using road-based surveys with systematic listening stops (e.g., Kansas; Van Pelt et al. 2013). Other states lack standardized survey protocols even within their jurisdictions. Non-standardized approaches make the comparison of common monitoring metrics (e.g., average number of males per lek) across administrative jurisdictions inappropriate and necessitate the use of more complex estimators (Garton et al. 2016). Regardless of survey platform (i.e., air- or ground-based), the development of a standardized and robust population monitoring protocol for prairie grouse should be prioritized so that regional and range-wide evaluations of population trends are possible (Runia et al. 2021).

9.8.4 Research Needs

In general, information regarding chick (i.e., survival from hatch to fledging) and juvenile survival (i.e., survival from fledging to first breeding) is lacking for prairie grouse. Most previous studies have been limited to flushing broods and have not monitored individual chicks with telemetry to evaluate survival, brood amalgamations, or survival to first breeding (Pitman et al. 2006b). Additionally, factors associated with stable populations are poorly understood, including minimum viable populations, dispersal and filters/barriers to movement, and minimum landscapes (i.e., patch size and connectivity) necessary to support viable populations. Modeling the effect of climate change projections on future distributions and estimated changes in lamba are also needed. Although multiple studies have evaluated the effects of energy development on space use and vital rates during the breeding season, research during the non-breeding season is generally lacking. Further evaluation is needed across species and landscapes before management recommendations are made. Additionally, although there are multiple publications that mention the effects of grazing on prairie grouse, most are either speculative, lack sufficient controls, or were not vetted by peer review (Dettenmaier et al. 2017; Table 9.1). We encourage future assessments of the effects of grazing on prairie grouse to be rigorous, include authors with expertise in rangeland ecology, and be peer reviewed. If site- and management-specific parameters are not considered empirically as covariates, then detailed descriptions of the study systems (e.g., soil or ecological sites, annual precipitation; timing, duration, and intensity of livestock grazing) should be provided so that reported effects can be put into context. Finally, most studies have been limited to short durations with inconsistent designs among studies. Multiple concurrent studies across variable landscapes using standardized approaches are needed to sort out effects of interest from site-specific variability (Lloyd et al. 2022).

References

Aldrich JW (1963) Geographic orientation of American Tetraonidae. J Wildl Manage 27:529–545

Apa A (1998) Habitat use and movements of sympatric sage and Columbian sharp-tailed grouse in South Eastern Idaho. University of Idaho, Moscow, ID, USA

Archer S, Schimel DS, Holland EA (1995) Mechanisms of shrubland expansion: land use, climate or CO_2? Clim Change 29:91–99

Atwood EL (1956) Validity of mail survey data on bagged waterfowl. J Wildl Manage 20:1–16

Aubry P, Guillemain M (2019) Attenuating the nonresponse bias in hunting bag surveys: the multiphase sampling strategy. PLoS ONE 14(3):e0213670. https://doi.org/10.1371/journal.pone.0213670

Augustine DJ, Derner JD (2015) Patch-burn grazing management, vegetation heterogeneity, and avian responses in a semi-arid grassland. J Wildl Manage 79:927–936. https://doi.org/10.1002/jwmg.909

Bain MR, Farley GH (2002) Display by apparent hybrid prairie-chickens in a zone of geographic overlap. Condor 104:683–687

Baldwin-Blocksome CE (2006) Sericea lespedeza: seed dispersal, monitoring, and effect on species richness. Dissertation, Kansas State University, Manhattan, Kansas

Beaman JJ, Vaske JJ, Miller CA (2005) Hunting activity record-cards and the accuracy of survey estimates. Hum Dimens Wildl 10:285–292. https://doi.org/10.2193/0022-541X(2005)069[0967:CPIHRO]2.0.CO;2

Beehler BM, and Foster MS (1988) Hotshots, hotspots, and female preference in the organization of lek mating systems. Am Nat 131:203–219. https://www.jstor.org/stable/2461845

Bell LA, Fuhlendorf SD, Patten MA et al (2010) Lesser prairie-chicken hen and brood habitat use on sand shinnery oak. Rangel Ecol Manag 63:478–486. https://doi.org/10.2111/08-245.1

Behney AC, Grisham BA, Boal CW et al (2012) Sexual selection and mating chronology of lesser prairie-chickens. The Wilson J Ornithol 124:96–105. https://doi.org/10.1676/11-079.1

Berger RP, Baydack RK (1992) Effects of aspen succession on sharp-tailed grouse, *Tympanuchus phasianellus*, in the Interlake region of Manitoba. Can Field Nat. 106:185–1991

Bergerud AT, Davies RG, Gardarsson A et al (1988) Population ecology of North American grouse. In: Bergerud AT, Gratson MW (eds) Adaptive strategies and population ecology of northern grouse. University of Minnesota Press, Minneapolis, pp 578–648

Best TL, Geluso K, Hunt JL, McWilliams LA (2003) The lesser prairie chicken (*Tympanuchus pallidicinctus*) in southeastern New Mexico: a population survey. Tex J Sci 55:225–234

Bibby CJ, Burgess ND, Hill DA, Mustoe S (2000) Bird census techniques, 2nd edn. Academic Press, San Diego

BirdLife International (2012) *Tympanuchus phasianellus*. In: The IUCN red list of threatened species 2012. International Union for Conservation of Nature and Natural Resources

Blomberg EJ, Hagen CA (2020) How many leks does it take? Minimum sample sizes for measuring local-scale conservation outcomes in Greater Sage-grouse. Avian Conserv Ecol 15(1):9. https://doi.org/10.5751/ACE-01517-150109

Boddicker ML, Huggins AJ (1965) Some parasites of sharp-tailed grouse in South Dakota. Transactions of the Annual Summer Conference, Central Mountains and Plains Section of the Wildlife Society 10:19

Boisvert JH (2002) Ecology of Columbian sharp-tailed grouse associated with Conservation Reserve Program and reclaimed surface mine lands in northwestern Colorado. M.S. thesis, University of Idaho, Moscow, ID

Boisvert JH, Hoffman RW, Reese KP (2005) Home range and seasonal movements of Columbian sharp-tailed grouse associated with Conservation Reserve Program and mine reclamation. West North Am Nat 65:36–44

Braun CE, Martin K, Remington TE, Young JR (1994) North American grouse: issues and strategies for the 21st century. In: McCabe RE, Wadsworth KG (eds) Trans 59th N Am Wildl Nat Resour Conf, pp 428–438

Breisjøberget JI, Odden M, Storaas T et al (2018) Harvesting a red-listed species: determinant factors for willow ptarmigan harvest rates, bag sizes, and hunting efforts in Norway. Eur J Wildl Res 64:54. https://doi.org/10.1007/s10344-018-1208-8

Broms K, Skalski JR, Millspaugh JJ et al (2010) Using statistical population reconstruction to estimate demographic trends in small game populations. J Wildl Manage 74:310–317. https://doi.org/10.2193/2008-469

Briske DD, Derner JD, Brown JR et al (2008) Rotational grazing on rangelands: reconciliation of perception and experimental evidence. Rangel Ecol Manag 61:3–17. https://doi.org/10.2111/06-159R.1

Brookshire EN, Stoy JPC, Currey B, Finney B (2020) The greening of the Northern Great Plains and its biogeochemical precursors. Glob Change Biol 26:5404–5413. https://doi.org/10.1111/gcb.15115

Buhnerkempe JE, Edwards WR, Vance DR, Westemeier RL (1984) Effects of residual vegetation on prairie-chicken nest placement and success. Wildl Soc Bull 12:382–386

Burr PC, Robinson AC, Larsen RT et al (2017) Sharp-tailed grouse nest survival and nest predator habitat use in North Dakota's Bakken Oil Field. PLoS ONE 12(1):e0170177. https://doi.org/10.1371/journal.pone.0170177

Burrell CJ, Howard CR, Murphy FA (2016) Fenner and White's medical virology, 5th edn. Academic Press, London, UK

Cannon RW, Knopf FL (1981) Lek numbers as a trend index to prairie grouse populations. The J Wildl Manage 45:776–778

Caudill D, Guttery MR, Terhune TM et al (2017) Individual heterogeneity and effects of harvest on greater sage-grouse populations. J Wildl Manage 81:754–765. https://doi.org/10.1002/jwmg.21241

Coates PS, Halstead BJ, Blomberg EJ et al (2014) A hierarchical integrated population model for greater sage-grouse (*Centrocercus urophasianus*) in the Bi-State Distinct Population Segment, California and Nevada: U.S. Geological Survey Open-File Report 2014–1165. https://doi.org/10.3133/ofr20141165

Coates PS, Prochazka BG, Ricca MA et al (2018) The relative importance of intrinsic and extrinsic drivers to population growth vary among local populations of Greater Sage-Grouse: an integrated population modeling approach. Auk 135(2):240–261. https://doi.org/10.1642/AUK-17-137.1

Collins CP (2004) Ecology of Columbian sharp-tailed grouse breeding in coal mine reclamation and native upland cover types in North Western Colorado. University of Idaho, Moscow, ID, USA

Connelly JW, Gratson MW, Reese KP (1998) Sharp-tailed grouse (*Tympanuchus phasianellus*). In: Poole A, Gill F (eds) The birds of North America. No. 354. Academy of Natural Sciences, Philadelphia

Coppedge BR, Fuhlendorf SD, Harrell WC, Engle DM (2008) Avian community response to vegetation and structural features in grasslands managed with fire and grazing. Biol Conserv 141, 1196e1203. https://doi.org/10.1016/j.biocon.2008.02.015

Dahlgren DK, Rodgers RD, Elmore RD, Bain MR (2016) Grasslands of western Kansas, north of the Arkansas River. In: Haukos DA, Boal CW (eds) Ecology and conservation of lesser prairie-chickens. Studies in avian biology (no. 48), CRC Press, Boca Raton, FL USA, pp 259–279

Dahlgren DK, Blomberg EJ, Hagen CA, Elmore RD (2021) Upland game bird harvest management. In: Pope KL, Powell LA (eds) Harvest of fish and wildlife: new paradigms for sustainable management. CRC Press, Boca Raton, FL, USA, pp 307–326

Dail D, Madsen L (2011) Models for estimating abundance from repeated counts of an open metapopulation. Biometrics 67:577–587. https://doi.org/10.1111/j.1541-0420.2010.01465.x

Davis SR, Mangelinckx BJ, Allen RB et al (2018) Survival and harvest of ruffed grouse in central Maine, USA. J Wildl Manage 82:1263–1272. https://doi.org/10.1002/jwmg.21483

Deeble BD (1996) Conservation of Columbian sharp-tailed grouse, with special emphasis on the upper Blackfoot Valley, Montana. University of Montana, Missoula, MT, USA

Dettenmaier SJ, Messmer TA, Hovick TJ et al (2017) Effects of livestock grazing on rangeland biodiversity: a meta-analysis of grouse populations. Ecol Evol 7:7620–7627.https://doi.org/10.1002/ece3.3287

DeYoung RW, Williford DL (2016) Genetic variation and population structure in the prairie grouse: Implications for conservation of the lesser prairie-chicken. In: Haukos DA, Boal CW (eds) Ecology and conservation of lesser prairie-chicken. Studies in avian biology (no. 48), CRC Press, Boca Raton, FL, pp 77–97

Earl JE, Fuhlendorf SD, Haukos D et al (2016) Characteristics of lesser prairie-chicken (*Tympanuchus pallidicinctus*) long-distance movements across their distribution. Ecosphere 7:e01441. https://doi.org/10.1002/ecs2.1441

Elmore RD, Dahlgren DK (2016) Public and private land conservation dichotomy, vol 48, pp 187–203. In: Haukos DA, Boal CW (eds) Ecology and conservation of lesser prairie-chickens. Studies in avian biology (no. 48), CRC Press, Boca Raton, FL, pp 187–203

Emlen ST, Oring LW (1977) Ecology, sexual selection, and the evolution of mating systems. Science 197(4300):215–223

Evans KE, Dietz DR (1974) Nutritional energetics of sharp-tailed grouse during winter. J Wildl Manage 38:622–629

Falkowski MJ, Evans J, Naugle D et al (2017) Mapping tree canopy cover in support of proactive prairie grouse conservation in western North America. Rangel Ecol Manag 70:15–24. https://doi.org/10.1016/j.rama.2016.08.002

Fandel SG, Hull S (2011) Wisconsin sharp-tailed grouse: a comprehensive management and conservation strategy. Wisconsin Department of Natural Resources. Madison WI

Fedy BC, Aldridge CL (2011) The importance of within-year repeated counts and the influence of scale on long-term monitoring of sage-grouse. J Wildl Manage 75:1022–1033. https://doi.org/10.1002/jwmg.155

Fedy BC, Doherty KE (2011) Population cycles are highly correlated over long time series and large spatial scales in two unrelated species: greater sage-grouse and cottontail rabbits. Oecologia 165:915–924. https://doi.org/10.1007/s00442-010-1768-0

Fischer CA, Keith LB (1974) Population response of central Alberta ruffed grouse to hunting. J Wildl Manage 35:585–600

Flanders-Wanner BL, White GC, McDaniel LL (2004) Weather and prairie grouse: dealing with effects beyond our control. Wildl Soc Bull 32:22–34. https://doi.org/10.2193/0091-7648(2004)32[22:WAPGDW]2.0.CO;2

Flanders-Wanner BL, White CC, McDaniel LL (2010) Validity of prairie grouse harvest age-ratios as production indices. J Wildl Manage 68:1088–1094. https://doi.org/10.2193/0022-541X(2004)068[1088:VOPGHR]2.0.CO;2

Fuhlendorf SD, Harrell WC, Engle DM, Hamilton RG, Davis CA, Leslie DM (2006) Should heterogeneity be the basis for conservation? Grassland bird response to fire and grazing. Ecol Appl 16:1706–1716. https://doi.org/10.1890/1051-0761(2006)016[1706:SHBTBF]2.0.CO;2

Fuhlendorf SD, Hovick TJ, Elmore RD et al (2017) A hierarchical perspective to woody plant encroachment for conservation of prairie-chickens. Rangel Ecol Manag 70:9–14. https://doi.org/10.1016/j.rama.2016.08.010

Fritts S, Grisham BA, Haukos DA, Boal CW et al (2016) Long-term lesser prairie-chicken nest ecology in response to grassland management. J Wildl Manage 80:527–539. https://doi.org/10.1002/jwmg.1042

Fritts S, Grisham B, Cox R et al (2018) Interactive effects of severe drought and grazing on the life history cycle of a bioindicator speceis. Ecol and Evol 8:9550–9562. https://doi.org/10.1002/ece3.4432

Garton EO, Hagen CA, Beauprez GM et al (2016) Population dynamics of the lesser prairie-chicken. In: Haukos DA, Boal C (eds) Ecology and conservation of lesser prairie-chicken. CRC Press, Boca Raton, pp 49–76

Geaumont BA, Graham DL (2020) Factors affecting sharp-tailed grouse brood habitat selection and survival. Wildlife Biol 2020(2). https://doi.org/10.2981/wlb.00633

Gerber LR (2016) Conservation triage or injurious neglect in endangered species recovery. Proc Nat Acad Sci 113:3563–3566. https://doi.org/10.1073/pnas.1525085113

Gillette GL (2014) Ecology and management of Columbian sharp-tailed grouse in southern Idaho: evaluating infrared technology, the Conservation Reserve Program, statistical population reconstruction, and the olfactory concealment theory. Dissertation, University of Idaho

Girvetz EH, Zganjar C, Raber GT et al (2009) Applied climate-change analysis: the climate wizard tool. PLoS ONE 4(12):e8320. https://doi.org/10.1371/journal.pone.0008320

Goddard AD, Dawson RD, Gillingham MP (2009) Habitat selection by nesting and brood-rearing sharp-tailed grouse. Can J Zool 87:326–336. https://doi.org/10.1139/Z09-016

Gratson MW (1988) Spatial patterns, movements, and cover selection by sharp-tailed grouse. University of Minnesota Press

Grisham BA, Boal CW, Haukos DA et al (2013) The predicted influence of climate change on lesser prairie-chicken reproductive parameters. PLoS ONE 8(7):e68225. https://doi.org/10.1371/journal.pone.0068225

Grisham BA, Borsdorf PK, Boal CW, Boydston KK (2014) Nesting ecology and nest survival of lesser prairiechickens on the Southern High Plains of Texas. J Wildl Manag (78):857–866. https://doi.org/10.1002/jwmg.716

Hagen CA (2010) Impacts of energy development on prairie grouse ecology: a research synthesis. Tran 75th N Amer Wildl Nat Res Conf 75:98–105

Hagen CA, Giesen KM (2005) Lesser prairie-chicken (*Tympanuchus pallidicinctus*) version 1.0. In: Poole AF (ed) Birds of the World. Cornell Lab of Ornithology, Ithaca. USA. [online] URL: http://bna.birds.cornell.edu/review/species/364

Hagen CA, Jamison BE, Giesen KM, Riley TZ (2004) Guidelines for managing lesser prairie-chicken populations and their habitats. Wildl Soc Bull 32:69–82. https://doi.org/10.2193/0091-7648(2004)32[69:GFMLPP]2.0.CO;2

Hagen C, Elmore RD (2016) Synthesis, conclusions, and a path forward. In: Haukos DA, Boal CS (eds) Ecology and conservation of lesser prairie-chickens. Cooper Ornithological Society, University of California Press, Los Angeles, pp 345–351

Hagen CA, Pitman JC, Sandercock BK et al (2005) Age-specific variation in apparent survival rates of male lesser prairie-chickens. Condor 107:78–86. https://doi.org/10.1093/condor/107.1.78

Hagen CA, Sandercock BK, Pitman JC et al (2006) Radiotelemetry survival estimates of lesser prairie-chickens in Kansas: are there transmitter biases? Wildl Soc Bull 34:1064–1069. https://doi.org/10.2193/0091-7648(2006)34[1064:RSEOLP]2.0.CO;2

Hagen CA, Pitman JC, Sandercock BK et al (2007a) Age-specific survival and probable causes of mortality in female lesser-prairie-chickens. J Wildl Manage 71:518–525. https://doi.org/10.2193/2005-778

Hagen CA, Pitman JC, Robel RJ et al (2007b) Niche partitioning in lesser prairie chickens *Tympanuchus pallidicinctus* and ring-necked pheasants *Phasianus colchicus* in southwestern Kansas. Wildl Biol 13(suppl 1):33–41. https://doi.org/10.2981/0909-6396(2007)13[34:NPBLPT]2.0.CO;2

Hagen CA, Sandercock BK, Pitman JC et al (2009) Spatial variation in lesser prairie-chicken demography: a sensitivity analysis of population dynamics and management alternatives. J Wildl Manage 73:1325–1332. https://doi.org/10.2193/2008-225

Hagen CA, Grisham BA, Boal CS et al (2013) A meta-analysis of lesser prairie-chicken nesting and brood-rearing habitats: Implications for habitat management. Wildl Soc Bull 37:750–758. https://doi.org/10.1002/wsb.313

Hagen CA, Garton EO, Beauprez G et al (2017) Lesser prairie-chicken population forecasts and extinction risks. Wildl Soc Bull 41:624–638. https://doi.org/10.1002/wsb.836

Hagen CA, Sedinger JE, Braun CE (2018) Estimating sex-ratio, survival and harvest susceptibility in greater sage-grouse: making the most of hunter harvests. Wildl Biol 2018(1). https://doi.org/10.2981/wlb.00362

Hagen CA et al (2020) Multi-scale occupancy of the lesser prairie-chicken: the role of private lands in conservation of an imperiled bird. Avian Conserv Ecol 15:17. https://doi.org/10.5751/ACE-01672-150217

Harrison JO, Bomberger Brown M, Powell LA et al (2017) Nest site selection and nest survival of Greater Prairie-Chickens near a wind energy facility. Condor 119:659–672. https://doi.org/10.1650/CONDOR-17-51.1

Hart CM, Lee OS, Low JB (1950) The sharp-tailed grouse in Utah: its life history, status and Management. Utah Department Fish and Game, Federal Aid Division Publication 3, 79 pp

Haukos DA, Smith LM (1999) Lek longevity and age structure of attending lesser prairie-chickens. Am Midl Nat 142:415–420. https://doi.org/10.1674/0003-0031(1999)142[0415:EOLAOA]2.0.CO;2

Hiller TL, McFadden JE, Powell LA et al (2019) Seasonal and interspecific landscape use of sympatric greater prairie-chickens and plains sharp-tailed grouse. Wildl Soc Bull 43:244–255. https://doi.org/10.1002/wsb.966

Hoffman RW (1981) Volunteer collection station use for obtaining grouse wing samples. Wildl Soc Bull 9:180–184

Hoffman RW, Griffin KA, Knetter JM et al (2015) Guidelines for the management of Columbian sharp-tailed grouse populations and their habitats. Western Association of Fish and Wildlife Agencies, Cheyenne

Holechek J, Pieper RD, Berbel CH (2011) Range management: principles and practices, 6th edn. Prentice Hall, Boston, MA

Horak GJ (1985) Kansas prairie chickens. Wildlife Bulletin No. 3. Kansas Fish and Game Commission, Pratt, Kansas 65 p

Hormay AL, Evanko AB (1958) Rest-rotation grazing: a management system for bunchgrass ranges. California Forest and Range Experiment Station, 14 pp

Hovick TJ, Elmore RD, Allred BW, Fuhlendorf SD, Dahlgren DK (2014a) Landscapes as a thermal moderator of thermal extremes: a case study from an imperiled grouse. Ecosphere 5:35. https://doi.org/10.1890/ES13-00340.1

Hovick TJ, Elmore RD, Fuhlendorf SD et al (2014b) Evidence of negative effects of anthropogenic structures on wildlife: a review of grouse survival and behavior. J Appl Ecol. https://doi.org/10.1111/1365-2664.12331

Hovick TJ, Allred BW, Elmore RD et al (2015a) Dynamic disturbance processes create dynamic lek site selection in a prairie grouse. PLoS ONE 10(9):e0137882. https://doi.org/10.1371/journal.pone.0137882

Hovick TJ, Elmore RD, Fuhlendorf SD et al (2015b) Weather constrains the influence of fire and grazing on nesting greater prairie-chickens. Rangel Ecol Manag 68:186–193. https://doi.org/10.1016/j.rama.2015.01.009

Hovick TJ, Dahlgren DK, Papeş M et al (2015c) Predicting greater prairie-chicken lek site suitability to inform conservation actions. PLoS ONE 10(8):e0137021. https://doi.org/10.1371/journal.pone.0137021

Hudson PJ (1986) The effect of a parasitic nematode on the breeding production of red grouse. J Anim Ecol 55:85–92

Jones RE (1963) Identification and analysis of lesser and greater prairie-chicken habitat. J Wildl Manage 27:757–778

Johnsgard PA (1973) Grouse and quails of North America. University of Nebraska Press, Lincoln, NE

Johnsgard P (2002) Grassland grouse and their conservation. Smithsonian Institution Press, Washington, DC

Johnsgard PA (2016) The North American grouse: their biology and behavior. Zea Books, Lincoln, NE, USA

Johnsgard PA, Wood RE (1968) Distributional changes and interaction between prairie chickens and sharp-tailed grouse in the Midwest. Papers in Ornithology 7

Johnson DH (1980) The comparison of usage and availability measurements for evaluating resource preference. Ecol 61:65–71

Johnson MD, Joseph K (1989) Feathers from the prairie: a short history of upland game birds. North Dakota Game and Fish Department, Bismarck

Johnson JA, Schroeder MA, Robb LA (2020) Greater prairie-chicken (*Tympanuchus cupido*), version 1.0. In: Poole AF (ed) Birds of the World. Cornell Lab of Ornithology, Ithaca. https://doi.org/10.2173/bow.grpchi.01

Kessler WB, Bosch RP (1982) Sharp-tailed grouse and range management practices in western rangelands In: Peek JM, Dalke PD (eds) Proc Wildlife-livestock Relationships Sympos, Coeur d'Alene, p 133–146

Kirby DR, Grosz KL (1995) Cattle grazing and sharp-tailed grouse nesting success. Rangelands 17:124–126

Klott JH, Lindzey FG (1990) Brood habitats of sympatric sage grouse and Columbian Sharp-tailed Grouse in Wyoming. J Wildl Manage 54:84–88

Kobriger GD (1965) Status, movements, habitats, and foods of prairie grouse on a Sandhills refuge. J Wildl Manage 29:788–800

Korschgen LJ (1962) Food habits of greater prairie chickens in Missouri. The Am Midl Nat 68(2):307–318

Kraft JD, Haukos DA, Bain MR, Rice MB, Robinson S, Sullins DS, Hagen CA, Pitman J, Lautenbach J, Plumb R, Lautenbach J (2021) Using grazing to manage herbaceous structure for a heterogeneity-dependent bird. J Wildl Manage 85:354–368. https://doi.org/10.1002/jwmg. 21984

Krausman PR, Naugle DE, Frisina MR et al (2009) Livestock grazing, wildlife habitat, and rangeland values. Rangelands 31:15–19. https://doi.org/10.2111/1551-501X-31.5.15

Lautenbach JM, Plumb RT, Robinson SG et al (2017) Lesser prairie-chicken avoidance of trees in a grassland landscape. Rangel Ecol Manag 70:78–86. https://doi.org/10.1016/j.rama.2016.07.008

Lautenbach JD, Haukos DA, Hagen CA et al (2021) Ecological disturbance through patch burn brazing influences lesser prairie-chicken space use. J Wildl Manage 85:1699–1710. https://doi.org/10.1002/jwmg.22118

LeBeau CW, Kauffman M, Smith KJ et al (2020). Placement of wind energy infrastructure matters: a quantitative study evaluating response of lesser prairie-chicken to a wind energy facility. Report to the American Wind Wildlife Institute. https://awwi.org/resources/wwrf-lesser-prairie-chicken-2019. Accessed 20 June 2021

Leupin EE, Chutter MJ (2007) Status of the sharp-tailed grouse, Columbianus subspecies in British Columbia. B.C. Ministry of Water, Land, and Air Protection, Wildlife Bulletin

Lloyd JD, Aldridge C, Allison T et al (2022) Prairie grouse and wind energy: the state of the science and implications for risk assessment. Wildl Soc Bull e1305. https://doi.org/10.1002/wsb.1305

Londe DW, Fuhlendorf SD, Elmore RD et al (2019) Female greater prairie-chicken response to energy development and rangeland management. Ecosphere 10(12):e02982. https://doi.org/10.1002/ecs2.2982

Londe DW, Elmore RD, Davis CA et al (2021a) Fine scale habitat selection limits tradeoffs between foraging and temperature in a grassland bird. Behav Ecol 32:625–637. https://doi.org/10.1093/beheco/arab012

Londe DW, Elmore RD, Davis CA et al (2021b) Weather influences multiple components of greater prairie-chicken reproduction. J Wildl Manage 85:121–134. https://doi.org/10.1002/jwmg.21957

Luukkonen DR, Minzey T, Maples TE et al (2009) Evaluation of population monitoring procedures for sharp-tailed grouse in the eastern upper peninsula of Michigan. Department of Natural Resources, Report 3503

Lwiwski TC, Koper N, Henderson DC (2015) Stocking rates and vegetation structure, heterogeneity, and community in a northern mixed-grass prairie. Rangel Ecol Manag 68:322–331. https://doi.org/10.1016/j.rama.2015.05.002

Manzer DL, Hannon SJ (2008) Survival of sharp-tailed grouse (*Tympanuchus phasianellus*) chicks and hens in a fragmented prairie landscape. Wildl Biol 14:16–25. https://doi.org/10.2981/0909-6396(2008)14[16:SOSGTP]2.0.CO;2

Marks JS, Marks VS (1988) Winter habitat use by Columbian sharp-tailed grouse in western Idaho. J Wildl Manage 52:743–746

Martinson RK, Whitesell DE (1964) Biases in a mail questionnaire survey of upland game hunters. Trans N Amer Wildl Nat Res Conf 29:287–294

Matthews SR, Coates PS, Delehanty, (2016) Survival of translocated sharp-tailed grouse: temporal and age effects. Wildl Res 43:220–227. https://doi.org/10.1071/WR15158

Matthews TW, Tyre AJ, Taylor JS et al (2011) Habitat selection and brood survival of greater prairie-chickens In: Sandercock BK, Martin K, Segelbacher G (eds) Ecology, conservation, and management of grouse. Studies in avian biology 39, University of California Press, Berkeley, pp 179–191

Matthews TW, Tyre AJ, Taylor JS et al (2013) Greater prairie-chicken nest success and habitat selection in southeastern Nebraska. J Wildl Manage 77:1202–1212. https://doi.org/10.1002/jwmg.564

McCaffery R, Lukacs PM (2016) A generalized integrated population model to estimate greater sage-grouse population dynamics. Ecosphere 7(11):e01585. https://doi.org/10.1002/ecs2.1585

McDonald L, Beauprez G, Gardner G, Griswold J, Hagen C, Klute D, Kyle S, Pitman J, Rintz T, Schoeling D, Timmer J, Van Pelt B (2014) Range-wide population size of the lesser prairie-chicken: 2012 and 2013. Wildl Soc Bull 38:536–546. https://doi.org/10.1002/wsb.417

McNew LB, Gregory AJ, Wisely SM et al (2011a) Human-mediated selection on life-history traits of greater prairie-chickens. In: Sandercock BK, Martin K, Segelbacher G (eds) Ecology, conservation, and management of grouse (SAB 39). University of California Press, Berkeley, pp 255–266

McNew LB, Gregory AJ, Wisely SM et al (2011b) Reproductive biology of a southern population of greater prairie-chickens. In: Sandercock BK, Martin K, Segelbacher G (eds) Ecology, conservation, and management of grouse (SAB 39). University of California Press, Berkeley, pp 209–221

McNew LB, Prebyl TJ, Sandercock BK (2012a) Effects of rangeland management on the site occupancy dynamics of prairie-chickens in a protected prairie preserve. J Wildl Manage 76:38–47. https://doi.org/10.1002/jwmg.237

McNew LB, Gregory AJ, Wisely SM et al (2012b) Demography of greater prairie-chickens: regional variation in vital rates, sensitivity values, and population dynamics. J Wildl Manage 76:987–1000. https://doi.org/10.1002/jwmg.369

McNew LB, Gregory AJ, Sandercock BK (2013) Spatial heterogeneity in habitat selection: nest site selection by prairie-chickens. J Wildl Manage 77:791–801. https://doi.org/10.1002/jwmg.493

McNew LB, Hunt LM, Gregory AJ et al (2014) Effects of wind energy development on the nesting ecology of greater prairie-chickens. Cons Biol 28:1089–1099. https://doi.org/10.1111/cobi.12258

McNew LB, Winder VL, Pitman JC et al (2015) Alternative rangeland management and the nesting ecology of greater prairie-chickens. Rangel Ecol Manag 68:298–304. https://doi.org/10.1016/j.rama.2015.03.009

McNew LB, Cascaddan B, Hicks-Lynch A, Milligan M, Netter A, Otto S, Payne J, Vold S, Wells S, Wyffels S (2017) Restoration plan for sharp-tailed grouse recovery in western Montana. Montana State University, Bozeman. https://doi.org/10.13140/RG.2.2.26189.95204.

Meints DR (1991) Seasonal movements, habitat use, and productivity of Columbian Sharp-tailed Grouse in southeastern Idaho. M.S. thesis, University of Idaho, Moscow

Merrill MD, Chapman KA, Poiani KA, Winter B (1999) Land-use patterns surrounding greater prairie-chicken leks in northwestern Minnesota. J Wildl Manag 63:189–198

Milligan MC, McNew LB (2022) Evaluating the cumulative effects of livestock grazing management on wildlife with an integrated population model. Front Ecol Evol 10:818050. https://doi.org/10.3389/fevo.2022.818050

Milligan MC, Wells SL, McNew LB (2018) A population viability analysis of sharp-tailed grouse to inform reintroductions. J Fish Wildl Manag 9:565–581. https://doi.org/10.3996/112017-JFWM-090

Milligan MC, Berkeley LI, McNew LB (2020a) Effects of rangeland management on the nesting ecology of sharp-tailed grouse. Rangel Ecol Manag 73:128–137. https://doi.org/10.1016/j.rama.2019.08.020

Milligan MC, Berkeley LI, McNew LB (2020b) Survival of sharp-tailed grouse under variable livestock grazing management. J Wildl Manage 84:1296–1305. https://doi.org/10.1002/jwmg.21909

Milligan MC, Berkeley LI, McNew LB (2020c) Habitat use of sharp-tailed grouse in rangelands managed for livestock. PloSOne 15(6):e0233756. https://doi.org/10.1371/journal.pone.0233756

Nasman K, Rints T, Pham D, McDonald L et al (2020) Range-wide population size of the lesser prairie-chicken: 2012–2020. WEST Inc., Fort Collins, CO, USA

Nielsen LS (1982) The effects of rest-rotation grazing on the distribution of sharp-tailed grouse. Dissertation, Montana State University, Bozeman

Niemuth ND (2000) Land use and vegetation associated with greater prairie-chicken leks in an agricultural landscape. J Wildl Manage 64:278–286. https://doi.org/10.2307/3803000

Niemuth ND (2003) Identifying landscapes for greater prairie-chicken translocation using habitat models and GIS—a case study. Wildl Soc Bull 31:145–155. https://doi.org/10.2307/3784368

Niemuth ND (2011) Spatially explicit habitat models for prairie grouse. In: Sandercock BK, Martin K, Segelbacher G (eds) Ecology, conservation, and management of grouse (SAB 39). University of California Press, Berkeley, pp 3–20

Niemuth ND, Boyce MS (2004) Influence of landscape composition on sharp-tailed grouse lek location and attendance in Wisconsin pine barrens. Ecoscience 11:209–217. https://doi.org/10.1080/11956860.2004.11682826

Northrup RD (1991) Sharp-tailed grouse habitat use during fall and winter on the Charles M. Russell National Wildlife Refuge. MS thesis, Montana State University, Bozeman

Norton MA, Jensen KC, Leif AP, Kirschenman TR, Wolbrink GA (2010) Resource selection of greater prairie-chicken and sharp-tailed grouse broods in central South Dakota. Prairie Nat 42:100–108

Oregon Department of Fish and Wildlife (2020) 2019–20 upland game bird harvest by harvest area. Available online: https://www.dfw.state.or.us/resources/hunting/upland_bird/harvest/docs/2019-20%20Upland%20Harvest%20Results.pdf. Accessed Dec 2020

Panjabi AO, Blancher PJ, Dettmers R, Rosenberg KV (2012) Partners in flight technical series No. 3. Rocky Mountain Bird Observatory website: http://www.rmbo.org/pubs/downloads/Handbook2012.pdf

Patten MA, Pruett C, Wolfe D (2011) Home range size and movements of greater prairie-chickens. In: Sandercock BK, Martin K, Segelbacher G (eds) Ecology, conservation, and management of grouse (SAB 39). University of California Press, Berkeley, pp 51–62

Pepper GW (1972) The ecology of sharp-tailed grouse during the spring and summer in the Aspen Parkland of Saskatchewan. Saskatchewan Department of Natural Resources, Wildlife Report # 1, 56 p

Peterson MJ (2004) Parasites and infectious diseases of prairie grouse: should managers be concerned? Wildl Soc Bull 32:35–55. https://doi.org/10.2193/0091-7648(2004)32[35:PAIDOP]2.0.CO;2

Plumb RT, Lautenbach JM, Robinson SG, Haukos DA, Winder VL, Hagen CA, Sullins DS, Pitman JC, Dahlgren DK (2019) Lesser prairie-chicken space use in relation to anthropogenic structures. J Wildl Manag 83:216–230. https://doi.org/10.1002/jwmg.21561

Pirius NE, Boal CW, Haukos DA et al (2013) Winter habitat use and survival of lesser prairie-chickens in West Texas. Wildl Soc Bull 37:759–765. https://doi.org/10.1002/wsb.354

Pitman JC, Hagen CA, Robel RJ et al (2005) Location and success of lesser prairie-chicken nests in relation to vegetation and human disturbance. J Wildl Manage 69:1259–1269. https://doi.org/10.2193/0022-541X(2005)069[1259:LASOLP]2.0.CO;2

Pitman JC, Hagen CA, Robel RJ et al (2006a) Nesting ecology of the lesser prairie-chicken in sand sagebrush prairie of Kansas. Wilson J Ornithol 118:23–35. https://doi.org/10.1676/1559-4491(2006)118[0023:neolpi]2.0.co;2

Pitman JC, Jamison BE, Hagen CA et al (2006b) Brood break-up and dispersal of juvenile lesser prairie-chickens in Kansas. Prairie Nat 38:86–99

Pollock KH, Winterstein SR, Bunck CM et al (1989) Survival analysis in telemetry studies: the staggered entry design. J Wildl Manage 53:7–15

Powell LA, Shacht WH, Ewald JP, McCollum KR (2020) Greater prairie-chickens and sharp-tailed grouse have similarly high nest survival in the Nebraska Sandhills. Prairie Nat 52:58–75

Prose BL (1987) Habitat suitability index models: plains sharp-tailed grouse. National Ecology Center, United States Department of the Interior Fish and Wildlife Service. https://pubs.er.usgs.gov/publication/fwsobs82_10_142

Raynor EJ, Harrison JO, Whalen CE et al (2019) Anthropogenic noise does not surpass land cover in explaining habitat selection of greater prairie-chicken (*Tympanuchus cupido*). Condor 121.https://doi.org/10.1093/condor/duz044

Reese KP, Bowyer RT (2007) Monitoring populations of sage-grouse. College of Natural Resources Experiment Station Bulletin 88

Riley TZ, Davis CA, Smith RA (1993) Autumn and winter foods of the lesser prairie-chicken (*Tympanuchus pallidicinctus*). Great Basin Nat 53:186–189

Robel RJ (1966) Booming territory size and mating success of the greater prairie chicken (*Tympanuchus cupido pinnatus*). Anim Behav 14:328–331

Robel RJ (1970) Possible role of behavior in regulating greater prairie-chickens. J Wildl Manage 34:306–312

Robel R, Briggs J, Dayton A et al (1970) Relationships between visual obstruction measurements and weight of grassland vegetation. J Range Manage 23:295–297

Robel RJ, Henderson FR, Jackson W (1972) Some sharp-tailed grouse population statistics from South Dakota. J Wildl Manage 36:87–98

Robinson SG, Haukos DA, Plumb RT et al (2018) Non-breeding home-range size and survival of lesser prairie-chickens. J Wildl Manage 82:413–423. https://doi.org/10.1002/jwmg.21390

Roseberry JL, Klimstra WD (1992) Further evidence of differential harvest rates among bobwhite sex-age groups. Wildl Soc Bull 20:91–94

Rosenquist EL, Toepfer JE (1995) A preliminary report on the winter ecology of the greater prairie-chicken in northwest Minnesota P 15–16 In: Kobriger GD (ed) Proceedings of the Twenty-first Prairie Grouse Technical Council Conference. North Dakota Game and Fish Department, Dickinson, North Dakota

Ross B, Haukos DA, Hagen CA et al (2016) The relative contribution of variation in climate to changes in lesser prairie-chicken abundance. Ecosphere 7.https://doi.org/10.1002/ecs2.1323

Ross B, Haukos DA, Hagen CA et al (2018) Combining multiple sources of data to inform conservation of lesser prairie-chicken populations. Auk 135:228–239. https://doi.org/10.1642/AUK-17-113.1

Royle JA, Dorazio RM (2008) Hierarchical modeling and inference in ecology: the analysis of data from populations, metapopulations and communities, 1st edn. Academic Press, Cambridge

Runia TJ, Solem AJ, Niemuth ND, Barnes KW (2021) Spatially explicit habitat models for prairie grouse: implications for improved population monitoring and targeted conservation. Wildl Soc Bull 45:36–54. https://doi.org/10.1002/wsb.1164

Sands JP, Pope MD (2010) A survey of galliform monitoring programs and methods in the United States and Canada. Wildl Biol 16:342–356. https://doi.org/10.2981/09-066

Santo AR, Donlan CJ, Hagen CA et al (2020) Characteristics and motivations of participants and nonparticipants in an at-risk species conservation program. Hum Dimens of Wildl 25:1–9. https://doi.org/10.1080/10871209.2020.1817631

Schaub M, Abadi F (2011) Integrated population models: a novel analysis framework for deeper insights into population dynamics. J Ornithol 152:227–237. https://doi.org/10.1007/s10336-010-0632-7

Schieltz JM, Rubenstein DI (2016) Evidence based review: positive versus negative effects of livestock grazing on wildlife. What do we really know? Environ Res Lett 11(11):113003. https://doi.org/10.1088/1748-9326/11/11/113003

Schindler AR, Haukos DA, Hagen CA, Ross BE (2020) A decision support tool to prioritize candidate landscapes for lesser prairie-chicken conservation. Landsc Ecol 35:1417–1434. https://doi.org/10.1007/s10980-020-01024-6

Schroeder MA (1991) Movements and lek visitation by female greater prairie-chickens in relation to Bradbury's female preference hypothesis of lek evolution. Auk 108:896–903

Schroeder MA, Braun CE (1992) Greater prairie-chicken attendance at leks and stability of leks in Colorado. Wilson Bull 104:273–284

Schroeder MA, White GC (1993) Dispersion of greater prairie chicken nests in relation to lek location: evaluation of the hot-spot hypothesis of lek evolution. Behav Ecol 4:266–270

Schroeder MA, Baydack RK, Harmon SA et al (2004) The North American grouse management plan. North American Grouse Partnership, Williamsport, Maryland

Schroeder MA, Ashley PR, Vander Haegen, M (2008) Terrestrial wildlife and habitat assessment on Bonneville Power Administration-funded wildlife areas in Washington: monitoring and evaluation activities. Washington Department of Fish and Wildlife

Schwilling MD (1955) A study of the lesser prairie-chicken in Kansas. Kansas Forestry, Fish and Game Commission, Pratt, KS, USA

Silvy NJ, Griffin CP, Lockwood MA et al (1999) The Attwater's prairie chicken—a lesson in conservation biology research. In: Svedarsky WD, Hier RH, Silvy NJ (eds) The greater prairie chicken: a national look. University of Minnesota Miscellaneous Publication 99–1999, St. Paul, MN

Sliwinski M, Burbach M, Powell A et al (2018) Factors influencing ranchers' intentions to manage for vegetation heterogeneity and promote cross-boundary management in the northern Great Plains. Ecol and Soc 23:45–62. https://doi.org/10.5751/ES-10660-230445

Smeins FE, Diamond DD, Hanselka CW (1991) Coastal prairie. In: Coupland RT (ed) Ecosystems of the world 8. Elsevier Press, New York, A-natural grasslands-introduction and western hemisphere, pp 269–290

Smith JT, Allred B, Boyd CS et al (2020) Sage-grouse: fine scale specialist or shrub steppe generalist? J Wildl Manage 84:759–774. https://doi.org/10.1002/jwmg.21837

South Dakota Department of Fish and Game (2010) Prairie grouse management plan for South Dakota 2011–2015. South Dakota Department of Game, Fish and Parks

Stinson DW, Schroeder MA (2012) Washington state recovery plan for the Columbian sharptailed grouse. Washington Department of Fish and Wildlife, Olympia. https://wdfw.wa.gov/publications/00882

Sullins DS, Kraft JD, Haukos DA, Robinson SG, Reitz JH, Plumb RT, Lautenbach JM, Lautenbach JD, Sandercock BK, Hagen CA (2018) Demographic consequences of conservation reserve program grasslands for lesser prairie-chickens. J Wildl Manage 82:1617–1632. https://doi.org/10.1002/jwmg.21553

Sullins DS, Haukos DA, Lautenbach JM, Lautenbach JD, Robinson SG, Rice MB, Sandercock BK, Kraft JD, Plumb RT, Reitz JH, Hutchinson J, Hagen CA (2019) Strategic conservation for lesser prairie-chickens among landscapes of varying anthropogenic influence. Biol Conserv (238) https://doi.org/10.1016/j.biocon.2019.108213

Svedarsky WD (1979) Spring and summer ecology of female greater prairie-chickens in northwestern Minnesota. PhD dissertation. University of North Dakota, Grand Forks, North Dakota. 166 p

Svedarsky WD (1988) Reproductive ecology of female greater prairie-chickens in Minnesota. In: Bergerud AT, Gratson MW (eds) Adaptive strategies and population ecology of northern grouse. University of Minnesota Press, Minneapolis, pp 193–267

Svedarsky WD, Westemeier RL, Robel RJ, Gough S, Toepher JE (2000) Status and management of the greater prairie-chicken Tympanuchus cupido pinnatus in North America. Wildl Biol 6:277–284. https://doi.org/10.2981/wlb.2000.027

Svedarsky WD, Toepfer JE, Westemeier RL, Robel RJ, Igl LD, and Shaffer JA (2022) The effects of management practices on grassland birds—greater prairie-chicken (Tympanuchus cupido pinnatus), chap. C of Johnson DH, Igl LD, Shaffer JA, DeLong JP (eds) The effects of management practices on grassland birds: U.S. Geological Survey Professional Paper 1842, 53 p. https://doi.org/10.3133/pp1842C

Swenson JE (1985) Seasonal habitat use by sharp-tailed grouse, Tympanuchus phasianellus, on mixed-grass prairie in Montana. Can Field Nat 99:40–46

Symstad AJ, Leis SA (2017) Woody encroachment in Northern Great Plains grasslands: perceptions, actions, and seeds. Nat Areas J 37:118–127. https://doi.org/10.3375/043.037.0114

Temple SA (1992) Population viability analysis of a sharp-tailed grouse metapopulation in Wisconsin. In: McCullough DR, Barrett RH (eds) Wildlife 2001: populations. Springer Publishing, New York City, NY, pp 750–758

Tews J, Brose U, Grimm V, Tielborger K, Wichmann MC, Schwager M, Jeltsch F (2004) Animal species diversity driven by habitat heterogeneity/diversity: the importance of keystone structures. J Biograph 31:79–92. https://doi.org/10.1046/j.0305-0270.2003.00994.x

Timmer, JM, Butler MJ, Ballard WB et al (2014) Spatially explicit modeling of lesser prairie-chicken lek density in Texas. J Wildl Manage 78:142–152.https://doi.org/10.1002/jwmg.646

Toombs TP, Derner JD, Augustine DJ, Krueger B, Gallagher S (2010) Managing biodiversity and livestock: a scale-dependent approach for promoting vegetation heterogeneity in western Great Plains grasslands. Rangelands 32:10–15. https://doi.org/10.2111/RANGELANDS-D-10-00006.1

Ulliman MJ (1995) Winter habitat ecology of Columbian sharp-tailed grouse in southeastern Idaho. MS thesis, University of Idaho, Moscow

U.S. Fish and Wildlife Service (2016) U.S. Fish and Wildlife Service removes lesser prairie-chicken from list of threatened and endangered species in accordance with Court Order USFWS, Washington, D.C., USA [online]. URL: https://www.fws.gov/news/ShowNews.cfm?ref=u.s.-fish-and-wildlife-serviceremoves-lesser-prairie-chicken-from-list-o&_ID=35739

U.S. Fish and Wildlife Service (2000) Endangered and threatened wildlife and plants; 12-month finding for a petition to list the Columbian sharp-tailed grouse as threatened. 65 Fed Reg 60391

U.S. Fish and Wildlife Service (2006) Endangered and threatened wildlife and plants; 90-day finding for a petition to list the Columbian sharp-tailed grouse as threatened. 71 Fed Reg 67318

Utah Department of Natural Resources (2002) Strategic management plan for Columbian sharp-tailed grouse. State of Utah, Department of Natural Resources Publication 01–19. https://wildlife.utah.gov/pdf/upland/02sharptail.pdf

Van Pelt WE, Kyle S, Pitman J et al (2013) The lesser prairie-chicken range-wide conservation plan. Western Association of Wildlife Agencies, Cheyenne, WY p 367

Vodehnal WL, Haufler JB (2007). A grassland conservation plan for prairie grouse. North American Grouse Partnership. Fruita, CO

Walsh DP, White GC, Remington TE et al (2004) Evaluation of the lek-count index for greater sage-grouse. Wildl Soc Bull 32:56–68. https://doi.org/10.2193/0091-7648(2004)32[56:EOTLIF]2.0.CO;2

Warheit KI, Dean CA (2009) Subspecific identification of sharp-tailed grouse (*Tympanuchus phasianellus*) samples from Montana. Washington Department of Fish and Wildlife, Molecular Genetics Laboratory Technical Report 09-1011

Wiens JA (1976) Population responses to patchy environments. Annu Rev Ecol Syst 7:81–120

Williamson RM (2009) Impacts of oil and gas development on sharp-tailed grouse on the Little Missouri National Grasslands. South Dakota State University, Brookings, South Dakota, USA, North Dakota

Winder VL, McNew LB, Hunt LM et al (2014a) Effects of wind energy development on seasonal survival of greater prairie-chickens. J Appl Ecol 51:395–405. https://doi.org/10.1111/1365-2664.12184

Winder VL, McNew LB, Gregory AJ et al (2014b) Space use by female greater prairie-chickens in response to wind energy development. Ecosphere 5:3. https://doi.org/10.1890/ES13-00206.1

Winder VL, Carrlson KM, Gregory AJ et al (2015) Factors affecting female space use in ten populations of prairie chickens. Ecosphere 6:166. https://doi.org/10.1890/ES14-00536.1

Winder VL, McNew LB, Pitman JC, Sandercock BK (2017) Space use of greater prairie-chickens in response to fire and grazing interactions. Rangel Ecol Manag 70:165–174. https://doi.org/10.1016/j.rama.2016.08.004

Winder VL, McNew LB, Pitman JC et al (2018) Effects of rangeland management on survival of female greater prairie-chickens. J Wildl Manage 82:113–122. https://doi.org/10.1002/jwmg.21331

Wisdom MJ, Mills LS (1997) Sensitivity analysis to guide population recovery: prairie-chicken as an example. J Wildl Manage 61:302–312. https://doi.org/10.2307/3802585

Wolfe DH, Patten MA, Shochat E et al (2007) Causes and patterns of mortality in lesser prairie-chickens *Tympanuchus pallidicinctus* and implications for management. Wildl Biol 13:95–104. https://doi.org/10.2981/0909-6396(2007)13[95:CAPOMI]2.0.CO;2

Chapter 10
Sage-Grouse

**Jeffrey L. Beck, Thomas J. Christiansen, Kirk W. Davies,
Jonathan B. Dinkins, Adrian P. Monroe, David E. Naugle,
and Michael A. Schroeder**

Abstract In this chapter, we summarize the ecology and conservation issues affecting greater (*Centrocercus urophasianus*) and Gunnison (*C. minimus*) sage-grouse, iconic and obligate species of rangelands in the sagebrush (*Artemisia* spp.) biome in western North America. Greater sage-grouse are noted for their ability to migrate, whereas Gunnison sage-grouse localize near leks year-round. Seasonal habitats include breeding habitat where males display at communal leks, nesting habitat composed of dense sagebrush and herbaceous plants to conceal nests, mesic summer habitats where broods are reared, and winter habitat, characterized by access to sagebrush for cover and forage. While two-thirds of sage-grouse habitat occurs on public lands, private land conservation is the focus of national groups including the USDA-NRCS Sage-Grouse Initiative. Sage-grouse are a species of great conservation concern due to population declines associated with loss and fragmentation of more than half of the sagebrush biome. Wildlife and land management agencies have been increasingly proactive in monitoring trends in sage-grouse populations

J. L. Beck (✉)
Department of Ecosystem Science and Management, University of Wyoming, Laramie,
WY 82071, USA
e-mail: jlbeck@uwyo.edu

T. J. Christiansen
Wyoming Game and Fish Department (Retired), Green River, WY 82935, USA

K. W. Davies
U.S. Department of Agriculture—Agricultural Research Service, Eastern Oregon Agricultural
Research Center, Burns, OR 97720, USA

J. B. Dinkins
Department of Animal and Rangeland Sciences, Oregon State University, Corvallis, OR 97331,
USA

A. P. Monroe
U.S. Geological Survey, Fort Collins Science Center, Fort Collins, CO 80526, USA

D. E. Naugle
Wildlife Biology Program, University of Montana, Missoula, MT 59812, USA

M. A. Schroeder
Washington Department of Fish and Wildlife, Bridgeport, WA 98813, USA

© The Author(s) 2023   295
L. B. McNew et al. (eds.), *Rangeland Wildlife Ecology and Conservation*,
https://doi.org/10.1007/978-3-031-34037-6_10

(e.g., lek count index), adapting regulations to reduce harvest on declining populations, and in designing and implementing conservation policies such as core areas to conserve sage-grouse habitats and populations. Much of the remaining sagebrush habitat is threatened by altered fire regimes, invasive annual grasses and noxious weeds, encroaching piñon (*Pinus edulis* and *monophylla*)-juniper (*Juniperus* spp.) woodlands, sagebrush conversion, anthropogenic development, and climate change. Several diseases affect sage-grouse, but to date, disease has not been a widespread cause of declines. Proper livestock grazing and limited hunting appear to be sustainable with sage-grouse, whereas improper grazing, increasing free-roaming equid populations, and sagebrush conversion are primary concerns for future conservation. Research has identified additional concerns for sage-grouse including effects from fence collisions, predation from common ravens (*Corvus corax*), and reduced habitat effectiveness resulting from grouse avoidance of anthropogenic infrastructure. There is a need for future research evaluating sage-grouse habitat restoration practices following improper rangeland management, habitat alteration from invasive species and fire, effects on small and isolated populations, and effects from diseases.

Keywords *Centrocercus urophasianus* · *Centrocercus minimus* · Ecosystem threats · Greater sage-grouse · Gunnison sage-grouse · Private and public land conservation · Rangeland management · Sagebrush

10.1 General Life History and Population Dynamics

Greater (*Centrocercus urophasianus*) and Gunnison (*C. minimus*) sage-grouse are icons of the sagebrush (*Artemisia* spp.) biome in western North America. Sage-grouse are world renowned for their spectacular lek breeding system where males congregate at traditional locations to display for and breed with females (Fig. 10.1). Competition among males at leks is intense and relatively few males breed with most of the females. Females are ground-nesting Galliformes that produce a maximum of 1 successful clutch with 7–9 chicks per year after a 4-week incubation period, with no help from the males (Schroeder et al. 2020; Young et al. 2020). Because nest success, chick survival, and rates of renesting are generally low, the relatively high survival of breeding-aged birds helps to maintain their populations (Crawford et al. 2004; Connelly et al. 2011a; Taylor et al. 2012; Blomberg et al. 2013c; Davis et al. 2014b; Dahlgren et al. 2016). Sage-grouse are long-lived relative to other game birds, with an observed maximum survival for greater sage-grouse of 9 years in females and 7 years in males (Zablan et al. 2003), though maximum longevity is likely higher.

Although greater sage-grouse often migrate seasonally, migrations are typically within the same general regions and ecosystems (Connelly et al. 1988; Fischer et al. 1996; Fedy et al. 2012). However, they stand out among upland gamebirds in their capability for long-distance migrations—the longest migration by a greater sage-grouse was documented at 240 km for a female between southern Saskatchewan and north-central Montana (Newton et al. 2017). Telemetry studies indicate most

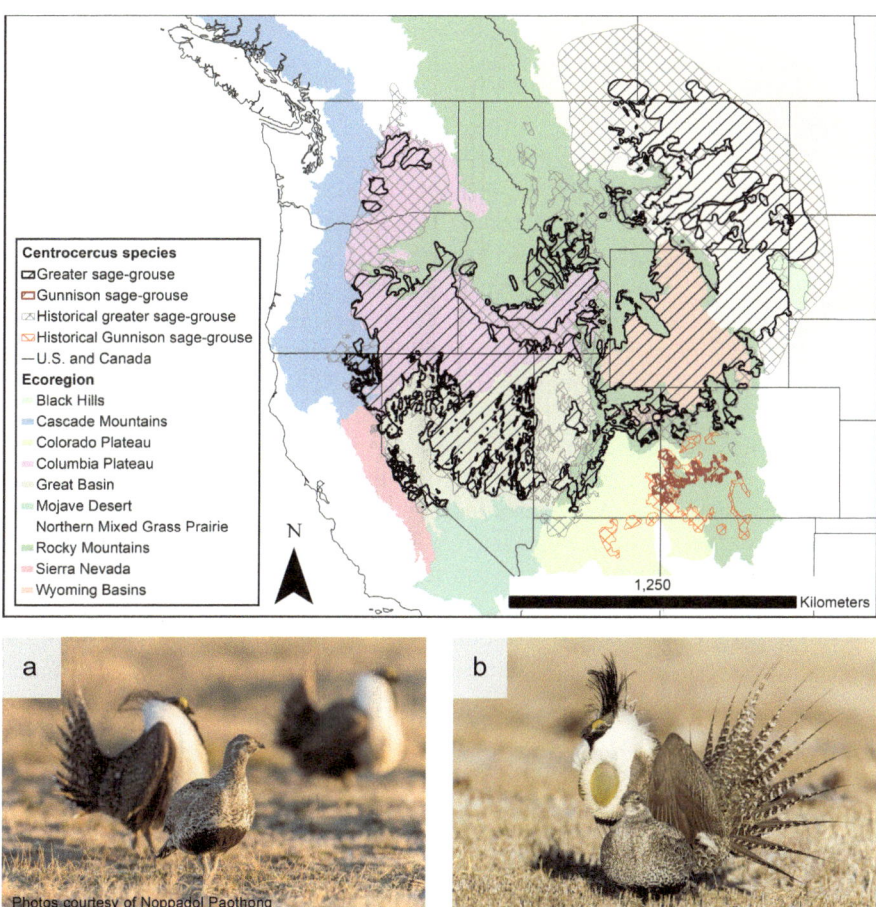

Fig. 10.1 Greater (Photo a) and Gunnison (Photo b) sage-grouse distributions in the western United States and Canada based on Schroeder et al. (2004), Environment Canada (2014), and USFWS (2015). Ecoregions represent EPA Level III ecoregions across North America (Wilken et al. 2011). Single hatched area represents current distribution of greater (black) and Gunnison (dark orange) sage-grouse, and cross-hatched area represents additional historical distribution of greater (grey) and Gunnison (light orange) sage-grouse

Gunnison sage-grouse spend their annual life cycles within 5 km of their lek of capture (Aldridge et al. 2012; Young et al. 2020). Greater sage-grouse populations are often partially migratory, with some individuals migrating and others remaining within the same areas year-round (Fedy et al. 2012; Pratt et al. 2017, 2019), which can complicate conservation efforts (Dinkins et al. 2017; Pratt et al. 2019). Greater sage-grouse in many populations migrate altitudinally from xeric lower elevation Wyoming big sagebrush (*A. tridentata wyomingensis*) in spring to more mesic higher elevation mountain big sagebrush (*A. t. vaseyana*) during summer, and then to lower elevations for winter and the following spring (Beck et al. 2006; Pratt et al. 2017,

2019). In some areas, greater sage-grouse may move to irrigated agricultural fields, in many cases alfalfa (*Medicago sativa*), in response to summer desiccation of native forbs (Fischer et al. 1996).

Sage-grouse select specific structural aspects of sagebrush as well as herbaceous understory components throughout life stages including dense sagebrush and grass cover for successful nesting, increased forb cover at early (through 2 weeks; Thompson et al. 2006), and late brood-rearing/summer (> 2 weeks; Smith et al. 2018c), and dense sagebrush for winter habitat (Fig. 10.2). Sage-grouse depend on sagebrush as a primary food source, particularly during the late autumn, winter, and early spring (Schroeder et al. 2020). They digest the leaves of sagebrush, while also tolerating high levels of monoterpenoids and other plant secondary metabolites, toxins most species cannot consume (Sauls 2006; Kohl et al. 2016; Oh et al. 2019). Sage-grouse also depend on the cover provided by sagebrush during nesting and brood-rearing and during all times of year for protection from potential predators (Schroeder et al. 2020; Young et al. 2020). Although sage-grouse are adept at snow burrowing to escape severe winter conditions (Back et al. 1987), winter landscapes that support sage-grouse are characterized by south and west aspects and patches of taller sagebrush that sage-grouse can access following deep snow accumulations (Hupp and Braun 1989).

The importance of sagebrush within sage-grouse habitat is clear, but additional vegetation components of these habitats provide necessary function as well (Connelly

Fig. 10.2 Life stages of sage-grouse within seasonal habitats. Successful nest and early brood photographs from N. Paothang. Nesting, early brood-rearing, and late brood-rearing/summer photos from T. Christiansen. Winter photograph from J. Lautenbach

et al. 2011b; Dumroese et al. 2015; Pennington et al. 2016). Herbaceous cover (grasses and forbs) provides essential concealment of nests, often explaining differences in nest site selection (Hagen et al. 2007). However, only one study has reported a weak effect of grass cover and height positively influencing nest success in greater sage-grouse (Holloran et al. 2005). A rangewide meta-analysis indicated common fine-scale herbaceous and shrub structural characteristics do not consistently influence nest success in greater sage-grouse (Smith et al. 2020). This study also found shrub characteristics such as sagebrush cover had moderate, yet context-dependent, effects, and herbaceous vegetation characteristics had weak effects, on sage-grouse nest site selection (Smith et al. 2020). Herbaceous cover, especially forbs, provides critical food during the late spring to early autumn period (Drut et al. 1994; Huwer et al. 2008), and provides habitat for a variety of invertebrates. Invertebrates such as ants, beetles, and grasshoppers are consumed by all sage-grouse (Klebenow and Gray 1968; Peterson 1970) and are essential for survival of young chicks (Johnson and Boyce 1990).

10.2   Species and Population Status

10.2.1   Historical Versus Current Distributions, Conservation Status

Sage-grouse were historically found throughout most of the vast sagebrush-dominated landscape in western North America (Bendire 1892; McClanahan 1940; Aldrich and Duvall 1955; Aldrich 1963; Zwickel and Schroeder 2003; Schroeder et al. 2004). Although the original distribution of both species (Fig. 10.1) was established by comparing historical observations (Aldrich and Duvall 1955) with the distribution of potentially suitable habitat (Küchler 1985), precise observations of distribution and abundance during the period prior to settlement by people of European descent were not possible (Schroeder et al. 2004). Examination of bones from the late Pleistocene and early Holocene shows that sage-grouse were present in western North America (Braun and Williams 2015; Wolfe and Broughton 2016), but delineation of a distribution from those data is not possible, especially because many of the bones were found outside the documented historical distribution (Braun and Williams 2015). Even with inevitable uncertainty, the association between sage-grouse and sagebrush-dominated habitat is undeniable (Schroeder et al. 2020; Young et al. 2020).

Gunnison sage-grouse were historically found in a relatively small area, including southern Colorado, northern New Mexico, northeastern Arizona, and southeastern Utah (Beck et al. 2003; Schroeder et al. 2004; Braun et al. 2014; Braun and Williams 2015). Greater sage-grouse were distributed more widely and historically found in portions of Arizona, Alberta, British Columbia, California, Colorado, Idaho, Montana, Nebraska, Nevada, North Dakota, Oregon, Saskatchewan, South Dakota,

Utah, Washington, and Wyoming (Schroeder et al. 2004; Fig. 10.1). Delineations of historical and current distributions of both species is complicated by the transloca-tions of grouse throughout the range, sometimes even including the translocation of greater sage-grouse into the historical range of Gunnison sage-grouse (Reese and Connelly 1997).

Long-term changes in the distribution of sage-grouse have followed trends in the distribution and quality of sagebrush habitat, typically dominated by big sagebrush (*A. tridentata*). However, in northeastern portions of the range including Alberta, eastern Montana, Saskatchewan, and North and South Dakota, silver sagebrush (*A. cana*) is extremely important or the sole source of sagebrush (Aldridge and Brigham 2002; Connelly et al. 2004; Carpenter et al. 2010). Because of declines in habitat quantity and quality, greater sage-grouse no longer occur in British Columbia, Arizona, and Nebraska and Gunnison sage-grouse no longer occur in Arizona and New Mexico (Schroeder et al. 2004; Fig. 10.1). Populations of greater sage-grouse are also dramatically reduced in most states and provinces, but especially Alberta, California, North and South Dakota, Saskatchewan, and Washington (Garton et al. 2011, 2015) where hunting is no longer permitted (Dinkins et al. 2021a). Research using rangewide greater sage-grouse data estimated > 50% extirpation probability for 45.7, 60.1, and 78.0% of leks based on 19, 38, and 56-year projections of popu-lation growth from 2019, respectively (Coates et al. 2021b: 3). Most extirpated leks were predicted to be on the periphery of greater sage-grouse range. This study also predicted > 50% extirpation probability for 12.3, 19.2, and 29.6% of populations, defined as neighboring clustered leks, over the same time frames (Coates et al. 2021b: 3).

Gunnison sage-grouse have not been legally hunted since 1999 (Dinkins et al. 2021a). Gunnison sage-grouse are almost extirpated from Utah and greatly reduced in Colorado; their populations are currently so small they have been federally listed as a Threatened species (USFWS 2014). In Canada, greater sage-grouse are federally listed as Endangered under the Alberta Wildlife Act, Saskatchewan Wildlife Act, and under the Canada's Species at Risk Act (Government of Canada 2021). Greater sage-grouse are not federally listed in the United States (USFWS 2015). Hunting is currently permitted in Colorado, Idaho, Montana, Nevada, Oregon, Utah, and Wyoming (Dinkins et al. 2021a). Sage-grouse have often been viewed as an umbrella species, protecting habitat for up to 350 vertebrates through conservation of their sagebrush habitats (Rowland et al. 2006; Gamo et al. 2013). However, research designed to address this question indicates sage-grouse likely do not serve well in a surrogate role due to mismatches between temporal and spatial scales of seasonal distributions for other species (Carlisle et al. 2018).

The historical and current distribution of greater sage-grouse falls within 9 EPA Level III ecoregions (Wilken et al. 2011; Fig. 10.1). Gunnison sage-grouse were historically found only in the Colorado Plateau and Rocky Mountains ecoregions, with most of their current distribution lying within the Rocky Mountains ecoregion within Utah and Colorado (Fig. 10.1). Greater sage-grouse occur within 720,141 km^2 across 10 western states and 2 Canadian provinces and Gunnison sage-grouse occur within 10,036 km^2 in Colorado and Utah (Table 10.1). The current distribution

Table 10.1 Land ownership (km^2 [%]) for greater and Gunnison sage-grouse within Western Association of Fish and Wildlife Agencies Sage-Grouse Management Zones (MZs)

Management zone	Federal km^2 (%)	Private	State	Tribal	Other	Total area
Greater sage-grouse						720,141
MZ I[a]	34,952 (17.9)	127,403 (65.1)	13,922 (7.1)	9820 (5.0)	9576 (4.9)	195,673[b]
MZ II	78,577 (52.4)	54,510 (36.4)	9835 (6.6)	6095 (4.1)	803 (0.5)	149,820
MZ III	103,616 (83.5)	16,024 (12.9)	2676 (2.2)	1290 (1.0)	451 (0.4)	124,057
MZ IV	99,339 (63.5)	46,026 (29.4)	7922 (5.1)	2156 (1.4)	917 (0.6)	156,360
MZ V	58,270 (74.4)	17,954 (22.9)	1298 (1.7)	512 (0.7)	259 (0.3)	78,293
MZ VI	2088 (18.7)	6963 (62.4)	823 (7.4)	1284 (11.5)	3 (0.0)	11,161
MZ VII	1740 (36.4)	1174 (24.6)	622 (13.0)	1207 (25.3)	34 (0.7)	4777
Gunnison sage-grouse						10,036
MZ VII	4738 (47.2)	5084 (50.7)	214 (2.1)	0 (0.0)	0 (0.0)	10,036

Gunnison sage-grouse only occur within Management Zone VII. Unspecified land ownership is denoted as other, which includes local government and unknown ownerships. Ownership in Canada is also unknown
[a] Area totals and percentages for federal, private, state, and tribal were not available for Canada; thus, these were only quantified for greater sage-grouse distribution within the U.S. Canada represented 9,166 km^2 (4.7% of MZ I)
[b] Total area for MZ I includes km^2 for Canada

of greater sage-grouse has been estimated at 56% of the historical distribution and the current distribution for Gunnison sage-grouse at 10% of its historical distribution (Schroeder et al. 2004).

10.2.2 Monitoring

The most common and widespread index for greater sage-grouse is based on annual counts of males attending leks (Connelly et al. 2003a, b). Seasonal and daily variation in lek attendance is well-documented (Emmons and Braun 1984; Walsh et al. 2004; Fremgen et al. 2016; Wann et al. 2019) and therefore using the maximum of repeated counts within a standardized period is recommended for more reliable inferences (Connelly et al. 2003a, b; Monroe et al. 2016). However, multiple factors can influence detection including weather (Baumgardt et al. 2017; Fremgen et al. 2019), vegetation and topography (Fremgen et al. 2016), males not attending leks (Blomberg et al. 2013b; Gibson et al. 2014), and age of grouse (Jenni and Hartzler 1978; Walsh et al. 2004; Wann et al. 2019), leading to concerns over the relevance of lek counts to the true population status. Any given count is likely an underesti-mate of the true population associated with a lek and does not indicate age and sex ratios or males attending unknown leks (Shyvers et al. 2018). Monitoring through mark-resight of leg bands or telemetry can estimate the attendance process (Walsh et al. 2004; Gibson et al. 2014; Fremgen et al. 2016; Wann et al. 2019), but also are more intensive. More recently, researchers have examined use of infrared imagery from aerial surveys to estimate sightability of sage-grouse, given their presence at the lek (Coates et al. 2019). Alternatively, researchers may take advantage of informa-tion from repeated counts to estimate the detection process using N-mixture models (McCaffery et al. 2016; Monroe et al. 2019). Due to the open nature of sage-grouse lek attendance, inferences from this modeling approach can be extended at most to the number of males attending a lek at least once in a season (Nichols et al. 2009). Indices may still offer useful inferences of population trends when detectability is constant or random over time (Johnson 2008; Monroe et al. 2016), but fail if detection covaries with population trends (Monroe et al. 2019; Blomberg and Hagen 2020). Additionally, the area used by sage-grouse attending leks is unknown and can vary, and not all active leks in a landscape are known to observers. A dual-frame approach may help jointly estimate both the number of active leks and the number of males attending leks, providing a more accurate estimate of the number of males across a landscape (Shyvers et al. 2018). Genetic data from non-invasive sources can be used to estimate winter (pre-breeding) population size with a mark-recapture approach (Shyvers et al. 2019).

Due to the cryptic nature of this species, use of radio telemetry and global posi-tioning systems (GPS) to monitor habitat use and survival is common (e.g., Aldridge and Boyce 2007; Smith et al. 2018a). However, this type of monitoring can incur substantial costs to researchers and management agencies, both financially and in potential bias from monitoring units themselves (Fremgen et al. 2017; Severson et al. 2019). Indeed, stage-specific data on demographic parameters for this species are relatively limited, particularly for chick and juvenile survival (Taylor et al. 2012), and other data sources may be needed. For example, hunter-harvest data could be used to estimate survival rates, sex ratios, and recruitment metrics (Braun and Schroeder 2015; Hagen et al. 2018; Wann et al. 2020), and annual counts of sage-grouse broods may provide an index of productivity (Connelly et al. 2003a, b). Integrated population models (IPMs) which combine demographic information with lek count data may improve estimates of sage-grouse population size and other demographic parameters (Davis et al. 2014a; McCaffery and Lukacs 2016; Coates et al. 2018).

10.3 Habitat Associations

10.3.1 Historical/Evolutionary

Natural variation in the historical distribution of sage-grouse has occurred because of climate fluctuations, specifically in relation to elevational gradients. For example, warmer and drier sagebrush habitats at lower elevations tend to have a smaller component of native herbaceous cover when compared with higher elevation habitats. As a result, some of these lower elevations have reduced capacity to support sage-grouse. This can be illustrated by the decline in sage-grouse in the Bonneville Basin in Utah during a 4000-year period as the climate warmed and dried (Wolfe and Broughton 2016). Furthermore, declines in sage-grouse abundance (Connelly et al. 2004; Garton et al. 2011) and occupancy (Schroeder et al. 2004) increases fragmentation and isolation of populations. This loss of connectivity has dramatic demographic and genetic consequences that can reduce population viability (Oyler-McCance et al. 2005; Oyler-McCance and Quinn 2011), thus further decreasing populations and distribution.

10.3.2 Contemporary

Declines in sage-grouse populations indicate that identification of important habitats, especially breeding habitats, is critical for long-term conservation of sage-grouse, as most year-round activity for greater sage-grouse populations occurs within 8 km of active leks (Fedy et al. 2012; Coates et al. 2013). Doherty et al. (2016) delineated breeding habitat for greater sage-grouse and Doherty et al. (2018) delineated breeding habitat for Gunnison sage-grouse. For both species of sage-grouse, breeding habitat probabilities $\geq 65\%$ provided a threshold to predict areas where leks occur (predicted breeding habitat; Doherty et al. 2016, 2018; Fig. 10.3). Predicted breeding habitat overlapped more (range = 15.6–24.7%) with federal lands than on other ownerships including private, state, and tribal in the U. S. within Western Association of Fish and Wildlife Agencies (WAFWA) Management Zones (MZs) II (Wyoming Basin), III (Southern Great Basin), IV (Snake River Plain), and V (Northern Great Basin; Doherty et al. 2016; Table 10.2). Predicted breeding habitat for greater sage-grouse overlapped more with private compared to public or tribal lands in MZs I (Northern Great Plains), VI (Columbia Basin), and VII (Colorado Plateau; Table 10.2). Only 14.8% (8.4% federal, 5.9% private, and 0.5% of state) of MZ VII was predicted breeding habitat for Gunnison sage-grouse in MZ VII (Table 10.2), with the majority located in the Gunnison Basin of Colorado (Doherty et al. 2018; Fig. 10.3).

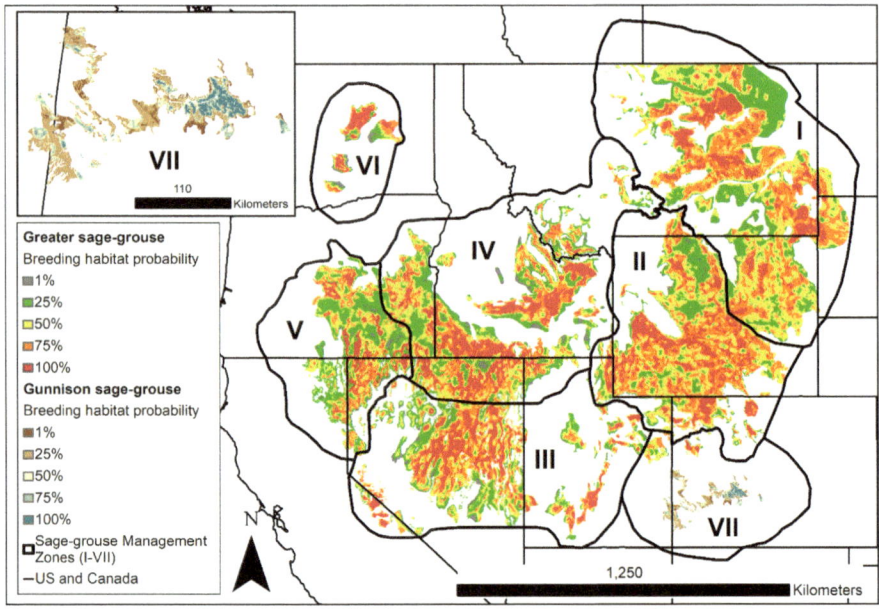

Fig. 10.3 Greater and Gunnison sage-grouse breeding habitat probabilities in the western United States. Inset map represents a detail of breeding habitat probabilities for distribution of Gunnison sage-grouse. Areas with > 65% breeding habitat probability indicate predicted breeding habitat suitable to support lek formation. The models that generated these predictions were originally created from separate analyses focused on greater (Doherty et al. 2016) and Gunnison (Doherty et al. 2018) sage-grouse. Models used data from 2010–2014 for greater and 2015 for Gunnison sage-grouse

10.4   Rangeland Management

We do not provide specific prescriptions for practitioners to manage or restore rangelands for sage-grouse. However, we do synthesize important issues related to Rangeland Management (Sect. 10.4; livestock grazing and grass height, improper grazing, mesic resources, fencing, habitat alteration treatments, feral equids, and ravens) and Ecosystem Threats (Sect. 10.6; fire, invasion from exotic annual grasses, conifer encroachment, sagebrush conversion and seeding introduced grasses, exurban development, energy development, and climate change). Topics listed under both sections highlight issues, describe concepts, and provide insights from relevant literature to assist in providing knowledge applicable to managing and restoring sage-grouse habitats at appropriate local scales. We encourage practitioners to access publications or other resources specifically developed to guide efforts to restore sage-grouse habitats such as Pyke et al. (2015a, b, 2017).

Table 10.2 Land ownership (km² [%]) for greater and Gunnison sage-grouse within areas with predicted breeding habitat sub-stratified by Western Association of Fish and Wildlife Agencies Sage-Grouse Management Zones (MZs)

Management zone	Federal	Private	State	Tribal	Other	Predicted breeding habitat	Total area
	km² (%)						
Greater sage-grouse							710,975
MZ I	12,387 (6.6)	26,087 (14.1)	3017 (1.6)	219 (0.1)	22 (0.0)	41,732 (22.4)	186,507
MZ II	28,578 (19.1)	16,084 (10.7)	3077 (2.1)	321 (0.2)	128 (0.1)	48,188 (32.2)	149,820
MZ III	30,614 (24.7)	3986 (3.2)	1096 (0.9)	705 (0.6)	241 (0.2)	36,642 (29.6)	124,057
MZ IV	32,770 (21.0)	10,519 (6.7)	2282 (1.5)	981 (0.6)	151 (0.1)	46,703 (29.9)	156,360
MZ V	12,206 (15.6)	1603 (2.0)	149 (0.2)	47 (0.1)	11 (0.0)	14,016 (17.9)	78,293
MZ VI	961 (8.6)	2923 (26.2)	424 (3.8)	154 (1.4)	1 (0.0)	4463 (40.0)	11,161
MZ VII	117 (2.5)	444 (9.3)	0 (0.0)	81 (1.7)	11 (0.2)	653 (13.7)	4777
Gunnison sage-grouse							10,036
MZ VII	847 (8.4)	594 (5.9)	50 (0.5)	0 (0.0)	0 (0.0)	1491 (14.8)	10,036

Predicted breeding habitat was based on breeding habitat probabilities $\geq 65\%$, indicating areas with adequate breeding habitat to support lek formation for greater (Doherty et al. 2016) and Gunnison (Doherty et al. 2018) sage-grouse (Fig. 10.3). For MZ I, areas with predicted breeding habitat were only quantified for greater sage-grouse distribution within the U.S., as there was no breeding habitat probability data available for Canada. Gunnison sage-grouse only occur within Management Zone VII. Unspecified land ownership was denoted as other, which included local government and unknown ownerships. Ownership in Canada was also unknown

10.4.1 Livestock Grazing and Grass Height

Potential responses from sage-grouse to livestock grazing management have been hypothesized, because positive correlations between sage-grouse nest site selection and survival and grass height have been documented (Holloran et al. 2005; Doherty et al. 2014; Gibson et al. 2016a). Indeed, the 2014 Threatened listing decision for Gunnison sage-grouse suspected that failure of multiple allotments meeting relevant land health standards might have negatively affected the species (USFWS 2014). One such habitat metric is grass height, which is incorporated in habitat management guidelines (Connelly et al. 2000; Stiver et al. 2015), but the relevance of this parameter is increasingly questioned. The first criticism is methodological; due to phenology and vegetation growth during a season, researchers measuring grass height at nesting outcome (fail or fledge) inherently induce a bias in their measurements, where grass height is often lower for failed nests (Gibson et al. 2016b; Smith et al. 2018b). Correcting for this discrepancy resulted in more modest, and variable, relationships between sage-grouse nest survival and grass height (Gibson et al. 2016b; Smith et al. 2018b). Secondly, habitat associations are often context-dependent, and there is risk in extrapolating results from individual, localized studies to other parts of the sage-grouse range (Smith et al. 2020). Finally, while attention has been focused on potential effects of grazing on sage-grouse nest habitat and survival, livestock may influence other components of sage-grouse life history such as brood-rearing habitat. For example, habitat selected for nesting may correlate with subsequent brood survival (Gibson et al. 2016a), and a diverse diet of abundant plants and invertebrates may benefit sage-grouse productivity (Blomberg et al. 2013a; Smith et al. 2019). However, there is a lack of empirical support for the hypothesis that livestock grazing manipulates food availability for sage-grouse, and, in turn influences sage-grouse vital rates. This is a compelling hypothesis, though, as research from central Montana found greater insect diversity and activity-density of arthropods eaten by sage-grouse in sagebrush grazed by livestock under rest-rotation as compared to lands ungrazed for over a decade (Goosey et al. 2019).

Studies examining responses of sage-grouse populations to grazing are also still limited, likely because of the large areas sage-grouse use for their life history (Fedy et al. 2014). In one of the few grazing studies, based on public lands records from across Wyoming, Monroe et al. (2017) estimated a negative response to higher levels of reported grazing early in the growing season among male sage-grouse attending leks and a positive response when grazing occurred later. This relationship was apparent in areas where vegetation productivity was low, but not in high productivity areas. A field experiment study based on research conducted on private ranches in central Montana indicated a positive trend in daily survival rates for greater sage-grouse nests on ranches that implemented rotational livestock grazing, yet support for this effect was weak (Smith et al. 2018a). This study did not find grazing rest of 1 year or greater increased daily survival rates for nests and rotational grazing and rest compared to other grazing strategies in the area had negligible effects on herbaceous vegetation height and cover. These authors concluded that grazing strategies played a minor role in sage-grouse nest success relative to other factors such as climate and predators in the northern Great Plains (Smith et al. 2018a).

10.4.2 Improper Grazing

Improper livestock grazing is generally heavy, repeated grazing, particularly in the spring, of sagebrush communities and can be detrimental to large perennial bunch-grasses and favor exotic annual grasses, particularly in sites with lower resilience and resistance (Stewart and Hull 1949; Daubenmire 1970; Mack 1981; Knapp 1996).

Improper grazing can negatively affect sagebrush communities by altering vegetation composition and structure. Some of the most significant effects of improper livestock grazing are related to its interaction with other factors. For example, improper grazing often depletes the perennial herbaceous component and increases sagebrush cover, but these communities may not convert to exotic annual grasslands until a wildfire removes the sagebrush and creates a pulse in resource availability (Davies et al. 2016a). In contrast to repeated heavy grazing, moderate levels of grazing with periods of growing season deferment and rest may not negatively impact sagebrush communities (West et al. 1984; Davies et al. 2018; Copeland et al. 2021) and may even decrease their risk of converting to exotic annual grasslands after burning by reducing fire severity (Davies et al. 2009, 2015a, 2016b).

10.4.3   Mesic Resources

Mesic resources include moist areas near springs, creeks, ponds, reservoirs, and wet meadows, which promote herbaceous production near water, attracting livestock and sage-grouse during summer as upland sites dry and herbaceous plants senesce (Connelly et al. 2011b; Swanson et al. 2015). Late brood-rearing habitat is more limited than early brood-rearing habitat, composing an estimated 5% of sage-grouse habitat in a study area in southern Alberta (Aldridge and Boyce 2007), and 2.4% of sage-grouse habitat in California, northwest Nevada, and Oregon (Donnelly et al. 2016). Furthermore, late brood-rearing habitat where broods survived composed only 2.8% of a study area in central Nevada (Atamian et al. 2010). Private lands are particularly important in providing mesic resources to sage-grouse, because 60% of sage-grouse habitat occurs on public lands, yet 68% of mesic sites available to sage-grouse occur on private lands (Donnelly et al. 2016, 2018). Sage-grouse population productivity, based on lek distribution and attendance data, increased with proximity to mesic sites for greater sage-grouse in California, northwest Nevada, and Oregon (Donnelly et al. 2016). Anecdotal evidence from earlier Nevada studies indicated moderate cattle grazing in mesic meadows induced use by sage-grouse because cattle herbivory exposed preferred forbs (Neel 1980; Evans 1986). Recent research from Idaho indicates use of early-season, high-intensity cattle grazing increases cover and biomass of high-value forbs used by sage-grouse in mesic meadows (Randall et al. 2022). Practices to restore meadows and riparian areas within the sagebrush biome include Zeedyk structures, beaver dam analogs, and grazing management—these practices may increase productivity of vegetation in sage-grouse brood-rearing habitats by 25% (Silverman et al. 2019). Conservation and management of mesic resources is imperative to maintain sage-grouse populations in the face of climate-driven variability in vegetation conditions (Donnelly et al. 2018).

10.4.4 Fencing

Structures associated with livestock production can be detrimental to sage-grouse. For example, fencing located near leks may increase collisions (Stevens et al. 2012b; Van Lanen et al. 2017), particularly in relatively flat topography (Stevens et al. 2012a; Fig. 10.4). Use of fence markers and wooden fence posts spaced < 4 m (13.1 feet) apart may reduce collisions (Stevens et al. 2012a; Van Lanen et al. 2017). Another option to consider is to move or remove fences in high-risk areas (Stevens et al. 2012b). Managers may also consider marking fences in sage-grouse breeding habitats when fence densities exceed 1 km/km^2 (0.62 mile/mile^2) within 2 km (1.2 miles) of active leks in areas with flat to gently rolling terrain (Stevens et al. 2012a). Marked compared to unmarked fences in Idaho reduced risk of fence collision by greater sage-grouse approximately 83% (Stevens et al. 2012b), and, in Wyoming, marking fences, regardless of marker type, reduced collisions approximately 57% (Van Lanen et al. 2017).

10.4.5 Habitat Alteration Treatments

Improper vegetation treatments are also a concern in sagebrush communities, especially brush management in lower elevation sagebrush communities, which are hotter and drier than those at higher elevations are. Attempts to improve sage-grouse and other wildlife habitat by reducing sagebrush cover in Wyoming big sagebrush communities has not achieved desired results of enhancing habitat conditions that bolster populations (Beck et al. 2012). Reducing sagebrush in these communities also often substantially increases exotic annual grass and forb abundance and cover (Davies et al. 2012; Davies and Bates 2014). Thus, sagebrush control in Wyoming big sagebrush communities may facilitate conversion to an exotic annual grassland. Manipulating sagebrush to benefit sage-grouse populations is largely untested (but see Dahlgren et al. 2006, 2015; Smith and Beck 2018; Smith et al. 2023), and cumulative effects of treatments can be detrimental to populations (Dahlgren et al. 2015). Effects likely vary by sagebrush treatment type and scale, with negative responses for mechanical and prescribed fire but neutral-to-positive long-term responses to chemical reduction, where many structural components of sagebrush are retained (Smith and Beck 2018).

10.4.6 Feral Equids

Negative ecological effects, mainly from increasing feral horse (*Equus ferus caballus*), and, in limited areas, burro (*E. asinus*) populations, on federal public rangelands are a growing concern for sage-grouse habitats (see Chap. 21: Feral

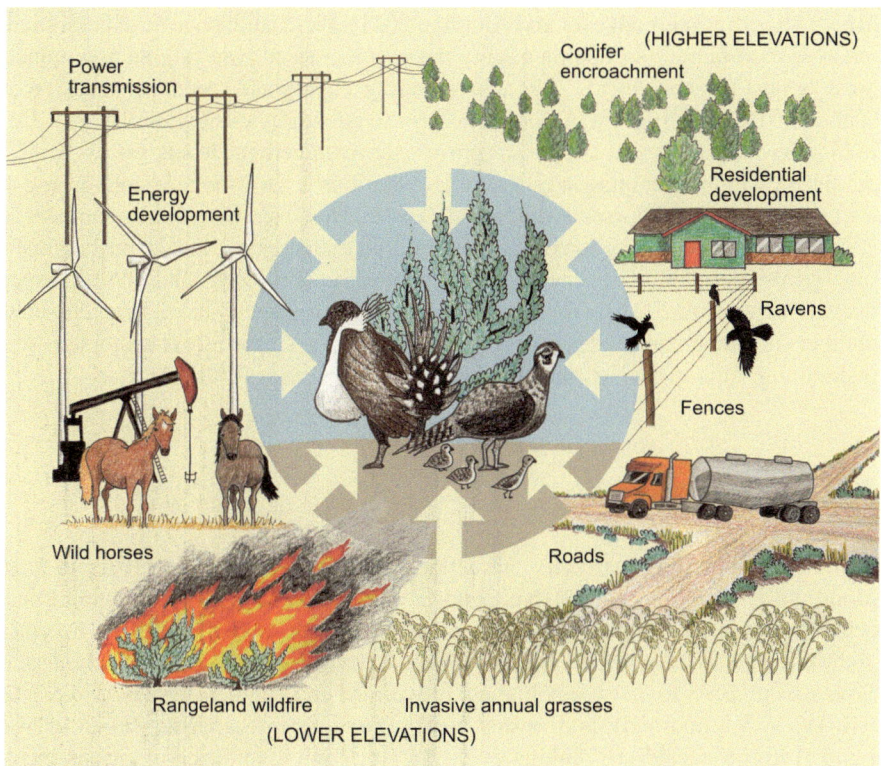

Fig. 10.4 Illustration depicting major threats to sage-grouse and their habitat. Multiple threats that may constrain sage-grouse populations are indicated by arrows pointing at an adult male, adult female, and chick sage-grouse in the center of the illustration. Land use and development threats can occur at any elevation, may occur solely, or in combination and are not ubiquitous across the range of sage-grouse. Land use and development threats include effects from renewable energy (e.g., wind energy), non-renewable energy (e.g., oil and gas development), linear features (e.g., fences, roads, and transmission lines), exurban residential development, increasing predators (e.g., common ravens that utilize developed areas to a higher degree), and feral horses. Persistent ecosystem threats related to exotic annual grasses, conifer encroachment and fire occur rangewide. Resistance to invasive annual grasses and resilience to disturbance increases with elevation owing to increased precipitation and cooler soil temperatures (Chambers et al. 2014, 2017). Wyoming big sagebrush dominates lower elevations where invasion by exotic annual grasses create continuous fuel beds, leading to increased fire frequency and reducing Wyoming big sagebrush. Encroaching conifers at higher elevations outcompete mountain big sagebrush and herbaceous plants (see Davies et al. 2011). Figure created by Emilene Ostlind, University of Wyoming

Equids; U. S. Fish and Wildlife Service 2013; Beever et al. 2018; Scasta et al. 2018). Rangewide, about 12% of sage-grouse habitat is also managed for feral equids. However, the amount of overlap varies among states, reaching a high of > 99% overlap where feral horses occur on Bureau of Land Management (BLM) lands in Wyoming (Beever and Aldridge 2011). Multiple direct and indirect effects have been hypothesized regarding how feral equids may influence sage-grouse life stages and habitat characteristics (Beever and Aldridge 2011). Field studies on feral horses have reported modification of habitat quality through increased bare ground and reduced grass height (Hennig et al. 2021), propagating cheatgrass seeds via feces (King et al. 2019), and disruption of male sage-grouse attending leks (Muñoz et al. 2021). Managing feral horses at or below appropriate management levels set by BLM is consistent with maintaining sage-grouse populations at levels where sage-grouse do not overlap with feral horses (Coates et al. 2021a). However, when horse numbers are 2, 2.5, and ≥ 3 times over maximum appropriate management levels, probability of sage-grouse population decline relative to controls is 76%, 97%, and > 99%, respectively (Coates et al. 2021a). These predictions indicate properly managing feral horse numbers that do not exceed appropriate management levels is most harmonious with maintaining sage-grouse populations where the two species share habitat.

10.4.7 Ravens

Common ravens (*Corvus corax* 'hereafter, ravens') are a major predator of sage-grouse nests (Dinkins et al. 2016a; Conover and Roberts 2017). During the last half century, ravens have expanded their distribution and increased in abundance in central and western North America (Dinkins et al. 2021b; Harju et al. 2021). Abundance of ravens from 1995 to 2014 was highest in western and southeastern WAFWA MZs (III, IV, V, VI, and VII), and ravens were expanding into and increasing in MZs I and II in the northeast (Dinkins et al. 2021b). High abundance of ravens in MZ VII indicates the Threatened Gunnison sage-grouse has been exposed to high raven numbers for a few decades (Dinkins et al. 2021b; Harju et al. 2021). Occurrence of ravens increases with presence of livestock and associated infrastructure, such as buildings and water sources (Coates et al. 2016a; Fig. 10.4). Increases in raven populations are related to their ability to exploit anthropogenic resource subsidies such as food, perches, and nesting structure in rangelands, uncoupling them from the availability of local indigenous resources (Boarman 2003). Adult female sage-grouse avoid nesting in habitat with higher densities of ravens (Dinkins et al. 2012). Predictive modeling across rangewide sage-grouse habitat indicates higher growth rates for ravens in landscapes with greater transmission line density. Dinkins et al. (2021b) found carrying capacity for ravens was higher with increasing proportion of urban land cover within 25 km and burned area within 3 km, and negatively correlated with greater distance from landfills and proportion of forest cover within 15 km. Management actions to reduce effects of ravens on sage-grouse populations include removing nesting structures and food sources, eliminating or covering landfills, and restoring burned sagebrush (Dinkins et al. 2021b).

10.5 Effects of Disease

10.5.1 General Concerns for Populations

Sage-grouse host a variety of potentially pathogenic organisms including macroparasitic arthropods (e.g., lice, ticks), helminths (e.g., nematodes, cestodes, trematodes), and microparasites (protozoa, bacteria, fungi and viruses; Christiansen and Tate 2011; Sinai et al. 2017). Various non-parasitic diseases or disease-like conditions can also affect sage-grouse (Christiansen and Tate 2011). Most macro- and microparasites, and infectious diseases documented in sage-grouse have not resulted in widespread population level effects to sage-grouse (Christiansen and Tate 2011). However, researchers have conducted few systematic surveys for parasites or pathogens in sage-grouse.

West Nile virus (WNV; Flaviviridae, *Flavivirus*) has emerged as a threat to sage-grouse (Walker and Naugle 2011). In addition to WNV, avian infectious bronchitis virus and other avian coronaviruses, avian retroviruses, *Mycoplasma* spp., and the *Eimeria* coccidians and associated enteric bacteria may be subject to amplification by climate change or anthropogenic disturbance or have a history of impacting sage-grouse. These historic and emerging risks indicate further monitoring and research of diseases and parasites in sage-grouse is warranted as suggested by Peterson (2004) for prairie grouse.

10.5.2 Diseases as Associated with Livestock

There are few direct disease relationships known to occur between rangeland livestock and sage-grouse. However, surface water, an important component of rangeland management, plays a role in the exposure of sage-grouse to WNV. Coccidiosis, tapeworms, and toxicosis are also discussed due to their local significance in the past and potential, though unlikely, risks to sage-grouse populations in the future.

West Nile Virus. West Nile virus, a mosquito-borne flavivirus, is recognized as an important source of mortality to sage-grouse in elevations below 1500 m, which is approximately 40% of current sage-grouse range (Walker and Naugle 2011). Few live sage-grouse have tested seropositive for WNV antibodies indicating sage-grouse rarely survive infection (Walker and Naugle 2011; Dusek et al. 2014). Population viability analyses indicate that local populations may be vulnerable to extirpation from even a single stressor, such as WNV (Taylor et al. 2013). Additionally, West Nile virus can have cascading effects as evidenced by a quadrupling of lek inactivity when populations already impacted by energy development are exposed to a West Nile virus outbreak (Taylor et al. 2013).

Livestock reservoirs, even water-filled hoof prints, can serve as breeding habitat for vector mosquitoes (Doherty 2007). The mesic areas created by the reservoirs attract sage-grouse in mid to late summer (Connelly et al. 2011b), during the peak of WNV transmission. Eliminating mosquito-breeding habitat or controlling mosquito larval populations in anthropogenic water sources can reduce effects of WNV on sage-grouse populations (Zou et al. 2006; Walker and Naugle 2011; Watchorn et al. 2018).

Summer temperatures may affect WNV viremia (Naugle et al. 2005; Walker and Naugle 2011) and vector mosquito autogeny (Brust 1991) indicating increasing temperatures associated with climate change may increase WNV risk to sage-grouse. Mammals, particularly humans and horses, can become infected through mosquito bites and represent dead-end hosts of WNV infection (Ahlers and Goodman 2018).

Coccidiosis. Prior to the emergence of West Nile virus, coccidiosis (*Eimeria* spp.) was the most important known disease of sage-grouse. Losses of young sage-grouse were documented in several states from 1932 to 1953, typically in areas where up to 2000 (estimated) birds congregated, resulting in fecal contamination of soil and water (Honess and Post 1968). Sporadic occurrence of coccidiosis-associated morbidity and mortality in individual birds is reported; however, notable mortality events attributed to coccidiosis in sage-grouse have not been documented since the early 1960s. This change in disease dynamic may be the result of decreased sage-grouse densities. Livestock are also susceptible to coccidiosis, but the infectious species of *Eimeria* are host specific (López-Osorio et al. 2020).

Tapeworms. The most visually apparent parasites of sage-grouse are tapeworms (Order: *cestoda*), which are commonly reported by hunters and field personnel. Sage-grouse show no apparent clinical signs of infection and may reflect an almost perfect adjustment between the host and its parasite (Honess 1982). However, "heavy" burdens of tapeworms could have direct and/or indirect adverse effects on individual birds, such as intestinal occlusion, reduction in vigor, and increased susceptibility to other parasites (Cole and Friend 1999). Livestock are also susceptible to tapeworms, but not the species that infect sage-grouse.

Toxicosis. During 1949–50, 1.7 million ha of Wyoming rangeland were aerially treated with Toxaphene and Chlordane bran bait to control grasshoppers (Family: *Acrididae*; Post 1951). Post (1951) reported game bird mortality and toxemia on treated areas. The scale and toxicity of grasshopper control efforts during this time indicate widespread but unquantified negative effects to sage-grouse populations. Neither Chlordane nor Toxaphene have been registered for grasshopper control since the early 1980s. Modern pesticides applied via Reduced Agent and Area Treatments approach (Lockwood et al. 2002) likely minimize effects.

Ivermectin, a broad-spectrum antiparasitic, is routinely and globally used to control parasitic worms of ruminant animals. Field studies have demonstrated the dung of animals treated with ivermectin influences abundance and ecology of invertebrates (Martinez et al. 2017; Finch et al. 2020). The direct or indirect effects of ivermectin (or similar compounds) to sage-grouse is unknown, but research in central Montana showed dramatically lower dung beetle activity-density on lands with managed grazing than on idled land, which were hypothesized to be due to anti-parasitic drugs like ivermectin (Goosey et al. 2019). However, sage-grouse food arthropods were still collected in higher numbers, overall, on managed than on idled land (Goosey et al. 2019).

10.6 Ecosystem Threats

Sagebrush ecosystems are experiencing numerous landscape-scale threats that decrease the quantity and quality of sagebrush communities (Davies et al. 2011). Historically, the sagebrush ecosystem occupied over 62 million hectares in western North America. By the early 2000s, it was estimated that sagebrush occupied less than 60% of its historical range and many remaining sagebrush communities were fragmented and degraded (Knick et al. 2003; Schroeder et al. 2004). The loss, degradation, and fragmentation of sagebrush occupied rangeland has continued with altered fire regimes, anthropogenic development, conifer encroachment, exotic annual grass invasion, climate change, and conversion to croplands and introduced grasslands. Many of these stressors are interrelated, with one threat exacerbating another, creating feedback loops. Habitat loss is viewed as the primary reason for declines in sage-grouse populations. For example, greater sage-grouse in Canada now only inhabit 7% of their historical distribution, primarily due to habitat loss from agricultural development and placement of anthropogenic features such as oil and gas wells in remaining habitats (Government of Canada 2021).

10.6.1 Altered Fire Regimes

Sagebrush ecosystems are experiencing widespread alteration to historical fire regimes. Fire frequency has increased in many hotter and drier sagebrush communities, driven mainly by exotic annual grass invasion (Balch et al. 2013) as well as increased human-caused ignitions (Bradley et al. 2018). Frequent fire prevents reestablishment of sagebrush and is detrimental to many native perennial species. In contrast, fire frequency has decreased in cooler and wetter sagebrush communities substantially because of fire suppression and historical heavy grazing (Miller and Wigand 1994; Miller and Rose 1999). Decreased fire frequency has allowed conifer woodlands to establish across many sagebrush communities, which further decreases fire frequency due to reduced fine fuel loads in woodland understories. Periodic fire in these cooler and wetter sagebrush communities is also important for promoting spatial and temporal heterogeneity of habitats (Davies and Bates 2020).

10.6.2 Invasion from Exotic Annual Grasses

Sagebrush plant communities are experiencing undesirable shifts in vegetation composition and structure from invasive species. Exotic annual grasses are the primary threat in hotter and drier sagebrush communities (Davies et al. 2011; Chambers et al. 2014) and native conifers are encroaching into cooler and wetter sagebrush communities (Miller and Wigand 1994; Miller et al. 2005). These two undesirable vegetation shifts largely occur in different sagebrush community types, but conifer encroachment and exotic annual grass invasion appear to be overlapping more frequently in recent decades (Davies et al. 2011) as annual grasses spread to higher elevations and north-facing aspects consistent with predictions of climate warming (Smith et al. 2022).

Exotic annual grasses, primarily cheatgrass (*Bromus tectorum*) and medusahead (*Taeniatherum caput-medusae*) have invaded tens of millions of hectares of the sagebrush ecosystem, particularly in the western portion of sage-grouse range (Meinke et al. 2009; Bradley et al. 2018). Cheatgrass, alone is estimated to be present in high abundance across almost a third (21 million ha) of the Great Basin (Bradley et al. 2018), primarily in sagebrush communities. Exotic annual grass invasion often triggers an exotic annual grass-fire cycle, where abundant exotic annual grass fuel promotes frequent fire, further decreasing native perennial vegetation (D'Antonio and Vitousek 1992; Balch et al. 2013). This cycle results in exponential declines in native plants and biodiversity as exotic annual grasses increase in abundance (Davies 2011). Currently there are no cost-effective treatments to control exotic annual grasses across the vast area of invasion (Stohlgren and Schnase 2006).

10.6.3 Conifer Encroachment

Conifers, primarily juniper (J. *osteosperma*, J. *occidentalis*, J. *scopulorum*) and piñon pine, in the sagebrush ecosystem were historically confined to fire-safe sites (i.e., rocky slopes and ridges with insufficient understory to carry surface fires) or occurred in open savannah-like stands (Romme et al. 2009). Decreased fire frequency has allowed conifers to expand into more productive sagebrush communities and increase in density, particularly in the western and southern regions of the sagebrush biome (Miller and Wigand 1994; Miller et al. 2005). Juniper and piñon woodlands currently occupy ~ 19 million ha in the Intermountain West; up to 90% of this area was historically sagebrush communities (Tausch et al. 1981; Miller et al. 2008). Conifer encroachment eliminates sagebrush and decreases the herbaceous understory, which can result in substantial erosion risk (Miller et al. 2000; Pierson et al. 2007). Conifer encroachment into sagebrush communities is especially detrimental to sage-grouse populations for several reasons. Greater sage-grouse will avoid sagebrush communities with as little as 4% tree cover (Baruch-Mordo et al. 2013; Severson et al. 2016). Conifer cover as low as 1.5% negatively influences survival for adult female greater sage-grouse (Coates et al. 2017) and increases risk of daily mortality, especially in juvenile and yearling birds, when navigating conifer-invaded sagebrush habitats (Prochazka et al. 2017). Population growth rates for greater sage-grouse are lower in conifer-invaded areas than adjacent areas where conifers have been removed (Olsen et al. 2021). In addition, simulated removal of conifer coverage up to 30% within a 0.56-km^2 scale was predicted to increase high-quality Gunnison sage-grouse breeding habitat fourfold (Doherty et al. 2018). Despite unprecedented efforts in the sagebrush biome, conifer removal is barely keeping pace with its rate of expansion (Reinhardt et al. 2020).

10.6.4   Sagebrush Conversion and Seeding Introduced Grasses

The most arable sagebrush communities in the United States have been converted to cropland, thus continued cultivation of intact landscapes produces marginal yields at high costs to wildlife. Crested wheatgrass (*Agropyron cristatum* and *A. desertorum*) has been seeded on 6–11 million hectares of rangelands in western North America, much of which was originally native sagebrush communities (Lesica and DeLuca 1996; Hansen and Wilson 2006). Crested wheatgrass is often seeded after wildfire in former sagebrush communities largely because of its ability to effectively compete with exotic annual grasses (Arredondo et al. 1998; Davies et al. 2010), but also because it is less expensive and often establishes more successfully than native bunchgrasses in hotter and drier sagebrush communities (Asay et al. 2003; James et al. 2012; Davies et al. 2015b). However, due to its ability to outcompete other species, crested wheatgrass can form monotypical plant communities, resulting in less value as wildlife habitat and lower overall biodiversity (Christian and Wilson 1999; Heidinga and Wilson 2002; Hamerlynck and Davies 2019).

Currently there exists a management conundrum that introduced grasslands are less desirable than native sagebrush rangelands, but native seedings using conventional seeding techniques frequently fail within sagebrush communities that have low resilience and resistance to exotic annual grass invasion (Knutson et al. 2014). Furthermore, exotic annual grasslands have lower habitat quality than introduced grasslands and the development of exotic annual grasslands increases the probability that surrounding areas will convert to exotic annual grasslands. For these reasons, introduced grasses are likely to continue to be seeded after fires in sagebrush communities with substantial risk of exotic annual grass dominance until native species establishment is improved. However, introduced grasses should not be seeded in more resilient and resistant sagebrush communities where seeded native species can successfully establish and persist (e.g. Davies et al. 2019; Urza et al. 2019).

10.6.5 Exurban Development

Exurban development is the process of dividing large parcels of mostly undeveloped lands into residential lots and is the fastest growing form of land conversion across the United States (Brown et al. 2005). As human population growth has continued in the western states and provinces, large expanses of the sagebrush ecosystem have seen negative effects from human development and subsequent habitat fragmentation. Exurban development directly converts native habitat, fragments habitat, and degrades remaining habitat for native species. This results in declines in native plant and animal diversity, increases in exotic species, and greatly limits the use of ecosystem management tools to achieve landscape level effects (Knight et al. 1995; Maestas et al. 2003; Hansen et al. 2005). Larger human populations are also tied to increased fire frequency in Mediterranean-climate ecosystems (Syphard et al. 2009), thus exurban development of sagebrush communities will likely increase fire frequency in some areas, further promoting exotic annual grass invasion and dominance.

10.6.6 Energy Development

Energy development has fragmented and degraded sagebrush communities in many western states and provinces (Bergquist et al. 2007; Lyon and Anderson, 2003; Naugle et al. 2011). This threat to the sagebrush ecosystem has increased during the twenty-first century with greater demand for renewable energy sources like wind and solar energy (Kiesecker and Naugle 2017). Influences of energy development are expected to rise as the United States continues to increase its domestic energy production (Doherty et al. 2010). For example, the land occupied by well pads, roads, and other facilities from recent (2000–2012) expansion of oil and gas extraction in North America is estimated at 3 million ha (Allred et al. 2015). This is not exclusive to sagebrush communities but highlights the threat of further energy development to these communities.

Energy extraction and development can cause high levels of fragmentation of sagebrush landscapes. In areas of northeastern Wyoming, every 1 km^2 was bisected by a powerline and bounded by a road where energy development and agricultural production occurred (Naugle et al. 2011). Infrastructure for energy extraction and development including roads, pipelines, earthen dams, and well pads create substantial surface disturbance and are vectors for exotic plant invasions within sagebrush communities, contributing to further degradation (Gelbard and Belnap 2003; Bergquist et al. 2007). Sage-grouse typically respond to fragmented landscapes by avoiding human infrastructure (e.g., Aldridge and Boyce 2007; Dinkins et al. 2014; Kirol et al. 2015). For example, translocated female sage-grouse in Alberta were more likely to select habitat with increasing distance from infrastructure, up to 2.5 km from roads, 3 km from trees and gas wells, 10 km from buildings, 15 km from settlements, and at least 23 km from power lines (Balderson 2017).

10.6.7 Climate Change

Climate change is expected to negatively affect the sagebrush ecosystem. Reduced snowpack and earlier snowmelt (Klos et al. 2014; Harte et al. 2015) and longer, more arid summer conditions leading to drier soils (Palmquist et al. 2016) are factors predicted to lead to future reduction in coverage of sagebrush and associated plant species, altering sage-grouse habitat (Homer et al. 2015). Cultivation of sagebrush habitats in the eastern range of sage-grouse has completely altered carbon storage (Sanderson et al. 2020). Retaining and restoring sagebrush habitats represents the single largest natural opportunity to maintain carbon storage in rangelands (Fargione et al. 2018). Farther west, most models of climate change predict warmer winters and altered precipitation patterns, as well as an earlier onset of fire season and more wild-fires, all conditions that perpetuate exotic annual grasses (Abatzoglou and Kolden 2011; Creutzburg et al. 2015). When these models are applied to the landscape, cheatgrass cover is predicted to remain stable or increase in much of the Great Basin for the next 50 years (Boyte et al. 2016). In addition, increasing atmospheric CO_2 concentrations may increase exotic annual grass productivity and litter retention, leading to increased fuel loads, which may increase fire intensity and frequency (Ziska et al. 2005). Ongoing increases in cheatgrass and other exotic annual grasses and associated increased fire frequency will likely further reduce the area occupied by sagebrush. Furthermore, climate change predictions indicate sagebrush may not be viable to its entire historical range and may experience a distribution shift (Schlaepfer et al. 2015). Future climate scenarios indicate loss of sagebrush, and ultimately sage-grouse habitat, in some areas (Schlaepfer et al. 2012; Homer et al. 2015; Palmquist et al. 2016; Renne et al. 2019), as was historically documented with a warmer, drier climate in the Bonneville Basin of Utah (Wolfe and Broughton 2016). Increased drought conditions will likely also be detrimental to sage-grouse demographics (Blomberg et al. 2014).

10.7 Conservation and Management Actions

10.7.1 Private Lands

Approximately one-third of western rangelands are privately owned and managed, encompassing some of the most productive sage-grouse habitats (Donnelly et al. 2016). Grazing by domestic livestock is the common thread that maintains open spaces of intact rangelands at ecosystem scales, hereafter referred to as working rangelands. Working rangelands connect a checkerboard of public lands that together provide ecological footprints large enough to sustain sage-grouse populations and rural communities. In the 1930s the father of modern conservation, Aldo Leopold, said that 'conservation will ultimately boil down to rewarding the private landowner who conserves the public interest (Flader et al. 1992: 202),' a prognosis still true today. Partnerships are the cornerstone of private lands conservation, with many like-minded, landowner-led groups coalescing into umbrella organizations to offer their shared vision of wildlife conservation through sustainable ranching. State, federal, and non-governmental partners commonly provide human and financial resources to implement beneficial conservation practices on ranchers' private operations.

The largest partnership is the Sage Grouse Initiative (SGI), which launched by the USDA Natural Resources Conservation Service (NRCS) in 2010, assists agricultural producers who volunteer to reduce threats facing sage-grouse on working rangelands. In 2012, the SGI served as the flagship for the establishment of Working Lands for Wildlife (WLFW), an effort to conserve other at-risk ecosystems and associated species. The NRCS employs the $60 billion conservation title of the federal Farm Bill legislation (title II of the Agriculture Improvement Act of 2018, FY2019–FY2023) to help landowners voluntarily implement conservation practices on private farms, ranches, and forestlands. Quantifying the outcomes of resulting sage-grouse conservation (Naugle et al. 2019; NRCS 2021) and the iterative use of emerging science to improve delivery are integral components of this effort (Naugle et al. 2020). Over the last decade, the SGI has become the primary catalyst for science-driven sagebrush conservation by using Farm Bill resources to restore or enhance more than 29,300 km^2 of sage-grouse habitat on more than 1850 ranches, while supporting sustainable agricultural productivity on these working lands. Private lands conservation was featured prominently in the most recent listing decision for greater sage-grouse, placing voluntary conservation on par with regulatory mechanisms on public lands.

With WLFW codified nationally in the 2018 Farm Bill, NRCS continues to contribute to conservation of working rangelands as part of the Sagebrush Conservation Strategy administered by WAFWA. Ongoing contributions strategically target removal of expanding conifer, restoration of riparian areas and wet meadows, and reduction of cultivation, exurban sprawl, and cheatgrass invasion. Recent removal of expanding conifer in > 3300 square kilometers of priority sage-grouse habitats in places like southern Oregon have increased the population growth rate of sage-grouse by + 12% (Olsen et al. 2021) and doubled the abundance of sagebrush songbirds (Holmes et al. 2017). Riparian and wet meadow restorations on working lands using Zeedyk structures, beaver dam analogs, and grazing management have increased by 25% the productivity of sage-grouse brood habitats (Silverman et al. 2019). The strategic placement of conservation easements to alleviate in perpetuity the threat of cultivation and subdivision in places like northcentral Montana also have conserved the longest known migration corridors of sage-grouse and pronghorn (*Antilocapra americana*) in North America (Tack et al. 2019). Targeting tools created from the new NRCS-sponsored Rangelands Analysis Platform (https://rangeland s.app) provide an integrated approach for reducing effects of cheatgrass (Western Governors' Association 2020).

10.7.2 Public Lands

Approximately two-thirds of sage-grouse habitat in the United States occurs on public lands, including lands managed by the BLM (51%), the U.S. Forest Service (USFS; 8%), and States (5%; Knick 2011). The majority of land inhabited by sage-grouse in Alberta and Saskatchewan is provincially owned and leased for grazing or contained within Grasslands National Park. Both active leks (as of 2021) in Saskatchewan lie within Grasslands National Park and of the 3 active leks in Alberta, 2 occur on provincial and 1 on private land (J. T. Nicholson, written communication, Alberta Environment and Parks, 2022). The Western Association of Fish and Wildlife Agencies delineated seven Management Zones (MZs) across the sage-grouse range (Stiver et al. 2006; Tables 10.1 and 10.2; Fig. 10.3), and these are distinguished as floristic provinces based on climatic, elevational, topographic, and edaphic characteristics (Connelly et al. 2004). Within each Management Zone in the U.S., areas are prioritized for sage-grouse conservation through coordination of local working groups and by multiple agencies including at the federal and state levels with habitat issues being primarily managed by the BLM and USFS on federal lands and each state on state-owned lands. Federal lands in the United States provide the majority of habitat for greater sage-grouse in MZs II (Wyoming Basin), III (Southern Great Basin), IV (Snake River Plain), and V (Northern Great Basin), whereas private lands provide the majority of habitat in MZs I (Northern Great Plains), VI (Columbia Basin), and VII (Colorado Plateau; Table 10.1). Gunnison sage-grouse only inhabit portions of MZ VII, where land ownership for this species is nearly evenly split between public (federal [47.2%] and state ([2.1%]) and private (50.7%; Table 10.1).

State and provincial wildlife management agencies have the primary role of managing sage-grouse populations, inclusive of providing sustainable hunting opportunities. To implement and prioritize coordinated conservation efforts, numerous conservation plans have been drafted by federal, state, provincial, and local working groups that detail specific threats and necessary conservation practices, which typically address disturbance (sagebrush conversion, energy and mining development, fire), habitat (conifer encroachment, invasive plants, grazing), and predation. State and local plans generally delineated priority areas based on breeding habitat surrounding sage-grouse leks, and these delineations were used to inform national Priority Areas for Conservation (PACs) and Priority Habitat Management Areas in the U.S (PHMAs; USFWS 2013, 2015). Examples of state priority areas and associated conservation policies include Core Areas in Oregon, Montana, and Wyoming, state Management Zones and Conservation Areas in Idaho, and Sage-Grouse Management Areas in Utah. Because federal and state designations of priority habitat were both informed by state assessments, their shape and extent tend to be similar (USFWS 2013), and they generally aim to maintain stable and reverse declining sage-grouse populations by minimizing threats and restoring degraded habitat. In Wyoming, the rate of energy development was restricted in core areas (Gamo and Beck 2017), and sage-grouse populations tend to perform better within core areas (Spence et al. 2017; Dinkins and Beck 2019; Heinrichs et al. 2019). However, priority areas tend to conserve breeding habitat over winter habitat (Smith et al. 2016; Dinkins et al. 2017), some plans for priority areas do not explicitly consider livestock management (Dinkins et al. 2016b), and high levels of human development adjacent to priority areas may still negatively affect sage-grouse populations within these priority areas (Spence et al. 2017; Heinrichs et al. 2019). Development activities within core areas, as allowed by specific policies, may negatively affect sage-grouse populations if adjacent to large leks or high-quality sage-grouse habitats.

With the 2015 sage-grouse listing decision by USFWS, the BLM and USFS committed to monitoring and reviewing grazing authorizations (permits and leases) in sagebrush focal areas (SFAs; USFWS 2015), lands deemed highest priority for conserving sage-grouse within PHMAs (USFWS 2014). Monitoring and evaluation based on Land Health Standards would be prioritized in these areas to determine whether habitat objectives for greater sage-grouse are met, whether modifications and management are needed, and to ensure compliance (USFWS 2015). Additionally, a conceptual framework was recently formalized for understanding the ability of sagebrush ecosystems to recover following stressors and disturbance such as drought and fire (resilience) and retain their original state such as by resisting invasion by annual grasses (Chambers et al. 2014). Soil temperature and moisture can largely determine resilience and resistance of sagebrush ecosystems (for example, faster recovery under cool and moist conditions), and resilience and resistance can be predicted across the landscape using maps of soils data (Maestas et al. 2016). Federal agencies currently use resilience and resistance concepts to evaluate threats and risk to sagebrush ecosystems and prioritize resources for sage-grouse conservation (Chambers et al. 2017; Crist et al. 2019). At finer scales, federal agencies are committed to using the Habitat Assessment Framework (HAF; Stiver et al. 2015) to determine management objectives for sage-grouse habitat on public lands (USFWS 2015). The BLM, USGS, and WAFWA are currently working to refine habitat objectives at multiple scales using rangewide datasets of sage-grouse lek locations, movement, and remote sensing products (C. L. Aldridge, U.S. Geological Survey, written communication, 2021).

Hunting regulations for sage-grouse were instituted among states starting around the turn of the twentieth century. Restrictions to hunting regulations, such as season length, bag limits, and hunting season closures, have generally been implemented in response to declines in sage-grouse numbers (Wambolt et al. 2002; Dinkins et al. 2021a). Although hunting can negatively impact sage-grouse populations (Connelly et al. 2003a, b; Blomberg 2015; Caudill et al. 2017), it is generally unclear that current harvest limits are detrimental (Sedinger et al. 2010; Taylor et al. 2012), and states and provinces have adjusted harvest regulations to reduce hunting exposure to declining populations or in response to management challenges such as extreme fire events (Dinkins et al. 2021a).

10.8   Research/Management Needs

As iconic species dependent on sagebrush-dominated rangelands across western North America, we have summarized the myriad conservation challenges facing greater and Gunnison sage-grouse, yet gaps remain in both our knowledge base and potential solutions. Livestock production is one of the oldest and most widespread land use types overlapping sage-grouse range; however, experiments directly evaluating effects of livestock on sage-grouse demography and population trends are limited (Beck and Mitchell 2000) but have been performed in Montana (Smith et al. 2018a) and are currently being conducted in Idaho (Conway et al. 2021). Despite uncertainty over the importance of grass height for sage-grouse nest success, herbaceous cover, particularly forbs and associated arthropods, may still be important for brood-rearing sage-grouse, yet little research exists for livestock management in brood-rearing habitat (but see Street 2020). Given variability in environmental conditions over space and time (such as precipitation; Blomberg et al. 2012; Guttery et al. 2013), replicating grazing experiments and observational studies over long temporal periods (> 2 years) is needed. Further, greater sage-grouse respond and use habitat at multiple scales, including for nest sites, brood-rearing, and wintering habitat, so examining how rangeland management affects these different components at relevant scales is warranted (for example, see Smith et al. 2023). Broad-scale studies are needed to anticipate how policies and land use alternatives may affect sage-grouse populations. Records of livestock on public lands may indicate levels of use across these vast landscapes (Veblen et al. 2014; Monroe et al. 2017), but these data are often coarse (reported annually at the scale of the allotment). An evaluation of how these records relate to forage consumption, and therefore to the structure and composition of vegetation, is warranted. Federal agencies have committed to implementing management standards for sage-grouse in priority areas (USFWS 2015), but consistent, long-term monitoring will be critical to detect deviations from desired management outcomes (Veblen et al. 2014). Effects of increasing feral equid populations on habitat alteration and sage-grouse populations also merit further study.

Studying each of the ecosystem threats listed above in the context of other threats will be necessary, given the potential for interactions among multiple stressors. Sustainable management of livestock is compatible with sage-grouse (Beck and Mitchell 2000; Smith et al. 2018a), but levels of forage production that define sustainable grazing could change under future climate conditions (Reeves et al. 2014). While resilience and resistance concepts can be used to anticipate effects of climate change (Bradford et al. 2019; Crist et al. 2019), current consideration for this stressor by federal agencies may be limited (Brice et al. 2020). Given the effects of increasing wildfire and exotic annual grass spread on sage-grouse populations (Coates et al. 2016b; Smith et al. 2022), additional research and management directed at controlling exotic annual grasses and restoring degraded habitat will benefit sage-grouse. Finally, greater study of effects to small and isolated populations including constrained movements, predation, genetic issues, and emerging infectious diseases could identify interactions with other environmental and anthropogenic threats, potentially revealing additional mechanisms driving sage-grouse population trends.

References

Abatzoglou JT, Kolden CA (2011) Climate change in western US deserts: potential for increased wildfire and invasive annual grasses. Rangeland Ecol Manag 64:471–478. https://doi.org/10.20111/REM-D-09-00151.1

Ahlers LRH, Goodman AG (2018) The immune responses of the animal hosts of West Nile virus: a comparison of insects, birds, and mammals. Front Cell Infect Mi 8:96. https://doi.org/10.3389/fcimb.2018.00096

Aldrich JW (1963) Geographic orientation of American Tetraonidae. J Wildl Manage 27:528–545. https://doi.org/10.2307/3798463

Aldrich JW, Duvall AJ (1955) Distribution of American gallinaceous game birds. United States Fish and Wildlife Service, Circular 34, Washington, DC

Aldridge CA, Boyce MS (2007) Linking occurrence and fitness to persistence: habitat-based approach for endangered greater sage-grouse. Ecol Appl 17:508–526. https://www-jstor-org.libproxy.uwyo.edu/stable/40061874

Aldridge CL, Brigham RM (2002) Sage-grouse nesting and brood habitat use in Southern Canada. J Wildl Manage 66:433–444. https://doi.org/10.2307/3803176

Aldridge CL, Saher DJ, Childers TM, Stahlnecker KE, Bowen ZH (2012) Crucial nesting habitat for Gunnison Sage-Grouse: a spatially explicit hierarchical approach. J Wildl Manage 76:391–406. https://doi.org/10.1002/jwmg.268

Allred BW, Smith WK, Twidwell D, Haggerty JH, Running SW, Naugle DE, Fuhlendorf SD (2015) Ecosystem services lost to oil and gas in North America. Science 348:401–402. https://doi.org/10.1126/science.aaa4785

Arredondo JT, Jones TA, Johnson DA (1998) Seedling growth of intermountain perennial and weedy annual grasses. J Range Manage 51:584–589. https://doi.org/10.2307/4003380

Asay KH, Chatterton NJ, Jensen KB, Jones TA, Waldron BL, Horton WH (2003) Breeding improved grasses for semiarid rangelands. Arid Land Res Manag 17:469–478. https://doi.org/10.1080/713936115

Atamian MT, Sedinger JS, Heaton JS, Blomberg EJ (2010) Landscape-level assessment of brood rearing habitat for greater sage-grouse in Nevada. J Wildl Manage 74:1533–1543. https://doi.org/10.2193/2009-226

Back GN, Barrington MR, McAdoo JK (1987) Sage grouse use of snow burrows in northeastern Nevada. Wilson Bull 99:488–490. https://www.jstor.org/stable/4162435

Balch JK, Bradley BA, D'Antonio CM, Gómez-Dans J (2013) Introduced annual grass increases regional fire activity across the arid western USA (1980–2009). Glob Change Biol 19:173–183. https://doi.org/10.1111/gcb.12046

Balderson KL (2017) Habitat selection and nesting ecology of translocated greater sage-grouse. University of Regina, Regina, Saskatchewan, Thesis

Baruch-Mordo S, Evans JS, Severson JP, Naugle DE, Maestas JD, Kiesecker JM, Falkowski MJ, Hagen CA, Reese KP (2013) Saving sage-grouse from the trees: a proactive solution to reducing a key threat to a candidate species. Biol Conserv 167:233–241. https://doi.org/10.1016/j.biocon.2013.08.017

Baumgardt JA, Reese KP, Connelly NW, Garton EO (2017) Visibility bias for sage-grouse lek counts. Wildl Soc Bull 41:461–470. https://doi.org/10.1002/wsb.800

Beck JL, Mitchell DL (2000) Influences of livestock grazing on sage grouse habitat. Wildlife Soc B 28:993–1002. https://www.jstor.org/stable/3783858

Beck JL, Mitchell DL, Maxfield BD (2003) Changes in the distribution and status of sage-grouse in Utah. West N Am Nat 63:203–214. https://www.jstor.org/stable/41717283

Beck JL, Reese KP, Connelly JW, Lucia MB (2006) Movements and survival of juvenile greater sage-grouse in southeastern Idaho. Wildlife Soc B 34:1070–1078. https://doi.org/10.2193/0091-7648(2006)34[1070:MASOJG]2.0.CO;2

Beck JL, Connelly JW, Wambolt CL (2012) Consequences of treating Wyoming big sagebrush to enhance wildlife habitats. Rangeland Ecol Manag 65:444–455. https://doi.org/10.2111/REM-D-10-00123.1

Beever EA, Aldridge CL (2011) Influences of free-roaming equids on sagebrush ecosystems, with a focus on greater sage-grouse. In: Knick ST, Connelly JW (eds) Greater sage-grouse: ecology and conservation of a landscape species and its habitats. Studies in Avian Biology 38, University of California Press, Berkeley, California, pp 273–290. https://doi.org/10.1525/California/978052 0267114.003.0015

Beever EA, Huntsinger L, Petersen SL (2018) Conservation challenges emerging from free-roaming horse management: a vexing social-ecological mismatch. Biol Conserv 226:321–328. https://doi.org/10.1016/j.biocon.2018.07.015

Bergquist E, Evangelista P, Stohlgren TJ, Alley N (2007) Invasive species and coal bed methane development in the Powder River Basin, Wyoming. Environ Monit Assess 128:381–394. https://doi.org/10.1007/s10661-006-9321-7

Bendire CE (1892) Life histories of North American birds. United States National Museum, Special Bulletin 1, Washington, DC

Blomberg EJ (2015) The influence of harvest timing on greater sage-grouse survival: a cautionary perspective. J Wildl Manage 79:695–703. https://doi.org/10.1002/jwmg.887

Blomberg EJ, Hagen CA (2020) How many leks does it take? Minimum samples sizes for measuring local-scale conservation outcomes in greater sage-grouse. Avian Conserv Ecol 15:9. https://doi.org/10.5751/ACE-01517-150109

Blomberg EJ, Sedinger JS, Atamian MT, Nonne DV (2012) Characteristics of climate and landscape disturbance influence the dynamics of greater sage-grouse populations. Ecosphere 3:55. https://doi.org/10.1890/ES11-00304.1

Blomberg EJ, Poulson SR, Sedinger JS, Gibson D (2013a) Prefledging diet is correlated with individual growth in greater sage-grouse (*Centrocercus urophasianus*). Auk 130:715–724. https://doi.org/10.1525/auk.2013.12188

Blomberg EJ, Sedinger JS, Nonne DV, Atamian MT (2013b) Annual male lek attendance influences count-based population indices of greater sage-grouse. J Wildl Manage 77:1583–1592. https://doi.org/10.1002/jwmg.615

Blomberg EJ, Sedinger JS, Nonne DV, Atamian MT (2013c) Seasonal reproductive costs contribute to reduced survival of female greater sage-grouse. J Avian Biol 44:149–158. https://doi.org/10.1111/j.1600-048X.2012.00013.x

Blomberg EJ, Sedinger JS, Gibson D, Coates PS, Casazza ML (2014) Carryover effects and climatic conditions influence postfledging survival of greater sage-grouse. Ecol Evol 4:4488–4499. https://doi.org/10.1002/ece3.1139

Boarman WI (2003) Managing a subsidized predator population: reducing common raven predation on desert tortoises. Environ Manage 32:205–217. https://doi.org/10.1007/s00267-003-2982-x

Boyte SP, Wylie BK, Major DJ (2016) Cheatgrass percent cover change: comparing recent estimates to climate change—driven predictions in the northern Great Basin. Rangeland Ecol Manag 69:265–279. https://doi.org/10.1016/j.rama.2016.03.002

Bradford JB, Schlaepfer DR, Lauenroth WK, Palmquist KA, Chambers JC, Maestas JD, Campbell SB (2019) Climate-driven shifts in soil temperature and moisture regimes suggest opportunities to enhance assessments of dryland resilience and resistance. Front Ecol Evol 7:358. https://doi.org/10.3389/fevo.2019.00358

Bradley BA, Curtis CA, Fusco EJ, Abatzoglou JT, Balch JK, Dadashi S, Tuanmu M-N (2018) Cheatgrass (*Bromus tectorum*) distribution in the intermountain Western United States and its relationship to fire frequency, seasonality, and ignitions. Biol Invasions 20:1493–1506. https://doi.org/10.1007/s10530-017-1641-8

Braun CE, Oyler-McCance SJ, Nehring JA, Commons ML, Young JR, Potter KM (2014) The historical distribution of Gunnison sage-grouse in Colorado. Wilson J Ornithol 126:207–217. https://doi.org/10.1676/13-184.1

Braun CE, Schroeder MA (2015) Age and sex identification from wings of sage-grouse. Wildl Soc Bull 39:182–187. https://doi.org/10.1002/wsb.517

Braun CE, Williams SO III (2015) History of sage-grouse (*Centrocercus* spp.) in New Mexico. Southwest Nat 60:207–213. https://doi.org/10.1894/MCG-141

Brice EM, Miller BA, Zhang H, Goldstein K, Zimmer SN, Grosklos GJ, Belmon P, Flint CG, Givens JE, Adler PB, Brunson MW, Smith JW (2020) Impacts of climate change on multiple use management of Bureau of Land Management land in the Intermountain West, USA. Ecosphere 11:e03286. https://doi.org/10.1002/ecs2.3286

Brown DG, Johnson KM, Loveland TR, Theobald DM (2005) Rural land-use trends in the conterminous United States, 1950–2000. Ecol Appl 15:1851–1863. https://doi.org/10.1890/03-5220

Brust RA (1991) Environmental regulation of autogeny in *Culex tarsalis* (Diptera: Culicidae) from Manitoba, Canada. J Med Entomol 28:847–853. https://doi.org/10.1093/jmedent/28.6.847

Carlisle JD, Keinath DA, Albeke SE, Chalfoun AD (2018) Identifying holes in the greater sage-grouse conservation umbrella. J Wildl Manage 82:948–957. https://doi.org/10.1002/jwmg.21460

Carpenter J, Aldridge C, Boyce MS (2010) Sage-grouse habitat selection during winter in Alberta. J Wildl Manage 74:1806–1814. https://doi.org/10.2193/2009-368

Caudill D, Guttery MR, Terhune TM II, Martin JA, Caudill G, Dahlgren DK, Messmer TA (2017) Individual heterogeneity and effects of harvest on greater sage-grouse populations. J Wildl Manage 81:754–765. https://doi.org/10.1002/jwmg.21241

Chambers JC, Bradley BA, Brown CS, D'Antonio C, Germino MJ, Grace JB, Hardegree SP, Miller RF, Pyke DA (2014) Resilience to stress and disturbance, and resistance to *Bromus tectorum* L. invasion in cold desert shrublands of western North America. Ecosystems 17:360–375. https://doi.org/10.1007/s10021-013-9725-5

Chambers JC, Maestas JD, Pyke DA, Boyd CS, Pellant M, Wuenschel A (2017) Using resilience and resistance concepts to manage persistent threats to sagebrush ecosystems and greater sage-grouse. Rangeland Ecol Manag 70:149–164. https://doi.org/10.1016/j.rama.2016.08.005

Christian JM, Wilson SD (1999) Long-term ecosystem impacts of an introduced grass in the northern Great Plains. Ecology 80:2397–2407. https://doi.org/10.1890/0012-9658(1999)080[2397:LTEIOA]2.0.CO;2

Christiansen TJ, Tate CM (2011) Parasites and diseases of greater sage-grouse. In: Knick ST, Connelly JW (eds) Greater sage-grouse: ecology and conservation of a landscape species and its habitats. Studies in avian biology 38, University of California Press, Berkeley, California, pp 113–126. https://doi.org/10.1525/California/9780520267114.003.0015

Coates PS, Casazza ML, Blomberg EJ, Gardner SC, Espinosa SP, Yee JL, Wiechman L, Halstead BJ (2013) Evaluating greater sage-grouse seasonal space use relative to leks: implications for surface use designations in sagebrush ecosystems. J Wildl Manage 77:1598–1609. https://doi.org/10.1002/jwmg.618

Coates PS, Brussee BE, Howe KB, Gustafson KB, Casazza ML, Delehanty DJ (2016a) Landscape characteristics and livestock presence influence common ravens: relevance to greater sage-grouse conservation. Ecosphere 7:e01203. https://doi.org/10.1002/ecs2.1203

Coates PS, Ricca MA, Prochazka BG, Brooks ML, Doherty KE, Kroger T, Blomberg EJ, Hagen CA, Casazza ML (2016b) Wildfire, climate, and invasive grass interactions negatively impact an indicator species by reshaping sagebrush ecosystems. Proc Natl A Sci-Biol 113:12745–12750. https://doi.org/10.1073/pnas.1606898113

Coates PS, Prochazka BG, Ricca MA, Gustafson KB, Ziegler P, Casazza ML (2017) Pinyon and juniper encroachment into sagebrush ecosystems impacts distribution and survival of greater sage-grouse. Rangeland Ecol Manag 70:25–38. https://doi.org/10.1016/j.rama.2016.09.001

Coates PS, Prochazka BG, Ricca MA, Halstead BJ, Casazza ML, Blomberg EJ, Brussee BE, Wiechman L, Tebbenkamp J, Gardner SC, Reese KP (2018) The relative importance of intrinsic and extrinsic drivers to population growth vary among local populations of greater sage-grouse: an integrated population modeling approach. Auk 135:240–261. https://doi.org/10.1642/AUK-17-137.1

Coates PS, Wann GT, Gillette GL, Ricca MA, Prochazka BG, Severson JP, Andrle KM, Espinosa SP, Casazza ML, Delehanty DJ (2019) Estimating sightability of greater sage-grouse at leks using an aerial infrared system and N-mixture models. Wildl Biol 1:1–11. https://doi.org/10.2981/wlb.00552

Coates PS, O'Neil ST, MuÑoz DA, Dwight IA, Tull JC (2021a) Sage-grouse population dynamics are adversely affected by overabundant feral horses. J Wildl Manage 85:1132–1149. https://doi.org/10.1002/jwmg.22089

Coates PS, Prochazka BG, O'Donnell MS, Aldridge CL, Edmunds DR, Monroe AP, Ricca MA, Wann GT, Hanser SE, Wiechman LA, Chenaille MP (2021b) Range-wide greater sage-grouse hierarchical monitoring framework—implications for defining population boundaries, trend estimation, and a targeted annual warning system: U.S. Geological Survey Open-File Report 2020-1154, 243 p. https://doi.org/10.3133/ofr20201154

Cole RA, Friend M (1999) Miscellaneous parasitic diseases. In: Friend M, Franson JC (eds) Field manual of wildlife diseases: general field procedures and diseases of birds. U.S. Geological Survey Biological Resources Division Information and Technical Report 1999–001, pp 249–258. https://pubs.er.usgs.gov/publication/itr19990001

Connelly JW, Browers HW, Gates RJ (1988) Seasonal movements of sage grouse in southeastern Idaho. J Wildl Manage 52:116–122. https://doi.org/10.2307/3801070

Connelly JW, Hagen CA, Schroeder MA (2011a) Characteristics and dynamics of greater sage-grouse populations. In: Knick ST, Connelly JW (eds) Greater sage-grouse: ecology and conservation of a landscape species and its habitats. Studies in avian biology 38, University of California Press, Berkeley, California, pp 53–67. https://doi.org/10.1525/California/9780520267114.003.0004

Connelly JW, Knick ST, Schroeder MA, Stiver SJ (2004) Conservation assessment of greater sage-grouse and sagebrush habitats. Unpublished report, Western Association of Fish and Wildlife Agencies, Cheyenne, Wyoming

Connelly JW, Reese KP, Garton EO, Commons-Kemner ML (2003a) Response of greater sage-grouse *Centrocercus urophasianus* populations to different levels of exploitation in Idaho, USA. Widl Biol 9:335–340. https://doi.org/10.2981/wlb.2003.022

Connelly JW, Reese KP, Schroeder MA (2003b) Monitoring of greater sage-grouse habitats and populations. Station Bulletin 80. University of Idaho, Moscow, USA. https://doi.org/10.5962/BHL.TITLE.153828

Connelly JW, Rinkes ET, Braun CE (2011b) Characteristics of greater sage-grouse habitats: a landscape species at micro- and macro scales. In: Knick ST, Connelly JW (eds) Greater sage-grouse: ecology and conservation of a landscape species and its habitats. Studies in avian biology 38, University of California Press, Berkeley, pp 69–83. https://doi.org/10.1525/California/9780520267114.003.0006

Connelly JW, Schroeder MA, Sands AR, Braun CE (2000) Guidelines to manage sage grouse populations and their habitats. Wildl Soc Bull 28:967–985. https://www.jstor.org/stable/3783856

Conover MR, Roberts AJ (2017) Predators, predator removal, and sage-grouse: a review. J Wildl Manage 81:7–15. https://doi.org/10.1002/jwmg.21168

Conway CJ, Tisdale CA, Launchbaugh K, Musil D, Makela P, Roberts S (2021) The grouse & grazing project: effects of cattle grazing on sage-grouse demographic traits—2021 annual report.

College of Natural Resources, University of Idaho. https://idahogrousegrazing.files.wordpress. com/2021/12/2021_annualreport.pdf

Copeland SM, Davies KW, Boyd CS, Bates JD (2021) Recovery of the herbaceous component of sagebrush steppe unimpeded by 75 years of moderate cattle grazing. Ecosphere 12:e03445. https://doi.org/10.1002/ecs2.3445

Crawford JA, Olson RA, West NE, Mosley JC, Schroeder MA, Whitson TD, Miller RF, Gregg MA, Boyd CS (2004) Ecology and management of sage-grouse and sage-grouse habitat. J Range Manage 57:2–19. https://doi.org/10.2111/1551-5028(2004)057[0002:EAMOSA]2.0.CO;2

Creutzburg MK, Halofsky JE, Halofsky JS, Christopher TA (2015) Climate change and land management in the rangelands of central Oregon. Environ Manage 55:43–55. https://doi.org/10.1007/s00267-014-0362-3

Crist MR, Chambers JC, Phillips SL, Prentice KL, Wiechman LA (eds) (2019) Science framework for conservation and restoration of the sagebrush biome: linking the department of the interior's integrated rangeland fire management strategy to long-term strategic conservation actions. Part 2. Management applications. Gen. Tech. Rep. RMRS-GTR-389. Department of Agriculture, Forest Service, Rocky Mountain Research Station, Fort Collins, Colorado, 237 pp. https://doi.org/10.2737/RMRS-GTR-389

D'Antonio CM, Vitousek PM (1992) Biological invasions by exotic grasses, the grass/fire cycle, and global change. Annu Rev Ecol Syst 23:63–87. https://www.jstor.org/stable/2097282

Dahlgren DK, Chi R, Messmer TA (2006) Greater sage-grouse response to sagebrush management in Utah. Wildl Soc Bull 34:975–985. https://www-jstor-org.libproxy.uwyo.edu/stable/4134306

Dahlgren DK, Larsen RT, Danvir R, Wilson G, Thacker ET, Black TA, Naugle DE, Connelly JW, Messmer TA (2015) Greater sage-grouse and range management: insights from a 25-year case study in Utah and Wyoming. Rangeland Ecol Manag 68:375–382. https://doi.org/10.1016/j.rama.2015.07.003

Dahlgren DK, Guttery MR, Messmer TA, Caudill D, Elmore RD, Chi R, Koons DN (2016) Evaluating vital rate contributions to greater sage-grouse population dynamics to inform conservation. Ecosphere 7:e01249. https://doi.org/10.1002/ecs2.1249

Daubenmire R (1970) Steppe vegetation of Washington. Washington Agricultural Experiment Station Technical Bulletin 62, 131 p

Davies KW (2011) Plant community diversity and native plant abundance decline with increasing abundance of an exotic annual grass. Oecologia 167:481–491. https://doi.org/10.1007/s00442-011-1992-2

Davies KW, Bates JD (2014) Attempting to restore herbaceous understories in Wyoming big sagebrush communities with mowing and seeding. Restor Ecol 22:608–615. https://doi.org/10.1111/rec.12110

Davies KW, Bates JD (2020) Re-introducing fire in sagebrush-steppe experiencing decreased fire frequency: does burning promote spatial and temporal heterogeneity? Int J Wildland Fire 29:686–695. https://doi.org/10.1071/WF20018

Davies KW, Svejcar TJ, Bates JD (2009) Interaction of historical and nonhistorical disturbances maintains native plant communities. Ecol Appl 19:1536–1545. https://doi.org/10.1890/09-0111.1

Davies KW, Nafus AM, Sheley RL (2010) Non-native competitive perennial grass impedes the spread of an invasive annual grass. Biol Invasions 12:3187–3194. https://doi.org/10.1007/s10530-010-9710-2

Davies KW, Boyd CS, Beck JL, Bates JD, Svejcar TJ, Gregg MA (2011) Saving the sagebrush sea: an ecosystem conservation plan for big sagebrush plant communities. Biol Conserv 144:2573–2584. https://doi.org/10.1016/j.biocon.2011.07.016

Davies KW, Bates JD, Nafus AM (2012) Mowing Wyoming big sagebrush communities with degraded herbaceous understories: has a threshold been crossed? Rangeland Ecol Manag 65:498–505. https://doi.org/10.2111/REM-D-12-00026.1

Davis AJ, Hooten MB, Phillips ML, Doherty PF Jr (2014a) An integrated modeling approach to estimating Gunnison sage-grouse population dynamics: combining index and demographic data. Ecol Evol 4:4247–4257. https://doi.org/10.1002/ece3.1290

Davis DM, Reese KP, Gardner SC (2014b) Demography, reproductive ecology, and variation in survival of greater sage-grouse in northeastern California. J Wildl Manage 78:1343–1355. https://doi.org/10.1002/jwmg.797

Davies KW, Boyd CS, Bates JD, Hulet A (2015a) Winter grazing can reduce wildfire size, intensity, and behaviour in a shrub-grassland. Int J Wildland Fire 25:191–199. https://doi.org/10.1071/WF15055

Davies KW, Boyd CS, Johnson DD, Nafus AM, Madsen MD (2015b) Success of seeding native compared to introduced perennial vegetation for revegetating medusahead-invaded sagebrush rangeland. Rangeland Ecol Manag 68:224–230. https://doi.org/10.1016/j.rama.2015.03.004

Davies KW, Bates JD, Boyd CS (2016a) Effects of intermediate-term grazing rest on sagebrush communities with depleted understories: evidence of a threshold. Rangeland Ecol Manag 69:173–178. https://doi.org/10.1016/j.rama.2016.01.002

Davies KW, Bates JD, Boyd CS, Svejcar TJ (2016b) Prefire grazing by cattle increases postfire resistance to exotic annual grass (*Bromus tectorum*) invasion and dominance for decades. Ecol Evol 6:3356–3366. https://doi.org/10.1002/ece3.2127

Davies KW, Boyd CS, Bates JD (2018) Eighty years of grazing by cattle modifies sagebrush and bunchgrass structure. Rangeland Ecol Manag 71:275–280. https://doi.org/10.1016/j.rama.2018.01.002

Davies KW, Bates JD, Boyd CS (2019) Postwildfire seeding to restore native vegetation and limit exotic annuals: an evaluation in juniper-dominated sagebrush steppe. Restor Ecol 27:120–127. https://doi.org/10.1111/rec.12848

Dinkins JB, Beck JL (2019) Comparison of conservation policy benefits for an umbrella and related sagebrush-obligate species. Hum Wildl Interact 13:447–458. https://doi.org/10.26077/4ypp-vj89

Dinkins JB, Conover MR, Kirol CP, Beck JL (2012) Greater sage-grouse (*Centrocercus urophasianus*) select nest-sites and brood-sites away from avian predators. Auk 129:600–610. https://doi.org/10.1525/auk.2012.12009

Dinkins JB, Conover MR, Kirol CP, Beck JL, Frey SN (2014). Greater sage-grouse (*Centrocercus urophasianus*) select habitat based on avian predators, landscape composition, and anthropogenic features. Condor: Ornithol Appl 116:629–642. https://doi.org/10.1650/CONDOR-13-163.1

Dinkins JB, Conover MR, Kirol CP, Beck JL, Frey SN (2016a) Effects of common raven and coyote removal and temporal variation in climate on greater sage-grouse nesting success. Biol Conserv 202:50–58. https://doi.org/10.1016/j.biocon.2016.08.011

Dinkins JB, Lawson KJ, Smith KT, Beck JL, Kirol CP, Pratt AC, Conover MR, Blomquist FC (2017) Quantifying overlap and fitness consequences of migration strategy with seasonal habitat use and a conservation policy. Ecosphere 8:e01991. https://doi.org/10.1002/ecs2.1991

Dinkins JB, Duchardt CJ, Hennig JD, Beck JL (2021a) Changes in hunting season regulations (1870s–2019) reduce harvest exposure on greater and Gunnison sage-grouse. PLoS ONE 16:e0253635. https://doi.org/10.1371/journal.pone.0253635

Dinkins JB, Perry LR, Beck JL, Taylor JD (2021b) Increased abundance of the common raven within the ranges of greater and Gunnison sage-grouse: influence of anthropogenic subsidies and fire. Hum Wildl Interact 15:270–288. https://doi.org/10.26077/mv59-jy24

Dinkins JB, Smith KT, Beck JL, Kirol CP, Pratt AC, Conover MR (2016b) Microhabitat conditions in Wyoming's Sage-Grouse Core Areas: effects on nest site selection and success. PLoS ONE 11(3):e0150798. https://doi.org/10.1371/journal.pone.0150798

Doherty MK (2007) Mosquito populations in the Powder River Basin, Wyoming: a comparison of natural, agricultural and effluent coal-bed natural gas aquatic habitats. Montana State University, Bozeman, Montana, Thesis

Doherty KE, Naugle DE, Evans JS (2010) A currency for offsetting energy development impacts: horse-trading sage-grouse on the open market. PLoS ONE 5:e10339. https://doi.org/10.1371/journal.pone.0010339

Doherty KE, Naugle DE, Tack JD, Walker BL, Graham JM, Beck JL (2014) Linking conservation actions to demography: grass height explains variation in greater sage-grouse nest survival. Wildl Biol 20:320–325. https://doi.org/10.2981/wlb.00004

Doherty KE, Evans JS, Coates PS, Juliusson LM, Fedy BC (2016) Importance of regional variation in conservation planning: a rangewide example of the greater sage-grouse. Ecosphere 7:e01462. https://doi.org/10.1002/ecs2.1462

Doherty KE, Hennig JD, Dinkins JB, Griffin KA, Cook AA, Maestas JD, Naugle DE, Beck JL (2018) Understanding biological effectiveness before scaling up rangewide restoration investments for Gunnison sage-grouse. Ecosphere 9:e02144. https://doi.org/10.1002/ecs2.2144

Donnelly JP, Naugle DE, Hagen CA, Maestas JD (2016) Public lands and private waters: scarce mesic resources structure land tenure and sage-grouse distributions. Ecosphere 7:e01208. https://doi.org/10.1002/ecs2.1208

Donnelly JP, Allred BW, Perret D, Silverman NL, Tack JD, Dreitz VJ, Maestas JD, Naugle DE (2018) Seasonal drought in North America's sagebrush biome structures dynamic mesic resources for sage-grouse. Ecol Evol 24:12492–12505. https://doi.org/10.1002/ece3.4614

Drut MS, Crawford JA, Gregg MA (1994) Brood habitat use by sage grouse in Oregon. Great Basin Nat 54:170–176. https://www.jstor.org/stable/41712827

Dumroese RK, Luna T, Richardson BA, Kilkenny FF, Runyon JB (2015) Conserving and restoring habitat for greater sage-grouse and other sagebrush-obligate wildlife: the crucial link of forbs and sagebrush diversity. Native Plants J 16:277–299. https://doi.org/10.3368/npj.16.3.276

Dusek RJ, Hagen CA, Franson JC, Budeau DA, Hofmeister EK (2014) Utilizing hunter harvest effort to survey for wildlife disease: a case study of West Nile virus in greater sage-grouse. Wildl Soc Bull 38:721–727. https://doi.org/10.1002/wsb.472

Emmons SR, Braun CE (1984) Lek attendance of male sage grouse. J Wildl Manage 48:1023–1028. https://doi.org/10.2307/3801461

Environment Canada (2014) Amended recovery strategy for the greater sage-grouse (*Centrocercus urophasianus urophasianus*) in Canada. Species at Risk Act Recovery Strategy Series, Environment Canada, Ottawa, Ontario, p 53

Evans CC (1986) The relationship of cattle grazing to sage grouse use of meadow habitat on the Sheldon National Wildlife Refuge. University of Nevada, Reno, Thesis

Fargione JE, Bassett S, Boucher T, Bridgham SD, Conant RT, Cook-Patton SC, Ellis PW, Falcucci A, Fourqurean JW, Gopalakrishna T, Gu H, Henderson B, Hurteau MD, Kroeger MD, Kroeger T, Lark TJ, Leavitt SM, Lomax G, McDonald RI, Megonigal JP, Miteva DA, Richardson CJ, Sanderman J, Shoch D, Spawn SA, Veldman JW, Williams CA, Woodbury PB, Zganjar C, Baranski M, Elias P, Houghton RA, Landis E, McGlynn E, Schlesinger WH, Siikamaki JV, Sutton-Grier AE, Griscom BW (2018) Natural climate solutions for the United States. Sci Adv 4:eaat1869. https://doi.org/10.1126/sciadv.aat1869

Fedy BC, Aldridge CL, Doherty KE, O'Donnell M, Beck JL, Bedrosian B, Holloran MJ, Johnson GD, Kaczor NW, Kirol CP, Mandich CA, Marshall D, McKee G, Olson C, Swanson CC, Walker BL (2012) Interseasonal movements of greater sage-grouse, migratory behavior, and an assessment of the core regions concept in Wyoming. J Wildl Manage 76:1062–1071. https://doi.org/10.1002/jwmg.337

Fedy BC, Doherty KE, Aldridge CL, O'Donnell M, Beck JL, Bedrosian B, Gummer D, Holloran MJ, Johnson GD, Kaczor NW, Kirol CP, Mandich CA, Marshall D, Mckee G, Olson C, Pratt AC, Swanson CC, Walker BL (2014) Habitat prioritization across large landscapes, multiple seasons, and novel areas: an example using greater sage-grouse in Wyoming. Wildl Monogr 190:1–39. https://doi.org/10.1002/wmon.1014

Finch D, Schofield H, Floate KD, Kubasiewicz LM, Mathews F (2020) Implications of endectocide residues on the survival of aphodiine dung beetles: a meta-analysis. Environ Toxicol Chem 39:863–872. https://doi.org/10.1002/etc.4671

Fischer RA, Reese KP, Connelly JW (1996) Influence of vegetal moisture content and nest fate on timing of female sage grouse migration. Condor 98:868–872. https://doi.org/10.2307/1369875

Flader SL, Callicott JB, Leopold A (1992) The river of the mother of god: and other essays by Aldo Leopold. University of Wisconsin Press, Madison, Wisconsin

Fremgen AL, Hansen CP, Rumble MA, Gamo RS, Millspaugh JJ (2016) Male greater sage-grouse detectability on leks. J Wildl Manage 80:266–274. https://doi.org/10.1002/jwmg.1001

Fremgen MR, Gibson D, Ehrlich RL, Krakauer AH, Forbey JS, Blomberg EJ, Sedinger JS, Patricelli GL (2017) Necklace-style radio-transmitters are associated with changes in display vocalizations of male greater sage-grouse. Wildl Biol 2017:wlb.00236. https://doi.org/10.2981/wlb.00236

Fremgen AL, Hansen CP, Rumble MA, Gamo RS, Millspaugh JJ (2019) Weather conditions and date influence male sage grouse attendance rates at leks. Ibis 161:35–49. https://doi.org/10.1111/ibi.12598

Gamo RS, Beck JL (2017) Effectiveness of Wyoming's sage-grouse core areas: influences on energy development and male lek attendance. Environ Manage 59:189–203. https://doi.org/10.1007/s00267-016-0789-9

Gamo RS, Carlisle JD, Beck JL, Bernard JAC, Herget ME (2013) Greater sage-grouse in Wyoming: an umbrella species for sagebrush-dependent wildlife. Wildlife Professional 7:56–59

Garton EO, Connelly JW, Horne JS, Hagen CA, Moser A, Schroeder MA (2011) Greater sage-grouse population dynamics and probability of persistence. In: Knick ST, Connelly JW (eds) Greater sage-grouse: ecology and conservation of a landscape species and its habitats. Studies in avian biology 38, University of California Press, Berkeley, pp 293–381. https://doi.org/10.1525/California/9780520267114.003.0016

Garton EO, Wells AG, Baumgardt JA, Connelly JW (2015) Greater sage-grouse population dynamics and probability of persistence. Final report to Pew Charitable Trusts. Available at https://www.pewtrusts.org/~/media/assets/2015/04/garton-et-al-2015-greater-sagegrouse-population-dynamics-and-persistence-31815.pdf

Gelbard JL, Belnap J (2003) Roads as conduits for exotic plant invasions in a semiarid landscape. Conserv Biol 17:420–432. https://doi.org/10.1046/j.1523-1739.2003.01408.x

Gibson D, Blomberg EJ, Atamian MT, Sedinger JS (2014) Lek fidelity and movement among leks by male greater sage-grouse Centrocercus urophasianus: a capture-mark-recapture approach. Ibis 156:729–740. https://doi.org/10.1111/ibi.12192

Gibson D, Blomberg EJ, Atamian MT, Sedinger JS (2016a) Nesting habitat selection influences nest and early offspring survival in greater sage-grouse. Condor: Ornithol Appl 118:689–702. https://doi.org/10.1650/CONDOR-16-62.1

Gibson D, Blomberg EJ, Sedinger JS (2016b) Evaluating vegetation effects on animal demographics: the role of plant phenology and sampling bias. Ecol Evol 6:3621–3631. https://doi.org/10.1002/ece3.2148

Goosey HB, Smith JT, O'Neill KM, Naugle DE (2019) Ground-dwelling arthropod community response to livestock grazing: implications for avian conservation. Environ Entomol 48:856–866. https://doi.org/10.1093/ee/nvz074

Government of Canada (2021) Greater sage-grouse. https://www.canada.ca/en/environment-climate-change/services/species-risk-education-centre/greater-sage-grouse.html. Accessed 27 Aug 2021

Guttery MR, Dahlgren DK, Messmer TA, Connelly JW, Reese KP, Terletzky PA, Burkepile N, Koons DN (2013) Effects of landscape-scale environmental variation on greater sage-grouse chick survival. PLoS ONE 8:e65582. https://doi.org/10.1371/journal.pone.0065582

Hagen CA, Connelly JW, Schroeder MA (2007) A meta-analysis of greater sage-grouse Centrocercus urophasianus nesting and brood-rearing habitats. Wildl Biol 13:42–50. https://doi.org/10.2981/0909-6396(2007)13[42:AMOGSC]2.0.CO;2

Hagen CA, Sedinger JE, Braun CE (2018) Estimating sex-ratio, survival, and harvest susceptibility in greater sage-grouse: making the most of hunter harvests. Wildl Biol 2018:wlb.00362. https://doi.org/10.2981/wlb.00362.

Hamerlynck EP, Davies KW (2019) Changes in the abundance of eight sagebrush-steppe bunchgrass species 13 yr after coplanting. Rangeland Ecol Manag 72:23–27. https://doi.org/10.1016/j.rama. 2018.07.001

Hansen AJ, Knight RL, Marzluff JM, Powell S, Brown K, Gude PH, Jones K (2005) Effects of exurban development on biodiversity: patterns, mechanisms, and research needs. Ecol Appl 15:1893–1905. https://doi.org/10.1890/05-5221

Hansen MJ, Wilson SD (2006) Is management of an invasive grass *Agropyron cristatum* contingent on environmental variation? J Appl Ecol 43:269–280. https://doi.org/10.1111/j.1365-2664.2006. 01145.x

Harju SM, Coates PS, Dettenmaier SJ, Dinkins JB, Jackson PJ, Chenaille MP (2021) Estimating trends of common raven populations in North America, 1966–2018. Hum Wildl Interact 15:In press. https://doi.org/10.26077/c27f-e335

Harte J, Saleska SR, Levy C (2015) Convergent ecosystem responses to 23-year ambient and manipulated warming link advancing snowmelt and shrub encroachment to transient and long-term climate- soil carbon feedback. Glob Change Biol 21:2349–2356. https://doi.org/10.1111/ gcb.12831

Heidinga L, Wilson SD (2002) The impact of an invading alien grass (*Agropyron cristatum*) on species turnover in native prairie. Divers Distrib 8:249–258. https://doi.org/10.1046/j.1472-4642.2002.00154.x

Heinrichs JA, O'Donnell MS, Aldridge CL, Garman SL, Homer CG (2019) Influences of potential oil and gas development and future climate on sage-grouse declines and redistribution. Ecol Appl 29:e01912. https://doi.org/10.1002/eap.1912

Hennig JD, Beck JL, Duchardt CJ, Scasta JD (2021) Variation in sage-grouse habitat quality metrics across a gradient of feral horse use. J Arid Environ 192:104550. https://doi.org/10.1016/j.aridenv. 2021.104550

Holloran MJ, Heath BJ, Lyon AG, Slater SJ, Kuipers JL, Anderson SH (2005) Greater sage-grouse nesting habitat selection and success in Wyoming. J Wildl Manage 69:638–649. https://doi.org/ 10.2193/0022-541X(2005)069[0638:GSNHSA]2.0.CO;2

Holmes AL, Maestas JD, Naugle DE (2017) Bird responses to removal of western juniper in sagebrush-steppe. Rangeland Ecol Manag 70:87–94. https://doi.org/10.1016/j.rama.2016. 10.006

Homer CG, Xian G, Aldridge CL, Meyer DK, Loveland TR, O'Donnell MS (2015) Forecasting sagebrush ecosystem components and greater sage-grouse habitat for 2050: learning from past climate patterns and Landsat imagery to predict the future. Ecol Indic 55:131–145. https://doi. org/10.1016/j.ecolind.2015.03.002

Honess RF (1982) Cestodes of grouse. In: Thorne ET, Kingston N, Jolley WR, Bergstrom RC (eds) Diseases of wildlife in Wyoming, second edition. Wyoming Game and Fish Department, Cheyenne, Wyoming, pp 161–164

Honess RF, Post G (1968) History of an epizootic in sage grouse. University of Wyoming Agricultural Experiment Station, Science Monograph 14, 32 p

Hupp JW, Braun CE (1989) Topographic distribution of sage grouse foraging in winter. J Wildl Manage 53:823–829. https://doi.org/10.2307/3809220

Huwer SL, Anderson DR, Remington TE, White GC (2008) Using human-imprinted chicks to evaluate the importance of forbs to sage-grouse. J Wildl Manage 72:1622–1627. https://doi.org/ 10.2193/2004-340

James JJ, Rinella MJ, Svejcar T (2012) Grass seedling demography and sagebrush steppe restoration. Rangeland Ecol Manag 65:409–417. https://doi.org/10.2111/REM-D-11-00138.1

Jenni DA, Hartzler JE (1978) Attendance at a sage grouse lek: implications for spring censuses. J Wildl Manage 42:46–52. https://doi.org/10.2307/3800688

Johnson DH (2008) In defense of indices: the case of bird surveys. J Wildl Manage 72:857–868. https://doi.org/10.2193/2007-294

Johnson GD, Boyce MS (1990) Feeding trials with insects in the diet of sage grouse chicks. J Wildl Manage 54:89–91. https://doi.org/10.2307/3808906

Kiesecker JM, Naugle DE (eds) (2017) Energy sprawl solutions: balancing global development and conservation. Island Press, Washington, D.C

King SR, Schoenecker KA, Manier DJ (2019) Potential spread of cheatgrass and other invasive species by feral horses in western Colorado. Rangeland Ecol Manag 72:706–710. https://doi.org/10.1016/j.rama.2019.02.006

Kirol CP, Beck JL, Huzurbazar SV, Holloran MJ, Miller SN (2015) Identifying greater sage-grouse source and sink habitats for conservation planning in an energy development landscape. Ecol Appl 25:968–990. https://doi.org/10.6084/m9.figshare.c.3296837.v1

Klebenow DA, Gray GM (1968) Food habits of juvenile sage grouse. J Range Manage 21:80–83. https://doi.org/10.2307/3896359

Klos PZ, Link TE, Abatzoglou JT (2014) Extent of the rain-snow transition zone in the western U.S. under historic and projected climate. Geophys Res Lett 41:4560–4568. https://doi.org/10.1002/2014GL060500

Knapp PA (1996) Cheatgrass (*Bromus tectorum* L) dominance in the Great Basin Desert: history, persistence, and influences to human activities. Glob Environ Change 6:37–52. https://doi.org/10.1016/0959-3780(95)00112-3

Knick ST (2011) Historical development, principal federal legislation, and current management of sagebrush habitats: implications for conservation. In: Knick ST, Connelly JW (eds) Greater sage-grouse: ecology and conservation of a landscape species and its habitats. Studies in avian biology 38, University of California Press, Berkeley, pp 13–31. https://doi.org/10.1525/California/9780520267114.003.0002

Knick ST, Dobkin DS, Rotenberry JT, Schroeder MA, Vander Haegen WM, van Riper IIIC (2003) Teetering on the edge or too late? Conservation and research issues for avifauna of sagebrush habitats. Condor 105:611–634. https://doi.org/10.1093/condor/105.4.611

Knight RL, Wallace GN, Riebsame WE (1995) Ranching the view: subdivisions versus agriculture. Conserv Biol 9:459–461. https://www.jstor.org/stable/2386971

Knutson KC, Pyke DA, Wirth TA, Arkle RS, Pilliod DS, Brooks ML, Chambers JC, Grace JB (2014) Long-term effects of seeding after wildfire on vegetation in Great Basin shrubland ecosystems. J Appl Ecol 51:1414–1424. https://doi.org/10.1111/1365-2664.12309

Kohl KD, Connelly JW, Dearing MD, Forbey JS (2016) Microbial detoxification in the gut of a specialist avian herbivore, the greater sage-grouse. FEMS Microbiol Lett 363:fnw144. https://doi.org/10.1093/femsle/fnw144

Küchler AW (1985) Potential natural vegetation. US Geological Survey. https://doi.org/10.3133/32574

Lesica P, DeLuca TH (1996) Long-term harmful effects of crested wheatgrass on Great Plains grassland ecosystems. J Soil Water Conserv 51:408–409

Lockwood JA, Anderson-Sprecher R, Schell SP (2002) When less is more: optimization of reduced agent-area treatments (RAATs) for management of rangeland grasshoppers. Crop Prot 21:551–562. https://doi.org/10.1016/S0261-2194(01)00145-4

López-Osorio S, Chaparro-Gutiérrez JJ, Gómez-Osorio LM (2020) Overview of poultry *Eimeria* life cycle and host-parasite interactions. Front Vet Sci 7:384. https://doi.org/10.3389/fvets.2020.00384

Lyon AG, Anderson SH (2003) Potential gas development impacts on sage grouse nest initiation and movement. Wildl Soc Bull 31:486–491. https://www.jstor.org/stable/3784329

Mack RN (1981) Invasion of *Bromus tectorum* L. into western North America: an ecological chronicle. Agro-Ecosystems 7:145–165. https://doi.org/10.1016/0304-3746(81)90027-5

Maestas JD, Knight RL, Gilgert WC (2003) Biodiversity across a rural land-use gradient. Conserv Biol 17:1425–1434. https://doi.org/10.1046/j.1523-1739.2003.02371.x

Maestas JD, Campbell SB, Chambers JC, Pellant M, Miller RF (2016) Tapping soil survey information for rapid assessment of sagebrush ecosystem resilience and resistance. Rangelands 38:120–128. https://doi.org/10.1016/j.rala.2016.02.002

Martínez MI, Lumaret J-P, Ortiz Zayas R, Kadiri N (2017) The effects of sublethal and lethal doses of ivermectin on the reproductive physiology and larval development of the dung beetle

Euoniticellus intermedius (Coleoptera: Scarabaeidae). Can Entomol 149:461–472. https://doi. org/10.4039/tce.2017.11

McCaffery R, Lukacs PM (2016) A generalized integrated population model to estimate greater sage-grouse population dynamics. Ecosphere 7:e01585. https://doi.org/10.1002/ecs2.1585

McCaffery R, Nowak JJ, Lukacs PM (2016) Improved analysis of lek count data using *N*-mixture models. J Wildl Manage 80:1011–1021. https://doi.org/10.1002/jwmg.21094

McClanahan RC (1940) Original and present breeding ranges of certain game birds in the United States. United States Bureau of Biological Survey, Wildlife Leaflet BS-158, Washington, DC

Meinke CW, Knick ST, Pyke DA (2009) A spatial model to prioritize sagebrush landscapes in the Intermountain West (U.S.A.) for restoration. Restor Ecol 17:652–659. https://doi.org/10.1111/j.1526-100X.2008.00400.x

Miller RF, Rose JA (1999) Fire history and western juniper encroachment in sagebrush steppe. J Range Manage 52:550–559. https://doi.org/10.2307/4003623

Miller RF, Wigand PE (1994) Holocene changes in semiarid pinyon-juniper woodlands: response to climate, fire, and human activities in the US Great Basin. Bioscience 44:465–474. https://doi. org/10.2307/1312298

Miller RF, Svejcar TJ, Rose JA (2000) Impacts of western juniper on plant community composition and structure. J Range Manage 53:574–585. https://doi.org/10.2458/azu_jrm_v53i6_miller

Miller RF, Bates JD, Svejcar TJ, Pierson FB, Eddleman LE (2005) Biology, ecology, and management of western juniper. Tech Bull 152. Oregon State University, Corvallis, Oregon. Available at https://ir.library.oregonstate.edu/concern/technical_reports/cz30pv075. Accessed 7 April 2022

Miller RF, Tausch RJ, MacArthur D, Johnson DD, Sanderson SC (2008) Development of post settlement piñon-juniper woodlands in the Intermountain West: a regional perspective. Research Paper Report RMRS-RP-69. USDA Forest Service, Rocky Mountain Research Station, Ft. Collins, Colorado

Monroe AP, Edmunds DR, Aldridge CL (2016) Effects of lek count protocols on greater sage-grouse population trend estimates. J Wildl Manage 80:667–678. https://doi.org/10.1002/jwmg.1050

Monroe AP, Aldridge CL, Assal TJ, Veblen KE, Pyke DA, Casazza ML (2017) Patterns in greater sage-grouse population dynamics correspond with public grazing records at broad scales. Ecol Appl 27:1096–1107. https://doi.org/10.1002/eap.1512

Monroe AP, Wann GT, Aldridge CL, Coates PS (2019) The importance of simulation assumptions when evaluating detectability in population models. Ecosphere 10:e02791. https://doi.org/10. 1002/ecs2.2791

Muñoz DA, Coates PS, Ricca MA (2021) Free-roaming horses disrupt greater sage-grouse lekking activity in the Great Basin. J Arid Environ 184:104304. https://doi.org/10.1016/j.jaridenv.2020. 104304

Natural Resources Conservation Service [NRCS] (2021) A decade of science support in the sagebrush biome. United States Department of Agriculture, Washington, D.C. 32 p. https://doi.org/ 10.32747/2021.7488985

Naugle DE, Aldridge CL, Walker BL, Doherty KE, Matchett MR, McIntosh J, Cornish TE, Boyce MS (2005) West Nile virus and sage-grouse: what more have we learned? Wildl Soc Bull 33:616–623. https://doi.org/10.2193/0091-7648(2005)33[616:WNVASW]2.0.CO;2

Naugle DE, Allred BW, Jones MO, Twidwell D, Maestas JD (2020) Coproducing science to inform working lands: the next frontier in nature conservation. Bioscience 70:90–96. https://doi.org/ 10.1093/biosci/biz144

Naugle DE, Doherty KE, Walker BL, Holloran MJ, Copeland HE (2011) Energy development and greater sage-grouse. In: Knick ST, Connelly JW (eds) Greater sage-grouse: ecology and conservation of a landscape species and its habitats. Studies in avian biology 38, University of California Press, Berkeley, pp 489–503. https://doi.org/10.1525/California/9780520267114. 003.0021

Naugle DE, Maestas JD, Allred BW, Hagen CA, Jones MO, Falkowski MJ, Randall B, Rewa CA (2019) CEAP quantifies conservation outcomes for wildlife and people on western grazing lands. Rangelands 41:211–217. https://doi.org/10.1016/j.rala.2019.07.004

Neel LA (1980) Sage grouse response to grazing management in Nevada, Thesis, University of Nevada, Reno

Newton RE, Tack JD, Carlson JC, Matchett MR, Fargey PJ, Naugle DE (2017) Longest sage-grouse migratory behavior sustained by intact pathways. J Wildl Manage 81:962–972. https://doi.org/10.1002/jwmg.21274

Nichols JD, Thomas L, Conn PB (2009) Inferences about landbird abundance from count data: recent advances and future directions. In: Thomson DL, Cooch EG, Conroy MJ (eds) Modeling demographic processes in marked populations. Springer, New York, New York, USA, pp 201–235

Oh KP, Aldridge CL, Forbey JS, Dadabay CY, Oyler-McCance SJ (2019) Conservation genomics in the sagebrush sea: population divergence, demographic history, and local adaptation in sage-grouse (*Centrocercus* spp.). Genome Biol Evol 11:2023–2034. https://doi.org/10.1093/gbe/evz112

Olsen AC, Severson JP, Maestas JD, Naugle DE, Smith JT, Tack JD, Yates KH, Hagen CA (2021) Reversing tree expansion in sagebrush steppe yields population-level benefit for imperiled grouse. Ecosphere 12:e03551. https://doi.org/10.1002/ecs2.3551

Oyler-McCance SJ, Quinn TW (2011) Molecular insights into the biology of greater sage-grouse. In: Knick ST, Connelly JW (eds) Greater sage-grouse: ecology and conservation of a landscape species and its habitats. Studies in avian biology 38, University of California Press, Berkeley, pp 85–94. https://doi.org/10.1525/california/9780520267114.003.0006

Oyler-McCance SJ, St. John J, Taylor SE, Apa AD, Quinn TW (2005) Population genetics of Gunnison sage-grouse: implications for management. J Wildl Manage 69:630–637.https://doi.org/10.2193/0022-541X(2005)069[0630:PGOGSI]2.0.CO;2

Palmquist KA, Schlaepfer DR, Bradford JB, Lauenroth WK (2016) Mid-latitude shrub steppe plant communities: climate change consequences for soil water resources. Ecology 97:2342–2354. https://doi.org/10.1002/ecy.1457

Pennington VE, Schlaepfer DR, Beck JL, Bradford JB, Palmquist KA, Lauenroth WK (2016) Sagebrush, greater sage-grouse, and the occurrence and importance of forbs. West N Am Nat 76:298–312. https://doi.org/10.3398/064.076.0307

Peterson JG (1970) The food habits and summer distribution of juvenile sage grouse in central Montana. J Wildl Manage 34:147–155. https://doi.org/10.2307/3799502

Peterson MJ (2004) Parasites and infectious diseases of prairie grouse: should managers be concerned? Wildl Soc Bull 32:35–55. https://doi.org/10.2193/0091-7648(2004)32[35:PAIDOP]2.0.CO;2

Pierson FB, Bates JD, Svejcar TJ, Hardegree SP (2007) Runoff and erosion after cutting western juniper. Rangeland Ecol Manag 60:285–292. https://doi.org/10.2111/1551-5028(2007)60[285:RAEAECW]2.0.CO;2

Pratt AC, Smith KT, Beck JL (2017) Environmental cues used by greater sage-grouse to initiate altitudinal migration. Auk: Ornithol Adv 134:628–643. https://doi.org/10.1642/AUK-16-192.1

Pratt AC, Smith KT, Beck JL (2019) Prioritizing seasonal habitats for comprehensive conservation of a partially migratory species. Glob Ecol Conserv 17:e00594. https://doi.org/10.1016/j.gecco.2019.e00594

Post G (1951) Effects of toxaphene and chlordane on certain game birds. J Wildl Manage 15:381–386. https://doi.org/10.2307/3796581

Prochazka BG, Coates PS, Ricca MA, Casazza ML, Gustafson KB, Hull JM (2017) Encounters with pinyon-juniper influence riskier movements in greater sage-grouse across the Great Basin. Rangel Ecol Manag 70:39–49. https://doi.org/10.1016/j.rama.2016.07.004

Pyke DA, Knick ST, Chambers JC, Pellant M, Miller RF, Beck JL, Doescher PS, Schupp EW, Roundy BA, Brunson M, McIver JD (2015b) Restoration handbook for sagebrush steppe ecosystems with emphasis on greater sage-grouse habitat—Part 2. Landscape level restoration decisions. U.S. Geological Survey Circular 1418, 21 p. https://doi.org/10.3133/cir1418.

Pyke DA, Chambers JC, Pellant M, Knick ST, Miller RF, Beck JL, Doescher PS, Schupp EW, Roundy BA, Brunson M, McIver JD (2015a) Restoration handbook for sagebrush steppe ecosystems with emphasis on greater sage-grouse habitat—Part 1. Concepts for understanding and applying restoration: U.S. Geological Survey Circular 1416, 44 p. https://doi.org/10.3133/cir 1416

Pyke DA, Chambers JC, Pellant M, Miller RF, Beck JL, Doescher PS, Roundy BA, Schupp EW, Knick ST, Brunson M, McIver JD (2017) Restoration handbook for sagebrush steppe ecosystems with emphasis on greater sage-grouse habitat—Part 3. Site level restoration decisions. U.S. Geological Survey Circular 1426, 62 p. https://doi.org/10.3133/cir1426

Randall KJ, Ellison MJ, Yelich JV, Price WJ, Johnson TN (2022) Managing forbs preferred by greater sage-grouse and soil moisture in mesic meadows with short-duration grazing. Rangeland Ecol Manag 82:66–75. https://doi.org/10.1016/j.rama.2022.02.008

Reese KP, Connelly JW (1997) Translocation of sage grouse Centrocercus urophasianus in North America. Wildlife Biol 3:235–241

Reeves MC, Moreno AL, Bagne KE, Running SW (2014) Estimating climate change effects on net primary production of rangelands in the United States. Clim Change 126:429–442. https://doi.org/10.1007/s10584-014-1235-8

Reinhart JR, Filippelli S, Falkowski M, Allred B, Maestas JD, Carlson JC, Naugle DE (2020) Quantifying pinyon-juniper reduction within North America's sagebrush ecosystem. Rangeland Ecol Manag 73:420–432. https://doi.org/10.1016/j.rama.2020.01.002

Renne RR, Schlaepfer DR, Palmquist KA, Bradford JB, Burke IC, Lauenroth WK (2019) Soil and stand structure explain shrub mortality patterns following global change–type drought and extreme precipitation. Ecology 100:e02889. https://doi.org/10.1002/ecy.2889

Romme WH, Allen CD, Bailey JD, Baker WL, Bestelmeyer BT, Brown PM, Eisenhart KS, Floyd ML, Huffman DW, Jacobs BF, Miller RF (2009) Historical and modern disturbance regimes, stand structures, and landscape dynamics in pinon–juniper vegetation of the western United States. Rangeland Ecol Manag 62:203–222. https://doi.org/10.2111/08-188R1.1

Rowland MM, Wisdom MJ, Suring LH, Meinke CW (2006) Greater sage-grouse as an umbrella species for sagebrush-associated vertebrates. Biol Conserv 129:323–335. https://doi.org/10.1016/j.biocon.2005.10.048

Sanderson JS, Beutler C, Brown JR, Burke I, Chapman T, Conant RT, Derner JD, Easter M, Fuhlendorf SD, Grissom G, Herrick JE, Liptzin D, Morgan JA, Murph R, Pague C, Rangwala I, Ray D, Rondeau R, Schulz T, Sullivan T (2020) Cattle, conservation, and carbon in the western Great Plains. J Soil Water Conserv 75:5A-12A. https://doi.org/10.2489/jswc.75.1.5A

Sauls HS (2006) The role of selective foraging and cecal microflora in sage-grouse nutritional ecology. Thesis, University of Montana, Missoula, Montana

Scasta JD, Hennig JD, Beck JL (2018) Framing contemporary U.S. wild horse and burro management processes in a dynamic ecological, sociological, and political environment. Hum Wildl Interact 12:31–45. https://doi.org/10.26077/2fhw-fz24

Schlaepfer DR, Lauenroth WK, Bradford JB (2012) Consequences of declining snow accumulation for water balance of mid-latitude dry regions. Glob Change Biol 18:1988–1997. https://doi.org/10.1111/j.1365-2486.2012.02642.x

Schlaepfer DR, Taylor KA, Pennington VE, Nelson KN, Martyn TE, Rottler CM, Lauenroth WK, Bradford JB (2015) Simulated big sagebrush regeneration supports predicted changes at the trailing and leading edges of distribution shifts. Ecosphere 6:1–31. https://doi.org/10.1890/ES14-00208.1

Schroeder MA, Aldridge CL, Apa AD, Bohne JR, Braun CE, Bunnell SD, Connelly JW, Deibert PA, Gardner SC, Hilliard MA, Kobriger GD, McAdam SM, McCarthy CW, McCarthy JJ, Mitchell DL, Rickerson EV, Stiver SJ (2004) Distribution of sage-grouse in North America. Condor 106:363–376. https://doi.org/10.1093/condor/106.2.363

Schroeder MA, Young JR, Braun CE (2020) Greater Sage-Grouse (Centrocercus urophasianus), version 1.0. In: Poole AF, Gill FB (eds) Birds of the world. Cornell Lab of Ornithology, Ithaca, New York. https://doi.org/10.2173/bow.saggro.01

Sedinger JS, White GC, Espinosa S, Partee ET, Braun CE (2010) Assessing compensatory versus additive harvest mortality: an example using greater sage-grouse. J Wildl Manage 74:326–332. https://doi.org/10.2193/2009-071

Severson JP, Hagen CA, Maestas JD, Naugle DE, Forbes JT, Reese KP (2016) Effects of conifer expansion on greater sage-grouse nesting habitat selection. J Wildl Manage 81:86–95. https://doi.org/10.1002/jwmg.21183

Severson JP, Coates PS, Prochazka BG, Ricca MA, Casazza ML, Delehanty DJ (2019) Global positioning system tracking devices can decrease greater sage-grouse survival. Condor: Ornithol Appl 121:1–15. https://doi.org/10.1093/condor/duz032

Shyvers JE, Walker BL, Noon BR (2018) Dual-frame lek surveys for estimating greater sage-grouse populations. J Wildl Manage 82:1689–1700. https://doi.org/10.1002/jwmg.21540

Shyvers JE, Walker BL, Oyler-McCance SJ, Fike JA, Noon BR (2019) Genetic mark-recapture analysis of winter faecal pellets allow estimation of population size in sage grouse *Centrocercus urophasianus*. Ibis 162:749–765. https://doi.org/10.1111/ibi.12768

Silverman NL, Allred BW, Donnelly JP, Chapman TB, Maestas JD, Wheaton JM, White J, Naugle DE (2019) Low-tech riparian and wet meadow restoration increases vegetation productivity and resilience across semiarid rangelands. Restor Ecol 27:269–278. https://doi.org/10.1111/rec.12869

Sinai NL, Coates PS, Andrle KM, Jefferis C, Sentíes-Cué CG, Pitesky ME (2017) A serosurvey of greater sage-grouse (*Centrocercus urophasianus*) in Nevada, USA. J Wildl Dis 53:136–139. https://doi.org/10.7589/2015-10-285

Smith KT, Beck JL (2018) Sagebrush treatments influence annual population change for greater sage-grouse. Restor Ecol 26:497–505. https://doi.org/10.1111/rec.12589

Smith KT, Beck JL, Pratt AC (2016) Does Wyoming's Core Area Policy protect winter habitats for greater sage-grouse? Environ Manage 58:585–596. https://doi.org/10.1007/s00267-016-0745-8

Smith JT, Tack JD, Berkeley LI, Szczypinski M, Naugle DE (2018a) Effects of rotational grazing management on nesting greater sage-grouse. J Wildl Manage 82:103–112. https://doi.org/10.1002/jwmg.21344

Smith JT, Tack JD, Doherty KE, Allred BW, Maestas JD, Berkeley LI, Dettenmaier SJ, Messmer TA, Naugle DE (2018b) Phenology largely explains taller grass at successful nests in greater sage-grouse. Ecol Evol 8:356–364. https://doi.org/10.1002/ece3.3679

Smith KT, Beck JL, Kirol CP (2018c) Reproductive state leads to intraspecific habitat partitioning and survival differences in greater sage-grouse: implications for conservation. Wildl Res 45:119–131. https://doi.org/10.1071/WR17123

Smith KT, Pratt AC, LeVan JR, Rhea AM, Beck JL (2019) Reconstructing greater sage-grouse chick diets: diet selection, body condition, and food availability at brood-rearing sites. Condor: Ornithol Appl 121:duy012. https://doi.org/10.1093/condor/duy012

Smith JT, Allred BW, Boyd CS, Carlson JC, Davies KW, Hagen CA, Naugle DE, Olsen AC, Tack JD (2020) Are sage-grouse fine-scale specialists or shrub-steppe generalists? J Wildl Manag 84:759–774. https://doi.org/10.1002/jwmg.21837

Smith JT, Allred BW, Boyd CS, Davies KW, Jones MO, Kleinhesselink AR, Maestas JD, Morford SL, Naugle DE (2022) The elevational ascent and spread of exotic annual grass dominance in the Great Basin, USA. Divers Distrib 28:83–96. https://doi.org/10.1111/ddi.13440

Smith KT, LeVan JR, Chalfoun AD, Christiansen TJ, Harter SR, Oberlie S, Beck, JL (2023) Response of greater sage-grouse to sagebrush reduction treatments in Wyoming big sagebrush. Wildl Monogr 212:e1075. https://doi.org/10.1002/wmon.1075

Spence ES, Beck JL, Gregory AJ (2017) Probability of lek collapse is lower inside sage-grouse Core Areas: effectiveness of conservation policy for a landscape species. PLoS ONE 12:e0185885. https://doi.org/10.1371/journal.pone.0185885

Stevens BS, Connelly JW, Reese KP (2012a) Multi-scale assessment of greater sage-grouse fence collision as a function of site and broad scale factors. J Wildl Manage 76:1370–1380. https://doi.org/10.1002/jwmg.397

Stevens BS, Reese KP, Connelly JW, Musil DD (2012b) Greater sage-grouse and fences: does marking reduce collisions? Wildl Soc Bull 36:297–303. https://doi.org/10.1002/wsb.142

Stewart G, Hull AC (1949) Cheatgrass (*Bromus tectorum* L.)—an ecologic intruder in southern Idaho. Ecology 30:58–74. https://doi.org/10.2307/193227

Stiver SJ, Apa AD, Bohne J, Bunnell SD, Deibert P, Gardner S, Hilliard M, McCarthy C, Schroeder MA (2006) Greater sage-grouse comprehensive conservation strategy. Unpublished Report, Western Association of Fish and Wildlife Agencies, Cheyenne, Wyoming, 442 pp

Stiver SJ, Rinkes ET, Naugle DE, Makela PD, Nance DA, Karl JW (eds) (2015) Sage-grouse habitat assessment framework: a multiscale assessment tool. Technical reference 6710-1. Bureau of Land Management and Western Association of Fish and Wildlife Agencies, Denver, Colorado

Stohlgren TJ, Schnase JL (2006) Risk analysis for biological hazards: what we need to know about invasive species. Risk Anal 26:163–173. https://doi.org/10.1111/j.1539-6924.2006.00707.x

Street PA (2020) Greater sage-grouse habitat and demographic responses to grazing by non-native ungulates. Dissertation, University of Nevada, Reno. http://hdl.handle.net/11714/7709

Swanson S, Wyman S, Evans C (2015) Practical grazing management to maintain or restore riparian functions and values on rangelands. J Rangeland Appl 2:1–28. https://thejra.nkn.uidaho.edu/index.php/jra/article/view/20/39

Syphard AD, Radeloff VC, Hawbaker TJ, Stewart SI (2009) Conservation threats due to human-caused increases in fire frequency in Mediterranean-climate ecosystems. Conserv Biol 23:758–769. https://doi.org/10.1111/j.1523-1739.2009.01223.x

Tack JD, Jakes AF, Jones PF, Smith JT, Newton RE, Martin BH, Hebblewhite M, Naugle DE (2019) Beyond protected areas: private lands and public policy anchor intact pathways for multi-species wildlife migration. Biol Cons 234:18–27. https://doi.org/10.1016/j.biocon.2019.03.017

Tausch RJ, West NE, Nabi AA (1981) Tree age and dominance patterns in Great Basin pinyon-juniper woodlands. J Range Manage 34:259–264. https://doi.org/10.2307/3897846

Taylor RL, Walker BL, Naugle DE, Mills LS (2012) Managing multiple vital rates to maximize greater sage-grouse population growth. J Wildl Manage 76:336–347. https://doi.org/10.1002/jwmg.267

Taylor RL, Tack JD, Naugle DE, Mills LS (2013) Combined effects of energy development and disease on greater sage-grouse. PLoS ONE 8:e71256. https://doi.org/10.1371/journal.pone.0071256

Thompson KM, Holloran MJ, Slater SJ, Kuipers JL, Anderson SH (2006) Early brood-rearing habitat use and productivity of greater sage-grouse in Wyoming. West N Am Nat 66:332–342. https://doi.org/10.3398/1527-0904(2006)66[332:EBHUAP]2.0.CO;2

Urza AK, Weisberg PJ, Chambers JC, Board D, Flake SW (2019) Seeding native species increases resistance to annual grass invasion following prescribed burning of semiarid woodlands. Biol Invasions 21:1993–2007. https://doi.org/10.1007/s10530-019-01951-9

U.S. Fish and Wildlife Service [USFWS] (2013) Greater Sage-grouse (*Centrocercus urophasianus*) Conservation Objectives: Final Report. U.S. Fish and Wildlife Service, Denver Colorado

U.S. Fish and Wildlife Service [USFWS] (2014) Endangered and threatened wildlife and plants: threatened status for Gunnison sage-grouse. Fed Reg 79:69191–69310

U.S. Fish and Wildlife Service [USFWS] (2015) Endangered and threatened wildlife and plants; 12-month finding on a petition to list the Greater sage-grouse (*Centrocercus urophasianus*) as an endangered or threatened species. Proposed Rule. Federal Register 80:59858–59942

Van Lanen NJ, Green AW, Gorman TR, Quattrini LA, Pavlacky DC Jr (2017) Evaluating efficacy of fence markers in reducing greater sage-grouse collisions with fencing. Biol Conserv 213:70–83. https://doi.org/10.1016/j.biocon.2017.06.030

Veblen KE, Pyke DA, Aldridge CL, Casazza ML, Assal TJ, Farinha MA (2014) Monitoring of livestock grazing effects on Bureau of Land Management land. Rangeland Ecol Manage 67:68–77. https://doi.org/10.2111/REM-D-12-00178.1

Walker BL, Naugle DE (2011) West Nile virus ecology in sagebrush habitat and impacts on greater sage-grouse populations. In: Knick ST, Connelly JW (eds) Greater sage-grouse: ecology and conservation of a landscape species and its habitats. Studies in Avian Biology 38, University of California Press, Berkeley, pp 127–142. https://doi.org/10.1525/California/9780520267114. 003.0010

Walsh DP, White GC, Remington TE, Bowden DC (2004) Evaluation of the lek-count index for greater sage-grouse. Wildl Soc Bull 32:56–68. https://doi.org/10.2193/0091-7648(2004)32[56: EOTLIF]2.0.CO;2

Wambolt CL, Harp AJ, Welch BL, Shaw N, Connelly JW, Reese KP, Braun CE, Klebenow DA, McArthur ED, Thompson JG, Torrell LA, Tanaka JA (2002) Conservation of greater sage-grouse on public lands in the western U.S.: implications of recovery and management policies. Policy Analysis Center for Western Public Lands Policy Paper SG-02-02, Caldwell, Idaho, 41 p

Wann GT, Coates PS, Prochazka BG, Severson JP, Monroe AP, Aldridge CL (2019) Assessing lek attendance of male greater sage-grouse using fine-resolution GPS data: implications for population monitoring of lek mating grouse. Popul Ecol 61:183–197. https://doi.org/10.1002/1438-390X.1019

Wann GT, Braun CE, Aldridge CL, Schroeder MA (2020) Rates of ovulation and reproductive success estimated from hunter-harvested greater sage-grouse in Colorado. J Fish Wildl Manag 11:151–163. https://doi.org/10.3996/072019-JFWM-063

Watchorn RT, Maechtle T, Fedy BC (2018) Assessing the efficacy of fathead minnows (*Pimephales promelas*) for mosquito control. PLoS ONE 13:e0194304. https://doi.org/10.1371/journal.pone. 0194304

West NE, Provenza FD, Johnson PS, Owens MK (1984) Vegetation change after 13 years of livestock grazing exclusion on sagebrush semidesert in west central Utah. J Range Manage 37:262–264. https://doi.org/10.2307/3899152

Western Governors' Association (2020) A toolkit for invasive annual grass management in the West. Available at: https://westgov.org/images/editor/FINAL_Cheatgrass_Toolkit_July_2020. pdf. Accessed 7 April 2022

Wilken E, Nava FJ, Griffith G (2011) North American terrestrial ecoregions—Level III. Commission for Environmental Cooperation, Montreal, Canada

Wolfe AL, Broughton JM (2016) Chapter 14 - Bonneville Basin avifaunal change at the Pleistocene/Holocene transition: evidence from Homestead Cave. In: Oviatt CG, Shroder JF (eds) Lake Bonneville: a scientific update. Developments in earth surface processes 20, pp 371–419. https://doi.org/10.1016/B978-0-444-63590-7.00014-7

Young JR, Braun CE, Oyler-McCance SJ, Aldridge CL, Magee PA, Schroeder MA (2020) Gunnison sage-grouse (*Centrocercus minimus*), version 1.0. In: Rodewald PG (ed) Birds of the World. Cornell Lab of Ornithology, Ithaca, New York. https://doi.org/10.2173/bow.gusgro.01

Zablan MA, Braun CE, White GC (2003) Estimation of greater sage-grouse survival in North Park, Colorado. J Wildl Manage 67:144–154. https://doi.org/10.2307/3803070

Ziska LH, Reeves JB III, Blank B (2005) The impact of recent increases in atmospheric CO_2 on biomass production and vegetative retention of cheatgrass (*Bromus tectorum*): implications for fire disturbance. Glob Change Biol 11:1325–1332. https://doi.org/10.1111/j.1365-2486.2005. 00992.x

Zou L, Miller SN, Schmidtmann ET (2006) Mosquito larval habitat mapping using remote sensing and GIS: implications of coalbed methane development and West Nile virus. Med Entomol 43:1034–1041. https://doi.org/10.1093/jmedent/43.5.1034

Zwickel FC, Schroeder MA (2003) Grouse of the Lewis and Clark expedition, 1803 to 1806. Northwest Nat 84:1–19. https://doi.org/10.2307/3536717

Chapter 11
Quails

Michelle C. Downey, Fidel Hernández, Kirby D. Bristow, Casey J. Cardinal, Mikal L. Cline, William P. Kuvlesky Jr., Katherine S. Miller, and Andrea B. Montalvo

Abstract Six species of quails occur on western United States (U.S.) rangelands: northern bobwhite, scaled quail, Gambel's quail, California quail, Montezuma quail, and mountain quail. These quails are found across a variety of vegetation types ranging from grasslands to mountain shrublands to coniferous woodlands. Given their ecological importance and gamebird status, there is considerable conservation, management, and research interest by ecologists and the public. Western quails in general are *r*-selected species whose populations are strongly influenced by weather. Based on Breeding Bird Survey data, 3 species are declining (northern bobwhite, scaled quail, and mountain quail), 2 species have inconclusive data (Gambel's quail

M. C. Downey (✉)
Yale School of the Environment, Yale University, New Haven, CT 06511, USA
e-mail: Michelle.Downey@yale.edu

F. Hernández · W. P. Kuvlesky Jr.
Caesar Kleberg Wildlife Research Institute, Texas A&M University-Kingsville, Kingsville, TX 78363, USA
e-mail: fidel.hernandez@tamuk.edu

W. P. Kuvlesky Jr.
e-mail: William.Kuvlesky@tamuk.edu

K. D. Bristow
Arizona Game and Fish Department, Tucson, AZ 85745, USA
e-mail: kbristow@azgfd.gov

C. J. Cardinal
New Mexico Department of Game and Fish, Santa Fe, NM 87507, USA
e-mail: Casey.Cardinal@dgf.nm.gov

M. L. Cline
Oregon Department of Fish and Wildlife, Salem, OR 97302, USA
e-mail: Mikal.L.Cline@odfw.oregon.gov

K. S. Miller
California Department of Fish and Wildlife, West Sacramento, CA 94244, USA
e-mail: Katherine.Miller@wildlife.ca.gov

A. B. Montalvo
East Foundation, Hebbronville, TX 78361, USA
e-mail: amontalvo@eastfoundation.net

© The Author(s) 2023
L. B. McNew et al. (eds.), *Rangeland Wildlife Ecology and Conservation*,
https://doi.org/10.1007/978-3-031-34037-6_11

339

and Montezuma quail), and 1 species is increasing (California quail). Grazing represents a valuable practice that can be used to create or maintain quail habitat on western rangelands if applied appropriately for a given species, site productivity, and prevailing climate. Invasive, nonnative grasses represent a notable threat to quails and their habitat given the negative influence that nonnative grasses have on the taxon. Numerous conservation programs exist for public and privately-owned rangelands with potential to create thousands of hectares of habitat for western quails. Although the taxon is relatively well-studied as a group, additional research is needed to quantify the cumulative impact of climate change, landscape alterations, and demographic processes on quail-population viability. In addition, research on quail response to rangeland-management practices is limited in scope (only 1–2 species) and geographic extent (mostly Texas, Oklahoma, and New Mexico) and warrants further investigation.

Keywords California quail · Gambel's quail · Grazing · Montezuma quail · Mountain quail · Nonnative grasses · Northern bobwhite · Quails · Rainfall · Scaled quail

11.1 General Life History and Population Dynamics

Quails and quail hunting represent an important component of the culture and economy of rural communities throughout the western United States (U.S.). Each year, thousands of quail hunters venture onto western rangelands for the opportunity to hunt wild quails. The popularity of quail hunting in western states extends not only from the beautiful landscapes that western rangelands provide for upland gamebird hunting but also from the rich diversity of quails. Six quail species occur in the U.S., and all 6 species are found on western rangelands. The 6 species of quail occur in 4 genera (*Colinus, Callipepla, Cyrtonyx,* and *Oreortyx*) and are classified within the order Galliformes, family Odontophoridae, and sub-family Odontophorinae. These quails are found across a variety of vegetation types in the U.S. ranging from grasslands to mountain shrublands to coniferous woodlands and consist of the northern bobwhite (*Colinus virginianus*), scaled quail (*Callipepla squamata*), Gambel's quail (*Callipepla gambelii*), California quail (*Callipepla californica*), Montezuma quail (*Cyrtonyx montezumae*), and mountain quail (*Oreortyx pictus*; Fig. 11.1a–f). Western quails are *r*-selected species whose populations are strongly influenced by weather, particularly rainfall (Brennan 2007).

Given the diversity of quails that occur on western rangelands, it is impractical to discuss each species' life history, ecology, and management. Consequently, we synthesize the literature on quails and provide generalizations of life history, ecology, and management for this taxon, acknowledging that individual species may show deviations from generalizations. In cases where such deviations are notable, we

Fig. 11.1 Six quail species inhabit the western rangelands of the United States. These quails are **a** northern bobwhite, **b** scaled quail, **c** Gambel's quail, **d** California quail, **e** Montezuma quail, and **f** mountain quail. Photographs by Larry Ditto (northern bobwhite, scaled quail, Gambel's quail, and Montezuma quail) and Brian Small (California quail and mountain quail)

reference the species. In addition, of the 6 quail species, northern bobwhite is the only species that also occurs in the eastern U.S. In this chapter, we focus on the ecology and management of northern bobwhite as it pertains to the western portion of its geographic distribution.

11.1.1 Nesting

Nesting season for quails generally begins shortly after covey break-up in the spring when males leave winter coveys and begin seeking female mates from other coveys (Gullion 1962; Gee et al. 2020; Table 11.1). Pair formation takes place generally 2–3 weeks prior to nesting but can occur much earlier (Gullion 1962; Wallmo 1954). Nests are usually built on the ground beneath herbaceous, succulent, or shrubby vegetation providing both security and thermal cover (Pope 2002; Stromberg et al. 2020). Although herbaceous cover is an important component of nest concealment, Gambel's quail have adapted to desert environments lacking such cover (Gee et al. 2020) and instead rely on cryptic coloration of the eggshells to reduce the probability of detection (Brennan 2007). Quails also select nesting structure depending on annual availability. For example, mountain quail in west-central Idaho relied more on woody cover for nesting and brood-rearing during a drier-than-average year but used more herbaceous cover in a wetter-than-normal year (Reese et al. 2005). Nest success varies greatly among species and within populations through time and space (Table 11.1).

11.1.2 Brood-Rearing

Female quail generally lay one egg per day to every other day until the clutch is complete ($\approx$ 12–14 eggs), with nest incubation initiating soon thereafter and lasting 21–26 days (Table 11.1). Both parents tend to incubate the clutch and care for the chicks, but the degree of care varies by species (Brennan 2007; Gutiérrez 1980). Quails traditionally have been considered monogamous and, of the 6 species, mountain quail likely are the most monogamous (Beck et al. 2005). However, ambisexual polyandry (i.e., one female mating with more than one male) is common and has been documented in several species. Both males and females are known to incubate and raise broods with more than one mate during the breeding season (Curtis et al. 1993; Brennan 2007; Davis et al. 2017). In addition, a small portion of the breeding population often produces multiple broods (i.e., individuals raising more than 1 brood per nesting season), at least in California quail (Francis 1965), Gambel's quail (Gullion 1956), and northern bobwhite (Guthery and Kuvlesky 1998). However, the influence of multiple broods on annual populations is likely insignificant because second and third broods contribute little to age ratios under a typical probability of nest success (Guthery and Kuvlesky 1998). In contrast to an ambisexual polyandry approach, female mountain quail lay two simultaneous clutches, incubated separately by the male and female in each monogamous pair and thereby optimize breeding success in mountainous areas typified by short growing seasons (Beck et al. 2005).

11.1.3 Brood Success and Chick Survival

Brood success and chick survival vary among quails and likely is related to habitat and weather conditions (Brennan 2007). Chicks of all quail species are precocial and susceptible to a variety of mortality sources such as predation and exposure to inclement weather. In mesic environments, exposure to rain during the first weeks of life has been associated with chick mortality (Terhune et al. 2019). In xeric environments, Heffelfinger et al. (1999) documented that hot, dry summer weather reduced the percent of juveniles in Gambel's quail populations in Arizona compared to cool, wet weather and speculated that reduced food availability reduced juvenile survival. Chick survival can have a significant impact on quail population dynamics, although less so than adult survival (Guthery and Kuvlesky 1998; Sandercock et al. 2008).

Table 11.1 General life history characteristics of 6 quail species inhabiting rangelands of the western United States

Common name	Scientific name	Critical precipitation	Covey breakup	Nesting	Clutch size	Nest success	Annual survival	Average life span
Northern bobwhite[a]	*Colinus virginianus*	Variable by region	Feb–Mar	May–Sep	7–28 eggs	35–45%	18–30%	< 1 year
Scaled quail[b]	*Callipepla squamata*	Jan–Jul	Feb–Mar	Apr–Sep	5–22 eggs	16–83%	14–17%	1–2 years
Gambel's quail[c]	*Callipepla gambelii*	Oct–Mar	Feb–Mar	Apr–Jun	5–20 eggs	No data available	10–60%	1.5 years
California quail[d]	*Callipepla californica*	Sep–Apr	Feb	May–Jul	1–26 eggs	5–30%	8–50%	No data available
Montezuma quail[e]	*Cyrtonyx montezumae*	Jul–Sep	Feb	Jul–Sep	2–15 eggs	12–75%	18–59%	No data available
Mountain quail[f]	*Oreortyx pictus*	Jan–Mar	Jan	Apr–Jun	6–14 eggs	70–76%	17–42%	No data available

[a]Brennan et al. (2020)
[b]Dabbert et al. (2020)
[c]Gee et al. (2020)
[d]Calkins et al. (2020)
[e]Stromberg et al. (2020)
[f]Gutiérrez and Delehanty (2020), Stephenson et al. (2011)

Reliable estimates of chick survival generally are lacking due to the difficulties in capturing and monitoring juvenile quail of all species; however, research on chick survival has increased during recent years given advances in technology (e.g., Orange et al. 2016; Terhune et al. 2019).

11.1.4 Non-breeding

Quails are gregarious species, and the covey is the primary social unit during much of the year.

Covey sizes generally are largest after brooding season (autumn). Depending on the species, autumn coveys are composed of 1 or more adult pairs and their broods, and covey sizes may range from 8 to 30 individuals. Covey sizes of Montezuma and mountain quail occur at the lower end of this range, whereas Gambel's and scaled quail occur at the upper end (Brennan 2007; Gutiérrez and Delehanty 2020). Whether in coveys or not, quails roost together at night. Quails most often roost on the ground in grass or shrubby ground cover, although Gambel's and California quail prefer to roost above ground in dense shrubs or trees (Gee et al. 2020; Calkins et al. 2020). Quails generally leave the roost shortly after sunrise to begin feeding (Gutiérrez and Delehanty 2020; Stromberg et al. 2020). Communal roosting and feeding presumably provides both thermal protection and enhanced predator detection (Anderson 1974).

11.1.5 Survival and Sources of Mortality

Annual survival of quails generally is low (< 20%) but varies among and within species (≈ 10–70%) and is considered a primary driver of populations (Guthery and Kuvlesky 1998; Sandercock et al. 2008; Table 11.1). Sources of adult quail mortality may include predation, exposure to weather and extreme temperature, disease, parasites, and starvation. Habitat quality and availability can exacerbate or ameliorate the effects of each of these (Brennan 2007). Mammalian predators are the primary predators of nests, whereas raptors pose the greatest threat to adults (Brennan 2007; Turner et al. 2014).

Similar to other Galliformes, quails tend to walk or run more often than fly and usually respond to potential predators with some variation of a "run and hide" escape strategy. For example, scaled quail will often run from potential predators and then, when pressured, fly long distances to hide (Dabbert et al. 2020). In contrast, Montezuma quail tend to crouch and hide in response to danger, relying on their cryptic coloration to prevent detection. Montezuma quail flush only when approached closely and fly short distances to again hide in the relatively dense oak (*Quercus*)-juniper (*Juniperus*) savanna they inhabit (Stromberg 1990). The other quails exhibit some variation between these two extremes, and the escape strategies they exhibit appear adapted to the habitat in which they evolved. For example, Montezuma quail will crouch and hide rather than fly even when found in areas lacking cover (Brown 1982; Stromberg 1990).

11.1.6 Seasonal Movements and Dispersal

Quails tend to be less mobile than other gallinaceous birds. Maximum annual movements of coveys < 4 km have been reported for several species (Stromberg 1990; Gee et al. 2020). Although quails are not known to migrate in a strict sense, mountain quail move seasonally between winter and breeding habitat presumably to avoid snow accumulation at higher elevations (Gutiérrez and Delehanty 2020). Similarly, scaled quail in the northern portions of their distribution are reported to make short (< 4 km) movements between summer and winter ranges (Dabbert et al. 2020). Information on movements from nesting to brood-rearing cover is limited. Large movements (e.g., > 20 km) by quails have been reported and may be associated with dispersing males (Campbell and Harris 1965 but see Townsend et al. 2003).

11.1.7 Population Dynamics

Quails are r-selected species (Guthery and Brennan 2007), and their population fluctuations are largely determined by weather (Brennan 2007). Variations in demographic parameters such as percent hens nesting, nesting rate, and nest success, combined with low annual survival, create conditions for fluctuating quail populations that are subject to the vagaries of habitat and weather conditions (Table 11.1). Given their low survival, quail population fluctuations largely are the result of varying reproductive success. For example, Swank and Gallizioli (1954) reported that 90% of the variation in Gambel's quail population indices were attributed to nesting success. Hernández et al. (2005) documented a lower percentage of northern bobwhite hens nesting, lower nesting rates, and shorter nesting seasons during drought compared to wet years. Consequently, in years of poor environmental conditions, quail numbers drop significantly only to rebound when conditions improve, resulting in "boom and bust" population dynamics (Hernández and Peterson 2007).

The reproductive success of quails that inhabit semiarid environments has been positively correlated with rainfall (Bridges et al. 2001; Hernández et al. 2005; Brennan 2007). The ideal timing for rainfall varies by species but generally occurs 1–3 months prior to the nesting season (Table 11.1). For example, northern bobwhite occurs over a wide range of vegetation types, and the months of critical rainfall as well as the relative influence of rainfall varies by region (Bridges et al. 2001; Hernández and Peterson 2007). Other researchers have explored the relationship between quail reproductive success and heat indices (Francis 1970; Heffelfinger et al. 1999) and have documented that cooler summer temperatures can have an ameliorating effect on drought with respect to quail reproduction (Heffelfinger et al. 1999).

The mechanism by which weather exerts its influence on quail reproduction presently is unknown (Hernández et al. 2002) but often attributed to the materialized effects of rainfall (e.g., increased food, nesting cover, etc.; Brennan 2007). For Gambel's quail, forb growth that proliferates after favorable winter rains is presumed to provide higher levels of Vitamin A, which is thought to stimulate reproductive organ development and positively influence reproductive success (Hungerford 1960, 1964). However, this relationship has not been empirically established in quails (Lehmann 1953; Guthery 2002). Investigations into other factors that may enhance (e.g., phosphorus) or possibly inhibit (e.g., phytoestrogens) quail reproduction have failed to provide conclusive evidence to explain the boom-and-bust population phenomenon (Cain et al. 1982, 1987). Research that has focused on food and water supplementation also has failed to provide explanatory evidence (Koerth and Guthery 1991; Harveson 1995; Lusk et al. 2002). More recently, thermal stress has been explored as a possible cause of decreased reproductive performance during dry conditions (Guthery et al. 2005) and, of all the proposed mechanisms, this heat-stress hypothesis presently appears the most plausible (Hernández et al. 2002).

11.2   Current Species and Population Status

There is considerable conservation concern among ecologists and the public regarding the population status of quails (Brennan 1991; Church et al. 1993; Hernández et al. 2013). Of the 6 western quails, 3 species are declining (northern bobwhite, scaled quail, and mountain quail), 2 species have inconclusive data (Gambel's quail and Montezuma quail), and 1 species is increasing (California quail; Table 11.2). Currently, none of the western quails are federally listed as endangered or threatened at the species level (Table 11.2). Some species, however, receive special protections at the state level given that most states have their own system for listing species beyond the federal Endangered Species Act. For example, California quail and mountain quail have received focused attention from state agencies due to their popularity (California quail is the official state bird of California) or limited scientific knowledge of their management (mountain quail).

Table 11.2 Conservation status and population trends of quails in the U.S.

Common name	Status	BBS trend (1966–2019)	CBC trend (1993–2019)	Federal status	State status
Northern bobwhite	Declining	− 3.1 (− 3.3, − 2.9)	− 5.25 (− 6.38, − 3.81)	*C. v. ridgwayi* is federally listed	No special status
Scaled quail	Declining	− 0.7 (− 1.6, 0.1)	− 8.11 (− 13.62, − 4.33)	No special status	No special status
Gambel's quail	Inconclusive	0.6 (− 1.8, 2.3)	− 0.88 (− 1.50, − 0.19)	No special status	No special status
California quail	Increasing	0.8 (0.2, 1.4)	1.71 (0.96, 2.51)	No special status	State wildlife action species (CA). *C. c. catalinensis* species of special concern
Montezuma quail	Inconclusive	Sample size too small for trends	3.82 (0.65, 6.98)	No special status	No special status
Mountain quail	Declining	0.0 (− 1.7, 1.3)	− 2.97 (− 5.02, − 0.83)	USFWS[a] determined eastern populations were not threatened (2003)	Species of greatest conservation concern (ID); sensitive species in northern basin (OR); state wildlife action species (NV)

Trends are percent annual change and 95% credible intervals (in parenthesis) as reported by Breeding Bird Surveys (BBS) and Christmas Bird Counts (CBC)

[a]United States Fish and Wildlife Service (USFWS)

11.2.1   Northern Bobwhite

Northern bobwhite have the largest geographic distribution of the 6 quail species. They can be found from the eastern U.S. west to the Great Plains, and from northern U.S. south to southern Mexico (Fig. 11.2A). Northern bobwhite have been declining at least since the early 1900s (Hernández et al. 2013), but ecologists did not take notice and become broadly aware of the continental decline of the species until the end of the century (Brennan 1991). According to data from the North American Breeding Bird Survey (BBS; Sauer et al. 2018), northern bobwhite declined 3.1% per year during 1966–2019 and have become extirpated (i.e., no longer documented during surveys) in the wild in New England states and functionally extirpated in surrounding states (e.g., New York, Pennsylvania, New Jersey; Table 11.2).

The masked bobwhite (*C. v. ridgwayi*), an endangered subspecies of northern bobwhite, possessed a historical geographic distribution that spanned southern

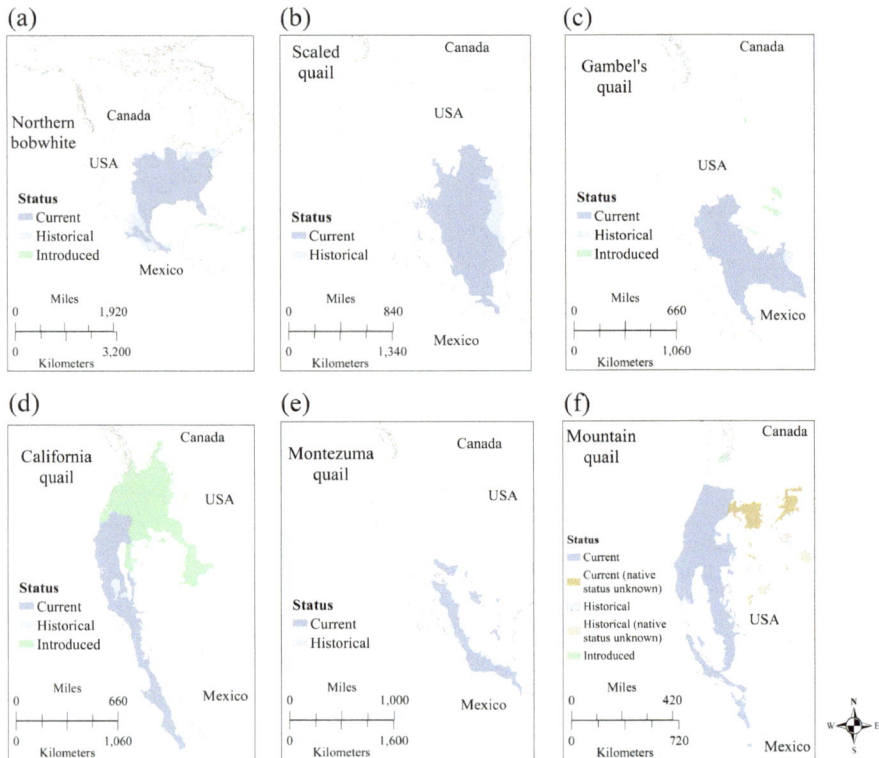

Fig. 11.2 Geographic distribution for **a** northern bobwhite, **b** scaled quail, **c** Gambel's quail, **d** California quail, **e** Montezuma quail, and **f** mountain quail. Historical and current geographic distributions are based on data from the North American Breeding Bird Survey except for Montezuma quail and mountain quail. For Montezuma quail, eBird data were used given the species is not detected during Breeding Bird Surveys. For mountain quail, in addition to data from Breeding Bird Survey, we used eBird data, state agency data (Idaho, Nevada, and Oregon), and Linsdale (1936). Breeding Bird Survey routes where ≥ 1 quail individual was detected were used to define historical (1967–1980) and current (2010–2019) distributions. Introduced geographic distributions represent areas where species have been introduced and formed a sustained population based on Breeding Bird Survey, eBird, and species accounts in the Birds of the World. Additional references consulted for geographic distributions included Guillon and Christiansen (1957), Brown (1989), Brennan (2007), Kamees et al. (2008), California Department of Fish and Wildlife (2017), and Idaho Department of Fish and Game (2019)

Arizona and northern Mexico (Hernández et al. 2006a, b). Today, the masked bobwhite is essentially "extinct" in the wild in the U.S., where populations consist of released captive-raised individuals. The species has not been detected during the BBS and rarely is documented during Christmas Bird Counts. Surveys from Buenos Aires National Wildlife Refuge—the only site where the subspecies is known to occur in the U.S.—indicated a declining trend during 1999–2011 (U.S. Fish and Wildlife Service 2014). Masked bobwhite may still exist in Sonora, Mexico (Hernández et al. 2006a, b).

11.2.2 Scaled Quail

The geographic distribution of scaled quail generally is associated with the Chihuahuan Desert and surrounding desert grasslands and chaparral of the southwestern U.S. (Fig. 11.2b). This species is found from southwestern Kansas and western Texas west to southeastern Arizona, and from southeastern Colorado south to central Mexico. Scaled quail declined 0.7% per year during 1966–2019, according to BBS data (Table 11.2). The chestnut-bellied scaled quail (*C. s. castanogastris*), a subspecies found in southern Texas, has been experiencing notable population declines in recent decades (Hernández et al. In Press).

11.2.3 Gambel's Quail

Gambel's quail possess a geographic distribution that may be described as centered in the Sonoran Desert of Arizona and northern Mexico and radiating from there into the surrounding contiguous states (Fig. 11.2c). Gambel's quail can be found from western Texas along the riparian areas of the Rio Grande River west to southeastern California, and from southwestern Utah south into northern Mexico. The population trend for Gambel's quail is inconclusive based on BBS data (Table 11.2). However, the species faces challenges associated with increased urban development (Zornes and Bishop 2009; Gee et al. 2020), especially solar energy development, the impacts of which are unknown.

11.2.4 California Quail

California quail possess a geographic distribution located along the western coast of the U.S. (Fig. 11.2d). The native geographic distribution of California quail is along the West Coast from southern Oregon, a small portion of western Nevada, south to California, and into Baja California, Mexico (Leopold 1985). However, California quail has been widely introduced throughout much of western North America and now occurs over most of Washington and Oregon, Idaho, Utah, and British Columbia. In contrast to other quail species, California quail increased 0.8% during 1966–2019 according to BBS data (Table 11.2). It is unknown why the species is increasing but may be related to the species' adaptability to human presence, often inhabiting cover adjacent to agricultural lands, riparian corridors, wooded suburbs, and even urban parks.

Similar to northern bobwhite, California quail have a subspecies (Catalina California quail, *C. c. catalinesis*) that receives special protection. The Catalina California quail is an insular subspecies believed to have been introduced to Santa Catalina

Island by Indigenous peoples about 12,000 years ago (Collins 2008; Calkins et al. 2020). This subspecies faces challenges endured by all small, isolated populations (e.g., threat of genetic inbreeding). The current population status of the Catalina California quail is unclear, given that the subspecies has been counted irregularly. Data from BBS indicated a decline from 191 quail in 2013 to 46 quail in 2017.

11.2.5   Montezuma Quail

Montezuma quail may be considered a Mexican species whose northern extent of its geographic distribution extends into southwestern U.S. Most of the Montezuma quail geographic distribution occurs in Mexico, but the species may be found from central and western Texas west to southwestern Arizona (Fig. 11.2e). Similar to Gambel's quail, the Montezuma quail population trend is inconclusive (Table 11.2). However, the species faces challenges, such as genetic erosion, in the easternmost portion of its distribution where the species occurs in relatively isolated populations (Mathur et al. 2019).

11.2.6   Mountain Quail

Of the 6 quail species, mountain quail are the least studied. Mountain quail occur primarily in the Sierra Nevada, Cascade, and Coast Ranges, but disjunct populations also occur in the Intermountain West of Idaho and Nevada as well as the Baja Peninsula (Fig. 11.2f). The species may be found from southern Washington south through western Oregon and western California. According to BBS data, mountain quail declined 0.01% during 1966–2019 (Table 11.2). Mountain quail have received focused attention from state agencies due to the limited scientific knowledge of the species (Pope and Crawford 2004; Reese et al. 2005; Stephenson et al. 2011).

11.3   Population Monitoring

11.3.1   National and Regional Level

Given the wide distribution of quails across the U.S., ecologists have relied on broad-scale datasets such as the BBS and the Audubon Christmas Bird Count (CBC) to monitor their populations. These monitoring programs analyze long-term datasets to estimate bird population trends at various spatial extents (e.g., state, national, Bird Conservation Region, geographic distribution). The BBS was initiated in 1966 to monitor North American bird populations (Sauer et al. 2018) and presently is coordinated by the U.S. Geological Survey. Surveys are conducted annually during the summer along thousands of 39.2-km routes that are distributed across North America. The CBC is coordinated by the National Audubon Society and was initiated in 1900 (Meehan et al. 2018). The CBC is conducted during winter (Dec–Jan) and involves observers counting birds within a 24.1-km diameter "count circle". The BBS and CBC provide complementary sources of information because the former occurs during the breeding season (summer), whereas the latter occurs during the non-breeding season (winter). Because these surveys are collected annually throughout the quails' geographic distributions, ecologists have used these data to understand quail-population response to changes in land use and weather patterns (e.g., Peterson et al. 2002; Murphy 2003; Veech 2006; Janke et al. 2017; Miller et al. 2018).

11.3.2 Ecoregion and Site Level

The BBS and CBC are designed to provide measures of bird populations at large spatial extents (e.g., statewide, multi-state, national). However, the need also exists to monitor quail populations at smaller spatial extents such as within a state or at a site level. The social nature of quails facilitates the monitoring of their populations at these smaller spatial extents. At the state level, state wildlife agencies have used rural mail carrier surveys (Robinson et al. 2000) and roadside surveys (DeMaso et al. 2002) to monitor quail populations. In Kansas, surveys are conducted annually throughout the state during specific weeks of the year by rural mail carriers making deliveries. These volunteers record their observations of quail and distance traveled for five consecutive days, and these data are used to obtain measures of relative abundance (Robinson et al. 2000). This method is very similar to roadside surveys. In Texas and Oklahoma, state agency personnel conduct annual roadside surveys whereby biologists drive along established roadside routes of known length and record the number of quail observed to estimate quail relative abundance for regions within the states (DeMaso et al. 2002).

Methods also exist for monitoring quail populations at a site level. These methods include techniques to obtain measures of relative abundance such as whistle counts (number of males calling per point), covey-call counts (number of calling coveys per point), and roadside counts (number of quail observed per distance traveled), as well as methods to obtain estimates of density or abundance such as distance sampling and mark-recapture, respectively. Recently, helicopter surveys within a distance sampling framework have been used to estimate quail density (Rusk et al. 2007; Schnupp et al. 2013). This recent development has permitted the monitoring of quail populations over relatively larger spatial extents (e.g., 20,000 ha) while reducing the survey effort that would be required with traditional walking transects. For more information on quail surveys and their protocols, we refer the reader to Brennan (2007) and Hernández and Guthery (2012).

11.4 Habitat Associations

Western quails occur across a variety of vegetation types (Fig. 11.3a–f). These include savannas and shrublands (northern bobwhite; Fig. 11.3a), desert grasslands or shrubland (scaled quail; Fig. 11.3b, Gambel's quail: Fig. 11.3c, California quail; Fig. 11.3d), oak-juniper woodlands (Montezuma quail; Fig. 11.3e), and mountain shrubland and regenerating forest (mountain quail; Fig. 11.3f). Because climate largely determines vegetation communities at broad scales, quails occur across a range of environmental and topographic gradients.

Quails are relatively sedentary in nature and therefore occur within plant communities that offer satisfactory food and cover in relatively close proximity (Wallmo 1956; Guthery 1999, Dabbert et al. 2020). Woody cover is a critical habitat component for all quails because it provides both food (e.g., seeds, mast, and leaves) and structure (e.g., roosting, escape, and loafing cover). In extreme climes, woody cover provides a retreat from inclement weather such as blizzards or extreme heat and provides access to food during snow accumulation (Lepper 1978; Reese et al. 2005; Palmer et al. 2021). Generally, quails prefer some mosaic of woody and herbaceous cover to support their daily and seasonal needs, but the specific amount of woody cover used by quails varies by species and scale (Hernández 2020). In addition, the optimal configuration of woody and herbaceous patches possesses "slack" in their arrangement such that a single optimal arrangement does not exist (Guthery 1999),

Fig. 11.3 Quail species inhabiting the western rangelands of the United States occur across a variety of vegetation communities as illustrated by typical habitat for **a** northern bobwhite in Texas, **b** scaled quail in Texas, **c** Gambel's quail in Arizona, **d** California quail in California, **e** Montezuma quail in Arizona, and **f** mountain quail in Oregon. Photographs by Fidel Hernández (northern bobwhite), Eric Grahmann (scaled quail), Arizona Game and Fish Department (Gambel's quail), Katherine Miller (California quail), Kirby Bristow (Montezuma quail), and Oregon Department of Fish and Wildlife (mountain quail)

at least for species such as northern bobwhite, scaled quail, and Gambel's quail (Guthery et al. 2001).

Despite these broad habitat commonalities, quail species possess unique habitat affinities and preferences. Following we provide brief descriptions for each species but refer the reader to Brennan et al. (2020), Calkins et al. (2020), Dabbert et al. (2020), Gee et al. (2020), Gutiérrez and Delehanty (2020), and Stromberg et al. (2020) for detailed descriptions.

11.4.1 Northern Bobwhite

Northern bobwhite extend into western rangelands only along the westernmost edge of their geographic distribution. Here, northern bobwhite occur in grasslands, shrublands, and savannas (Fig. 11.3A; Brennan et al. 2020). Northern bobwhite use open ground for travel, herbaceous plants for food and nesting cover, and woody plants for thermal cover and predator protection, as well as nesting (Lehmann 1984; Hernández et al. 2007). Woody cover is important as thermal cover for northern bobwhite in semiarid rangelands, given the regular occurrence of high temperatures and drought (Guthery et al. 2005; Parent et al. 2016).

11.4.2 Scaled Quail, Gambel's Quail, and Masked Bobwhite

Quails of the semiarid southwestern U.S. (Gambel's quail, scaled quail, and masked bobwhite) inhabit desert grasslands, shrublands, brushy arroyos, pinyon (*Pinus* spp.)-juniper woodlands, and chaparral (Anderson 1974; Silvy et al. 2007). These sympatric quails appear to partition available habitat and thereby minimize interspecific competition (Guthery et al. 2001). For example, in Arizona, scaled quail have a stronger grassland association, if a patchy shrub component with minimal tree cover and open bare ground is available (Fig. 11.3b; Bristow and Ockenfels 2006, Dabbert et al. 2020). Gambel's quail evolved in association with thorny legumes, succulents, and scrub-shrub grasslands of the desert (Fig. 11.3c; Brown 1989; Kuvlesky et al. 2007; Gee et al. 2020). This species tends to inhabit areas with more woody cover than either scaled quail or masked bobwhite and prefers mesquite-rimmed riparian areas, particularly along the southern limits of its geographic distribution (Guthery et al. 2001; Ortega-Sánchez 2006; Kuvlesky et al. 2007). Masked bobwhite habitat is characterized by more herbaceous cover and less bare ground relative to Gambel's and scaled quail (Goodwin and Hungerford 1977; Guthery et al. 2001).

11.4.3 California Quail

California quail is an adaptable species that is associated with brushy cover such as riparian edges, foothill woodlands, chaparral, sagebrush (*Artemisia* spp.), grassland oak, and recently disturbed or converted forest (Fig. 11.3d; Leopold 1985; Calkins et al. 2020). California quail also occur along the edges of urban areas such as suburban neighborhoods and apparently do well in such environments (Iknayan et al. 2021); however, the species has been harmed by certain levels of urbanization (Crooks et al. 2004). California quail need access to early successional habitat for foraging, but these early seral stages must be intermixed with woody cover (Koford 1987; Calkins et al. 2020). In the rangelands of the Great Basin, California quail rely on areas of dense shrub such as willows (*Salix* spp.), thorny shrub thickets, saltbush (*Atriplex* spp.), and junipers for protection from snowfall (Nielson 1952; Jewett et al. 1953; Brown 1989).

11.4.4 Montezuma Quail

Montezuma quail occur at higher elevations than other quail species of the south-western U.S. The species is strongly associated with oak and pine (*Pinus* spp.) wood-lands possessing an understory of tall, perennial bunchgrasses and typically inhabit steep, rugged slopes (Fig. 11.3e; Leopold and McCabe 1957; Harveson et al. 2007). An important component of Montezuma quail habitat is the availability of corms, tubers, bulbs, and rhizomes that primarily compose their diet (Hernández et al. 2006a, b; Harveson et al. 2007).

11.4.5 Mountain Quail

Mountain quail prefer steep, shrub-dominated slopes and generally avoid grassland habitats (Fig. 11.3f; Brennan et al. 1987; Gutiérrez and Delehanty 2020). Exam-ples of shrub-dominated communities include chaparral, mixed desert scrub, and early-successional-stage shrub vegetation following disturbance (e.g., fire, logging) (Gutiérrez and Delehanty 2020). Mountain quail may also be found in mixed evergreen-hardwood forests and montane conifer forests (Gutiérrez and Delehanty 2020). Although this species may not strictly inhabit what may be considered typical rangeland environments, the eastern extent of its geographic distribution includes rangelands in the Great Basin (Pope 2002) and western Idaho (Beck et al. 2005; Reese et al. 2005). Here, mountain quail can be found in association with pinyon-juniper, aspen (*Populus* spp.)-sagebrush, shrub-steppe, and riparian areas that are generally steep, rugged, and brushy (Gutiérrez 1980; Brennan et al. 1987; Gutiérrez and Delehanty 2020).

11.5   Rangeland Management

11.5.1   Livestock Grazing

Livestock grazing can be a useful tool for managing quail habitat. Like other range-land management practices such as prescribed fire or brush management, how the practice is applied and where it is applied will determine whether the effect is positive or negative for quails. In xeric environments, excessive grazing can reduce critical cover (e.g., nesting, escape, thermal, etc.) for quails (Ortega-S and Bryant 2005). Conversely, in more mesic environments, livestock grazing can be a valuable tool for reducing dense, rank vegetation while increasing forb abundance and diversity (Holechek 1981; Grahmann et al. 2018). Overall, livestock grazing can be a useful tool to manage quail habitat, but the impact it will have on quail habitat depends on factors such as grazing intensity, rangeland site productivity, and climate regimes.

The perceived impact of livestock grazing on wildlife habitat has tradition-ally differed between areas dominated by private lands and areas comprised of mostly public lands. For example, Texas is 95% privately owned and possesses large contiguous tracts of native rangelands where northern bobwhite, scaled quail, Montezuma quail, and Gambel's quail occur. Privately owned ranches in areas such as the Rio Grande Plains and the Rolling Plains of Texas benefit from fee-lease hunting for quails (Hernández et al. 2002) and applying grazing strategies that benefit quail habitat therefore directly contributes to their financial success. These grazing strategies include reduced stocking rates (number of animal units per area per time) and grazing stockers (weaned, yearling cattle) rather than cow-calf pairs. Grazing with stockers is a more quail-friendly strategy because stockers generally are grazed during spring–summer and sold during autumn but can be sold any time during the grazing period should drought occur and forage become limited. Consequently, adjustments in stocking rates can be made more promptly and easily when grazing stockers than cow-calf pairs because the latter involves consideration of the reproduc-tive phase (gestation, weaning, etc.) of the cattle among other logistical and financial considerations. In Texas, grazing is an important habitat management tool for quails that supports privately-operated hunting operations (Brennan 2007; Hernández and Guthery 2012). In contrast, the effects of livestock grazing on wildlife habitat in western states dominated by public land has been contentious. This is likely due to public land agencies in the West being charged to manage lands for multiple uses such as recreation, oil-and-gas production, mining, timber, and wildlife (Brown et al. 1993; Krausman 1996).

The impact of grazing on quails varies by species given their unique ecology and environment they inhabit. Of the 6 quails, masked bobwhite and Montezuma quail exhibit the highest sensitivity to grazing, whereas Gambel's quail exhibits the least sensitivity. Overgrazing has been attributed to the near extinction of masked bobwhite (Kuvlesky et al. 2000; Hernández et al. 2006a, b). The effect of grazing on masked bobwhites likely is exacerbated by the arid climate the subspecies inhabits in Arizona and Sonora, Mexico. These areas experience drought and low herbaceous productivity. The floodplains and drainages that support herbaceous vegetation are preferred by masked bobwhites and cattle, thereby creating conflicts in use between the two (Kuvlesky et al. 2000). Consequently, grazing is prohibited in the Buenos Aries National Wildlife Refuge, the only location in the U.S. where the masked bobwhite occurs (USFWS 2014). Grazing also can negatively affect Montezuma quail because grazing may result in the loss of herbaceous cover, which is critical for this species for nesting, thermal, and hiding cover (Stromberg 1990). If herbaceous cover is severely reduced by livestock, local extirpations may occur (Brown 1982). Similarly, grazing has been cited as a contributing factor to the loss of mountain quail in Idaho resulting from the loss of herbaceous cover and plant diversity in the low-elevation riparian areas inhabited by mountain quail during winter (Brennan 1994).

Although grazing livestock has the ability to negative impact quail habitat and their populations, it also has the ability to have a positive impact if applied appropriately for the climate and site productivity present. Leopold (1985) noted that livestock grazing was necessary to reduce herbaceous cover and increase forb abundance for California quail in the coastal ranges and Sacramento Valley foothills of California where precipitation was higher (Leopold 1985). In southern Texas, livestock grazing also may be beneficial to scaled quail and northern bobwhite, particularly in rangelands dominated by nonnative grasses. Scaled quail strongly avoid dense monocultures of nonnative grasses, and grazing can be used to increase bare ground and forb diversity for both scaled quail (Fulbright et al. 2019; Kline et al. 2019) and northern bobwhite (Grahmann et al. 2018). It is important to note that, even in native rangeland, grazing and quail presence can be compatible if properly managed. For example, northern bobwhite has persisted for decades in huntable numbers over millions of hectares in Texas ecoregions (i.e., the Rolling Plains and Rio Grande Plains) where grazing is a dominant land use (Hernández et al. 2002).

Proper grazing management for quails depends on applying the appropriate grazing pressure to match a site's productivity. Higher grazing pressure may be possible in more mesic and productive sites whereas lower or no grazing pressure may be appropriate for more xeric and lower productivity sites (Spears et al. 1993). Balancing quail habitat and livestock use is possible by using appropriate and flexible stocking rates to always ensure sufficient herbaceous cover for quails across space and time, including during drought (Hernández and Guthery 2012; Bruno 2018).

11.5.2   Other Rangeland Management Practices

Except for northern bobwhite, little research exists on the use of rangeland management practices such as prescribed fire, mechanical treatments, and chemical treatments to manage quail habitat. This research focus on northern bobwhite likely is due to its inhabiting primarily private lands (in the western portion of its geographic distribution) where its long history as an important gamebird provides strong economic, cultural, and ecological incentives for landowners, state agencies, and non-governmental organizations to purposefully manage the species. The other five western quails occur mostly in states dominated by public land where users are the general public and therefore the incentives for active management are considerably fewer. Consequently, management for most western quails besides northern bobwhite tends to be accidental rather than purposeful (Brennan 2007).

Regarding northern bobwhite, research on the impacts of rangeland-management practices has been limited in geographic extent (mostly Texas, Oklahoma, New Mexico; Hernández et al. 2002) and has been discussed in detail elsewhere (Guthery 2000; Brennan 2007; Hernández and Guthery 2012). Brennan (2007) includes the sparse research that exists on the impacts of rangeland management on some of the other western quails, and Hernández and Guthery (2012) provides detailed discussion on the use of prescribed fire, mechanical treatments (e.g., root-plowing, roller-chopping, chaining, grubbing, etc.), and chemical treatments (herbicides, equipment, patterns of application, etc.) for northern bobwhite. We refer the reader to these publications for such information but provide the following general recommendations regarding the use of these or any other rangeland management practice for quail-habitat management.

Rangeland management practices for quails should be implemented in a manner that (1) preserves uncommon or rare vegetation community types present on the site, (2) treats smaller portions (e.g., 120 ha) of more pastures rather than larger portions (e.g., 500 ha) of fewer pastures, (3) treats areas of the same pasture with different but appropriate methods, and (4) treats different areas in different years (Hernández and Guthery 2012). The general goal of such a rangeland-management approach is the promotion of rangeland heterogeneity. Regarding determination of the appropriate rangeland-management practice for a given situation, the decision requires (1) an understanding of plant-community response based on soils and management techniques, (2) knowledge of the amount of the target cover present on the rangeland relative to quail requirements, and (3) some reasonable prediction of the desired outcome (Hernández et al. In Press).

We conclude this section with a brief discussion of a management practice that has generated perennial interest in the management of western quails: water provision. This long-time interest in water provision likely is the result of the semiarid and desert environments that western quails inhabit and the common observation of quails at watering sources. Guzzlers generally are means through which water is provided to western quails, and their use has been evaluated in several species including scaled quail (Rollins et al. 2009), Gambel's quail (Campbell 1960), and mountain quail (Delehanty et al. 2004). Research suggests that, despite the common use of guzzlers by quails, guzzlers do not influence quail vital rates (i.e., adult survival, nest survival) and therefore a practice that likely is of limited value for western quails from a population-response perspective (Campbell 1960; Tanner et al. 2015).

11.6   Effects of Disease

There is no direct association involving livestock as a causative agent for disease in quails. However, parasitic infections and disease research has made a resurgence in the past decade, particularly in Texas for northern bobwhite (Dunham et al. 2014; Bruno et al. 2018) and to a lesser extent scaled quail (Fedynich et al. 2019). Beyond this regional emphasis, quail disease research is scattered across the West with some focus on Gambel's quail in Arizona and New Mexico, and mountain and California quail in California, Oregon, and Washington. However, none of these species has been investigated for parasites and disease in the last 2–3 decades. Given the recent documentation of parasites and diseases in northern bobwhite (Dunham et al. 2014; Bruno et al. 2018), we provide a brief overview of quail parasites and their documented impact on quails.

11.6.1   Microparasites

Parasites can be categorized into microparasites (bacteria, viruses, and fungi) and macroparasites (helminths and arthropods; Peterson 2007). Microparasitic infections that could potentially cause population decline in quails include avian pox and avian malaria (Peterson 2007). Avian pox (*Avipoxvirus spp.*) cases have been reported for northern bobwhite in the southeastern U.S. (Davidson et al. 1982), scaled quail in Texas (Wilson and Crawford 1988), and Gambel's quail in Arizona (Blankenship et al. 1966). Avian malaria has been documented in northern bobwhite in Colorado (Stabler and Kitzmiller 1976); California quail (O'Roke 1930), scaled quail, and Gambel's quail in New Mexico (Campbell and Lee 1953); scaled and Gambel's quail in Arizona (Wood and Herman 1943; Hungerford 1955); and Gambel's quail in Nevada (Gullion 1957). O'Roke (1930) observed California quail infected with avian malaria that were weakened and anorexic, which can lead to death in rare instances, whereas others (Campbell and Lee 1953; Hungerford 1955) noted that malaria is likely not a significant disease for Gambel's quail.

11.6.2   Macroparasites

Helminths are well documented in northern bobwhite and scaled quail. However, information is limited for other species possibly due to the lack of helminth presence in arid and semiarid conditions such as occur in the western U.S. (Moore et al. 1989). Of the helminth species documented, some cause morbidity and mortality in pen-raised quail and potentially wild quails, but their impact on wild quail populations is unknown. *Dispharynx nasuta*, a nematode inhabiting the proventriculus, can cause mortality in chicks of pen-raised northern bobwhite (Kellogg and Prestwood 1968) and has been reported in wild northern bobwhite, California quail, and Gambel's quail in the western U.S. (Table 11.3). Perhaps the most cited example of helminth population regulation in Galliformes is the cecal worm (*Trichostrongulus tenuis*) in red grouse (*Lagopus lagopus scoticus*; Hudson et al. 1998), which causes internal inflammation and bleeding in the ceca of grouse that can decrease grouse survival. The larvae of *T. tenuis* typically favor mesic habitats so its occurrence in western quails is relatively low. Moore et al. (1988) found *T. tenuis* occurring in mountain quail in Oregon that inhabited high-elevation mesic areas (Table 11.3). The cecal worm *T. cramae* is more commonly found in northern bobwhite in Texas (Demarais et al. 1987; Purvis et al. 1998) and is not known to be pathogenic, that is, able to cause disease (Freehling and Moore 1993).

Research from Texas has identified two helminths as potentially pathogenic: the eyeworm (*Oxyspirura petrowi*) and the cecal worm (*Aulonocephalus pennula*). The eyeworm was first reported in Texas in scaled quail and northern bobwhite in the Rolling Plains ecoregion (Table 11.3) and has been a central topic of study in the

Table 11.3 Literature review for four helminth species occurring in quails inhabiting rangelands of western United States

Parasite	Host	State	First reported	Highest reported prevalence	Prevalence N (%)
Aulnocephalus pennula	Bobwhite	TX	Webster and Addis (1945)	Dunham et al. 2017	123 (99.2)[a]
	Gambel's	NV	Gullion (1957)	Gullion (1957)	110 (24.0)
	Scaled	AZ	Canavan (1929)	Canavan (1929)	–
	Scaled	NM	Campbell and Lee (1953)	Campbell and Lee (1953)	–
	Scaled	TX	Canavan (1929)	Howard 1981	240 (100.0)
Dispharynx nasuta	Bobwhite	TX	Purvis et al. (1998)	Purvis et al. (1998)	5 (62.0)
	California	OR	Moore et al. (1989)	Moore et al. (1989)	80 (38.0)
	Gambel's	AZ	Gorsuch (1934)	Gorsuch (1934)	–
Oxyspirura petrowi	Bobwhite	TX	Jackson and Greene (1965)	Dunham et al. 2017	125 (95.2)[a]
	Gambel's	AZ	Dunham and Kendall (2017)	Dunham and Kendall (2017)	59 (1.7)
	Montezuma	TX	Pence (1975)	Pence (1975)	3 (67.0)
	Scaled	NM	Dunham and Kendall (2017)	Dunham and Kendall (2017)	53 (28.3)
	Scaled	TX	Wallmo (1956)	Dunham et al. 2017	33 (72.7)[b]
Trichostrongulus tenuis	Mountain	OR	Moore et al. (1989)	Moore et al. (1989)	2 (100.0)

past decade (Bruno et al. 2015; Dunham et al. 2016a, b; Kalyanasundaram et al. 2019; Henry et al. 2020). Concern about the eyeworm arose with the identification of a higher prevalence (95%; Dunham et al. 2016a) and a greater intensity of infection (i.e., 90–100 individuals) in northern bobwhite in the Rolling Plains of Texas than previously reported (30 individuals, Jackson and Greene 1965). Surveys have reported eye worms in scaled quail (Wallmo 1956; Dancak et al. 1982; Landgrebe et al. 2007; Fedynich et al. 2019), Gambel's quail (Dunham and Kendall 2017), and Montezuma quail (Pence 1975) in western Texas, although in lower intensities of infection (Table 11.3).

The cecal worm has garnered similar attention for its high prevalence and intensity of infection. Over 500 worms in an individual host have been reported from northern bobwhite (Dunham et al. 2016a; Bruno et al. 2018) and scaled quail (Fedynich et al. 2019) from Texas. The cecal worm is free floating and does not appear to attach to the cecal wall; however, a disruption in regular feed intake or digestion could negatively impact the host, particularly during times of increased stress. Cecal worms have been reported in scaled quail and Gambel's quail from Nevada and Arizona, but in lower prevalence and intensities (Table 11.3).

11.7 Ecosystem Threats

11.7.1 Habitat Loss

Habitat loss and fragmentation are considered leading causes of global declines and extinctions of species, and these factors also threaten quails on western rangelands (Brennan 1991; Church et al. 1993; Hernández et al. 2013). Habitat loss for quails can occur in at least two forms: (1) actual habitat loss due to factors such as urbanization where the total amount of habitat is reduced and (2) habitat loss via degradation of rangelands due to factors such as establishment of nonnative grasses where the total amount of habitat may remain the same, but the suitability in portions of the existing habitat declines. Habitat loss due to degradation may involve processes such as establishment of nonnative grasses, encroachment of woody plants, and overgrazing and often is amendable by management, albeit sometimes costly. Here we focus on habitat loss due to degradation, the second type.

Quail populations decline when components of important vegetation communities are altered or degraded. The endangered masked bobwhite is thought to have been extirpated in Arizona because of overgrazing by livestock and the accompanying invasion of shrubs (Engel-Wilson and Kuvlesky 2002). California quail and Montezuma quail declines also have been attributed to habitat loss due to overgrazing of herbaceous cover on rangelands and forested savannas, respectively (Brennan 1994). Brennan (1994) believed that intensive agriculture and the construction of hydroelectric reservoirs in the region where the Snake River and Columbia River meet (southeastern Washington, northwestern Idaho, northeastern Oregon), along with overgrazing of secondary riparian corridors, reduced important habitat for mountain quail sufficiently to cause population declines. Even northern bobwhite, which have broader habitat requirements than most other western quails, have experienced significant population declines due to habitat loss in the form of the proliferation of clean farming practices, high-density pine silviculture, and forest succession in the southeastern U.S. (Brennan 1991, 1994).

11.7.2   Invasive Species

Nonnative grass invasions have become a significant form of habitat loss for western quails. Grasses such as coastal Bermuda grass (*Cynodon dactylon*), buffelgrass (*Cenchrus ciliaris*), yellow bluestem (*Bothriochloa ischaemum*), Lehmann lovegrass (*Eragrostis lehmanniana*), and cheatgrass (*Bromus tectorum*) are all nonnative grass species that were introduced either intentionally or unintentionally to native-plant communities in western rangelands. The significance of nonnative grass invasions for quails is that they degrade quail habitat and negatively impact their abundance (Kuvlesky et al. 2012). Nonnative grasses can form dense monocultures that result in reduced forb diversity, grass diversity, arthropod abundance, and bare ground, thereby negatively impacting quail foraging, movements, and space use (Fulbright et al. 2019). Quail abundance therefore tends to be higher in rangelands dominated by native grasses, which provide higher quality habitat than nonnative grasses. For example, northern bobwhite were twice as abundant on areas dominated by native grass compared to areas dominated by buffelgrass or Lehmann lovegrass in southern Texas (Flanders et al. 2006). DeMaso and Dillard (2007) believed that the disappearance of northern bobwhite from the Cross Timbers and Prairies, Post Oak Savanna, and Blackland Prairie ecoregions of Texas partly could be attributed to the introduction and accompanying invasions of coastal Bermudagrass to tens of thousands of hectares. Additionally, Fulbright et al. (2019) reported that scaled quail avoided areas dominated by nonnative grasses and concluded that nonnative grasses could be responsible for declines in scaled-quail populations in southern Texas.

However, nonnative invasive grasses can be of use for quails in certain situations. Kuvlesky et al. (2012) noted that nonnative grasses provide quails with important escape, thermal, nesting, and brood cover, particularly in vegetation communities where these cover types are limited. They noted that the endangered masked bobwhite in Sonora, Mexico likely would not have persisted on the grazed rangelands of this state without buffelgrass, which provided essentially the only cover available to the subspecies. In Texas, northern bobwhite nest in buffelgrass (Buelow 2009; Sands et al. 2012) and use guineagrass (*Urochloa maxima*), another nonnative species, as loafing cover (Moore 2010). Nonnative grasses also do not appear to negatively impact Gambel's quail in Arizona given adequate shrub cover and bare ground (King 1998), and introduced California quail that were successfully established in Washington heavily relied on nonnative plants for food and cover (Crawford 1993). The impact that nonnative grasses have on quails likely depends on the species' life history and the degree by which the nonnative grass has established dominance in an area. For northern bobwhite, the threshold beyond which nonnative grasses such as buffelgrass and Lehmann lovegrass begin to negatively impact their habitat use appears to be ≥ 20% cover (Edwards 2019).

11.7.3 Climate Change

Climate models project that the Southwest and Central Plains of the U.S. will become drier during the twenty-first century, a transition that already appears underway (Archer and Predick 2008; Cook et al. 2015). These regions are projected to experience warmer temperatures and higher frequency of extreme weather events (e.g., droughts, heat waves, and floods; Archer and Predick 2008). For both the Southwest and Central Plains, the risk of multidecadal drought is expected to increase from < 12% (1950–2000) to ≥ 80% (2050–2099), a level of aridity that exceeds even the persistent megadroughts of the Medieval era (1100–1300 CE) (Cook et al. 2015). This projected change in climate may negatively impact western quails, particularly those species inhabiting semiarid and arid environments. The primary impacts likely will involve how quails respond to increasing temperatures and aridity, as well as accompanying distributional and compositional changes in vegetation communities resulting from climate change and projected increases in wildfire frequency (Heidari et al. 2021).

Quails inhabiting arid and semiarid environments live near their physiological limits. For example, the thermal neutral zone for northern bobwhite is estimated at 30–35 °C (Lustick et al. 1972; Forrester et al. 1998), with gular flutter occurring at 35.0–38.5 °C (Case and Robel 1974) and death at 40 °C if individuals are exposed to this temperature for a prolonged period of time (Case and Robel 1974). The thermal environment therefore strongly influences quail life history and ecology, and minor changes in climate can substantially influence their performance (Guthery et al. 2000; Burger et al. 2017). High temperatures are known to cause embryonic mortality (Reyna and Burggren 2012), reduce food intake (Case and Robel 1974), reduce egg laying (Case and Robel 1974), decrease productivity (Heffelfinger et al. 1999), and shorten the nesting season (Guthery et al. 1988). Quails can partly minimize the risk of thermal stress via modifications in space use. For example, northern bobwhite and scaled quail in Oklahoma and New Mexico nest in sites with temperatures that are 6–8 °C cooler than the available landscape (Carroll et al. 2018; Kauffman et al. 2021). However, such behavioral adjustments depend on the availability of thermally suitable sites, which can be limited even in the present climate (Kline et al. 2019; Palmer et al. 2021). The proportion of thermally suitable areas on a landscape may be as little as 40–60% during the hottest time of the day (Forrester et al. 1998) and may become even more limited in the future.

In addition to demographic responses of quails to climate change, quails also can respond by adjusting their geographic distribution because of compositional or distributional changes in vegetation communities. The National Audubon Society used their large-scale, bird-observation database and climate models to project how climate change may affect the geographic distributions of birds (www.audubon.org/climate/survivalbydegrees). Assuming a 3 °C increase in temperature as projected by climate models, 1 quail species is considered to possess high vulnerability (Montezuma quail), 1 moderate vulnerability (scaled quail), 2 low vulnerability (California quail and mountain quail), and 2 stable (northern bobwhite and Gambel's quail) relative to changes in their respective geographic distribution (Table 11.4). These projections agree in general with those of Tanner et al. (2017) who modeled changes in geographic distribution of western quails using an ensemble approach of four general circulation models. They documented that 4 of the 6 species (scaled quail, California quail, Montezuma quail, and mountain quail) are projected to have a net loss in area of geographic distribution. The geographic distributions of Montezuma quail and mountain quail are projected to shift higher in elevation as potential distribution contractions occur in lower latitudes and gains occur in higher latitudes. The net change in the geographic distribution of northern bobwhite is projected to be minimal; however, the species is projected to lose population strongholds. Gambel's quail is the only species projected to experience an increase in area of geographic distribution. Collectively, the geographic distributions of western quails are projected to be displaced northward and eastward, with losses in their southernmost extents (Tanner et al. 2017).

Table 11.4 Projected changes in the geographic distribution of western quails as reported by the National Audubon Society (www.audubon.org/climate/survivalbydegrees) based on a 3 °C increase in temperature

Common name	Species vulnerability	Geographic distribution gained (%)	Geographic distribution maintained (%)	Geographic distribution lost (%)
Northern bobwhite	Stable	37	90	11
Scaled quail	Moderate	28	72	28
Gambel's quail	Stable	56	92	8
California quail	Low	49	57	43
Montezuma quail	High	6	26	74
Mountain quail	Low	56	52	48

11.8   Conservation and Management Actions

The rangelands that western quails inhabit represent a mix of ownerships including federal government, state governments, local municipalities, tribes, corporations, and private individuals (USGS GAP 2018). The differing management authorities among these entities can create a disconnect in conservation objectives for quails. Additionally, wildlife species do not recognize jurisdictional boundaries, further complicating management of western quails. Collaborative efforts among these managing entities have had, and will continue to have, the greatest potential for quail conservation and management in western rangelands.

11.8.1   *Conservation Programs for Public Rangelands*

The federal government manages a substantial proportion of western lands, and some federal agencies operate under directives to manage lands for multiple uses including the provision of fish and wildlife habitat (Vincent et al. 2020). It is estimated that the Bureau of Land Management (BLM) alone contains more than 8.1 million hectares of quail habitat: 4.9 million hectares (Gambel's quail), 1.6 million hectares (scaled quail), 1.2 million hectares (California quail), 1.1 million acres (mountain quail), 0.5 million hectares (northern bobwhite), and 110,000 ha (Montezuma quail; Sands et al. 1992). This large holding of quail habitat represents great potential for management and opportunities for federal and state agency collaboration on quail management and conservation. The Sikes Act of 1974 (Public Law 93-452) provides one avenue for collaborative funding for wildlife habitat on federal lands by requiring people who hunt, fish, or trap on certain federal lands to purchase a stamp that provides funding for the conservation and restoration of these lands (Public Law 93-452). New Mexico created the Habitat Stamp Program in 1986 under the federal Sikes Act and since then has raised more than $26 million dollars and completed more than 2000 projects, some of which have benefitted quails (NMDGF 2017).

In addition to routine habitat management on federal lands within the geographic distributions of quails, federal agencies also are able to create initiatives aimed at specific species or habitats. "Answer the Call" was one such initiative directed at managing habitats for quails on federal lands. Started in 1988 as part of the U.S. Forest Service's (USFS) Get Wild program, "Answer the Call" was directed to make improvements to quail habitat on National Forest System lands (USDA 1991). The USFS collaborated with Quail Unlimited (a former non-government organization), BLM, and National Fish and Wildlife Foundation to implement this program and improve over 80,000 hectares of quail and associated wildlife habitat on National Forests across the U.S. (USDA 2004). "Answer the Call" is still available through the USFS but Quail Unlimited disbanded in 2013 thereby slowing the implementation of the program.

11.8.2 Conservation Programs for Private Rangelands

Despite the fact that a smaller proportion of western rangelands is privately owned (Vincent et al. 2020), private lands have conservation value for western quails. Much of rural, private land is used for agricultural purposes (Robertson and Swinton 2005), and land-use decisions generally are made by landowners to support their livelihoods and families (Heard 2000). Such heavy reliance of these private rangelands on agricultural use has earned them the name of "working lands" (i.e., privately owned land in agricultural production) (Naugle et al. 2020).

Conservation of wildlife species on working lands, specifically grassland birds such as quails, can be achieved through voluntary conservation efforts by private landowners that are supported by strong partnerships between landowners and resource professionals (Drum et al. 2015). Conservation programs or initiatives for private lands must consider socioeconomic factors and how they impact landowner decisions-making (Drum et al. 2015). Conservation practices that are cost-effective, sustainable, and compatible with agricultural systems are often attractive to landowners (Burger et al. 2006, 2019). For example, private landowners in Texas placed great importance on minimizing out-of-pocket costs and labor input when making decisions about whether and how to restore northern bobwhite habitat (Valdez et al. 2019).

The 1985 Food Security Act (Farm Bill) is "an omnibus, multiyear law that governs an array of agricultural and food programs", including conservation incentive programs (Stubbs 2019). The U.S. Farm Bill provides private landowners cost-share payments for implementing United States Department of Agriculture (USDA) conservation practices (Briske et al. 2017). Thus, Farm Bill programs are a primary vehicle for implementing quail conservation on private lands (Burger et al. 2006), and the primary land conservation program of the Farm Bill is the Conservation Reserve Program (CRP).

The Conservation Reserve Program provides compensation to private landowners who voluntarily remove lands from agricultural production to improve soil and water quality (Stubbs 2019). The initial impact of CRP on quails has varied by region and method of implementation (Burger 2006). In the Midwest, CRP lands planted to native grasses were extremely beneficial to quails, but CRP lands planted to nonnative grasses or enrolled in tree planting practices produced minimal benefits for quails (Burger 2000, 2006). In addition, the disturbance frequency and intensity of mid-contract management that CRP requires may not provide the level of disturbance needed to create the greater habitat heterogeneity that species such as the northern bobwhite require (Pavlacky et al. 2021). However, the Continuous CRP provides an option to create a more species-directed approach to the program and, in 2004, a new continuous CRP practice (CP33–Habitat Buffers for Upland Birds) was announced (Burger et al. 2006). In these 10-year contracts, field buffers are planted with native grass, forb, and shrub mixes, or re-established through natural succession (USDA FSA 2010) and followed up with site disturbance (mid-contract management) to maintain early successional habitat (Burger et al. 2006). The CP33 practice has provided habitat for quails while compensating landowners for removing hard-to-farm lands from production (Burger et al. 2006).

The two largest working lands programs of the Farm Bill are the Environmental Quality Incentives Program (EQIP) and the Conservation Stewardship Program (CSP; Stubbs 2019). These programs financially incentivize landowners to adopt conservation practices on their privately owned lands (Burger et al. 2019), and research indicates that northern bobwhite have responded positively to buffers, creation of early succession habitat, and restoration of native grasslands when managed to maintain appropriate vegetative structure (USDA NRCS 2009).

Another important collaborative program created by the 2014 Farm Bill is the Regional Conservation Partnership Program (RCPP) whereby conservation partners select an area of concern, determine conservations goals, and implement conservation practices using funding provided by Farm Bill and partners (Stubbs 2019). The RCPP has potential for large-scale conservation of western quail habitat. For example, the Oaks and Prairies Joint Venture received RCPP funding to implement its Grassland Restoration Incentive Program in Texas and Oklahoma that has potential to positively impact northern bobwhite (NBCI 2018).

Factors that may limit the effectiveness of Farm Bill conservation efforts on private lands are the lack of documented outcomes and staff capacity at USDA offices. Briske et al. (2017) concluded that the existing conservation practice standards are insufficient to conserve rangelands at a large scale and recommends that USDA-NRCS modify conservation programs to incorporate evidence-based conservation, including collaborative monitoring of conservation practices to understand environmental outcomes. To address the staff capacity issue, local, state, private, and federal partners have created partner biologist positions to provide technical assistance and work with private landowners to promote USDA conservation programs (PLJV 2019). These partner positions often work in local USDA service centers and provide technical and financial assistance to private landowners for habitat improvements (PLJV 2019). The non-government organization Pheasants Forever/Quail Forever has created 188 positions in 30 states to maximize implementation of USDA conservation programs (Burger et al. 2019), thereby indicating that non-government organizations will be increasingly important in the future.

11.8.3 Conservation Partnerships

The Association of Fish and Wildlife Agencies (AFWA) and their regional affiliates (WAFWA, MAFWA, SEAFWA, NEAFWA) agencies have been critical in facilitating meetings among wildlife managers, funding collaborative efforts, and providing staff to assist in multijurisdictional management. As conservation issues arise, the associations create working groups or technical committees comprised of state biologists or other wildlife professionals. The development of such groups provides collaborative opportunities for biologists working throughout the geographic distributions of quails.

11.8.3.1   National Bobwhite Conservation Initiative

During 1980–1999, northern bobwhite populations declined by an estimated 65.8% across their geographic distribution (Dimmick et al. 2002). This decline led the Southeast Association of Fish and Wildlife Agencies to task the Southeast Quail Study Group with creating a plan for the recovery of northern bobwhite (Dimmick et al. 2002) resulting in the National Bobwhite Conservation Initiative (NBCI) in 2002 (Dimmick et al. 2002). The NBCI was the first collaborative effort to create a range-wide management plan for northern bobwhite (NBTC 2011), and the NBCI now partners with a variety of federal, non-governmental, and academic organizations to carry out its mission.

11.8.3.2   Western Quail Working Group

Following the successes from the NBCI, the Resident Game Bird Working Group of the AFWA directed the creation of the Western Quail Management Plan (Zornes and Bishop 2009). This plan was a collaborative effort of biologists across the West to compile and evaluate information on western quails at both their geographic distributions and individual Bird Conservation Regions (BCR). Information provided for each BCR included population size, habitat abundance, current threats, management recommendations, and research needs. Following the finalization of the Western Quail Management Plan in 2009, the WAFWA signed a memorandum of understanding to create the Western Quail Working Group (WQWG; WAFWA 2011) and help foster cooperation across state lines to effectively manage species at regional scales (WAFWA 2011).

11.8.3.3   Joint Ventures

Bird Habitat Joint Ventures were established in the late-1980s to provide coordinated conservation planning for migratory birds at regional scales (USFWS 2005). There are currently 18 Bird Habitat Joint Ventures that encompass most of the U.S. and are comprised of self-directed partnerships between government and non-government organizations, corporations, and private individuals (Faaborg et al. 2010; Giocomo et al. 2012). Joint Venture administrative boundaries are primarily defined by Bird Conservation Regions boundaries (Giocomo et al. 2012). Given that both the NBCI and Western Quail Management Plan delineate quail management objectives by Bird Conservation Regions, Joint Ventures are well positioned to aid in quail conservation efforts. Since their inception, Joint Ventures have facilitated collaboration among > 5700 partners and assisted in habitat conservation on 10.9 million acres (USFWS 2018). Although created to focus on migratory birds, many regional Joint Ventures include non-migratory species such as northern bobwhite as priority species. In 2017, 7 of the 12 Joint Ventures that occur within the geographic distribution of northern bobwhite listed it as a priority species (DeMaso 2017).

11.9 Research Needs

Although game species tend to be well studied, the dynamic nature of western range-lands and the increasing human footprint create a perennial need to address emerging issues. We provide general research and management priorities for quails as a taxon and at the scale of their geographic distributions. From a demographic perspective, the need exists to quantify the cumulative impact of climate change, landscape alterations (e.g., habitat loss and fragmentation, non-native grasses, large wildfires), and demographic processes (e.g., dispersal, predation, disease) on quail-population viability. Investigations on population genetics of quails also are necessary to develop a more thorough understanding of genetic relatedness, taxonomy, and evolutionary history of quails to aid in their conservation efforts. From a management perspective, research on quail response to rangeland-management practices is limited in scope (1–2 species) and geographic extent (mostly Texas, Oklahoma, and New Mexico) and warrants investigation. In addition, the need exists to develop effective management strategies for invasive, nonnative grasses. Reliable monitoring techniques also are needed for quails that can be applied at both small and large spatial extents, especially for species such as Montezuma quail and mountain quail that have low detection probabilities. In recent years, the translocation of wild quails to restore declining populations of western quails has received research attention (Troy et al. 2013; Downey et al. 2017; Ruzicka et al. 2017) but warrants further evaluation to determine the viability of the technique as an effective conservation tool.

References

Anderson WL (1974) Scaled quail: social organization and movements. Master's thesis, University of Arizona

Archer SR, Predick KI (2008) Climate change and ecosystems of the southwestern United States. Rangelands 30:23–28. https://doi.org/10.2111/1551-501X(2008)30[23:CCAEOT]2.0.CO;2

Beck JL, Reese KP, Zager P et al (2005) Simultaneous multiple clutches and female breeding success in mountain quail. Condor 107:891–899. https://doi.org/10.1093/condor/107.4.889

Blankenship LH, Reed RE, Irby HD (1966) Pox in mourning doves and Gambel's quail in southern Arizona. J Wildl Manage 30:253–257

Brennan LA (1991) How can we reverse the northern bobwhite population decline? Wildl Soc Bull 19:544–555

Brennan LA (1994) Broad–scale population declines in four species of North American quail: an examination of possible causes. In: Covington WW, Debano LF (eds) Sustainable ecological systems: implementing an ecological approach to land management. USDA Forest Service General Technical Report RM–247, USDA Forest Service, Fort Collins, Colorado, pp 45–50

Brennan LA (2007) Texas quails; ecology and management. Texas A&M University Press, College Station

Brennan LA, Block WM, Gutiérrez RJ (1987) Habitat use by mountain quail in northern California. Condor 89:66–74. https://doi.org/10.2307/1368760

Brennan LA, Hernández F, Williford D (2020) Northern Bobwhite (*Colinus virginianus*). In: Poole AF (ed) Birds of the world, vers. 1.0. Cornell Lab of Ornithology, Ithaca, New York. https://doi.org/10.2173/bow.norbob.01

Bridges AS, Peterson MJ, Silvy NJ et al (2001) Differential influence of weather on regional quail abundance in Texas. J Wildl Manage 65:10–18

Briske DD, Bestelmeyer BT, Brown JR et al (2017) Assessment of USDA-NRCS rangeland conservation programs: recommendation for an evidence-based conservation platform. Ecol Appl 27:94–104. https://doi.org/10.1002/eap.1414

Bristow KD, Ockenfels RA (2006) Fall and winter habitat use by scaled quail in southeastern Arizona. Rangel Ecol Manag 59:308–313. https://doi.org/10.2111/04-117R2.1

Brown RL (1982) Effects of livestock grazing on Mearns' quail in southeastern Arizona. J Range Manage 35:727–732

Brown DE (1989) Arizona game birds. University of Arizona Press, Tucson

Brown DE, Sands A, Clubine S et al (1993) Appendix A: grazing and range management. Proc Natl Quail Symp 3:176–177

Bruno A (2018) Monitoring vegetation and northern bobwhite density in a grazing demonstration project in South Texas. Dissertation, Texas A&M University, Kingsville

Bruno A, Fedynich AM, Smith-Herron A et al (2015) Pathological response of northern bobwhites to *Oxyspirura petrowi* infections. J Parasitol 101:364–368.https://doi.org/10.1645/14-526.1

Bruno A, Fedynich AM, Rollins D et al (2018) Helminth community and host dynamics in northern bobwhites from the Rolling Plains ecoregion, U.S.A. J Helminthol, 1–7. https://doi.org/10.1017/S0022149X18000494

Buelow MC (2009) Effects of tanglehead on northern bobwhite habitat use. Master's thesis, Texas A&M University-Kingsville

Burger LW Jr (2000) Wildlife responses to the Conservation Reserve Program in the Southeast. In: Hohman WL (ed) A comprehensive review of Farm Bill contributions to wildlife conservation 1985–2000. Technical report, USDA/NRCS/WHMI-2000, U.S. Department of Agriculture, Natural Resources Conservation Service, Wildlife Habitat Institute, Madison, Mississippi, pp 55–74

Burger LW Jr (2006) Creating wildlife habitat through federal farm programs: an objective-driven approach. Wildl Soc Bull 34:994–999. https://doi.org/10.2193/0091-7648(2006)34[994:CWHTFF]2.0.CO;2

Burger LW Jr, McKenzie D, Thackston R et al (2006) The role of farm policy in achieving large-scale conservation: bobwhite and buffers. Wildl Soc Bull 34:986–993. https://doi.org/10.2193/0091-7648(2006)34[986:TROFPI]2.0.CO;2

Burger LW Jr, Dailey TV, Ryan MR et al (2017) Effect of temperature and wind on metabolism of northern bobwhite in winter. Proc Natl Quail Symp 8:300–307

Burger LW Jr, Evans KO, McConnell MD et al (2019) Private lands conservation: a vision for the future. Wildl Soc Bull 43:1–10. https://doi.org/10.1002/wsb.1001

Cain JR, Beasom SL, Rowland LO et al (1982) The effects of varying dietary phosphorus on breeding bobwhites. J Wildl Manage 46:1061–1065

Cain JR, Lien RJ, Beasom SL (1987) Phytoestrogen effects on reproductive performance of scaled quail. J Wildl Manage 51:198–201

California Department of Fish and Wildlife. 2017. California Wildlife Habitat Relationships. https://wildlife.ca.gov/Data/CWHR

Calkins JD, Gee JM, Hagelin JC et al (2020) California quail (*Callipepla californica*). In: Poole A (ed) Birds of the world, vers. 1.0. Cornell Lab of Ornithology, Ithaca, New York. https://doi.org/10.2173/bow.calqua.01

Campbell H (1960) An evaluation of gallinaceous guzzlers for quail in New Mexico. J Wildl Manage 24:21–26

Campbell H, Lee L (1953) Studies on quail malaria in New Mexico and notes on other aspects of quail populations (No. 3). New Mexico Department of Game and Fish, Santa Fe, New Mexico

Campbell H, Harris BK (1965) Mass population dispersal and long-distance movements: scaled quail. J Wildl Manage 29:801–805

Canavan WP (1929) Nematode parasites of vertebrates in the Philadelphia Zoological Garden and vicinity. Parasitology 21:63–102. https://doi.org/10.1017/S0031182000022794

Carroll RL, Davis CA, Fuhlendorf SD et al (2018) Avian parental behavior and nest success influenced by temperature fluctuations. J Therm Biol 74:140–148. https://doi.org/10.1016/j.jtherbio.2018.03.020

Case RM, Robel RJ (1974) Bioenergetics of the bobwhite. J Wildl Manage 38:638–652

Church KE, Sauer JR, Droege S (1993) Population trends of quails in North America. Proc Natl Quail Symp 3:44–54

Collins PW (2008) Catalina California quail (*Callipepla californica catalinensis*). In: Shuford WD and Gardali T (eds) California bird species of special concern. California Department of Fish and Game, Studies of Western Birds No. 1, Sacramento, California, pp 107–111

Cook BI, Ault TR, Smerdon JE (2015) Unprecedented 21st century drought risk in the American Southwest and Central Plains. Sci Adv 1:e1400082. https://doi.org/10.1126/sciadv.1400082

Crawford JA (1993) California quail in western Oregon: a review. Proc Natl Quail Symp 3:1–7

Crooks KR, Suarez AV, Bolger DT (2004) Avian assemblages along a gradient of urbanization in a highly fragmented landscape. Biol Conserv 115:451–462. https://doi.org/10.1016/S0006-3207(03)00162-9

Curtis PD, Mueller BS, Doerr PD et al (1993) Potential polygamous breeding behavior in northern bobwhite. Proc Natl Quail Symp 3:55–63

Dabbert CB, Pleasant G, and Schemnitz SD (2020) Scaled quail (*Callipepla squamata*). In: Poole AF (ed) Birds of the world, vers.1.0. Cornell Lab of ornithology, Ithaca, New York. doi–org.proxy.osl.state.or.us/https://doi.org/10.2173/bow.scaqua.01

Dancak K, Pence DB, Stormer FA et al (1982) Helminths of the scaled quail, *Callipepla squamata*, from northwest Texas. Proc Helminthol Soc Wash 49:144–146

Davidson WR, Kellogg FE, Doster GL (1982) Avian pox infections in southeastern bobwhites: historical and recent information. Proc Natl Quail Symp 2:64–68

Davis CA, Orange JP, Van Den Bussche RA et al (2017) Extrapair paternity and nest parasitism in two sympatric quail. Auk 134:811–820. https://doi.org/10.1642/AUK-16-162.1

Delehanty DJ, Eaton SS, Campbell TG (2004) Mountain quail fidelity to guzzlers in the Mojave Desert. Wildl Soc Bull 32:588–593. https://doi.org/10.2193/0091-7648(2004)32[588:FTFMQF]2.0.CO;2

Demarais S, Everett DD, Pons ML (1987) Seasonal comparison of endoparasites of northern bobwhites from two types of habitat in southern Texas. J Wildl Dis 23:256–260. https://doi.org/10.7589/0090-3558-23.2.256

DeMaso SJ (2017) The role of Joint Ventures in northern bobwhite conservation. Proc Natl Quail Symp 8:117

DeMaso SJ, Dillard J (2007) Bobwhites on the cross timbers and prairies. In: Brennan LA (ed) Texas quails: ecology and management. Texas A&M University Press, College Station, pp 142–155

DeMaso SJ, Peterson MJ, Purvis JR et al (2002) A comparison of two quail abundance indices and their relationship to quail harvest in Texas. Proc Natl Quail Symp 5:206–2012

Dimmick RW, Gudlin MJ, McKenzie DF (2002) The northern bobwhite conservation initiative. Publication of the Southeastern Association of Fish and Wildlife Agencies, Columbia, South Carolina, p 96

Downey MC, Rollins D, Hernández F et al (2017) An evaluation of northern bobwhite translocation to restore populations. J Wildl Manage 81:800–813. https://doi.org/10.1002/jwmg.21245

Drum RG, Ribic CA, Koch K et al (2015) Strategic grassland bird conservation throughout the annual cycle: linking policy alternative, landowner decisions, and biological population outcomes. PLoS ONE 10:e0142525. https://doi.org/10.1371/journal.pone.0142525

Dunham NR, Kendall RJ (2017) Eyeworm infections of *Oxyspirura petrowi*, Skrjabin, 1929 (Spirurida: Thelaziidae), in species of quail from Texas, New Mexico and Arizona, USA. J Helminthol 91:491–496. https://doi.org/10.1017/S0022149X16000468

Dunham NR, Soliz LA, Fedynich AM et al (2014) Evidence of an *Oxyspirura petrowi* epizootic in northern bobwhites (*Colinus virginianus*). J Wildl Dis 50:552–558. https://doi.org/10.7589/2013-10-275

Dunham NR, Bruno A, Almas S et al (2016a) Eyeworms (*Oxyspirura petrowi*) in northern bobwhites (*Colinus virginianus*) from the Rolling Plains ecoregion of Texas and Oklahoma, 2011–2013. J Wildl Dis 52:562–567. https://doi.org/10.7589/2015-04-103

Dunham NR, Reed S, Rollins D et al (2016b) *Oxyspirura petrowi* infection leads to pathological consequences in northern bobwhite (*Colinus virginianus*). Int J Parasitol Parasites Wildl 5:273–276. https://doi.org/10.1016/j.ijppaw.2016.09.004

Dunham NR, Heny C, Brym M et al (2017) Caecal worm, *Aulonocephalus pennula*, infection in the northern bobwhite quail, *Colinus virginianus*. Int J Parasitol Parasites Wildl 6:35–38. https://doi.org/10.1016/j.ijppaw.2017.02.001

Edwards JT (2019) Habitat, weather, and raptors as factors in the northern-bobwhite and scaled-quail population declines. Dissertation, Texas A&M University-Kingsville

Engel-Wilson R, Kuvlesky WP Jr (2002) Arizona quail: species in jeopardy? Proc Natl Quail Symp 5:1–7

Faaborg J, Holmes RT, Anders AD et al (2010) Conserving migratory land birds in the New World: do we know enough? Ecol Appl 20:398–418. https://doi.org/10.1890/09-0397.1

Fedynich AM, Bedford K, Rollins D et al (2019) Helminth fauna in a semi-arid host species–scaled quail (*Callipepla squamata*). J Helminthol 1–5.https://doi.org/10.1017/S0022149X19000580

Flanders AA, Kuvlesky WP Jr, Ruthven DC III et al (2006) Effects of invasive exotic grasses on South Texas rangeland breeding birds. Auk 123:171–182. https://doi.org/10.1093/auk/123.1.171

Forrester ND, Guthery FS, Kopp SD et al (1998) Operative temperature reduces habitat space for northern bobwhites. J Wildl Manage 62:1505–1510

Francis WJ (1965) Double broods in California quail. Condor 67:541–542. https://doi.org/10.1093/condor/67.6.541

Francis WJ (1970) The influence of weather on population fluctuations in California quail. J Wildl Manage 34:249–266

Freehling M, Moore J (1993) Host specificity of *Trichostrongylus tenuis* from red grouse and northern bobwhites in experimental infections of northern bobwhites. J Parasitol 79:538–541. https://doi.org/10.2307/3283379

Fulbright TE, Kline HN, Wester DB et al (2019) Non-native grasses reduce scaled quail habitat. J Wildl Manage 83:1581–1591. https://doi.org/10.1002/jwmg.21731

Gee JM, Brown DE, Hagelin JC et al (2020) Gambel's quail (*Callipepla gambelii*). In: Poole A (ed) Birds of the world, vers. 1.0. Cornell Lab of Ornithology, Ithaca, New York. https://doi.org/10.2173/bow.gamqua.01

Giocomo JJ, Gustafson M, Duberstein JN, Boyd C (2012) The role of joint ventures in bridging the gap between research and management. In: Sands JP, DeMaso SJ, Schnupp MJ et al (eds) Wildlife science connecting research with management. CRC Press, Boca Raton, Florida, pp 239–252

Goodwin JG Jr, Hungerford CR (1977) Habitat use by native Gambel's and scaled quail and released masked bobwhite quail in southern Arizona. USDA Forest Service Research Paper RM–197, Fort Collins, Colorado

Gorsuch DM (1934) Life history of the Gambel's quail in Arizona. Biol Sci Bull 5(4), 2:1–89

Grahmann ED, Fulbright TE, Hernández F et al (2018) Demographic and density response of northern bobwhites to pyric herbivory of non-native grasslands. Rangel Ecol Manag 71:458–469. https://doi.org/10.1016/j.rama.2018.02.008

Gullion GW (1956) Evidence of double-brooding in Gambel quail. Condor 58:232–234. https://doi.org/10.2307/1364678

Gullion GW (1957) Gambel's quail disease and parasite investigations in Nevada. Am Midl Nat 57:414–420. https://doi.org/10.2307/2422407

Gullion GW (1962) Organization and movements of coveys of a Gambel's quail population. Condor 64:402–415. https://doi.org/10.2307/1365548

Guillon GW, Christiansen GC (1957) A review of the distribution of gallinaceous birds of Nevada. Condor 59:128–138. https://doi.org/10.2307/1364574

Guthery FS (1999) Slack in the configuration of habitat patches for northern bobwhites. J Wildl Manage 63:245–250

Guthery FS (2000) On bobwhites. Texas A&M University Press, College Station

Guthery FS (2002) The technology of bobwhite management: the theory behind the practice. Iowa State University Press, Ames

Guthery FS, Kuvlesky WP Jr (1998) The effect of multiple-brooding on age ratios of quail. J Wildl Manage 62:540–549. https://doi.org/10.2307/3801075

Guthery FS, Brennan LA (2007) The science of quail management and the management of quail science. In: Brennan LA (ed) Texas Quails Book. Texas A&M University Press, College Station, pp 407–420

Guthery FS, Koerth NE, Smith DS (1988) Reproduction of northern bobwhites in semiarid environments. J Wildl Manage 52:144–149

Guthery FS, Forrester ND, Nolte KR et al (2000) Potential effects of global warming on quail populations. Proc Natl Quail Symp 4:198–204

Guthery FS, King NM, Kuvlesky WP et al (2001) Comparative habitat use by three quails in desert grassland. J Wildl Manage 65:850–860. https://doi.org/10.2307/3803034

Guthery FS, Rybak AR, Fuhlendorf SD et al (2005) Aspects of the thermal ecology of bobwhites in North Texas. Wildl Monogr 159.https://doi.org/10.2193/0084-0173(2004)159[1:AOTTEO]2.0. CO;2

Gutiérrez RJ (1980) Comparative ecology of the Mountain and California Quail in Carmel Valley, California. Living Bird 18:71–93

Gutiérrez RJ, Delehanty DJ (2020) Mountain Quail (*Oreortyx pictus*). In: Poole A, Gill FB (eds) Birds of the world, vers. 1.0. Cornell Lab of Ornithology, Ithaca, New York. https://doi.org/10. 2173/bow.mouqua.01

Harveson LA (1995) Nutritional and physiological ecology of reproducing northern bobwhites in southern Texas. Master's thesis, Texas A&M University-Kingsville

Harveson LA, Allen TH, Hernández F et al (2007) Montezuma quail ecology and life history. In: Brennan LA (ed) Texas quails: ecology and management, 1st edn. Texas A&M University Press, College Station, pp 23–39

Heard LP (2000) Introduction. In: Hohman WL, Halloum DJ (eds) A comprehensive review of Farm Bill contributions to wildlife conservation, 1985–2000. U.S. Department of Agriculture, Washington, DC, pp 1–4

Heffelfinger JR, Guthery FS, Olding RJ et al (1999) Influence of precipitation timing and summer temperatures on reproduction of Gambel's quail. J Wildl Manage 63:154–161. https://doi.org/10.2307/3802496

Heidari H, Arabi M, Warziniack T (2021) Effects of climate change on natural-caused fire activity in western US national forests. Atmosphere 12:981. https://doi.org/10.3390/atmos12080981

Henry C, Kalyanasundaram A, Brym MZ et al (2020) Molecular identification of *Oxyspirura petrowi* intermediate hosts by nested PCR using internal transcribed Spacer 1 (ITS1). J Parasitol 106:46–52. https://doi.org/10.1645/19-135

Hernández F (2020) Ecological discord and the importance of scale in scientific inquiry. J Wildl Manage 84:1427–1434. https://doi.org/10.1002/jwmg.21942

Hernández F, Peterson MJ (2007) Northern bobwhite ecology and life history. In: Brennan LA (ed) Texas quails: ecology and management. Texas A&M University Press, College Station, pp 40–64

Hernández F, Guthery FS (2012) Beef, brush, and bobwhites: quail management in cattle country. Texas A&M University Press, College Station

Hernández F, Guthery FS, Kuvlesky WP Jr (2002) The legacy of bobwhite research in South Texas. J Wildl Manage 66:1–18. https://doi.org/10.2307/3802866

Hernández F, Arredondo JA, Hernández F et al (2005) Influence of weather on population dynamics of northern bobwhite. Wildl Soc Bull 33:1071–1079

Hernández F, Kuvlesky WP Jr, DeYoung RW et al (2006a) Recovery of a rare species: case study of the masked bobwhite. J Wildl Manage 70:617–631. https://doi.org/10.2193/0022-541X(2006)70[617:RORSCS]2.0.CO;2

Hernández F, Harveson LA, Hernández FCE et al (2006b) Habitat characteristics of Montezuma quail foraging areas in Trans-Pecos Texas. Wildl Soc Bull 34:856–860. https://doi.org/10.2193/0091-7648(2006)34[856:HCOMQF]2.0.CO;2

Hernández F, Perez RM, Guthery FS (2007) Bobwhites on the South Texas Plains. In: Brennan LA (ed) Texas quails: ecology and management, 1st edn. Texas A&M University Press, College Station, pp 273–298

Hernández F, Brennan LA, DeMaso SJ et al (2013) On reversing the northern bobwhite population decline: 20 years later. Wildl Soc Bull 37:177–188. https://doi.org/10.1002/wsb.223

Hernández F, Perez RM, Guthery FS et al (In Press) Quails on the South Texas Plains. In: Brennan LA, Hernández F (eds) Texas quails: ecology and management, second edition. Texas A&M University Press, College Station

Holechek JL (1981) Livestock grazing impacts on public lands: a viewpoint. J Range Manage 34:251–254

Howard MO (1981) Food habits and parasites of scaled quail in southeastern Pecos County, Texas. Dissertation, Sul Ross State University

Hudson PJ, Dobson AP, Newborn D (1998) Prevention of population cycles by parasite removal. Science 282:2256–2258. https://doi.org/10.1126/science.282.5397.2256

Hungerford CR (1955) A preliminary evaluation of quail malaria in southern Arizona in relation to habitat and quail mortality. Trans N Am Wildl Conf 20:209–219

Hungerford CR (1960) The factors affecting the breeding of Gambel's quail *Lophortix gambelii* in Arizona. Dissertation, University of Arizona

Hungerford CR (1964) Vitamin A and productivity in Gambel's quail. J Wildl Manage 28:141–147

Idaho Department of Fish and Game (2019) Idaho upland game management plan, 2019–2025, Idaho

Iknayan KJ, Wheeler MM, Safran SM et al (2021) What makes urban parks good for California quail? Evaluating park suitability, species persistence, and the potential for reintroduction into a large urban national park. J Appl Ecol. https://doi.org/10.1111/1365-2664.14045

Jackson AS, Green H (1965) Dynamics of bobwhite quail in the west Texas Rolling Plains: parasitism in bobwhite quail. Texas Parks and Wildlife Department, Federal Aid Project No. W–88–R–4, Austin, Texas

Janke AK, Terhune TM, Gates RJ et al (2017) Northern bobwhite population responses to winter weather along their northern range. Wildl Soc Bull 41:479–488. https://doi.org/10.1002/wsb.779

Jewett SG, Taylor WP, Aldrich JW (1953) Birds of Washington State. University of Washington Press, Seattle

Kalyanasundaram A, Brym MZ, Blanchard KR et al (2019) Life-cycle of *Oxyspirura petrowi* (*Spirurida: Thelaziidae*), an eyeworm of the northern bobwhite quail (*Colinus virginianus*). Parasit Vectors 12:2–10. https://doi.org/10.1186/s13071-019-3802-3

Kamees L, Mitchusson T, Gruber M (2008) New Mexico's quail: biology, distribution, and management recommendations. New Mexico Department of Game and Fish

Kauffman KL, Elmore RD, Davis CA et al (2021) Role of the thermal environment in scaled quail (*Callipepla squamata*) nest site selection and survival. J Therm Biol 95:102791. https://doi.org/10.1016/j.jtherbio.2020.102791

Kellogg FE, Prestwood AK (1968) Gastrointestinal helminths from wild and pen-raised bobwhites. J Wildl Manage 32:468–475

King NM (1998) Habitat use by endangered masked bobwhites and other quail on the Buenos Aires National Wildlife Refuge. Master's thesis, University of Arizona

Kline HN, Fulbright TE, Grahmann ED et al (2019) Temperature influences resource use by chestnut-bellied scaled quail. Ecosphere 10(2):e02599. https://doi.org/10.1002/ecs2.2599.10.1002/ecs2.2599

Koerth NE, Guthery FS (1991) Water restriction effects on northern bobwhite reproduction. J Wildl Manage 55:132–137. https://doi.org/10.2307/3809250

Koford EJ (1987) Variations in California Quail productivity in relation to precipitation in Baja California Norte. Master's thesis, University of California, Davis

Krausman PR (ed) (1996) Rangeland widlife. Society for Range Management, Denver, Colorado

Kuvlesky WP Jr, Gall SA, Dobrott SJ et al (2000) The status of masked bobwhite recovery in the United States and Mexico. Proc Natl Quail Symp 4:42–57

Kuvlesky WP Jr, DeMaso SJ, Hobson MD (2007) Gambel's quail ecology and life history. In: Brennan LA (ed) Texas quails: ecology and management, 1st edn. Texas A&M University Press, College Station, pp 6–22

Kuvlesky WP Jr, Brennan LA, Fulbright TE et al (2012) Impacts of invasive, exotic grasses on quail of southwestern rangelands: a decade of progress? Proc Natl Quail Symp 7:25–33

Landgrebe JN, Vasquez B, Bradley RG et al (2007) Helminth community of scaled quail (*Callipepla squamata*) from western Texas. J Parasitol 93:204–208. https://doi.org/10.1645/GE-3578RN.1

Lehmann VW (1953) Bobwhite population fluctuations and vitamin A. Trans N Am Wildl Conf 18:199–246

Lehmann VW (1984) Bobwhites in the Rio Grande Plain of Texas. Texas A&M University Press, College Station

Leopold AS (1985) The California quail. University of California Press, Berkeley

Leopold AS, McCabe RA (1957) Natural history of the Montezuma quail in Mexico. Condor 59:3–26. https://doi.org/10.2307/1364613

Lepper MG (1978) Covey behavior in California quail (*Lophortyx californicus* Shaw) in Nevada. Sociobiology 3:107–124

Lusk JM, Guthery FS, George RR et al (2002) Relative abundance of bobwhites in relation to weather and land use. J Wildl Manage 66:1040–1051. https://doi.org/10.2307/3802936

Lustick S, Voss T, Peterle TJ (1972) Effects of DDT on steroid metabolism and energetics in bobwhite quail (*Colinus virginianus*). Proc Natl Quail Symp 1:213–233

Mathur S, Tomeček JM, Heniff A et al (2019) Evidence of genetic erosion in a peripheral population of a North American game bird: the Montezuma quail (*Cyrtonyx montezumae*). Conserv Genet 20:1369–1381. https://doi.org/10.1007/s10592-019-01218-9

Meehan TD, LeBaron GS, Dale K et al (2018) Abundance trends of birds wintering in the USA and Canada, from Audubon Christmas Bird Counts, 1966–2017, version 2.1. National Audubon Society, New York

Miller KS, Brennan LA, Perotto-Baldivieso HL et al (2018) Correlates of habitat fragmentation and northern bobwhite abundance in the Gulf Coast Prairie Landscape Conservation Cooperative. J Fish Wildl Manag 10:3–18. https://doi.org/10.3996/112017-JFWM-094

Moore SF (2010) Effects of guineagrass on northern bobwhite habitat use. Master's thesis, Texas A&M University-Kingsville

Moore J, Freehling M, Crawford JA et al (1988) *Dispharynx nasuta* (Nematoda) California Quail (*Callipepla califomica*) in Western Oregon. J Wildl Dis 24:564–567. https://doi.org/10.7589/0090-3558-24.3.564

Moore J, Freehling M, Platenberg R et al (1989) Helminths of California quail (*Callipepla califomica*) and Mountain quail (*Oreortyx pictus*) in Western Oregon. J Wildl Dis 25:422–424. https://doi.org/10.7589/0090-3558-25.3.422

Murphy MT (2003) Avian population trends within then evolving agricultural landscape of Eastern and Central United States. Auk 120:20–34. https://doi.org/10.1093/auk/120.1.20

National Bobwhite Quail Conservation Initiative (2018) State of the bobwhite 2018. National Bobwhite Technical Committee Publication, Knoxville, Tennessee, 68 p

National Bobwhite Technical Committee (2011) The national bobwhite conservation initiative: A range–wide plan for recovering bobwhites. In: Palmer WE, Terhune TM, and McKenzie DF (eds) National Bobwhite Technical Committee Publication, ver. 2.0. Knoxville, Tennessee, 212 pp

Naugle DE, Allred BW, Jones MO et al (2020) Coproducing science to inform working lands: the next frontier in nature conservation. Bioscience 70:90–96. https://doi.org/10.1093/biosci/biz144

New Mexico Department of Game and Fish (2017) Habitat stamp. New Mexico Department of Game and Fish publication, Santa Fe, New Mexico, p 2

Nielson RL (1952) Factors affecting the California quail populations of Uintah County, Utah. Master's thesis, Utah State University

Orange JP, Davis CA, Elmore RD et al (2016) Evaluating the efficacy of brood flush counts: a case study in two quail species. West N Am Nat 76: 485–492.https://doi.org/10.3398/064.076.0409

O'Roke EC (1930) The morphology, transmission, and life-history of *Haemoproteus lophortyx* O'Roke, a blood parasite of the California valley quail. Univ California Pub in Zool 36:1–50

Ortega-S JA, Bryant FC (2005) Cattle management to enhance wildlife habitat in South Texas. 2005. Wildlife Management Bulletin of the Caesar Kleberg Wildlife Research Institute No. 6. Texas A&M University–Kingsville

Ortega-Sánchez A (2006) Delineation of habitats and a comparison of density estimators for Gambel's quail in the Trans-Pecos, Texas. Master's thesis, Sul Ross State University

Palmer BJ, Fulbright TE, Grahmann ED et al (2021) Vegetation structural attributes providing thermal refugia for northern bobwhites. J Wildl Manage 85:543–555. https://doi.org/10.1002/jwmg.22006

Parent CJ, Hernández F, Brennan LA et al (2016) Northern bobwhite abundance in relation to precipitation and landscape structure. J Wildl Manage 80:7–18. https://doi.org/10.1002/jwmg.992

Pavlacky DC Jr, Hagen CA, Bartuszevige AM et al (2021) Scaling up private land conservation to meet recovery goals for grassland birds. Conserv Biol 35:1564–1574. https://doi.org/10.1111/cobi.13731

Pence DB (1975) Eyeworms (Nematoda: *Thelaziidae*) from west Texas quail. Proc Helminthol Soc Wash 42:181–183

Peterson MJ (2007) Diseases and parasites of Texas quails. In: Brennan LA (ed) Texas quails: ecology and management, 1st edn. Texas A&M University Press, College Station, pp 89–114

Peterson MJ, Wu XB, Rho P (2002) Rangewide trends in land use and northern bobwhite abundance: a preliminary analysis. Proc Natl Quail Symp 5:35–44

Playa Lakes Joint Venture (PLJV) (2019) Private lands biologists. Playa lakes joint venture website. https://pljv.org/for-landowners/private-lands-biologists. Accessed 30 Nov 2020

Pope M (2002) The ecology of mountain quail in Oregon. Dissertation, Oregon State University

Pope MD, Crawford JA (2004) Survival rates of translocated and native mountain quail in Oregon. West N Am Nat 64:331–337

Purvis JR, Peterson MJ, Dronen NO et al (1998) Northern bobwhites as disease indicators for the endangered Attwater's prairie chicken. J Wildl Dis 34:348–354. https://doi.org/10.7589/0090-3558-34.2.348

Reese KP, Beck JL, Zager P et al (2005) Nest and brood site characteristics of mountain quail in west-central Idaho. Northwest Sci 79:254–264

Reyna KS, Burggren WW (2012) Upper lethal temperatures of Northern Bobwhite embryos and the thermal properties of their eggs. Poult Sci 91:41–46. https://doi.org/10.3382/ps.2011-01676

Robertson GP, Swinton SM (2005) Reconciling agricultural productivity and environmental integrity: a grand challenge for agriculture. Front Ecol Environ 3:38–46. https://doi.org/10.1890/1540-9295(2005)003[0038:RAPAEI]2.0.CO;2

Robinson DA Jr, Jensen WE, Applegate RD (2000) Observer effect on a rural mail carrier survey population index. Wildl Soc Bull 28:330–332

Rollins D, Taylor BD, Sparks TD et al (2009) Species visitation at quail feeders and guzzlers in southern New Mexico. Proc Natl Quail Symp 6:210–219

Rusk JP, Hernández F, Arredondo JA et al (2007) An evaluation of survey methods for estimating northern bobwhite abundance in southern Texas. J Wildl Manage 71:1336–1343. https://doi.org/10.2193/2006-071

Ruzicka RE, Campbell KB, Downey MC et al (2017) Efficacy of a soft release strategy for translocating scaled quail in the Rolling Plains of Texas. Proc Natl Quail Symp 8:389–394

Sandercock BK, Jensen WE, Williams CK et al (2008) Demographic sensitivity of population change in northern bobwhite. J Wildl Manage 72:970–982. https://doi.org/10.2193/2007-124

Sands A, Braun CE, Brubaker R et al (1992) Upland game bird habitat management on the rise. United States Department of the Interior, Bureau of Land Management, Washington, DC, USA 40 p

Sands JP, Brennan LA, Hernández F et al (2012) Impacts of introduced grasses on breeding season habitat use of northern bobwhite in the South Texas Plains. J Wildl Manage 76:608–618. https://doi.org/10.1002/jwmg.305

Sauer JR, Link WA, Niven DK et al (2018) The North American breeding bird survey, analysis results 1966–2017. Version 20180924. United States Geological Survey. https://doi.org/10.5066/P9A4OAEH

Schnupp MJ, Hernández F, Redeker EJ et al (2013) An electronic system to collect distance-sampling data during helicopter surveys of northern bobwhite. Wildl Soc Bull 37:236–245. https://doi.org/10.1002/wsb.232

Silvy NJ, Rollins D, Whisenant SW (2007) Scale quail ecology and life history. In: Brennan LA (ed) Texas quails: ecology and management. Texas A&M University Press, First edition, pp 65–88

Spears GS, Guthery FS, Rice SM et al (1993) Optimum seral stage for northern bobwhites as influenced by site productivity. J Wildl Manage 57:805–811. https://doi.org/10.2307/3809083

Stabler RM, Kitzmiller NJ (1976) Plasmodium (*Giovannolaia pedioecetii*) from gallinaceous birds of Colorado. J Parasitol 62:539–544. https://doi.org/10.2307/3279408

Stephenson JA, Reese KP, Zager P et al (2011) Factors influencing survival of native and translocated mountain quail in Idaho and Washington. J Wildl Manage 75:1350–1373. https://doi.org/10.1002/jwmg.189

Stromberg MR (1990) Habitat, movements, and roost characteristics of Montezuma quail in southeastern Arizona. Condor 92:229–236. https://doi.org/10.2307/1368404

Stromberg MR, Montoya AB, Holdermann D (2020) Montezuma quail (*Cyrtonyx montezumae*). In: Rodewald PG (ed) Birds of the world, vers. 1.0. Cornell Lab of Ornithology, Ithaca, New York. https://doi.org/10.2173/bow.monqua.01

Stubbs M (2019) Agricultural Conservation in the 2018 Farm Bill. Congressional Research Service R45698. Washington DC, USA, 44 p

Swank WG, Gallizioli S (1954) The influence of hunting and rainfall on Gambel's quail populations. Trans N Am Wildl Natural Res Conf 19:283–296

Tanner EP, Elmore RD, Fuhlendorf SD et al (2015) Behavioral responses at distribution extremes: how artificial surface water can affect quail movement patterns. Rangel Ecol Manag 68:476–484. https://doi.org/10.1016/j.rama.2015.07.008

Tanner EP, Papeş M, Elmore RD et al (2017) Incorporating abundance information and guiding variable selection for climate-based ensemble forecasting of species' distributional shifts. PLoS ONE 12:e0184316. https://doi.org/10.1371/journal.pone.0184316

Terhune TM, Palmer WE, Wellendorf SD (2019) Northern bobwhite chick survival effects of weather. J Wildl Manage 83:963–974. https://doi.org/10.1002/jwmg.21655

Townsend DE II, Leslie DM Jr, Lochmiller RL et al (2003) Fitness costs and benefits associated with dispersal in northern bobwhites (*Colinus virginianus*). Am Midl Nat 150:73–82. https://doi.org/10.1674/0003-0031(2003)150[0073:FCABAW]2.0.CO;2

Troy RJ, Coates PS, Connelly JW et al (2013) Survival of mountain quail translocated from two distinct source populations. J Wildl Manage 77:1031–1037. https://doi.org/10.1002/jwmg.549

Turner JW, Hernández F, Boal CW et al (2014) Raptor abundance and northern bobwhite survival and habitat use. Wildl Soc Bull 38:689–696. https://doi.org/10.1002/wsb.476

United State Department of Agriculture, Farm Service Agency (2010) Fact sheet. Conservation Reserve Program: Northern bobwhite quail habitat initiative. United States Department of Agriculture, Washington DC, USA, 2 p

United States Fish and Wildlife Service (2005) Population management series: migratory bird conservation. United States Department of the Interior, Fish and Wildlife Service Manual Part 721, Chapter 6

United States Fish and Wildlife Service (2014) Masked bobwhite (Colinus virginianus ridwayi), 5-year review: summary and evaluation. Buenos Aires National Wildlife Refuge, Sasabe, AZ, USA, 37 p

United States Fish and Wildlife Service (2018) Migratory bird joint ventures. United States Fish and Wildlife Service Website. https://www.fws.gov/birds/management/bird-conservation-par tnership-and-initiatives/migratory-bird-joint-ventures.php. Accessed 30 Nov 2020

United States Geological Survey Gap Analysis Project (2018) Protected areas database of the United States (PAD–US). U.S. Geological Survey data release.https://doi.org/10.5066/P955KPLE. Accessed 30 Nov 2020

United States Department of Agriculture (1991) Sharing the Commitment: Partnerships for wildlife, fish and rare plants on the National Forests. Publication FS–491. U.S. Government Printing Office: 1991–295–809. Washington DC, USA, 18p

United State Department of Agriculture (2004) Forest service and Quail Unlimited renew partnership with new MOU. On The Wild Side: Wildlife program newsletter

United State Department of Agriculture, Natural Resources Conservation Service (2009) Managing working lands for northern bobwhite: the USDA NRCS Bobwhite Restoration Project. In: Burger LW Jr, Evans KO (eds) United States Department of Agriculture, Washington DC, USA, 209 p

Valdez RX, Peterson MJ, Peterson TR et al (2019) Multi-attribute preferences for northern bobwhite habitat restoration among Texas landowners. Wildl Soc Bull 43:272–281. https://doi.org/10.1002/wsb.975

Veech J (2006) Increasing and declining populations of northern bobwhites inhabit different types of landscapes. J Wildl Manage 70:922–930. https://doi.org/10.2193/0022-541X(2006)70[922:IADPON]2.0.CO;2

Vincent CH, Hanson LA, Bermejo LF (2020) Federal land ownership: Overview and data. Congressional Research Service R42346. Washington DC, 25 p

Wallmo OC (1954) Nesting of Mearns' quail in southeastern Arizona. Condor 56:125–128. https://doi.org/10.2307/1364778

Wallmo OC (1956) Ecology of scaled quail in west Texas. Texas Game and Fish Commission, Austin, Texas

Webster JD, Addis CJ (1945) Helminths from the bob-white quail in Texas. J Parasitol 31:286–287

Western Association of Fish and Wildlife Agencies (2011) Memorandum of Understanding among members of the Western Association of Fish and Wildlife Agencies western quail management plan implementation. Approved July 2011. Big Sky, Montana, USA

Wilson MH, Crawford JA (1988) Poxvirus in scaled quail and prevalences of poxvirus-like lesions in northern bobwhites and scaled quail from Texas. J Wildl Dis 24:360–363. https://doi.org/10.7589/0090-3558-24.2.360

Wood SF, Herman CM (1943) The occurrence of blood parasites in birds from southwestern United States. J Parasitol 29:187–196. https://doi.org/10.2307/3273097

Zornes M, Bishop RA (2009) Western quail conservation plan. Association of Fish and Wildlife Agencies, Washington, D.C., p 92

Chapter 12
Rangeland Songbirds

Anna D. Chalfoun, Tracey N. Johnson, and Jill A. Shaffer

Abstract Songbirds that occur across the diverse types of North American range-lands constitute many families within the Order Passeriformes, and hundreds of species. Most are declining, and many are considered potential indicator species for rangeland ecosystems. We synthesized information on the natural and life history, habitat requirements, conservation status, and responses to management of song-birds associated with North American grasslands and sagebrush steppe, two of the most geographically extensive types of rangelands. We provide a more targeted examination of the habitat associations and management considerations for two focal species, the grassland-obligate grasshopper sparrow (*Ammodramus savannarum*) and sagebrush-obligate Brewer's sparrow (*Spizella breweri*). Grassland- and sagebrush-obligate species rely on expansive stands of grasslands and sagebrush, respectively, and we discuss how key ecological processes and rangeland management approaches—grazing, fire, and mechanical treatments—influence rangeland songbirds. Rangeland management practices can affect breeding songbirds considerably, primarily through the resultant structure and composition of vegetation, which influences the availability of preferred nesting substrates, refugia from predators, and foraging success. Optimal management strategies to limit negative consequences to rangeland songbirds will depend on the target species and local topoedaphic and climatic conditions. The maintenance of large, contiguous patches of native habitats and restoration of previously degraded areas will help facilitate the population persistence of rangeland-associated songbirds. Maintaining structural heterogeneity of habitats within landscapes, moreover, can facilitate local species diversity. Information pertaining to periods outside of the nesting stage is severely lacking for

A. D. Chalfoun (✉)
USGS Wyoming Cooperative Fish and Wildlife Research Unit, Dept. of Zoology and Physiology, University of Wyoming, 1000 East University Ave., Dept. 3166, Laramie, WY 82071, USA
e-mail: Achalfou@uwyo.edu

T. N. Johnson
Department of Fish & Wildlife Sciences, University of Idaho, 322 E. Front St., Boise ID 83702, USA

J. A. Shaffer
US Geological Survey, Northern Prairie Wildlife Research Center, 8711 37th St. SE, Jamestown ND 58401, USA

© The Author(s) 2023
L. B. McNew et al. (eds.), *Rangeland Wildlife Ecology and Conservation*,
https://doi.org/10.1007/978-3-031-34037-6_12

most species, which is concerning because effective management necessitates understanding of threats and limiting factors across the full annual life cycle. Moreover, information on disease effects and prevalence, the effects of a changing climate, and how both may interact with management strategies, also comprise key gaps in knowledge.

Keywords Brewer's sparrow · Conservation · Grasshopper sparrow · Grassland songbirds · Habitat · Management · Sagebrush songbirds

12.1   Life/Natural History and Population Dynamics

The songbird species that inhabit North American rangelands have relatively fast life histories, with first breeding attempts typically occurring in the first year of adulthood. The distributions of some species are restricted (e.g., Baird's sparrow [*Centronyx bairdii*]; Green et al. 2020), whereas the distributions of other species span multiple continents (e.g., horned larks [*Eremophila alpestris*]; Beason 2020). Most rangeland songbirds are migratory, and territorial on breeding grounds. Primary foods include arthropods during the breeding season and seeds during the winter. The annual life cycle of rangeland-associated songbirds can be classified as nesting, post-fledging, fall migration, over-wintering, and spring migration.

12.1.1   Nesting

Songbird males establish breeding territories shortly after arriving on breeding grounds in spring. Males often have elaborate courtship songs, and many combine songs with aerial displays. Song dialects can vary regionally, and males of some species (e.g., Brewer's sparrow [*Spizella breweri*] and grasshopper sparrow [*Ammodramus savannarum*]) have different song types for pre- and post-pairing. Most species are socially monogamous at least within a breeding season, though some such as the bobolink (*Dolichonyx oryzivorus*) and dickcissel (*Spiza americana*) are polygynous, a mating system in which the desirable males will pair with more than one female (Renfrew et al. 2020; Temple 2020). Even for socially monogamous populations, extra-pair paternity can be common (e.g., Danner et al. 2018).

Nest placement is variable, with some species nesting on the ground amidst vegetation, and others within shrubs or trees. Nest structures typically are open or domed cups, constructed with sticks, grasses, forbs, and/or sedges, and lined with finer material such as rootlets, mammal hair or feathers of other species. Females lay one egg per day until clutch completion, and clutch sizes vary from approximately 2–7 eggs. Eggs develop and remain viable within a specific range of temperatures regulated by incubation (Deeming 2001). Incubation is conducted primarily by females, although males contribute in some species, and females are sometimes provisioned with food on the nest by their mates. Incubation periods typically range from 10 to 13 days for open-cup nesters, whereas cavity nesters incubate for longer periods. The incubation for the juniper titmouse (*Baeolophus ridgwayi*), for example, is approximately 17 days (Cicero et al. 2020). Nestlings are altricial and highly dependent upon parental care for food and thermoregulation. Nestling periods range from 8 to 14 days for most species, and young are almost always fed by both parents.

Songbird nest survival varies across habitat conditions, sites, and years. The primary source of nesting mortality is predation from a wide variety of species including snakes, rodents, mustelids, canids, domesticated or feral cats (*Felis catus*), raccoons (*Procyon lotor*), raptors, shrikes (*Lanius* spp.) and even ungulates including deer (*Odocoileus* spp.,) and elk (*Cervus canadensis*) (Pietz and Granfors 2000; Renfrew and Ribic 2003; Hethcoat and Chalfoun 2015a; Lyons et al. 2015). Many songbirds also experience brood parasitism by the brown-headed cowbird (*Molothrus ater*) (Shaffer et al. 2019a), though some species such as sage thrashers (*Oreoscoptes montanus*) remove cowbird eggs from their nests (Reynolds et al. 2020). Other causes of nesting failures include extreme weather events, such as snowstorms or hail (Hightower et al. 2018), and anthropogenic activities.

12.1.2 Post-fledging

Songbird nestlings typically depart nests before they are fully capable of flight. Mortality from predation or inclement weather during the early post-fledging period can therefore be high for most if not all species (e.g., Fisher and Davis 2011; Hovick et al. 2011). Fledglings are fed by parents for at two least weeks after leaving the nest, achieving adult body mass within about a month (Jones et al. 2018). Family groups likely rely on habitats with sufficient cover to shelter young from predators and the elements (Fisher and Davis 2011). Unfortunately, the post-fledging period for many songbirds rarely is studied (Davis and Fisher 2009; Ribic et al. 2018, 2019), and estimates of post-fledging habitat use and survival are lacking. Where studied, estimates of fledgling survival range from 26 to 36% (Yackel Adams et al. 2006; Berkeley et al. 2007; Hovick et al. 2011; Young et al. 2019). Nestling body condition and wing development, which vary with food availability and provisioning rates, tend to be positively related to post-fledging survival (Yackel Adams et al. 2006; Jones et al. 2017; Jones and Ward 2020).

12.1.3 Non-breeding

Most songbirds inhabiting North American rangelands during the breeding season are migratory, although some populations inhabiting southern areas are year-round residents. Adults typically complete a full molt of their feathers towards the end of the nesting season, and migrants often form large, single, or mixed-species flocks for southward migration. The length of migration distances ranges from short to long, with many species over-wintering in the southwestern U.S. and northern Mexico (e.g., chestnut-collared longspur [*Calcarius ornatus*]), and others that migrate to South America (e.g., bobolink and dickcissel). Flocks periodically use migratory stopover habitats to forage and rest. Over-wintering migrants tend to use habitats similar in structure to their breeding habitats (Igl and Ballard 1999; Hovick et al. 2014).

12.1.4 Survival and Sources of Mortality

Songbird nests are depredated by a wide variety of species (see Nesting section). For most songbird species, much less is known about predator species and rates of predation during the post-fledging, migratory, and over-wintering periods, though many species of raptors (e.g., accipters, falcons) are known to kill adult songbirds (Lima 2009). Fledglings are consumed by raptors, corvids, shrikes, snakes, and mammals (Yackel Adams et al. 2006; Berkeley et al. 2007; Hovick et al. 2011; Young et al. 2019). Free-ranging domestic and feral cats kill billions of songbirds in North America each year (Loss et al. 2013a). Other sources of adult mortality of songbirds include collisions with buildings, vehicles, guy wires extending from communication towers, and wind turbines (Longcore et al. 2012; Loss et al. 2013b; Erickson et al. 2014).

12.1.5 Seasonal Movements and Dispersal

Movement and dispersal data are rare for most rangeland songbirds. Historically, the logistical challenges of safely radio-tracking very small birds were an impediment. Recent technological advances, however, have enabled the manufacture of smaller, lighter transmitters and light-level geolocators that record the movements and locations of small birds across time upon recapture. Soon after independence, immature birds join post-breeding flocks of adults, leave their natal area, and begin moving with pre-migratory flocks (e.g., Temple 2020). An understanding of the connectivity between the breeding grounds and particular migration routes or over-wintering areas is lacking for most grassland and sagebrush songbird species. Site fidelity, or the repeated return, to breeding sites varies across species, habitats and locations. Juveniles sometimes return to the general area where they were born (natal philopatry; e.g., Renfrew et al. 2020). Some species appear to be facultatively nomadic on breeding grounds between years (e.g., chestnut-collared longspur and lark bunting [*Calamospiza melanocorys*]), likely as an evolved response to shifting habitat suitability associated with the unpredictable influences of fire, drought, and the movements and grazing of bison (*Bison bison*) herds (Green et al. 2019).

12.1.6 Population Dynamics

Whereas offspring mortality during the nesting (e.g., Kerns et al. 2010; Hethcoat and Chalfoun 2015b; Verheijen et al. 2022) and post-fledging (e.g., Young et al. 2019) periods can be high, a lack of research encompassing the full annual life cycle constrains an understanding of which life stages tend to be most limiting for rangeland songbirds (Marra et al. 2015). Because most songbirds have relatively fast life histories, the influence of reproductive success (clutch size, nest survival, post-fledging survival) on population growth likely is high (Saether and Bakke 1997). Nest density also may influence population growth, as avian productivity within an area is the product of per capita nest survival and density (Pulliam et al. 2021). Moreover, carryover effects from over-wintering grounds and migratory stopover sites can affect the timing and reproductive success of songbirds via the interaction between arrival times and food availability, and the condition of adults at the onset of nesting (Bayly et al. 2016).

12.2 Current Species and Population Status

Most populations of songbirds that breed within North American rangelands are declining, some drastically, concomitant with broad-scale habitat loss and alteration (Table 12.1; Rosenberg et al. 2019; Sauer et al. 2020). For example, of 34 species of New World sparrows, which include scrub-successional, aridland, and grassland species, 17 exhibited significant declines and 27 had negative trend estimates (Sauer et al. 2013). Moreover, numbers of grassland and aridland birds declined by an estimated 55% and 23%, respectively, during 1970–2017 (Rosenberg et al. 2019).

12.3 Population Monitoring

There is no monitoring program devoted specifically to rangeland songbirds, though many populations are monitored as part of broader efforts. The North American Breeding Bird Survey (BBS) and Christmas Bird Count (see Chap. 11) are used frequently to assess the status and general trends of rangeland songbirds (Table 12.1). Laurent et al. (2012) provide details on these and other national and regional programs, such as the Strategic Multi-scale Grassland Bird Population Monitoring Protocol (SMGBPM) and Monitoring Avian Productivity and Survivorship (MAPS). The SMGBPM uses counties as management units and was developed because of the concern that BBS may underestimate grassland bird numbers in some areas. MAPS utilizes a network of mist-netting efforts and mark-recapture analyses to assess demographic parameters including annual survival and productivity of North American birds over time. Citizen-science programs include eBird, which is an online database

Table 12.1 Representative songbird species inhabiting the major (but not all) vegetation types (Barbour and Billings 2000) composing North American rangelands

Vegetation type	Typical songbird species	Conservation status	
		PIF score	BBS trend (%)
Grasslands			
Tallgrass prairie	Bobolink (*Dolichonyx oryzivorus*)	14	− 1.5
	Eastern meadowlark (*Sturnella magna*)	11	− 2.6
	Henslow's sparrow (*Centronyx henslowii*)	15	− 1.9
N. mixed-grass prairie	Baird's sparrow (*Centronyx bairdii*)	15	− 0.9
	Chestnut-collared longspur (*Calcarius ornatus*)	15	− 2.5
	Sprague's pipit (*Anthus spragueii*)	14	− 3.2
S. mixed-grass prairie	Cassin's sparrow (*Peucaea cassinii*)	11	− 0.4
	Dickcissel (*Spiza americana*)	11	− 0.6
	Lark sparrow (*Chondestes grammacus*)	10	− 1.2
Shortgrass prairie	Horned lark (*Eremophila alpestris*)	9	− 1.9
	Lark bunting (*Calamospiza melanocorys*)	12	− 3.7
	Thick-billed longspur (*Rhynchophanes mccownii*)	15	− 2.1
Palouse prairie	Grasshopper sparrow (*Ammodramus savannarum*)	12	− 2.5
	Vesper sparrow (*Pooecetes gramineus*)	11	− 0.8
	Western meadowlark (*Sturnella neglecta*)	10	− 0.9
Warm deserts and grasslands	Loggerhead shrike (*Lanius ludovicianus*)	11	− 2.6
	Cactus wren (*Campylorhynchus brunneicapillus*)	12	− 1.3
	Rufous-crowned sparrow (*Aimophila ruficeps*)	11	0.4
Shrublands			
Sagebrush[a]	Brewer's sparrow (*Spizella breweri*)	11	− 0.9
	Sagebrush sparrow (*Artemisiospiza nevadensis*)	10	− 1.2
	Sage thrasher (*Oreoscoptes montanus*)	11	− 0.4
Juniper-pinyon	Juniper titmouse (*Baeolophus ridgwayi*)	11	+ 0.1
	Pinyon jay (*Gymnorhinus cyanocephalus*)	14	− 2.1

(continued)

Table 12.1 (continued)

Vegetation type	Typical songbird species	Conservation status	
		PIF score	BBS trend (%)
	Black-throated gray warbler (*Setophaga nigrescens*)	13	− 0.7

Species are listed only once across types even though they may be found in several. See Vickery et al. (1999) for a more complete list of obligate and facultative grassland and arid shrubland birds. Conservation status is indexed by the Partner's in Flight (PIF) Avian Conservation Assessment Database maximum continental combined score (Partners in Flight 2021) and the range-wide Breeding Bird Survey trend during 1966–2019 (% population change per year; Sauer et al. 2020). The PIF score integrates information about the global population size, distribution, threats, and trends. Scores range from 4 to 20, with higher values associated with greater concern. Species with scores of 14 or higher, or with a concern score of 13 and a steeply declining population trend, are those most at risk of extinction without significant conservation actions to reverse declines and reduce threats
[a]Sagebrush steppe and Great Basin sagebrush types combined

of bird observations, and NestWatch, which focuses on reproductive success. Finally, several facilitated databases, including the Avian Knowledge Network, store data that land managers, scientists, and others can access for research and conservation (Laurent et al. 2012).

Monitoring programs vary in their degree of statistical rigor, spatial inference, and limitations. Selection of monitoring data on which to base research or management decisions should therefore depend on the desired metrics (e.g., occupancy, distribution, abundance trends over time, productivity, species richness) and precision. The BBS, for example, was established in 1966 and has been valuable for documenting general population trends of over 400 North American bird species. Surveys, however, are conducted as annual roadside routes, which may under-sample species sensitive to human infrastructure. Current protocols also do not account for potential spatiotemporal differences in the probability of detecting birds. The Integrated Monitoring in Bird Conservation Regions, coordinated by the Bird Conservancy of the Rockies, incorporates randomized sampling and was designed to provide robust estimates of avian occupancy and density across time and multiple spatial scales.

12.4 Habitat Associations

Songbird species inhabiting the grasslands and arid shrublands of North America (see Table 12.1 for representative species) include habitat specialists (or "obligates") and those that are open-country generalists. The thick-billed longspur (*Rhynchophanes mccownii*), for example, is a shortgrass-prairie specialist with a restricted distribution (With 2021). By contrast, the western meadowlark (*Sturnella neglecta*) inhabits a wide variety of open habitat types and agricultural fields throughout the entire western

Juniper titmouse	Sage thrasher	Thick-billed longspur	Chestnut-collared longspur	Bobolink

Juniper-pinyon	Sagebrush steppe	Shortgrass prairie	Mixed-grass prairie	Tallgrass prairie

Fig. 12.1 Artistic rendering of representative songbird and plant species within five of the major rangeland types in North America. Plant species from left to right include western juniper (*Juniperus occidentalis*), big sagebrush (*Artemisia tridentata*), buffalo grass (*Bouteloua dactyloides*), blue grama (*Bouteloua gracilis*), and big bluestem (*Andropogon gerardii*). Assemblages are organized by relative longitude. Artwork by Bethann Merkle. *Source* photograph credits include Jack Parlapiano (titmouse), Tayler Scherr (thrasher), Rick Bohn (chestnut-collared longspur), Dan Casey (thick-billed longspur), Dave Lambeth (bobolink), Sarah McIntire (juniper), and Anna Chalfoun (sagebrush). Other plant photos drawn from open sources

portion of North America (Davis and Lanyon 2020). The assemblage of songbird species that occupies a given site varies by geographic location, vegetation type, habitat structure, and extent of habitat degradation (Fig. 12.1).

The habitat preferences of rangeland songbirds evolved based on the conditions most associated with successful survival and reproduction (Nelson et al. 2020). Such preferences often are scale-dependent (Chalfoun and Martin 2007; Lipsey et al. 2017; Box 12.2). Several species may inhabit the same area but primarily nest or forage within more differentiated niches (Grinnell 1917). Such differences likely arose to limit competition for nest sites and food. Within mixed-grass prairie, for example, the nest sites of sympatric songbirds are distributed across microhabitat gradients ranging from shorter, sparser vegetation (e.g., thick-billed longspur) to taller, denser grasses (e.g., western meadowlark; Fig. 12.1). The vertical and horizontal partitioning of nest sites within an area may benefit the reproductive success of co-occurring songbird species by reducing predator search efficiency (Martin 1993). The maintenance of microhabitat heterogeneity within landscapes is therefore a key management consideration. Boxes 12.1 and 12.2 provide more in-depth descriptions of habitat associations for two focal species, a grassland-obligate songbird (grasshopper sparrow) and a sagebrush-steppe obligate (Brewer's sparrow).

12.5 Rangeland Management

The primary ecological processes and management practices that influence rangeland songbirds are grazing, fire, and mowing, whereas mechanical management practices include the application of herbicides and pesticides, mowing, chaining, and discing (reviewed in Shaffer and DeLong 2019). Some management interventions, such as bison grazing and fire, often are geared towards mimicking historical disturbance regimes (see Chaps. 6 and 8), whereas mechanical management practices may be used to produce similar outcomes but within faster time frames. These management approaches may be used singly or in combination (e.g., patch-burn grazing). Ecological processes and management practices influence local avian biodiversity primarily through their effect on vegetation structure and composition. A management approach will have variable outcomes depending on timing, intensity, and frequency. Timing, or seasonality, refers to when during the year a management approach is applied. Intensity refers to the degree to which a management approach is applied. In terms of fire, intensity is the amount of heat produced (Chap. 6), whereas for grazing, intensity refers to the number of grazing animals and length of time grazing occurs, or how much biomass is removed. Frequency refers to how often ecological processes or management practices have been applied, either within or among seasons (Chap. 4).

Management approaches depend on goals, and outcomes often are site- or species-specific. Management guidance for individual species is summarized in the accounts constituting Johnson et al. (2019). Thorough coverage of management approaches for grasslands bird species can be found in Sample and Mossman (1997), whereas management considerations pertaining to sagebrush species can be found in Paige and Ritter (1999) and Walker et al. (2020). The two case studies in this chapter (Boxes 12.1 and 12.2) illustrate the complexity of the decisions involved in the application of ecological processes and management practices that maintain and create habitat for specific songbird species.

12.5.1 Grazing

Direct effects of livestock grazing on rangeland songbird species are rare and include trampling of eggs, nestlings, or adults, and in some cases apparent predation (Nack and Ribic 2005; Bleho et al. 2014). Nest destruction by livestock generally increases with grazing intensity during the nesting season, though for some species of songbirds, the creation of habitat via grazing may offset the minimal nest losses (Owens and Myres 1973; Bleho et al. 2014). Indirect effects of livestock grazing include alteration in vegetation structure (e.g., decreased litter cover, increased bare ground) and composition (e.g., dominance of some plant species over others). Such changes can lead to altered insect food availability or nest predation risk (Johnson et al. 2012). Indirect effects of grazing therefore tend to be more impactful than direct effects in influencing whether grazed rangelands comprise high-quality nesting habitat (Cody 1985; Martin and Possingham 2005). Additional indirect effects of grazing may include increased nest parasitism by brown-headed cowbirds or increased nest predator populations, via the addition of water, feed, and carcasses (Goguen and Mathews 1999, 2000; Coates et al. 2016). However, parasitism rates within grasslands also are influenced by the landscape matrix within which pastures are located. Brown-headed cowbirds may be less likely to parasitize nests of grassland songbirds where tree cover on the landscape is greater and nests of woodland species are readily available as alternative cowbird hosts (Pietz et al. 2009; Hovick and Miller 2013).

Livestock grazing can be used to manipulate vegetation to create desired conditions for rangeland songbirds (Derner et al. 2009; Bleho et al. 2014). The effects of grazing on vegetation, however, can be highly variable and affected by grazing regime (Chap. 4), livestock characteristics (species, breed, sex, age, and genetic factors), precipitation (amount, seasonality), current vegetation structure and composition, soil characteristics, historical land use, and presence and types of other disturbances (Briske et al. 2008; Sliwinski and Koper 2015; Lipsey and Naugle 2017). Thus, prior to implementing a grazing system within any given year to obtain a desired vegetation structure (e.g., habitat outcome; Pulliam et al. 2020, 2021), the characteristics of that grazing system may need to be modified based on the region's expected precipitation and other aforementioned factors.

Long- and short-term monitoring of the effects of grazing on vegetation structure is important in terms of assessing the effects of grazing on avian abundance, community composition, and reproduction (Pulliam et al. 2021). Short-term effects include the reduction of herbaceous cover or height of vegetation, which can affect songbird species dependent on litter and grass cover for nest concealment. Long-term effects can manifest as altered composition of plant species or reduced vegetation productivity of a site (Briske et al. 2008). For example, repeated livestock grazing can affect shrub and tree establishment, thereby affecting songbird species dependent on non-herbaceous vegetation (Bock et al. 1993). Accordingly, rangeland songbird species may respond differently to grazing-induced changes over time (e.g., Johnson et al. 2011; Sliwinski and Koper 2015). Most species that are affected negatively by grazing are those that are dependent on relatively dense herbaceous ground cover or heavy shrub cover for nesting and foraging. The responses of species in sagebrush and montane coniferous habitats to livestock grazing, however, remain understudied.

Albeit not yet well understood, native grazers such as bison and prairie dogs (*Cynomys* spp.) may influence songbirds and their habitats differently than livestock (Allred et al. 2011a). The abundance of vesper and grasshopper sparrows in Montana were more abundant in pastures grazed year-round by bison compared with those that were grazed seasonally by cattle, although the abundance of 7 other songbird species and diversity measures did not vary by grazing type (Boyce et al. 2021). The grazing and fossorial activities of prairie dogs have played an influential role in the maintenance and composition of grassland and arid shrubland communities and can facilitate co-occurring bird species (Duchardt et al. 2019, 2021; Chap. 15).

12.5.2 Fire

As with grazing, the direct effects of fire on songbird species include the destruction of nests and young. Indirect effects involve altered vegetation characteristics as influenced by the timing, intensity, and frequency of fire applications (Chap. 6). In most rangeland systems, fire will reduce the biomass of live and dead herbaceous vegetation and shrub or tree cover, depending on fire intensity, and stimulate regrowth of herbaceous species through nutrient recycling (Sample and Mossman 1997). Responses of rangeland songbirds to fire are a function of each species' preferences for the resultant post-burn vegetation conditions, and responses may change with time since fire as vegetation recovers. Grassland songbirds occur in fire-evolved ecosystems that historically had more frequent fires than sagebrush ecosystems (Chap. 6). Prescribed fire, therefore, is applied more frequently in grasslands to maintain songbird habitat than in shrubsteppe.

The timing of prescribed burning is an important consideration. Prescribed fire applied outside of the breeding season precludes the destruction of nests and allows for vegetation regrowth before the nesting season (Higgins 1986; Sample and Mossman 1997). Spring burns, however, can be most effective at suppressing the spread of invasive plant species by damaging plants during a vulnerable growth stage (Shaffer and DeLong 2019).

The consideration of historical fire-return intervals within regions and rangeland types is critical for the maintenance of songbird habitat. For example, in low-elevation, xeric sagebrush habitats, the invasion of nonnative annual grasses such as cheatgrass (*Bromus tectorum*) increases fuel loads, fire frequency, extent, and severity; and reduces shrub cover, which affects habitat suitability for shrub-dependent birds (Knick et al. 2005; Pilliod et al. 2017). Even in fire-dependent grasslands, fire-return intervals shortened relative to historical regimes can result in changes in the composition and structure of vegetation, with resultant reduced habitat quality for some grassland songbird species (Zimmerman 1997; Reinking 2005; With et al. 2008). For example, annual fires can eliminate the residual cover used as avian nesting substrates. Conversely, lengthened fire-return intervals, and especially the suppression of wildfires, may cause the expansion of woody vegetation into previously vast expanses of grassland and high-elevation sagebrush steppe (Grant and Murphy 2005; Noson et al. 2006; Anderson and Steidl 2019).

12.5.3 Mowing

Mowing uniformly reduces vegetation height, woody vegetation, and litter (Herkert et al. 1996; Sample and Mossman 1997). Mowing can therefore be implemented as a management tool for some grassland songbirds that prefer such conditions, both within the current harvest year (Mabry and Harms 2020) and occasionally the subsequent year (Igl and Johnson 2016). However, mowing can have direct negative effects on ground-nesting songbirds if conducted during the breeding season because nests may be abandoned or destroyed, or incubating adults, eggs, nestlings, and recently fledged young may be killed (Bollinger et al. 1990). Indirect effects of mowing include the reduction of invertebrate populations that serve as important prey for breeding birds (Zalik and Strong 2008). Plant species composition also can be affected over longer time scales with repeated mowing (Sample and Mossman 1997; Allen et al. 2001).

Effects of haying on songbirds depend on the timing and frequency of disturbance. Traditional hayland practices employed by agricultural producers aim to maximize the amount and quality of forage and typically involve an early initial cutting and one or more subsequent harvests that coincide with the avian breeding cycle, which can negatively affect avian reproductive success. Fields that are mowed multiple times within a breeding season and with short intervals between mowing may therefore cause complete avian reproductive failure (Rodenhouse et al. 1995). Conservation-focused haying strategies aim to avoid negative effects on birds by conducting operations after the nesting period (after mid-July or August, depending on location), haying periodically but not annually, and leaving portions of fields un-mowed (Shaffer and DeLong 2019).

12.5.4   Managing for Heterogeneity

A primary goal of livestock producers is to facilitate livestock growth via the maximal consumption of vegetation, which depending on management can decrease vegetation heterogeneity (variation in plant species composition and structure; Chaps. 8 and 9). Such practices can promote the dominance of a few plant species that are valuable to domesticated livestock but do not necessarily facilitate biological diversity. Traditional grazing systems (Chap. 4) wherein beef production is a primary objective, and without the use of fire, can therefore be insufficient in providing the vegetation heterogeneity required to support a diverse local suite of grassland birds (Sliwinski et al. 2019, 2020). In some situations, increasing the habitat heterogeneity within the overall landscape or region for biodiversity may entail managing for conditions that are rare or absent in surrounding areas.

In some grassland ecosystems of the Great Plains, patch-burn grazing, also known as pyric-herbivory, has been promoted as an alternative rangeland management strategy that aims to increase vegetation heterogeneity and avian and vegetation biodiversity while maintaining profitability for livestock producers (Fuhlendorf and Engle 2001; Allred et al. 2011b; Neilly et al. 2016). Patch-burn grazing entails shifting mosaics of burned patches designed to influence grazing distribution and increase vegetation heterogeneity (Fuhlendorf et al. 2006). Where fire is not a feasible management option, local habitat heterogeneity can be enhanced by herding, strategic placement of salt, minerals, or fencing, or alteration of stocking rates and season of use (Scasta et al. 2015; Sliwinski et al. 2019). The extent to which grazing may be used to increase local habitat heterogeneity will depend also on the spatial and habitat use of cattle, which tend to vary with factors such as topography, soils, water, and stocking rate (Bailey 2005; Rivero et al. 2021; Chap. 4). Cattle tend to decrease habitat selectivity under high stocking rates, which can increase habitat homogeneity (Rivero et al. 2021).

In sagebrush steppe, habitat heterogeneity within a landscape that provides for the entire suite of songbirds may be facilitated through shifts in the relative dominance of woody versus herbaceous vegetation, and promoting both sagebrush and other shrub species in patches of various heights (Knick et al. 2008; Hanser and Knick 2011; Miller et al. 2017). Heterogeneity in plant structure and composition in sagebrush-dominated systems can be influenced by grazing management (Veblen et al. 2014) or reintroduction of fire into communities (e.g., mountain big sagebrush) that have experienced prolonged fire exclusion both of which can help maintain plant diversity (Manier and Hobbs 2006; Davies and Bates 2020). If sagebrush-obligate songbirds (Brewer's sparrow; sagebrush sparrow, *Artemisiospiza nevadensis*; sage thrasher, *Oreoscoptes montanus*) are of primary management interest, the prioritization of areas with relatively tall shrubs (50–200 cm) and high (greater than > 25%) shrub cover is paramount (Chalfoun and Martin 2007; Martin and Carlson 2020; Reynolds et al. 2020). Given the extensive loss of sagebrush habitat range-wide, and the agricultural value of areas within the sagebrush steppe consisting of more mesic, well-drained soils, such conditions have become rare (Knick et al. 2008).

12.6 Disease

The effects of disease on rangeland songbirds are poorly studied. West Nile Virus has been detected in several rangeland-inhabiting species including the bobolink, brown-headed cowbird, black-chinned sparrow (*Spizella atrogularis*), field sparrow (*Spizella pusilla*), lark sparrow (*Chondestes grammacus*), pinyon jay (*Gymnorhinus cyanocephalus*), and Savannah sparrow (*Passerculus sandwichensis*) (Centers for Disease Control 2016). Avian pox viruses have been recorded for sagebrush sparrow and Savannah sparrow (Martin and Carlson 2020; Wheelwright and Rising 2020). Songbirds are affected by outbreaks of salmonellosis, which has a high mortality rate; however, the extent to which this disease affects rangeland songbirds in particular is largely unknown. Species that congregate in flocks and are exposed to contaminated feces appear to be most at risk. Some rangeland songbirds may therefore be vulnerable, including those that use feeders or roost in groups. Brown-headed cowbirds seem to be particularly at risk and may serve as a reservoir for salmonellosis, possibly influenced by their association with cattle (Tizard 2004).

Parasites, such as bird blowflies (*Protocalliphora* spp. and *Trypocalliphora braueri*), are widespread in songbirds and can inflict serious harm. Effects of blowflies have included reduced nestling survival and fledging success for sage thrashers (Howe 1992), reduced tarsi length for sagebrush sparrow nestlings (Peterson et al. 1986), and retarded feather growth for Savannah sparrow nestlings (Bedard and McNeil 1979). Detrimental effects of ectoparasites on songbird nestlings can be ameliorated by increased food availability and feeding rates by adults, but exacerbated by environmental conditions that decrease foraging opportunities (e.g., adverse weather; Howe 1992; De Lope et al. 1993; Tripet and Richner 1997). Finally, rangeland songbirds also may experience anemia from haematophagous parasites, to a largely unknown extent (Boyd 1951).

12.7 Ecosystem Threats

12.7.1 Habitat Conversion and Alteration

The biggest collective threat to rangeland songbird species is habitat loss, fragmentation, and degradation. Large and rapid declines in grassland and aridland species often are linked to the loss and alteration of habitat on breeding grounds (Sauer et al. 2013; Rosenberg et al. 2019). Historically, agricultural practices, and particularly cropland agriculture, have been the greatest causes of native grassland and sagebrush loss in North America (Knick et al. 2003; Rosenberg et al. 2019). Urban development and sprawl in exurban areas, and development for energy resources, have caused further habitat loss and fragmentation (Marzluff and Ewing 2001; Northrup and Wittemyer 2013). The spread of invasive plant species and woody encroachment also causes degradation in habitat quality for songbird species (Archer et al. 2017).

12.7.2 Energy Development

Portions of North American rangelands coincide with on-going energy extraction, including oil, natural gas, and wind (Northrup and Wittemyer 2013). Effects of oil and gas development on rangeland songbirds include reduced abundance, altered habitat use, and reduced reproductive success (Gilbert and Chalfoun 2011; Kalyn Bogard and Davis 2014; Thompson et al. 2015; Chalfoun 2021 and references therein). Habitat alteration associated with energy development activities can alter trophic dynamics among wildlife species and result in decreased reproduction or survival. In Wyoming's sagebrush steppe, for example, the nest success of three sagebrush-obligate songbird species decreased with adjacent surface disturbance from natural gas development (Hethcoat and Chalfoun 2015a, b). Nest failures were attributed primarily to increased abundance of rodent nest predators that were attracted to the re-seeded areas surrounding well pads, pipelines and roads (Sanders and Chalfoun 2018).

Activities associated with energy development simultaneously alter many characteristics within landscapes in addition to the footprint, including human activity, noise, and lighting. Yet, the specific mechanisms underlying avian responses are rarely tested or understood (Jones et al. 2015; Chalfoun 2021; but see Bernath-Plaistad and Koper 2016; Mejia et al. 2019). Wind facilities can cause both direct (mortality due to turbine strikes; Allison et al. 2019) and indirect (reduced reproductive success, avoidance of suitable habitat; Mahoney and Chalfoun 2016; Shaffer and Buhl 2016; Shaffer et al. 2019b) effects on rangeland songbirds. Solar installations are increasing in parts of the western U.S. and may pose additional management challenges (Loss 2016).

12.7.3 Invasive Species

Invasive plant species can affect rangeland songbird habitat in a myriad of ways. Many species of rangeland songbirds occur in areas that contain non-native plants, and use them for various activities including nesting or perching (e.g., Ruehmann et al. 2011; Nelson et al. 2017). Evaluation of the extent to which such use has negative consequences for songbirds, however, has implications for the growth of avian populations (e.g., Ruehmann et al. 2011; Nelson et al. 2018). Moreover, a few species of invasive plants, including cheatgrass, can exert such influence that they change the overall functioning of ecosystems and substantially eliminate or alter songbird habitat (Brooks et al. 2004; Knick et al. 2005; Coffman et al. 2014; Bestelmeyer et al. 2018).

The effects of invasive plants on songbirds include the alteration of habitat structure or composition that can influence habitat use, movements, abundance, survival, or reproductive success in a context- and species-specific manner (Stoleson and Finch 2001; Hovick and Miller 2013; Nelson et al. 2017; Stinson and Pejchar 2018). The abundance of songbirds in the northern mixed-grass prairie, for example, decreased slightly or remained the same with exotic grass encroachment (Pulliam et al. 2020). The cover of exotic grass, however, co-varied with herbaceous biomass. Areas with high leafy spurge (*Euphorbia esula*) in North Dakota decreased the breeding densities of some, but not all, species of grassland songbirds (Scheiman et al. 2003). Similarly, patterns of occurrence of songbirds in Saskatchewan between native pastures and those partially comprised of crested wheatgrass (*Agropyron cristatum*) were mixed (Davis and Duncan 1999). Relationships between songbird reproductive success and invasive plants generally have been neutral or positive (Stinson and Pejchar 2018, but see Lloyd and Martin 2005). Other indirect effects include altered prey availability, because native plants typically support more abundant and diverse invertebrate assemblages (Hickman et al. 2006; Litt et al 2014) which can influence reproductive parameters such as nestling growth (Lloyd and Martin 2005). However, the nestling mass of Botteri's sparrows (*Peucaea botterii*) and several other species of grassland songbirds was unaffected by invasive grasses (Jones and Bock 2005; Kennedy et al. 2009).

Examples of invasive woody plant species include eastern redcedar (*Juniperus virginiana*) in southern grasslands (Archer et al. 2017), *Pinus* spp. and *Juniperus* spp. in sagebrush communities (Knick et al. 2014), and willow (*Salix* spp.) and aspen

(*Populous tremuloides*) in northern grasslands (Grant et al. 2004). Woody encroachment alters both the vertical and horizontal characteristics of vegetation communities, and in some cases results in monocultures with little to no understory (Frost and Powell 2011; Archer et al. 2017; Nackley et al. 2017). These vegetation changes often cause avian species turnover and shifts in avian community composition (Grant et al. 2004; Anderson and Steidl 2019). Changes in habitat quality can occur within grasslands with woody encroachment via altered nest predation and brood parasitism rates, and decreased food availability and quality (Archer et al. 2017). In the Great Plains, for example, eastern red cedar encroachment has increased the habitat fragmentation of remnant grassland patches, with resultant decreases in the abundance of rangeland songbirds, at least partially to increased rates of nest predation (Coppedge et al. 2001; Engle et al. 2003). Similar fragmentation effects and reduced avian abundances have occurred in areas where western juniper (*Juniperus occidentalis*) has expanded into sagebrush steppe (Noson et al. 2006).

12.7.4 Climate Change

By one estimate, 53% of North American bird species are projected to lose more than half of their current geographic range across three scenarios of climate change by the end of the century (Langham et al. 2015). Grassland habitats and birds are expected to be particularly affected by climate change. Nearly half (42%) of grassland breeding bird species were deemed highly vulnerable under a scenario of a 3.0 °C increase in global mean temperature (Wilsey et al. 2019). Sagebrush songbirds similarly have been deemed threatened with respect to changing climate (Fleishman et al. 2014; National Audubon Society 2014; Nixon et al. 2016).

Spatial and temporal variation in precipitation and temperature influence the occurrence, distribution, and reproductive success of rangeland songbird species (Rotenberry and Wiens 1991; Shaffer and DeLong 2019). Years with moderate moisture and temperatures tend to lead to the highest reproductive output for rangeland songbirds, with implications for increasing variation in precipitation regimes (Ludlow et al. 2014; Conrey et al. 2016; Ruth and Skagen 2018). Increasing intensity of storms, such as those producing hail, can result in local mortality of young and adults tending nests (Carver et al. 2017; Hightower et al. 2018). Moreover, increasing temperatures and drought frequency in the western U.S. will likely decrease the productivity of nesting birds (Skagen and Yackel Adams 2012), especially in areas with higher habitat loss (Zuckerberg et al. 2018). Mismatches between the timing of peak availability of invertebrate prey and peak nesting activity also are likely to continue to become more common with a changing climate, which can lower reproductive success (Lany et al. 2016).

12.8   Conservation and Management Actions

12.8.1   Reversing the Loss and Fragmentation of Native Grasslands and Shrublands

One aspect central to all wildlife conservation is the necessity to maintain large and relatively intact landscapes, most of which are at least partially composed of private lands. Landscape protection therefore necessitates broad coalitions and partnerships (e.g., Chap. 27).

12.8.2   Habitat Management

Given the complexities of the short- and long-term effects of management activities on vegetation and birds in rangelands, and differences in preferred habitat across species, a universal approach to managing rangelands for songbirds does not exist (Duchardt et al. 2019, Shaffer and DeLong 2019). The management practices that facilitate the habitat needs of one species will not necessarily meet the needs of others. Ideal management prescriptions will therefore depend upon specific goals. Because some songbird species are more imperiled than others, a focus on managing for the species of highest conservation concern may be warranted in some scenarios (Herkert et al.1996). Alternatively, management might focus on sensitive species with limited breeding ranges, and whose core breeding ranges occur within the land manager's jurisdiction. Management suggestions pertaining to individual species can be found in Shaffer and DeLong (2019) and Boxes 12.1 and 12.2 herein. The maintenance of heterogeneity within landscapes can provide the requisite microhabitat diversity for the success of individual songbird species and support a variety of species (Engle et al. 2003; Powell 2006). Patches (e.g., sandy draws) within landscapes consisting of tall shrubs and/or higher shrub cover, for example, support the highest breeding densities of sagebrush-obligate songbird species (Chalfoun and Martin 2007; Williams et al. 2011), and other declining species such as the loggerhead shrike (*Lanius ludovicianus*).

Other factors that influence the effectiveness of management for songbird habitat are regional differences in dominant vegetation types (e.g., warm-season or cool-season grasses), rangeland health (degree of degradation and level of biotic diversity), microclimate, and soil type (Shaffer and DeLong 2019). The previous and current land uses of a management unit also warrant consideration. Rangeland management for the conservation of birds may include ongoing maintenance of extant or degraded native grasslands or shrublands, and restoration of areas that had been converted for another use (e.g., agricultural production) to a more native state. Emulating historic, natural disturbances that resulted in a mosaic of habitats and vegetation structure can facilitate habitat heterogeneity and avian diversity. Resource managers may need to experiment with combinations of management tools at different sites with varying soil moisture conditions to maintain the array of habitats required to facilitate the biotic diversity of rangeland ecosystems (Ryan 1990).

Given limited resources for conservation, the premise is that management geared towards a single habitat specialist with large home-range requirements, such as the greater sage-grouse (*Centrocercus urophasianus*), can simultaneously protect other co-occurring species of concern often is appealing. The efficacy of relying on such "umbrella species" (Caro 2010), however, partly depends upon the spatial scale at which management is implemented. At broad spatial scales, the reduction of habitat loss and fragmentation certainly may benefit some co-occurring species (Carlisle et al. 2018a). At finer scales, however, the specific resource needs of the umbrella and sympatric species can diverge, and targeted management actions for the umbrella species may be detrimental to other species (Hanser and Knick 2011; Carlisle and Chalfoun 2020). For example, the experimental reduction of sagebrush cover to benefit sage-grouse during the brood-rearing stage led to complete loss of nesting habitat for sagebrush-obligate songbirds (Carlisle et al. 2018b).

Finally, the need to consider the on-going influence of shifting climatic regimes on vegetation and songbird species will be critical for the long-term success of management actions. Adaptive management strategies that accommodate the shrinking and shifting distributions of climate-sensitive species may be one effective mechanism (Langham et al. 2015).

12.9   Research and Management Needs

The remaining informational gaps and research needs for rangeland songbirds are extensive, as most have not received the same level of prioritization as many game species. Experimental and longer-term studies would help clarify the specific habitat factors, disturbances, and management interventions that most affect songbird responses and the underlying mechanisms (Chalfoun 2021). The further development of tools to mitigate the effects of energy development on songbirds is merited (Sanders and Chalfoun 2018; Shaffer et al. 2019b), which will necessitate mechanistic understanding of the effects of different types of energy development on songbird species. Efforts to restore habitats to pre-disturbance conditions and protect native ecosystems most at risk of conversion for new energy extraction will be paramount.

A better understanding of how and why songbird abundance and community composition change in areas affected by invasive plant species and woody-plant encroachment would be useful. The development of statistically rigorous (e.g., Before-After Control-Impact) studies of rangeland songbird species in relation to specific management prescriptions within rangeland types would clarify optimal management approaches. Experimental designs that account for the independent contributions of potentially confounding variables, such as the effects of burning versus grazing, also would be fruitful. Improved understanding of the effects of ecological processes and mechanical management practices on avian abundance and productivity at scales relevant to management (e.g., grazing allotments) would further clarify optimal management approaches for songbird management (Pulliam et al. 2021). Potential carryover effects of management activities, such as grazing across years, also would provide more holistic understanding (Johnson et al. 2011). In addition, a better understanding of the influence of multiple stressors, including interactions between changing climatic conditions and their effects on songbird habitats, will be critical for the effective management of rangeland songbirds into the future.

A lack of information about the demography of most rangeland songbird species across the full annual life cycle (i.e., outside of the nesting period), and which life stage(s) are most affected by habitat changes and the most limiting to population growth greatly hampers understanding of ideal management allocation (Marra et al. 2015). The post-fledging survival, migratory routes, key stopover areas, overwintering locations, and annual survival of most grassland and sagebrush songbird populations remain unknown, partly because of historic limitations on tracking technologies that could be deployed safely on small birds. Because most rangeland-associated songbirds leave the nest prior to being capable of sustained flight (Yackel Adams et al. 2006), habitat requirements, and rates and causes of mortality during the post-fledgling period may be particularly important to understand for threat and population assessments (Yackel Adams et al. 2006; Davis and Fisher 2009; Hovick et al. 2011). Lack of knowledge about the movements and cause-specific mortality of many grassland and shrubsteppe songbird species during migration and winter also inhibits understanding of the relative influence of the breeding versus non-breeding periods on annual survival and therefore population growth (Fletcher et al. 2006). Finally, conditions and processes during particular life stages can carry-over into subsequent stages (Akresh et al. 2021), albeit to an unknown extent for most songbirds inhabiting North American rangelands.

Study of the prevalence and effects of disease (e.g., salmonellosis), endoparasites, and ectoparasites (e.g., blowflies) on the condition and vital rates of rangeland songbirds is in its infancy. Fairly high blow fly loads have been observed on some nestlings in Montana and Wyoming, which can result in partial or complete mortality of the brood (A. Chalfoun, personal observation). Whether particular conditions such as energy development, livestock grazing, or weather influence the susceptibility of songbirds to disease or parasites, and whether such changes scale up to influence populations, remains unclear.

Finally, the importance of understanding and acknowledging the contribution of native peoples' role in wildlife management, and the incorporation of indigenous and local knowledge into management policies, has been emphasized recently by scholars and U.S. legislators (Lam et al. 2020). Such information is rarely incorporated into rangeland management plans, yet such knowledge offers historical insights that may complement and enrich contemporary approaches to sustainable use of landscapes and encourage practices that are more culturally inclusive and holistic (Lam et al. 2020).

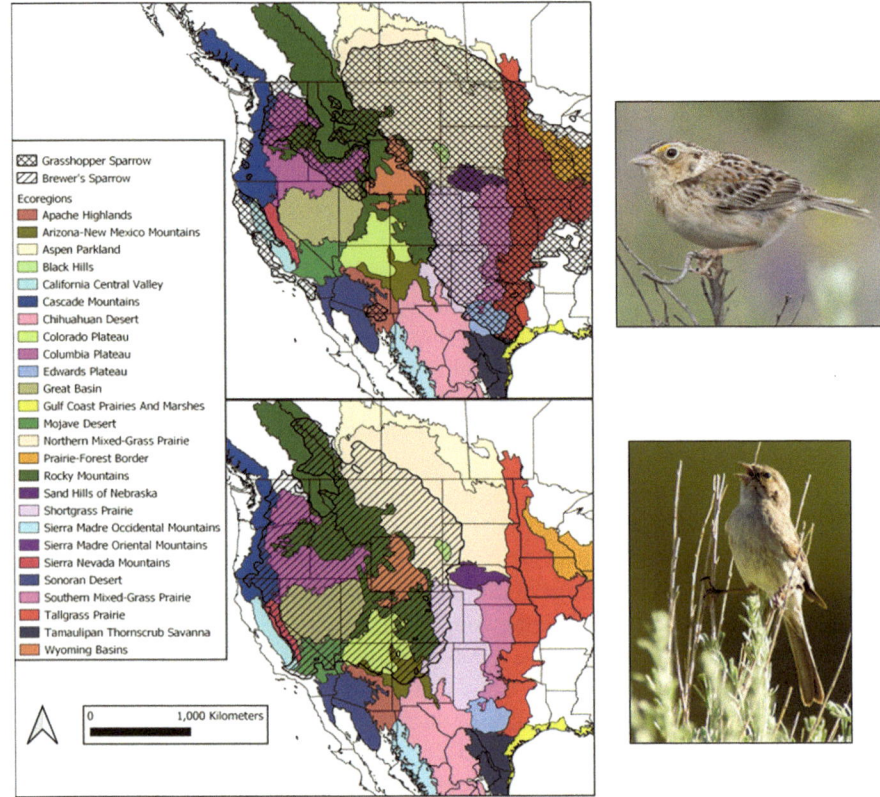

Fig. 12.2 Grasshopper sparrow (*Ammodramus savannarum*) and Brewer's sparrow (*Spizella breweri*) breeding distributions in relation to rangeland ecoregions of North America. *Photo credits* Blair Dudeck and Dave Showalter

---

Box 12.1. Grasshopper Sparrow (Ammodramus savannarum)

Habitat Associations

The grasshopper sparrow is a grassland-obligate songbird with a trans-coastal distribution, whose highest breeding densities occur throughout the Great Plains (Vickery 2020; Fig. 12.2). Grasshopper sparrows breed in a variety of native habitats including prairies, desert grasslands, savannahs, and sagebrush steppe, and in nonnative habitats such as planted grasslands (Shaffer et al. 2021). Throughout the grasslands of the Great Plains, grasshopper sparrows tend to avoid areas with woody vegetation (Bakker et al. 2002; Patten et al. 2006; Herse et al. 2018), where the species is reported to be area sensitive [that is, to show a preference for larger extents of grassland; reviewed in Ribic et al. (2009) and Shaffer et al. (2021)]. Within sagebrush steppe, however, the species is more tolerant of shrubby habitats that include native bunchgrasses in the understory (Holmes and Miller 2010; Earnst and Holmes 2012). Vegetation structure likely influences the habitat decisions of grasshopper sparrows to a greater extent than plant species' composition (Henderson and Davis 2014; Shaffer et al. 2021), as grasshopper sparrows will nest within areas dominated by nonnative grasses such as Kentucky bluegrass *(Poa pratensis)* and smooth brome *(Bromus inermis)* (Grant et al. 2004; Shaffer et al. 2021). Structural attributes of vegetation associated with grasshopper sparrow occupancy include intermediate height (8–166 cm), moderate density (4–80 cm visual obstruction reading), moderately deep litter (≤ 9 cm), low-to-moderate bare ground ($\leq 38\%$ bare ground), and low shrub cover ($< 35\%$) (Shaffer et al. 2021). This narrative will focus on management approaches to benefit grasshopper sparrows breeding in grassland habitats and will not discuss management approaches for other ecosystems such as sagebrush.

Rangeland Management

Management approaches that maintain open expanses of grassland benefit the grasshopper sparrow. Typical management approaches for the grasshopper sparrow involve grazing-only or a combination of burning and grazing (Shaffer et al. 2021). Optimal management strategies vary across the species' range depending on how the resultant habitat structure and composition comports with the species' habitat requirements.

The effects of grazing on grasshopper sparrows depend on local factors such as rangeland type, climate, topoedaphic conditions, and landscape composition (Lipsey and Naugle 2017; Vold et al. 2019; Davis et al. 2021). Further considerations include the timing of grazing, grazing intensity (e.g., stocking rate and duration), and livestock type, all of which may influence the abundance and reproductive success of grasshopper sparrows (Shaffer and DeLong 2019; Shaffer et al. 2021). Appropriate intervals between management treatments depend on rangeland type; for example, mesic prairies regenerate litter more rapidly (1–3 years) than dry prairies (4–6 years) and sooner in southern than northern prairies (Swengel 1996). In tallgrass prairies, moderate-to-heavy grazing reduces vegetation biomass and curtails woody encroachment (Ahlering and Merkord 2016). In mixed-grass prairies, light-to-moderate grazing that reduces vegetation height and density and creates patchy areas is compatible with the habitat needs of the grasshopper sparrow, however, heavy grazing can reduce litter depth and cover to unsuitable levels (Shaffer et al. 2021). Nest densities in Palouse prairie decreased with cattle stocking rates, and no nests were found in pastures with the highest stocking rates of 43.2 animal unit months (46% vegetation utilization) (Johnson et al. 2011). Grazing studies within desert grasslands have been very localized (Ruth 2015), and even light grazing can be detrimental if vegetation becomes too short and open for grasshopper sparrow use (Bock and Webb 1984; Bock et al. 1984, 1993). Heavy grazing in arid grasslands tends to reduce already sparse bunchgrass cover and exclude grasshopper sparrows (Bock and Webb 1984).

In contrast to earlier studies in mixed-grass prairies that suggested that rotational grazing systems may achieve desired vegetation heterogeneity for birds (Shaffer et al. 2021), rest-rotation grazing in northern mixed-grass prairies failed to increase grasshopper sparrow abundance, relative to traditional season-long or summer-rotation grazing (Vold et al. 2019). Similarly, rotational grazing systems in the Nebraska Sandhills (Sliwinski et al. 2019, 2020), shortgrass prairies (Davis et al. 2020), and tallgrass prairies (Temple et al. 1999) did not convey population benefits to the grasshopper sparrow. Rotational grazing systems may fail to create sufficient vegetation heterogeneity to be ecologically relevant to bird communities (Sliwinski et al. 2020), especially when other factors such as year, ecological site (Davis et al. 2020), and stocking rate (Sliwinski et al. 2019; Vold et al. 2019) can have greater effects on bird abundance than grazing system. Livestock type also may influence avian abundance, as grasshopper sparrows were more abundant in Montana pastures grazed by bison than in pastures grazed by cattle (Boyce et al. 2021) and equally as abundant in cattle- and bison-grazed pastures in Colorado (Wilkins et al. 2019).

Fire as a management strategy for grasshopper sparrows is more common within Great Plains grasslands than desert grasslands, where burns usually are the result of wildfires (Shaffer et al. 2021). Similar to grazing, the response of grasshopper sparrows to burning will depend on how vegetation structure is affected, which may vary locally by climate, ecosystem or habitat type, type of burn (e.g., prescribed burn versus wildfire), season, frequency, and intensity (Shaffer and DeLong 2019; Shaffer et al. 2021). The effects of burning-only management on the abundance of grasshopper sparrows in mixed-grass and tallgrass prairies have been varied (Madden et al. 1999; Grant et al. 2010; Byers et al. 2017). In desert grasslands, fires that destroy shrubs may be detrimental, as small shrubs are used by grasshopper sparrows as thermal refugia during extremely hot temperatures (Ruth et al. 2020).

The combination of burning and grazing is a common management approach implemented throughout the Great Plains. Geographic variation in management objectives and approaches, study designs, and timing of the application of burning and grazing, however, make a statement of broad generalizations about the effect on grasshopper sparrow abundance or success difficult (Shaffer et al. 2021). Burning and grazing approaches predominate in northern mixed-grass prairies (e.g., Richardson et al. 2014), shortgrass prairies (e.g., Augustine and Derner 2015), tallgrass prairies (e.g., Fuhlendorf et al. 2006), and sand sagebrush grasslands (Holcomb et al 2014). The patch-burn grazing strategy explained in this chapter's main section is advocated primarily for tallgrass prairies, with a focus on the Flint Hills (Fuhlendorf and Engle 2001). Examinations between the effects of the traditional burning and grazing approach in the Flint Hills (that is, annual early-spring burns followed immediately by grazing) and the patch-burn grazing approach have yielded variable results on grasshopper sparrow abundance and productivity (Shaffer et al. 2021).

Box 12.2. Brewer's Sparrow (Spizella breweri)

Habitat Associations

The Brewer's sparrow often is referred to as a sagebrush-obligate (Rich et al. 2005), along with the sagebrush sparrow (*Artemisiospiza nevadensis*) and sage thrasher (*Orescoptes montanus*), although Brewer's sparrows occasionally inhabit other shrubby habitats. Brewer's sparrows are migratory, and overwinter in the southwestern U.S. and Mexico (Knick et al. 2014; Valencia-Herverth et al. 2018). The breeding range of Brewer's sparrows covers most of the extent of North American shrubsteppe, from southern British Columbia, Alberta and Saskatchewan in Canada, south to southern California, southern Nevada, and northern Arizona and New Mexico (Walker et al. 2020; Fig. 12.2).

Brewer's sparrows prefer and are found in the highest breeding densities within landscapes composed of relatively high cover of sagebrush (> 30%) and taller shrubs (Rotenberry and Wiens 1980; Chalfoun and Martin 2007; Walker et al. 2020). Because Brewer's sparrows primarily sing, forage and nest within the shrub layer (e.g., Rotenberry and Wiens 1998; Fig. 12.3), the attributes of the shrub layer are paramount. Brewer's sparrows may be entirely absent from areas with shrub cover ≤ 8% (Walker et al. 2020). Habitat preferences and quality, moreover, are scale-dependent. At smaller spatial scales, Brewer's sparrows preferred and had higher reproductive success in territories and nest patches (5-m radius) with higher shrub density, and particularly densities of potentially suitable nest shrubs (Chalfoun and Martin 2007, 2009).

Fig. 12.3 Brewer's sparrow adult incubating eggs, a clutch of eggs, and a nest in the process of hatching in western Wyoming, USA. *Photo credits* Anna Chalfoun

Brewer's sparrows prefer to forage and place nests in shrubs with live canopies (Peterson and Best 1985; Rotenberry and Wiens 1998; Chalfoun and Martin 2009; Fig. 12.3).

Rangeland Management

As a near sagebrush-obligate, management actions that remove or decrease big sagebrush (*Artemisia tridentata*) cover, including burning (Bock and Bock 1987; Knick et al. 2005; Noson et al. 2006), the application of herbicides (Best 1972; Schroeder and Sturges 1975; Kerley and Anderson 1995), and mechanical treatments (Castrale 1982; Carlisle et al. 2018b) tend to eliminate or reduce the local abundance of the Brewer's sparrow (Walker et al. 2020). Because big sagebrush does not re-sprout after fire, frequent fires increase the cover of annual grasses and reduce sagebrush cover which decreases habitat for sagebrush-obligates including the Brewer's sparrow (Knick et al. 2003). Burning also can negatively affect sagebrush songbirds by promoting the spread of nonnative weeds and the subsequent conversion of shrubsteppe habitats to nonnative annual grasslands. The planting of nonnative grasses following sagebrush removal hinders recolonization by sagebrush and delays or prohibits the recovery of Brewer's sparrow habitat (Reynolds and Trost 1980; McAdoo et al. 1989). Insecticide treatments during the nesting period have the potential to reduce arthropod prey and thereby alter Brewer's sparrow habitat use and productivity (Howe et al. 1996).

Management activities that reduce coniferous encroachment into sagebrush habitats have shown positive effects on sagebrush songbirds (Crow and van Riper 2010). However, habitat treatments traditionally geared towards the reduction of the sagebrush canopy and enhancement of herbaceous understories, thought to benefit the greater sage-grouse (*Centrocercus urophasianus*) and the Gunnison sage-grouse (*Centrocercus minimus*) during the brood rearing period, usually negatively affect sagebrush-obligate songbirds. Mechanical treatments (e.g., roller chopping, disking, mowing,) in Colorado significantly decreased densities of Brewer's sparrows (Lukacs et al. 2015). Moreover, experimental evaluation of mowing effects in central Wyoming resulted in the complete loss of nesting habitat for Brewer's sparrows and sage thrashers (Carlisle et al. 2018b).

Rigorous investigations of the effects of grazing regimes on the Brewer's sparrows have been limited. The abundance of Brewer's sparrows did not differ between rest-rotation versus season-long grazing treatments in Montana (Golding and Dreitz 2017). However, Brewer's sparrow abundance decreased with the highest grazing treatment during a study in southern Idaho and northern Utah, which corresponded with lower shrub cover and higher cover of exotic annuals (Bradford et al. 1998). Brewer's sparrows tend to be less affected by moderate grazing compared with grassland songbirds that are more reliant on the herbaceous understory (Bock et al. 1993).

References

Ahlering MA, Merkord CL (2016) Cattle grazing and grassland birds in northern tallgrass prairie. J Wildl Manage 80:643–654. https://doi.org/10.1002/jwmg.1049

Akresh ME, King DI, Marra PP (2021) Hatching date influences winter habitat occupancy: examining seasonal interactions across the full annual cycle in a migratory songbird. Ecol Evol 11:9241–9253. https://doi.org/10.1002/ece3.750

Allen AW, Cade BS, Vandever MW (2001) Effects of emergency haying on vegetative characteristics within selected Conservation Reserve Program fields in the northern Great Plains. J Soil Water Conserv 56:120–125. https://www.jswconline.org/content/56/2/120

Allison TD, Diffendorfer JE, Baerwald EF, Beston J, Drake D, Hale A, Hein C, Huso MM, Loss S, Lovich JE, Strickland D, Williams K, Winder V (2019) Impacts to wildlife of wind energy siting and operation in the United States. Issues in Ecol, Report No. 21. https://www.esa.org/wp-content/uploads/2019/09/Issues-in-Ecology_Fall-2019.pdf

Allred BW, Fuhlendorf SD, Hamilton RG (2011) The role of herbivores in Great Plains conservation: comparative ecology of bison and cattle. Ecosphere 2:1–17. https://doi.org/10.1890/ES10-00152.1

Allred BW, Fuhlendorf SD, Engle DM, Elmore RD (2011) Ungulate preference for burned patches reveals strength of fire-grazing interaction. Ecol Evol 1:132–144. https://doi.org/10.1002/ece3.12

Anderson EM, Steidl RJ (2019) Woody plant encroachment restructures bird communities in semiarid grasslands. Biol Cons 240:108276. https://doi.org/10.1016/j.biocon.2019.108276

Archer SR, Anderson EM, Predick KI, Schwinning S, Steidl RJ, Woods SR (2017) Woody plant encroachment: causes and consequences. In: Briske D (ed) Rangeland systems: processes, management, and challenges. Springer Series on Environmental Management, pp 25–84. https://doi.org/10.1007/978-3-319-46709-2

Augustine DJ, Derner JD (2015) Patch-burn grazing management, vegetation heterogeneity, and avian responses in a semi-arid grassland. J Wildl Manage 79:927–936. https://doi.org/10.1002/jwmg.909

Bailey DW (2005) Identification and creation of optimum habitat conditions for livestock. Rangel Ecol Manag 58:109–118. https://doi.org/10.2111/03-147.1

Bakker KK, Naugle DE, Higgins KF (2002) Incorporating landscape attributes into models for migratory grassland bird conservation. Conserv Biol 16:1638–1646. https://doi.org/10.1046/j.1523-1739.2002.01328.x

Barbour MG, Billings WD (2000) North American terrestrial vegetation. Cambridge University Press, Cambridge

Bayly NJ, Gomez C, Hobson KA, Rosenberg KV (2016) Prioritizing tropical habitats for long-distance migratory songbirds: an assessment of habitat quality at a stopover site in Colombia. Avian Conserv Ecol 11:article 5. https://doi.org/10.5751/ACE-00873-110205

Beason RC (2020) Horned lark (*Eremophila alpestris*). In: Billerman SM (ed) Birds of the World, vers.1.0. Cornell Lab of Ornithology, Ithaca, New York. https://doi.org/10.2173/bow.horlar.01

Bedard J, McNeil JN (1979) *Protocalliphora hirudo* (Diptera: Calliphoridae) infesting Savannah sparrow, *Passerculus sandwichensis* (Aves: Fringillidae), in eastern Quebec. Can Entomol 111:111–112. https://doi.org/10.4039/Ent111111-1

Berkeley LI, McCarty JP, Wolfenbarger LL (2007) Postfledging survival and movement in dickcissels (*Spiza americana*): implications for habitat management and conservation. Auk 124:396–409. https://doi.org/10.1093/auk/124.2.396

Bernath-Plaistad J, Koper N (2016) Physical footprint of oil and gas infrastructure, not anthropogenic noise, reduces nesting success of some grassland songbirds. Biol Conserv 204:434–441. https://doi.org/10.1016/j.biocon.2016.11.002

Best LB (1972) First-year effects of sagebrush control on two sparrows. J Wildl Manage 36:534–544

Bestelmeyer BT, Peters DPC, Archer SR, Browning DM, Okin GS, Schooley RL, Webb NP (2018) The grassland-shrubland regime shift in the southwestern United States: misconceptions and their implications for management. Bioscience 68:678–690. https://doi.org/10.1093/biosci/biy065

Bleho BI, Koper N, Machtans CS (2014) Direct effects of cattle on grassland birds in Canada. Conserv Biol 28:724–734. https://doi.org/10.1111/cobi.12259

Bock CE, Webb B (1984) Birds as grazing indicator species in southeastern Arizona. J Wildl Manage 48:1045–1049. https://doi.org/10.2307/3801466

Bock CE, Bock JH (1987) Avian habitat occupancy following fire in a Montana shrubsteppe. Prairie Naturalist 19:153–158

Bock CE, Bock JH, Kenney WR, Hawthorne VM (1984) Responses of birds, rodents, and vegetation to livestock exclosure in a semidesert grassland site. J Range Manage 37:239–242. https://doi.org/10.2307/3899146

Bock CE, Saab VA, Rich TD, Dobkin DS (1993) Effects of livestock grazing on Neotropical migratory landbirds in western North America. In: Finch DM, Stangel PW (eds) Status and management of Neotropical migratory birds. U.S. Department of Agriculture, Forest Service, Rocky Mountain Forest and Range Experiment Station, General Technical Report RM-229. Fort Collins, Colorado, pp 296–309

Bollinger EK, Bollinger PB, Gavin TA (1990) Effects of hay-cropping on eastern populations of the bobolink. Wilson Bull 18:142–150. https://www.jstor.org/stable/3782128

Boyce AM, Shamon H, Kunkel KE, McShea WJ (2021) Grassland bird diversity and abundance in the presence of native and non-native grazers. Avian Cons Ecol 16:article 13. https://doi.org/10.5751/ACE-01944-160213

Boyd EM (1951) The external parasites of birds: a review. Wilson Bull 63:363–369

Bradford DF, Franson SE, Neale AC, Heggem DT, Miller GR, Canterbury GE (1998) Bird species assemblages as indicators of biological integrity in Great Basin rangeland. Environ Monit Assess 49:1–22. https://doi.org/10.1023/A:1005712405487

Briske DD, Derner JD, Brown JR, Fuhlendorf SD, Teague WR, Havstad KM, Gillen RL, Ash AJ, Willms WD (2008) Rotational grazing on rangelands: reconciliation of perception and experimental evidence. Rangel Ecol Manag 61:3–17. https://doi.org/10.2111/06-159R.1

Brooks ML, D'Antonio CM, Richardson DM, Grace JB, Keeley JE, DiTomaso JM, Hobbs RJ, Pellant M, Pyke D (2004) Effects of invasive alien plants on fire regimes. Bioscience 54:677–683. https://doi.org/10.1641/0006-3568(2004)054[0677:EOIAPO]2.0.CO;2

Byers CM, Ribic CA, Sample DW, Dadisman JD, Guttery MR (2017) Grassland bird productivity in warm season grass fields in southwest Wisconsin. Am Midl Nat 178:47–63. https://doi.org/10.1674/0003-0031-178.1.47

Carlisle JD, Chalfoun AD (2020) The abundance of greater sage-grouse as a proxy for the abundance of sagebrush-associated songbirds in Wyoming, USA. Avian Cons Ecol 15:article 16. https://doi.org/10.5751/ACE-01702-150216

Carlisle JD, Keinath DA, Albeke SE, Chalfoun AD (2018a) Identifying holes in the greater sage-grouse conservation umbrella. J Wildl Manage 82:948–957. https://doi.org/10.1002/jwmg.21460

Carlisle JD, Chalfoun AD, Smith KT, Beck JL (2018b) Nontarget effects on songbirds from habitat manipulation for greater sage-grouse: implications for the umbrella species concept. Condor Ornithol Appl 120:439–455. https://doi.org/10.1650/CONDOR-17-200.1

Caro TM (2010) Conservation by proxy: indicator, umbrella, keystone, flagship, and other surrogate species. Island Press, Washington, DC

Carver AR, Ross JD, Augustine DJ et al (2017) Weather radar data correlate to hail-induced mortality in grassland birds. Remote Sens Ecol Conserv 3:90–101. https://doi.org/10.1002/rse2.41

Castrale JS (1982) Effects of two sagebrush control methods on nongame birds. J Wildl Manage 46:945–952. https://doi.org/10.2307/3808227

Centers for Disease Control (2016) Species of dead birds in which West Nile virus has been detected, United States, 1999–2016. https://www.cdc.gov/westnile/resources/pdfs/birdspecies1999-2016.pdf

Chalfoun AD (2021) Responses of vertebrate wildlife to oil and natural gas development: patterns and frontiers. Curr Landscape Ecol Rep 6:71–84. https://doi.org/10.1007/s40823-021-00065-0

Chalfoun AD, Martin TE (2007) Assessments of habitat preferences and quality depend on spatial scale and metrics of fitness. J Appl Ecol 44:983–992. https://doi.org/10.1111/j.1365-2664.2007.01352.x

Chalfoun AD, Martin TE (2009) Habitat structure mediates predation risk for sedentary prey: experimental tests of alternative hypotheses. J Anim Ecol 78:497–503. https://doi.org/10.1111/j.1365-2656.2008.01506.x

Cicero C, Pyle P, Patten MA (2020) Juniper titmouse (*Baeolophus ridgwayi*). In: Rodewald PG (ed) Birds of the world, vers. 1.0. Cornell Lab of Ornithology, Ithaca, New York. https://doi.org/10.2173/bow.juntit1.01

Coates PS, Brussee BE, Howe KB et al (2016) Landscape characteristics and livestock presence influence common ravens: relevance to greater sage-grouse conservation. Ecosphere 7:e01203. https://doi.org/10.1002/ecs2.1203

Cody ML (1985) Habitat selection in grassland and open country birds. In: Cody ML (ed) Habitat selection in birds. Academic Press, New York, pp 191–226

Coffman JM, Bestelmeyer BT, Kelly JF, Wright TF, Schooley RL (2014) Restoration practices have positive effects on breeding bird species of concern in the Chihuahuan desert. Restor Ecol 22:336–344. https://doi.org/10.1111/rec.12081

Conrey RY, Skagen SK, Yackel Adams AA, Panjabi AO (2016) Extremes of heat, drought and precipitation depress reproductive performance in shortgrass prairie passerines. Ibis 158:614–629. https://doi.org/10.1111/ibi.12373

Coppedge BR, Engle DM, Masters RE, Gregory MS (2001) Avian response to landscape change in fragmented southern Great Plains grasslands. Ecol Appl 11:47–59. https://doi.org/10.1890/1051-0761(2001)011[0047:ARTLCI]2.0.CO;2

Crow C, van Riper IIIC (2010) Avian community responses to mechanical thinning of a pinyon-juniper woodland: specialist sensitivity to tree reduction. Nat Areas J 30:191–201. https://doi.org/10.3375/043.030.0206

Danner JE, Small DM, Ryder TB, Lohr B, Masters BS, Gill DE, Fleischer RC (2018) Temporal patterns of extra-pair paternity in a population of grasshopper sparrows (*Ammodramus savannarum*) in Maryland. Wilson J Ornithol 130:40–51. https://doi.org/10.1676/16-049.1

Davies KW, Bates JD (2020) Re-introducing fire in sagebrush steppe experiencing decreased fire frequency: does burning promote spatial and temporal heterogeneity? Int J Wildland Fire 29:686–695. https://doi.org/10.1071/WF20018

Davis SK, Duncan DC (1999) Grassland songbird occurrence in native and crested wheatgrass pastures of southern Saskatchewan. In: Vickery PD, Herkert JR (eds) Ecology and conservation of grassland birds of the Western Hemisphere. Stud Avian Biol 19:211–218

Davis SK, Lanyon WE (2020) Western meadowlark (*Sturnella neglecta*). In: Poole AF (ed) Birds of the world, vers. 1.0. Cornell Lab of Ornithology, Ithaca, New York. https://doi.org/10.2173/bow.wesmea.01

Davis SK, Fisher RJ (2009) Post-fledging movements of Sprague's pipit. Wilson J Ornithol 121:198–202. https://doi.org/10.1676/08-025.1

Davis KP, Augustine DJ, Monroe AP, Derner JD, Aldridge CL (2020) Adaptive rangeland management benefits grassland birds utilizing opposing vegetation structure in the shortgrass steppe. Ecol Appl 30:e02020. https://doi.org/10.1002/eap.2020

Davis KP, Augustine DJ, Monroe AP, Aldridge CL (2021) Vegetation characteristics and precipitation jointly influence grassland bird abundance beyond the effects of grazing management. Ornithol Appl 123:1–15. https://doi.org/10.1093/ornithapp/duab041

de Lope F, González G, Pérez JJ, Møller AP (1993) Increased detrimental effects of ectoparasites on their bird hosts during adverse environmental conditions. Oecologia 95:234–240. https://doi.org/10.1007/BF00323495

Deeming DC (2001) Avian incubation: environment, behavior and evolution. Oxford University Press, London

Derner JD, Lauenroth WK, Stapp P, Augustine DJ (2009) Livestock as ecosystem engineers for grassland bird habitat in the western Great Plains of North America. Rangel Ecol Manag 62:111–118. https://doi.org/10.2111/08-008.1

Duchardt CJ, Augustine DJ, Beck JL (2019) Threshold responses of grassland and sagebrush birds to patterns of disturbance created by an ecosystem engineer. Landsc Ecol 34:895–909. https://doi.org/10.1007/s10980-019-00813-y

Duchardt CJ, Porensky LM, Pearse IS (2021) Direct and indirect effects of a keystone engineer on a shrubland-prairie food web. Ecology 102:e03195. https://doi.org/10.1002/ecy.3195

Earnst SL, Holmes AL (2012) Bird-habitat relationships in Interior Columbia Basin shrubsteppe. Condor 14:15–29. https://doi.org/10.1525/cond.2009.080109

Engle DM, Fuhlendorf SD, Coppedge BR (2003) Conservation priorities on fragmented, homogenized Great Plains landscapes. In: Fore S (ed) Proceeding of the 18th North American Prairie Conference, pp 1–6

Erickson WP, Wolfe MM, Bay KJ, Johnson DH, Gehring JL (2014) A comprehensive analysis of small-passerine fatalities from collision with turbines at wind energy facilities. PLoS ONE 9:e107491. https://doi.org/10.1371/journal.pone.0107491

Fisher RJ, Davis SK (2011) Post-fledging dispersal, habitat use, and survival of Sprague's pipits: are planted grasslands a good substitute for native? Biol Conserv 144:263–271. https://doi.org/10.1016/j.biocon.2010.08.024

Fleishman E, Thomson JR, Kalies EL, Dickson BG, Dobkin DS, Leu M (2014) Projecting current and future location, quality, and connectivity of habitat for breeding birds in the Great Basin. Ecosphere 5: article 82. https://doi.org/10.1890/ES13-00387.1

Fletcher RJ Jr, Koford RR, Seaman DA (2006) Critical demographic parameters for declining songbirds breeding in restored grasslands. J Wildl Manage 70:145–157. https://doi.org/10.2193/0022-541X(2006)70[145:CDPFDS]2.0.CO;2

Frost JS, Powell LA (2011) Cedar infestation impacts avian communities along the Niobrara River Valley, Nebraska. Restor Ecol 19:529–536. https://doi.org/10.1111/j.1526-100X.2009.00618.x

Fuhlendorf SD, Engle DM (2001) Restoring heterogeneity on rangelands: ecosystem management based on evolutionary grazing patterns. Bioscience 51:625–632. https://doi.org/10.1641/0006-3568(2001)051[0625:RHOREM]2.0.CO;2

Fuhlendorf SD, Harrell WC, Engle DM, Hamilton RG, Davis CA, Leslie DM Jr (2006) Should heterogeneity be the basis for conservation? Grassland bird response to fire and grazing. EcolAppl 16:1706–1716. https://doi.org/10.1890/10510761(2006)016%5B1706:SHBTBF%5D2.0.CO;2

Gilbert MG, Chalfoun AD (2011) Energy development affects populations of sagebrush songbirds in Wyoming. J Wildl Manage 75:816–824. https://doi.org/10.1002/jwmg.123

Goguen CB, Mathews NE (1999) Review of the causes and implications of the association between cowbirds and livestock. In: Hahn DC, Hall LS, Morrison ML et al (eds) Research and management of the brown-headed cowbird in western landscapes. Stud Avian Biol 18:10–17

Goguen CB, Mathews NE (2000) Local gradients of cowbird abundance and parasitism relative to livestock grazing in a western landscape. Conserv Biol 14:1862–1869. https://doi.org/10.1111/j.1523-1739.2000.99313.x

Golding JD, Dreitz VJ (2017) Songbird response to rest-rotation and season-long cattle grazing in a grassland sagebrush ecosystem. J Environ Manage 204:605–612. https://doi.org/10.1016/j.jenvman.2017.09.044

Grant TA, Murphy RK (2005) Changes in woodland cover on prairie refuges in North Dakota, USA. Nat Areas J 25:359–368

Grant TA, Madden E, Berkey GB (2004) Tree and shrub invasion in northern mixed-grass prairie: implications for breeding grassland birds. Wildl Soc Bull 32:807–818. https://doi.org/10.2193/0091-7648(2004)032[0807:TASIIN]2.0.CO;2

Grant TA, Madden E, Shaffer TL, Dockens JS (2010) Effects of prescribed fire on vegetation and passerine birds in northern mixed-grass prairie. J Wildl Manage 74:1841–1851. https://doi.org/10.2193/2010-006

Green AW, Pavlacky DC Jr, George TL (2019) A dynamic multi-scale occupancy model to estimate temporal dynamics and hierarchical habitat use for nomadic species. Ecol Evol 9:793–803. https://doi.org/10.1002/ece3.4822

Green MT, Lowther PE, Jones SL, Davis SK, Dale BC (2020) Baird's sparrow (*Centronyx bairdii*). In: Poole AF, Gill FB (eds) Birds of the world, vers. 1.0. Cornell Lab of Ornithology, Ithaca, New York. https://doi.org/10.2173/bow.baispa.01

Grinnell J (1917) The niche-relationships of the California thrasher. Auk 34:427–433. https://doi.org/10.2307/4072271

Hanser SE, Knick ST (2011) Greater sage-grouse as an umbrella species for shrubland passerine birds: a multiscale assessment. In: Knick S, Connelly JW (eds) Ecology, conservation, and management of grouse. Stud Avian Biol 38:475–488

Henderson AE, Davis SK (2014) Rangeland health assessment: a useful tool for linking range management and grassland bird conservation? Rangel Ecol Manag 67:88–98. https://doi.org/10.2111/REM-D-12-00140.1

Herkert JR, Sample DW, Warner RE (1996) Management of midwestern grassland landscapes for the conservation of migratory birds. In: Thompson III FR (ed) Management of midwestern landscapes for the conservation of Neotropical migratory birds. U.S. Department of Agriculture, Forest Service, North Central Forest Experiment Station, General Technical Report GTR-NC-187. St. Paul, Minnesota, pp 89–116

Herse MR, With KA, Boyle WA (2018) The importance of core habitat for a threatened species in changing landscapes. J Appl Ecol 55:2241–2252. https://doi.org/10.1111/1365-2664.13234

Hethcoat MG, Chalfoun AD (2015) Toward a mechanistic understanding of human-induced rapid environmental change: a case study linking energy development, nest predation, and predators. J Appl Ecol 52:1492–1499. https://doi.org/10.1016/j.biocon.2015.02.009

Hethcoat MG, Chalfoun AD (2015) Energy development and avian nest survival in Wyoming, USA: a test of a common disturbance index. Biol Conserv 184:327–334. https://doi.org/10.1016/j.biocon.2015.02.009

Hickman KR, Farley GH, Channell R, Steier JE (2006) Effects of old world bluestem (*Bothriochloa ischaemum*) on food availability and avian community composition within the mixed-grass prairie. Southwest Nat 51:524–530. https://doi.org/10.1894/0038-4909(2006)51[524:EOOWBB]2.0.CO;2

Higgins KF (1986) A comparison of burn season effects on nesting birds in North Dakota mixed-grass prairie. Prairie Nat 18:219–228

Hightower JN, Carlisle JD, Chalfoun AD (2018) Nest mortality of sagebrush songbirds due to a severe hailstorm. Wilson J Ornithol 130:561–567. https://doi.org/10.1676/17-025.1

Holcomb ED, Davis CA, Fuhlendorf SD (2014) Patch-burn management: implications for conservation of avian communities in fire-dependent sagebrush ecosystems. J Wildl Manage 78:848–856. https://doi.org/10.1002/jwmg.723

Holmes AL, Miller RF (2010) State-and-transition models for assessing grasshopper sparrow habitat use. J Wildl Manage 74:1834–1840. https://doi.org/10.2193/2009-417

Hovick TJ, Miller JR (2013) Broad-scale heterogeneity influences nest selection by brown-headed cowbirds. Landsc Ecol 28:1493–1503. https://doi.org/10.1007/s10980-013-9896-7

Hovick TJ, Miller JR, Koford RR, Engle DM, Debinski DM (2011) Post-fledging survival of grasshopper sparrows in grasslands managed with fire and grazing. Condor 113:429–437. https://doi.org/10.1525/cond.2011.100135

Hovick TJ, Elmore RD, Fuhlendorf SD (2014) Structural heterogeneity increases diversity of non-breeding grassland birds. Ecosphere 5:1–13. https://doi.org/10.1890/ES14-00062.1

Howe FP (1992) Effects of *Protocalliphora braueri* (Diptera: Calliphoridae) parasitism and inclement weather on nestling sage thrashers. J Wildl Dis 28:141–143. https://doi.org/10.7589/0090-3558-28.1.141

Howe FP, Knight RL, McEwen LC, George TL (1996) Direct and indirect effects of insecticide applications on growth and survival of nestling passerines. Ecol Appl 6:1314–1324. https://doi.org/10.2307/2269609

Igl LD, Ballard BM (1999) Habitat associations of migrating and overwintering grassland birds in southern Texas. Condor 101:771–782. https://doi.org/10.2307/1370064

Igl LD, Johnson DH (2016) Effects of haying on breeding birds in CRP grasslands. J Wildl Manage 80:1189–1204. https://doi.org/10.1002/jwmg.21119

Johnson TN, Kennedy PL, DelCurto T, Taylor RV (2011) Bird community responses to cattle stocking rates in a Pacific Northwest bunchgrass prairie. Agric Ecosyst Environ 144:338–346. https://doi.org/10.1016/j.agee.2011.10.003

Johnson TN, Kennedy PL, Etterson MA (2012) Nest success and cause-specific nest failure of grassland passerines breeding in prairie grazed by livestock. J Wildl Manage 76:1607–1616. https://doi.org/10.1002/jwmg.437

Johnson DH, Igl LD, Shaffer JA, DeLong JP (2019) The effects of management practices on grassland birds. U.S. Geological Survey Professional Paper 1842. https://doi.org/10.3133/pp1842

Jones ZF, Bock CE (2005) The Botteri's sparrow and exotic Arizona grasslands: an ecological trap or habitat regained? Condor 107:731–741. https://doi.org/10.1093/condor/107.4.731

Jones TM, Ward MP (2020) Pre- to post-fledging carryover effects and the adaptive significance of variation in wing development for juvenile songbirds. J Anim Ecol 89:2235–2245. https://doi.org/10.1111/1365-2656.13285

Jones NF, Pejchar L, Kiesecker JM (2015) The energy footprint: how oil, natural gas, and wind energy affect land for biodiversity and the flow of ecosystem services. Bioscience 65:290–301. https://doi.org/10.1093/biosci/biu224

Jones TM, Ward MP, Benson TJ, Brawn JD (2017) Variation in nestling body condition and wing development predict cause-specific mortality in fledgling dickcissels. J Avian Biol 48:439–447. https://doi.org/10.1111/jav.01143

Jones TM, Brawn JD, Ward MP (2018) Development of activity rates in fledgling songbirds: when do young birds begin to behave like adults? Behaviour 155:337–350. https://doi.org/10.1163/1568539X-00003492

Kalyn Bogard HJ, Davis SK (2014) Grassland songbirds exhibit variable responses to the proximity and density of natural gas wells. J Wildl Manage 78:471–482. https://doi.org/10.1002/jwmg.684

Kennedy PL, DeBano SJ, Bartuszevige AM, Lueders AS (2009) Effects of native and non-native grassland plant communities on breeding passerine birds: implications for restoration of northwest bunchgrass prairie. Restor Ecol 17:515–525. https://doi.org/10.1111/j.1526-100X.2008.00402.x

Kerley LL, Anderson SH (1995) Songbird responses to sagebrush removal in a high elevation sagebrush steppe ecosystem. Prairie Naturalist 27:129–146

Kerns CK, Ryan MR, Murphy RK, Thompson FR III, Rubin CS (2010) Factors affecting songbird nest survival in northern mixed-grass prairie. J Wildl Manage 74:257–264. https://doi.org/10.2193/2008-249

Knick ST, Dobkin DS, Rotenberry JT, Schroeder MA, Vander Haegen WM, van Riper IIIC (2003) Teetering on the edge or too late? Conservation and research issues for avifauna of sagebrush habitats. Condor 105:611–634. https://doi.org/10.1650/7329

Knick ST, Holmes AL, Miller RF (2005) The role of fire in structuring sagebrush habitats and bird communities. In: Saab VA, Powell HDW (eds) Fire and avian ecology in North America. Stud Avian Biol 30:63–75

Knick ST, Rotenberry JT, Leu M (2008) Habitat, topographical, and geographical components structuring shrubsteppe bird communities. Ecography 31:389–400. https://doi.org/10.1111/j.0906-7590.2008.05391.x

Knick ST, Leu M, Rotenberry JT, Hanser SE, Fesenmyer KA (2014) Diffuse migratory connectivity in two species of shrubland birds: evidence from stable isotopes. Oecologia 174:595–608. https://doi.org/10.1007/s00442-013-2791-8

Lam DPM, Hinz E, Lang DJ, Tengo M, von Wehrden H, Martin-López B (2020) Indigenous and local knowledge in sustainability transformations research: a literature review. Ecol Soc 25: article 3. https://doi.org/10.5751/ES-11305-250103

Langham GM, Schuetz JG, Distler T, Soykan CU, Wilsey C (2015) Conservation status of North American birds in the face of future climate change. PLoS ONE 10:e0135350. https://doi.org/10.1371/journal.pone.0135350

Lany NK, Ayres MP, Stange EE, Sillett TS, Rodenhouse NL, Holmes RT (2016) Breeding timed to maximize reproductive success for a migratory songbird: the importance of phenological asynchrony. Oikos 125:656–666. https://doi.org/10.1111/oik.02412

Laurent EJ, Bart J, Giocomo J, Harding S, Koch K, Moore-Barnhill L, Mordecai R, Sachs E, Wilson T (2012) A field guide to southeast bird monitoring programs and protocols. Southeast Partners in Flight. http://SEmonitoringguide.sepif.org

Lima SL (2009) Predators and the breeding bird: behavioral and reproductive flexibility under the risk of predation. Biol Rev 84:485–513. https://doi.org/10.1111/j.1469-185X.2009.00085.x

Lipsey MK, Naugle DE (2017) Precipitation and soil productivity explain effects of grazing on grassland songbirds. Rangel Ecol Manag 70:331–340. https://doi.org/10.1016/j.rama.2016.10.010

Lipsey MK, Naugle DE, Nowak J, Lukacs PM (2017) Extending utility of hierarchical models to multi-scale habitat selection. Divers Distrib 23:783–793. https://doi.org/10.1111/ddi.12567

Litt AR, Cord EE, Fulbright TE, Schuster GL (2014) Effects of invasive plants on arthropods. Conserv Biol 28:1532–1549. https://doi.org/10.1111/cobi.12350

Lloyd JD, Martin TE (2005) Reproductive success of chestnut collared longspurs in native and exotic grassland. Condor 107:363–374. https://doi.org/10.1650/7701

Longcore T, Rich C, Mineau P, MacDonald B, Bert DG, Sullivan LM, Mutrie E, Gauthreaux SA Jr, Avery ML, Crawford RL, Manville AM II, Travis ER, Drake D (2012) An estimate of avian mortality at communication towers in the United States and Canada. PLoS ONE 7:e34025. https://doi.org/10.31371/journal.pone.0034025

Loss SR (2016) Avian interactions with energy infrastructure in the context of other anthropogenic threats. Condor 118:424–432. https://doi.org/10.1650/condor-16-12.1

Loss SR, Will T, Marra PP (2013a) The impact of free-ranging domestic cats on wildlife of the United States. Nat Commun 4:article 1396. https://doi.org/10.1038/ncomms2380

Loss SR, Will T, Marra PP (2013b) Estimates of bird collision mortality at wind facilities in the contiguous United States. Biol Conser 168:201–209. https://doi.org/10.1016/j.biocon.2013.10.007

Ludlow SM, Brigham RM, Davis SK (2014) Nesting ecology of grassland songbirds: effects of predation, parasitism, and weather. Wilson J Ornithol 126:686–699. https://doi.org/10.1676/13-176.1

Lukacs P, Seglund A, Boyle S (2015) Effects of Gunnison sage-grouse habitat treatment efforts on associated avifauna and vegetation structure. Avian Conserv Ecol 10:article 7. https://doi.org/10.5751/ACE-00799-100207

Lyons TP, Miller JR, Debinski DM, Engle DM (2015) Predator identity influences the effect of habitat management on nest predation. Ecol Appl 25:1596–1605. https://doi.org/10.1890/14-1641

Mabry CM, Harms TM (2020) Impact of delayed mowing on restoring populations of grassland birds of conservation concern. Ecol Restor 38:77–82

Madden E, Hansen AJ, Murphy RK (1999) Influence of prescribed fire history on habitat and abundance of passerine birds in northern mixed-grass prairie. Can Field Nat 113:627–640

Mahoney A, Chalfoun AD (2016) Reproductive success of horned lark and McCown's longspur in relation to wind energy infrastructure. Condor: Ornithol Appl 118:360–375. https://doi.org/10.1650/CONDOR-15-25.1

Manier DJ, Hobbs NT (2006) Large herbivores influence the composition and diversity of shrub-steppe communities in the Rocky Mountains, USA. Oecologia 146:641–651. https://doi.org/10.1007/s00442-005-0065-9

Marra PP, Cohen EB, Loss SR, Rutter JE, Tonra CM (2015) A call for full annual cycle research in animal ecology. Biol Lett 11:e20150552. https://doi.org/10.1098/rsbl.2015.0552

Martin TG (1993) Nest predation and nest sites: new perspectives on old patterns. Bioscience 43:523–532

Martin TG, Possingham HP (2005) Predicting the impact of livestock grazing on birds using foraging height data. J Appl Ecol 42:400–408. https://doi.org/10.1111/j.1365-2664.2005.01012.x

Martin JW, Carlson BA (2020) Sagebrush sparrow (*Artemisiospiza nevadensis*). In: Poole AF (ed) Birds of the world, vers. 1.0. Cornell Lab of Ornithology, Ithaca, New York. https://doi.org/10.2173/bow.sagspa1.01

Marzluff J, Ewing K (2001) Restoration of fragmented landscapes for the conservation of birds: a general framework and specific recommendations for urbanizing landscapes. Restor Ecol 9:280–292. https://doi.org/10.1046/j.1526-100X.2001.009003280.x

McAdoo JK, Longland WS, Evans RA (1989) Nongame bird community responses to sagebrush invasion of crested wheatgrass seedings. J Wildl Manage 53:494–502. https://doi.org/10.2307/3801155

Mejia EC, McClure CJW, Barber JR (2019) Large-scale manipulation of the acoustic environment can alter the abundance of breeding birds: evidence from a phantom natural gas field. J Appl Ecol 56:2091–2101. https://doi.org/10.1111/1365-2664.13449

Miller RA, Bond L, Migas PN, Carlisle JD, Kaltenecker GS (2017) Contrasting habitat associations of sagebrush-steppe songbirds in the intermountain west. Western Birds 48:35–55

Nack JL, Ribic CA (2005) Apparent predation by cattle at grassland bird nests. Wilson Bull 117:56–62

Nackley LL, West AG, Skowno AL, Bond WJ (2017) The nebulous ecology of native invasions. Trends Ecol Evol 32:814–824. https://doi.org/10.1016/j.tree.2017.08.003

National Audubon Society (2014) Audubon's birds and climate change report: a primer for practitioners. Vers. 1.2. National Audubon Society, New York. http://climate.audubon.org/sites/default/files/Audubon-Birds-Climate-Report-v1.2.pdf

Neilly H, Vanderwal J, Schwarzkopf L (2016) Balancing biodiversity and food production: a better understanding of wildlife response to grazing will inform off-reserve conservation on rangelands. Rangel Ecol Manag 69:430–436. https://doi.org/10.1016/j.rama.2016.07.007

Nelson SB, Coon JJ, Duchardt CJ, Fischer JD, Halsey SJ, Kranz AJ, Parker CM, Schneider SC, Swartz TM, Miller JR (2017) Patterns and mechanisms of invasive plant impacts on North American birds: a systematic review. Biol Invasions 19:1547–1563. https://doi.org/10.1007/s10530-017-1377-5

Nelson SB, Coon JJ, Duchardt CJ, Miller JM, Debinski DM, Schacht WH (2018) Contrasting impacts of invasive plants and human-altered landscape context on nest survival and brood parasitism of a grassland bird. Landsc Ecol 33:1799–1813. https://doi.org/10.1007/s10980-018-0703-3

Nelson SB, Coon JJ, Miller RJ (2020) Do habitat preferences improve fitness? Context-specific adaptive habitat selection by a grassland songbird. Oecologia 193:15–26. https://doi.org/10.1007/s00442-020-04626-8

Nixon AE, Fisher RJ, Stralberg D, Bayne EM, Farr DR (2016) Projected responses of North American grassland songbirds to climate change and habitat availability at their northern range limits in Alberta, Canada. Avian Conserv Ecol 11:1–14. https://doi.org/10.5751/ACE-00866-110202

Northrup JM, Wittemyer G (2013) Characterising the impacts of emerging energy development on wildlife, with an eye towards mitigation. Ecol Lett 16:112–125. https://doi.org/10.1111/ele.12009

Noson AC, Schmitz RA, Miller RF (2006) Influence of fire and juniper encroachment on birds in high-elevation sagebrush steppe. West N Am Nat 66:343–353. https://doi.org/10.3398/1527-0904(2006)66[343:IOFAJE]2.0.CO;2

Owens RA, Myres MT (1973) Effects of agriculture upon populations of native passerine birds of an Alberta fescue grassland. Can J Zool 51:697–713

Paige C, Ritter SA (1999) Birds in a sagebrush sea: managing sagebrush habitats for bird communities. Partners in Flight Western Working Group, Boise, Idaho. https://partnersinflight.org/res ources/birds-in-a-sagebrush-sea/. Accessed 1 Aug 2022

Partners in Flight (2021) Avian conservation assessment database, vers. 2021. https://partnersinfl ight.org/what-we-do/science/databases/

Patten MA, Shochat E, Reinking DL, Wolfe DH, Sherrod SK (2006) Habitat edge, land management, and rates of brood parasitism in tallgrass prairie. Ecol Appl 16:687–695. https://doi.org/10.1890/ 10510761(2006)016%5B0687:HELMAR%5D2.0.CO;2

Peterson KL, Best LB (1985) Brewer's sparrow nest-site characteristics in a sagebrush community. J Field Ornithol 56:23–27

Peterson KL, Best LB, Winter BM (1986) Growth of nestling sage sparrows and Brewer's sparrows. Wilson Bull 98:535–546

Pietz PJ, Granfors DA (2000) Identifying predators and fates of grassland passerine nests using miniature video cameras. J Wildl Manage 64:71–87. https://doi.org/10.2307/3802976

Pietz PJ, Buhl DA, Shaffer JA, Winter M, Johnson DH (2009) Influence of trees in the landscape on parasitism rates of grassland passerine nests in southeastern North Dakota. Condor 111:36–42. https://doi.org/10.1525/cond.2009.080012

Pilliod DS, Welty JL, Arkle RS (2017) Refining the cheatgrass-fire cycle in the Great Basin: precipitation timing and fine fuel composition predict wildfire trends. Ecol Evol 7:8126–8151. https:/ /doi.org/10.1002/ece3.3414

Powell AFLA (2006) Effects of prescribed burns and bison (Bos bison) grazing on breeding bird abundance in tallgrass prairie. Auk 123:183–197. https://doi.org/10.1642/00048038(2006)123% 5B0183:EOPBAB%5D2.0.CO;2

Pulliam JP, Somershoe S, Sather M, McNew LM (2020) Habitat targets for imperiled grassland birds in northern mixed-grass prairie. Rangel Ecol Manag 73:511–519. https://doi.org/10.1016/ j.rama.2020.02.006

Pulliam JP, Somershoe S, Sather M, McNew LM (2021) Nest density drives productivity in chestnut-collared longspurs: implications for grassland bird conservation. PLoS ONE 16:e0256346. https://doi.org/10.1371/journal.pone.0256346

Reinking DL (2005) Fire regimes and avian responses in the central tallgrass prairie. In: Saab VA, Powell HDW (eds) Fire and avian ecology in North America. Stud Avian Biol 30:116–126

Renfrew R, Strong AM, Perlut NG, Martin SG, Gavin TA (2020) Bobolink (Dolichonyx oryzivorus). In: Rodewald PG (ed) Birds of the world, vers. 1.0. Cornell Lab of Ornithology, Ithaca, New York. https://doi.org/10.2173/bow.boboli.01

Renfrew RB, Ribic CA (2003) Grassland passerine nest predators near pasture edges identified on videotape. Auk 120:371–383. https://doi.org/10.1642/0004-8038(2003)120[0371:GPNPNP]2. 0.CO;2

Reynolds TD, Trost CH (1980) The response of native vertebrate populations to crested wheatgrass planting and grazing by sheep. J Range Manag 33:122–125. https://doi.org/10.2307/3898425

Reynolds TD, Rich TD, Stephens DA (2020) Sage thrasher (Oreoscoptes montanus). In: Poole AF, Gill FB (eds) Birds of the world, vers. 1.0. Cornell Lab of Ornithology, Ithaca, New York. https:/ /doi.org/10.2173/bow.sagthr.01

Ribic CA, Koford RR, Herkert JR (2009) Area sensitivity in North American grassland birds: patterns and processes. Auk 126:233–244. https://doi.org/10.1525/auk.2009.1409

Ribic CA, Ng CS, Koper N et al (2018) Diel fledging patterns among grassland passerines: relative impacts of energetics and predation risk. Auk 135:1100–1112. https://doi.org/10.1642/AUK-17-213.1

Ribic CA, Rugg DJ, Koper N et al (2019) Behavior of adult and young grassland songbirds at fledging. J Field Ornithol 90:143–153. https://doi.org/10.1111/jofo.12289

Rich TD, Wisdom M J, Saab VA (2005) Conservation of priority birds in sagebrush ecosystems. In: Ralph CJ, Rich TD (eds) Bird conservation implementation and integration in the Americas:

Proceedings of the 3rd international partners in flight conference, 2002 March 20–24, Asilomar, California, vol 1. U.S. Department of Agriculture, Forest Service, Pacific Southwest Research Station, General Technical Report PSW-GTR-191. Albany, California, pp 589–606

Richardson AN, Koper N, White KA (2014) Interactions between ecological disturbances—burning and grazing and their effects on songbird communities in northern mixed-grass prairies. Avian Conserv Ecol 9:article 5. https://doi.org/10.5751/ACE-00692-090205

Rivero MJ, Grau-Campanario P, Mullan S et al (2021) Factors affecting site use preference of grazing cattle Studied from 2000 to 2020 through GPS tracking: a review. Sensors 21:article 2696. https://doi.org/10.3390/s21082696

Rodenhouse NL, Best LB, O'Connor RJ et al (1995) Effects of agricultural practices and farmland structures. In: Martin TE, Finch DM (eds) Ecology and management of neotropical migratory birds. Oxford University Press, New York, pp 269–293

Rosenberg KV, Dokter AM, Blancher PJ et al (2019) Decline of the North American avifauna. Science 366:120–124. https://doi.org/10.1126/science.aaw1313

Rotenberry JT, Wiens JA (1980) Habitat structure, patchiness, and avian communities in North American steppe vegetation—a multivariate analysis. Ecology 61:1228–1250. https://doi.org/10.2307/1936840

Rotenberry JT, Wiens JA (1991) Weather and reproductive variation in shrubsteppe sparrows—a hierarchical analysis. Ecology 72:1325–1335. https://doi.org/10.2307/1941105

Rotenberry JT, Wiens JA (1998) Foraging patch selection by shrubsteppe sparrows. Ecology 79:1160–1173. https://doi.org/10.1890/0012-9658(1998)079[1160:FPSBSS]2.0.CO;2

Ruehmann MB, Desmond MJ, Gould WR (2011) Effects of smooth brome on Brewer's sparrow nest survival in sagebrush steppe. Condor 113:419–428. https://doi.org/10.1525/cond.2011.100022

Ruth JM (2015) Status Assessment and Conservation Plan for the grasshopper sparrow (*Ammodramus savannarum*). U.S. Fish and Wildlife Service, Lakewood. Vers. 1.0. https://digitalcommons.unl.edu/usfwspubs/471/

Ruth JM, Skagen SK (2018) Reproductive response of Arizona grasshopper sparrows to weather patterns and habitat structure. Condor 120:596–616. https://doi.org/10.1650/CONDOR-17-128.1

Ruth JM, Talbot WA, Smith EK (2020) Behavioral response to high temperatures in a desert grassland bird: use of shrubs as thermal refugia. West N Am Nat 80:265–275. https://doi.org/10.3398/064.080.0215

Ryan MR (1990) A dynamic approach to the conservation of the prairie ecosystem in the Midwest. In: Sweeney JM (ed) Management of dynamic ecosystems. The Wildlife Society, North Central Section, West Lafayette, Indiana

Saether BE, Bakke O (1997) Avian life history variation and contribution of demographic traits to the population growth rate. Ecology 81:642–653. https://doi.org/10.2307/177366

Sample DW, Mossman MJ (1997) Managing habitat for grassland birds—A guide for Wisconsin Wisconsin Department of Natural Resources, Madison, Wisconsin

Sanders LE, Chalfoun AD (2018) Novel landscape elements within natural gas fields increase densities but not fitness of an important songbird nest predator. Biol Conser 228:132–141. https://doi.org/10.1016/j.biocon.2018.10.020

Sauer JR, Link WA, Fallon JE, Pardieck DL, Ziolkowski DJ (2013) The North American breeding bird survey 1966–2011: summary analysis and species accounts. North American Fauna 79:1–32. https://doi.org/10.3996/nafa.79.0001

Sauer JR, Link WA, Hines JE (2020) The North American breeding bird survey, analysis results 1966–2019: U.S. Geological Survey data release. https://doi.org/10.5066/P96A7675

Scasta JD, Thacker ET, Hovick TJ, Engle DM, Allred BW, Fuhlendorf SD, Weir JR (2015) Patch-burn grazing (PBG) as a livestock management alternative for fire-prone ecosystems of North America. Renew Agric Food Syst 31:550–567. https://doi.org/10.1017/S1742170515000411

Scheiman DM, Bollinger EK, Johnson DH (2003) Effects of leafy spurge infestation on grassland birds. J Wildl Manage 67:115–121. https://doi.org/10.2307/3803067

Schroeder MH, Sturges DL (1975) The effects on the Brewer's sparrow of spraying big sagebrush. J Range Manag 28:294–297

Shaffer JA, Buhl DA (2016) Effects of wind-energy facilities on breeding grassland bird distributions. Conserv Biol 30:59–71. https://doi.org/10.1111/cobi.12569

Shaffer JA, DeLong JP (2019) The effects of management practices on grassland birds: an introduction to North American grasslands and the practices used to manage grasslands and grassland birds. In: Johnson DH, Igl LD, Shaffer JA, DeLong JP (eds) The effects of management practices on grassland birds, chap. A. U.S. Geological Survey Professional Paper 1842. https://doi.org/10.3133/pp1842A

Shaffer JA, Igl LD, Johnson DH (2019a) The effects of management practices on grassland birds—rates of brown-headed cowbird (*Molothrus ater*) parasitism in nests of North American grassland birds. In: Johnson DH, Igl LD, Shaffer JA, DeLong JP (eds) The effects of management practices on grassland birds, chap. PP. U.S. Geological Survey Professional Paper 1842. https://doi.org/10.3133/pp1842PP

Shaffer JA, Loesch CR, Buhl DA (2019) Estimating offsets for avian displacement effects of anthropogenic impacts. Ecol Appl 29:e01983. https://doi.org/10.1002/eap.1983

Shaffer JA, Igl LD, Johnson DH, Sondreal ML, Goldade CM, Nenneman MP, Wooten TL, Euliss BR (2021) The effects of management practices on grassland birds—grasshopper sparrow (*Ammodramus savannarum*). In: Johnson DH, Igl LD, Shaffer JA, DeLong JP (eds) The effects of management practices on grassland birds, chap. GG. U.S. Geological Survey Professional Paper 1842. https://doi.org/10.3133/pp1842GG

Skagen SK, Yackel Adams A (2012) Weather effects on avian breeding performance and implications of climate change. Ecol Appl 22:1131–1145. https://doi.org/10.1890/11-0291.1

Sliwinski MS, Powell LA, Schacht WH (2019) Grazing systems do not affect bird habitat on a Sandhills landscape. Rangel Ecol Manag 72:136–144. https://doi.org/10.1016/j.rama.2018.07.006

Sliwinski MS, Powell LA, Schacht WH (2020) Similar bird communities across grazing systems in the Nebraska Sandhills. J Wildl Manage 84:802–812. https://doi.org/10.1002/jwmg.21825

Sliwinski MS, Koper N (2015) Managing mixed-grass prairies for songbirds using variable cattle stocking rates. Rangel Ecol Manag 68:470–475.https://doi.org/10.1016/j.rama.2015.07.010

Stinson LT, Pejchar L (2018) The effects of introduced plants on songbird reproductive success. Biol Invasions 20:1403–1416. https://doi.org/10.1007/s10530-017-1633-8

Stoleson SH, Finch DM (2001) Breeding bird use of and nesting success in exotic Russian olive in New Mexico. Wilson Bull 113:452–455

Swengel SR (1996) Management responses of three species of declining sparrows in tallgrass prairie. Bird Conserv Int 6:241–253. https://doi.org/10.1017/S0959270900003130

Temple SA (2020) Dickcissel (*Spiza americana*). In: Poole AF, Gill FB (eds) Birds of the world, vers. 1.0. Cornell Lab of Ornithology, Ithaca, New York. https://doi.org/10.2173/bow.dickci.01

Temple SA, Fevold BM, Paine LK, Undersander DJ, Sample DW (1999) Nesting birds and grazing cattle—accommodating both on midwestern pastures. In: Vickery PD, Herkert JR (eds) Ecology and conservation of grassland birds of the Western Hemisphere. Stud Avian Biol 19:196–202

Thompson SJ, Johnson DH, Niemuth ND, Ribic CA (2015) Avoidance of unconventional oil wells and roads exacerbates habitat loss for grassland birds in the North American Great Plains. Biol Conser 192:82–90. https://doi.org/10.1016/j.biocon.2015.08.040

Tizard I (2004) Salmonellosis in wild birds. Semin Avian Exotic Pet Med 13:50–66. https://doi.org/10.1053/j.saep.2004.01.008

Tripet F, Richner H (1997) Host responses to ectoparasites: food compensation by parent blue tits. Oikos 78:557–561. https://doi.org/10.2307/3545617

Valencia-Herverth J, Garrido D, Valencia-Herverth R (2018) Contributions to the knowledge of the distribution of birds in the state of Hidalgo, Mexico. Southwest Nat 63:83–87. https://doi.org/10.1894/0038-4909.63.83

Veblen KE, Pyke DA, Aldridge CL, Casazza ML, Assal TJ, Farinha MA (2014) Monitoring of livestock grazing effects on Bureau of Land Management land. Rangel Ecol Manag 67:68–77. https://doi.org/10.2111/REM-D-12-00178.1

Verheijen BHF, Erickson AN, Boyle WA, Leveritte KS, Sojka JL, Spahr LA, Williams EJ, Winnicki SK, Sandercock BK (2022) Predation, parasitism, and drought counteract the benefits of patch-burn grazing for the reproductive success of grassland songbirds. Ornithol Appl 124:1–22. https://doi.org/10.1093/ornithapp/duab066

Vickery PD (2020) Grasshopper sparrow (*Ammodramus savannarum*) In: Poole AF, Gill FB (eds) Birds of the world, vers 1.0. Cornell Lab of Ornithology, Ithaca, New York. https://doi.org/10.2173/bow.graspa.01

Vickery PD, Tubaro PL, da Silva JMC, Peterjohn BG, Herkert JR, Cavalcanti RB (1999) Conservation of grassland birds in the Western Hemisphere. In: Vickery PD, Herkert JR (eds) Ecology and conservation of grassland birds of the Western Hemisphere. Stud Avian Biol 19:2–26

Vold ST, Berkeley LI, McNew LB (2019) Effects of livestock grazing management on grassland birds in a northern mixed-grass prairie ecosystem. Rangel Ecol Manag 72:933–945. https://doi.org/10.1016/j.rama.2019.08.005

Walker BL, Igl LD, Shaffer JA (2020) The effects of management practices on grassland birds—Brewer's sparrow (*Spizella breweri breweri*). In: Johnson DH, Igl LD, Shaffer JA, DeLong JP (eds) The effects of management practices on grassland birds, chap. AA. U.S. Geological Survey Professional Paper 1842. https://doi.org/10.3133/pp1842AA

Wheelwright NT, Rising JD (2020) Savannah sparrow (*Passerculus sandwichensis*). In: Poole AF (ed) Birds of the world, vers 1.0. Cornell Lab of Ornithology, Ithaca, New York. https://doi.org/10.2173/bow.savspa.01

Wilkins K, Pejchar L, Garvoille R (2019) Ecological and social consequences of bison reintroduction in Colorado. Conserv Sci Pract 1:e9. https://doi.org/10.1111/csp2.9

Williams MI, Paige GB, Thurow TL, Hild AL, Gerow KG (2011) Songbird relationships to shrub-steppe ecological site characteristics. Rangel Ecol Manage 64:109–118. https://doi.org/10.2111/REM-D-10-00076.1

Wilsey C, Taylor L, Bateman B, Jensen C, Michel N, Panjabi A, Langham G (2019) Climate policy action needed to reduce vulnerability of conservation-reliant grassland birds in North America. Conserv Sci Pract 1:e21. https://doi.org/10.1111/csp2.21

With KA (2021) Thick-billed longspur (*Rhynchophanes mccownii*). In: Poole AF (ed) Birds of the world, vers. 1.1. Cornell Lab of Ornithology, Ithaca, New York. https://doi.org/10.2173/bow.mcclon.01.1

With KA, King AW, Jensen WE (2008) Remaining large grasslands may not be sufficient to prevent grassland bird declines. Biol Conser 141:3152–3167. https://doi.org/10.1016/j.biocon.2008.09.025

Yackel Adams A, Skagen SK, Savidge JA (2006) Modeling post-fledging survival of lark buntings in response to ecological and biological factors. Ecology 87:178–188. https://doi.org/10.1890/04-1922

Young AC, Cox W, McCarty JP, Wolfenbarger L (2019) Postfledging habitat selection and survival of Henslow's sparrow: management implications for a critical life stage. Avian Conserv Ecol 14:article 10. https://doi.org/10.5751/ACE-01418-1402

Zalik NJ, Strong AM (2008) Effects of hay cropping on invertebrate biomass and the breeding ecology of Savannah sparrows (*Passerculus sandwichensis*). Auk 125:700–710. https://doi.org/10.1525/auk.2008.07106

Zimmerman JL (1997) Avian community responses to fire, grazing, and drought in the tallgrass prairie. In: Knopf FL, Samson FB (eds) Ecology and conservation of Great Plains vertebrates. Springer-Verlag, New York, pp 167–180

Zuckerberg B, Ribic CA, McCauley LA (2018) Effects of temperature and precipitation on grassland bird nesting success as mediated by patch size. Conserv Biol 32:872–882. https://doi.org/10.1111/cobi.13089

Chapter 13
Waterfowl and Wetland Birds

Josh L. Vest, David A. Haukos, Neal D. Niemuth, Casey M. Setash, James H. Gammonley, James H. Devries, and David K. Dahlgren

Abstract The future of wetland bird habitat and populations is intrinsically connected with the conservation of rangelands in North America. Many rangeland watersheds are source drainage for some of the highest functioning extant wetlands. The Central and Pacific Flyways have significant overlap with available rangelands in western North America. Within these flyways, the importance of rangeland management has become increasingly recognized by those involved in wetland bird conservation. Within the array of wetland bird species, seasonal habitat needs are highly

J. L. Vest (✉)
U.S. Fish and Wildlife Service, Prairie Pothole Joint Venture, 922 Bootlegger Trail, Great Falls, MT 59404, USA
e-mail: josh_vest@fws.gov

D. A. Haukos
U.S. Geological Survey, Kansas Cooperative Fish and Wildlife Research Unit, 1128 N. 7th Street, Manhattan, KS 66506, USA
e-mail: dhaukos@ksu.edu

N. D. Niemuth
U.S. Fish and Wildlife Service, Habitat and Population Evaluation Team, 3425 Miriam Avenue, Bismarck, ND 58501, USA
e-mail: neal_niemuth@fws.gov

C. M. Setash
Department of Fish, Wildlife, and Conservation Biology, Colorado State University, 1474 Campus Delivery, Fort Collins, CO 80523, USA
e-mail: csetash@rams.colostate.edu

J. H. Gammonley
Colorado Parks and Wildlife, 317 West Prospect, Fort Collins, CO 80526, USA
e-mail: jim.gammonley@state.co.us

J. H. Devries
Ducks Unlimited Canada, Institute for Wetland and Waterfowl Research, P.O. Box 1160, Stonewall, MB R0C 2Z0, Canada
e-mail: j_devries@ducks.ca

D. K. Dahlgren
Department of Wildland Resources, Utah State University, 5230 Old Main Hill, Logan, UT 84322, USA
e-mail: dave.dahlgren@usu.edu

L. B. McNew et al. (eds.), *Rangeland Wildlife Ecology and Conservation*,
https://doi.org/10.1007/978-3-031-34037-6_13

417

variable. During the breeding period, nest survival is one of the most important drivers of population growth for many wetland bird species and rangelands often provide quality nesting cover. Throughout spring and fall, rangeland wetlands provide key forage resources that support energetic demands needed for migration. In some areas, stock ponds developed for livestock water provide migration stopover and wintering habitat, especially in times of water scarcity. In the Intermountain West, drought combined with water demands from agriculture and human population growth are likely headed to an ecological tipping point for wetland birds and their habitat in the region. In the Prairie Pothole Region, conversion of rangeland and draining of wetlands for increased crop production remains a significant conservation issue for wetland birds and other wildlife. In landscapes dominated by agricultural production, rangelands provide some of the highest value ecosystem services, including water quality and wetland function. Recent research has shown livestock grazing, if managed properly, is compatible and at times beneficial to wetland bird habitat needs. Either directly, or indirectly, wetland bird populations and their habitat needs are supported by healthy rangelands. In the future, rangeland and wetland bird managers will benefit from increased collaboration to aid in meeting ultimate conservation objectives.

Keywords Conservation · Livestock grazing · Management · Rangeland · Shorebirds · Waterbirds · Waterfowl · Wetland birds

13.1 Introduction

Rangeland systems and the wetland birds using them vary across western North America. This chapter addresses three groups of birds dependent on wetlands: waterfowl, shorebirds, and waterbirds. Many wetland bird conservation plans recognize the significant influence rangelands have on associated wetlands, including the North American Waterfowl Management Plan (NAWMP 2018), U.S. Shorebird Conservation Plan (Brown et al. 2001), Canadian Shorebird Conservation Plan (Donaldson et al. 2000), and North American Waterbird Conservation Plan (Kushlan et al. 2002). Wetland birds typically exhibit large-scale mobility, including seasonal migration across North America capitalizing on ecoregional resources to meet their annual cycle needs. Wetland birds breeding in northern latitudes take advantage of primary productivity associated with extended summer daylight but as winter nears, they seek resources at more southerly latitudes. Wetland bird migrations have heralded seasonal change for societies over human history. Connected wetland networks sustain migrations by providing rest and food resources and have demographic consequences for populations.

Within seasonal home-ranges, wetlands birds can be highly mobile, and a single wetland or wetland type can rarely meet daily, seasonal, or annual needs. Seasonal wetlands tend to have high biological productivity, whereas wetlands with stable water levels typically have reduced biological productivity. Because wetlands are dynamic, their availability and quality as habitat can be highly variable. Consequently, wetland birds generally select landscapes with a diversity of wetlands to maximize resources. Diversity within a complex of wetlands is a key strategy for resource managers throughout North America (Baldassarre and Bolen 2006). Wetland bird conservation has been coordinated across migration corridors (i.e., flyways) and regions. Rangelands cover significant areas of the Great Plains and the West (Fig. 13.1; Table 13.1 [avian scientific names presented]). Throughout this chapter ecoregional terminology is used consistent with wetland bird conservation and management plans.

Previous reviews have provided important information on the ecology and management of waterfowl (e.g., Smith et al. 1989; Batt et al. 1992; Baldassarre and Bolen 2006; Baldassarre 2014), shorebirds (Helmers 1992; Iglecia and Winn 2021),

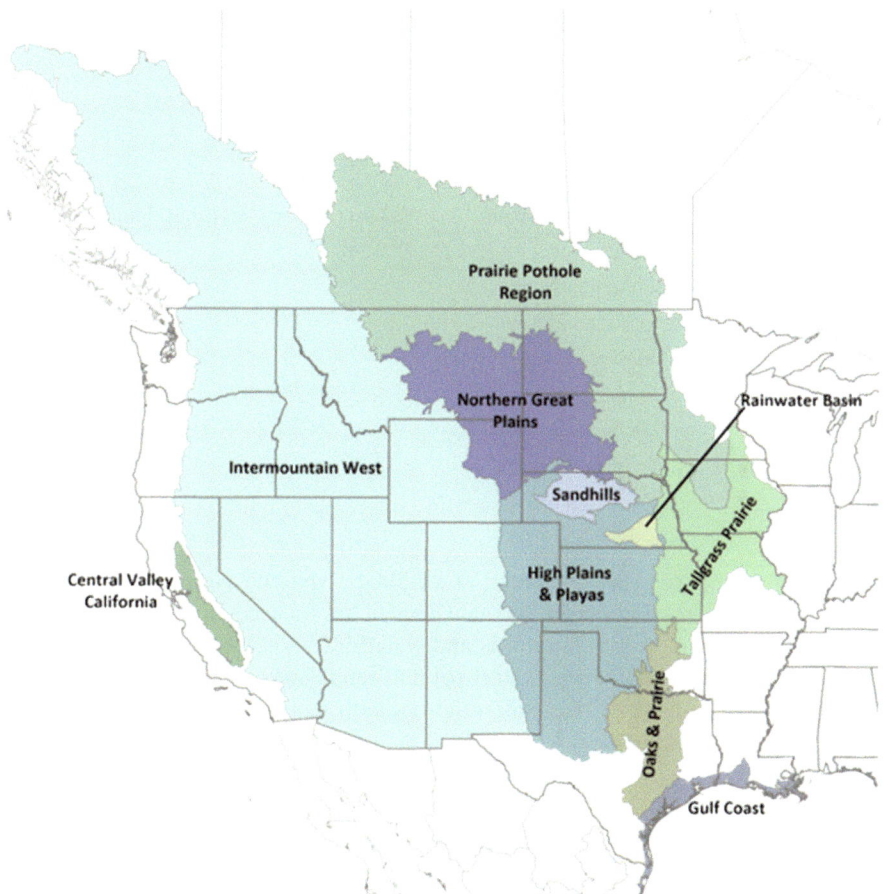

Fig. 13.1 Major wetland bird ecoregions of rangeland systems in North America

Table 13.1 Major wetland bird ecoregions within central and western North America with subregions and regions of western rangelands

Major Wetland Bird Region	Rangeland	
	Subregion	Region
Prairie Pothole	Aspen Parkland	Great Plains
	Northern Mixed Grass Prairie	
Northern Great Plains		
Sandhills of Nebraska	Sand Hills of Nebraska	
Rainwater Basin	Tallgrass and Southern Mixed Grass Prairie	
Oaks and Prairie		
High Plains & Playas	Southern Mixed Grass and Shortgrass Prairie	
Gulf Coast	Gulf Coast Prairies and Marshes	Gulf Coastal Prairie
Intermountain West	Rocky Mountains	Rocky Mountains
	Sierra Nevada Mountains	Sierra Nevada Mountains
	Cascade Mountains	Western Deserts, Grasslands, Shrublands, and Woodlands
	Great Basin	
	Columbia Plateau	
	Colorado Plateau	
	Mojave Desert	
	Sonoran Desert	
	Chihuahuan Desert	
Central Valley of California	California Central Valley	

and other waterbirds (Beyersbergen et al. 2004; Ivey and Herziger 2006). Available research addressing rangeland management has focused on waterfowl, with less empirical information for shorebirds and waterbirds. Therefore, this chapter relies heavily on science addressing waterfowl and rangeland relationships. We provide overviews of life history, regional variation, and population dynamics of wetland birds that may be influenced by rangeland management and conservation.

13.2 Wetland Systems

Wetlands occupy a relatively small footprint in many rangelands. However, wetlands, riparian systems, and mesic habitats are often vital to the productivity, function, and biodiversity of rangeland systems (Johnson 2019, Chap. 7). Wetlands provide substantial ecosystem services and structure biological communities well beyond their immediate footprint (Mitsch and Gosselink 2015; Donnelly et al. 2016; Johnson 2019). Wetlands are transitional areas with characteristics of both aquatic and terrestrial ecosystems in addition to their own unique ecological conditions. Wetlands typically occur where groundwater is at or near the surface or land is covered by collection of water through runoff of surface water within a watershed (Cowardin et al. 1979; Mitsch and Gosselink 2015). Wetlands have dynamic hydrology resulting in conditions ranging from near-terrestrial to fully aquatic. Availability of habitat can vary temporally and is subject to variation in response to climate patterns. Identifying jurisdictional (i.e., subject to legal authority) wetlands includes a combination of key factors: (1) presence of shallow water or moist soil for 14–21 days during the growing season, (2) water-adapted plants (i.e., hydrophytic vegetation), and (3) hydric soils influenced by anaerobic conditions of saturation (Cowardin et al. 1979; Weller 1999). Not all wetlands are considered jurisdictional or subject to legal protections. For example, some wetlands in more arid environments are ephemeral, in some cases inundated only once or twice over years.

Hydrology and water budget determine wetland type and associated ecological processes (Mitsch and Gosselink 2015). Wetlands are commonly classified by hydroperiod (Cowardin et al. 1979; Table 13.2). Hydrologic conditions such as water depth, flow patterns, and flood frequency and duration (i.e., hydroperiod) influence abiotic and biotic components. The hydroperiod is determined by water inflows and outflows. Hydroperiod, largely dictates resource availability for wetland birds, other wildlife, and livestock. For example, recharge wetlands are solely dependent upon surface runoff linking hydroperiod to precipitation patterns. Conversely, discharge wetlands have hydroperiods based on groundwater. Hydroperiod is more dynamic in recharge versus discharge wetlands. Small hydrologic fluctuations can lead to significant changes in plant and animal composition (Mitsch and Gosselink 2015). Wetlands referenced in this chapter are either palustrine (i.e., marshy fresh or inland saline waters or vegetated margins of large water bodies; Cowardin et al. 1979), or lacustrine wetlands (i.e., relatively shallow, open, freshwater lakes or their sparsely vegetated margins).

Wetland bird use tends to vary by water depth, vegetation characteristics, and size (Laubhan and Gammonley 2000; Weller 1999; Ma et al. 2010). There are five general types of wetland plant associations: submerged plants, floating-leaved plants, emergent plants, moist-soil plants, and woody plants. Submerged aquatic vegetation (SAV) communities provide important food sources for wetland birds—especially waterfowl—through their seeds, tubers, and leafy materials as well as associated aquatic macroinvertebrates. Light penetration and turbidity affect subsurface photosynthesis and influence establishment and productivity of SAV. Floating-leaved

Table 13.2 Definitions of inland wetland hydroperiods

Permanently flooded—flooded throughout the year in all years
Intermittently exposed—flooded throughout the year except in years of extreme drought
Semipermanently flooded—flooded during the growing season in most years
Seasonally flooded—flooded for extended periods during the growing season, but usually no surface water by end of the growing season
Saturated—substrate is saturated for extended periods during the growing season, but standing water is rarely present
Temporarily flooded—flooded for brief periods during the growing season but the water table is otherwise well below surface
Intermittently flooded—surface is usually exposed with surface water present for variable periods without detectable seasonal pattern

Source Cowardin et al. (1979)

communities include both rooted and free-floating aquatic plants and provide little value to most wetland birds. Function and productivity within rangeland wetlands is primarily provided by emergent plants. These plants range from dense, robust emergents such as cattail (*Typha* spp.), and bulrushes (*Scirpus* spp.), to relatively shorter emergents with varying flood tolerances including sedges (*Carex* spp.), rushes (*Juncus* spp.), spike-rushes (*Eleocharis* spp.), and water-tolerant grasses such as cordgrass (*Spartina* spp.), panic grasses (*Panicum* spp.), and whitetop (*Scholochloa festucacea)*. Many emergent wetland species can be common livestock forages (Kirby et al. 2002). Moist-soil plants include annuals or perennials that germinate following drying events on exposed mudflats and provide abundant food via seeds and aquatic invertebrates (Fredrickson and Taylor 1982; Haukos and Smith 1993; Anderson and Smith 2000). Common moist-soil plants in western rangelands include smartweeds (*Polygnum* spp.), barnyardgrass (*Echinochloa crus-galli*), spike-rushes, curly dock (*Rumex crispus*), goosefoots and Lamb's quarters (*Chenopodium spp.*), and alkali bulrush (*Scheonoplectus maritumus*; Kadlec and Smith 1989; Haukos and Smith 1993; Dugger et al. 2007). Management of moist-soil habitats has been extensively applied to wetland complexes providing forage for waterfowl (Fredrickson and Taylor 1982; Baldassarre and Bolen 2006). Periodic drying can temporarily reduce wetland bird use, but is essential for cycling nutrients, succession of plant communities, and maintaining productivity (Harris and Marshall 1963; Murkin et al. 1997).

13.2.1 Flyway Wetlands

Flyways including Atlantic, Central, Mississippi, and Pacific are useful constructs for the administration of migratory bird management (Anderson et al. 2018; Roberts et al. 2023), with rangelands primarily overlapping the Central and Pacific Flyways. The Central Flyway includes prairie potholes, playas, and coastal marshes. Central

Mixed-Grass Prairie and Tallgrass Prairie regions in the southern Central Flyway provide key wetland habitats during migration (Smith et al. 1989; DU 2021; Hagy et al. *in review*). Millions of pothole wetlands occur in the Northern Mixed-Grass Prairie, northwestern Tallgrass Prairie, and Aspen Parklands within the Prairie Pothole Region (PPR; Fig. 13.1). High wetland density with associated grasslands makes the PPR unique and ecologically important in North America, and globally, for breeding and migrating wetland birds (Baldassarre and Bolen 2006; Niemuth et al. 2010). The PPR is known as the "Duck Factory" producing between half to two-thirds of all ducks in North America (Smith et al. 1964; Batt et al. 1989; Baldassarre and Bolen 2006) along with important water and forage resources for livestock (Johnson 2019). Playas are shallow, ephemeral, recharge wetlands abundant on the High Plains of Central and Southern Shortgrass and Mixed-Grass prairies. Playas' hydroperiods are highly variable and inundation can range from days to years. Playas are drivers of biodiversity in the region and the primary source of Ogallala Aquifer recharge (Haukos and Smith 1994; Smith et al. 2012; Gitz and Brauer 2016). Millions of wetland birds use playas during migration and winter (Haukos and Smith 1994; Moon and Haukos 2008; Smith et al. 2012). Coastal marshes, tidal freshwater swamps, and adjacent lagoons are defining features of the Gulf Coast. Freshwater and brackish marshes generally support the most valuable habitats for wetland birds, particularly waterfowl (Chabreck et al. 1989; Davis 2012). Coastal wetlands of Louisiana and Texas are wintering grounds for millions of wetland birds (Baldassarre and Bolen 2006; Vermillion 2012; Henkel and Taylor 2015).

The Pacific Flyway includes the Intermountain West with a variety of wetlands comprising < 10% of the area (McKinstry et al. 2004; Donnelly and Vest 2012). Many seasonal wetlands have been converted to irrigated pastures and hay meadows for production agriculture (McKinstry et al. 2004) and water management is generally complex and controversial (Downard and Endter-Wada 2013; Donnelly et al. 2020; Lovvorn and Crozier 2022). Within the region, wetlands are critical to sustaining wetland birds, other wildlife, and agricultural-based economies (Sketch et al. 2020; Donnelly et al. 2021, 2022; King et al. 2021). For most wildlife species, wetlands are part of their annual life cycle (McKinstry et al. 2004). The region provides migration, breeding, and wintering habitats for > 10 million wetland birds (Donnelly and Vest 2012; IWJV 2013). Due to precipitation patterns, wetlands experience high annual variability in availability and productivity maintaining a network of functional wetlands is critical to wetland bird conservation (Haig et al. 1998; Mackell et al. 2021; Donnelly et al. 2020, 2021, 2022).

13.3 Life History, Annual Cycle, and Population Dynamics

The diverse taxa comprising wetland birds span a continuum of life-history strategies that prioritize different fitness components (e.g., fecundity versus survival). However, management occurs primarily at population levels and key vital rates that shape population dynamics allow for some generalizations (Koons et al. 2014). Life histories vary from short-lived and high reproductive rates (i.e., more R-selected) to long-lived and lower reproductive (i.e., more K-selected) strategies (Stearns 1992). Accordingly, adult survival will have more influence on population growth rate for species with moderate-to-long generation times, like geese, compared to species with faster life histories, like teal, where reproductive success is more impactful (Koons et al. 2014). Overall, both reproduction and survival of wetland bird populations are influenced by environmental and habitat conditions. Sustaining functional wetland networks, especially within rangelands, across flyways provides resiliency against environmental stressors for wetland bird populations (Albanese and Haukos 2017; Haig et al. 2019; Donnelly et al. 2020).

13.3.1 Nest and Female Survival

Nest survival, the probability that $\geq$ 1 egg hatches, is one of the primary drivers of duck population growth rate and often the focus of management (Hoekman et al. 2002; Reynolds et al. 2006). Duck population growth rates can also be sensitive to adult female survival with increased predation risk for nesting females (Hoekman et al. 2002). Nest survival is generally higher for larger species like geese and swans averaging $\geq$ 70%, whereas ducks average 15–20% (Hoekman et al. 2002; Baldassarre and Bolen 2006; Baldassarre 2014). Clutch sizes range from 4 to 6 eggs for geese and swans and 8–12 eggs for ducks, whereas shorebirds typically lay 4 eggs and some other waterbird clutches may only have 1 egg (e.g., sandhill cranes). Waterfowl and shorebirds that commonly nest in rangelands tend to be solitary nesters, but semi-colonial behavior may occur where nest densities are high (e.g., islands). Nest initiation starts in mid-April for early nesters like mallards and northern pintail, to late June for late nesters like gadwall in high elevation systems (Baldassarre 2014). Growing season interacts with environmental conditions dictating nesting phenology and the propensity for renesting (Baldassarre 2014; Raquel et al. 2016).

Some wetland birds are generalists (e.g., mallards) that will nest in uplands, emergent vegetation in wetland margins, artificial nest structures, or woody vegetation along riparian areas (Baldassarre 2014). Others, like inland populations of snowy plovers, nest exclusively in specialized habitat (e.g., unvegetated shorelines and sandbars; Anteau et al. 2012). Agricultural lands can become ecological traps, such as when northern pintail select cropland resulting in low nest survival (Buderman et al. 2020). Waterfowl nesting habitat has three broad categories: (1) uplands including grasslands, shrublands, and agriculture lands, (2) overwater vegetation such as cattails and bulrushes or man-made platforms, and (3) cavities in trees or nest boxes.

Ducks select nesting cover based on species, local conditions, and availability. For example, mallards tend to select denser cover whereas northern pintail typically select shorter, less dense vegetation (Baldassarre 2014). Proximity to wetlands is important for upland nesting ducks but varies by species. Blue-winged teal have relatively small home ranges and nest closer to wetlands. Mallard and northern pintail can nest > 2 km from a wetland (Reynolds et al. 2006). Lesser scaup have limited mobility in uplands and nest very close to wetlands. When uplands lack cover, upland nesters tend to seek cover in dry wetlands at the emergent fringe (Lovvorn and Crozier 2022).

Most adult female mortality (i.e., 65–80%) of ducks occurs during the breeding season where nesting females are vulnerable to predators (Hoekman et al. 2002; Arnold et al. 2012). Providing quality nesting habitat helps increase both nest and female survival (Reynolds et al. 1995; Arnold et al. 2012). At the population level, nest survival is impacted by large-scale environmental factors and local nest-site characteristics; vegetation structure is more important than composition (Ringelman et al. 2018; Sherfy et al. 2018; Bortolotti et al. 2022). Nest survival generally increases with larger patch size and more perennial vegetation (Baldassarre and Bolen 2006; Bortolotti et al. 2022). The relationship of habitat and nest survival is complex, varies regionally, and difficult to differentiate among confounding factors like landscape characteristics, environmental changes, and predator communities (Clark and Nudds 1991; Horn et al. 2005; Walker et al. 2013a; Ringelman et al. 2018; Bortolotti et al. 2022; Pearse et al. 2022). Rangelands, with associated wetlands, generally provide extensive areas of perennial cover and reliably have high duck nest survival (Stephens et al. 2005; Walker et al. 2013a; Bortolotti et al. 2022). Increased nest survival in rangelands, compared to cropland landscapes, is likely due to reduced predator efficiency within large intact habitat and/or lower predator densities (Ball et al. 1995; Phillips et al. 2003; Horn et al. 2005). Large areas of intact rangelands may also support a greater abundance and diversity of other prey, reducing predation pressure on duck nests (Ackerman 2002). Although not fully understood at continental and population scales, intact rangelands are likely important in sustaining waterfowl in North America due to the potential for high nesting productivity (Higgins et al. 2002; PHJV 2021; PPJV 2017).

Nearly all shorebirds are ground nesters, but habitats and breeding behavior vary widely by species (Iglecia and Winn 2021). Before Euro-American settlement, breeding shorebirds in the Great Plains specialized in exploiting the diverse grass-land mosaics left by bison (*Bison bison*) and fire (Eldridge 1992). Shorebird breeding habitat includes unvegetated beaches and salt/alkali flats to moderately tall and dense grasslands (Eldridge 1992; Iglecia and Winn 2021). Long-billed curlew, marbled godwit, willet, killdeer, and mountain plover all nest and forage in short (< 15 cm) grassland vegetation often far from wetlands. Wilson's phalarope and upland sand-piper typically use taller (10–30 cm) and denser vegetation (Eldridge 1992). For species that rely on wetland invertebrates, proximity to wetlands is important when selecting nesting habitat (e.g., Wilson's phalarope, American avocet, piping plover, snowy plover, marbled godwit, willet; Eldridge 1992; Specht et al. 2020). Drivers of shorebird nest survival may be similar to those of waterfowl due to shared nest predators (Specht et al. 2020).

Diving ducks and swans (Table 13.3) primarily build overwater nests from emergent vegetation such as bulrush, cattail, and sedges (Baldassarre 2014). These over-water nesters often have limited available nesting cover and are generally associated with semi-permanent and permanent wetlands (Baldassarre 2014). Overwater nests are more protected, and survival tends to be higher than upland nests, although predation rates can increase with decreasing water levels (Baldassarre and Bolen 2006). Across the PPR, mallards nest in emergent wetland vegetation and experience relatively higher nest survival rates compared to upland nests (Baldassarre and Bolen 2006; Baldassarre 2014). Other waterbird species also nest over water either in dense emergent vegetation (e.g., sandhill crane) or on floating mats of vegetation (e.g., grebes). Some waterbirds nest on islands (e.g., pelicans) and in trees (e.g., herons; Beyersbergen et al. 2004).

13.3.2 Juvenile Survival

Juvenile survival can also strongly influence population growth rate for wetland birds, especially dabbling ducks (Hoekman et al. 2002). Chick survival is lowest within the first two weeks post-hatch. Small size and lack of thermoregulation during this time makes chicks vulnerable to exposure (Bloom et al. 2012; Iglecia and Winn 2021) and a wide range of predators (Sargeant and Raveling 1992; Baldassarre and Bolen 2006). Females can move their brood long distances to find quality habitat, which includes abundant invertebrates for food and security cover. Brood occurrence and survival has been shown to correlate with the availability of perennial herbaceous vegetation and wetland area (Krapu et al. 2000; Walker et al. 2013b). Rangelands with abundant and diverse wetlands in both size and hydroperiod are essential to sustaining wetland bird populations in North America (Helmers 1992; Beyersbergen et al. 2004; Walker et al. 2013b).

Waterbird species have chicks that range from precocial to altricial. Sandhill crane colts leave the nest directly after hatching whereas loons, grebes, most rails, and coots rely on parental feeding at the nest for several days. Gull and tern chicks may quickly leave the nest but remain close to the nest site for several days. Ibis, pelicans, cormorants, and herons feed chicks in nests until mobility develops, which varies from 2 to 11 weeks (Weller 1999). Sandhill crane parents feed young for the first few weeks and colt mortality can be high at this time (Gerber et al. 2015). Sandhill cranes have the lowest recruitment of hunted avian species in North America (Drewien et al. 1995).

13.3.3 Post-breeding Survival and Migration

Post-breeding is bracketed by the reproductive and fall migration periods (Hohman et al. 1992). Most waterfowl molt flight feathers rendering birds flightless for 3–5 weeks (Baldassarre and Bolen 2006; Fox et al. 2014). Post-breeding waterfowl are vulnerable to habitat changes (e.g., drying, or de-watering of wetlands) that increase predation risk or limit access to food resources (Hohman et al. 1992). Molting has

Table 13.3 Common waterfowl species in North America and their primary occurrence in rangelands, population size, trend, and conservation or management status in the United States

Common name	Scientific name	Rangeland overlap[a]	Population Estimate: LTA-TSA[b]	Estimate: PIF (US, CA)[c]	Trend (%/yr)[d]	Status[e]
Northern Pintail[DA]	*Anas acuta*	B, NB	3,866,300	3,200,000	− 1.2	BMC
Green-winged Teal[DA]	*Anas crecca*	B, NB	2,179,200	3,900,000	1.7	BMC
Mexican Duck*[DA]	*Anas diazi*	B, NB		55,000		
Mottled Duck[DA]	*Anas fulvigula*	B, NB		180,000	− 2.5	BMC, BCC
Mallard[DA]	*Anas platyrhynchos*	B, NB	7,930,400	11,000,000	0.7	BMC
American Black Duck[DA]	*Anas rubripes*			700,000	− 1	BMC
Muscovy Duck[DA]	*Cairina moschata*	B, NB				
American Wigeon[DA]	*Mareca americana*	B, NB	2,618,100	2,700,000	− 0.2	BMC
Gadwall[DA]	*Mareca strepera*	B, NB	2,057,300	3,400,000	2.4	BMC
Northern Shoveler[DA]	*Spatula clypeata*	B, NB	2,643,900	4,400,000	2.3	BMC
Cinnamon Teal[DA]	*Spatula cyanoptera*	B, NB		440,000	− 2.2	BMC, BCC
Blue-winged Teal[DA]	*Spatula discors*	B, NB	5,127,600	7,800,000	1.5	BMC
Wood Duck[DC]	*Aix sponsa*	B, NB		4,600,000	1.8	BMC
Fulvous Whistling-Duck[WD]	*Dendrocygna bicolor*	B, NB		120,000		
Black-bellied Whistling-Duck[WD]	*Dendrocygna autumnalis*	B, NB			7.4	
Lesser Scaup[DI]	*Aythya affinis*	B, NB	4,947,300[c]	3,700,000	− 1.2	BMC
Redhead[DI]	*Aythya americana*	B, NB	732,700	1,200,000	1.6	BMC
Ring-necked Duck[DI]	*Aythya collaris*	B, NB		2,000,000	3.3	BMC
Greater Scaup[DI]	*Aythya marila*	NB	c	720,000	− 1.5	BMC
Canvasback[DI]	*Aythya valisineria*	B, NB	591,300	690,000	0.8	BMC

(continued)

Table 13.3 (continued)

Common name	Scientific name	Rangeland overlap[a]	Population			
			Estimate: LTA-TSA[b]	Estimate: PIF (US, CA)[c]	Trend (%/yr)[d]	Status[e]
Ruddy Duck[DO]	*Oxyura jamaicensis*	B, NB		1,300,000	1.7	BMC
Bufflehead[SM]	*Bucephala albeola*	b, NB		1,300,000	3.5	
Common Goldeneye[SM]	*Bucephala clangula*	B, NB		1,200,000	0.7	BMC
Barrow's Goldeneye[SM]	*Bucephala islandica*	B, NB		180,000	− 1.4	
Long-tailed Duck[SM]	*Clangula hyemalis*			1,000,000	− 4.8	BMC
Harlequin Duck[SM]	*Histrionicus histrionicus*	b		170,000	− 0.5	BMC
Hooded Merganser[SM]	*Lophodytes cucullatus*	B, NB		1,100,000	4.7	
Black Scoter[SM]	*Melanitta americana*			500,000	− 2.3	BMC
White-winged Scoter[SM]	*Melanitta deglandi*	b, nb		400,000	− 0.6	BMC
Surf Scoter[SM]	*Melanitta perspicillata*	nb		470,000	0.2	BMC
Common Merganser[SM]	*Mergus merganser*	B, NB		1,200,000	− 0.4	
Red-breasted Merganser[SM]	*Mergus serrator*	NB		400,000	− 3.3	
Steller's Eider[SM]	*Polysticta stelleri*			660	− 4	ESA
Spectacled Eider[SM]	*Somateria fischeri*			20,000		ESA
Common Eider[SM]	*Somateria mollissima*			750,000	1	BMC
King Eider[SM]	*Somateria spectabilis*			600,000	− 6.4	BMC
Greater White-fronted Goose[GA]	*Anser albifrons*	b, NB		4,300,000	4.9	BMC
Snow Goose[GA]	*Anser caerulescens*	NB		15,000,000	6.1	BMC
Emperor Goose[GA]	*Anser canagicus*			98,000	0.4	BMC, BCC
Ross's Goose[GA]	*Anser rossii*	NB		1,600,000	11.7	BMC

(continued)

Table 13.3 (continued)

Common name	Scientific name	Rangeland overlap[a]	Population			
			Estimate: LTA-TSA[b]	Estimate: PIF (US, CA)[c]	Trend (%/ yr)[d]	Status[e]
Brant[GA]	*Branta bernicla*	nb		340,000	0.2	BMC, BCC
Canada Goose[GA]	*Branta canadensis*	B, NB		7,500,000	10.3	BMC
Cackling Goose[GA]	*Branta hutchinsii*	nb		4,100,000	6.2	BMC
Trumpeter Swan[SC]	*Cygnus buccinator*	B, NB		63,000	6.6	BMC
Tundra Swan[SC]	*Cygnus columbianus*	NB		190,000	0	BMC
Mute Swan[SC]	*Cygnus olor*			31,000	3.6	

[a]Species occurrence in central and western rangeland regions of North America and annual cycle importance. B (b) = breeding, NB (nb) = non-breeding; capital letters indicate common or abundant, lowercase letters indicate uncommon, rare, or minimal rangeland overlap

[b]Population estimate based on the long-term average (LTA[1955–2022]) from the traditional survey area (TSA) of the Breeding Waterfowl and Habitat Survey conducted by U.S. Fish & Wildlife Service and Canadian Wildlife Service (USFWS 2022). Population estimates for lesser and greater scaup combined

[c]Population estimate from Partners in Flight (2021; PIF)

[d]Population trend (% change per year) from Partners in Flight (2021)

[e]Conservation and management status identified by U.S. Fish and Wildlife Service. BMC = birds of management concern, BCC = birds of conservation concern, ESA = threatened or endangered status under the Endangered Species Act

Guild and (Tribe): [DA]Dabbler (Anatani), [DC]Dabbler (Cairinini), [WD]Whistling Duck (Dendrocygnini), [DI]Diver (Aythini), [DO]Diver (Oxyurini), [SM]Sea Duck (Mergini), [GA]Goose (Anserini), [SC]Swan (Cygnini)

high nutrient demands like protein-rich foods (e.g., aquatic insects). Post-breeding waterfowl select habitats that lower predation risk and offer abundant food resources (Fox et al. 2014). Semi-permanent or permanent wetlands with emergent vegetation and open water are often selected post-breeding (Hohman et al. 1992; Fleskes et al. 2010). Such habitats also offer key migratory stopover areas when energetic demands increase and wetland bird diets transition to more carbohydrate-rich food sources such as wetland plant seeds, tubers, rhizomes, and agricultural grains (Baldasarre and Bolen 2006; Donnelly et al. 2021). Shorebirds will consume small amounts of plant material, but they primarily consume invertebrates for energy and some species may double their body mass prior to migration (Baker et al. 2014; Iglecia and Winn 2021).

Across rangelands, wetland availability is lowest during late summer and early fall (Johnson et al. 2010; Donnelly et al. 2019). Habitat availability is typically lowest during the post-breeding period when birds have high nutrient demands. Low nutrient reserves may negatively affect autumn survival (Sedinger and Alisauskas 2014). Additionally, diseases such as botulism, avian cholera, and avian influenza virus increase mortality risk, particularly for waterfowl, especially with decreased wetland availability (Friend et al. 2001; Baldassarre and Bolen 2006; Kent et al. 2022). Changes in land and water use, often in combination with drought, decrease wetland availability resulting in bird concentrations and recurring disease issues (Fleskes et al. 2010; Donnelly et al. 2022; Kahara et al. 2021). Similar to the breeding period, rangelands that provide wetland habitat during the post-breeding period are vital to wetland birds (Johnson et al. 2010; Gerber et al. 2015; Kemink et al. 2021; Donnelly et al. 2022).

Ideally, migration and wintering habitat provide key nutrients and energy (i.e., lipids) sources during migration and highlight the importance of available wetland complexes (Moon and Haukos 2006, 2009; Davis et al. 2014; Yetter et al. 2018). Selected food resources may change based on physiology and behavior and in response to environmental conditions. Narrow migration windows may or may not align with food availability. Donnelly et al. (2019) found that most seasonal wetlands were available during spring migration, whereas ≤ 20% were available for fall migration. In winter, freezing and snow accumulation can decrease food (e.g., grains) availability inhibiting migration.

Some wetland birds (e.g., waterfowl, coots, sandhill cranes and other rails) are hunted during fall and winter. Hunters, through harvest reporting (e.g., band returns, wing collections, surveys) and funding (e.g., duck stamp), have increased our understanding of population dynamics, movements, and conservation (Anderson et al. 2018). For example, adult female mallard survival during the non-breeding season has little impact on population growth rates relative to the breeding period and males have low natural mortality making them even more available for sustainable harvest (Hoekman et al. 2002). Consequently, hunting harvest is the primary mortality cause for male ducks (Hoekman et al. 2002; Riecke et al. 2022a). Female ducks generally experience lower harvest rates than males (Riecke et al. 2022a, b). Waterfowl harvest is carefully managed across flyways and represents one of the most successful examples of adaptive management in the world (Nichols et al. 2019).

Non-breeding habitat conditions can have carry-over effects to breeding success (Sedinger and Alisauskas 2014; Swift et al. 2020). Generally, birds in better nutritional state (i.e., body condition) during winter and spring may arrive in breeding areas earlier, nest earlier, and experience greater breeding success (Devries et al. 2008; Sedinger and Alisauskas 2014; Swift et al. 2020). Management that enhances nutritive resources in non-breeding habitats can also increase vital rates (Davis et al. 2014; Stafford et al. 2014). More information is needed to better understand shorebird vital rates and population dynamics, along with the impacts of migration and winter habitat. However, adult annual survival sustains populations for several arctic-nesting shorebirds during migration (Weiser et al. 2020). Like other wetland birds, wetland networks are critical (Albanese and Davis 2015). Wetlands within rangelands generally have less functional impairment than in croplands and are critical to wetland bird survival (Tsai et al. 2012; Collins et al. 2014; Albanese and Davis 2015; McCauley et al. 2015; Tangen et al. 2022).

13.3.4   Spring Migration

Spring migration is another critical time for wetland birds and includes additional energetic demands, like molt and courtship (Anteau et al. 2011; Stafford et al. 2014). Survival is usually high (Moon and Haukos 2006; Osnas et al. 2021), and habitat availability remains important (Anteau and Afton 2011; Sedinger and Alisauskas 2014). For many species, early arrival to breeding areas correlates with increased reproductive success. Shallow flooded wetlands are often the first to thaw and provide important food resources in spring. Overall, the timing, stop-over frequency, and duration of spring migration is influenced by weather conditions, habitat availability (i.e., food abundance), and initial body condition (Miller et al. 2005; Haukos et al. 2006; Stafford et al. 2014). Wetland networks are therefore needed to support migration survival and breeding success (Devries et al. 2008; Zarzycki 2017; Osnas et al. 2021).

13.4   Current Species Population Status and Monitoring

More than 200 million individuals of 280 species of wetland birds occur in North America (PIF 2021). Over half of these species have seasonal distributions that overlap with rangelands, comprising > 160 million wetland birds (Tables 13.4, 13.5 and 13.6). Wetland bird populations have increased between 1970 and 2017, primarily from waterfowl and geese, but other wetland birds have declined (Rosenberg et al. 2019).

13.4.1   Monitoring Programs

Large-scale programs have been developed to monitor population status. Several are agency-led, particularly for hunted species, while some rely on citizen science efforts. Since 1955, the U.S. Fish and Wildlife Service (USFWS) and Canadian Wildlife Service (CWS) have conducted the Waterfowl Breeding Population and Habitat Survey (WBPHS) to estimate breeding populations in Alaska, Canada, and north-central United States (USFWS 2022). The WBPHS is used for estimates of multiple waterfowl species populations and wetland abundance. The USFWS, in coordination with state wildlife agencies, conducts an annual mid-winter waterfowl survey within each flyway to index waterfowl populations (USFWS 2023). Large-scale monitoring for shorebirds has been proposed, with implementation of some periodic, regional surveys (Cavitt et al. 2014). Secretive marsh bird surveys have been implemented in multiple regions (Johnson et al. 2009).

Publicly-sourced data collection has become increasingly important. The Breeding Bird Survey (BBS) is the main source of avian population status in North America and provides representative sampling of wetlands (Sauer et al. 2003;

Table 13.4 Common shorebird species in North America and their primary occurrence in rangelands, population size, trend, and conservation or management status in the United States

Common name	Scientific name	Rangeland overlap[a]	Population		
			Estimate: PIF (US, CA)[b]	Trend (%/ yr)[c]	Status[d]
Piping Plover[C]	*Charadrius melodus*	B, NB	8400	− 1.9	ESA
Mountain Plover[C]	*Charadrius montanus*	B, NB	20,000	− 3.1	BMC, BCC
Snowy Plover[C]	*Charadrius nivosus*	B, NB	24,000	0.4	BMC, BCC
Semipalmated Plover[C]	*Charadrius semipalmatus*	NB	200,000	− 0.4	
Killdeer[C]	*Charadrius vociferus*	B, NB	1,800,000	− 1	
Wilson's Plover[C]	*Charadrius wilsonia*	B	8600	− 1.9	BMC, BCC
American Golden-Plover[C]	*Pluvialis dominica*	NB	500,000	− 1.9	BCC
Pacific Golden-Plover[C]	*Pluvialis fulva*		43,000	− 1.7	
Black-bellied Plover[C]	*Pluvialis squatarola*	NB	360,000	− 1.6	
Black Oystercatcher[H]	*Haematopus bachmani*		10,000	3.5	BMC, BCC
American Oystercatcher[H]	*Haematopus palliatus*	B	12,000	1	BMC, BCC
Northern Jacana[J]	*Jacana spinosa*	B, NB			
Black-necked Stilt[R]	*Himantopus mexicanus*	B, NB	180,000	2.4	
American Avocet[R]	*Recurvirostra americana*	B, NB	450,000	0.5	BCC
Spotted Sandpiper[S]	*Actitis macularius*	B, NB	660,000	− 1.4	
Ruddy Turnstone[S]	*Arenaria interpres*	NB	250,000	− 4.7	BCC
Black Turnstone[S]	*Arenaria melanocephala*	NB	95,000	− 0.4	BCC
Upland Sandpiper[S]	*Bartramia longicauda*	B, NB	750,000	0.5	BMC, BCC
Sanderling[S]	*Calidris alba*	NB	300,000	− 3.3	

(continued)

Table 13.4 (continued)

Common name	Scientific name	Rangeland overlap[a]	Population		
			Estimate: PIF (US, CA)[b]	Trend (%/yr)[c]	Status[d]
Dunlin[S]	*Calidris alpina*	nb	1,500,000	− 2.7	BMC, BCC
Baird's Sandpiper[S]	*Calidris bairdii*	NB	280,000	1.3	
Red Knot[S]	*Calidris canutus*	MB	140,000	− 5.7	BMC, BCC
White-rumped Sandpiper[S]	*Calidris fuscicollis*	NB	1,700,000	1.3	
Stilt Sandpiper[S]	*Calidris himantopus*	NB	1,200,000	− 1.5	
Purple Sandpiper[S]	*Calidris maritima*		25,000	− 1.2	BMC, BCC
Western Sandpiper[S]	*Calidris mauri*	NB	3,500,000	− 0.4	
Pectoral Sandpiper[S]	*Calidris melanotos*		1,500,000	− 2	BCC
Least Sandpiper[S]	*Calidris minutilla*	NB	700,000	− 0.2	
Rock Sandpiper[S]	*Calidris ptilocnemis*		140,000	− 2.8	BMC, BCC
Semipalmated Sandpiper[S]	*Calidris pusilla*	NB	2,300,000	− 3.1	BMC, BCC
Buff-breasted Sandpiper[S]	*Calidris subruficollis*	NB	56,000	1.8	BMC, BCC
Surfbird[S]	*Calidris virgata*	NB	70,000	− 1.7	
Wilson's Snipe[S]	*Gallinago delicata*	B, NB	2,000,000	0.3	BMC
Short-billed Dowitcher[S]	*Limnodromus griseus*	NB	150,000	− 2.9	BMC, BCC
Long-billed Dowitcher[S]	*Limnodromus scolopaceus*	NB	520,000	− 0.3	
Marbled Godwit[S]	*Limosa fedoa*	B, NB	170,000	− 0.7	BCC
Hudsonian Godwit[S]	*Limosa haemastica*	NB	77,000	− 3.4	BMC, BCC
Bar-tailed Godwit[S]	*Limosa lapponica*		90,000		BMC, BCC
Long-billed Curlew[S]	*Numenius americanus*	B, NB	140,000	0	BMC, BCC

(continued)

Table 13.4 (continued)

Common name	Scientific name	Rangeland overlap[a]	Population		
			Estimate: PIF (US, CA)[b]	Trend (%/ yr)[c]	Status[d]
Eskimo Curlew[S]	*Numenius borealis*		50		ESA
Whimbrel[S]	*Numenius phaeopus*	MNB	80,000	− 2.1	BMC
Bristle-thighed Curlew[S]	*Numenius tahitiensis*		10,000		BMC, BCC
Red Phalarope[S]	*Phalaropus fulicarius*		1,600,000		
Red-necked Phalarope[S]	*Phalaropus lobatus*	NB	2,500,000		
Wilson's Phalarope[S]	*Phalaropus tricolor*	B, NB	1,500,000	− 0.2	
American Woodcock[S]	*Scolopax minor*		3,500,000	− 0.8	BMC
Lesser Yellowlegs[S]	*Tringa flavipes*	NB	660,000	− 2.8	BMC, BCC
Wandering Tattler[S]	*Tringa incana*	NB	16,000	− 4.3	BCC
Greater Yellowlegs[S]	*Tringa melanoleuca*	NB	140,000	0.5	
Willet[S]	*Tringa semipalmata*	B, NB	250,000	− 0.6	BCC
Solitary Sandpiper[S]	*Tringa solitaria*	B, NB	190,000	0.7	BMC, BCC

[a]Species occurrence in central and western rangeland regions of North America and annual cycle importance. B (b) = breeding, NB (nb) = non-breeding; capital letters indicate common or abundant, lowercase letters indicate uncommon, rare, or minimal rangeland overlap
[b]Population estimate from Partners in Flight (2021; PIF)
[c]Population trend (% change per year) from Partners in Flight (2021)
[d]Conservation and management status identified by U.S. Fish and Wildlife Service. BMC = birds of management concern, BCC = birds of conservation concern, ESA = threatened or endangered status under the Endangered Species Act
Family names: [C]Charadriidae, [H]Haematopodidae, [J]Jacanidae, [R]Recurvirostridae, [S]Scolopacidae

Niemuth et al. 2007; Veech et al. 2017). The BBS may not suffice for all species and formal evaluations are needed concerning wetland birds (Hudson et al. 2017). For many wetland birds, BBS data could be more useful if wetland habitat availability were included (Niemuth and Solberg 2003; Niemuth et al. 2009). For other species, targeted monitoring may be necessary. eBird, a global online database launched in 2002 (Cornell Lab of Ornithology), compiles public records of avian species detections with location and date (Sullivan et al. 2014). Biologists have used eBird data to

Table 13.5 Common waterbird species in North America and their primary occurrence in rangelands, population size, trend, and conservation or management status in the United States

Common name	Scientific name	Rangeland overlap[a]	Population		
			Estimate: PIF (US, CA)[b]	Trend (%/yr)[c]	Status[f]
Black Tern[CL]	*Chlidonias niger*	B, NB	2,300,000	− 1.9	BMC, BCC
Bonaparte's Gull[CL]	*Chroicocephalus philadelphia*	NB	790,000	1.9	
Gull-billed Tern[CL]	*Gelochelidon nilotica*	B, NB	8,000	1.3	BMC, BCC
Caspian Tern[CL]	*Hydroprogne caspia*	B, NB	78,000	0.9	
Herring Gull[CL]	*Larus argentatus*	B, NB	2,900,000	− 3.9	
California Gull[CL]	*Larus californicus*	B, NB	1,100,000	− 1.6	BCC
Ring-billed Gull[CL]	*Larus delawarensis*	B, NB	3,700,000	1.5	
Glaucous-winged Gull[CL]	*Larus glaucescens*	B, NB	440,000	− 0.6	
Iceland Gull[CL]	*Larus glaucoides*	NB	84,000		
Heermann's Gull[CL]	*Larus heermanni*	NB			BCC
Yellow-footed Gull[CL]	*Larus livens*	B, NB			BCC
Laughing Gull[CL]	*Leucophaeus atricilla*	B, NB	680,000	2	
Franklin's Gull[CL]	*Leucophaeus pipixcan*	B, NB	2,300,000	− 1.9	BCC
Sooty Tern[CL]	*Onychoprion fuscatus*	B			
Black Skimmer[CL]	*Rynchops niger*	B, NB	60,000	− 3.1	BCC
Forster's Tern[CL]	*Sterna forsteri*	B, NB	130,000	− 1.4	BCC
Common Tern[CL]	*Sterna hirundo*	B	470,000	− 2.1	
Least Tern[CL]	*Sternula antillarum*	B	52,000	− 3.6	ESA
Elegant Tern[CL]	*Thalasseus elegans*	B, NB			BCC
Royal Tern[CL]	*Thalasseus maximus*	B	35,000	0.5	
Sandwich Tern[CL]	*Thalasseus sandvicensis*	B, NB	94,000	1.4	BMC, BCC
Wood Stork[CC]	*Mycteria americana*	NB	16,000	1.6	ESA (SE pop)

(continued)

Table 13.5 (continued)

Common name	Scientific name	Rangeland overlap[a]	Population		
			Estimate: PIF (US, CA)[b]	Trend (%/yr)[c]	Status[f]
Common Loon[GA]	*Gavia immer*	B, NB	1,100,000	0.8	
Sandhill Crane[GU]	*Antigone canadensis*	B, NB	500,000	5.1	BMC
Whooping Crane[GU]	*Grus americana*	B, NB	370		ESA
Yellow Rail[GR]	*Coturnicops noveboracensis*	B, NB	12,000		BMC, BCC
American Coot[GR]	*Fulica americana*	B, NB	5,500,000	0.2	BMC
Common Gallinule[GR]	*Gallinula galeata*	B, NB	500,000	− 1.1	
Black Rail[GR]	*Laterallus jamaicensis*	B, NB			BMC
Purple Gallinule[GR]	*Porphyrio martinicus*	B, NB	20,000	− 1.9	
Sora[GR]	*Porzana carolina*	B, NB	4,400,000	0.5	BMC
Clapper Rail[GR]	*Rallus crepitans*	B, NB	170,000	− 0.8	BMC
King Rail[GR]	*Rallus elegans*	B, NB	63,000	− 4.5	BMC, BCC
Virginia Rail[GR]	*Rallus limicola*	B, NB	230,000	1.5	BMC
Great Egret[PA]	*Ardea alba*	B, NB	710,000	2.5	
Great Blue Heron[PA]	*Ardea herodias*	B, NB	620,000	0.7	
American Bittern[PA]	*Botaurus lentiginosus*	B, NB	2,500,000	− 0.6	BMC
Cattle Egret[PA]	*Bubulcus ibis*	B, NB	2,800,000	− 1.4	
Green Heron[PA]	*Butorides virescens*	B, NB	770,000	− 1.9	
Little Blue Heron[PA]	*Egretta caerulea*	B, NB	270,000	− 1.4	BMC, BCC
Reddish Egret[PA]	*Egretta rufescens*	B, NB	2400	1.3	BMC
Snowy Egret[PA]	*Egretta thula*	B, NB	220,000	2.2	
Tricolored Heron[PA]	*Egretta tricolor*	B, NB	58,000	− 0.5	
Least Bittern[PA]	*Ixobrychus exilis*	B, NB	130,000	0.7	BMC
Yellow-crowned Night-Heron[PA]	*Nyctanassa violacea*	B, NB	130,000	− 0.5	

(continued)

Table 13.5 (continued)

Common name	Scientific name	Rangeland overlap[a]	Population		
			Estimate: PIF (US, CA)[b]	Trend (%/yr)[c]	Status[f]
Black-crowned Night-Heron[PA]	*Nycticorax nycticorax*	B, NB	420,000	− 0.4	BMC
American White Pelican[PP]	*Pelecanus erythrorhynchos*	B, NB	410,000	6.3	BCC
Brown Pelican[PP]	*Pelecanus occidentalis*	B, NB	100,000	3.6	
White Ibis[PT]	*Eudocimus albus*	B, NB	1,200,000	4.4	
Roseate Spoonbill[PT]	*Platalea ajaja*	B, NB	11,000	7.3	
White-faced Ibis[PT]	*Plegadis chihi*	B, NB	1,300,000	3.5	
Glossy Ibis[PT]	*Plegadis falcinellus*		36,000	5.3	
Clark's Grebe[PO]	*Aechmophorus clarkii*	B, NB	72,000	− 2.9	BCC
Western Grebe[PO]	*Aechmophorus occidentalis*	B, NB	990,000	− 3.6	BMC, BCC
Horned Grebe[PO]	*Podiceps auritus*	B, NB	250,000	− 1.4	BMC
Red-necked Grebe[PO]	*Podiceps grisegena*	B	740,000	0.9	
Eared Grebe[PO]	*Podiceps nigricollis*	B, NB	2,000,000	1.1	BMC
Pied-billed Grebe[PO]	*Podilymbus podiceps*	B, NB	1,100,000	0.9	
Least Grebe[PO]	*Tachybaptus dominicus*	B, NB			
Anhinga[SA]	*Anhinga anhinga*	B, NB	27,000	1.6	
Double-crested Cormorant[SP]	*Phalacrocorax auritus*	B, NB	560,000	4	BMC (OA)
Neotropic Cormorant[SP]	*Phalacrocorax brasilianus*	B, NB		7.6	

[a]Species occurrence in central and western rangeland regions of North America and annual cycle importance. B (b) = breeding, NB (nb) = non-breeding; capital letters indicate common or abundant, lowercase letters indicate uncommon, rare, or minimal rangeland overlap

[b]Population estimate from Partners in Flight (2021; PIF)

[c]Population trend (% change per year) Partners in Flight (2021)

[d]Conservation and management status identified by U.S. Fish and Wildlife Service. BMC = birds of management concern, BCC = birds of conservation concern, ESA = threatened or endangered status under the Endangered Species Act 1973

Order and Family: [CL]Ciconiiformes Laridae, [CC]Ciconiiformes Ciconiidae, [GA]Gaviiformes Gaviidae, [GU]Gruiformes Gruidae, [GR]Gruiformes Rallidae, [PA]Pelecaniformes Ardeidae, [PP]Pelecaniformes Pelecanidae, [PT]Pelecaniformes Threskiornithidae, [PO]Podicipediformes Podidipedidae, [SA]Suliformes Anhingidae, [SP]Suliformes Phalacrocoracidae

assess migration chronology, distribution, abundance, and population trends (Walker and Taylor 2017; Horns et al. 2018; Fink et al. 2020).

International banding programs provide information on movements and demographics of many wetland bird populations. The U.S. Geological Survey Bird Banding Laboratory distributes about one million aluminum leg bands to managers and researchers in the U.S. and Canada each year and manages an archive of over 77 million banding records and 5 million band encounters (U.S. Geological Survey, Bird Banding Laboratory 2020). Hunter participation in waterfowl band reporting has been one of the longest and most significant information sources for waterfowl research and conservation. Mark-recovery methods (Brownie et al. 1985; Williams et al. 2002) are used to identify harvest distribution and associated breeding areas, estimate harvest rates, and survival rates for species, age, and sex (Smith et al. 1989). The USFWS and CWS conduct annual hunter surveys for hunted wetland birds (Martin and Carney 1977; Cooch et al. 1978; Martin et al. 1979). Harvest estimates can provide an index to population trends when other data sources are limited for some species. Banding and harvest survey data can be combined to estimate population abundance in some cases (Lincoln 1930; Alisauskas et al. 2009, 2014).

13.4.2 Waterfowl

As a group, there are fewer waterfowl species (n = 46) than either shorebird (n = 51) or waterbirds (n = 62), but waterfowl populations are more abundant. Waterfowl have also had more monitoring due to their gamebird status and associated socio-economic values (Anderson et al. 2018). Indices of breeding ducks from the WBPHS have fluctuated from lows of 25 million in the early 1960s and 1990s to nearly 50 million in 2014–2015. Duck populations show a cyclical pattern over time influenced largely by conditions in the PPR (Fig. 13.2; Baldassarre and Bolen 2006; USFWS 2022).

Duck species that rely primarily on rangeland wetlands tend to have small populations or be in decline. Cinnamon teal are widely distributed across western rangelands. Mottled ducks occur primarily along the Gulf Coast. Cinnamon teal, mottled duck, and Mexican duck are among the least studied species with lower abundance and are identified as Species of Conservation Concern (Table 13.3; Baldassarre 2014). Northern pintails have declined since the early 1970s and remain below population objectives (NAWMP 2018; USFWS 2022). Rangeland conversion to row-crop production, especially in the PPR, has contributed to pintail declines (Baldassarre 2014; Buderman et al. 2020). Lesser and greater scaup have had similar population declines and status as pintails (NAWMP 2018; USFWS 2022). Most lesser scaup breed in the Western Boreal Forest, but at least 25% breed in rangeland wetlands where livestock grazing is a prominent land-use (Baldassarre 2014).

Goose and swan populations have generally increased since the 1970s, with overabundance of some goose populations (USFWS BMC 2011; Baldassarre 2014; USFWS 2022). The dramatic increase in snow geese is largely due to 4 factors: (1) increased food availability due to crop-conversion and enhanced fertilizer-based

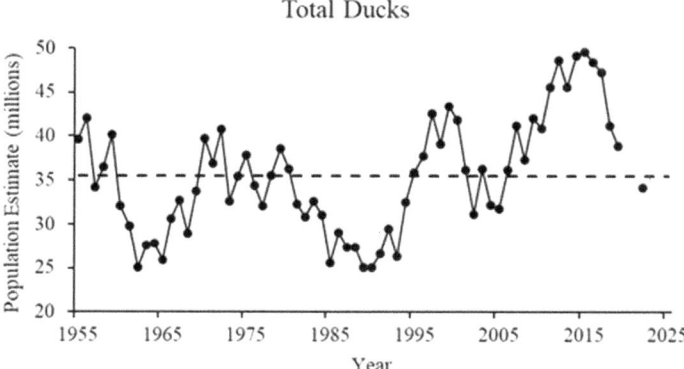

Fig. 13.2 Total duck population change 1955–2022 from the Traditional Survey Area of the Waterfowl Breeding Population and Habitat Survey (USFWS 2022)

yields, (2) establishment of staging and wintering areas on refuges, (3) declines in harvest rates, and (4) climate change (Jefferies et al. 2003; Baldassarre 2014). Several Canada goose populations breed extensively in rangelands (Baldassarre 2014). The Hi-Line population in north-central Montana increased tenfold following their 1960s reintroductions concomitant with reservoir and stock pond development (Nieman et al. 2000; Baldassarre 2014).

13.4.3 Shorebirds

Shorebirds have experienced significant declines (i.e., 37%) since the 1970s (Fig. 13.2; Rosenberg et al. 2019; Smith et al. 2023). Most shorebirds are considered species of high conservation concern with 5 listed under the Endangered Species Act (1973; Table 13.4; U.S. Shorebird Conservation Plan Partnership 2016). Over 80% of breeding shorebirds migrate to Mexico, Central, and South America (Iglecia and Winn 2021). At least 16 species have significant breeding range overlap with rangelands and ~ 50% exhibit declining population trends (Table 13.4). Causes of shorebird declines are poorly understood.

13.4.4 Waterbirds

Waterbirds include > 180 species across 7 taxonomic orders that use marine and inland aquatic habitats (PIF 2021). More than 60 waterbird species inhabit rangelands (Table 13.5). Status of some waterbirds, especially secretive species, is poorly understood (Sauer 1999; Johnson et al. 2009), but survey information suggests variation in trends. Rosenberg et al. (2019) estimated a 22% decrease across 77 species. Sandhill crane populations have increased in recent decades (Seamans 2022).

13.5 Habitat Associations

A broad overview of functional habitat relationships across groups of wetland birds is provided herein. Habitat use varies by species, season, and time of day and habitat associations are available for most species (e.g., eBird, Birds of the World). Breeding and foraging characteristics are highly varied among waterbirds, shorebirds, and waterfowl. Heterogeneous habitat with assorted wetland types, water depths, vegetation density, and food support a diversity of wetland birds (Ma et al. 2010).

13.5.1 *Waterfowl*

Puddle ducks, or dabblers, are associated with shallow wetlands foraging near the water surface by "tipping-up" to reach food items (Table 13.3; Fig. 13.3). However, they can perform shallow dives to avoid predators or reach food. Dabblers use seasonal and perennial wetlands with emergent vegetation for foraging and escape cover, particularly important during the brooding period (Walker et al. 2013b; Fig. 13.3). Seasonal wetlands are in overall decline (Collins et al. 2014; McCauley et al. 2015; Donnelly et al. 2022). In the Intermountain West, rangeland seasonal wetlands have been converted to flood-irrigated fields but can still provide important habitat (Fleskes and Gregory 2010; Donnelly et al. 2019; Mackell et al. 2021). Semipermanent wetlands are habitat for dabblers and may be especially important during periods of water scarcity (McCauley et al. 2015; Donnelly et al. 2022). Dabblers also use open water as roosting habitat, especially when foraging habitat is nearby. During the breeding period, grass-dominated upland habitats are vital for dabbler nesting habitat. For example, nearly 90% of waterfowl in the PPR nest in uplands (PPJV 2017).

Pochards, or diving ducks, are adapted to deeper aquatic systems where they forage in the water column or benthic substrate (Figs. 13.3 and 13.4; Table 13.3). Common benthic forage includes bivalves, worms, and insect larvae (Baldassarre and Bolen 2006). Divers often use wetlands with SAV for foraging (e.g., pondweeds [Potamogetonaceae]). Sea ducks generally have high salinity tolerance, forage deeper, and are uncommon in rangelands (Table 13.3), though bufflehead and common goldeneye use rangeland wetlands (Baldassarre and Bolen 2006). Mergansers also use rangeland aquatic habitats and forage on small fish, often in deeper water systems.

Swans use wetland habitat similar to diving ducks (Fig. 13.4). Swans use their long necks to access SAV. Breeding trumpeter swans use freshwater marshes, ponds, lakes,

Table 13.6 General waterbird habitat associations based on amount of emergent vegetation, open water, and nesting habitat

Group A	Group B	Group C	Group D	Group E
Wetland with:	Wetland with:	Wetland with:	Wetland with:	Lake or River with:
• Substantial emergent vegetation	• Emergent vegetation	• Emergent vegetation	• Emergent vegetation	• Open water
• Variable open water	• Partial open water	• Extensive open water	• Open water	• Barren ground
			• Nesting trees	• Islands
American Bittern	Sandhill Crane	Common Loon	Great Blue Heron	American White Pelican
Least Bittern	White-faced Ibis	Pied-billed Grebe	Great Egret	Double-crested Cormorant
Black-crowned Night-Heron	Franklin's Gull	Horned Grebe	Snowy Egret	Ring-billed Gull
Yellow Rail	Bonaparte's Gull	Red-necked Grebe	Tricolored Heron	California Gull
Black Rail	Forster's Tern	Eared Grebe	Little Blue Heron	Herring Gull
King Rail	Black Tern	Western Grebe	Cattle Egret	Caspian Tern
Virgina Rail		Clark's Grebe	Green Heron	Common Tern
Sora		White-faced Ibis	Yellow-crowned Night Heron	Least Tern
		American Coot		
		Common Moorhen		

Adapted from Beyersbergen et al. (2004)

and occasionally slowly moving streams. Basic breeding habitat features include sufficient open water to take flight (about 100 m), SAV, stable water levels, structure for nest sites, and low human disturbance (Baldassarre 2014; Mitchell and Eicholz 2020). Both tundra and trumpeter swans can forage in upland agricultural areas during the non-breeding season. Migrating tundra swans show strong selection for wetlands with sago pondweed (*Stuckenia pectinata*) while nonforaging swans selected large open water areas (Earnst 1994).

Canada geese commonly occur in rangelands (Baldassarre 2014) and are primarily grazers of grasses and sedges, though non-breeding geese can be dependent on crops. Canada geese use a greater diversity of nest sites than other waterfowl (Baldassarre 2014). Common brood-rearing habitat includes gradually sloping ponds or river shorelines, abundant graminoids, and mudflats (Mowbray et al. 2020).

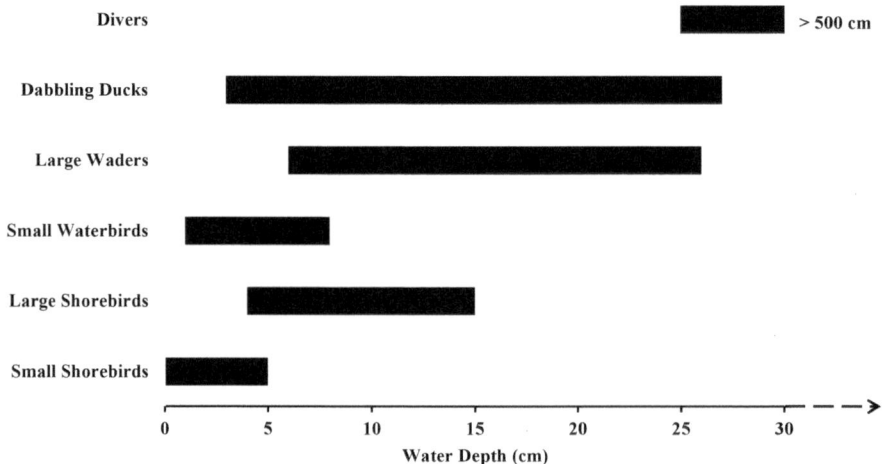

Fig. 13.3 Preferred foraging depths of select wetland birds. Modified from Tori et al. (2002), Helmers (1992), and Richmond et al. (2012)

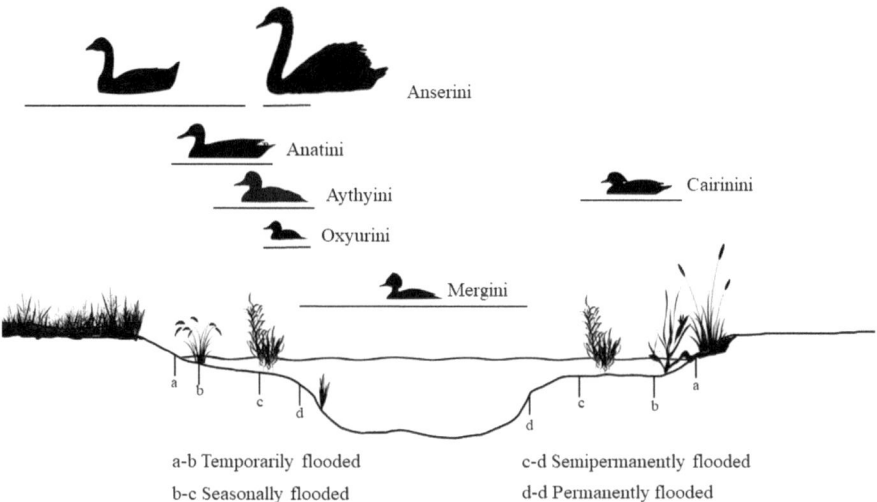

Fig. 13.4 Principal foraging habitats of various waterfowl groups with respect to water depth, plant communities, and wetland hydroperiod. Modified from Krapu and Reinecke (1992)

13.5.2 Shorebirds

Shorebirds use a variety of wetland and upland habitats throughout the year. Most shorebirds select shallow wetlands, wet meadows, shorelines, and open mud flats for foraging and avoid tall and dense vegetation (Iglecia and Winn 2021). For example, marbled godwits and willets select short sparse upland vegetation and wetland complexes for nesting and foraging (Niemuth et al. 2012; Shaffer et al. 2019a, b; Specht et al. 2020). However, these species can use taller and denser vegetation when brooding (Shaffer et al. 2019a, b). Some shorebirds use uplands for breeding, but shift to wetlands later (Shaffer et al. 2019a, c; Niemuth et al. 2012). Shorebirds exhibit varied wetland salinity tolerances. Some breeding shorebirds solely use uplands (Shaffer et al. 2019d, e, f; Iglecia and Winn 2021). During migration, shorebirds select shallow, sparsely vegetated wetlands often with mudflats. For example, shorebirds in the fall correlate positively with grazing pressure, and negatively with denser vegetation (Albanese and Davis 2015). Aquatic and terrestrial invertebrates are common shorebird foods, although seeds, vegetation, algae, and small fish are consumed opportunistically. Dominant invertebrate prey items include chironomids, flies (Diptera), beetles (Coleoptera), true bugs (Hemiptera), amphipods (Amphipoda), snails (Gastropoda), and clams and mussels (Bivalvia). Water depth, in combination with leg and bill length, determines food availability and habitat types used by different shorebirds (Fig. 13.3).

13.5.3 Waterbirds

Waterbirds exhibit diversity in morphology, life history, and habitat use. Species range from large and conspicuous cranes to secretive marsh birds such as bitterns and rails (Table 13.5). Waterbirds use various wetland types with assorted amounts of emergent vegetation, open water, water depth, and woody vegetation (Fig. 13.3; Table 13.6; Beyersbergen et al. 2004). Species use different areas within a wetland. For example, white-faced ibis nest on emergent vegetation in colonies and use shallow flooded areas to forage (Coons 2021; Moulton et al. 2022). Similarly, sandhill cranes nest on mounds in shallow water and use adjacent uplands for foraging (Austin et al. 2007; Ivey and Dugger 2008). Many other waterbirds use flooded areas in rangelands where management often mimics natural hydroperiods (Ivey and Herziger 2006).

13.6 Rangeland Management

Livestock production and wetland bird populations are linked by their dependence on rangelands and surface water (Bue et al. 1964; Richmond et al. 2012; Brasher et al. 2019). Grazing, burning, haying, and water management in wetlands and uplands often increase resources for wetland birds (Kadlec and Smith 1992; Naugle et al. 2000; Baldassarre and Bolen 2006). Response to management practices vary among species, spatial scales, biological parameters, season, and locale. Managers should consider objectives, seasonal habitat needs, and potential tradeoffs. Generally, management that provides a mosaic of upland and wetland habitat is best (Naugle et al. 2000; Baldassarre and Bolen 2006; Krausman et al. 2009; Ma et al. 2010).

13.6.1 Grazing

Upland nesting ducks generally favor dense cover within 4 km (~ 2.5 miles) of wetlands (Reynolds et al. 2006). Nest survival correlates positively with vegetation height (Baldassarre and Bolen 2006; Bloom et al. 2013) and the amount of adjacent grassland (Greenwood et al. 1995; Reynolds et al. 2001; Stephens et al. 2005), reinforcing the need to conserve rangelands. Lack of disturbance can negatively impact grassland, and duck productivity (Naugle et al. 2000; Dixon et al. 2019; Grant et al. 2009). Recent literature has demonstrated the compatibility of livestock grazing with waterfowl habitat (Naugle et al. 2000; Ignatiuk and Duncan 2001; Warren et al. 2008; Bloom et al. 2013; Rischette et al. 2021). Livestock grazing is a land-use that can ultimately support wetland birds (PPJV 2017; Brasher et al. 2019; PHJV 2021). However, localized impacts of grazing can depend on timing, intensity, duration, bird species, and demographics (Briske et al. 2011; Lipsey and Naugle 2017).

Managing grazing for residual cover (> 28 cm; Bloom et al. 2013) will enhance waterfowl nest survival (Warren et al. 2008; Rischette et al. 2021), which is highest when cover provides a physical barrier to predators. Grazing timing and intensity has complex interactions with nest density and survival, along with local and landscape conditions such as precipitation, site quality, and predator dynamics (Herkert et al. 2003; Stephens et al. 2005; Warren et al. 2008; Bloom et al. 2013; Ringelman et al. 2018). Mismanagement leading to overgrazing is detrimental to wetland birds and rangeland health (Kadlec and Smith 1992; Krausman et al. 2009). To maximize productivity, disturbances (e.g., grazing) should occur after or late in the nesting period (Barker et al. 1990; Naugle et al. 2000). From an operational viewpoint, when areas must be grazed, moderate to low stocking rates are preferred for waterfowl nesting cover (Bloom et al. 2013; Rischette et al. 2021).

Multiple grazing systems can support wetland birds while meeting rangeland health and producer objectives. Generally, systems that emphasize residual and dense grass cover are beneficial for waterfowl nesting habitat (Chap. 4; Table 4.2; Holechek et al. 1982; Barker et al. 1990; Ignatiuk and Duncan 2001; Murphy et al. 2004; West and Messmer 2006; Krausman et al. 2009). Studies have indicated that grazing systems with deferment, rotation, and rest (e.g., deferred rotation, rest rotation, deferred rest rotation, and high-intensity low-frequency) can increase residual cover and support wetland bird productivity (Gjersin 1975; Mundinger 1976; Barker et al. 1990; Ignatiuk and Duncan 2001; Murphy et al. 2004; Carroll et al. 2007; Emery et al. 2005; Shaffer et al. 2019a, b, d, e). Resting or deferring grazing in wetlands during the non-breeding season can maintain plant-based foods for waterfowl. Conversely, for many shorebird species, abundance correlates positively with increased grazing pressure, particularly in the non-breeding season (Holechek et al. 1982; Powers and Glimp 1996; Albanese and Davis 2015). In areas with longer growing seasons (e.g., Central Valley of California), grazing July–October supported forage for wintering geese and cranes along with nesting cover (Carroll et al. 2007). However, fall and winter grazing within shorter growing seasons may reduce initial residual cover, albeit with less influence on later nests due to vegetation growth. Evaluating contributions of local-scale management over the short-term (2–3 years) is challenging because productivity can also be influenced by large-scale and carry-over effects (Ringelman et al. 2018; Bortolotti et al. 2022).

Maintaining wetland vegetation structure and availability (Murkin et al. 1997; Masto et al. 2022) is key to nesting, foraging, and brood-rearing habitat for most species (Harrison et al. 2017). In wetlands dominated by robust and monotypic perennials (e.g., cattail) or invasives (e.g., reed canary grass [*Phalaris arundinacea*]), grazing can improve habitat structural diversity, especially in conjunction with practices such as fire, herbicides, and water-level manipulation (Stutzenbaker and Weller 1989; Schultz et al. 1994; Anderson et al. 2019; Bansal et al. 2019; Hillhouse 2019). Maintaining emergent vegetation is important for escape cover and food (Walker et al. 2013b). While reducing vegetation structure along shorelines may be better for shorebirds, excessive grazing can reduce habitat quality for other species (Hoffman and Stanley 1978; Harrison et al. 2017; Iglecia and Winn 2021). For nests along shorelines (e.g., snowy plover), restricting livestock access, or delaying grazing, can increase productivity (Iglecia and Winn 2021). Many wetland plants have high nutrition value and forage production generally exceeds uplands sites (Johnson 2019). Graminoids in mesic areas usually provide high forage quality for livestock (Hubbard 1988; Kirby et al. 2002).

In regions where available water is limited, livestock disproportionately select wet areas increasing the risk of habitat degradation. Historically, improper grazing has led to deterioration of wetlands and negatively impacted wetland birds (Tessman 2004). However, there is a paucity of research concerning grazing impacts on wetland bird survival and productivity, especially in the Intermountain West (Gilbert et al. 1996; Powers and Glimp 1996; Ivey and Dugger 2008; McWethy and Austin 2009). Risk of nest failure due to predation or trampling is generally associated with increased stocking rates (Littlefield and Paullin 1990; Bleho et al. 2014; Harrison et al. 2017; Shaffer et al. 2019d, e). Increases in water scarcity will likely exacerbate grazing impacts on wetland birds.

During the non-breeding period, moist-soil vegetation and seasonally flooded areas should be the focus of resource managers (Fredrickson and Taylor 1982; Smith et al. 1989; Haukos and Smith 1993; Hillhouse 2019). Moist-soil communities dominated by annual plants such as smartweed, common ragweed (*Ambrosia artemisiifolia*), and barnyardgrass, as well as perennials such as sedges, spike-rushes, giant bur-reed (*Sparganium eurycarpum*), and dock (*Rumex* spp.) offer high quality forage (Chabreck et al. 1989; Haukos and Smith 1993; Anderson et al. 2019). Consequently, factors that decrease seed production reduce food availability and carrying capacity. Grazing late summer can reduce seed production, whereas grazing until mid-summer may allow plants and seed production to recover (Chabreck et al. 1989; Anderson et al. 2019; Hillhouse 2019).

At landscape scales, livestock grazing helps maintain rangeland and wetland habitat, but negative effects can occur at smaller scales, although most issues can be addressed through management. For example, rotation and cross fencing can be used to control when and where grazing occurs and help maintain economic and ecological viability (Fynn and Jackson 2022). However, fencing can facilitate meso-predator movements and cause collisions for wetland birds, especially for species that fly close to the water surface or take flight by running across the water surface (Cornwell and Hochbaum 1971; Allen and Ramirez 1990).

13.6.2 Haying/mowing

Delaying haying until late nesting season helps minimize adult mortality and nest failure. However, optimal hay quality in some areas may occur earlier creating a challenge for livestock operations (Epperson et al. 1999; Gruntorad et al. 2021). Flushing bars mounted to haying equipment may help prevent adult mortality, but nests are still destroyed. Haying patterns that move concentrically out from the middle of the field may provide more opportunity for young birds to escape (Ivey 2011). Haying reduces residual vegetation the following nesting season and generally results in lower nest densities and productivity (Renner et al. 1995; Naugle et al. 2000; Rischette et al. 2021). Early nesting species (e.g., mallard, northern pintail) are impacted more by haying than later nesting species (Luttschwager et al. 1994; Renner et al. 1995). Ideally, haying should be late enough to minimize disturbance to nesting birds but early enough for precipitation and regrowth late in the growing season (Rischette et al. 2021).

Many wetland resources depend on irrigation with haying and grazing (Lovvorn and Hart 2004; Copeland et al. 2010; Donnelly et al. 2021). Early haying (e.g., mid-June) may cause nest failure and reduced foraging as well as mortality of unfledged waterbirds (Littlefield 1999; Ivey and Herziger 2006). Concomitantly, irrigated hayfields provide productive breeding habitat for species that select shorter and sparse vegetation (Hartman and Oring 2009; Shaffer et al. 2019d). The short-stature vegetation from haying (or grazing) can provide foraging habitat the following spring and summer when these areas are flooded (Fleskes and Gregory 2010; Donnelly et al. 2019).

13.6.3 Fire

Historically, fire was a principal driver of ecosystem structure throughout the Great Plains (Chap. 6). Burns reset succession to more productive states providing improved nesting and foraging habitat. Prescribed burns can be used to provide desired plant communities for wetland birds (Smith et al. 1989; Kadlec and Smith 1992; Anderson et al. 2019). Fire in marshes and prairie wetlands can reduce dense vegetation, increase food resources, promote desirable plants, provide new growth, and increase plant nutrition (Smith and Kadlec 1985, 1992; Chabreck et al. 1989; Stutzenbaker and Weller 1989; Naugle et al. 2000; Brennan et al. 2005; Venne and Frederick 2013; Anderson et al. 2019). Fire effects vary by location, season, and species needs.

Because seasonal habitat requirements vary widely across wetland bird species, providing a mosaic of burned and unburned areas at multiple scales is likely ideal (Gray et al. 2013). If well-managed, fire can support broad ecological functioning (Hovick et al. 2017).

13.6.4 Water Management

Livestock operations in semi-arid rangelands have long used surface water developments. Inadequate water can lead to poor livestock distribution and utilization issues (Bue et al. 1964; Holechek et al. 2011). Water developments (e.g., stock ponds) for livestock can provide habitat for wetland birds (Forman et al. 1996; Pederson et al. 1989; May et al. 2002; Baldassarre and Bolen 2006). Stock ponds are dammed watercourses, excavated areas, or a combination of both. Excavated stock ponds in seasonal wetlands provide additional water accumulation, causing altered hydroperiods and less-preferred vegetation (Gray et al. 2013; Smith 2003; Baldassarre and Bolen 2006). Constructing terraces can provide shallow water and emergent vegetation (Gray and Bolen 1987). Selection of stock ponds is influenced by multiple factors including size, water depth, emergent and submergent vegetation, proximity to other wetlands, and adjacent nesting cover (Austin and Buhl 2009).

Stock ponds that provide various water depths and diverse vegetation will be attractive to multiple species (Ma et al. 2010). Surface area, shoreline complexity, and vegetation composition are key characteristics for breeding season selection (Flake et al. 1977; Austin and Buhl 2009). Shorebirds may benefit from grazed pond margins and adjacent uplands (Laubhan and Gammonley 2000; May et al. 2002). Irregular shorelines, improved water quality, and SAV are attractive to breeding ducks (Hudson 1983; Svingen and Anderson 1998; Austin and Buhl 2009). Ponds, and natural wetlands, that approximate a 50:50 ratio of emergent vegetation and open water (i.e., hemi-marsh) provide ideal conditions for many wetland birds, particularly waterfowl (Murkin et al. 1997; Smith et al. 2004). Stock ponds are common in areas with limited water availability. Rumble and Flake (1983) recommend ponds for waterfowl broods that have: (1) larger surface area, (2) shallow water supporting submersed and emergent vegetation, (3) grazing management fostering emergent vegetation, (4) adjacent upland cover, and (5) undrained nearby wetlands. Exclusion fencing in shallows may promote emergent and moist-soil plants for food and cover.

Water developments for livestock are also used during non-breeding periods. Approximately half of the ducks during 1997–2014 mid-winter surveys in Texas were detected on stock ponds (Texas Parks and Wildlife Department unpublished data; DU 2021). Medium-sized ponds (0.81–16.2 ha) had higher occupancy (32–51%) compared to smaller ponds (< 0.81 ha; 11–26%; Texas Parks and Wildlife Department unpublished data; Mason et al. 2013). Evidence suggests stock ponds may help offset reduced habitat availability during drought (DU 2021). Along the Gulf Coast, stock ponds provided wintering habitat and freshwater sources for waterfowl, shaping distribution, abundance, and foraging patterns (Adair et al. 1996; Ballard et al. 2010).

Forage availability in stock ponds is currently not well understood, but likely highly variable (Kraai 2003; Clark 2016). Stock ponds may provide important refugia during non-breeding periods (Kraai 2003; K. Kraai, Texas Parks and Wildlife Department, personal communication).

The relationship between irrigation and wetlands is complex (Bolen et al. 1989; Lovvorn and Hart 2004; Bishop and Vrtiska 2008; Moore 2016; Donnelly et al. 2020; King et al. 2021). Donnelly et al. (2022) indicate rapid wetland decline in western North America may be approaching an ecological tipping point for wetland bird populations. In the West, most surface water rights are agricultural and used in irrigation systems (Kendy 2006; Downard and Endter-Wada 2013; Donnelly et al. 2020; King et al. 2021). In many areas, availability of wetland habitat follows irrigation schedules and further research is needed to better understand benefits and relative tradeoffs for wetland birds throughout the annual cycle (Copeland et al. 2010; Donnelly et al. 2019, 2020, 2021; Lovvorn and Crozier 2022). Water scarcity is intensifying socio-political pressures, including "use it or lose it" policies, to improve efficiency (Grafton et al. 2018; Sketch et al. 2020). However, more efficient irrigation practices (e.g., pressurized sprinklers) could lead to significant declines in flood irrigation and negatively impact wetland bird habitat and other ecosystem services (Baker et al. 2014; Moulton et al. 2016; Donnelly et al. 2020, 2021). Rapid wetland declines in the West may be approaching an ecological tipping point for wetland bird populations (Donnelly et al. 2020, 2022).

13.7   Ecosystem Threats

Wetland loss has been extensive, with > 50% declines in the western U.S. and Great Plains (Dahl 1990, 2014) and comparable losses (40–70%) in western Canada (Doherty et al. 2013), and Mexico (25–98%; Landgrave and Mereno-Casasola 2012). In the Great Plains, the most significant driver has been wetland drainage (e.g., tiling) tied to row-crop expansion, and loss of wetland legal protections (Dahl 1990, 2011; Lark et al. 2020). In the West, agricultural development and large-scale overexploitation of beavers in the 1800s led to widespread wetland losses (Dahl 1990; Lemly et al. 2000; McKinstry et al. 2001; Chap. 7). Intact wetlands and rangelands tend to be associated with livestock production and land owned by public agencies. Wetland bird conservation is therefore intrinsically linked to livestock production (Higgins et al. 2002; Anderson et al. 2018; Brasher et al. 2019). Climate change is predicted to exacerbate threats (Niemuth et al. 2014; Haig et al. 2019; Lark et al. 2020; Donnelly et al. 2021; Moon et al. 2021). Conservation of remaining wetlands, especially in rangelands, will be important to sustain wetland birds (Bartuszevige et al. 2012; Tsai et al. 2012; PPJV 2017; PHJV 2021; Donnelly et al. 2021).

13.7.1 Habitat Conversion and Alteration

Recent changes in row-crop agriculture, such as the development of drought-resistant crop varieties and increased farming efficiencies, provide incentives to convert rangeland and other marginal areas into crop production (Higgins et al. 2002; Doherty et al. 2013; Lark et al. 2020). Recently, the most extensive conversion has occurred in the PPR and High Plains (RWBJV 2013; Fields and Barnes 2019; Lark et al. 2020). Lark et al. (2020) found recently converted grasslands and wetlands in the PPR had 37% less nesting accessibility for ducks than non-converted areas, demonstrating the significant risk of agricultural conversion to wetland bird productivity. Along the Gulf Coast, human development, crop conversion, non-native grass pastures, and wetland draining has led to < 1% of native prairie remaining and significant loss to wetland bird nesting habitat (Smeins et al. 1991; Wilson and Esslinger 2002; Vermillion et al. 2008). The loss of ranching operations and subdivision of land ownership has contributed to habitat declines. Future development is expected to increase 72% over the next 80 years putting remaining rangeland and wetlands at further risk (Moon et al. 2021). In the Intermountain West, human population growth and water scarcity have intensified competition for water resources driving substantial land-use changes that impact wetland bird habitat (Hansen et al. 2002; Baker et al. 2014; Donnelly et al. 2021; King et al. 2021). Water-use is increasingly transferred from agricultural to municipal holdings for growing urban water demands, increasing the challenge of maintaining regional wetland networks (Brewer et al. 2007; Dilling et al. 2019; Donnelly et al. 2021). The accumulation of increasing threats within the Intermountain West has potential negative population-level impacts (Haig et al. 1998, 2019; Donnelly et al. 2020, 2021; Mackell et al. 2021).

13.7.2 Energy Development

Energy development continues to increase across rangelands (Ott et al. 2021). Collisions, habitat loss and degradation, and displacement are common impacts from energy development that threaten wetland bird populations (Shaffer et al. 2019g). Oil field wastewater developments in semi-arid rangelands are commonly mistaken as habitat by wetland birds resulting in mortality (Flickinger 1981; Flickinger and Bunck 1987; Trail 2006; Ramirez 2010). Oil spills and flowback water from fracking occur regularly and can contaminate wetlands. Brine contamination has been frequently reported in wetlands in the Bakken Formation and can negatively affect local aquatic invertebrates (Preston and Ray 2017; Blewett et al. 2017). The demand for biofuel, particularly corn ethanol, has accelerated grassland and wetland conversion of > 400,000 ha per year (Wright and Wimberly 2013; Lark et al. 2015, 2020). This conversion leads to increases in land values, affecting livestock operation sustainability, and portends challenges for ranching economies and associated ecosystem services (Johnson and Stephens 2011).

Indirect losses from energy development include fragmentation and displacement, which significantly increases the footprint of habitat loss (Johnson and Stephens 2011; Loesch et al. 2013). Indirect effects vary by species, seasons, and spatial scale of habitat (Shaffer et al. 2019g; Pearse et al. 2021). Lower breeding (Loesch et al. 2013) and wintering abundance (Lange et al. 2018) of ducks have been documented near wind energy facilities as well as avoidance during migration by whooping cranes (Pearse et al. 2021). Fragmentation and displacement from wind development are of greater conservation concern compared to direct mortality (Shaffer et al. 2019g; Hise et al. 2020). Larger wetland birds, such as sandhill cranes, are at greater risk of collision (Brown and Drewien 1995; Navarrete and Griffis-Kyle 2014; Murphy et al. 2016; Pearse et al. 2016; Hays et al. 2021).

13.7.3   Invasive Species

Invasive flora and fauna affect wetland birds in rangelands. Invasive aquatic plants reduce overall biodiversity and habitat quality for waterbirds. Native and non-native plant species such as cattail, common reed, reed canary grass, and creeping foxtail (*Alopecurus arundinaceus*) form dense monotypic stands that outcompete more desirable vegetation (Baldassarre and Bolen 2006; Hillhouse 2019; Johnson 2019). Cattail species have proliferated in the absence of natural disturbances (e.g., grazing and fire) and row-crop agriculture provides conditions that promote cattail establishment and vigor (i.e., nutrient runoff, sediment accumulation; Bansal et al. 2019). Dense stands of cattail can dominate wetlands, eliminate open-water, replace emergents and SAV, and preclude wetland bird species (Bansal et al. 2019). Similarly, common reed (i.e., phragmites) is a growing problem in the Intermountain West (Duncan et al. 2019; Rohal et al. 2019). Reed canary grass is widely used as livestock forage but can quickly form dense unproductive monotypic stands (Paveglio and Kilbride 2000; Evans-Peters et al. 2012; Hillhouse 2019).

Invasive animals also negatively affect waterfowl either directly through predation (e.g., northern pike (*Esox lucius*), or indirectly through habitat degradation. Common carp (*Cyprinus carpio*) are pervasive and degrade habitat quality and waterfowl productivity by consuming SAV and increasing turbidity that reduces forage availability (Ivey et al. 1998; Bajer et al. 2009). However, carp control is challenging, and success is often short-lived (Pearson et al. 2019). PPR wetlands evolved under isolated and intermittent drying conditions with only temporary surface-hydrologic connections. Wetland drainage has resulted in deeper, more stabilized hydrology, with interconnected basins that permit fish to persist (McLean et al. 2022). Similarly, fish are not endemic to playas in the High Plains but excavated ponds for irrigation support introduced fish, causing similar issues as above (Bolen et al. 1989; Smith et al. 2012).

13.7.4 Climate Change

The availability and function of wetlands are balanced by precipitation and evapo-transpiration, making them sensitive to changes in climate (McKenna et al. 2021a). Climate change will likely have variable effects on wetland function and productivity throughout North American rangelands. Indirect climate change impacts on land-use are also conservation concerns for wetland birds (McKenna et al. 2019). In response to climate change, water availability and land-use patterns will increasingly challenge agricultural-based economies and wetland bird populations.

The PPR has received considerable attention for evaluating potential climate change effects on wetland birds. Recommendations for waterfowl conservation strategies have shifted as climate change has been increasingly understood. Areas that currently support the largest densities of intact wetlands and breeding populations will likely be most critical to future continental waterfowl populations (Loesch et al. 2012; Niemuth et al. 2014; Sofaer et al. 2016; McKenna et al. 2021a). Many of these wetlands overlap rangeland areas with ranch and livestock-based economies (PPJV 2017). In the southern PPR, a shift from winter to summer and fall precipitation-driven hydrology has occurred in recent decades (McKenna et al. 2017). More precipitation may initially seem beneficial, but wetland productivity and function can decline with less periodic drying (Euliss et al. 2004; McCauley et al. 2015). Under wetter conditions, wetlands would deepen and have more stable water levels promoting fish persistence and cattail domination (Anteau et al. 2016). Shorebirds that require exposed shorelines and mudflats would be less likely to find habitat (Anteau et al. 2016). Alternatively, prolonged dry periods can result in loss of seasonal wetlands and shrinking wetlands alter plant and invertebrate communities. Upland management, such as grazing and burning, adjacent to wetlands can help increase runoff into wetlands and reduce ponding loss during the breeding season (McKenna et al. 2021b).

In the Southern Great Plains, spring and summer are expected to become hotter and drier with fewer, but more intense and unpredictable, precipitation events (Londe et al. 2022). Recent models indicate a high likelihood that wetland networks will exhibit reduced connectivity, with playas especially at risk (Uden et al. 2015; Albanese and Haukos 2017; McIntyre et al. 2018; Verheijen et al. 2020; Londe et al. 2022). Loss of stopover habitat and forage can reduce survival during the non-breeding period (Moon and Haukos 2006) and subsequent reproductive success (Sedinger and Alisauskas 2014). Reduced summer wetland inundation also means less available water for livestock. Opportunities to introduce rangeland management practices, such as fire and/or grazing in Conservation Reserve Program lands, (Cariveau et al. 2011; Smith et al. 2011) may become increasingly important to address climate impacts.

The West is experiencing rising temperatures, reduced snowpack, and earlier runoff resulting in water scarcity (Kapnick and Hall 2012; Mote et al. 2018; Snyder et al. 2019). Snowpack runoff drives availability and function for most western wetlands. In recent decades, water surface area in wetlands have declined by 47% or more while important aquatic systems like the Great Salt Lake have declined by 27% (Donnelly et al. 2020). Terminal basins and lower portions of watersheds in the Great Basin (Kadlec and Smith 1989; Donnelly et al. 2020) are strongly influenced by upstream water management decisions (Moore 2016; Null and Wurtsbaugh 2020; King et al. 2021; Donnelly et al. 2022). Climate change brings increasing temperatures and evapotranspiration rates intensifying water scarcity and ultimately impacting wetland bird habitat in the region (Downard and Endter-wada 2013; Moore 2016; Haig et al. 2019; Donnelly et al. 2020, 2021, 2022).

Climate change has potential to affect Gulf Coast habitat through sea-level rise and intensification of tropical storms. Coastal wetlands are vulnerable to increasing salinity, which decreases primary production, altering habitat quality (Battaglia et al. 2012; Moon et al. 2021). Freshwater and irregularly flooded marshes (Chabreck et al. 1989; Wilson and Esslinger 2002), are projected to dramatically decrease (Moon et al. 2021). Inland prairie and agricultural wetlands are also at risk (Battaglia et al. 2012; Moon et al. 2021) but may continue to provide vital habitat to species like mottled duck (Moon et al. 2021).

13.8 Conservation and Management Actions

13.8.1 Addressing Loss and Fragmentation of Wetlands and Rangelands

Minimizing the conversion of wetlands and rangelands to cultivated agricultural production is one of the greatest conservation challenges and priorities for wetland birds. Unfortunately, increases in commodity prices and the slow pace of conservation actions are unlikely to reverse wetland bird habitat losses in rangelands or offset anticipated future losses (Higgins et al. 2002; Doherty et al. 2013; Lark et al. 2020). However, maintaining livestock production on rangelands decreases the likelihood of cropland conversion and other land use changes (Higgins et al. 2002). Therefore, sustaining grazing as part of the region's socio-economic fabric will be vital for conserving wetland bird habitats (Higgins et al. 2002). Where grasslands have been lost, the maintenance and conservation of wetland basins supports wetland bird persistence (Reynolds et al. 2006; Niemuth et al. 2009). Nevertheless, keeping rangelands "green side up" and wetlands intact are primary conservation goals to sustain wetland bird populations.

Flood irrigation, beaver restoration, and low-tech riparian and wet meadow restoration (e.g., beaver dam mimicry or analogs, Zeedyk structures) offer opportunities to enhance natural water storage (Blevins 2015; Silverman et al. 2018; Moore and McEvoy 2022). Enhanced soil water storage capacity from such practices can increase watershed resilience to climate drivers, enhance wetland wildlife habitat, and increase livestock forage production (Silverman et al. 2018). Financial incentives, access to technical assistance, and local partnerships help managers implement restoration as well as maintain or upgrade flood irrigation infrastructure (Sketch et al. 2020; Donnelly et al. 2021; Moore and McEvoy 2022). Watershed and state-based partnerships will help managers navigate water management, water rights, and restoration techniques within the social-ecological systems of western watersheds (Downard and Endter-Wada 2013; Moore and McEvoy 2022).

13.8.2 *Partnerships and Programs*

Conserving wetland birds requires effective public–private partnerships at local, regional, and international scales (Anderson et al. 2018; Brasher et al. 2019). For example, the NAWMP acknowledges sustaining waterfowl populations is impossible without conservation on private lands and no single entity can solely address habitat loss. Conservation partnerships for wetlands and grasslands have historically focused on voluntary incentive programs such as those available through the federal Farm Bill (Hohman et al. 2014) for sustaining and growing wetland bird populations (e.g., Gray and Teels 2006; Reynolds et al. 2006; Bishop and Vrtiska 2008; Drum et al. 2015). More recent partnerships have focused on adaptive conservation projects in working rangelands, including the creation and maintenance of water sources that concurrently improve livestock grazing management and wildlife habitat. Effective wetland bird conservation includes a broad suite of short-term and long-term stewardship programs and incentives for livestock operations (Higgins et al. 2002; Brasher et al. 2019).

Numerous agency programs are available to assist with range improvements, grazing infrastructure, and wetland restoration and protection (Brasher et al. 2019). Prominent federal examples include Natural Resource Conservation Service (e.g., EQIP, WRE—including reserved grazing rights option) and the U.S. Fish and Wildlife Service Partners for Fish and Wildlife Program which can provide technical expertise and funding for wetland conservation projects that align with supporting producer objectives. Community-based conservation efforts can foster productive dialogue among stakeholders for meaningful conservation actions (Neudecker et al. 2011; Bennett et al. 2021). Voluntary conservation easements, including NRCS's Agricultural Conservation Easement program, limit sub-division, development, and conversion of rangelands to other land-uses (Brasher et al. 2019; Bennett et al. 2021). Prioritization is needed to help distribute limited resources (Niemuth et al. 2022). Facilitating land-use changes, like the transition of expiring Conservation Reserve Program lands into grazed rangeland will sustain or improve habitat conditions for wetland birds, expand grazing opportunities, and improve landscape resilience by supporting sustainable ranching economies that keep grasslands and wetlands on the landscape (Higgins et al. 2002; PPJV 2017; NRCS 2021).

The growing awareness of ecosystem services provided to society through wetlands and rangelands are likely to generate additional public–private partnership opportunities and funding sources for conservation. Ecosystem services from rangelands and wetlands include flood control, water quality, groundwater recharge and discharge, carbon storage, and ecological resilience (Mitsch and Gosselink 2015). Focus on improved wildlife resources has been a primary message for conservation groups to date; however, helping people understand the life-sustaining ecosystem services provided by rangeland and wetlands may increase stakeholder interest and funding available for conservation (Bartuszevige et al. 2016; Humburg et al. 2018; Brasher et al. 2019).

References

Ackerman JT (2002) Of mice and mallards: positive indirect effects of coexisting prey on waterfowl nest success. Oikos 99:469–480

Adair SE, Moore JL, Kiel WH (1996) Wintering diving duck use of coastal ponds: an analysis of alternative hypotheses. J Wildl Manage 60:83–93

Albanese G, Davis CA (2015) Characteristics within and around stopover wetlands used by migratory shorebirds: is the neighborhood important? Condor 117:328–340

Albanese G, Haukos DA (2017) A network model framework for prioritizing wetland conservation in the Great Plains. Landsc Ecol 32:115–130

Alisauskas RT, Drake KL, Nichols JD (2009) Filling a void: abundance estimation of North American populations of Arctic geese using hunter recoveries. In: Thomson DL, Cooch EG, Conroy MJ (eds) Modeling demographic processes in marked populations. Environmental and ecological statistics vol 3. Springer Science, New York, pp 463–489

Alisauskas RT, Arnold TW, Leafloor JO, Otis DL, Sedinger JS (2014) Lincoln estimates of mallard (*Anas platyrhynchos*) abundance in North America. Ecol Evol 4:132–143

Allen GT, Ramirez P (1990) A review of bird deaths on barbed-wire fences. Wilson Bull 102:553–558

Anderson JT, Smith LM (2000) Invertebrate response to moist-soil management of playa wetlands. Ecol Appl 550–558

Anderson MG, Alisauskas RT, Batt BDJ et al (2018) The migratory bird treaty and a century of waterfowl conservation. J Wild Manage 82:247–259

Anderson BE, Hillhouse HL, Bishop AA et al (2019) Grazing Rainwater Basin wetlands. Extension circular EC3040, Institute of Agriculture and Natural Resources. University of Nebraska, Lincoln, NE, USA. https://extensionpubs.unl.edu/publication/9000020939194/grazing-rainwater-basin-wetlands/

Anteau MH, Afton AD (2011) Lipid catabolism of invertebrate predator indicates widespread wetland ecosystem degradation. PLoS ONE 6:e16029

Anteau MJ, Afton AD, Anteau ACE et al (2011) Fish and land use influence *Gammarus lacustris* and *Hyalella Azteca* (Amphipoda) densities in large wetlands across the upper Midwest. Hydrobiologia 664:69–80

Anteau MJ, Shaffer TL, Sherfy MH et al (2012) Nest survival of piping plovers at a dynamic reservoir indicates an ecological trap for a threatened population. Oecologia 170:1167–1179

Anteau MJ, Wiltermuth MT, van der Burg MP et al (2016) Prerequisites for understanding climate-change impacts on northern prairie wetlands. Wetlands 36:299–307

Arnold TW, Roche EA, Devries JH et al (2012) Costs of reproduction in breeding female mallards: predation risk during incubation drives annual mortality. Avian Conserv Ecol 7:1. https://doi.org/10.5751/ACE-00504-070101

Austin JE, Buhl DA (2009) Factors associated with duck use of impounded and natural wetlands in western South Dakota. Prairie Nat 41:1–27

Austin JE, Henry AR, Ball IJ (2007) Sandhill crane abundance and nesting ecology at Grays Lake, Idaho. J Wild Manage 71:1067–1079

Bajer PG, Sullivan G, Sorensen PW (2009) Effects of a rapidly increasing population of common carp on vegetative cover and waterfowl in a recently restored Midwestern shallow lake. Hydrobiologia 632:235–245

Baker JM, Everett Y, Liegel L, Van Kirk R (2014) Pattern of irrigated agricultural land conversion in a Western U.S. watershed: implication for landscape-level water management and land-use planning. Soc Nat Resour 27(11):1145–1160. https://doi.org/10.1080/08941920.2014.918231

Baldassarre GA (ed) (2014) Ducks, geese, and swans of North America. Johns Hopkins University Press, Baltimore, Maryland

Baldassarre GA, Bolen EG (2006) Waterfowl ecology and management, 2nd edn. Krieger Publishing, Malabar, Florida, USA

Ball IJ, Eng RL, Ball SK (1995) Population density and productivity of ducks on large grassland tracts in northcentral Montana. Wild Soc Bull 23:767–773

Ballard BM, James JD, Bingham RL et al (2010) Coastal pond use by redhead wintering in the Laguna madre, Texas. Wetlands 30:669–674

Bansal S, Lishawa SC, Newman S et al (2019) Typha (Cattail) invasion in North American wetlands: biology, regional problems, impacts, ecosystem services, and management. Wetlands 39:645–684

Barker WT, Sedivec KK, Messmer TA et al (1990) Effects of specialized grazing systems on waterfowl production in south-central North Dakota. Trans North Am Wildl Nat Resour Conf 55:462–474

Bartuszevige AM, Pavlacky DC Jr, Burris L et al (2012) Inundation of playa wetlands in the western Great Plains relative to landcover context. Wetlands 32:1103–1113

Bartuszevige AM, Taylor K, Daniels A et al (2016) Landscape design: integrating ecological, social, and economic considerations into conservation planning. Wildl Soc Bull 40:411–422

Batt BDJ, Anderson MG, Anderson CD et al (1989) The use of prairie potholes by North American ducks. In: van der valk AG (ed) Northern Prairie wetlands. Iowa State University Press, Ames, Iowa, pp 2–14

Batt BD, Afton AD, Anderson MG et al (eds) (1992) Ecology and management of breeding waterfowl. University of Minnesota Press, Minneapolis, Minnesota

Battaglia LL, Woodrey MS, Peterson MS et al (2012) Wetlands of the northern Gulf Coast. In: Batzer DP, Baldwin AH (eds) Wetland habitats of North America. University of California Press, Berkely and Los Angeles, California, pp 75–88

Bennett DE, Knapp CN, Knight RL (2021) The evolution of rangeland trusts network as a catalyst for community-based conservation in the American West. Conserv Sci Pract 3:e257

Beyersbergen GW, Niemuth ND, Norton MR (eds) (2004) Northern Prairie and Parkland Waterbird conservation plan. Waterbird Conservation for the Americas Initiative, Prairie Pothole Joint Venture, Denver, Colorado, USA

Bishop AA, Vrtiska M (2008) Effects of the wetland reserve program on waterfowl carrying capacity in the Rainwater Basin Region of south-central Nebraska. A Conservation Effects Assessment Project, Wildlife component assessment. U.S. Department of Agriculture Natural Resources Conservation Service, Washington, D.C. 51 pp. https://nrcs.usda.gov/publications/ceap-wildlife-2008-wrp-waterfowl-rainwater-nebraska.pdf

Bleho BI, Koper N, Machtans CS (2014) Direct effects of cattle on grassland birds in Canada. Conserv Biol 28:724–734

Blevins SC (2015) Valuing the non-agricultural benefits of flood irrigation in the Upper Green River Basin. Master's Thesis, Department of Agriculture and Applied Economics, University of Wyoming, Laramie, Wyoming, USA, 79 pp

Blewett TA, Delompré PLM, He Y et al (2017) Sublethal and reproductive effects of acute and chronic exposure to flowback and produced water from hydraulic fracturing on the water flea Daphnia magna. Environ Sci Technol 51:3032–3039. https://doi.org/10.1021/acs.est.6b05179

Bloom PM, Clark RG, Howerter DW (2012) Landscape-level correlates of mallard duckling survival: implications for conservation programs. J Wildl Manage 76:813–823

Bloom PM, Howerter DW, Emery RE et al (2013) Relationships between grazing and waterfowl production in the Canadian prairies. J Wildl Manage 77:534–544

Bolen EG, Smith LM, Schramm HL Jr (1989) Playa lakes: prairie wetlands of the Southern high plains. Bioscience 39:615–623

Bortolotti LE, Emery RB, Armstrong LM et al (2022) Landscape composition, climate variability, and their interaction drive waterfowl nest survival in the Canadian prairies. Ecosphere 13:e3908

Brasher MG, Giocomo JJ, Azure DA et al (2019) The history and importance of private lands for North American waterfowl conservation. Wildl Soc Bull 43:338–354

Brennan EK, Smith LM, Haukos DA et al (2005) Short-term response of wetland birds to prescribed burning in Rainwater Basin wetlands. Wetlands 25:667–674

Brewer J, Glennon R, Ker A et al (2007) Water markets in the West: prices, trading, and contractual forms. NBER Working Paper No. 13002. National Bureau of Economic Research, Cambridge, MA, USA. http://www.nber.org/papers/w13002

Briske DD, Derner JD, Milchunas DG et al (2011) An evidence-based assessment of prescribed grazing practices. Rotational grazing on rangelands: reconciliation of perception and experimental evidence. Rangel Ecol Manag 61:3–17

Brown WM, Drewien RC (1995) Evaluation of two power line markers to reduce crane and waterfowl collision mortality. Wildl Soc Bull 23(2):217–227. https://www.jstor.org/stable/3782794

Brown S, Hickey C, Harrington B, Gill R (eds) (2001) The U.S. shorebird conservation plan, 2nd edn. Manomet Center for Conservation Sciences, Manomet, MA. https://www.shorebirdplan.org/

Brownie C, Anderson Dr, Burnham KP et al (1985) Statistical inference from band recovery data—a handbook. U. S. Department of the Interior, Fish and Wildlife Service, Resource Publication 156

Buderman FE, Devries JH, Koons DN (2020) Changes in climate and land use interact to create an ecological trap in a migratory species. J Anim Ecol 89:1961–1977

Bue IG, Uhlig HG, Smith JD (1964) Stock ponds and dugouts. In: Linduska JP, Nelson AL (eds) Waterfowl tomorrow. US Department of Interior, Washington, DC, pp 391–398

Cariveau AB, Pavlacky DC Jr, Bishop AA et al (2011) Effects of surrounding land use on playa inundation following intense rainfall. Wetlands 31:65–73

Carroll LC, Arnold TW, Beam JA (2007) Effects of rotational grazing on nesting ducks in California. J Wildl Manage 71:681–1021

Cavitt JF, Jones SL, Wilson NM et al (2014) Atlas of breeding colonial waterbirds in the interior western United States. Research Report, U.S. Department of the Interior, Fish and Wildlife Service, Denver, Colorado

Chabreck RH, Joanen T, Paulus SL (1989) Southern coastal marshes and lakes. In: Smith LM, Pederson RL, Kaminski RM (eds) Habitat management for migrating and wintering waterfowl in North America. Texas Tech University Press, Lubbock, Texas, pp 249–277

Clark L (2016) Stock pond forage resource relationships for nonbreeding ducks in the Rolling Plains of Texas. Thesis, Texas Tech University, Lubbock, Texas

Clark RG, Nudds TD (1991) Habitat patch size and duck nest success: the crucial experiments have not been performed. Wildl Soc Bull 19:534–543

Collins SD, Heintzman LJ, Starr SM et al (2014) Hydrologic dynamics of temporary wetlands in the southern Great Plains as a function of surrounding land use. J Arid Environ 109:6–14

Cooch FG, Wendt S, Smith GEJ et al (1978) The Canada migratory game bird hunting permit and associated surveys. In: Boyd H, Finney GH (eds) Migratory game bird hunters and hunting in Canada. Canadian Wildlife Service Report Series vol 43, pp 8–39

Coons SP (2021) Monitoring the wetland landscape: white-faced ibis (*Plegadis chihi*) breeding habitat as a model assemblage. Graduate student theses, dissertations, and professional papers. 11836. University of Montana, Missoula, Montana. https://scholarworks.umt.edu/etd/11838

Copeland HE, Tessman SA, Girvetz EH et al (2010) A geospatial assessment on the distribution, condition, and vulnerability of Wyoming's wetlands. Ecol Indic 10:869–879

Cornwell G, Hochbaum HA (1971) Collisions with wires: a source of Anatid mortality. Wilson Bull 83:305–306

Cowardin LM, Carter V, Golet FC et al (1979) Classification of wetlands and deepwater habitats of the United States. US Fish and Wildlife Service, Washington, DC, USA. FWS/OBS-79/31

Dahl TE (1990) Wetland losses in the United States, 1780s to 1980s. U.S. Fish and Wildlife Service, Washington, D.C. USA. https://www.fws.gov/wetlands/documents/Wetlands-Status-and-Trends-in-the-Conterminous-United-States-Mid-1970s-to-Mid-1980s.pdf

Dahl TE (2011) Status and trends of wetlands in the conterminous United States 2004–2009. U.S. Department of the Interior, Fish and Wildlife Service, Washington, D.C. 108 pp

Dahl TE (2014) Status and trends of prairie wetlands in the United States 1997–2009. U.S. Department of the Interior, Fish and Wildlife Service, Ecological Services, Washington, D.C. USA, 67pp

Davis BE (2012) Habitat use, movements, and ecology of female mottled ducks in the Gulf Coast of Louisiana and Texas. Dissertation, Louisiana State University, Baton Rouge, Louisiana

Davis JB, Guillemain M, Kaminski RM (2014) Habitat and resource use by waterfowl in the northern hemisphere in autumn and winter. Wildfowl (2014 Special Issue) 4:17–69

Devries JH, Brook RW, Howerter DW et al (2008) Effects of spring body condition and age on reproduction in mallards (*Anas platyrhynchos*). Auk 125:618–628

Dilling L, Berggren J, Henderson J, Kenney D (2019) Savior of rural landscapes or Solomon's choice? Colorado's experiment with alternative transfer methods for water (ATMs). Water Secur 6:100027. https://doi.org/10.1016/j.wasec.2019.100027

Dixon C, Vacek S, Grant T (2019) Evolving management paradigms on the U.S. Fish and Wildlife Service lands in the Prairie Pothole Region. Rangelands 41:36–43

Doherty KE, Ryba AJ, Stemler CL, Niemuth ND, Meeks WA (2013) Conservation planning in an era of change: state of the U.S. Prairie Pothole Region. Wildl Soc Bull 37(3):546–563. https://doi.org/10.1002/wsb.284

Donaldson G, Hyslop C, Morrison G, Dickson L, Davidson I (eds) (2000) Canadian shorebird conservation plan. Canadian Wildlife Service, Environment Canada, Ottaway, Ontario, Canada

Donnelly JP, Vest JL (2012) Identifying science priorities: 2013–2018 wetland focal strategies. Intermountain West Joint Venture technical series 2012-3. Intermountain West Joint Venture, Missoula, Montana. https://iwjv.org/wp-content/uploads/2019/08/iwjv_3_science_wetlands_2013-2018.pdf

Donnelly JP, Naugle DE, Hagen CE et al (2016) Public lands and private waters: scarce mesic resources structure land tenure and sage-grouse distributions. Ecosphere 7:1. https://doi.org/10.1002/ecs2.1208

Donnelly JP, Naugle DE, Collins DP et al (2019) Synchronizing conservation to seasonal wetland hydrology and waterbird migration in semi-arid landscapes. Ecosphere 10(6):e02758. https://doi.org/10.1002/ecs2.2758

Donnelly JP, King S, Silverman NL et al (2020) Climate and human water use diminish wetland networks supporting continental waterbird migration. Glob Chang Biol 26(4):2042–2059. https://doi.org/10.1111/gcb.15010

Donnelly JP, King SL, Knetter J et al (2021) Migration efficiency sustains connectivity across agroecological networks supporting sandhill crane migration. Ecosphere 12:6. https://doi.org/10.1002/ecs2.3543

Donnelly JP, Moore JN, Casazza ML et al (2022) Functional wetland loss drives emerging risks to waterbird migration networks. Front Ecol Evol 10:844278. https://doi.org/10.3389/fevo.2022.844278

Downard R, Endter-Wada J (2013) Keeping wetlands wet in the western United States: adaptations to drought in agriculture-dominated human-natural systems. J Environ Manag 131:394–406. https://doi.org/10.1016/j.jenvman.2013.10.008

Drewien RC, Brown WM, Kendall WL (1995) Recruitment of rocky mountain greater sandhill cranes and comparison with other crane populations. J Wildl Manage 59:339–356

Drum RG, Loesch CR, Carrlson K et al (2015) Assessing the biological benefits of the USDA-Conservation Reserve Program (CRP) for waterfowl and grassland passerines in the Prairie Pothole Region of the United States: spatial analyses for targeting CRP to maximize benefits for migratory birds. Final Report for USDA-FSA Agreement: 12-IA-MRE-CRP-TA

DU (2021) Ducks unlimited conservation priorities. Ducks Unlimited, Inc., Memphis, Tennessee

Dugger BD, Moore ML, Finger RS et al (2007) True metabolizable energy for seeds of common moist-soil plant species. J Wildl Manage 71:1964–1967

Duncan BL, Hansen R, Hambrecth K et al (2019) Cattle grazing for invasive *Phragmites australis* (common reed) management in Northern Utah wetlands. Utah State University Extension Fact Sheet NR/Wildlands/2019-01pr, Logan, Utah, USA

Earnst SL (1994) Tundra swan habitat preferences during migration in North Dakota. J Wildl Manage 58:546–551

Eldridge J (1992) 13.2.14 management of habitat for breeding and migrating shorebirds in the Midwest. Waterfowl Management Handbook, US Fish and Wildlife Service Leaflet vol 11, p 7

Emery RB, Howerter DW, Armstrong LM et al (2005) Seasonal variation in waterfowl nesting success and its relation to cover management in the Canadian prairies. J Wildl Manage 69:1181–1193

Epperson WL, Eadie JM, Marcum DB et al (1999) Late season hay harvest provides habitat for marshland birds. Calif Agric 53:12–17

Euliss NH, LaBaugh JW, Fredrickson LH et al (2004) The wetland continuum: a conceptual framework for interpreting biological studies. Wetlands 24:448–458

Evans-Peters GR, Dugger BD, Petrie MJ (2012) Plant community composition and waterfowl food production on wetland reserve program easements compared to those on managed public lands in western Oregon and Washington. Wetlands 32:391–399

Fields S, Barnes K (2019) Grassland assessment of North American Great Plains Migratory Bird Joint Ventures. Unpublished report. Prairie Pothole Joint Venture, Great Falls, MT USA. http://ppjv.org/assets/docs/Great_Plains_Grassland_Assessment_Final_Report.pdf

Fink D, Auer T, Johnston A et al (2020) Modeling avian full annual cycle distribution and population trends with citizen science data. Ecol Appl 30:02056. https://doi.org/10.1002/eap.2056

Flake LD, Petersen GL, Tucker WL (1977) Habitat relationships of breeding waterfowl on stock ponds in northwestern South Dakota. Proc South Dakota Acad Sci 56:135–151

Fleskes JP, Gregory CJ (2010) Distribution and dynamic of waterbird habitat during spring in Southern Oregon-Northeastern California. West North Am Nat 70(1):26–38. https://doi.org/10.3398/064.070.0104

Fleskes JP, Mauser DM, Yee JL, Blehert D, Yarris GS (2010) Flightless and post-molt survival and movements of female mallards molting in Klamath Basin. Waterbirds 33(2):208–220. https://doi.org/10.1675/063.033.0209

Flickinger EL (1981) Wildlife mortality at petroleum pits in Texas. J Wildl Manage 45:560–564

Flickinger EL, Bunck CM (1987) Number of oil-killed birds and fate of bird carcasses at crude oil pits in Texas. Southwestern Nat 32(3):377–381. https://doi.org/10.2307/3671456

Forman KJ, Madsen CR, Hogan MJ (1996) Creating multiple purpose wetlands to enhance livestock grazing distribution, range condition, and waterfowl production in western South Dakota. In: Schaack J, Anderson SS (eds) Water for agriculture and wildlife and the environment: win-win opportunities. U.S. Committee on Irrigation and Drainage, Denver, CO, USA, pp185–192

Fox AD, Flint PL, Hohman WL et al (2014) Waterfowl habitat use and selection during the remigial moult period in the northern hemisphere. Wildfowl 4:131–168

Fredrickson LH, Taylor TS (1982) Management of seasonally flooded impoundments for wildlife. Resource Publication 148. US Department of Interior, Fish and Wildlife Service, Washington, DC, USA

Friend M, McLean RG, Dein FJ (2001) Disease emergence in birds: challenges of the twenty-first century. Auk 118:290–303

Fynn R, Jackson J (2022) Grazing management on commercial cattle ranches: incorporating foraging ecology and biodiversity conservation principles. Rangelands 44:136–147

Gerber BD, Kendall WL, Hooten MB et al (2015) Optimal population prediction of sandhill crane recruitment based on climate-mediated habitat limitations. J Anim Ecol 84:1299–1310

Gilbert DW, Anderson DR, Ringleman JK et al (1996) Response of nesting ducks to habitat management on the Monte Vista National Wildlife Refuge, Colorado. Wildl Monogr 131

Gitz D, Brauer D (2016) Trends in playa inundation and water storage in the Ogallala Aquifer on the Texas High Plains. Hydrology 3:31. https://doi.org/10.3390/hydrology3030031

Gjersin FM (1975) Waterfowl production in relation to rest-rotation grazing. Rangel Ecol Manag 28:37–42

Grafton RQ, Williams J, Perry CJ et al (2018) The paradox of irrigation efficiency. Science 361:748–750

Grant TA, Flanders-Wanner B, Shaffer TL et al (2009) An emerging crisis across Northern Prairie refuges: prevalence of invasive plants and a plan for adaptive management. Ecol Restor 27:58–65

Gray PN, Bolen EG (1987) Seed reserves in tailwater pits of playa lakes in relation to waterfowl management. Wetlands 7:11–23

Gray RL, Teels BM (2006) Wildlife and fish conservation through the farm bill. Wildl Soc Bull 34:906–913

Gray MJ, Hagy HM, Nyman JA et al (2013) Management of wetlands for wildlife. In: Anderson JT, Davis CA (eds) Wetland techniques. Springer, Dordrecht, pp 121–180

Greenwood RJ, Sargent AB, Johnson DG et al (1995) Factors associated with duck nest success in the Prairie Pothole Regin of Canada. Wildl Monogr 128:57

Gruntorad MP, Graham KA, Arcilla N et al (2021) Is hay for the birds? Investigating landowner willingness to time hay harvests for grassland bird conservation. Animals 11:1030. https://doi.org/10.3390/ani11041030

Hagy HM, Brasher MG, Flesks JP et al (in review) Important geographies for migrating and wintering waterfowl. In: Ballard, BM, Brasher MG, Fleskes JP (eds) Waterfowl in winter. Texas A&M University Press, College Station, USA

Haig SM, Mehlman DW, Oring LW (1998) Avian movements and wetland connectivity in landscape conservation. Conserv Biol 12:749–758

Haig SM, Murphy SP, Matthews JH et al (2019) Climate-altered wetlands challenge waterbird use and migratory connectivity in arid landscapes. Sci Rep 9:4666. https://doi.org/10.1038/s41598-019-41135-y

Hansen AJ, Rasker R, Maxwell B et al (2002) Ecological causes and consequences of demographic change in the New West. Bioscience 52:151–162. https://doi.org/10.1641/0006-3568(2002)052[0151:ECACOD]2.0.CO;2

Harris SW, Marshall WH (1963) Ecology of water-level manipulations on a northern marsh. Ecology 44:331–343

Harrison RB, Jones WM, Clark D et al (2017) Livestock grazing in intermountain depressional wetlands: effects on breeding waterfowl. Wetl Ecol Manag 25:471–484

Hartman CA, Oring LW (2009) Reproductive success of long-billed curlews (Numenius americanus) in northeastern Nevada hay fields. Auk 126:420–430

Haukos DA, Smith LM (1993) Moist-soil management of playa lakes for migrating and wintering ducks. Wildl Soc Bull 21:288–298

Haukos DA, Smith LM (1994) The importance of playa wetlands to biodiversity of the Southern High Plains. Landsc Urban Plan 28:83–98

Haukos DA, Miller MR, Orthmeyer DL et al (2006) Spring migration of northern pintails from Texas and New Mexico, USA. Waterbirds 29:127–241

Hays QR, Tredennick AT, Carlisle JD et al (2021) Spatially explicit assessment of sandhill crane exposure to potential transmission line collision risk. J Wildl Manage 85(7):1440–1449. https://doi.org/10.1002/jwmg.22100

Helmers DL (1992) Shorebird management manual. Western Hemisphere Shorebird Reserve Network, Manomet, MA, 58pp

Henkel JR, Taylor CM (2015) Migration strategy predicts stopover ecology in shorebird on the northern Gulf of Mexico. Anim Migr 2:63–75

Herkert JR, Reinking DL, Wiedenfeld DA et al (2003) Effects of prairie fragmentation on the nest success of breeding birds in the midcontinental United States. Conserv Biol 17:587–594

Higgins KF, Naugle DE, Forman, KJ (2002) A case study of changing land use practices in the Northern Great Plains, U.S.A.: an uncertain future for waterbird conservation. Waterbirds 25(Special Publication 2):42–50. https://www.jstor.org/stable/1522450

Hillhouse HL (2019) Impacts of cattle grazing on seed production in Rainwater Basin wetlands. Wetl Ecol Manag 27:141–147

Hise C, Obermeyer B, Ahlering M et al (2020) Site wind right: identifying low-impact wind development areas in the central United States. Unpublished manuscript. https://doi.org/10.1101/2020.02.11.943613

Hoekman ST, Mills LS, Howerter DW et al (2002) Sensitivity analyses of the life cycle of midcontinent mallards. J Wildl Manage 66:883–900

Hoffman GR, Stanley LD (1978) Effects of cattle grazing on shore vegetation of fluctuating water level reservoirs. J Range Manag 31:412–416

Hohman WL, Ankney CD, Gordon DH (1992) Ecology and management of postbreeding waterfowl. In: Batt BDJ, Afton AD, Anderson MG et al (eds) Ecology and management of breeding waterfowl. University of Minnesota Press Minneapolis, Minnesota, pp 128–189

Hohman WL, Lindstrom EB, Rashford BS et al (2014) Opportunities and challenges to waterfowl habitat conservation on private land. Wildfowl (special Issue) 4:368–406

Holechek JL, Pieper RD, Herbel CH (2011) Range management: principles and practices, 6th edition. Pearson Education, Inc. New York, NewYork, USA

Holechek JL, Valdez R, Schemnitz SD et al (1982) Manipulation of grazing to improve or maintain wildlife habitat. Wildl Soc Bull 10:204–210

Horn DJ, Phillips ML, Koford RR et al (2005) Landscape composition, patch size, and distance to edges: interactions affecting duck reproductive success. Ecol Appl 15:1367–1376

Horns JJ, Adler FR, Şekercioğlu CH (2018) Using opportunistic citizen science data to estimate avian population trends. Biol Conserv 221:151–159

Hovick TJ, Carroll JM, Elmore RD et al (2017) Restoring fire to grasslands is critical for migrating shorebird populations. Ecol Appl 27:1805–1814

Hubbard DE (1988) Using your wetland as forage. South Dakota State University, Cooperative Extension, Extension Fact Sheets 1023. https://openprairie.sdstate.edu/extension_fact/1023

Hudson M (1983) Waterfowl production on three age-classes of stock ponds in Montana. J Wildl Manage 47:112–117

Hudson M, Francis CM, Campbell KJ et al (2017) The role of the North American breeding bird survey in conservation. Condor 119:526–545

Humburg DD, Anderson MG, Brasher MG (2018) Implementing the 2012 North American Waterfowl Management Plan revision: populations, habitat, and people. J Wildl Manage 82:275–286

Iglecia M, Winn B (2021) A shorebird management manual. Manomet, Massachusetts, USA. https://www.manomet.org/wp-content/uploads/2021/01/Iglecia_and_Winn_2021_AShorebirdManagementManual-012021-web.pdf

Ignatiuk JB, Duncan DC (2001) Nest success of ducks on rotational and season-long grazing systems in Saskatchewan. Wildl Soc Bull 29:211–217

Ivey G (2011) Making a home for Greater Sandhill Cranes on private lands. Audubon California. Working Land Series, Summer 2011. Audubon California, Oakland, CA, USA. https://www.academia.edu/535531/Audubon_Californias_Working_Lands_Series_Making_a_home_for_Greater_Sandhill_Cranes_on_private_lands

Ivey GL, Dugger BD (2008) Factors influencing nest success of Greater Sandhill Cranes at Malheur National Wildlife Refuge, Oregon. Waterbirds 31:52–61

Ivey GL, Herziger CP (2006) Intermountain West waterbird conservation plan, version 1.2. Waterbird Conservation for the Americas Initiative, US Fish and Wildlife Service, Portland, Oregon, USA

Ivey GL, Cornely JE, Ehlers BD (1998) Carp impacts on waterfowl at Malheur National Wildlife Refuge, Oregon. In: Transactions of the 63rd North American wildlife and natural resources conference vol. 63. pp 66–74

IWJV (2013) Intermountain West Joint Venture 2013 Implementation plan. Intermountain West Joint Venture, Missoula, Montana. https://iwjv.org/resource/iwjv-2013-implementation-plan-entire-plan/

Jefferies RL, Rockwell RF, Abraham KF (2003) The embarrassment of riches: agricultural food subsidies, high goose numbers, and loss of arctic wetlands—continuing saga. Environ Rev 11:193–232

Johnson WC (2019) Ecosystem services provided by prairie wetlands in northern rangelands. Rangeland 41:44–48

Johnson GD, Stephens SE (2011) Wind power and biofuels: a green dilemma for wildlife conservation. In: Naugle DE (ed) Energy development and wildlife conservation in Western North America. Island Press, Washington, D.C., USA, pp 131–155

Johnson DH, Gibbs JP, Herzog M (2009) A sampling design framework for monitoring secretive marshbirds. Waterbirds 32:203–215

Johnson WC, Werner B, Guntenspergen GR et al (2010) Prairie wetland complexes as landscape functional units in a changing climate. Bioscience 60:128–140

Kadlec JA, Smith LM (1989) The Great Basin marshes. In: Smith LM, Pederson RL, Kaminski RM (eds) Habitat management for migrating and wintering waterfowl in North America. Texas Tech University Press, Lubbock, Texas, pp 451–474

Kadlec JA, Smith LM (1992) Habitat management for breeding areas. In: Batt BDJ, Afton AD, Anderson MG et al (eds) Ecology and management of breeding waterfowl. University of Minnesota Press, Minneapolis, MN, USA

Kahara SN, Skalos D, Madurapperuma B, Hernandez K (2021) Habitat quality and drought effects on breeding mallard and other water populations in California. J Wildl Manage. https://doi.org/10.1002/jwmg.22133

Kapnick S, Hall A (2012) Causes of recent changes in western North American snowpack. Clim Dyn 38:1885–1899

Kemink KM, Adams VM, Pressey RL (2021) Integrating dynamic processes into waterfowl conservation prioritization tools. Divers Distrib 27:585–601

Kendy E (2006) Impacts of changing land use and irrigation practices on western wetlands. Nat Wetlands Newsletter 28:12–27

Kent CM, Ramey AM, Ackerman JT (2022) Spatiotemporal changes in influenza A virus prevalence among wild waterfowl inhabiting the continental United States throughout the annual cycle. Sci Rep 12:13083. https://doi.org/10.1038/s41598-022-17396-5

King SL, Laubhan MK, Tashjian P et al (2021) Wetland conservation: challenges related to water law and farm policy. Wetlands 41:54. https://doi.org/10.1007/s13157-021-01449-y

Kirby DR, Krabbenhoft KD, Sedivec KK et al (2002) Wetlands in Northern Plains prairies: benefitting wildlife and livestock. Rangelands 24:22–25

Koons DN, Gunnarson G, Schmutz JA et al (2014) Drivers of waterfowl population dynamics: from teal to swans. Wildfowl (2014 Special Issue) 4:169–191

Kraai KJ (2003) Late winter feeding habits, body condition, and feather molt intensity of female mallards utilizing livestock ponds in northeast Texas. Thesis, Texas A&M University, Commerce, Texas

Krapu GL, Pietz PJ, Brandt DA et al (2000) Factors limiting mallard brood survival in Prairie Pothole landscapes. J Wildl Manage 64:553–561

Krapu GL, Reinecke KJ (1989) Foraging ecology and nutrition. In: Batt BDJ, Afton AD, Anderson MG et al (ed) Ecology and Management of Breeding Waterfowl. University of Minnesota Press, Minneapolis, Minnesota pp 1–30

Krausman PR, Naugle DE, Frisina MR et al (2009) Livestock grazing, wildlife habitat, and rangeland values. Rangelands 31:15–19

Kushlan JA, Steinkamp MJ, Parson KC et al (2002) Waterbird conservation for the Americas: the Northern American waterbird conservation plan, version 1. Waterbird Conservation for the Americas, Washington, DC, USA, 78 pp. https://www.fws.gov/partner/north-american-waterbird-conservation-plan

Landgrave R, Moreno-Casasola P (2012) Evaluación cuantitativa de la pérdida de humedales en México. Inv Ambiental 4(1):19–35. https://proyectopuente.com.mx/wp-content/uploads/2019/05/121-707-1-pb.pdf

Lange CJ, Ballard BM, Collins DP (2018) Impacts of wind turbines on redheads in the Laguna Madre. J Wildl Manage 82(3):531–537. https://doi.org/10.1002/jwmg.21415

Lark TJ, Salmon M, Gibbs HK (2015) Cropland expansion outpaces agricultural and biofuel polices in the United States. Environ Res Lett 10:044003. https://doi.org/10.1088/1748-9326/10/4/044003

Lark TJ, Spawn SA, Bougie M, Gibbs HK (2020) Cropland expansion in the United states produces marginal yields at high costs to wildlife. Nat Commun 11:4295. https://doi.org/10.1038/s41467-020-18045-z

Laubhan MK, Gammonley JH (2000) Density and foraging habitat selection of waterbirds breeding in the San Luis Valley of Colorado. J Wildl Manage 64:808–819

Lemly AD, Kingsford RT, Thompson JR (2000) Irrigated agriculture and wildlife conservation: conflict on a global sale. Environ Manag 25(5):485–512. https://doi.org/10.1007/s002679910039.pdf

Lincoln FC (1930) Calculating waterfowl abundance on the basis of banding returns. U.S. Department of Agriculture Circular No. 118

Lipsey MK, Naugle DE (2017) Precipitation and soil productivity explain effects of grazing on grassland songbirds. Rangel Ecol Manag 70:331–340

Littlefield CD (1999) Greater Sandhill Crane productivity on privately owned wetlands in Eastern Oregon. Western Birds 30:206–210

Littlefield CD, Paullin DG (1990) Effects of land management on nesting success of Sandhill Cranes in Oregon. Wildl Soc Bull 18:63–65

Loesch CR, Reynolds RE, Hansen LT (2012) An assessment of re-directing breeding waterfowl conservation relative to predictions of climate change. J Fish Wildl Manag 3:1–22

Loesch CR, Walker JA, Reynolds RE et al (2013) Effect of wind energy development on breeding duck densities in the Prairie Pothole Region. J Wildl Manage 77(3):587–598

Londe DW, Dvorett D, Davis CA et al (2022) Inundation of depressional wetland declines under a changing climate. Clim Change 172:27

Lovvorn JR, Crozier ML (2022) Duck use of saline wetlands created by irrigation in a semiarid landscape. Wetlands 42:4. https://doi.org/10.1007/s13157-021-01525-3

Lovvorn JR, Hart EA (2004) Irrigation, salinity, and landscape patterns of natural palustrine wetlands. In: McKinstry MC, Hubert WA, Anderson SH (eds) Wetland and riparian areas of the Intermountain West: ecology and management. University of Texas Press, Austin, pp 105–129

Luttschwager KA, Higgins KF, Jenks JA (1994) Effects of emergency haying on duck nesting in conservation reserve program fields, South Dakota. Wildl Soc Bull 22:403–408

Ma Z, Cai Y, Li B et al (2010) Managing wetland habitats for waterbirds: an international perspective. Wetlands 30:15–27

Mackell DA, Casazza ML, Overton CT et al (2021) Migration stopover ecology of Cinnamon Teal in western North America. Ecol Evol 1–14.https://doi.org/10.1002/ece3.8115

Martin EM, Carney SM (1977) Population ecology of the mallard: IV. A review of duck hunting regulations, activity, and success, with special reference to the mallard. U.S. Fish and Wildlife Service, Resource Publication 130, Washington, D.C., USA

Martin FW, Pospahala RS, Nichols JD (1979) Assessment and population management of North American migratory birds. In: Cairns J, Patil GP, Walters WE (eds) Environmental biomonitoring, assessment, prediction, and management-certain case studies and related quantitative issues. International Cooperative Publishing House, Fairland, Maryland, pp 187–239

Mason CD, Whiting RM Jr, Conway WC (2013) Time-activity budgets of waterfowl wintering on livestock ponds in northeast Texas. Southeast Nat 12:757–768

Masto NM, Kaminski RM, Prince HH (2022) Hemi-marsh concept prevails? Kaminski and Prince (1981) revisited. J Wildl Manage 86:e22301

May SM, Naugle DE, Higgins KF (2002) Effects of land use on nongame wetland birds in western South Dakota stock ponds, U.S.A. Waterbirds 25(Special Publication):51–55

McCauley LA, Anteau MH, van der Burg MP et al (2015) Land use and wetland drainage affect water levels and dynamics of remaining wetlands. Ecosphere 6:92. https://doi.org/10.1890/ES14-0094.1

Mckenna OP, Kucia SR, Mushet DM et al. (2019) Synergistic interaction of climate and land-use drivers alter function of North American, Prairie Pothole Wetlands. Sustain 11:6581. https://doi.org/10.3390/su11236581

McLean K, Mushet D, Sweetman J (2022) Climate and land use driven ecosystem homogenization in the Prairie Pothole Region. Water 14:3106. https://doi.org/10.3390/w14193106

McIntyre NE, Collins SD, Heintzman LJ et al (2018) The challenge of assaying landscape connectivity in a changing world: a 27-year case study in the southern Great Plains (USA) playa network. Ecol Indic 91:607–616

McKenna OP, Mushet DM, Rosenberry DO et al (2017) Evidence for a climate-induced ecohydrological state shift in wetland systems of the southern Prairie Pothole Region. Clim Change 145:273–287

McKenna OP, Mushet DM, Kucia SR et al (2021a) Limited shifts in the distribution of migratory bird breeding habitat density in response to future changes in climate. Ecol Appl 31:e02428

McKenna OP, Renton DA, Mushet DM et al (2021b) Upland burning and grazing as strategies to offset climate-change effects on wetlands. Wetl Ecol Manag 29:193–208

McKinstry MC, Caffrey P, Anderson SH (2001) The importance of beaver to wetland habitats and waterfowl in Wyoming. J Am Water Resour Assoc 37(6):1571–1577

McKinstry MC, Hubert WA, Anderson SH (eds) (2004) Wetland and riparian areas of the Intermountain West. University of Texas Press, Austin, Texas, USA

McWethy DB, Austin JE (2009) Nesting ecology of Greater Sandhill Cranes (Grus canadensis tabida) in riparian and palustrine wetlands of Eastern Idaho. Waterbirds 32:106–115

Miller MR, Takekawa JY, Fleskes JP et al (2005) Spring migration of northern pintails from California's Central Valley wintering area tracked with satellite telemetry: routes, timing, and destinations. Can J Zool 83:1314–1332

Mitchell CD, Eicholz MW (2020) Trumpeter swan (Cygnus buccinator), version 1.0. In: Rodewald PG (ed) Birds of the world. Cornell Lab of Ornithology, Ithaca, NY, USA. https://doi.org/10.2173/bow.truswa.01

Mitsch WJ, Gosselink JG (2015) Wetlands, 5th edn. Wiley, Hoboken, New Jersey, USA

Moon JA, Haukos DA (2006) Survival of female northern pintails wintering in the Playa Lakes Region of northwestern Texas. J Wildl Manage 70:777–783

Moon JA, Haukos DA (2008) Habitat use by northern pintails wintering in the Playa Lakes Region. Proc Southeast Assoc Fish Wildlife Agencies 62:82–87

Moon JA, Haukos DA (2009) Factors affecting body condition of Northern Pintails wintering in the Playa Lakes Region. Waterbirds 32:87–95

Moon JA, Lehnen SE, Metzger KL et al (2021) Projected impact of sea-level rise and urbanization on mottled duck (Anas fulvigula) habitat along the Gulf Coast of Louisiana and Texas through 2100. Ecol Ind 132:108276. https://doi.org/10.1016/j.ecolind.2021.108276

Moore JM (2016) Recent desiccation of western Great Basin saline lakes: lessons from Lake Abert, Oregon, U.S.A. Sci Total Environ 554–555:142–154

Moore MA, McEvoy J (2022) "In Montana, you're only a week away from a drought": Rancher's perspectives on flood irrigation and beaver mimicry as drought mitigation strategies. Rangelands 44:258–269

Mote PW, Li S, Lettenmaier DP et al (2018) Dramatic declines in snowpack in the western US. NPJ Clim Atmos Sci 1:2

Moulton CE, Carlisle JD, Knetter SJ et al (2022) Importance of flood irrigation for foraging colonial waterbirds. J Wildl Manage 86:e22288

Mowbray TB, Ely CR, Sedinger JS et al (2020) Canada goose (*Branta canadensis*), version 1.0. In Rodewald PG (ed) Birds of the world. Cornell Lab of Ornithology, NY, USA. https://doi.org/10.2173/bow.cangoo.01

Mundinger JG (1976) Waterfowl response to rest-rotation grazing. J Wildl Manage 40:60–68

Murkin HR, Murkin EJ, Ball JP (1997) Avian habitat selection and prairie wetland dynamics: a 10-year experiment. Ecol Appl 7:1144–1159

Murphy RK, Schindler DJ, Crawford RD (2004) Duck nesting on rotational and continuous grazed pastures in North Dakota. Prairie Nat 36:83–94

Murphy RK, Dwyer JF, Mojica EK et al (2016) Reactions of sandhill cranes approaching a marked transmission power line. J Fish Wildl Manag 7(2):480–489. https://doi.org/10.3996/052016-JFWM-037

Naugle DE, Higgins KF, Bakker KK (2000) A synthesis of the effects of upland management practices on waterfowl and other birds in the Northern Great Plains of the U.S. and Canada. College of Natural Resources, University of Wisconsin-Stevens Point, WI. Wildlife Technical Report 1, 28p

Navarrete L, Griffis-Kyle KL (2014) Sandhill crane collisions with wind turbines in Texas. In: Proceedings of the North Americna Crane workshop vol 12. pp 65–67. https://digitalcommons.unl.edu/nacwgproc/380/

NAWMP (2018) North American Waterfowl Management Plan (NAWMP) Update: connecting people, waterfowl, and wetlands. Canadian Wildlife Service, U.S. Fish and Wildlife Service, and Secretaria de Medio Ambiente y Recursos Naturales. https://www.fws.gov/partner/north-american-waterfowl-management-plan

Neudecker GA, Duvall AL, Stutzman JW (2011) Community-based landscape conservation: a roadmap for the future. In: Naugle DE (ed) Energy development and wildlife conservation in western North America. Island Press, Washington, DC, USA

Nichols JD, Kendall WL, Boomer GS (2019) Accumulating evidence in ecology: once is not enough. Ecol Evol 9:13991–14004

Nieman DJ, Didiuk AB, Smith JR (2000) Status of Canada Geese of the Canadian prairies. In: Dickson KM (ed) Towards conservation of the diversity of Canada geese (*Branta canadensis*). Occasional Paper 103. Canada Wildlife Service, Ottawa, Ontario, pp 139–150

Niemuth ND, Solberg SW (2003) Response of waterbirds to number of wetlands in the Prairie Pothole Region of North Dakota, U.S.A. Waterbirds 26:233–238

Niemuth ND, Dahl AL, Estey ME et al (2007) Representation of landcover along breeding bird survey routes in the Northern Plains. J Wildl Manage 71:2258–2265

Niemuth ND, Reynolds RE, Granfors DE et al (2009) Landscape-level planning for conservation of wetland birds in the U.S. Prairie Pothole Region. In: Millspaugh JJ, Thompson FR III (eds) Models for planning wildlife conservation in large landscapes. Elsevier Science, pp 533–560

Niemuth ND, Wangler B, Reynolds RE (2010) Spatial and temporal variation in wet area of wetlands in the Prairie Pothole Region of North Dakota and South Dakota. Wetlands 30:1053–1064

Niemuth ND, Estey ME, Reynolds RE (2012) Factors influencing presence and detection of breeding shorebirds in the Prairie Pothole Region of North Dakota, South Dakota, and Montana, USA

Niemuth ND, Fleming KK, Reynolds RE (2014) Waterfowl conservation in the US Prairie Pothole Region: confronting the complexities of climate change. PLoS ONE 9(6):e100034. https://doi.org/10.1371/journal.pone.0100034

Niemuth ND, Barnes KW, Tack JD et al (2022) Past is prologue: historic landcover patterns predict contemporary grassland loss in the U.S Northern Great Plains. Landsc Ecol 37:3011–3027

NRCS (2021) A framework for conservation action in the Great Plains grassland biome. USDA Natural Resources Conservation Service, Working Lands for Wildlife

Null SE, Wurtsbaugh WA (2020) Water development, consumptive water uses, and Great Salt Lake. In: Baxter B, Butler J (eds) Great Salt Lake biology a terminal lake in a time of change. Springer, Cham, Switzerland, pp 1–21

Osnas EE, Boomer GS, Devries JH et al (2021) Decision-support framework for linking regional-scale management actions to continental-scale conservation of wide-ranging species. US Geological Survey Open-File Report 2020–1084.https://doi.org/10.3133/ofr20201084

Ott JP, Hanberry BB, Khalil M, Paschke MW, van der Burg MP, Prenni AJ (2021) Energy development and production in the Great Plains: implications and mitigation opportunities. Rangel Ecol Manag 78:257–272. https://doi.org/10.1016/j.rama.2020.05.003

Paveglio FL, Kilbride K (2000) Response of vegetation to control of reed canarygrass in seasonally managed wetlands of southwestern Washington. Wildl Soc Bull 28:730–740

Pearse AT, Brandt DA, Krapu GL (2016) Wintering sandhill crane exposure to wind energy development in the central and southern Great Plains, USA. Condor 118:391–401. https://doi.org/10.1650/CONDOR-15-99.1

Pearse AT, Metzger KL, Brandt DA et al (2021) Migrating whooping cranes avoid wind-energy infrastructure when selecting stopover habitat. Ecol Appl 30(5):e02324. https://doi.org/10.1002/eap.2324

Pearse AT, Anteau MJ, van der Burg MP et al (2022) Reassessing perennial cover as a driver of duck nest survival in the Prairie Pothole Region. J Wildl Manage 86:e22227. https://doi.org/10.1002/jwmg.22227

Pearson J, Dunham J, Bellmore JR et al (2019) Modeling control of common carp (Cyprinus carpio) in a shallow lake-wetland system. Wetl Ecol Manag 27:663–682

Pederson RL, Jorde DG, Simpson SG (1989) Northern Great Plains. In: Smith LM, Pederson RL, Kaminski RM (eds) Habitat management for migrating and wintering waterfowl in North America. Texas Tech University Press, Lubbock, Texas

Phillips ML, Clark WR, Sovada MA et al (2003) Predator selection of prairie landscape features and its relation to duck nest success. J Wildl Manage 67:104–114

PHJV (2021) Prairie Habitat Joint Venture implementation plan 2021–2025: the Prairie Parklands. Report to the Prairie Habitat Joint Venture. Environment Canada, Edmonton, Alberta

PIF (2021) Avian conservation assessment database, version 2021. Partners in Flight, Available at http://pif.birdconservancy.org/ACAD

Powers LC, Glimp HA (1996) Impacts of livestock on shorebirds: a review and application to shorebirds of the western Great Basin. Int Wader Stud 9:55–63

PPJV (2017) Implementation plan. Prairie Pothole Joint Venture, Lakewood, Colorado. https://ppjv.org/assets/pdf/2017-PPJV-implementation-plan.zip

Preston TM, Ray AM (2017) Effects of energy development on wetland plants and macroinvertebrate communities in Prairie Pothole Region wetlands. J Freshw Ecol 32(1):29–34. https://doi.org/10.1080/02705060.2016.1231137

Ramirez P Jr (2010) Bird mortality in oil field wastewater disposal facilities. Environ Manage 46:820–826. https://doi.org/10.1007/s00267-010-9557-4

Raquel AJ, Devries JH, Howerter DW et al (2016) Timing of nesting of upland-nesting ducks in the Canadian prairies and its relation to spring wetland conditions. Can J Zool 94:575–581

Renner RW, Reynolds RE, Batt BDJ (1995) The impact of haying conservation reserve program lands on productivity of ducks nesting in the Prairie Pothole Region of North Dakota and South Dakota. In: Transactions of the 60th North American wildlife and natural resources conference, vol 60. pp 221–229

Reynolds RE, Blohm RJ, Nichols JD et al (1995) Spring-summer survival rates of yearlings versus adult mallard females. J Wildl Manage 59:691–696

Reynolds RE, Shaffer TL, Renner RW et al (2001) Impact of the conservation reserve program on duck recruitment in the U.S. Prairie Pothole Region. J Wildl Manage 65:765–780

Reynolds RE, Shaffer TL, Loesch CR (2006) The farm bill and duck production in the Prairie Pothole Region: increasing the benefits. Wildl Soc Bull 34:963–974

Richmond OMW, Tecklin J, Beissinger SR (2012) Impact of cattle grazing in the occupancy of a cryptic, threatened rail. Ecol Appl 22:1655–1664

Riecke TV, Sedinger BS, Arnold TW et al (2022a) A hierarchical model for jointly assessing ecological and anthropogenic impacts on animal demography. J Anim Ecol 91:1612–1626

Riecke TV, Lohman MG, Sedinger BS (2022b) Density-dependence produces spurious relationships among demographic parameters in a harvested species. J Anim Ecol 00:1–12. https://doi.org/10.1111/1365-2656.13807

Ringelman KM, Walker J, Ringelman JK et al (2018) Temporal and multi-spatial environmental drivers of duck nest survival. Auk 135:486–494

Rischette AC, Geaumont BA, Elmore RD et al (2021) Duck nest density and survival in post-conservation reserve program lands. Wildl Soc Bull 45:630–637

Roberts A, Scarpignato AL, Huysman A et al (2023) Migratory connectivity of North American waterfowl across administrative flyways. Ecol Appl 2023:e2788. https://doi.org/10.1002/eap.2788

Rohal CB, Cranney C, Hazelton ELG et al (2019) Invasive Phragmites australis management outcomes and native plant recovery are context dependent. Ecol Evol 9:13835–13849

Rosenberg KV, Dokter AM, Blancher PJ et al (2019) Decline of North American avifauna. Science 366:120–124

Rumble MA, Flake LD (1983) Management considerations to enhance use of stock ponds by waterfowl broods. Rangel Ecol Manag 36:691–694

RWBJV (2013) Rainwater Basin Joint Venture waterfowl plan: a regional contribution to the North American Waterfowl Management Plan and the Rainwater Basin Joint Venture Implementation Plan. Rainwater Basin Joint Venture, Grand Island, Nebraska, USA. https://www.rwbjv.org/wp-content/uploads/Rainwater-Basin-Joint-Venture-Waterfowl-Plan-2013.pdf

Sargeant AB, Raveling DG (1992) Mortality during the breeding season. In: Batt BDJ, Afton AD, Anderson MG et al (eds) Ecology and management of breeding waterfowl. University of Minnesota Press Minneapolis, Minnesota, pp 396–422

Sauer JR (1999) Marsh birds and the North American breeding bird survey: judging the value of a landscape level survey for habitat specialist species with low detection rates. In: Ribic CA, Lewis SJ, Melvin S et al (eds) Proceedings of the marsh bird monitoring workshop. U.S. Fish and Wildlife Service and U.S. Geological Survey, Laurel, MD, USA. p 43

Sauer JR, Fallon JE, Johnson R (2003) Use of North American breeding bird survey data to estimate population change for bird conservation regions. J Wildl Manage 67:372–389

Schultz BD, Hubbard DE, Jens JA et al (1994) Plant and waterfowl responses to cattle grazing in two South Dakota semipermanent wetlands. Proc South Dakota Acad Sci 73:121–134

Seamans ME (2022) Status and harvests of sandhill cranes: mid-continent, Rocky Mountain, Lower Colorado River Valley and Eastern Populations. Administrative Report, U.S. Fish and Wildlife Service, Lakewood, Colorado

Sedinger JS, Alisauskas RT (2014) Cross-seasonal effects and the dynamics of waterfowl populations. Wildfowl 4:277–304

Shaffer JA, Igl LD, Johnson DH et al (2019a) The effects of management practices on grassland birds–Marbled Godwit (*Limosa fedora*), Chapter H. In: Johnson DH, Igl LD, Shaffer JA et al (eds) The effects of management practices on grassland birds. U.S. Geological Survey Professional Paper 1842, 9p. https://doi.org/10.3133/pp1842H

Shaffer JA, Igl LD, Johnson DH et al (2019b) The effects of management practices on grassland birds–Willet (*Tringa semipalmata inornata*), Chapter I. In: Johnson DH, Igl LD, Shaffer JA et al (eds) The effects of management practices on grassland birds. U.S. Geological Survey Professional Paper 1842, 9p. https://doi.org/10.3133/pp1842I

Shaffer JA, Igl LD, Johnson DH et al (2019c) The effects of management practices on grassland birds–Wilson's Phalarope (*Phalaropus tricolor*), Chapter J. In: Johnson DH, Igl LD, Shaffer

JA et al (eds) The effects of management practices on grassland birds. U.S. Geological Survey Professional Paper 1842, 10p. https://doi.org/10.3133/pp1842J

Shaffer JA, Igl LD, Johnson DH et al (2019d) The effects of management practices on grassland birds–Long-billed Curlew (*Numenius americanus*), Chapter G. In: Johnson DH, Igl LD, Shaffer JA et al (eds) The effects of management practices on grassland birds. U.S. Geological Survey Professional Paper 1842, 12p. https://doi.org/10.3133/pp1842G

Shaffer JA, Igl LD, Johnson DH et al (2019e) The effects of management practices on grassland birds–Upland Sandpiper (*Bartramia longicauda*), Chapter F. In: Johnson DH, Igl LD, Shaffer JA et al (eds) The effects of management practices on grassland birds. U.S. Geological Survey Professional Paper 1842, 20p. https://doi.org/10.3133/pp1842F

Shaffer JA, Igl LD, Johnson DH et al (2019f) The effects of management practices on grassland birds–Mountain Plover (*Charadrius montanus*), Chapter E. In: Johnson DH, Igl LD, Shaffer JA et al (eds) The effects of management practices on grassland birds. U.S. Geological Survey Professional Paper 1842, 10p. https://doi.org/10.3133/pp1842E

Shaffer JA, Loesch CR, Buhl DA (2019g) Estimating offsets for avian displacement effects of anthropogenic impacts. Ecol Appl 28(8):e01983. https://doi.org/10.1002/eap.1983

Sherfy MH, Anteau MJ, Shaffer TL et al (2018) Density and success of upland duck nests in native- and tame-seeded conservation fields. Wildl Soc Bull 42:204–212

Silverman NJ, Allred BW, Donnelly JP (2018) Low-tech riparian and wet meadow restoration increases vegetation productivity and resilience across semiarid rangelands. Restor Ecol 27:269–278

Sketch M, Dayer AA, Metcalf AL (2020) Western ranchers' perspectives on enablers and constraints to flood irrigation. Rangel Ecol Manag 73(2):285–296

Smeins FE, Diamond DD, Hanselka CW (1991) Coastal prairie. In: Coupland RT (ed) Ecosystems of the world 8. A natural grasslands-introduction and western hemisphere. Elsevier Press, New York. pp 269–290

Smith LM (2003) Playas of the Great Plains. University of Texas Press, Austin, Texas, USA

Smith LM, Kadlec JA (1985) Fire and herbivory in a Great Salt Lake marsh. Ecology 66:259–265

Smith AG, Stoudt JH, Gollp JB (1964) Prairie potholes and marshes. In: Linduska JP, Nelson AL (eds) Waterfowl tomorrow. US Department of Interior, Washington, DC, pp 39–50

Smith LM, Pederson RL, Kadlec JA (1989) Habitat management for migrating and wintering waterfowl. Texas Tech University Press, Lubbock, Texas, USA

Smith LM, Haukos DA, Prather RM (2004) Avian response to vegetative pattern in playa wetlands during winter. Wildl Soc Bull 32:474–480

Smith LM, Haukos DA, McMurry ST et al (2011) Ecosystem services provided by playas in the High Plains; potential influences of USDA conservation programs. Ecol Appl 21:82–92

Smith LM, Haukos DA, McMurry ST (2012) High Plains playas. In: Batzer DP, Baldwin AH (eds) Wetland habitats of North America: ecology and conservation concerns. University of California Press, Berkely and Los Angeles, California, pp 299–311

Smith PA, Smith AC, Andres B et al (2023) Accelerating declines of North America's shorebirds signal the need for urgent conservation action. Ornithol Appl 125:1–14

Snyder KA, Evers L, Chambers JC et al (2019) Effects of changing climate on the hydrological cycle in cold desert ecosystems of the Great Basin and Columbia Plateau. Rangel Ecol Manag 72:1–12

Sofaer HR, Skagen SK, Barsugli JJ et al (2016) Projected wetland densities under climate change: habitat loss but little geographic shift in conservation strategy. Ecol Appl 26:1677–1692

Specht H, St-Louis V, Gratto-Trevor CL et al (2020) Habitat selection and nest survival in two Great Plains shorebirds. Avian Conserv Ecol 15:3. https://doi.org/10.5751/ACE-01487-150103

Stafford JD, Janke AK, Anteau MJ et al (2014) Spring migration of waterfowl in the northern hemisphere: a conservation perspective. Wildfowl 4:70–85

Stearns SC (1992) The evolution of life histories. Oxford University Press, Oxford, England

Stephens SE, Rotella JJ, Lindberg MS et al (2005) Duck nest survival in the Missouri Coteau of North Dakota: landscape effects at multiple spatial scales. Ecol Appl 15:2137–2149

Stutzenbaker CD, Weller MW (1989) The Texas Coast. In: Smith LM, Pederson RL, Kaminski RM (eds) Habitat management for migrating and wintering waterfowl in North America. Texas Tech University Press, Lubbock, Texas, USA

Sullivan BL, Aycrigg JL, Barry JH et al (2014) The eBird enterprise: an integrated approach to development and application of citizen science. Biol Conserv 169:31–40

Svingen D, Anderson SH (1998) Waterfowl management on grass-sage stock ponds. Wetlands 18:84–89

Swift RJ, Rodewald AD, Johnson JA, Andres BA, Senner NR (2020) Seasonal survival and reversible state effects in a long-distance migratory shorebird. J Anim Ecol 89:2043–2055

Tangen BA, Bansal S, Jones S et al (2022) Using a vegetation index to assess wetland condition in the Prairie Pothole Region of North America. Front Environ Sci 10:889170. https://doi.org/10.3389/fenvs.2022.889170

Tessman SA (2004) Management of created palustrine wetlands. In: McKinstry MC, Hubert WA, Anderson SH (eds) Wetland and riparian areas of the Intermountain West. University of Texas Press, Austin, Texas, pp 154–184

Tori GM, McLeod S, McKnight K et al (2002) Wetland conservation and Ducks Unlimited: real world approaches to multispecies management. Waterbirds 25 (Special Publication 2):115–121

Trail PW (2006) Avian mortality at oil pits in the United States: a review of the problem and efforts for its solution. Environ Manage 38:532–544. https://doi.org/10.1007/s00267-005-0201-7

Tsai J, Venne LS, Smith LM et al (2012) Influence of local and landscape characteristics on avian richness and density of wet playas of the Southern Great Plains, USA. Wetlands 32:605–618

U.S. Geological Survey Bird Banding Laboratory (2020) USGS celebrates 100 years of bird banding lab: A century of advancing bird conservation science. https://www.usgs.gov/news/featured-story/

U.S. Shorebird Conservation Plan Partnership (2016) U.S. shorebirds of conservation concern–2016. U.S. Fish and Wildlife Service. http://www.shorebirdplan.org/science/assessment-conservation-status-shorebirds/

Uden DR, Allen CR, Bishop AA et al (2015) Predictions of future ephemeral springtime waterbird stopover habitat availability under global change. Ecosphere 6:1–26

USFWS (2011) Birds of management concern and focal species. U.S. Fish and Wildlife Service, Migratory Bird Program. U.S. Department of the Interior, Washington, DC, USA

USFWS (2022) Waterfowl population status, 2022. U.S. Fish and Wildlife Service, U.S. Department of the Interior, Washington, DC, USA

USFWS (2023) Fish and Wildlife Service. Mid-winter waterfowl survey. U.S. Department of the Interior, Washington, DC, USA. https://migbirdapps.fws.gov/mbdc/databases/mwi/mwidb.asp

Veech JA, Pardieck KL, Ziolkowski DJ Jr (2017) How well do route survey areas represent landscapes at larger spatial extents? An analysis of land cover composition along Breeding Bird Survey routes. Condor 119:607–615

Venne LS, Frederick PC (2013) Foraging wading birds (CICONIIFORMES) attraction to prescribed burns in an oligotrophic wetland. Fire Ecol 9:78–95

Verheijen BHF, Varner DM, Haukos DA (2020) Future loss of playa wetlands decrease network structure and connectivity of the Rainwater Basin, Nebraska. Landsc Ecol 35:453–467

Vermillion WG (2012) Fall habitat objectives for priority Gulf Coast Joint Venture shorebird species using managed wetlands and grasslands, version 4.0. Gulf Coast Joint Venture, Lafayette, Louisiana

Vermillion W, Eley JW, Wilson B et al (2008) Partners in Flight bird conservation plan, Gulf Coastal Prairie, bird conservation region 37, version 1.3. Gulf Coast Bird Observatory, Lake Jackson, Texas, USA

Walker J, Taylor PD (2017) Using eBird data to model population change of migratory bird species. Avian Conserv Ecol 12(1):4. https://doi.org/10.5751/ACE-00960-120104

Walker J, Rotella JJ, Stephens SE et al (2013a) Time-lagged variation in pond density and primary productivity affects duck nest survival in the Prairie Pothole Region. Ecol Appl 23:1061–1074

Walker J, Rotella JJ, Schmidt JH et al (2013b) Distribution of duck broods relative to habitat characteristics in the Prairie Pothole Region. J Wildl Manage 77:392–404

Warren J, Rotella J, Thompson J (2008) Contrasting effects of cattle grazing intensity on upland-nesting duck production at nest and field scales in the Aspen Parkland. Avian Conserv Ecol 3:6

Weiser EL, Lanctot RB, Brown SC et al (2020) Annual adult survival drives trends in Arctic-breeding shorebirds but knowledge gaps in other vital rates remain. Condor 122:1–14

Weller M (1999) Wetland birds: habitat resources and conservation implications. Cambridge University Press, Cambridge. https://doi.org/10.1017/CBO9780511541919

West BC, Messmer TA (2006) Effects of livestock grazing on duck nesting habitat in Utah. Rangel Ecol Manag 59:208–211

Williams BK, Nichols JD, Conroy MJ (2002) Analysis and management of animal populations. Academic Press, San Diego, California, USA

Wilson BC, Esslinger CG (2002) North American waterfowl management plan, Gulf Coast Joint Venture: Texas Mid-Coast initiative. North American Waterfowl Management Plan, Albuquerque, New Mexico, USA, 28pp

Wright CK, Wimberly MC (2013) Recent land use change in the Western Corn Belt threatens grasslands and wetlands. PNAS 110(10):4134–4139. https://doi.org/10.1073/pnas.1215404110

Yetter AP, Hagy HM, Horath MM et al (2018) Mallard survival, movements, and habitat use during autumn in Illinois. J Wildl Manage 82:182–191

Zarzycki M (2017) Evidence for cross-seasonal effects: insights from long-term data on northern pintail. Thesis, Oregon State University, Corvallis, Oregon, USA

Chapter 14
Avian Predators in Rangelands

Bryan Bedrosian

Abstract Management of avian predators in western rangelands is uniquely challenging due to differences in managing for/against particular species, management of sensitive prey species, long-standing human/wildlife conflicts, and the unique legal protections within this ecological group. In general, many avian predator species considered rangeland specialists have been declining due to habitat loss, fragmentation, human sensitivity, and direct persecution. Conversely, avian predators that are more human-tolerant and/or are subsidized by human activities are significantly increasing across rangelands. The complicated nature of inter- and intra-species guilds, coupled with human dynamics has created a challenging scenario for both management for avian predators, as well as their prey. Human-mediated population control, both legal and illegal, continues for avian predators to reduce livestock conflict, aid sensitive prey populations, and/or because of general predator persecution. Conversion of rangeland to development for energy, cultivation, and urbanization remains the largest impediment to maintaining viable, historical assemblages of avian predators. Large-scale habitat protections, reduction of invasive plants, and reducing wildfire will continue to enhance at-risk populations of predators and their prey. Further, mediating human-induced mortality risks will also aid at-risk predator populations, such as reducing direct killing (poisoning and shooting), secondary poisoning from varmint control and lead ammunition use, electrocutions, and vehicle strikes, while reducing anthropogenic subsidies can help curtail population expansion of corvids. Additional understanding of long-term, successful predator control efforts for corvids and mitigation options for declining raptors is needed to help balance the avian predator–prey dynamic in western rangelands.

Keywords Anthropogenic · Avian predators · Buteo · Corvid · Eagle · Raptor · Rangeland · Raven

B. Bedrosian (✉)
Teton Raptor Center, 5450 W HWY 22, Wilson, WY 83014, USA
e-mail: bryan@tetonraptorcenter.org

© The Author(s) 2023
L. B. McNew et al. (eds.), *Rangeland Wildlife Ecology and Conservation*,
https://doi.org/10.1007/978-3-031-34037-6_14

14.1   General Life History and Population Dynamics

Avian predators have often been considered flagship or umbrella species due to their large home ranges and unique legal protections (Sergio et al. 2006; Donazar et al. 2016). However, the complex anthropogenic and ecological relationships of avian predators in rangeland habitats have varied drastically through history and continue to shift. The diverse migratory strategies and intra- and inter-specific competition among avian predators can impact community dynamics of raptors (hawks, eagles, and owls), corvids (ravens, crows, and magpies), and gulls in rangeland systems. Further, raptors and corvids are unique among rangeland wildlife because of their varied and complicated relationship with humans, largely due to their influence on economic interests, development, historical and contemporary persecution, conflicting multi-species management goals, and multiple legal protections.

Community composition and abundance of avian predators across rangelands are affected by inter- and intra-specific competition (Craighead and Craighead 1969), habitat quality (e.g. Dunk et al. 2019), and species-specific habitat associations. Most raptors of rangelands have evolved behaviors or traits to help facilitate hunting, movements, and breeding in open landscapes. Specialist species like the semi-fossorial burrowing owl (*Athene cunicularia*) are unique by relying on burrows excavated by mammals, such as prairie dogs and ground squirrels for nesting. As such, they have co-evolved with those species to the point where they use and rely on prairie dog alarm calls to alert them to potential predators (Bryan and Wunder 2014). Other raptors relying on open rangelands, like ferruginous hawks (*Buteo regalis*) and golden eagles (*Aquila chrysaetos*) have evolved traits including ground nesting, morphology and flight dynamics for aerial foraging, and increased sensitivity to human disturbance. There is a wide range of population trends for various species, largely based on tolerance and reliance on anthropogenic features across rangelands. Species like common ravens (*Corvus corax*) have largely increasing populations due to reliance on anthropogenic subsidies for nesting and foraging while other species like golden eagles are becoming increasingly at risk. There is also a large seasonal component to these dynamics driven by differences in migration strategies among species. Some groups exhibit prey-based partitioning, with different species occupying the same habitats but selecting differential prey (e.g., American kestrels *Falco sparverius* and red-tailed hawks *Buteo jamaicensis*). Others avoid competition by timing, either diurnal/nocturnal or by season (e.g., great horned owls *Bubo virginianus* and red-tailed hawks).

14.1.1   Nesting

Almost all avian predators are highly territorial during nesting. Those species that cannot avoid competition through prey or temporal niche partitioning can exhibit significant territoriality, or habitat partitioning (Restani 1991; Kennedy et al. 2014).

This is classically apparent within raptor species, where territorial pairs aggressively defend breeding territories from conspecifics (Newton 2010). Notably, there are some exceptions with corvids. Territorial common ravens are similar in that they aggressively defend breeding areas from conspecifics but differ by allowing non-breeding conspecifics within a territory when a food bonanza (e.g., large carcass) occurs (Webb et al. 2012). Because larger raptors can also prey on smaller raptor species, habitat partitioning in rangelands not only is a result of prey availability but also intra-guild predator–prey dynamics. This regular territoriality in avian predators, coupled with the relative ease of locating large raptor stick nests, has led to largely nest-centric management practices across rangelands, with various sized protection buffers placed around most raptor nests for disturbances during the breeding season (USFWS 2022).

Generally, raptors and corvids are k-selected species, with relatively slow reproductive rates (Newton 2010). For example, golden eagles typically do not begin reproducing until $\geq$ 5 years of age and breeding pairs produce an average of < 1 fledgling per year (Katzner et al. 2020a, b). This slow reproductive rate, coupled with prey or habitat specialization, has led to many raptor species' declines. While popular perception is often that breeding raptors "mate for life," this theory is a bit misleading. Many avian predators do have high territory and mate fidelity, which can often lead to increased lifetime reproductive success (Leon-Ortega et al. 2017), but some studies suggest that individual quality is a better driver of reproductive success than territory quality in long-lived raptor species (Zabala and Zuberogoitia 2014). Most avian predators do maintain mates across years (e.g., golden eagles, common ravens) but will regularly, and quickly, replace a mate that dies or does not return to the breeding territory following migration (Watson 2010; Webb et al. 2012). Further, some individuals of these species regularly switch both mates and territories between years (Steenhof and Peterson 2009; author, unpublished data), while other species like northern harriers (*Circus cyaneus*) and Harris's hawks (*Parabuteo unicinctus)* commonly practice alternative breeding strategies like polygyny (Simmons et al. 1986) and cooperative breeding (Bednarz and Ligon 1988).

Population dynamics of many avian predators are strongly influenced by inter-annual fluctuations in prey population abundance and human presence on the landscape (Newton 2010). Several well-adapted prey species, such as leporids, prairie dogs, and prairie grouse species were abundant in rangelands until recent history (Bedrosian et al. 2019). Changes in prey, land use, and anthropogenic influences has generally led to a reduced diversity and overall population sizes of many historic raptor species and an increase in corvids and gulls in rangelands today. While many factors have influenced the decline of some species, declining prey abundance has also had significant further effects on avian predators, guild dynamics, and management actions (Newton 2010).

14.1.2 Post-fledging

The most vulnerable period for avian predator survival is after the post-fledging dependence period, or after the young disperse from their natal territory (Newton 2010; Millsap et al. 2022). Typically, mortality is the highest during this time and often a result of starvation or predation (Millsap et al. 2022). The post-fledging dependence period (after fledging but before dispersal) can range from several weeks up to > 1 year for large species like golden eagles and California condors (Gymnogyps californianus). Young golden eagles tracked with transmitters have been documented to have home ranges that include their natal territory for several years (Murphy et al. 2017). Habitat associations also can affect survival in various ways, depending on the species. For example, common raven post-fledging survival increased as the nest distance decreased from the nearest human settlement and subsidies (Webb et al. 2004; Bedrosian 2004), but the causes of mortality switched from natural causes to anthropogenic as ravens nested nearer to towns (author, unpublished data). Further, in desert rangelands with limited resources, raven post-fledging survival is drastically lower (38%; Webb et al. 2004) than more mesic, diverse habitats (83%; Bedrosian 2004), further indicating the importance of habitat quality on predator survival.

14.1.3 Non-breeding

Because many avian predators do not breed in their first year of life (e.g., most large-bodied raptors and ravens), the non-breeding component of the population can be large and differ in habitat use, prey use, and survival than breeding adults. Popularly referred to as "sub-adults" or "pre-breeding," these individuals can represent a significant portion of the population within specific areas of rangelands. For example, sub-adult golden eagles occupy habitats more often associated with wintering habitat than typical breeding habitat, even in the summer months (author, unpublished data). Pre-breeding and non-breeding adult ravens also occur at anthropogenic point subsidies (e.g., landfills) to a much higher degree than breeding adults (Harju et al. 2018). The differentiation and understanding of varying habitat and space use among different age-classes of avian predators can have significant impact on management of these species across rangelands, particularly for the benefit of prey species, like greater sage-grouse (Centrocercus urophasianus) (Harju et al. 2018).

14.1.4 Survival

An avian predator species' ability to adapt and evolve in response to changing habitat conditions in rangelands is a function of their reproductive rate, diet breadth, and tolerance for anthropogenic features. However, several confounding factors also contribute to a species' persistence, including sensitivity to chemicals or toxic elements and plasticity in habitat selection. California condors are at one extreme of that spectrum, with little ability to adapt quickly to changing landscapes and a low tolerance of toxic elements in their environment (Finkelstein et al. 2020). Historically, carcasses of American bison (*Bison bison*) and other big-game sustained their populations but as bison and large carnivores were eventually replaced with livestock across the range, the abundance of carcasses available to scavengers dwindled. The species shifted to other available carrion, like hunter-harvested big game and euthanized livestock. The decreased abundance of food, coupled with their increased mortality from ingesting lead ammunition fragments in gut piles and non-steroidal anti-inflammatory drugs in deceased livestock, further exacerbated their decline (Finkelstein et al. 2012). At the other end of the spectrum are species like common ravens, great-horned owls, and red-tailed hawks. Unlike many other predators, these species' evolutionary history has led to a greater tolerance of human activities, wide diet breadth, and the ability to nest in a wide range of habitats and climates; all of which has led to their expansion in many human-altered habitats, including rangelands (Boarman and Heinrich 2020).

Post-fledging, most avian predators have high survival with species like golden eagles nearing 90% annual survival rates for adults and 70% for first-year golden eagles in western rangelands (Millsap et al. 2022). Cause specific mortality for young avian predators is mostly due to natural causes (e.g., starvation and predation) but eventually switch to primarily human-caused mortality in older-age classes (see Sect. 14.4).

14.1.5   Seasonal Movements and Dispersal

The diverse migratory behaviors of avian predators have led to large seasonal shifts in abundance and distribution across rangelands. Some species exhibit complete migration, like Swainson's hawks (*Buteo swainsoni*), which occupy rangelands only during the breeding season, then migrate to South America during the non-breeding season. Similarly, rough-legged hawks (*Buteo lagopus*) breed in the arctic tundra and migrate south to winter in rangelands. Other species that make nomadic or irruptive migratory movements, like short-eared owls (*Asio flammeus*), can have different breeding territories each year, sometimes hundreds of kilometers apart (Shaffer et al. 2021). Corvids typically do not exhibit migratory movements in the classical sense but can drastically increase their home ranges during the non-breeding season and occur more often in areas of anthropogenic subsidies compared to the breeding season (Harju et al. 2018). Species of gulls that occupy rangelands, like California Gulls (*Larus californicus*) and Ring-billed Gulls (*Larus delawarensis*) typically migrate during the winter to western coasts, but some small populations overwinter along the Snake River corridor and near Great Salt Lake (Pollet et al. 2020; Winkler 2020). Snowy owls (*Bubo scandiacus*), which breed in the arctic, make irruptive migrations to the coterminous US in years of high prey abundance in northern rangelands

(Robillard et al. 2016). Other species employ a mixture of these strategies, like ferruginous hawks (*Buteo regalis*), which maintain disparate breeding and wintering ranges across years (both typically in rangelands) but make nomadic (typically northern) movements after breeding during late summer, followed by a typical migration to more southern latitudes (Watson et al. 2018; Watson and Keren 2019). Finally, some species exhibit diverse migratory patterns that vary across their range and life stages. For example, golden eagles can be (1) year-round residents in much of the coterminous US (Crandall et al. 2015), (2) complete, long-distance migrants from the arctic tundra and grasslands of Canada (Bedrosian et al. 2019), (3) migrate north from the arid southwest (Murphy et al. 2018), or (4) have very large but no regular seasonal ranges across multiple states (Poessel et al. 2022). All these sub-groups generally converge in the grassland and sage-steppe rangelands of the U.S. during the winter months. The diversity of migratory patterns within and among species results in dynamic variation in the assemblage of rangeland avian predators in space and time.

14.1.6 Population Dynamics

Species composition of avian predator communities in rangelands has shifted in response to alterations in habitat composition, prey abundance, and anthropogenic use (Donazar et al. 2016). Changes in habitat are largely driven by anthropogenic causes, ranging from increased fragmentation, conversion, invasive plants, fire, and combinations of these factors. Predators that have evolved in rangelands typically occupy large seasonal ranges and have reproductive strategies to accommodate fluctuating prey populations and dispersed resources (Johnson et al. 2022). This reliance on large home ranges in rangelands can lead to negative population consequences as fragmentation of these landscapes increase.

Some human-tolerant species have significantly increased in rangeland habitats due, in part, to increased anthropogenic use and alteration of rangelands (Coates et al. 2016; Boarman and Heinrich 2020). While several species of raptors and gulls are included in this group, corvids are the most significant example of this across the West. Historically occurring in low densities across the deserts, sage-steppe, and grasslands, corvids have had unparalleled expansion into rangelands due to several compounding factors (Bui et al. 2010). First, their plasticity in both habitat selection and foraging strategies has allowed ravens to occupy nearly every type of habitat in the West. Second, declines in other raptor species can decrease inter-specific competition that may have otherwise excluded corvids. Finally, anthropogenic food, water, and nesting subsidies in rangelands are more readily used by corvids and gulls due to their human tolerance (Harju et al. 2018; Winkler 2020) and these subsidies create nesting territories where they historically would not have occurred (e.g., ravens do not nest on the ground). However, some species considered obligates of native rangelands may also benefit from some degree of habitat heterogeneity resulting from agriculture and human infrastructure, such as ferruginous hawks nesting on anthropogenic structures (Wallace et al. 2016b). These population shifts comes with significant management challenges from declining raptor populations to native prey species management.

The complexities of intra- and inter-specific habitat use, seasonality, competition, and human tolerance are among the many factors that make avian predator management difficult in changing rangelands across North America. Avian predator habitat selection typically occurs at larger scales in rangelands than more heterogeneous or productive landscapes, which can complicate management actions. Management efforts must address multiple ecosystem-level processes, spatial scales, trophic cascades, and multiple species to be effective.

14.2 Current Species and Population Status

Numerous avian predator species occupy North American rangelands from the northern arctic tundra to the southern deserts. The large and diverse types of rangelands host both specialized and generalist avian predators. While very few raptor species rely solely on rangelands for year-round habitat needs, rangelands provide important seasonal habitat with a wide and vast array of avian predators. Rangelands occur in most avian predator home-ranges across the West and these habitats are essential for large portions of many populations.

Herein, the focus is on avian predators of rangeland habitats south of the boreal forests in North America and exclude forest-obligate species (Table 14.1). However, forested habitats may be used for livestock grazing and forest-obligate raptors may also be affected by rangeland management practices that are adjacent to forests or woodlands. Several distinct groups of avian predators occur within western rangelands, including raptors (i.e., eagles, hawks, owls), corvids (i.e., ravens, crows, magpies), vultures, and gulls. Each group and species have unique habitat and management needs and may occupy different rangelands in different seasons. Most avian predators occupy various rangeland types year-round and populations of predators typically increase in winter as northern migrants flood into the habitats occupied by year-round residents in the coterminous US. The best example of this phenomenon is golden eagles (see above), which may pose additional management complications for both predator and prey species.

14.2.1 Golden Eagles

One of the largest raptors in North America, the golden eagle regularly occurs in and largely relies on western rangelands. Golden eagles are year-round residents in much of the western North America and their breeding range extends from Alaska and the Canadian arctic to Mexico (Katzner et al. 2020a, b). Golden eagle populations are

Table 14.1 List of avian predator species occupying rangeland in western North America, typical rangeland habitat association, and population status (IUCN; www.iucn.org)

General avian predator class				
Order	*Genus Species*	Season	Rangeland	Status
Common name			Type	
Dirunal Raptors				
Accipitridae				
Golden Eagle	*Aquila Chrysaetos*	Year-round	Shrublands, Grasslands	Stable/ Declining
Bald Eagle	*Haliaeetus leucocephalus*	Year-round	Grasslands	Increasing
Ferruginous Hawk	*Buteo regalis*	Breeding	All	Stable/ Declining
Swainson's Hawk	*Buteo swainsoni*	Breeding	Grasslands	Unknown
Rough-legged Hawk	*Buteo lagopus*	Winter	Grasslands, Shrublands	Unknown
Red-tailed Hawk	*Buteo jamacensis*	Year-round	All	Stable
Harris's Hawk	*Parabuteo unicinctus*	Year-round	Desert, shrublands	Stable/ Increasing
Northern Harrier	*Circus hudsonius*	Year-round	Grasslands, shrublands	Stable
Mississippi Kite	*Ictinia mississippiensis*	Breeding	Grasslands	Stable/ Increasing
White-tailed Kite	*Elanus leucurus*	Breeding	Grasslands	Declining
Falconidae				
Prairie Falcon	*Falco mexicanus*	Year-round	All	Unknown
Aplomado Falcon	*Falco femoralis*	Year-round	Desert	Critically endangered
American Kestrel	*Falco sparverius*	Year-round	All	Declining
Merlin	*Falco columbarius*	Year-round	Grasslands	Unknown
Peregrine Falcon	*Falco peregrinus*	Year-round	Shrublands	Increasing
Gyrfalcon	*Falco rusticolus*	Winter	Grasslands	Stable/ Increasing
Crested Caracara	*Caracara cheriway*	Year-round		Unknown
Nocturnal Raptors				
Strigidae				
Burrowing Owl	*Athene cunicularia*	Breeding	Grasslands	Declining
Short-eared Owl	*Asio flammeus*	Year-round	Shrublands, grasslands	Unknown
Elf Owl	*Micrathene whitneyi*	Year-round	Desert	Unknown

(continued)

Table 14.1 (continued)

General avian predator class				
Order	Genus Species	Season	Rangeland	Status
Common name			Type	
Ferruginous Pygmy Owl	Glaucidium brasilianum	Year-round	Desert	Declining
Long-eared Owl	Asio otus	Year-round	Shrublands, grasslands	Unknown
Great-horned Owl	Bubo virginianus	Year-round	All	Stable
Snowy Owl	Bubo scandiacus	Winter	Grasslands	Declining
Tytonidae				
Barn Owl	Tyto alba	Year-round	Shrublands, grasslands	Unknown
Corvids				
Corvidae				
Common Raven	Corvus corax	Year-round	All	Increasing
Chihauhuan Raven	Corvus cryptoleucus	Year-round	Desert	Stable
American Crow	Corvus brachyrhynchos	Year-round	All	Increasing
Black-Billed Magpie	Pica hudsonia	Year-round	Shurblands, grasslands	Stable
Vultures				
Cathartidae				
Turkey Vulture	Cathartes aura	Breeding	All	Increasing
Black Vulture	Coragyps atratus	Year-round	All	Increasing
California Condor	Gymnogyps californianus	Year-round	Shurblands, desert	Critically endangered
Gulls				
Laridae				
Franklin's Gull	Leucophaeus pipixcan	Year-round		Stable
Ring-billed Gull	Larus delawarensis	Year-round		Stable
California Gull	Larus californicus	Year-round		Stable

believed to be stable in North America in recent decades (Millsap et al. 2013). Despite apparent stability of golden eagle populations in the western US from 1968 to 2014 (Millsap et al. 2013), population projections suggest current rates of human-caused mortality are sufficient to cause a decline in the future (Millsap et al. 2022). Golden eagles use a wide range of open habitats where they prey primarily on mammals and nest on cliffs and/or trees. This contrasts with significantly increasing populations of bald eagles (*Haliaeetus leucocephalus*), whose habitat is more strongly associated with lakes and rivers that provide foraging habitat and nesting trees. The long-lives and delayed reproduction of golden eagles, coupled with diverse migratory strategies, results in multiple population segments co-occurring and potentially competing in western rangelands. These include resident breeders that hold territories year-round, sub-adult residents that may occupy larger yearly ranges or wander, and a vast number of migrants that breed across Canada and Alaska and migrate long distances into the conterminous U.S. every winter.

Eagles have been targets of widespread human persecution across rangelands (Bedrosian et al. 2019) even though they receive a special degree of legal protection under the Bald and Golden Eagle Protection Act (BGEPA; 16 U.S.C. 668-668c). The largest source of mortality for golden eagles is natural starvation or disease for first-year eagles, anthropogenic poisoning for sub-adults, and shooting for adults (Millsap et al. 2022). This underscores the historic and continued, contemporary persecution of eagles. In recent years, both eagle species have been the subject of concern because of their vulnerability to mortality from collisions with wind turbines, with golden eagles especially at risk (Pagel et al. 2013). Eagles are also affected by other common risks to large raptors, including electrocution, vehicle collisions while foraging on road-kill, and lead poisoning. Golden eagles are particularly vulnerable to electrocution because of their broad wingspan and frequent use of power poles for perching in open habitats (Mojica et al. 2017).

Golden eagles are powerful and regular predators in most rangelands, preying mainly on rabbits, hares, ground squirrels, and prairie dogs but also taking larger prey such as antelope, deer, sheep, and young livestock (Bedrosian et al. 2017). However, they are also facultative scavengers, which exposes them to risks like lead poisoning from eating lead fragments in hunter-harvested game and gut piles (Bedrosian et al. 2012; Langner et al. 2016; Slabe et al. 2022) and vehicle collision when feeding on road-killed ungulates (Slater et al. 2022). Golden eagles may have increased mortality risk during winter when scavenging increases and secondary factors like sub-lethal lead intoxication occurs. Scavenging can also result in eagles congregating at ranches during lambing and calving to feed on afterbirth or stillborn livestock. This behavior can be associated with opportunistic predation on young livestock, but more often to the perception of livestock predation risk (Bedrosian et al. 2019).

Golden eagles rely on native rangelands for nesting and foraging habitat. Reproductive output is tied to inter-annual fluctuations in prey abundance, while occupancy of territories is consistent across long time periods (Kochert and Steenhof 2012). Fluctuations in productivity of golden eagles can be dramatic in areas where prey populations fluctuate cyclically or from epizootics (C. Preston, personal communication). Sage-steppe and grassland habitats host most breeding eagles in the coterminous US, which likely a function of where their main prey occur (Nielson et al. 2016). The proportion of golden eagles nesting in trees (both deciduous and coniferous) can be near 50% (Crandall et al. 2016) in heterogenous habitats and closer to 100% in the Great Plains, where loss of older-aged cottonwoods that provide some of the only nesting substrate is a conservation concern for golden eagles (Bedrosian et al. 2019). Cliff nests are often reused and as many as 39 years has been recorded between uses (Kochert and Steenhof 2012). Some nests are used regularly and for many generations. One golden nest on a basalt cliff in Montana was estimated to be more than seven meters tall and measured at > 500 years-old based on carbon dating of a stick from within the base of the nest (Ellis et al. 2009). It is suggested that nests be protected for at least 10 years after their last confirmed use (Kochert and Steenhof 2012). The longevity of nest sites highlights the importance of conserving raptor nesting sites and territories for generations.

14.2.2 Buteo Hawks

As large-bodied hawks, most Buteo species world-wide use a perch-and-pounce hunting strategy. However, perches have historically been limited across range-lands and the wing-shape, flight dynamics, and foraging strategy of Buteos that have evolved on rangelands is notable. The Buteos most associated with rangelands in the U.S. either employ a more aerial-based or ground-based hunting strategy than congenerics in forested and mixed habitat types. Buteo species co-occurring in rangeland systems partition habitat based on both nesting substrates (Restani 1991; Kennedy et al. 2014) and diet (MacLaren et al. 1988).

The most rangeland-specialist Buteo is arguably the ferruginous hawk, which is the largest hawk species in North America by mass and wingspan (Ng et al. 2020). This species has the greatest conservation need of Buteos regularly occurring in rangelands but other species, like Swainson's and Harris's hawks, may also have local conservation concerns. The broad wingspan of this grassland, shrub-steppe, and desert raptor allows for efficient long-term soaring to augment its perch and pounce hunting strategy. The large body size of ferruginous hawks is likely a reflection of the larger size of its main prey: jackrabbits, prairie dogs, and ground squirrels (Ng et al. 2020). Historically associated with open habitats in western rangelands, ferruginous hawks have adapted to nesting on the ground where trees, cliffs, or other elevated substrates are absent.

Like most raptors, ferruginous hawk populations are assumed to have significantly declined from historical abundance. While listed as a federally threatened species in Canada between 1980 and 1995 and again since 2010 due to declining numbers, most breeding populations are generally considered stable in the US with some evidence of recent declines in grassland habitats (Sauer et al. 2017). Like other raptors of grasslands and shrub-steppe, productivity of ferruginous hawks can fluctuate with prey populations and some regional declines have been linked to declines in prey species, like jackrabbits (Smith et al. 1981). There is some evidence to suggest this species is particularly sensitive to human disturbance (White and Thurow 1985; Keeley and Bechard 2011). In experimental trials in Canada, disturbance of nesting hawks was greater from foot traffic than vehicles and in more remote areas (Nordell et al. 2017). Loss and fragmentation of native habitats (Coates et al. 2014) by tillage agriculture (Schmutz 1987) has negatively affected breeding success of ferruginous hawks, possibly because of the sensitivity of this species to disturbance at nest sites.

Ferruginous hawks may, however, benefit from other anthropogenic modifications of their habitat, including some types of roads (Gilmer and Stewart 1983; MacLaren et al. 1988), prey in edge habitats (Zelenak and Rotella 1997; Keough and Conover 2012), and anthropogenic structures for perching and nesting (Steenhof et al. 1993; Keough and Conover 2012), including nest platforms installed for habitat enhancement and mitigation (Tigner et al. 1996; Wallace et al. 2016a, b). Previous studies on effects of roads and oil and gas well pads on ferruginous hawks are equivocal: some document positive relationships of productivity and roads (Zelenak and Rotella 1997), occupancy and roads (Neal et al. 2010; Wallace et al. 2016a; Squires et al. 2020), and occupancy and well pads (Keough and Conover 2012), whereas others report negative relationships of occupancy (Wiggins et al. 2017) and productivity with well pads (Harmata 1991; Keough 2006), and no apparent response of occupancy (Wallace et al. 2016a) or breeding success (Van Horn 1993; Wallace et al. 2016b) to well pads.

Habitat selection and breeding performance of ferruginous hawks are also influenced by natural factors, including vegetative cover (Wallace et al. 2016a; Squires et al. 2020), prey abundance (Smith et al. 1981; Schmutz et al. 2008), congeneric competition (Restani 1991), spring weather (Gilmer and Stewart 1983; Wallace et al. 2016b), and availability of nesting substrates (Kennedy et al. 2014). Somewhat unique to ferruginous hawks, individuals engage in wide-ranging nomadism post-breeding before their fall migration (Watson and Keren 2019; author unpublished data). The late-summer nomadic movements can be > 800 km in the opposite direction of their winter range (Watson et al. 2018). All seasonal habitats are typically associated with prey habitat (i.e., prairie dogs) but may be more tied to agriculture during the winter months compared to breeding habitats.

14.2.3   Burrowing Owls

Owls are largely nocturnal predators that occur in diverse habitats, where they take a wide variety of prey from insects to medium-sized mammals. Species in North American rangelands vary from rare habitat specialists, like the burrowing owl, to widespread generalists, like the great-horned owl, and their distributions span the continent from the Arctic tundra breeding grounds of the snowy owl to the preferred nesting habitat of the elf owl (*Micrathene whitneyi*) in cacti of the Sonoran Desert.

The western subspecies of the burrowing owl can be considered one of the few raptor species that are reliant on rangelands, with breeding habitat in open grasslands, shrub steppe, and deserts (Shaffer et al. 2022; Poulin et al. 2020). This semi-fossorial owl is dependent on burrowing mammals, such as prairie dogs and ground squirrels, to excavate tunnels that the owl uses as nest sites. Burrowing owls prefer short-grass prairies or areas where vegetation has been grazed to short heights, and most often select burrows in active prairie dog colonies. Colonies abandoned for even one year can be unsuitable for owls (Shaffer et al. 2022). This species generally avoids agricultural areas in native rangelands but does occur in both agricultural and urban habitats in some portions of its range. For example, in Canada, burrowing owls are almost exclusively found in native grasslands (Poulin et al. 2005), while owls in southern Idaho have benefited from being near irrigated agriculture due to increased prey densities there (King and Belthoff 2001). Native pastures may be used more readily for nesting than re-seeded, historical croplands (e.g., Conservation Reserve Program) (Shaffer et al. 2022). Burrowing owls are listed as endangered in Canada, threatened in Mexico and Colorado, and a species of concern for most other states. The species faces primary threats such as native habitat conversion and loss of prairie dog and ground squirrel colonies from control measures and plague. As a migrant species, they also have added habitat needs and management challenges across seasons. Because invertebrates comprise the most frequent prey for burrowing owls, insecticides can significantly reduce reproductive success or lead to direct and indirect mortality (James and Fox 1987). Similarly, pesticides applied to control mammals may also affect survival and reproduction of owls (James et al. 1990).

14.2.4 Corvids

The well-documented intelligence of ravens, crows, and magpies make them adaptable, effective predators and opportunists in North American rangelands. Currently, corvids present some of the greatest conservation challenges to other sensitive species across rangelands since they, themselves, are native species. Corvids are often considered "invasive species" by the public and some managers since their populations are rapidly expanding across many rangeland habitats, primarily because of human alterations of the landscape (Boarman and Heinrich 2020). Ravens typically nest on trees and cliffs but take advantage of many other available nesting structures. In many areas with energy development, ravens have adapted to nesting on oil and gas infrastructure or other anthropogenic substrates, including abandoned buildings, windmills, power lines, billboards, and virtually any elevated structure in rangeland habitats where food and water subsidies exist. This species now presents conservation challenges by predating various native wildlife species of concern, including greater sage-grouse and desert tortoise (*Gopherus agassizii*) (Boarman and Heinrich 2020). Similarly, the distribution of the American crow (*Corvus brachyrhynchos*) has expanded since European settlement as clearing of forests, expansion of agriculture, planting of trees in the Great Plains, and urban sprawl have created more open, human-altered habitats. Crows are efficient avian nest predators and are also considered agricultural pests in some areas, where flocks damage crops, like grains and tree fruits (Verbeek and Caffrey 2020).

14.3 Population Monitoring

Because avian predators typically occur in low densities and have large home ranges, monitoring population trends and status can be difficult at large scales. While citizen-science counts, such as the Breeding Bird Survey, Christmas Bird Count, and Mid-winter Bald Eagle Count, can inform long-term trends for many species, data for rarely encountered species, like most raptors, are difficult to interpret and do not capture variation away from roadways. Raptor space use does not conform to typical bird conservation administrative boundaries that many broad monitoring efforts have been based on, e.g., Bird Conservation Regions, flyways, Migratory Bird Joint Ventures, Landscape Conservation Cooperatives (Brown et al. 2017). Standardized data for nesting raptors and information sharing among agencies, industry, and non-governmental organizations has been lacking but would significantly help management actions across rangelands. There have been several publications dedicated to terminology used in nest monitoring (Steenhof 2017), some states have raptor working groups (UT, WY, CA) and Wyoming recently developed state-wide monitoring protocols and datasheets to address this concern. More widespread coordination of population monitoring efforts across the West would be beneficial to many agencies and conservation actions.

Recently, the USFWS developed a range-wide monitoring program to assess the status of Golden Eagles in the western U.S. (Millsap et al. 2013; Nielson et al. 2014; Nielson et al. 2016). This significant effort uses a combination of distance sampling (Thomas et al. 2010) and mark-recapture methods (Borchers et al. 2006) to estimate the population size detection probability for golden eagles observed during standardized aerial transect surveys. The method can be used to infer population trends at large scales (e.g., bird conservation regions). While this monitoring program is the first to undertake such a large-scale effort to estimate population trends for raptors at the population-level, the data are likely insufficient to detect trends at smaller spatial scales (e.g., State of Wyoming) without additional transects added. The difference in nesting timing between avian predator species in rangelands also contributes to inefficiencies in monitoring efforts. For example, golden eagles begin nesting in February–March while ferruginous hawks are not reliably back from their wintering grounds until May. Attempting to survey for multiple species simultaneously may miss early eagle nest failures or late arriving raptors without careful coordination.

14.4 Habitat Associations

14.4.1 Historical Habitat Use

Species composition was likely very different prior to European settlement of the western rangelands for several reasons. Most open-habitat raptors have evolved in varying landscapes of prairie and shrub-steppe ecosystems. While each ecological sub-region has varying conditions, several historical habitat features, such as fire regimes and mammal assemblages, likely influenced the composition and abundance of avian predator species. Most raptor species prefer foraging in shorter grasslands for easier visual access to prey, which were more abundant in fire-rich and/or ungulate-grazed prairie habitats. Native American and First Nations peoples may have helped maintain fire within some grassland systems to hunt bison *en masse* (Roos et al. 2018), both of which would have benefited avian predators and scavengers. The ecology of the grasslands of the plains and prairie states was largely different than current conditions due to the historical presence and abundance of ungulates and large carnivores. It is suspected that species like common ravens and California condors were more abundant across the grasslands in the nineteenth century as a result (Boarman and Heinrich 2020). Similarly, the historical widespread abundance and distribution of prey, such as large prairie dog and ground squirrel colonies and prairie grouse populations likely supported large raptor populations.

Another important factor influencing avian predator distribution and abundance is the large negative effect of direct human persecution. Virtually all avian predators were actively persecuted throughout the late nineteenth and the majority of the twentieth century. Government bounties existed for most raptors and shooting, trapping, and poisoning of predators was encouraged and practiced for generations (Madden et al. 2019). Secondary poisoning and trapping of raptors from mammalian predator control was also widespread across rangelands.

Finally, habitat conversion from native grasslands and shrub-steppe to agriculture was a significant driver of species abundance and composition that remains a management challenge today. Much of middle North America was historically prairie habitat that likely supported large populations of nesting raptors that currently are considered rangeland species. As native habitats were lost to urbanization or converted to agriculture, the species relying on these habitats either adapted to new habitat types or experienced range contraction and population declines.

14.4.2 Contemporary Habitat Use

As with most wildlife species, the ability to quickly alter behaviors with changing habitats and perturbations has been a major driver of species' abundance and distribution in rangelands. While some habitats and resources have remained intact, like large swaths of sage-steppe in Wyoming, all ecosystems have been altered in one way or another. Bison and other native large ungulates have largely been replaced with livestock and large mammalian predators have been removed or significantly reduced in abundance in almost all systems. Many habitats have been converted from native grasslands and shrub-steppe to agriculture or development. This conversion both reduces nesting habitat, particularly for ground-nesting raptors, and can reduce and alter prey populations. The remaining grasslands and shrublands provide extant rangelands but are threatened by invasive plant species like cheatgrass (*Bromus tectorum*), which has significantly altered the severity and frequency of fire and is subsequently affecting raptor productivity (Slater et al. 2013).

Undoubtedly, the avian predator guild species composition has been drastically shifting over the last century and has been a continual management challenge for conservation-reliant prey species like greater sage-grouse and desert tortoise. Specialized species historically associated with native habitats, but now associated with remaining rangelands, are generally struggling to maintain population viability and more human-tolerant species are replacing those specialized species. Similarly, specialized prey populations (e.g., sage-grouse) have been declining for similar reasons but are further exacerbated by this shift in the avian predator community. Without exception, all avian predator species currently in North America's rangelands have been historically present, albeit in different densities over time. Increased human presence and alteration of the landscape has negatively influenced the occurrence of sensitive species, fragmented habitat that influenced prey populations, directly reduced both prey and predator populations, increased anthropogenic subsidies of water, food, and nesting structures. As each species reacts differently to these factors, the guild dynamics of avian predators continues to shift towards those species tolerant of and subsidized by human activities.

14.5   Rangeland Management

The management of each species is unique, and recommendations required for successful avian predator ecosystem management include the species-specific management, in addition to management of prey populations and anthropogenic use of rangelands. Management of grazing, fire, invasive species, habitat patch size, and many other factors can be unique for many species, but generalizations can be applied to two basic sub-groups of avian predators in rangelands: human-intolerant and human-subsidized species. Across the West, the general trend over time has been increasingly challenging for conserving the former while reducing the latter.

14.5.1   Livestock Grazing

The largest ecological connection between livestock grazing and avian predator species abundance and richness is the interaction of grazing with prey species. While little work has been dedicated directly to the link between grazing and raptors, Johnson and Horn (2008) found that raptor abundance decreases in grazed pastures of mesic coastal grasslands in California because of lower rodent density. In Mediterranean grasslands, an experiment that increased European kestrel (*Falco tinnunculus*) populations in grazed pastures did not additively decrease small mammal abundance or richness, meaning grazing was the driving factor in small mammal declines, not avian predators. World-wide, high- and medium-intensity grazing has been shown to decrease small mammal abundance and diversity by reducing available forage and increasing soil compaction (e.g., Eccard et al. 2000; Saetnan and Skarpe 2006; Torre et al. 2007; Cao et al. 2016). In any system with decreased small mammals, the abundance of raptors dependent on those prey will follow similar trends.

Often, confounding factors of annual precipitation and wildfire have been interwoven with studies assessing grazing effects on small mammal communities (Yarnell et al. 2007; Bock et al. 2011), with low-intensity grazing and fires appearing to have some benefit to small mammal diversity in some areas. However, it appears that grazing has a larger impact on small mammal communities than both wildfire and meadow wetness (Horncastle et al. 2019).

The simple presence of livestock on rangelands can also be directly correlated with abundance of human-subsidized predators, like common ravens. The odds of raven occurrence can increase as much as 45% when free-range livestock are present compared to similar habitats without cattle (Coates et al. 2016). This association is not clear but may be linked to water provided to cattle or increased insect availability for foraging around and under cattle fecal piles.

14.5.2 Predator Control

Beyond the ecological connection, humans and avian predators have a long, adversarial history on rangelands. Because raptors can, and do, prey on livestock, gamebirds, and sensitive wildlife species, there have been illegal and legal control actions taken against avian predators. In a survey of 274 ranchers in Wyoming, Scasta et al. (2017) found that avian predators accounted for 19% of all livestock losses in the year prior to the survey. However, this survey also included turkey vultures (*Cathartes aura*) as an avian predator. While turkey vultures are technically considered raptors (McClure et al. 2019), they are an obligate scavenger incapable of killing livestock (Kirk and Mossman 2020) and its inclusion in the study highlights the inaccurate assumption that avian scavengers (e.g., eagles, ravens, magpies) feeding on a carcass or afterbirth predated that animal. The inability to distinguish scavenging from predation can lead to both artificially inflated estimates of avian predation on livestock and to continued negative perceptions of avian predators (Scasta et al. 2017).

Eagles are federally protected species by both the Migratory Bird Treaty Act (MBTA) and the BGEPA. Corvids and other raptors are all protected by the MBTA, but American crows are also considered a game and/or varmint species in many states and can be legally harvested during particular seasons, often times without a hunting licenses or bag limits (e.g., Wyoming). Common ravens are not a game species in any state and misidentification from hunters between crows and ravens can be a problem in areas where the species ranges overlap. However, regardless of their legal protections, many avian predators are illegally poisoned, shot, and trapped. Additionally,

USDA Wildlife Services and local animal control boards have permissions to control tens-of-thousands corvids across the West suspected of impacting both livestock and game species, such as greater sage-grouse.

Golden eagles can predate young calves, lambs, ewes, and rams and are greatest threats in open country lambing operations and predation events typically involve young lambs or goats (Phillips and Blom 1988; Matchett and O'Gara 1991; Avery and Cummings 2004). In just one year of low leporid (i.e., rabbit and hare) abundance in South Dakota, golden eagles were verified to depredate at least 142 lambs from seven ranches (Waite and Phillips 1994). Legal action for ranchers is typically very onerous and time consuming, which can involves hiring local, state, or federal control officers for species other than eagles, and all means of avian predator abatement have shown very little effectiveness (Scasta et al. 2017). Several historical attempts to relocate golden eagles to reduce lamb predation have occurred, but have not been successful, with most adults (12 of 14) returning within 11–316 days, even after being moved > 400 km (Miner 1975; Phillips and Blom 1988; O'Gara and Rightmire 1987; Phillips et al. 1991). One study relocated 432 golden eagles from ranchlands near Butte, Montana, but the effort resulted in little to no effect on lamb depredation rates from 1975 to 1983 (Avery and Cummings 2004). This lack of ability for producers to deal with predation from raptors can lead to animosity and/or illegal killing of these avian predators. Increased research and experimentation to determine how to control depredation more efficiently will help alleviate this issue.

14.5.3   Fire

Given the diverse habitat associations of avian predators across rangelands, fire can have varied effects for different species. In general, both prescribed and wild fire can have negative effects on sensitive species of nesting raptors (e.g., Marzluff et al. 1997; Kochert et al. 1999). Fires during the nesting season can destroy cover and active ground nests for species like northern harriers, short-eared owls, and ferruginous hawks (Johnson et al. 2019). Large wildfires can also destroy the few nesting trees that occur across rangelands thereby eliminating nesting territories if no other nesting structure is present.

Fires not only affect nesting structures, but also prey populations that can indirectly affect raptors. Because fires in scrublands significantly change the habitat type, fires in those ecotypes may affect raptors to a greater degree than in prairies and grasslands which are more adapted to fires. For example, golden eagles in shrubland habitats had significantly reduced productivity post-fire (Kochert et al. 1999). The increase in annual invasive grasses (i.e., cheatgrass) further exacerbated this issue by both reducing prey habitat and increasing future fire risk. In a 44-year study of golden eagle productivity and diet relative to wildfires, Heath et al. (2021) found that eagles were able to shift their diets from typical, preferred scrub prey species (i.e., lagomorphs) to a more diverse diet, but at the expense of productivity due to novel diseases which may result in negative population-level effects.

14.5.4 Water Subsidies

Providing artificial water sources in rangelands for livestock also can impact avian predators, both in positive and negative ways. Anthropogenic sources of water can be very beneficial in many arid and semi-arid landscapes. However, most raptor species obtain their water through their food and do not often directly drink water. Corvids and gulls, however, can significantly benefit from these water sources in an otherwise inhospitable environment (Kristin and Boarman 2007). This type of anthropogenic subsidy may affect abundance of these species in habitats they otherwise would not occur in. Raptors and other wildlife have been observed drowning in stock tanks when escape ladders are not provided. Raptors likely use these stock tanks for bathing and cooling but can have a difficult time escaping after becoming wet. A variety of simple ladders and ramps are readily available and increasingly used to mitigate this source of mortality (Rocky Mountain Bird Observatory 2006).

14.6 Impacts of Disease

Raptors have been found to be particularly susceptible to strains of avian influenza (Shearn-Bochsler et al. 2019) and West Nile virus (Nemeth et al. 2006) but transmission in rangelands is likely limited. Avian influenza can be transmitted from exposure between poultry farms and waterfowl. West Nile virus is likely more of a concern for both raptors and corvids in rangelands, with transmission occurring from foraging on infected prey and from mosquito transmission while nesting. Walker and Naugle (2011) provide an overview of West Nile virus ecology in sagebrush habitats and Bedrosian et al. (2019) and Wallace et al. (2019) provide reviews of transmission incidence in golden eagles on rangelands that can likely be applied to other raptor species. West Nile virus occurrence has also been documented in ferruginous hawks (Datta et al. 2015), burrowing owls and American kestrels (Dusek et al. 2010), and likely occurs in most raptor species. The larval habitats of the main mosquito vector for West Nile virus (*Culex tarsalis*) are small areas of standing water (< 4 ha) with high organic matter (Beehler and Mulla 1995). Most mosquito breeding areas in rangelands are created by human activities, including livestock watering ponds, water-storage areas, and discharge watering ponds in coalbed methane extraction regions (Denke and Spackman 1990). Irrigated agricultural sources of larval ponds produce significantly less mosquitos than coalbed methane ponds and outlets, and for a shorter duration (Doherty 2007). Further, Zou et al. (2006) mapped potential mosquito breeding areas through remote sensing in the Powder River Basin and found a 75% increase in area of potential larval habitats from 1999 to 2004, particularly in coalbed methane extraction areas. Drought may exacerbate outbreaks by concentrating mosquitos in restricted water sources from anthropogenic sources. Prevalence of outbreaks is predicted to increase by 2050 in the West (Harrigan et al. 2014) with substantial increases because of climate change (Schrag et al. 2011).

14.7 Ecosystem Threats

Anthropogenic changes to the North American rangelands can have drastically varying effects on avian predator species. Habitat fragmentation and/or loss through conversion to agriculture or development can lead to declines in species typically associated with native habitats, such as ferruginous hawks and golden eagles, while simultaneously increasing more human-tolerant and dependent species like common ravens, California gulls, great horned owls, and red-tailed hawks. The interconnected nature of these shifts is difficult to tease apart into direct, cause-effect relationships (see above). As rangelands and their predator composition change, the ecosystem-level changes further threaten sensitive species reliant on native rangeland.

14.7.1 Human-Persecution

Avian predators have a long-standing, typically negative, association with humans across all rangelands. Raptors have had long-standing direct conflict with people because they are predators of livestock (economic loss), wildlife people hunt (direct competition), and sensitive species (management conflicts). Corvids have a long-standing negative association within many European cultures dating back to medieval times when ravens and crows fed on human corpses after battles and during the bubonic plague (Król and Hernik 2020).

Indirectly, many raptors are affected by persecution of their prey too, which are often suppressed by humans due to competition for limited forage with livestock. Further, a minority of recreational shooters of small game and varmints illegally shoot non-game animals, often avian predators, in rangelands (Katzner et al. 2020a, b). Because of the link between avian predators and sensitive species that have cascading management implications and restrictions to private lands and industry, this increases pressure to reduce avian predator populations. Increased management and raptor-specific restrictions on extractive industries like mining, oil, and gas development has further created a negative sentiment for raptors across rangelands due to concerns over economic losses. All these reasons contribute to the ongoing direct human persecution through illegal shooting, poisoning, and trapping of avian predators, even with additional federal protections in place.

Raptors are most vulnerable to persecution while incubating or activities otherwise associated with nesting behavior. Most raptors nesting in rangelands have large, conspicuous nests because there are few places to hide a large stick nest. This conspicuousness and increased human persecution have likely led to behavior responses that avoid humans. As such, some raptors exhibit a large degree of sensitivity and increase flushing rates from their nests when associated with anthropogenic disturbances (Keeley and Bechard 2011). As human occurrence increases in rangelands, these sensitivities become exacerbated and may lead to population declines for human-intolerant species.

14.7.2 Habitat Conversion and Invasive Species

Conversion of native grassland and shrubland habitats to cultivated crops is a large driving factor in predator species composition in rangelands. Habitat conversion to agriculture tends to reduce foraging habitat for most raptors by reducing native prey habitat (e.g., scurids and lagomorphs). The Great Plains have already experienced significant habitat loss since European settlement and is expected to increase in the next 100 years due to climate change making grasslands more suitable for agriculture and increasing demand for biofuels (Sleeter et al. 2012; Sohl et al. 2012). Further, livestock grazing, which can reduce prey populations (see 14.5.1), and hay production is predicted to increase by 270% in the northwestern Great Plains by 2050 (Sleeter et al. 2012), which will likely alter raptor distribution and abundance. Expansion of invasive plant species, like cheatgrass, also creates a monoculture not conducive to prey habitat needs and increases wildfire frequency (Vilà et al. 2011; Bachen et al. 2018), which in turn affects abundance and reproductive rates of raptors.

Habitat conversion in arid rangelands can change the complexity of species richness, ecological diversity, and functional diversity of raptors (Tinajero et al. 2017). Increasing agricultural area can decrease functional diversity, or the component of biodiversity that influences ecosystem dynamics, stability, and ecosystem functioning (Tilman 2001). Because most raptor territories are in areas of complex habitat structure, shifts towards increasing agriculture reduce this complexity and favors more generalist species (Tinajero et al. 2017). Like other birds, raptor size and diet specificity appear to be the most important factors tied to tolerance of modified environments, with larger and more specialized species being more at risk (Sekercioglu 2012; Tinajero et al. 2017).

14.7.3 Energy Development

Unlike agricultural conversion that directly replaces habitat, energy development poses different challenges for avian predators through increased fragmentation and human presence (Shaffer et al. 2019). In areas of oil and gas development where vertical structures become available, species with increased tolerance of human disturbance at nest sites can sustain, or increase, in abundance (e.g., common ravens). Further, as fragmentation and human traffic increases, non-native plant species like cheatgrass can increase in abundance and further exacerbate these changes to prey habitat and subsequent raptor communities (see above).

Wind development across rangelands has the potential for significant impacts on raptor species, particularly golden eagles (Katzner et al. 2017; Millsap et al. 2022) due to the direct mortality of raptor collisions with turbine blades (Pagel et al. 2013). This new energy sector is poised to rapidly increase across rangelands due to a variety of factors, including increases in direct mortality, habitat fragmentation, human presence/vehicles, and power distribution.

Power distribution (e.g., power and transmission lines) infrastructure is another significant anthropogenic feature affecting occurrence and survival of avian predators in rangelands (Bedrosian et al. 2020). Distribution poles for water pumps and other power needs provide both perch and nesting sites in landscapes that would otherwise be devoid of vertical structure. Legacy distribution poles that were constructed before the *Avian Power Line Interaction Committee* guidelines (APLIC and USFWS 2005) have higher risk for electrocutions for large raptors. Similarly, transformers at dead-end poles used for water pumps that are unprotected also pose significant risk of electrocution due to exposed wiring. A bird, while landing or taking off, is at risk of touching two exposed energized parts that will cause electrocution. Further, a raptor may catch fire during electrocution and fall to the ground below the pole, causing a wildfire.

14.8 Conservation and Management Actions

14.8.1 Loss and Fragmentation of Rangeland

Similar to most rangeland wildlife, habitat change is the underlying force for most avian predator conservation issues in rangelands. Conversion of native habitats (e.g., to agriculture or invasive plant monocultures) and fragmentation from anthropogenic development both alter prey population dynamics and avian predator occupancy and abundance. The shift from low intensity use, such as livestock production, to higher intensity uses like oil and gas development have cumulative negative impacts on raptor species that require large territories to meet their survival and demographic requirements. Most shifts in habitat and anthropogenic use are trending to favor more generalist avian predators with a tolerance for human alterations across the landscape. More specialized raptors that require large expanses of habitat with little human use are becoming increasingly at risk. The most cost-efficient management is to protect the highest priority habitats, in other words, conserve large intact contiguous range-lands. Although restoration activities are commendable, for many avian predators that rely on rangelands the large spatial scale needed often makes full habitat restoration cost prohibitive.

14.8.2 Predator Management

Predator communities in rangelands have been altered due to human-induced habitat change and anthropogenic subsidies pose the greatest management challenge for sensitive prey species. For example, in the Chihuahua Desert, human-provided subsidies of food, water, and nest sites has caused a significant increase of common ravens (Kristan and Boarman 2007). Concurrently, desert tortoise populations have been declining from habitat loss, disease, and other perturbations at such a rate that has caused them to become a federally listed endangered species. Because common ravens are very successful generalists, their increased abundance in tortoise habitat has led to greater predation on young tortoises (Kristan and Boarman 2003). These two independent factors have now led to a significant ecosystem conflict for two federally protected species on opposite population trajectories. Similar conflicts with ravens are widespread in greater sage-grouse sagebrush rangelands and has resulted in raven control efforts to mitigate increased predation pressure on grouse nests and chicks (Dinkins et al. 2016). Across the West, USDA Wildlife Services objectives are to control > 11,000 and displace > 125,000 ravens a year for livestock and sensitive species conflicts (https://www.aphis.usda.gov/aphis/ourfocus/wildlifedamage/SA_Reports/SA_PDRs). While these control actions may help local-level issues temporarily, they fail to address the ultimate cause of increasing raven populations: habitat alterations by humans.

Predator control has been a common management action to limit human-subsidized avian predator species and may seem less challenging compared to habitat conservation or restoration. However, studies have shown that direct control may reduce raven abundance for short periods in localized areas (Coates et al. 2007; Dinkins 2013), but the effectiveness of long-term suppression on management objectives (i.e., benefits to livestock and/or prey species) has yet to be demonstrated. Moreover, adult breeding ravens are the cohort most often occupying native rangelands with at-risk species and tend not to use subsidies during the breeding period (Bui et al. 2010; Harju et al. 2018). On the other hand, lethal control efforts are largely conducted in areas with high raven concentrations, such as roosts, landfills, and areas with other subsidies, to increase efficiency and effectiveness of the control efforts. However, breeding individuals may rarely be targeted due to their wariness of humans and their relatively reduced use of these subsidies (Harju et al. 2018).

Similar management efforts have occurred, and are continuing, to mitigate sheep predation by golden eagles (see 14.5.2). While eagle relocations have not been successful, some management actions have been, including installation of netting over lambing pens, using "scarecrows" on ridges where lambs bed for the night, removing dead livestock and other potential eagle attractants, and the use of guard dogs all have helped minimize or curtail eagle-lamb depredations (O'Gara and Rightmire 1987).

14.8.3   Management of Direct and Indirect Mortality

A major management objective to benefit raptor, corvid, and vulture populations in rangelands should include reducing anthropogenic-caused mortalities, including illegal persecution, vehicle strikes, turbine collisions, electrocutions, and poisoning. Road-killed ungulates should be moved at least 12 m from roadways to reduce raptor-vehicle collision risk when scavengers are feeding on roadkill (Slater et al. 2022). Lead-free ammunition should be used for both big-game, upland and varmint hunting to reduce secondary lead poisoning in raptors (Haig et al. 2014). Recent models of risk have been completed across the western rangelands as a function of power pole density (Dwyer et al. 2020) to help inform mitigation efforts to retrofit power poles in areas of high eagle breeding density. Retrofitting of power poles and transmission equipment can significantly reduce risk of electrocution can significantly reduce mortality risk if done correctly (Dwyer et al. 2015; Dwyer et al. 2017). Impacts from rodenticide poisoning can be avoided by discontinuing use in important habitats of raptors (Herring et al. 2017). Additionally, chemicals used to euthanize livestock are known to kill eagles (Viner et al. 2016). Poisoning by euthanasia agents can be avoided by burying, cremating, covering, or otherwise disposing of carcasses such that they are not available to scavengers. Management to benefit raptor populations in rangelands should include conserving and reducing disturbance to nesting and roosting habitats, following best practices energy development and other infrastructure, education to reduce human persecution, and conservation of native vegetation communities that support populations of prey species.

14.8.4   Habitat Management

Artificial nesting structures have been regularly used to increase nesting density and success for ground-nesting raptors, like ferruginous hawks and burrowing owls (Fig. 14.1). Because of the vulnerability of ground nests, increased traffic, human presence, noise from hunting/shooting activities, and land alterations have caused lower reproductive success. Nesting on anthropogenic structures has been linked to increased nest success and offers a mitigation tool in areas with at-risk populations (Wallace et al. 2016a, b). Artificial nesting platforms may also serve to reduce potential electrocutions of raptors when nesting on power poles. This management technique has been used successfully to relocate at risk ferruginous hawks and golden eagles (Kemper et al. 2020; G. McKee, personal communication). Similarly, providing artificial burrows for burrowing owls may help maintain and bolster populations in and adjacent to areas with habitat conversion or loss of burrowing mammals (Moulton et al. 2006; Menzel 2018). Conversely, nesting on anthropogenic structures has been an increasing problem for predators in conflict with sensitive species, like common ravens. Ravens regularly nest on human structures but will not nest on the

Fig. 14.1 Example of a ground nest built by a ferruginous hawk in western Wyoming (left) and an elevated artificial nesting structure in the same area (right). Photo credits: author

ground. While managing to reduce raven occurrence and abundance in rangelands through direct control the same managers could simply eliminate raven nest materials before eggs are laid, which would not violate the MBTA.

Recent conservation efforts on rangelands have been shifting to focus on identification and protection of priority habitats. The best example of this is not for raptors, but for greater sage-grouse. The Core Area Policy in Wyoming, for example, is designed to restrict development in the areas that host the largest number of breeding birds (Wyoming Executive Order 2019-3). The heart of this strategy is to identify and conserve the smallest areas that protect the largest number of birds. Similar efforts have been underway for golden eagles across their western range due to at-risk populations and the novel threat of increasing wind development in key eagle habitat (e.g., Dunk et al. 2019). Like the core area concept, protecting areas that host dense populations of breeding and wintering eagles will have disproportionately larger conservation benefits. For example, if 50% of all golden eagle nests in a state are located on only 10% of the landscape, then protecting that 10% will have greater benefit than conserving the other 90% of the state. With the priority area concept, it is extremely important to understand the entirety of a species' habitat requirements prior to delineating specific areas. For example, for multiple rangeland wildlife species managers have focused on breeding habitats while largely ignoring winter habitat or areas necessary for seasonal movement or genetic connectivity. For raptors, understanding migratory routes and seasonal ranges has been nearly impossible. But the contemporary advancement in tracking technologies for birds has largely filled this knowledge gap and we can now accurately assess habitat use and needs of migratory species. Because many rangeland raptors have intercontinental migratory and seasonal habitats, continued collaboration among countries, agencies, and other appropriate entities is essential to conserve the year-round habitat needs. A key example of this connectivity and need for international conservation of rangeland raptors is the Swainson's hawk, where pesticide use in South America in the

1990s threatened the population persistence of this species in North America and international efforts were successful in managing and largely eliminating this threat (Goldstein et al. 1999).

Prioritizing habitat conservation and protection prior to disturbance, especially direct habitat loss and fragmentation, will be key to conserving the long-term ecology of rangelands and is much simpler and less expensive than trying to restore degraded and altered habitats. Identifying and prioritizing areas of largely undisturbed intact habitat is critical for rangeland species due to limited remaining resources and the need of many species for large intact home ranges. Continuing low-density human use of rangelands, from a mix of livestock production, large ranches, and public lands, instead of increasing fragmentation from energy development and other anthropogenic uses, will be vital to maintaining biodiversity and ecological function within western rangelands.

14.9 Research Needs

There are many aspects of avian predator ecology that remain understudied. Most population status and trend estimates of raptors could be significantly improved. More information is needed on prey populations across rangelands and how to increase their populations. The complex and compounding relationship between changing prey density, climate change, and the interaction with avian predators needs further research. There is an increasing need for research concerning plague outbreaks in prairie dog populations and other raptor disease concerns like avian influenza and West Nile virus. The effects of both livestock grazing and vegetation treatments meant to support livestock production on raptor abundance and productivity on rangelands needs further attention. Research focusing on the interaction of grazing intensity, human presence, small mammal abundance, and the avian predator guild is currently lacking.

As in most regions, climate change has significant potential to alter rangeland systems, including avian predators that are associated with rangelands. As rangelands dry out with continued droughts, anthropogenic subsidies will become increasingly important to manage. Increasing fire frequency may hinder some rangelands (e.g., sage-steppe), while potentially benefiting others (e.g., grasslands). Heat-stress has also been shown to directly affect home range size and productivity of some rangeland raptors (Braham et al. 2015; Kochert et al. 2019). Secondary effects on seasonal shifts of prey (e.g., hibernation emergence or reproduction) and those effects on nesting raptors remain unknown.

Management actions centered on predator reduction still need more critical evaluation on their success, cost–benefit, scalability, and long-term success. More alternative actions, particularly non-lethal techniques like nesting deterrence options, need to be developed. Some ideas could be assessed, such as taste-aversion in corvids for grouse management, reducing anthropogenic nesting substrate and subsidies for generalist species, livestock herd protections (e.g., scarecrows for eagles), roadkill removal to reduce winter eagle abundance, and non-lethal control options. While there are situations where lethal control methods need to be employed, there is a need to understand the efficacy and efficiency of lethal control methods and non-target impacts.

Finally, new and emerging threats to rangelands will continue to increase, further reducing and fragmenting native rangelands. Understanding and mitigating effects prior to these disturbances will be critical in maintaining raptor populations in rangelands. For example, the renewable energy demands in the U.S. is likely to lead to an increase in wind power development and biofuels. Wind development is increasingly more prominent in western rangelands and can be a significant risk to raptors from direct collisions and habitat alteration. Understanding and prioritizing the entire landscape for raptors is essential for long-term management through identification and protection of critical habitats. Further, more compensatory mitigation options are needed to offset any losses from this development since power pole retrofitting to reduce electrocutions is the only currently accepted management action to offset eagle mortalities. Other options, such as using lead-free ammunition for hunting, road-kill removal, and breeding habitat enhancements will all benefit rangeland raptor management. Prioritizing and conserving critical and key habitats for all sensitive species, including, but not limited to raptors, and all life history phases for each species will greatly enhance management decisions for the multiple uses and threats the future rangelands will face.

References

Avery ML, Cummings JL (2004) Livestock depredations by black vultures and golden eagles. Sheep Goat Res J 19:58–63

Avian Power Line Interaction Committee [APLIC] and U.S. Fish and Wildlife Service [USFWS] (2005) Avian protection plan (APP) guide-lines. APLIC, Washington, D.C., USA

Bachen DA, Litt AR, Gower CN (2018) Simulating cheatgrass (Bromus tectorum) invasion decreases access to food resources for small mammals in sagebrush steppe. Biol Invasions. 20:2301–2311

Bednarz JC, Ligon JD (1988) A study of the ecological bases of cooperative breeding in the Harris' Hawk. Ecology 69:1176–1187

Bedrosian B (2004) Nesting and post-fledging ecology of the common raven in Grand Teton National Park, Wyoming. Master's Thesis, Arkansas State University

Bedrosian B, Craighead D, Crandall R (2012) Lead exposure in bald eagles from big game hunting, the continental implications and successful mitigation efforts. PLoS ONE 7(12):e51978

Bedrosian BE, Wallace Z, Bedrosian G et al (2019) Northwestern Plains Golden Eagle conservation strategy. Unpublished report prepared for the U.S. Fish and Wildlife Service Western Golden Eagle Team by Teton Raptor Center. Available online at https://ecos.fws.gov/ServCat/Reference/Profile/98141

Bedrosian G, Carlisle JD, Woodbridge B et al (2020) A spatially explicit model to predict the relative risk of Golden Eagle electrocutions in the Northwestern Plains, USA. J Raptor Res 54:110–125

Bedrosian G, Watson JW, Steenhof K, Kochert MN, Preston CR, Woodbridge B, Williams GE, Keller KR, Crandall RH (2017) Spatial and temporal patterns in golden eagle diets in the western United States, with implications for conservation planning. J Raptor Res 3:347–67

Beehler JW, Mulla MS (1995) Effects of organic enrichment on temporal distribution and abundance of culicine egg rafts. J Am Mosq Control Assoc 11(2 Pt 1):167–171

Boarman WI, Heinrich B (2020) Common Raven (Corvus corax), version 1.0. In: Birds of the world. Cornell Lab of Ornithology. https://doi.org/10.2173/bow.comrav.01

Bock CE, Jones ZF, Kennedy LJ, Bock JH (2011) Response of rodents to wildfire and livestock grazing in an Arizona desert grassland. Am Midl Nat 166:126–138

Borchers DL, Laake JL, Southwell C, Paxton CG (2006) Accommodating unmodeled heterogeneity in double-observer distance sampling surveys. Biometrics. 62(2):372–378

Braham M, Miller T, Duerr AE et al (2015) Home in the heat: dramatic seasonal variation in home range of desert golden eagles informs management for renewable energy development. Biol Cons 186:225–232

Brown JL, Bedrosian B, Bell DA et al (2017) Patterns of spatial distribution of Golden Eagles across North America: how do they fit into existing landscape-scale mapping systems? J Raptor Res 51:197–215. https://doi.org/10.3356/JRR-16-72.1

Bryan RD, Wunder MB (2014) Western burrowing owls (*Athene cunicularia hypugaea*) eavesdrop on alarm calls of black-tailed prairie dogs (*Cynomys ludovicianus*). Ethology 20:180–188

Bui TV, Marzluff JM, Bedrosian B (2010) Common raven activity in relation to land use in western Wyoming: implications for greater sage-grouse reproductive success. Condor 112(1):65–78

Cao C, Shuai LY, Xin XP et al (2016) Effects of cattle grazing on small mammal communities in the Hulunber meadow steppe. PeerJ 4:e2349

Coates PS, Howe KB, Casazza ML et al (2014) Landscape alterations influence differential habitat use of nesting buteos and ravens within sagebrush ecosystem: implications for transmission line development. Condor: Ornithol Appl 116:341–356

Coates PE, Brussee BE, Howe KB et al (2016) Landscape characteristics and livestock presence influence common ravens: relevance to greater sage-grouse conservation. https://doi.org/10.1002/ecs2.1203

Coates PS, Spencer Jr JO, Delehanty DJ (2007) Efficacy of CPTH-treated egg baits for removing ravens. Hum Wildl Conflicts 1(2):224–234

Craighead JJ and FC Craighead (1969) Hawks, Owls, and Wildlife. Stackpole Co., Harrisburg, PA and Wildlife Management Institute, Washington, DC. p 443

Crandall RH, Bedrosian BE, Craighead D (2015) Habitat selection and factors influencing nest survival of Golden Eagles in south-central Montana. J Raptor Res 4:413–428

Crandall RH, Craighead DJ, Bedrosian BE (2016) A comparison of nest survival between cliff-and tree-nesting Golden Eagles. J Raptor Res 50(3):295–300

Datta S, Jenks JA, Knudsen D, Jensen K, Inselman WM, Swanson CC, Grovenburg TW (2015) West Nile Virus and ferruginous hawks (Buteo regalis) in the northern Great Plains. Prairie Naturalist 47:38

Denke PM, Spackman EW (1990) The mosquitoes of Wyoming. Bulletin-Wyoming University, Coop Extension Serv (USA)

Dinkins JB (2013) Common raven density and greater sage-grouse nesting success in southern Wyoming: potential conservation and management implications. Doctoral Dissertation. Utah State University

Dinkins JB, Conover MR, Kirol CP et al (2016) Effects of common raven and coyote removal and temporal variation in climate on greater sage-grouse nesting success. Biol Conserv 202:50–58. ISSN 0006-3207. https://doi.org/10.1016/j.biocon.2016.08.011

Doherty MK (2007) Mosquito populations in the Powder River Basin, Wyoming: A comparison of natural, agricultural and effluent coal bed natural gas aquatic habitats. Doctoral dissertation, Montana State University-Bozeman, College of Agriculture

Donázar JA, Cortes-Avizanda A, Fargallo JA et al (2016) Roles of raptors in a changing world: from flagships to providers of key ecosystem services. Ardeola 63:181–234

Dunk JR, Woodbridge B, Lickfett TM, Bedrosian G, Noon BR, LaPlante DW, Brown JL, Tack JD (2019) Modeling spatial variation in density of Golden Eagle nest sites in the western United States. PLoS One 14(9):e0223143

Dusek RJ, Iko WM, Hofmeister EK (2010) Occurrence of West Nile virus infection in raptors at the Salton Sea, California. J Wildl Dis 46(3):889–897

Dwyer JF, Harness RE, Eccleston D (2017) Avian electrocutions on incorrectly retrofitted power poles. J Raptor Res 51(3):293–304

Dwyer JF, Kratz GE, Harness RE, Little SS (2015) Critical dimensions of raptors on electric utility poles. J Raptor Res 49(2):210–216

Dwyer JF, Bednarz JC, Raitt RJ (2020) Chihuahuan Raven (*Corvus cryptoleucus*), version 1.0. In: Poole AF (ed) Birds of the world. Cornell Lab of Ornithology, Ithaca, NY, USA. https://doi.org/10.2173/bow.chirav.01

Eccard JA, Walther RB, Milton SJ (2000) How livestock grazing affects vegetation structures and small mammal distribution in the semi-arid Karoo. J Arid Environ 46:103–106

Ellis DH, Craig T, Craig E, Postupalsky S, LaRue CT, Nelson RW, Anderson DW, Henny CJ, Watson J, Millsap BA, Dawson JW (2009). Unusual raptor nests around the world. J Raptor Res 43(3):175–198

Finkelstein ME, Doak DF, George D et al (2012) Lead poisoning and the deceptive recovery of the critically endangered California condor. P Natl Acad Sci USA 28:11449–11454

Finkelstein M, Kuspa Z, Snyder NF, Schmitt NJ (2020) California condor (*Gymnogyps californianus*), version 1.0. In: Rodewald PG (ed) Birds of the world. Cornell Lab of Ornithology, Ithaca, NY, USA. https://doi.org/10.2173/bow.calcon.01

Gilmer DS, Stewart RE (1983) Ferruginous hawk populations and habitat use in North Dakota. J Wildl Manag 47:146–157

Goldstein MI, Lacher TE, Zaccagnini ME et al (1999) Monitoring and assessment of Swainson's Hawks in Argentina following restrictions on monocrotophos use, 1996–97. Ecotoxicology 8:215–224. https://doi.org/10.1023/A:1026448415467

Haig SM, D'Elia J, Eagles-Smith C et al (2014) The persistent problem of lead poisoning in birds from ammunition and fishing tackle. Condor: Ornithol Appl 116(3):408–428

Harju SM, Olson CV, Hess JE, Bedrosian B (2018) Common raven movement and space use: influence of anthropogenic subsidies within greater sage grouse nesting habitat. Ecosphere (7):e02348

Harmata AR (1991) Impacts of oil and gas development on raptors associated with Kevin Rim. Monitoring report prepared for the Bureau of Land Management, Great Falls, Montana, USA

Harrigan RJ, Thomassen HA, Buermann W, Smith TB (2014) A continental risk assessment of West Nile virus under climate change. Glob Change Biol 20(8):2417–2425

Heath JA, Kochert MN, Steenhof K (2021) Golden Eagle dietary shifts following wildfire and shrub loss have negative consequences for nestling survivorship. Condor 123(4):duab034

Herring G, Eagles-Smith CA, Buck J (2017) Characterizing golden eagle risk to lead and anticoagulant rodenticide exposure: a review. J Raptor Res 51:273–292

Horncastle VJ, Chambers CL, Dickson BG (2019) Grazing and wildfire effects on small mammals inhabiting montane meadows. J Wildl Manag 83:534–543

James PC, Fox GA (1987) Effects of some insecticides on productivity of Burrowing Owls. Blue Jay 45(2)

James PC, Fox GA, Ethier TJ (1990) Is the operational use of strychnine to control ground squirrels detrimental to Burrowing Owls. J Raptor Res 24(4):120–123

Johnson DL, Henderson MT, Anderson DL, Booms TL, Williams CT (2022) Isotopic niche partitioning and individual specialization in an Arctic raptor guild. Oecologia 198(4):1073–1084

Johnson TN, Nasman K, Wallace ZP, Olson LE, Squires JR, Nielson RM, Kennedy PL (2019) Survey design for broad-scale, territory-based occupancy monitoring of a raptor: Ferruginous hawk (Buteo regalis) as a case study. Plos one 14(3):e0213654

Johnson MD, Horn CM (2008) Effects of rotational grazing on rodents and raptors in a coastal grassland. Western North Am Nat 68:444–452

Katzner TE, Carlisle JD, Poessel SA et al (2020a) Illegal killing of nongame wildlife and recreational shooting in conservation areas. Conserv Sci Pract 2(11):e279

Katzner TE, Kochert MN, Steenhof K et al (2020b) Golden Eagle (*Aquila chrysaetos*), version 2.0. In: Birds of the world. Cornell Lab of Ornithology. https://doi.org/10.2173/bow.goleag.02

Katzner TE, Nelson DM, Braham MA, Doyle JM, Fernandez NB, Duerr AE, Bloom PH, Fitzpatrick MC, Miller TA, Culver RC, Braswell L (2017) Golden Eagle fatalities and the continental-scale consequences of local wind-energy generation. Conserv Biol 31(2):406–415

Keeley WH, Bechard MJ (2011) Flushing distances of ferruginous hawks nesting in rural and exurban New Mexico. J Wildl Manag 75:1034–1039

Kemper CM, Wellicome TI, Andre DG, McWilliams BE, Nordell CJ (2020) The use of mobile nesting platforms to reduce electrocution risk to Ferruginous Hawks. J Raptor Res 54(2):177–185

Kennedy PL, Bartuszevige AM, Houle M et al (2014) Stable occupancy by breeding hawks (Buteo spp.) over 25 years on a privately managed bunchgrass prairie in northeastern Oregon, USA. Condor: Ornithol Appl 116:435–445

Keough H (2006) Factors influencing breeding ferruginous hawks (*Buteo regalis*) in the Uintah Basin, Utah. Dissertation Utah State University, Logan, USA

Keough HL, Conover MR (2012) Breeding-site selection by ferruginous hawks within Utah's Uintah basin. J Raptor Res 46:378–388

King AR, Belthoff JR (2001) Post-fledging dispersal of burrowing owls in southwestern Idaho: characterization of movements and use of satellite burrows. Condor 103(1):118–126

Kirk DA, Mossman MJ (2020) Turkey vulture (*Cathartes aura*), version 1.0. In: Birds of the world. Cornell Lab of Ornithology. https://doi.org/10.2173/bow.turvul.01

Kochert MN and Steenhof K (2012) Frequency of nest use by Golden Eagles in southwestern Idaho. J Raptor Res 46(3):239–247

Kochert MN, Steenhof K, Carpenter LB, Marzluff JM (1999) Effects of fire on golden eagle territory occupancy and reproductive success. J Wildl Manag 63:773–780. https://doi.org/10.2307/380 2790

Kristan WB III, Boarman WI (2003) Spatial pattern of risk of common raven predation on desert tortoises. Ecology 84:2432–2443. https://doi.org/10.1890/02-0448

Kristan WB, Boarman WI (2007) Effects of anthropogenic developments on common ravens nesting in the west Mojave Desert. Ecol Appl 17:1703–1713. https://doi.org/10.1890/06-1114.1

Król K, Hernik J (2020) Crows and ravens as indicators of socioeconomic and cultural changes in urban areas. Sustainability 12:10231. https://doi.org/10.3390/su122410231

Langner HW, Domenech R, Slabe VA et al (2016) Lead and mercury in fall migrant golden eagles from western North America. Arch Environ Contam Toxicol 69:54–61

León-Ortega M, Jiménez-Franco MV, Martínez JE, Calvo JF (2017) Factors influencing territorial occupancy and reproductive success in a Eurasian eagle-owl (Bubo bubo) population. PLoS ONE. https://doi.org/10.1371/journal.pone.0175597

MacLaren PA, Anderson SH, Runde DE (1988) Food habits and nest characteristics of breeding raptors in southwestern Wyoming. Great Basin Nat 48:548–553

Madden KK, Rozhon GC, Dwyer JF (2019) Conservation letter: raptor persecution. J Raptor Res 53:230–233. https://doi.org/10.3356/JRR-18-37

Marzluff JM, Knick ST, Vekasy MS et al (1997) Spatial use and habitat selection of golden eagles in Southwestern Idaho. Auk 114:673–687. https://doi.org/10.2307/4089287

Matchett MR, O'Gara BW (1991) Golden eagles and the livestock industry: an emotionally charged issue. Western Wildlands 17:18–24

McClure JCW, Schulwitz SE, Anderson DL et al (2019) Commentary: defining raptors and birds of prey. J Raptor Res 53:419–430

Menzel S (2018) Artificial burrow use by burrowing owls in northern California. J Raptor Res 52(2):167–177

Millsap BA, Zimmerman GS, Kendall WL, Barnes JG, Braham MA, Bedrosian BE, Bell DA, Bloom PH, Crandall RH, Domenech R, Driscoll D (2022) Age—specific survival rates, causes of death, and allowable take of golden eagles in the western United States. Ecol Appl 32(3):e2544

Millsap BA, Zimmerman GS, Sauer JR et al (2013) Golden eagle population trends in the western United States: 1968–2010. J Wildl Manag 77:1436–1448

Miner NR (1975) Montana Golden Eagle Removal and Translocation Project. In: Great Plains Wildlife Damage Control Workshop Proceedings 201. https://digitalcommons.unl.edu/gpwdcw p/201

Mojica EK, Dwyer JF, Harness RE, Williams G et al (2017) Review and synthesis of research investigating golden eagle electrocutions: golden eagle electrocutions. J Wildl Manag. https:// doi.org/10.1002/jwmg.21412

Moulton CE, Brady RS, Belthoff JR (2006) Association between wildlife and agriculture: underlying mechanisms and implications in Burrowing Owls. J Wildl Manage 70(3):708–716

Murphy RK, Dunk JR, Woodbridge B et al (2017) First-year dispersal of golden eagles from natal areas in the southwestern United States and implications for second-year settling. J Raptor Res 51:216–233. https://doi.org/10.3356/JRR-16-80.1

Murphy RK, Stahlecker DW, Millsap BA et al (2018) Natal dispersal distance of golden eagles in the southwestern United States. J Fish Wildl Manag 10(213–218):e1944–e2687. https://doi.org/ 10.3996/052018-JFWM-039

Neal MC, Smith JP, Slater SJ (2010) Artificial nest structures as mitigation for natural-gas development impacts to ferruginous hawks (*Buteo regalis*) in south-central Wyoming. U.S. Department of the Interior, Bureau of Land Management, Washington, D.C., USA

Nemeth N, Gould D, Bowen R, Komar N (2006) Natural and experimental West Nile virus infection in five raptor species. J Wildl Dis 42(1):1–3

Newton I (2010) Population ecology of raptors. Bloomsbury Publishing, London, UK, A&C Black

Ng J, Giovanni MD, Bechard MJ et al (2020) Ferruginous hawk (*Buteo regalis*), version 1.0. In: Rodewald PG (ed) Birds of the world. Cornell Lab of Ornithology, Ithaca, NY, USA. https:// doi.org/10.2173/bow.ferhaw.01

Nielson RM, Mcmanus L, Rintz T, Mcdonald LL, Murphy RK, Howe WH, Good RE (2014) Monitoring abundance of golden eagles in the western United States. J Wildl Manage 78(4):721–730

Nielson RM, Murphy RK, Millsap BA et al (2016) Modeling late-summer distribution of golden eagles (*Aquila chrysaetos*) in the Western United States. PLoS ONE 11(8):e0159271. https:// doi.org/10.1371/journal.pone.0159271

Nordell CJ, Wellicome TI, Bayne EM (2017) Flight initiation by Ferruginous Hawks depends on disturbance type, experience, and the anthropogenic landscape. PLoS ONE 12(5):e0177584

O'Gara, BW and Rightmire W (1987) Wolf, Golden Eagle, and coyote problems in Montana. In: Third Eastern Wildlife Damage Control Conference 42. https://digitalcommons.unl.edu/ewd cc3/42

Pagel JE, Kritz KJ, Millsap BA et al (2013) Bald eagle and golden eagle mortalities at wind energy facilities in the contiguous United States. J Raptor Res 47:311–315

Phillips RL, Cummings JL, Berry JD (1991) Responses of breeding golden eagles to relocation. Wildl Soc Bull (1973–2006). 19(4):430–434

Phillips RL, Blom FS (1988) Distribution and magnitude of eagle/livestock conflicts in the western United States. In: Proceedings of the thirteenth vertebrate pest conference. University of California, Davis

Poessel SA, Woodbridge B, Smith et al (2022) Interpreting long-distance movements of non-migratory golden eagles: prospecting and nomadism? Ecosphere 13:e4072. https://doi.org/10. 1002/ecs2.4072

Pollet IL, Shutler D, Chardine JW, Ryder JP (2020) Ring-billed gull (*Larus delawarensis*), version 1.0. In: Poole AF (ed) Birds of the world. Cornell Lab of Ornithology, Ithaca, NY, USA. https:/ /doi.org/10.2173/bow.ribgul.01

Poulin RG, Todd LD, Dohms KM, Brigham RM, Wellicome TI (2005) Factors associated with nest- and roost-burrow selection by Burrowing Owls (Athene cunicularia) on the Canadian prairies. Can J Zool 10:1373–1380

Poulin RG, Todd LD, Haug EA, Millsap BA, Martell MS (2020) Burrowing Owl (Athene cunicularia), version 1.0. InBirds of the World (AF Poole, Editor). Cornell Lab Ornithol Ithaca NY, USA

Restani M (1991) Resource partitioning among three Buteo species in the Centennial Valley, Montana. Condor, pp 1007–1010

Robillard A, Therrien JF, Gauthier G et al (2016) Pulsed resources at tundra breeding sites affect irruptions at temperate latitudes of a top predator, the snowy owl. Oecologia 181:423–433

Rocky Mountain Bird Observatory (2006) Stock tank ladders from Rocky Mountain Bird Observatory. Internal Report. http://www.rmbo.org/dataentry/postingArticle/dataBox/WildlifeEscapeLadder[1].pdf. Accessed 11 July 2022

Roos CI, Zedeño MN, Hollenback KL, Erlick MM (2018) Indigenous impacts on North American Great Plains fire regimes of the past millennium. Proc Natl Acad Sci 115:8143–8148

Saetnan ER, Skarpe C (2006) The effect of ungulate grazing on a small mammal community in southeastern Botswana. Afr Zool 41:9–16

Sauer, JR, Niven DK, Hines JE, Ziolkowski DJ Jr, Pardieck KL, Fallon JE, Link, WA (2017) The North American Breeding Bird Survey, results and analysis 1966–2015. Version 2.07.2017. U.S. Geological Survey, Patuxent Wildlife Research Center, Laurel, Maryland, USA

Scasta JD, Stam B, Windh JL (2017) Rancher-reported efficacy of lethal and non-lethal livestock predation mitigation strategies for a suite of carnivores. Sci Rep 7(1):1–11

Schmutz JK (1987) The effect of agriculture of ferruginous and Swainson's hawks. J Range Manag 40:438–440

Schmutz JK, Flockhart DTT, Houston CS, McLoughlin PD (2008) Demography of ferruginous hawks breeding in western Canada. J Wildl Manag 72:1352–1360

Schrag A, Konrad S, Miller S, Walker B, Forrest S (2011) Climate-change impacts on sagebrush habitat and West Nile virus transmission risk and conservation implications for greater sage-grouse. GeoJ. 76:561–575

Sekercioglu CH (2012) Bird functional diversity and ecosystem services in tropical forests, agroforests and agricultural areas. J Ornithol 153(Suppl 1):153–161

Sergio F, Newton I, Marchesi L, Pedrini P (2006) Ecologically justified charisma: preservation of top predators delivers biodiversity conservation. J Appl Ecol 43:1049–1055

Shaffer JA, Igl LD, Johnson DH et al (2019) The effects of management practices on grassland birds—Ferruginous Hawk (Buteo regalis), chap. N. In: Johnson DH, Igl LD, Shaffer JA, DeLong JP (eds) The effects of management practices on grassland birds. U.S. Geological Survey Professional Paper 1842, 13p. https://doi.org/10.3133/pp1842N

Shaffer JA, Igl LD, Johnson et al (2021) The effects of management practices on grassland birds—short-eared Owl (Asio flammeus), chap. Q. In: Johnson DH, Igl LD, Shaffer JA, DeLong JP (eds) The effects of management practices on grassland birds. U.S. Geological Survey Professional Paper 1842, 12p. https://doi.org/10.3133/pp1842Q

Shaffer JA, Igl LD, Johnson DH, Sondreal ML, Goldade CM, Rabie PA, Thiele JP, Euliss BR (2022) The effects of management practices on grassland birds—Burrowing Owl (Athene cunicularia hypugaea) (ver. 1.1, May 2023), chap. P of Johnson DH, Igl LD, Shaffer JA, DeLong JP (eds) The effects of management practices on grassland birds: U.S. Geological Survey Professional Paper 1842, 35p

Shearn-Bochsler VI, Knowles S, Ip H (2019) Lethal infection of wild raptors with highly pathogenic avian influenza H5N8 and H5N2 viruses in the USA, 2014–15. J Wildl Dis 55(1):164–168

Simmons RE, Smith PC, MacWhirter RB (1986) Hierarchies among Northern Harrier (Circus cyaneus) harems and the costs of polygyny. J Anim Ecol 55:755–771

Slater SJ et al (2013) Utah Legacy Raptor Project: Great Basin bird species-at-risk an invasive species management partnership. HawkWatch International, Inc., Final Report—Phase 3, Salt Lake City, USA

Slater SJ, Maloney DM, Taylor JM (2022) Golden eagle use of winter roadkill and response to vehicles in the western United States. J Wildl Manag e22246. https://doi.org/10.1002/jwmg.22246

Slabe VA, Anderson JT, Millsap BA, Cooper JL, Harmata AR, Restani M, Crandall RH, Bodenstein B, Bloom PH, Booms T, Buchweitz J (2022) Demographic implications of lead poisoning for eagles across North America. Sci 6582:779–782

Sleeter BM, Sohl TL, Bouchard MA, Reker RR, Soulard CE, Acevedo W, Griffith GE, Sleeter RR, Auch RF, Sayler KL, Prisley S (2012) Scenarios of land use and land cover change in the conterminous United States: Utilizing the special report on emission scenarios at ecoregional scales. Glob Environ Change. 22(4):896–914

Smith DG, Murphy JR, Woffinden ND (1981) Relationships between jackrabbit abundance and ferruginous hawk reproduction. Condor 83:52–56

Sohl TL, Sleeter BM, Sayler KL, Bouchard MA, Reker RR, Bennett SL, Sleeter RR, Kanengieter RL, Zhu Z (2012) Spatially explicit land-use and land-cover scenarios for the Great Plains of the United States. Agric Ecosyst Environ 153:1–5

Squires JR, Olson LE, Wallace ZP et al (2020) Resource selection of apex raptors: implications for siting energy development in sagebrush and prairie ecosystems. Ecosphere 11:e03204

Steenhof K (2017) Coming to terms about describing golden eagle reproduction. J Raptor Res 51:378–390

Steenhof K, Kochert MN, Roppe JA (1993) Nesting by raptors and common ravens on electrical transmission line towers. J Wildl Manag 57:271–281

Steenhof K, Peterson B (2009) Site fidelity, mate fidelity, and breeding dispersal in American Kestrels. Wilson J Ornithol 121(1):12–21

Thomas L, Buckland ST, Rexstad EA, Laake JL, Strindberg S, Hedley SL, Bishop JR, Marques TA, Burnham KP (2010) Distance software: design and analysis of distance sampling surveys for estimating population size. J Appl Ecol 47(1):5–14

Tigner JR, Call MW, Kochert MN (1996) Effectiveness of artificial nesting structures for ferruginous hawks in Wyoming. In: Bird DM, Varland DE, Negro JJ (eds) Raptors in human landscapes: adaptation to built and cultivated environments. Academic Press, Waltham, Massachusetts, USA, pp 137–144

Torre I, Díaz M, Martínez-Padilla J et al (2007) Cattle grazing, raptor abundance and small mammal communities in Mediterranean grasslands. Basic Appl Ecol 8(6):565–575

Tilman D (2001) Functional diversity. Encycl Biodivers 3(1):109–120

Tinajero R, Barragán F, Chapa-Vargas L (2017) Raptor functional diversity in scrubland-agricultural landscapes of northern-central-Mexican dryland environments. Trop Conserv Sci. https://doi.org/10:1940082917712426

U.S. Fish and Wildlife Service (2022) Wyoming ecological services field office raptor guidelines. https://www.fws.gov/media/wyoming-ecological-services-field-office-raptor-guidelines-2022. Accessed 29 June 2022

Van Horn RC (1993) Ferruginous hawk and prairie falcon reproductive and behavioral responses to human activity near Kevin Rim, Montana. Thesis, Montana State University, Bozeman, USA

Verbeek NA, Caffrey C (2020) American crow (Corvus brachyrhynchos), version 1.0. In: Poole AF, Gill FB (ed) Birds of the world. Cornell Lab of Ornithology, Ithaca, NY, USA. https://doi.org/10.2173/bow.amecro.01

Vilà M, Espinar JL, Hejda M, Hulme PE, Jarošík V, Maron JL, Pergl J, Schaffner U, Sun Y, Pyšek P (2011) Ecological impacts of invasive alien plants: a meta-analysis of their effects on species, communities and ecosystems. Ecol Lett 14(7):702–708

Viner TC, Hamlin BC, McClure PJ, Yates BC (2016) Integrating the forensic sciences in wildlife case investigations: a case report of pentobarbital and phenytoin toxicosis in a bald eagle (Haliaeetus leucocephalus). Vet Pathol 53:1103–1106. https://doi.org/10.1177/0300985816641176

Waite BC, Phillips RL (1994) An approach to controlling golden eagle predation on lambs in South Dakota. In: Proceedings of the Vertebrate Pest Conference 16(16)

Walker BL, Naugle DE (2011) West Nile virus ecology in sagebrush habitat and impacts on greater sage-grouse populations. Studies in Avian Biology. 38:127–142.

Wallace, Z, Bedrosian G, Woodbridge B, Williams G, Bedrosian BE, and Dunk J (2019) Wyoming and Uinta Basins Golden Eagle Conservation Strategy. Unpublished report prepared for the U.S. Fish and Wildlife Service Western Golden Eagle Team by the Wyoming Natural Diversity Database and Eagle Environmental, Inc

Wallace ZP, Kennedy PL, Squires JR et al (2016a) Re-occupancy of breeding territories by ferruginous hawks in Wyoming: relationships to environmental and anthropogenic factors. PLoS ONE 11:e0152977

Wallace ZP, Kennedy PL, Squires JR et al (2016b) Human-made structures, vegetation, and weather influence ferruginous hawk breeding performance. J Wildl Manag 80:75–90

Watson J (2010) The golden eagle. Yale University Press, New Haven, CT. 448p

Watson JW, Keren IN (2019) Repeatability in migration of Ferruginous Hawks (*Buteo regalis*) and implications for nomadism. Wilson J Ornithol 131(3):561–570

Watson JW, Banasch U, Byer T et al (2018) Migration patterns, timing, and seasonal destinations of adult Ferruginous Hawks (*Buteo regalis*). J Raptor Res 52:267–281

Webb WC, Boarman WI, Rotenberry JT (2004) Common raven juvenile survivorship in a human augmented landscape. Condor 106:517–528

Webb WC, Marzluff JM, Hepinstall-Cymerman J (2012) Differences in space use by common ravens in relation to sex, breeding status, and kinship. Condor 114(3):584–594

White CM, Thurow TL (1985) Reproduction of ferruginous hawks exposed to controlled disturbance. Condor 87:14–22

Wiggins DA, Grzybowski JA, Schnell GD (2017) Ferruginous hawk demography in areas differing in energy extraction activity. J Wildl Manag 81(2):337–341

Winkler DW (2020) California gull (*Larus californicus*), version 1.0. In: Poole AF, Gill FB (eds) Birds of the world. Cornell Lab of Ornithology, Ithaca, NY, USA. https://doi.org/10.2173/bow.calgul.01

Yarnell RW, Scott DM, Chimimba CT, Metcalfe DJ (2007) Untangling the roles of fire, grazing and rainfall on small mammal communities in grassland ecosystems. Oecologia 154(2):387–402

Zabala J, Zuberogoitia I (2014) Individual quality explains variation in reproductive success better than territory quality in a long-lived territorial raptor. PLoS ONE 9(3):e90254. https://doi.org/10.1371/journal.pone.0090254

Zelenak JR, Rotella JJ (1997) Nest success and productivity of ferruginous hawks in northern Montana. Can J Zool 75:1035–1041

Zou L, Miller SN, Schmidtmann ET (2006) Mosquito larval habitat mapping using remote sensing and GIS: implications of coalbed methane development and West Nile virus. J Med Entomol 43(5):1034–1041

Chapter 15
Burrowing Rodents

David J. Augustine, Jennifer E. Smith, Ana D. Davidson, and Paul Stapp

Abstract Burrowing rodents have unusually disproportionate effects on rangeland ecosystems because they (1) engineer their environment through burrow construction and modification of vegetation structure, (2) influence ecosystem processes including aboveground plant production, nutrient cycling rates, and water infiltration patterns, (3) alter plant community composition, and (4) provide a prey base for a diverse array of predators. In some cases, engineering effects create habitat for certain faunal species that inhabit burrows or colonies of these rodents. We review the ecology and management of burrowing rodents that function as ecosystem engineers in western North America, which includes prairie dogs (five species in the genus *Cynomys*), ground squirrels (11 species in the genera *Otospermophilus*, *Poliocitellus*, and *Urocitellus*), pocket gophers (16 widespread species in the genera *Cratogeomys*, *Geomys*, and *Thomomys*), and kangaroo rats (eight widespread species in the genus *Dipodomys*). Effects of burrowing rodents on vegetation structure, species composition, and nutrient content vary with diet, degree of sociality, body size, and hibernation patterns, and potentially have significant effects on coexisting large grazers, including domestic livestock. Diets of prairie dogs overlap substantially with livestock. Impacts on ranching enterprises can vary with their abundance and seasonally, and may be

D. J. Augustine (✉)
Rangeland Resources and Systems Research Unit, USDA—Agricultural Research Service, 2150 Centre Ave, Building D, Fort Collins, CO 80526, USA
e-mail: David.Augustine@usda.gov

J. E. Smith
Biology Department, University of Wisconsin Eau Claire, 105 Garfield Avenue, Eau Claire, WI 54702, USA
e-mail: smitjenn@uwec.edu

A. D. Davidson
Colorado Natural Heritage Program and Department of Fish, Wildlife, and Conservation Biology, Colorado State University, 1475 Campus Delivery, Fort Collins, CO 80523, USA
e-mail: ana.davidson@colostate.edu

P. Stapp
Department of Biological Science, California State University Fullerton, Fullerton, CA 92834, USA
e-mail: pstapp@fullerton.edu

© The Author(s) 2023
L. B. McNew et al. (eds.), *Rangeland Wildlife Ecology and Conservation*,
https://doi.org/10.1007/978-3-031-34037-6_15

greatest when burrowing rodents reduce dormant-season forage availability. Ground squirrel, pocket gopher, and kangaroo rat interactions with livestock vary among species in relation to their diet, degree of coloniality, and population density. All prairie dog and ground squirrel species are affected by outbreaks of plague caused by *Yersinia pestis*, a non-native disease. Plague and population control via rodenticides are the primary factors determining the distribution and abundance of these species. In contrast, pocket gophers and kangaroo rats are unaffected by plague. Management and conservation efforts that enable burrowing rodents to coexist with livestock across broad landscapes will likely be essential for the conservation of a unique suite of bird, mammal, herpetofaunal and arthropod species that depend on them as prey or on their engineering activities for habitat.

Keywords Ecosystem engineers · Ground squirrel · Livestock competition · Kangaroo rat · Plague · Pocket gopher · Prairie dog

15.1   Introduction

Rangelands around the world are inhabited and shaped by a diverse array of fossorial and semi-fossorial (burrowing), herbivorous mammals (Davidson et al. 2012). Many of these species function as ecosystem engineers (Jones et al. 1994) because they construct burrow systems and alter the structure of vegetation and soils (e.g., Huntly and Inouye 1988; Reichman and Seabloom 2002; Lenihan 2007; Davidson and Lightfoot 2008; Prugh and Brashares 2012; Baker et al. 2013). These engineering activities alter the composition of plant communities, and create habitat features upon which other fauna depend (e.g., Davidson et al. 2012; Augustine and Baker 2013). In addition, burrowing mammals often serve as the prey base for a diverse array of predators, including raptors and mammalian carnivores. Here, we provide a review of the burrowing rodent species that function as ecosystem engineers in rangelands of western North America. In this review, we examine a group of ground squirrels that are social and colonial, which often concentrates their effects on rangelands in a spatially heterogeneous manner. These colonial species can be divided in terms of taxonomy and body size into the prairie dogs (five species in the genus *Cynomys*, which tend to be larger than other colonial ground squirrels) versus the somewhat smaller ground squirrels in the genera *Otospermophilus* (one species), *Poliocitellus* (one species), and *Urocitellus* (nine species; Table 15.1). Non-colonial, burrowing rodents that exert important engineering effects on western rangelands consist of pocket gophers in the genera *Cratogeomys*, *Geomys*, and *Thomomys* (16 widespread species, plus several restricted-range endemics) and kangaroo rats in the genus *Dipodomys* (eight widespread species, plus several restricted-range endemics; Table 15.2). We first describe the general life history and distribution of representative species in each of these groups, followed by a description of their ecosystem associations and the ways in which they influence the structure and function of rangelands. We follow with a discussion of our current state of knowledge on how burrowing rodents influence livestock management and production, the degree to which they are regulated by diseases (especially plague, caused by the introduced bacterium *Yersinia pestis*), and how interactions with disease and livestock fundamentally shape the management and conservation of burrowing rodents.

15.2 Life History, Ecology, and Distribution

15.2.1 Prairie Dogs and Ground Squirrels

The black-tailed prairie dog (*C. ludovicianus*) is the most widespread species, occurring from the prairies of Saskatchewan to the northern Chihuahuan Desert of Mexico (Fig. 15.1). The closely-related Mexican prairie dog (*C. mexicanus*) occurs in a disjunct, southern portion of the Chihuahuan Desert, and is likely derived from a population of *C. ludovicianus* that became isolated from the main population following the Wisconsin glaciation (Ceballos and Wilson 1985). Their relatively long, black-tipped tail distinguishes these two species from the three white-tailed species (*C. parvidens*, *C. gunnisoni*, and *C. leucurus*). Of these, the white-tailed prairie dog (*C. leucurus*) predominantly inhabits sagebrush (*Artemisia* spp.) steppe in the Wyoming Basin of Wyoming, Colorado, and Utah as well as northern portions of the Colorado Plateau. Gunnison's prairie dog (*C. gunnisoni*) occurs to the south, in sagebrush steppe and grasslands across the Colorado Plateau, Apache highlands, and Chihuahuan Desert, as well as high-elevation grasslands within the southern Rocky Mountains (Fig. 15.1). The Utah prairie dog (*C. parvidens*) inhabits grasslands and shrublands in a relatively restricted range in southwestern Utah (Fig. 15.1).

The extent to which prairie dogs affect ecosystem processes and interact with other taxa is largely influenced by degree of social organization (which influences animal density within a colony), body size, and hibernation patterns. All five species feed primarily on grasses and forbs, hence their diet overlaps substantially with that of livestock and native ungulates. *C. ludovicianus* additionally clip tall vegetation and girdle shrubs on their colonies to enhance visibility, whereas the three white-tailed species do not (Tileston and Lechleitner 1966). Body size of all species varies seasonally and is greater in males, with the range being ~ 700–1100 g for *C. ludovicianus* and *mexicanus*, ~ 650–1100 g for *C. leucurus* and *C. parvidens*, and ~ 400–800 g for *C. gunnisoni* (Hoogland 2003). Prairie dogs live in spatially discrete colonies composed of multiple family units (coteries or clans) and exhibit greater social complexity (as measured in terms of relative frequencies of different age and sex classes living within a single social group) than the other ground squirrels described below (Blumstein and Armitage 1997). *C. ludovicianus* and *C. parvidens* exhibit more complex social structure than *C. leucurus* and *C. gunnisoni* (Blumstein

Table 15.1 Prairie dog and ground squirrel species (all within the family Sciuridae) that occupy rangelands of western North America, along with estimates of two different measures of their degree of social organization (Social Grade[a], and Social Complexity index[b]), range of reported population densities (individuals ha^{-1}), mean body size (g), and mean litter size

Common name	Latin name	Social grade	Social complexity	Density (no ha^{-1})	Adult male mass (g)	Adult female mass (g)	Litter size
Black-tailed prairie dog	Cynomys ludovicianus	5	1.12	10–194	760 (660–1000)	700 (670–900)	4.6
Mexican prairie dog	Cynomys mexicanus	5	–	–	1000 (900–1100)	890 (880–1010)	–
Utah prairie dog	Cynomys parvidens	–	1.23	1.0–74	920 (670–1130)	750 (650–830)	4.8
Gunnison's prairie dog	Cynomys gunnisoni	4	1.03	8.0–60	670 (550–770)	570 (400–630)	4.6
White-tailed prairie dog	Cynomys leucurus	2	0.84	7.0–19	950 (530–1150)	660(550–740)	5.2
Uinta ground squirrel	Urocitellus armatus	2	0.44	23.0	333	266	6.0
Merriam's ground squirrel	Urocitellus canus	–	–	–	–	–	–
Columbian ground squirrel	Urocitellus columbianus	3	0.65	12.2	490	406	3.3
Belding's ground squirrel	Urocitellus beldingi	2	0.40	43.9	284	243	6.4
Wyoming ground squirrel	Urocitellus elegans	2	0.43	–	266	203	6.0
Piute ground squirrel	Urocitellus mollis	–	–	–	–	–	–
Richardson's ground squirrel	Urocitellus richardsonii	2	0.39	23.5	260	225	7.0
Townsends ground squirrel	Urocitellus townsendii	1–2	0.41	–	–	155	9.3
Washington ground squirrel	Urocitellus washingtoni	–	–	–	–	–	–
Idaho ground squirrel	Urocitellus brunneus	–	–	–	–	–	–
California ground squirrel	Otospermophilus beecheyi	2–3	0.26	20.1	625	480	6.3

(continued)

Table 15.1 (continued)

Common name	Latin name	Social grade	Social complexity	Density (no ha^{-1})	Adult male mass (g)	Adult female mass (g)	Litter size
Franklin's ground squirrel	*Poliocitellus franklinii*	1–2	–	–	–	–	–

Data summarized from Heaney (1984), Rayor (1988), Hoogland (2003), Magle et al. (2007). Nomenclature for ground squirrels follows Helgen et al. (2009) and Phuong et al. (2014)
[a] As defined by Holekamp (1984). This index varies from 1 to 5, with 1 representing solitary species and 5 representing species with the most complex social system
[b] As defined by Blumstein and Armitage (1997). This index can vary from 0 for species with no social groups to an unlimited upper value based on the complexity of the demographic structure within social groups of a given species

and Armitage 1997; Table 15.1), which is reflected to some extent in the density of individuals within colonies and degree of modification of vegetation structure. *C. ludovicianus* often occurs at densities on the order of 10–35 individuals per ha, and can even increase to > 100 individuals per ha in suburban landscapes where dispersal is curtailed (Magle et al. 2007; Table 15.1). Densities of the three white-tailed species can also vary widely (e.g., 1–60 individuals ha^{-1}; Table 15.1), but typically occur at lower densities than *C. ludovicianus*, on the order of 2–32 ha^{-1} (Menkens and Anderson 1991; Nelson and Theimer 2012; USFWS 2012).

Although the pre-settlement distribution and abundance of prairie dogs are difficult to assess from journals of early explorers, comprehensive reviews of cumulative accounts of prairie dog control efforts in the late 1800s and early 1900s show that the distribution and abundance of all five prairie dog species declined dramatically after European settlement (Knowles et al. 2002). Effects of human control efforts were compounded by the introduction of plague in the early 1900s. Because of the cost of monitoring the vast areas often surveyed for *C. ludovicianus*, surveys typically measure colony area rather than animal numbers, using methods that include aerial photography or high spatial resolution satellite imagery to detect colonies (Brennan et al. 2020), aerial surveys of colony intercepts by observers in fixed-wing aircraft (White et al. 2005), and ground-based mapping of colony boundaries (Sidle et al. 2012). Thus far, ground-based surveys have been required to accurately distinguish colonies actively occupied by prairie dogs from former colony-sites recently extirpated by poisoning or plague (Sidle et al. 2012). For Utah prairie dogs, populations are monitored annually using direct counts of aboveground individuals on known colonies (USFWS 2012).

Twelve species of ground squirrels inhabit western North America all of which are smaller in body size and exhibit less complex social organization compared to prairie dogs (Table 15.1; Figs. 15.2 and 15.3). We focus primarily on five of these because of their relatively large ecological roles, widespread distribution, and existing research on impacts to western rangelands. The California ground squirrel (*Otospermophilus beecheyi*) primarily resides in open grasslands, oak (*Quercus* spp.) savannah, oak

Table 15.2 Pocket gopher and kangaroo rat species that occupy rangelands of western North America

Common name	Latin name	Family	Distribution	Adult female mass (g)
Yellow-faced pocket gopher	*Cratogeomys castanops*	Geomyidae	Widespread (Fig. 15.4)	267
Desert pocket gopher	*Geomys arenarius*	Geomyidae	Restricted (Rio Grande River Valley)	206
Plains pocket gopher	*Geomys bursarius*	Geomyidae	Widespread (Fig. 15.4)	204
Hall's pocket gopher	*Geomys jugossicularis*	Geomyidae	Restricted (shortgrass prairie)	–
Jones's pocket gopher	*Geomys knoxjonesi*	Geomyidae	Restricted (shortgrass prairie)	–
Sand Hills pocket gopher	*Geomys lutescens*	Geomyidae	Restricted (NE Sand Hills)	–
Llano pocket gopher	*Geomys texensis*	Geomyidae	Restricted (Edwards plateau)	–
Botta's pocket gopher	*Thomomys bottae*	Geomyidae	Widespread (Fig. 15.4)	123
Camas pocket gopher	*Thomomys bulbivorus*	Geomyidae	Restricted (Willamette River Valley)	360
Wyoming pocket gopher	*Thomomys clusius*	Geomyidae	Restricted (Wyoming Basins)	58
Idaho pocket gopher	*Thomomys idahoensis*	Geomyidae	Restricted (Rocky Mountains)	67
Western pocket gopher	*Thomomys mazama*	Geomyidae	Restricted (Sierra Nevada)	93
Mountain pocket gopher	*Thomomys monticola*	Geomyidae	Restricted (Sierra Nevada)	81
Northern pocket gopher	*Thomomys talpoides*	Geomyidae	Widespread (Fig. 15.4)	105
Townsend's pocket gopher	*Thomomys townsendii*	Geomyidae	Restricted (Columbia Plateau)	263
Southern pocket gopher	*Thomomys umbrinus*	Geomyidae	Restricted (Sonoran, Chihuahuan Desert)	126
California kangaroo rat	*Dipodomys californicus*	Heteromyidae	Restricted (North Great Central Valley)	85

(continued)

Table 15.2 (continued)

Common name	Latin name	Family	Distribution	Adult female mass (g)
Texas kangaroo rat	*Dipodomys elator*	Heteromyidae	Restricted (Southern Mixed Grass Prairie)	106
Heermann's kangaroo rat	*Dipodomys heermanni*	Heteromyidae	Restricted (South Great Central Valley)	63
Giant kangaroo rat	*Dipodomys ingens*	Heteromyidae	Restricted (Great Central Valley)	114
Merriam's kangaroo rat	*Dipodomys merriami*	Heteromyidae	Widespread (not shown)	38
Great basin kangaroo rat	*Dipodomys microps*	Heteromyidae	Restricted (Great Basin)	56
San Joaquin kangaroo rat	*Dipodomys nitratoides*	Heteromyidae	Restricted (South Great Central Valley)	42
Ord's kangaroo rat	*Dipodomys ordii*	Heteromyidae	Widespread (Fig. 15.4)	50
Banner-tailed kangaroo rat	*Dipodomys spectabilis*	Heteromyidae	Widespread (Fig. 15.4)	125

Data for mean female body mass are adapted from Davidson et al. (2017)

woodland, nearshore rocky outcrops, and on agricultural lands where the openness of the habitat permits individuals to detect predators. Abundant throughout its range in Pacific Coast states, *O. beecheyi* densities typically vary between 8 and 92 animals ha^{-1}, mass of adults ranges from 280 to 738 g, and females usually produce a single litter each year of 4–11 young (Smith et al. 2016). Two species, the Richardson's ground squirrel (*Urocitellus richardsonii*) and Franklin's ground squirrel (*Poliocitellus franklinii*) occupy the northern Great Plains (Fig. 15.2). Adult mass of *U. richardsonii* varies greatly, with pre-hibernation masses for adult females of 350–435 g and males of 500–655 g (Michener and Koeppl 1985). *P. franklinii* is slightly smaller and is less social than the *Urocitellus* species, generally living alone or in pairs rather than in colonies (Ostroff and Finck 2003). The Wyoming ground squirrel (*Urocitellus elegans*) now includes three subspecies, each of which occurs in distinct geographic ranges (Thorington et al. 2012; Fig. 15.2). Here, we focus primarily on *U. e. elegans* due to their prevalence and impacts in sagebrush steppe (Zegers 1984; Thorington et al. 2012; Fig. 15.2). *U. e. elegans* select habitats with talus slopes or well-drained soils that facilitate burrow construction. The Uinta ground squirrel (*Urocitellus armatus*) is a large ground squirrel that resides mainly in the in or near Utah and Idaho in open meadows and sagebrush steppe (Fig. 15.3; Eshelman and Sonnemann 2000).

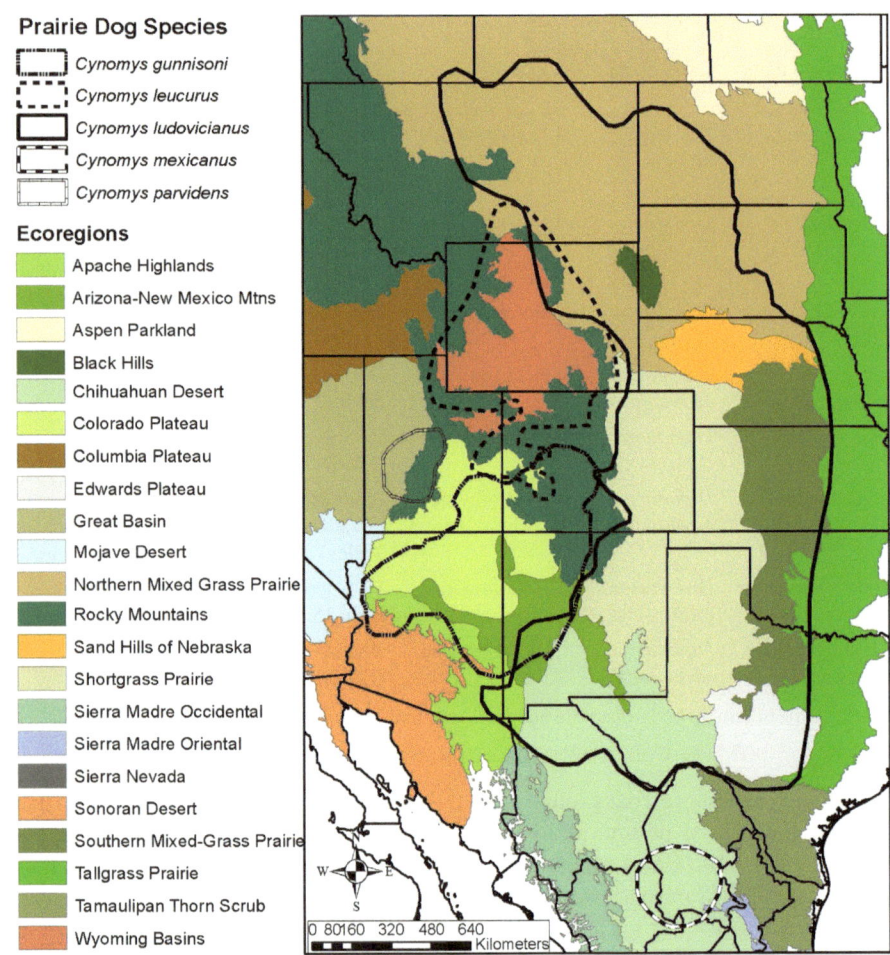

Fig. 15.1 Geographic ranges of the five species of prairie dogs that inhabit rangelands of western North America, overlaid with the distribution of rangeland ecoregions

Seven other ground squirrel species have more restricted distribution in western North America (Figs. 15.2 and 15.3). These include the Belding's (*U. beldingii*), Columbian (*U. columbianus*), Idaho (*U. brunneus*), Piute (*U. mollis*), Townsends (*U. townsendii*), and Washington (*U. washingtonii*) ground squirrels. Collectively, these seven species occupy a broad swath of the Intermountain Region extending from southern Alberta to southern Nevada (Figs. 15.2 and 15.3), and all of these species can be locally important in terms of effects on rangelands.

For most rangeland ground squirrels, population numbers are limited by food availability, as shown experimentally for *U. columbianus* (Dobson and Kjelgaard 1985). Population densities of *O. beecheyi*, for example, vary radically with typical densities of 1.2–11 adults/ha (Schitoskey and Woodmansee 1978). Moreover, *U. u.*

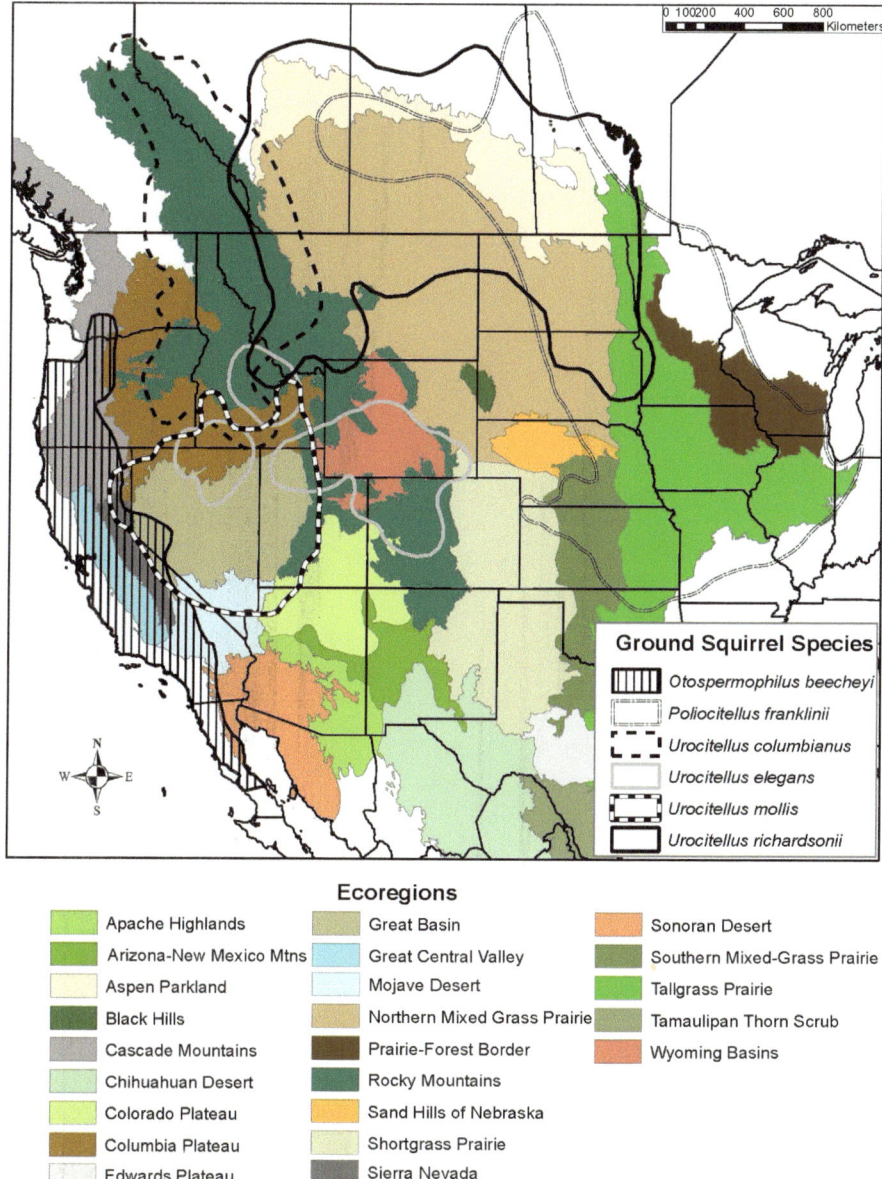

Fig. 15.2 Geographic range of six widely distributed species of ground squirrels that inhabit rangelands of western North America, overlaid with the distribution of rangeland ecoregions

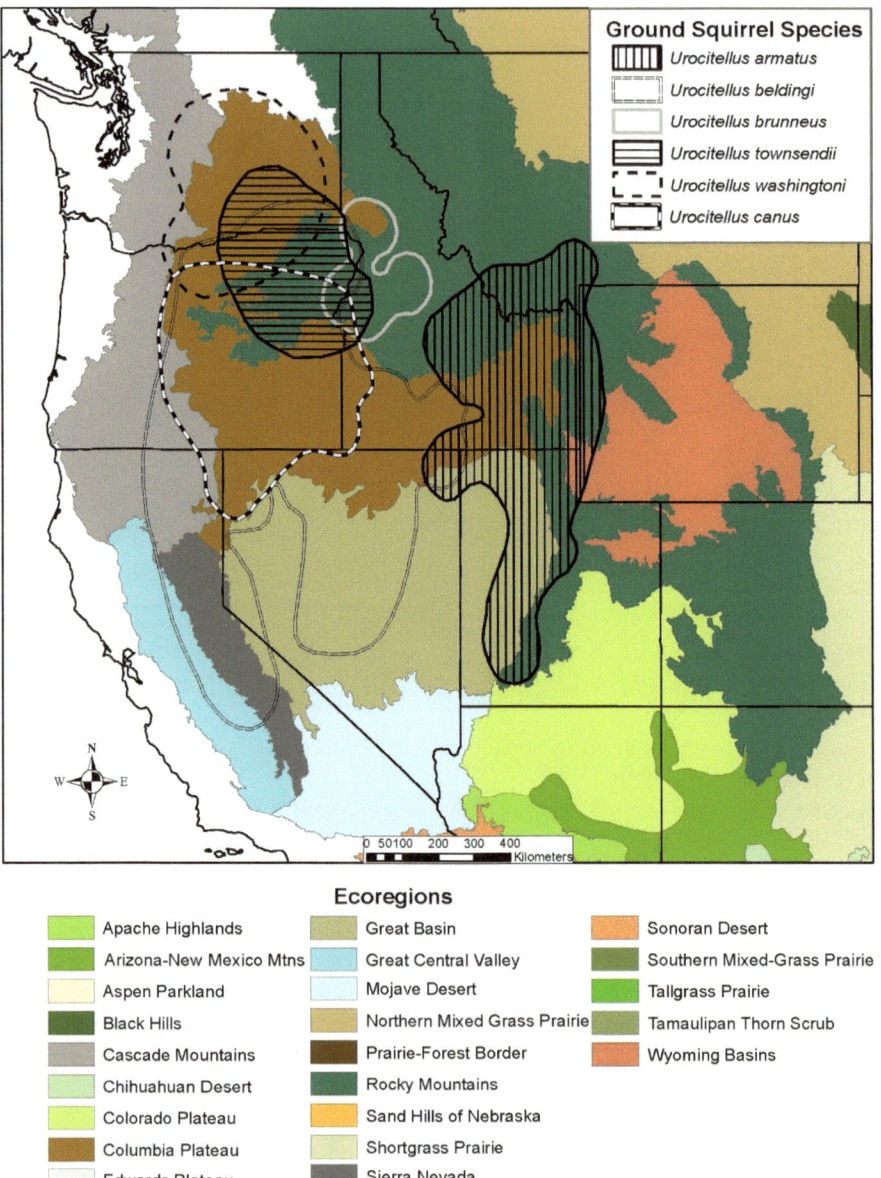

Fig. 15.3 Geographic ranges of six species of ground squirrels that occupy the Intermountain Region of western North America. Four of these have relatively restricted ranges within the Columbia Plateau and in grasslands and meadows of the Rocky Mountains

elegans densities vary seasonally with hibernation but densities, including juveniles, reach up to 44 ha^{-1} in shortgrass prairie and vary from 14 to 48 ha^{-1} in montane meadows (Zegers 1984). *P. franklinii* typically occurs at lower densities (1.5–2.5 adults ha^{-1}. Ostroff and Finck 2003).

15.2.2 Pocket Gophers and Kangaroo Rats

Pocket gophers of the family Geomyidae include 41 recognized species restricted to North and Central America. Pocket gophers are truly fossorial and possess multiple adaptations for a life spent mostly underground, including a stocky, fusiform body with stout forelimbs and enlarged claws, skin on the snout that grows behind the incisors to prevent soil from entering the mouth while digging, reduced eyes and ears, and a short, mostly naked tail (Baker et al. 2013). The pelage is short and fine, and dorsal coloration often matches the color of the specific soils in which they live, presumably as a mechanism for avoiding aerial predators (Krupa and Geluso 2000). They possess fur-lined, external cheek pouches that can be used to temporarily hold and transport food.

Three genera (*Cratogeomys*, *Geomys*, and *Thomomys*) inhabit arid, semiarid and montane rangelands of western North America (Fig. 15.4). Five widespread species (*C. castanops, G. bursarius, T. bottae, T. talpoides, and T. townsendii*) collectively occupy all of the rangeland ecoregions of western North America except the Nebraska sandhills (Fig. 15.4). Maximum adult body sizes vary from about 90 to 630 g; most species weigh less than 400 g (Reid 2006), and are similar morphologically and behaviorally. They are mostly active at night, and unlike prairie dogs and ground squirrels, are described as solitary, territorial, and asocial (Baker et al. 2013). In milder climates and irrigated farmlands, females can have multiple litters per year, whereas those in colder environments and shorter growing seasons tend to have a single litter of 4–6 pups. In highly seasonal environments, two peaks of burrowing activity correspond to the onsets of breeding and juvenile dispersal (Miller 1964). All species are strictly herbivores. In California, gophers move seasonally in response to drying of vegetation or flooding of burrows (Fitch and Bentley 1949).

Gopher species differ in soil affinities (Miller 1964), with larger species restricted to deeper, sandier soils, and widely distributed species such as *T. bottae* and *T. talpoides* inhabiting diverse soil types. Some species are specialized to specific ecoregions, such as *T. clusius* in the Wyoming Basins, *G. texensis* on the Edwards Plateau, and *G. knoxjonesi* in the southern shortgrass prairie (Fig. 15.4). These three species are shown as examples in Fig. 15.4, but nine additional pocket gopher species not displayed have more restricted ranges in western rangelands (Table 15.2). Soil moisture limits smaller *Thomomys* species. Typically, only one gopher occupies a burrow at a time and, in areas of high gopher activity, the distance between burrow systems is remarkably consistent, regardless of sex, age, or reproductive status (Reichman et al. 1982). Population densities of gophers are highly variable and biased by the size of the area studied; Smallwood and Morrison (1999) estimated an average density of 35

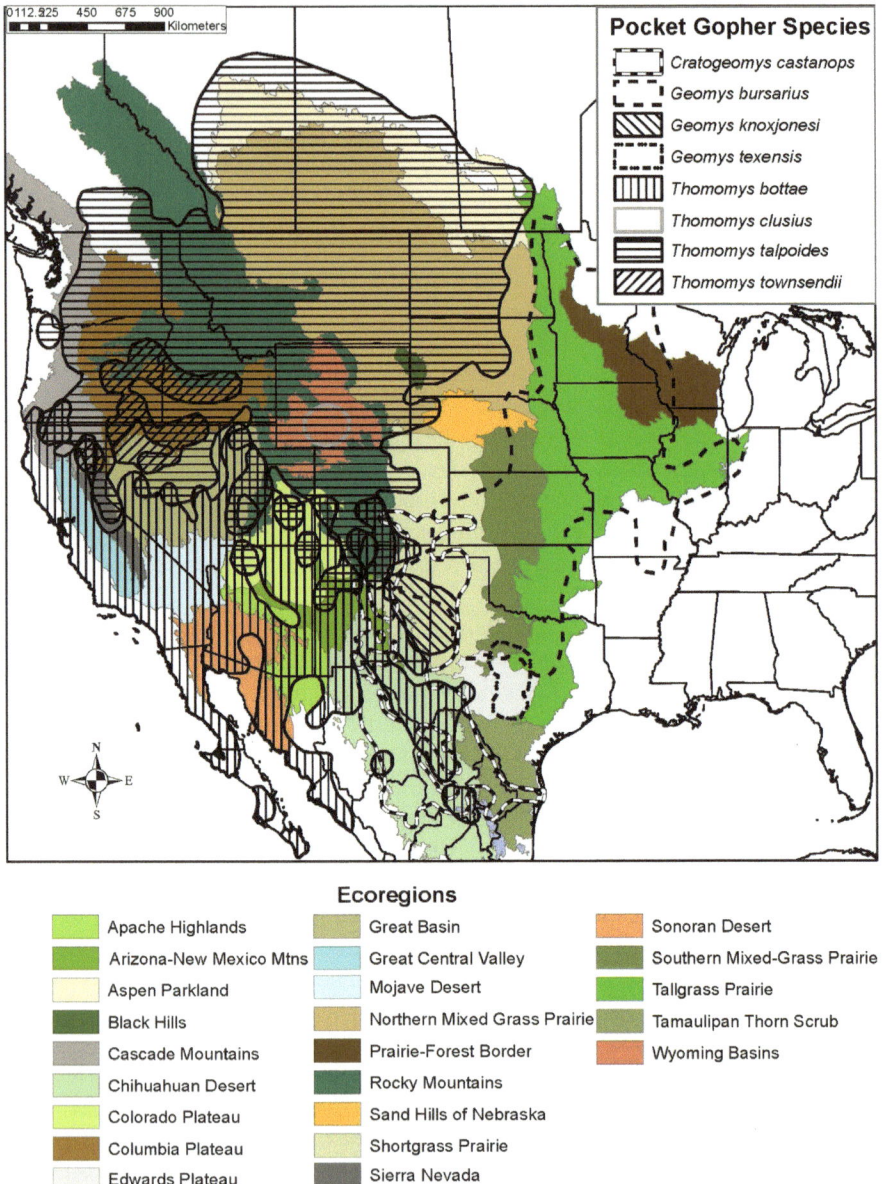

Fig. 15.4 Geographic ranges of eight species of pocket gophers that inhabit rangelands of western North America, overlaid with the distribution of rangeland ecoregions. The five most widespread species occupy all rangeland ecoregions with the exception of the Sand Hills of Nebraska. Three additional species with restricted ranges are also shown to illustrate their close association with specific ecoregions

gophers ha^{-1} for six common western species, although higher densities are possible (49–83 ha^{-1}; Hansen and Remmenga 1961). Population densities are usually estimated by kill- or live-trapping, although there have attempts to convert counts of mounds and burrows to densities (Smallwood and Morrison 1999).

Kangaroo rats are solitary, bipedal, granivorous rodents (Genoways and Brown 1993). Collectively, the seven most widespread species occur in all of the rangeland ecoregions in western North America, except for the tall-structured grasslands and savannas of the eastern Great Plains (Fig. 15.5). They dig for seeds in the soil and fill their external fur-lined cheek pouches with seeds that they scatter-hoard in superficial subsurface caches often near their mounds and within their burrows (Brown et al. 1979). They complement their granivorous diet with green grass and insects when available, and they have highly efficient kidneys that enable them to extract water from their food. Body sizes range from 30 to 200 g, with many species weighing around 50 g, while the three largest species (*D. deserti, D. ingens,* and *D. spectabilis*) weigh ~ 150 g. Typical litter sizes are 2–4, with some species, e.g., *D. merriami*, capable of producing larger litters and breeding multiple times in a year if food resources and environmental conditions permit (Kenagy and Bartholomew 1985). Different-sized species of kangaroo rats and other rodents often coexist in the same environment, partitioning seed resources based on seed size (Brown et al. 2000).

Because of their ability to respond numerically to pulses of production of seed resources, population densities of kangaroo rats are highly variable, among species and populations, and across years. Their populations are typically monitored with Sherman live traps in trapping grids or webs. Lima et al. (2008) reported densities in southern Arizona ranging from approximately 2–16 ha^{-1} for *D. ordii* and 18–50 ha^{-1} for *D. merriami*; average densities for three species at the same site ranged from 2 to 12 ha^{-1} (Brown and Zongyong 1989), which is similar to estimates in other habitats (e.g., Orland and Kelt 2007; Stapp et al. 2008). Kangaroo rats also have been extensively studied as part of long-term experiments that have revealed much about their ecologies and species co-existence at the community level (Brown and Heske 1990; Kelt 2011).

15.3 Role of Burrowing Rodents as Ecosystem Engineers

Burrowing rodents have unusually disproportionate effects on rangelands because they (1) engineer their environment through burrow construction and modification of vegetation structure, (2) influence ecosystem processes including above ground plant production, nutrient cycling rates, and water infiltration patterns, (3) alter plant community composition, and (4) provide a prey base for a diverse array of predators (Coggan et al. 2018). In some cases, engineering effects create or enhance habitat for certain faunal species that uniquely inhabit burrows or colonies of these rodents. Modifications to vegetation structure, species composition, and nutrient content potentially have significant effects on coexisting large grazers, including domestic livestock (Krueger 1986; Derner et al. 2006; Augustine and Springer 2013).

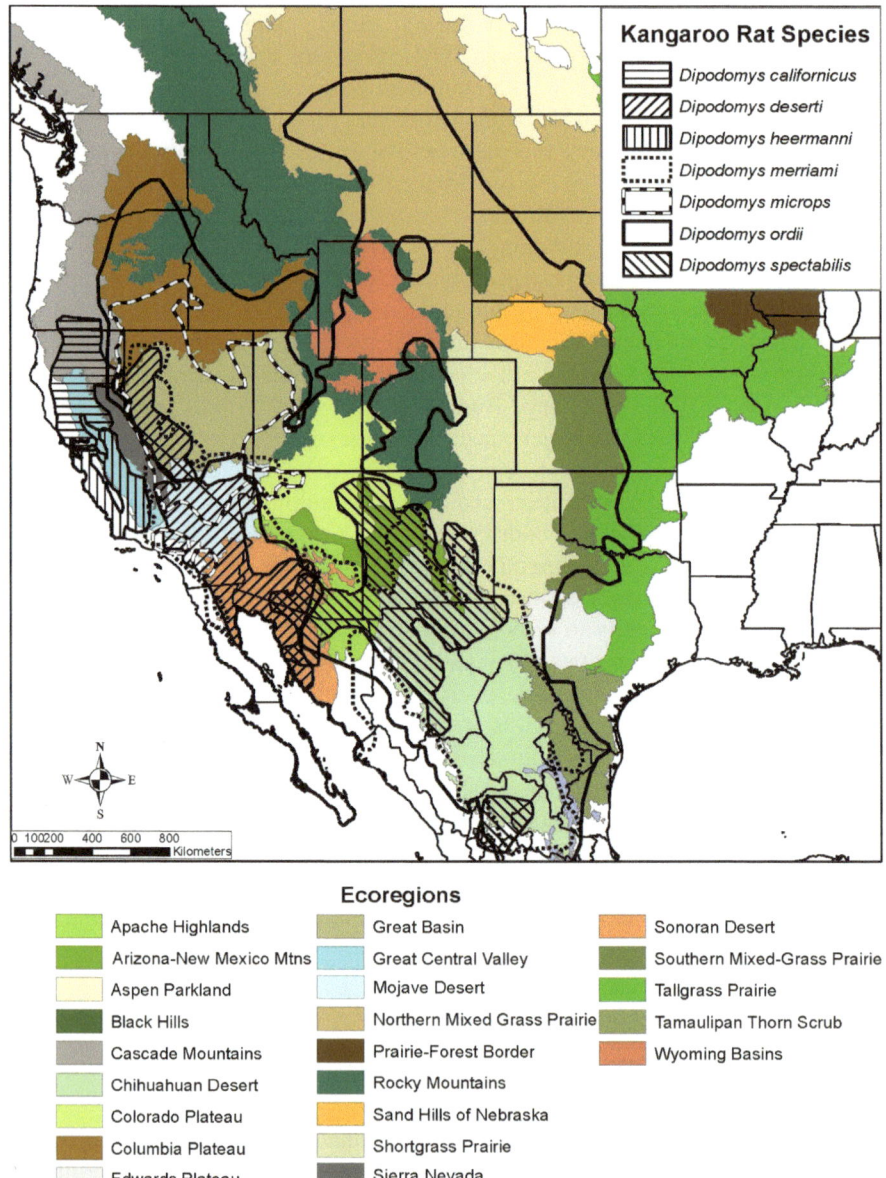

Fig. 15.5 Geographic ranges of seven species of kangaroo rats that inhabit rangelands of western North America, overlaid with the distribution of rangeland ecoregions. These seven species collectively occupy all rangeland ecoregions, with the exception of tall-structured grasslands in the Tallgrass Prairie, Prairie-Forest Border, and Aspen Parkland

While all of the burrowing rodents we discuss here serve as ecosystem engineers in rangelands, the strength and specific nature of these effects varies among the different taxa, and among different types of rangelands (Stapp 1998; Cully et al. 2010a, b; Baker et al. 2013; Fig. 15.6). Furthermore, the strength and nature of effects can be contingent on rainfall patterns (Augustine and Springer 2013) and the spatial extent of areas occupied by different species (Derner et al. 2006).

15.3.1 Prairie Dogs and Ground Squirrels

Large body size of *C. ludovicianus* relative to other rodents and high social complexity contribute to their ability to occur at high densities and exert dramatic effects on vegetation within their colonies. This includes a substantial increase in bare soil exposure, reduced vegetation height and biomass, and increased abundance of annual forbs, grazing-tolerant grasses, and some unpalatable subshrubs (e.g., Coppock et al. 1983; Cid et al. 1991; Hartley et al. 2009; Augustine et al. 2014; Fig. 15.6). Plant diversity is typically enhanced on versus off colonies, although diversity may decline with increasing years of occupancy as dominant mid-height grasses are lost (Archer et al. 1987; Fahnestock and Detling 2002). Mounds at burrow entrances are the most conspicuous aspect of prairie dog colonies, but typically only cover about 2% of the total colony area (Stapp et al. 2008). The unique habitats that colonies and mounds provide for plants and animals increases diversity across the landscape (Davidson et al. 2012). However, prairie dog grazing activity combined with mounds can increase total bare soil exposure to > 50% in some cases, even during the growing season (Augustine and Derner 2012). All prairie dog species construct extensive burrow systems that are typically 5–14 m in length and extend 1–2 m in depth below the ground's surface. Burrow construction results in substantial mixing of soil horizons, estimated to affect 200–225 kg of soil per burrow system for *C. ludovicianus* (Whicker and Detling 1988), and burrow mounds have increased soil nutrient concentrations and water infiltration rates (Barth et al. 2014). Soil disturbance and intense grazing by prairie dogs accelerates nitrogen mineralization in the soil and uptake by plants, thereby improving forage quality for large herbivores (Holland et al. 1992; Fahnestock and Detling 2002; Augustine and Springer 2013).

The extent of bare soil exposure created on colonies provides key nesting habitat for mountain plovers (*Charadrius montanus*) throughout the western Great Plains (Dinsmore et al. 2005; Augustine and Derner 2012; Duchardt et al. 2019; Table 15.3), and enhances habitat for other birds such as horned larks (*Eremophila alpestris*), mourning doves (*Zenaida macroura*) and upland sandpipers (*Bartramia longicauda*; Augustine and Baker 2013; Geaumont et al. 2019) and some non-fossorial rodents (Stapp 1997; Cully et al. 2010a). *C. ludovicianus* burrows additionally provide essential nest sites for burrowing owls (*Athene cunicularia*; Desmond et al. 2000) and winter hibernacula for prairie rattlesnakes (*Crotalus viridis*; Shipley and Reading 2006). *C. ludovicianus* frequently girdle sagebrush, which creates uniquely herbaceous-dominated patches in some northern portions of their range (Baker et al.

Fig. 15.6 Examples of burrow mounds created by prairie dogs and ground squirrels, and their effects on vegetation on and surrounding the mounds. Upper panels show the contrast between vegetation height and composition in the absence of prairie dogs **a** versus on an active black-tailed prairie dog (*C. ludovicianus*) colony **b** near the peak of the growing season in northern mixed-grass prairie on the Buffalo Gap National Grassland, South Dakota. Photos (**a**) and (**b**) were taken within 200 m of one another on the same day, and include a cow pat of approximately the same size in the lower right corner for scale. Panel **c** shows closely cropped vegetation on Gunnison's prairie dog (*C. gunnisoni*) colony during the growing season at the Sevilleta National Wildlife Refuge, New Mexico, and **d** shows the extent of bare soil on black-tailed prairie dog colony during the dormant season in the Chihuahuan Desert, near Janos, Mexico. Insets in (**c**) and (**d**) show Gunnison's and black-tailed prairie dogs respectively. Photo **e** illustrates the effect of a California ground squirrel (*O. beecheyi*) burrow mound on the plant community and bare soil exposure, and **f** illustrates the fan of soil left at the entrance of a recently excavated burrow. In (**f**), the burrow entrance is approximately 11 cm in diameter, and the inset illustrates a social group of California ground squirrels. *Photo credits* **a**, **b** David Augustine, **c**, **d** Ana Davidson, **e**, **f** Paul Stapp/Jennifer Smith

Fig. 15.6 (continued)

2013) and suppress invasion by undesired shrubs such as mesquite (*Prosopsis glandulosa*) in the south (Weltzin et al. 1997; Ponce-Guevara et al. 2016; Hale et al. 2020).

The influence of the three white-tailed species of prairie dog on rangelands in some cases can be similar to black-tailed prairie dogs, such as creating open grassland and burrows that provide habitat for grassland fauna (Keeley et al. 2016; Davidson et al. 2018). However, in portions of their range, *C. gunnisoni* and *C. lucurus* have less effect on bare soil exposure and vegetation height, relative to *C. ludovicianus*, and do not clip or girdle shrubs (Baker et al. 2013). As a result, the white-tailed species have fewer cascading effects on ground-nesting birds, and domestic and native large herbivores. For example, because some rangelands inhabited by *C. gunnisoni* already have substantial bare soil exposure due to aridity, their grazing effect on vegetation is not necessary to create breeding habitat for mountain plovers (Pierce et al. 2017). *C. gunnisoni* can have substantial effects in some rangelands (e.g., Chihuahuan desert grassland; Davidson and Lightfoot 2008), but lesser effects in others (e.g., Stapp 1998, Baker et al. 2013).

Like prairie dogs, gregarious species of ground squirrels alter ecosystems through burrow construction and effects on rangeland vegetation. In particular, *O. beecheyi* imposes disproportional effects on rangelands through construction of burrows used for shelter and breeding, which results in soil mixing and deposition of large, fan-shaped mounds at burrow entrances. A typical tunnel is roughly 5 m in length (Van Vuren and Ordeñana 2012), but soil type and squirrel density influence length and complexity of tunnels, varying from 0.9 to 70 m (Grinnell 1923). Most burrows have interconnected tunnels with multiple (e.g., 6–20) openings to the surface, each with an average diameter of 11 cm (Grinnell 1923). In contrast, *U. e. elegans'* burrow construction involves excavation of sticks, rocks, and sagebrush leaves to produce a pile of debris near each entrance (Andelt and Hopper 2016). Burrow construction and forage consumption by another ground squirrel, *U. richardsonii*, promotes overall plant community diversity and soil nitrate content on intensely grazed rangelands

Table 15.3 Vertebrate species and invertebrate taxa that are positively associated with burrowing rodents in western North American rangelands to varying degrees

Entirely dependent on burrowing rodents	Burrowing rodent species	Effect type	Supporting Citations
Black-footed ferret	Prairie dogs	Prey	Biggins and Eads (2018)
Burrowing owl	Prairie dogs, ground squirrels	Burrows	Desmond et al. (2000), Lenihan (2007), Augustine and Baker (2013), McCullough Hennessy et al. (2016)

Significant increase in populations throughout much of their range

American badger	Prairie dogs, ground squirrels	Prey	Minta et al. (1992), Goodrich and Buskirk (1998), Lomolino and Smith (2003), Lenihan (2007), Davidson et al. (2018)
Mountain plover	Black-tailed prairie dog	Bare soil	Dinsmore et al. (2005), Augustine and Derner (2012), Duchardt et al. (2019)
Ferruginous hawk	Prairie dogs, ground squirrels	Prey	Plumpton and Andersen (1997), Bak et al. (2001), Cook et al. (2003), Smith and Lomolino (2004)
Prairie rattlesnake	Prairie dogs, California ground squirrel	Burrows and prey	Kretzer and Cully (2001), Shipley and Reading (2006), Lenihan (2007)

Greater abundance documented on versus off-colony

Mexican kit fox, swift fox, coyotes	Black-tailed prairie dog	Burrows and Prey	Lomolino and Smith (2003), Moehrenschlager et al. (2007)
Grasshopper mouse	Black-tailed prairie dog; pocket gophers	Burrows, arthropods	Stapp (1997), Cully et al. (2010a), Kraft and Stapp (2013)
Golden eagle, American kestrel	California ground squirrel	Prey, arthropods	Lenihan (2007)
Horned lark, mourning dove, killdeer, thick-billed longspur	Black-tailed prairie dog		Smith and Lomolino (2004), Augustine and Baker (2013)
Lesser earless lizard	Black-tailed and Gunnison's prairie dog; banner-tailed kangaroo rat	Burrows, bare soil	Davidson et al. (2008), Kretzer and Cully (2001)
Striped and New Mexico whiptails	Banner-tailed kangaroo rat	Burrows	Davidson et al. (2008)
Gopher snake	California ground squirrel	Burrows and prey	Lenihan (2007)

(continued)

Table 15.3 (continued)

Entirely dependent on burrowing rodents	Burrowing rodent species	Effect type	Supporting Citations
Tiger salamanders	Black-tailed prairie dog; California ground squirrel; pocket gophers		Vaughan (1961), Kretzer and Cully (2001), Lomolino and Smith (2003)
Elodes spp., Gryllacridid crickets	Gunnison's prairie dog	Burrows, bare soil	Bangert and Slobodchikoff (2006)
Tenebrionidae and Anobiidae beetles; Rhaphidophoridae crickets	Gunnison's prairie dog, banner-tailed kangaroo rat	Burrows	Davidson and Lightfoot (2007)
Centipedes and ground beetles	California ground squirrel	Burrows, bare soil	Lenihan (2007)

The effect type column indicates the primary mechanism by which burrowing rodents benefit the associated species

(Newediuk et al. 2015). Burrows of most ground squirrel species are used by a diversity of commensal species, including amphibians, reptiles and mammals (Lenihan 2007; McCullough Hennessy et al. 2016; Conway 2018; Table 15.3).

15.3.2 Pocket Gophers and Kangaroo Rats

Because pocket gophers are small and asocial, their engineering effects on aboveground vegetation appear less intensive than that of prairie dogs, and their effects primarily occur via soil disturbance. Pocket gophers create mounds by pushing soil through burrows to the surface in inclined tunnels, and mounds can be distinguished from those of other rodents and moles by their crescent shape and visible soil plug, which will be quickly replaced if removed. Individual mounds typically range in size from 20 to 50 cm; however, in areas of high activity, mounds exist as irregular clusters and disturbed soils overlying burrows. The density and coverage of mounds varies with soil type, texture, and topography, ranging up to about 20% (Laycock and Richardson 1975; Grant et al. 1980; Carlson and Crist 1999; Stapp et al. 2008), or in unusual cases up to 50% (Stromberg and Griffin 1996). Burrows tend to be 10–70 cm belowground (Wilkins and Roberts 2007). Burrow diameter, depth and length vary across and even within species, and may be more related to plant distribution and soil characteristics than to body size (Romañach et al. 2005; Wilkins and Roberts 2007). Due to the highly clustered nature of gopher mounds and burrows at low and intermediate densities (Hansen and Remmenga 1961), the effects of gophers are likely to be more spatially heterogeneous than those of prairie dogs, and more widespread on the landscape. The density and dispersion of gopher populations ultimately reflect the availability and spatial patterning of preferred food plants and the friability of the soil, which are determined by soil type, topography, and land-use practices (Huntly and Inouye 1988). Gopher species differ in their affinities for particular soil types and textures (Miller 1964), with larger species restricted to deeper, sandier soils, and widely distributed species such as *T. talpoides* and *T. bottae* inhabiting a broad range of soil conditions.

Gophers are capable of transporting large amounts of soil: synthesizing studies of five common western gopher species, Smallwood and Morrison (1999) estimated soil excavation rates of 12.6–21.7 m^3 ha^{-1} yr^{-1} (mean 17.6 m^3 ha^{-1} yr^{-1}). A single gopher produces roughly 110 mounds per year (Romañach et al. 2007), mixing soil and nutrients between the surface and deeper soil layers (Huntly and Inouye 1988). Climate or land-use history of a given location can influence the degree to which mounds have more or less nutrients and moisture than off-mound sites. Regardless of location, mounds increase spatial heterogeneity and microtopographic variation, which affects primary productivity and plant communities (Reichman and Seabloom 2002). Mounds bury individual plants, creating small-scale openings where early seral species can establish. Increased availability of nitrogen adjacent to mounds enhances plant growth (Reichman 2007); in Colorado, this led to a 5.5% increase in primary productivity, which more than offset the loss of plants covered by mounds (Grant et al. 1980). By 3–4 years after appearance of a mound, cover of perennial and annual forbs increases (Foster and Stubbendieck 1980). In California, gopher mounds also favor establishment of exotic annual plants (Stromberg and Griffin 1996). As a consequence of the vertical mixing of the soils and increased spatial heterogeneity of resources and vegetative cover, gopher disturbances increase soil fertility, water flow, and plant species diversity (Reichman and Seabloom 2002), and in the some areas, alter soil development (Mielke 1977).

Vaughan (1961) reported at least 22 species of vertebrates using gopher mounds in Colorado and that the occurrence of tiger salamanders (*Ambystoma tigrinum*) was directly related to the presence of mounds. Connior (2011) documented 45 vertebrate species and numerous arthropods that were associated with *G. bursarius* habitat. The friable soils on mounds are used for dustbathing by grasshopper mice (*Onychomys leucogaster*) and the abundance of arthropods on mounds compared to other microhabitats may explain why foraging mice are oriented to gopher mounds, as they are with prairie dog burrows (Stapp 1997; Kraft and Stapp 2013). Gopher mounds enhance abundance of certain grasshopper species by providing favorable oviposition and sunning sites, in addition to a diversity of potential food plants (Huntly and Inouye 1988). Root foraging by gophers in Idaho resulted in higher damage to plants from mobile, chewing insects and decreased abundance of sedentary, sucking insects, such as aphids (Ostrow et al. 2002). Gophers also indirectly increase plant reproductive success by altering interactions between plants and their pollinators (Underwood and Inouye 2017).

Like pocket gophers, kangaroo rats are often described as ecosystem engineers, keystone species, or as a keystone guild (Davidson and Lightfoot 2008; Prugh and Brashares 2012). Experimental exclusion of multiple kangaroo rat species from grasslands has illuminated their effects on community structure and composition through selective harvesting of large seeds. For example, long-term removal of kangaroo rats from Chihuahuan Desert induced a transition from shrubland to a grassland dominated by an introduced, large-seeded grass (Brown and Heske 1990). Through a combination of selective foraging, burrowing, and other soil disturbances, *D. ingens* increased plant gamma diversity, biomass, and productivity, invertebrate biomass and diversity, and lizard and squirrel densities in the Carrizo Plain of California (Prugh and Brashares 2012). Similarly, removal of *D. stephensi* decreased overall plant species diversity and bare ground, causing dramatic increases in an introduced annual forb (Brock and Kelt 2004).

The larger species of kangaroo rats (*D. deserti, D. ingens, D. spectabilis*) create large mounds or precincts (e.g., 3–10 m in diameter) that dot rangeland landscapes (Davidson and Lightfoot 2008; Prugh and Brashares 2012; Fig. 15.7). Although mounds appear aggregated at landscape scales due to the spatial patterning of soils, plant communities, and livestock grazing, they are uniformly dispersed at finer scales, reflecting the intensity of intraspecific competition (Schooley and Wiens 2001). Mounds represent nutrient-rich patches with distinct soils and plant and animal assemblages compared to adjacent off-mound areas, increasing landscape heterogeneity and biodiversity (e.g. Davidson and Lightfoot 2008; Koontz and Simpson 2010; Fig. 15.7). In some cases, large kangaroo rat mounds can facilitate establishment of exotic weed species, because of the disturbed mound soil (Schiffman 1994). Grasshoppers that associate with bare soil and annual plants are abundant on the kangaroo rat mounds, as are insects that consume annual plant seeds (Davidson and Lightfoot 2007). Lizards use the mounds for basking and squirrels and prairie dogs use them as open, high points for viewing the landscape (Davidson et al. 2008; Prugh and Brashares 2012). Kangaroo rat mounds are honeycombed with shallow burrows, and provide homes and refugia for numerous species of arthropods, amphibians, lizards, snakes, other rodents, and rabbits (Hawkins and Nicoletto 1992; Hawkins 1996; Davidson and Lightfoot 2007; Prugh and Brashares 2012; Table 15.3). For example, Davidson and Lightfoot (2007) found that banner-tailed kangaroo rat (*D. spectabilis*) burrows enhanced abundance and species richness of multiple trophic and taxonomic groups of surface-active arthropods, as well as obligate arthropod burrow specialists. Burrows are also modified and used by larger vertebrates such as kit foxes (*Vulpes macrotis*), burrowing owls, American badgers (*Taxidea taxus*), and weasels (Hawkins and Nicoletto 1992; Conway 2018).

15.4 Predators of Burrowing Rodents

Burrowing rodents in rangelands of western North America also serve as a key prey source supporting a diverse array of predators. The endangered black-footed ferret (*Mustela nigripes*) is well-known to rely exclusively on prairie dogs as a prey source, and long-term conservation of this species depends upon landscape-scale prairie dog conservation (Biggins and Eads 2018). In portions of their range, Mexican kit foxes (*Vulpes macrotis zinseri*) also rely on prairie dogs and their burrows as a critical habitat (Moehrenschlager et al. 2007). The abundance of ferruginous hawks (*Buteo*

Fig. 15.7 Examples of mounds and surface disturbance created by the burrowing activities of pocket gophers in California annual grasslands (**a**) with a Botta's pocket gopher (*T. bottae*) shown in the inset, and by banner-tailed kangaroo rats (*D. spectabilis*) in the Janos grasslands of northern Chihuahua, Mexico, with a banner-tailed kangaroo rat in the inset. *Photo credits* **a** Paul Stapp, **b** Ana Davidson/David Lightfoot

regalis) and American badgers (*Taxidea taxus*) are both closely tied to the abundance of prairie dogs, ground squirrels, and pocket gophers across much of their range (Goodrich and Buskirk 1998; Cook et al. 2003; Lomolino and Smith 2003; Lenihan 2007). Most ground squirrels contribute to biodiversity as major prey items for a diverse assemblage of snakes, raptors, and predatory mammals, including *O. beecheyi* (Lenihan 2007; McCullough Hennessy et al. 2016), *U. richardsonii* (Michener 1979), *U. u. elegans* (Andelt and Hopper 2016), *U. columbianus* (Macwhirter 1991), and *U. armatus* (Minta et al. 1992). Although pocket gophers spend relatively little time above ground, they are surprisingly common prey in the diet of many raptors and owls, presumably because soil movement reveals their presence, and their vision is poor (Cartron et al. 2004).

15.5 Interactions with Livestock

15.5.1 Prairie Dogs and Ground Squirrels

Diets of prairie dogs overlap substantially with livestock, and all five species have experienced widespread efforts to control their populations, typically via rodenticides (Detling 2006; Miller et al. 2007). Prairie dogs significantly increase bare soil exposure and reduce vegetation height and biomass on their colonies across a diverse suite of rangeland ecosystems (Baker et al. 2013). This creates substantial concern for livestock producers because forage limitations during dry periods are a major determinant of long-term stocking rates on western rangelands. Traditional models of rangeland dynamics (reviewed by Briske et al. 2005) also frequently associate vegetation conditions on prairie dog colonies with a degraded or overgrazed state (e.g., Augustine et al. 2014). However, prairie dogs are also well-known to enhance protein content and digestibility of forage on their colonies, and cattle sometimes graze preferentially on colonies (Sierra-Corona et al. 2015), raising questions about when, where, and to what extent prairie dogs negatively affect livestock. Early research found minimal differences in cattle weight gains in pasture with versus without black-tailed prairie dogs (O'Meilia et al. 1982), but more recent work found that weight gains of yearling steers declined by 15% as prairie dog abundance increased from 0 to 60% of a pasture in shortgrass steppe (Derner et al. 2006).

Prairie dogs influence forage quality and quantity in both shortgrass and mixed grass rangelands, with the largest effects in mixed grass (Augustine and Springer 2013). This can lead to prairie dogs enhancing cattle gains in wet years, but suppressing cattle gains in dry years (Augustine and Springer 2013; Connell et al. 2019). Because prairie dog populations now fluctuate dramatically throughout their range due to plague outbreaks (even in the absence of human control efforts), long-term studies are needed to understand how colony expansion and contraction interact with varying precipitation to influence livestock operations beyond a single growing season. Additionally, suppression of forage quantity by *C. ludovicianus* during the dormant season (as they are typically active in this time period, with only occasional facultative episodes of torpor; Lehmer and Biggins 2005), can clearly impact livestock operations. These dormant-season effects need to be incorporated into assessments of multi-year effects on livestock operations.

C. ludovicianus also have unique effects on Chihuahuan Desert grasslands, which have undergone widespread transitions from grassland to shrubland encroached by mesquite and other desert shrubs (Van Auken 2009). Prairie dogs clip and girdle desert shrubs to increase their ability to detect predators, and they consume the seedlings of shrubs (Weltzin et al. 1997; Ceballos et al. 2010; Baker et al. 2013; Ponce et al. 2016). By doing so they help maintain grasslands that ranchers depend on for their cattle. Cattle facilitate the presence of prairie dogs through their grazing, keeping vegetation height low, which helps promote the ecological role of prairie dogs in controlling mesquite expansion (Ponce et al. 2016, Sierra-Corona et al. 2015). For example, Ponce et al. (2016) found that mesquite abundance in the Chihuahuan Desert grasslands was up to five times greater in plots without versus with prairie dogs. Mesquite canopy cover increased 61% over a 23-year period following prairie dog eradication (Weltzin et al. 1997). Ceballos et al. (2010) found shrub cover expanded by 34% into a desert grassland over seven years following prairie dog poisoning.

We are unaware of studies directly examining effects of the three white-tailed prairie dog species on livestock. Of the three, *C. parvidens* may have had the greatest localized effect on livestock during the twentieth century because of their affinity for productive swales, which is likely linked to intensive control efforts across their range, followed by eventual listing as an Endangered Species in 1973 (USFWS 2012). Recovery efforts have aimed to relocate populations in conflict with agriculture on private lands to historically occupied public lands (USFWS 2012). All three white-tailed species hibernate during the dormant season, which reduces effects on dormant-season livestock forage. In addition, growing-season effects of *C. gunnisoni* and *C. leucurus* on forage growth are likely to be less than that of *C. ludovicianus* due to lower population densities and per-animal forage requirements (Grant-Hoffman and Detling 2006; Table 15.1).

Livestock grazing also has reciprocal effects on prairie dog populations. In portions of their range with tall vegetation, *C. ludovicianus* often depend on livestock or native large grazers to maintain sufficiently short vegetation for them to persist or increase in abundance (Davidson et al. 2010). In these areas, reductions in live-stock grazing pressure can be an effective means of discouraging colony expansion or recolonization of former colonies (Cable and Timm 1987; Truett et al. 2001). In more arid rangelands, some prairie dog populations are strongly limited by forage availability during drought (e.g., *C. gunnisoni*; Davidson et al. 2018), and forage loss to simulated livestock grazing has been shown to suppress population growth (e.g. *C. parvidens*; Cheng and Ritchie 2006).

Ground squirrel interactions with livestock vary among species in relation to their diet, degree of coloniality, and population density. *O. beecheyi* often forage on seeds of grasses and oaks, which reduces direct competition with cattle for food (Linsdale 1946). However, during the growing season, *O. beecheyi* forage almost exclusively on herbaceous vegetation, giving rise to seasonal competition with livestock. In one direct test for cattle-ground squirrel competition, heifers gained more weight during the growing season where *O. beecheyi* were controlled with rodenticides compared to where they were not (Howard and Bentley 1959). A follow-up study showed that the overall energy requirements of *O. beecheyi* are minimal (e.g., 94 cal/g/d) and their dietary preferences generally differ from those of cattle (Schitoskey and Woodmansee 1978), but they may still reduce forage for livestock to a greater extent than indicated by energy requirements as a result of clipping and burrowing effects. *O. beecheyi* also benefits from foraging in open, grazed habitat alongside cattle (Ortiz et al. 2019; Hammond et al. 2019). *O. beecheyi* can also be highly opportunistic, frequently stealing feed grains, pellets, and molasses lick blocks from facilities with poultry and livestock (Baker 1984), making them a serious agricultural pest, responsible for at least 12–16 million dollars in annual losses in California (Marsh 1998).

Other ground squirrel species that focus heavily on green leaves and grasses, such as *U. richardsonii* (Michener and Koeppl 1985) and *U. armatus* (Eshelman and Sonnemann 2000), may overlap more with the foraging niche of grazing livestock. Competition between native rodents and cattle is particularly common when rangelands are heavily grazed by livestock. Drought can amplify the effects of poor management practices, favoring population irruptions such as those well documented for *U. richardsonii* in Canadian prairies (Proulx 2010). Much like prairie dogs, *U. richardsonii* population densities are favored in heavily grazed rangelands where squirrels benefit from foraging on leaves, flowers, and seeds in open areas with reduced predator risk (Michener and Koeppl 1985). In tallgrass prairie, potential impacts of *P. franklinii* on livestock are low compared to those of *Urocitellus* species because *P. franklinii* is omnivorous, burrows are dispersed at lower density, and the species declines in mowed or heavily grazed grasslands (Ostroff and Finck 2003).

Although ranchers often express concerns about livestock breaking legs in the holes of burrowing mammals (Minta and Marsh 1988), there is little empirical evidence that burrows are a significant source of injury, and none of the multiple long-term studies of livestock-prairie dog interactions have reported this as an issue. However, Weir et al. (2016) found that 16% of 131 ranchers surveyed from British Columbia, Canada, reported injuries to livestock from burrows during the previous five years. Although pastures were inhabited by ground squirrels (*U. columbianus*), pocket gophers, marmots (*Marmota flaviventris*), and badgers, most (79%) of the injuries were caused by rodent burrows, and were to horses (58%) rather than cattle or other livestock, with 25% of the injuries ultimately requiring euthanasia. Many (53%) ranchers also reported damage from burrows to agricultural machinery including swathers and balers, over that period.

15.5.2 Pocket Gophers and Kangaroo Rats

Interactions between gophers and livestock are not as well-studied as they are for prairie dogs and ground squirrels. Gophers affect livestock mostly through effects on the availability and nutritional quality of forage. Over the short term, the creation of mounds buries forage plants, reducing basal cover, and consumption of belowground parts increases mortality, although these impacts may be offset by higher plant growth of plants adjacent to mounds (Grant et al. 1980; Reichman and Seabloom 2002). Gopher mounds may facilitate the establishment of exotic species (Stromberg and Griffin 1996) and plant species that are unpalatable to livestock (Foster and Stubbendieck 1980). Over the longer term, gopher activity can improve deteriorated rangelands by breaking up compacted soils and redistributing belowground nutrients (Grinnell 1923). In turn, the activities of livestock and other large grazers may influence the abundance and dispersion of gophers via their effects on the productivity and spatial patterning of major food plants (Steuter et al 1995).

To estimate forage loss to pocket gophers, California ground squirrels, and kangaroo rats in rangeland of central California, Fitch and Bentley (1949) stocked each rodent species in separate enclosures, all of which lacked cattle. They estimated that during the wet season, these rodents destroyed 25, 35, and 16% respectively of the potential forage yields in those enclosures. Losses included consumption, trampling, burying live plants, and clipping or caching belowground. Rodents removed comparatively little vegetation during the dry season. They concluded that competition between livestock and rodents is much more significant during the growing season, but the degree of competition depended on rodent population densities and annual variation in herbage production. Turner (1969) estimated that gophers may reduce standing crop biomass in Colorado mountain rangeland by as much as 20%. In western Nebraska, Foster and Stubbendieck (1980) reported that forage production, especially of perennial grasses, in gopher-disturbed pastures was 21–49% lower than in undisturbed pastures.

High densities of gopher mounds traditionally have been considered an indicator of overgrazing (Laycock and Richardson 1975), but mounds also tend to be more conspicuous in disturbed, heavily grazed pastures. Studies quantifying relationships between livestock grazing intensity and mound coverage have not identified consistent relationships (Grant et al. 1980; Stromberg and Griffin 1996; Carlson and Crist 1999).

Kangaroo rats are generally not considered major competitors with livestock in rangelands because they are primarily granivores, but they have been subject to past extermination efforts for rangeland management (Reynolds 1958). In some desert grasslands with heavy livestock grazing, their clipping of perennial grasses, consumption of large-seeded perennial grasses, and dispersal of mesquite seed pods has been suggested to further reduce perennial grass cover and therefore play a role in desertification (Reynolds 1958). However, long-term exclosure experiments (21 years) in the Chihuahuan desert grassland revealed significant increases in mesquite establishment when kangaroo rats were removed, suggesting they actually help prevent shrub invasion and therefore desertification of grasslands (Valone and Thornhill 2001).

15.6 Impacts of Disease

Prairie dogs and ground squirrels are affected by a wide array of bacterial, viral, parasitic and fungal diseases, some of which are zoonotic and hence of concern for human health (reviewed by Donnelly et al. 2015). Most notably, all of the prairie dog and ground squirrel species in western North America are affected to some extent by enzootic and/or epizootic outbreaks of plague. In contrast, pocket gophers and kangaroo rats are not known to be regulated by plague, nor are they known to be hosts of plague or other major zoonotic diseases. Plague was first documented in *O. beecheyi* in 1908, and then spread eastward, reaching *C. ludovicianus* populations in the Great Plains by the 1940s (Cully and Williams 2001; Biggins and Eads 2019). This disease is one of the most important factors currently driving population fluctuations in prairie dogs and ground squirrels. Plague can also be transmitted from prairie dogs and ground squirrels to humans and their pets through fleabites and direct contact with infected animals. As a result, it is important for people managing prairie dogs or ground squirrels to avoid movement of fleas from carcasses or burrows onto themselves.

Epizootic outbreaks of plague periodically decimate prairie dog populations throughout their range, often causing > 95% decline in the size of individual colonies or entire colony complexes (distributed across landscapes of up to > 100,000 ha) within a single year (Cully and Williams 2001; Stapp et al. 2004; Augustine et al. 2008; Cully et al. 2010b). Because plague transmission is density dependent and its spread depends on colony dispersion across the landscape, the introduction of plague to the Great Plains altered the historical metapopulation dynamics of prairie dogs by increasing the likelihood of extinction of large colonies, as well as the smaller, neighboring satellite colonies (Stapp et al. 2004, 2008). In prairie dogs, plague can be transmitted via multiple flea species, but the most important plague vectors are *Oropsylla hirsuta* and *O. tuberculata cynomuris* (Salkeld et al. 2016). Rapid plague spread across a large colony complex is related to colony size, low inter-colony distances, and proximity of colonies to dry creek drainages, which prairie dogs use for dispersal (Roach et al. 2001; Stapp et al. 2004; Johnson et al. 2011). Less rapid spread is related to large inter-colony distances, as well as the distribution of roads, lakes and streams that reduce dispersal and plague transmission (Collinge et al. 2005). In addition, overall population response to plague has been less severe for *C. leucurus* compared to *C. gunnisoni* and *C. ludovicianus*, presumably because *C. leucurus* has lower population densities (up to 10 times lower than that of *C. gunnisoni* and *C. ludovicianus*), is less social, and has smaller colony areas (Cully and Williams 2001). Recent studies suggest some populations of *C. gunnisoni* and *C. ludovicianus* that experienced plague outbreaks since the 1940s may be less susceptible than populations with no history of plague (Rocke et al. 2012; Busch et al. 2013). Nevertheless, inter and intra-species differences in vulnerability to plague remain. For example, *C. gunnisoni* and *C. parvidens* populations from high elevations remain highly susceptible, despite historical exposure (Russell et al. 2018).

How the disease is maintained between epizootics and factors driving epizootics are the subject of substantial ongoing research (Cully et al. 2010b; Salkeld et al. 2016; Biggins and Eads 2019). Uncertainty on these subjects and variability in colony size/connectivity across rangelands makes it difficult to predict the precise location and timing of epizootics, although observed spatial patterns suggest plague is spread by dispersing prairie dogs or carnivores that can move between colonies carrying infected fleas, rather than separate local foci (Stapp et al. 2004; Salkeld et al. 2016). Epizootics typically occur at intervals of 5–15 years, with slow, steady regrowth of the colonies in between (Stapp et al. 2004; Augustine et al. 2008; Hartley et al. 2009; Cully et al. 2010b). The likelihood of epizootics is also influenced by complex interactions between precipitation and temperature and the bacterium, fleas that transmit it, prairie dog health, the presence of amplifying alternate hosts, and the movements of other species capable of moving the pathogen between colonies (Stapp et al. 2004; Salkeld et al. 2016; Eads and Biggins 2017).

Prior to European settlement and the introduction of plague, prairie dog colonies in the western Great Plains were typically large, stable features on the landscape that did not undergo periodic, dramatic plague-induced declines (Knowles et al. 2002). Dramatic expansions of prairie dog colonies between epizootics are not necessarily population increases above historical levels, but rather are population recoveries following population collapse from a non-native disease (Cully et al. 2010b). Plague also kills and is a major impediment to recovery of the endangered black-footed ferret (Matchett et al. 2010). Epizootics can also prevent prairie dog colonies from occupying the same location continuously over long time periods, which reduces their impact on vegetation composition and productivity in the western Great Plains (Hartley et al. 2009; Augustine et al. 2014).

Like prairie dogs, ground squirrels in western rangelands host fleas that can transmit plague (Gage and Kosoy 2005). In California, *O. beecheyi* in combination with the fleas *Orophsylla montana* and *Hoplopsyllus anomalous*, form the principal complex for amplifying plague (Barnes 1982; Lang 2004), which are often detected because outbreaks can result in large numbers of dead squirrels. However, some populations of *O. beecheyi* show considerable resistance to plague mortality (e.g., Williams et al. 1979), so that they may not experience the extreme die-offs seen in prairie dog populations. Moreover, *O. beecheyi* hosts have consistent individual differences in their flea abundance/community stability (Smith et al. 2021) and degree of sociality (Smith et al. 2018) across years. Individuals vary considerably in these traits such that host heterogeneity is likely a strong determinant of plague transmission. As with prairie dogs, plague-related fluctuations in *O. beecheyi* numbers can influence prey availability for multiple predators (Lenihan 2007).

15.7 Threats

Three primary threats affect prairie dogs and ground squirrels: (1) periodic outbreaks of epizootic plague, (2) direct control by humans via rodenticide and shooting, and (3) loss and fragmentation of habitat. Plague and poisoning can rapidly reduce colony size and density, thereby removing the functional role of prairie dogs and ground squirrels in rangelands for several years until populations recover, with ramifications for prairie-dog/ground squirrel-associated species during these low points. The combination of plague and poisoning can influence metapopulation dynamics, whereby colonies extirpated by plague and/or control rely on recolonization from other colonies, critical to long-term prairie dog population viability. As a result, fragmented land ownership patterns can also threaten populations where lands upon which prairie dogs are controlled are closely interspersed with lands where they are not (Augustine et al. 2021).

Loss of habitat due to expanding human development is affecting some portions of the range of *C. ludovicianus*, and can even lead to small, isolated populations with unusually high density (Magle et al. 2007), but vast areas of their range remain unaffected. Because *C. parvidens* has a restricted range and preferentially colonizes productive swales in valleys where grasslands have been converted to housing, golf courses, and hay production, land development continues to affect recovery efforts (USFWS 2012).

Prairie dogs are additionally threatened by drought and climate change. In the southern and far northern portion of their range, drought can greatly limit recruitment, causing population declines and preventing recovery of reintroduced populations (Ceballos et al. 2010; Facka et al. 2010; Davidson et al. 2014; Hayes et al. 2016; Stephens et al 2018). One of the formerly largest remaining colonies of *C. ludovicianus* in Chihuahuan Desert grasslands (in Janos, Chihuahua, Mexico) has collapsed by 90% (Ceballos et al. 2010; Ponce pers. Comm.). It is uncertain if plague caused the population collapse, but the collapse occurred during multiple years of extreme drought and little offspring recruitment (Ceballos et al. 2010; Ponce pers. comm.). The population still has not recovered, presumably because of increasing aridity and desertification (Ceballos et al. 2010; Ponce pers. Comm.). Prairie dogs are also vulnerable in the northern portion of their range to harsh winters, which could be ameliorated by climate warming (Stephens et al. 2018).

Kangaroo rats that inhabit desert grasslands, such as *D. spectabilis*, have experienced major population declines due to desertification (Waser and Ayers 2003). Other species of kangaroo rats with small, endemic ranges throughout California's grasslands and deserts have lost habitat and risk extinction due to expansive urban development, invasion of exotic annual grasses, and widespread agricultural conversion (Goldingay 1997; Longland and Dimitri 2020). Climate change is an increasing threat to kangaroo rats, with loss of habitat and range shifts predicted, as well as increased temperatures expected to exceed physiological tolerance of the desert-adapted rodents (Price et al. 2000; Widick and Bean 2019; Wilkening et al. 2019).

15.8 Management and Conservation Actions

15.8.1 Prairie Dogs and Ground Squirrels

As a result of widespread populations declines, four prairie dog species—*C. ludovicianus*, *C. parvidens*, *C. gunnisoni*, and *C. mexicanus*—have been proposed for listing under the U. S. Endangered Species Act. *C. ludovicianus* was proposed for listing in 1999. Following reviews, the species was removed as a Candidate for listing in 2004 on the basis that improved state agency surveys estimated 745,750 ha of occupied habitat in the United States (Federal Register Vol 69, No 159, 8/18/2004 pp 55,217–51,226). *C. ludovicianus* conservation in the United States is currently led by the 11 individual states where the species occurs, coordinated through a multistate conservation plan (Luce 2003). In Canada, where *C. ludovicianus* inhabits a restricted portion of Saskatchewan that includes Grasslands National Park, the species is listed as threatened. *C. ludovicianus* also occupies a portion of the northern Chihuahuan Desert in Mexico, where it is listed as endangered and populations have

contracted over the past three decades in response to severe droughts and shrub inva-
sion (Ceballos et al. 2010). In 2009, the Mexican government established the Janos
Biosphere Reserve to advance conservation of prairie dogs and associated species in
Mexico.

C. gunnisoni was removed as a Candidate for listing in 2013 on the basis that
occupancy surveys indicated that populations had stabilized, and dusting burrows
with insecticide effectively controls plague (Federal Register Vol 78, No 220, 11/
14/2013, ppl 68,660–68,665). *C. parvidens* was listed as an Endangered Species in
the U.S. in 1973, and downlisted to threatened in 1984. Populations are limited to
seven counties in southwest Utah, and have either increased or remained stable over
the past 30 years (USFWS 2012). Management to recover populations of both *C.
gunnisoni* and *C. parvidens* has focused primarily on translocations from agricultural fields and urban conflict areas onto public lands where they historically occurred
(USFWS 2012; Nelson and Theimer 2012; Curtis et al. 2014). Attempts to translo-
cate *C. gunnisoni* to historic sites in the southern portion of their range have been
compromised by severe drought (Davidson et al. 2014).

C. mexicanus is listed as Endangered by the U.S. and Mexican governments,
and occurs in six valleys within the south-central Chihuahuan Desert of Mexico,
primarily on low-productivity rangelands with gypsum-derived soils (Yeaton and
Flores-Flores 2006). Following decades of range contraction, populations may have
stabilized where rangelands cannot be converted to cropland due to lack of irrigation
water (Yeaton and Flores-Flores 2006); there is no formal conservation or recovery
plan.

Public lands with a focus on *C. ludovicianus* conservation include Charles
M. Russell and UL Bend National Wildlife Refuges (MT), Theodore Roosevelt,
Badlands and Wind Cave National Parks (ND and SD), and Grasslands National
Park (SK). Public lands with a multiple use focus that includes balancing prairie
dog conservation and livestock production include 14 National Grasslands in 9
states managed by the US Forest Service, and areas administered by the Bureau of
Land Management in New Mexico, Wyoming, and Montana. Because most of these
public lands are closely intermingled with private lands, cross-boundary management of prairie dogs to both conserve populations and minimize impacts on livestock
producers has emerged as a key management issue (e.g., Miller et al. 2007; Augustine
et al. 2021). Tribal lands in the northern Great Plains also host extensive populations
of *C. ludovicianus*, and in some cases are contiguous with public lands.

Management to enhance prairie dog populations primarily relies on plague mitigation and population translocations. The application of insecticides (e.g., deltamethrin)
directly to burrows to control the flea vector is effective in preventing epizootic plague
and maintaining prairie dog genetic diversity (Jones et al. 2012; Eads and Biggins
2019). Recent studies also show that application of flour-based baits containing
fipronil that are consumed by prairie dogs can suppress fleas for up to a year (Eads

et al. 2021). However, both external and internal insecticides still require labor-intensive application to colonies annually. A bait-based vaccine increased survivorship in field trials (Rocke et al. 2017), but it is still undergoing evaluation and refinement as a management tool. Best practices for translocation are discussed by Truett et al. (2001), Curtis et al. (2014) and Davidson et al. (2018).

Conversely, management to reduce or eliminate prairie dog populations primarily occurs through direct poisoning with rodenticides consisting of toxicants or anticoagulants, and recreational shooting. The latter does not as dramatically affect numbers as poisoning efforts, but shooting can significantly alter behavior and suppress reproduction in local populations (Pauli and Buskirk 2007a) and is of concern due to lead fragment impacts on mammals and birds that scavenge carcasses (Pauli and Buskirk 2007b). Similarly, anticoagulants are of concern due to their impact on scavengers (Witmer et al. 2016).

One example of how all these management tools can be combined with cross-jurisdictional collaboration to enhance prairie dog conservation is in a portion of South Dakota encompassing Badlands National Park, Buffalo Gap National Grassland, the Pine Ridge Indian Reservation, and intermingled private lands. Here, management to enhance prairie dog populations by controlling plague (primarily via insecticides) in conservation zones is implemented alongside management to control prairie dogs via poisoning within boundary zones to prevent dispersal onto privately owned rangelands. Management of both plague and prairie dogs across jurisdictions at this landscape scale has been essential in sustaining a wild population of black-footed ferrets at this site (Phillips et al. 2020).

In contrast to prairie dogs, ground squirrel species described in this chapter are of least concern and their robust numbers are often the focus of integrated pest management programs; these include: (i) monitoring, (ii) preventative practices (e.g., educating farmers), and (iii) implementation of a variety of control methods (e.g., mechanical, physical, biological and chemical; Andelt and Hopper 2016). Over the past century, extensive efforts have been undertaken to control most rangeland ground squirrel species (e.g., Gilson and Salmon 1990; Marsh 1994; Proulx 2010). The application of toxicants, and to a lesser extent the use of shooting or fumigants, continue to be the most widely used control methods (e.g., Marsh 1994; Baldwin et al. 2014). The state of California now bans the use lead bullets because secondary ingestion of *O. beecheyi* carcasses can be lethal for wildlife at higher trophic levels (Smith et al. 2016). Historical effects include the killing of endangered species such as California condors (*Gymnogyps californianus*) and native predators (Marsh et al. 1987; Pattee et al. 1990). Toxic fumigants applied to *O. beecheyi* burrows may also kill commensal species, such as the California tiger salamander (*Ambystoma californiense*) and burrowing owl. Recreational shooting of other ground squirrel species also occurs throughout their ranges; carcasses of shot squirrels are known to pose a lead poisoning hazard for scavengers of *U. richarsonii* (Knopper et al. 2006) and *U. beldingi* (Herring et al. 2016).

While there has been some success in using targeted poisons applied at times that minimize death of nontarget species (Whisson 1999), nonlethal controls (e.g., habitat modification, translocations) can also be effective alternatives (Gilson and

Salmon 1990; McCullough Hennessy et al. 2016; Swaisgood et al. 2019). Plague in ground squirrels can be suppressed effectively through application of insecticide at bait stations or burrow entrances (Barnes 1982). Despite this, lethal methods such as poisoning and gassing remain the most widely used method to reduce the size of the potential plague reservoir in rangelands (Wobeser 1994).

15.8.2 Pocket Gophers and Kangaroo Rats

For most species and throughout their ranges, pocket gophers are either considered innocuous or managed as pests. Control methods include flood irrigation, trapping, fumigants, or toxic baits (Baldwin et al. 2014), although most methods are effective only at small scales. There is increasing interest in biological control methods, such as erecting nest boxes for barn owls (*Tyto alba*; Browning et al. 2016), but degree of efficacy is not clear (Moore et al. 1998). In the past, kangaroo rats experienced some population control (Reynolds 1958); zinc and aluminum phosphide are approved to reduce kangaroo rat damage in some areas, but these animals are rarely considered pests and many species and populations are declining.

Some pocket gopher species (e.g., *T. clusius*) and subspecies (*T. bottae curtatus*) are of state or federal conservation concern. Six rangeland species of *Dipodomys* (*elator, ingens, nitratoides exilis, nitratoides nitratoides, spectabilis, and stephensi*) are considered Endangered, Vulnerable, or Near-Threatened by IUCN. Conservation efforts have primarily focused on establishing small parks and natural areas to prevent further habitat loss to urban and agricultural development, restoring habitat on these lands, and genetic studies to understand variation among disjunct populations (e.g., Price and Endo 1989; USFWS 2020).

15.9 Conclusion

Our review of the distribution and impact of ground squirrels, pocket gophers, and kangaroo rats in western North America highlights the strong and geographically widespread effects they have on rangeland ecosystems. Despite many differences among species and functional groups of burrowing rodents in terms of body size, diet, habitat ecology, and physiology, they all function as key agents of soil disturbance, engineers of belowground refugia, modifiers of vegetation structure and plant community composition, and a prey base for a suite of predators. Of these ecosystem functions, the extent of vegetation modification varies, with the more social, herbivorous, and larger-bodied prairie dogs having the strongest effects on vegetation structure and composition, and the non-social, granivorous kangaroo rats having comparatively more subtle effects. However, all four groups of burrowing rodents reviewed here can induce substantial movement and mixing of soils in rangelands, with important consequences for nutrient cycling, water infiltration, soil structure, the establishment of early-successional plants, and for associated species that use their burrows.

As the extent of rangelands contract in the face of human development and cropland conversion and as livestock production intensifies to meet the demands of our growing population, the management and conservation of burrowing rodents is likely to become both increasingly important and controversial on western rangelands. The conservation of burrowing rodents with relatively restricted ranges (e.g., *C. parvidens, D. ingens*) may be achieved through localized protection in parks and preserves, translocations to public lands where conflicts with livestock and human development are minimized, and the development of a plague vaccine. Yet, conservation of specific species at small scales on lands that do not support livestock grazing may not effectively conserve the role burrowing rodents play in creating habitat for associated species and as a prey base for predators across broad rangeland landscapes. Management and conservation efforts that enable burrowing rodents to coexist with livestock and native ungulates across broad landscapes will likely be essential for the conservation of black-footed ferrets, mountain plovers, burrowing owls, and ferruginous hawks, along with a diverse array of herpetofauna, other small mammals, and arthropods. Continued research is needed to develop creative management approaches that minimize impacts of burrowing rodents on livestock producers, while sustaining their ecologically important role as engineers. Such approaches may include both lethal control of burrowing rodent populations in some locations, while simultaneously mitigating plague and enhancing rodent populations in other locations at the spatial scales necessary to sustain associated species. As a result, cross-jurisdictional collaboration and coordination among managers of both public and private lands will be essential for achieving desired outcomes associated with burrowing rodents in rangelands.

References

Andelt WF, Hopper SN (2016) Managing Wyoming ground squirrels: fact sheet. Natural Resources Series: Wildlife, Colorado State University, Fort Collins, CO. https://extension.colostate.edu/topic-areas/natural-resources/managing-wyoming-ground-squirrels-6-505/

Archer S, Garrett MG, Detling JK (1987) Rates of vegetation change associated with prairie dog (*Cynomys ludovicianus*) grazing in North American mixed-grass prairie. Vegetatio 72:159–166. https://doi.org/10.1007/BF00039837

Augustine DJ, Baker BW (2013) Associations of grassland bird communities with black-tailed prairie dogs in the North American Great Plains. Conserv Biol 27(2):324–334. https://doi.org/10.1111/cobi.12013

Augustine DJ, Derner JD (2012) Disturbance regimes and mountain plover habitat in shortgrass steppe: large herbivore grazing does not substitute for prairie dog grazing or fire. J Wildl Manage 76(4):721–728. https://doi.org/10.1002/jwmg.334

Augustine DJ, Springer TL (2013) Competition and facilitation between a native and a domestic herbivore: trade-offs between forage quantity and quality. Ecol Appl 23(4):850–863

Augustine DJ, Matchett MR, Toombs TP, Cully JF Jr, Johnson TL, Sidle JG (2008) Spatiotemporal dynamics of black-tailed prairie dog colonies affected by plague. Landsc Ecol 23(3):255–267. https://doi.org/10.1007/s10980-007-9175-6

Augustine DJ, Derner JD, Detling JK (2014) Testing for thresholds in a semiarid grassland: the influence of prairie dogs and plague. Rangeland Ecol Manage 67(6):701–709. https://doi.org/10.2111/REM-D-14-00032.1

Augustine DJ, Davidson A, Dickinson K, Van Pelt B (2021) Thinking like a Grassland: challenges and opportunities for biodiversity conservation in the great plains of North America. Rangel Ecol Manag 78:281–295. https://doi.org/10.1016/j.rama.2019.09.001

Bak JM, Boykin KG, Thompson BC, Daniel D (2001) Distribution of wintering ferruginous hawks (*Buteo regalis*) in relation to black-tailed prairie dog (*Cynomys ludovicianus*) colonies in southern New Mexico and northern Chihuahua. J Raptor Res 35:124–129

Baker RO (1984) Comingling of Norway and roof rats with native rodents. In: Clark DO (ed) Proceedings of the 10th vertebrate pest conference. University of California, Davis, CA, USA, pp 103–111

Baker B, Augustine D, Sedgwick J, Lubow B (2013) Ecosystem engineering varies spatially: a test of the vegetation modification paradigm for prairie dogs. Ecography 36:230–239. https://doi.org/10.1111/j.1600-0587.2012.07614.x

Barnes A (1982) Surveillance and control of bubonic plague in the United States. In: Edwards M, McDonnell U (eds) Animal disease in relation to animal conservation. Academic Press, Zoological Society of London

Barnes AM (1982) Surveillance and control of plague in the United States. Animal disease in relation to animal conservation. Edwards MA, McDonnell U (eds) Symposia of the Zoological Society of London 50. Academic Press, New York, pp 237–270

Baldwin RA, Salmon TP, Schmidt RH, Timm RM (2014) Perceived damage and areas of needed research for wildlife pests of California agriculture. Integr Zool 9(3):265–279. https://doi.org/10.1111/1749-4877.12067

Bangert RK, Slobodchikoff CN (2006) Conservation of prairie dog ecosystem engineering may support arthropod beta and gamma diversity. J Arid Environ 67(1):100–115. https://doi.org/10.1016/j.jaridenv.2006.01.015

Barth CJ, Liebig MA, Hendrickson JR, Sedivec KK, Halvorson G (2014) Soil change induced by prairie dogs across three ecological sites. Soil Sci Soc Am J 78:2054–2060. https://doi.org/10.2136/sssaj2014.06.0263

Biggins DE, Eads DA (2018) Evolution, natural history, and conservation of black-footed ferrets. In: Biology and conservation of musteloids. pp 340–356. https://doi.org/10.1093/oso/9780198759805.003.0015

Biggins DE, Eads DA (2019) Prairie dogs, persistent plague, flocking fleas, and pernicious positive feedback. Front Vet Sci 6. https://doi.org/10.3389/fvets.2019.00075

Blumstein DT, Armitage KB (1997) Does sociality drive the evolution of communicative complexity? A comparative test with ground-dwelling sciurid alarm calls. Am Nat 150(2):179–200. https://doi.org/10.1086/286062

Brennan JR, Johnson PS, Hanan NP (2020) Comparing stability in random forest models to map Northern Great Plains plant communities in pastures occupied by prairie dogs using Pleiades imagery. Biogeosciences 17(5):1281–1291. https://doi.org/10.5194/bg-17-1281-2020

Briske DD, Fuhlendorf SD, Smeins FE (2005) State-and-transition models, thresholds, and rangeland health: a synthesis of ecological concepts and perspectives. Rangeland Ecol Manage 58:1–10. https://doi.org/10.2111/1551-5028(2005)58%3c1:SMTARH%3e2.0.CO;2

Brock RE, Kelt DA (2004) Keystone effects of the endangered Stephens' kangaroo rat (*Dipodomys stephensi*). Biol Conserv 116(1):131–139. https://doi.org/10.1016/S0006-3207(03)00184-8

Brown JH, Heske EJ (1990) Control of a desert-grassland transition by a keystone rodent guild. Science 250(4988):1705–1707. https://doi.org/10.1126/science.250.4988.1705

Brown JH, Zongyong Z (1989) Comparative population ecology of eleven species of rodents in the Chihuahuan Desert. Ecology 70(5):1507–1525. https://doi.org/10.2307/1938209

Brown JH, Reichman OJ, Davidson DW (1979) Granivory in desert ecosystems. Annu Rev Ecol Evol Syst 10(1):201–227. https://doi.org/10.1146/annurev.es.10.110179.001221

Brown JH, Fox BJ, Kelt DA (2000) Assembly rules: desert rodent communities are structured at scales from local to continental. Am Nat 156:314–321. https://doi.org/10.1086/303385

Browning M, Cleckler J, Knott K, Johnson M (2016) Prey consumption by a large aggregation of Barn Owls in an agricultural setting. In: Baldwin RA (ed) Proceedings of the 27th vertebrate pest conference. University of California, Davis, CA, USA, pp 337–344

Busch JD, Van Andel R, Stone NE, Cobble KR, Nottingham R, Lee J et al. (2013) J Wildl Dis 49:920–931. https://doi.org/10.7589/2012/08/209

Cable KA, Timm RM (1987) Efficacy of deferred grazing in reducing prairie dog reinfestation rates. In: Eighth Great Plains wildlife damage control workshop proceedings. Rapid City, South Dakota, USA

Carlson JM, Crist TO (1999) Plant responses to pocket-gopher disturbances across pastures and topography. J Range Manage 52(6):637–645. https://doi.org/10.2307/4003635

Cartron JLE, Polechla PJ, Cook RR (2004) Prey of nesting Ferruginous Hawks in New Mexico. Southwestern Nat 49:270–276

Ceballos G, Wilson DE (1985) *Cynomys Mexicanus*. Mamm Species 248:1–3. https://doi.org/10.2307/3503981

Ceballos G, Davidson A, List R, Pacheco J, Manzano-Fischer P, Santos-Barrera G, Cruzado J (2010) Rapid decline of a grassland system and its ecological and conservation implications. PLoS ONE 5(1):e8562. https://doi.org/10.1371/journal.pone.0008562

Cheng E, Ritchie ME (2006) Impacts of simulated livestock grazing on Utah prairie dogs (*Cynomys parvidens*) in a low productivity ecosystem. Oecologia 147(3):546–555. https://doi.org/10.1007/s00442-005-0286-y

Cid MS, Detling JK, Whicker AD, Brizuela MA (1991) Vegetational responses of a mixed-grass prairie site following exclusion of prairie dogs and bison. J Range Manage 44(2):100–105. https://doi.org/10.2307/4002305

Coggan NV, Hayward MW, Gibb H (2018) A global database and "state of the field" review of research into ecosystem engineering by land animals. J Anim Ecol 87(4):974–994. https://doi.org/10.1111/1365-2656.12819

Collinge SK, Johnson WC, Ray C, Matchett R, Grensten J, Cully Jr. JF, Gage KL, Kosoy MY, Loye JE, Martin AP (2005) Landscape structure and plague occurrence in black-tailed prairie dogs on grasslands of the Western USA. Landsc Ecol 20(8):941–955. https://doi.org/10.1007/s10980-005-4617-5

Connell LC, Porensky LM, Scasta JD (2019) Prairie dog (*Cynomys ludovicianus*) influence on forage quantity and quality in a grazed grassland-shrubland ecotone. Rangeland Ecol Manage 72:360–373. https://doi.org/10.1016/j.rama.2018.10.004

Connior MB (2011) *Geomys bursarius* (Rodentia: Geomyiae). Mamm Species 43(879). https://doi.org/10.1644/879.1

Conway CJ (2018) Spatial and temporal patterns in population trends and burrow usage of burrowing owls in North America. J Raptor Res 52(2):129–142. https://doi.org/10.3356/JRR-16-109.1

Cook RR, Cartron JLE, Polechla PJ Jr (2003) The importance of prairie dogs to nesting ferruginous hawks in grassland ecosystems. Wildl Soc Bull 31(4):1073–1082

Coppock DL, Detling JK, Ellis JE, Dyer MI (1983) Plant-herbivore interactions in a North American mixed-grass prairie—I. Effects of black-tailed prairie dogs on intraseasonal aboveground plant biomass and nutrient dynamics and plant species diversity. Oecologia 56(1):1–9. https://doi.org/10.1007/BF00378210

Cully JF Jr, Williams ES (2001) Interspecific comparisons of sylvatic plague in prairie dogs. J Mammal 82(4):894–905. https://doi.org/10.1644/1545-1542(2001)082%3c0894:ICOSPI%3e2.0.CO;2

Cully JF, Collinge SK, Van Nimwegen RE, Ray C, Johnson WC, Thiagarajan B, Conlin DB, Holmes BE (2010a) Spatial variation in keystone effects: small mammal diversity associated with black-tailed prairie dog colonies. Ecography 33(4):667–677. https://doi.org/10.1111/j.1600-0587.2009.05746.x

Cully JF Jr, Johnson TL, Collinge SK, Ray C (2010b) Disease limits populations: plague and black-tailed prairie dogs. Vector-Borne Zoonotic Dis 10(1):7–15. https://doi.org/10.1089/vbz.2009.0045

Curtis R, Frey SN, Brown NL (2014) The effect of coterie relocation on release-site retention and behavior of Utah prairie dogs. J Wildl Manage 78(6):1069–1077. https://doi.org/10.1002/jwmg.755

Davidson AD, Lightfoot DC (2007) Interactive effects of keystone rodents on the structure of desert grassland arthropod communities. Ecography 30(4):515–525. https://doi.org/10.1111/j.2007.0906-7590.05032.x

Davidson AD, Lightfoot DC (2008) Burrowing rodents increase landscape heterogeneity in a desert grassland. J Arid Environ 72(7):1133–1145. https://doi.org/10.1016/j.jaridenv.2007.12.015

Davidson AD, Lightfoot DC, McIntyre JL (2008) Engineering rodents create key habitat for lizards. J Arid Environ 72:2142–2149. https://doi.org/10.1016/j.jaridenv.2008.07.006

Davidson AD, Ponce E, Lightfoot DC, Fredrickson EL, Brown JH, Cruzado J, Brantley SL, Sierra-Corona R, List R, Toledo D, Ceballos G (2010) Rapid response of a grassland ecosystem to an experimental manipulation of a keystone rodent and domestic livestock. Ecology 91(11):3189–3200. https://doi.org/10.1890/09-1277.1

Davidson AD, Detling JK, Brown JH (2012) Ecological roles and conservation challenges of social, burrowing, herbivorous mammals in the world's grasslands. Front Ecol Environ 10(9):477–486. https://doi.org/10.1890/110054

Davidson AD, Friggens MT, Shoemaker KT, Hayes CL, Erz J, Duran R (2014) Population dynamics of reintroduced Gunnison's Prairie dogs in the southern portion of their range. J Wildl Manage 78(3):429–439. https://doi.org/10.1002/jwmg.755

Davidson AD, Shoemaker KT, Weinstein B, Costa G, Radeloff V, Rondinini C, Ceballos G, Graham C (2017) Geography of global mammal extinction risk. PLoS ONE 12(11):1–18. https://doi.org/10.1371/journal.pone.0186934

Davidson AD, Hunter EA, Erz J, Lightfoot DC, McCarthy AM, Mueller JK, Shoemaker KT (2018) Reintroducing a keystone burrowing rodent to restore an arid North American grassland: challenges and successes. Restor Ecol 26(5):909–920. https://doi.org/10.1890/110054

Derner JD, Detling JK, Antolin MF (2006) Are livestock weight gains affected by black-tailed prairie dogs? Front Ecol Environ 4(9):459–464. https://doi.org/10.1890/1540-9295(2006)4[459:ALWGAB]2.0.CO;2

Desmond MJ, Savidge JA, Eskridge KM (2000) Correlations between burrowing owl and black-tailed prairie dog declines: A 7-year analysis. J Wildl Manage 64(4):1067–1075. https://doi.org/10.2307/3803217

Detling JK (2006) Do prairie dogs compete with livestock? In: Hoogland JL (ed) Conservation of the black-tailed prairie dog. Island Press, Washington D.C., USA

Dinsmore SJ, White GC, Knopf FL (2005) Mountain plover population responses to black-tailed prairie dogs in Montana. J Wildl Manage 69(4):1546–1553. https://doi.org/10.2193/0022-541X(2005)69[1546:MPPRTB]2.0.CO;2

Dobson FS, Kjelgaard JD (1985) The influence of food resources on population dynamics in Columbian ground squirrels. Can J Zool 63(9):2095–2104. https://doi.org/10.1139/z85-308

Donnelly TM, Bergin I, Ihrig M (2015) Chapter 7—biology and diseases of other rodents. Laboratory animal medicine: 3rd edn. Academic Press, London, UK, pp 284–350

Duchardt CJ, Beck JL, Augustine DJ (2019) Mountain Plover habitat selection and nest survival in relation to weather variability and spatial attributes of black-tailed prairie dog disturbance. Condor 122(1):duz059. https://doi.org/10.1093/condor/duz059

Eads DA, Biggins DE (2017) Paltry past-precipitation: predisposing prairie dogs to plague? J Wildl Manage 81(6):990–998. https://doi.org/10.1002/jwmg.21281

Eads DA, Biggins DE (2019) Plague management of prairie dog colonies: degree and duration of deltamethrin flea control. J Vect Ecol 44(1):40–47. https://doi.org/10.1111/jvec.12327

Eads DA, Livieri TM, Dobesh P, Childers E, Noble LE, Vasquez MC, Biggins DE (2021) Fipronil pellets reduce flea abundance on black-tailed prairie dogs: potential tool for plague management and black-footed ferret conservation. J Wildl Dis 57:434–438. https://doi.org/10.7589/JWD-D-20-00161

Eshelman BD, Sonnemann CS (2000) *Spermophilus Armatus*. Mamm Species 637:1–6. https://doi.org/10.1644/1545-1410

Facka AN, Roemer GW, Mathis VL, Kam M, Geffen E (2010) Drought leads to collapse of black-tailed prairie dog populations reintroduced to the Chihuahuan Desert. J Wildl Manage 74(8):1752–1762. https://doi.org/10.2193/2009-208

Fahnestock J, Detling J (2002) Bison-prairie dog-plant interactions in a North American mixed-grass prairie. Oecologia 132:86–95. https://doi.org/10.1007/s00442-002-0930-8

Fitch HS, Bentley JJ (1949) Use of California annual-plant forage by rodents. Ecology 30:306–321. https://doi.org/10.2307/1932612

Foster MA, Stubbendieck J (1980) Effects of the plains pocket gopher (*Geomys bursarius*) on rangeland. J Range Manage 33:74–78. https://doi.org/10.2307/3898233

Gage KL, Kosoy MY (2005) Natural history of plague: perspectives from more than a century of research. Ann Rev Entomol 50:505–528. https://doi.org/10.1146/annurev.ento.50.071803.130337

Geaumont BA, Hovick TJ, Limb RF, Mack WM, Lipinski AR, Sedivec KK (2019) Plant and Bird community dynamics in mixed-grass prairie grazed by native and domestic herbivores. Rangeland Ecol Manage 72:374–384. https://doi.org/10.1016/j.rama.2018.10.002

Genoways HH, Brown JH (1993) Biology of the Heteromyidae, vol 10. American Society of Mammalogists, Provo, UT, USA

Gilson A, Salmon TP (1990) Ground squirrel burrow destruction: control implications. In: Davis LR, Marsh RE (eds) Proceedings of the 14th vertebrate pest conference. University of California, Davis, CA, USA, pp 97–98

Goldingay RL (1997) The kangaroo rats of California: endemism and conservation of keystone species. Pac Conserv Biol 3(1):47–60. https://doi.org/10.1071/pc970047

Goodrich JM, Buskirk SW (1998) Spacing and ecology of North American badgers (*Taxidea taxus*) in a prairie-dog (*Cynomys leucurus*) complex. J Mammal 79(1):171–179. https://doi.org/10.2307/1382852

Grant W, French N, Folse L (1980) Effects of pocket gopher mounds on plant production in shortgrass prairie ecosystems. Southwest Nat 25:215–224

Grant-Hoffman MN, Detling JK (2006) Vegetation on Gunnison's prairie dog colonies in southwestern Colorado. Rangeland Ecol Manage 59:73–79. https://doi.org/10.2111/1551-5028(2006)59[073:VOGPDC]2.0.CO;2

Grinnell J (1923) The burrowing rodents of California as agents in soil formation. J Mammal 4:137–149

Hale SL, Koprowski JL, Archer SR (2020) Black-tailed prairie dog (*Cynomys ludovicianus*) reintroduction can limit woody plant proliferation in grasslands. Front Ecol Evol 8:233. https://doi.org/10.3389/fevo.2020.00233

Hammond TT, Vo M, Burton CT, Surber LL, Lacey EA, Smith JE (2019) Physiological and behavioral responses to anthropogenic stressors in a human-tolerant mammal. J Mammal 100(6):1928–1940. https://doi.org/10.1093/jmammal/gyz134

Hansen R, Remmenga E (1961) Nearest neighbor concept applied to pocket gopher populations. Ecology 42:812–814. https://doi.org/10.2307/1933511

Hartley LM, Detling JK, Savage LT (2009) Introduced plague lessens the effects of an herbivorous rodent on grassland vegetation. J Appl Ecol 46:861–869. https://doi.org/10.1111/j.1365-2664.2009.01660.x

Hawkins LK (1996) Burrows of kangaroo rats are hotspots for desert soil fungi. J Arid Environ 32(3):239–249. https://doi.org/10.1006/jare.1996.0020

Hawkins LK, Nicoletto PF (1992) Kangaroo rat burrows structure the spatial organisation of ground-dwelling animals in a semiarid grassland. J Arid Environ 23(2):199–208. https://doi.org/10.1016/s0140-1963(18)30531-7

Hayes CL, Talbot WA, Wolf BO (2016) Abiotic limitation and the C3 hypothesis: isotopic evidence from Gunnison's prairie dog during persistent drought. Ecosphere 7(12):e01626. https://doi.org/10.1002/ecs2.1626

Heaney LR (1984) Climatic influences of life-history tactics and behavior of North American tree squirrels. In: Murie JO, Michener GR (eds) The biology of ground-dwelling squirrels. University of Nebraska Press, Lincoln, NE, pp 43–70

Helgen KM, Cole FR, Helgen LE, Wilson DE (2009) Generic revision in the Holartic ground squirrel genus *Spermophilus*. J Mammal 90(2):270–305. https://doi.org/10.1644/07-MAMM-A-309.1

Herring G, Eagles-Smith CA, Wagner MT (2016) Ground squirrel shooting and potential lead exposure in breeding avian scavengers. PLoS ONE 11(12):e0167926. https://doi.org/10.1371/journal.pone.0167926

Holekamp KE (1984) Dispersal in ground-dwelling sciurids. In: Murie JO, Michener GR (eds) The biology of ground-dwelling squirrels. University of Nebraska Press, Lincoln, NE, USA, pp 297–320

Holland EA, Parton WJ, Detling JK, Coppock DL (1992) Physiological responses of plant populations to herbivory and their consequences for ecosystem nutrient flow. Am Nat 140(4):685–706. https://doi.org/10.1086/285435

Hoogland JL (2003) Sexual dimorphism of prairie dogs. J Mammal 84(4):1254–1266. https://doi.org/10.1644/BME-008

Howard WE, Bentley JR (1959) Competition between ground squirrels and cattle for range forage. J Range Manage 12:110–115

Huntly N, Inouye RS (1988) Pocket gophers in ecosystems: patterns and mechanisms. Bioscience 38:786–793. https://doi.org/10.2307/1310788

Johnson TL, Cully JF, Collinge SK, Ray C, Frey CM, Sandercock BK (2011) Spread of plague among black-tailed prairie dogs is associated with colony spatial characteristics. J Wildl Manage 75(2):357–368. https://doi.org/10.1002/jwmg.v75.2

Jones CG, Lawton JH, Shachak M (1994) Organisms as ecosystem engineers. Oikos 69(3):373–386. https://doi.org/10.2307/3545850

Jones PH, Biggins DE, Eads DA, Eads SL, Britten HB (2012) Deltamethrin flea-control preserves genetic variability of black-tailed prairie dogs during a plague outbreak. Conserv Genet 13(1):183–195. https://doi.org/10.1007/s10592-011-0275-0

Keeley WH, Bechard MJ, Garber GL (2016) Prey use and productivity of ferruginous hawks in rural and exurban New Mexico. J Wildl Manage 80(8):1479–1487. https://doi.org/10.1002/jwmg.21130

Kelt DA (2011) Comparative ecology of desert small mammals: a selective review of the past 30 years. J Mammal 92(6):1158–1178. https://doi.org/10.1644/10-MAMM-S-238.1

Kenagy GJ, Bartholomew GA (1985) Seasonal reproductive patterns in five coexisting California desert rodent species. Ecol Monogr 55:371–396. https://doi.org/10.2307/2937128

Knopper LD, Mineau P, Scheuhammer AM, Bond DE, McKinnon DT (2006) Carcasses of shot Richardon's ground squirrels may pose lead hazards to scavenging hawks. J Wildl Manage 70:295–299. https://doi.org/10.2193/0022-541X(2006)70[295:COSRGS]2.0.CO;2

Knowles CJ, Proctor JD, Forrest SC (2002) Black-tailed prairie dog abundance and distribution in the Great Plains based on historic and contemporary information. Great Plains Res 12:219–254

Koontz TL, Simpson HL (2010) The composition of seed banks on kangaroo rat (*Dipodomys spectabilis*) mounds in a Chihuahuan Desert grassland. J Arid Environ 74(10):1156–1161. https://doi.org/10.1016/j.jaridenv.2010.03.008

Kraft JP, Stapp P (2013) Movements and burrow use by northern grasshopper mice as a possible mechanism of plague spread in prairie dog colonies. J Mammal 94(5):1087–1093. https://doi.org/10.1644/12-MAMM-A-197.1

Kretzer JE, Cully JF Jr (2001) Effects of black-tailed prairie dogs on reptiles and amphibians in Kansas shortgrass prairie. Southwest Nat 46(2):171–177. https://doi.org/10.2307/3672525

Krueger K (1986) Feeding relationships among bison, pronghorn, and prairie dogs: an experimental analysis. Ecology 67(3):760–770. https://doi.org/10.2307/1937699

Krupa JJ, Geluso KN (2000) Matching the color of excavated soil: cryptic coloration in the plains pocket gopher (Geomys bursarius). J Mammal 81(1):86–96. https://doi.org/10.1644/1545-154 2(2000)081%3c0086:MTCOES%3e2.0.CO;2

Lang JD (2004) Rodent-flea-plague relationships at the higher eleveations of San Diego County, California. J Vector Ecol 29(2):236–247

Laycock WA, Richardson BZ (1975) Long-term effects of pocket gopher control on vegetation and soils of a subslpine grassland. J Range Manage 28:458–462. https://doi.org/10.2307/3897222

Lehmer EM, Biggins DE (2005) Variation in torpor patterns of free-ranging black-tailed and Utah prairie dogs across gradients of elevation. J Mammal 86(1):15–21. https://doi.org/10.1644/1545-1542(2005)086%3c0015:VITPOF%3e2.0.CO;2

Lenihan CM (2007) The ecological role of the California ground squirrel (Spermophilus beecheyi). University of California Press, Davis, California, USA

Lima M, Ernest SKM, Brown JH, Belgrano A, Stenseth NC (2008) Chihuahuan desert kangaroo rats: nonlinear effects of population dynamics, competition, and rainfall. Ecology 89(9):2594–2603. https://doi.org/10.1890/07-1246.1

Linsdale JM (1946) The California ground squirrel: a record of observations made on the Hastings natural history reservation. University of California Press, Berkeley, California, USA

Lomolino M, Smith G (2003) Terrestrial vertebrate communities at black-tailed prairie dog (Cynomys ludovicianus) towns. Biol Conserv 115:89–100. https://doi.org/10.1016/S0006-320 7(03)00097-1

Longland WS, Dimitri LA (2020) Kangaroo rats: ecosystem engineers on western rangelands. Rangelands 2:1–9. https://doi.org/10.1016/j.rala.2020.10.004

Luce B (2003) A multi-state conservation plan for the black-tailed prairie dog, Cynomys ludovicianus, in the United States addendum to the black-tailed prairie dog conservation assessment and strategy of 3 Nov 1999. Available at https://wafwa.org/wpdm-package/a-multi-state-conser vation-plan-for-the-black-tailed-prairie-dog/. Accessed 29 June 2022

Macwhirter RB (1991) Effects of reproduction on activity and foraging behaviour of adult female Columbian ground squirrels. Can J Zool 69(8):2209–2216. https://doi.org/10.1139/z91-30

Magle SB, McClintock BT, Tripp DW, White GC, Antolin MF, Crooks KR (2007) Mark-resight methodology for estimating population densities for prairie dogs. J Wildl Manage 71(6):2067–2073. https://doi.org/10.2193/2006-138

Marsh RE (1994) Current ground squirrel control practices in California. In: Halverson WS, Crabb AC (eds) Proceedings of the 16th vertebrate pest conference. University of California, Davis, CA, USA, pp 61–65

Marsh RE (1998) Historical review of ground squirrel crop damage in California. Int Biodeterior Biodegradation 42:93–99

Marsh RE, Schmidt RH, Howard WE (1987) Secondary hazards to coyotes of ground squirrels poisoned with 1080 or strychnine. Wildl Soc Bull 15:380–385

Matchett MR, Biggins DE, Carlson V, Powell B, Rocke T (2010) Enzootic plague reduces black-footed ferret (Mustela nigripes) survival in montana. Vector-Borne Zoonotic Dis 10(1):27–35. https://doi.org/10.1089/vbz.2009.0053

McCullough Hennessy S, Deutschman DH, Shier DM, Nordstrom LA, Lenihan C, Jp M, Wisinski CL, Swaisgood RR, Sarah McCullough Hennessy C, Diego Zoo S (2016) Experimental habitat restoration for conserved species using ecosystem engineers and vegetation management. Anim Conserv 19:506–514. https://doi.org/10.1111/acv.12266

Menkens GE Jr, Anderson SH (1991) Population dynamics of white-tailed prairie dogs during an epizootic of sylvatic plague. J Mammal 72(2):328–331. https://doi.org/10.2307/1382103

Michener GR (1979) Yearly variations in the population dynamics of Richardson's ground squirrels. Can Field-Nat 93:363–370

Michener GR, Koeppl JW (1985) *Spermophilus Richardsonii*. Mamm Species 243:1–8. https://doi. org/10.2307/3503990

Mielke HW (1977) Mound building by pocket gophers (*Geomyidae*): their impact on soils and vegetation in North America. J Biogeogr 4:171–180. https://doi.org/10.2307/3038161

Miller RS (1964) Ecology and distribution of pocket gophers (*Geomyidae*) in Colorado. Ecology 45:256–272. https://doi.org/10.2307/1933839

Miller B, Reading R, Biggins D, Detling J, Forrest S, Hoogland J, Javersak J, Miller S, Proctor J, Truett J, Urest D (2007) Prairie dogs: an ecological review and current biopolitics. J Wildl Manage 71:2801–2810. https://doi.org/10.2193/2007-041

Minta SC, Marsh RE (1988) Badgers (*Taxidea taxus*) as occasional pests in agriculture. In: Crabb CA, Marsh RE (eds) Proceedings of the 13th vertebrate pest conference. University of California, Davis, CA, USA, pp 199–208

Minta SC, Minta KA, Lott DF (1992) Hunting associations between badgers (*Taxidea taxus*) and coyotes (*Canis latrans*). J Mammal 73(4):814–820. https://doi.org/10.2307/1382201

Moehrenschlager A, List R, Macdonald DW (2007) Escaping intraguild predation: Mexican kit foxes survive while coyotes and golden eagles kill Canadian swift foxes. J Mammal 88(4):1029–1039. https://doi.org/10.1644/06-MAMM-A-159R.1

Moore T, Van Vuren D, Ingels C (1998) Are Barn Owls a biological control for gophers? Evaluating effectiveness in vineyards and orchards. In: Baker RO, Crabb AC (eds) Proceedings of the 18th vertebrate pest conference. University of California, Davis, CA, USA

Nelson EJ, Theimer TC (2012) Translocation of Gunnison's prairie dogs from an urban and suburban colony to abandoned wildland colonies. J Wildl Manage 76(1):95–101. https://doi.org/10.1002/jwmg.281

Newediuk LJ, Waters I, Hare JF (2015) Aspen parkland pasture altered by Richardson's ground squirrel (*Urocitellus richardsonii* Sabine) activity: the good, the bad, and the not so ugly? Can Field-Nat 129(4):331–341. https://doi.org/10.22621/cfn.v129i4.1755

O'Meilia ME, Knopf FL, Lewis JC (1982) Some consequences of competition between prairie dogs and beef cattle. J Range Manage 5:580–585

Orland MC, Kelt DA (2007) Responses of a heteromyid rodent community to large- and small-scale resource pulses: diversity, abundance, and home-range dynamics. J Mammal 88(5):1280–1287. https://doi.org/10.1644/06-MAMM-A-408.1

Ortiz CA, Pendleton EL, Newcomb KL, Smith JE (2019) Conspecific presence and microhabitat features influence foraging decisions across ontogeny in a facultatively social mammal. Behav Ecol Sociobiol 73(4):42–42. https://doi.org/10.1007/s00265-019-2651-6

Ostroff AC, Finck EJ (2003) *Spermophilus Franklinii*. Mamm Species 724:1–5. https://doi.org/10. 1644/724

Ostrow DG, Huntly N, Inouye RS (2002) Plant-mediated interactions between the northern pocket gopher, *Thomomys talpoides*, and aboveground herbivorous insects. J Mammal 83(4):991–998. https://doi.org/10.1644/1545-1542(2002)083%3c0991:PMIBTN%3e2.0.CO;2

Pattee OH, Bloom PH, Scott JM, Smith MR (1990) Lead hazards within the range of the California condor. Condor 92:931–937. https://doi.org/10.2307/1368729

Pauli JN, Buskirk SW (2007a) Risk-disturbance overrides density dependence in a hunted colonial rodent, the black-tailed prairie dog *Cynomys ludovicianus*. J Appl Ecol 44(6):1219–1230. https://doi.org/10.1111/j.1365-2664.2007.01337.x

Pauli JN, Buskirk SW (2007b) Recreational shooting of prairie dogs: a portal for lead entering wildlife food chains. J Wildl Manage 71(1):103–108. https://doi.org/10.2193/2005-620

Phillips P, Livieri TM, Swanson BJ (2020) Genetic signature of disease epizootic and reintroduction history in an endangered carnivore. J Mammal 101(3):779–789. https://doi.org/10.1093/jmammal/gyaa043

Phuong MA, Lim MC, Wait DR, Rowe KC, Moritz C (2014) Delimiting species in the genus *Otospermophilus* (Rodentia: Sciuridae), using genetics, ecology, and morphology. Bological J Linnean Soc 113:1136–1151

Pierce AK, Dinsmore SJ, Jorgensen D, Wunder MB (2017) Migration routes and timing of Mountain Plovers revealed by geolocators. J Field Ornithol 88(1):30–38. https://doi.org/10.1111/jofo. 12184

Plumpton DL, Anderson DE (1997) Habitat use and time budgeting by wintering Ferruginous Hawks. Condor 99:888–893. https://doi.org/10.2307/1370139

Ponce-Guevara E, Davidson A, Sierra-Corona R, Ceballos G (2016) Interactive effects of black-tailed prairie dogs and cattle on shrub encroachment in a desert grassland ecosystem. PLoS ONE 11(5):e0154748. https://doi.org/10.1371/journal.pone.0154748

Price MV, Endo PR (1989) Estimating the distribution and abundance of a cryptic species, *Dipodomys stephensi* (Rodentia: Heteromyidae), and implications for management. Conserv Biol 3(3):293–301. https://doi.org/10.1111/j.1523-1739.1989.tb00089.x

Price MV, Waser NM, McDonald SA (2000) Elevational distributions of kangaroo rats (Genus *Dipodomys*): long-term trends at a Mojave Desert site. Am Midl Nat 144(2):352–361. https://doi.org/10.1674/0003-0031(2000)144[0352:EDOKRG]2.0.CO;2

Proulx G (2010) Factors contributing to the outbreak of Richardson's ground squirrel populations in the Canadian prairies. In: Proceedings of the 24th vertebrate pest conference. University of California, Davis, CA, USA, pp 213–217

Prugh LR, Brashares JS (2012) Partitioning the effects of an ecosystem engineer: Kangaroo rats control community structure via multiple pathways. J Anim Ecol 81(3):667–678. https://doi.org/10.1111/j.1365-2656.2011.01930.x

Rayor LS (1988) Social organization and space-use in Gunnison's prairie dog. Behav Ecol Sociobiol 22:69–78. https://doi.org/10.1007/BF00395699

Reichman OJ (2007) The influence of pocket gophers on the biotic and abiotic environment. In: Begall S, Burda H, Schleich CE (eds) Subterranean rodents: news from underground. Springer-Verlag, Berlin, pp 271–286

Reichman OJ, Seabloom EW (2002) The role of pocket gophers as subterranean ecosystem engineers. Trends Ecol Evol 17(1):44–49. https://doi.org/10.1016/S0169-5347(01)02329-1

Reichman OJ, Whitham TG, Ruffner GA (1982) Adaptive geometry of burrow spacing in two pocket gopher populations. Ecology 63(3):687–695. https://doi.org/10.2307/1936789

Reid F (2006) Peterson field guide to the Mammals of North America. Houghton Mifflin, New York, NY, USA

Reynolds HG (1958) The ecology of the Merriam kangaroo rat (*Dipodomys merriami* Mearns) on the grazing lands of southern Arizona. Ecol Monogr 28(2):111–127. https://doi.org/10.2307/1942205

Roach JL, Stapp P, Van Horne B, Antolin MF (2001) Genetic structure of a metapopulation of black-tailed prairie dogs. J Mammal 82(4): 946–959

Rocke TE, Williamson J, Cobble KR, Busch JD, Antolin MF, Wagner DM (2012) Resistance to plague among black-tailed prairie dog populations. Vector Borne Zoonot Dis 12(2):111–116. https://doi.org/10.1089/vbz.2011.0602

Rocke TE, Tripp DW, Russell RE, Abbott RC, Richgels KLD, Matchett MR, Biggins DE, Griebel R, Schroeder G, Grassel SM, Pipkin DR, Cordova J, Kavalunas A, Maxfield B, Boulerice J, Miller MW (2017) Sylvatic plague vaccine partially protects prairie dogs (*Cynomys* spp.) in field trials. EcoHealth 14(3):438–450. https://doi.org/10.1007/s10393-017-1253-x

Romañach SS, Seabloom EW, Reichman OJ, Rogers WE, Cameron GN (2005) Effects of species, sex, age, and habitat on geometry of pocket gopher foraging tunnels. J Mammal 86(4):750–756. https://doi.org/10.1644/1545-1542(2005)086[0750:EOSSAA]2.0.CO;2

Romañach SS, Seabloom EW, Reichman OJ (2007) Costs and benefits of pocket gopher foraging: linking behavior and physiology. Ecology 88(8):2047–2057. https://doi.org/10.1890/06-1461.1

Russell RE, Abbott RC, Tripp DW, Rocke TE (2018) Local factors associated with on-host flea distributions on prairie dog colonies. Ecol Evol 8(17):8951–8972. https://doi.org/10.1002/ece3. 4390

Salkeld DJ, Stapp P, Tripp DW, Gage KL, Lowell J, Webb CT, Brinkerhoff RJ, Antolin MF (2016) Ecological traits driving the outbreaks and emergence of zoonotic pathogens. Bioscience 66(2):118–129. https://doi.org/10.1093/biosci/biv179

Schiffman PM (1994) Promotion of exotic weed establishment by endangered giant kangaroo rats (*Dipodomys ingens*) in a California grassland. Biodiv Conserv 3(6):524–537. https://doi.org/10.1007/BF00115158

Schitoskey F, Woodmansee SR (1978) Energy requirements and diet of the California ground squirrel. J Wildl Manage 42:373–382. https://doi.org/10.2307/3800273

Schooley RL, Wiens JA (2001) Dispersion of kangaroo rat mounds at multiple scales in New Mexico, USA. Landsc Ecol 16:267–277. https://doi.org/10.1023/A:1011122218548

Shipley BK, Reading RP (2006) A comparison of herpetofauna and small mammal diversity on black-tailed prairie dog (*Cynomys ludovicianus*) colonies and non-colonized grasslands in Colorado. J Arid Environ 66(1):27–41. https://doi.org/10.1016/j.jaridenv.2005.10.013

Sidle JG, Augustine DJ, Johnson DH, Miller SD, Cully JF Jr, Reading RP (2012) Aerial surveys adjusted by ground surveys to estimate area occupied by black-tailed prairie dog colonies. Wildl Soc Bull 36(2):248–256. https://doi.org/10.1002/wsb.146

Sierra-Corona R, Davidson A, Fredrickson EL, Luna-Soria H, Suzan-Azpiri H, Ponce-Guevara E, Ceballos G (2015) Black-tailed prairie dogs, cattle, and the conservation of North America's Arid Grasslands. PLoS ONE 10(3):e0118602. https://doi.org/10.1371/journal.pone.0118602

Smallwood KS, Morrison ML (1999) Estimating burrow volume and excavation rate of pocket gophers (Geomyidae). Southwest Nat 44(2):173–183

Smith G, Lomolino M (2004) Black-tailed prairie dogs and the structure of avian communities on the shortgrass plains. Oecologia 138:592–602. https://doi.org/10.1007/s00442-003-1465-3

Smith JE, Long DJ, Russell ID, Newcomb KL, Muñoz VD (2016) *Otospermophilus beecheyi* (Rodentia: Sciuridae). Mammal Species 48(939):91–108. https://doi.org/10.1093/mspecies/sew010

Smith JE, Gamboa DA, Spencer JM, Travenick SJ, Ortiz CA, Hunter RD, Sih A (2018) Split between two worlds: automated sensing reveals links between above-and belowground social networks in a free-living mammal. Philos Trans R Soc Lond B: Biol Sci 373(1753):20170249. https://doi.org/10.1098/rstb.2017.0249

Smith JE, Smith IB, Working CL, Russell ID, Krout SA, Singh KS, Sih A (2021) Host traits, identity, and ecological conditions predict consistent flea abundance and prevalence on free-living California ground squirrels. Int J Parasitol 51:587–598. https://doi.org/10.1016/j.ijpara.2020.12.001

Stapp P (1997) Habitat selection by an insectivorous rodent: patterns and mechanisms across multiple scales. J Mammal 78(4):1128–1143. https://doi.org/10.2307/1383055

Stapp P (1998) A reevaluation of the role of prairie dogs in Great Plains grasslands. Conserv Biol 12(6):1253–1259. https://doi.org/10.1111/j.1523-1739.1998.97469.x

Stapp P, Antolin MF, Ball M (2004) Patterns of extinction in prairie dog metapopulations: plague outbreaks follow El Niño events. FrontEcol Environ 2(5):235–240. https://doi.org/10.1890/1540-9295(2004)002[0235:POEIPD]2.0.CO;2

Stapp P, Van Horne B, Lindquist M (2008) Ecology of mammals of the shortgrass steppe. In: Lauenroth WK, Burke IC (eds) Ecology of the shortgrass steppe: a long-term perspective. Oxford University Press, Oxford, UK, pp 132–180

Stephens T, Wilson SC, Cassidy F, Bender D, Gummer D, Smith DHV, Lloyd N, McPherson JM, Moehrenschlager A (2018) Climate change impacts on the conservation outlook of populations on the poleward periphery of species ranges: a case study of Canadian black-tailed prairie dogs (*Cynomys ludovicianus*). Glob Change Biol 24(2):836–847. https://doi.org/10.1111/gcb.13922

Steuter AA, Steinauer EM, Hill GL, Bowers PA, Tieszen LL (1995) Distribution and diet of bison and pocket gophers in a sandhills prairie. Ecol Appl 5(3):756–766. https://doi.org/10.2307/1941983

Stromberg MR, Griffin JR (1996) Long-term patterns in coastal California grasslands in relation to cultivation, gophers, and grazing. Ecol Appl 6(4):1189–1211. https://doi.org/10.2307/2269601

Swaisgood RR, Montagne JP, Lenihan CM, Wisinski CL, Nordstrom LA, Shier DM (2019) Capturing pests and releasing ecosystem engineers: translocation of common but diminished species to re-establish ecological roles. Anim Conserv 22(6):600–610. https://doi.org/10.1111/acv.12509

Thorington RW, Koprowski JL, Steele MA, Whatton JF (2012) Squirrels of the world. Johns Hopkins University Press Books, Baltimore, MD, USA

Tileston J, Lechleitner R (1966) Some comparisons of black-tailed and white-tailed prairie dogs in north-central Colorado. Am Midl Nat 75:292–316. https://doi.org/10.2307/2423393

Truett JC, Dullum JALD, Matchett MR, Owens E, Seery D (2001) Translocating prairie dogs: a review. Wildl Soc Bull 29(3):863–872

Turner GT (1969) Responses of mountain grassland vegetation to gopher control, reduced grazing, and herbicide. J Range Manage 22:377–383

Underwood N, Inouye BD (2017) Pathways for effects of small-scale disturbances on a rare plant: How *Mimulus angustatus* benefits from gopher mounds. Ecosphere 8(6):e01838. https://doi.org/10.1002/ecs2.1838

U.S. Fish and Wildlife Service (2012) Utah prairie dog (*Cynomys parvidens*) revised recovery plan. U.S. Fish and Wildlife Service, Denver, CO, USA

U.S. Fish and Wildlife Service (2020) Species status assessment report for the Giant Kangaroo Rat (*Dipodomys ingens*) U.S. Fish and Wildlife Service, Sacramento, CA plan. U.S. Fish and Wildlife Service, Denver, CO, USA

Valone TJ, Thornhill DJ (2001) Mesquite establishment in arid grasslands: an experimental investigation of the role of kangaroo rats. J Arid Environ 48(3):281–288. https://doi.org/10.1006/jare.2000.0757

Van Auken OW (2009) Causes and consequences of woody plant encroachment into western North American grasslands. J Environ Manage 90(10):2931–2941. https://doi.org/10.1016/j.jenvman.2009.04.023

Van Vuren DH, Ordeñana MA (2012) Factors influencing burrow length and depth of ground-dwelling squirrels. J Mammal 93:1240–1246. https://doi.org/10.1644/12-MAMM-A-049.1

Vaughan TA (1961) Vertebrates inhabiting pocket gopher burrows in Colorado. J Mammal 42:171–174. https://doi.org/10.2307/1376826

Waser PM, Ayers JM (2003) Microhabitat use and population decline in banner-tailed kangaroo rats. J Mammal 84(3):1031–1043. https://doi.org/10.1644/BBa-032

Weir RD, Davis H, Gayton DV, Lofroth EC (2016) Chapter 11. Fact or fantasy? Damage to livestock and agricultural machinery by American badgers and other burrowing mammals in British Columbia, Canada. In: Proulx G, Do Linh San E (eds) Badgers: systematics, biology, conservation and research techniques. Alpha Wildlife Publications, Sherwood Park, Alberta, Canada, pp 299–310

Weltzin J, Archer S, Heitschmidt R (1997) Small-mammal regulation of vegetation structure in a temperate savanna. Ecology 78(3):751–763. https://doi.org/10.1890/0012-9658(1997)078[0751:SMROVS]2.0.CO;2

Whicker A, Detling J (1988) Ecological consequences of prairie dog disturbances. Bioscience 38:778–785

Whisson DA (1999) Modified bait stations for California ground squirrel control in endangered kangaroo rat habitat. Wildl Soc Bull 27:172–177

White GC, Dennis JR, Pusateri FM (2005) Area of black-tailed prairie dog colonies in eastern Colorado. Wildl Soc Bull 33(1):265–272. https://doi.org/10.2193/0091-7648(2005)33[265:AOBPDC]2.0.CO;2

Widick IV, Bean WT (2019) Evaluating current and future range limits of an endangered, keystone rodent (*Dipodomys ingens*). Divers Distrib 25(7):1074–1087. https://doi.org/10.1111/ddi.12914

Wilkening J, Pearson-Prestera W, Mungi NA, Bhattacharyya S (2019) Endangered species management and climate change: when habitat conservation becomes a moving target. Wildl Soc Bull 43(1):11–20. https://doi.org/10.1002/wsb.944

Wilkins KT, Roberts HR (2007) Comparative analysis of burrow systems of seven species of pocket gophers (Rodentia: Geomyidae). Southwest Nat 52(1):83–88. https://doi.org/10.1894/0038-490 9(2007)52[83:CAOBSO]2.0.CO;2

Williams JE, Moussa MA, Cavanaugh DC (1979) Experimental plague in the California ground squirrel. J Infec Dis 140:618–621.https://doi.org/10.1093/infdis/140.4.618

Witmer GW, Snow NP, Moulton RS (2016) Retention time of chlorophacinone in black-tailed prairie dogs informs secondary hazards from a prairie dog rodenticide bait. Pest Manage Sci 72(4):725–730. https://doi.org/10.1002/ps.4045

Wobeser GA (1994) Disease management through manipulation of the host population. In: Wobeser GA (ed) Disease in wild animals. Springer, NY, USA, pp 217–245. https://doi.org/10.1007/978-3-540-48978-8_12

Yeaton RI, Flores-Flores JL (2006) Patterns of occurrence and abundance in colony complexes of the Mexican prairie dog (*Cynomys mexicanus*) in productive and unproductive grasslands. Acta Zool Mexicana 22(3):107–130

Zegers DA (1984) *Spermophilus Elegans*. Mammal Species 214:1–7. https://doi.org/10.2307/350 3955

Chapter 16
Mesocarnivores of Western Rangelands

Julie K. Young, Andrew R. Butler, Joseph D. Holbrook, Hila Shamon, and Robert C. Lonsinger

Abstract There are 22 species of mesocarnivores (carnivores weighing < 15 kg) belonging to five families that live in rangelands of the western United States. Mesocarnivores are understudied relative to large carnivores but can have significant impacts on ecosystems and human dimensions. In this chapter, we review the current state of knowledge about the biology, ecology, and human interactions of the mesocarnivores that occupy the rangelands of the central and western United States. In these two regions, mesocarnivores may serve as the apex predator in areas where large carnivores no longer occur, and can have profound impacts on endemic prey, disease ecology, and livestock production. Some mesocarnivore species are valued because they are harvested for food and fur, while others are considered nuisance species because they can have negative impacts on ranching. Many mesocarnivores have flexible life history strategies that make them well-suited for future population growth or range expansion as western landscapes change due to rapid human population growth, landscape development, and alterations to ecosystems from climate change; however other mesocarnivores continue to decline. More research on this important guild is needed to understand their role in western working landscapes.

J. K. Young (✉)
Department of Wildland Resources, Utah State University, 5230 Old Main Hill, Logan, UT 84322, USA
e-mail: julie.young@usu.edu

A. R. Butler
Prairie Ecology Lab, Department of Forestry and Environmental Conservation, Clemson University, Clemson, SC, USA

J. D. Holbrook
Haub School of Environment and Natural Resources, Department of Zoology and Physiology, University of Wyoming, Laramie, WY, USA

H. Shamon
Smithsonian Conservation Biology Institute, National Zoological Park, 1500 Remount Road, Front Royal, VA, USA

R. C. Lonsinger
U.S. Geological Survey, Oklahoma Cooperative Fish and Wildlife Research Unit, Stillwater, OK, USA

© The Author(s) 2023
L. B. McNew et al. (eds.), *Rangeland Wildlife Ecology and Conservation*,
https://doi.org/10.1007/978-3-031-34037-6_16

Keywords Canids · Felids · Fur harvest · Intraguild interactions · Mesopredator · Mustelids · Procyonids · Rangeland management · Skunks

16.1 General Natural History of Mesocarnivores

Western rangelands host 22 mesocarnivore species, belonging to five families (Fig. 16.1). Mesocarnivores are defined as mid-sized carnivores, weighing < 15 kg; they are typically more abundant than large carnivores and carry important ecological roles that may regulate trophic levels similar to their larger counterparts (Roemer et al. 2009). The high number of mesocarnivores is reflected by their diverse adaptations to their environment, which results in a diversity of dietary portfolios, and substantial variation in spatiotemporal patterns of habitat use. The diversity among mesocarnivores enables them to effectively partition resources (e.g., via spatial, temporal, or dietary partitioning) and coexist within similar habitats throughout their range (Roemer et al. 2009; Lesmeister et al. 2015).

16.1.1 Species and Population Statuses

Although there are many mesocarnivores in western landscapes, most are understudied, and information is therefore limited on their distribution and population status. In general, many are declining, notably up to 62% of the world's small carnivore species (Belant et al. 2009), and there is a need for improving conservation efforts that reduce population threats (Marneweck et al. 2021). Contrasting this plight, some populations are expanding and are considered to be nuisance species. We present this information collectively by their scientific Family to highlight similarities within groups and where information remains sparse.

16.1.2 Family: Canidae

Five medium-sized canids occur in western rangelands; coyotes (*Canis latrans*) and four fox species (gray fox [*Urocyon cinereoargenteus*], kit fox [*Vulpes macrotis*], red fox [*Vulpes vulpes*], and swift fox [*Vulpes velox*]). The most widespread and largest (7–20 kg) of the five is the coyote (Kays 2018). Coyotes historically occupied a considerable portion of the western third of North America, excluding parts of northern Canada and Alaska (Hody and Kays 2018). They mostly occurred in open habitats including grasslands and shrublands. During the last century, the exploratory tendencies and generalist diet of coyotes, combined with human modifications that opened forested habitats and the historical elimination of gray wolves (*Canis lupus*) and cougars (*Puma concolor*) from much of their respective ranges, have allowed

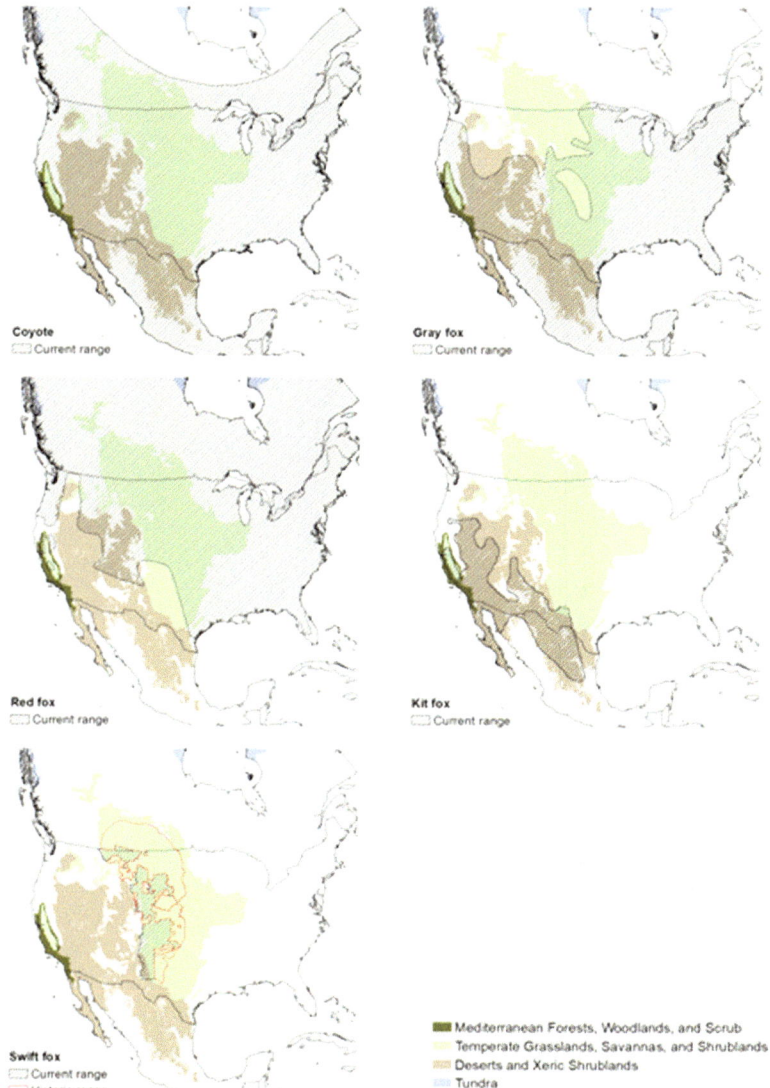

Fig. 16.1 Range of mesocarnivore species inhabiting rangelands of western North America and adjoining regions. Canidae: Coyote (Kays 2018); gray fox (Roemer et al. 2016); red fox (Moehren-schlager and Sovada 2016); kit fox (Cypher and List 2014); swift fox (Moehrenschlager and Sovada 2016). Mephitidae: Eastern spotted skunk (Gompper and Jachowski 2016); western spotted skunk (Cuarón et al. 2016c); hog-nosed skunk (Helgen 2016); hooded skunk (Cuarón et al. 2016a); striped skunk (Helgen and Reid 2016a). Mustelidae: mink (Reid et al. 2016b); American badger (Helgen and Reid 2016b); river otter (Serfass et al. 2015); black-footed ferret (Belant et al. 2015); least weasel (McDonald et al. 2019); short-tailed weasel (Reid et al. 2016a); long-tailed weasel (Helgen and Reid 2016c). Felidae: Bobcat (Kelly et al. 2016); ocelot (Paviolo et al. 2015). Procyonidae: Raccoon (Timm et al. 2016); ringtail (Reid et al. 2016c); white-nosed coati (Cuarón et al. 2016b). Rangelands landcover (Olson et al. 2001)

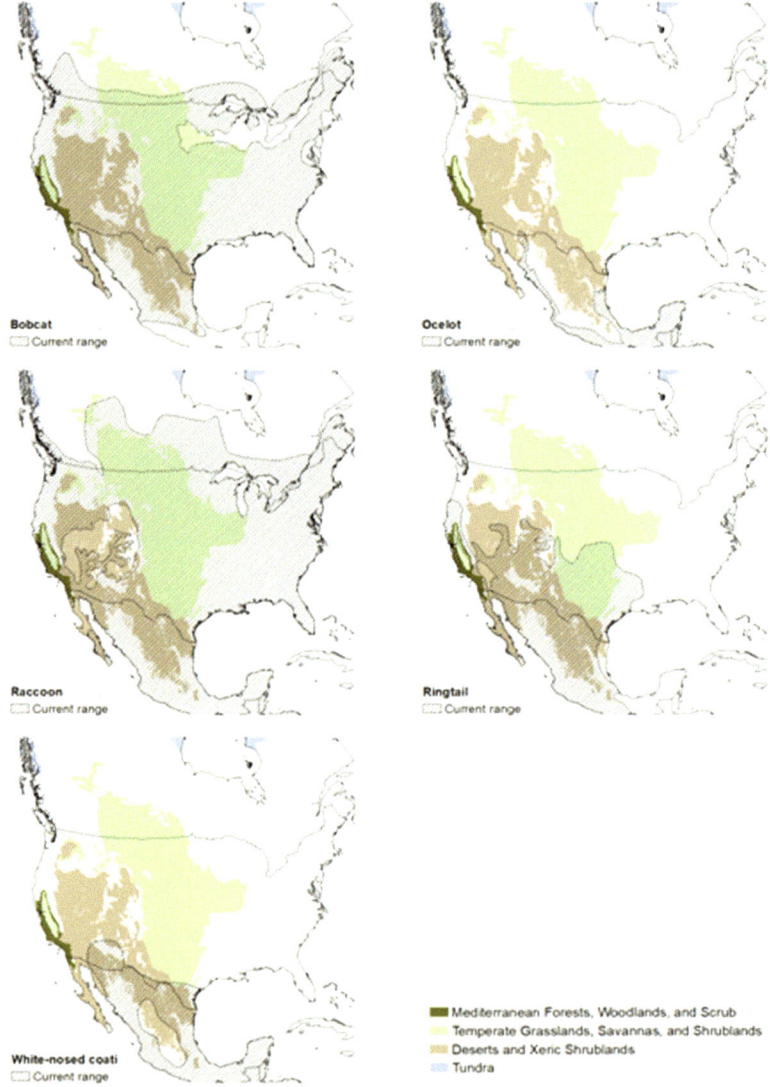

Fig. 16.1 (continued)

coyotes to expand their range by over 40% (Gompper 2002; Hody and Kays 2018). Coyote populations are rarely assessed even though some states have harvest seasons or bounty programs, but they are generally considered stable or increasing throughout their range.

The red fox is also widespread throughout North America; however, some populations are a mix of native and non-native individuals (e.g., Black et al. 2018). Native red foxes originated in boreal and mountainous habitats in northern North America

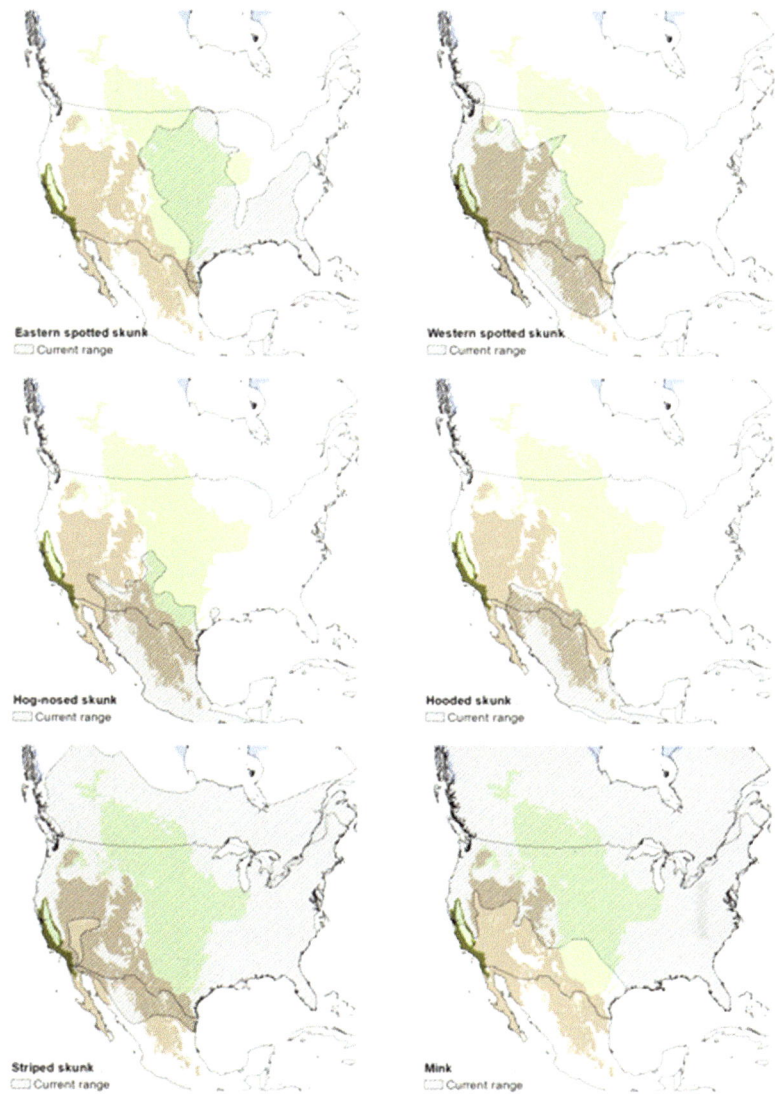

Fig. 16.1 (continued)

(Kamler and Ballard 2002; Aubry et al. 2009). Non-native red foxes were likely transported to the United States from Europe in the 1700s (Kasprowicz et al. 2016); however, it is likely interbreeding was limited to the eastern parts of the United States and is not of concern for western populations (Statham et al. 2012; Sacks et al. 2018; Kuo et al. 2019). Their omnivorous diet and ability to thrive near human habitation enabled red foxes to occupy much of North America (Kamler and Ballard 2002; Hoffmann and Sillero-Zubiri 2016). Red foxes occur in higher densities near

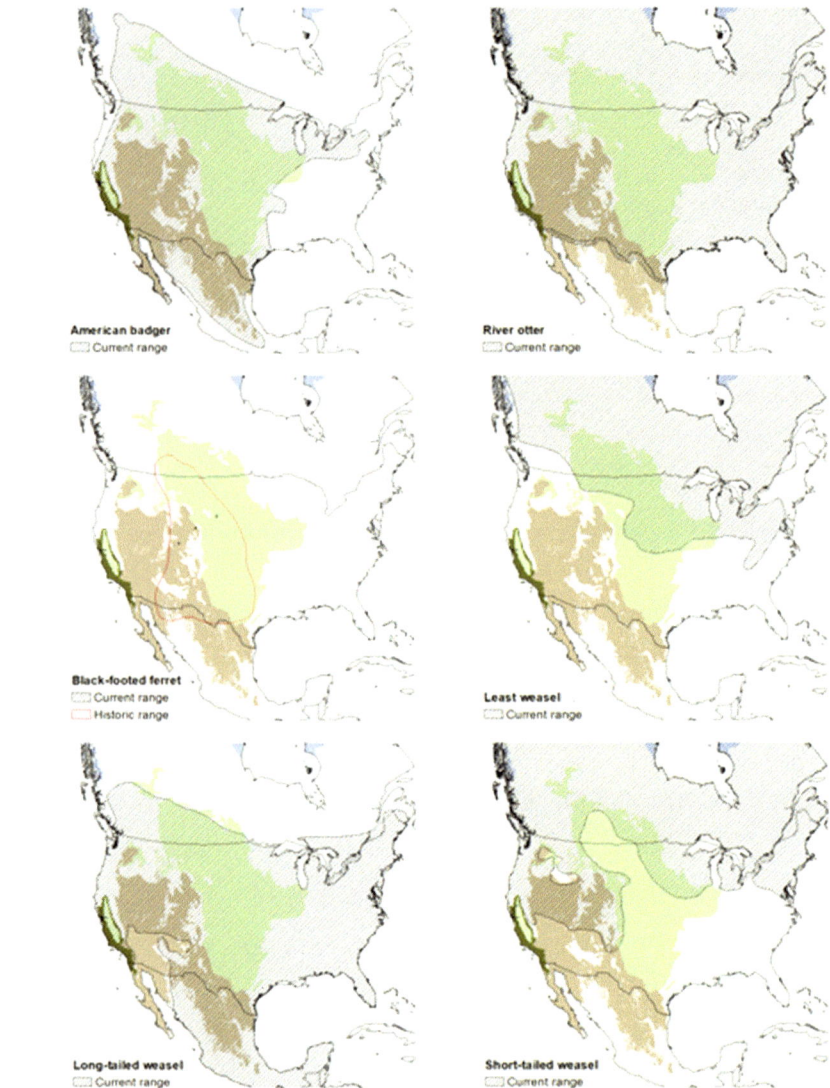

Fig. 16.1 (continued)

agriculture, towns, or dry and patchy landscapes with shrubs and woodlands, and occur at lower densities in grasslands (Hoffmann and Sillero-Zubiri 2016). Red fox populations are rarely assessed but there are current attempts related to threatened and endangered subpopulations, such as for the Sierra Nevada red fox (*V. v. necator*; e.g., Hatfield et al. 2021).

The gray fox overlaps with the red fox throughout most of its range and there is evidence that the two species can coexist with little competition, likely due to the ability of gray foxes to climb (Lesmeister et al. 2015). The range of the gray fox extends farther southwest, while the range of the red fox extends farther north. Similar to coyotes and red foxes, the diet of gray foxes is omnivorous and consists mostly of small mammals, fruits, and seeds (Larson et al. 2015). They occur in woodland, riparian forests, and dense shrublands but also at agricultural and urban edges (Roemer et al. 2016). Gray fox populations are rarely assessed but considered stable throughout their range.

The kit fox is found in arid deserts and grasslands of southwest North America, whereas the historical range of the swift fox included the Great Plains grasslands and shrublands. Kit foxes and swift foxes only overlap in the southern most extent of their ranges, in parts of Texas and New Mexico. They were considered conspecific until 2005, when they were separated into two species following genetic assessments (Cypher and List 2014; Moehrenschlager and Sovada 2016). They are similar in size (1.3–3.5 kg; swift foxes are slightly smaller) and have similar biology. The diet of both species is comprised by a high proportion of rodents and insects (Hines and Case 1991; Pechacek et al. 2000). Kit foxes are considered stable throughout the southern portions of their range, and declining in the northern third of their range (Lonsinger et al. 2020). Swift foxes were eliminated from ~ 90% of their range by the 1950s, and today occupy about 40% of their historical range, with lower densities in the northern portion of their range (Zimmerman 1998; Sovada et al. 2009). Intraguild predation by coyotes on swift foxes and kit foxes has been documented in many areas throughout the foxes' range and is linked to local extinctions (Nelson et al. 2007; Thompson and Gese 2007; Karki et al. 2007; Lonsinger et al. 2017).

16.1.3   Family: Felidae

There are three felid mesocarnivores found in North America's rangelands: bobcat (*Lynx rufus*), ocelot (*Leopardus pardalis*), and jaguarundi (*Herpailurus yagouaroundi*). Only bobcats and ocelots are found within the United States today. Bobcats are common and widespread across North America, whereas, the ocelot's range is primarily through Central and South America (Paviolo et al. 2015; Kelly et al. 2016). Both species can withstand certain degrees of anthropogenic disturbance, albeit ocelots are significantly less tolerant of human disturbances. In the United States, ocelots are considered endangered (Paviolo et al. 2015); they require unfragmented thornscrub habitat, which has limited availability (Jackson et al. 2005; Horne et al. 2009; Janečka et al. 2016). The majority of ocelots in the United States are currently on private rangelands in Texas (Lombardi et al. 2020b) and experience high mortality from road collisions (Haines et al. 2005; Blackburn et al. 2021). Bobcat populations are stable or increasing throughout most of their range (Roberts and Crimmins 2010). Even so, they are closely regulated due to inclusion in Appendix II of the Convention on International Trade in Endangered Species of Wild Flora and Fauna (CITES) as the only spotted cat legally traded worldwide. Local bobcat densities are dependent on prey availability (CITES 2021). Bobcats are generalists and adapt to changes in prey composition (Newbury and Hodges 2018). In southern rangelands, ocelot and bobcat diets overlap, consisting of small rodents, lagomorphs, and birds (Booth-Binczik et al. 2013). Overlap in diet and competition with bobcats may affect the ocelot's recovery, however, further research is needed to underline the ecological mechanisms of co-occurrence of these species (Lombardi et al. 2020a).

16.1.4 Family: Procyonidae

Three procyonids are found in North America's rangelands, raccoons (*Procyon lotor*), ringtail (*Bassariscus astutus*), and white-nosed coati (*Nasua narica*). All are generalist omnivores occupying diverse habitats. Raccoons are highly flexible and cohabitate with humans. They inhabit grasslands and shrublands and are known to prey on grassland birds and nests. Raccoons are limited by water resources and tend to select sites near streams and riparian forests (Timm et al. 2016; Berry et al. 2017). Their expansion to arid rangelands is partially attributed to anthropogenic water resources (Kamler et al. 2003b). Raccoons thrive on anthropogenic resources and have flexible social organization, from solitary in natural habitats to social in urban habitats. Interactions among raccoons and with other species are a concern for the transmission of pathogens (Hirsch et al. 2013). Raccoon populations are rarely assessed but they are considered to be stable or increasing.

Ringtail and the white-nosed coati are semi-arboreal species, although ringtails may be more appropriately described as scansorial because they primarily occur in areas with little or no tree cover by exploiting canyons and similar orographic features. Ringtails are found in diverse habitats that include forests, deserts, rocky cliffs, and tropical areas and withstand low levels of disturbance and human habitation (Reid et al. 2016c), from southern Oregon south into Mexico. There are no data on ringtail populations (Reid et al. 2016c), and their status varies from furbearer to fully protected in states within the United States where they are found.

White-nosed coatis occur in low densities in southwestern rangelands of the United States. They are common in tropical habitats, and are also found in hardwood riparian forests of deserts in the southwestern United States and Mexico (Cuarón et al. 2016b). White-nosed coati are decreasing globally, although abundant in some areas (Cuarón et al. 2016b).

16.1.5 Family: Mephitidae

Five mephitids occur in western rangelands. The striped skunk (*Mephitis mephitis*) is the most common and widespread (Helgen and Reid 2016a). Striped skunks are opportunistic feeders, have an omnivorous diet that consists mostly of insects, but rodents, birds, and fruits are also consumed (Greenwood et al 1999). Populations are rarely assessed, but they have expanded their range in Canada (Long 2003). Eastern spotted skunks (*Spilogale putorius*) and western spotted skunks (*Spilogale gracilis*) occupy the east and west portions of the United States, respectively, and there is some overlap across a portion of the Great Plains (Cuarón et al. 2016c; Gompper and Jachowski 2016). The population of western spotted skunks is unknown but thought to be decreasing alongside decreasing prairie habitat, while eastern spotted skunks are declining throughout their range (Gompper and Hackett 2005). Spotted skunks consume invertebrates, small mammals, snakes, amphibians, birds, and plants, along with scavenging large mammalian prey (Sprayberry and Edelman 2016). Hog-nosed skunks (*Conepatus leuconotus*) and hooded skunks (*Mephitis macroura*) range from the southwestern United States through much of Central America (Cuarón et al. 2016a; Helgen 2016). Hog-nosed skunks are insectivorous (Hall and Dalquest 1963), more so than other skunks, but also an opportunistic feeder that consumes a variety of small vertebrates and fruits (Dragoo and Honeycutt 1999). The population is declining (Dragoo and Sheffield 2009). There are almost no studies on the diet of hooded skunks, considered to be omnivores, and their populations are considered stable (Cuarón et al. 2016a). Mephitids are fairly opportunistic and adaptable to differing conditions.

16.1.6   Family: Mustelidae

The most diverse group of mesocarnivores are the mustelids. Comprised of seven species, North America's mustelids differ in size, specialization, habitat selection, diet composition, and activity patterns. The two semi-aquatic species, river otters (*Lontra canadensis*) and mink (*Neovison vison*) occur throughout much of North America. River otters occupy rivers and streams. Their populations were extremely reduced in the first half of the twentieth century. Thanks to habitat restoration, stricter regulations around harvest, and reintroduction programs, they now occupy about 90% of their historical range (Roberts et al. 2020). Despite their impressive recovery, otter densities and reproductive success are susceptible to heavy metals and polycyclic aromatic compound contamination in rivers and food resources (Thomas et al. 2021). River otters primarily consume fish and cetaceans (Melquist et al. 2003).

Mink are obligate carnivores that occupy areas by small streams, marshes, and dense vegetation (Reid et al. 2016b; Holland et al. 2019). Captive populations of mink are found throughout the United States and maintained for their fur. Farm minks may come into contact with wild animals through fence lines or by escaping confinement, which could spread diseases to wild animals; captive mink have transmitted COVID-19 to wildlife (e.g., Shriner et al. 2021). Mink are highly adaptable, considered generalist predators that eat fish, invertebrates, birds, amphibians, and small mammals. They primarily consume muskrats and lagomorphs in much of their native range (Dunstone 1993).

American badgers (*Taxidea taxus*) are widespread throughout western range-lands. Their diet mostly consists of small mammals but they also consume birds, reptiles, and insects (Helgen and Reid 2016b). Despite being characterized as gener-alists, badgers select for prairie dog (*Cynomys spp.*) colonies when available (Grassel and Rachlow 2018). Badgers are also considered ecosystem engineers where their den mounds contribute to soil nutrient patchiness that in turn affects vegetation composition (Eldridge and Whitford 2009). Additionally, badger burrows provide subterranean habitat to a wide diversity of species (Andersen et al. 2021).

Black-footed ferrets (*Mustela nigripes*) are one of the most endangered mammals in North America. Black-footed ferrets are obligate carnivores that feed mostly on prairie dogs (Brickner et al. 2014). The extermination of prairie dogs from North American rangelands caused the precipitous decline of black-footed ferrets (Knowles et al. 2002). Black-footed ferrets were believed to be extinct when the last known population died out in South Dakota until a small population was discovered in 1981 in Wyoming. Since then, several breeding facilities have been established. Ferrets have been reintroduced to 30 sites across the Great Plains, and as of 2019 occur in 23 sites. Reintroductions were successful as long as sufficient prairie dog acreage remained (Santymire and Graves 2019). Sylvatic plague, caused by the bacteria *Yersinia pestis,* is a significant threat to prairie dog persistence, can be contracted by black footed ferrets, and therefore threatens the existence of black-footed ferrets. Today, ~ 300 black-footed ferrets remain in the wild, though only 1–2 reintroduced populations are considered potentially viable (Belant et al. 2015). An additional ~ 320 captive individuals are still maintained and continue to be important to the population recovery program (Goldman 2021).

There are three weasel species in North America. The least weasel (*Mustela nivalis*) is the smallest mesocarnivore in the world. They are the most fossorial and subnivean in their hunting strategies of the three weasels. The least weasel, along with the short-tailed weasel (*Mustela erminea*), have a Circumboreal Holarctic distribution (Reid et al. 2016a; McDonald et al. 2019). They occupy diverse habitats that include grasslands, shrublands, riparian, tundra, and farmlands. The long-tailed weasel (*Mustela frenata*) is common throughout parts of Canada and the United States, and its range extends to northern South America (Helgen and Reid 2016c). All three weasel species feed predominantly on rodents and other small mammals. They also are all commonly regulated and harvested as a single group (i.e., weasel), despite little knowledge of population size or trends and some concern about population sizes of the least weasels, which are listed as a species of greatest concern in some states.

16.2   Intraguild Associations

The spatial organization of mesocarnivore communities is influenced not only by the distribution of resources (e.g., prey) and abiotic factors, but also intraguild interactions (Schoener 1974; Thompson and Gese 2007). The frequency and intensity of interspecific interactions among mesocarnivores often depends on multiple factors including dietary niche overlap, temporal activity patterns, and resource availability (Heithaus 2001; Donadio and Buskirk 2006; Atwood et al. 2011).

16.2.1   Dietary Overlap

Exploitative competition is is likely widespread among mesocarnivores as it is often inferred from patterns of dietary overlap. Sympatric canid species demonstrate high levels of dietary overlap and likely compete for resources. For example, red foxes have high dietary overlap with endangered San Joaquin kit foxes (*V. m. mutica*; Clark et al. 2005). Coyotes are considered a generalist and are the most widespread canid in western rangelands (Gompper 2002). Coyotes have high dietary overlap with sympatric red foxes (Azevedo et al. 2006), swift foxes, (Kitchen et al. 1999), kit foxes (Byerly et al. 2018), and gray foxes (Neale and Sacks 2001). Coyote diet overlaps with badgers for consumption of small mammals, but they may form hunting associations that compliment instead of compete for access to prey (Minta et al. 1992; Thornton et al. 2018). Striped skunks and raccoons are widespread generalists, and have more omnivorous diets that primarily include insects, plant materials, and eggs (rather than mammalian prey). Their diets likely put them in competition with one another but limits competition with canids, felids, and mustelids (Azevedo et al. 2006).

While many mesocarnivores have omnivorous diets, the felids are hypercarnivorous. This dietary specialization helps them secure taurine, an essential amino acid found in animal protein (Hedberg et al. 2007). Even so, bobcats and coyotes have also been shown to have high dietary overlap but may partition dietary resources. In Arizona, bobcats and coyotes both consumed rodents and lagomorphs, but bobcats consumed more rodents whereas coyotes consumed more lagomorphs; coyotes also supplemented their diet with larger prey (e.g., deer [*Odocoileus* spp.], javelina [*Tayassu tajacu*]) and plant material (e.g., fruit, seeds), which were consumed infrequently by bobcats (McKinney and Smith 2007). These examples illustrate how competition among species can vary in intensity among mesocarnivore dyads.

Dietary overlap among sympatric mesocarnivores often changes seasonally. For example, diets of gray foxes included fruit during summer and fall, creating relatively high dietary overlap with coyotes but little overlap with bobcats. In contrast, during winter and spring when fruits are not available, gray foxes shift to using rodents and therefore reduce dietary overlap with coyotes and increase dietary overlap with bobcats (Neale and Sacks 2001). Changes in the amount of dietary overlap across seasons is more likely to occur when at least one species practices seasonal shifts in their diet.

16.2.2 Intraguild Predation

Interference competition among heterospecific mesocarnivores often manifests as intraguild predation or interspecific killing. Intraguild predation occurs when two species compete for limited resources and also prey upon one another (Polis et al. 1989). Interspecific killing, often described as a form of intraguild predation, is an intense form of interference competition in which an individual of one species kills (but does not consume) a competitor (Lourenço et al. 2014). In mammalian predator guilds, intraguild predation is typically asymmetric with larger, and often more generalist predators (the intraguild predator), killing smaller, and often more specialized predators (the intraguild prey; Polis et al. 1989; Verdy and Amarasekare 2010). The resource-ratio hypothesis (Holt and Polis 1997; Miller et al. 2005) predicts three stable states for intraguild predation systems. Under resource-poor conditions, resources are insufficient to support the intraguild predator but are sufficient to support an intraguild prey that is a superior exploitative competitor. When resource levels are high, the intraguild predator is supported at sufficient levels to exclude the intraguild prey. Finally, only when resources are at intermediate levels can stable coexistence between the intraguild predator and prey be achieved, provided the intraguild predator benefits from the consumption of the intraguild prey. In western rangelands, coyotes and kit foxes have been used as a model system to evaluate patterns of intraguild predation. Studies found evidence for all three predicted stable states (Robinson et al. 2014; Lonsinger et al. 2017). In southern New Mexico, kit foxes persisted in resource-poor environments that excluded coyotes (Robinson et al. 2014). In western Utah, coyote abundance in resource-rich habitats was sufficiently high to competitively exclude kit foxes (Lonsinger et al. 2017). Habitats with intermediate resource levels supporting the coexistence of both species were identified in New Mexico shrublands (Robinson et al. 2014) and Utah grasslands (Lonsinger et al. 2017).

Despite predictions of the resource-ratio hypothesis (Holt and Polis 1997) and models of intraguild predation, which predict the coexistence of an intraguild predator and intraguild prey is unlikely when dietary overlap is high (Heithaus 2001), the coexistence of intraguild mesocarnivores is common. The coexistence of mesocarnivores may be facilitated by alternative prey (Holt and Huxel 2007), behavioral avoidance (Wilson et al. 2010), or increased vigilance of the intraguild prey (Kimbrell et al. 2007). Alternative prey available to the intraguild predator and outside of the handling capacity of the intraguild prey is common in mesocarnivores and likely contributes to coexistence in western rangelands. For example, coyotes kill ungulates, which are not common prey for sympatric fox species (Kitchen et al. 1999; Neale and Sacks

2001; Azevedo et al. 2006; Byerly et al. 2018). Patterns of avoidance that may stabilize coexistence include spatial or temporal resource partitioning. Spatial avoidance of intraguild predators by intraguild prey (i.e., safety matching) has reportedly facilitated the co-occurrence of coyotes with both swift foxes (Thompson and Gese 2007) and bobcats (Wilson et al. 2010). Patterns of predation risk and safety matching may also be influenced by sex. For example, avoidance of badgers (an intraguild predator) was stronger for female than male black-footed ferrets (an intraguild prey; Grassel et al. 2015). When mesocarnivores rely on similar prey resources, the temporal availability of prey may limit temporal partitioning. Even so, temporal partitioning may be important in facilitating co-occurrence between intraguild predator and prey. For example, gray foxes used water sources in Texas at times that minimized the potential for interactions with coyotes and bobcats (Atwood et al. 2011). Finally, the coexistence of intraguild predator and prey may be facilitated by increased vigilance by the intraguild prey, which decreases the foraging efficiency of the intraguild prey and may lead to decreased vigilance and increased susceptibility of shared prey (Kimbrell et al. 2007). Increased vigilance is likely common for intraguild prey, which must consider predation risk while foraging, leading to changes in behavior that can influence shared prey (Rosenheim 2004).

Patterns of higher intraguild predation may lead to reductions in mesocarnivore populations (i.e., mesocarnivore suppression), whereas lower intraguild predation may lead to increases (i.e., mesocarnivore release; Soulé et al. 1988; Crooks and Soulé 1999). Eradication of large carnivores from many rangeland systems has relaxed top-down pressures, allowing mesocarnivores to increase in abundance, ultimately increasing pressure on prey populations (Prugh et al. 2009). These increases may be further exacerbated by relaxed bottom-up constraints, particularly when synanthropic mesocarnivores (species ecologically associated with humans) benefit from anthropogenic subsidies (e.g., food via trash or crops, increased denning structures). In some cases, coyotes took the role of apex predators in areas where larger carnivores were extirpated (Cherry et al. 2016; Schuttler et al. 2017). In rangelands, coyotes limit prey species such as pronghorn (*Antilocapra americana*; Berger and Conner 2008). The implications of mesocarnivore suppression for species conservation and management can be complicated. For example, coyotes are a primary source of mortality across their range for smaller sympatric foxes, including swift and kit foxes (Nelson et al. 2007; Thompson and Gese 2007) and endangered San Joaquin kit foxes (Cypher and Spencer, 1998). Consequently, to induce mesocarnivore release by San Joaquin kit foxes, coyotes have been lethally controlled in some areas (Cypher and Scrivner 1992). However, despite the lethal removal of nearly 600 coyotes in five years from one site, indices of abundance and survival rates did not increase, nor did fewer foxes get killed by coyotes (Cypher and Scrivner 1992). Further, Clark et al. (2005) cautioned that coyote control could harm San Joaquin kit foxes by reducing competitive exclusion of nonnative red foxes by coyotes and increasing the potential negative interaction between red foxes and kit foxes, which have greater dietary overlap than kit foxes and coyotes.

16.3   Rangeland Management

16.3.1   Livestock Conflicts

Several mesocarnivores, especially the coyote, have been the focus of lethal and nonlethal tools to reduce livestock depredation in western rangelands for over a century. An abundant population of mesocarnivores may result in increased livestock loss through direct predation. Direct predation by abundant mesocarnivores has received the majority of attention concerning mesocarnivores on and around rangelands, and is generally discussed in terms of carnivore-livestock conflict (e.g., Scasta et al. 2017; Mosley et al. 2020). That is, mesocarnivores are suggested to negatively impact livestock through direct predation, which requires some type of lethal or non-lethal intervention to decrease losses (e.g., Andelt 1992; Knowlton et al. 1999; Mitchell et al. 2004; Scasta et al. 2017). However, the discussion and severity of mesocarnivore-livestock conflicts seem to vary across taxonomic groups and among species that are common and rare. Further, the scientific knowledge concerning mesocarnivore ecology related to livestock ranching and rangelands beyond direct predation is limited. For instance, evaluating how mesocarnivores may reduce potential forage competitors of livestock such as jackrabbits (*Lepus* spp.) or prairie dogs has received some mention, yet little empirical attention, despite evidence of jackrabbits consuming a substantial portion of livestock forage (Ranglack et al. 2015). The mesopredator release hypothesis (Soulé et al. 1988; Crooks and Soulé 1999) could apply to this context (e.g., Prugh et al. 2009; Prugh and Sivy 2020). We would expect abundant mesocarnivore populations to have a positive effect on livestock forage by reducing herbivore populations, which is a mechanism that has received empirical support (Henke and Bryant 1999).

Despite the possible strong role of mesocarnivores to rangelands, few studies have evaluated questions associated with how intact and working rangelands contribute to mesocarnivore habitat and conservation, or how mesocarnivores may provision ecosystem services on rangelands. As a contemporary demonstration, we performed a literature search over the last 20 years (i.e., 2000–2021) within the Web of Science[1] database. We searched for articles using search terms of the mesocarnivore species and the term ranch; an example for American badgers was *"badger" & "ranch."* This combination of terms returned all articles that included badger and ranch (e.g., could be ranching, ranchland, etc.), which we then examined to ensure the articles were relevant. Once all articles were gathered, we determined if the main theme of articles were associated with livestock conflict, which we used to classify articles into two groups: (1) was or (2) was not associated with conflict. Results from this assessment indicated that (1) there are very few articles assessing mesocarnivore ecology and ranching, and (2) that most articles discussing coyotes on rangelands are associated

---

[1] Any use of trade, firm, or product names is for descriptive purposes only and does not imply endorsement by the U.S. Government.

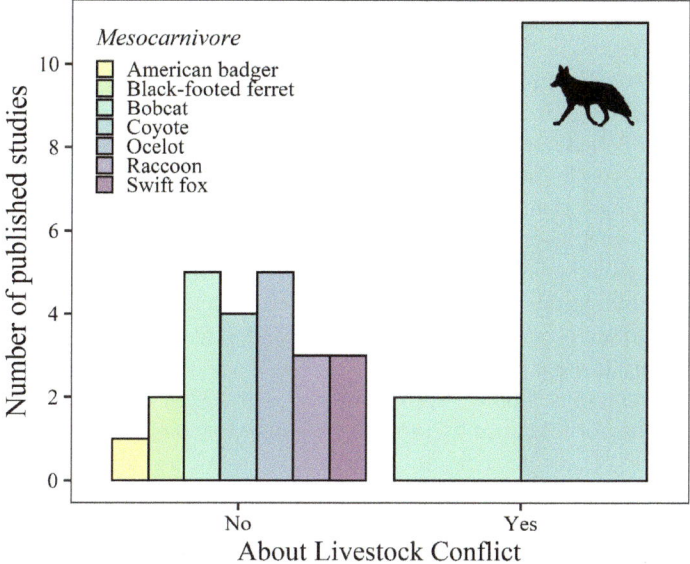

Fig. 16.2 Mesocarnivore studies referencing rangelands during 2000–2021. Data were gathered from Web of Science using search terms: mesocarnivore species* and ranch*. Thirty-seven relevant studies were gathered from search results. We determined if the main theme of the article was associated with livestock conflict or not. Studies of coyotes (*Canis latrans*) were generally about conflict with livestock

with livestock conflict (Fig. 16.2). Consequently, there is vast potential to advance the state of understanding in the future regarding the role of working rangelands in the ecology and conservation of mesocarnivores.

16.3.2   Canid Predation

Coyotes are the most discussed mesocarnivore concerning ranching and rangelands. Most assessments of coyotes on rangelands have been in the context of livestock predation and losses (Knowlton et al. 1999; Larson et al. 2019). Coyotes have been documented killing and consuming livestock (Knowlton et al. 1999; Blejwas et al. 2002; Sacks and Neale 2002; Palmer et al. 2010) and a suite of techniques, both lethal and non-lethal, have been implemented to reduce livestock losses. For instance, lethal removal in the form of trapping, calling and shooting, and aerial gunning have been employed to reduce livestock depredation (Knowlton et al. 1999; Blejwas et al. 2002). Non-lethal control in the form of livestock husbandry, fencing, electronic frightening devices, livestock guard dogs, and sterilization of breeding pairs have also been used to reduce livestock losses from coyotes (Andelt 1992; Bromley and Gese 2001; Knowlton et al. 1999; van Eeden et al. 2018a; Bromen et al. 2019; Mosley et al. 2020).

Despite differing approaches to limit, reduce, or eliminate livestock losses from coyotes, few rigorous studies exist where the consequence of treatment have been effectively evaluated in a controlled or semi-controlled context (Eklund et al. 2017; van Eeden et al. 2018b). This pattern has also been observed in wildlife studies, where the impacts of predator management or removals on measurable outcomes (e.g., recruitment of young) of prey species is absent, muddled, and/or context-dependent (Ballard et al. 2001, but see Mahoney et al. 2018 and Seidler et al. 2014). Rigorous experimental work over longer timeframes is needed to assess which techniques may be most effective in preventing livestock losses across different landscapes and ecological communities (van Eeden et al. 2018b). No technique will be universally effective and often a variety of approaches may need to be used in an integrated and complimentary fashion (Knowlton et al. 1999).

Despite numerous studies evaluating and discussing modifications to coyote density or behavior to reduce livestock losses, very few studies have evaluated questions related to coyote behavior and demography as a consequence of management efforts. An exception was a study in southeastern Colorado that demonstrated coyote diurnal activity increased with a reduction in human persecution (i.e., shooting, trapping, and intense aerial gunning; Kitchen et al. 2000). This alteration in activity may have had carry-over effects on coyote diet or other interspecific interactions, but this was not evaluated as part of the study. As researchers and managers seek to understand the effectiveness of different treatments to reduce livestock losses, there will be additional opportunities to assess the consequences of treatments on coyote ecology.

Beyond coyotes, few studies have assessed the relationship between livestock ranching and the ecology of other medium-sized canids. However, many studies on the ecology of foxes (e.g., space use, den site selection, and movement) have occurred on rangelands and may be useful to predicting relationships between livestock ranching and canid ecology. For example, studies of space use, den site selection, and movement of swift foxes has often occurred on rangelands (Nicholson et al. 2006, 2007; Sasmal et al. 2015; Butler et al. 2019, 2020). Even so, many additional questions remain concerning the ecology of medium-sized canids and livestock ranching.

16.3.3 Felid Predation

There is less research regarding livestock ranching and rangelands and medium-sized felids relative to that of canids. Some studies have assessed livestock losses by bobcats (e.g., Scasta et al. 2017; Bromen et al. 2019), while others have evaluated bobcat occupancy or density on rangelands (Greenspan et al. 2020; Lombardi et al. 2020a). Ocelots have received more attention with specific regard to rangelands and detailed questions about their ecology. This is likely because the majority of ocelot habitat in the United States occurs on private ranchlands (Lombardi et al. 2020b), and thus successful conservation efforts require ranchland connectivity. Beyond the importance of working ranchlands for ocelot conservation, additional questions remain regarding the influence of different ranching practices on both bobcats and ocelots, and the potential interactions between these species on private ranches.

16.3.4 Mustelid, Mephitid, and Procyonid Predation

The majority of species within the families *Mustelidae*, *Mephitidae*, and *Procyonidae* have received little-to-no attention concerning their ecology on rangelands. Some notable exceptions, however, include black-footed ferrets, American badgers, and raccoons. For example, research on black-footed ferrets has examined habitat use and resource selection of reintroduced individuals on rangelands in New Mexico (Chipault et al. 2012). In Wyoming, researchers have recently examined the role of habitat provisioning wherein American badgers make subterranean habitat accessible to non-fossorial species; this work assessed the diversity and frequency of species using abandoned badger burrows on rangelands (Andersen et al. 2021). Finally, researchers have evaluated the role of raccoon fecal deposition on *Escherichia coli* *(E. coli)* distribution and concentration within Texas floodplains (Parker et al. 2013). There remain many additional opportunities to better assess the ecology of these species on working rangelands.

16.4 Harvest of Mesocarnivores

The harvest of mesocarnivores has a long history in North America. Humans have harvested mesocarnivores in North America for thousands of years to use for clothing, food, and religious ceremonies (Wright 1987). The Indigenous peoples of western North America harvested mesocarnivores such as badger, mink, raccoon, skunks, weasels, and wildcats (*Lynx* spp.) for their pelts for ceremonial decoration, clothing, quivers, baskets, and occasionally for eating before the arrival of Europeans (McGee 1987). Systematic fur trading, primarily for American beaver (*Castor canadensis*) pelts, between Europeans and Indigenous peoples began in the 1580s in northeastern North America (Obbard et al 1987). In the late 1700s and early 1800s, European and American fur traders expanded into the southwest and Great Plains of the United States (Ray 1987). Across North America, European and American fur trading companies kept records that have provided insights into the scale at which fur trapping occurred (Obbard et al. 1987; Ray 1987).

Harvest levels of mesocarnivores before the 1930s are difficult to decipher as fur records are incomplete, and the most continuous data sets come from the Hudson Bay Company, which predominately operated in Canada (Obbard et al. 1987). Using data from the Hudson Bay Company and other fur trading companies, Obbard et al. (1987) found that the annual mean harvest of foxes (all species grouped), raccoon, mink, and bobcat (grouped with Canada lynx; *Lynx canadensis*) substantially increased from a few thousand pelts to tens or hundreds of thousands of pelts annually from the 1800s to the 1980s in North America. River otter annual mean harvest also increased from a few thousand to approximately 43,000 in the 1780s, then declined to less than 16,000 between 1900 and 1909, and finally rebounded again to historically high levels starting in 1980. Ringtail and badger harvests were lower than some other species, likely because of the low utility and quality of their pelts. Annual mean harvest of ringtails and badgers largely began after 1900 and reached approximately 70,000 and 38,000 in the 1980s, respectively. Like foxes, historical harvest records of skunks encompassed all species of skunks until the 1900s, and weasel records continue to group species together. Skunk annual mean harvest increased from 113,000 in the 1920s to 1.2 million by the 1930s and 1940s, and weasel annual mean harvest increased from 490,000 to 900,000 over the same time frame. Both declined, with annual mean skunk harvest of 38,000 in the 1960s and annual mean weasel harvest of 71,000 in the 1980s.

Up until the early 1900s, coyote annual mean harvest totals were likely combined with those of wolves (*C. lupus*) making it difficult to identify long-term trends. Coyote annual mean harvest fluctuated between 20,000 and 107,000 pelts between the 1920s and 1950s before increasing to 500,000 by the 1980s (Obbard et al. 1987). The totals are from a time when the harvest was largely unregulated, which began to change at the beginning of the twentieth century when some species became locally scarce or extirpated (Hubert 1982).

Furbearer management began to include trapping and fur buyer licenses, and annual trapper surveys in the early 1900s (Hubert 1982). Regulations on harvest such as duration of the season, bag limits, and trapping best management practices for each of the 23 species of furbearing animals in North America were put in place to maintain a sustainable yield for perpetuity (White et al. 2015; White et al. 2021). The implementation of regulations varied by state and species during the 1900s. For example, in 1971, bobcats were unprotected in 10 of 12 western states (Faulkner 1971) but by 1987, protection was given to bobcats in all western states (Melchior et al. 1987). In contrast, raccoons changed from regulated to less regulated harvests. They were harvested during distinct hunting seasons without bag limits in eight midwestern states and year-round in one state with no bag limit in 1982 (Melchior et al. 1987). In 1993, seven midwestern states had hunting seasons with no bag limits, one state had a hunting season with a bag limit of 20, and two states allowed hunting year-round with no bag limit (Rogers 1995). In 2017, harvest was widespread with 14 species or groups of mesocarnivore in 22 midwestern and western states totaling 1,049,994 individuals (Association of Fish and Wildlife Agencies 2015, unpublished data).

In addition to records during the fur trade, harvest surveys and harvest efforts have historically been used to track population levels of mesocarnivores (Clark and Andrew 1982; Roberts and Crimmins 2010; Roberts et al. 2020). Many factors influence furbearer harvest totals beyond population abundance such as trapper experience, trapping regulations, trapper effort, pelt prices, and winter weather (Elsken-Lacy et al. 1999; Ruette et al. 2003; DeVink et al. 2011). These factors potentially confound the ability of harvest data to indicate biological changes in the population (Allen et al. 2020). Recently, many studies have evaluated the influence of different combinations of these factors simultaneously and found that the direction and relative importance of these effects on harvest varied by species (Hiller et al. 2011; Kapfer and Potts 2012; Ahlers et al. 2016; Allen et al. 2018; Bauder et al. 2020a, b). For example, winter temperature had a strong negative effect on bobcat harvest in Minnesota (Kapfer and Potts 2012), but only a weak negative effect on raccoon harvest in Illinois (Bauder et al 2020b). Raccoon, red fox, and gray fox harvests in Illinois were positively influenced by gasoline prices, while gasoline prices had a negative influence on muskrat (*Ondatra zibethicus*) harvest in Illinois (Ahlers et al. 2016). While the harvest of some mesopredators is not influenced by pelt prices (Hiller et al. 2011), pelt prices influenced harvest of other species, including red fox, gray fox, and muskrat (Ahlers et al. 2016). Collectively, these studies indicate biological trends can be inferred from harvest data after accounting for confounding factors. However, as these studies were conducted in the midwestern United States, it is currently unknown if their findings apply to harvest data from the western United States and Canada, which could vary in management regulations and economic and social pressures. Therefore, future research should investigate how these factors, and others, influence the harvest of mesocarnivores in rangeland-dominated jurisdictions.

Harvest of mesocarnivores is not entirely for the fur but also a source of sport for some hunters, although hunting for sport is not mutually exclusive to harvest for fur. Coyotes are often the focal animal for sport hunters. In many western rangeland systems, local communities, counties, and other organizations host coyote hunting contests, also called coyote call contests. These contests typically provide a cash reward for the most animals killed within a set time. Foxes and other mesocarnivores may also be the focus of these contests. In recent years, some western states have banned hunting contests; for example, New Mexico and Arizona banned these contests in 2019 and 2021, respectively (NM S-B76 2019, AZ AC R12-4-303 2021).

16.5 Predator Control

In addition to legal harvest of mesocarnivores for fur, food, and sport, mesocarnivores have also been killed for centuries to reduce human-carnivore conflicts. These events occur in response to actual or perceived threats of mesocarnivores to domestic livestock or wildlife populations (Reynolds and Tapper 1996). The earliest form of formal predator control was the bounty system. Beginning in 1800, every state in the midwestern United States has had a bounty placed on a mesocarnivore, most often coyote, red fox, and gray fox (11 states), but also bobcat (8 states) and badger (1 state; Hubert 1982). Moreover, bounty programs across North America for species such as coyote, red fox, and bobcat remained in effect through the 1970s and 1980s (Hubert 1982; Slough et al. 1987) and into the twenty-first century, with coyote bounty programs in effect in South Dakota, Texas, and Utah through 2003 (Bartel and Brunson 2003). There is currently a coyote bounty program in Utah that began in 2012, where participants can receive up to 50 USD per coyote submitted, and a nest predator bounty program in South Dakota for raccoons, badgers, striped skunks, red foxes, and Virgina opossums (*Didelphis virginiana*) that began in 2019, where participants can receive up to 10 USD per tail submitted.

Impacts of bounty programs are largely unclear. There is no consolidation of records to determine how many mesopredators have been killed through bounty programs but there are reports from some bounty programs to indicate high numbers of animals are taken. One report from 1972 documented that 111,569 coyotes, 17,169 bobcats, and 494,635 foxes were taken in Michigan between 1935 and 1970, accounting for 4.46 million USD in payments (Cain et al. 1972). Although the number of trappers and hunters has declined across the years, the data for number of animals submitted to current bounty programs remains high. In South Dakota, 54,471 tails were submitted in 2019 and 26,390 tails were submitted in 2020; data are available at the agency's website: https://sdgfp.maps.arcgis.com/apps/opsdashboard/index.html#/e7bbbd6fa93b48c6a31985aa7c57c5ff. For the Utah bounty program, 7160 coyotes were submitted in its first year, 2013 (UDWR 2013). The number of coyotes submitted annually has ranged from the 7000s to 10,000s, although only 4109 coyotes were submitted in 2020. This low number may have been due to COVID-19 restrictions. Although the number of animals submitted to bounty programs appears high, it is not clear if bounty programs have the desired effect of reducing predation pressure on livestock or other wildlife. Research on the effects of removing several species of the mesocarnivore community (typically coyotes, raccoons, red foxes, and striped skunks) on waterfowl nesting success showed conflicting effects (Rohwer et al. 1995; Sargeant et al. 1995; Docken 2011; Blythe and Boyce 2020), no effect on upland bird populations (Guthery and Beasom 1977; Lawrence and Silvy 1995; Frey et al. 2003; Lyons et al. 2009; Docken 2011; Reid 2019), and no effect on rodent and lagomorph densities (Guthery and Beasom 1977; Henke and Bryant 1999). Similarly, studies on the effects of coyote removal on other mesocarnivores showed conflicting results (Henke and Bryant 1999; Kamler et al. 2003a; Karki et al. 2007). There is evidence that coyote removal has short-term positive effects on white-tailed deer (*O. virginianus*), mule deer (*O. hemionus*), and pronghorn (Guthery and Beasom 1977; Harrington and Conover 2007; Brown and Conover 2011; Mahoney et al. 2018).

In the United States, federal programs aimed at the prevention and control of wildlife damage emerged near the end of the nineteenth century and have been the primary entity responsible for reducing livestock depredation rates (Miller 2007). Known today as the United States Department of Agriculture—Animal and Plant Health Inspection Service—Wildlife Services (WS), WS became involved in predator control in 1915 when charged with reducing wolf and coyote livestock depredation. Subsequently, part of their mission was to research and develop new lethal predator control methods (Miller 2007; Feldman 2007) such as traps, M-44's, predacides, ground shooting, snares, denning, dogs, and aerial shooting (Evans and Pearson 1980). Coyote depredation continues to be the largest cause of cattle and sheep predator loss in the United States in 2015 (United States Department of Agriculture—National Agricultural Statistics Service 2015a, b), which has been confirmed by research on sheep (Palmer et al. 2010). WS killed or euthanized 68,905 coyotes in 2015 and killed or euthanized 62,002 coyotes during WS operations in 2019 (Wildlife Services 2015, 2019). These numbers are similar to the annual number killed during the 1970s (Evans and Pearson 1980) even though there are dramatically lower numbers of sheep in the United States today. Wagner and Conover (1999) found that aerial shooting of coyotes decreases the number of lambs lost to coyotes. However, this may not be a long-term solution given the financial costs and high immigration rates in exploited populations (Kilgo et al. 2017). Although an evaluation of the program's efficacy has been called for across decades (Wagner 1988; Shivik 2014), no such evaluation has occurred. This is important because recent research has investigated the efficacy of nonlethal methods to reduce coyote depredation such as sterilization (Seidler and Gese 2012; Young et al. 2019b), livestock guard dogs (Kinka and Young 2018; Saitone and Bruno 2020), and fladry (Young et al. 2019a), which showed promising results and should continue to be part of an integrated approach and a focus of future studies.

16.6 Impacts of Disease

16.6.1 Disease Concerns for Mesocarnivore Populations

Disease enzootics in mesocarnivores can have significant implications for the dynamics of mesocarnivore communities, disease management, and human perceptions of mesocarnivores. Mesocarnivores serve as a primary reservoir for generalist pathogens (i.e., those having a wide host range), which may be transmitted among native carnivore species or spread to domestic animals (Roemer et al. 2009). Generalist pathogens maintained by mesocarnivores may also be zoonotic (i.e., transmitted between animals and humans) and, therefore, represent a public health concern. Generalist viruses and pathogens of principal concern for mesocarnivores in western rangelands include canine distemper virus (*Morbillivirus sp.*), parvoviruses (including feline panleukopenia virus and canine parvovirus; genus *Parvovirus*), rabies (genus *Lyssavirus*), and mange (primarily sarcoptic *Sarcoptes scabiei* and notoedric *Notoedres cati*).

Canine distemper virus is an infectious disease that impacts all families of terrestrial carnivores and has high mortality rates in some species (Deem et al. 2000). Being highly transmissible, canine distemper infections occur through contact with or inhalation of aerosolized virus, which may be shed by an individual through respiratory droplets or bodily secretions for up to 90 days after infection, and can persist in the environment for up to 14 days (Deem et al. 2000; Anis et al. 2020). The impact of canine distemper varies among species. Canine distemper is enzootic in raccoons and gray foxes, which serve as reservoirs in North America (Deem et al. 2000). Raccoons support the coinfection of multiple canine distemper strains, potentially leading to new strains (Pope et al. 2016). Among mesocarnivores, mustelids appear to be the most susceptible to canine distemper with critically high mortality rates approaching 100% (Beineke et al. 2015). For this reason, canine distemper has been considered the "most significant infectious disease" limiting endangered black-footed ferrets, which are also susceptible to canine distemper induced through a modified-live vaccine (Williams et al. 1988, 1996).

Parvoviruses impacting mesocarnivores emerged in the 1970s and have continued to evolve (Steinel et al. 2001; Hueffer et al. 2003; Allison et al. 2012). Originally described as feline panleukopenia virus, cross-species transmission to canid hosts lead to canine parvovirus (CPV) type-2 and the subsequent establishment of two antigenic strains—CPV-2a and CPV-2—which regained the ability to infect felids and became widespread (Steinel et al. 2001; Allison et al. 2013; Stuetzer and Hartmann 2014). Parvoviruses are easily transmitted from hosts through contact with their feces or contaminated objects and can persist in the environment for extended periods (weeks to months; Steinel et al. 2001). Consequently, parvoviruses can be transmitted among individuals without direct contact and are highly contagious (Allison et al. 2013). The tendency of many mesocarnivores to use latrines, olfactory cues for territorial maintenance, and shared focal resources (e.g., water resources in arid rangelands; Atwood et al. 2011) likely contributes to the high transmission capacity of parvoviruses. CPV antibodies reported in coyotes (71–100%; McCue and O'Farrell 1988; Gese et al. 1991, 2004) and adult foxes (kit and swift foxes: 60–71%; Miller et al. 2000; Gese et al. 2004) were generally high, inidicating the virus may be enzootic in these species. Evidence of CPV antibodies was higher in adults than juveniles for coyotes and foxes (Miller et al. 2000; Gese et al. 2004). This is likely explained by the fact that parvoviruses impact juvenile survival significantly more than adults, which in turn may limit recruitment and threaten populations (Gese et al. 1997).

Rabies is a zoonotic, neurotrophic virus most commonly transmitted among conspecifics through a bite from an infected animal, but interspecific transmission may also occur (Hass and Dragoo 2006; Ma et al. 2020). Rabies initially evolved in bats and later spread to mesocarnivores (Kuzmin et al. 2012). Although direct evidence of transmission between bats and mesocarnivores is rare, multiple transmission events of a bat variant of rabies to striped skunks were detected in Arizona (Leslie et al. 2006; Kuzmin et al. 2012). Mesocarnivores are reservoirs for rabies, with striped skunks and, to a lesser extent, gray foxes maintaining enzootic levels of rabies in western rangelands (Hass and Dragoo 2006; DeYoung et al. 2009; Ma et al. 2020). Rabies has also been detected in hooded, hog-nosed, and western spotted skunks (Crawford-Miksza et al. 1999; Dragoo et al. 2004), but their role in maintaining the virus is unclear. Epizootics of the skunk variant of rabies have been associated with periods following increased precipitation and may be driven by increased density of skunks resulting from bottom-up processes (e.g., increased primary productivity and prey; Hass and Dragoo 2006). While most mesocarnivore rabies cases in western rangelands are attributable to skunks, foxes, and raccoons, rabies has been detected in bobcats and coyotes, and less frequently in mustelids and other procyonids. Management of rabies in wildlife populations in western rangelands has focused largely on surveillance of wildlife populations with targeted oral vaccination programs aimed at limiting the spread and reducing the risk of transmission to pets and humans. In Texas, oral vaccination programs guided by information on dispersal tendencies and population genetic structure of gray foxes (DeYoung et al. 2009) were successful in eliminating localized gray fox and coyote-dog variants of the virus (Slate et al. 2005).

Mange is a disease caused by mites that infest a host's epidermis leading to intense irritation and itching, hair loss, and callousing of the epidermis (Niedringhaus et al. 2019). Secondary cracking of the skin, combined with bacterial or yeast infections, may lead to severe emaciation and death (Niedringhaus et al. 2019). Sarcoptic and notoedric mange are among the most common in mesocarnivores, with sarcoptic mange being widespread (Niedringhaus et al. 2019). Transmission of mange among individuals may occur directly through contact or indirectly through shared environments, with mites being able to persist under some environmental conditions without a host (Niedringhaus et al. 2019). Species relying on dens may be at heightened risk of mange infestation, owing to the increased potential for indirect transmission (Montecino-Latorre et al. 2019). Den use, particularly when dens are used by more than one individual, family group, or species, may facilitate the spread and cross-species transmission of mange (Niedringhaus et al. 2019; Montecino-Latorre et al. 2019). In North American rangelands, coyotes are a primary host of sarcoptic mange, though other canid species may also sustain the disease (Pence et al. 1983; Niedringhaus et al. 2019). The impact of mange is likely influenced by host age and condition, with infections being more rapid in juveniles (Pence et al. 1983). Population-level impacts of mange likely vary among species and populations. For example, endangered San Joaquin kit foxes infected with mange are unlikely to recover, even with treatment, and mange can represent a significant threat to population persistence (Cypher et al. 2017). In contrast, while mange-infected coyotes had increased mortality risk, these mortalities were compensatory and did not influence population-level survival (Pence et al. 1983). Furthermore, there is evidence that some coyotes and red foxes can recover from mange infections (Chronert et al. 2007; Nimmervoll et al. 2013). Although less widespread, notoedric mange has caused acute population declines in bobcats (Serieys et al. 2013) and has been detected in endangered ocelots (Pence et al. 1995).

16.6.2   Disease Concerns for Other Rangeland Animals

In western rangelands, disease epizootics in mesocarnivores are of greatest concern due to the potential for transmission from the reservoir host to other species, which could include imperiled species, domestic animals, and humans. Endangered black-footed ferrets and San Joaquin kit foxes are at high risk of being impacted by diseases, and disease outbreaks in these species have demonstrated the population-level threat of pathogens. Recovery of endangered black-footed ferrets has been hampered by direct impacts of diseases that cycle in sympatric canid populations (e.g., canine distemper virus and rabies; Williams et al. 1988, 1994; Gese et al. 1991) and indirect impacts of sylvatic plague (caused by *Yersinia pestis*) on prairie dogs, the black-footed ferret's primary prey. The role of disease in regulating San Joaquin kit fox populations is less clear (Cypher et al. 2017). Antibodies of CPV have been detected in San Joaquin kit foxes, but it is suspected to be enzootic and has not been implicated in population declines (McCue and O'Farrell 1988). In contrast, canine distemper virus, rabies, and sarcoptic mange have all led to mortality events or substantial declines in at least one San Joaquin kit fox population (White et al. 2000; Cypher et al. 2017; Rudd et al. 2019). Mesocarnivores tend to occur in higher densities than large carnivores and disease spillover has the potential to influence sympatric populations of imperiled large carnivores. For example, the introduction of CPV to wolves on Isle Royale from a domestic dog (*C. familiaris*) resulted in a precipitous decline in wolves (Wilmers et al. 2006), and similar spillover events from mesocarnivore reservoirs could threaten recovery efforts of large carnivores (e.g., Mexican gray wolves [*C. l. baileyi*]) in western rangelands, but these community dynamics have not been well documented or studied (Roemer et al. 2009).

The complex mesocarnivore community offers primary and secondary reservoir species for most generalist pathogens, making disease management challenging. For small, isolated populations, such as those characterizing endangered mesocarnivores, the risk of pathogen-caused local extinctions is increased when generalist pathogens can persist in sympatric species and when transmission rates are artificially increased by anthropogenic habitat changes (Smith et al. 2009). Land management practices in western rangelands may influence disease dynamics by promoting wildlife aggregation (e.g., water developments, refuge habitats) and potentially increasing contact rates. Water catchments to support game populations and livestock have greatly expanded the availability of water (Rosenstock et al. 1999) and increased habitat for some mesocarnivores (e.g., raccoons; Kamler et al. 2003b). Catchments may increase contact rates among individuals and increase the potential for intra- and inter-specific disease transmission, but there is little empirical evidence for this hypothesis (Rosenstock et al. 1999) and mesocarnivores may be able to mitigate risks in some systems through spatial and temporal partitioning of catchment usage (Atwood et al. 2011). Similarly, many mesocarnivores may experience increased spatial and temporal overlap in refuge habitats within urban (e.g., parks) or agricultural (e.g., shelterbelts) landscapes (Sévêque et al. 2020).

Many mesocarnivores have shown a remarkable ability to exploit both natural and anthropogenic landscapes (Šálek et al. 2015). The potential for disease transmission from domestic animals to mesocarnivores, and vice versa, is of great concern. Domestic dogs and cats (*Felis catus*) are widespread, occur in high densities, and can serve as a host for many pathogens that impact mesocarnivores (Smith et al. 2009). The ability of some mesocarnivores (e.g., striped skunks and raccoons) to be synanthropic increases the risk of spillover of generalist pathogens between domestic animals and mesocarnivores, particularly if anthropogenic subsidies associated with agriculture or urban-suburban gradients increase the mesocarnivore densities, interactions among mesocarnivores and domestic animals, or both (Ordeñana et al. 2010; Tardy et al. 2014; Theimer et al. 2015).

16.7 Ecosystem Threats

There are several new and continuing threats to western ecosystems that are likely to have different impacts on mesocarnivores. For example, in the last several years extreme fire events and drought have become prevalent through much of the rangelands in the western United States, yet the impacts of such fires on mesocarnivores remains unclear (e.g., Holbrook et al. 2016). Coyotes in controlled burned areas prefer recently burned habitats, where their prey are also more abundant (Stevenson et al. 2019) but in areas where fires are more severe and uncontrolled, it is likely the same responses of prey and mesocarnivores are not possible due to slower recovery of vegetation and the larger extent of recent wildfires. Instead, extreme weather events and rapid development are likely to result in more homogenous habitats that may limit the ability of mesocarnivores to co-occur because of an inability to partition habitat that currently allows for co-occurrence (e.g., Mueller et al. 2018).

The two largest ecosystem threats to mesocarnivore populations are the use of biological resources and land-use changes (Marneweck et al. 2021), both of which are increasing (Bell et al. 2004; Willcox 2020). Rangelands are being transformed into housing for the growing human population or infrastructure to support humans, such as energy extraction and irrigated croplands (Ellis et al. 2010; DuToit et al. 2017). In the Prairie Pothole Region of the United States, where mesocarnivores have occurred with irrigated crops for a longer time, there are positive relationships between mesocarnivore abundance and crops (Crimmins et al. 2016), likely related to higher abundance of prey in this landscape. Thus, we may also expect to see increasing populations of mesocarnivores in rangelands that transition to irrigated crops.

Rapid oil and gas developments are carried out with limited data about environmental impacts (Allred et al. 2015). Surprisingly, there are few publications on the effects of gas and oil development on mesocarnivore populations even though gas and oil development has experienced substantial growth in western rangelands of the United States in recent decades. One study found that mesocarnivores have unique responses to increasing levels of oilfield development (Fiehler et al. 2017). Coyotes and badgers were active in areas with high levels of development, whereas, the San Joaquin kit fox selected fields with no or medium levels of development (Fiehler et al. 2017). Swift foxes are similar to the latter and may even prefer oilfields due to lower coyote densities (Butler et al. 2020). These findings illustrate the complexities of how mesocarnivores respond to changing landscapes and the need for further studies. Studies that currently focus on species at the edge of their range (e.g., Sacks et al. 2018) or in areas where they overlap with exotic or invasive species (e.g., Moreira-Arce et al. 2015) will likely provide the best metrics to forecast population ecology and management needs for mesocarnivores.

Development of alternative energies is another rapidly growing industry, with many wind and solar energy facilities being constructed in western rangelands (Agha et al. 2020). The southwest United States has been identified as having the largest potential for solar energy (Lovich and Ennen 2011; Kabir et al. 2018), and construction of new facilities in rangelands is expected to continue. While these alternative energies may provide net benefits to ecosystems, the immediate costs to local wildlife populations are only starting to be understood. Scavengers may benefit from carcass resources under wind turbines (Smallwood et al. 2010). However, to date, almost no information is available on the response of mesocarnivores to wind and solar facilities; most studies focus on impacts to threatened and endangered or aerial species (Allison et al. 2019; Chambers et al. 2017; Lovich et al. 2011). Agha et al. (2017) found that wind energy facilities influence mesocarnivore behavior, likely by creating access to new dirt roads as travel routes and changes in prey behavior occur. Studies from Europe have shown red foxes were tolerant of wind turbines (Łopucki et al. 2017), while European badgers (*Meles meles*) displayed elevated stress hormones in proximity to turbines (Agnew et al. 2016). In that and another study (Smith et al. 2017), mesocarnivores appeared to avoid wind turbines, which could create micro-refuges for prey. Further studies are needed to understand how the local environmental changes associated with wind farms, such as noise, presence of humans, traffic, construction and maintenance disturbance, visual alterations to the habitat, toxins, new smells, collision threats, and more, may have short and long-term effects on mesocarnivore behavior and population dynamics.

16.8 Research and Management Needs

As detailed in this chapter, mesocarnivores on rangeland systems are primarily either considered problematic for human endeavors, such as livestock grazing and disease risks, or ignored and understudied. In general, there is too little scientific information and research on mesocarnivores (Marneweck et al. 2021), even for those that have garnered more attention, like coyotes. Identifying ways to fill in research gaps are needed. For example, the formation of the Eastern Spotted Skunk Cooperative Study working group has advanced research and conservation of this understudied species (Jachowski and Edelman 2021). Their efforts to engage seemingly disparate researchers and agency personnel that previously worked independently of one another created more awareness and information than previous solo efforts. A similar model of harnessing the power of multiple research groups to broaden our understanding of a species is underway by one of the authors (J.K. Young) with 16 research teams studying coyotes. Today, the ease for which researchers and agency personnel can work collaboratively despite being at distant locations from one another has created new opportunities for cooperative research, management, and conservation groups to form.

Some mesocarnivore populations are stable or increasing but many are declining, and others are too data deficient to track effectively. Greater information on population statuses and reasons for population dynamics of mesocarnivores is needed. This is especially important because rangeland landscapes are rapidly changing due to climate change and anthropogenic impacts and this guild can serve as "sentinels for global change" (Marneweck et al. 2022) but only if we understand their current status and population dynamics.

As noted in 16.7, rangeland conditions are undergoing rapid changes, primarily through climate change and rapid human development, and how mesocarnivores adapt to new challenges is largely unclear. This is important to consider because mesocarnivores provide substantial ecosystem services that could be lost with shifting populations and behavior in response to accelerated global changes (Marneweck et al. 2021). Mesocarnivores that exhibit flexible dietary and habitat requirements, such as coyotes, raccoons, and skunks, will likely adapt to landscape modifications, others with stricter habitat or dietary requirements, like ocelots and black-footed ferrets, may suffer population declines. However, how populations that currently are considered adaptive actually cope with these changes remains to be seen too.

The rapid development of urban-wildland interfaces and the infrastructure to support burgeoning human populations are also indirectly impacting mesocarnivore populations via altering connectivity. It is likely that additional roads and other structures decrease connectivity which, in turn, can impact immigration, emigration and genetic diversity of populations (e.g., Butler et al. 2020). As much of the western United States rangeland system becomes more arid, mesocarnivores and their prey are likely to shift ranges to seek moister areas. In many cases, this shift will include seeking out artificial water sources created by humans, from water guzzlers to water fountains. Thus, as cities and towns expand, so does the potential for some mesocarnivore species to interact with feral and free-roaming dogs (Young et al. 2011). Combined, these form a human footprint that mesocarnivores are likely responding to in different ways behaviorally and spatially (e.g., Carricondo-Sanchez et al. 2019). Many opportunities exist to study these potential impacts.

16.9 Summary

Our chapter highlights that mesocarnivores are important to ecosystems (Roemer et al. 2009) but they remain relatively understudied, especially in rangelands. The historical and current distribution, population status (when available), and conservation status of the 22 mesocarnivores in western rangeland landscapes illustrates how some species have thrived under changing conditions in western rangeland landscapes—increasing their range and population size—while others have been negatively impacted and are declining or at least unable to recover from earlier declines. Even so, most mesocarnivores continue to be affected by direct management actions, such as persecution for real or perceived human- and livestock- conflict and fur harvest, or indirect actions associated with the ever-growing anthropogenic footprint, such as increased number of roads, traffic, and developed landscapes that are typically the cause of decreased prairie and rangeland habitats. At the same time mesocarnivores are affected by management decisions and ecosystem threats, they are also impacted by diseases and intraguild interactions.

It remains unclear how ecosystem threats will further shift mesocarnivore populations and their distributions. Mesocarnivores serve as apex predators in some western ecosystems, but as populations of large carnivores continue to recover in the West and landscapes are converted by humans or human-caused climate change, it will be important to continue to study this important guild to determine their impacts on the natural and working landscapes they occupy.

References

Agha M, Lovich JE, Ennen JR, Todd BD (2020) Wind, sun, and wildlife: do wind and solar energy development 'short-circuit' conservation in the western United States? Environ Res Lett 15(7):075004. https://doi.org/10.1088/1748-9326/ab8846

Agha M, Smith AL, Lovich JE, Delaney D, Ennen JR, Briggs J, Fleckenstein LJ, Tennant LA, Puffer SR, Walde A, Arundel TR (2017) Mammalian mesocarnivore visitation at tortoise burrows in a wind farm. J Wildl Mgmt 81:1117–1124

Agnew RC, Smith VJ, Fowkes RC (2016) Wind turbines cause chronic stress in badgers (Meles meles) in Great Britain. J Wildl Dis 52(3):459–467. https://doi.org/10.7589/2015-09-231

Ahlers AA, Heske EJ, Miller CA (2016) Economic influences on trapper participation and per capita harvest of muskrat. Wildl Soc Bull 40:548–553. https://doi.org/10.1002/wsb.696

Allen ML, Roberts NM, Van Deelen TR (2018) Hunter selection for larger and older male bobcats affects annual harvest demography. Roy Soc Open Sci 5:180668. https://doi.org/10.1098/rsos. 180668

Allen ML, Roberts NM, Bauder JM (2020) Relationships of catch-per-unit-effort metrics with abundance vary depending on sampling method and population trajectory. PLoS ONE 15:0233444. https://doi.org/10.1371/journal.pone.0233444

Allison AB, Harbison CE, Pagan I, Stucker KM, Kaelber JT, Brown JD, Ruder MG, Keel MK, Dubovi EJ, Holmes EC, Parrish CR (2012) Role of multiple hosts in the cross-species transmission and emergence of a pandemic parvovirus. J Virology 86:865–872. https://doi.org/10.1128/ JVI.06187-11

Allison AB, Kohler DJ, Fox KA, Brown JD, Gerhold RW, Shearn-Bochsler VI, Dubovi EJ, Parrish CR, Holmes EC (2013) Frequent cross-species transmission of parvoviruses among diverse carnivore hosts. J Virol 87:2342–2347. https://doi.org/10.1128/JVI.02428-12

Allison TD, Diffendorfer JE, Baerwald EF, Beston JA, Drake D, Hale AM, Hein CD, Huso MM, Loss SR, Lovich JE, Strickland MD (2019) Impacts to wildlife of wind energy siting and operation in the United States. Issues Ecol 21(1):2–18

Allred BW, Smith WK, Twidwell D, Haggerty JH, Running SW, Naugle DE, Fuhlendorf SD (2015) Ecosystem services lost to oil and gas in North America. Science 348(6233):401–402. https:// doi.org/10.1126/science.aaa4785

Andelt W (1992) Effectiveness of livestock guarding dogs for reducing predation on domestic sheep. Wildl Soc Bull 20:55–62

Andersen ML, Bennett DE, Holbrook JD (2021) Burrow webs: clawing the surface of interactions with burrows excavated by American badgers. Ecol Evol 11(17):11559–11568. https://doi.org/ 10.1002/ece3.7962

Anis E, Needle DB, Stevens B, Yan L, Wilkes RP (2020) Genetic characteristics of canine distemper viruses circulating in wildlife in the United States. J Zoo Wildl Med 50:790–797. https://doi. org/10.1638/2019-0052

Arizona Administrative Code R12-4-303 (2021) Unlawful devices, methods, and ammunition

Association of Fish and Wildlife Agencies (2015) National fur harvest database, 1970–2018. https:/ /www.fishwildlife.org/download_file/view/2896/1213

Atwood TC, Fry TL, Leland BR (2011) Partitioning of anthropogenic watering sites by desert carnivores. J Wildl Manage 75:1609–1615. https://doi.org/10.1002/jwmg.225

Aubry KB, Statham MJ, Sacks BN, Perrine JD, Wisely SM (2009) Phylogeography of the North American red fox: vicariance in Pleistocene forest refugia. Mol Ecol 18(12):2668–2686. https://doi.org/10.1111/j.1365-294X.2009.04222.x

Azevedo FCC, Lester V, Gorsuch W, Larivière S, Wirsing AJ, Murray DL (2006) Dietary breadth and overlap among five sympatric prairie carnivores. J Zool 269:127–135. https://doi.org/10.1111/j.1469-7998.2006.00075.x

Ballard WB, Lutz D, Keegan TW, Carpenter LH, De Vos J (2001) Deer-predator relationships: a review of recent North American studies with emphasis on mule and black-tailed deer. Wildl Soc Bull 29:99–115

Bartel RA, Brunson MW (2003) Effects of Utah's coyote bounty program on harvester behavior. Wildl Soc Bull 31:736–743. https://doi.org/10.2307/3784593

Bauder JM, Allen ML, Ahlers AA, Benson TJ, Miller CA, Stodola KW (2020a) Identifying and controlling for variation in canid harvest data. J Wildl Manage 84:1234–1245. https://doi.org/10.1002/jwmg.21919

Bauder JM, Stodola KW, Benson TJ, Miller CA, Allen ML (2020b) Raccoon pelt price and trapper harvest relationships are temporally inconsistent. J Wildl Manage 84:1601–1610. https://doi.org/10.1002/jwmg.21928

Beineke A, Baumgärtner W, Wohlsein P (2015) Cross-species transmission of canine distemper virus-an update. One Health 1:49–59. https://doi.org/10.1016/j.onehlt.2015.09.002

Belant JL, Schipper J, Conroy J (2009) The conservation status of small carnivores in the Americas. Small Carniv Conserv 41:3–8

Belant J, Biggins D, Garelle D et al (2015) *Mustela nigripes*. The IUCN Red List of Threatened Species 2015: e.T14020A45200314. https://doi.org/10.2305/IUCN.UK.2015-4.RLTS.T14020A45200314.en.8235

Bell D, Roberton S, Hunter PR (2004) Animal origins of SARS coronavirus: possible links with the international trade in small carnivores. Phil Trans R Soc B: Biol Sci 359:1107–1114. https://doi.org/10.1098/rstb.2004.1492

Berger KM, Conner MM (2008) Recolonizing wolves and mesopredator suppression of coyotes: impacts on pronghorn population dynamics. Ecol Appl 8(3):599–612

Berry B, Schooley RL, Ward MP (2017) Landscape context affects use of restored grasslands by mammals in a dynamic agroecosystem. Am Midland Nat 177:165–182. https://doi.org/10.1674/0003-0031-177.2.165

Black KL, Petty SK, Radeloff VC, Pauli JN (2018) The Great Lakes Region is a melting pot for vicariant red fox (*Vulpes vulpes*) populations. J Mamm 99(5):1229–1236. https://doi.org/10.1093/jmammal/gyy096

Blackburn AM, Anderson CJ, Veals AM, Tewes ME, Wester DB, Young JH, DeYoung RW, Perotto-Baldivieso HL (2021) Landscape patterns of ocelot–vehicle collision sites. Landscape Ecol 36:497–511. https://doi.org/10.1007/s10980-020-01153-y

Blejwas KM, Sacks BN, Jaeger MM, McCullough DR (2002) The effectiveness of selective removal of breeding coyotes in reducing sheep predation. J Wildl Manage 66:451. https://doi.org/10.2307/3803178

Blythe EM, Boyce MS (2020) Trappings of success: predator removal for duck nest survival in Alberta Parklands. Diversity 12:119. https://doi.org/10.3390/d12030119

Booth-Binczik SD, Bradley RD, Thompson CW et al (2013) Food habits of ocelots and potential for competition with bobcats in southern Texas. Southwest Nat 58:403–410. https://doi.org/10.1894/0038-4909-58.4.403

Brickner KM, Grenier MB, Crosier AE, Pauli JN (2014) Foraging plasticity in a highly specialized carnivore, the endangered black-footed ferret. Biol Conserv 169:1–5. https://doi.org/10.1016/j.biocon.2013.10.010

Bromen NA, French JT, Walker JW, Tomeček JM (2019) Spatial relationships between livestock guardian dogs and mesocarnivores in central Texas. Human-Wildl Interact 13:29–41. https://doi.org/10.26076/0d01-xz26

Bromley C, Gese EM (2001) Surgical sterilization as a method of reducing coyote predation on domestic sheep. J Wildl Manage 65:510–519. https://doi.org/10.2307/3803104

Brown DE, Conover MR (2011) Effects of large-scale removal of coyotes on pronghorn and mule deer productivity and abundance. J Wildl Manage 75:876–882. https://doi.org/10.1002/jwmg.126

Butler AR, Bly KLS, Harris H, Inman RM, Moehrenschlager A, Schwalm D, Jachowski DS (2019) Winter movement behavior by swift foxes (Vulpes velox) at the northern edge of their range. Can J Zool 97:922–930. https://doi.org/10.1139/cjz-2018-0272

Butler AR, Bly KL, Harris H, Inman RM, Moehrenschlager A, Schwalm D, Jachowski DS (2020) Home range size and resource use by swift foxes in northeastern Montana. J Mamm 101(3):684–696. https://doi.org/10.1093/jmammal/gyaa030

Byerly PA, Lonsinger RC, Gese EM, Kozlowski AJ, Waits LP (2018) Resource partitioning between kit foxes (Vulpes macrotis) and coyotes (Canis latrans): a comparison of historical and contemporary dietary overlap. Canadian J Zool 96:497–504. https://doi.org/10.1139/cjz-2017-0246

Cain SA, Kadlec JA, Allen DL, Cooley RA, Hornocker MH, Wagner FH (1972) Predator control–1971: report to the President's Council on Environmental Quality and the US Department of the Interior by the Advisory Committee of Predator Control. University of Michigan Press, Ann Arbor, MI, p 230

Carricondo-Sanchez D, Odden M, Kulkarni A, Vanak AT (2019) Scale-dependent strategies for coexistence of mesocarnivores in human-dominated landscapes. Biotropica 51(5):781–791. https://doi.org/10.1111/btp.12705

Chambers JC, Maestas JD, Pyke DA, Boyd CS, Pellant M, Wuenschel A (2017) Using resilience and resistance concepts to manage persistent threats to sagebrush ecosystems and greater sage-grouse. Rangeland Ecol Manage 70(2):149–164. https://doi.org/10.2737/RMRS-GTR-356

Chipault JG, Biggins DE, Detling JK, Long DH, Reich RM (2012) Fine-scale habitat use of reintroduced black-footed ferrets on prairie dog colonies in New Mexico. Western N Am Nat 72:216–227. https://doi.org/10.3398/064.072.0211

Cherry MJ, Morgan KE, Rutledge BT, Conner LM, Warren RJ (2016) Can coyote predation risk induce reproduction suppression in white-tailed deer? Ecosphere 7:e01481

Chronert JM, Jenks JA, Roddy DE, Wild MA, Powers JG (2007) Effects of sarcoptic mange on coyotes at Wind Cave National Park. J Wildl Manage 71:1987–1992. https://doi.org/10.2193/2006-225

Clark W, Andrew RD (1982) Review of population indices applied in furbearer management. In: Sanderson GC (ed) Proceedings of the Midwest furbearer management symposium. North Central Section, Central Mountains and Plains Section, and Kansas Chapter, The Wildlife Society, 7–8 Dec 1981, Wichita, KS, USA, pp 11–22

Clark HO, Warrick GD, Cypher BL, Kelly PA, Williams DF, Grubbs DE (2005) Competitive interactions between endangered kit foxes and nonnative red foxes. Western N Am Nat 65:153–163

Convention on International Trade in Endangered Species of Wild Fauna and Flora [CITES] (2021) Lynx rufus. In: CITES 2021 Appendix II. https://cites.org/eng

Crawford-Miksza LK, Wadford DA, Schnurr DP (1999) Molecular epidemiology of enzootic rabies in California. J Clinic Virol 14:207–219. https://doi.org/10.1016/S1386-6532(99)00054-2

Crimmins SM, Walleser LR, Hertel DR, McKann PC, Rohweder JJ, Thogmartin WE (2016) Relating mesocarnivore relative abundance to anthropogenic land-use with a hierarchical spatial count model. Ecography 39(6):524–532. https://doi.org/10.1111/ecog.01179

Crooks KR, Soulé ME (1999) Mesopredator release and avifaunal extinctions in a fragmented system. Nature 400:563–566. https://doi.org/10.1038/23028

Cuarón AD, González-Maya JF, Helgen K et al (2016a) *Mephitis macroura*. The IUCN Red List of Threatened Species 2016a: e.T41634A45211135. https://doi.org/10.2305/IUCN.UK.2016-1. RLTS.T41634A45211135.en. 8235

Cuarón AD, Helgen K, Reid F et al (2016b) *Nasua narica*. The IUCN Red List of Threatened Species 2016b: e.T41683A45216060. https://doi.org/10.2305/IUCN.UK.2016-1.RLTS.T41683A45216 060.en, https://doi.org/10.2305/IUCN.UK.2016-1.RLTS.T41683A45216060.en

Cuarón AD, Helgen K, Reid F (2016c) *Spilogale gracilis*. The IUCN Red List of Threatened Species 2016c: e.T136797A45221721. https://doi.org/10.2305/IUCN.UK.2016-1.RLTS. T136797A45221721.en

Cypher B, List R (2014) *Vulpes macrotis*. The IUCN Red List of Threatened Species 2014: e.T41587A62259374. https://doi.org/10.2305/IUCN.UK.2014-3.RLTS.T41587A62259374.en

Cypher BL, Spencer KA (1998) Competitive interactions between coyotes and San Joaquin kit foxes. J Mamm 79:204–214. https://doi.org/10.2307/1382855

Cypher BL, Scrivner JH (1992) Coyote control to protect endangered San Joaquin kit foxes at the Naval Petroleum Reserves, California. In: Proceedings of the 15th vert pest conference, pp 42–47

Cypher BL, Rudd JL, Westall TL, Woods LW, Stephenson N, Foley JE, Richardson D, Clifford DL (2017) Sarcoptic mange in endangered kit foxes (*Vulpes macrotis mutica*): case histories, diagnoses, and implications for conservation. J Wildl Dis 53:1–8. https://doi.org/10.7589/2016-05-098

Deem SL, Spelman LH, Yates RA, Montali RJ (2000) Canine distemper in terrestrial carnivores: a review. J Zoo Wildl Med 31:441–451. https://doi.org/10.1638/1042-7260(2000)031[0441:CDI TCA]2.0.CO;2

DeVink J, Berezanski D, Imrie D (2011) Comments on Brodie and Post: Harvest effort: the missing covariate in analyses of furbearer harvest data. Pop Ecol 53:261–262. https://doi.org/10.1007/s10144-010-0241-6

DeYoung RW, Zamorano A, Mesenbrink BT, Campbell TA, Leland BR, Moore GM, Honeycutt RL, Root JJ (2009) Landscape-genetic analysis of population structure in the Texas gray fox oral rabies vaccination zone. J Wildl Manage 73:1292–1299. https://doi.org/10.2193/2008-336

Docken NR (2011) Effects of block predator management on duck and pheasant nest success in eastern South Dakota. M.S. thesis, South Dakota State University, Brookings, SD, 92pp

Donadio E, Buskirk SW (2006) Diet, morphology, and interspecific killing in Carnivora. Am Nat 167:524–536. https://doi.org/10.1086/501033

Dragoo JW, Honeycutt RL (1999) Eastern hog-nosed skunk/Conepatus leuconotus. In: Wilson DE, Ruff S (eds) The smithsonian book of North American mammals. Smithsonian Institution Press. Washington, D.C., pp 190–191

Dragoo JW, Sheffield SR (2009) *Conepatus leuconotus* (Carnivora: Mephitidae). Mamm Spec 827:1–8

Dragoo JW, Matthes DK, Aragon A, Hass CC, Yates TL (2004) Identification of skunk species submitted for rabies testing in the desert southwest. J Wildl Dis 40:371–376. https://doi.org/10. 7589/0090-3558-40.2.371

du Toit JT, Cross PC, Valeix M (2017) Managing the livestock–wildlife interface on rangelands. In: Briske DD (ed) Rangeland systems, pp 395–425

Dunstone N (1993) The mink. UK, T. & A. D. Poyser

Eklund A, López-Bao JV, Tourani M, Chapron G, Frank J (2017) Limited evidence on the effectiveness of interventions to reduce livestock predation by large carnivores. Sci Rep 7:1–9. https://doi.org/10.1038/s41598-017-02323-w

Eldridge DJ, Whitford WG (2009) Badger (*Taxidea taxus*) disturbances increase soil heterogeneity in a degraded shrub-steppe ecosystem. J Arid Environ 73:66–73. https://doi.org/10.1016/j.jar idenv.2008.09.004

Ellis EC, Klein Goldewijk K, Siebert S, Lightman D, Ramankutty N (2010) Anthropogenic transformation of the biomes, 1700–2000. Global Ecol Biogeog 19(5):589–606. https://doi.org/10. 1111/j.1466-8238.2010.00540.x

Elsken-Lacy P, Wilson AM, Heidt GA, Peck JH (1999) Arkansas gray fox fur price-harvest model revisited. J Arkansas Acad Sci 53:50–54

Evans GD, Pearson EW (1980) Federal coyote control methods used in the western United States, 1971–77. Wildl Soc Bull 1:34–39

Faulkner CE (1971) The legal status of wildcats in the United States. In: Jorgensen RM, Mech LD (ed) Proceedings of a symposium on the native cats of North America, their status and management. 36th American wildlife and natural resources conference. U.S. Fish and Wildlife Service, Twin Cities, Minnesota, pp 124–125

Feldman JW (2007) Public opinion, the Leopold Report, and the reform of federal predator control policy. Human-Wildl Confl 1:112–124. https://doi.org/10.26077/eh4w-3894

Fiehler CM, Cypher BL, Saslaw LR (2017) Effects of oil and gas development on vertebrate community composition in the southern San Joaquin Valley, California. Global Ecol Conserv 9:131–141. https://doi.org/10.1016/j.gecco.2017.01.001

Frey SN, Majors S, Conover MR, Messmer TA, Mitchell DL (2003) Effect of predator control on ring-necked pheasant populations. Wildl Soc Bull 31:727–735. https://doi.org/10.2307/378 4592

Gese EM, Schultz RD, Rongstad OJ, Andersen DE (1991) Prevalence of antibodies against canine parvovirus and canine distemper virus in wild coyotes in southeastern Colorado. J Wildl Dis 27:320–323. https://doi.org/10.7589/0090-3558-27.2.320

Gese EM, Schultz RD, Johnson MR, Williams ES, Crabtree RL, Ruff RL (1997) Serological survey for diseases in free-ranging coyotes (*Canis latrans*) in Yellowstone National Park, Wyoming. J Wildl Dis 33:47–56. https://doi.org/10.7589/0090-3558-33.1.47

Gese EM, Karki SM, Klavetter ML, Schauster ER, Kitchen AM (2004) Serologic survey for canine infectious diseases among sympatric swift foxes (*Vulpes velox*) and coyotes (*Canis latrans*) in Southeastern Colorado. J Wildl Dis 40:741–748. https://doi.org/10.7589/0090-3558-40.4.741

Goldman HV (2021) Nearly extirpated by plague and distemper in the 1980s, black-footed ferrets now vaccinated for COVID-19. Small Carnivore Conserv 59. https://smallcarnivoreconserv ation.com/index.php/sccg/article/view/3463

Gompper ME (2002) Top carnivores in the suburbs? Ecological and conservation issues raised by colonization of north eastern North America by coyotes. Bioscience 52:185. https://doi.org/10. 1641/0006-3568(2002)052[0185:TCITSE]2.0.CO;2

Gompper ME, Hackett HM (2005) The long-term, range-wide decline of a once common carnivore: the eastern spotted skunk (*Spilogale putorius*). Anim Conserv 8(2):195–201. https://doi.org/10. 1017/S1367943005001964

Gompper M, Jachowski D (2016) *Spilogale putorius*. The IUCN Red List of Threatened Species 2016: e.T41636A45211474. https://doi.org/10.2305/IUCN.UK.2016-1.RLTS.T41636A45211 474.en

Grassel SM, Rachlow JL (2018) When generalists behave as specialists: local specialization by American badgers (*Taxidea taxus*). Canadian J Zool 96:592–599. https://doi.org/10.1139/cjz-2017-0125

Grassel SM, Rachlow JL, Williams CJ (2015) Spatial interactions between sympatric carnivores: asymmetric avoidance of an intraguild predator. Ecol Evol 5:2762–2773. https://doi.org/10. 1002/ece3.1561

Greenspan E, Anile S, Nielsen CK (2020) Density of wild felids in Sonora, Mexico: a comparison of spatially explicit capture-recapture methods. Eur J Wildl Res 66:1–12. https://doi.org/10.1007/ s10344-020-01401-1

Greenwood RJ, Sargeant AB, Piehl JL et al (1999) Foods and foraging of prairie striped skunks during the avian nesting season. Wildl Soc Bull 27:823–832

Guthery FS, Beasom SL (1977) Responses of game and nongame wildlife to predator control in south Texas. J Range Manage Arch 30:404–409

Haines AM, Tewes ME, Laack LL (2005) Survival and sources of mortality in ocelots. J Wildl Manage 69:255–263. https://doi.org/10.2193/0022-541X(2005)069%3c0255:SASOMI%3e2. 0.CO;2

Haines AM, Janecka JE, Tewes ME, Grassman LI, Morton P (2006) The importance of private lands for ocelot *Leopardus pardalis* conservation in the United States. Oryx 40:90–94. https://doi.org/10.1017/S0030605306000044

Hall ER, Dalquest WW (1963) The mammals of Veracruz. Univ Kansas Publ Mus Nat Hist 14:165–362

Harrington JL, Conover MR (2007) Does removing coyotes for livestock protection benefit free-ranging ungulates? J Wildl Manage 71:1555–1560

Hass CC, Dragoo JW (2006) Rabies in hooded and striped skunks in Arizona. J Wildl Dis 42:825–829

Hatfield BE, Runcie JM, Siemion EA, Quinn CB, Stephenson TR (2021) New detections extend the known range of the state-threatened Sierra Nevada red fox. Calif Fish Game 107:438–443

Hedberg GE, Dierenfeld ES, Rogers QR (2007) Taurine and zoo felids: considerations of dietary and biological tissue concentrations. Zoo Biol 26:517–531

Heithaus MR (2001) Habitat selection by predators and prey in communities with asymmetrical intraguild. Oikos 92:542–554

Helgen K (2016) *Conepatus leuconotus*. The IUCN Red List of Threatened Species 2016: e.T41632A45210809. https://doi.org/10.2305/IUCN.UK.2016-1.RLTS.T41632A45210809.en

Helgen K, Reid F (2016a) *Mephitis mephitis*. The IUCN Red List of Threatened Species 2016a: e.T41635A45211301. https://doi.org/10.2305/IUCN.UK.2016-1.RLTS.T41635A45211301.en. https://doi.org/10.1007/978-1-4615-6422-5_26

Helgen K, Reid F (2016b) *Taxidea taxus*. The IUCN Red List of Threatened Species 2016b: e.T41663A45215410. https://doi.org/10.2305/IUCN.UK.2016-1.RLTS.T41663A45215410.en

Helgen K, Reid F (2016c) *Mustela frenata*. The IUCN Red List of Threatened Species 2016c: e.T41654A45213820. https://doi.org/10.2305/IUCN.UK.2016-1.RLTS.T41654A45213820.en

Henke S, Bryant F (1999) Effects of coyote removal on the faunal community in western Texas. J Wildl Manage 63:1066–1081

Hiller TL, Etter DR, Belant JL, Tyre AJ (2011) Factors affecting harvests of fishers and American martens in northern Michigan. J Wildl Manage 75:1399–1405

Hines TD, Case RM (1991) Diet, home range, movements, and activity periods of swift fox in Nebraska. Prairie Nat 23:131–138

Hirsch BT, Prange S, Hauver SA, Gehrt SD (2013) Raccoon social networks and the potential for disease transmission. PLoS ONE 8:e75830

Hody JW, Kays R (2018) Mapping the expansion of coyotes (*Canis latrans*) across North and Central America. Zookeys 81–97.https://doi.org/10.3897/zookeys.759.15149

Hoffmann M, Sillero-Zubiri C (2016) *Vulpes vulpes*. The IUCN Red List of Threatened Species 2016: e.T23062A46190249. https://doi.org/10.2305/IUCN.UK.2016-1.RLTS.T23062A46190249.en. IUCN Red List Threat Species 2016 8235:e.T23062A4

Holbrook JD, Arkle RS, Rachlow JL, Vierling KT, Pilliod DS, Wiest MM (2016) Occupancy and abundance of predator and prey: implications of the fire-cheatgrass cycle in sagebrush ecosystems. Ecosphere 7(6):e01307

Holland AM, Schauber EM, Nielsen CK, Hellgren EC (2019) River otter and mink occupancy dynamics in riparian systems. J Wildl Manage 83:1552–1564. https://doi.org/10.1002/jwmg.21745

Holt RD, Huxel GR (2007) Alternative prey and the dynamics of intraguild predation: theoretical perspectives. Ecology 88:2706–2712

Holt RD, Polis GA (1997) A theoretical framework for intraguild predation. Am Nat 149:745–764

Horne JS, Haines AM, Tewes ME, Laack LL (2009) Habitat partitioning by sympatric ocelots and bobcats: implications for recovery of ocelots in southern Texas. Southwest Nat 54:119–126. https://doi.org/10.1894/PS-49.1

Hubert GF (1982) History of Midwestern furbearer management and a look to the future. In: Sanderson GC (ed) Midwest furbearer management: proceedings of the symposium of the 43rd Midwest fish and wildlife conference. Kansas Chapter of the Wildlife Society, Wichita, KS, pp 175–190

Hueffer K, Parker JSL, Weichert WS, Geisel RE, Sgro J-Y, Parrish CR (2003) The natural host range shift and subsequent evolution of canine parvovirus resulted from virus-specific binding to the canine transferrin receptor. J Virol 77:1718–1726. https://doi.org/10.1128/jvi.77.3.1718-1726.2003

Jachowski DS, Edelman AJ (2021) Advancing small carnivore research and conservation: the Eastern Spotted Skunk Cooperative Study Group model. Southeast Nat 20:1–12. https://doi.org/10.1656/058.020.0sp1102

Jackson VL, Laack LL, Zimmerman EG (2005) Landscape metrics associated with habitat use by ocelots in south Texas. J Wildl Manage 69:733–738

Janečka JE, Tewes ME, Davis IA, Haines AM, Caso A, Blankenship TL, Honeycutt RL (2016) Genetic differences in the response to landscape fragmentation by a habitat generalist, the bobcat, and a habitat specialist, the ocelot. Conserv Genet 17(5):1093–1108. https://doi.org/10.1007/s10592-016-0846-1

Kabir E, Kumar P, Kumar S, Adelodun AA, Kim KH (2018) Solar energy: potential and future prospects. Renew Sustain Energy Rev 82:894–900. https://doi.org/10.1016/j.rser.2017.09.094

Kamler JF, Ballard WB (2002) A review of native and nonnative red foxes in North America. Wildl Soc Bull 30:370–379. https://doi.org/10.2307/3784493

Kamler JF, Ballard WB, Gilliland RL, Lemons PR, Mote K (2003a) Impacts of coyotes on swift foxes in northwestern Texas. J Wildl Manage 67:317–323. https://doi.org/10.2307/3802773

Kamler JF, Ballard WB, Helliker BR, Stiver S (2003b) Range expansion of raccoons in western Utah and central Nevada. West North Am Nat 63:406–408

Kapfer PM, Potts KB (2012) Socioeconomic and ecological correlates of bobcat harvest in Minnesota. J Wildl Manage 76:237–242. https://doi.org/10.2307/41418264

Karki SM, Gese EM, Klavetter ML (2007) Effects of coyote population reduction on swift fox demographics in southeastern Colorado. J Wildl Manage 71:2707–2718. https://doi.org/10.2193/2006-275

Kasprowicz AE, Statham MJ, Sacks BN (2016) The fate of the other red coat: remnants of colonial British red foxes in the Eastern United States. J Mamm 97:298–309. https://doi.org/10.1093/jmammal/gyv179

Kays R (2018) Canis latrans (errata version published in 2020). The IUCN Red List of Threatened Species 2018: e.T3745A163508579. https://doi.org/10.2305/IUCN.UK.2018-2.RLTS.T3745A163508579.en

Kelly M, Morin D, López-González CA (2016). Lynx rufus. The IUCN Red List of Threatened Species 2016: e.T12521A50655874. https://doi.org/10.2305/IUCN.UK.2016-1.RLTS.T12521A50655874.en

Kilgo JC, Shaw CE, Vukovich M, Conroy MJ, Ruth C (2017) Reproductive characteristics of a coyote population before and during exploitation. J Wildl Manage 81:1386–1393. https://doi.org/10.1002/jwmg.21329

Kimbrell T, Holt RD, Lundberg P (2007) The influence of vigilance on intraguild predation. J Theor Biol 249:218–234. https://doi.org/10.1016/j.jtbi.2007.07.031

Kinka D, Young JK (2018) A livestock guardian dog by any other name: similar response to wolves across livestock guardian dog breeds. Rangeland Ecol Manage 71(4):509–517. https://doi.org/10.1016/j.rama.2018.03.004

Kitchen AM, Gese EM, Schauster ER (1999) Resource partitioning between coyotes and swift foxes: space, time, and diet. Can J Zool 77:1645–1656. https://doi.org/10.1139/cjz-77-10-1645

Kitchen AM, Gese EM, Schauster ER (2000) Changes in coyote activity patterns due to reduced exposure to human persecution. Can J Zool 78:853–857. https://doi.org/10.1139/cjz-78-5-853

Knowles CJ, Proctor JD, Forrest SC (2002) Black-tailed prairie dog abundance and distribution in the great plains based on historic and contemporary information. Great Plains Res 12:219–254

Knowlton FF, Gese EM, Jaeger MM (1999) Coyote depredation control: an interface between biology and management. Soc Range Manage 52:398–412. https://doi.org/10.2307/4003765

Kuo YH, Vanderzwan SL, Kasprowicz AE, Sacks BN (2019) Using ancestry-informative SNPs to quantify introgression of European alleles into North American red foxes. J Heredity 110(7):782–792. https://doi.org/10.1093/jhered/esz053

Kuzmin IV, Shi M, Orciari LA, Yager PA, Velasco-Villa A, Kuzmina NA, Streicker DG, Bergman DG, Rupprecht CE (2012) Molecular inferences suggest multiple host shifts of rabies viruses from bats to mesocarnivores in Arizona during 2001–2009. PLoS Pathogens 8.https://doi.org/10.1371/journal.ppat.1002786

Larson RN, Morin DJ, Wierzbowska IA, Crooks KR (2015) Food habits of coyotes, gray foxes, and bobcats in a coastal Southern California urban landscape. West North Am Nat 75:339–347. https://doi.org/10.3398/064.075.0311

Larson S, McGranahan DA, Timm RM (2019) The Marin County livestock protection program: 15 years in review. Human-Wildl Interact 13:63–78. https://doi.org/10.5070/V427110695

Lawrence JS, Silvy NJ (1995) Effect of predator control on reproductive success and hen survival of Attwater's prairie-chicken. In: Proceedings Southeast Association of Fish Wildlife Agencies, vol 49, pp 275–282

Leslie MJ, Messenger S, Rohde RE, Smith J, Cheshier R, Hanlon C, Rupprecht CE (2006) Bat-associated rabies virus in skunks. Emerg Infect Dis 12:1274–1277.https://doi.org/10.3201/eid 1208.051526

Lesmeister DB, Nielsen CK, Schauber EM, Hellgren EC (2015) Spatial and temporal structure of a mesocarnivore guild in midwestern north America. Wild Mon 191:1–61. https://doi.org/10.1002/wmon.1015

Lombardi JV, MacKenzie DI, Tewes ME et al (2020a) Co-occurrence of bobcats, coyotes, and ocelots in Texas. Ecol Evol 10:4903–4917. https://doi.org/10.1002/ece3.6242

Lombardi JV, Tewes ME, Perotto-Baldivieso HL, Mata JM, Campbell TA (2020b) Spatial structure of woody cover affects habitat use patterns of ocelots in Texas. Mamm Res 65:555–563. https://doi.org/10.1007/s13364-020-00501-2

Long JL (2003) Introduced mammals of the world: their history, distribution and influence. CABI Publishing, Wallingford, UK, xxi + 589 pp

Lonsinger RC, Gese EM, Bailey LL, Waits LP (2017) The roles of habitat and intraguild predation by coyotes on the spatial dynamics of kit foxes. Ecosphere 8:e01749. https://doi.org/10.1002/ecs2.1749

Lonsinger RC, Kluever BM, Hall LK, Larsen RT, Gese EM, Waits LP, Knight RN (2020) Conservation of kit foxes in the Great Basin Desert: Review and recommendations. J Fish Wildl Mgmt 11:679–698

Łopucki R, Klich D, Gielarek S (2017) Do terrestrial animals avoid areas close to turbines in functioning wind farms in agricultural landscapes?. Environ Monit Assess 189:1–1.

Lourenço R, Penteriani V, Rabaça JE, Korpimäki E (2014) Lethal interactions among vertebrate top predators: a review of concepts, assumptions and terminology. Biol Rev 89:270–283. https://doi.org/10.1111/brv.12054

Lovich JE, Ennen JR (2011) Wildlife conservation and solar energy development in the desert southwest, United States. Bioscience 61:982–992. https://doi.org/10.1525/bio.2011.61.12.8

Lovich JE, Ennen JR, Madrak S, Grover B (2011) Turtles, culverts, and alternative energy development: an unreported but potentially significant mortality threat to the desert tortoise (*Gopherus agassizii*). Chelonian Conserv Biol 10:124–129. https://doi.org/10.2744/CCB-0864.1

Lyons EK, Frost J, Rollins D, Scott C (2009) An evaluation of short-term mesocarnivore control for increasing hatch rate in Northern Bobwhites. Nat Quail Symp Proc 6:447–455

Ma X, Monroe BP, Cleaton JP, Orciari LA, Gigante CM, Kirby JD, Chipman RB, Fehlner-Gardiner C, Cedillo VG, Petersen BW, Olson V, Wallace RM (2020) Rabies surveillance in the United States during 2018. J Am Vet Med Assoc 256:195–208. https://doi.org/10.2460/javma.256.2.195

Mahoney PJ, Young JK, Hersey KR, Larsen RT, McMillan BR, Stoner DC (2018) Spatial processes decouple management from objectives in a heterogeneous landscape: predator control as a case study. Ecol Applic 28(3):786–797. https://doi.org/10.1002/eap.1686

Marneweck C, Butler AR, Gigliotti LC, Harris SN, Jensen AJ, Muthersbaugh M, Jachowski DS (2021) Shining the spotlight on small mammalian carnivores: global status and threats. Biol Conserv 255:109005. https://doi.org/10.1016/j.biocon.2021.109005

Marneweck CJ, Allen BL, Butler AR, Do Linh San E, Harris SN, Jensen AJ, Jachowski DS (2022) Middle-out ecology: small carnivores as sentinels of global change. Mamm Rev. https://doi.org/10.1111/mam.12300

McCue PM, O'Farrell TP (1988) Serological survey for selected diseases in the endangered San Joaquin kit fox (*Vulpes macrotis mutica*). J Wildl Dis 24:274–281. https://doi.org/10.7589/0090-3558-24.2.274

McDonald RA, Abramov AV, Stubbe M et al (2019) *Mustela nivalis* (amended version of 2016 assessment). The IUCN Red List of Threatened Species 2019: e.T70207409A147993366. https://doi.org/10.2305/IUCN.UK.2016-1.RLTS.T70207409A147993366.en

McGee Jr HF (1987) The use of furbearers by native North Americans after 1500. In: Novak M, Baker JA, Obbard M, Malloch B (eds) Wild furbearer management and conservation in North America. Ontario Ministry of Natural Resources, Toronto, Ontario, pp 13–20

McKinney T, Smith TW (2007) Diets of sympatric bobcats and coyotes during years of varying rainfall in central Arizona. West North Am Nat 67:8–15

Melchior HR, Johnson NF, Phelps JS (1987) Wild furbearer management in the western United States and Alaska In: Novak M, Baker JA, Obbard ME, Malloch B (eds) Wild furbearer management and conservation in North America. Ontario Ministry of Natural Resources, Toronto, Ontario, pp 1117–1128

Melquist WE, Polechla PJ, Toweill D (2003) River otter, *Lontra canadensis*. In: Feldhamer GA, Thompson BC, Chapman JA (eds) Wild mammals of North America: biology, management, and conservation. Johns Hopkins University Press, Baltimore, Md, pp 708–734

Miller JE (2007) Evolution of the field of wildlife damage management in the United States and future challenges. Human-Wildl Confl 1:13–20. https://doi.org/10.26077/f1df-nm52

Miller DS, Covell DF, McLean RG, Adrian WJ, Niezgoda M, Gustafson JM, Rongstad OJ, Schultz RD, Kirk LJ, Quan TJ (2000) Serologic Survey for selected infectious disease agents in swift and kit foxes from the western United States. J Wildl Dis 36:798–805. https://doi.org/10.7589/0090-3558-36.4.798

Miller TE, Burns JH, Munguia P, Walters EL, Kneitel M, Richards PM, Mouquet N, Buckley HL (2005) A critical review of twenty years' use of the resource-ratio theory. Am Nat 165:439–448. https://doi.org/10.1086/428681

Minta SC, Minta KA, Lott DF (1992) Hunting associations between badgers (*Taxidea taxus*) and coyotes (*Canis latrans*). J Mamm 73(4):814–820. https://doi.org/10.1139/cjz-2017-0234

Mitchell BR, Jaeger MM, Barrett RH (2004) Coyote depredation management: current methods and research needs. Wildl Soc Bull 32:1209–1218. https://doi.org/10.2193/0091-7648(2004)032[1209:CDMCMA]2.0.CO;2

Moehrenschlager A, Sovada M (2016) *Vulpes velox*. The IUCN Red List of Threatened Species 2016: e.T23059A57629306. https://doi.org/10.2305/IUCN.UK.2016-3.RLTS.T23059A57629306.en

Montecino-Latorre D, Cypher BL, Rudd JL, Clifford DL, Mazet JAK, Foley JE (2019) Assessing the role of dens in the spread, establishment and persistence of sarcoptic mange in an endangered canid. Epidemics 27:28–40. https://doi.org/10.1016/j.epidem.2019.01.001

Moreira-Arce D, Vergara PM, Boutin S, Simonetti JA, Briceño C, Acosta-Jamett G (2015) Native forest replacement by exotic plantations triggers changes in prey selection of mesocarnivores. Biol Conserv 192:258–267. https://doi.org/10.1016/j.biocon.2015.09.015

Mosley JC, Roeder BL, Frost RA, Wells SL, McNew LB, Clark PE (2020) Mitigating human conflicts with livestock guardian dogs in extensive sheep grazing systems. Range Ecol Manage 73:724–732. https://doi.org/10.1016/j.rama.2020.04.009

Mueller MA, Drake D, Allen ML (2018) Coexistence of coyotes (*Canis latrans*) and red foxes (*Vulpes vulpes*) in an urban landscape. PLoS ONE 13(1):e0190971. https://doi.org/10.1371/journal.pone.0190971

Neale JCC, Sacks BN (2001) Food habits and space use of gray foxes in relation to sympatric coyotes and bobcats. Can J Zool 79:1794–1800. https://doi.org/10.1139/cjz-79-10-1794

Nelson JL, Cypher BL, Bjurlin CD, Creel S (2007) Effects of habitat on competition between kit foxes and coyotes. J Wildl Manage 71:1467–1475. https://doi.org/10.2193/2006-234

Newbury RK, Hodges KE (2018) Regional differences in winter diets of bobcats in their northern range. Ecol Evol 8:11100–11110. https://doi.org/10.1002/ece3.4576

New Mexico Senate Bill 76 (2019) Prohibit coyote killing contests

Nicholson KL, Ballard WB, Mcgee BK, Surles J, Kamler JF, Lemons PR (2006) Swift fox use of black-tailed prairie dog towns in northwest Texas. J Wildl Manage 70:1659–1666

Nicholson KL, Ballard WB, McGee BK, Whitlaw HA (2007) Dispersal and extraterritorial movements of swift foxes (*Vulpes velox*) in Northwestern Texas. West North Am Nat 67:102–108. https://doi.org/10.3398/1527-0904(2007)67[102:DAEMOS]2.0.CO;2

Niedringhaus KD, Brown JD, Sweeley KM, Yabsley MJ (2019) A review of sarcoptic mange in North American wildlife. Internat J Parasit: Parasites Wildl 9:285–297. https://doi.org/10.1016/j.ijppaw.2019.06.003

Nimmervoll H, Hoby S, Robert N, Lommano E, Welle M, Ryser-Degiorgis MP (2013) Pathology of sarcoptic mange in red foxes (*Vulpes vulpes*): macroscopic and histologic characterization of three disease stages. J Wildl Disease 49:91–102. https://doi.org/10.7589/2010-11-316

Obbard ME, Jones JG, Newman R, Booth A, Satterthwaite AJ, Linscombe G (1987) Furbearer harvests in North America. In: Novak M, Baker JA, Obbard ME, Malloch B (eds) Wild furbearer management and conservation in North America. Ontario Ministry of Natural Resources, Toronto, Ontario, pp 1007–1034

Olson DM, Dinerstein E, Wikramanayake ED, Burgess ND, Powell GV, Underwood EC, D'amico JA, Itoua I, Strand HE, Morrison JC, Loucks CJ (2001) Terrestrial ecoregions of the world: A new map of life on Earth: A new global map of terrestrial ecoregions provides an innovative tool for conserving biodiversity. BioScience 51:933–938

Ordeñana MA, Crooks KR, Boydston EE, Fisher RN, Lyren LM, Siudyla S, Haas CD, Harris S, Hathaway SA, Turschak GM, Miles AK, Van Vuren DH (2010) Effects of urbanization on carnivore species distribution and richness. J Mamm 91:1322–1331. https://doi.org/10.1644/09-MAMM-A-312.1

Palmer BC, Conover MR, Frey SN (2010) Replication of a 1970s study on domestic sheep losses to predators on Utah's summer rangelands. Range Ecol Manage 63:689–695. https://doi.org/10.2111/REM-D-09-00190.1

Parker ID, Lopez RR, Padia R, Gallagher M, Karthikeyan R, Cathey JC, Silvy NJ, Davis DS (2013) Role of free-ranging mammals in the deposition of *Escherichia coli* into a Texas floodplain. Wildl Res 40:570–577. https://doi.org/10.1071/WR13082

Paviolo A, Crawshaw P, Caso A et al (2015) *Leopardus pardalis*. The IUCN Red List of Threatened Species 2015: e.T11509A97212355. https://doi.org/10.2305/IUCN.UK.2015-4.RLTS.T11509A50653476.en

Pechacek PE, Lindzey FRG, Anderson STH (2000) Autumn and winter diet of the swift fox (*Vulpes velox*) in south-eastern Wyoming. Hystrixthe Ital J Mammal 11:83–87. https://doi.org/10.4404/hystrix-11.2-4153

Pence DB, Windberg LA, Pence BC, Sprowls R (1983) The epizootiology and pathology of sarcoptic mange in coyotes, *Canis latrans*, from south Texas. J Parasit 69:1100–1115

Pence DB, Tewes ME, Shindle DB, Dunn DM (1995) Notoedric mange in an ocelot (*Felis pardalis*) from southern Texas. J Wildl Dis 31:558–561. https://doi.org/10.7589/0090-3558-31.4.558

Polis GA, Myers CA, Holt RD (1989) The ecology and evolution of intraguild predation: potential competitors that eat each other. Ann Rev Ecol System 20:297–330. https://doi.org/10.1146/annurev.es.20.110189.001501

Pope JP, Miller DL, Riley MC, Anis E, Wilkes RP (2016) Characterization of a novel Canine distemper virus causing disease in wildlife. J Vet Diagn Invest 28:506–513

Prugh LR, Sivy KJ (2020) Enemies with benefits: integrating positive and negative interactions among terrestrial carnivores. Ecol Lett 23:902–918. https://doi.org/10.1111/ele.13489

Prugh LR, Stoner CJ, Epps CW, Bean WT, Ripple WJ, Laliberte AS, Brashares JS (2009) The rise of the mesopredator. Bioscience 59:779–791. https://doi.org/10.1525/bio.2009.59.9.9

Ranglack DH, Durham S, du Toit JT (2015) Competition on the range: science versus perception in a bison-cattle conflict in the western USA. J Appl Ecol 52:467–474. https://doi.org/10.1111/1365-2664.12386

Ray AJ (1987) The fur trade in North America: an overview from a geographical perspective. In: Novak M, Baker JA, Obbard ME, Malloch B (eds) Wild furbearer management and conservation in North America. Ontario Ministry of Natural Resources, Toronto, Ontario, pp 21–30

Reid JM (2019) Effects of predator reduction on northern bobwhite (Colinus virginianus) in the Rolling Plains of Texas. M.S. Thesis, Texas Tech University, Lubbock, TX, 64pp

Reid F, Helgen K, Kranz A (2016a) Mustela erminea. The IUCN Red List of Threatened Species 2016a: e.T29674A45203335. http://dx.doi.org/https://doi.org/10.2305/IUCN.UK.2016-1.RLTS.T29674A45203335.en Copyright: Mamm Species 8235:1. https://doi.org/10.2307/3503967

Reid F, Schiaffini M, Schipper J (2016b) Neovison vison. The IUCN Red List of Threatened Species 2016b: e.T41661A45214988. https://doi.org/10.2305/IUCN.UK.2016-1.RLTS.T41661A45214988.en

Reid F, Schipper J, Timm R (2016c) Bassariscus astutus. The IUCN Red List of Threatened Species 2016c: e.T41680A45215881. https://doi.org/10.2305/IUCN.UK.2016-1.RLTS.T41680A45215881.en

Reynolds JC, Tapper SC (1996) Control of mammalian predators in game management and conservation. Mamm Rev 26:127–155. https://doi.org/10.1111/j.1365-2907.1996.tb00150.x

Roberts NM, Crimmins SM (2010) Bobcat population status and management in North America: evidence of large-scale population increase. J Fish Wildl Manage 1:169–174. https://doi.org/10.3996/122009-JFWM-026

Roberts NM, Lovallo MJ, Crimmins SM (2020) River otter status, management, and distribution in the United States: evidence of large-scale population increase and range expansion. J Fish Wildl Manage 11:279–286. https://doi.org/10.3996/102018-JFWM-093

Robinson QH, Bustos D, Roemer GW (2014) The application of occupancy modeling to evaluate intraguild predation in a model carnivore system. Ecology 95:3112–3123. https://doi.org/10.1890/13-1546.1

Roemer GW, Gompper ME, Van Valkengurgh B (2009) The ecological role of the mammalian mesocarnivore. Bioscience 59:165–173. https://doi.org/10.1525/bio.2009.59.2.9

Roemer G, Cypher B, List R (2016) Urocyon cinereoargenteus. The IUCN Red List of Threatened Species 2016: e.T22780A46178068. https://doi.org/10.2305/IUCN.UK.2016-1.RLTS.T22780A46178068.en. Mamm Species 8235:1. https://doi.org/10.2307/3503957

Rogers RE (1995) A review of state raccoon hunting and coondog training regulations. Wildl Soc Bull 23:398–406

Rohwer FC, Garrettson PR, Mense BJ (1995) Can predator trapping improve waterfowl recruitment in the Prairie Pothole region? In: Proceedings of the 7th Eastern wildlife damage management conference vol 7, pp 12–22

Rosenheim JA (2004) Top predators constrain the habitat selection games played by intermediate predators and their prey. Israel J Zool 50:129–138. https://doi.org/10.1560/K796-DMB2-546Q-Y4AQ

Rosenstock SS, Ballard WB, Devos JC (1999) Viewpoint: benefits and impacts of wildlife water developments. J Range Manage 52:302–311

Rudd J, Clifford D, Richardson D, Cypher B, Westall T, Kelly E, Foley J (2019) Hematologic and serum chemistry values of endangered San Joaquin kit foxes (Vulpes macrotis mutica) with sarcoptic mange. J Wildl Dis 55:410–415. https://doi.org/10.7589/2017-10-270

Ruette S, Stahl P, Albaret M (2003) Factors affecting trapping success of red fox Vulpes, stone marten Martes foina and pine marten M. martes in France. Wildl Biol 9:11–19. https://doi.org/10.2981/wlb.2003.003

Sacks BN, Neale JCC (2002) Foraging strategy of a generalist predator toward a special prey: coyote predation on sheep. Ecol Appl 12:299–306. https://doi.org/10.1890/1051-0761(2002)012[0299: FSOAGP]2.0.CO;2

Sacks BN, Lounsberry ZT, Statham MJ (2018) Nuclear genetic analysis of the red fox across its Trans-Pacific Range. J Heredity 109(5):573–584. https://doi.org/10.1093/jhered/esy028

Saitone TL, Bruno EM (2020) Cost effectiveness of livestock guardian dogs for predator control. Wildl Soc Bull 44:101–109. https://doi.org/10.1002/wsb.1063

Šálek M, Drahníková L, Tkadlec E (2015) Changes in home range sizes and population densities of carnivore species along the natural to urban habitat gradient. Mamm Rev 45:1–14. https:// doi.org/10.1111/mam.12027

Santymire R, Graves G (2019) Black-footed ferret SAFE program action plan 2019–2021. 26

Sargeant AB, Sovada MA, Shaffer TL (1995) Seasonal predator removal relative to hatch rate of duck nests in waterfowl production areas. Wildl Soc Bull 23:507–513

Sasmal I, Honness K, Bly K, McCaffery M, Kunkel K, Jenks JA, Phillips M (2015) Release method evaluation for swift fox reintroduction at Bad River Ranches in South Dakota. Restor Ecol 23:491–498.https://doi.org/10.1111/rec.12211

Scasta JD, Stam B, Windh JL (2017) Rancher-reported efficacy of lethal and non-lethal livestock predation mitigation strategies for a suite of carnivores. Sci Rep 7:1–11. https://doi.org/10.1038/ s41598-017-14462-1

Schoener TW (1974) Resource partitioning in ecological communities. Science 185:27–39. https:/ /doi.org/10.1126/science.185.4145.27

Schuttler SG, Parsons AW, Forrester TD, Baker MC, McShea WJ, Costello R, Kays R (2017) Deer on the lookout: how hunting, hiking and coyotes affect white-tailed deer vigilance. J Zool 301:320–327

Seidler RG, Gese EM (2012) Territory fidelity, space use, and survival rates of wild coyotes following surgical sterilization. J Ethol 30:345–354. https://doi.org/10.1007/s10164-012-0330-4

Seidler RG, Gese EM, Conner MM (2014) Using sterilization to change predation rates of wild coyotes: a test case involving pronghorn fawns. Appl Anim Behav Sci 154:83–92. https://doi. org/10.1016/j.applanim.2014.02.006

Serfass T, Evans S, Polechla P (2015) Lontra canadensis. The IUCN Red List of Threatened Species 2015: e.T12302A21936349. https://doi.org/10.2305/IUCN.UK.2015-2.RLTS.T12302 A21936349

Serieys LEK, Foley J, Owens S, Woods L, Boydston EE, Lyren LM, Poppenga RH, Clifford DL, Stephenson N, Rudd J, Riley SPD (2013) Serum chemistry, hematologic, and post-mortem findings in free-ranging bobcats (Lynx rufus) with notoedric mange. J Parasit 99:989–996. https:/ /doi.org/10.1645/12-175.1

Sévêque A, Gentle LK, López-Bao JV, Yarnell RW, Uzal A (2020) Human disturbance has contrasting effects on niche partitioning within carnivore communities. Biolog Rev 95:1689–1705. https://doi.org/10.1111/brv.12635

Shivik J (2014) The predator paradox: ending the war with wolves, bears, cougars, and coyotes. Beacon Press

Shriner SA, Ellis JW, Root JJ, Roug A, Stopak SR, Wiscomb GW, Zierenberg JR, Ip HS, Torchetti MK, DeLiberto TJ (2021) SARS-CoV-2 exposure in escaped mink, Utah, USA. Emerg Infect Dis 27(3):988. https://doi.org/10.3201/eid2703.204444

Slate D, Rupprecht CE, Rooney JA, Donovan D, Lein DH, Chipman RB (2005) Status of oral rabies vaccination in wild carnivores in the United States. Virus Res 111:68–76. https://doi.org/ 10.1016/j.virusres.2005.03.012

Slough BG, Jessup RH, McKay DI, Stephenson AB (1987) Wild furbearer management in western and northern Canada. In: Novak M, Baker JA, Obbard ME, Malloch B (eds) Wild furbearer management and conservation in North America. Ontario Ministry of Natural Resources, Toronto, Ontario, pp 1062–1976

Smallwood KS, Bell DA, Snyder SA, DiDonato JE (2010) Novel scavenger removal trials increase wind turbine—caused avian fatality estimates. J Wildl Manage 74(5):1089–1096. https://doi.org/10.2193/2009-266

Smith KF, Acevedo-Whitehouse K, Pedersen AB (2009) The role of infectious diseases in biological conservation. Anim Conserv 12:1–12. https://doi.org/10.1111/j.1469-1795.2008.00228.x

Smith JA, Brown MB, Harrison JO, Powell LA (2017) Predation risk: a potential mechanism for effects of a wind energy facility on Greater Prairie-Chicken survival. Ecosphere 8(6):e01835. https://doi.org/10.1002/ecs2.1835

Soulé M, Bolger D, Alberts A, Wright J, Sorice M, Hill S (1988) Reconstructed dynamics of rapid extinctions of chaparral-requiring birds in urban habitat islands. Conserv Biol 2:75–92. https://doi.org/10.1111/j.1523-1739.1988.tb00337.x

Sovada MA, Woodward RO, IGL LD (2009) Historical range, current distribution, and conservation status of the swift fox, *Vulpes velox*, in North America. Can Field-Nat 123:346–367. https://doi.org/10.22621/cfn.v123i4.1004

Sprayberry TR, Edelman AJ (2016) Food provisioning of kits by a female eastern spotted skunk. Southeast Nat 15(4). https://doi.org/10.1656/058.015.0417

Statham MJ, Sacks BN, Aubry KB, Perrine JD, Wisely SM (2012) The origin of recently established red fox populations in the United States: translocations or natural range expansions? J Mamm 93(1):52–65. https://doi.org/10.1644/11-MAMM-A-033.1

Steinel A, Parrish CR, Bloom ME, Truyen U (2001) Parvovirus infections in wild carnivores. J Wildl Dis 37:594–607. https://doi.org/10.7589/0090-3558-37.3.594

Stevenson ER, Lashley MA, Chitwood MC, Garabedian JE, Swingen MB, DePerno CS, Moorman CE (2019) Resource selection by coyotes (*Canis latrans*) in a longleaf pine (*Pinus palustris*) ecosystem: effects of anthropogenic fires and landscape features. Can J Zool 97(2):165–171. https://doi.org/10.1139/cjz-2018-0150

Stuetzer B, Hartmann K (2014) Feline parvovirus infection and associated diseases. Vet J 201:150–155. https://doi.org/10.1016/j.tvjl.2014.05.027

Tardy O, Massé A, Pelletier F, Mainguy J, Fortin D (2014) Density-dependent functional responses in habitat selection by two hosts of the raccoon rabies virus variant. Ecosphere 5:1–16. https://doi.org/10.1890/ES14-00197.1

Theimer TC, Clayton AC, Martinez A, Peterson DL, Bergman DL (2015) Visitation rate and behavior of urban mesocarnivores differs in the presence of two common anthropogenic food sources. Urban Ecosys 18:895–906. https://doi.org/10.1007/s11252-015-0436-x

Thomas PJ, Newell EE, Eccles K et al (2021) Co-exposures to trace elements and polycyclic aromatic compounds (PACs) impacts North American river otter (*Lontra canadensis*) baculum. Chemosphere 265:128920. https://doi.org/10.1016/j.chemosphere.2020.128920

Thompson CM, Gese EM (2007) Food webs and intraguild predation: community interactions of a native mesocarnivore. Ecology 88:334–346. https://doi.org/10.1890/0012-9658(2007)88[334:FWAIPC]2.0.CO;2

Thornton D, Scully A, King T, Fisher S, Fitkin S, Rohrer J (2018) Hunting associations of American badgers (*Taxidea taxus*) and coyotes (*Canis latrans*) revealed by camera trapping. Can J Zool 96(7):769–773. https://doi.org/10.1139/cjz-2017-0234

Timm R, Cuarón AD, Reid F et al (2016) *Procyon lotor*. The IUCN Red List of Threatened Species 2016: e.T41686A45216638. https://doi.org/10.2305/IUCN.UK.2016-1.RLTS.T41686A45216638.en

UDWR (2013) Utah's predator control program summary. Final Report

United States Department of Agriculture—National Agricultural Statistics Service (2015a) Sheep and lamb predator and nonpredator death loss in the United States, 2015a. USDA–APHIS–VS–CEAH–NAHMS Fort Collins, CO #721.0915. www.nass.usda.gov. Accessed 25 Feb 2021

United States Department of Agriculture—National Agricultural Statistics Service (2015b) Cattle and calves death loss in the United States due to predator and nonpredator causes, 2015b. USDA–APHIS–VS–CEAH. Fort Collins, CO #745.1217. www.nass.usda.gov. Accessed 25 Feb 2021

van Eeden LM, Crowther MS, Dickman CR, Macdonald DW, Ripple WJ, Ritchie EG, Newsome TM (2018a) Managing conflict between large carnivores and livestock. Conserv Biol 32:26–34. https://doi.org/10.1111/cobi.12959

van Eeden LM, Eklund A, Miller JRB, López-Bao JV, Chapron G, Cejtin MR, Crowther MS, Dickman CR, Frank J, Krofel M, Macdonald DW, McManus J, Meyer TK, Middleton AD, Newsome TM, Ripple WJ, Ritchie EG, Schmitz OJ, Stoner KJ, Tourani M, Treves A (2018b) Carnivore conservation needs evidence-based livestock protection. PLoS Biol 16:1–8. https://doi.org/10.1371/journal.pbio.2005577

Verdy A, Amarasekare P (2010) Alternative stable states in communities with intraguild predation. J Theor Biol 262:116–128. https://doi.org/10.1016/j.jtbi.2009.09.011

Wagner FH (1988) Predator control and the sheep industry. Regina Books, Claremont, California

Wagner KK, Conover MR (1999) Effect of preventive coyote hunting on sheep losses to coyote predation. J Wildl Manage 63:606–612. https://doi.org/10.2307/3802649

White PJ, Berry WH, Eliason JJ, Hanson MT (2000) Catastrophic decrease in an isolated population of kit foxes. Southwest Nat 45:204–211. https://doi.org/10.2307/3672462

White HB, Decker T, O'Brien MJ, Organ JF, Roberts NM (2015) Trapping and furbearer management in North American wildlife conservation. Internat J Environ Stud 5:756–769. https://doi.org/10.1080/00207233.2015.1019297

White HB, Batcheller GR, Boggess EK, Brown CL, Butfiloski JW, Decker TA, Erb JD, Fall MW, Hamilton DA, Hiller TL, Hubert GF Jr (2021) Best management practices for trapping furbearers in the United States. Wildl Monog 207(1):3–59. https://doi.org/10.1002/wmon.1057

Wildlife Services (2015) Program Data Report G—animals dispersed/killed or euthanized/removed or destroyed/freed or relocated. https://www.aphis.usda.gov/aphis/ourfocus/wildlifedamage/SA_Reports/SA_PDRs. Accessed 24 Feb 2021

Wildlife Services (2019) Program Data Report G—animals dispersed/killed or euthanized/removed or destroyed/freed or relocated. https://www.aphis.usda.gov/aphis/ourfocus/wildlifedamage/SA_Reports/SA_PDRs. Accessed 24 Feb 2021

Willcox D (2020) Conservation status, ex situ priorities and emerging threats to small carnivores. Int Zoo Yb 54:1–16. https://doi.org/10.1111/izy.12275

Williams ES, Thorne ET, Appel MJG, Belitsky DW (1988) Canine distemper in black-footed ferrets (*Mustela nigripes*) from Wyoming. J Wildl Dis 24:385–398. https://doi.org/10.7589/0090-3558-24.3.385

Williams ES, MillsK KDR, Thorne ET, Boerger-Fields A (1994) Plague in a black-footed ferret (*Mustela nigripes*). J Wildl Dis 30:581–585. https://doi.org/10.7589/0090-3558-30.4.581

Williams ES, Erson SL, Cavender J, Lynn C, List K, Hearn C, Appel MJG (1996) Vaccination of black-footed ferrets (*Mustela nigripes*) x Siberian polecat (*M. eversmanni*) hybrids against domestic ferrets (*M. putorius furo*) against canine distemper. J Wildl Dis 32:417–423

Wilmers CC, Post E, Peterson RO, Vucetich JA (2006) Predator disease out-break modulates top-down, bottom-up and climatic effects on herbivore population dynamics. Ecol Lett 9:383–389. https://doi.org/10.1111/j.1461-0248.2006.00890.x

Wilson RR, Blankenship TL, Hooten MB, Shivik JA (2010) Prey-mediated avoidance of an intraguild predator by its intraguild prey. Oecologia 164:921–929. https://doi.org/10.1007/s00442-010-1797-8

Wright JV (1987) Archeological evidence for the use of furbearers in North America. In: Novak M, Baker JA, Obbard ME, Malloch B (eds) Wild furbearer management and conservation in North America. Ontario Ministry of Natural Resources, Toronto, Ontario, pp 3–12

Young JK, Olson KA, Reading RP, Amgalanbaatar S, Berger J (2011) Is wildlife going to the dogs? Impacts of feral and free-roaming dogs on wildlife populations. Bioscience 61(2):125–132. https://doi.org/10.1525/bio.2011.61.2.7

Young JK, Draper JP, Breck S (2019a) Mind the gap: experimental tests to improve efficacy of fladry for nonlethal management of coyotes. Wildl Soc Bull 43:265–271. https://doi.org/10.1002/wsb.970

Young JK, Draper JP, Kinka D (2019b) Spatial associations of livestock guardian dogs and domestic sheep. Human–Wildl Interact 13:6. https://doi.org/10.26076/frv4-jx12

Zimmerman A (1998) Reestablishment of swift fox in north central Montana. M.S. Thesis, Montana State University

Chapter 17
Black-Tailed and Mule Deer

Randy T. Larsen and Brock R. McMillan

Abstract Black-tailed and mule deer (both designated as *Odocoileus hemionus*; hereafter referred to as "deer" or "mule deer") comprise an iconic species that is broadly distributed across western North America. This species occurs in all rangeland types including grasslands, desert shrublands, forests, savannah woodlands, and even portions of tundra. The distribution of mule deer has changed little since Euro-American settlement, but abundance has fluctuated in response to environmental variation and rangeland management practices. These deer are medium-sized, polygynous mammals classified as generalist herbivores (foregut fermenters). Population growth in this species is strongly influenced by survival of adult females and recruitment of young. The management of rangelands has direct influence on deer populations given the wide distribution of this species and measurable responses to rangeland management practices. Rangeland management practices including development of water, grazing by domestic livestock, prescribed fire, energy extraction, vegetation alteration, and others can have positive or negative influences or both on this species. Although mule deer are widely distributed and relatively abundant, conservation of this species is challenged by rapid changes currently occurring on rangelands of western North America. Altered fire regimes due to climate change and invasive plants, competition (with feral horses [*Equus ferus caballus*], livestock, and other wild ungulates), development of energy, ex-urban and urban expansion, and many other challenges threaten continued abundance of this species. Rangelands and their associated management will continue to play a disproportionally large role in the conservation of mule deer in the future.

Keywords Black-tailed deer · Mule deer · Conservation · Habitat · *Odocoileus hemionus* · Rangelands · Wildlife · Management

R. T. Larsen (✉) · B. R. McMillan
Department of Plant and Wildlife Sciences, Brigham Young University, Provo, UT, USA 84602
e-mail: randy_larsen@byu.edu

© The Author(s) 2023
L. B. McNew et al. (eds.), *Rangeland Wildlife Ecology and Conservation*,
https://doi.org/10.1007/978-3-031-34037-6_17

591

17.1 General Life History and Population Dynamics

Black-tailed and mule deer (both designated as *Odocoileus hemionus* and hereafter referred to as "deer" or "mule deer") are polygynous breeders with the breeding season occurring during the late fall or early winter and a gestation period of 199–208 days or approximately 7 months (Anderson 1981). Pregnancy rates for prime-aged (between 2 and 6 years of age) females typically exceed 90% with limited variation in response to environmental or rangeland conditions (Freeman et al. 2014; Montieth et al. 2014). Pregnancy rates for yearling females (1.5-year-olds during fall/winter breeding season), however, are generally lower and more variable across rangelands and in response to environmental conditions (Lawrence et al. 2004; Montieth et al. 2014). Likewise, pregnancy rates for older (≥ 6.5) females are also more variable than prime-aged females and have been reported as low as 73% in arid rangelands (Lawrence et al. 2004). Timing of parturition varies across latitudes with earlier parturition at northern latitudes and later parturition at more southern latitudes, ostensibly in response to the increased influence of monsoon moisture during summer on rangelands in western North America from north to south (Freeman et al. 2014; Stoner et al. 2016).

Typically, one or two offspring are born during the spring or summer following gestation. Number of offspring varies with age (average litter size reduced for yearling females) and condition of parturient females (Montieth et al. 2014). Mean body mass of neonatal mule deer at birth was estimated at 3.4 kg, but varies across rangelands from 2.7 to 4.0 kg depending on condition of females, presence of a twin, location, and year (Lomas and Bender 2006). Neonatal mule deer are weaned over the summer and fall months prior to the subsequent breeding season and most females have offspring each year (Bowyer 1991).

Survival of mule deer varies by age, sex, and across rangelands in western North America. Mean estimates of survival for neonates (from 0 to 6 months of age) range between zero and 62% depending on location and year with high rates of predation commonly observed (Pojar and Bowden 2004; Lomas and Bender 2006; Montieth et al. 2014; Shallow et al. 2015). A weighted average for mean survival to 6 months of age using data collected from across the range of mule deer was 44% (95% CI: 33% to 55%) (Forrester and Wittmer 2013). Survival for fawns (from 6 months of age through first winter) was also averaged at 44% (SE = 3%), but considerable variation was found with estimates ranging between 4% (SE = 3%) and 81% (SE = 7%) depending on location and year (Unsworth et al. 1999). Observed rates of survival for both neonates and fawns are lower and more variable than commonly observed with other ungulates (Forrester and Wittmer 2013). Survival of adult females is less variable than survival of neonates or fawns with a weighted mean estimated from across the range at 84% (95% CI 75–94%) (Forrester and Wittmer 2013). For most populations, annual survival of adult males is strongly dependent on harvest by humans with estimates of survival ranging from 60 to 92% (Pac and White 2005; Bender et al. 2012).

Population dynamics for ungulates are typically most strongly influenced by survival of adult females which is often relatively high and stable (Gaillard et al. 2000). Some estimates suggest adult survival for mule deer is nearly four times more influential than other demographic rates on population growth (Lukacs and Nowak 2023). In other research however, survival of fawns and recruitment were identified as most influential on population growth as 4 of 5 studies found this demographic rate more influential than adult survival (Forrester and Wittmer 2013). This discrepancy may occur because survival of juvenile deer was lower and more variable than observed with other ungulates (Forrester and Wittmer 2013). Moreover, unlike many ungulates, prime-aged deer typically give birth to twins and consequently may rely on relatively high fecundity rates as a driver of population growth (Forrester and Wittmer 2013).

As with other ungulates, predation is the most commonly identified source of mortality for neonates between 0 and 6 months of age and typically accounts for between 50 and 100% of all mortalities (Linnell et al. 1995). Fawns (6 months to 1.5 years of age) are also vulnerable to predation, but high mortality rates are also regularly observed at northern latitudes during winter due to starvation. A diverse group of predators have been reported to prey on mule deer during their first year of life with coyotes (*Canis latrans*), mountain lions (*Puma concolor*), bobcats (*Lynx rufus*), wolves (*Canis lupus*), and black bears (*Ursus americanus*) reported as the primary predator depending on years evaluated and location (Forrester and Wittmer 2013). Predation is also the most cited cause of mortality for adult deer with between 22 and 66% of all mortalities due to predation from primarily mountain lions and wolves with secondary sources of mortality identified as disease, malnutrition, vehicle strikes, and miscellaneous other causes (Forrester and Wittmer 2013).

Nonetheless, mule deer do show evidence of density dependence (Bergman et al. 2015). Adult male to adult female ratios, for example, are negatively correlated with ratios of production suggesting intraspecific competition occurs and harvest strategies designed to increase the proportion of males in a population may have a regulating effect on population growth (Bishop et al. 2005a; Bergman et al. 2011, 2015). Likewise, a severe (76%) reduction in density of a mule deer population in Colorado was associated with a more than two-fold increase (31% to 77% in control and treatment areas, respectively) in fawn survival (White and Bartmann 1998). Interestingly, in the same area, reductions in density of 16% and 22% were not associated with increases in overwinter fawn survival (Bartmann et al. 1992).

Mule deer are of tremendous interest to the public for hunting, collection of shed antlers, and wildlife viewing. For many western states and provinces, demand for deer hunting permits greatly exceeds available supply. Sales of hunting permits and excise taxes on purchases of hunting equipment provide millions of dollars annually to state and provincial agencies to be used for conservation and management.

Because mule deer follow a polygynous breeding system, harvest management strategies are often focused on removal of males (less likely to influence population dynamics) and most exploited populations have male/female ratios that are heavily skewed towards females. Selective harvest of males has mixed effects on mule deer populations.

Rates of pregnancy (99% vs. 97%) and timing of parturition at comparable latitudes, for example, were similar for populations with relatively low (14 males per 100 females) and relatively high (26 males per 100 females) male/female ratios, respectively (Freeman et al. 2014). This finding suggests that male/female ratios as low as 14 males per 100 females did not limit breeding opportunity or influence dates of conception. Moreover, increases in male/female ratios have been associated with decreased fawn/adult female ratios suggesting that annual production is actually reduced in the presence of high male/female ratios (Bishop et al. 2005a). This effect may be caused by competition for resources between adult females and adult males resulting in poor condition and reduced productivity for adult females.

Hunting can influence other aspects of the ecology of mule deer. Number of males harvested and timing of harvest, for example, were correlated with reduced prevalence of chronic wasting disease, an emerging conservation challenge for this species (Conner et al. 2021). Additionally, selective harvest of large males over the last century has been proposed as the most likely cause of an approximate 3% decline in antler size (Monteith et al. 2013). Further, disturbance associated with hunting can influence movement rates and habitat selection. Female mule deer demonstrated increased movement rates during the daytime during hunting season, but similar movement rates during the night while maintaining high site fidelity (Brown et al. 2020). Similarly, at relatively fine scales (within summer range areas and at stopover sites along migration routes), female mule deer used habitats that retained high-quality forage consistently throughout the hunting season whereas male mule deer selected more secure habitats away from motorized routes (Rodgers et al. 2021). Nonetheless, conservation of mule deer has benefitted from the interest and revenue associated with pursuit and harvest of this species.

17.2 Species and Population Status

The genus *Odocoileus*, which contains mule deer along with white-tailed deer (*O. virginianus*) is one of 18 genera found across the world in the family Cervidae (Wilson and Mittermeier 2011). This genus first evolved in North America approximately 3.5 million years ago with a form similar to white-tailed deer present in North America at least 3 million years ago (Miller et al. 2003). Eleven subspecies of *O. hemionus* are recognized, but little genetic or morphological variation is found between many of them and they are subsequently grouped into two morphological types: black-tailed deer (*O. h. columbianus* and *O. h. sitkensis*) and mule deer (*O. h. hemionus, O. h. fulginatus, O. h. californicus, O. h. inyoensis, O. h. eremicus, O. h. crooki, O. h.peninsulae, O. h. sheldoni, and O. h. cerrosensis*) (Cronin 1991; Latch et al. 2009). Black-tailed deer are a relatively recent species that likely diverged from white-tailed deer within the last 500,000 years (Polzhein and Strobek 1998). Mule deer (the morphological type that includes all of the subspecies excluding *O. h. columbianus* and *O. h. sitkensis*) are even younger and likely evolved within the last

12,000 years following extinction of Pleistocene megafauna, glacial retreat, and putative hybridization between black-tailed deer and white-tailed deer in western North America (Geist 1998). Hereafter we refer to black-tailed deer (*O. h. columbianus* and *O. h. sitkensis*) and mule deer (remaining subspecies) as mule deer unless referring to an individual subspecies.

17.2.1 Historical Versus Current Distribution

Climatic oscillations and the subsequent expansion and retreat of glaciers in North America influenced the distribution of this species (Latch et al. 2009). Genetic analyses suggest that black-tailed deer persisted in a single refugium in the Pacific Northwest near the coast prior to expanding north and inland following the last glacial maximum (Latch et al. 2009). Conversely, mule deer likely persisted in multiple refugia in the southern portion of western North America from which they expanded north following the last glacial maximum (Latch et al. 2009).

More recently, mule deer were first encountered by Lewis and Clark while traveling up the Missouri River (Hays 1869). Lewis and Clark also noted the presence of mule deer as they journeyed across the northern portion of the United States with black-tailed deer (*O. h. columbianus*) observed west of the Cascade Mountains (Kay 2007). Rock art from multiple Native American cultures also provides evidence that mule deer occurred in locations consistent with their current distribution in the southern portion of their range (Murray 2013). Moreover, mule deer were also noted in the journals of early explorers of the southwestern United States including Dominguez and Escalante (Rawley 1985).

17.2.2 Distribution Map

Black-tailed deer and mule deer currently occupy portions of 6 Canadian provinces, 6 states in Mexico, and 18 western states in the United States of America (Fig. 17.1; Heffelfinger and Latch 2023). Comparison of the current distribution with early accounts of this species from Euro-American explorers suggest only a few differences since the seventeenth and eighteenth centuries. Lewis and Clark, for example, first encountered mule deer along the Missouri River (Hays 1869) which is near the eastern boundary currently recognized for mule deer (Fig. 17.1; Heffelfinger and Latch 2023). Moreover, pictographs and other archaeological records coupled with early records from Dominguez and Escalante (Rawley 1985) also suggest little change in distribution in recent centuries.

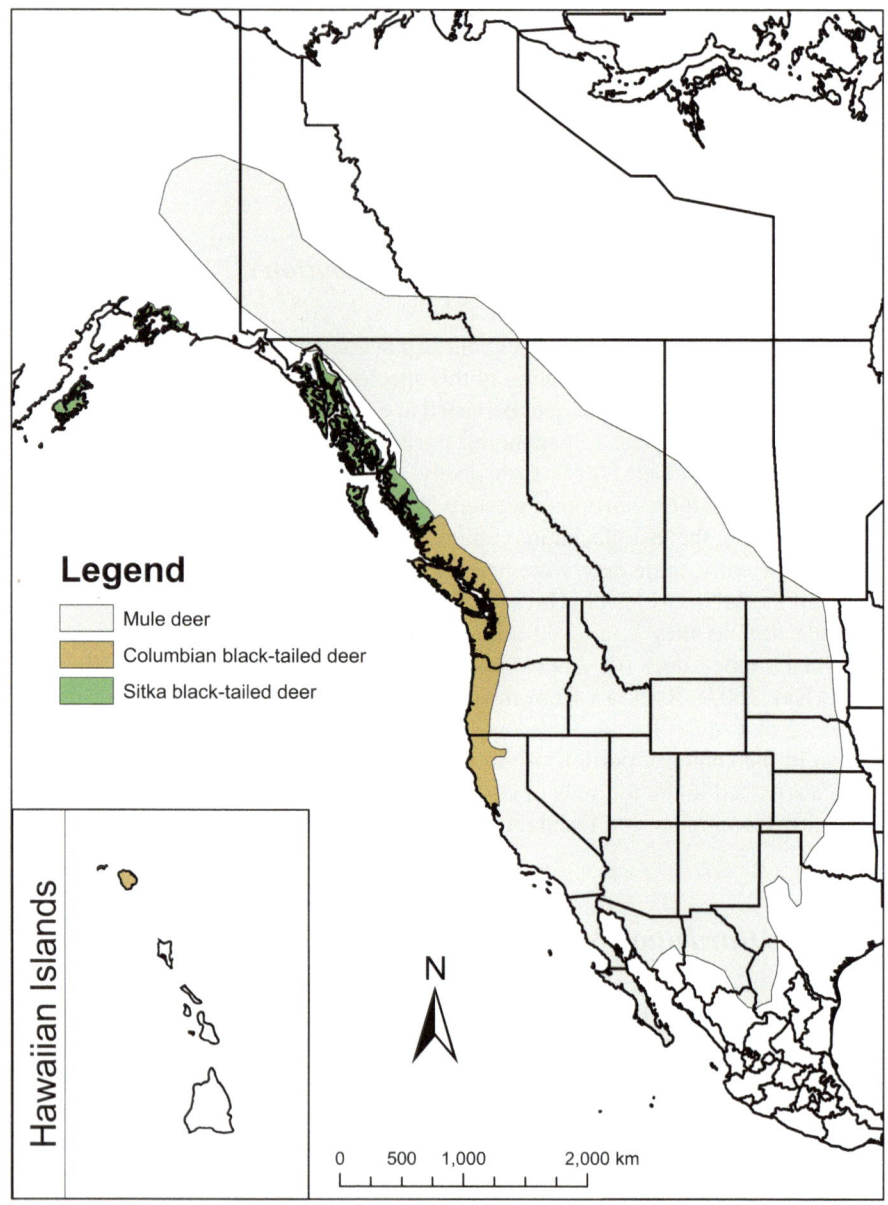

Fig. 17.1 Distribution map depicting the current distribution of mule deer (*Odocoileus hemionus*) and black-tailed deer (subspecies *O. h. sitkensis* and *O. h. columbianus*) modified from Heffelfinger and Latch (2023). The distribution of this species remains largely unchanged since Euro-American settlement of western North America

17.2.3   *Historical Versus Current Abundance*

Despite a broad distribution during the period of exploration by Euro-Americans in western North America, mule deer may have persisted at relatively low abundance in many parts of their range. On their journey up the Missouri River and across the northern portion of the western United States, Lewis and Clark harvested more white-tailed deer than all other large mammals combined and mule deer were rare (Kay 2007). This discrepancy may have resulted from differences in habitat selection as the Lewis and Clark expedition navigated along rivers where white-tailed deer were perhaps more abundant. Alternatively, behavioral differences between mule deer and white-tailed deer may also explain the difference as mule deer were thought to have elevated risk of harvest by Native Americans because some of the highest encounter rates for this species occurred along tribal boundary zones (Geist 1998; Kay 2007). In the southern portion of their range, mule deer are outnumbered at least 2 to 1 by bighorn sheep (*Ovis canadensis*) in rock art suggesting reduced abundance histor-ically relative to that ungulate (Castleton 2002). Additionally, journals from many early Euro-American settlers in areas of western North America where mule deer are now abundant suggest they were only encountered rarely during the late 18th and early nineteenth centuries (Rawley 1985).

Mule deer populations undoubtedly fluctuated in response to environmental condi-tions and landscape changes (e.g., fire). Some records suggest this species was at least locally abundant in some portions of its range during the nineteenth century. Market hunter Frank Mayer, for example, harvested over 250 ungulates between late August and early November 1878 in Middle Park, Colorado and 89 of these were mule deer (Gill et al. 1999). Expanding settlement by humans and unregulated harvest including market hunting, however, led to severe declines for most ungulate species in western North America during the late 19th and early twentieth centuries (Krausman and Bleich 2013).

Following these severe declines, regulations on harvest coupled with human-induced changes to landscapes and predator communities led to population growth for mule deer during the twentieth century. The famous Kaibab mule deer population in Northern Arizona, for example, was estimated at 4,000 in 1906, but at least 7 times that number in 1924 less than two decades later (Caughley 1970). Moreover, between 1930 and 1960, mule deer populations irrupted in the Intermountain West and likely reached all-time high abundance in recorded history between 1940 and 1950 (Gruell 1986). By the late 1940s, mule deer populations in at least portions of each of the western United States were recognized as over-populated and a cause of rangeland degradation (Leopold et al. 1947; Binkley et al. 2006). Four competing hypotheses have been proposed for this irruption. These hypotheses include: (1) conversion of rangelands to shrub-dominated systems favored by mule deer following overgrazing by domestic animals during the early twentieth century, (2) logging of forested land-scapes and subsequent succession of those plant communities to forbs and shrubs, (3) widespread predator control including use of poison, and (4) reductions in live-stock grazing following implementation of the Taylor Grazing Act (Gruell 1986).

Although some authors favor conversion of rangelands to shrub-dominated systems as a leading hypothesis (Gruell 1986), it is impossible to disentangle that idea from the other concepts as landscape change, predator control, and reductions in grazing occurred simultaneously and may have been synergistic.

Following the historic highs of the 1950s and 1960s, mule deer populations appeared to decline across their range, although good estimates of abundance are not available until the latter part of the twentieth century. During the last half of the twentieth century and early part of the twenty-first century, estimates of abundance for mule deer populations fluctuated in both space and time with periods of general decline (late 1970s, early 1990s) and growth (early 1980s, early 2010s) noted across their range (Mule Deer Working Group 2015). The most recent estimates from governmental entities that participate in the Western Association of Fish and Wildlife Agencies suggest increasing populations in 6 jurisdictions (provinces or states), stable populations in 10, and declining populations in 7 (Mule Deer Working Group 2021). Current estimates of abundance suggest approximately 4 million mule deer occur across their range making this species the most abundant ungulate in western North America (Mule Deer Working Group 2021). Mule deer are listed as "stable" and designated a species of "least concern" by the International Union for Conservation of Nature (IUCN) (Sanchez-Rojas and Tessaro 2016).

17.2.4 Monitoring

Declines in mule deer populations following historic highs in the 1940s and 1950s prompted state and federal agencies to develop methods to estimate abundance. Mule deer are now monitored extensively on rangelands across western North America using a variety of methods (Keegan et al. 2011). Because management of deer resides with individual agencies within provincial and state governments, there are a myriad of approaches and strategies used to monitor populations (Rabe et al. 2002).

Monitoring efforts often include surveying populations from the ground or by fixed-wing aircraft or helicopter and classification of age and sex composition (Freddy et al. 2004; Keegan et al. 2011). Some agencies use distance sampling (often using fixed-wing or rotary aircraft although this can also be done from the ground) which allows for estimation of density after first estimating detection probability (Koenen et al. 2002). This method assumes perfect detection of deer at the survey point or along the survey transect, limited movement of animals in response to the surveyor prior to detection, accurate measurement of distance from the survey point or perpendicular distance from the transect to each individual deer, and no double counting (Buckland et al. 1993). Consequently, this technique is most often used on rangelands with relatively high visibility where statistical assumptions can be reasonably attained. Detection probabilities can also be estimated on surveys if deer are uniquely marked (e.g., ear tags, collars). Detection probabilities vary from 19 to 86% and are associated with many factors including deer activity, group size, topography, vegetation, and weather conditions (Bartmann et al. 1986; Zabransky et al. 2016).

Harvest surveys are also commonly used to estimate abundance across the range where mule deer occur (Rupp et al. 2000). More recently, advancements with remote cameras have allowed estimation of abundance of deer using instantaneous sampling, space-to-event, and time-to-event models, which may prove particularly helpful on rangelands with reduced visibility (Moeller et al. 2018). Likewise, advancements with unmanned aerial systems and thermal imaging cameras show some ability to estimate parameters such as abundance and survival (Williams et al. 2020). Some agencies have also used DNA collected from fecal pellets (often referred to as fecal DNA or fDNA) to identify individuals and help with estimation of abundance (Furnas et al. 2018). Finally, deer are also commonly monitored via mark-recapture techniques and either GPS or VHF collars, which can provide estimates of fecundity, recruitment, and survival when the sample of marked individuals is monitored closely.

Most provincial or state agencies employ dozens of biologists to collect monitoring data for deer within their jurisdictions. Advancements in statistical analyses and computing power now allow for incorporation of multiple sources of data with varying degrees of uncertainty into estimates of abundance or density (Furnas et al. 2018). Data obtained on populations of deer (e.g., flight surveys, remote camera surveys, mark-recapture data from collared animals, etc.) are now commonly integrated with harvest information into population models to help with decision making (e.g., integrated population models; Riecke et al. 2019). Population models are then commonly used to adjust harvest permits or quotas on an annual or every few years' basis in an adaptive management framework (Nagy-Reis et al. 2021). These efforts create a robust data set to monitor mule deer populations across their range.

17.2.5 Migration Ecology and Overcoming Barriers to Movement

Many mule deer populations are migratory including some individuals and populations with long distance (> 100 km) migrations (Sawyer et al. 2005). Nonetheless, tremendous variation in migratory behavior occurs including some individuals and populations that do not migrate (McCorquodale 1999; Van de Kerk et al. 2021). Migratory behavior allows some mule deer populations to maximize intake of nutritious forage on rangelands during spring green-up by prolonging the period when they can consume high-quality forage (Merkle et al. 2016). Conversely, migratory behavior is energetically costly and can be associated with increased risk of mortality as animals navigate migratory routes. Thus, a tradeoff exists for deer between prolonged access to nutritious forage and the ability to escape deep snows (both of which likely have fitness benefits) versus risks associated with migration. Long-distance migrants in one population of mule deer, for example, were exposed to increased risk of mortality from fences and highways, whereas animals that did not migrate as far experienced fewer of those risks (Sawyer et al. 2016). Migratory behavior for another population, however, was associated with higher survival for mule deer that migrated compared to non-migratory animals (Schuyler et al. 2019).

Knowledge of migration routes and timing is thought to be transmitted culturally for ungulates, but has not been specifically tested in mule deer (Jesmer et al. 2018). This cultural transmission of knowledge likely includes learning of both routes and stopover areas, the latter of which are increasingly recognized as important. Mule deer in one population, for example, took an average of 3 weeks to complete migrations, but spent 95% of that time in stopover areas (Sawyer and Kauffman 2011). These animals averaged use of a stopover every 5.3 and 6.7 km during spring and fall migrations, respectively (Sawyer and Kauffman 2011). Stopovers likely act as physiological refugia (forage and rest) and were associated with lower measures of stress hormones (fecal glucocorticoid metabolites) along a lengthy migration on rangelands in Wyoming (Jachowski et al. 2018). Although plasticity on whether to migrate and where to go during migration has been documented (Van de Kerk et al. 2021), many populations show strong (> 80%) fidelity to migration routes and stopover areas across years with limited evidence of plasticity (Sawyer et al. 2019).

Consequently, preservation of migratory routes and stopover areas is critical to conservation of mule deer populations. Terrestrial migrations in general are imperiled around the globe including on rangelands in western North America where deer occur (Middleton et al. 2019). Migratory behavior in mule deer populations is now threatened by rapidly changing landscapes on western rangelands. Anthropogenic activities and structures including fences, highways, homes, and extraction of natural resources can influence migratory behavior and ecology of deer. On a lengthy migratory route on rangelands in Wyoming, for example, it was estimated that mule deer crossed an average of five highways and 171 fences per year during migration (Sawyer et al. 2016). Mule deer have also been shown to increase rate of movement, decrease time in stopover areas, and shift the location of stopover areas in response to energy and residential development (Wyckoff et al. 2018). Likewise, housing development and roadways reduced the effective width of a bottleneck along a migratory route for over 2,500 migratory mule deer to < 0.8 km (Sawyer et al. 2005). Additionally, anthropogenic disturbance associated with development of natural gas resources was associated with delayed departure, but earlier arrival suggesting more rapid transit of migratory routes along with shorter migration distances (Lendrum et al. 2013).

Increased understanding of the importance of migratory routes and stopover areas for mule deer has led to conservation efforts to improve permeability along migration routes by removing movement barriers (e.g., fences) or constructing passage ways such as highway underpasses to facilitate movement. Fences, can alter movement and pose a risk of entanglement for migratory animals. Removal of fencing or replacement of fencing with designs that include a smooth lower wire that is raised (46 cm) and a shorter (107 cm) top wire can increase permeability and mitigate risk of entanglement (Segar and Keane 2020). Likewise, highway underpasses coupled with exclusionary fencing facilitated safe movement of nearly 50,000 mule deer over 3 years under highway 30 in southwestern Wyoming while reducing collisions with vehicles by 81% (Sawyer et al. 2012). In general, passage structures with high "openness ratios" (width multiplied by height/length) are considered most effective (Clevenger and Waltho 2000). Collaborative efforts between governmental agencies and vested stakeholders coupled with improved data from GPS collars have helped identify barriers to movement for migratory mule deer (Middleton et al. 2019). This information should be incorporated into landscape planning so that crossing structures and mitigation efforts preserve migratory routes and stopover areas for mule deer.

17.3 Habitat Associations

Mule deer occupy grasslands, desert shrublands, savannah woodlands, forests, and even portions of tundra. Deer are medium-sized, generalist herbivores that are foregut fermenters with small mouth parts (concentrate selectors) (Hofmann 1989). These characteristics assist mule deer in selecting the most nutritious parts of the many different plant species they consume across the varied rangelands where they occur.

Mule deer consume a variety of different plants throughout the year including forbs, grasses, shrubs, and trees (Stewart et al. 2003; Berry et al. 2019). Across their range, the relative importance of these functional groups, however, varies seasonally. During spring and summer on rangelands at northern latitudes, mule deer primarily consume forbs, grasses, and deciduous shrubs whereas during fall and winter when herbaceous plants senesce or are covered in snow, evergreen shrubs and trees constitute the majority of consumed forage (Scasta et al. 2016; Berry et al. 2019). On rangelands at southern latitudes, seasonal use of forbs and grasses varies in response to precipitation patterns that can be highly variable and shrubs often constitute a majority of consumed forage annually (Krausman et al. 1997; Marshal et al. 2012).

Although classified as concentrate selectors, mule deer have some characteristics consistent with intermediate foragers. Composition of volatile fatty acids in the rumen, for example, was similar between red deer (*Cervus elaphus*; intermediate forager) and mule deer (Prins and Geelen 1971). Likewise, papillae density, dry weight of rumen digesta, and intestinal length are all greater in mule deer compared to white-tailed deer allowing them to make use of less nutritious forage (Zimmerman et al. 2006). Relative to white-tailed deer, for example, mule deer required 54% less digestible protein and 21% less digestible energy per day to maintain body mass and nitrogen balance (Staudenmaier et al. 2021).

Moreover, mule deer have adaptations that allow them to process plant secondary compounds such as tannins and terpenes common in shrubs. Mule deer commonly consume multiple different plant species which often allows for overall greater forage intake than consumption of a single species when secondary compounds are present (Freeland and Janzen 1974). Additionally, mule deer, have relatively large parotid salivary glands that produce proteins that bind to tannins and ameliorate the impact of those secondary compounds on digestibility (Hagerman and Robbins 1993).

Mule deer were also better able to process forage containing the monoterpene α-pinene compared to white-tailed deer, suggesting a more efficient and less energetically costly method of detoxifying this compound (Staudenmaier et al. 2021). Consequently, mule deer may have an advantage over white-tailed deer on rangelands dominated by low-quality forages that are chemically defended (Staudenmaier et al. 2021).

17.4 Rangeland Management

The rangelands where mule deer occur (Fig. 17.1) differ greatly in precipitation patterns, plant community composition, and soils. Management of these disparate rangelands also differs regionally and across jurisdictions. Consequently, it is difficult to make definitive statements concerning the influence of rangeland management activities on mule deer, or their habitat. Mule deer, like all species, require cover, food, space, and water in an arrangement where all are accessible (sensu Leopold 1933). When rangeland management activities promote these elements, populations benefit. Conversely, when rangeland management activities eliminate access to or degrade these essential components to habitat, populations decline.

17.4.1 Livestock Grazing

Dietary overlap between domestic livestock and mule deer covaries with species of livestock, rangeland type, and annual variation in availability of forage. Dietary overlap with cattle (*Bos taurus*) is often low (e.g., Stewart et al. 2003; Beck and Peek 2005), but can increase during years of low precipitation or with high deer density and heavy stocking rates (Campbell and Johnson 1983; Hansen and Reid 1975). Overlap between mule deer and domestic sheep (*Ovis aries*) also varies depending on plant community composition and season of grazing. Spring grazing by domestic sheep, for example, resulted in low (15%) dietary overlap with mule deer on a sagebrush (*Artemisia* sp.) rangeland in Colorado (MacCracken and Hansen 1981). Moderate overlap (22–65%, depending on year), however, was observed in summer in aspen (*Populus tremuloides*)-sagebrush communities in northeastern Nevada (Beck and Peek 2005).

Grazing by domestic livestock can have both positive and negative influences or both depending on the species of livestock, stocking rate, timing of grazing, and response of plant communities to grazing (Krausman et al. 2011). Influences of grazing by domestic livestock can be both direct and indirect (Chaikina and Ruckstuhl 2006). Presence of livestock, for example, can influence habitat selection leading to avoidance of areas used by livestock (Ragotzkie and Bailey 1991; Loft et al. 1991; Stewart et al. 2002). Avoidance of grazed pastures may occur even beyond the time when livestock are removed (Clegg 1994; Kinka et al. 2021). This interaction and the strength of selection against areas used by livestock, however, are influenced by stocking rate and density of deer with preference for ungrazed pastures diminished at low stocking rates or high deer densities (Austin and Urness 1986).

Mule deer altered their foraging behavior in relation to stocking rates of cattle by feeding for longer durations in areas grazed at high stocking rates when forage was limited, but not when herbaceous plants were abundant (Kie 1996). Removal of vegetation under moderate and high stocking rates can also lead to decreased hiding cover—particularly for neonates (Loft et al. 1987). Additionally, mule deer are more likely to compete with livestock for forage during dry years when forage production on rangelands is limited (Kie et al. 1991). Domestic livestock are potential vectors for exotic and invasive plants through both endozoochory (passage of viable seed through the digestive tract) and epizoochory (transport of seeds on skin and fur) (Chuong et al. 2016). On rangelands in the Northwestern portion of North America, cattle were estimated to disperse (via endozoochory) an order of magnitude more seeds from exotic grasses than either elk (*Cervus canadensis*) or deer (Bartuszevige and Endress 2008).

Conversely, grazing by livestock can also result in positive outcomes for range-lands. For rangelands dominated by grasses, grazing by cattle has been associated with removal of standing dead vegetation leading to more nutritious forage during the subsequent growing season (Short and Knight 2003; Taylor et al. 2004). Moreover, grazing by cattle produced more nutritious forage during the subsequent growing season than mowing in a rough fescue (*Festuca scabrella*) rangeland (Taylor et al. 2004). Similarly, grazing by domestic sheep can increase nutritional quality (e.g., protein content and digestibility) of plants—particularly shrubs—during fall and winter when those plants are grazed during spring at moderate (less than 55%) utilization (Rhodes and Sharrow 1990; Alpe et al. 1999).

Changes in vegetation due to grazing by domestic livestock, however, have not been directly linked to increased abundance, condition, or production in mule deer populations and caution is warranted. Nonetheless, suggested guidelines for manage-ment of grazing by domestic livestock on rangelands important to deer during winter include grazing during the spring to balance utilization of shrubs by deer during winter with consumption of forbs and grasses by livestock (Austin 2000). Alter-nating species of livestock along with moderate (50%) utilization in a rest-rotation system have also been suggested as best practices to maintain or enhance rangelands for mule deer (Jensen et al. 1972; Austin 2000).

Grazing of rangelands by domestic livestock can also be used prescriptively to enhance habitat for mule deer by reducing abundance of undesirable species. Targeted grazing of cheatgrass (*Bromus tectorum*) by cattle during spring, for example, was associated with reduced risk of catastrophic fire in the fall (Diamond et al. 2009). Use of grazing by domestic livestock to reduce invasive plants, however, is challenging across rangelands with large spatial extents because intensive management (e.g., electric fencing, supplemental protein and energy to overcome secondary compounds in targeted plant species) is typically required to reach high utilization rates (Popay and Field 1996; Dziba et al. 2007).

Fig. 17.2 Mule deer
(*Odocoileus hemionus*)
caught in a fence used to
manage livestock on federal
rangelands. Photo credit to
Jason Nicholes with the Utah
Division of Wildlife
Resources

Fencing associated with management of domestic livestock can also influence deer. In western Wyoming, density of fences within mule deer range was estimated at 0.59 km/km^2 and mule deer encountered fences an average of 119 times per year (Xu et al. 2020). Fences altered normal movement patterns on nearly 40% of encounters (Xu et al. 2020). Moreover, mule deer can become entangled in fences resulting in injury or death (Fig. 17.2). Up to 0.08 mortalities per km of fencing were estimated annually for mule deer in Utah and Wyoming (Harrington and Conover 2006). Fencing mortalities peaked in August and juveniles were more likely to be entangled in fences than adult animals (Harrington and Conover 2006). Increased height of the bottom wire was associated with increased probability of successfully crossing fences (Jones et al. 2020). Woven wire fences with a single strand of barbed wire were the most lethal compared to 4-strand barbed wire fences or woven wire with two strands above the mesh (Harrington and Conover 2006). Replacement of 4-strand barbed wire fences with wildlife friendly fencing where the bottom wire was smooth and raised (46 cm) along with a shorter top wire (107 cm) was associated with an increase of over 18% in successful crossings indicating some ability to mitigate effects of fences with design modifications (Segar and Keane 2020).

17.4.2 *Interactions with Coexisting Feral and Wild Ungulates*

Mule deer coexist with several species of native and non-native, free-ranging ungulates throughout portions of their geographic range (Fig. 17.1). Free-ranging native ungulates with potential for interspecific interactions include American elk, white-tailed deer, moose (*Alces alces*), pronghorn (*Antilocapra americana*), bighorn sheep, mountain goat (*Oreamnos americanus*), bison (*Bison bison*), and collared peccary (*Pecari tajuca*). In addition, there are several species of free-ranging, non-native animals including feral horses (*Equus caballus*), feral burros (*E. asinus*), feral pigs (*Sus scrofa*), feral sheep (*Ovis aries*), and feral goats (*Capra hircus*) that occur on rangelands with mule deer. Based on literature examining interactions between deer and many of these species, it appears that species of greatest interest and potential to influence mule deer include American elk, white-tailed deer, and feral equids (horses and burros). There is also significant risk for future interactions with feral pigs if that species continues to expand and increase in abundance (O'Brien et al. 2019).

American elk have generally increased in population size across western North America in recent decades. Concurrent with this increase in abundance of elk has been a general decrease in abundance of mule deer in many areas. This concurrent and inverse relationship in abundance has led many to postulate that elk may be responsible for the decrease in abundance of deer. Both species occupy the same rangelands across much of western North America. However, results from many studies examining the potential for competition between mule deer and elk are inconsistent. Because of their smaller body size, deer require higher-quality forage than elk (Wickstrom et al. 1984). Elk are considered more generalist foragers and their diet is typically comprised of more low-quality food such as grasses, except during spring when diet is more similar—likely due to both species needing higher quality forage to recover from winter and reproduce (Stewart et al. 2003; Sandoval et al. 2005; Torstenson et al. 2006). In addition, elk are much larger than deer and appear to be socially dominant and capable of physically displacing deer, which provides some evidence for interference competition between these two species (Stewart et al. 2002). However, the evidence for displacement is not universal (Sallee et al. 2023).

White-tailed deer are the most widespread ungulate in North America and co-occur with mule deer across much of western North America. Similar to patterns observed with elk over recent decades, there appears to be an overall decline in abundance of mule deer that is concurrent with an increase in abundance of white-tailed deer. Among sympatric populations, mule deer typically demonstrated population decline concurrent with population increase by white-tailed deer (Robinson et al. 2002) even though mule deer appear to be competitively dominant (Anthony and Smith 1977). Genomic analyses suggest this pattern has a much older origin with effective population size for white-tailed deer increasing over the last 500,000 years while that same metric has declined for mule deer (Lamb et al. 2021). Because of the similarity in body size and digestive systems between both species, there is potential for forage competition. Indeed, studies have generally demonstrated a high degree of dietary overlap between these two species, but there is evidence for partitioning of food resources (Berry et al. 2019). Mule deer, for example, were better able to process forage containing the monoterpene α-pinene compared to white-tailed deer which may broaden forage options relative to white-tailed deer (Staudenmaier et al. 2021).

Mule deer also coexist on rangelands with feral equids in large portions of western North America (Stoner et al. 2021). Over 131,000 km² of mule deer habitat in the western United States is occupied by feral equids and 97% of management units for feral equids contain mule deer (Stoner et al. 2021). Moreover, over 80% of federally managed herds of feral equids exceed population objectives (BLM 2018). Overabundance of feral equids has been associated with habitat degradation and loss of biodiversity in some areas where mule deer occur (Zeigenfuss et al. 2014; Davies and Boyd 2019). Dietary overlap between feral equids and mule deer is limited in most seasons, but varies regionally and seasonally (Scasta et al. 2016). In a Sonoran Desert rangeland where both feral asses and mule deer consumed primarily browse species, overlap was highest and biologically significant during periods of abundant forage (summer and early fall seasons) compared to periods of relatively low forage availability when each species focused on forage that maximized physiological differences (Marshal et al. 2012). Feral horses appear to interfere with mule deer access to drinking water when limited availability of water occurs on arid and semi-arid landscapes (Hall et al. 2018). Conversely, feral burros may improve access to drinking water for mule deer on some rangelands in the southern portion of their range by digging wells in dry washes (Lundgren et al. 2021). Because of the ongoing range expansion and increase in abundance of feral equids, there is considerable potential for competition.

17.4.3 Fire

Historically, fire played a large role in structuring plant communities on rangelands where mule deer occur (Block et al. 2016). Historical fire-return intervals vary across rangelands in western North America from every few years to more than 300 years (Rollins 2009; Stevens et al. 2020). Tremendous variation (10–200 years) in return intervals can even occur within the same rangeland type (Miller and Tausch 2001). Variation in extent, severity, and return intervals associated with fire would have created a shifting mosaic on rangelands with associated plant communities in various stages of succession. This historical heterogeneity, however, has been reduced since Euro-American settlement. Suppression of fires coupled with changes in rangelands due to climate change and plant invasions (e.g., annual grasses) has led to increased size and severity of wildfires and an overall reduction in rangeland heterogeneity where mule deer occur (Dennison et al. 2014; Jolly et al. 2015).

The influence of fire (both prescribed and wild) on mule deer populations depends on the responses of rangeland plant communities to this disturbance which can be both positive and negative or both (Block et al. 2016). Both above-ground biomass and nutrients in plants can increase following fire (Rau et al. 2008; Roerick et al. 2019). Prescribed fire, for example, increased both crude protein and in vitro digestible organic matter in plants available to mule deer (Hobbs and Spowart 1984). Likewise, nitrogen in forbs and grasses regularly consumed by mule deer also increased following fire (Rau et al. 2008).

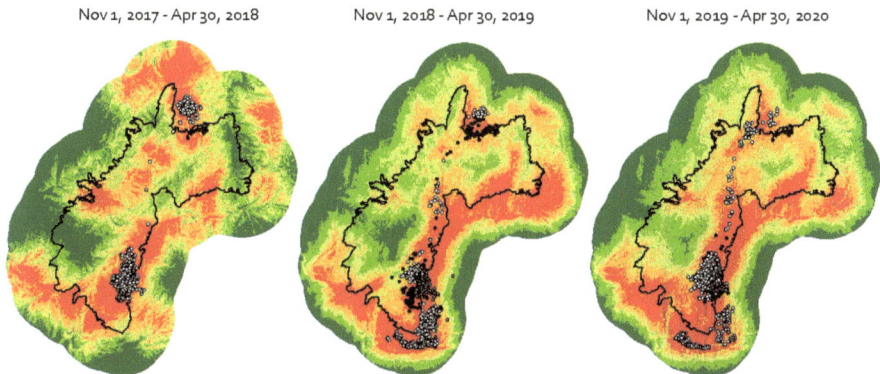

Nov 1, 2017 - Apr 30, 2018   Nov 1, 2018 - Apr 30, 2019   Nov 1, 2019 - Apr 30, 2020

Fig. 17.3 Habitat selection during winter (3rd-order selection or habitat patches within a home range) for an adult female mule deer (*Odocoileus hemionus*) in relation to fire with the black polygon representing the burn boundary for a portion of the Pole Creek fire in central Utah which occurred during fall of 2018. Figure shows preference (warm colors) for edges and avoidance of areas on the interior of the fire polygon during the initial two winters following the fire. Red to green colors represent high, medium–high, medium, medium–low, and low probabilities of selection, respectively

Changes in availability of plants and nutrition within plants following fire can influence habitat selection by mule deer (Fig. 17.3). On southern rangelands, mule deer selected for burned habitats unless they had been impacted recently (< 5 years) by high-severity fire and then they were avoided (Roerick et al. 2019; Bristow et al. 2020). Increased availability of nutritious plants on rangelands following fire should lead to increased health of adult females with cascading effects on reproduction including litter size and birthweight of neonates (Shallow et al. 2015). Indeed, recruitment rates over 13 years were positively correlated with acreage burned when precipitation patterns were also favorable (Holl and Bleich 2010). Selection for rangelands following fire, however, does not always translate into population growth (Klinger et al. 1989).

Moreover, fire has the potential to negatively impact deer populations when rangeland plant communities respond negatively. Where cheatgrass or red brome (*Bromus rubens*) occur in combination with nonsprouting shrubs such as sagebrush (*Artemisia* spp.), for example, fire can alter rangelands in a way that is detrimental to deer. When these invasive grasses become dominant, they can create a negative feedback loop that leads to increased frequency of fires and eventual elimination of shrubs (Pilliod et al. 2017). Conversion of shrublands to grasslands across large acreages of western North America has been facilitated by invasive annual grasses and fire (D'Antonio and Vitousek 1992). These changes have the potential to negatively impact mule deer populations. Consequently, managers should carefully consider potential responses of rangeland plant communities prior to use of prescribed fire (Block et al. 2016).

17.4.4   Vegetation Management—Chaining and Mastication of Conifers

Rangeland managers have a long history of vegetation treatment to improve rangelands using a variety of methods. Mechanical methods such as chaining, lop and scatter, mastication (also sometimes referred to as shredding), and mowing have been used across the range where mule deer occur. Likewise, treatment of vegetation with herbicide has also occurred, often in conjunction with post-fire restoration or mechanical treatment. Each of these methods is designed to improve rangelands by reducing risk of catastrophic fire, increasing forage for domestic livestock, improving habitat for wildlife including mule deer, and general promotion of rangeland health.

Chaining and mastication are most frequently used to reduce cover of pinyon (*Pinus* spp.)-juniper (*Juniperus* spp.) and promote forbs, grasses, and shrubs in this cover type (Monaco and Gunnell 2020). Unlike chaining where woody debris is moved into piles, mastication allows rangeland managers to turn woody material into mulch and spread it onto the soil where it can facilitate positive responses from the plant community (Bybee et al. 2016; Havrilla et al. 2017; Monaco and Gunnell 2020). In recent years, efforts to reduce conifers from sagebrush rangelands have matched estimates of expansion (1.5% per year) for this cover type (Sankey and Germino 2008; Reinhart et al. 2020).

Mule deer respond to conifer removal positively when forage plants respond well and adequate concealment and thermal cover remains within or adjacent to treatment areas. Positive outcomes for mule deer were noted in some treated areas, whereas others showed no increase or even declines depending on response of plant communities and presence of adequate cover (Bombaci and Pejchar 2016). Because reduced use of chained habitats by mule deer has been noted beyond 120 m from cover, suggested guidelines include interspersing food and cover such that areas where chaining or mastication has occurred are no more than 200 m from cover (Fairchild 1999). Managers also need to provide mule deer with a diversity of forage options for consumption across seasons. Mule deer selected for mastication treatments in New Mexico during summer 1–4 years after treatment, but switched to patches > 4 years old in winter, presumably due to preference for herbaceous plants in the summer and browse species in the winter (Sorensen et al. 2020).

When deer populations are resource limited, vegetation treatments including mastication can increase population growth rates if plant communities respond favorably. Over-winter survival of 6-month old fawns in pinyon-juniper habitats treated with mastication and herbicide to control invasive grasses averaged 77% (SE = 8%) compared to 68% (SE = 11%) for areas without treatment or with treatment and no control of invasive grasses (Bergman et al. 2014a). Likewise, measures of condition for adult females (ingesta free body fat; Cook et al. 2007, 2010) were higher in areas with these same vegetation treatments (Bergman et al. 2014b). Similarly, mechanical removal of pinyon-juniper trees on a rangeland composed of perennial grasses interspersed with pinyon-juniper stands and limited shrublands influenced space-use patterns (smaller) and metrics of condition (greater) suggesting treatments improved forage for mule deer (Bender et al. 2013).

Conversely, many studies involving mule deer response to reduction of pinyon-juniper woodlands on rangelands have failed to identify a positive response (Bombaci and Pejchar 2016). Evidence that mule deer selected for juniper trees on sagebrush rangelands in the northwestern United States sparked a lively debate about the value of mastication projects for this species (Coe et al. 2018; Clark et al. 2019; Maestas et al. 2019). Coe et al. (2018) found selection for trees by mule deer at multiple spatial and temporal scales in a sagebrush-dominated rangeland and concluded that mastication of western juniper (*J. occidentalis*) may not improve habitat for this species. Maestas et al. (2019) countered that habitat selection was imprecise and unreliable, and response of mule deer to mastication could only be evaluated by looking at demographic responses (see Clark et al. 2019 for a rebuttal).

17.4.5 Vegetation Management—Mowing of Shrubs

Managers also treat rangelands dominated by shrubs where deer occur with mechanical implements such as a mower or Lawson aerator. Treatment of shrubs is often intended to reduce shrub cover in favor of herbaceous plants and create increased availability of forage for livestock and wildlife. Shrub response to vegetation treatments varies with species and environmental conditions. Positive outcomes have more potential in mountain big sagebrush (*A. tridentata vaseyana*) communities but have rarely been shown in Wyoming big sagebrush (*A. tridentata wyomingensis*) leading to calls for caution and more research (Beck et al. 2012). Consideration of seasonal use of rangelands by mule deer should also be considered in relation to treatment of shrubs. Mowing alters the structure and height of sagebrush with effects that can persist for more than 20 years (Davies et al. 2009). Consequently, mowing could have detrimental effects for mule deer if done on winter ranges where treated shrubs could then be covered in snow and unavailable (Davies et al. 2009).

17.4.6 Vegetation Treatment—Herbicide

Herbicides such as imazapic or Plateau® (BASF, Ludwigshafen, Germany), indaziflam or Rejuvra® (Bayer, Cary, NC, USA), and tebuthiuron or Spike® (Dow Agro-Sciences, Indianapolis, IN, USA) are also used for vegetation treatment where deer occur. Tebuthiuron has been used to thin browse species and can be applied aerially resulting in highly variable mortality rates for shrubs (Scifres et al. 1979). Imazapic and indaziflam are specific herbicides developed for control of annual grasses (Mealor et al. 2013). These two herbicides can reduce biomass of undesirable annual plants such as cheatgrass, medusahead (*Taeniatherum caput-medusae*), or red brome and they are often used in combination with mechanical methods to restore rangelands (Elseroad and Rudd 2011; Burnett and Mealor 2015).

Responses of mule deer to vegetation treatments with herbicide are variable and dependent on the response of preferred forage species. Treatment of sagebrush with Tebuthiuron resulted in greater crude protein in leaves compared to plants in both control plots and sagebrush that was mowed during the initial year following treatment, but increases were modest (Smith et al. 2022). These treatments may have improved palatability of sagebrush for mule deer because plant secondary metabolites (e.g., terpenes) were unchanged in relation to treatment while crude protein increased, but marginal increases likely do not compensate for loss of cover or density of plants in sagebrush systems (Smith et al. 2022). Herbicide treatments in forested habitats were associated with reduced bite sizes and reduced digestible energy, but increased digestible protein for black-tailed deer (Ulappa et al. 2020). Similarly, mule deer in the Great Plains selected for sites treated with tebuthiuron five years earlier, ostensibly due to improved quality of forage (Gage 2011). Mastication of pinyon-juniper followed by control of annual grasses with imazapic was associated with increased condition of adult female mule deer and increased survival of mule deer fawns compared to untreated areas and areas treated with mastication alone where invasive grasses were not controlled (Bergman et al. 2014a, b).

17.4.7 Water Development

Water is an essential element for all life on earth including deer. Water, however, is available in forms other than drinking water and this species does not always need to drink. Three forms of water are recognized including metabolic water, preformed water, and free or drinking water. Metabolic water is produced when compounds such as carbohydrates, fats, and proteins are oxidized (Robbins 1983). A single gram of carbohydrate, for example, produces over half (0.56 g) a gram of metabolic water and the conversion ratio is essentially one to one for fatty compounds (Gill 1994). Deer also consume preformed water which is available in plants. Average moisture content of plants consumed by mule deer can vary from < 10% for some seeds and senescent grasses to > 80% for forbs and succulent plants (Cain et al. 2008). When metabolic and preformed water are adequate to meet the needs of deer, they do not need to drink. When metabolic and preformed water are inadequate to meet physiological needs, mule deer access free or drinking water from natural sources such as springs or streams and anthropogenic sources such as water developments (Larsen et al. 2012).

Although some have called for more research on the influence of water developments on deer populations (Simpson et al. 2011), water development has the potential to benefit deer—particularly on arid rangelands. On arid rangelands, mule deer regularly use water developments including troughs and wells (Fig. 17.4; Krausman 2002). Mule deer visited water sources every 1–4 days and consumed between 1 and 6 L of water per visit with higher consumption rates occurring during the summer months (Hazam and Krausman 1988; Shields et al. 2012). Frequency of visits to water sources was higher for females compared to males—particularly during summer when females were lactating (Fig. 17.5; Hervert and Krausman 1986; Shields et al. 2012).

Fig. 17.4 Image of a male mule deer (*Odocoileus hemionus*) at a livestock water development designed as a rangeland improvement practice to increase availability and distribution of free or drinking water

Fig. 17.5 Image of a female mule deer (*Odocoileus hemionus*) with two neonates at a wildlife water development designed to increase density and distribution of this species. Note the fence posts, but lack of wire as it has been removed to facilitate access by this species

During summer months, mule deer in arid rangelands are often located closer (typically within 5 km) to sources of water than they are during other seasons (Ordway and Krausman 1986; Krausman and Etchberger 1995). Moreover, availability of water was a factor associated with migration in arid environments; mule deer migrated to areas with available water during the summer months (Rautenstrauch and Krausman 1989). During summer months, availability of water was also associated with reduced movements in arid rangelands suggesting that mule deer were able to meet their resource needs in smaller areas when water was available (McKee et al. 2015). Access to drinking water may allow mule deer in arid rangelands to consume a wider variety of forage plants including species with low pre-formed water content. Moreover, higher densities of mule deer in the arid Chihuahuan Desert of Mexico were noted near available drinking water, although those densities may reflect habitat selection as opposed to increased abundance (Sánchez-Rojas and Gallina 2000a, b). Nonetheless, many of the arid rangelands where mule deer occur are considered water-limited (Cox et al. 2009).

Water developments may receive very little use by mule deer in arid rangelands— particularly if they are newly constructed, fenced, or heavily used by livestock or feral equids. Water developments available for longer than 3 years received more use than those built more recently suggesting mule deer required some time to find and acclimate to newly constructed developments (Marshal et al. 2006a, b). Fencing that successfully deterred feral horses from water developments was also associated with reduced use by mule deer—particularly when the area fenced around water sources was relatively small (Larsen et al. 2011). When water developments are not fenced, feral equids may outcompete mule deer and other wildlife for access to drinking water (Fig. 17.6; Hall et al. 2016, 2018). Competition for water can be particularly acute at relatively small water sources where feral horses can monopolize access for most of a 24 h period (Hall et al. 2018).

Water developments have also been used to mitigate loss of water resources due to anthropogenic effects (e.g., urban, agricultural, transportation, industrial development; Rosenstock et al. 1999; Krausman et al. 2006). In Joshua Tree National Park, both the number of available springs and the volume of water flowing from those springs have declined over the last 50 years (Longshore et al. 2009). Wildlife water developments, however, were able to partially offset the predicted loss of suitable habitat for bighorn sheep over those same years (Longshore et al. 2009). Projected loss of naturally occurring water sources will make water development an increasingly important rangeland management practice for mule deer (Seager et al. 2013). To maximize value to mule deer in arid rangelands, water developments should be spaced at 3.2–4.8 km from other sources of water (Heffelfinger 2006; Krausman et al. 2006).

Fig. 17.6 Feral horses (*Equus caballus*) at a livestock water development. Feral horses have been shown to limit access to water for mule deer (*Odocoileus hemionus*) and other species at water sources

17.4.8 Predator Management

Mule deer are vulnerable to a suite of predators and predation is typically the most common identified cause of mortality across all age and sex classes, except during severe winters and in the presence of high harvest by hunters (Linnell et al. 1995; Unsworth et al. 1999; Bishop et al. 2005b; Forrester and Wittmer 2013). The relationship between deer and their predators, however, is complex and nuanced with predation rates that vary across age and sex categories, in relation to animal condition, with availability of alternative prey for predators, in response to habitat conditions (e.g., availability of hiding cover), and by type of predator. Coyote predation on neonates during the summer, for example, was lowest when abundance of microtine rodents was high (Hamlin et al. 1984). Predation can also be considered compensatory or additive depending on where deer populations are in relation to carrying capacity and condition of rangelands (Ballard et al. 2001; Forrester and Wittmer 2013).

Given the complex ecological relationships between deer and their predators, it isn't surprising that the results of predator control on population growth is highly variable and the literature equivocal. Removal of coyotes over a large area (> 10,500 km^2), for example, was not associated with increased production of mule deer as measured by fawn/adult female ratios (Brown and Conover 2011). Likewise, although removal of coyotes and mountain lions showed some short-term increases in survival rates depending on age class, it did not appreciably change the long-term dynamics for a population of mule deer on rangelands in southeastern Idaho (Hurley et al. 2011). Conversely, removal of wolves on Vancouver Island led to increased survival, higher production, and positive population growth rates for black-tailed deer (Hatter and Janz 1994). Similarly, removal of coyotes was associated with increased survival of neonates and population growth when that predator was removed over consecutive years from fawning habitat when the deer population had room to grow (McMillan et al.2023).

Variation in outcomes associated with deer populations (along with ungulates in general) in response to predator removal highlight uncertainty surrounding this management action, the need for better science, and the requisite nature of information on limiting factors affecting populations targeted for increase following predator removal (Clark and Hebblewhite 2020). Conditions under which predator control was effective at influencing population growth in deer include deer populations below carrying capacity, predation identified as limiting and additive, control efforts adequate to significantly reduce predator densities for the species exerting "top-down" control, and control efforts conducted at optimal spatial and temporal scales (Ballard et al. 2001; Forrester and Wittmer 2013; Mahoney et al. 2018). When these conditions are not present, reductions in predators are unlikely to influence population growth.

17.5   Impacts of Disease

Similar to other species, deer are susceptible to numerous pathogens including bacteria, fungi, parasites, prions, and viruses. Diseases associated with a bacterial vector include rain rot (often associated with ticks and flies), necrobacillosis, gangrene, keratoconjunctivitisrosis, and others (Mule Deer Working Group 2014). Many different parasites including bot flies, fleas, lice, round worms, tapeworms, ticks, and others also occur in mule deer (Mule Deer Working Group 2014). Chronic wasting disease (CWD) is an emerging disease in mule deer caused by an infectious prion (modified protein). This disease is similar to bovine spongiform encephalopathy (sometimes referred to as "mad cow disease") in cattle and scrapie in domestic sheep and is spreading rapidly in mule deer populations (Haley and Hoover 2015). Diseases caused by viral pathogens include hemorraghic diseases (blue tongue and epizootic hemorrhagic disease or EHD), and fibroma tumors caused by the papilloma virus (Mule Deer Working Group 2014). Up to 40% of sampled white-tailed deer were found to have antibodies for Covid-19 and mule deer are likely also susceptible (Chandler et al. 2021).

17.5.1   Impacts of Disease on Populations

Most of the diseases listed above affect individuals with little influence on populations. Exceptions with potential to influence population growth include the hemmoraghic diseases and chronic wasting disease. Hemmoraghic diseases result from a viral infection spread by an insect vector (several species of *Culicoides* midges). Infection usually occurs in late summer or early fall and can result in significant mortality—particularly at northern latitudes (Howerth et al. 2001). In one outbreak on rangelands in California, over 1,000 mule deer were estimated to have perished with pathology for a sample of these animals consistent with bluetongue or EHD (Woods et al. 1996).

Impacts to mule deer populations have also been observed with CWD which is a relatively new disease currently found in mule deer and spreading rapidly (Haley and Hoover 2015). Chronic wasting disease presents as degenerative because it affects the central nervous system of host animals. The source of CWD in mule deer is not completely understood, however, it seems likely it may have originated in north-central Colorado and southeastern Wyoming because original diagnosis in captive herds occurred there in the late 1970s (Williams et al. 2002). Prevalence of CWD in cervids has increased exponentially over the last 6 decades and it has now been detected in 4 Canadian provinces and 29 US states including many where mule deer occur (Otero et al. 2021).

Although no differences in susceptibility to this disease have been identified between female and male mule deer, prevalence of CWD is higher for males and increases with age, ostensibly due to behavioral differences and higher contact rates for mature males in this species (Miller and Conner 2005). In areas with high prevalence rates (> 20%), population-level impacts have been noted (DeVivo et al. 2017). Mule deer infected with CWD are more susceptible to predation and vehicle strikes and large differences in annual survival (e.g., 32% compared to 76%) and population growth rate (e.g., $\lambda = 0.79$ compared to $\lambda = 1.00$) of CWD positive versus CWD-negative animals has been noted (Miller et al. 2008; DeVivo et al. 2017).

17.5.2 Disease Interactions with Livestock

Livestock including cattle and domestic sheep can be infected with hemorrhagic diseases, but the disease is usually subclinical for these species (Howerth et al. 2001). The *Culicoides* midges that spread the virus associated with hemmoraghic diseases reproduce in water and some species (e.g., *C. sonorensis*) appear to thrive in stale waters or mud enriched by fecal material and urine from domestic livestock or wild animals (Pfannenstiel et al. 2015). Congregation of both livestock and wildlife including mule deer at water sources in the summer may help create favorable conditions for midges and spread of this disease (Pfannenstiel et al. 2015). Consequently management of livestock (e.g., stocking rate, group sizes) and water resources (e.g., protecting spring heads) may provide an option to reduce outbreaks of hemmoraghic diseases, but the etiology of the diseases are not completely understood and specific recommendations are not available (Pfannenstiel et al. 2015).

The prions associated with CWD can persist in the environment for years or even decades and contraction of the disease by mule deer from the environment is documented and perhaps more frequent than transmission between animals (Miller et al. 2004, 2006). Although natural transmission of CWD from mule deer to livestock has not been observed, passage of CWD to sheep can be induced via intracranial inoculation suggesting some potential to cross species barriers to domestic livestock (Cassman et al. 2021). Nonetheless, understanding of CWD and the potential for interactions with livestock are limited. Monitoring programs for CWD can be found in most states and provincial agencies. These programs often include collection of samples from harvested animals and are typically "hunter-based" (Smolko et al. 2021). Diagnostic tests for CWD have evolved and improved rapidly over the last few decades, but tests with high accuracy remain relegated to those where tissues (e.g., tonsil, lymph nodes) are collected postmortem and limitations persist for effective testing of live animals (Haley and Richt 2017). Despite several decades of research on CWD, much remains to be learned about the etiology of this disease and how to successfully manage it in mule deer populations on rangelands in western North America.

17.6   Ecosystem Threats

Despite a broad distribution (Fig. 17.1) and relatively high abundance, mule deer populations face many threats. The human population in western North America has grown rapidly over the last century with concomitant changes to landscapes including habitat loss, degradation, and fragmentation. In 2008, the estimated human footprint (physical area occupied by humans as housing, roads, intensively managed agricultural lands, and other infrastructure) covered 13% (402,000 km^2) of the land area in the Western United States (Leu et al. 2008). The size of this footprint has increased rapidly in recent decades. Between 1980 and 2010, for example, residential land-use increased by 37% and impacted more than 1,000,000 ha of western Colorado (Johnson et al. 2017). Likewise, the estimated land area occupied by energy extraction infrastructure (wells, well pads, roads, storage facilities) in the central portion of North America including many areas where mule deer occur on the eastern portion of their range increased by 3 million ha between 2000 and 2012 leading to an estimated loss of 10 Tg of plant biomass (Allred et al. 2015). Changes in land use and increases in habitat loss in western Colorado were associated with an average decrease of 0.5 fawns per 100 adult females per year with a strong negative trend noted in relation to increased residential development (Johnson et al. 2017).

Direct loss of habitat can reduce availability of forage for deer, but these effects are often magnified by avoidance of areas near human structures creating indirect effects. Mule deer, for example, avoided areas within 2.7 and 3.7 km of well pads during winter suggesting that indirect effects of energy extraction were greater than direct effects (Sawyer et al. 2010). Avoidance of areas near roads and well pads magnifies the impact of surface disturbance with recent estimates suggesting a multiplier of 4.6 should be applied for this species to account for indirect effects associated with avoidance of energy infrastructure (Dwinnell et al. 2019). Avoidance of wells and well pads was most pronounced during the active drilling phase when presence of humans, vehicles, noise, and artificial light was greatest (Northrup et al. 2021). Nonetheless, even after 15 years this species failed to habituate to energy infrastructure in some areas suggesting effects can be long-term and persistent (Sawyer et al. 2017).

Moreover, fragmentation of landscapes used by mule deer can disrupt migration routes and timing (Lendrum et al. 2013). Avoidance of human structures when selecting stopover areas, for example, has been documented for migratory mule deer (Wyckoff et al. 2018). Additionally, decreased movement efficiency and increased energy expenditure were noted for mule deer while migrating through areas impacted by surface mining potentially leading to fitness consequences (Blum et al. 2015). Likewise, there can be genetic consequences of habitat fragmentation evidenced by genetic structure that corresponded to highway boundaries (Fraser et al. 2019). Impacts including avoidance of areas disturbed by energy development have been observed for migratory mule deer after surface disturbance reached only 3% of rangelands (Sawyer et al. 2020; Lambert et al. 2022). Disruptions to migratory routes and reductions in fitness for deer migrating across disturbed and fragmented landscapes could lead to complete loss of migratory knowledge by populations if information on routes and timing is transmitted culturally in this species as it is with other ungulates (Jesmer et al. 2018).

Additional indirect effects associated with growth in human populations in western North America include altered dynamics between deer and their predators and responses to increased recreation by humans on rangelands. Interactions between mule deer and mountain lions, for example, were influenced by urbanization and presence of artificial light (Benson et al. 2016; Ditmer et al. 2021). These altered interactions may trigger trophic cascades that lead to changes in plant composition on rangelands (Waser et al. 2014). Additionally, hikers and in particular—hikers off established trails with dogs—have been shown to influence mule deer with increased vigilance and energy expenditure common responses (Miller et al. 2001). Likewise, approximately 1/3rd of mule deer with GPS collars responded to people searching for and collecting shed antlers by leaving established home ranges which increases energy expenditure and may lead to increased predation risk (Bates et al. 2021).

17.6.1 Climate Change

Deer demonstrate a wide thermal tolerance zone which is reflected in their broad distribution (Wallmo 1981). The thermo-neutral zone for mule deer has been estimated at operative temperatures (temperatures experienced by the animal after accounting for wind and radiation) between $-20\ °C$ and $5\ °C$ ($-4\ °F$ to $41\ °F$) in winter and $< 25\ °C$ ($77\ °F$) in summer (Parker and Robbins 1994). Outside of these operative temperatures, mule deer must alter behavior (i.e., seek shade) or use water and energy to maintain homeostasis (Parker and Robbins 1984). As temperatures warm under predicted climate change scenarios, deer may be forced to adjust behavior and alter resource selection. Because operative temperatures experienced by deer are strongly influenced by cover, changes to rangelands (e.g., conversion from shrubland to grasslands due to increased fire frequency) that result in reduced cover may exacerbate the effect of increased temperatures (Parker and Gillingham 1990). Consequently, the relative role of thermal cover and water to resource selection, for example, may increase for this species and rangelands with limited thermal cover may require mule deer to expend more energy to meet their needs, which may influence fitness.

Nonetheless, the biggest impacts to this species from climate change will likely be indirect. Massive conifer mortality across much of western North America, for example, is predicted due to increased temperatures (McDowell et al. 2016). Increased temperatures will likely result in more frequent and larger wildfires (Schoennagel et al. 2017). Increased frequency and severity of wildfires, in turn, will lead to loss of forests and shrublands in favor of grasslands—particularly where annual grasses such as cheatgrass have invaded. Once established, annual grasses such as cheatgrass prolong the fire season and increase availability of fine fuels in a negative feedback loop that leads to more frequent fire that furthers conversion of shrublands and forests to rangelands dominated by annual grasses. Conversion of rangelands dominated by forests and shrubs to grass-dominated systems will have far-reaching and cascading consequences for mule deer. These changes are likely to favor more generalist ungulates (e.g., elk) which may exacerbate potential competition with mule deer.

Moreover, long-standing benefits associated with migration for mule deer including prolonged access to nutritious forage for migrating animals are likely to be reduced or eliminated (Aikens et al. 2020). Massive losses in surface water are also predicted for western North America which will reduce availability of free or drinking water for this species and likely increase competition at remaining water sources where mule deer are at a competitive disadvantage to feral horses (Hall et al. 2018). Increased aridity and drought on rangelands in western North America may also influence disease dynamics associated with hemorrhagic diseases as outbreaks typically occur in summer and are often most severe during drought years. Increased temperatures may congregate animals at water sources and favor conditions for the midges associated with transmission of this disease. Indirect effects including ecosystem change associated with climate change represent a serious threat to mule deer populations across their range.

17.7 Conservation and Management Actions

Deer have a long history of management by provincial and state agencies. Agencies typically develop conservation plans to guide management decisions with spatial resolution for specific plans often at the population level (i.e., individual conservation or management plans for specific herd units). Conservation and management planning, however, can be challenged by migratory mule deer that regularly cross administrative boundaries (Middleton et al. 2019). In many jurisdictions, habitat restoration efforts are also coordinated between provincial or state management agencies and land owners. Mule deer are of tremendous interest to the public and many non-governmental organizations work closely with provincial, state, tribal, and federal agencies to conserve and manage this species. Most state and provincial agencies have active programs to conserve and restore habitat for mule deer.

In recent decades, many of these programs have focused on preservation and restoration of winter ranges. The working hypothesis for many mule deer populations is that they are nutritionally limited and many in particular across the northern portion of their distribution are thought to be limited by quality of winter range (Bergman et al. 2015). Consequently, millions of dollars have been spent to conserve and restore rangelands used by mule deer during winter. Likewise, large investments in restoration of natural springs and construction or maintenance of wildlife water developments for mule deer have occurred in more arid portions of their range. Total expenditures for development and maintenance of wildlife water developments across western states in the US exceeded $1,000,000 in 1999 (Rosenstock et al. 1999). Recent work has also highlighted the importance of maternal condition on population growth suggesting that populations may also be limited by quality of summer range (Lamb et al. 2023).

Most recently, advances with tracking technology including GPS collars have allowed for unprecedented understanding of migration ecology for this species including the ability to identify migration routes, stopover areas, and barriers to migration. State and provincial agencies have been able to use this information to construct highway crossing structures (e.g., underpasses and overpasses) which have successfully facilitated crossing of major roadways (Sawyer et al. 2012). Likewise, information derived from GPS collars has been valuable for conservation and management of populations that use rangelands on multiple land ownerships throughout the year (Middleton et al. 2019).

Mule deer are of tremendous interest to the public and many non-governmental organizations (e.g., Mule Deer Foundation). These conservation organizations provide resources including funding, equipment, and volunteers that can be leveraged with state or federal resources to complete more conservation projects at the larger scales needed to benefit this species. Engagement with the public and these organizations including local stake holders is crucial to sustain conservation actions (Middleton et al. 2019). Furthermore, production and sharing of information with the public in a format that is easily accessible to a diverse group of constituents and stakeholders can facilitate conservation planning and implementation of management actions that benefit mule deer (Middleton et al. 2019). Although the challenges to long-term conservation of mule deer are immense, so is the interest amongst the public associated with this species.

17.8 Research/Management Needs

Despite a long history of management and research on mule deer in North America, much remains to be learned about this species. Tremendous variation in habitat selection and migratory behavior exists across populations of mule deer and thus information at relatively fine spatial scales is often needed for effective conservation and management actions—particularly in the face of rapid growth of human populations in western North America. Information on the response of mule deer to landscape changes currently influencing rangelands is urgently needed for conservation and management of this species. Recent research on response of mule deer to oil and gas development, for example, suggests impacts when as little as 3% of the landscape is disturbed (Sawyer et al. 2020), but no information is available regarding the applicability of that threshold to renewable energy. Thus, information on response of mule deer to solar or wind energy developments would help with conservation planning for this species. Likewise, we lack clear understanding of the impacts to mule deer from recreational activities occurring on rangelands. Increased vigilance and some displacement of individuals has been noted (Miller et al. 2001; Bates et al. 2021), but these effects have not been tied to maternal condition, demographics, or population growth.

It is also clear that there is potential for the management of co-occuring species and the rangelands where deer occur to influence this species. There is a need to better understand the potential for apparent competition between co-occuring ungulates and mule deer. Apparent competition occurs when top-down effects on a population are mediated by a common predator. For example, mule deer are a preferred food source by a generalist predator (e.g., mountain lion, Cooley et al. 2008). If co-occurring species of ungulates such as elk or feral equids provide a supplemental food source such that mountain lions can maintain a higher population size or even maintain a consistent population size during times of declining abundance for mule deer, then there is potential for coexisting ungulates to have an indirect effect on mule deer. Indeed, there is some evidence for such an interaction between mule deer and white-tailed deer (Wielgus 2017), but no information on mule deer with elk or feral equids.

Additionally, aggressive measures to reduce prevalence of CWD have been implemented or proposed in many areas including altered hunt structures to impose increased harvest on mature males. Many of the proposed mitigation measures, however, are untested and deeply unpopular with consumptive users. Spread of CWD has increased exponentially since it was discovered and this disease shows little sign of abating (Otero et al. 2021). Chronic wasting disease has the potential to disrupt management of deer populations along with funding to provincial and state governments. Thus, there is an urgent need to identify measures to mitigate the spread of CWD.

Land managers across western North America are faced with increasing challenges to rangelands due to invasive plants, increased frequency and severity of wildfires, and climate change with all of the interacting and cascading effects. Improved understanding of how mule deer, and their habitats are likely to respond to these disturbances including climate change would help with conservation planning. More importantly, however, are solutions that can be employed on rangelands to mitigate and restore these ecosystems. Land managers would also benefit from help with prioritization so that resources can be maximized for conservation value. In recent decades, for example, substantial energy, effort, and money have been invested in restoration of winter ranges for mule deer, but it is unclear if those efforts represent the best use of limited resources. Some argument can be made that restoration of summer habitats may impact population growth of mule deer as much or more than restoration of winter habitat (Clements and Young 1997). Consequently land managers would benefit from increased understanding of the relative value of summer versus winter habitat quality on population growth in deer.

References

Aikens EO, Monteith KL, Merkle JA, Dwinnell SPH, Fralick GL, Kauffman MJ (2020) Drought reshuffles plant phenology and reduces the foraging benefit of green-wave surfing for a migratory ungulate. Glob Chang Biol 26:4215–4225. https://doi.org/10.1111/gcb.15169

Allred BW, Smith WK, Twidwell D, Haggerty JH, Running SW, Naugle DE, Fuhlendorf SD (2015) Ecosystem services lost to oil and gas in North America. Science 348:401–402. https://doi.org/10.1126/science.aaa4785

Alpe MJ, Kingery JL, Mosley JC (1999) Effects of summer sheep grazing on browse nutritive quality in autumn and winter. J Wildl Manage 63:346–354. https://doi.org/10.2307/3802518

Anderson AE (1981) Morphology and physiological characteristics. In Wallmo OC (ed) Mule Deer and Black-tailed Deer of North America. University of Nebraska Press, Lincoln, Nebraska, USA pp 27–97.

Anthony RG, Smith NS (1977) Ecological relationships between mule deer and white-tailed deer in southeastern Arizona. Ecol Monogr 47:255–277. https://doi.org/10.2307/1942517

Austin DD (2000) Managing livestock grazing for mule deer (*Odocoileus hemionus*) on winter range in the Great Basin. West N Am Nat 60:198–203. https://www.jstor.org/stable/41717030

Austin DD, Urness PJ (1986) Effects of cattle grazing on mule deer diet and area selection. J Range Manage 39:18–21. https://doi.org/10.2307/3899678

Ballard WB, Lutz D, Keegan TW, Carpenter LH, deVos Jr JC (2001) Deer-predator relationships: a review of recent North American studies with emphasis on mule and black-tailed deer. Wildl Soc Bull 29:99–115. https://www.jstor.org/stable/3783986

Bartmann RM, Carpenter LH, Garrott RA, Bowden DC (1986) Accuracy of helicopter counts of mule deer in pinyon-juniper woodland. Wildl Soc Bull 14:356–363. https://www.jstor.org/stable/3782266

Bartmann RM, White GC, Carpenter LH (1992) Compensatory mortality in a Colorado mule deer population. Wildl Monogr 121:3–39. https://www.jstor.org/stable/pdf/3830602.pdf

Bartuszevige AM, Endress BA (2008) Do ungulates facilitate native and exotic plant spread? Seed dispersal by cattle, elk, and deer in northeastern Oregon. J Arid Environ 72:904–913. https://doi.org/10.1016/j.jaridenv.2007.11.007

Bates SB, Whiting JC, Larsen RT (2021) Comparison of effects of shed antler hunting and helicopter surveys on ungulate movements and space use. J Wildl Manage 85:437–448. https://doi.org/10.1002/jwmg.22008

Beck JL, Peek JM (2005) Diet composition, forage selection, and potential for forage competition among elk, deer, and livestock on aspen-sagebrush summer range. Rangel Ecol Manag 58:135–147. https://doi.org/10.2111/03-13.1

Beck JL, Connelly JW, Wambolt CL (2012) Consequences of treating Wyoming big sagebrush to enhance wildlife habitats. Rangel Ecol Manag 65:444–455. https://doi.org/10.2111/REM-D-10-00123.1

Bender LC, Hoenes BD, Rodden CL (2012) Factors influencing survival of desert mule deer in the greater San Andres Mountains, New Mexico. Hum Wildl Interact 6:245–260. https://doi.org/10.26077/h5bg-1829

Bender LC, Boren JC, Halbritter H, Cox S (2013) Effects of site characteristics, pinyon-juniper management, and precipitation on habitat quality for mule deer in New Mexico. Hum Wildl Interact 7:47–59. https://doi.org/10.26077/4kq0-y179

Benson JF, Sikich JA, Riley SPD (2016) Individual and population level resource selection patterns of mountain lions preying on mule deer along an urban-wildland gradient. PLoS ONE 11(7):e0158006. https://doi.org/10.1371/journal.pone.0158006

Bergman EJ, Watkins BE, Bishop CJ, Lukacs PM, Lloyd M (2011) Biological and socio-economic effects of statewide limitation of deer licenses in Colorado. J Wildl Manage 75:1443–1452. https://doi.org/10.1002/jwmg.168

Bergman EJ, Bishop CJ, Freddy DJ, White GC, Doherty PF Jr (2014a) Habitat management influences overwinter survival of mule deer fawns in Colorado. J Wildl Manage 78:448–455. https://doi.org/10.1002/jwmg.683

Bergman EJ, Doherty PF Jr, Bishop CJ, Wolfe LL, Banulis BA (2014b) Herbivore body condition response in altered environments: mule deer and habitat management. PLoS ONE 9:e106374. https://doi.org/10.1371/journal.pone.0106374

Bergman EJ, Doherty PF Jr, White GC, Holland AA (2015) Density dependence in mule deer: a review of the evidence. Wildl Biol 21:18–29. https://doi.org/10.2981/wlb.00012

Berry SL, Shipley LA, Long RA, Loggers C (2019) Differences in dietary niche and foraging behavior of sympatric mule and white-tailed deer. Ecosphere 10(7):e02815. https://doi.org/10.1002/ecs2.2815

Binkley D, Moore MM, Romme WH, Brown PM (2006) Was Aldo Leopold right about the Kaibab deer herd? Ecosystems 9:227–241. https://doi.org/10.1007/s10021-005-0100-z

Bishop CJ, White GC, Freddy DJ, Watkins BE (2005a) Effect of limited antlered harvest on mule deer sex and age ratios. Wildl Soc Bull 33:662–668. https://www.jstor.org/stable/3785094

Bishop CJ, Unsworth JW, Garton EO (2005b) Mule deer survival among adjacent populations in southwest Idaho. J Wildl Manage 69:311–321. https://www.jstor.org/stable/3803607

Block WM, Conner LM, Brewer PA, Ford P, Haufler J, Litt A, Masters RE, Mitchell LR, Park J (2016) Effects of prescribed fire on wildlife and wildlife habitat in selected ecosystems of North America. The Wildlife Society Technical Review 16–01. The Wildlife Society, Bethesda, Maryland, USA. https://wildlife.org/wp-content/uploads/2014/05/TechManual16-01FINAL.pdf

Blum ME, Stewart KM, Schroeder C (2015) Effects of large-scale gold mining on migratory behavior of a large herbivore. Ecosphere 6:74. https://doi.org/10.1890/ES14-00421.1

Bombaci S, Pejchar L (2016) Consequences of pinyon and juniper woodland reduction for wildlife in North America. Rangel Ecol Manag 365:34–50. https://doi.org/10.1016/j.foreco.2016.01.018

Bowyer RT (1991) Timing of parturition and lactation in southern mule deer. J Mammal 72:138–145. https://doi.org/10.2307/1381988

Bristow KD, Harding LE, Lucas RW, McCall TC (2020) Influence of fire severity and vegetation treatments on mule deer (*Odocoileus hemionus*) winter habitat use on the Kaibab Plateau, Arizona. Anim Prod Sci 60:1292–1302. https://doi.org/10.1071/AN19373

Brown DE, Conover MR (2011) Effects of large-scale removal of coyotes on pronghorn and mule deer productivity and abundance. J Wildl Manage 75:876–882. https://doi.org/10.1002/jwmg.126

Brown CL, Smith JB, Wisdom MJ, Rowland MM, Spitz DB, Clark DA (2020) Evaluating indirect effects of hunting on mule deer spatial behavior. J Wildl Manage 84:1246–1255. https://doi.org/10.1002/jwmg.21916

Buckland ST, Anderson DR, Burnham KP, Laake JL (1993) Distance sampling: estimating abundance of biological populations. Chapman and Hall, London, United Kingdom, p 446

Burnett SA, Mealor BA (2015) Imazapic effects on competition dynamics between native perennial grasses and downy brome (*Bromus tectorum*). Invasive Plant Sci Manag 8:72–80. https://doi.org/10.1614/IPSM-D-14-00032.1

Bybee J, Roundy BA, Young KR, Hulet A, Roundy DB, Crook L, Aanderud Z, Eggett DL, Cline NL (2016) Vegetation response to pinon and juniper tree shredding. Rangel Ecol Manag 69:224–234. https://doi.org/10.1016/j.rama.2016.01.007

Cain JW, Krausman PR, Morgart JR, Jansen BD, Pepper MP (2008) Responses of desert bighorn sheep to removal of water sources. Wildl Monogr 171:1–32. https://doi.org/10.2193/2007-209

Campbell EG, Johnson RL (1983) Food habits of mountain goats, mule deer, and cattle on Chopaka Mountain, Washington, 1977–1980. J Range Manage 36:488–491. https://doi.org/10.2307/3897949

Cassman ED, Frese RD, Greenlee JJ (2021) Second passage of chronic wasting disease of mule deer to sheep by intracranial inoculation compared to classical scrapie. J Vet Diagn Invest 33:711–720. https://doi.org/10.1177/10406387211017615

Castleton KB (2002) Petroglyphs and pictographs of Utah, Utah Museum of Natural History, Salt Lake City, Utah, p 232

Caughley G (1970) Eruption of ungulate populations, with emphasis on Himalayan Thar in New Zealand. Ecology 51:53–72. https://doi.org/10.2307/1933599

Chaikina NA, Ruckstuhl KE (2006) The effect of cattle on native ungulates: the good, the bad, and the ugly. Rangelands 28:8–14. https://doi.org/10.2111/1551-501X(2006)28[8:TEOCGO]2.0.CO;2

Chandler JC, Bevins SN, Ellis JW, Linder TJ, Tell RM, Jenkins-Moore M, Root JJ, Lenoch JB, Robbe-Austerman S, DeLiberto TJ, Gidlewski T, Torchetti MK, Shriner SA (2021) SARS-CoV-2 exposure in wild white-tailed deer (*Odocoileus virginianus*). PNAS 118(47):e2114828118. https://doi.org/10.1073/pnas.211482811

Chuong J, Huxley J, Spotswood EN, Nichols L, Mariotte P, Suding KN (2016) Cattle as dispersal vectors of invasive and introduced plants in a California annual grassland. Rangel Ecol Manag 69:52–58. https://doi.org/10.1016/j.rama.2015.10.009

Clark TJ, Hebblewhite M (2020) Predator control may not increase ungulate populations in the future: a formal meta-analysis. J Appl Ecol 58:812–824. https://doi.org/10.1111/1365-2664.13810

Clark DA, Coe PK, Gregory SC, Hedrick MJ, Johnson BK, Jackson DH (2019) Habitat use informs species needs and management: a reply to Maestas et al. J Wildl Manage 83:762–766. https://www.jstor.org/stable/26689577

Clegg K (1994) Density and feeding habits of elk and deer in relation to livestock disturbance. Utah State University, Logan, USA, Thesis

Clements CD, Young JA (1997) A viewpoint: rangeland health and mule deer habitat. J Range Manage 50:129–138. https://doi.org/10.2307/4002369

Clevenger AP, Waltho N (2000) Factors influencing the effectiveness of wildlife underpasses in Banff National Park, Alberta, Canada. Conserv Bio 14:47–56. https://doi.org/10.1046/j.1523-1739.2000.00099-085.x

Coe PK, Clark DA, Nielson RM, Gregory SC, Cupples JB, Hedrick MJ, Johnson BK, Jackson DH (2018) Multiscale models of habitat use by mule deer in winter. J Wildl Manage 82:1285–1299. https://doi.org/10.1002/jwmg.21484

Conner MM, Wood ME, Hubbs A, Binfet J, Holland AA, Meduna LR, Roug A, Runge JP, Nordeen TD, Pybus MJ, Miller MW (2021) The relationship between harvest management and chronic wasting disease prevalence trends in western mule deer (*Odocoileus hemionus*) herds. J Wildl Dis 57:831–843. https://doi.org/10.7589/JWD-D-20-00226

Cook RC, Stephenson TR, Myers WL, Cook JB, Shipley LA (2007) Validating predictive models of nutritional condition for mule deer. J Wildl Manage 71:1934–1943. https://doi.org/10.2193/2006-262

Cook RC, Cook JG, Stephenson TR, Myers WL, McCorquodale SM, Vales DJ, Irwin LI, Hall PB, Spencer RD, Murphie SL, Schoenecker KA, Miller PJ (2010) Revisions of rump fat and body scoring indices for deer, elk, and moose. J Wildl Manage 74:880–896. https://doi.org/10.2193/2009-031

Cooley HS, Robinson HS, Wielgus RB, Lambert CS (2008) Cougar prey selection in a white-tailed deer and mule deer community. J Wildl Manage 72:99–106. https://doi.org/10.2193/2007-060

Cox M, Lutz DW, Wasley T, Fleming M, Compton BB, Keegan T, Stroud D, Kilpatrick S, Gray K, Carlson J, Carpenter L, Urquhart K, Johnson B, McLaughlin C (2009) Habitat guidelines for mule deer: intermountain West Ecoregion. Mule Deer Working Group, Western Association of Fish and Wildlife Agencies, Boise, Idaho, USA

Cronin MA (1991) Mitochondrial and nuclear genetic relationships of deer (Odocoileus spp.) in western North America. Can J Zool 69:1270–1279. https://doi.org/10.1139/z91-179

D'Antonio CM, Vitousek PM (1992) Biological invasions by exotic grasses, the grass/fire cycle and global change. Annu Rev Ecol Syst 23:63–87. https://www.jstor.org/stable/2097282

Davies KW, Bates JD, Johnson DD, Nafus AM (2009) Influence of mowing Artemisia tridentata ssp. wyomingensis on winter habitat for wildlife. Env Manage 44:84–92. https://doi.org/10.1007/s00267-008-9258-4

Davies KW, Boyd CS (2019) Ecological effects of free-roaming horses in North American rangelands. Bioscience 69:558–565. https://doi.org/10.1093/biosci/biz060

Dennison PE, Brewer SC, Arnold JD, Moritz MA (2014) Large wildfire trends in the western United States, 1984–2011. Geophys Res Lett 41:2928–2933. https://doi.org/10.1002/2014GL059576

DeVivo MT, Edmunds DR, Kauffman MJ, Schumaker BA, Binfet J, Kreeger TJ, Richards BJ, Schatzl HM, Cornish TE (2017) Endemic chronic wasting disease causes mule deer population decline in Wyoming. PLoS ONE 12:e0186512. https://doi.org/10.1371/journal.pone.0186512

Diamond JM, Call CA, Devoe N (2009) Effects of targeted cattle grazing on fire behavior of cheatgrass-dominated rangeland in the northern Great Basin, USA. Int J Wildland Fire 18:944–950. https://doi.org/10.1071/WF08075

Ditmer MA, Stoner DC, Francis CD, Barber JR, Forester JD, Choate DM, Ironside KE, Longshore KM, Hersey KR, Larsen RT, McMillan BR, Olson DD, Andreasen AM, Beckmann JP, Holton PB, Messmer TA, Carter NH (2021) Artificial nightlight alters the predator-prey dynamics of an apex carnivore. Ecography 44:149–161. https://doi.org/10.1111/ecog.05251

Dwinnell SP, Sawyer H, Randall JE, Beck JL, Forbey JS, Fralick GL, Monteith KL (2019) Where to forage when afraid: Does perceived risk impair use of the foodscape? Ecol Appl 29:9(7):e01972. https://doi.org/10.1002/eap.1972

Dziba LE, Provenza FD, Villalba JJ, Atwood SB (2007) Supplemental energy and protein increase use of sagebrush by sheep. Small Rumin Res 69:203–207. https://doi.org/10.1016/j.smallrumres.2005.12.013

Elseroad AC, Rudd NT (2011) Can imazapic increase native species abundance in cheatgrass (Bromus tectorum) invaded native plant communities? Rangel Ecol Manag 64:641–648. https://doi.org/10.2111/REM-D-10-00163.1

Fairchild JA (1999) Pinyon-juniper chaining design guidelines for big game winter range enhancement projects. In: Monsen SB, Stevens R (compilers) Proceedings: ecology and management of pinyon-juniper communities within the Interior West, RMRS-P-9. United States Department of Agriculture, Forest Service, Rocky Mountain Research Station, Ogden, Utah, USA, pp 278–280

Forrester TD, Wittmer HU (2013) A review of the population dynamics of mule deer and black-tailed deer Odocoileus hemionus in North America. Mamm Rev 43:292–308. https://doi.org/10.1111/mam.12002

Fraser DL, Ironside K, Wayne RK, Boydston EE (2019) Connectivity of mule deer (Odocoileus hemionus) populations in a highly fragmented urban landscape. Landsc Ecol 34:1097–1115. https://doi.org/10.1007/s10980-019-00824-9

Freddy DJ, White GC, Kneeland MC, Kahn RH, Unsworth JW, deVergie WJ, Graham VK, Ellenberger JH, Wagner CH (2004) How many mule deer are there? Challenges of credibility in Colorado. Wildl Soc Bull 32:916–927. https://www.jstor.org/stable/3784816

Freeland WJ, Janzen DH (1974) Strategies in herbivory by mammals: the role of plant secondary compounds. Am Nat 108:269–289. https://doi.org/10.1086/282907

Freeman ED, Larsen RT, Peterson ME, Anderson CR, Hersey KR, McMillan BR (2014) Effects of male-biased harvest on mule deer: Implications for rates of pregnancy, synchrony, and timing of parturition. Wildl Soc Bull 38:806–811. https://doi.org/10.1002/wsb.450

Furnas BJ, Landers RH, Hill S, Itoga SS, Sacks BN (2018) Integrated modeling to estimate population size and composition of mule deer. J Wildl Manage 82:1429–1441. https://doi.org/10.1002/jwmg.21507

Gage RT (2011) Effects of spike 20p on habitat use and movement of mule deer and other wildlife in Trans-Pecos. Thesis, Sul Ross State University, Alpine, Texas, USA, Texas

Gaillard J-M, Festa-Bianchet M, Yoccoz NG, Loison A, Toïgo C (2000) Temporal variation in fitness components and population dynamics of large herbivores. Annu Rev Ecol Syst 31:367–393. https://doi.org/10.1146/annurev.ecolsys.31.1.367

Geist V (1998) Deer of world: their evolution, behavior, and ecology. Stackpole Books, Mechanicsburg, Pennsylvania USA, p 421

Gill RB, Beck TDI, Bishop CJ, Freddy DJ, Hobbs NT, Kahn RH, Miller MW, Pojar TM, White GC (1999) Declining mule deer populations in Colorado: reasons and responses. pp. 54

Gill FB (1994) Ornithology, Second edition. W.H. Freeman, New York, New York, USA, pp 763

Gruell GE (1986) Post-1900 mule deer irruptions in the Intermountain West: principle cause and influences. General Technical Report INT-206. United States Department of Agriculture, Forest Service, p 44

Hagerman AE, Robbins CT (1993) Specificity of tannin-binding salivary proteins relative to diet selection by mammals. Can J Zool 71:628–633. https://doi.org/10.1139/z93-085

Haley NJ, Hoover EA (2015) Chronic wasting disease of cervids: current knowledge and future perspectives. Annu Rev Anim Biosci 3:305–325. https://doi.org/10.1146/annurev-animal-022 114-111001

Haley NJ, Richt JA (2017) Evolution of diagnostic tests for chronic wasting disease, a naturally occurring prion disease of cervids. Pathogens 6:35. https://doi.org/10.3390/pathogens6030035

Hall LK, Larsen RT, Westover MD, Day CC, Knight RN, McMillan BR (2016) Influence of exotic horses on the use of water by communities of native wildlife in a semi-arid environment. J Arid Environ 127:100–105. https://doi.org/10.1016/j.jaridenv.2015.11.008

Hall LK, Larsen RT, Knight RN, McMillan BR (2018) Feral horses influence both spatial and temporal patterns of water use by native ungulates in a semi-arid environment. Ecosphere 9:e02096. https://doi.org/10.1002/ecs2.2096

Hamlin KL, Riley SJ, Pyrah D, Dood AR, Mackie RJ (1984) Relationships among mule deer fawn mortality, coyotes, and alternate prey species during summer. J Wildl Manage 48:489–499. https://doi.org/10.2307/3801181

Hansen RM, Reid LD (1975) Diet overlap of deer, elk, and cattle in southern Colorado. J Range Manag 28:43–47. https://doi.org/10.2307/3897577

Harrington JL, Conover MR (2006) Characteristics of ungulate behavior and mortality associated with wire fences. Wildlife Soc Bull 34:1295–1305. https://doi.org/10.2193/0091-7648(200 6)34[1295:COUBAM]2.0.CO;2

Havrilla CA, Faist AM, Barger NN (2017) Understory plant community responses to fuel-reduction treatments and seeding in an upland Piñon-Juniper woodland. Rangel Ecol Manag 70:609–620. https://doi.org/10.1016/j.rama.2017.04.002

Hatter IW, Janz DW (1994) Apparent demographic changes in black-tailed deer associated with wolf control on northern Vancouver Island. Can J Zool 72:878–884. https://doi.org/10.1139/z94-119

Hays WJ (1869) The mule deer. Am Nat 3:180–181. https://doi.org/10.1086/270403

Hazam JE, Krausman PR (1988) Measuring water consumption of desert mule deer. J Wildl Manage 52:528–534. https://doi.org/10.2307/3801605

Heffelfinger JR (2006) Deer of the Southwest. Texas A&M University Press, College Station, USA, p 282

Heffelfinger JR, Latch EK (2023) Origin, Classification, and Distribution. In: Heffelfinger JR, Krausman PR (eds) Ecology and Management of Black-tailed and Mule Deer in North America. In Press. CRC Press, Boca Raton, Florida, USA

Heffelfinger JR, Brewer C, Alcalá-Galván CH, Hale B, Weybright DL, Wakeling BF, Carpenter LH, Dodd NL (2006) Habitat guidelines for mule deer: Southwest Deserts Ecoregion. Mule Deer Working Group, Western Association of Fish and Wildlife Agencies, Boise, Idaho, USA

Hervert JJ, Krausman PR (1986) Desert mule deer use of water developments in Arizona. J Wildl Manage 50:670–676. https://doi.org/10.2307/3800979

Hobbs NT, Spowart RA (1984) Effects of prescribed fire on nutrition of mountain sheep and mule deer during winter and spring. J Wildl Manage 48:551–560. https://doi.org/10.2307/3801188

Hofmann RR (1989) Evolutionary steps of ecophysiological adaptation and diversification of ruminants: a comparative view of their digestive system. Oecologia 78:443–457. https://doi.org/10.1007/BF00378733

Holl SA, Bleich VC (2010) Responses of bighorn sheep and mule deer to fire and rain in the San Gabriel Mountains, California. Biennial Symposium of the Northern Wild Sheep and Goat Council 17:139–157

Howerth EW, Stallknecht DE, Kirkland PD (2001) Blue-tongue, epizootic hemorrhagic disease, and other orbivirus-related diseases. In: Williams ES, Barker IK (eds) Infectious diseases of wild mammals. Iowa State University Press, Ames, Iowa, pp 77–97

Hurley MA, Unsworth JW, Zager P, Hebblewhite M, Garton EO, Montgomery DM, Skalski JR, Maycock CL (2011) Demographic response of mule deer to experimental reduction of coyotes and mountain lions in southeastern Idaho. Wildl Monogr 178:1–33. https://doi.org/10.1002/wmon.4

Jachowski DS, Kauffman MJ, Jesmer BR, Sawyer H, Millspaugh JJ (2018) Integrating physiological stress into the movement ecology of migratory ungulates: a spatial analysis with mule deer. Conserv Physiol 6(1). https://doi.org/10.1093/conphys/coy054

Jensen CH, Smith AD, Scotter GW (1972) Guidelines for grazing sheep on rangelands used by big game in winter. J Range Manag 25:346–352. https://doi.org/10.2307/3896543

Jesmer BR, Merkle JA, Goheen JR, Aikens EO, Beck JL, Courtemanch AB, Hurley MA, McWhirter DE, Miyasaki HM, Monteith KL, Kauffman MJ (2018) Is ungulate migration culturally transmitted? Evidence of social learning from translocated animals. Science 361:1023–1025. https://doi.org/10.1126/science.aat098

Johnson HE, Sushinsky JR, Holland A, Bergman EJ, Balzer T, Garner J, Reed SE (2017) Increases in residential and energy development are associated with reductions in recruitment for a large ungulate. Glob Change Biol 23(2) 578–591. https://doi.org/10.1111/gcb.13385

Jolly WM, Cochrane MA, Freeborn PH, Holden ZA, Brown TJ, Williamson GJ, Bowman DMJS (2015) Climate-induced variations in global wildfire danger from 1979 to 2013. Nat Commun 6:7537. https://doi.org/10.1038/ncomms8537

Jones PF, Jakes AF, MacDonald AM, Hanlon JA, Eacker DR, Martin BH, Hebblewhite M (2020) Evaluating responses by sympatric ungulates to fence modifications across the Northern Great Plains. Wildl Soc Bull 44:130–141. https://doi.org/10.1002/wsb.1067

Kay CE (2007) Were native people keystone predators? A continuous-time analysis of wildlife observations made by Lewis and Clark in 1804–1806. Can Field Nat 121:1–16. https://doi.org/10.22621/cfn.v121i1.386

Keegan TW, Ackerman BB, Aoude AN, Bender LC, Boudreau T, Carpenter LH, Compton BB, Elmer M, Heffelfinger JR, Lutz DW, Trindle BD, Wakeling BF, Watkins BE (2011) Methods for monitoring mule deer populations. Mule Deer Working Group, Western Association of Fish and Wildlife Agencies, USA

Kie JG, Evans CJ, Loft ER, Menke JW (1991) Foraging behavior by mule deer: the influence of cattle grazing. J Wildl Manage 55:665–674. https://doi.org/10.2307/3809516

Kie JG (1996) The effects of cattle grazing on optimal foraging in mule deer (*Odocoileus hemionus*). For Ecol Manage 88:131–138. https://doi.org/10.1016/S0378-1127(96)03818-2

Kinka D, Schultz JT, Young JK (2021) Wildlife responses to livestock guard dogs and domestic sheep on open range. Glob Ecol Conserv 31:e01823. https://doi.org/10.1016/j.gecco.2021.e01823

Klinger RC, Kutilek MJ, Shellhammer HS (1989) Population response of black-tailed deer to prescribed burning. J Wildl Manage 53:863–871. https://doi.org/10.2307/3809578

Koenen KKG, DeStefano S, Krausman PR (2002) Using distance sampling to estimate seasonal densities of desert mule deer in a semidesert grassland. Wildl Soc Bull 30:53–63. https://www.jstor.org/stable/3784635

Krausman PR (2002) Introduction to wildlife management: the basics. Prentice Hall, Upper Saddle River, New Jersey, USA, p 478

Krausman PR, Etchberger RC (1995) Response of desert ungulates to a water project in Arizona. J Wildl Manage 59:292–300. https://doi.org/10.2307/3808942

Krausman PR, Bleich VC (2013) Conservation and management of ungulates in North America. Int J Environ Stud 70:372–382. https://doi.org/10.1080/00207233.2013.804748

Krausman PR, Kuenzi AJ, Etchberger RC, Rautenstrauch KR, Ordway LL, Hervert JJ (1997) Diets of desert mule deer. J Range Manag 50:513–522. https://doi.org/10.2307/4003707

Krausman PR, Rosenstock SS, Cain JW (2006) Developed waters for wildlife: science, perception, values, and controversy. Wildl Soc Bull 34:563–569. https://www.jstor.org/stable/3784681

Krausman PR, Bleich VC, Block WM, Naugle DE, Wallace MC (2011) An assessment of rangeland practices on wildlife populations and habitat. In: Briske DD (ed) Conservation benefits of rangeland practices: assessment, recommendations, and knowledge gaps. U.S. Department of Agriculture, Natural Resources Conservation Service, Washington, D.C., USA, pp 253–290

Lamb S, McMillan BR, van de Kerk M, Frandsen PB, Hersey KR, Larsen RT (2023) From conception to recruitment: Nutritional condition of the dam dictates the likelihood of success in a temperate ungulate. Front Ecol Evolution. https://doi.org/1110.3389/fevo.2023.1090116

Lamb S, Taylor AM, Hughes TA, McMillan BR, Larsen RT, Khan R, Weisz D, Dudchenko O, Aiden EL, Edelman NB, Frandsen PB (2021) De novo chromosome-length assembly of the mule deer (Odocoileus hemionus) genome. Gigabyte 1:2021. https://doi.org/10.46471/gigabyte.34

Larsen RT, Bissonette JA, Flinders JT, Robinson AC (2011) Does small-perimeter fencing inhibit mule deer or pronghorn use of water developments? J Wildl Manage 75:1417–1425. https://doi.org/10.1002/jwmg.163

Larsen RT, Bissonette JA, Flinders JT, Whiting JC (2012) Framework for understanding the influences of wildlife water developments in the western United States. Calif Fish Game 98:148–163

Latch EK, Heffelfinger JR, Fike JA, Rhodes OE Jr (2009) Species-wide phylogeography of North American mule deer (Odocoileus hemionus): cryptic glacial refugia and postglacial recolonization. Mol Ecol 18:1730–1745. https://doi.org/10.1111/j.1365-294X.2009.04153.x

Lambert MS, Sawyer H, Merkle JA (2022) Responses to natural gas development differ by season for two migratory ungulates. Ecol Appl. https://doi.org/10.1002/eap.2652

Lawrence RK, Demarais S, Relyea RA, Haskell SP, Ballard WB, Clark TL (2004) Desert mule deer survival in southwest Texas. J Wildl Manage 68:561–569. https://www.jstor.org/stable/3803389

Lendrum PE, Anderson CR Jr, Monteith KL, Jenks JA, Bowyer RT (2013) Migrating mule deer: effects of anthropogenically altered landscapes. PLoS ONE 8(5):e64548. https://doi.org/10.1371/journal.pone.0064548

Leopold A (1933) Game management. Charles Scribner's Sons, New York, New York, USA, p 520

Leopold A, Sowls LK, Spencer DL (1947) A survey of over-populated deer ranges in the United States. J Wildl Manage 11:162–177. https://doi.org/10.2307/3795561

Leu M, Hanser SE, Knick ST (2008) The human footprint in the west: a large-scale analysis of anthropogenic impacts. Ecol Appl 18:1119–1139. https://doi.org/10.1890/07-0480.1

Linnell JDC, Aanes R, Andersen R (1995) Who killed bambi? The role of predation in the neonatal mortality of temperate ungulates. Wildlife Biol 1:209–223. https://doi.org/10.2981/wlb.1995.0026

Loft ER, Menke JW, Kie JG, Bertrom RC (1987) Influence of cattle stocking rate on the structural profile of deer hiding cover. J Wildl Manage 51:655–664. https://doi.org/10.2307/3801285

Loft ER, Menke JW, Kie JG (1991) Habitat shifts by mule deer: the influence of cattle grazing. J Wildl Manage 55:16–26. https://doi.org/10.2307/3809236

Lomas LA, Bender LC (2006) Survival and cause-specific mortality of neonatal mule deer fawns, north-central New Mexico. J Wildl Manage 71:884–894. https://doi.org/10.2193/2006-203

Longshore KM, Lowrey C, Thompson DB (2009) Compensating for diminishing natural water: predicting the impacts of water development on summer habitat of desert bighorn sheep. J Arid Environ 73:280–286. https://doi.org/10.1016/j.jaridenv.2008.09.021

Lukacs PM, Nowak JJ (2023) Modeling Population Dynamics of Black-Tailed and Mule Deer. In Heffelfinger JR, Krausman PR (eds) Ecology and Management of Black-tailed and Mule Deer in North America. In Press. CRC Press, Boca Raton, Florida, USA.

Lundgren EJ, Ramp D, Stromberg JC, Wu J, Nieto NC, Sluk M, Moeller KT, Wallach AD (2021) Equids engineer desert water availability. Science 372:491–495. https://doi.org/10.1126/science.abd6775

MacCracken JG, Hansen RM (1981) Diets of domestic sheep and other large herbivores in southcentral Colorado. J Rang Manage 34:242–243. https://doi.org/10.2307/3898054

Maestas JD, Hagen CA, Smith JT, Tack JD, Allred BW, Griffiths T, Bishop CJ, Stewart KM, Naugle DE (2019) Mule deer juniper use is an unreliable indicator of habitat quality: comments on Coe et al. (2018). J Wildl Manage 83:755–761. https://doi.org/10.1002/jwmg.21614

Mahoney PJ, Young JK, Hersey KR, Larsen RT, McMillan BR, Stoner DC (2018) Spatial processes decouple management from objectives in a heterogenous landscape: predator control as a case study. Ecol Appl 28:786–797. https://doi.org/10.1002/eap.1686

Marshal JP, Bleich VC, Krausman PR, Reed ML, Andrew NG (2006a) Factors affecting habitat use and distribution of desert mule deer in an arid environment. Wildl Soc Bull 34:609–619. https://doi.org/10.2193/0091-7648(2006)34[609:FAHUAD]2.0.CO;2

Marshal JP, Krausman PR, Bleich VC, Rosenstock SS, Ballard WB (2006b) Gradients of forage biomass and ungulate use near wildlife water developments. Wildl Soc Bull 34:620–626. https://doi.org/10.2193/0091-7648(2006)34[620:GOFBAU]2.0.CO;2

Marshal JP, Bleich VC, Krausman PR, Reed M-L, Neibergs A (2012) Overlap in diet and habitat between the mule deer (*Odocoileus hemionus*) and feral ass (Equus asinus) in the Sonoran Desert. Southwest Nat 57:16–25. https://doi.org/10.1894/0038-4909-57.1.16

McCorquodale SM (1999) Movements, survival, and mortality of black-tailed deer in the Klickitat Basin of Washington. J Wildl Manage 63:861–871. https://doi.org/10.2307/3802799

McDowell NG, Williams AP, Xu C, Pockman WT, Dickman LT, Sevanto S, Pangle R, Limousin J, Plaut J, Mackay DS, Ogee J, Domec JC, Allen CD, Fisher RA, Jiang X, Muss JD, Breshears DD, Rauscher SA, Koven C (2016) Multi-scale predictions of massive conifer mortality due to chronic temperature rise. Nat Clim Chang 6:1048. https://doi.org/10.1038/NCLIMATE2873

McKee CJ, Stewart KM, Sedinger JS, Bush AP, Darby NW, Hughson DL, Bleich VC (2015) Spatial distributions and resource selection by mule deer in an arid environment: responses to provision of water. J Arid Environ 122:76–84. https://doi.org/10.1016/j.jaridenv.2015.06.008

McMillan BR, Hall JT, Freeman ED, Hersey KR, Larsen RT (2023) Both temporal and spatial aspects of predator management influence survival of a temperate ungulate through early life. Front Ecol Evolution. https://doi.org/1110.3389/fevo.2023.1087063

Mealor BA, Mealor RD, Kelley WK, Bergman DL, Burnett SA, Decker TW, Fowers B, Herget ME, Noseworthy CE, Richards JL, Brown CS, Beck KG, Fernandez-Gimenez MF, Frasier M, Munis M, Lupis S, Roath R, Coghenour M, Verdone M, Brumbaugh M (2013) Cheatgrass management handbook: managing an invasive annual grass in the Rocky Mountain Region. Laramie, and Colorado State University, Fort Collins, USA, University of Wyoming, p 136

Merkle JA, Monteith KL, Aikens EO, Hayes MM, Hersey KR, Middleton AD, Oates BA, Sawyer H, Scurlock BM, Kauffman MJ (2016) Large herbivores surf waves of green-up during spring. Proc R Soc B 283:20160456. https://doi.org/10.1098/rspb.2016.0456

Middleton AD, Sawyer H, Merkle JA, Kauffman MJ, Cole EK, Dewey SR, Gude JA, Gustine DD, McWhirter DE, Proffitt KM, White PJ (2019) Conserving transboundary wildlife migrations:

recent insights from the Greater Yellowstone Ecosystem. Front Ecol Environ 18:83–91. https://doi.org/10.1002/fee.2145

Miller RF, Tausch RJ (2001) The role of fire in pinyon and juniper woodlands: a descriptive analysis. In: Galley K, Wilson T (eds) Fire conference 2000: the first national congress on fire ecology, prevention and management. Tall Timbers Research Station, Tallahassee, Florida, USA, pp 15–30

Miller MW, Conner MM (2005) Epidemiology of chronic wasting disease in free-ranging mule deer: spatial, temporal, and demographic influences on observed prevalence patterns. J Wildl Dis 41:275–290. https://doi.org/10.7589/0090-3558-41.2.275

Miller SG, Knight RL, Miller CK (2001) Wildlife responses to pedestrians and dogs. Wildl Soc Bull 29:124–132. https://www.jstor.org/stable/3783988

Miller KV, Muller LI, Demarais S (2003) White-tailed Deer. In: Feldhammer GA, Thompson BC, Chapman JA (eds) Wild mammals of North America: Biology, management, and conservation. The John Hopkins University Press, London, United Kingdom, pp 906–930

Miller MW, Williams ES, Hobbs NT, Wolfe LL (2004) Environmental sources of prion transmission in mule deer. Emerg Infect Dis 10:1003–1006. https://doi.org/10.3201/eid1006.040010

Miller MW, Hobbs NT, Tavener SJ (2006) Dynamics of prion disease transmission in mule deer. Ecol Appl 16:2208–2214. https://doi.org/10.1890/1051-0761(2006)016[2208:dopdti]2.0.co;2

Miller MW, Swanson HM, Wolfe LL, Quartarone FG, Huwer SL, Southwick CH, Lukacs PM (2008) Lions and prions and deer demise. PLoS ONE 3:e4019. https://doi.org/10.1371/journal.pone.0004019

Moeller AK, Lukacs PM, Horne JS (2018) Three novel methods to estimate abundance of unmarked animals using remote cameras. Ecosphere 9:e02331. https://doi.org/10.1002/ecs2.2331

Monaco TA, Gunnell KL (2020) Understory vegetation change following woodland reduction varies by plant community type and seeding status: a region-wide assessment of the ecological benefits and risks. Plants 9:1113. https://doi.org/10.3390/plants9091113

Monteith KL, Long RA, Bleich VC, Heffelfinger JR, Krausman PR, and Bowyer RT (2013) Effects of harvest, culture, and climate on trends in size of horn-like structures in trophy ungulates. Wildl Monogr 183:1–28.

Montieth KL, Bleich VC, Stephenson TR, Pierce BM, Conner MM, Kie JG, Bowyer RT (2014) Life-history characteristics of mule deer: effects of nutrition in a variable environment. Wildl Monogr 186:1–62. https://doi.org/10.1002/wmon.1011

Mule Deer Working Group (2014) Diseases and parasites of mule deer. Fact Sheet 11, Western Association of Fish and Wildlife Agencies, 2 p

Mule Deer Working Group (2015) Historical and current mule deer abundance. Fact Sheet 32, Western Association of Fish and Wildlife Agencies, 2 p

Mule Deer Working Group (2021) 2021 range-wide status of black-tailed and mule deer. Western Association of Fish and Wildlife Agencies, 46 p

Murray WB (2013) Deer: sacred and profane. In: Gillette DL, Greer M, Hayward MH, Murray B, Murray W, Breen W (eds) Rock art and sacred landscapes. Springer, New York, pp 195–206. https://doi.org/10.1007/978-1-4614-8406-6

Nagy-Reis M, Reimer JR, Lewis MA, Jensen WF, Boyce MS (2021) Aligning population models with data: adaptive management for big game harvests. Glob Ecol Conserv 26:e01501. https://doi.org/10.1016/j.gecco.2021.e01501

Northrup JM, Anderson CR Jr, Gerber BD, Wittemeyer G (2021) Behavioral and demographic responses of mule deer to energy development on winter range. Wildl Monogr 208:1–37. https://doi.org/10.1002/wmon.1060

O'Brien P, Vander Wal E, Koen EL, Brown CD, Guy J, van Beest FM, Brook RK (2019) Understanding habitat co-occurrence and the potential for competition between native mammals and invasive wild pigs (Sus scrofa) at the norther edge of their range. Can J Zool 97:537–546. https://doi.org/10.1139/cjz-2018-0156

Ordway LL, Krausman PR (1986) Habitat use by desert mule deer. J Wildl Manage 50:677–683. https://doi.org/10.2307/3800980

Otero A, Velasquez CD, Aiken J, McKenzie D (2021) Chronic wasting disease: a cervid prion infection looming to spillover. Vet Res 52:115. https://doi.org/10.1186/s13567-021-00986-y

Pac DF, White GC (2005) Survival and cause-specific mortality of male mule deer under different hunting regulations in the Bridger Mountains, Montana. J Wildl Manag 71:816–827. https://doi.org/10.2193/2005-713

Parker KL, Robbins CT (1984) Thermoregulation in deer and elk. Can J Zool 62:1409–1422. https://doi.org/10.1139/z84-202

Parker KL, Gillingham MP (1990) Estimates of critical thermal environments for mule deer. J Rang Manage 43:73–81. https://doi.org/10.2307/3899126

Pfannenstiel RS, Mullens BA, Ruder MG, Zurek L, Cohnstaedt LW, Nayduch D (2015) Management of North American Culicoides biting midges: current knowledge and research needs. Vector Borne Zoonotic Dis 15:374–384. https://doi.org/10.1089/vbz.2014.1705

Pilliod DS, Welty JL, Arkle RS (2017) Refining the cheatgrass-fire cycle in the Great Basin: precipitation timing and fine fuel composition predict wildfire trends. Ecol Evol 7:8126–8151. https://doi.org/10.1002/ece3.3414

Popay I, Field R (1996) Grazing animals as weed control agents. Weed Technol 10:217–231. https://www.jstor.org/stable/3987805

Pojar TM, Bowden DC (2004) Neonatal mule deer fawn survival in west-central Colorado. J Wildl Manage 68:550–560. https://www.jstor.org/stable/3803388

Polzhein RO, Strobek C (1998) Phylogeny of wapiti, red deer, sika deer, and other North American cervids as determined from mitochondrial DNA. Mol Phylogenet Evol 10:249–258. https://doi.org/10.1006/mpev.1998.0527

Prins RA, Geelen MJH (1971) Rumen characteristics of red deer, fallow deer, and roe deer. J Wildl Manage 35:673–680. https://doi.org/10.2307/3799772

Rabe MJ, Rosenstock SS, deVos Jr JC (2002) Review of big-game survey methods used by wildlife agencies of the western United States. Wildl Soc Bull 30:46–52. https://www.jstor.org/stable/3784634

Ragotzkie KE, Bailey JA (1991) Desert mule deer use of grazed and ungrazed habitats. J Rang Manage 44:487–490. https://doi.org/10.2307/4002750

Rau BM, Chambers JC, Blank RR, Johnson DW (2008) Prescribed fire, soil, and plants: burn effects and interactions in the central Great Basin. Rangel Ecol Manag 61:169–181. https://doi.org/10.2111/07-037.1

Rautenstrauch KR, Krausman PR (1989) Influence of water availability and rainfall on movements of desert mule deer. J Mammal 70:197–201. https://doi.org/10.2307/1381689

Rawley EV (1985) Early records of wildlife in Utah. Utah Division of Wildlife Resources, Salt Lake City, Utah, 102 p

Reinhart JR, Filippelli S, Falkowski M, Allred B, Maestas JD, Carlson JC, Naugle DE (2020) Quantifying pinyon-juniper reduction within North America's sagebrush ecosystem. Rangel Ecol Manag 73:420–432. https://doi.org/10.1016/j.rama.2020.01.002

Rhodes BD, Sharrow SH (1990) Effect of grazing by sheep on the quantity and quality of forage available to big game in Oregon's Coast Range. J Rang Manage 43:235–237. https://doi.org/10.2307/3898680

Riecke TV, Williams PJ, Behnke TL, Gibson D, Leach AG, Sedinger BS, Street PA, Sedinger JS (2019) Integrated population models: model assumptions and inference. Methods Ecol Evol 10:1072–1082. https://doi.org/10.1111/2041-210X.13195

Robinson HS, Wielgus RB, Gwilliam JC (2002) Cougar predation and population growth of sympatric mule deer and white-tailed deer. Can J Zool 80:556–568. https://doi.org/10.1139/Z02-025

Robbins CT (1983) Wildlife feeding and nutrition. Academic Press, Orlando, Florida, 343 p. https://doi.org/10.1016/C2013-0-11381-5

Rodgers PA, Sawyer H, Mong TW, Stephens S, Kauffman MJ (2021) Sex-specific behaviors of hunted mule deer during rifle season. J Wildl Manage 85:215–227. https://doi.org/10.1002/jwmg.21988

Roerick TM, Cain JW, Gedir JV (2019) Forest restoration, wildfire, and habitat selection by female mule deer. For Ecol Manage 447:169–179. https://doi.org/10.1016/j.foreco.2019.05.067

Rollins M (2009) Landfire: a nationally consistent vegetation, wildland fire, and fuel assessment. Int J Wildland Fire 18:235–249. https://doi.org/10.1071/WF08088

Rosenstock SS, Ballard WB, deVos Jr JC (1999) Viewpoint: benefits and impacts of wildlife water developments. J Rang Manage 52:302–311. https://doi.org/10.2307/4003538

Rupp SP, Ballard WB, Wallace MC (2000) A nationwide evaluation of deer hunter harvest survey techniques. Wildl Soc Bull 28:570–578. https://www.jstor.org/stable/3783605

Sallee DW, McMillan BR, Hersey KR, Petersen SL, Larsen RT (2023) Influence of interspecific competition on mule deer birthing and rearing site selection. J Wildlife Manag 87(1). https://doi.org/10.1002/jwmg.v87.1, https://doi.org/10.1002/jwmg.22318

Sánchez-Rojas G, Gallina S (2000a) Factors affecting habitat use by mule deer (*Odocoileus hemionus*) in the central part of the Chihuahuan Desert, Mexico: an assessment with univariate and multivariate methods. Ethol Ecol Evol 12:405–417. https://doi.org/10.1080/08927014.2000.9522795

Sánchez-Rojas G, Gallina S (2000b) Mule deer (*Odocoileus hemionus*) density in a landscape element of the Chihuahuan Desert, Mexico. J Arid Environ 44:357–368. https://doi.org/10.1006/jare.1999.0605

Sanchez-Rojas G, Tessaro SG (2016) Odocoileus hemionus. The IUCN Red List of Threatened Species 2016: e.T42393A22162113. https://doi.org/10.2305/IUCN.UK.2016-1.RLTS.T42393A22162113.en. Accessed 11 Jan 2022

Sandoval L, Holechek J, Biggs J, Valdez R, VanLeeuwen D (2005) Elk and mule deer diets in north-central New Mexico. Rangel Ecol Manage 58:366–372. https://doi.org/10.2111/1551-5028(2005)058[0366:EAMDDI]2.0.CO;2

Sankey TT, Germino MJ (2008) Assessment of juniper encroachment with the use of satellite imagery and geospatial data. Rangel Ecol Manage 61:412–418. https://doi.org/10.2111/07-141.1

Sawyer H, Kauffman MJ (2011) Stopover ecology of a migratory ungulate. J Anim Ecol 80:1078–1087. https://doi.org/10.1111/j.1365-2656.2011.01845.x

Sawyer H, Lindzey F, McWhirter D (2005) Mule deer and pronghorn migration in western Wyoming. Wildl Soc Bull 33:1266–1273. https://doi.org/10.2193/0091-7648(2005)33[1266:MDAPMI]2.0.CO;2

Sawyer H, Nielson RM, Lindzey F, McDonald LL (2010) Winter habitat selection of mule deer before and after development of a natural gas field. J Wildl Manage 70:396–403. https://doi.org/10.2193/0022-541X(2006)70[396:WHSOMD]2.0.CO;2

Sawyer H, Lebeau C, Hart T (2012) Mitigating roadway impacts to migratory mule deer—a case study with underpasses and continuous fencing. Wildl Soc Bull 36:492–498. https://doi.org/10.1002/wsb.166

Sawyer H, Middleton AD, Hayes MH, Kauffman MJ, Monteith KL (2016) The extra mile: ungulate migration distance alters the use of seasonal range and exposure to anthropogenic risk. Ecosphere 7(10):e01534. https://doi.org/10.1002/ecs2.1534.10.1002/ecs2.1534

Sawyer H, Korfanta NM, Nielson RM, Monteith KL, Strickland D (2017) Mule deer and energy development—long-term trends of habituation and abundance. Glob Chang Biol 23:4521–4529. https://doi.org/10.1111/gcb.13711

Sawyer H, Merkle JA, Middleton AD, Dwinnell SPH, Monteith KL (2019) Migratory plasticity is not ubiquitous among large herbivores. J Anim Ecol 88:450–460. https://doi.org/10.1111/1365-2656.12926

Sawyer H, Lambert MS, Merkle JA (2020) Migratory disturbance thresholds with mule deer and energy development. J Wildl Manage 84:930–937. https://doi.org/10.1002/jwmg.21847

Scasta JD, Beck JL, Angwin CJ (2016) Meta-analysis of diet composition and potential conflict of wild horses with livestock and wild ungulates on western rangelands of North America. Rangel Ecol Manage 69:310–318. https://doi.org/10.1016/j.rama.2016.01.001

Schoennagel T, Balch JK, Brenkert-Smith H, Dennison PE, Harvey BJ, Krawchuk MA, Mietkiewicz N, Morgan P, Moritz MA, Rasker R, Turner MG, Whitlock C (2017) Adapt to more wildfire in western North American forests as climate changes. PNAS 114:4582–4590. https://doi.org/10.1073/pnas.161746411

Schuyler EM, Dugger KM, Jackson DH (2019) Effects of distribution, behavior, and climate on mule deer survival. J Wildl Manage 83:89–99. https://doi.org/10.1002/jwmg.21558

Scifres CJ, Mutz JL, Hamilton WT (1979) Control of mixed brush with tebuthiuron. J Rang Manage 32:155–158. https://doi.org/10.2307/3897563

Seager R, Ting M, Li C, Naik N, Cook B, Nakamura J, Liu H (2013) Projections of declining surface-water availability for the southwestern United States. Nat Clim Chang 3:482–486. https://doi.org/10.1038/nclimate1787

Segar J, Keane A (2020) Species and demographic responses to wildlife-friendly fencing on ungulate crossing success and behavior. Conserv Sci Pract 2:e285. https://doi.org/10.1111/csp2.285

Shallow JRT, Hurley MA, Monteith KL, Bowyer RT (2015) Cascading effects of habitat on maternal condition and life-history characteristics of neonatal mule deer. J Mammal 96:194–205. https://doi.org/10.1093/jmammal/gyu024

Shields AV, Larsen RT, Whiting JC (2012) Summer watering patterns of mule deer in the Great Basin Desert, USA: implications of differential use by individuals and the sexes for management of water resources. Sci World J 9:846218. https://doi.org/10.1100/2012/846218

Short JJ, Knight JE (2003) Fall grazing affects big game forage on rough fescue grasslands. J Rang Manage 56:213–217. https://doi.org/10.2307/4003809

Simpson NO, Stewart KM, Bleich VC (2011) What have we learned about water developments? Not enough. Calif Fish Game 97:190–209

Smith JT, Allred BW, Boyd CS, Davies KW, Jones MO, Kleinhesselink AR, Maestas JD, Morford SL, Naugle DE (2022) The elevational ascent and spread of exotic annual grass dominance in the Great Basin, USA. Divers Distrib 28:83–96. https://doi.org/10.1111/ddi.13440

Smolko P, Seidel D, Pybus M, Hubbs A, Ball M, Merrill E (2021) Spatio-temporal changes in chronic wasting disease risk in wild deer during 14 years of surveillance in Alberta. Canada. Prev Vet Med 197:105512. https://doi.org/10.1016/j.prevetmed.2021.105512

Sorensen GE, Kramer DW, Cain JW, Taylor CA, Gipson PS, Wallace MC, Cox RD, Ballard WB (2020) Mule deer habitat selection following vegetation thinning treatments in New Mexico. Wildl Soc Bull 44:122–129. https://doi.org/10.1002/wsb.1062

Staudenmaier AR, Shipley LA, Camp MJ, Forbey JS, Hagerman AE, Brandt AE, Thornton DH (2021) Mule deer do more with less: comparing their nutritional requirements and tolerances with white-tailed deer. J Mammal 103:178–195. https://doi.org/10.1093/jmammal/gyab116

Stevens JT, Kling MM, Schwilk DW, Varner JM, Kane JM (2020) Biogeography of fire regimes in western U.S. conifer forests: a trait-based approach. Glob Ecol Biogeogr 29:944–955. https://doi.org/10.1111/geb.13079

Stewart KM, Bowyer RT, Kie JG, Cimon NJ, Johnson BK (2002) Temporospatial distributions of elk, mule deer, and cattle: resource partitioning and competitive displacement. J Mammal 83:229–244. https://doi.org/10.1644/1545-1542(2002)083%3c0229:TDOEMD%3e2.0.CO;2

Stewart KM, Bowyer RT, Kie JG, Dick BL, Ben-David M (2003) Niche partitioning among mule deer, elk, and cattle: do stable isotopes reflect dietary niche? Ecoscience 10:297–302. https://www.jstor.org/stable/42902531

Stoner DC, Sexton JO, Nagol J, Bernales HH, Edwards TC Jr (2016) Ungulate reproductive parameters track satellite observations of plant phenology across latitude and climatological regimes. PLoS ONE 11(2):e0148780. https://doi.org/10.1371/journal.pone.0148780

Stoner DC, Anderson MT, Schroeder CA, Bleke CA, Thacker ET (2021) Distribution of competition potential between native ungulates and free-roaming equids on western rangelands. J Wildl Manage 85:1062–1073. https://doi.org/10.1002/jwmg.21993

Taylor N, Knight JE, Short JJ (2004) Fall cattle grazing versus mowing to increase big game forage. Wildl Soc Bull 32:449–455. https://www.jstor.org/stable/3784984

Torstenson WLF, Mosely JC, Brewer TK, Tess MW, Knight JE (2006) Elk, mule deer, and cattle foraging relationships on foothill and mountain rangeland. Rangel Ecol Manage 59:80–87. https://doi.org/10.2111/05-001R1.1

Ulappa AC, Shipley LA, Cook RC, Cook JG, Swanson ME (2020) Silvicultural herbicides and forest succession influence understory vegetation and nutritional ecology of black-tailed deer in managed forests. For Ecol Manage 470–471:118216. https://doi.org/10.1016/j.foreco.2020. 118216

Unsworth JW, Pac DF, White GC, Bartmann RM (1999) Mule deer survival in Colorado, Idaho, and Montana. J Wildl Manage 63:315–326. https://doi.org/10.2307/3802515

Van de Kerk M, Larsen RT, Olson DD, Hersey KR, McMillan BR (2021) Variation in movement patterns of mule deer: have we oversimplified migration? Mov Ecol 9:44. https://doi.org/10. 1186/s40462-021-00281-7

Wallmo OC (1981) Mule and black-tailed deer distribution and habitats. In: Wallmo OC (ed) Mule Deer and black-tailed deer of North America. University of Nebraska Press, Lincoln Nebraska, pp 1–25

Waser NM, Price MV, Blumstein DT, Arozqueta SR, Escobar BDC, Pickens R, Pistoia A (2014) Coyotes, deer, and wildflowers: diverse evidence points to a trophic cascade. Naturwissenschaften 101:427–436. https://doi.org/10.1007/s00114-014-1172-4

Wielgus RB (2017) Resource competition and apparent competition in declining mule deer (*Odocoileus hemionus*). Can J Zool 95:499–504. https://doi.org/10.1139/cjz-2016-0109

White GC, Bartmann RM (1998) Effect of density reduction on overwinter survival of free-ranging mule deer fawns. J Wildl Manage 62:214–225. https://doi.org/10.2307/3802281

Wickstrom ML, Robbins CT, Hanley TA, Spalinger DE, Parish SM (1984) Food intake and foraging energetics of elk and mule deer. J Wildl Manage 48:1285–1301. https://doi.org/10.2307/3801789

Williams ES, Miller MW, Kreeger TJ, Kahn RH, Thorne ET (2002) Chronic wasting disease of deer and elk: a review with recommendations for management. J Wildl Manage 66:551–563. https://doi.org/10.2307/3803123

Williams PJ, Schroeder C, Jackson P (2020) Estimating reproduction and survival of unmarked juveniles using aerial images and marked adults. J Agr Bio Environ Stat 25:133–147. https:// doi.org/10.1007/s13253-020-00384-5

Wilson DE, Mittermeier AR (eds) (2011) Handbook of the mammals of the world, vol 2 (Hoofed Mammals). Lynx Edicions. ISBN 978-84-96553-77-4

Woods LW, Swift PK, Barr BC, Horzinek MC, Nordhausen RW, Stillian MH, Patton JF, Oliver MN, Jones KR, MacLachlan NJ (1996) Systemic adenovirus infection associated with high mortality in mule deer (*Odocoileus hemionus*) in California. Vet Path 33:125–132. https://doi. org/10.1177/030098589603300201

Wyckoff TB, Sawyer H, Albeke SE, Garman SL, Kauffman MJ (2018) Evaluating the influence of energy and residential development on the migratory behavior of mule deer. Ecosphere e02113. https://doi.org/10.1002/ecs2.2113

Xu W, Dejid N, Herrman V, Sawyer H, Middleton AD (2020) Barrier behavior analysis (BaBA) reveals extensive effects of fencing on wide-ranging ungulates. J Appl Ecol 58:690–698. https:/ /doi.org/10.1111/1365-2664.13806

Zabransky CJ, Hewitt DG, Deyoung RW, Gray SS, Richardson C, Litt AR, Deyoung CA (2016) A detection probability model for aerial surveys of mule deer. J Wildl Manage 80:1379–1389. https://doi.org/10.1002/jwmg.21143

Zeigenfuss LC, Schoenecker KA, Ransom JI, Ignizio DA, Mask T (2014) Influence of nonnative and native ungulate biomass and seasonal precipitation on vegetation production in a Great Basin eco-system. West N Am Nat 74:286–298. https://doi.org/10.3398/064.074.0304

Zimmerman TS, Jenks JA, Leslie DM Jr (2006) Gastrointestinal morphology of female white-tailed and mule deer: effects of fire, reproduction, and feeding type. J Mammal 87:598–605. https:// doi.org/10.1644/05-mamm-A-356R1.1

Chapter 18
White-Tailed Deer

Timothy E. Fulbright

Abstract White-tailed deer are geographically widespread and occupy a variety of ecosystems from semi-desert shrubland and grasslands to forests. They have a relatively high reproductive potential but recruitment may be limited in semiarid rangelands where annual variation in precipitation is high. They eat browse and forbs but mast may seasonally comprise most of the diet. White-tailed deer select areas with a mixture of woody vegetation and areas dominated by herbaceous vegetation. They use woody vegetation for cover and often forage in adjacent herbaceous-dominated areas. They are highly adaptable and can adjust to changes in vegetation resulting from rangeland management practices; however, excessive grazing reduces habitat quality. Brush management minimally affects white-tailed deer and their habitat when adequate resources such as thermal cover, hiding cover, and browse-and-mast-producing vegetation remain on the landscape. Empirical evidence that creating mosaics of herbaceous-dominated foraging patches and woody cover improves demographics or productivity is equivocal; however, managing for increased spatial heterogeneity in vegetation may increase fawn survival. Chronic wasting disease is a major threat to white-tailed deer populations. White-tailed deer use behavioral adaptations to reduce excessive heat loads resulting from climate change in the southern part of their range. Paradoxically, populations are expanding in the northern part of their range in part because of milder winters. Hunting is the primary tool to manage white-tailed deer populations. Combining recreational hunting with livestock production increases revenue for ranchers. Ironically, white-tailed deer are often a nuisance in eastern forests, but they can be an economically important asset on rangelands.

Keywords *Odocoileus virginianus* · Brush management · Climate change · Density dependence · Grazing · Livestock

T. E. Fulbright (✉)
Ceasar Kleberg Wildlife Research Institute, Texas A&M University-Kingsville, 700 University Blvd., Kingsville, TX 78363, USA
e-mail: Timothy.Fulbright@tamuk.edu

635

18.1 Introduction

A remarkable ability to adapt to different environments and to human presence is a hallmark trait of white-tailed deer (*Odocoileus virginianus*; Fig. 18.1). White-tailed deer populations persist and reproduce in environments ranging from the temperate forests of the eastern United States to the western deserts, grasslands, shrublands, and woodlands and range from Alaska to Mexico in North America. They occupy landscapes ranging from relatively undisturbed National Parks to the suburbs of major cities (Potratz et al. 2019). Although white-tailed deer is among the most studied and managed wildlife species in North America, less research has been conducted in western rangelands than in other parts of their distribution. The majority of research on rangeland-associated populations has occurred in the Southern Great Plains and Tamaulipan Vegetation Region of Texas. Throughout this chapter, I have included information from other regions whenever possible and examples from non-rangeland areas when appropriate.

18.2 General Life History and Population Dynamics

White-tailed deer have a relatively high reproductive potential. Number of offspring produced by females depends on many factors including age, nutritional status, environment, deer density, and behavioral interactions (Verme 1969; DeYoung 2011; DeYoung et al 2019). Under favorable conditions, adult female white-tailed deer may produce an average of two or more offspring/year. Under extremely favorable conditions, female fawns may breed, and some adult females may have triplets (Ozoga 1987; DeYoung 2011). Age at which fecundity declines is unclear; some reports suggest it does not decline until well into maturity (DelGiudice et al. 2007).

Males reach sexual maturity at 1.5 years of age (Sauer 1984; DeYoung and Miller 2011). White-tailed deer populations commonly lack older males because of heavy harvest of yearlings and younger males by hunters. In populations with an older age structure, males 3.5 years and older sire about 70% of the fawns in a given year with 1.5- and 2.5-year-old males siring the remainder (DeYoung et al. 2009; DeYoung and Miller 2011).

Fawns use bed sites in grassland and areas dominated by woody vegetation (Grovenburg et al. 2010; Michel et al. 2020; Fulbright et al. 2023). Bed sites tend to have more grass cover and taller grasses than the surrounding habitat (Uresk et al. 1999). Woody plant cover is important along with grass cover at daytime fawn bed sites in the Tamaulipan Vegetation Region (Hyde et al. 1987; Fulbright et al. 2023; Fig. 18.2). Bed site cover may help fawns to avoid predation; however, the importance of bed site cover in evading predators such as coyotes (*Canis latrans*) is unclear. For example, increasing bed site cover was only weakly related to fawn survival in a recent study not conducted on rangeland (Chitwood et al. 2015). In contrast, white-tailed deer in a separate study avoided recently burned areas during fawn-rearing

Fig. 18.1   White-tailed deer occupy a variety of plant communities and environments. Photograph © Timothy E. Fulbright

possibly because hiding cover was lacking (Cherry et al. 2017). Bed site cover may be important for thermoregulation during midday in warm environments (Fulbright et al. 2023). Woody cover that supplies shade and cooler temperatures during fawning and fawn-rearing could be particularly important in hot environments. In cooler environments, thermal cover may be important to help fawns avoid hypothermia. In the northern Great Plains bed sites in Conservation Reserve Program grasslands provided more cover and were warmer during summer than bed sites in wheat fields where mortality of fawns may have resulted from hypothermia (Grovenburg et al. 2012b).

Growth and development of white-tailed deer depend on sex and environmental factors including latitude and habitat quality (Ditchkoff 2011). In general, deer in

Fig. 18.2 Daytime fawn bed sites often consist of an overhead shrub canopy and tall grass. Photograph © Timothy E. Fulbright

environments with good nutrition reach maximum body mass at older ages than deer in resource-limited environments (Strickland and Demarais 2000; Monteith et al. 2009). Further illustrating the effect of environment, adult white-tailed deer from the Black Hills of South Dakota are smaller than those from more productive rangelands in eastern South Dakota (Monteith et al. 2009). Body mass of adult white-tailed deer along the Gulf Coastal Prairie is smaller and males have smaller antlers than adult white-tailed deer in the western Tamaulipan Vegetation Region (Rankins et al. 2021). One of the possible reasons deer in the western Tamaulipan Vegetation Region are larger is that digestible energy in browse and mast is greater than along the coast.

Photoperiod regulates seasonal timing of antler growth (Demarais and Strickland 2011). Yearling bucks may be spikes (unbranched antlers) or they may have branched antlers. The percentage of spikes in the yearling cohort strongly depends on nutrition (DeYoung et al. 2019). Antler size increases with age and reaches an asymptote at around five years old (Monteith et al. 2009; Hewitt et al. 2014).

Mortality of white-tailed deer follows a U-shaped curve with highest mortality in fawns and old (> 5 years) adults with higher survival rates in between (DeYoung 2011). Predation is the primary cause of fawn mortality on rangelands with coyotes being the primary predator (Bartush and Lewis 1981; Kie and White 1985; Whittaker and Lindzey 1999; Grovenburg et al. 2012b). Mule deer (*O. hemionus*)—white-tailed deer hybrids may be more susceptible to predation than nonhybrids because their gait is slower and mechanically less efficient (Lingle 1993).

Annual natural mortality of adult females tends to be low in the northern Great Plains (Dusek et al. 1992; Grovenburg et al. 2011). Annual survival of adult female white-tailed deer was 88% in in the northern Great Plains during the winters of 2014 and 2015, with mortality increasing as winter progressed (Moratz et al. 2018). Predation and hunting are the major causes of mortality of adult female white-tailed deer in the northern Great Plains (Dusek et al. 1992; Moratz et al. 2018). Enhancing nutrition by providing pelleted feed high in protein and energy increased survival of adult males and females in the Tamaulipan Vegetation Region, demonstrating that nutrition is a limiting factor for white-tailed deer populations in that region (DeYoung et al. 2019).

White-tailed deer population dynamics are linked to vegetation dynamics (DeYoung et al. 2019). The conventional paradigm in rangeland vegetation dynamics, the equilibrium model, assumes heavy grazing results in a shift in plant community composition to less palatable plants or those more tolerant of herbivory. The non-equilibrium model of vegetation dynamics is an alternative paradigm where abiotic factors such as variable precipitation drive plant community and ecosystem characteristics with plant–herbivore interactions weakly linked (Briske et al. 2003). Consequently, herbivore populations are density-independent, particularly in regions where the coefficient of variation in annual rainfall exceeds 30–33% (Briske et al. 2003; Derry and Boone 2010). Rangeland ecosystems can exhibit both equilibrium and non-equilibrium vegetation dynamics (Briske et al. 2003; Derry and Boone 2010). The conventional rangeland model of vegetation dynamics parallels density dependence theory in that loss of palatable plants results in a decline in forage quality and availability (DeYoung et al. 2019).

Biologists often use deer management models that assume populations act in a density dependent fashion. However, density-dependent population behavior is often difficult to detect (McCullough 1999). Research on non-equilibrium models of vegetation dynamics has focused on livestock and the effects of environmental stochasticity on white-tailed deer population dynamics has received little attention (DeYoung et al. 2019). DeYoung et al. (2008) hypothesized that high annual variation in precipitation, low soil fertility, and severe winters that limit populations may obscure density dependence in white-tailed deer and predicted that simple density-dependent models may not be useful in more than half of the range of white-tailed deer in the United States. In the Tamaulipan Vegetation Region, DeYoung et al. (2019) examined the effects of three different white-tailed deer densities on population growth rates, fawn and adult survival, and deer morphometrics where the coefficient of variation in annual precipitation exceeded 30%. They concluded that white-tailed deer in the region were only weakly density dependent and that in the absence of several consecutive wet years harvest of females would be additive, not compensatory mortality (DeYoung et al. 2019). One reason for weak density dependence in the region was that cycles of drought followed by periods of high precipitation had a much stronger effect on vegetation than increasing white-tailed deer density from 10 deer 81 ha^{-2} to 40 deer 81 ha^{-2} (DeYoung et al. 2019). Environmental stochasticity is a characteristic of many rangeland ecosystems; however, research linking models of vegetation dynamics with white-tailed deer population dynamics in systems other than the Tamaulipan Vegetation Region is lacking. Additional research on white-tailed deer in different rangeland ecosystems is needed to determine the utility of simple density dependent models in management.

Home range sizes of white-tailed deer vary from < 100 ha to > 1000 ha depending on a variety of factors including, sex, age, season, and population density (DeYoung and Miller 2011). Home range size is generally larger in drier, unproductive areas than in more productive, mesic environments (Stewart et al. 2011). Males typically have larger home ranges than females (DeYoung and Miller 2011; Stewart et al. 2011) and home range sizes of males tend to be larger during the breeding season than at other times of the year in low-density populations. Home range size declines when white-tailed deer population density increases. In the Tamaulipan Vegetation Region, for example, home ranges during late gestation, summer lactation, and early rut were 2.4 times larger when deer density was 10 deer 81 ha^{-2} than when density was 40 deer 81 ha^{-2} (Fulbright et al. 2023).

White-tailed deer form relatively small groups and rarely come together in large herds (DeYoung and Miller 2011). Females form groups that typically consist of an older matriarch and several generations of her offspring. Males 1.5-years and older form bachelor groups during the non-breeding season but are solitary during the breeding season. Males and females often separate spatially and use areas with different habitat characteristics during the non-breeding season (Stewart et al. 2011). In contrast to mule deer, groups of white-tailed deer do not consistently change group size or formation when confronted with predators (Lingle 2001).

White-tailed deer are crepuscular and are usually most active during early morning and late evening (Wiemers et al. 2014). Most of their active time, except during the breeding season, is spent foraging and searching for food. White-tailed deer consume forage amounting to 2–4% of their live body weight on a dry-matter basis (Halls 1978). Deer have a small rumen to body mass ratio relative to other ruminants. Consequently, they are concentrate feeders that select the most nutritious plants and plant parts (Hewitt 2011). They depend on a relatively short retention time of plant parts in the rumen so they can process the readily digestible nutrients and then quickly pass the undigested material making space for additional forage. White-tailed deer forages are typically classified as browse, forbs, grass, and mast (Hewitt 2011). However, they also consume flowers, dead leaves, and fungi (Darr et al. 2019). They select forbs over browse and grasses when forbs are available (Fulbright and Ortega-S 2013). On rangelands, forbs are often ephemeral. Browse often composes a major part of white-tailed deer diets when forbs are unavailable. Mast may be the dominant dietary component during certain seasons, particularly when acorns or honey mesquite (*Prosopis glandulosa*) mast are available. For example, mast including honey mesquite pods and prickly pear fruits formed up to 90% of deer diets during summer in a study in the Tamaulipan Vegetation Region (Fulbright et al. 2023). Flowers may also be an important dietary component when they are available composing up to 48% of deer diets based on a study in the Tamaulipan Vegetation Region (Darr et al. 2019).

White-tailed deer eat a variety of plant species and plant parts (Fulbright and Ortega-S 2013). Plant species vary in mineral content and in protein and energy.

Consuming a diverse diet may aid in optimizing the nutrient content of their diet (Provenza et al. 2003). Many rangeland shrub genera and species are high in secondary compounds that have anti-nutrition effects on herbivores. Examples of genera with species high in secondary compounds include oaks (*Quercus* spp.), sage-brush (*Artemisia* spp.), sumac (*Rhus* spp.), and acacias (*Acacia* spp.). Consuming a diverse diet may help to neutralize anti-nutrition effects of secondary compounds and improve the nutrient content of white-tailed deer diets (Provenza et al. 2003, 2009).

18.3   Species and Population Status

White-tailed deer occupy a large geographic area from Alaska and Canada to South America (Heffelfinger 2011; Fig. 18.3). About 38 subspecies of white-tailed deer occur within this geographic range. In contrast to range contractions for elk (*Cervus canadensis*; Chap. 20), the geographic range of white-tailed deer has expanded, particularly along the northern fringe of its range (Heffelfinger 2011). Climate change may be involved in range expansion along with human-imposed changes in the landscape such as forest cutting and cultivated agriculture expansion.

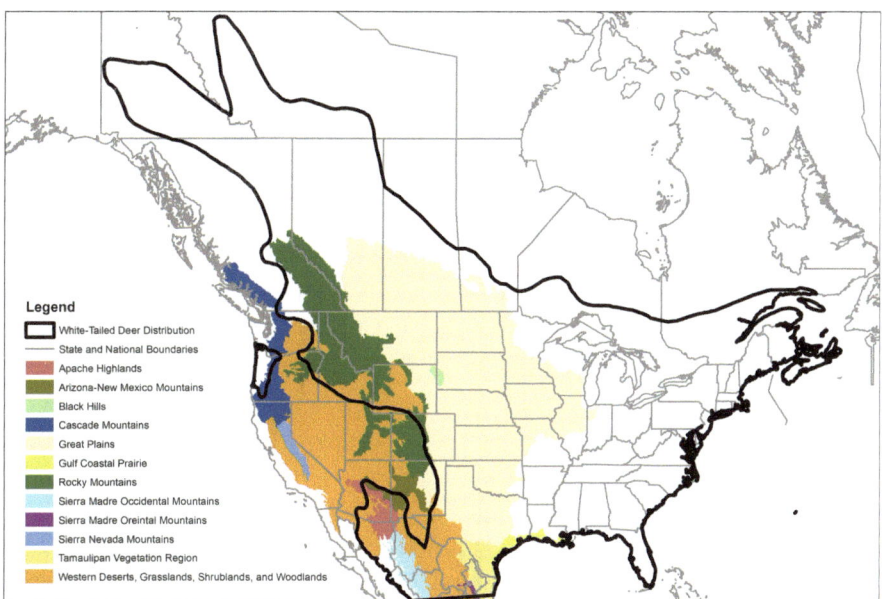

Fig. 18.3 Geographic distribution of white-tailed deer. Map created by H. Perotto

Several techniques are used to monitor populations of white-tailed deer. Older methods include pellet-group counts, track counts, night spotlighting, and mark-resight (DeYoung 2011). Helicopter surveys are commonly used in rangelands. Infrared or motion-triggered cameras also are used to estimate population density and can be used in combination with mark-resight techniques (Moore et al. 2014). Camera surveys and N-mixture modeling have been shown to be highly effective methods of estimating white-tailed deer populations (Keever et al. 2017). N-mixture modeling does not require capturing and marking individual deer. The procedure estimates detection probability and abundance with covariates that vary in time and space. Infrared thermal imaging is a technology with potential use for monitoring white-tailed deer populations (DeYoung 2011). Distance sampling can be used to estimate deer densities corrected from imperfect detection using a variety of survey methods including night spotlighting, helicopter surveys, and surveys using infrared thermal imaging (Montague et al. 2017; Peterson et al. 2020). However, conventional distance sampling assumes a monotonically decreasing detection probability with distance from the survey route, which may be violated for deer surveys occurring in areas with variable topography or vegetation cover. Hierarchical distance sampling models that allow and adjust for site-specific covariates on detectability may produce similar parameter estimates to N-mixture modeling (Christensen et al. 2021a).

White-tailed deer are typically undercounted in part because of visual obstruction from vegetation or rough topography. Sightability models are commonly used for aerial surveys from fixed-wing aircraft and helicopters to account for visibility bias. Sightability models use logistic regressions to model detections and non-detections of deer and develop correction factors to account for behavioral and environmental factors that influence rate of detection of animals (Anderson et al. 1998). Precision of sightability models for white-tailed deer declines with increased distance of deer from transects and vegetation obstruction (Dyal et al. 2021). Surveys using cameras and infrared thermal imaging (FLIR, Forward Looking Infrared) produced similar point estimates and detection probabilities (Haus et al. 2019). Use of infrared thermal imaging from unmanned aerial systems (drones) is a promising technology for monitoring white-tailed deer (Chrétien et al. 2016). Current limitations of the technology are limited flight distance of unmanned aerial systems and regulations.

18.4 Habitat Associations

White-tailed deer occupy a variety of different plant communities and ecosystems ranging from grasslands to forests, and from semi-desert shrubland to suburbs (Figs. 18.1 and 18.3). The fossil record of *Odocoileus* goes back four million years and fossils occur throughout most of their contemporary range (Heffelfinger 2011). Their success in a variety of settings over a long period of time is testimony to the adaptability of the species. In rangelands, white-tailed deer are most abundant in areas where woody vegetation dominates part of the landscape (Fulbright 2011). In the Great Plains, for example, white-tailed deer are strongly associated with wooded riparian corridors or bottomland areas (Compton et al. 1988). Similarly, Columbian white-tailed deer (*O. v. leucurus*) in Oregon were generally associated with riparian systems (Smith 1987). In western Texas, white-tailed deer densities increased with increasing woody plant cover (Wiggers and Beasom 1986). Highest densities of Columbian white-tailed deer occurred in areas with ≥ 50% woody vegetation (Smith 1987). In the Tamaulipan Vegetation Region, areas most heavily used by white-tailed deer had ≥ 85% woody canopy cover (Pollock et al. 1994). White-tailed deer typically bed in patches dominated by woody cover and forage where herbaceous vegetation dominates (Volk et al. 2007).

Because they forage in areas with herbaceous vegetation, white-tailed deer typically select vegetation communities that have a mixture of woody-plant-dominated and herbaceous-dominated patches (van der Hoek et al. 2002; Volk et al. 2007). Use of areas dominated by woody vegetation by white-tailed deer varies seasonally and with time of day. In the Tamaulipan Vegetation Region, for example, white-tailed deer used areas with 60–97% woody canopy cover during summer (Steuter and Wright 1980). In January, however, woody plant canopy cover did not influence use by white-tailed deer. In another study, vegetation height was strongly and positively related to relative probability of use during the day (Wiemers et al. 2014). Conversely, vegetation height was negatively related to relative probability of use at night. The negative relationship at night occurred because white-tailed deer were feeding in areas of herbaceous vegetation. A similar temporal and spatial pattern of habitat use was reported in the Great Plains in Kansas (Volk et al. 2007). In the Kansas study, white-tailed deer used areas dominated by woody vegetation at a course (6.25 ha) scale and avoided open grasslands. However, at a fine spatial scale deer that were foraging used open grassland. In the Tamaulipan Vegetation Region, vegetation height was less positively related to relative probability of use in the morning than at midday and was unrelated to relative probability of use in the evening (Wiemers et al. 2014). Selection of woody vegetation during summer and during midday is driven in part by the need for thermal cover to reduce heat loads. In the Great Plains of Colorado, Whittaker and Lindzey (2004) suggested that security cover was the primary driver of white-tailed habitat use. Woody vegetation provides both thermal and security cover and disentangling the two uses is difficult.

18.5 Rangeland Management

Grazing by domestic livestock is the dominant land use on rangelands. Consequently, responses of white-tailed deer habitat and populations to livestock and associated management practices including brush management, fencing, and water development are important considerations when managing white-tailed deer on rangelands.

18.5.1 Livestock Grazing

White-tailed deer management on rangeland is prone to error if the influence of domestic livestock is ignored. Cattle grazing and foraging by white-tailed deer are sometimes viewed as complimentary land uses because cattle primarily consume grass whereas white-tailed deer consume primarily forbs and browse (Fulbright and Ortega-S 2013). In fact, cattle grazing has been suggested as a tool to reduce grasses and increase forbs for wildlife (Lyons and Wright 2003). Nevertheless, the effect of cattle grazing on white-tailed deer and their habitat depends on factors such as season, management decisions regarding grazing intensity and stocking rate, and environment.

Livestock grazing may affect white-tailed deer through (1) competition resulting from diet overlap, (2) modifying species composition of plant communities, (3) social interactions, and (4) negative impacts on fawn production and survival. Diet overlap between cattle and white-tailed deer is greater during stress periods (e. g., winter and drought) and on overgrazed rangelands (Fulbright and Ortega-S 2013). Based on a quantitative review of literature on cattle grazing and white-tailed deer and mule deer, cattle and deer diet overlap ranged from 0.6 to 65% (Hines et al. 2021). Diet overlap was greatest during winter and spring. Diet overlap between cattle and deer increased 0.5% with every 0.1 AUY (Animal Unit Year) ha^{-1} increase in cattle stocking rate.

Domestic sheep (*Ovis aries*) and, to a lesser degree, goats (*Capra aegagrus*) compete with white-tailed deer for forbs (Bryant et al. 1979) and dietary overlap varies seasonally. For example, potential competition for forbs among goats, sheep, and white-tailed deer in the southern Great Plains (Edward's Plateau of Texas) is greatest during winter and early spring (Bryant et al. 1979). The effects of competition on population dynamics of white-tailed deer are unknown but likely vary depending on timing, stocking rates, and local conditions. In the Tamaulipan Vegetation Region, white-tailed deer were able to shift diet composition to less palatable shrubs when Angora goats depleted shrubs that were more palatable to deer (Ekblad et al. 1993). Consequently, both Angora goats and white-tailed deer were able to stabilize the nutrient content of their diet regardless of diet overlap. Indices of diet overlap between white-tailed deer and Angora goats in the Tamaulipan Vegetation Region study ranged from 0.75 to 0.88 when goats were stocked at 0, 2, 4, and 6 goats ha^{-1}.

Long-term, heavy livestock grazing can shift composition of plant communities from palatable plant species toward greater abundance of less palatable plant species. In general, a shift from grassland to dominance of woody vegetation may favor occupancy by white-tailed deer. Rangeland in the southern Great Plains (Edward's Plateau of Texas) with a history of heavy grazing had larger standing crop of browse than less heavily grazed rangeland (Bryant et al. 1981). White-tailed deer spent more time foraging, however, on heavily grazed than on lightly grazed rangeland suggesting that palatable forage was scarcer on heavily grazed rangeland. In addition, diet samples from lightly grazed rangeland were higher in crude protein and phosphorus, and, except for winter, in digestible energy. Forb diversity is greater on lightly than on heavily grazed rangeland in the southern Great Plains (Edward's Plateau; Warren and Krysl 1983).

Season, soil properties, and geographic location influence how cattle grazing affects standing crop or percent canopy cover of forbs (Hines et al. 2021). Forbs were more likely to decrease than increase in response to cattle grazing going from south to north across North America, possibly because the likelihood of grazing reducing forbs increased with cooler temperatures and shorter growing seasons. Forbs were more likely to increase in response to grazing going east and south across North America. In the drier ecosystems of western North America, variation in precipitation and amount of annual precipitation likely influenced forb response more than cattle grazing.

In theory, managing rangelands for increased plant species richness should benefit white-tailed deer because of the importance of plant diversity in optimizing the nutrient content of ruminant diets. Based on the intermediate disturbance hypothesis, plant species diversity may peak under moderate grazing intensities (Gao and Carmel 2020). On semiarid rangelands, however, plant species diversity may decline with increasing grazing intensity with the shape of the relationship depending on evolutionary history of grazing (Milchunas et al. 1988). Based on published literature, using livestock to increase species richness appears to be less applicable in semiarid rangeland ecosystems than in subhumid and humid parts of the distribution of white-tailed deer. Based on a meta-analysis of published papers, Gao and Carmel (2020) found that moderate grazing caused a slight increase in plant species richness in subhumid and humid areas, but plant species richness declined in arid and semiarid areas. Response of plant species richness to grazing intensity depended on the type of livestock. For example, in arid and semiarid areas species richness declined with grazing intensity with a mix of sheep and goats but grazing by sheep alone did not influence plant species richness.

The influence of grazing systems on white-tailed deer is unclear due to a lack of replicated research. In the southern Great Plains (Texas Edward's Plateau), white-tailed deer densities were greater under a seven-pasture, short duration system than under a Merrill three-herd, four-pasture system (Reardon et al. 1978). Results of several studies have suggested that continuous year-long grazing benefitted deer more than rotational grazing systems (Cohen et al. 1989; Martinez et al. 1997; Ortega et al. 1997a, b). In the Tamaulipan Vegetation Region, white-tailed deer avoided intense concentrations of cattle in short-duration grazing cells (Cohen et al. 1989). Precipitation and topo-edaphic conditions mediate the effects of livestock grazing management on wildlife responses (e.g., Lipsey and Naugle 2017) and additional research is needed to evaluate the effects of livestock management (e.g., grazing system, stocking rates and timing) on white-tailed deer in other rangeland ecosystems.

White-tailed deer avoid areas grazed by livestock if areas that are not grazed are available to them. In a review of 70 published papers on cattle-deer interactions which included mule deer, Hines et al. (2021) found that in two-thirds of the papers deer either increased home-range size or used an alternative vegetation community if cattle were present. In the Tamaulipan Vegetation Region, spatial distribution of deer and cattle overlapped in productive areas, but the two species used the areas at different times (Cooper et al. 2008). White-tailed deer tended to move if a cow approached to within 46 m. In Oregon, white-tailed deer avoided pastures grazed by cattle but used the grazed pastures two or three months after cattle removal (Gavin et al. 1984). Possibly, white-tailed deer used the previously grazed pastures because of greater plant species richness. Cattle and white-tailed deer heavily use riparian areas; however, white-tailed deer avoid these areas when cattle are present (Compton et al. 1988; Cooper et al. 2008). There is little spatial overlap between white-tailed deer and cattle in rocky areas, dense shrub communities with little herbaceous vegetation, and areas distant from water because these areas are avoided by cattle (Owens et al. 1991; Cooper et al. 2008).

White-tailed deer in Oregon selected areas with little or no use by cattle or sheep for fawning (Smith and Coblentz 2010). Females in areas with livestock made large shifts in their activity center for fawning; three of seven females established home ranges geographically separate from their annual home range. Females in areas with little or no livestock made small activity center shifts for fawning and used sites within their annual home range. Possibly, females shifted home ranges in areas with livestock because of reductions in height and cover of vegetation resulting from livestock grazing.

Cattle stocking rate had little influence on white-tailed deer densities under average precipitation and temperature conditions based on computer-simulation models of trends in white-tailed deer densities (Glasscock 2001). Combined effects of low winter temperatures, low precipitation, and heavy stocking rates caused rapid declines in deer densities. In contrast, number of fawns surviving to a year old declined with increasing stocking rates consisting of a combination of cattle, sheep, and goats in the southern Great Plains (Edward's Plateau of Texas; McMahan and Ramsey 1965). In Oklahoma and Arkansas fetuses female^{-1} declined with increasing cattle stocking rate (Jenks and Leslie 2003). White-tailed deer had 2 fetuses female^{-1} with no grazing, compared to 1.4 and 1.2 fetuses female^{-1} with moderate and heavy grazing, respectively.

18.5.2 *Brush Management and Vegetation Manipulation*

Brush management includes removal, reduction, or manipulation of woody vegetation (Hamilton et al. 2004). Brush management methods can be grouped broadly as fire, mechanical, and chemical approaches. Brush management has traditionally been applied to meet livestock needs such as increasing herbaceous forage (Fulbright et al. 2018). More recently, approaches to brush management have taken wildlife responses into account or have included improving white-tailed deer habitat as a goal. Efficacy of brush management in reducing woodland expansion in the Great Plains was questioned by Scholtz et al. (2021); brush management treatments were generally short-lived and woody cover showed little reduction at regional scales meaningful to population management.

Documenting whether or not brush management improved habitat quality for white-tailed deer is difficult. Vegetation metrics such as an increase in food plants, particularly forbs, or forage quality are often used to infer improved habitat quality (Fulbright et al. 2018). Basing inferences on forbs is inadequate because mast of woody plants such as mesquite is the primary item in the diet during summer and during drought when forbs are sparse. Ironically, killing mesquite is often a primary goal of brush management. Further, food plants are but one part of white-tailed deer habitat. Other habitat characteristics such as thermal and hiding cover are critically important. Consequently, animal metrics such as demographic characteristics or productivity metrics such as fecundity or body mass are more reliable indicators of changes in habitat quality than vegetation metrics (Van Horne 1983; Fulbright et al. 2018). Unfortunately, vegetation metrics are used more often to infer changes in habitat quality because animal data are expensive and time-consuming to collect.

18.5.3   Fire

Controlled and prescribed fire may alter food resources, cover, and patterns of habitat use by white-tailed deer. In the Southern Great Plains and Tamaulipan Vegetation Region, prescribed fire was the brush management approach that most consistently resulted in an increase in white-tailed deer food plants based on a review of literature published between 1966 and 2011 (Fulbright and Ortega-S 2013; Fig. 18.4). White-tailed deer are likely attracted to burned areas because forage quality (e.g., crude protein) of vegetation recovering post-fire is typically higher than mature, unburned vegetation (Fulbright and Ortega-S 2013). White-tailed deer may temporarily concentrate in burned patches. For example, use of resprouting shrubs in burned patches relative to use of shrubs in unburned areas peaked 12 to 20 weeks post-fire and remained higher up to 30 weeks post-fire (Fulbright et al. 2011). Although white-tailed deer are attracted to burned areas, they may maintain portions of their home range in unburned areas based on research in non-rangeland environments (Cherry et al. 2018).

Burning may alter predator–prey relationships. For example, although burning increased high-quality forage in a forested area of Georgia, white-tailed deer avoided recently burned areas (Cherry et al. 2017). Similar results have been reported on rangeland. For example, white-tailed deer in the northern Great Plains also avoided burned areas during the first winter after fire (Dubreuil 2003). In both studies, researchers attributed avoidance of recently burned areas to a lack of cover increasing susceptibility to predation. Predators may be attracted to burned areas where prey concentrate. Although not documented on rangeland, panthers (*Felis concolor*) in Florida are attracted to prescribed burns < 1 year old possibly because of higher numbers of white-tailed deer and other prey (Dees et al. 2001).

Males may use burned areas differently than females. Differential use between sexes may influence how beneficial burning is in managing white-tailed deer habitat; however, research on this topic is minimal on rangelands. Female white-tailed deer

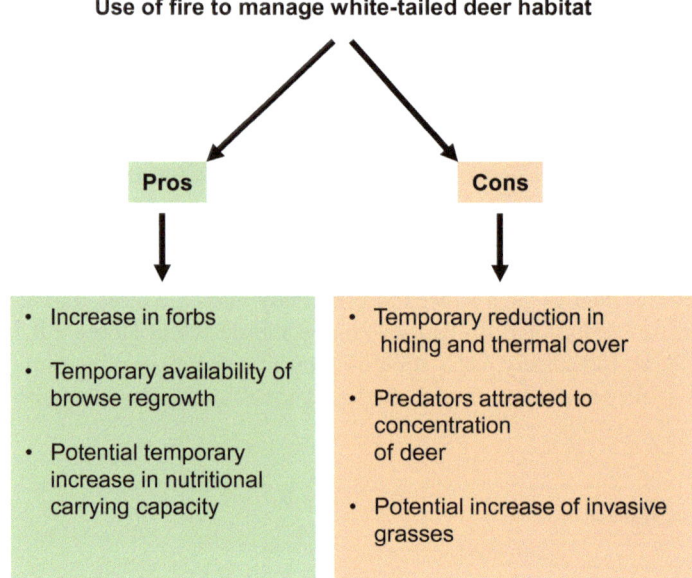

Fig. 18.4 Pros and cons associated with the use of fire to manage white-tailed deer habitat

in Georgia and North Carolina avoided recently burned patches possibly because of the lack of cover (Lashley et al. 2015; Cherry et al. 2017). Males may be more prone to take advantage of the improved forage quality after fire (Lashley et al. 2015). In the eastern Great Plains of Oklahoma, male and female white-tailed deer exhibited differential selection for fire and herbicide treatments (Leslie et al. 1996). For example, based on pooling two years of data, male deer avoided a treatment with fire and no herbicides during spring and autumn. In comparison, females used the fire with no herbicide treatment in proportion to availability during all seasons. Males selected a treatment with a combination of triclopyr application and fire during summer and autumn. Females avoided the treatment during autumn and winter and selected the triclopyr and fire treatment during spring.

Evidence that fire temporarily improves habitat quality for white-tailed deer is largely based on vegetation metrics and from a restricted geography outside of rangelands. For example, a one-year fire return interval in pine-hardwood forests in Alabama increased estimated nutritional carrying capacity (Glow et al. 2019). There is limited evidence that fire may benefit measures of white-tailed deer productivity such as fawn survival or antler size. In one of the few studies on fire and deer productivity, fawn biomass, and antler size of two-year-old males were greater during the initial year after fire (Springer 1977).

18.5.4 Mechanical

Mechanical treatments on rangeland include brush management and activities such as haying and mowing. Brush management ranges from removal of individual woody plants by hand-grubbing to use of heavy equipment to uproot plants (Hamilton et al. 2004). Selective removal of individual plants to reduce woody plant density has been referred to as "brush sculpting" (Ansley et al. 2003). Forms of brush management such as roller chopping remove top growth of woody plants leaving the crowns and roots of the plants intact. Re-sprouting woody plant species such as honey mesquite produce sprouts from buds in the crowns and quickly produce new sprouts following top removal.

White-tailed deer can shift diet composition and maintain diet quality when brush management has altered vegetation composition. For example, woody vegetation re-establishes after root-plowing in a decade or two, but the re-established woody plant community may lack woody plant species important for browse (Fulbright and Beasom 1987; Ruthven et al. 1993, 1994). Seventeen years after root plowing in the eastern Rio Grande Plains of Texas, root-plowed sites were dominated by huisache (*Vachellia farnesiana*) compared to mesquite-mixed brush in untreated areas. White-tailed deer tended to eat more browse and less huisache mast and forbs in untreated than in root-plowed sites (Ruthven et al. 1994). However, reproductive measures and population status of white-tailed deer were similar in untreated and root-plowed areas. The temporal scales of mechanical treatment effects likely vary across rangeland types in relation to a variety of local conditions (e.g., soil properties, precipitation, production potential), but information in most rangeland systems are lacking.

Effects of mechanical brush management on white-tailed deer vary depending on scale and pattern of application. Clearing large tracts of brushland to create extensive grassland with no woody cover reduces white-tailed deer densities (McMahan and Inglis 1974; Darr and Klebenow 1975). Conversely, reducing woody vegetation canopy cover to 50–70% may not reduce white-tailed deer densities (Rollins et al. 1988), although effects of treatment scale and proximity of untreated woody vegetation were not taken into account.

Brush management is often done in strips or other patterns to create mosaics of woody plant-dominated patches and interspersed herbaceous-dominated patches (Archer et al. 2011). Root-plowing to create a mosaic consisting of an alternating sequence of 85-m-wide woody-plant-dominated strips separated by 95-m-wide root plowed strips in the Tamaulipan Vegetation Region had little effect on white-tailed deer home range size or placement (Dykes 2022).

As with fire, males may use brush management treatments differently than females (Stewart et al. 2003). In the Tamaulipan Vegetation Region, for example, adult females used roller chopped strips more than untreated strips regardless of season. In contrast, adult males used roller chopped strips more than untreated strips during autumn but not during spring.

There is no clear scale or pattern of clearing woody plants and creating woody-plant strips or clusters that is optimal for white-tailed deer. A wide range of scales and patterns may exist across which demographic or measures of white-tailed deer productivity are similar. Selection of scales and patterns by range and wildlife managers is based on economics and aesthetics more than knowledge of optimum treatment designs (Fulbright and Ortega-S 2013; Fulbright et al. 2018). From an economic perspective, clearing linear strips is more cost-effective than creating shrub clusters. In regard to aesthetics, humans perceive savanna-like landscapes as more pleasing than woody-plant dominated landscapes (Ulrich et al. 1991). The human-value orientation influencing woody plant management decisions is well illustrated by use of terms such as "brush sculpting." It is difficult to disentangle what is beneficial to white-tailed deer from what is perceived as aesthetically pleasing by humans because of our incomplete knowledge of the effects of brush management on white-tailed deer at the population level.

As with fire, documentation that mechanical brush management improves habitat quality for white-tailed deer is largely based on vegetation metrics. A knowledge gap exists regarding the question of whether or not mechanical brush management can be used to improve habitat quality based on demographic or productivity metrics of white-tailed deer.

Mowing of grassland to reduce vegetation height has been used to increase use by white-tailed deer (Washburn and Seamans 2007). However, hayed grassland may be avoided by fawns until vegetation regrows enough to provide cover (Grovenburg et al. 2012a).

18.5.5 Chemical

Chemical treatments are usually applied on rangeland to manage woody vegetation, but they have also been used to manipulate grassland structure and composition for wildlife (Washburn and Seamans 2007). Herbicides can be broadly grouped as soil applied or foliar applied. Methods of application range from applying herbicide to the trunks or canopy of individual woody plants to large-scale broadcast applications from aircraft. Herbicides can be applied in a brush sculpting fashion to kill individual woody plants. Herbicides can also be applied in patterns of alternating treated and untreated strips (Fulbright and Garza 1991). A slightly more complex mosaic pattern created with herbicides is the variable rate pattern (Scifres et al. 1988). Variable rate patterns of herbicide application were developed to increase grass for livestock while leaving adequate woody vegetation for wildlife. In the variable rate pattern, a checkerboard of patches of woody vegetation receives different rates of herbicides. The result is that only a portion of woody plants are killed in some patches and woody plants are totally killed in others.

White-tailed deer may temporarily leave areas aerially treated with broadcast herbicides. One of the possible reasons for this is that some herbicides kill forbs; in addition, herbicides may reduce browse and mast. White-tailed deer densities may return to pre-treatment densities once forbs reestablish (Beasom and Scifres 1977). In South Texas, treating 100% of an area resulted in a 40% reduction in white-tailed deer densities between 15 and 27 months after treatment. However, densities returned to pre-treatment levels 27 months post-treatment.

Herbicide treatments applied in a mosaic fashion may have a neutral to potentially beneficial effect on habitat quality. For example, applying tebuthiuron in alternating treated and untreated strips had little overall effect on white-tailed deer nutritional status in live-oak (*Quercus fusiformis*) dominated rangeland in South Texas (Fulbright and Garza 1991). In Oklahoma, treating Cross Timbers and Prairies vegetation in a mosaic of herbicide treatments and annual spring burning resulted in greater white-tailed deer body mass and dietary nitrogen concentrations (Soper et al. 1993).

18.5.6 Managing for Heterogeneity

Standing crop and species composition of herbaceous vegetation in rangeland systems varies in space and time. A traditional paradigm in rangeland management has been focused on stabilizing or increasing livestock productivity by increasing rangeland homogeneity (Fuhlendorf and Engle 2001; Wilcox et al. 2021). A paradigm shift in contemporary rangeland management is to increase spatial heterogeneity (Fuhlendorf et al. 2017). One approach is use of pyric-herbivory to increase spatial heterogeneity by incorporating a combination of fire and grazing to create a mosaic of patches differing in grazing intensity and time since fire (McGranahan et al. 2012). In the southern Great Plains, livestock productivity decreased with declining precipitation in more homogeneous environments (Allred et al. 2014). In heterogeneous environments, livestock productivity was unrelated to precipitation.

White-tailed deer may also benefit from increased landscape heterogeneity resulting from pryric-herbivory. For example, fire and grazing have been used to reduce grass canopy cover and increase forbs (Ramirez-Yanez et al. 2007). Theoretically, creating a mosaic of different forb guilds and successional states increases the diversity of foods available to deer, perhaps conferring nutritional benefits. Fawn survival may be higher in heterogeneous landscapes than in more homogeneous landscapes (Rohm et al. 2007; Grovenburg et al. 2012c; Gulsby et al. 2017; Kilburn 2018). Reasons for higher fawn survival in heterogeneous landscapes are unclear. Several explanations have been proposed, such as higher quality of food in heterogeneous areas allowing females to have smaller home ranges and additional time for defense and nursing of fawns, high availability of food for predators in heterogeneous areas buffering predation, and reduced susceptibility of fawns to predation because females travel more rapidly and further in homogeneous environments (Rohm et al. 2007; Gulsby et al. 2017; Kilburn 2018).

Much of the research on pyric-herbivory has been done in the tallgrass prairies of the Great Plains. Rangelands that are more arid or semiarid often have more inherent heterogeneous spatial structure consisting of patches of bare ground or sparse herbaceous vegetation and patches of perennial grasses or woody plants (Aguiar and Sala 1999, van de Koppel et al. 2002; Segoli et al. 2012). Spatial redistribution of surface water or nutrients is an important ecosystem process to maintain productivity in these systems. The focus of management in more arid and semiarid systems may be maintaining heterogeneity and ecosystem function rather than trying to create it.

Habitat heterogeneity in rangelands is temporal as well as spatial. High variation in precipitation drives variation in the composition and structure of vegetation and, as a result, the abundance of deer foods. In the Tamaulipan Vegetation Region, for example, annual precipitation across six study sites during 2012–2019 varied from 28.9 to 84.5 cm (Fulbright et al. 2021). Standing crop of forbs selected by white-tailed deer during that period varied more than four-fold, from 82 to 442 kg ha^{-1}. On rangelands with highly variable precipitation, season, soil texture, and precipitation may have a greater impact on standing crop of forbs than grazing by ungulates (Fulbright et al. 2021).

Most studies of the influence of white-tailed deer on vegetation have been conducted in the mid-western and eastern United States. Herbivory by white-tailed deer in the eastern portion of their range strongly influences composition of understory vegetation (Frerker et al. 2014, Habeck and Schultz 2015). On rangelands, effects of foraging by white-tailed deer on plant community composition appears to be less dramatic (DeYoung et al. 2019; Bloodworth et al. 2020). Research on rangelands is limited, however, and additional research is needed to clarify how deer impact vegetation in different rangeland plant communities and ecosystems where vegetation dynamics may follow equilibrium or non-equlibrium dynamics, or a combination.

18.5.7   Habitat Restoration

In the Northern Great Plains, restoration of herbaceous vegetation through the Conservation Reserve Program (CRP) has increased white-tailed deer occurrence and abundance (Nagy-Reis et al. 2019). Fawns in the northern Great Plains selected CRP over other vegetation types (Grovenburg et al. 2012a). Interestingly, revenues from hunting as a result of Conservation Reserve Program plantings override the net economic effect of losses in crop production revenues (Bangsund et al. 2004).

18.5.8   Water Development

White-tailed deer drink from earthen ponds and concrete water troughs associated with water wells constructed for livestock (Prasad and Guthery 1986; Fulbright et al. 2023). However, the importance of these water sources to white-tailed deer is unclear. We do not know for sure if white-tailed deer require free-standing water or if they can meet their needs with preformed water (dietary moisture) in forage. However, we do know that white-tailed deer use free-standing water when it is available.

White-tailed deer drank free-standing water from concrete troughs in South Texas in exclosures with no livestock (Fulbright et al. 2023). Male white-tailed deer drank an average of 1.57 gallons of water month^{-1}. Females consumed an average of 1.33 gallons month^{-1} with a minimum and maximum monthly average of 0.08 and 4.8 gallons, respectively. In the Tamaulipan Vegetation Region, white-tailed deer avoided concrete water troughs at the center of short-duration grazing cells possibly because of heavy use by livestock and increased human presence (Prasad and Guthery 1986).

18.5.9 Fencing

Livestock fencing is a semipermeable barrier to white-tailed deer (Burkholder et al. 2018). White-tailed deer in the northern Great Plains preferred to crawl under fences rather than jumping over them. In the northern Great Plains, odds of a white-tailed deer successfully crossing a fence increased with increasing height of the bottom wire (Jones et al. 2020). Increasing height reduces the number of deer jumping over the fence; 14% fewer deer jump fences 1.8 m tall compared to 1.5 m tall (VerCauteren et al. 2010). White-tailed deer can become entangled in wire fences but the relative importance of fences as a cause of mortality on rangeland is unclear. Entanglement in fences was a minor cause of mortality in the Tamaulipan Vegetation region (Webb et al. 2007). Webb et al. (2007) tracked 48 mature male white-tailed deer for two years; out of 21 mortalities they recorded one that resulted from fence entanglement.

18.6 Impacts of Disease

Diseases are an important management concern for deer because outbreaks of some diseases can reduce populations; whereas, others are transmissible to humans and livestock. Epizootic hermorrhagic disease is the most important cause of viral-related mortality in white-tailed deer (Christiansen et al. 2021b); however, direct population-level effects are poorly documented (Gaydos et al. 2004). Epizootic hemorrhagic disease is transmitted by biting midges (*Culicoides* spp.; Stevens et al. 2015). Losses of white-tailed deer in the northern Great Plains from epizootic hermorrhagic disease are normally minor but can be large (South Dakota Game, Fish, and Parks 2022). The disease was implicated in a population decline in the northern Great Plains during the late 1970s (Dusek et al. 1989). Epizootic hemorrhagic disease also affects cattle; however, they rarely exhibit clinical signs of the disease (Campbell and VerCauteren 2011; Stevens et al. 2015). Baiting and feeding of deer increase the probability of direct transmission of the disease from infected animals (Rivera et al. 2021). In addition, white-tailed deer kept in breeding pens have increased prevalence of epizootic hemorrhagic disease (Rivera 2021). In one study, presence of captive white-tailed deer resulted in higher infection rates of epizootic hemorrhagic disease among cattle (Becker et al. 2020).

Bluetongue is a viral disease closely related to epizootic hemorrhagic disease and is also transmitted by biting midges in the genus *Culicoides* (Campbell and VerCauteren 2011). All ruminants are susceptible to being infected by bluetongue but the disease is most common in sheep (Sperlova and Zendulkova 2011). Although bluetongue is less common in white-tailed deer than epizootic hemorrhagic disease, serious outbreaks sometimes occur. For example, up to 10,000 white-tails perished from the disease in an outbreak in Idaho in 2011 (Phillips 2015).

Bacterial diseases in white-tailed deer include anthrax, dermatophilosis, brain abscesses, bovine tuberculosis, paratuberculosis, leptospirosis, salmonella, and lyme disease (Campbell and VerCauteren 2011). Of these, anthrax is the deadliest. Anthrax is relatively uncommon with the most frequent outbreaks in white-tailed deer occurring in southwestern Texas (Blackburn and Goodin 2013; Mullins et al. 2015). Population-level effects of dermatophilosis and bacterial dermatologic diseases in free-ranging white-tailed deer are probably minimal (Nemeth et al. 2014).

Bovine tuberculosis primarily affects cattle; however, white-tailed deer can contract the disease and are the primary maintenance host of the disease in North America (Carstensen et al. 2008; Campbell and VerCauteren 2011). The disease is endemic to a five-county area in Michigan and an area around Riding Mountain National Park in Manitoba, Canada (Atwood et al. 2007; Brook et al. 2013). Potential for transmission of the disease from white-tailed deer to cattle can be reduced by protecting cattle feeders from white-tailed deer, reducing deer densities, and other strategies than minimize contact between deer and cattle (Campbell and VerCauteren 2011; Brook et al. 2013). White-tailed deer are not an important reservoir for paratuberculosis, which is uncommon in wild ruminants (Campbell and VerCauteren 2011).

In Mexico, 5.6% of white-tailed deer tested had antibodies against leptospirosis (Cantu-C et al. 2008). Probability of deer testing positive for leptospirosis was 3.6 times greater where cattle were continuously grazed than if they were rotationally grazed. In a survey in the United States, about 40% of white-tailed deer tested had titers to the serovars of *Leptospira*; however, only 3% of the animals tested demonstrated recent infection (Pedersen et al. 2018). Pedersen et al. (2018) concluded that white-tailed deer could be important contributors to the cycle of infection of leoptospirosis and may be involved in transmission of the disease to livestock.

Chronic wasting disease (CWD) is a transmissible spongiform encephalopathy. It is not caused by a bacterium or a virus, but rather a prion which is a misfolded form of a protein. First discovered in captive deer in the 1960s, the first case of CWD in free-ranging wildlife was discovered in mule deer in Colorado in 1980. The disease has spread widely since 1980 and affects white-tailed deer from southern Canada to Texas. Chronic wasting disease is of particular concern on rangelands because white-tailed deer in semiarid environments depend on high adult survivorship to counter low fawn recruitment. Using simulation models, Foley et al. (2016) found that CWD increases additive mortality. Annually, white-tailed deer with CWD are 4.5 times more likely to die than those testing negative for the disease (Edmunds et al. 2016). Chronic wasting disease has the potential to limit white-tailed deer populations if the disease becomes endemic (Edmunds et al. 2016).

White-tailed deer serve as hosts for cattle fever ticks (Fulbright and Ortega-S 2013). This creates a challenge for tick eradication because white-tailed deer are free ranging and highly mobile (Currie et al. 2020). Cattle fever ticks were considered eradicated from the United States in 1943 (Thomas et al. 2020). However, since 2008 fever tick infestations in Texas near the border with Mexico have increased. Researchers in the region have been developing ways to reduce fever tick infestations in white-tailed deer. For example, consumption of ivermectin-medicated corn reduces the probability of infestation with cattle fever ticks in white-tailed deer (Currie et al., 2020).

White-tailed deer are highly susceptible to acute respiratory syndrome coronavirus 2 (SARS-Co-V-2; Palmer et al. 2021). Consequently, they are a potential reservoir of the disease that could be transmitted to humans (Palermo et al. 2022).

18.7 Ecosystem Threats

Continued spread of chronic wasting disease is a major threat to white-tailed deer. The disease is spreading rapidly in North America (Escobar et al. 2020). In the United States, the disease has been reported in wild cervids from Idaho, Montana, and South Dakota south to Texas (Centers for Disease Control 2022). Where the disease is well established, infection rates in free-ranging deer and elk may exceed 10% and cases with infection rates > 25% have been reported. Infection rates are highest in captive herds, reaching 80–90% in certain cases (Haley and Hoover 2015). Chronic wasting disease has been detected in > 175 captive cervid facilities (Carlson et al. 2018). Spread of chronic wasting disease occurs through natural movements of infected animals, movement of infected captive cervids by humans, and escape of infected animals from captive facilities (Carlson et al. 2018; Rivera et al. 2019).

Energy development also alters white-tailed deer habitat on rangelands. Published results of research on the effects of energy development on white-tailed deer is limited. In North and South Dakota, oil and gas development did not appear to alter survival and health of white-tailed deer (Moratz 2016). However, oil and gas development did alter distribution of white-tailed deer (Gullikson 2019); white-tailed deer avoided well pads and avoided areas with oil field development at the population level during summer. For similar reasons, renewable energy development has the potential to reduce white-tailed deer habitat on rangelands. However, the effects on white-tailed deer of wind farms and solar parks and their associated road networks and infrastructure are unknown.

A variety of exotic ungulates have been introduced in areas occupied by white-tailed deer. Potential negative interactions between exotics and white-tailed deer on western US rangelands is primarily restricted to Texas. There were more than two million exotic animals of about 135 species in Texas in the early twenty-first century

(Gill 2020). Axis deer (*Axis axis*), fallow deer (*Dama dama*), and sika deer (*Cervus nippon*) are among the most abundant exotics that potentially compete directly with white-tailed deer. These exotic deer species can consume a diet high in grass (Henke et al. 1988). White-tailed deer, in contrast, cannot digest grass as efficiently so exotic deer species have a competitive advantage when forbs and browse are limited in availability. Competition between sika deer and white-tailed deer in Maryland resulted in white-tailed deer consuming lower quality forage (Kalb et al. 2018). As a result of their competitive ability, exotic deer species have the potential to reduce productivity of white-tailed deer and displace them from higher quality habitat (Faas and Weckerly 2010).

Climate change presents a paradox for white-tailed deer with differing effects in the southern and northern parts of their range. In the southern part of their geographic distribution, warming may cause changes in white-tailed deer behavior to cope with higher temperatures. White-tailed deer have few physiological adaptations to reduce heat loads. They can reduce heat loads by panting, but panting results in water loss, which may be maladaptive in dry rangeland environments. White-tailed deer therefore rely primarily on behavioral adaptations such as reducing activity and seeking shade to deal with excessive heat. In Minnesota, white-tailed deer were active at temperatures between 6 and 16 °C but became less active as temperatures warmed above 16 °C (Beier and McCullough 1990). Although white-tailed deer are crepuscular, warming temperatures may cause them to be more active at night when temperatures are cooler. White-tailed deer increased the amount of time they fed at night to avoid hot daytime temperatures in Mississippi (Wolff et al. 2020). In the hot, dry rangelands of the Tamaulipan Vegetation Region, white-tailed deer selected taller vegetation with the lowest operative temperature during morning and midday (Wiemers et al. 2014). Higher temperatures resulting from climate change may make availability of thermal cover even more important to white-tailed deer. White-tailed deer may also alter their behavior to deal with effects of extreme climatic events such as hurricanes that are predicted by climatologists to increase in strength and frequency with climate change (Abernathy et al. 2019).

Climate change may be at least partly responsible for expansion of the range of white-tailed deer in the northern part of their geographic distribution. Dawe and Boutin (2016) modeled the effect of climate change on white-tailed deer distribution in the boreal forest of North America. Their model predicted that during the first half of the twenty-first century the range of white-tailed deer will expand 100 km further north in northeastern Alberta. In Ontario, Kennedy-Slaney et al. (2018) used simulation models to predict that northward expansion of white-tailed deer will not be limited by severe winters by 2100. Weiskopf et al. (2019) predicted that white-tailed deer will become more abundant in the midwestern United States as a result of climate change. Factors facilitating greater abundance included increased survival because warmer temperatures reduced snowpacks.

18.8   Conservation and Management Actions

Chronic wasting disease is an insidious threat to white-tailed deer throughout the United States. The disease has been reported in 27 states (Centers for Disease Control and Prevention 2022). The primary mechanism of spread for CWD is movement of live animals by humans (Miller and Fischer 2016). Approaches to containing the spread of the disease include local population reduction, regulating the translocation of white-tailed deer and other cervids by humans, and bans on baiting and feeding (Campbell and VerCauteren 2011). Culling of host animals and restrictions on export of meat are also recommended (Mysterud et al. 2021).

Hunting is the primary tool for managing white-tailed deer populations throughout their range (Woolf and Roseberry 1998; Brown et al. 2000; McShea 2012). About 10 million hunters pursued white-tailed deer annually during 2010–2013 (Hewitt 2015). Hunting white-tailed deer provides significant economic benefit to ranchers and other landowners in rangelands of the western United States. In the United States, about 33% of the private land is leased or owned for wildlife-related recreation (Macaulay 2016). Ranching enterprises with a combination of livestock production and hunting have a higher internal rate of return than enterprises with only livestock or only hunting (Genho et al. 2003) and fee hunting will likely compose a larger component of the diverse economies that maintain private ranching operations in the future (Chap. 27). In addition, potential for wildlife-related recreation adds more to real estate values than potential for agricultural production (Baen 1997; Haggerty et al. 2018).

Management of white-tailed deer in the western United States is typically of low intensity on public and private land (Jacobson et al. 2011). However, popularity of more intensive white-tailed deer management is growing. Intensive management typically occurs on private lands and is directed at increasing antler size and managing for older males (Jacobson et al. 2011). Tools of intensive management including high fences that restrict deer ingress and egress and supplemental feeding (Knox 2011). Intensive management has evolved into a deer breeding industry in which captive deer are bred for large antlers. Objections to intensive deer management including privatization of deer, which are considered a publically-owned resource in the United States; lack of fair-chase hunting; reducing the wildness of deer; and exacerbating the spread of chronic wasting disease and other diseases.

18.9   Research and Management Needs

Comparisons of the effect of different white-tailed deer population densities on population dynamics is needed in rangeland ecosystems with highly variable precipitation and low soil fertility to determine the usefulness of simple density dependent population models for management. Demographic responses of white-tailed deer to different livestock grazing intensities and grazing strategies represent a gap in our knowledge of white-tailed deer-livestock interactions. In particular, livestock grazing may reduce hiding cover for fawns, making them more susceptible to predation. Greater understanding of these interactions is important because dual white-tailed deer hunting and livestock production offer greater returns to ranchers than livestock alone in areas with huntable white-tailed deer populations.

Differential use by males and females may dictate benefits of prescribed burning in managing white-tailed deer habitat. Research on differential use of burned areas and the effects of fire on survival, productivity, and population growth of white-tailed deer is lacking in rangeland environments. Demographic and productivity responses of white-tailed deer to mechanical and chemical brush management on rangelands are also a gap in our knowledge.

Non-native grasses are often planted following brush management. Non-native grasses also have invaded large areas of white-tailed deer habitat on rangelands (Fulbright et al. 2013). White-tailed deer have been implicated in the spread of non-native plants in the eastern United States (Averill et al. 2018). Research is needed to determine if white-tailed deer have a role in dissemination of non-native plant seeds on rangelands.

There is a knowledge gap regarding spatial heterogeneity of vegetation and white-tailed deer nutrition and population ecology on rangelands, especially on fawn survival. In addition, we need to develop a better understanding of the influence of white-tailed deer foraging on vegetation dynamics in different rangeland ecosystems.

There is reason for concern about the impacts of renewable energy development because there is evidence of negative effects of wind farms on other deer species. For example, roe deer (*Capreolus capreolus*) in Poland have elevated stress levels in response to large wind farms (Klich et al. 2020). They avoid the interior part of wind farms and avoid proximity to wind turbines (Łopucki et al. 2017). Extensive road networks associated with wind farms increase potential for invasion of non-native plants (Keehn and Feldman 2018) that may degrade white-tailed deer habitat.

Acknowledgements I thank David Hewitt and Mike Cherry for reviewing early drafts of the manuscript. This chapter is CKWRI manuscript number 22-104.

References

Abernathy HN, Crawford DA, Garrison EP, Chandler RB, Conner ML, Miller KV, Cherry MJ (2019) Deer movement and resource selection during hurricane Irma: implications for extreme climatic events and wildlife. Proc Royal Soc B 286. https://doi.org/10.1098/rspb.2019.2230

Aguiar MR, Sala OE (1999) Patch structure, dynamics and implications for the functioning of arid ecosystems. Trends Ecol Evol 14:273–277

Anderson C Jr, Moody D, Smith B, Lindzey F, Lanka R (1998) Development of sightability models for summer elk surveys. J Wildl Manag 62:1055–1066. https://doi.org/10.2307/380255

Allred BW, Scasta JD, Hovick T, Fuhlendorf SD, Hamilton RG (2014) Spatial heterogeneity stabilizes livestock productivity in a changing climate. Agric Ecosyst Environ 193:37–41. https://doi.org/10.1016/j.agee.2014.04.020

Ansley RJ, Kramp BA, Jones DL (2003) Converting mesquite thickets to savanna through foliage modification with clopyralid. J Range Manag 56:72–80

Archer SR, Davies KW, Fulbright TE, McDaniel KC, Wilcox BP, Predick KI (2011) Brush management as a rangeland conservation strategy: a critical evaluation. In: Briske DD (ed) Conservation benefits of rangeland practices: assessment, recommendations, and knowledge gaps. USDA Natural Resources Conservation Service, Washington, DC, pp 105–170

Atwood TC, VerCauteren KC, DeLiberto TJ, Smith HJ, Stevenson JS (2007) Coyotes as sentinels for monitoring bovine tuberculosis prevalence in white-tailed deer. J Wildl Manag 71:1545–1554. https://doi.org/10.2193/2006-441

Averill KM, Mortensen DA, Smithwick EAH, Kalisz S, McShea WJ, Bourg NA, Parker JD, Royo AA, Abrams MD, Apsley DK, Blossey B, Boucher DH, Caraher KL, DiTommaso A, Johnson SE, Masson R, Nuzzo VA (2018) A regional assessment of white-tailed deer effects on plant invasion. AoB PLANTS 10. https://doi.org/10.1093/aobpla/plx047

Baen J (1997) The growing importance and value implications of recreational hunting leases to agricultural land investors. J Real Estate Res 14:399–414. https://doi.org/10.1080/1083557. 1997.12090909

Bangsund DA, Hodur NM, Leistritz FL (2004) Agricultural and recreational impacts of the conservation reserve program in Noarth Dakota, USA. J Environ Manage 71:293–303. https://doi.org/ 10.1016/j.jenvman.2003.12.017

Bartush WS, Lewis JC (1981) Mortality of white-tailed deer fawns in the Wichita Mountains. Proc Oklahoma Acad Sci 61:23–27

Beasom SL, Scifres CJ (1977) Population reactions of selected game species to aerial herbicide applications in South Texas. J Range Manag 30:138–143. https://doi.org/10.2307/3897757

Becker ME, Roberts J, Schroeder ME, Gentry G, Foil LD (2020) Prospective study of epizootic hemorrhagic disease virus and bluetongue virus transmission in captive ruminants. J Med Entomol 57:1277–1285. https://doi.org/10.1093/jme/tjaa027

Beier P, McCullough DR (1990) Factors influencing white-tailed deer activity patterns and habitat use. Wildl Monogr 109:3–51

Blackburn JK, Goodin DG (2013) Differentiation of springtime vegetation indices associated with summer anthrax epizootics in West Texas, USA, deer. J Wildl Dis 49:699–703. https://doi.org/ 10.7589/2012-10-253

Bloodworth KJ, Ritchie ME, Komatsu KJ (2020) Effects of white-tailed deer exclusion on the plant community composition of an upland tallgrass prairie ecosystem. J Veg Sci 31:899–907. https:/ /doi.org/10.1111/jvs.12910

Briske DD, Fuhlendorf SD, Smeins FE (2003) Vegetation dynamics on rangelands: a critique of the current paradigms. J Appl Ecol 40:601–614

Brook RK, Vander Wal E, van Beest FM, McLachlan S (2013) Evaluating use of cattle winter feeding areas by elk and white-tailed deer: implications for managing bovine tuberculosis transmission risk from the ground up. Preventative Vet Med 108:137–147. https://doi.org/10.1016/j.prevet med.2012.07.017

Brown TL, Decker DJ, Riley SJ, Enck JW, Lauber TB, Curtis PD, Mattfeld GF (2000) The future of hunting as a mechanism to control white-tailed deer populations. Wildl Soc Bull 28:797–807

Bryant FC, Kothmann MM, Merrill LB (1979) Diets of sheep, angora goats, Spanish goats, and white-tailed deer under excellent range conditions. J Range Manag 32:412–417

Bryant FC, Taylor CA, Merrill LB (1981) White-tailed deer diets from pastures in excellent and poor range condition. J Range Manag 34:193–200

Burkholder EN, Jakes AF, Jones PF, Hebblewhite M, Bishop CJ (2018) To jump or not to jump: mule deer and white-tailed deer fence crossing decisions. Wildl Soc Bull 42:420–429. https:// doi.org/10.1002/wsb.898

Campbell TA, VerCauteren KC (2011) Diseases and parasites. In: Hewitt DG (ed) Biology and management of white-tailed deer. CRC Press, Boca Raton, FL, pp 219–249

Cantu-C A, Ortega-S JA, Mosqueda J, Garcia-Vasquez Z, Henke SE, George JE (2008) Prevalence of infectious agents in free-ranging white-tailed deer in northeastern Mexico. J Wildl Dis 44:1002–1007. https://doi.org/10.7589/0090-3558-44.4.1002

Carlson CM, Hopkins MC, Nguyen NT, Richards BJ, Walsh DP, Walter WD (2018) Chronic wasting disease: status, science, and management support by the U. S. Geological Survey. US Department of the Interior. https://pubs.usgs.gov

Carstensen M, Butler E, DonCarlos M, Cornicelli L (2008) Managing bovine tuberculosis in white-tailed deer in northwestern Minnesota: a 2008 progress report. Mich Bovine Tuberc Bibliography Database 18. https://digitalcommons.unl.edu/michbovinetb/18

Centers for Disease Control and Prevention (2022) Chronic wasting disease. https://www.cdc.gov/prions/cwd/index.html

Cherry MC, Warren RJ, Conner LM (2017) Fire-mediated foraging tradeoffs in white-tailed deer. Ecosphere 8:e01784. https://doi.org/10.1002/ecs2.1784

Cherry MC, Chandler RB, Garrison EP, Crawford DA, Kelly BD, Shindle DB, Godsea KG, Miller KV, Conner LM (2018) Wildfire affects space use and movement of white-tailed deer in a tropical pyric landscape. For Ecol Manage 409:161–169. https://doi.org/10.1016/j.foreco.2017.11.007

Chitwood MC, Lashley MA, Kilgo JC, Pollock KH, Moorman CE, DePerno CS (2015) Do biological and bedsite characteristics influence survival of neonatal white-tailed deer? PLoS ONE 10:e0119070. https://doi.org/10.1371/journal.pone.0119070

Chrétien L, Théau J, Ménard P (2016) Visible and thermal infrared remote sensing for the detection of white-tailed deer using an unmanned aerial system. Wildl Soc Bull 40:181–191. https://doi.org/10.1002/wsb.629

Christensen S, Farr M, Williams D (2021a) Assessment and novel application of N-mixture models for aerial surveys of wildlife. Ecosphere 12:e03725. https://doi.org/10.1002/ecs2.3725

Christensen SJ, Williams DM, Rudolph BA, Porter WF (2021b) Spatial variation of white-tailed deer (*Odocoileus virginianus*) population impacts and recovery from epizootic hemorrhagic disease. J Wildl Dis 57:82–93. https://doi.org/10.7589/JWD-D-20-00030

Cohen WE, Drawe DL, Bryant FC, Bradley LC (1989) Observations on white-tailed deer and habitat response to livestock grazing in South Texas. J Range Manag 42:361–365

Compton BB, Mackie RJ, Dusek GL (1988) Factors influencing distribution of white-tailed deer in riparian habitats. J Wildl Manag 52:544–548. https://doi.org/10.2307/3801607

Cooper SM, Perotto-Baldivieso HL, Owens MK, Meek MG, Figueroa-Pagán (2008) Distribution and interaction of white-tailed deer and cattle in a semi-arid grazing system. Agric Ecosyst Environ 127:85–92. https://doi.org/10.1016/j.agee.2008.03.004

Currie CR, Hewitt DG, Ortega-S JA, Shuster GL, Campbell TA, Lohmeyer KH, Wester DG, Pérez de León A (2020) Efficacy of white-tailed deer (Odocoileus virginianus) treatment for cattle fever ticks in southern Texas. J Wildl Dis 56:588–596. https://doi.org/10.7589/2015-11-304

Darr GW, Klebenow DA (1975) Deer, brush control, and livestock on the Texas Rollings Plains. J Range Manag 25:115–119

Darr RL, Williamson KM, Garver LW, Hewitt DG, DeYoung CA, Fulbright TE, Gann KR, Wester DB, Draeger DA (2019) Effects of enhanced nutrition on white-tailed deer foraging behavior. Wildl Monogr 202:27–34. https://doi.org/10.1002/wmon.1040

Dawe KL, Boutin S (2016) Climate change is the primary driver of white-tailed deer (*Odocoileus virginianus*) range expansion at the northern extent of its range; land use is secondary. Ecol Evol 6:6435–6451. https://doi.org/10.1002/ece3.2316

Dees CS, Clark JD, Van Manen FT (2001) Florida panther habitat use in response to prescribed fire. J Wildl Manag 65:141–147. https://doi.org/10.2307/3803287

DelGiudice GD, Lenarz MS, Powell MC (2007) Age-specific fertility and fecundity in northern free-ranging white-tailed deer: evidence for reproductive senescence? J Mammal 88:427–435. https://doi.org/10.1644/06-MAMM-A-164R.1

Demarais S, Strickland BK (2011) Antlers. In: Hewitt DG (ed) Biology and management of white-tailed deer. CRC Press, Boca Raton, FL, pp 107–145

Derry JF, Boone RB (2010) Grazing systems are a result of equilibrium and non-equilibrium dynamics. J Arid Environ 74:307–309. https://doi.org/10.1016/j.jaridenv.2009.07.010

DeYoung CD (2011) Population dynamics. In: Hewitt DG (ed) Biology and management of white-tailed deer. CRC Press, Boca Raton, FL, pp 147–180

DeYoung RW, Miller KV (2011) White-tailed deer behavior. In: Hewitt DG (ed) Biology and management of white-tailed deer. CRC Press, Boca Raton, FL, pp 311–351

DeYoung CD, Draw DL, Fulbright TE, Hewitt DG, Stedman SW, Synatzske DR, Teer JG (2008) Density dependence in deer populations: relevance for management in variable environments. In: Fulbright TE, Hewitt DG (eds) Wildlife science: linking ecological theory and management applications. CRC Press, Boca Raton, FL, pp 203–222

DeYoung RW, Demarais S, Gee KL, Honeycutt RL, Hellickson MW, Gonzales RA (2009) Molecular evaluation of the white-tailed deer (*Odocoileus virginianus*) mating system. J Mammal 90:946–953. https://doi.org/10.1644/08-MAMM-A-227.1

DeYoung CD, Fulbright TE, Hewitt DG, Wester DB, Draeger DA (2019) Linking white-tailed deer density, nutrition, and vegetation in a stochastic environment. Wildl Monogr 202:1–63. https://doi.org/10.1002/wmon.1040

Ditchkoff SS (2011) Anatomy and physiology. In: Hewitt DG (ed) Biology and management of white-tailed deer. CRC Press, Boca Raton, FL, pp 43–79

Dubreuil RP (2003) Habitat selection of white-tailed deer and mule deer in the southern Black Hills, South Dakota. MS Thesis, South Dakota State University, Brookings, USA

Dusek GL, MacKie RJ, Herriges JD Jr, Compton BB (1989) Population ecology of white-tailed deer along the lower Yellowstone River. Wildl Monogr 104:3–68

Dusek GL, Wood AK, Stewart ST (1992) Spatial and temporal patterns of mortality among female white-tailed deer. J Wildl Manag 56:645–650. https://doi.org/10.2307/3809455

Dyal JR, Miller KV, Cherry MJ, D'Angelo GJ (2021) Estimating sightability for helicopter surveys using surrogates of white-tailed deer. J Wildl Manag 85:887–896. https://doi.org/10.1002/jwmg.22040

Dykes JL (2022) Thermal ecology of white-tailed deer on southwestern rangelands. PhD dissertation, Texas A&M University-Kingsville

Edmunds DR, Kauffman MJ, Schumaker BA, Lindzey FG, Cook WE, Kreeger TJ, Grogan RG, Cornish TE (2016) Chronic wasting disease drives population decline of white-tailed deer. PLoS ONE 11:e0161127. https://doi.org/10.1371/journal.pone.0161127

Ekblad RL, Stuth JW, Owens MK (1993) Grazing pressure impacts on potential foraging competition between Angora goats and white-tailed deer. Small Rumin Res 11:195–208. https://doi.org/10.1016/0921-4488(93)90045-J

Escobar LE, Pritzkow S, Winter SN, Grear DA, Kirchgessner MS, Dominguez-Villegas E, Machado G, Peterson A, Soto C (2020) The ecology of chronic wasting disease in wildlife. Biol Rev Camb Philos Society 95:393–408. https://doi.org/10.1111/brv.12568

Faas CJ, Weckerly FW (2010) Habitat interference by axis deer on white-tailed deer. J Wildl Manag 74:698–706. https://doi.org/10.2193/2009-135

Foley AM, Hewitt DG, DeYoung CA, DeYoung RW, Schnupp MJ (2016) Modeled impacts of chronic wasting disease on white-tailed deer in a semi-arid environment. PLoS ONE. https://doi.org/10.1371/journal.pone.016359

Frerker K, Sabo A, Waller D (2014) Long-term regional shifts in plant community composition are largely explained by local deer impact experiments. PLoS ONE 9:e115843. https://doi.org/10.1371/journal.pone.0115843

Fulbright TE (2011) Managing white-tailed deer: western North America. In: Hewitt DG (ed) Biology and management of white-tailed deer. CRC Press, Boca Raton, FL, pp 537–563

Fulbright TE, Beasom SL (1987) Long term effects of mechanical treatments on white tailed deer browse. Wildl Soc Bull 15:560–564

Fulbright TE, Garza A Jr (1991) Forage yield and white-tailed deer diets following live oak control. J Range Manag 44:451–455

Fulbright TE, Ortega-S JA (2013) White-tailed deer habitat: ecology and management on rangelands, 2nd edn. Texas A&M University Press, College Station, TX

Fulbright TE, Dacy EC, Drawe DL (2011) Does browsing reduce shrub survival and vigor following summer fires? Acta Oecologica 37:10–15. https://doi.org/10.1016/j.actao.2010.10.007

Fulbright TE, Hickman KR, Hewitt DG (2013) Exotic grass invasion and wildlife abundance and diversity, south-central United States. Wildl Soc Bull 37:503–509. https://doi.org/10.1002/wsb.312

Fulbright TE, Davies KW, Archer SR (2018) Wildlife responses to brush management: a contemporary evaluation. Rangel Ecol Manage 71:35–44. https://doi.org/10.1016/j.rama.2017.07.001

Fulbright TE, Drabek DJ, Ortega-S JA, Hines SL, Saenz R III, Campbell TA, Hewitt DG, Wester DB (2021) Forb standing crop response to grazing and precipitation. Rangel Ecol Manage 79:175–185. https://doi.org/10.1016/j.rama.2021.08.007

Fulbright TE, DeYoung CA, Hewitt DG, Draeger DA (2023) Advanced white-tailed deer management: the nutrition—density sweet spot. Texas A&M University Press, College Station, TX

Fuhlendorf SD, Engle DM (2001) Restoring heterogeneity on rangelands: ecosystem management based on evolutionary grazing patterns: we propose a paradigm than enhances heterogeneity instead of homogeneity to promote biological diversity and wildlife habitat on rangelands grazed by livestock. Bioscience 51:625–632. https://doi.org/10.1641/0006-3568(2001)051[0625:RHOREM]2.0.CO;2

Fuhlendorf SD, Fynn RWS, McGranahan A, Twidwell D (2017) Heterogeneity as the basis for rangeland management. In: Briske DD (ed) Rangeland systems: processes, management and challenges. Springer Series on Environmental Management, Cham, Switzerland, pp 169–196. https://doi.org/10.1007/978-3-319-46709-2

Gavin TA, Suring H, Vohs PA Jr, Meslow EC (1984) Population characteristics, spatial organization, and natural mortality in Columbian white-tailed deer. Wildl Monogr 91:1–41

Gao J, Carmel Y (2020) Can the intermediate disturbance hypothesis explain grazing-diversity relations at a global scale? Oikos 129:493–502. https://doi.org/10.1111/oik.06338

Gaydos JK, Crum JM, Davidson WR, Cross SS, Owen SF, Stallknecht DE (2004) Epizootiology of an epizootic hemorrhagic disease outbreak in West Virginia. J Wildl Dis 40:383–393. https://doi.org/10.7589/0090-3558-40.3.383

Genho P, Hunt J, Rhyne M (2003) Managing for the long term while surviving the short term. In: Forgason CA, Bryant FC, Genho P (eds) Ranch management: integrating cattle, wildlife, and range. King Ranch, Kingsville, Texas, USA, pp 81–107

Gill C (2020) Escaped exotic animals are changing the Texas landscape. https://pitchstonewaters.com/escaped-exotics-animals-are-changing-the-texas-landscape/. Accessed 14 Jan 2021

Glasscock SN (2001) Analysis of vegetation dynamics, wildlife interactions, and management strategies in a semi-arid rangeland system: the Welder Wildlife Refuge model (Texas). PhD dissertation, Texas A&M University, College Station

Glow MP, Ditchkoff SS, Smith MD (2019) Annual fire return interval influences nutritional carrying capacity of white-tailed deer in pine-hardwood forests. Forest Sci 65:483–491. https://doi.org/10.1093/forsci/fxy063

Grovenburg TW, Jacques CN, Klaver RW, Jenks JA (2010) Bed site selection by neonate deer in grassland habitats on the Northern Great Plains. J Wildl Manag 74:1250–1256. https://doi.org/10.1111/j.1937-2817.2010.tb01245.x

Grovenburg TW, Swanson CC, Jacques CN, Deperno CS, Klaver RW, Jenks JA (2011) Female white-tailed deer survival across ecoregions in Minnesota and South Dakota. Am Midl Nat 165:426–435. https://doi.org/10.1674/0003-0031-165.2.426

Grovenburg TW, Klaver RW, Jenks JA (2012a) Spatial ecology of white-tailed deer fawns in the Northern Great Plains: implications of loss of conservation reserve program grasslands. J Wildl Manag 76:632–644. https://doi.org/10.1002/jwmg.288

Grovenburg TW, Klaver RW, Jenks JA (2012b) Survival of white-tailed deer fawns in the grasslands of the Northern Great Plains. J Wildl Manag 76:944–956. https://doi.org/10.1002/jwmg.339

Grovenburg TW, Monteith KL, Klaver RW, Jenks JA (2012c) Predator evasion by white-tailed deer fawns. Anim Behav 84:59–65. https://doi.org/10.1016/j.anbehav.2012.04.005

Gullikson BS (2019) Effects of energy development on movements, home ranges, and resource selection of white-tailed deer in the western Dakotas. PhD dissertation, South Dakota State University

Gulsby WD, Kilgo JC, Vukovich M, Martin JA (2017) Landscape heterogeneity reduces coyote predation on white-tailed deer fawns. J Wildl Manag 81:601–609. https://doi.org/10.1005/jwmg.21240

Habeck CW, Schultz AK (2015) Community-level impacts of white-tailed deer on understory plants in North American forests: a meta-analysis. AoB Plants 7:plv119. https://doi.org/10.1093/aobpla/plv119

Haggerty JH et al (2018) Land use diversification and intensification on Elk winter range in greater Yellowstone: framework and agenda for social-ecological research. Rangeland Ecol Manage 71:171–174. https://doi.org/10.1016/j.rama.2017.11.002

Haley NJ, Hoover EA (2015) Chronic wasting disease of cervids: current knowledge and future perspectives. Ann Rev Animal Biosci 3:305–325. https://doi.org/10.1146/annurev-animal-022114-111001

Hamilton WT, McGinty A, Ueckert DN, Hanselka CW, Lee MR (2004) Brush management: past, present, future. Texas A&M University Press, College Station

Halls LK (1978) White-tailed deer. In: Schmidt JL, Gilber DL (eds) Big game of North America. Stackpole Books, Harrisburg, PA, pp 43–65

Haus J, Eyler T, Bowman J (2019) A spatially and temporally concurrent comparison of popular abundance estimators for white-tailed deer. Northeast Nat 26:305–324. https://doi.org/10.1656/045.026.0207

Heffelfinger JR (2011) Deer of the Southwest: a complete guide to the natural history, biology, and management of southwestern mule deer and white-tailed deer. Texas A&M University Press, College Station

Henke SE, Demarais S, Pfister JA (1988) Digestive capacity of white-tailed deer and exotic ruminants. J Wildl Manag 52:595–598. https://doi.org/10.23007/3800913

Hewitt DG (2011) Nutrition. In: Hewitt DG (ed) Biology and management of white-tailed deer. CRC Press, Boca Raton, FL, pp 75–105

Hewitt DG (2015) Hunters and the conservation and management of white-tailed deer (*Odocoileus virginianus*). Int J Environ Stud 72:839–849. https://doi.org/10.1080/00207233.2015.1073473

Hewitt DG, Hellickson MW, Lewis JS, Wester DB, Bryant FS (2014) Age-related patterns of antler development in free-ranging white-tailed deer. J Wildl Manag 78:979–984. https://doi.org/10.1002/jwmg.741

Hines SL, Fulbright TE, Ortega-S AJ, Webb SL, Hewitt DG, Boutton TW (2021) Compatibility of dual enterprises for cattle and deer in North America: a quantitative review. Rangeland Ecol Manag 74:21–31 https://doi.org/10.1016/j.rama.2020.10.005

Hyde KJ, DeYoung CA, Garza A Jr (1987) Bed sites of white-tailed deer fawns in South Texas. In: Proceedings of the annual conference of the southeastern association of fish and wildlife agencies, vol 41, pp 288–293

Jacobson HA, DeYoung CA, DeYoung RW, Fulbright TE, Hewitt DG (2011) Management on private property. In: Hewitt DG (ed) Biology and management of white-tailed deer. CRC Press, Boca Raton, FL, pp 453–479

Jenks JA, Leslie DM (2003) Effect of domestic cattle on the condition of female white-tailed deer in southern pine-bluestem forests, USA. Acta Theriol 48:131–144

Jones PF, Jakes AF, MacDonald AM, Hanlon JA, Eacker DR, Martin BH, Hebblewhite M (2020) Evaluating responses by sympatric ungulates to fence modifications across the northern Great Plains. Wildl Soc Bull 44:130–131. https://doi.org/10.1002/wsb.1067

Kalb DM, Bowman JL, DeYoung RW (2018) Dietary resource use and competition between white-tailed deer and introduced sika deer. Wildl Res 45:457–472. https://doi.org/10.1071/WR17125

Keehn JE, Feldman CR (2018) Disturbance affects biotic community composition at desert wind farms. Wildl Res 45:383–396. https://doi.org/10.1071/WR17059

Keever A, McGowan C, Ditchkoff S, Acker P, Grand J, Newbolt C (2017) Efficacy of N-mixture models for surveying and monitoring white-tailed deer populations. Mammal Res 62. https://doi.org/10.1007/s13364-017-0319-z

Kennedy-Slaney L, Bowman J, Walpole AA, Pond BA (2018) Northward bound: the distribution of white-tailed deer in Ontario under a changing climate. Wildl Res 45:220–228. https://doi.org/10.1071/WR17106

Kie JG, White M (1985) Population dynamics of white-tailed deer (*Odocoileus virginianus*) on the Welder Wildlife Refuge, Texas. Southwest Nat 30:105–118. https://doi.org/10.2307/3670664

Kilburn J (2018) Spatial variability in abundance, detectability, and survival of white-tailed deer across a heterogenous landscape of fear. MS Thesis, University of Connecticut

Klich D, Łopucki R, Ścibior A, Gołębiowska D, Wojciechowska M (2020) Roe deer stress response to a wind farms: methodological and practical implications. Ecol Ind 117. https://doi.org/10.1016/j.ecolind.2020.106658

Knox WM (2011) The antler religion. Wildl Soc Bull 35:45–48. https://doi.org/10.1002/wsb.5

Lashley MA, Chitwood MC, Kays R, Harper CA, DePerno CS, Moorman CE (2015) Prescribed fire affects female white-tailed deer habitat use during lactation. For Ecol Manage 348:220–225. https://doi.org/10.1016/j.foreco.2015.03.041

Leslie DM Jr, Soper RB, Lochmiller RL, Engle DM (1996) Habitat use by white-tailed deer on cross timbers rangeland following brush management. J Range Manag 49:401–406

Lingle S (1993) Escape gaits of white-tailed deer, mule deer, and their hybrids: body configuration, biomechanics, and function. Can J Zool 71:708–724. https://doi.org/10.1139/z93-095

Lingle S (2001) Anti-predator strategies and grouping patterns in white-tailed and mule deer. Ethology 107:295–314. https://doi.org/10.1046/j.1439-0310.2001.00664.x

Lipsey MK, Naugle DE (2017) Precipitation and soil productivity explain effects of grazing on grassland songbirds. Rangel Ecol Manage 70:331–340. https://doi.org/10.1016/j.rama.2016.10.010

Łopucki R, Klich D, Gielarek S (2017) Do terrestrial animals avoid areas close to turbines in functioning wind farms in agricultural landscapes? Environ Monit Assess 189:343. https://doi.org/10.1007/s10661-017-6018-z

Lyons RK, Wright BD (2003) Using livestock to manage wildlife habitat. AgriLife Extension Publication B-6136, Texas A&M University, College Station

Macaulay L (2016) The role of wildlife-associated recreation in private land use and conservation: providing the missing baseline. Land Use Policy 58:218–233. https://doi.org/10.1016/j.landusepol.2016.06.024

Martinez MA, Molina V, Gonzalez-S F, Marroquin JS, Navar Ch J (1997) Observations of white-tailed deer and cattle diets in Mexico. J Range Manag 50:253–257. https://doi.org/10.2307/4003725

McCullough DR (1999) Density dependence and life-history strategies of ungulates. J Mammal 80:1130–1146. https://doi.org/10.2307/1383164

McGranahan DA, Engle DM, Fuhlendorf SD, Winter SJ, Miller JR, Debinski DM (2012) Spatial heterogeneity across five rangelands managed with pyric-herbivory. J Appl Ecol 49:903–910. https://doi.org/10.1111/j.1365-2664.2012.02168.x

McMahan CA, Inglis JM (1974) Use of rio grande plain brush types by white-tailed deer. J Range Manag 27:369–374. https://doi.org/10.2307/3896494

McMahan CA, Ramsey CW (1965) Response of deer and livestock to controlled grazing in Central Texas. J Range Manag 18:1–7

McShea WJ (2012) Ecology and management of white-tailed deer in a changing world. Ann N Y Acad Sci 1249:45–56. https://doi.org/10.1111/j.1749-6632.2011.06376.x

Michel ES, Gullikson BS, Brackel KL, Schaffer BA, Jenks JA, Jensen WF (2020) Habitat selection of white-tailed deer fawns and their dams in the Northern Great Plains. Mammal Res 68:825–833. https://doi.org/10.1007/s13364-020-00519-6

Milchunas DG, Sala OE, Lauenroth WK (1988) A generalized model of the effects of grazing by large herbivores on grassland community structure. Am Nat 132:87–106. https://doi.org/10.1086/284839

Miller MW, Fischer JR (2016) The first five (or more) decades of chronic wasting disease: lessons for the five decades to come. In: Transactions of the North American wildlife and natural resources conference, vol 81, pp 1–12

Montague DM, Montague RD, Fies ML, Kelly MJ (2017) Using distance-sampling to estimate density of white-tailed deer in forested, mountainous landscapes in Virginia. Northeast Nat 24:505–519. https://doi.org/10.1656/045.024.0409

Monteith KL, Schmitz LE, Jenks JA, Delger JA, Bowyer RT (2009) Growth of male white-tailed deer: consequences of maternal effects. J Mammal 90:651–660. https://doi.org/10.1644/08-MAMM-A-191R1.1

Moore MT, Foley AM, DeYoung CD, Hewitt DG, Fulbright TE, Draeger DA (2014) Evaluation of population estimates of white-tailed deer from camera survey. J Southeast Assoc Fish Wildl Agencies 1:127–132

Moratz KL (2016) Effect of oil and gas development on survival and health of white-tailed deer in the western Dakotas. PhD Dissertation, South Dakota State University

Moratz KL, Gullikson BS, Michel ES, Jenks JA, Grove DM, Jensen WF (2018) Assessing factors affecting adult female white-tailed deer survival in the northern Great Plains. Wildl Res 45:679–684. https://doi.org/10.1071/WR18032

Mullins JC, Van Ert M, Hadfield T, Nikolich MP, Hugh-Jones ME, Blackburn JK (2015) Spatio-temporal patterns of an anthrax outbreak in white-tailed deer, *Odocoileus virginianus*, and associated genetic diversity of *Bacillus anthracis*. BMC Ecol 15:23. https://doi.org/10.1186/s12898-015-0054-8

Mysterud A, Benestad SL, Rolandsen CM, Våge J (2021) Policy implications of an expanded chronic wasting disease universe. J Appl Ecol 58:281–285. https://doi.org/10.1111/1365-2664.13783

Nagy-Reis MB, Lewis MA, Jensen WF, Boyce MS (2019) Conservation reserve program is a key element for managing white-tailed deer populations at multiple spatial scales. J Environ Manage 248:109299. https://doi.org/10.1016/j.jenvman.2019.109299

Nemeth NM, Ruder MG, Gerhold RW, Brown JD, Munk BA, Oesterle PT, Kubiski SV, Keel MK (2014) Demodectic mange, dermatophilosis, and other parasitic and bacterial dermatologic diseases in free-ranging white-tailed deer (*Odocoileus virginianus*) in the United States from 1975 to 2012. Vet Pathol 51:633–640

Ortega IM, Soltero-Gardea S, Bryant FC, Drawe DL (1997a) Evaluating grazing strategies for cattle: deer and cattle food partitioning. J Range Manag 50:622–630

Ortega IM, Soltero-Gardea S, Bryant FC, Drawe DL (1997b) Evaluating grazing strategies for cattle: nutrition of cattle and deer. J Range Manag 50:631–637

Owens ML, Launchbaugh KL, Holloway JW (1991) Pasture characteristics affecting spatial distribution of utilization by cattle in mixed brush communities. J Range Manag 44:118–123

Ozoga JJ (1987) Maximum fecundity in supplementally-fed northern Michigan white-tailed deer. J Mammal 68:878–879. https://doi.org/10.2307/1381573

Palermo PM, Orbegozo J, Watts DM, Morrill JC (2022) SARS-CoV-2 neutralizing antibodies in white-tailed deer from Texas. Vector-Borne Zoonotic Dis 22:62–64. https://doi.org/10.1089/vbz.2021.0094

Palmer MV, Martins M, Falkenberg S, Buckley A, Caserta LC, Mitchell PK, Cassmann ED, Rollins A, Zylich NC, Renshaw RW, Guarino C (2021) Susceptibility of white-tailed deer (*Odocoileus virginianus*) to SARS-CoV-2. J Virol 95:e00083–21.17.6. https://doi.org/10.1128/JVI.00083-21

Pedersen K, Anderson TD, Maison RM, Wiscomb GW, Pipas MJ, Sinnett DR, Baroch JA, Gidlewski T (2018) Leptospira antibodies detected in wildlife in the USA and the US Virgin Islands. J Wildl Dis 54:450–459. https://doi.org/10.7589/2017-10-269

Peterson MK, Foley AM, Tri AN, Hewitt DG, DeYoung RW, DeYoung CA, Campbell TA (2020) Mark-recapture distance sampling for aerial surveys of ungulates on rangelands. Wildl Soc Bull 44:713–723. https://doi.org/10.1002/wsb.1144

Phillips R (2015) Fish and Game confirms outbreak of bluetongue disease in whitetails. Idaho Fish and Game. https://idfg.idaho.gov/

Pollock MT, Whittaker DG, Demarais S, Zaiglan RE (1994) Vegetation characteristics influencing site selection by male white-tailed deer in Texas. J Range Manag 47:235–239

Potratz EJ, Brown JS, Gallo T, Anchor C, Santymire RM (2019) Effects of demography and urbanization on stress and body condition in urban white-tailed deer. Urban Ecosystems. https://doi.org/10.1007/s11252-019-00856-8

Prasad NLNS, Guthery FS (1986) Wildlife use of water under short duration and continuous grazing. Wildl Soc Bull 14:450–454

Provenza FD, Villalba JJ, Dziba LE, Atwood SB, Banner RE (2003) Linking herbivore experience, varied diets, and plant biochemical diversity. Small Rumin Res 49:257–274. https://doi.org/10.1016/S0921-4488(03)00143-3

Provenza FD, Villalba JJ, Wiemeier RW, Lyman T, Owens J, Lisonbee L, Clemensen A, Welch KD, Gardner DR, Lee ST (2009) Value of plant diversity for diet mixing and sequencing in herbivores. Rangelands 31:45–49. https://doi.org/10.2111/1551-501X-31.1.45

Ramirez-Yanez LE, Ortega-S, Brennan LA, Rasmussen GA (2007) Use of prescribed fire and cattle grazing to control guineagrass. In: Proceedings of the 23rd tall timbers fire ecology conference: fire in grassland and shrubland ecosystems. Tall Timbers Research Station Tallahassee, FL, USA, pp 240–245

Rankins ST, DeYoung RW, Foley AM, Ortega-S JA, Fulbright TE, Hewitt DG, Schofield LR, Campbell TA (2021) Energy content of browse: a regional driver of white-tailed deer size. In: Abstracts of the 44th annual meeting of the Southeast Deer Study Group

Reardon PO, Merrill PA, Taylor CA Jr (1978) White-tailed deer preferences and hunter success under various grazing systems. J Range Manag 31:40–42

Rivera NA, Brandt AL, Novakofski JE, Mateus-Pinilla NE (2019) Veterinary medicene: research and reports 10:123–139. https://doi.org/10.2147/VMRR.S197404

Rivera NA, Varga C, Ruder MG, Dorak SJ, Roca AL, Novakofski JE, Mateus-Pinilla NE (2021) Bluetongue and epizootic hemorrhagic disease in the United States of America at the wildlife-livestock interface. Pathogens 10:915. https://doi.org/10.3390/pathogens10080915

Rohm JH, Nielsen CK, Woolf A (2007) Survival of neonatal white-tailed deer in an exurban population. J Wildl Manag 71:940–944. https://doi.org/10.2193/2006-116

Rollins D, Bryant FC, Waid DD, Bradley LC (1988) Deer response to brush management in central Texas. Wildl Soc Bull 16:277–284

Ruthven DC III, Fulbright TE, Beasom SL, Hellgren EC (1993) Long-term effects of root plowing on vegetation in the eastern South Texas Plains. J Range Manag 46:351–354

Ruthven DC III, Hellgren EC, Beasom SL (1994) Effects of root plowing on white-tailed deer condition, population status, and diet. J Wildl Manag 58:59–70. https://doi.org/10.2307/3809549

Sauer PR (1984) Physical characteristics. In: Halls LK (ed) White-tailed deer: ecology and management. Stackpole Books, Harrisburg, PA, pp 73–90

Scifres CJ, Hamilton WT, Koerth BH, Flinn RC, Crane RA (1988) Bionomics of patterned herbicide application for wildlife enhancement. J Range Manag 41:317–321

Scholtz R, Fuhlendorf SD, Ulden DR, Allred BW, Jones MO, Naugle DE, Twidwell D (2021) Challenges of brush management treatment effectiveness in southern Great Plains, United States. Rangel Ecol Manage 77:57–63. https://doi.org/10.1016/j.rama.2021.03.007

Segoli M, Ungar ED, Shachak M (2012) Fine-scale spatial heterogeneity of resource modulation in semi-arid "Islands of Fertility." Arid Land Res Manag 26:344–354. https://doi.org/10.1080/15324982.2012.694397

Smith WP (1987) Dispersion and habitat use by sympatric Columbian white-tailed deer and Columbian black-tailed deer. J Mammal 68:337–347. https://doi.org/10.2307/1381473

Smith WP, Coblentz BE (2010) Cattle or sheep reduce fawning habitat available to Columbian white-tailed deer in western Oregon. Northwest Sci 84:315–326. https://doi.org/10.3955/046.084.0401

Soper RB, Lochmiller RL, Leslie DM, Jr Engle DM (1993) Condition and diet quality of white-tailed deer in response to vegetation management in central Oklahoma. In: Proceedings of the Oklahoma academy of science, vol 73, pp 53–61

South Dakota Game, Fish, and Parks (2022) Epizootic hermorrhagic disease (EHD). https://gfp.sd.gov

Sperlova A, Zendulkova D (2011) Bluetongue: a review. Vet Med 9:430–452

Springer MD (1977) The influence of prescribed burning on nutrition in white-tailed deer in the Coastal Plain of Texas. PhD dissertation, Texas A&M University, College Station, USA

Steuter AA, Wright HA (1980) White-tailed deer densities and brush cover on the Rio Grande Plain. J Range Manag 33:328–331

Stevens G, McCluskey B, King A, O'Hearn E, Mayr G (2015) Review of the 2012 epizootic hemorrhagic disease outbreak in domestic ruminants in the United States. PLoS ONE 10:e0133359. https://doi.org/10.1371/journal.pone.0133359

Stewart KM, Fulbright TE, Drawe DL, Bowyer RT (2003) Sexual segretation in white-tailed deer: responses to habitat manipulations. Wildl Soc Bull 31:1210–1217

Stewart KM, Bowyer RT, Weisburg PJ (2011) Spatial use of landscapes. In: Hewitt DG (ed) Biology and management of white-tailed deer. CRC Press, Boca Raton, FL, pp 181–217

Strickland BK, Demarais S (2000) Age and regional differences in antlers and mass of white-tailed deer. J Wildl Manag 64:903–911. https://doi.org/10.2307/3803198

Thomas DB, Klafke G, Busch JD, Olafson PU, Miller RA, Mosqueda J, Stone NE, Scoles G, Wagner DM, Perez-De-Leon A (2020) Tracking the increase of acaricide resistance in an invasive population of cattle fever ticks (Acari:Ixodidae) and implementation of real-time PCR assays to rapidly genotype resistance mutations. Ann Entomol Soc Am 113:298–309. https://doi.org/10.1093/aesa/saz053

Ulrich RS, Simons RF, Losito BD, Fiorito E, Miles MA, Zelson M (1991) Stress recovery during exposure to natural and urban environments. J Environ Psychol 11:201–230. https://doi.org/10.1016/S0272-4944(05)80184-7

Uresk DW, Benzon TA, Severson KE, Benkobi L (1999) Characteristics of white-tailed deer fawn beds, Black Hills, South Dakota. Great Basin Naturalist 59:348–354

Van der Hoek D, Knapp AK, Briggs JM, Bokdam J (2002) White-tailed deer browsing on six shrub species of tallgrass prairie. Great Plains Res 12:141–156

Van de Koppel J, Rietkerk M, van Langevelde F, Kumar L, Klausmeier CA, Fryxell JM, Hearne JW, van Andel J, de Ridder N, Skidmore A, Stroosnijder L, Prins HHT (2002) Spatial heterogeneity and irreversible vegetation change in semiarid grazing systems. Am Nat 159:209–218. https://doi.org/10.1086/324791

Van Horne B (1983) Density as a misleading indicator of habitat quality. J Wildl Manag 47:893–901. https://doi.org/10.2307/3808148

VerCauteren KC, VanDeelen, Lavelle MJ, Hall WH (2010) Assessment of abilities of white-tailed deer to jump fences. J Wildl Manag 74:1378–1381. https://doi.org/10.1111/j.1937-2817.2010.tb01260.x

Verme LJ (1969) Reproductive patterns of white-tailed deer related to nutritional plane. J Wildl Manag 33:881–887. https://doi.org/10.2307/3799320

Volk MD, Kaufman DW, Kaufman GA (2007) Diurnal activity and habitat associations of white-tailed deer in tallgrass prairie of eastern Kansas. Trans Kans Acad Sci 110:145–154. https://doi.org/10.1660/0022-8443(2007)110(145:DAAHAO12.0.CO:2

Warren RJ, Krysl LJ (1983) White-tailed deer food habitats and nutritional status as affected by grazing and deer-harvest management. J Range Manag 36:104–109

Washburn BE, Seamans TW (2007) Wildlife responses to vegetation height management in cool-season grasslands. Rangel Ecol Manage 60:319–323. https://doi.org/10.2111/1551-5028(2007)60[319:WRTVHM]2.0.CO;2

Webb SL, Hewitt DG, Hellickson MW (2007) Survival and cause-specific mortality of mature male white-tailed deer. J Wildl Manag 71:555–558. https://doi.org/10.2193/2006-189

Weiskopf SR, Ledee OE, Thompson LM (2019) Climate change effects on deer and moose in the Midwest. J Wildl Manag 83:769–781. https://doi.org/10.1002/jwmg.21649

Whittaker DG, Lindzey FG (1999) Effect of coyote predation on early fawn survival in sympatric deer species. Wildl Soc Bull 27:256–262

Whittaker DG, Lindzey FG (2004) Habitat use patterns of sympatric deer species on Rocky Mountain Arsenal, Colorado. Wildl Soc Bull 32:1114–1123. https://doi.org/10.2193/0091-7648(2004)032[1114:HUPOSD]2.0CO;2

Wiemers DW, Fulbright TE, Wester DB, Ortega-S JA, Rasmussen GA, Hewitt DG, Hellickson MW (2014) Role of thermal environment in habitat selection by male white-tailed deer during summer in Texas, USA. Wildl Biol 20:47–56. https://doi.org/10.2981/wlb.13029

Wiggers EP, Beasom SL (1986) Characterization of sympatric or adjacent habitats of 2 deer species in west Texas. J Wildl Manag 50:129–134. https://doi.org/10.2307/3801502

Wilcox BP, Fuhlendorf SD, Walker JW, Twidwell D, Wu XB (2021) Saving imperiled grassland biomes by recoupling fire and grazing: a case study from the Great Plains. Front Ecol Environ. https://doi.org/10.1002/fee.2448

Woolf A, Roseberry JL (1998) Deer management: our profession's symbol of success or failure? Wildl Soc Bull 26:515–521

Wolff CL, Demarais S, Brooks CP, Barton BT (2020) Behavioral plasticity mitigates the effect of warming on white-tailed deer. Ecol Evol 10:2579–2587. https://doi.org/10.1002/ece3.6087

Chapter 19
Pronghorn

Paul F. Jones, Adele K. Reinking, Andrew F. Jakes, Myrna M. Miller,
Terry Creekmore, and Rich Guenzel

Abstract Pronghorn (*Antilocapra americana*) are an endemic ungulate in western
North America and occupy rangelands concurrently with domestic livestock. When
rangelands are in healthy condition, there is little-to-no competition between
pronghorn and domestic livestock. When rangeland health deteriorates, direct competition occurs when both compete for limited resources. Pronghorn are a highly
mobile species that cope with challenging environmental conditions (both natural and
human-imposed) through daily and seasonal movements to more favorable habitats.
Maintaining healthy rangelands and rangeland connectivity will allow pronghorn to
move freely and adapt to increased human disturbance. In addition, understanding the
cumulative effects and identifying mitigation strategies of deleterious anthropogenic
effects (i.e., habitat conversion, linear features, energy development, and climate
changes) will help to ensure long-term persistence of pronghorn populations. Mitigation will be critical, in conjunction with expanded research efforts, to help gain a
greater knowledge of the role of environmental conditions and anthropogenic disturbances on pronghorn fitness, persistence, and their ability to move across the land in
response to an ever-changing landscape.

P. F. Jones (✉)
Alberta Conservation Association, Lethbridge, AB T1J 0P3, Canada
e-mail: paul.jones@ab-conservation.com

A. K. Reinking
Cooperative Institute for Research in the Atmosphere, Colorado State University, Fort Collins,
CO 80523, USA
e-mail: adele.reinking@colostate.edu

A. F. Jakes
Smithsonian's National Zoo and Conservation Biology Institute, Front Royal, VA 22630, USA
e-mail: JakesAF@si.edu

M. M. Miller
Wyoming State Veterinary Laboratory (Retired), University of Wyoming, Laramie, WY 82070,
USA
e-mail: MillerMM@uwyo.edu

T. Creekmore · R. Guenzel
Wyoming Game and Fish Department (Retired), Laramie, WY 82070, USA

© The Author(s) 2023
L. B. McNew et al. (eds.), *Rangeland Wildlife Ecology and Conservation*,
https://doi.org/10.1007/978-3-031-34037-6_19

Keywords *Antilocapra americana* · Connectivity · Habitat · Pronghorn · Rangeland management

19.1 General Life History and Population Dynamics

Pronghorn (*Antilocapra americana*), commonly called antelope, are an endemic western North American (Fig. 19.1) ungulate that are found nowhere else in the world. The unique pelage of pronghorn makes them readily identifiable, with a large white rump, white underbelly, white bands on the neck, and a dark nose (Fig. 19.1). Both males and females possess horns, but when present in females, they tend to be shorter than their ears (O'Gara 2004a). The name pronghorn comes from the front prong on the horns of mature males (Fig. 19.1). Pronghorn are the world's second fastest land mammal, able to reach speeds between 70 and 100 km hr^{-1} [40–60 miles hr^{-1} (O'Gara 2004a)].

Pronghorn are the last remaining species of their taxonomic family (*Antilocapridae*), which roamed North America during the Pleistocene epoch (O'Gara and Janis 2004a). The current form of pronghorn has evolved over the last 20 million years (O'Gara and Janis 2004a). The most common of five subspecies is *A. a. americana* which is widely distributed from Texas, north into Alberta and Saskatchewan. The Sonoran pronghorn (*A. a. sonoriensis*) is the smallest subspecies, is classified as endangered, and can be found only in southwestern Arizona and northwestern Sonora, Mexico (USFWS 2015). The Mexican pronghorn (*A. a. mexicana*) is an endangered subspecies found in Mexico and the Marathon Basin of Texas (O'Gara and Janis 2004b). The peninsular pronghorn (*A. a. peninsularis*) is also an endangered subspecies found in the Vizcaino Desert, Mexico (O'Gara and Janis 2004b). The Oregon pronghorn (*A. a. oregona*) is a subspecies found in Oregon, Idaho, California, and Nevada (O'Gara and Janis 2004b). However, Lee Jr (1992) analyzed mitochondrial DNA and concluded that pronghorn in the range of *A. a. oregona* were not dissimilar to *A. a. americana* and therefore should not be treated as a subspecies. While 3 subspecies are currently federally listed as endangered, the most recognized subspecies *A. a. americana* are common on the rangelands of western North America, though local populations vary in their conservation status (Jakes 2021).

19.2 Distribution and Population Status

19.2.1 Distribution

The current distribution of pronghorn spans 23 jurisdictions including 17 US states, 2 Canadian provinces, and 4 Mexican states (Yoakum et al. 2014). The range of pronghorn in 2000 is depicted in Fig. 19.1 (Jensen et al. 2004). Missing from the

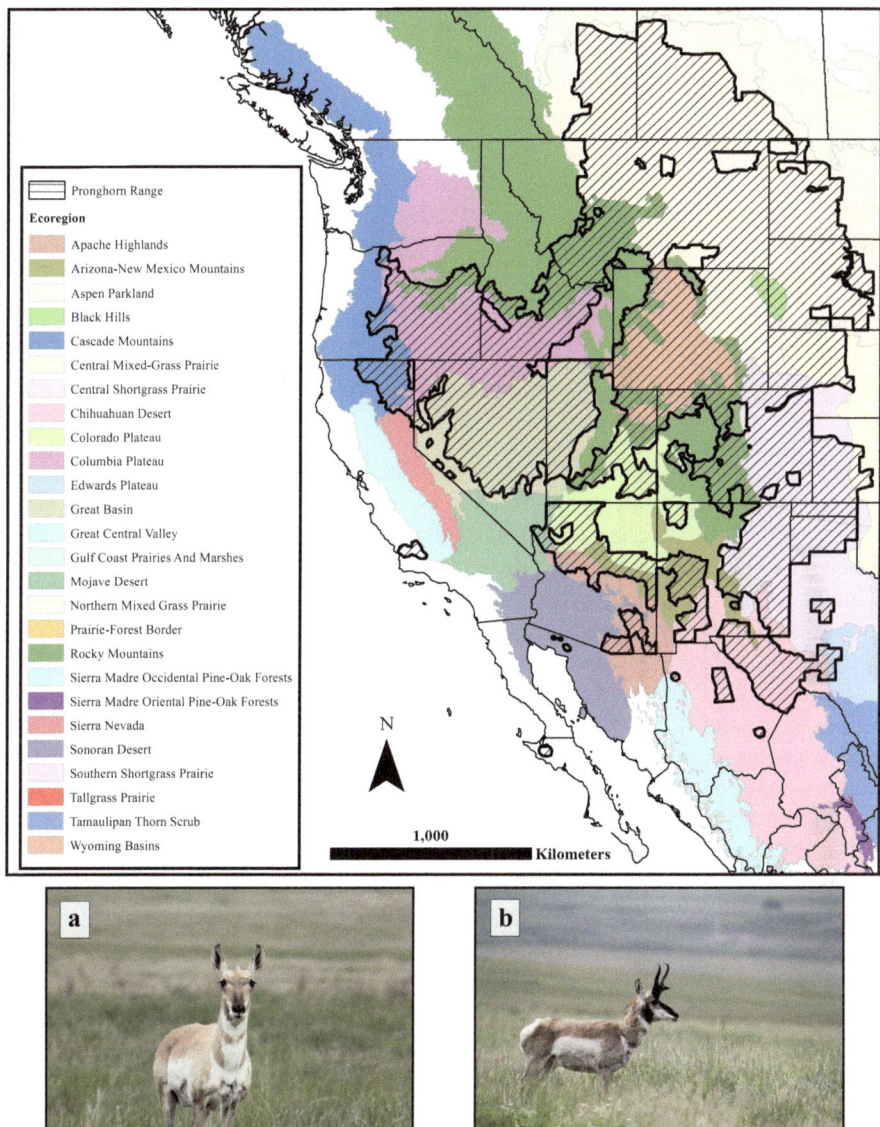

Fig. 19.1 The 2000 geographical distribution of pronghorn across North America. Photos are of a female (**a**) and a male (**b**) pronghorn in the grasslands of Alberta, Canada. Photos: P. Jones, Alberta Conservation Association. Spatial data source: Jensen et al. (2004)

figure are populations in Washington that are the result of recent re-introductions (Jakes 2021). Almost half of all pronghorn are found in Wyoming, and approximately 80% of the population occurs within Colorado, Montana, New Mexico, and Wyoming, with these 4 states being considered the "core area" of suitable pronghorn habitat (Schroeder 2018; Jakes 2021). While pronghorn occupy most of their historic range, their numbers are drastically lower than prior to European settlement (Yoakum 2004a). The 2017 population estimate was just under 1 million pronghorn, compared with historical estimates of 30–40 million (Yoakum 2004a; Schroeder 2018; Jakes 2021).

19.2.2   Monitoring

Pronghorn are surveyed to determine population estimates and demographic data for setting harvest rates by each jurisdiction or for assessing species status across their range. Population surveys for pronghorn are dependent upon the survey objective(s), local habitat, population density, and the distribution (e.g., evenly distributed, clumped, etc.) of animals across the landscape (Yoakum et al. 2014). Most jurisdictions that survey pronghorn use aerial surveys (via fixed-wing aircraft or helicopter), with a few still using ground surveys (Schroeder 2018). Surveys to detect animals using fixed-wing aircraft disturb pronghorn less than helicopters due to being flown at higher altitudes with lower noise levels (Yoakum et al. 2014). Survey protocol and coverage is often dictated by available financial resources and human safety requirements. Most surveys are conducted between May and August when pronghorn are most widely distributed, in smaller groups, with mobile and detectable fawns allowing for the classification of both sex and age structure (Yoakum et al. 2014).

   A variety of survey protocols have been employed to estimate pronghorn population size including: (1) strip transects, (2) line transects with distance sampling, and (3) quadrats or area sampling (Pojar and Guenzel 1999; Pojar 2004). While the ideal survey would produce a population estimate with an associated confidence interval, this is not always achievable. Some jurisdictions have used the strip transect method and relied on trend counts to assess annual differences in relative population estimates. The detection of a change using trend data is contingent upon the assumption that survey conditions (e.g., weather, time of survey, habitat, observer, etc.) are consistent and that the percentage of animals detected is similar between surveys (Nichols 1992). Recent developments in survey methodology and statistical analysis allows for more precise population estimates. For example, the use of line transects with distance sampling allows for the correction of population estimates based on the detection probability of observing animals on the transect (Ward 2016). Whichever survey protocol is used, one should strive to minimize bias (e.g., observer, survey assumptions), produce the most precise estimate possible, and validate visibility bias for the geographic region and survey protocol to which they will be applied (Guenzel 1997; Ward 2016).

In addition to estimating population size, most surveys assess the ratio of fawns and males (bucks) to females (does; i.e., ff:dd and bb:dd ratios). Late summer is the optimal time to conduct classification surveys, especially to estimate ff:dd ratios as postnatal fawn mortality has subsided and fawns are still easily distinguishable from females, which is not the case come fall or winter (Yoakum et al. 2014). The ff:dd ratios can be used to estimate recruitment in population models. Fall surveys are not ideal for estimating bb:dd ratios because fawns can be mistaken for adult females, which inflates the female count and widens the bb:dd ratio (Yoakum et al 2014). Winter surveys are not ideal as males lose their horn sheaths after October which could result in younger males being classified as females and would underestimate the bb:dd ratios. The bb:dd ratios are used as sex ratios in population models. Linking demography data (sex, age) with spatiotemporal variables can help forecast and classify populations based on current structure, as well as current and future landscape conditions (Arnold et al. 2018). Pronghorn are considered to have ecological and economic value across their range and therefore, State and Provincial agencies are responsible for setting pronghorn harvest rates (Jakes 2021; Stoner et al. 2021). Demography data combined with population estimates form the foundation for which decisions on pronghorn harvest levels are made by wildlife managers. Most pronghorn tag numbers are based on limited-quota or limited-entry licences due to the low number of animals in most states and provinces (O'Gara and Morrison 2004), making pronghorn one of the most sought-after harvestable species.

19.3 Habitat Associations

19.3.1 Historical/Evolutionary

Pronghorn are largely found in the same habitats that they occupied historically, including Grasslands (i.e., southern mid-grass prairie, northern mid-grass prairie, short grass prairie; hereafter grasslands), Intermountain Valleys and Lower Mountain Slopes (e.g., Great Basin Sagebrush [*Artemisia*], Sagebrush Steppe [*Artemisia*-Perennial Bunchgrasses]; hereafter shrub-steppe), and Warm Deserts and Grasslands (i.e., Chihuahuan Desert, including chaparral in Mexico; Sonoran Desert, including chaparral in Arizona; hereafter desert; Yoakum 2004a). Collectively we refer to grasslands, shrub-steppe, and desert as rangelands. Pronghorn, with their excellent long-distance vision and speed are uniquely adapted to these relatively flat, rolling landscapes (O'Gara 2004a). Many of these adaptations are relics of the predator species with which pronghorn coexisted millions of years ago. Their ability to reach tremendous speeds, for example, is attributed to the ancient predation threat of the now-extinct American cheetah (*Miracinonyx spp.;* Byers 1997). Pronghorn once roamed alongside ungulates including camels (*Paracamelus spp.*) and tapirs (*Tapiris spp.*) and faced predation threats from saber-toothed cats (genera *Megantereon*, *Smilodon*, and *Homotherium*), giant short-faced bears (*Arctodus simus*), and dire wolves (*Canis dirus;* Byers 1997; McCabe et al. 2004). More recently pronghorn share habitat with bison (*Bison bison*), elk (*Cervus canadensis*), deer (*Odocoileus spp.*), gray wolves (*Canis lupus*), golden eagles (*Aquila chrysaetos*), mountain lions (*Puma concolor*), bobcats (*Lynx rufus*), and coyotes (*Canis latrans*) (Byers 1997). While pronghorn still occupy rangelands with other ungulate species, many predators that were previously common in pronghorn habitats are often absent today or occur at lower densities than they did historically (Byers 1997). Therefore, predation is typically not a limiting factor for most pronghorn populations. However, predation of fawns can be significant, and, in some populations, adult predation can be high (O'Gara 2004b; Keller et al. 2013).

Historically, fires were the chief disturbance in the grassland, shrub-steppe, and desert regions that pronghorn occupy (Yoakum 2004b). It has been suggested that reduced shrub density and increased forb availability resulting from periodic burns are likely to benefit pronghorn populations (Greenquist 1983; Augustine and Derner 2015). As Europeans settled in North America in the early 1800s, such natural disturbance regimes were altered, and new sources of habitat changes ensued, resulting in habitat conversion, loss, and fragmentation (Greenquist 1983; O'Gara and McCabe 2004). Across much of the current pronghorn range, vast networks of wire fencing associated with nineteenth century property delineation and livestock production are still present. These fences currently represent a source of direct and indirect pronghorn mortality (Oakley 1973; Harrington and Conover 2006; Jones 2014; Jones et al. 2019) and alter behavior and movement (Jakes et al. 2018a; Seidler et al. 2018; Reinking et al. 2019; Smith et al. 2020). The conversion of rangelands that began in the 1800s, coupled with additional anthropogenic development since, has reduced native habitat availability to pronghorn and caused deterioration of rangelands through erosion, weeds, conifer encroachment, and brush removal (O'Gara and McCabe 2004). This habitat loss and degradation continue to present issues across much of the current pronghorn range and are further described in Sect. 19.7.

19.3.2 Contemporary

Current pronghorn habitat is characterized by low, rolling hills with limited visual barriers, and ranges in elevation from roughly 0 to 3000 m (0–9850 ft) above sea level (Yoakum 2004b). Vegetation in pronghorn habitats mainly consists of grasses, forbs, and shrubs, with vegetation height typically ranging from 13 to 76 cm (5–30 in), though use at the upper end of this height range is minimal (Yoakum 2004c). The usage of vegetation types for forage varies by location, availability, and season, and is described in Sect. 19.5.1.

Annual precipitation varies widely across pronghorn range, but most animals occur in areas receiving 20–40 cm (8–16 in) annually (Yoakum 2004b). Population persistence depends on both the amount and timing of annual precipitation (Brown et al. 2006; Simpson et al. 2007). Precipitation during late gestation and lactation may be especially important, particularly for animals in the arid southwestern United

States (Gedir et al. 2015). During colder seasons, most of the current pronghorn range (70%) typically experiences precipitation in the form of snow (Yoakum 2004b). Pronghorn mainly rely on snow and free water (Yoakum et al. 2014), but succulent forage may also be used as a water resource in drier areas or drought years (Büechner 1950; Beale and Smith 1970; Clemente et al. 1995).

Pronghorn habitat requirements also include topographic and vegetative features (e.g., taller shrubs) that provide protection (i.e., cover) from both the elements and predators. Thermal cover can include shade-providing features (e.g., tall trees and shrubs) to help keep animals cool when air temperature is high (Yoakum 2004b; Wilson and Krausman 2008). However, pronghorn have a high heat tolerance and are typically able to mitigate high temperatures through unique morphological and physiological adaptations. Topographic and vegetative features can provide refuge from high wind speeds by minimizing wind chill in low temperatures, while offering areas of shallower snow (Bruns 1977; Ryder and Irwin 1987). Security cover that provides protection from predators is also required, but mainly as fawn hiding habitat (Barrett 1982; Jacques et al. 2015).

19.3.3 Seasonal

Suitable pronghorn habitat must provide adequate seasonal ranges, as well as functional landscapes connecting seasonal ranges (see Sect. 19.4). These varied habitats allow pronghorn to maintain access to forage, minimize energetic demands, and maximize fitness (i.e., survival and reproductive success) as resources fluctuate annually (Dalton 2009; Yoakum et al. 2014). In winter, pronghorn seasonal ranges are generally larger than in summer (Sheldon 2005; Reinking et al. 2019). Winter range is largely selected to avoid deep snow and maximize the period of exposure to high quality forage and can be either lower in elevation or latitude than summer range, fawning areas, or migration habitat (Yoakum 2004b). Snow depths < 15 cm (< 6 in) are preferable in pronghorn winter range both to maintain forage accessibility above the snowpack and mitigate the energetic costs of locomotion through snow (Yoakum et al. 2014). Snow depths become particularly detrimental at roughly 30 cm (12 in), limiting access to forage, and when at mid-limb height on an individual, inhibiting their movement (Telfer and Kelsall 1984; Yoakum et al. 2014). The interaction of snow conditions and anthropogenic features like railroads, highways, and fences can also present extreme challenges (Jones et al. 2020a). Deep snow can force animals onto snow-cleared railroads and highways that offer easier movement (O'Gara 2004b) but increase the risk of collision and energy expenditure (Seidler et al. 2018; Jones et al. 2020a). Moreover, deep snows can reduce the open space beneath wire fences, eliminating the ability of pronghorn to pass underneath and move to more suitable habitats during winter when resources are already limited (Bruns 1977; Sheldon 2005; Yoakum et al. 2014; Seidler et al. 2018).

Summer range requirements are largely synonymous with ideal fawning habitat. These areas provide high quality herbaceous vegetation for does and fawns, offer sufficient vegetative cover to protect fawns and vulnerable birthing females from predators, and usually have higher temperatures with little to no snow (Yoakum 2004b). Unlike other ungulates that largely rely on previously acquired fat stores to fuel reproduction and survival (i.e., capital breeders; Jönsson 1997), pronghorn are thought to be income breeders, meaning that they mainly meet energetic demands as they arise with the immediate intake of resources (Smyser et al. 2005; Reinking et al. 2018). Therefore, fawn survival and the survival of adult females facing the high energetic costs of reproduction are dependent on high forage quality and availability on summer range (Smyser et al. 2005; Reinking et al. 2018; Panting et al. 2020; Bender and Rosas-Rosas 2021).

19.4 Movement, Migration, and Dispersal

Pronghorn move amongst and between habitats or to completely new suitable habitats for population maintenance (Dingle and Drake 2007). Movements undertaken by pronghorn provide connections between suitable habitats across spatiotemporal scales, which include daily movements amid vegetation patch types, annual migrations between seasonal ranges, or dispersal events to seek out appropriate habitat in new areas, thus providing functional connections between herds and populations (Sawyer et al. 2005; Jacques and Jenks 2007; Kolar et al. 2011; Collins 2016; Jakes et al. 2018b). Because migration is an annually repeated phenomenon, it can be a useful focus for identifying and maintaining landscape connectivity to sustain pronghorn populations. Pronghorn use such movements to maximize access to high-nutrition vegetation, improve physical condition to increase reproductive success, find mates, decrease intraspecific competition, and respond to changing environmental conditions (Hoskinson and Tester 1980; Bolger et al. 2007; Barnowe-Meyer et al. 2017). Across North America, caribou (*Rangifer tarandus*) and mule deer (*Odocoileus hemionus*) are the only ungulates reported to have made greater annual long-distance movements than pronghorn (Joly et al. 2019).

Pronghorn populations are often partially migratory (White et al. 2007; Jacques et al. 2009; Kolar et al. 2011; Jakes et al. 2018b), meaning that some individuals migrate, and others do not (Dingle and Drake 2007). At the northern range, pronghorn that migrated were found to have a 7% increase in survival probability, compared to individuals that remained residential (Jones et al. 2020a). Some pronghorn individuals switched movement tactics from one year to the next (Jakes et al. 2018b), suggesting that pronghorn exhibit plasticity in movement decisions. Indeed, factors such as demography and learning through social interactions, may also influence the strategy employed, indicating that migration may not be a fixed behavior (Bauer et al. 2011; Barnowe-Meyer et al. 2013; Jesmer et al. 2018).

Depending on the distance and duration of migration, pronghorn may use stopover sites to energetically recover and amass fat and protein reserves to complete their journey (Bolger et al. 2007; Sawyer et al. 2009). Stopover sites are typically areas of higher forage productivity with lower densities of anthropogenic features relative to migratory pathways (Jakes 2015). However, pronghorn may stopover along suboptimal areas such as roads and fences (Seidler et al. 2015). These human-induced stopovers can delay migration and deplete important energy reserves needed to navigate terrain successfully or detect alternative locations to traverse these features. In some instances, linear features become an impermeable barrier and deter pronghorn crossing opportunities altogether.

Other long-distance movements by pronghorn have been observed at various times of year. Across their range, pronghorn may display unpredictable movements to apparently follow forage maturation and availability (e.g., nomadism) as opposed to exhibiting fidelity to any one area, although this is not well understood (Milligan et al. 2021; Morrison et al. 2021). Alternatively, long-distance movements may occur as a survival tactic in response to stochastic events such as fire, drought, or extreme snowfall. For example, at the northern periphery of pronghorn range, movements from one winter range to another in response to extreme environmental conditions (i.e., facultative winter migration), as well as movements from an initial distinct fawning range during known parturition dates to a separate summer range (i.e., potential post-fawning migration), have been reported (Jakes et al. 2018b). In general, facultative winter migrations made by pronghorn occurred from winter range, where sagebrush and other forage was unavailable, to winter range where sagebrush was accessible (Jakes et al. 2018b).

Pronghorn seasonal and daily movements are influenced by environmental gradients and anthropogenic factors. In general, pronghorn spring migrations follow the 'green-wave' of available forage to acquire protein-rich resources while avoiding heavily used or high densities of human development (Mysterud 2013; Jakes et al. 2020). For pronghorn, anthropogenic disturbances include features such as roads, fences, energy infrastructure, and other developments such as houses (Sheldon 2005; Jones et al. 2019; Jakes et al. 2020). During fall migration, pronghorn tend to select for native grasslands and avoid roads, with some populations also following large stream and river systems, to quickly arrive onto winter grounds (Jakes et al. 2020). Unfragmented rangelands offer the best areas for pronghorn to move through during these succinct, yet important migratory periods. Alternatively, examination of daily movement rates can identify spatiotemporal factors that are significant to pronghorn movements, including migration and dispersal (Jones et al. 2017). Increased movement rates were observed following periods where migrations were protracted by linear features such as roads and fences, which may act as semi- or complete barriers to movement (Seidler et al. 2015). While spatiotemporal components are extremely important in understanding pronghorn movements, cognitive learning, as well as individual and group memory, likely influence pronghorn movements, though these are not fully understood (Barnowe-Meyer et al. 2013).

19.5   Interaction with Livestock Grazing Management

It is estimated that 99% of pronghorn populations share their distribution with domestic or feral livestock (Yoakum 2004d; Stoner et al. 2021) including domestic cattle (*Bos taurus*), sheep (*Ovis aries*), and domestic and feral horses (*Equus ferus caballus*), with low co-occurrence with pigs (*Sus domesticus*), goats (*Capra hircus*), and burros (*E. asinus*). With such a large overlap in distribution, interactions between pronghorn and livestock are inevitable. These interactions may be direct (i.e., diet overlap or competition for forage/water) or indirect (i.e., management practices for livestock affect habitat selection by pronghorn). The following subsections will focus on the direct and indirect interactions between pronghorn with domestic cattle, sheep, and feral horses.

19.5.1   Forage Competition and Diet Overlap

Pronghorn have physiological traits similar to other concentrate feeders (Van Soest 1994) and intermediate feeders (Hofmann 1989), suggesting they are adapted to feed on diets high in cell solubles, such as forbs and higher quality shrubs. Showing preference for forbs and shrubs during all seasons and having a digestive system engineered to pass food through the system relatively quickly is consistent with the intermediate (Hofmann 1989) or mixed feeder category (Kauffman et al. 2021). Indeed Yoakum (2004c) called pronghorn "forage switchers" because of their ability to switch forage preference to take advantage of succulent vegetation resulting from seasonal phenological changes. To demonstrate, pronghorn forage on grasses that tend to green up before forbs during spring, then switch to predominantly forbs during summer months, then switch to shrubs in fall and winter (Mitchell and Smoliak 1971; Pyrah 1987; Yoakum 2004c). In grassland diet studies, the vegetation composition was predominately grass (74%), followed by forbs (16%), and shrubs (9%) with pronghorn diet selection being predominately forbs (62%), followed by grasses (19%) and shrubs (17%; Yoakum et al. 2014). In contrast, the vegetation composition in shrub-steppe studies was predominately shrubs (46%), followed by grasses (37%), and forbs (15%), with pronghorn diet selection being predominately shrubs (62%), followed by forbs (30%), and grasses (7%; Yoakum et al. 2014). The diet preference between forbs and shrubs in the desert biome is regulated by sporadic precipitation, with forbs being preferred when adequate rainfall provides succulent forbs (Cancino 1994; Yoakum 2004c). The diets of desert-dwelling pronghorn likely include more succulent and cacti species than are consumed by populations in grasslands and shrub-steppe (Yoakum 2004c). Pronghorn in extremely arid environments utilize succulents not only to meet their nutritional requirements, but also as a major water source (Büechner 1950; Beale and Smith 1970; Clemente et al. 1995). In years of particularly severe drought, succulents may be crucial for pronghorn survival.

The documented breadth of forage species selected by pronghorn is tremendous with the use of 124 different species (96 forbs, 14 shrubs and 14 grasses; Mitchell and Smoliak 1971; Pyrah 1987). Of the plants identified as being consumed by pronghorn, 21 were considered poor forage, and 51 were unpalatable to livestock (Büechner 1950). Indeed, pronghorn consume many plants considered toxic or poisonous to livestock, including locoweed (*Astragalus* spp.), larkspur (*Delphinium* spp.), lupine (*Lupinus* spp.), and death camas (*Toxicoscordion* spp.), to name a few (Einarsen 1948; Büechner 1950; Yoakum 2004c).

Pronghorn propensity and variety of plant species consumed results in little to no competition for forage with cattle and horses, but competition can be extensive with domestic sheep. There is little dietary overlap between pronghorn, which prefer forbs, and domestic cattle and horses, which prefer grasses (Yoakum et al. 2014; Scasta et al. 2016). Yoakum (2004d) determined the annual diet overlap was less than 25% between cattle and pronghorn, and less than 36% between horses and pronghorn. Domestic sheep prefer forbs, which results in intense competition for forage with pronghorn, and diet overlap can range between 33% (moderate overlap) and 66% (high overlap; Yoakum 2004d). Yet, diet overlap with livestock in general is based on rangeland conditions being in good health, and when rangeland conditions deteriorate, competition for remaining forage intensifies (Yoakum 2004d). In addition, indirect competition may occur in areas where habitat quality is decreased through soil compaction and increased erosion (Eldridge et al. 2020). Lastly, there are specific instances when competition can be prevalent. For example, feral horses can compete directly with pronghorn in arid environments for water resources (Gooch et al. 2017). While Hennig et al. (2021) found significant temporal overlap in the use of watering sources between pronghorn and feral horses in Wyoming they could not conclude that interference was occurring between the 2 species. However, they did note the infrequent occurrence of both species being observed together.

19.5.2 Rangeland Management Practices

Western rangeland management has historically been for the benefit of livestock production, but recently, specific management actions have been completed with wildlife solely in mind. Management actions fall into two categories: (1) livestock grazing management, and (2) rangeland improvements for livestock and/or wildlife. Actions associated with livestock management include type and breed of livestock grazed, grazing intensity (i.e., stocking rate), timing of grazing (e.g., year-long, spring, etc.), and grazing system (e.g., rest-rotation, deferred, etc.). Pronghorn occupy an assortment of rangeland types; therefore, it is not our intent to evaluate and/or recommend prescriptive livestock management actions. However, we provide general livestock grazing management principles that can be practiced across a diversity of vegetation communities to improve or maintain pronghorn habitat. Rangelands that are maintained in good ecological condition and provide ecological resiliency will benefit both pronghorn and livestock (Yoakum 2004d). Livestock managers should consult with local rangeland specialists and wildlife managers when designing their grazing system. The following general recommendations are adapted from Yoakum (2004d) and Yoakum et al. (2014) and are intended as guidelines for livestock managers to enhance pronghorn habitat while maintaining high-quality livestock grazing:

- Livestock grazing systems should be designed around the local ecosystem and vegetation community and should account for the forage needs of pronghorn. Grazing systems that result in seral vegetation conditions and closely resemble the ecological potential of the local area will provide the greatest benefit to pronghorn. Grazing systems that restrict, alter, limit, or deleteriously affect the native vegetation community will negatively impact pronghorn habitat and should therefore include mitigation and alternative procedures for enhancing pronghorn habitat. As part of the grazing system, adequate amounts of preferred forage should be allocated for pronghorn and should include a variety of forbs, shrubs, and grasses identified as key forage species for pronghorn.
- Grazing capacity should be designed around the local ecosystem and vegetation community and should account for the forage needs of pronghorn. Grazing capacity should be modified based on annual precipitation levels (e.g., reduced during drought). Livestock should be restricted from key pronghorn fawning areas during the fawning season to ensure adequate forage and hiding cover.
- Livestock mangers should consider developing a ranch or allotment management plan that accounts for the needs of their livestock as well as local wildlife populations, including pronghorn.

Rangeland improvement and wildlife enhancement projects are used by livestock managers to either improve existing forage or change the utilization of existing forage by redistributing livestock (Yoakum 2004d). Improvements focused on enhancing existing forage include seeding, brush control, and burning. Seeding projects can be beneficial or detrimental to pronghorn, depending on the species used to seed the area. If the seed mixture includes forb and shrub seeds, then the project can enhance pronghorn habitat (Yoakum 2004d) but comes at a higher monetary cost than just seeding a monoculture of grasses (Yoakum et al. 2014; Downey et al. 2013). Historically, seeding projects, in which the goal was to increase forage for livestock or establish permanent vegetative cover, used seed mixtures limited to either a single or a few grass species (Yoakum 2004d). The lack of vegetation diversity established on these seeded sites generally made them poor pronghorn habitat (Yoakum et al. 2014). Recently, seeding projects have begun to use mixtures of native species to re-establish rangelands for both wildlife and livestock utilization (Downey et al. 2013; Espeland 2014). Areas that have entered late successional stages and are dominated by shrubs and shrubby trees provide poor pronghorn habitat and are of limited value to livestock. For pronghorn, once an area becomes composed of 25% or greater shrub cover, with shrubs that are $\geq$ 76 cm ($\geq$ 30 in) tall, the area provides poor pronghorn habitat because of limited forage availability and the resulting reduction in predator detection capacity (Yoakum 2004e; Yoakum et al. 2014). Areas with high shrub cover and height can be treated, either mechanically, chemically, or through prescribed fire; however, prior to any management the habitat needs of other sensitive species (e.g., sage-grouse, *Centrocercus spp.*) in the area should be considered. Yoakum (2004e) recommended shrub treatment projects be no larger than 405 ha (1000 ac) and implemented in a mosaic fashion so not all shrubs (especially those palatable to pronghorn) are removed; 5–20% retention of shrubs is ideal to maintain winter forage and fawn hiding habitat (Bayless 1969; Bruns 1977; Barrett 1981). Fire has the potential to benefit pronghorn if it returns climatic vegetation communities back to early successional stages of forbs and grasses (Yoakum 2004e). Pronghorn quickly move into areas following a fire and readily forage on newly sprouted forbs and cacti that have had their spines burned off (Courtney 1989; Van Dyke 1990; Payne and Bryant 1998; Augustine and Derner 2015). In areas with silver sagebrush (*A. cana*) burning resulted in low plant kill rates and vigorous resprouting (White and Currie 1983). However, other species of sagebrush (e.g., *A. tridentata subsp. wyomingensis*) when burned can create conditions were invasive species such as cheatgrass (*Bromus tectorum*) become dominant (Davies et al. 2007; Crist et al. 2021). Therefore, caution should be exercised before using fire in sagebrush habitat as the impacts to pronghorn habitat can be detrimental.

Improvement practices associated with livestock distribution are frequently employed on western rangelands. These practices, such as fencing, water development, salting/mineral supplementation, and in the case of domestic sheep and goats, herders, are implemented to enhance livestock distribution to maximize the use of available forage. Fencing has historically impacted pronghorn negatively and is discussed in Sect. 19.5.3. Pronghorn will readily use natural and artificial water sources (Einarsen 1948; Beale and Smith 1970; Gooch et al. 2017). Water developments allow greater pronghorn distribution, particularly during dry seasons or periods of drought (Beale and Smith 1970). However, Yoakum (2004e) noted that water developments have the potential to cause competition by allowing livestock to move to previously under-utilized areas; therefore, new water developments need to be assessed in terms of their benefit or disadvantages for pronghorn. Pronghorn will utilize salt and mineral blocks placed on the landscape to improve the distribution of livestock, but the nutritional benefits to pronghorn are poorly understood (Yoakum et al. 2014).

19.5.3 Fencing and Pronghorn

Fencing is a ubiquitous feature on rangeland landscapes (Jakes et al. 2018a; Mcinturff et al. 2020), and as far back as 1877 has been documented as a detriment to pronghorn (Caton 1877: 48 in Yoakum et al. 2014). Having evolved on treeless landscapes, pronghorn have not developed an instinct to jump over vertical obstacles, including fences (O'Gara 2004c), although they can physically jump (Harrington and Conover 2006; Jones et al. 2018, 2020b). Fences can cause mortality when pronghorn get caught in the wires (Harrington and Conover 2006). In addition, fences indirectly impact pronghorn when wounds are inflicted by barbs when crawling underneath the bottom wire or between wires, when the fence reduces access to resources (e.g., prime habitat, water, etc.), or when fences alter a pronghorn's ability to freely move across the landscape, trapping them in inhospitable habitat during environmental extremes (Jones 2014; Jones et al. 2019; Reinking et al. 2019; Xu et al. 2020).

The primary purposes of fencing on the landscape are to delineate ownership boundaries, control the distribution of livestock, and keep livestock and wildlife off roads to reduce wildlife-vehicle collisions (Jakes et al. 2018a). While there are a variety of fence designs used on western rangelands (e.g., 4-strand barbed-wire, woven wire, etc.), it is the height of the bottom wire that determines if pronghorn are able to cross the fence successfully. The predominant recommendation (Fig. 19.2) is to raise or set the bottom wire height to a minimum of 46 cm (18 in) above the ground to allow ample room for pronghorn to crawl under (Jones et al. 2018, 2020b). In addition, it is recommended that a double stranded smooth wire be used on the bottom to reduce potential injuries to pronghorn from crawling under a fence with a barbed bottom wire (Jones 2014). Enhancements to existing sheep fences (i.e., woven wire) are more problematic for livestock producers because of the requirement for a low bottom wire to contain sheep and goats. Woven wire fences can be replaced with a 4-strand barbed-wire fence with a bottom wire 25 cm (10 in) above the ground (Paige 2020). While not an ideal bottom wire height, using a barbed-wire fence (with a smooth wire on bottom), as opposed to a woven wire fence, does create some opportunity for pronghorn to pass underneath. Ideally sections of woven wire fence could be dropped when small livestock are not present. Other mitigations include leaving gates open, virtual fencing, or using lay down fence designs when livestock are not present (Paige 2020).

A FRIENDLIER FENCE FOR WILDLIFE

The friendliest fences are very visible and allow wild animals to easily jump over or slip under the wires or rails.

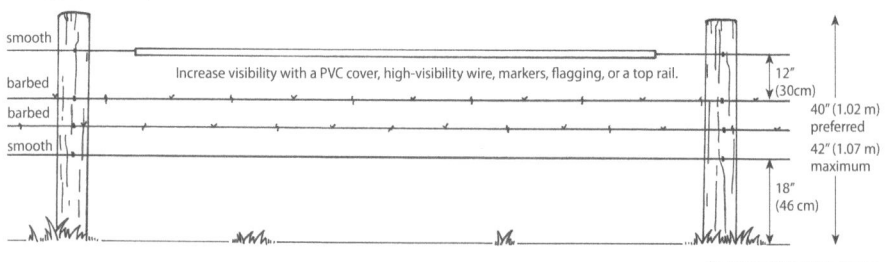

Illustration: Kristen Rumbolt Miller

Fig. 19.2 A pronghorn friendly alternative fence design with a double stranded bottom wire at 18″ (46 cm) from the ground. Adopted from Paige 2020. Illustration by K. Rumbolt Miller

19.6 Impacts of Disease

Infectious diseases can cause locally extensive mortality, but seldom produce the population level impacts that are associated with severe weather, habitat degradation, and barriers to movement. Diseases affecting pronghorn caused by viruses, bacteria, or parasites are typically shared with other wild or domestic ruminants, and frequently occur in partnership with other stressors.

Respiratory diseases of pronghorn are a frequent cause of death and are typically present as adhesions of the lung to the surface of the chest cavity, pneumonia, and fluid or hemorrhages in the lungs. Bacterial pathogens that are identical or related to those of cattle and sheep are often found. Viruses are infrequently identified, but transient infections are thought to make individuals susceptible to secondary bacterial pneumonia.

The list of bacterial pathogens which impact pronghorn is extensive, and many of these bacterial diseases are exacerbated by poor rangeland quality and over-crowding, both among pronghorn and with domestic livestock (O'Gara 2004d). Significant bacterial diseases that infect pronghorn include Anaplasmosis, Campy-lobacter, leptospirosis, *Mycoplasma bovis*, and necrobacillosis. Additional bacterial diseases which impact pronghorn to varying degrees include Actinobacillosis, Acti-nomycosis, *Escherichia coli* infections, Pasteurellosis, and Vibriosis (Jaworski et al. 1998; Kreeger et al. 2011). *Mycoplasma bovis* is a bacterial disease of cattle causing pneumonia, mastitis, and arthritis. Mycoplasma pneumonia has recently been iden-tified as the cause-of-death for hundreds of pronghorn in northern Wyoming (Malm-berg et al. 2020). These mortality events occurred in the late winter to early spring. The pronghorn died quickly, even though they were in good body condition, and upon necropsy, results indicated massive pneumonia with yellow fibrin covering the surface of the lungs. At this point, it is unclear whether mycoplasma pneumonia in pronghorn is a localized problem, or if the organism is established in pronghorn, but only infrequently causes disease.

The most significant viral pathogens of pronghorn are those causing hemorrhagic disease, including epizootic hemorrhagic disease virus (EHDV) and bluetongue virus (BTV). Hemorrhagic disease outbreaks in pronghorn can produce significant die-offs but occur in four- to seven-year cycles. Typically, there is minimal or no mortality between large-scale hemorrhagic disease events. Outbreak years often correspond with exceptionally hot and dry summers, which favor large vector populations and increased animal density around limited water sources. These seasonal variations, combined with waning population immunity, contribute to the risk for outbreaks. Bluetongue virus and EHDV also affect deer, elk, cattle, and domestic sheep.

Scours, or diarrhea, can be caused by a rapid change in diet, particularly during the spring green up, but it also occurs during the summer months. Animals are seen with a soiled hind end and may be listless and appear unkempt. Scours is more prevalent in young animals, but all ages and both sexes can be affected, and pronghorn mortality can be locally extensive. Scours frequently occurs in animals feeding on alfalfa (*Medicago sativa*), but a variety of bacteria, viruses, and parasites have also been identified as potential causes. Extensive research has failed to identify a definitive origin for this condition, which is frequently a cause of concern for livestock producers whose animals share rangeland with affected pronghorn.

Pronghorn harbor a number of parasites also present in domestic ruminants, but health impacts are usually restricted to crowded situations or overgrazed rangelands shared with livestock. Increased transmission occurs near water sources such as stock ponds and water tanks that are heavily used by livestock and wildlife. The large stomach worm or barber pole worm (*Haemonchus contortus*), is the most significant parasite of pronghorn (Kreeger et al. 2011). This parasite attaches to the mucosa of the fourth stomach and feeds on blood. Heavy infection results in anemia and may contribute to mortality in animals already in poor nutritional condition. This parasite is well recognized in domestic livestock, especially sheep, goats, and cattle, and parasite burdens may increase on rangelands shared by susceptible ruminant species. Although infrequently found, round worms (e.g., *Ostertagia* sp., *Nematodirus sp.*, and *Cooperia sp.*), lung worm *(Protostrongylus macrotis),* and tapeworms *(Monezia sp.)* infect pronghorn, cattle, and sheep (Goldsby and Eveleth 1954; Greiner et al. 1974).

Foot rot is the common term used to describe the disease caused by the bacterium *Fusobacterium necrophorum*. Animals often show signs of lameness, with swollen feet and fetlocks, but may also have ulcers in their mouths. Mortality occurs during the spring when snowmelt produces muddy conditions, or during the summer when pronghorn congregate around ponds or stock tanks where the muddy substrate has been contaminated with feces containing the bacteria. Mortality events are usually localized with most animals recovering from infection.

Management practices for both pronghorn and livestock influence the transmission of diseases and parasites between individuals and among species. Good nutrition and maintaining animals at or below the carrying capacity of their summer range are the basic tenets of healthy populations, both wild and domestic.

19.7   Ecosystem Threats

During the nineteenth century, pronghorn populations range-wide were decimated from market hunting by European settlers, and by the 1920s, the species was nearly extinct across their range (Grinnell 1929; Greenquist 1983; O'Gara and McCabe 2004; McCabe et al. 2004). As twentieth century regulation of pronghorn hunting was implemented, initiating the species' recovery (Greenquist 1983; O'Gara and McCabe 2004), multiple factors impacting pronghorn sustainability on the landscape began to shift.

Today, long-term pronghorn population persistence is chiefly threatened by human-caused habitat conversion, fragmentation, and loss (O'Gara and McCabe 2004). Additionally, anthropogenic development and activity have impacted pronghorn populations by producing behavioral changes (Sawyer et al. 2002; Beckmann et al. 2012; Seidler et al. 2015; Reinking et al. 2019; Jones et al. 2019) and dramatically altered weather and climate regimes related to global climate change (Christie et al. 2015; Gedir et al. 2015; McKelvey and Buotte 2018). The variety of ecosystem threats facing pronghorn populations today and into the future are explored in greater detail below.

19.7.1 Farming and Ranching

Habitat alteration associated with farming and ranching began in the nineteenth century with the arrival of European settlers. Agricultural production equated to the conversion of native pronghorn range, particularly in grassland habitats, where nutrient-rich soils are ideal for the growth of staple crops like corn and wheat (O'Gara and McCabe 2004). In the 1930s, the dust-filled winds and economic recession of the Great Depression frequently caused farming families to abandon property located on marginal prairie lands. Despite the drought conditions of that time, many of these uninhabited areas reverted to native vegetation, creating short-term benefits for pronghorn populations (O'Gara and McCabe 2004). However, reprieve was temporary; as rampant drought abated, farming expanded and became increasingly mechanized, and practices that caused rapid deterioration of rangelands were further employed. Clearing of native vegetation on highly erodible grasslands (i.e., sodbusting), was common into the 1990s, despite legislation designed to discourage the practice (e.g., the Sodbuster Provision of the 1985 Food Security Act; O'Gara and McCabe 2004). Although federal, state, and non-government organization's programs and partnerships look to curb habitat conversion, the threat of losing additional native habitats to agricultural land still exists (Smith et al. 2016).

Improper livestock management represents a source of pronghorn habitat alteration. Heavy livestock stocking rates can negatively impact pronghorn habitat quality through overgrazing and trampling of native vegetation, compaction of soil, and damage to riparian areas (O'Gara and McCabe 2004). Such imprudent management of rangelands, exacerbated by the historical prioritization of livestock grazing over wildlife management, resulted in degraded landscapes and was particularly problematic in the arid, desert portions of pronghorn range (O'Gara and McCabe 2004). Legislation, including the Public Rangelands Improvement Act of 1978, and interventions like the removal of cattle from crucial Sonoran pronghorn habitat in the 1980s, helped to bring livestock production into greater equilibrium with pronghorn conservation and management (O'Gara and McCabe 2004). While improper grazing still occurs, but to a lesser extent than the last several decades, today the largest impact of ranching practices on pronghorn populations is the fragmentation caused by fences erected to exclude wildlife or contain livestock (see Sect. 19.5.3).

19.7.2   Habitat Alteration

One prevalent threat to pronghorn persistence is the alteration of their habitat through increased invasion of rangelands by non-desirable species (i.e., non-native grasses and shrubs or trees) and the associated changes in fire regimes. Many non-native grasses (e.g., cheatgrass, smooth brome [*B*. inermis], and crested wheatgrass [*Agropyron cristatum*]) are capable of out-competing native vegetation, resulting in critical habitat changes (Boyd et al. 2021; Gaskin et al. 2021). Many invasive grasses can dominate native species because they quickly colonize disturbed areas, mature early, have short root systems for absorbing water quickly in soil, and are prolific seed producers (Boyd et al. 2021). In addition, many invasive species, and especially cheatgrass, respond positively to and can alter fire regimes, catalyzing a detrimental invasive plant-fire regime cycle (Brooks et al. 2004). These characteristics result in pronghorn habitat being altered from diverse mosaics of grasses, forbs, and shrubs to monocultures of invasive grasses (Boyd et al. 2021; Gaskin et al. 2021). Moreover, climate change has the potential to exacerbate these changes (Adler et al. 2021).

Fire suppression on western rangelands has caused the transition of vegetative communities from early to late succession. Associated with this change is the encroachment and expansion of coniferous trees, especially pinyon pine (*Pinus sp.*) and juniper (*Juniperus sp.*), into western rangelands (Maestas et al. 2021). This increase in coniferous trees results in declines in perennial grasses, forbs, and more generally, productivity (Maestas et al. 2021). Such encroachment has also resulted in changes to sagebrush communities as they become increasingly susceptible to invasive species (e.g., cheatgrass) because of increases in overstory crown fires (Chambers et al. 2014; Maestas et al. 2021). With these changes, the availability of forage and cover provided by native rangeland species declines considerably, resulting in a dramatic shift to rangelands composed of more coniferous species (Maestas et al. 2021).

19.7.3   Residential and Urban Development

In addition to the expansion of agriculture and livestock production, European settlement of the North American West spurred residential and urban growth (O'Gara and McCabe 2004). As was common in farming and ranching practices, fences were frequently used to delineate property boundaries, posing risks to pronghorn survival, and presenting direct barriers to movement (Harrington and Conover 2006; Jones et al. 2018). For example, in the 1960s in Arizona, the sectioning of large swathes of native rangeland for residential plots (i.e., ranchettes), allowed people to feel closer to wildlife in relatively rural settings but ultimately fragmented large portions of the landscape at a detriment to pronghorn (O'Gara and McCabe 2004). In addition to habitat fragmentation, residential expansion can result in direct and indirect habitat loss and mortality for pronghorn because of greater traffic levels and higher road/

fence densities. In rare instances, residential and urban development can also cause direct mortality for pronghorn; for example, animals have been known to forage on toxic, ornamental vegetation in landscaped yards or even trash at city waste facilities, ultimately dying from inflamed stomachs and toxicity (O'Gara 2004b).

19.7.4 Energy Development

Much of pronghorn range is conducive to energy development (wind, solar, oil and natural gas, mining) due to the impressive wind speeds, high incoming solar radiation, and sizable underlying fossil fuel and mineral depositions found in these rangelands (Yoakum 2004b; Copeland et al. 2009). A growing body of literature indicates that the infrastructure and activity associated with these land uses can directly eliminate portions of pronghorn habitat and indirectly cause habitat loss by altering pronghorn behavior (Sawyer et al. 2002; Beckmann et al. 2012; Christie et al. 2015; Reinking et al. 2019; Jakes et al. 2020; Smith et al. 2020). Notably, the infrastructure of these developments also typically includes high densities of roads and fencing, which influence pronghorn behavior (Gavin and Komers 2006; Seidler et al. 2015; Jones et al. 2018) and present direct and indirect mortality risks (Harrington and Conover 2006; Jones 2014).

The risk-avoidance hypothesis suggests that animals can perceive human-induced landscape disturbance similarly to predators, and that similar risk-avoidance responses may result when they are exposed to anthropogenic activity and infrastructure (Frid and Dill 2002; Gavin and Komers 2006). These responses include behavioral alterations, such as spending an increased proportion of time in a vigilant state and less time foraging (Gavin and Komers 2006; Seidler et al. 2015, 2018), and avoidance of developed areas (Beckmann et al. 2012; Reinking et al. 2019; Smith et al. 2020). For pronghorn, these responses may be amplified in winter, when individuals are generally in reduced physical condition and already stressed by limited forage and increased energetic requirements (Yoakum et al. 2014). For example, pronghorn in the Shirley Basin, Wyoming were found to avoid wind turbines after installment in their winter home range, and this effect was stronger in more severe winters (Taylor et al. 2016; Smith et al. 2020). Within the same study area, Milligan et al. (2021) found pronghorn were displaced when selecting a home range by existing turbines in both summer and winter, but there was little evidence of avoidance behaviour within the home range at the population level. A similar trend has been observed in multiple studies of the impacts of oil and natural gas development and associated infrastructure (Beckmann et al. 2012; Reinking et al. 2019). Moreover, energy development within winter range can cause cumulative changes in pronghorn habitat use over time including initial and continued avoidance that can ultimately result in increasing abandonment of these seasonally crucial areas (Sawyer et al. 2019). While studies have yet to make the mechanistic connection between energy development and its potential influence on pronghorn survival and reproductive success, it is likely that the habitat loss and behavioral changes these disturbances produce negatively impact pronghorn fitness (Sawyer et al. 2002; Beckmann et al. 2016). This is supported by research linking long-term pronghorn population declines to the density of energy development on the landscape (Christie et al. 2015).

Given its relatively low cost (e.g., seeding to non-natives) and simple implementation, reclamation of rangelands altered by energy extraction efforts is often more preferable for industry stakeholders than modifying their procedures, such as directional drilling of oil and gas wells to minimize the number of required well pads (O'Gara and McCabe 2004). However, reclamation has largely been proven to be inadequate in the biomes pronghorn inhabit, as much of the landscape damage that results from energy development is irreversible (Rottler et al. 2018). Reclamation efforts frequently fail to restore habitat to its former condition, and can result in the establishment of invasive, noxious weeds (Padgett 2020). Additionally, the mitigation requirements for energy development stakeholders are often vague, with little to no post-reclamation monitoring or land use management required (Zimmerman 1983; O'Gara and McCabe 2004).

19.7.5 Climate Change

Given the spatially expansive nature of pronghorn range, and the variety of habitats it includes, the expected alterations to weather and climate regimes resulting from global climate change are myriad. In general, pronghorn range-wide are likely to experience increased air temperatures year-round, causing warmer winters, more extreme summer heat, and greater frequency of drought conditions (McKelvey and Buotte 2018; Adler et al. 2021). Other climate alterations, such as changes in precipitation patterns, are likely to be more influential at the periphery of pronghorn range and will vary both latitudinally and longitudinally. Overall, high quality pronghorn seasonal and connectivity habitat is likely to be reduced because of climate change (Zeller et al. 2021).

In the northern portion of their range, pronghorn will likely experience more stochastic winter conditions, which will include some years with reduced winter precipitation as well as years with dramatically increased winter precipitation (McKelvey and Buotte 2018; Adler et al. 2021). This will result in winters with severely reduced snowpack, and some with extremely deep snow accumulations. Years of limited winter precipitation will produce drier conditions in the following summer, which in-turn, can lead to increased frequency and severity of wildfire (Halofsky et al. 2018). Increased frequency and severity of wildfires can reduce overall shrub density and allow non-native, noxious weeds to flourish (Yoakum et al. 2014; Adler et al. 2021; Boyd et al. 2021). Years with greater winter precipitation and lower temperatures could result in population declines (Barrett 1982; Christie et al. 2015; Jones et al. 2020a). The physical properties of future snowpacks, such as their ability to support pronghorn on the surface or the wetness of the snow, are also likely to be altered (Berteaux et al. 2017; Boelman et al. 2019). These properties can influence the energetic expense of moving through snow (Parker et al. 1984) and therefore have implications for pronghorn fitness.

The southern portions of pronghorn range are expected to receive less moisture and average higher air temperatures (Gedir et al. 2015; Adler et al. 2021). Studies that project the impacts of such reduced precipitation in southwestern regions anticipate decreased pronghorn abundance and local extirpations resulting from these hotter, drier conditions (Gedir et al. 2015). In arid and semi-arid areas, precipitation is crucial for maintaining adequate forage and water resources on the landscape (Beale and Smith 1970; Yoakum 2004b). Deficiencies in these resources resulting from drought conditions have been linked to reduced pronghorn reproductive success, lower survival, and ultimately, population declines (Brown et al. 2006; Simpson et al. 2007; McKinney et al. 2008).

19.8   Conservation and Management Actions

Across the extent of pronghorn range, the landscape is a matrix of habitats (i.e., grass-lands, sagebrush, agricultural crops) and ownership (public and private). In addition, pronghorn individuals and populations currently move between jurisdictions (e.g., between Colorado—Wyoming, Montana—Idaho, and Alberta—Saskatchewan) and even countries (e.g., between Saskatchewan, Canada and Montana, USA). Continued prospects to travel throughout and between habitats, independent of jurisdictional boundaries, is particularly important to pronghorn as movement is one of their key adaptations to maintain populations and genetic diversity. Landscape connectivity for pronghorn allows them to track spatiotemporal shifts in vegetation condition and availability, adapt to anthropogenic influences, and move to landscapes that may become more suitable for pronghorn over time (e.g., as a result of climate change) while maintaining genetic diversity (Hilty et al. 2006).

19.8.1   Barriers to Movement and Functional Connectivity

In general, natural landscapes are more connected, functioning, and resilient ecosystems than those inundated by human-made features and development. Subsequently, pronghorn need specified areas and/or identified locales to navigate anthropogenic impediments and sustain movements across fragmented landscapes (Beier and Noss 1998; Hilty et al. 2006). Simple and cost-effective measures can be taken to allow for continued daily and seasonal use by pronghorn. Solutions exist for providing pronghorn safe passage across linear anthropogenic features, such as roads and fences that fragment the landscape.

Roads (paved and unpaved and with or without fences) typically have a major influence on pronghorn, presenting barriers to movement and in many cases causing avoidance behaviors, increased vigilance, and reduced foraging opportunities (Gavin and Komers 2006; Dodd et al. 2011; Jones et al. 2019; Jakes et al. 2020). In concert with roadside fencing, direct mortalities along roads occur to pronghorn by being caught in fencing, fawns being separated from does and predated upon, or individuals being trapped within the road right-of-way and struck by vehicles (Sawyer and Rudd 2005; Harrington and Conover 2006; Seidler et al. 2018). While mitigation opportunities do exist, the risk of wildlife-vehicle collisions, both in terms of the safety of vehicular passengers and risk of property damage, must be considered (Dodd et al. 2011; Lee et al. 2021). One mitigation measure is wildlife crossing structures. In Wyoming, pronghorn have been observed to use highway underpasses (Plumb et al. 2003) to navigate roads, but given a choice, pronghorn preferred to use highway overpasses 93% of the time, rather than underpasses (Sawyer et al. 2016). The construction of wildlife crossings, particularly overpasses, has been effective in allowing for continued seasonal migrations of pronghorn and provides an additional option to communities and jurisdictions to allow for wildlife movement in a safe manner for both people and wildlife (Seidler et al. 2018). While the up-front costs of planning and constructing these features can be significant, they are offset by the long-term savings in costs associated with insurance claims and the value of increased human safety (Huijser et al. 2009).

Fencing can similarly be modified to allow for continued pronghorn daily and seasonal movements while also addressing human needs. Fences along roadways can be modified to create an opportunity for pronghorn to cross at a specific location (accounting for pronghorn use, traffic levels, and proper fence design), or can provide a funneling mechanism to direct animals towards a crossing structure (i.e., underpass, overpass; O'Gara and McCabe 2004; Sawyer and Rudd 2005; Yoakum et al. 2014). Paired right-of-way fencing gates and lay-down fences have been installed more across the West in the last ten years and are considered important conservation measures benefiting pronghorn movement and landscape connectivity, in general (Paige 2020). In addition, several sportsman and conservation groups (e.g., Alberta Fish and Game Association, Arizona Antelope Foundation, Jackson Hole Wildlife Foundation, etc.) hold volunteer events that modify fences for the benefit of pronghorn and other wildlife species. Pasture fence design and modifications are discussed in detail in Sect. 19.5.3.

19.8.2 Managing Pronghorn on the Private–Public Landscape Matrix

Habitat management and enhancement within the private–public landscape matrix is important for maintaining pronghorn populations. For example, pronghorn in the Northern Great Plains were found to migrate through a greater percentage of private lands than public lands (Tack et al. 2019). Therefore, listening to, understanding, and accounting for private landowner perspectives is essential to properly manage wildlife populations. For example, landowners require fences that contain livestock in appropriate pastures and at the same time, they spend time and money on fixing fences that are damaged by the wildlife navigating them. The solution is to find 'win–win' approaches for both the landowner and wildlife to minimize fence damage and keep livestock contained. Installation of wildlife-friendlier fencing, the use of fence modifications on existing fences, or the installment of gates will result in win–win opportunities (Jones et al. 2018, 2020b). Similarly, water is in limited supply across pronghorn range, and water development, design, and placement will influence its use by pronghorn and domestic livestock (Larsen et al. 2011). Wildlife managers can work with landowners to design stock tanks that most effectively facilitate pronghorn use. Finally, working with landowners to identify priority native habitat for wildlife is of utmost importance. For example, conservation easements have been used as an effective tool to conserve greater sage-grouse habitat that also protected pronghorn habitat (Tack et al. 2019). Many state, provincial, federal, and non-government organizations' programs provide funding for conservation easements, fence modifications, vegetation treatments, management of annual invasive grass, and water developments which provide benefits to a suite of rangeland wildlife species.

Public land across the range of pronghorn is managed for multi-use including livestock grazing, energy development, recreational use (e.g., hunting and viewing), and wildlife habitat. Within the USA most of the public land falls within the jurisdiction of the Bureau of Land Management (BLM) and the Forest Service and represents 50% of total pronghorn habitat (Yoakum 2004f). Historically public land has been managed with livestock in mind. This fact is exemplified with 95% of the BLM expenditures for rangeland improvements being for the benefit of livestock (Donahue 1999). More recently there has been a decrease in priority for grazing as a balance been livestock and wildlife needs has been struck. For example, at the Hart Mountain National Antelope Refuge sheep, cattle, and feral horses have been removed resulting in improved habitat and pronghorn numbers on the refuge (Yoakum 2004f). In addition, rangeland improvement projects are now completed with wildlife in mind, such as modifying fences to wildlife friendly designs and installation of water developments.

19.8.3 Genetic Diversity

One area lacking research is the analysis of pronghorn genetics (see Yoakum et al 2014 for a review). Initial genetic work has focused on endangered subspecies, with more recent work focused on the use of genetics to estimate populations and determine if natural and anthropogenetic landscape features are barriers to movement. Both Sonoran and Peninsular subspecies of pronghorn are endangered and continue to severely lose genetic diversity, and if they are to persist, careful genetic management is required through continued captive breeding (Stephen et al. 2005; Klimova et al. 2014). Recently, the use of noninvasively collected fecal DNA and capture-recapture designs at watering holes have been evaluated to determine if the use of genetics can improve population estimates (Woodruff et al. 2016). Except the work of Lee Jr et al. (1994), few genetic studies have been conducted on *A. americana* that characterize genetic variation between populations. Using mitochondrial DNA, Lee Jr et al. (1994) found differences in allozyme variation in 29 populations across the West. More recently, pronghorn populations in Wyoming were found to be genetically connected throughout the core of their range (LaCava et al. 2020), which is encouraging, given naturally occurring landscape barriers (e.g., mountain ranges) and anthropogenic fragmentation across the state (Copeland et al. 2009).

19.9 Research and Management Needs

The rangelands of western North America are under increasing pressure from anthropogenic disturbances (e.g., roads, fences, houses/residential development, agricultural conversion, energy development), and climate change. Understanding the effects of a changing landscape will be key to conserving pronghorn populations across their range. We suggest addressing the following research and management needs to ensure healthy, sustainable pronghorn populations, though we do not consider this list exhaustive:

- While continued understanding of each impact is warranted (e.g., wind and solar energy), the real need is to understand the cumulative effects of these factors on pronghorn population persistence (i.e., fitness). To understand the cumulative effects requires long-term datasets and intrinsic information (e.g., body condition, reproductive status, recruitment, and survival). Lastly, understanding cumulative effects of impacts should identify threshold levels that result in population declines or local extirpation.
- The potential effects of climate change on pronghorn population persistence need further exploration, particularly as climate regime alterations continue. Large-scale connectivity modelling that accounts for future climate scenarios should occur. Long-distance movements may be a vital adaptation for pronghorn at the periphery of their range, because these movements offer escape from extreme environmental conditions, stochastic weather and disturbance events, and habitat alterations. Future research is required to identify long-distance movements more clearly and understand the mechanisms driving them.
- A greater understanding of the effects of linear features (i.e., fences and transportation infrastructure) on pronghorn movement and fitness is required. For example, in the management of fences the first step is to develop tools and designate resources to map fences including design specifications across broad spatial scales. For transportation infrastructure, citizen science programs (e.g., Wildlife Xing (www.pronghornxing.org)) can be implemented and promoted. These programs will allow us to better understand where pronghorn interact (e.g., killed, cross, stage) within transportation corridors to assist in identifying key areas for mitigation (e.g., overpasses). Then, these datasets should be coupled with long-term movement datasets and intrinsic fitness information.

- Associated with gaining a greater understanding of the effects of linear features is the evaluation of whether these features, as well as natural features, are acting as barriers to gene flow. In addition, a genetic analysis of the populations across the range of pronghorn is warranted to determine relatedness and to confirm the number of distinct subspecies.
- The development of integrated population models is needed to account for the influence of spatiotemporal factors (e.g., seasonally variable environmental conditions) on pronghorn

Competing Interest
All coauthors do not have any competing interest associated with this chapter.

References

Adler PB, Bradford JB, Chalfoun A et al (2021) Climate adaptation. In: Remington TE, Deibert PA, Hanser SE et al (eds) Sagebrush conservation strategy—challenges to sagebrush conservation. U.S. Geological Survey Open-File Report 2020–1125, Reston, Virginia, pp 121–137. https://doi.org/10.3133/ofr20201125

Arnold TW, Clark RG, Koons DN et al (2018) Integrated population models facilitate ecological understanding and improved management decisions. J Wildl Manage 82:266–274. https://doi.org/10.1002/jwmg.21404

Augustine DJ, Derner JD (2015) Patch burn grazing management in a semiarid grassland: Consequences for pronghorn, plains prickly pear, and wind erosion. Rangel Ecol Manag 68:40–47. https://doi.org/10.1016/j.rama.2014.12.010

Barnowe-Meyer KK, White PJ, Davis TL et al (2017) Seasonal foraging strategies of migrant and non-migrant pronghorn In Yellowstone National Park. Northwest Nat 98:82–90. https://doi.org/10.1898/NWN16-10.1

Barnowe-Meyer KK, White PJ, Waits LP et al (2013) Social and genetic structure associated with migration in pronghorn. Biol Conserv 168:108–115. https://doi.org/10.1016/j.biocon.2013.09.022

Barrett MW (1981) Environmental characteristics and functional significance of pronghorn fawn bedding sites in Alberta. J Wildl Manage 45:120–131. https://doi.org/10.2307/3807880

Barrett MW (1982) Ranges, habitat, and mortality of pronghorns at the northern limits of their range. Doctor of Philosophy Dissertation, University of Alberta. https://doi.org/10.7939/R3C J87X4R

Bauer S, Nolet BA, Giske J et al (2011) Cues and decision rules in animal migration. In: Milner-Gulland EJ, Fryxell JM, Sinclair ARE (eds) Animal migration: a synthesis. Oxford University Press, Oxford, pp 68–87. https://doi.org/10.1093/acprof:oso/9780199568994.001.0001

Bayless SR (1969) Winter food habits, range use, and home range of antelope in Montana. J Wildl Manage 33:538–551. https://doi.org/10.2307/3799376

Beale DM, Smith AD (1970) Forage use, water consumption, and productivity of pronghorn antelope in Western Utah. J Wildl Manage 34:570–582. https://doi.org/10.2307/3798865

Beckmann JP, Murray K, Seidler RG et al (2012) Human-mediated shifts in animal habitat use: sequential changes in pronghorn use of a natural gas field in Greater Yellowstone. Biol Conserv 147:222–233. https://doi.org/10.1016/j.biocon.2012.01.003

Beckmann JP, Olson SH, Seidler RG et al (2016) Sub-lethal effects of energy development on a migratory mammal—The enigma of North American pronghorn. Glob Ecol Conserv 6:36–47. https://doi.org/10.1016/j.gecco.2016.02.001

Beier P, Noss R (1998) Do habitat corridors provide connectivity? Conserv Biol 12:1241–1252. https://doi.org/10.1111/j.1523-1739.1998.98036.x

Bender LC, Rosas-Rosas OC (2021) Actual precipitation, predicted precipitation, and large herbivore condition in arid and semi-arid southern New Mexico. J Arid Environ 185: article 104378. https://doi.org/10.1016/j.jaridenv.2020.104378

Berteaux D, Gauthier G, Domine F et al (2017) Effects of changing permafrost and snow conditions on tundra wildlife: critical places and times. Arct Sci 3:65–90. https://doi.org/10.1139/as-2016-0023

Boelman NT, Liston GE, Gurarie E et al (2019) Integrating snow science and wildlife ecology in Arctic-boreal North America. Environ Res Lett 14: article 1. https://doi.org/10.1088/1748-9326/aaeec1

Bolger DT, Newmark WD, Morrison TA et al (2007) The need for integrative approaches to understand and conserve migratory ungulates. Ecol Lett 11:63–77. https://doi.org/10.1111/j.1461-0248.2007.01109.x

Boyd CS, Davis DM, Germino MJ et al (2021) Invasive plant species. In: Remington TE, Deibert PA, Hanser SE et al (eds) Sagebrush conservation strategy—challenges to sagebrush conservation. U.S. Geological Survey Open-File Report 2020–1125, Reston, Virginia, pp 99–119. https://doi.org/10.3133/ofr20201125

Brooks ML, D'Antonio CM, Richardson DM et al (2004) Effects of invasive alien plants on fire regimes. Biosci 54:677–688. https://doi.org/10.1641/0006-3568(2004)054[0677:EOIAPO]2.0.CO;2

Brown DE, Warnecke D, McKinney T et al (2006) Effects of midsummer drought on mortality of doe pronghorn (*Antilocapra americana*). Southwest Nat 51:220–225. https://doi.org/10.1894/0038-4909

Bruns EH (1977) Winter behavior of pronghorns in relation to habitat. J Wildl Manage 41:560–571. https://doi.org/10.2307/3800530

Büechner HK (1950) Life history, ecology, and range use of the pronghorn antelope in Trans-Pecos Texas. Am Midl Nat 43:257–354. https://doi.org/10.2307/2421904

Byers JA (1997) American pronghorn: social adaptations and the ghosts of predators past. The University of Chicago Press, Chicago

Cancino J (1994) Food habits of the peninsular pronghorn. Proc Prong Ant Work 16:176–185

Caton JD (1877) The antelope and deer of America. Hurd and Houghton, New York

Chambers JC, Miller RF, Board DI et al (2014) Resilience and resistance of sagebrush ecosystems—implications for state and transition models and management treatments. Rangel Ecol Manag 67:440–454. https://doi.org/10.2111/REM-D-13-00074.1.]

Christie KS, Jensen WF, Schmidt JH et al (2015) Long-term changes in pronghorn abundance index linked to climate and oil development in North Dakota. Biol Conserv 192:445–453. https://doi.org/10.1016/j.biocon.2015.11.007

Clemente F, Valdez R, Holechek JL et al (1995) Pronghorn home range relative to permanent water in Southern New Mexico. Southwest Nat 40:38–41. https://doi.org/10.2307/30054391

Collins GH (2016) Seasonal distribution and routes of pronghorn in the Northern Great Basin. West N Am Nat 76:101–112. https://doi.org/10.3398/064.076.0111

Copeland HE, Doherty KE, Naugle DE et al (2009) Mapping oil and gas development potential in the US Intermountain West and estimating impacts to species. PLoS ONE 4:e7400. https://doi.org/10.1371/journal.pone.0007400

Courtney RF (1989) Pronghorn use of recently burned mixed prairie in Alberta. J Wildl Manage 53:302–305. https://doi.org/10.2307/3801127

Crist MR, Belger R, Davies KW et al (2021) Altered fire regimes. In: Remington TE, Deibert PA, Hanser SE et al (eds) Sagebrush conservation strategy—Challenges to sagebrush conservation. U.S. Geological Survey Open-File Report 2020–1125, Reston, Virginia, pp 79–98. https://doi.org/10.3133/ofr20201125

Dalton KA (2009) Pronghorn: migration triggers and resource selection in southeastern Oregon. Master's Thesis, Washington State University

Davies KW, Bates JD, Miller RF (2007) Short-term effects of burning Wyoming Big Sagebrush Steppe in southeast Oregon. Rangel Ecol Manag 60:515–522. https://doi.org/10.2111/1551-502 8(2007)60[515:SEOBWB]2.0.CO;2

Dingle H, Drake VA (2007) What is migration? Biosci 57:113–121. https://doi.org/10.1641/B57 0206

Dodd NL, Gagnon JW, Sprague S et al (2011) Assessment of pronghorn movements and strategies to promote highway permeability: US Highway 89. Arizona Department of Transportation, Phoenix, Arizona

Donahue D (1999) The western range revisited: removing livestock from public lands to conserve native biodiversity. University of Oklahoma Press, Norman

Downey BA, Blouin F, Richman JD, Downey BL, Jones PF (2013) Restoring mixed grass prairie in southeastern Alberta, Canada. Rangelands 35:16–20. https://doi.org/10.2111/RANGELANDS-D-12-00082.1

Einarsen AS (1948) The pronghorn antelope and its management. Monument Printing Press, Baltimore

Eldridge DJ, Ding J, Travers SK (2020) Feral horse activity reduces environmental quality in ecosystems globally. Biol Conserv 241:108367. https://doi.org/10.1016/j.biocon.2019.108367

Espeland EK (2014) Choosing a reclamation seed mix to maintain rangelands during energy development in the Bakken. Rangelands 36:25–28. https://doi.org/10.2111/RANGELANDS-D-13-00056.1

Frid A, Dill LM (2002) Human-caused disturbance stimuli as a form of predation risk. Conserv Ecol 6: article 11. https://doi.org/10.5751/ES-00404-060111

Gaskin JF, Espeland E, Johnson CD et al (2021) Managing invasive plants on Great Plains grasslands: a discussion of current challenges. Rangel Ecol Manag 78:235–249. https://doi.org/10.1016/j.rama.2020.04.003

Gavin SD, Komers PE (2006) Do pronghorn (*Antilocapra americana*) perceive roads as a predation risk? Can J Zool 84:1775–1780. https://doi.org/10.1139/z06-175

Gedir JV, Cain JW, Harris G et al (2015) Effects of climate change on long-term population growth of pronghorn in an arid environment. Ecosphere 6: article 189. https://doi.org/10.1890/ES15-00266.1

Goldsby AI, Eveleth DF (1954) Internal parasites in North Dakota antelope. J Parasitol 40:637–648. https://doi.org/10.2307/3273702

Gooch AMJ, Petersen SL, Collins GH et al (2017) The impact of feral horses on pronghorn behavior at water sources. J Arid Environ 138:38–43. https://doi.org/10.1016/j.jaridenv.2016.11.012

Greenquist CM (1983) The American pronghorn antelope in Wyoming: a history of human influences and management. Doctor of Philosophy Dissertation, University of Oregon

Greiner EC, Worley DE, O'Gara BW (1974) Protostrongylus macrotis (Nematoda: Metastrongyloidea) in pronghorn antelope from Montana and Wyoming. J Wildl Dis 10:70–73. https://doi.org/10.7589/0090-3558-10.1.70

Grinnell GB (1929) Pronghorn antelope. J Mammal 10:135–141. https://doi.org/10.2307/1373835

Guenzel RJ (1997) Estimating pronghorn abundance using aerial line-transect sampling. Wyoming Game and Fish Department, Cheyenne, Wyoming. https://doi.org/10.13140/RG.2.2.15682.94407

Halofsky JE, Peterson DL, Dante-Wood SK et al (2018) Climate change vulnerability and adaptation in the Northern Rocky Mountains: Part 2. Rocky Mountain Research Station, General Technical Report 374, Fort Collins

Harrington JL, Conover MR (2006) Characteristics of ungulate behavior and mortality associated with wire fences. Wildl Soc Bull 34:1295–1305. https://doi.org/10.2193/0091-7648(200 6)34[1295:COUBAM]2.0.CO;2

Hennig JD, Beck JL, Gray CJ, Scasta JD (2021) Temporal overlap among feral horses, cattle, and native ungulates at water sources. J Wildl Manage 85:1084–1090. https://doi.org/10.1002/jwmg.21959

Hilty JA, Lidicker WZ Jr, Merenlender AM (2006) Corridor ecology: the science and practice of linking landscapes for biodiversity conservation. Island Press, New York

Hofmann RR (1989) Evolutionary steps of ecophysiological adaptation and diversification of ruminants: a comparative view of their digestive system. Oecologia 78:443–457. https://doi.org/10.1007/BF00378733

Hoskinson RL, Tester JR (1980) Migration behavior of pronghorn in southeastern Idaho. J Wildl Manage 44:132–144. https://doi.org/10.2307/3808359

Huijser MPM, Duffield JWJ, Clevenger AP et al (2009) Cost-benefit analyses of mitigation measures aimed at reducing collisions with large ungulates in the United States and Canada: A decision support tool. Ecol Soc 14: article 15

Jacques CN, Jenks JA (2007) Dispersal of yearling pronghorns in Western South Dakota. J Wildl Manage 71:177–182. https://doi.org/10.2193/2005-704

Jacques CN, Jenks JA, Klaver RW (2009) Seasonal movements and home-range use by female pronghorns in sagebrush-steppe communities of western South Dakota. J Mammal 90:433–441. https://doi.org/10.1644/07-MAMM-A-395.1

Jacques CN, Jenks JA, Grovenburg TW et al (2015) Influence of habitat and intrinsic characteristics on survival of neonatal pronghorn. PLoS ONE 10:e0144026. https://doi.org/10.1371/journal.pone.0144026

Jakes AF (2015) Factors influencing seasonal migrations of pronghorn across the Northern Sagebrush Steppe. Doctorate of Philosophy Dissertation, University of Calgary, Calgary. https://doi.org/10.11575/PRISM/26150

Jakes AF (2021) Chapter F: Pronghorn. In: Remington TE, Deibert PA, Hanser SE et al (eds) Sagebrush conservation strategy—challenges to sagebrush conservation. U.S. Geological Survey open-file report 2020–1125. Fort Collins, Colorado, pp 37–42. https://doi.org/10.3133/ofr202001125

Jakes AF, Jones PF, Paige C et al (2018a) A fence runs through it: a call for greater attention to the influence of fences on wildlife and ecosystems. Biol Conserv 227:310–318. https://doi.org/10.1016/j.biocon.2018.09.026

Jakes AF, Gates CC, DeCesare NJ et al (2018b) Classifying the migration behaviors of pronghorn on their northern range. J Wildl Manage 82:1229–1242. https://doi.org/10.1002/jwmg.21485

Jakes AF, DeCesare NJ, Jones PF et al (2020) Multi-scale habitat assessment of pronghorn migration routes. PLoS ONE 15:e0241042. https://doi.org/10.1371/journal.pone.0241042

Jaworski MD, Hunter DL, Ward AC (1998) Biovariants of isolates of Pasteurella from domestic and wild ruminants. J Vet Diagn Invest 10:49–55. https://doi.org/10.1177/104063879801000109

Jensen WF, Hosek BM, Rudd WJ (2004) Mapping continental range distribution of pronghorn using geographic information systems technology. Bien Prong Work 21:18–36

Jesmer BR, Merkle JA, Goheen JR et al (2018) Is ungulate migration culturally transmitted? Evidence of social learning from translocated animals. Science 361:1023–1025. https://doi.org/10.1126/science.aat0985

Joly K, Gurarie E, Sorum MS et al (2019) Longest terrestrial migrations and movements around the world. Sci Rep 9:article 15333. https://doi.org/10.1038/s41598-019-51884-5

Jones PF (2014) Scarred for life; the other side of the fence debate. Hum-Wild Int 8:150–154. https://doi.org/10.26077/mppv-tt76

Jones PF, Hurly JA, Jensen C et al (2017) Diel and monthly movement rates by migratory and resident female pronghorn. TPN 46:3–12

Jones PF, Jakes AF, Eacker DR et al (2018) Evaluating responses by pronghorn to fence modifications across the Northern Great Plains. Wildl Soc Bull 42:225–236. https://doi.org/10.1002/wsb.869

Jones PF, Jakes AF, Telander AC et al (2019) Fences reduce habitat for a partially migratory ungulate in the Northern Sagebrush Steppe. Ecosphere 10:e02782. https://doi.org/10.1002/ecs2.2782

Jones PF, Jakes AF, Eacker DR et al (2020a) Annual pronghorn survival of a partially migratory population. J Wildl Manage 84:1114–1126. https://doi.org/10.1002/jwmg.21886

Jones PF, Jakes AF, MacDonald AM et al (2020b) Evaluating responses by sympatric ungulates to fence modifications across the Northern Great Plains. Wildl Soc Bull 44:130–141. https://doi.org/10.1002/wsb.1067

Jönsson KI (1997) Capital and income breeding as alternative tactics of resource use in reproduction. Oikos 78:57–66. https://doi.org/10.2307/3545800

Kauffman MJ, Aikens EO, Esmaeili S et al (2021) Causes, consequences, and conservation of ungulate migration. Annu Rev Ecol Evol Syst 52:453–478. https://doi.org/10.1146/annurev-ecolsys-012021-011516)

Keller BJ, Millspaugh JJ, Lehman CP et al (2013) Adult pronghorn (*Antilocapra americana*) survival and cause-specific mortality in Custer State Park, S.D. Am Midl Nat 170:311–322. https://doi.org/10.1674/0003-0031-170.2.311

Klimova A, Munguia-Vega A, Hoffman JL et al (2014) Genetic diversity and demography of two endangered captive pronghorn subspecies from the Sonoran Desert. J Mammal 95:1263–1277. https://doi.org/10.1644/13-MAMM-A-321

Kolar JL, Millspaugh JJ, Stillings BA (2011) Migration patterns of pronghorn in southwestern North Dakota. J Wildl Manage 75:198–203. https://doi.org/10.1002/jwmg.32

Kreeger TJ, Cornish T, Creekmore TE et al (2011) Antilopcapridae, pronghorn. In: Field guide to diseases of Wyoming wildlife. Wyoming Game and Fish Department, Cheyenne, pp 57–69

LaCava MEF, Gagne RB, Love Stowell SM et al (2020) Pronghorn population genomics show connectivity in the core of their range. J Mammal 101:1061–1071. https://doi.org/10.1093/jmammal/gyaa054

Larsen RT, Bissonette JA, Flinders JT et al (2011) Does small-perimeter fencing inhibit mule deer or pronghorn use of water developments? J Wildl Manage 75:1417–1425. https://doi.org/10.1002/jwmg.163

Lee Jr. TE (1992) Mitochondrial DNA and allozyme analysis of pronghorn populations in North America. Doctorate of Philosphy Dissertation, Texas A&M University, College Station

Lee TE Jr, Bickham JW, Scott MD (1994) Mitochondrial DNA and allozyme analysis of North American pronghorn populations. J Wildl Manage 58:307–318. https://doi.org/10.2307/3809396

Lee TS, Creech TG, Martinson A et al (2021) Prioritizing human safety and multispecies connectivity across a regional road network. Conserv Sci Pract 3:e327. https://doi.org/10.1111/csp2.327

Maestas JD, Naugle DE, Chambers JC et al (2021) Conifer expansion. In: Remington TE, Deibert PA, Hanser SE et al (eds) Sagebrush conservation strategy—challenges to sagebrush conservation. U.S. Geological Survey Open-File Report 2020–1125, Reston, Virginia, pp 139–152. https://doi.org/10.3133/ofr20201125

Malmberg JL, O'Toole D, Creekmore T et al (2020) Mycoplasma bovis infections in free-ranging pronghorn, Wyoming, USA. Emerg Infect Dis 26:2807–2814. https://doi.org/10.3201/eid2612.191375

McCabe RE, O'Gara BW, Reeves HM (2004) Prairie Ghost: pronghorn and human interaction in early America. University Press of Colorado, Boulder

Mcinturff A, Xu W, Wilkinson CE et al (2020) Fence ecology: framework for understanding the ecological effects of fences. Biosci 70:971–985. https://doi.org/10.1093/biosci/biaa103

McKelvey KS, Buotte PC (2018) Effects of climate change on wildlife in the Northern Rockies. In: Halofsky JE, Peterson DL (eds) Climate change and rocky mountain ecosystems. Springer International Publishing, New York, pp 143–167. https://doi.org/10.1007/978-3-319-56928-4

McKinney T, Brown DE, Allison L (2008) Winter precipitation and recruitment of pronghorns in Arizona. Southwest Nat 53:319–325. https://doi.org/10.1894/CJ-147.1

Milligan MC, Johnston AN, Beck JL et al (2021) Variable effects of wind-energy development on seasonal habitat selection of pronghorn. Ecosphere 12:e03850. https://doi.org/10.1002/ecs2.3850

Mitchell GJ, Smoliak S (1971) Pronghorn antelope range characteristics and food habits in Alberta. J Wildl Manage 35:238–250. https://doi.org/10.2307/3799597

Morrison TA, Merkle JA, Hopcraft JGC et al (2021) Drivers of site fidelity in ungulates. J Anim Ecol 90:955–966. https://doi.org/10.1111/1365-2656.13425

Mysterud A (2013) Ungulate migration, plant phenology, and large carnivores: the times they are a-changin'. Ecol 94:1257–1261. https://doi.org/10.1890/12-0505.1

Nichols JD (1992) Capture-recapture models using marked animals to study population dynamics. Biosci 42:94–102. https://doi.org/10.2307/1311650

O'Gara BW (2004a) Physical characteristics. In: O'Gara BW, Yoakum JD (eds) Pronghorn: ecology and management. University Press of Colorado, Boulder, pp 109–143

O'Gara BW (2004b) Mortality factors. In: O'Gara BW, Yoakum JD (eds) Pronghorn: ecology and management. University Press of Colorado, Boulder, pp 379–408

O'Gara BW (2004c) Behavior. In: O'Gara BW, Yoakum JD (eds) Pronghorn: ecology and management. University Press of Colorado, Boulder, pp 145–194

O'Gara BW (2004d) Disease and parasites. In: O'Gara BW, Yoakum JD (eds) Pronghorn: ecology and management. University Press of Colorado, Boulder, pp 299–336

O'Gara BW, Janis CM (2004a) The fossil record. In: O'Gara BW, Yoakum JD (eds) Pronghorn: ecology and management. University Press of Colorado, Boulder, pp 27–39

O'Gara BW, Janis CM (2004b) Scientific classification. In: O'Gara BW, Yoakum JD (eds) Pronghorn: ecology and management. University Press of Colorado, Boulder, pp 3–25

O'Gara BW, McCabe RE (2004) From exploitation to conservation. In: O'Gara BW, Yoakum JD (eds) Pronghorn: ecology and management. University Press of Colorado, Boulder, pp 41–73

O'Gara BW, Morrison B (2004) Managing the harvest. In: O'Gara BW, Yoakum JD (eds) Pronghorn: ecology and management. University Press of Colorado, Boulder, pp 675–704

Oakley C (1973) Effects of livestock fencing on antelope. In: Wyoming wildlife. Wyoming Game and Fish Department, Cheyenne, pp 26–29

Padgett PE (2020) Weeds, wheels, fire, and juniper: threats to sagebrush steppe. U.S. In: Dumroese RK, Moser WK (eds) Northeastern California plateaus bioregion science synthesis. Gen. Tech. Rep. RMRS-GTR-409, Department of Agriculture, Forest Service, Rocky Mountain Research Station, Department of Agriculture, U.S. Forest Service, Rocky Mountain Research Station, Fort Collins, Colorado, pp 64–76. https://doi.org/10.2737/RMRS-GTR-409

Paige C (2020) Alberta Landholder's guide to wildlife friendly fencing. Alberta Conservation Association, Sherwood Park

Panting BR, Gese EM, Conner MM et al (2020) Factors influencing survival rates of pronghorn fawns in Idaho. J Wildl Manage 85:97–108. https://doi.org/10.1002/jwmg.21956

Parker KL, Robbins CT, Hanley TA (1984) Energy expenditures for locomotion by mule deer and elk. J Wildl Manage 48:474–488. https://doi.org/10.2307/3801180

Payne NF, Bryant FC (1998) Wildlife habitat management of forestlands, rangelands, and farmlands. Krieger Publishing Company, Malabar

Plumb RE, Gordon KM, Anderson SH (2003) Pronghorn use of a wildlife underpass. Wildl Soc Bull 31:1244–1245. https://doi.org/10.2307/3784474

Pojar TM (2004) Survey methods to estimate populations. In: O'Gara BW, Yoakum JD (eds) Pronghorn: ecology and management. University Press of Colorado, Boulder, pp 631–694

Pojar TM, Guenzel RJ (1999) Comparison of fixed-wing line-transect and helicopter quadrant pronghorn surveys. Pro Prong Work 18:64–68

Pyrah DB (1987) American pronghorn antelope in the Yellow Water Triangle, Montana: A study of social distribution, population dynamics, and habitat use. Montana Department of Fish, Wildlife and Parks and the U.S. Bureau of Land Management, Helena. https://doi.org/10.5962/bhl.title.117229

Reinking AK, Smith KT, Monteith KL et al (2018) Intrinsic, environmental, and anthropogenic factors related to pronghorn summer mortality. J Wildl Manage 82:608–617. https://doi.org/10.1002/jwmg.21414

Reinking AK, Smith KT, Mong TW et al (2019) Across scales, pronghorn select sagebrush, avoid fences, and show negative responses to anthropogenic features in winter. Ecosphere 10:e02722. https://doi.org/10.1002/ecs2.2722

Rottler CM, Burke IC, Palmquist KA et al (2018) Current reclamation practices after oil and gas development do not speed up succession or plant community recovery in big sagebrush ecosystems in Wyoming. Restor Ecol 26:114–123. https://doi.org/10.1111/rec.12543

Ryder TJ, Irwin LL (1987) Winter habitat relationships of pronghorns in southcentral Wyoming. J Wildl Manage 51:79–85. https://doi.org/10.2307/3801635

Sawyer H, Rudd B (2005) Pronghorn roadway crossings: a review of available information and potential options. Federal Highway Administration, Wyoming Department of Transportation and Wyoming Game and Fish Department, Cheyenne

Sawyer H, Lindzey F, McWhirter D et al (2002) Potential effects of oil and gas development on mule deer and pronghorn populations in western Wyoming. U.S. Bureau of Land Management, Cheyenne

Sawyer H, Lindzey F, McWhirter D (2005) Mule deer and pronghorn migration in western Wyoming. Wildl Soc Bull 33:1266–1273. https://doi.org/10.2193/0091-7648(2005)33[1266:MDAPMI]2.0.CO;2

Sawyer H, Kauffman MJ, Nielson RM et al (2009) Identifying and prioritizing ungulate migration routes for landscape level conservation. Ecol Appl 19:2016–2025. https://doi.org/10.1890/08-2034.1

Sawyer H, Rodgers PA, Hart T (2016) Pronghorn and mule deer use of underpasses and overpasses along U.S. Highway 191. Wildl Soc Bull 40:211–216. https://doi.org/10.1002/wsb.650

Sawyer H, Beckmann JP, Seidler RG et al (2019) Long-term effects of energy development on winter distribution and residency of pronghorn in the Greater Yellowstone Ecosystem. Conserv Sci Pract 1:e83. https://doi.org/10.1111/csp2.83

Scasta JD, Beck JL, Angwin CJ (2016) Meta-analysis of diet composition and potential conflict of wild horses with livestock and wild ungulates on western rangelands of North America. Rangel Ecol Manag 69:310–318. https://doi.org/10.1016/j.rama.2016.01.001

Schroeder C (2018) Western state and province pronghorn status report, 2018. Bien Prong Work 28:29–35

Seidler RG, Long RA, Berger J et al (2015) Identifying impediments to long-distance mammal migrations. Conserv Biol 29:99–109. https://doi.org/10.1111/cobi.12376

Seidler RG, Green DS, Beckmann JP (2018) Highways, crossing structures and risk: Behaviors of greater Yellowstone pronghorn elucidate efficacy of road mitigation. Glob Ecol Conserv 15:e00416. https://doi.org/10.1016/j.gecco.2018.e00416

Sheldon D (2005) Pronghorn movement and distribution patterns in relation to roads and fences in southwestern Wyoming. Master's Thesis, University of Wyoming

Simpson DC, Harveson LA, Brewer CE et al (2007) Influence of precipitation on pronghorn demography in Texas. J Wildl Manage 71:906–910. https://doi.org/10.2193/2005-753

Smith JT, Evans JS, Martin BH et al (2016) Reducing cultivation risk for at-risk species: predicting outcomes of conservation easements for sage-grouse. Biol Conserv 201:10–19. https://doi.org/10.1016/j.biocon.2016.06.006

Smith KT, Taylor KL, Albeke SE et al (2020) Pronghorn winter resource selection before and after wind energy development in South-Central Wyoming. Rangel Ecol Manag 73:227–233. https://doi.org/10.1016/j.rama.2019.12.004

Smyser TJ, Garton EO, Zager P (2005) The influence of habitat variables on pronghorn recruitment. Idaho Department of Fish and Game, Boise

Stephen CL, Devos JC Jr, Lee TE Jr et al (2005) Population genetic analysis of Sonoran pronghorn (Antilocapra americana sonoriensis). J Mammal 86:782–792. https://doi.org/10.1644/1545-1542(2005)086[0782:PGAOSP]2.0.CO;2

Stoner DC, Anderson MT, Schroeder CA et al (2021) Distribution of competition potential between native ungulates and free-roaming equids on western rangelands. J Wildl Manage 85:1062–1073. https://doi.org/10.1002/jwmg.21993

Tack J, Jakes AF, Jones PF et al (2019) Beyond protected areas: private lands and public policy anchor intact pathways for multi-species wildlife migration. Biol Conserv 234:18–27. https://doi.org/10.1016/j.biocon.2019.03.017

Taylor KL, Beck JL, Huzurbazar SV (2016) Factors influencing winter mortality risk for pronghorn exposed to wind energy development. Rangel Ecol Manag 69:108–116. https://doi.org/10.1016/j.rama.2015.12.003

Telfer ES, Kelsall JP (1984) Adaptation of some large North American mammals for survival in snow. Ecol 65:1828–1834. https://doi.org/10.2307/1937779

U. S. Fish and Wildlife Service (2015) Draft Recovery Plan for the Sonoran pronghorn (*Antilocapra americana sonoriensis*), Second Revision. U.S. Fish and Wildlife Service, Albuquerque

Van Dyke W (1990) Oregon pronghorn status report: 1990. Proc Prong Ant Work 14:14–16

Van Soest PJ (1994) Chapter 2: nutritional concepts. In: Van Soest PF (ed) Nutritional ecology of the ruminant. Cornell University Press, Ithaca, pp 7–21

Ward CL (2016) Evaluation of survey techniques and sightability for pronghorn antelope (*Antilocapra americana*) in Texas. Master's Thesis, University of Georgia

White RS, Currie PO (1983) The effects of prescribed burning on silver sagebrush. J Range Manage 36:611–613. https://doi.org/10.2307/3898352

White PJ, Davis TL, Barnowe-Meyer KK et al (2007) Partial migration and philopatry of Yellowstone pronghorn. Biol Conserv 135:502–510. https://doi.org/10.1016/j.biocon.2006.10.041Get rights and content

Wilson RR, Krausman PR (2008) Possibility of heat-related mortality in desert ungulates. J Ariz-Nev Acad Sci 40:12–15. https://doi.org/10.2181/1533-6085(2008)40[12:POHMID]2.0.CO;2

Woodruff SP, Lukacs PM, Christianson D et al (2016) Estimating Sonoran pronghorn abundance and survival with fecal DNA and capture-recapture methods. Conserv Biol 30:1102–1111. https://doi.org/10.1111/cobi.12710

Xu W, Dejid N, Herrmann V et al (2020) Barrier behaviour analysis (BaBA) reveals extensive effects of fencing on wide-ranging ungulates. J Appl Ecol 58:690–698. https://doi.org/10.1111/1365-2664.13806

Yoakum JD (2004a) Distribution and abundance. In: O'Gara BW, Yoakum JD (eds) Pronghorn: ecology and management. University Press of Colorado, Boulder, pp 75–105

Yoakum JD (2004b) Habitat characteristics and requirements. In: O'Gara BW, Yoakum JD (eds) Pronghorn: ecology and management. University Press of Colorado, Boulder, pp 409–445

Yoakum JD (2004c) Foraging ecology, diet studies and nutrient values. In: O'Gara BW, Yoakum JD (eds) Pronghorn: ecology and management. University Press of Colorado, Boulder, pp 447–502

Yoakum JD (2004d) Relationships with other herbivores. In: O'Gara BW, Yoakum JD (eds) Pronghorn: ecology and management. University Press of Colorado, Boulder, pp 503–538

Yoakum JD (2004e) Habitat conservation. In: O'Gara BW, Yoakum JD (eds) Pronghorn: ecology and management. University Press of Colorado, Boulder, pp 571–630

Yoakum JD (2004f) Relationship with other herbivores. In: O'Gara BW, Yoakum JD (eds) Pronghorn: ecology and management. University Press of Colorado, Boulder, pp 503–538

Yoakum JD, Jones PF, Cancino J et al (2014) Pronghorn management guides, 5th edn. Western Association of Fish and Wildlife Agencies' Pronghorn Workshop and New Mexico Department of Game and Fish, Santa Ana Pueblo

Zeller KA, Schroeder CA, Wan HY, Collins G et al (2021) Forecasting habitat and connectivity for pronghorn across the Great Basin ecoregion. Divers Distrib. https://doi.org/10.1111/ddi.13402

Zimmerman GM (1983) Rehabilitation of pronghorn habitat on surface mines of the Northern Great Plains. Master's Thesis, Montana State University

Chapter 20
Elk and Rangelands

Michel T. Kohl, Shawn M. Cleveland, Calvin C. Ellis, Ashlyn N. Halseth, Jerod A. Merkle, Kelly M. Proffitt, Mary M. Rowland, and Michael J. Wisdom

Abstract Elk (*Cervus canadensis*) are the second largest member of the deer family that reside in North America. Historically, the species occupied most of North America, however, today, they occupy only a small proportion of that range. Across their historical and contemporary distribution, they occupied diverse vegetation communities including both rangelands and forest ecosystems. Given this broad distribution, elk face numerous conservation and management threats including competition with wild and domestic ungulates, disease considerations, and human-elk conflicts. This chapter highlights these and other conservation and management concerns, especially as they pertain to rangelands. In closing, we identify current and future research needs that will be important for the continued persistence and expansion of elk populations across their range.

Keywords Carnivores · Climate change · Conservation · Disease · Elk-livestock · Habitat · Management

M. T. Kohl (✉) · C. C. Ellis · A. N. Halseth
Warnell School of Forestry and Natural Resources, University of Georgia, Athens, GA, USA
e-mail: michel.kohl@uga.edu

S. M. Cleveland
Colorado State University, Pueblo, CO, USA

J. A. Merkle
Department of Zoology and Physiology, University of Wyoming, Laramie, WY, USA

K. M. Proffitt
Montana Department of Fish, Wildlife, and Parks, Bozeman, MT, USA

M. M. Rowland · M. J. Wisdom
Pacific Northwest Research Station, USDA Forest Service, La Grande, OR, USA

© The Author(s) 2023
L. B. McNew et al. (eds.), *Rangeland Wildlife Ecology and Conservation*,
https://doi.org/10.1007/978-3-031-34037-6_20

20.1   General Life History and Population Dynamics

Elk (*Cervus canadensis*), comprised of six subspecies, were one of the most widely distributed deer species in North America (Fig. 20.1). However, since European settlement, the Eastern elk (*C. c. canadensis*) and Merriam's elk (*C. c. merriami*), have been driven to extinction. Of the remaining subspecies, the Tule (*C. c. nannodes*) and Manitoban (*C. c. manitobensis*) elk only occupy a small fraction of their historical range. In contrast, the Roosevelt (*C. c. roosevelti*), and Rocky Mountain elk (*C. c. nelsoni*) subspecies occur across much of their historical range. Of these, the Rocky Mountain subspecies is the most widely distributed and the subspecies found across most of North America rangelands today.

Elk ecology and behavior of both males (hereafter bulls) and females (hereafter cows) are driven by the energetic requirements associated with breeding and calving periods (Geist 2002; Cook et al. 2013). The courtship and breeding activity, termed the "rut", occurs in late September. For cows, the rut is followed by an ~ 250-day gestation period, with the peak of calving occurring in early June to coincide with the high nutritional quality provided by vegetative green-up (Cook et al. 2013). After birthing, cows will track forage availability to increase fat reserves for winter while also continuing energetically expensive lactation activities into winter. Like cows, bulls will track vegetation conditions during summer to replenish fat reserves for the

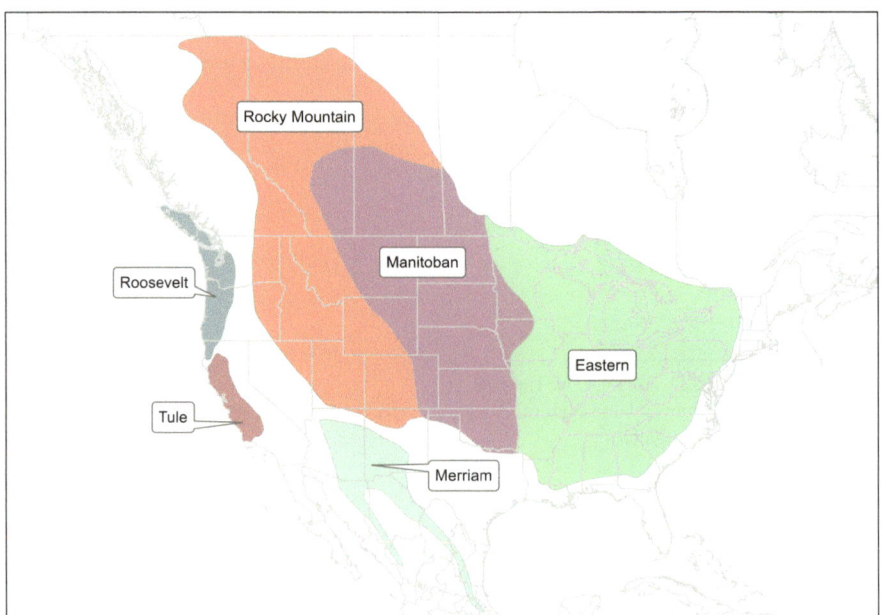

Fig. 20.1 Historical distribution of elk in North America. Adapted from figures available in Wisdom and Thomas (1996) and from Historic Elk Range Map Geographic Information System Polygon files compiled by Rocky Mountain Elk Foundation

next breeding season. Because antler shape and size are important for establishing dominance among bulls, and maintaining a harem of cows during the rut, bulls will also seek out mineral resources during the antler growth period (~ March–July). For bulls, the rut is a period of intense fat loss driven by reduced feeding and high energy expenditure. Following the rut, bulls will maximize energy conservation during winter to maintain fat reserves. Across most western rangelands, elk migrate seasonally to meet these nutritional requirements. Elk will typically move up in elevation to forested summer ranges during May and move down in elevation to lower-elevation rangelands between September and December where they winter. However, the spatial and temporal variability of these forage and mineral resources across populations and geographic regions results in significant variability in when and how elk use rangelands across the U.S.

When elk can appropriately track changing forage conditions, populations can undergo strong population growth. However, stochasticity in environmental conditions and predator communities likely shape annual variation in calf survival (Lukacs et al. 2018), and thus, overall population size. A review of 37 studies reported annual calf survival estimates ranged from 6 to 72% suggesting that calf survival is a prominent driver of elk population growth rate (Raithel et al. 2007). In contrast, adult female survival was relatively constant across populations (Raithel et al. 2007). Moreover, human harvest is the primary source of mortality in most hunted populations, for both cows and bulls, suggesting that management objectives strongly influence elk abundance (Keller et al. 2015). In a study of 45 different elk populations across the western U.S. and Canada, cow elk survival was 85 and 95% (Brodie et al. 2013) and bull elk survival was 56 and 79% in harvested and unharvested populations, respectively (Unsworth et al. 1993; Lubow et al. 2002).

20.2   Current Species and Population Status

Following the market hunting period, only 60,000 elk, distributed across 7 western states, remained in North America (Jackson 1944). However, by 2021, large-scale reintroductions and conservation actions have led to an estimated 1.18–1.22 million wild elk distributed across 27 U.S. states and five Canadian provinces, based on a collation of state and provincial elk management reports and media statements (Fig. 20.2; Table 20.1). Because of continued restoration and management efforts, elk continue to increase in abundance and distribution in most portions of their range. The species *Cervus canadensis* is ranked "least concern" by the International Union for the Conservation of Nature and "globally secure" by NatureServe. State- and province-level ranks in North America also indicate secure populations except in Ontario (critically imperiled).

In areas of established elk populations, surveys are common. Surveys have predominantly been conducted via fecal pellets (Rowland et al. 1984) and road- or aerial-based sampling surveys (Samuel et al. 1987), collecting data on the observed number of bulls and cows and approximate age class (juvenile, sub-adult, adult).

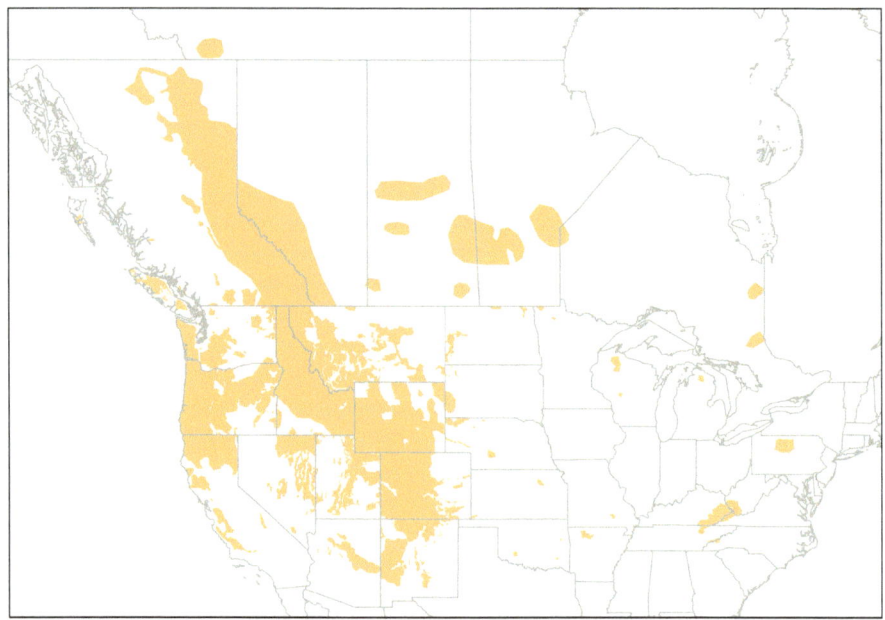

Fig. 20.2 Current distribution of elk in North America. State- or province-specific distribution data were obtained from state websites, by georeferencing habitat mapping documents (e.g., state elk management plans), or replicated from data compiled by Rocky Mountain Elk Foundation. Distribution in Alaska not shown. Elk residing in National Parks or within Tribal or First Nations boundaries are not shown as they are not included in Table 20.1

Additionally, information on age of harvest obtained via check stations and phone surveys has been used to determine population structure (Bender and Spencer 1999). Traditional survey methodology is comprehensively reviewed in Toweill and Thomas (2002).

In recent years, thermal imaging has been used to increase detection probabilities during rangeland aerial surveys (Dunn et al. 2002) and to assess vegetation impacts by elk (Biederbeck et al. 2016). Unmanned aerial surveys are also seeing increased use, providing both thermal imaging and real-time, spatially explicit population information (Witczuk et al. 2018; Graves et al. 2022). Remote cameras are also now being used to estimate elk abundance (Moeller et al. 2018). The vast data collected using traditional and contemporary survey methods, and advances in computer processing capabilities have facilitated corresponding improvements for estimating elk abundance and assessing harvest management scenarios (Eacker et al. 2017; Bender and Spencer 1999) and habitat use (Sawyer et al. 2007; Boyce et al. 2003).

The significance of diseases in elk has emerged in the last half century as game-farming and winter-feeding grounds have exposed elk and facilitated the spread of a variety of diseases. More recently, the emergence of management-oriented epidemiological investigations has expanded our understanding of these diseases within free-ranging elk populations and their consequences on population performance. Thus,

Table 20.1 Elk population estimates for U.S. states and Canadian provinces. Estimates are unlikely to include elk abundance within national parks or on Tribal or First Nations lands

Location	Elk estimate	Source	Date
U.S.A			
AK	1300–1500	S	2021
AZ	35,000–45,000	W, M[1]	2021
AR	450	S	2021
CA	12,900	P	2018
CO	292,760	P	2019
ID	> 120,000	W	2021
KS	450–500	M[1]	2018
KY	15,876	P	2020
MI	930–1462	M[1]	2019
MN	259	M[1]	2020
MO	> 200	M[1]	2020
MT	136,151	P	2020
NE	2500–3000	S	2021
NV	13,500	P	2018
NM	70,000–90,000	M[1]	2021
NC	150–200	W	2021
ND	700–1000	U[2]	2021
OK	> 5000	M[1]	2019
OR	126,646	P	2019
PA	1350	W	2020
SD	7682–10,076	P	2020
TN	> 400	W	2020
TX	3500	O3	2014
UT	81,000	P	2020
WA	46,150–53,150	W	2021
WI	400	W	2021
WV	85	W	2021
WY	112,900	M[1]	2020
Canada			
AB	26,000	U[2]	2021
BC	43,000–43,500	W	2021
MB	6500	W	2021

(continued)

Table 20.1 (continued)

Location	Elk estimate	Source	Date
ON	600–1000	M[1]	2020
SK	7350	O[4]	2014
Total	1,183,939–1,225,865		

Source M—Media reports (refers to agency biologist statements in news articles), O—Other, P—Elk management plan or population estimate, S—Wildlife agency staff, U—Unconfirmed, W—Wildlife agency website
[1] "Media reports" refer to agency biologist statements within news articles
[2] Third party websites (e.g., state-specific hunting organizations)
[3] Texas Tech University fact sheet
[4] University of Saskatchewan thesis

disease surveillance and management are now important components of many elk management programs.

20.3 Habitat Associations

Historically, North American elk occupied a diverse mix of habitats to meet basic ecological needs for forage, water, and security from weather and predators (i.e., both human and nonhuman). This diversity reflected the broad distribution of elk across the continent prior to European settlement (Murie 1951). Nearly all major vegetation types were occupied by elk other than the humid forests of the southeastern U.S. and hot desert communities of the Southwest (Murie 1951; Skovlin et al. 2002). Different subspecies of elk exploited this habitat mix (Fig. 20.1). For example, the extirpated Merriam's elk occurred in dry forests and chaparral of the Southwest (Skovlin et al. 2002), where its distribution may have been limited by water. By contrast, Roosevelt elk were the most forest-associated taxon, occurring primarily in coastal rainforests but with seasonal use of open meadows (Murie 1951). The Tule elk of California were the subspecies most adapted to plains environments, occasionally inhabiting even chaparral or woodlands (Murie 1951; McCullough 1969); however, their habitat associations were little studied before their near extirpation. Populations of Eastern elk were widely distributed in the plains of the midwestern U.S., (e.g., Iowa and Illinois), where they lived year-round and co-occurred with bison (Murie 1951). These habitat associations continue today across the geographic range of elk, apart from those elk extirpated in plains states (Murie 1951).

The broad range of habitat associations for elk reflects their status as mixed feeders of intermediate selectivity, consuming a wide variety of graminoids, forbs, and woody plants (Cook 2002). As a result, rangelands provide key seasonal habitats for elk, primarily through their provisioning of non-woody forage (Cook 2002). Winter diets of elk are also dominated by graminoids (Christianson and Creel 2007), but some elk herds rely on woody browse more frequently in winter than in other seasons

(Rowland et al. 1983; Cook 2002). Today, rangelands remain an important habitat component for many elk populations (Sawyer et al. 2007), particularly those that migrate between forested summer range to more open rangeland in winter. The seasonal use of rangelands has become increasingly common in the Western U.S. with elk selecting bedding sites in sage-steppe biomes in Washington (McCorquodale et al. 1986), foraging sites in agricultural lands in Manitoba, Canada and Montana, USA (Proffitt et al. 2013; Brook 2010), and spatial refugia in urbanized landscapes (Polfus and Krausman 2012).

Forested stands, especially with low-moderate canopy cover, offer additional foraging opportunities for elk in late summer and early fall as vegetation senesces in open habitats (Cook 2002). Additionally, forested communities may benefit elk by reducing predation risk due to increased visual obstruction (Skovlin et al. 2002; Lowrey et al. 2020). However, forests are not a required habitat component for elk if rugged topography offers visual obstruction (Lehman et al. 2016) and sufficient forage. Land ownership and its associated management may also influence habitat selection such that elk behaviors diverge from traditional habitat associations as elk attempt to minimize predation risk from human hunting (Proffitt et al. 2010).

As human populations and their footprint have expanded into historical elk winter range, some of these habitat associations have been altered due to changes in forage characteristics and predation risk. For example, conversion of winter range to crop-lands has led to elk selecting for areas of increased forage potential (Brook 2010). In cases of high forage availability, elk may select forage over the thermal cover provided in adjacent forested landscapes that they would have historically used (Long et al. 2014). This transition has altered forage preferences to more non-traditional food sources available in urban settings, such as golf courses, subdivisions, and the wildland-urban interface (WUI; Tucker et al. 2004; Skovlin 2002). Further, the spread of exotic species, such as spotted knapweed (*Centaurea stoebe*), into rangelands has altered elk diets and forage availability (Kohl et al. 2012).

The attractiveness of higher forage quality and availability from urbanization of elk winter range has been exacerbated by reduced predation risk as predators may avoid humans, and because hunter harvest is usually not allowed (Berger 2007). Thus, elk in these situations have ample, high-quality forage with reduced or absent preda-tion risk, increasing residency time on adjacent rangelands (Cleveland et al. 2012) and potential loss of traditional migratory behavior (Hebblewhite et al. 2006). For example, elk migratory behavior has been altered or lost in Banff, Alberta, Canada; Estes Park, CO; Sequim, WA; and Jackson Hole, WY. The resulting, burgeoning elk populations in urban and WUI settings have negative impacts on rangeland condition that are largely beyond managerial control (Haggerty and Travis 2006) and have led to corresponding increases in human/elk conflict.

Although conflicts between ranchers and elk have developed on rangelands and croplands adjacent to forest preserves and urbanized rangelands (Tucker et al. 2004), parallel opportunities have emerged to improve range condition for elk. For example, in cooperation with cattle ranchers, wildlife managers can "pre-condition" rangelands for elk via cattle grazing, thus improving winter range condition (Clark et al. 2000). Additionally, advanced, interactive tools are being developed to provide site specific recommendations for range management on agricultural lands (https://tinyurl.com/54fu93cb). Further work on habitat associations of urbanizing landscapes, predation risk, migration and crop depredation remain important contemporary issues for elk management and research.

20.4 Rangeland Management

Shared grazing management of elk with other wild and domestic ungulates has been controversial for over a century (Miller 2002). This controversy centers on how the perceived competitive advantages of elk (e.g., their broad diet, large body size, tendency to form large herds, aggressive behavior toward smaller ungulates) may negatively impact other ungulates on co-occupied rangeland (Miller 2002). However, the shared use of rangelands by elk with other ungulates does not automatically indicate competition. Multi-ungulate grazing systems can be complementary, indifferent, or beneficial. Thus, many factors must be considered (Ager et al. 2004; Hughey et al. 2021), with effects that are season- and area-specific and often requiring formal monitoring or research.

20.4.1 Elk-Cattle Competition

It is unlikely that elk will compete with other native wild ungulates, however elk may compete with cattle because of the following reasons (Wisdom and Thomas 1996; Clark et al. 2017): (1) the two ungulates co-occupy millions of ha of rangelands across the western U.S. and Canada, among the largest areas of shared range of ungulates in North America; (2) their diets can converge when either or both graze at high population density during seasonal forage limitations, such as senescence of herbaceous forage during late summer-fall; and (3) some seasonal ranges (e.g., winter) that are co-occupied by elk and cattle may have limited space and forage availability. Under each scenario, there is potential for elk and cattle to compete directly for food. Furthermore, elk-cattle dietary overlap can be high within and across seasons (Torstenson et al. 2006), and both ungulates readily adapt to available forages seasonally (Scasta et al. 2016). In turn, both ungulates can substantially reduce available biomass of nutritious forages, altering the abundance, composition, and structure of plant communities under moderate to high grazing use (Endress et al. 2016; Rhodes et al. 2018), affecting each other and potentially other rangeland species.

20.4.2 Cattle Grazing Prescriptions that Benefit Elk

In many cases, livestock grazing systems are likely compatible with elk. However, it may also be either positive or negative. For example, grazing prescriptions can be designed and used to condition grasses for nutritional benefit of elk, and to maintain desired elk distributions. This was demonstrated by research from Montana in which grazing intensity, duration, rotation, and rest periods was manipulated to maximize foraging efficiency and dietary quality for elk (Alt 1992; Frisina 1992). Rest-rotation cattle-grazing systems in Oregon were also documented to support extended elk use of grazing lands by providing desired nutritional benefits (Anderson et al. 1990). Similar benefits to elk are possible from traditional, deferred and rest-rotation cattle grazing systems in more productive rangelands (Vavra and Sheehy 1996; Crane et al. 2016). However, the benefits are not always realized when beef cattle production is the primary goal (Chaikina and Ruckstuhl 2006; Tolleson et al. 2012). Cattle grazing in many arid and semi-arid rangelands also is not likely to provide nutritional benefits to the other ungulates under moderate or high stocking rates (Hobbs et al. 1996; Krausman et al. 2009; Damiran et al. 2019).

20.4.3 Elk Competition with Other Ungulates

The potential for feral equid-elk competition has become an increasing management concern, and yet, elk and feral equids share less than 4.5% of their distributions across the western U.S. (Stoner et al. 2021). It should be noted that where these species overlap, feral equids can have substantial impacts on elk habitat. Beyond equids, a notable and emerging grazing management controversy is the potential for elk to compete with mule deer. Elk may displace mule deer (Johnson et al. 2000; Stewart et al. 2002) and can substantially reduce biomass of high-quality mule deer forages, particularly during late-summer and fall (Findholt et al. 2004). In addition to mule deer, concerns regarding the potential for elk-bison competition are increasing as bison reintroduction efforts continue throughout North America. To date, research on bison-elk competition has been limited and inconclusive. In Wind Cave National Park, bison and elk had moderate spatial overlap, however differences in food habitats limited overall competition (Wydeven and Dahlgren 1985). In contrast, Coughenour (2005) suggested that at high densities, bison and elk may compete for available forage.

20.4.4 Estimating Elk and Ungulate Competition

We summarized conditions under which the potential for elk-ungulate competition may be low, high, or uncertain on co-occupied rangelands, based on 12 generalizations developed and reviewed extensively in Wisdom and Thomas (1996; Fig. 20.3). These generalizations have been supported by contemporary research and still apply today (Clark et al. 2017). Although specific to elk-cattle competition, similar approaches could be applied to better understand competition between elk and other ungulates. Our summary is not intended to replace field assessments, monitoring, or research, but could be used as a first step to prioritize areas and times in which greater attention may be warranted to address the potential for competitive interactions. Figure 20.3 could be used, for example, to identify grazing periods when formal methods of "forage allocation" may help mitigate undesired grazing impacts and potential competitive interactions. In this context, forage allocation represents the desired proportional availability of nutritional resources across space and time for each type of ungulate grazer.

Traditional methods for estimating forage allocation, based on stocking rate of cattle and elk, are reviewed in Wisdom and Thomas (1996). More recent applications (Riggs et al. 2015) further documented the challenges of applying these methods at landscape scales, owing to the diversity and data accuracy that must be considered (Clark et al. 2017). Fine-scale, spatially- and temporally-dynamic forage allocation methods such as use of linear programming (e.g., Cooperrider and Bailey 1984;

Competition Potential	Specific Conditions	Grazing Context	Specific Conditions	Competition Potential
LOW	Late-spring and early summer ranges support high forage biomass and quality (peak productivity of forbs and grasses) over large areas.	Competition varies by season and extent of spatial overlap.	Winter and spring-fall ranges have limited grazing area and low forage biomass and quality.	HIGH
LOW	Cattle and elk are temporally separate on more productive, mesic rangelands where spring-summer grazing by one ungulate facilitates regrowth of grasses to benefit the other.	Grazing effects on forage can be beneficial or competitive.	Grazing on less productive, xeric rangelands does not facilitate regrowth of grasses, increasing competition for limited forage following peak forage production.	HIGH
LOW	Cattle and elk distribute themselves separately based on different responses to environmental conditions, with independent spatial forage use.	Elk and cattle are spatially separate.	Elk avoid cattle, which may or may not restrict elk grazing choices or limit cattle conditioning of grasses to benefit elk after cattle are moved from a pasture.	UNCERTAIN
LOW	High precipitation results in abundant forage of sufficient quality, with full expression of dietary differences and low dietary overlap, regardless of spatial overlap.	Forage quality and quantity vary by timing and extent of drought.	Prolonged drought, often during late summer and fall, results in low forage quantity and quality and convergence of elk-cattle diets when using same areas.	HIGH
LOW	Productive, mesic rangelands historically grazed by multiple ungulate species, such as tall- and mid-grass prairie systems, have high resilience to fire and grazing.	Range site productivity drives competition.	Less productive, xeric rangelands have low resilience to fire and grazing, especially for areas with a declining range trend and high use by cattle or elk.	HIGH
UNCERTAIN	Rangelands have highly variable seasonal climate regimes, resulting in low predictability of herbaceous forage dynamics.	High number of unknowns due to insufficient data.	Perceptions are the basis for inferring competitive interactions.	UNCERTAIN

Fig. 20.3 Potential for competition between elk and cattle (high, low, or uncertain) based on grazing context on western rangelands, summarized from Wisdom and Thomas (1996), as estimated for arid and semi-arid rangelands (areas receiving < 50 cm of annual precipitation)

Johnson et al. 1996) and foraging simulation models (Ager et al. 2004; Riggs et al. 2015) have been used successfully in research but required data are often lacking for effective management applications. Regardless of method, estimating forage allocation requires four major inputs (Ager et al. 2004): (1) biomass of key forages available to each ungulate; (2) allowable use of those forages for each ungulate; (3) percent spatial overlap between ungulates; and (4) percent dietary overlap between ungulates as offset by degree of spatial overlap.

Reasonable estimates of biomass of key forage species available to each ungulate often can be obtained from past monitoring conducted in an area or from published sources for the associated ecoregion (Ager et al. 2004) or using remote-sensed products (Garroutte et al. 2016; Allred et al. 2021). Establishing the allowable use of key forages for each ungulate is a major management decision best made in relation to the ecological resilience (i.e., capacity to survive and recover) of key forages under a specified level of grazing use by each ungulate. Data on range condition and trend can often be used as the basis for making decisions about grazing use for each ungulate, which are typically available on most public grazing allotments. Estimating spatial overlap involves mapping the expected spatial distributions of each ungulate in relation to the environmental features shown in the literature to have consistent and measurable influence on each ungulate type's use of a landscape, independent of the other. For example, geographic information systems and extensive, widely-available, spatial data could provide efficient mapping of elk and cattle spatial distributions to evaluate spatial overlap (Stewart et al. 2002). Lastly, the main dietary items (i.e., key forages) must be identified to estimate biomass available to each ungulate within areas of shared spatial overlap (Wisdom and Thomas 1996), which, in some cases, can now be done at landscape scales with DNA metabarcoding (Nichols et al. 2016).

20.5 Impacts of Disease

Elk can host a suite of viral, bacterial, prion, and nutritional diseases that have varying levels of impact on elk populations (Table 20.2). While disease surveillance and management often focus on the diseases that influence elk population performance, much of the focus of elk diseases relate to their consequences for livestock. The transmission of disease from elk to livestock or livestock to elk, which we refer to as spillover, can have important consequences for both elk and domestic livestock health. For example, bacterial diseases such as bovine tuberculosis or anthrax can be transmitted between livestock and elk and result in animal health concerns and/ or death. The risk of spillover to livestock can create conflict, as disease spillover has the potential to adversely affect livestock health, economic activity, and support for elk conservation. Below we detail two of the most important and contemporary diseases affecting elk, brucellosis (Cross et al. 2010a) and chronic wasting disease (CWD; Williams and Young 1980).

Brucellosis is a global zoonotic disease caused by the bacteria *Brucella abortus* that infects cattle, elk, and bison (Olsen 2010). Brucellosis was nearly eradicated

Table 20.2 Bacterial, viral, prion, and parasitic diseases that affect elk

Diseases or Condition	Description	Primary references
Bacterial		
Anaplasmosis (gall sickness)	Disease of blood cells primarily affecting domestic cattle that is caused by *Anaplasma* bacteria and transmitted by ectoparasite	Kuttler (1984), Zaugg et al. (1996)
Anthrax	Zoonosis afflicting animals and humans globally, caused by the spore-forming, environmentally-maintained bacterium *Bacillus anthracis*	Turnbull (2008), Blackburn et al. (2014)
Bovine Tuberculosis	A zoonotic disease, due to *Mycobacterium bovis,* classically carried by cattle and spilling over into wildlife reservoirs	Rhyan and Saari (1995), Brook et al. (2013)
Brucellosis	A highly infectious zoonosis affecting animals and humans worldwide caused by *Brucella abortus*	Cheville et al. (1998), Rhyan et al. (2013)
Leptospirosis spp.	An infective serological group of bacteria that can infect nearly all mammals	Bender and Hall (1996), Lilenbaum and Martins (2014)
Necrotic Stomatitis	A bacterial infection in the mouth which causes abscesses, necrosis, and loss of teeth	Murie (1930), Murray et al. (1996)
Paratuberculosis (Johne's Disease)	An infectious granulomatous enteritis caused by *Mycobacterium avium* paratuberculosis causing significant economic losses in livestock	Williams et al. (1983), Carta et al. (2013)
Viral		
Hemorrhagic Disease	Transmitted by biting midges in the *Culicoides* spp. and other arthropods, causing acute and frequently fatal hemorrhagic disease in domestic and wild ungulates	Howerth et al. (2001), Ruder et al. (2015)
Prion		
Chronic Wasting Disease	Transmissible spongiform encephalopathy caused by prion-induced folding of proteins in the brain	Williams and Young (1980), Williams et al. (2002)
Parasites and Parasitic Diseases		
External parasites: Mange Mites and Winter tick	Common part of the normal biology of most animals; however, some external parasites including mange mites (*Psoroptic* spp.) and winter tick (*Dermacentor albipictus*) may cause morbidity and mortality in elk	Samuel et al. (1991), Corn and Nettles (2001)
Internal parasites: Lungworm, Ecchinoccus granulosis	Common part of the normal biology of most animals; however, some internal parasites, e.g., lungworm (*Dictyocaulus viviparus*) or *Ecchinoccus granulosis*, may cause morbidity and mortality in elk	Foreyt et al. (2000), Thompson (2008)

from the U.S. in the early 2000's, but the disease persists in elk and bison populations in the Greater Yellowstone Area (GYA; Rhyan et al. 2013). Brucellosis prevalence within GYA elk populations ranges from 0–53% (Rayl et al. 2019; NASEM 2020), and is increasing in many herds (Cross et al. 2010b; Brennan et al. 2017). *B. abortus* concentrates in the reproductive system and typically causes abortion during the third trimester (Cheville et al. 1998). Transmission occurs when individuals ingest *B. abortus* bacteria from infected fetuses or birthing fluids on tissues, soil or vegetation which may persist for 21–81 days depending on conditions (Aune et al. 2012).

Elk are responsible for transmitting the disease to livestock in multiple recent outbreaks (Kamath et al. 2016). Because transmission from elk to livestock occurs where livestock may contact and ingest the elk-aborted fetus or birthing fluids, disease management programs focus on maintaining spatial separation between elk and livestock during this transmission risk period. In addition, the controversial elk winter feed grounds in Wyoming are used in part to reduce comingling between elk and livestock. Management to reduce transmission risk may also include non-lethal and lethal actions aimed at redistributing elk away from livestock (Jones et al. 2021).

CWD is a transmissible spongiform encephalopathy caused by abnormal folding of proteins that accumulate in brains of infected animals and eventually lead to central nervous system failure and death. As CWD may infect members of the cervid family, including elk, deer, and moose, it has critical implications for elk population performance and management on rangelands and elsewhere. CWD was first recognized in the 1970's at a captive deer research facility in northwestern Colorado (Williams and Young 1980), and since has spread throughout captive and free-ranging elk, deer, and moose populations across North America (Mysterud and Edmunds 2019). Control efforts to date have largely been unsuccessful at reducing the spatial spread and prevalence of CWD in free-ranging elk populations (Uehlinger et al. 2016).

Clinical signs of CWD include severe weight loss, and behavioral changes such as stumbling, tremors, and teeth grinding (Miller et al. 1998). CWD has an incubation period lasting for several months to several years, during which an infected elk may show few signs of illness but still shed prions in urine, feces, and saliva. Transmission to susceptible animals may occur directly through contact with an infected animal or indirectly through environmental contamination (Williams et al. 2002). There is no evidence that infected elk can transmit CWD to domestic livestock (Williams 2005).

The effect of CWD on elk population demography is primarily due to reduced adult female survival rates, as infected individuals will continue to reproduce (Mysterud and Edmunds 2019). Depending on the prevalence and other factors interacting to influence adult female survival rate, CWD may have variable effects on elk population performance (Monello et al. 2014; Mysterud and Edmunds 2019). For example, in the Rocky Mountain elk population, CWD prevalence rates exceeding 13% have the potential to decrease population growth (Monello et al. 2014). It should be noted that recent work suggests that natural resistance to CWD may be increasing in wild cervid populations (Monello et al. 2017).

20.6 Ecosystem Threats

A primary threat to elk across western rangelands stems from continued human population growth and its contribution to urban, suburban, and exurban growth. The result has been increased development and associated infrastructure leading to habitat alteration, fragmentation, and destruction. Beyond direct reduction in available habitat, residential expansion and development also contributes to shifts in elk behavior that contribute to changing elk use of rangelands (Polfus and Krausman 2012). In some places, human development has led to fragmented and diminished habitat quality with elk demonstrating avoidance of small ownership parcels (Wait and McNally 2004) and faster movements in areas close to houses (Cleveland et al. 2012). In other cases, elk may select for developed areas because of increased forage opportunities (e.g., manicured lawns, irrigated fields), reduced snow depth, and reduced predator densities that may potentially contribute to increased human-wildlife conflicts (Thompson and Henderson 1998).

Beyond residential development, human population growth requires substantial infrastructure (Soulard 2006) including road and energy development, much of which is occurring on elk winter and transitional ranges. Generally, elk avoid roads open to public motorized use (Rowland et al. 2000; Sawyer et al. 2007; Frair et al. 2008), a behavior that is particularly evident for hunted populations during fall and winter (Beck et al. 2013) and during daylight and twilight hours (Prokopenko et al. 2017). It should be noted, however, that elk response to roads in refuge areas are less predictable (Wisdom et al. 2018). While there is little evidence that elk–vehicle collisions are a significant influence on elk survival, roads provide a means of incidental mortality from legal and illegal harvest by humans (McCorquodale et al. 2003; Frair et al. 2007). Through such mortality, the road network has the possibility to influence elk population dynamics (Frair et al. 2008).

As human population continues to grow, there is an ever-expanding energy development network that is required (Kiesecker and Naugle 2017), much of which overlaps with ungulate winter range in North America (Hebblewhite 2011). Previous research has demonstrated that these surface disturbances (e.g., wells, access roads, etc.) negatively affect elk, however the magnitude of those effects has varied across studies. For example, elk in northeastern Wyoming altered their behavior in response to the development of a coalbed natural gas field (Buchanan et al. 2014). Compared to pre-development years, elk selected areas with greater cover, increased terrain ruggedness, and farther from roads post-development, leading to a decrease in preferred habitat use for the population (Buchanan et al. 2014). Despite these displacement behaviors, the size of this elk population has remained stable (Buchanan et al. 2014) or increased since the development (Bureau of Land Management 2015). In contrast, the risk of mortality for elk in New Mexico and Southern Colorado decreased for elk in proximity to energy development disturbance (Dzialak et al. 2011). More research is necessary to clarify how elk respond to energy development, and studies in North Dakota and Wyoming are underway.

Concomitant to human population growth, interest in recreational activities is also increasing leading to concerns about its impact on elk behavior and demography. For example, increased recreational use (e.g., all-terrain vehicle use, hiking, horseback riding, mountain biking) has been shown to increase movement rates (Wisdom et al. 2004) and reduce feeding time (Naylor et al. 2009) by elk, translating to higher energetic costs which may contribute to lower vital rates (Phillips and Alldredge 2000).

Land ownership changes are becoming increasingly common, and in turn, posing significant challenges to elk populations and to wildlife managers (Haggerty and Travis 2006). When land ownership changes result in a shift away from traditional ranching activities, and toward restricted hunting access, elk use of private lands may intensify due to increases in high quality forage (Barker et al. 2019a) or due to enhanced security relative to neighboring hunted areas (Conner et al. 2001; Vieira et al. 2003). This has contributed to overabundant elk populations in states such as Wyoming and Montana. Importantly, reactivation of hunting on these private lands can quickly reverse elk behaviors (Sergeyev et al. 2022).

Long-term climate change poses one of the most significant threats to elk populations across western rangelands. Through its alteration of the environment, climate change has the potential to influence elk distribution, migration, and population sizes via changes in seasonal forage availability and quality. The West is experiencing lower snowpack, earlier snowmelt, and an increase in both drought frequency and the rate of spring green-up (Marshall et al. 2019). Temperature and other weather patterns are less predictable, with increased frequency of large storms, and the dry periods between them (Groisman and Knight 2008). These changes in climate are also affecting broad scale forest disturbance (e.g., bark beetle outbreak) and fire regimes, both with impacts on elk habitat use and management (Lamont et al. 2020; Spitz et al. 2018; Proffitt et al. 2019). Each of these factors will continue to influence availability of high-quality forage on both summer and winter range, and during migration (Rowland et al. 2018). Moreover, these changes may particularly impact elk spring migrations, which are closely tied to the phenology of snow melt and plant green-up (Hebblewhite et al. 2008) and fall migrations that are tied to the timing and amount of snow (Rickbeil et al. 2019). Significant mismatches in timing between migration and plant phenology can negatively influence reproductive rates and overwinter survival (Middleton et al. 2018; Cook et al. 2004).

20.7 Conservation Actions

Generally, elk populations are stable or increasing, precluding the need for explicit conservation actions. One exception is Tule elk which are the focus of diverse conservation actions following its near extirpation in the 1800s. Despite this history some herds have grown exponentially resulting in conflict with livestock producers, especially in protected areas (Watt 2015). Given their high genetic diversity and relatively low allelic diversity and heterozygosity, transplants among existing Tule elk herds may be the best strategy to conserve this subspecies (Williams et al. 2004). However, removal of elk or contraception may be needed where populations remain above objective (e.g., Howell et al. 2002). Due to the location of their habitat, future conservation of Tule elk will require management in a socioecological context (Ciriacy-Wantrup et al. 2019; Denryter and Fischer 2022).

Transplants can restore elk populations extirpated or dramatically reduced by market hunting, habitat loss, or other stressors, with successful reintroductions occurring in many locales (Sargeant and Oehler 2007; O'Neil and Bump 2014). This is particularly evident in eastern North America, where elk reintroductions have led to establishment of ~ 20,000 elk across nine states and two Canadian provinces (Table 20.1). Most reintroductions have been aimed at providing hunting opportunity, but conflicts with landowners must also be addressed when reintroduced elk extend onto private lands. In rare cases, translocations may also help alleviate problems associated with overabundant elk (Walter et al. 2010). Translocation options are increasingly limited by state policies due to the potential of spreading diseases such as CWD (Corn and Nettles 2001).

Large-scale habitat alteration from wildfires, leading to habitat loss and fragmentation, can pose special conservation challenges for elk, especially in more arid regions. Habitat restoration in such sites is challenging given high spatial and temporal variability in precipitation patterns and forage resources (Chambers et al. 2014). Although fire can improve the nutritional landscape for elk, especially in higher elevations (Proffitt et al. 2019), realized benefits depend on fire intensity, size, pattern, and affected vegetation community. Moreover, forage may be limited in the short term as fires can damage shrublands that are used seasonally by elk (McCorquodale et al. 1986) and may require decades to become reestablished (Davies et al. 2012). Additionally, exotic herbaceous plants often predominate following fire in shrublands and supplant native species, diminishing resources for elk and often require active restoration through seeding and other practices (Chambers et al. 2014).

Climate change may also alter elk habitat and migratory patterns and these relationships have primarily been studied in forested and alpine habitats, although Denryter and Fischer (2022) assessed movements in non-forested habitat. The complexities of climate change interactions with migration, disease ecology, and harvest will challenge elk conservation into the future, although elk have exhibited plastic behaviors that can partially compensate for these changes (Rickbeil et al. 2019). Where elk are considered vulnerable or below objective, protecting special habitat features such as calving areas, migratory corridors, or security areas from human disturbance or outright loss, while engaging collaboration across all relevant stakeholders, is recommended (Shively et al. 2005; Middleton et al. 2019).

20.8   Management Actions

Elk receive significant management attention in most areas they occur, from harvest to habitat management, and in some cases human conflict issues can become prominent. In some areas predators that prey on elk are also considered an integral part of elk management. While state and provincial wildlife agencies often take the lead on elk management, many other entities have significant involvement including federal land management agencies, research institutions, conservation groups, sportsman interests, livestock producers, and others.

20.8.1   Harvest Management

In general, most elk management programs seek to maintain the size and distribution of elk populations at socially acceptable levels compatible with other land uses while meeting recreational demands, including harvest. Harvest is an important management tool for manipulating the distribution and abundance of elk, given they are among the most iconic species of the American West and are highly valued by sportsmen and women. Harvest management is complex, integrating biological objectives and reflecting political, economic, and social considerations. Elk have important influences on vegetation (Wisdom et al. 2006), which can create conflicts with private landowners (Hobbs et al. 1996; Walter et al. 2010). After objectives for managing elk populations are determined for a population and/or management unit, specific harvest regulations can be designed to achieve those objectives. Because population management objectives vary widely across western populations, harvest strategies also vary widely (Stalling et al. 2002).

Elk population management objectives and associated harvest regulations are usually defined in state Elk Management Plans. These plans guide annual regulations that determine the allocation of hunting licenses and specify the number, age, and sex of animals allowable for harvest, the hunting period and areas of harvest, and allowable weapon types. Tribal harvest of elk, whether on lands ceded as treaty hunting areas or on reservations, is regulated by each tribe and is primarily "need-driven" (McCorquodale 1997). Tribal harvest likely contributes only marginally to the total harvest of elk in rangelands given the relative paucity of tribal members participating in elk hunts (McCorquodale 1997); nonetheless, elk remain an important cultural resource in much of the US, where they are considered a "First Food" (Long and Lake 2018).

As elk populations increased and recovered in the mid-1900s, conservative hunting regulations only allowed for harvest of males. Elk harvest management has changed as populations have largely recovered, and in some cases exceeded socially tolerable levels. For overabundant elk populations, regulations that allow for antlerless harvest and/or prolonged hunting periods may be implemented to reduce cow survival rates and corresponding population growth rates. Conflicts on private lands related to crop depredation are increasing and may be related to the abundance or distribution of elk (Walter et al. 2010). To alleviate these conflicts, harvest regulations that apply to specific parcels of land and/or outside of traditional hunting season dates may also be implemented. Conversely, to manage small or declining elk populations, restrictive regulations may be applied to increase cow survival rates and population growth rates.

20.8.2 Habitat Management

The overarching management objective for elk habitat is to provide for a mix of seasonal habitats that include adequate nutrition while minimizing risk of disturbance from predators and humans. By emphasizing both vegetation conditions and disturbance risk, managers will be best equipped to optimize distribution and abundance of elk. Within this context, habitat management is often targeted toward shifting elk distributions from private to public lands to reduce damage on the former and increase hunting and viewing opportunities on the latter. Enhancing forage production for elk in natural systems can help meet these goals (Barker et al. 2019a, b). A special consideration of elk habitat is vulnerability to harvest, given the gentle topography and lower tree canopy cover common in rangeland systems (Edge and Marcum 1991; Wisdom and Thomas 1996).

In habitats with abundant non-native plants or encroachment of shrublands or woodlands, active restoration through seeding and/or prescribed fire can reduce invasive plant abundance and improve nutritional conditions, assuming precipitation is adequate and competition from invasive plant species is reduced (Chambers et al. 2014). For elk, this situation is most common on winter ranges, where invasive plant species like cheatgrass (*Bromus tectorum*) and knapweeds (*Centaurea* spp.) can be common. However, elk may consume these species in moderate amounts, especially during winter and spring (Kohl et al. 2012).

Elk feed grounds constitute a unique management tool, typically occupying traditional elk winter ranges. They can be effective in alleviating damage on private lands thus improving social tolerance of elk and landowner-wildlife agency relations, however, they can lead to degraded ranges and disease transmission, such as brucellosis or CWD (Maichak et al. 2009; Thorne et al. 1991).

20.8.3 Carnivore Management

Declines in some elk populations have been attributed to the restoration and recovery of large carnivore populations such as mountain lions (*Puma concolor*), grizzly bears (*Ursus arctos horribilis*), and wolves (*Canis lupus*) (Lehman et al. 2018;

Horne et al. 2019; Proffitt et al. 2020). Management tools to limit carnivore abundances and impacts on elk populations may be limited by state or federal legislation, lack of public support for carnivore harvest or carnivore population reductions (Mitchell et al. 2018), and disagreements within the scientific community regarding the effectiveness of carnivore control. As carnivore populations expand and increase, wildlife managers will need to employ integrated programs to effectively achieve both carnivore and elk population objectives (Proffitt et al. 2020).

20.9 Research/Management Needs

Future research to support management of elk can be summarized in four priority topics: competitive interactions with other ungulates, ongoing and emerging diseases and pathogens, effects of climate change, and socio-ecological effects of changing population distributions.

20.9.1 Competitive Interactions with Other Ungulates

Elk exhibit a variety of perceived competitive advantages over other ungulates, particularly mule deer and cattle, and all three co-occupy vast areas of western rangelands. Competitive interactions remain highly controversial and major sources of uncertainty remain. Study priorities include the need for:

- Manipulative landscape experiments to evaluate the behavioral, distributional, dietary, and population responses of mule deer to reductions in elk density in areas of historical mule deer range where populations have declined while elk populations have increased. Designs such as before-after-control impact (BACI) studies would be optimal.
- Observational landscape studies of mule deer diet, habitat-use, distribution, and individual and population performance across a gradient of elk densities (zero to high) under similar background environmental conditions. Data from these descriptive landscape studies could be used to validate predictions developed from the manipulative landscape experiments.
- Diet and spatial overlap of elk with cattle, and associated levels of biomass reduction and changes in nutritional quality of diets of both ungulates under high or moderate densities of both ungulates to assess potential for exploitative competition and effects on animal performance of each ungulate at specified densities of each.
- New decision support tools to accurately assess and model the spatio-temporal interactions of elk, deer, and cattle and the potential for interference and exploitative competition under varying densities of each ungulate, and under different

cattle grazing systems and practices across the xeric to mesic gradient of grassland, shrubland, woodland, and forested rangelands.

20.9.2   Diseases and Pathogens

Much has been learned about diseases and pathogens that affect elk, with new diseases emerging as climate change and other factors affect elk and their habitats (Rayl et al. 2019). Today, the expansion of CWD across elk populations is a primary concern of state wildlife agencies, as CWD is being found in new herds annually (Galloway et al. 2021). Spillover of brucellosis between elk and livestock remains a special challenge as transmission from elk to livestock continues (Kamath et al. 2016). Research to further investigate the spatial and temporal extent of disease transmission risk would inform management approaches to reduce disease spread in elk. For farmed elk, further research on disease outbreaks in confined herds is needed to broaden understanding of the risk of transmission to wild elk. Synthesizing this knowledge to develop comprehensive and systematic disease surveillance protocols would help state agencies decide when and what management actions (e.g., increased harvest, culling) to implement.

20.9.3   Climate Change

Studies that document changes in forage phenology under changing temperature and precipitation regimes are among the highest of elk research and management needs, given their cascading effects on all other reproductive phases of elk life history and potential for dramatic shifts in population distributions, including migration (Middleton et al. 2013; Rickbeil et al. 2019). Effects of climate change include those during winter, such as from diminished snowpack and more rapid snowmelt, and during summer, such as from more extended drought and higher temperatures. Similarly, research is needed to understand how altered fire regimes, facilitated by climate change, fire management, and land uses (Loehman et al. 2018; Chambers et al. 2019), affect forage phenology, dynamics, and nutrition, and subsequently affect spatial distributions and performance of elk populations. Study topics of high priority to evaluate changing climate and fire regimes include:

- Effects on herbaceous forage phenology related to timing and rate of green up and brown down, and duration of high photosynthetic activity.
- Resultant changes in timing and duration of forage biomass and quality in relation to potential mismatches with calf birth dates and their lactation needs.
- Potential cascading effects on body fat dynamics of lactating females from mismatched forage dynamics with birth dates and subsequent increase in alternate-year productivity.

- Possible shifts in population distribution from increasingly xeric habitats, often associated with public rangelands, to more mesic habitats on private lands.
- Changes in timing, routes, and predictability of migration and demographic consequences.
- Effects of altered fire regimes (intensity, scale, frequency) on forage phenology and nutritional resource dynamics in arid and semi-arid rangelands, and subsequent effects on spatial distributions and performance of elk populations.
- Restoration of desired native forages of high nutritional value in the face of increasing competition with exotic plants following wildfires in arid and semi-arid rangelands.

20.9.4 Socio-Ecological Research

Elk are one of the most widely studied wildlife species in the world, but management conflicts persist. Thus, integrated socio-ecological solutions to issues such as elk distributions on private versus public lands are required (Carter et al. 2014). All stakeholders should be at the table, including state wildlife, tribal, and public land management agencies, private landowners, the public, and local governing bodies (White and Ward 2010). Knowledge gaps exist about the social tolerance of stakeholders for elk, which could be addressed through structured, qualitative interviews and listening sessions designed by social scientists in a knowledge co-production framework. This process could elucidate why landowners do or do not desire elk on their properties and help explain apparent contradictions in the mutual desire of some for-hunter opportunity on public lands coupled with opposition to road closures. In addition, the economic value of elk has not been well-documented and most research on this topic is decades old, generally not from rangeland systems, and primarily based on consumptive use (Bolon 1994; Fried et al. 1995; Chapagain et al. 2020).

Although a broad understanding of how elk respond to management actions such as road closures or timber thinning has been gained by a wealth of published studies (Spitz et al. 2018; Wertz et al. 1996), a formal meta-analysis could help answer questions such as: how large an area must be treated to attract elk to a seasonal range and hold them there? What habitat features do elk seek on private versus public lands and how does that differ across rangeland systems? How does human disturbance interact with elk response to vegetation treatments? Finally, more adaptive management experiments employing hunting regulations, e.g., general season antlerless elk damage tags as recently piloted by Oregon Department of Fish and Wildlife (https://myodfw.com/articles/general-season-antlerless-elk-damage-tag) or multiple, targeted hunts (Cleveland et al. 2012; Sergeyev et al. 2022), will advance knowledge about best practices to mitigate elk depredation issues on private lands.

20.10 Summary

Human interest in elk is well documented in indigenous culture and after European settlement. With settlement came the extirpation of Eastern and Merriam's elk, the near extirpation of Tule elk, and significant reduction of Roosevelt and Rocky Mountain elk populations. Through expansive wildlife conservation and management efforts, Roosevelt and Rocky Mountain elk populations have largely recovered. Initial regulations focused on bull harvest to increase calf and cow survival, leading to burgeoning Rocky Mountain and Roosevelt elk populations across the U.S. and Canadian rangelands. With little exception, notably in New Mexico and Ontario, elk populations are at, or above objectives established by state and provincial agencies, leading to a shift in strategy toward cow harvest to curtail population growth.

As elk populations have recovered, habitat associations closely tied to rangelands have emerged, and research demonstrates the importance of seasonal (e.g., summer and winter) ranges to maximize calf recruitment and cow and bull overwinter survival (Murie 1951; Cook et al. 2013). There has also been an increased appreciation for the importance of balancing the cumulative impacts of human development, specifically roads, which can increase elk harvest rates (Polfus and Krausman 2012). Further, as the human population grows and development of historic winter ranges expands, the need for conservation easements and cooperative grazing plans has emerged as an important management strategy. The realization that human development has disrupted migratory elk behavior and altered historical habitat associations continues to be a point of conservation concern for the long-term management of elk populations.

These alternations contribute to changes in forage quality and availability, predator–prey interactions, and at times, a reduction or loss of migratory corridors that all facilitate increased human/elk conflict, specifically in rangelands near urbanized areas or areas in which hunter access is restricted (Brook 2010; Proffitt et al. 2013). To mitigate these conflicts, cooperative grazing management has been implemented (Wisdom and Thomas 1996) that can benefit both elk and livestock, thus reducing competition for forage resources. The forage allocation strategy can also lead to improved rangeland management for a variety of sympatric rangeland species of concern, such as mule deer. Still, other rangeland management challenges remain. Feral horses and burros lack sufficient predators and population control measures and can lead to deleterious impacts on rangeland resources and potential competition between ungulate species, both domestic and wild (Stoner et al. 2021).

The spatial overlap of elk, other wild ungulates, and domestic livestock may also pose disease transmission concerns. Elk transmission of brucellosis to domestic livestock has resulted in increased conflict and at times reduced support for elk conservation. In addition, the emergence of CWD and its increasing spread throughout elk populations is a rapidly growing concern across occupied elk range (Uehlinger et al. 2016).

Energy development has become an area of increasing concern in rangelands. The development of oil and gas fields and their corresponding road networks can alter elk habitat use, potentially increasing vulnerability and possibly impacting population dynamics (Hebblewhite 2011). Further, "Green Energy" expansion of wind farms and solar energy arrays in rangelands will likely lead to additional habitat alterations and impacts on population distribution and dynamics. However, this relationship is poorly understood and proper siting guidelines and best management practices for wind and solar energy development are currently limited or lacking.

Climate change is altering the timing of spring green-up and the duration and accumulation of snow in winter, both of which impact elk recruitment and survival (Cook et al. 2004). Further, the expanse and intensity of fire in rangelands, as was evident in the unprecedented duration and longevity of fires in 2020 across the Western US, is attributable to climate change. Beyond direct habitat alterations in the immediate aftermath of these fires, invasive species, such as cheatgrass and spotted knapweed, are expanding into rangelands through habitat disturbance such as fire, further altering historical elk-forage relationships.

Given this breadth of challenges facing elk populations, additional elk conservation and management actions are warranted. The focus on connected, unaltered rangelands to preserve existing habitat associations and migratory behaviors must continue. These large, connected and intact areas help bolster elk populations against the impacts of climate change and slow energy development-induced land alterations. Given their high behavioral plasticity, elk will likely be able to adapt to these stressors if proper rangeland management and conservation efforts continue into the future (Rickbeil et al. 2019).

As elk have been and remain the focus of both indigenous cultural traditions and recreational harvest and viewing, continued support for elk conservation in rangelands is needed. Despite significant challenges such as invasive species, landscape alteration, disease emergence, and climate change, continued focus is critical for managing rangelands for multiple use and multiple species. Research and management are needed that focus on competitive interactions of elk with sympatric ungulates, emerging diseases and pathogens, interactive effects of climate change, and the socio-ecological effects of shifts in population distribution. Addressing these contemporary issues is a pressing management need and will require broad and diverse partnerships to ensure the viability of elk populations across North American rangelands into the future.

References

Ager AA, Johnson BK, Coe PK, et al (2004) Landscape simulation of foraging by elk, mule deer, and cattle on summer range. In: Transactions, North American Wildlife and Natural Resource Conference, vol 69, pp 687–707

Allred BW, Bestelmeyer BT, Boyd CS et al (2021) Improving Landsat predictions of rangeland fractional cover with multitask learning and uncertainty. Methods Ecol Evol 12:841–849

Alt KL, Frisina MR, King FJ (1992) Coordinated management of elk and cattle, a perspective-Wall Creek Wildlife Management Area. Rangelands 14:12–15

Anderson EW, Franzen DL, Melland JE (1990) Forage quality as influenced by prescribed grazing. In: Severson KE (ed) Can livestock be used be used as a tool to enhance wildlife habitat? RM-194. US Department of Agriculture, Forest Service, Rocky Mountain Research Station, Fort Collins, CO, pp 56–70

Aune K, Rhyan JC, Russell R et al (2012) Environmental persistence of Brucella abortus in the Greater Yellowstone Area. J Wildlife Manage 76:253–261

Barker KJ, Mitchell MS, Proffitt KM (2019a) Native forage mediates influence of irrigated agriculture on migratory behaviour of elk. J Anim Ecol 88:1100–1110

Barker KJ, Mitchell MS, Proffitt KM et al (2019b) Land management alters traditional nutritional benefits of migration for elk. J Wildl Manage 83:167–174

Beck JL, Smith KT, Flinders JT et al (2013) Seasonal habitat selection by elk in north central Utah. West N Am Naturalist 73:442–456

Bender LC, Hall PB (1996) Leptospira interrogans exposure in free-ranging elk in Washington. J Wildl Dis 32:121–124

Bender LC, Spencer RD (1999) Estimating elk population size by reconstruction from harvest data and herd ratios. Wildl Soc B 27:636–645

Berger J (2007) Fear, human shields and the redistribution of prey and predators in protected areas. Biol Lett. https://doi.org/10.1098/rsbl.2007.0415

Biederbeck HH, Jackson DH, VandeBergh DJ (2016) Aerial high resolution digital imagery elk survey. Wildlife Technical Report, vol 006. Oregon Department of Fish and Wildlife, Corvallis, OR

Blackburn JK, Van Ert M, Mullins JC et al (2014) The necrophagous fly anthrax transmission pathway: empirical and genetic evidence from wildlife epizootics. Vector-Borne Zoonotic Dis 14:576–583

Bolon NA (1994) Estimates of the values of elk in the Blue Mountains of Oregon and Washington: evidence from the existing literature. General Technical Report. U.S. Forest Service, Pacific Northwest Research Station, PNW-GTR-316, Portland, Oregon, USA

Boyce MS, Mao JS, Merrill EH et al (2003) Scale and heterogeneity in habitat selection by elk in Yellowstone National Park. Ecoscience 10:421–431

Brennan A, Cross PC, Portacci K et al (2017) Shifting brucellosis risk in livestock coincides with spreading seroprevalence in elk. PLoS ONE 12(6):e0178780

Brodie J, Johnson H, Mitchell M et al (2013) Relative influence of human harvest, carnivores, and weather on adult female elk survival across western North America. J Appl Ecol 50:295–305

Brook RK (2010) Habitat selection by parturient elk (Cervus elaphus) in agricultural and forested landscapes. Can J Zool 88:968–976

Brook RK, Vander Wal E, van Beest FM et al (2013) Evaluating use of cattle winter feeding areas by elk and white-tailed deer: implications for managing bovine tuberculosis transmission risk from the ground up. Prev Vet Med 108:137–147

Buchanan CB, Beck JL, Bills TE et al (2014) Seasonal resource selection and distributional response by elk to development of a natural gas field. Rangeland Ecol Manag 67:369–379

Bureau of Land Management (2015) Fortification creek planning area annual monitoring report 2015. Buffalo, Wyoming

Carta T, Álvarez J, de la Lastra JP et al (2013) Wildlife and paratuberculosis: a review. Res Vet Sci 94:191–197

Carter NH, Viña A, Hull V et al (2014) Coupled human and natural systems approach to wildlife research and conservation. Ecol Soc 19:43

Chaikina NA, Ruckstuhl KE (2006) Native ungulates: the good, the bad, and the ugly. Rangelands 28:8–14

Chambers JC, Bradley A, Brown CS et al (2014) Resilience to stress and disturbance, and resistance to Bromus tectorum L. invasion in the cold desert shrublands of western North America. Ecosyst 7:360–375

Chambers JC, Brooks ML, Germino MJ et al (2019) Operationalizing resilience and resistance concepts to address invasive grass-fire cycles. Front Ecol Evol 7:185

Chapagain BP, Poudyal NC, Warkins C (2020) A travel cost analysis of elk-viewing opportunity generated from an elk reintroduction project in Tennessee. Hum Dimens Wildlife. https://doi.org/10.1080/10871209.2020.1864067

Cheville NF, McCullough DR, Paulson LR et al (1998) Brucellosis in the greater Yellowstone area. National Academies Press

Christianson DA, Creel S (2007) A review of environmental factors affecting elk winter diets. J Wildl Manage 71:164–176

Ciriacy-Wantrup SV, Bishop RC, Andersen SO (2019) Conservation of the California Tule Elk: a socioeconomic study of a survival problem. Natural Resource Economics. Selected Papers, pp 231–246. Routledge

Clark PE, Krueger WC, Bryant LD et al (2000) Livestock grazing effects on forage quality of elk winter range. J Range Manag 53:97–105

Clark PE, Johnson DE, Ganskopp DC et al (2017) Contrasting daily and seasonal activity and movement of sympatric elk and cattle. Rangel Ecol Manage 70:183–191

Cleveland SM, Hebblewhite M, Thompson M et al (2012) Linking elk movement and resource selection to hunting pressure in a heterogeneous landscape. Wildl Soc B 36:658–668

Conner MM, White GC, Freddy DJ (2001) Elk movement in response to early-season hunting in northwest Colorado. J Wildl Manage 926–940

Cook JG (2002) Nutrition and food. In: Toweill D, Thomas JW (eds) North American elk: ecology and management. Smithsonian Institution Press, Washington, DC, pp 259–350

Cook JG, Johnson BK, Cook RC et al (2004) Effects of summer-autumn nutrition and parturition date on reproduction and survival of elk. Wildl Monogr 155(1):1–61

Cook RC, Cook JG, Vales DJ et al (2013) Regional and seasonal patterns of nutritional condition and reproduction in elk. Wild Mon 184:1–45

Cooperrider AY, Bailey JA (1984) A simulation approach to forage allocation. Developing strategies for rangeland management. Westview Press, Boulder, CO, USA, pp 525–560

Corn JL, Nettles VF (2001) Health protocol for translocation of free-ranging elk. J Wildlife Dis 37:413–426

Coughenour MB (2005) Bison and elk in Yellowstone National Park—linking ecosystem, animal nutrition, and population processes. Part 3 of a Final Report to U.S. Geological Survey, Bozeman, MT

Crane KK, Mosley JC, Mosley TK et al (2016) Elk foraging site selection on foothill and mountain rangeland in spring. Rangeland Ecol Manage 69:319–325

Cross P, Ebinger MR, Patrek V et al (2010a) Brucellosis in cattle, bison, and elk: management conflicts in a society with diverse values. Knowing Yellowstone Taylor Trade Publishing, Lanham, Maryland, USA, pp 80–93

Cross PC, Cole E, Dobson AP et al (2010b) Probable causes of increasing brucellosis in free-ranging elk of the Greater Yellowstone Ecosystem. Ecol Appl 20:278–288

Damiran D, DelCurto T, Findholt SL et al (2019) The effects of previous grazing on the subsequent nutrient supply of ungulates grazing late-summer mixed-conifer rangelands. Sustain Agric Res 8. https://doi.org/10.5539/sar.v8n4p13

Davies GM, Bakker JD, Dettweiler-Robinson E et al (2012) Trajectories of change in sagebrush steppe vegetation communities in relation to multiple wildfires. Ecol Appl 22:1562–1577

Denryter K, Fischer K (2022) Mitigating anthropogenic barriers to facilitate distributional shifts helps reduce vulnerability of a large herbivore to climate change. Animal Conserv. https://doi.org/10.1111/acv.12776

Dunn WC, Donnelly JP, Krausmann WJ (2002) Using thermal infrared sensing to count Elk in the Southwestern United States. Wildl Soc B 30:963–967

Dzialak MR, Webb SL, Harju SM et al (2011) The spatial pattern of demographic performance as a component of sustainable landscape management and planning. Landscape Ecol 26:775–790

Eacker DR, Lukacs PM, Proffitt KM et al (2017) Assessing the importance of demographic parameters for population dynamics using Bayesian integrated population modeling. Ecol Appl 27:1280–1293

Edge WD, Marcum C (1991) Topography ameliorates the effects of roads and human disturbance on elk. In: Proceedings of a symposium on elk vulnerability. Montana State University, Bozeman, MT, pp 132–137

Endress BA, Naylor BJ, Pekin BK et al (2016) Aboveground and belowground mammalian herbivores regulate the demography of deciduous woody species in conifer forests. Ecosphere 7:e01520

Findholt SL, Johnson BK, Damiran D et al (2004) Diet composition, dry matter intake, and diet overlap of mule deer, elk, and cattle. In: Transactions, North American wildlife and natural resources conference, vol 69, pp 670–686

Foreyt WJ, Hunter D, Cook JG et al (2000) Susceptibility of elk to lungworms from cattle. J Wildl Dis 36:729–733

Frair JL, Merrill EH, Allen JR et al (2007) Know thy enemy: experience affects elk translocation success in risky landscapes. J Wildl Manage 71:541–554

Frair JL, Merrill EH, Beyer HL et al (2008) Thresholds in landscape connectivity and mortality risks in response to growing road networks. J Appl Ecol 45:1504–1513

Fried BM, Adams RM, Berrens RP et al (1995) Willingness to pay for a change in elk hunting quality. Wildl Soc B 23:680–686

Frisina MR (1992) Elk habitat use within a rest-rotation grazing system. Rangelands 14:93–96

Galloway NL, Monello RJ, Brimeyer D et al (2021) Supporting adaptive management with ecological forecasting: chronic wasting disease in the Jackson Elk herd. Ecosphere 12:e03776

Garroutte EL, Hansen AJ, Lawrence RI (2016) Using NDVI and EVI to map spatiotemporal variation in the biomass and quality of forage for migratory elk in the Greater Yellowstone Ecosystem. Remote Sensing 8(404):1–25

Geist V (2002) Adaptive Behavioral Strategies. In: Toweill DE, Thomas JW (eds) North American elk: ecology and management, vol 1. Smithsonian Institution Press. Washington D.C, USA, pp 389–433

Groisman PY, Knight RW (2008) Prolonged dry episodes over the conterminous United States: New tendencies emerging during the last 40 years. J Climate 21:1850–1862

Graves TA, Yarnall MJ, Johnston AN et al (2022) Eyes on the herd: quantifying ungulate density from satellite, unmanned aerial systems, and GPS collar data. Ecol Appl e2600

Haggerty JH, Travis WR (2006) Out of administrative control: absentee owners, resident elk and the shifting nature of wildlife management in southwestern Montana. Geoforum 37:816–830

Hebblewhite M (2011) Effects of energy development on ungulates. In: Naugle DE (ed) Energy development and wildlife conservation in Western North America. Island Press/Center for Resource Economics, Washington, DC, pp 71–94

Hebblewhite M, Merrill EH, McDermid G (2008) A multi-scale test of the forage maturation hypothesis in a partially migratory ungulate population. Ecol Monogr 78:141–166

Hebblewhite M, Merrill EH, Morgantini LE et al (2006) Is the migratory behavior of montane elk herds in peril? The case of Alberta's Ya Ha Tinda elk herd. Wildlife Soc B 34:1280–1294

Hobbs NT, Baker DL, Bear GD et al (1996) Ungulate grazing in sagebrush grassland: mechanisms of resource competition. Ecol Appl 6:200–217

Horne JS, Hurley MA, White CG et al (2019) Effects of wolf pack size and winter conditions on elk mortality. J Wildl Manage 83:1103–1116

Howell JA, Brooks GC, Semenoff-Irving M et al (2002) Population dynamics of Tule Elk at point Reyes national seashore, California. J Wildl Manage 66:478–490

Howerth EW, Stallknecht DE, Kirkland PD (2001) Bluetongue, epizootic hemorrhagic disease, and other orbivirus-related diseases. Infect Dis Wild Mamm 3:77–97

Hughey LF, Shoemaker KT, Stewart KM et al (2021) Effects of human-altered landscapes on a reintroduced ungulate: Patterns of habitat selection at the rangeland-wildland interface. Biol Cons 257:109073

Jackson HHT (1944) Big game resources of the United States. PNW-325. U.S. Dept Inter Bur Sport Fish Wildlife Res

Johnson BK, Ager, Crim SA et al (1996) Allocating forage among wild and domestic ungulates-a new approach. In: Edge WD, Olson-Edge SL (eds) Proceedings of the sustaining rangeland ecosystems symposium, Oregon State University, Corvallis. OR, pp 166–169

Johnson BK, Kern JW, Wisdom MJ et al (2000) Resource selection of mule deer and elk during spring. J Wildl Manage 64:685–697

Jones JD, Proffitt KM, Paterson JT et al (2021) Elk responses to management hunting and hazing. J Wildl Manag 85:1721–1738

Kamath PL, Foster JT, Drees KP et al (2016) Genomics reveals historic and contemporary transmission dynamics of a bacterial disease among wildlife and livestock. Nat Comm 7:1–10

Keller BJ, Montgomery RA, Campa HR III et al (2015) A review of vital rates and cause-specific mortality of elk Cervus elaphus populations in eastern North America. Mamm Rev 45:146–159

Kiesecker JM, Naugle DE (2017) Energy sprawl solutions: balancing global development and conservation. Island Press, Washington DC

Kohl MT, Hebblewhite M, Cleveland SM et al (2012) Forage value of invasive species to the diet of Rocky Mountain elk. Rangelands 34:24–28

Krausman PR, Naugle DE, Frisina MR et al (2009) Livestock grazing, wildlife habitat, and rangeland values. Rangelands 31:15–19

Kuttler K (1984) Anaplasma infections in wild and domestic ruminants: a review. J Wildl Dis 20:12–20

Lamont BG, Kauffman MJ, Merkle JA et al (2020) Bark beetle-affected forests provide elk only a marginal refuge from hunters. J Wildl Manage 84:413–424

Lehman CP, Rumble MA, Rota CT et al (2016) Elk resource selection at parturition sites, Black Hills, South Dakota. J Wildl Manage 80:465–478

Lehman CP, Rota CT, Raithel JD et al (2018) Pumas affect elk dynamics in absence of other large carnivores. J Wild Manage 82:344–353

Lilenbaum W, Martins G (2014) Leptospirosis in cattle: a challenging scenario for the understanding of the epidemiology. Transbound Emerg Dis 61:63–68

Loehman R, Flatley W, Holsinger L et al (2018) Can land management buffer impacts of climate changes and altered fire regimes on ecosystems of the southwestern United States? Forests 9:192

Long JW, Lake FK (2018) Escaping social-ecological traps through tribal stewardship on national forest lands in the Pacific Northwest, United States of America. Ecol Soc 23. https://doi.org/10.2307/26799109

Long RA, Bowyer RT, Porter WP et al (2014) Behavior and nutritional condition buffer a large-bodied endotherm against direct and indirect effects of climate. Ecol Monogr 84:513–532

Lowrey B, Devoe J, Proffitt KM et al (2020) Hiding without cover? Defining elk security in a beetle-killed forest. J Wildl Manage 84:138–149

Lubow BC, Singer FJ, Johnson TL et al (2002) Dynamics of interacting elk populations within and adjacent to Rocky Mountain National Park. J Wildl Manage 66:757–775

Lukacs PM, Mitchell MS, Hebblewhite M et al (2018) Factors influencing elk recruitment across ecotypes in the Western United States. J Wildl Manage 82:698–710

Maichak EJ, Scurlock BM, Rogerson JD et al (2009) Effects of management, behavior, and scavegning on risk of Brucellosis transmission in elk of western Wyoming. J Wildl Dis 45:398–410

Marshall AM, Abatzoglou JT, Link TE et al (2019) Projected changes in interannual variability of peak snowpack amount and timing in the Western United States. Geophys Res Lett 46:8882–8892

McCorquodale SM (1997) Cultural contexts of recreational hunting and native subsistence and ceremonial hunting: their significance for wildlife management. Wildl Soc B 25:568–573

McCorquodale SM, Raedeke KJ, Taber RD (1986) Elk habitat use patterns in the shrub-steppe of Washington. J Wildl Manage 50:664–669

McCorquodale SM, Wiseman R, Marcum CL (2003) Survival and harvest vulnerability of elk in the Cascade Range of Washington. J Wildlife Manage 67:248–257

McCullough D (1969) The tule elk: its history, behavior, and ecology, vol 88. University of California Press, Berkeley

Middleton AD, Kauffman MJ, McWirther DE et al (2013) Animal migration amid shifting patterns of phenology and predation: lessons from a Yellowstone elk herd. Ecol 94:1245–1256

Middleton AD, Merkle JA, McWhirter DE et al (2018) Green-wave surfing increases fat gain in a migratory ungulate. Oikos 127:1060–1068

Middleton AD, Sawyer H, Merkle JA et al (2019) Conserving transboundary wildlife migrations: recent insights from the greater yellowstone ecosystem. Front Ecol Environ 18:83–91

Miller W (2002) Elk interactions with other ungulates. In: Toweill DE, Thomas JW (eds) North American elk: ecology and management. Smithsonian Institution Press, Washington, D.C., USA, pp 435–447

Miller MW, Wild MA, Williams ES (1998) Epidemiology of chronic wasting disease in captive Rocky Mountain elk. J Wildl Dis 34:532–538

Mitchell MS, Cooley H, Goode JA, et al (2018) Distinguishing values from science in decision making: setting harvest quotas for mountain lions in Montana. Wildlife Soc B 42:13–21

Moeller AK, Lukacs PM, Horne JS (2018) Three novel methods to estimate abundance of unmarked animals using remote cameras. Ecosphere 9:e02331

Monello RJ, Powers JG, Hobbs NT et al (2014) Survival and population growth of a free-ranging elk population with a long history of exposure to chronic wasting disease. J Wildl Manage 78:214–223

Monello RJ, Galloway NL, Powers JG et al (2017) Pathogen-mediated selection in free-ranging elk populations infected by chronic wasting disease. P Natl Acad Sci USA 114:12208–12212

Murie OJ (1930) An epizootic disease of elk. J Mammal 11:214–222

Murie OJ (1951) The elk of North America. Stackpole Books

Murray R, Downham D, Clarkson M et al (1996) Epidemiology of lameness in dairy cattle: description and analysis of foot lesions. Vet Rec 138:586–591

Mysterud A, Edmunds DR (2019) A review of chronic wasting disease in North America with implications for Europe. Eur J Wildl Res 65:26

National Academies of Sciences, Engineering, and Medicine [NASEM] (2020) Revisiting brucellosis in the greater Yellowstone area. National Academies Press

Naylor LM, Wisdom MJ, Anthony RG (2009) Behavioral responses of North American elk to recreational activity. J Wildl Manag 37:328–338

Nichols RV, Åkesson M, Kjellander P (2016) Diet assessment based on rumen contents: a comparison between DNA metabarcoding and macroscopy. PLoS ONE 11:e0157977

O'Neil S, Bump J (2014) Modeling habitat potential for elk expansion in Michigan, USA. Wildl Bio in Pract 10:111–131

Olsen S (2010) Brucellosis in the United States: role and significance of wildlife reservoirs. Vaccine 28:F73–F76

Phillips GE, Alldredge AW (2000) Reproductive success of elk following disturbance by humans during calving season. J Wildl Manage 64:521–530

Polfus JL, Krausman PR (2012) Impacts of residential development on ungulates in the Rocky Mountain West. Wildl Soc B 36:647–657

Proffitt KM, Grigg JL, Garrott RA et al (2010) Changes in elk resource selection and distributions associated with a late-season elk hunt. J Wildl Manage 74:210–218

Proffitt KM, Gude JA, Hamlin KL et al (2013) Effects of hunter access and habitat security on elk habitat selection in landscapes with a public and private land matrix. J Wildl Manage 77:514–524

Proffitt KM, DeVoe J, Barker K et al (2019) A century of changing fire management alters ungulate forage in a wildfire-dominated landscape. For Int J For Res 92:523–537

Proffitt KM, Garrott R, Gude JA et al (2020) Integrated carnivore-ungulate management: a case study in west-central Montana. Wildl Monogr 206:1–28

Prokopenko CM, Boyce MS, Avgar T (2017) Characterizing wildlife behavioural responses to roads using integrated step selection analysis. J Appl Ecol 54:470–479

Raithel JD, Kauffman MJ, Pletscher DH (2007) Impact of spatial and temporal variation in calf survival on the growth of elk populations. J Wildl Manage 71:795–803

Rayl ND, Proffitt KM, Almberg ES et al (2019) Modeling elk-to-livestock transmission risk to predict hotspots of brucellosis spillover. J Wildl Manage 83:817–829

Rhyan J, Saari D (1995) A comparative study of the histopathologic features of bovine tuberculosis in cattle, fallow deer (Dama dama), sika deer (Cervus nippon), and red deer and elk (Cervus elaphus). Vet Pathol 32:215–220

Rhyan JC, Nol P, Quance C et al (2013) Transmission of brucellosis from elk to cattle and bison, Greater Yellowstone Area, USA, 2002–2012. Emerg Inf Dis 19:1992

Rickbeil GJ, Merkle JA, Anderson G et al (2019) Plasticity in elk migration timing is a response to changing environmental conditions. Glob Change Biol 25:2368–2381

Riggs RA, Keane RE, Cimon N et al (2015) Biomass and fire dynamics in a temperate forest-grassland mosaic: Integrating multi-species herbivory, climate, and fire with the FireBGCv2/GrazeBGC system. Ecol Model 296:57–78

Rhodes AC, Larsen RT, Clair SBS (2018) Differential effects of cattle, mule deer, and elk herbivory on aspen forest regeneration and recruitment. Forest Ecol Manag 422:273–280

Rowland MM, Alldredge AW, Ellis JE et al (1983) Comparative winter diets of elk in New Mexico. J Wildl Manage 47:924–932

Rowland M, White G, Karlen E (1984) Use of pellet-group plots to measure trends in deer and elk populations. Wildl Soc B 12:147–155

Rowland MM, Wisdom MJ, Johnson BK et al (2000) Elk distribution and modeling in relation to roads. J Wildl Manage 64:672–684

Rowland MM, Wisdom MJ, Nielson RM et al (2018) Modeling elk nutrition and habitat use in western Oregon and Washington. Wildl Monogr 199:1–102

Ruder MG, Lysyk TJ, Stallknecht DE et al (2015) Transmission and epidemiology of bluetongue and epizootic hemorrhagic disease in North America: current perspectives, research gaps, and future directions. Vector-Borne Zoonot 15:348–363

Samuel MD, Garton EO, Schlegel MW et al (1987) Visibility Bias during Aerial Surveys of Elk in Northcentral Idaho. J Wildl Manage 51:622–630

Samuel W, Welch D, Smith B (1991) Ectoparasites from elk (Cervus elaphus nelsoni) from Wyoming. J Wildl Dis 27:446–451

Sargeant GA, Oehler MW Sr (2007) Dynamics of newly established elk populations. J Wildl Manage 71:1141–1148

Sawyer H, Nielson RM, Lindzey FG et al (2007) Habitat selection of Rocky Mountain elk in a nonforested environment. J Wildl Manage 71:868–874

Scasta JD, Beck JL, Angwin CJ (2016) Meta-analysis of diet composition and potential conflict of wild horses with livestock and wild ungulates on western rangelands of North America. Rangeland Ecol Manage 69:310–318

Sergeyev M, McMillan BR, Hall LK et al (2022) Reducing the refuge effect: using private-land hunting to mitigate issues with hunter access. J Wildl Manage 86:e22148

Shively KJ, Alldredge AW, Phillips GE (2005) Elk reproductive response to removal of calving season disturbance by humans. J Wildl Manage 69:1073–1080

Skovlin JM, Zager P, Johnson PK (2002) Elk habitat selection and evaluation. In: Toweill DE, Thomas JW (eds) North American elk: ecology and management. Smithsonian Institution Press, Washington DC, p 531–556

Soulard CE (2006) Land cover trends of the central basin and range ecoregion. U.S. Geological Survey, Reston, Virginia

Spitz DB, Clark DA, Wisdom MJ et al (2018) Fire history influences large-herbivore behavior at circadian, seasonal, and successional scales. Ecol Appl 28:2082–2091

Stalling DH, Wolfe GJ, Crockett DK (2002) Regulating the hunt. North American elk: eclogy and management. Smithsonian Insitituion Press, London, U.K, pp 749–791

Stewart KM, Bowyer RT, Kie JG et al (2002) Temporospatial distributions of elk, mule deer, and cattle: resource partitioning and competitive displacement. J Mammal 83:229–244

Stoner DC, Anderson MT, Schroeder CA et al (2021) Distribution of competition potential between native ungulates and free-roaming equids on western rangelands. J Wildl Manage 85:1062–1073

Thompson R (2008) The taxonomy, phylogeny and transmission of Echinococcus. Exp Parasitol 119:439–446

Thompson MJ, Henderson RE (1998) Elk habituation as a credibility challenge for wildlife professionals. Wildl Soc B 26:477–483

Thorne E, Herriges Jr JD, Reese AD (1991) A Bovine brucellosis in elk: conflicts in the greater Yellowstone area. In: Proceedings of the elk vulnerability symposium, Bozeman, Montana, pp 296–291

Tolleson D, Halstead L, Howery L et al (2012) The effects of a rotational grazing system on elk diets in Arizona piñon–juniper rangeland. Rangelands 34:19–25

Torstenson WLF, Mosley JC, Brewer TK et al (2006) Elk, mule deer, and cattle foraging relationships on foothill and mountain rangeland. Rangeland Ecol Manage 59:80–87

Toweill DE, Thomas JW (2002) North American Elk: Ecology and Management. Smithsonian Institution Press, Washington and London

Tucker DG, Gardner ES, Wakeling BF (2004) Elk habitat use in relation to residential development in the Hualapai Mountains, Arizona. In: Van Riper C, Cole KL (eds) The Colorado Plateau: cultural, biological, and physical research. University of Arizona Press, Tucson, AZ, pp 89–96

Turnbull PCB (2008) Anthrax in humans and animals. World Health Organization

Uehlinger F, Johnston A, Bollinger T et al (2016) Systematic review of management strategies to control chronic wasting disease in wild deer populations in North America. BMC Vet Res 12:1–16

Unsworth JW, Kuck L, Scott MD et al (1993) Elk mortality in the Clearwater drainage of northcentral Idaho. J Wildl Manage 57:495–502

Vavra M, Sheehy DP (1996) Improving elk habitat characteristics with livestock grazing. Rangelands 18:182–185

Vieira ME, Conner MM, White GC et al (2003) Effects of archery hunter numbers and opening dates on elk movement. J Wildl Manage 67:717–728

Wait S, McNally H (2004) Selection of habitats by wintering elk in a rapidly subdividing area of La Plata County, Colorado. In: Proceedings of the fourth international symposium on urban wildlife conservation. University of Arizona, Tucson, Arizona, USA, pp 200–209

Walter WD, Lavelle MJ, Fischer JW et al (2010) Management of damage by elk (Cervus elaphus) in North America: a review. Wildl Res 37:630–646

Watt LA (2015) The continuously managed wild: Tule Elk at point Reyes national seashore. J Int Wildl Law Pol 18:289–308

Wertz TL, Blumton A, Erickson LE et al (1996) Strategies to keep wildlife where you want them - do they work? INT-GTR-343 In: Evans KE (ed) Sharing common ground on western rangelands: Proceedings of a livestock/big game symposium. U.S. Forest Service, Intermountain Research Station, Ogden, UT, pp 70–72

White PCL, Ward AI (2010) Interdisciplinary approaches for the management of existing and emerging human–wildlife conflicts. Wildl Res 37:623–629

Williams E (2005) Chronic wasting disease. Vet Pathol 42:530–549

Williams E, Young S (1980) Chronic wasting disease of captive mule deer: a spongiform encephalopathy. J Wildl Dis 16:89–98

Williams ES, Snyder SP, Martin KL (1983) Experimental infection of some North American wild ruminants and domestic sheep with Mycobacterium paratuberculosis: clinical and bacteriological findings. J Wildl Dis 19:185–191

Williams ES, Miller MW, Kreeger TJ et al (2002) Chronic wasting disease of deer and elk: a review with recommendations for management. J Wildl Manage 66:551–563

Williams CL, Lundrigan B, Rhodes OE Jr (2004) Microsatellite DNA variation in tule elk. J Wildl Manage 68:109–119

Wisdom MJ, Thomas JW (1996) Elk. In: Krausman PR (ed) Rangeland wildlife. Society for Range Management, Denver, CO, pp 157–181

Wisdom MJ, Ager AA, Preisler HK et al (2004) Effects of off-road recreation on mule deer and elk. In: Transactions of the 69th North American wildlife and natural resources conference 2004, pp 531–550

Wisdom MJ, Vavra M, Boyd JM et al (2006) Understanding ungulate herbivory—episodic disturbance effects on vegetation dynamics: knowledge gaps and management needs. Wildl Soc B 34:283–292

Wisdom MJ, Preisler HK, Naylor LM et al (2018) Elk responses to trail-baised recreation on public forests. For Ecol Manage 411:223–233

Witczuk J, Pagacz S, Zmarz A (2018) Exploring the feasibility of unmanned aerial vehicles and thermal imaging for ungulate surveys in forests - preliminary results. Int J Remote Sens 39:5503–5520

Wydeven AP, Dahlgren RB (1985) Ungulate habitat relationships in Wind Cave National Park. J Wildl Manage 49:805–813

Zaugg JL, Goff WL, Foreyt W et al (1996) Susceptibility of elk (Cervus elaphus) to experimental infection with Anaplasma marginale and A. ovis. J Wildl Dis 32:62–66

Chapter 21
Feral Equids

Steven L. Petersen, J. Derek Scasta, Kathryn A. Schoenecker,
and Jacob D. Hennig

Abstract Feral horses (*Equus ferus caballus*) and burros (*E. asinus*) in North
America, often referred to as **free-roaming, free-ranging, or wild horses and
burros**, are introduced species that are currently increasing in arid and semi-arid
rangelands. They differ from all other North American mammals by being the only
feral species protected by federal law. These equids inhabit areas featuring rough
topography, limited net primary productivity, and extreme weather conditions, and
have potential to cause long-term ecosystem impacts. In this chapter, we review
the historical and modern context of feral equids on North American rangelands
including their evolutionary past and introduction to the continent, their relationships to the environment, and challenges associated with their management. The
management of feral equids is perhaps more scrutinized than any other species
because their legal status, body size, physiology, foraging patterns, and local abundance directly interacts and competes with rangeland resource quality, impacts native
wildlife populations, and conflicts with the multiple-uses of the land that they inhabit.

Keywords Feral · Free-roaming · *Equus ferus caballus* · Bureau of Land
Management · U.S. Forest Service · Social science · Population estimation ·
Public opinion · Population growth · Domestication · Cecal digestion

S. L. Petersen (✉)
Department of Plant and Wildlife Sciences, Brigham Young University, Provo, UT, USA
e-mail: steven_petersen@byu.edu

J. D. Scasta · J. D. Hennig
Department of Ecosystem Science and Management, University of Wyoming, Laramie, WY, USA

K. A. Schoenecker
Fort Collins Science Center, United States Geological Survey (USGS), Fort Collins, CO, USA

735

21.1 General Life History

21.1.1 Feral Equid Species

Feral equids of North America, referred to as free-roaming, free-ranging, or wild equids, include horses (*Equus ferus caballus*) and burros (*E. asinus*), and are the only federally-protected feral species in North America. The term feral constitutes "species that have been established from intentional or accidental release of domestic stock that results in a self-sustaining population(s)" and "are generally non-indigenous and often invasive" (The Wildlife Society 2021). Feral animals are wild descendants of a domesticated species. To better understand how feral equids became federally-protected, we must consider the evolutionary and domestication history of these animals and their relationship to humans. The socio-ecological mismatch of protecting a feral species translates into great potential for feral equids to negatively affect the ecosystems they inhabit. Together, these aspects frame the controversy surrounding the contemporary management of feral equids on western North American rangelands (Beever et al. 2018; Scasta et al. 2018). In this chapter, we provide greater content and focus on feral horses because they are more numerous and more widely researched than burros in North America. In contrast, the lack of research on burros has resulted in a general gap in our knowledge of this species.

21.1.2 Evolutionary and Domestication History

Equidae, the family containing horses and burros, originated in North America approximately 50 million years ago (Hurlbert Jr. 1993). Ancient equids included a diverse assemblage of species possessing a variety of physiological and morphological features. *Hypohippus* was a three-toed browsing species while *Dinohippus* was a single-toed grazing species (Fig. 21.1). All equid species in North America ultimately became extinct during the late-Pleistocene epoch due to a combination of environmental change, disease, and the arrival of humans and hunting (Buck and Bard 2007). Prior to their North American extinction, equids crossed the Bering Strait and dispersed into Eurasia 20 million years ago (Kelekna 2009). The horses that radiated across the steppes of Eurasia eventually were domesticated by humans approximately 6,000 years ago (Outram et al. 2009). Burros, meanwhile, originated from African wild asses (*E. africanus*), and were likely domesticated in Egypt and Mesopotamia over 5,000 years ago.

The earliest records of horse domestication were from the Botai people of north-central Kazakhstan whose horse-centric cultures were highly influential (Outram et al. 2009). Early cultures hunted horses and likely captured orphaned foals leading to breeding horses and keeping them for milk and meat in an intimate association where horse and human survival were closely intertwined (Levine 1999). Horse domestication was a critical component of human history and provided a valuable utility for

Fig. 21.1 Equidae evolved in North America and ancient horses were physiologically and morphologically diverse. Depicted here by the American Natural History Museum, as example, are the large single-toed grazing species *Dinohippus* (left), the small three-toed mixed feeder *Nannippus* (center), and the three-toed browsing species *Hypohippus*. Picture provided by the American Natural History Museum (ANHM)

many cultures. Domestic horses were transported across the globe and their distribution generally tracks the expansion and distribution of humans. Today, the emotional attachment of humans to horses helps explain the ubiquity of feral equids worldwide. It was the horse, and it's raw "horse-power", that enabled cultures to disperse and advance agriculture, transportation, industry, commerce, and warfare (Ransom and Kaczensky 2016). Domestication included artificial selection for certain traits over many years leading to horses that were optimized for particular size, color, and reproduction characteristics. All domesticated and feral horses today differ genetically and phenotypically from their non-domesticated ancestors (Fages et al. 2019) and they are morphologically different from their only extant wild relative, the Przewalski's horse (*Equus ferus przewalskii*; Groves 1994).

21.1.3 Feralization and Protection of Equids in North America

Italian explorer Christopher Columbus first transported domestic horses to North America on his second voyage to the continent in 1493 (Kelekna 2009). The best evidence suggests that burros were brought to North America around the same time (Antonius 1938; McKnight 1958). A half-century later, an estimated 10,000 horses roamed central Mexico with both Pueblo and Apache peoples possessing equestrian skills (Kelekna 2009). In 1680, the Pueblo peoples revolted against Spanish conquistadors, facilitating the release of several thousand horses which served as the "nucleus" of mustang horse herds in North America (Kelekna 2009). Ever since, such horses have become a fundamental aspect of North American human cultural evolution (Berger 1986). Additional escapes along with intentional releases by Native Americans, European settlers, and the military during the 1700s provided more sources of horses that enhanced genetic diversity and boosted population densities (Mitchell 2015). With the advent of the industrial age in the nineteenth and twentieth centuries, demand for horses and burros declined due to a combination of a rapidly urbanizing and mechanized society and high costs of equid care and land (Garrott 2018; Scasta et al. 2018). Consequently, the post-industrial period in the mid-twentieth century saw an increase in intentional horse releases. Feral horses became more abundant across western rangelands, until they were captured by mustangers and others who sold them for slaughter, re-sale, or other economic purposes (Danvir 2018). Spurred by citizens concerned about the dwindling population of horses and burros in the West, the U.S. government enacted a law called the Wild Free-roaming Horses and Burros Act (WFRHBA) in 1971 to protect the remaining populations of feral equids on federally-owned land (Public Law 92-195, see Rangeland Management section).

21.2 Distribution and Population Dynamics

21.2.1 Distribution of Feral Equids in the United States

Feral equids are generally found in areas where they escaped after humans no longer needed them or were released on public lands during stark economic times (to avoid feeding costs, e.g.). The areas where horses have been allowed to remain typically have low human population densities, minimal human use, and are of little economic value for row-crop agriculture or commercial development. Feral equids can be found across the United States with most populations occurring on rangelands in western states (Fig. 21.2). Small populations also exist on barrier islands off the Atlantic coast, along with isolated populations in eastern forests. Feral equids inhabit federally-owned land managed by the Bureau of Land Management (BLM), the U.S. Forest Service (USFS), the U.S. Fish and Wildlife Service (USFWS), the National Park Service (NPS), and the Department of Defense (DOD). Horses and burros can also be found on private, municipal, state, and sovereign tribal lands. The feral equids that occur on BLM and USFS lands are protected by the WFHBA (Public Law 92-195). These populations are managed in the areas where they occurred at the time of the Act's passing. On BLM land, these areas are called Herd Management Areas (HMA). There are also populations on BLM land where they are not specifically managed for, and these areas are known as Herd Areas (HA). On USFS land, management areas are termed Wild Horse and Burro Territories (WHBT). In total, there are 177 HMAs and 53 WHBTs spread across 10 western U.S. states (BLM 2022; USFS 2022).

21.2.2 Global Distribution of Feral Equids

Feral equids inhabit a wide range of habitats throughout the world, with many populations existing in ecosystems characterized by rugged topography, limited net primary

Fig. 21.2 Feral horses in western North America arid and semi-arid regions characterized by complex topography and extreme temperatures. (Top) Southwestern Wyoming, January 2017, photo credit: J. D. Scasta. (Bottom) Southern Nevada, September 2015

production, and extreme weather patterns (Fig. 21.3). We do not present an exhaustive list of all global feral equid populations here; rather, we list select populations to highlight that they are widespread across the globe. In North America, feral equids also occur in Canada and Mexico, in addition to the U.S. (Schoenecker et al. 2021). In South America, populations occur in Ecuador and Argentina (Scorolli 2018). Australia is thought to have the greatest abundance of feral equids of any country (Schoenecker et al. 2021), and New Zealand also contains feral equids (the Kaimanawas). A small population also occurs in French Polynesia in the South

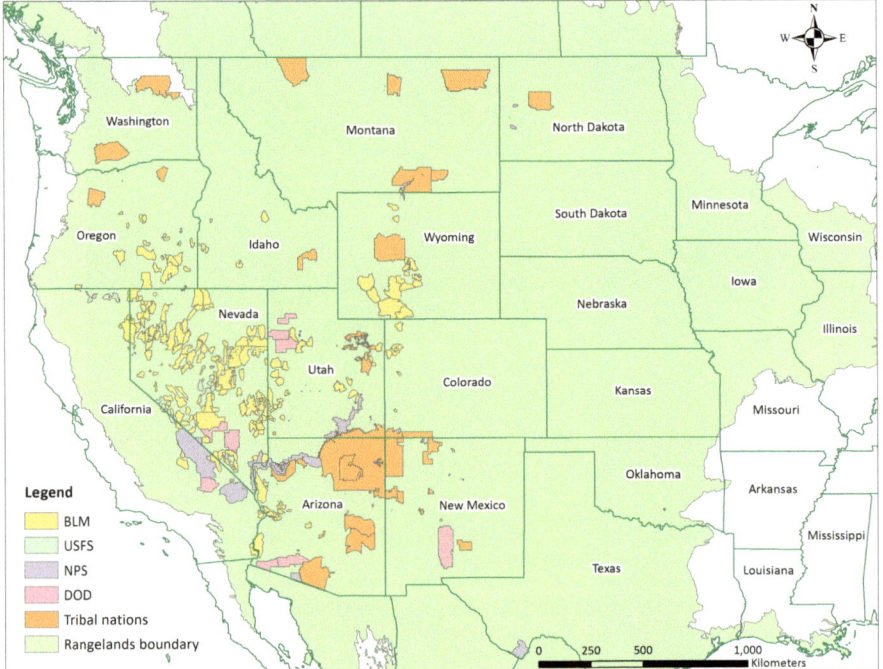

Fig. 21.3 Approximate range of known feral horse and burro populations on western United States rangelands. *Note* not all areas within each polygon are occupied by horses or burros, and there are likely feral equid populations not represented here

Pacific. In Africa, feral horses and burros are known to inhabit the Namib desert (Cothran et al. 2001). In Europe, some populations have been introduced as part of rewilding efforts (Linnartz and Meissner 2014), while others are managed extensively (i.e., handled annually). Populations are present in France (Camargue), in the United Kingdom (e.g. Dartmoor; Exmoor, New Forest, and Welsh Mountain ponies), in the Danube Delta region of Romania, in the Pyrenees Mountains of France and Spain (Galacia ponies, Pottoka horses), and in Portugal (Sorraia horses and Garrano ponies). In Asia there are some Misaki-uma horses occurring within the designated National Monument on Cape Toi, Japan.

21.2.3 Population Estimates of Feral Equids in the United States

The nationwide estimate of feral free-ranging equids across all land jurisdictions is approximately 275,000 (Table 21.1). The majority of feral equids are thought to occur on tribal nations, with 75,000 horses estimated on the Navajo nation alone (Schoenecker et al. 2021; Wallace et al. 2021). There were roughly 72,000 horses and 14,500 burros on BLM land in 2021 (BLM 2022), and approximately 9,000 feral equids on USFS land (T. Drotar, pers. comm.). These estimates far exceed maximum appropriate management levels (AML) which are population ranges set to balance equid populations with the other uses of public rangelands (see *Rangeland Management* for more details). The nationwide AML for feral equids is 26,785 on BLM land and 2,253 on USFS land (BLM 2022; USFS 2014). Feral equid population growth rates range from 11% to over 25% (Roelle et al. 2010), but the protected status of feral equids on BLM and USFS lands makes them a challenge for management (Messmer et al. 2021). In addition, there were an estimated 59,749 horses and 862 burros in 2021 living in 'off-range' BLM facilities consisting of corrals and pastures (BLM 2022).

Table 21.1 Population estimates of feral horses (*Equus ferus caballus*) on different land jurisdictions in the United States

Land jurisdiction or entity	Feral horse estimate	Date(s) of estimates used for tally	Source(s)
Tribal Nations	103,654	2020	Pers comm. with tribes, Beever et al. (2019), Wallace et al. (2021), Schoenecker USGS unpublished data (Navajo Nation)
DOI BLM On-range	86,189	2021	BLM Wild Horse and Burro Program
DOI BLM Off-range	59,007	2021	BLM Wild Horse and Burro Program
DOI National Park Service	1606	2021	Direct pers comm. with 20 NPS units; Powers J.E. 2014; Cumberland Island horse—Wikipedia https://www.nps.gov/calo/learn/management/upload/Annual-Horse-Findings-Report-2020-final.pdf
DOI Fish and Wildlife Service	150	2020	https://www.fws.gov/refuge/Chincoteague/wildlife_and_habitat/ponies.html
USDA Forest Service	9000	2020	Teresa Drotar, USFS WH&B Program Lead; pers commun. 2021
Department of Defense	1295	2020	Pers comm. with military installations; Trespass Horse Working Group, LA (Fort Polk)
Municipal and State Lands (AZ, NV, FL, CO)	13,950	2020, 2021	Jim French Humboldt County Commissioner NV; Science and Conservation Center MT, TJ Holmes volunteer CO, Paynes Prairie State Park FL; B. Lubow estimates
Total	274,851		

Estimates were compiled April 2021, and obtained from various sources for each category of land management jurisdiction

21.2.4   Population Monitoring

The BLM, USFS, and NPS conduct regular population surveys for feral equids following established methods (Lubow and Ransom 2016, 2009; Griffin et al. 2020). Feral equid populations on other land jurisdictions are surveyed less regularly. Survey methods differ among populations but include simultaneous double-observer aerial surveys (Lubow and Ransom 2016; Griffin et al. 2020; Hennig et al. 2022), photo mark-resight surveys (Lubow and Ransom 2009), genetic capture-recapture models using fecal DNA (Schoenecker et al. 2021), employing distance sampling within aerial infrared surveys (Schoenecker et al. 2018) and direct visual counts by ground observers (Friends of a Legacy, Little Book Cliffs HMA, Colorado).

21.3   Habitat Associations and Impacts

21.3.1   Habitat Selection, Home Range Sizes, and Movement Patterns

Because feral equids did not co-evolve within the areas they reside in, generalizing habitat selection across populations is inherently difficult. While habitat selection is context dependent, there are a few patterns that are common across studies. Terrain strongly influences the habitat selection of feral horses, and they are much more likely to utilize relatively flat topography or gently sloping ridgetops (Ganskopp and Vavra 1986; Henning 2022; Schoenecker et al. 2022a, b) than steep slopes. Habitat selection by feral horses is also strongly linked to forage availability (Schoenecker et al. 2016, 2022a, b). Horses are large-bodied grazers (Van Soest 1994) that consume large quantities of graminoids (King 2002; King and Gurnell 2005; Girard et al. 2013); therefore they tend to select for grassland or shrubland landcover types (Smith 1986; Crane et al. 1997; King 2002; King and Gurnell 2005; Schoenecker et al. 2022a, b). Horses that inhabit heavily forested environments select for disturbed areas, such as roadside edges, where grass production is higher (Irving 2001; Girard et al. 2013). Equids are relatively inefficient in water retention, compared to ruminants, owing to their cecal digestion (Janis 1976). Consequently, equids select for closer proximity to water sources during the growing season and foaling season (Arandhara et al. 2020; Esmaeili et al. 2021; Schoenecker et al. 2022a, b; Girard et al. 2013). Horses can eat snow for hydration, and are therefore less reliant on open water during the winter (Mejdell and Boe 2005; Kaczensky et al. 2008; Salter and Hudson 1979). The social status of individuals can also affect habitat selection. Different male social classes vary in their use of the landscape: harem-holding stallions are constrained by the habitat selection of their mares who need to remain closer to surface water during foaling and lactation, whereas bachelors are free to travel longer distances to access prime forage (Schoenecker et al. 2022a, b).

Few studies have evaluated the movement patterns of feral horses, but variation in resources across space and time seem to drive their movements. Berger (1986) found that a horse population in the Great Basin exhibited altitudinal migration to enhance their access to forage availability, while a population in the Red Desert of Wyoming, where spatiotemporal variation was less extreme, exhibited relatively stable, year-long home ranges (Hennig 2021). Movements of equids are strongly influenced by seasonal vegetation biomass and availability (Salter and Hudson 1982; Kaczensky et al. 2008), which subsequently influences home range size (McLoughlin and Ferguson 2000). Older studies in North America that relied on visual observations reported wide variation in horse home range size, between 2.6 and 48 km^2 (Pellegrini 1971; Feist and McCullough 1976; Berger 1977, 1986; Salter and Hudson 1982; Miller 1983). Home range size from these earlier studies are smaller than what has been found in studies using global positioning system (GPS) telemetry data. Home ranges sizes reported for feral horses living in forested areas in Alberta and open shrublands in Wyoming were 48.4 km^2 and 40.4 km^2, respectively (Girard et al. 2013; Hennig et al. 2018). In Utah, average home range size for mares was 110.3 km^2 (Schoenecker et al. 2022a, b). Mares in Alberta and Wyoming inhabited areas with abundant water sources; whereas mares in Utah had larger home range sizes most likely to accommodate larger distances to water (Schoenecker et al. 2022a, b).

21.3.2 Feral Equid Effects on Rangeland Ecosystems

Equids are cecal digestors with agile lips and upper sets of canines and incisors (Janis 1976; Scasta et al. 2016). Cecal digestion is comparatively less efficient at nutrient extraction than rumination, meaning that equids need to consume more plant biomass relative to a comparatively-sized ruminant (Hanley 1982; Menard et al. 2002). Their agile lips and upper teeth allow equids to crop plants closer to the ground, compared to cattle, when grazing (Menard et al. 2002). Together, and along with their relatively large body size, poorly-managed feral equid populations can have severe negative effects on the rangeland systems they inhabit (Boyd et al. 2017; Eldridge et al. 2020). Studies have linked feral horse grazing with decreased vegetation biomass, lower plant height, decreased plant species richness, increased cover of exotic and invasive species, reduced seed banks, increased soil penetration resistance, and increased bare ground cover (Baur et al. 2018; Beever 2003; Beever and Brussard 2004; Beever and Herrick 2006; Beever et al. 2008; Beever and Aldridge 2011; Boyd et al. 2017; Davies and Boyd 2019; King et al. 2019; Loydi et al. 2012; Stoppelaire et al. 2004; Zeigenfuss et al. 2014; Hennig 2021). These effects contribute to decreased overall rangeland health, less forage for livestock and native herbivores, and degraded wildlife habitat (Jones 2000; Beever 2003; Scasta et al. 2018). Indeed, research has documented lower small mammal, reptile, and invertebrate densities in horse-occupied versus un-occupied sites (Beever and Brussard 2004; Beever and Herrick 2006). Moreover, increasing populations of feral horses was correlated with population declines of the greater sage-grouse (*Centrocercus urophasianus*; Coates et al. 2021).

In arid rangelands, feral equid effects extend to interference competition at limited water sources. Feral horses are large and often aggressive, which can translate into subordinate species altering their behavior at water. Bighorn sheep (*Ovis canadensis*) have been shown to avoid water sites when horses are present (Osterman-Kelm et al. 2008), and pronghorn (*Antilocapra americana*) show increased vigilant activity around horses (Gooch et al. 2017). Both pronghorn and mule deer (*Odocoileus hemionus*) have been documented to shift their temporal or spatial watering activity in response to horses, and watering sites with horses tend to have fewer vertebrate species richness (Hall et al. 2016, 2018). Equid grazing and trampling at watering sites influences plant communities, particularly during the critical growing period. Impacts can include reduced vegetation cover, greater percent bare ground, and less litter (Boyd et al. 2017). In combination with other grazers, forage species and soils become highly vulnerable to grazing impacts when they are in close proximity to these water sources. Agencies and land owners that limit equid access to riparian areas experience increased vegetation cover and greater soil protection from compaction and erosion. For example, following 3 years of exclusion, Boyd et al. (2017) found that plant cover and litter increased by as much as 40% and the extent of bare ground decreased by 30%. Higher vegetation cover and reduced bare ground can reduce erosion potential and decrease the vulnerability of these sites to invasive species.

21.4   Rangeland Management

21.4.1   Guiding Federal Policies

The complexity of rangeland management of feral equids on federally-owned public land in the United States is better understood when considering the laws that govern feral equid protection and public land use. The first law dealing with protection and management of horses and burros was the Wild Horse Protection Act of 1959 (WHPA; Public Law 86-234). This act prevents the use of aircraft or motor vehicles to hunt and capture unbranded horses or burros on public lands. It also prohibits the pollution or poisoning of water holes on public land for the purpose of trapping or killing horses or burros. Congress next implemented the Wild Free-Roaming Horses and Burros Act in 1971 (WFRHBA; Public Law 92-195), which is the sentinel law concerning horse and burro protection and management. This act protects any unbranded or unclaimed horse or burro on public lands from capture, branding, harassment, or death (Public Law 92-195). It also mandates that the BLM and USFS provide habitat for horses and burros in areas where they existed at the time of enactment. These agencies were granted permission to conduct management actions to maintain a natural ecological balance between equid populations and the capacity for public lands to offer other ecosystem services, including livestock grazing, wildlife habitat, and recreation. The WFRHBA gives authority to the BLM and USFS to remove excess horses and burros for private adoption or to humanely destroy individuals if it was deemed necessary to preserve rangeland condition for multiple uses.

The Federal Land Policy and Management Act of 1976 (FLPMA; Public Law 94-579) amended the WFRHBA by authorizing the BLM and USFS to use helicopters for transporting captured horses and burros and the Omnibus Parks and Public Lands Management Act of 1996 (Public Law 104-333) extended the use of helicopters for gathering. FLPMA further defines the concept of multiple uses as the managing of public lands so that they best meet the present and future needs of citizens. This means protecting the ecological, scenic, and historical values and preserving habitat for wildlife and livestock. The WFRHBA was additionally amended through the Public Rangelands Improvement Act in 1978 (PRIA; Public Law 95-514). This act required inventories of horse and burro populations on federal lands and directed the BLM and USFS to determine appropriate management levels (AML) within horse and burro herd management areas (HMA). PRIA gave BLM or USFS the authority to determine whether AML should be achieved by removal or destruction of excess animals, or through non-lethal methods such as sterilization.

When equid populations in HMAs are found to be above the maximum AML, PRIA directs the BLM or USFS to decide which population control method (removal, destruction, sterilization, or other) is most appropriate to implement. Their decisions must be approved by the general public and are often legally challenged and successfully overturned (see Scasta et al. 2018). When removals do occur, excess healthy animals are put up for adoption, but the WFRHBA states that if excess animals are not adopted after three attempts, then they shall be humanely destroyed; however, due to annual riders (amendments) attached to federal appropriations bills, destruction of healthy animals is currently prohibited (Garrott and Oli 2013).

21.4.2   Livestock Grazing Management in the Feral Equid Context

Administration of livestock grazing on public lands in the western U.S. was prompted by the Taylor Grazing Act of 1934 (Public Law 73-482). This Act ended open grazing on public rangelands and created the Division of Grazing in the Department of Interior (DOI), which has been used to regulate the entry and practice of grazing on approximately 80 million acres of unreserved federal lands (excluding Alaska). This resulted in a highly regulated process that includes permitting, fees, and multi-year leasing. In addition, livestock numbers (i.e., animal unit months or AUMs) and timing of grazing are explicitly stipulated within a permit that is reviewed by specialists from the BLM and USFS in the context of rangeland monitoring data. Adjustments over time are made through collaborative dialogue with permittees. Violations of livestock grazing stipulations, deteriorating rangeland condition concerns, or weather patterns such as drought can manifest in a reduction of AUMs and grazing duration.

Compared to livestock grazing on public lands, feral equid use is much less regulated. In addition to controlling the numbers of livestock and timing of grazing, the areas that livestock can graze are often managed using fencing, deferred grazing rotation, herding, and salt and water distribution (Beever 2003). Contrastingly, feral equids graze year-round in largely unfenced areas that permit free movement across the landscape. Livestock grazing is annually assessed in the context of rangeland monitoring data and adaptively managed to alleviate problems, as compared to equid grazing which is managed with gathers and removals (Fig. 21.4) to move numbers closer to AML (Hurwitt 2017).

Fig. 21.4 Helicopter gathering of horses in southern Wyoming (above) and in Utah (below). Note the handler and Judas horse in the foreground in the Utah roundup. The Judas horse is a trained horse that is released as horses are led into the corrals, subsequently leading the group of feral horses into the trap (not shown). Photo credit USGS

21.4.3 *Feral Equid Population Management Tools*

Management of feral equid populations involves different approaches to reduce total population on western rangelands and/or growth rates (Scasta et al. 2018; Hendrickson 2018). Non-lethal approaches are the primary strategy, particularly in the most recent report to Congress (BLM 2018) and include several options:

(1) *Reproduction management* where animals are gathered, chemical immunocontraceptive or surgical sterilization are administered, and animals released back 'on-range'. Some immunocontraceptives can be delivered through darting in the field and do not require gathering animals (Kirkpatrick and Turner 2008; Kane 2018; Bechert et al. 2021).
(2) *Removal and Adoption* where animals are gathered and then adopted to private individuals (Bender and Stowe 2020; Fig. 21.5).

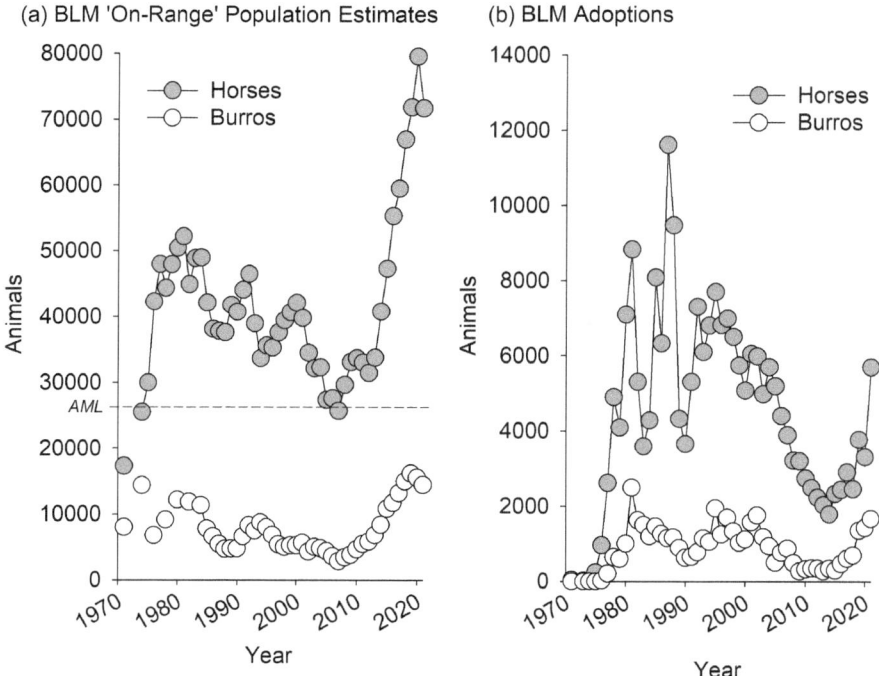

Fig. 21.5 (Left) On-range population estimates of feral horses and burros within Bureau of Land Management (BLM) Herd Management Areas from 1970 to 2020. The dotted line signifies the nationwide maximum Appropriate Management Level (AML). In areas where equid populations are above maximum AML, the BLM may conduct gathers to remove excess individuals. These individuals are either put up for adoption or housed in long-term holding facilities. (Right) The number of adopted feral horses and burros by private citizens between 1970 and 2020. All data were acquired via the BLM Wild Horse and Burro Program website (https://www.blm.gov/whb)

(3) *Relocation to off-range facilities* where unadopted animals are transferred to long-term pastures in the central U.S. that are privately owned and a per head payment is provided by the BLM (Elizondo et al. 2016).

Lethal strategies are not currently allowed but do need mention here and include:

(1) *Capture and euthanasia* where an animal is in stress and/or pain due to age, injury, or other condition inhibiting horse welfare. This is in adherence to Instruction Memorandum (IM) 2015-070 for BLM Animal Health, Maintenance, Evaluation, and Response and established the policy and procedures for proactive and preventative medical care (BLM 2015).
(2) *Slaughter* where animals are gathered and killed off-site and the meat is utilized (either human or non-human purposes). While WFRHBA (Public Law 92-195) does provide the authority for "destroying" either excess horses for which there is no adoption demand [see §1333. Powers and Duties of Secretary (a)(2)(C)]; this is not used in the United States currently because the U.S. Congress has prohibited slaughter since 2007 with the Agriculture, Rural Development, Food and Drug Administration, and Related Agencies Appropriations Act (Public Law 109-97) that prohibits use of federal funds for horse inspection, followed by subsequent amendments and ultimately a 2014 federal budget which explicitly prohibited horse slaughter (Norris 2018).

21.5 Threats to Feral Equid Populations

21.5.1 Disease

Domestic and feral equids are affected by a variety of maladies (Table 21.2). There is the potential for wild populations to act as a disease reservoir (Gilchrist and Sergeant 2011), with a difference in potential for spread depending on whether they are on-range, or in holding facilities. Additionally, disease is more likely to be expressed and spread in holding facilities due to high density of horses from various HMAs and high stress levels in captive equids. Gastrointestinal parasites can be common among feral equids, which can impair gastrointestinal function, reduce body condition, lower reproductive success, and decrease overall health and longevity (Debaffe et al. 2016; Pihl et al. 2018). In south-east Australia, Harvey et al. (2019) found that the parasite *Strongylus vulgaris* had infection rates as high as 97%, with symptoms that included fever, elevated heart rate, pain, and gastric reflux. This parasite was transmissible to domestic herds through direct contact with wild horse herds.

Blindness, lameness and hoof disorders or damage (i.e. laminitis) all occur to feral equids. Blindness may result from trauma (fighting), impact trauma from branches or grass stems, or disease (i.e. Equine recurrent uveitis, also known as moon blindness, which is the most common cause of blindness in horses). Common causes of lameness include trauma, infection, acquired disorders, metabolic disorders, and nervous and circulatory system disease (Adams 2015). Horses evolved and were

Table 21.2 A non-exhaustive list of diseases, infections, and disorders that may affect both domestic and feral equids

Disease/Disorder	Health concern/Risk	Treatment
Diseases		
Brucellosis	Reproductive issues, discharge	Antibiotics
Equine encephalomyelitis	Impaired vision, weakness, convulsions, death	
Equine infectious anemia	Fever, hemorrhage, weight loss	None, quarantine
Equine influenza	Fever, respiratory issues	Rest
Equine papillomavirus	Skin tumors, warts	Lye, formaldehyde, iodine
Equine protozal myeloencephalitis	Brian damage, spinal issues, atrophy	Anitprotozal, SDZ/PYR
Equine rabies	Depression, lameness, tremors, death	None
Potomac horse fever	Anorexia, fever, diarrhea	Antibiotics, fluid treatment
Rhinopneumonitis herpesvirus	Fever, nasal, inflammation	Rest
Ringworm	Blistering, scabbing	Anti-fungal
Streptococcus equi (stangles)	Severe inflammation, discharge, death	Antibiotics
Tetanus	Muscle stiffness, spasms, death	Antibiotics, antitoxin
West Nile virus	Ataxia, fever, weakness, paralysis	None
Disorders		
Blindness	Locomotion	
Lameness	Abnormal stance, locomotor difficulties	
Lamanitis	Extreme pain, hoof rotation	

artificially selected to travel long distances with repeated low-load concussive conditions, typical of hard terrain. However, they are subsequently predisposed to hoof and leg abnormalities (Hampson et al. 2013). These can also lead to issues such as osteoarthritis, joint pain, foot irregularities, and laminitis. Laminitis is a hoof ailment that has been commonly observed in Australian feral horses than can cause severe pain and difficulty during travel (Hampson et al. 2010a, b).

21.5.2 Climate Change

Effective management of feral equids will require an understanding of the current and future threats from a changing climate (Tietjen and Jeltsch 2007). Forecasted global climate change suggests western North America will be warmer and experience greater variability of extreme events including droughts (Pokhrel et al. 2021). The effects of climate change could be exacerbated in xeric climates. Data suggests that impacts can include high variability in precipitation levels, with xeric areas becoming dryer (Dore 2005). These changes may subsequently impact vegetation and forage production as intensity in precipitation increases but total quantity remains the same, creating more variable soil moisture conditions. If forage production decreases, carrying capacity will also decrease leading to potential overgrazing by herbivores (Tietjen and Jeltsch 2007). Impacts to feral equids may include death and sickness caused by starvation, greater conflicts in urban areas, and increased intraspecific competition. The use of wildlands for grazing are at risk because of unpredictable trends in climate and vegetation dynamics and therefore require careful monitoring and planning to prevent overgrazing and negative impacts by feral equid and other ungulate grazers.

21.6 Conservation and Management Challenges

21.6.1 Social Challenges

The management of feral equids is a contentious issue to say the least. While federal protection is stipulated by the WFRHBA, so is the proper management of the broader suite of natural resources (Public Law 92-195). The federal government's role has been characterized as "a national injustice" and "systematic removal and eradication of an American icon". Generally, the situation has pitted those who advocate for horses against those who advocate for multiple use and healthy rangelands. Yet, these two groups may not be mutually exclusive because as the population of feral equids increases, there may be negative consequences for horses due to degraded rangelands. In other words, an overabundance of horses and burros leads to overgrazing and potentially health issues for horses and burros as well as a cascade of other issues for soils, water, plants, wildlife, and other user groups. Increasing equid populations, especially in arid landscapes, may lead to decreased body condition, reduced access to forage and water, and an increase in emergency gathers conducted by BLM (Fuller et al. 2016). Further exacerbating the problem is the financial cost of gathering, removing, and maintaining horses in off-range facilities. Off-range care and feeding that are primary costs covered by the BLM Wild Horse and Burro program and these costs exceeded $65.5 million in FY 2020. These off-range costs are projected to be approximately $360 million annually in the next 15–18 years if on-range populations are reduced to AML (BLM 2020b). Future progress on the issue will require finding common ground among different stakeholder groups that enhances the health of the land and the horses and burros.

21.6.2 Antithetical Litigation

Aside from financial constraints, a major impediment to feral equid management is the prevalence of litigation. Scasta et al. (2018) provided examples of cases filed against the BLM for both managing and not managing equid populations. For example, one lawsuit attempted to bar the BLM from implementing a plan to gather approximately 2,700 wild horses in western Nevada. In a contrasting case, the BLM was sued for allowing too many free-ranging horses in Nevada. This antithetical litigation dynamic creates a very difficult situation for the federal government to effectively manage horse populations, ultimately leading to instances of management stasis while horse populations continue to grow and ecological problems continue to intensify.

21.7 Research and Management Needs

Feral equids inhabit a vast area of the western North American landscape but their ecology is less understood compared to native ungulates. Only a handful of recent studies have characterized habitat use of feral equids (Edouard et al. 2009; Girard et al. 2013; van Beest et al. 2014; Leverkus et al. 2018; Hennig 2021; Schoenecker et al. 2022a, b). There is a dearth of information regarding feral equids for several reasons. Little funding has been available to study feral equids since the inception of the WFRHBA. Further, feral species ecology was of little interest to basic science (Boyce et al. 2021). Feral equids are both domesticated and introduced; thus their ecology isn't studied within the context of prevailing evolutionary theory. Instead, their abundances and distributions are a product of human introductions and land use decisions. Consequently, there is a critical need for research examining topics including resource selection, niche overlap and interspecific competition, and density-dependence to better understand the role of how feral species interact with novel environments. In a management context, specific questions that require further research attention include understanding the comparative effects of feral equids versus livestock on rangelands, quantifying competition between equids and both wild and domestic herbivores, assessing if feral equids decrease the fitness or survival of sympatric wildlife species, and better understanding of social issues such as how the general public perceives the feral equid issue. More information on all of these topics will help natural resources managers with sustaining healthy lands and healthy herds into the future.

References

Adams SB (2015) Overview of lameness in horses. Merck Veterinary Manual https://www.merckv etmanual.com/musculoskeletal-system/lameness-in-horses/overview-of-lameness-in-horses

Antonius O (1938) On the geographical distribution in former times and today, of the recent Equidae. Proc Zool Soc 107:557–564

Baur LE, Schoenecker KA, Smith MD (2018) Effects of feral horse herds on rangeland plant communities across a precipitation gradient. West N Am Nat 77:526–539. https://doi.org/10. 3398/064.077.0412

Bechert US, Turner JW, Baker DL, Eckery DC, Bruemmer J, Lyman CC, Prado T, King SRB, Fraker MA (2021b) Fertility control options for management of free-ranging horse populations. Hum Wildl Interact (in review)

Beever EA (2003) Management implications of the ecology of free-roaming horses in semi-arid ecosystems of the western United States. Wildl Soc Bull 3:887–895. https://www.jstor.org/sta ble/3784615

Beever EA, Brussard PF (2004) Community- and landscape-level responses of reptiles and small mammals to feral horse grazing in the Great Basin. J Arid Environ 59:271–297. https://doi.org/ 10.1016/j.jaridenv.2003.12.008

Beever EA, Herrick JE (2006) Effects of feral horses in Great Basin landscapes on soils and ants: direct and indirect mechanisms. J Arid Environ 66:96–112. https://doi.org/10.1016/j.jaridenv. 2005.11.006

Beever EA, Huntsinger L, Petersen SL (2018) Conservation challenges emerging from free-roaming horse management: A vexing social-ecological mismatch. Biol Conserv 226321–328. https:// doi.org/10.1016/j.biocon.2018.07.015

Beever EA, Taush RJ, Thogmartin WE (2008) Multi-scale responses of vegetation to removal of horse grazing from the Great Basin (USA) mountain ranges. Plant Ecol 196:163–184. https:// doi.org/10.1007/s11258-007-9342-5

Beever EA, Simberloff D, Crowley SL, Al-Chokhachy R, Jackson HA, Petersen SL (2019) Social–ecological mismatches create conservation challenges in introduced species management. Front Ecol Environ 17(2):117–125. https://doi.org/10.1002/fee.2019.17.issue-2, https://doi.org/10. 1002/fee.2000

Berger JC (1986) Wild horses of the Great Basin: social competition and population size. University of Chicago Press, Chicago

Boyce PN, Hennig JD, Brook RK, McLoughlin PD (2021) Causes and consequences of lags in basic and applied research into feral wildlife ecology: the case for feral horses. Basic Appl Ecol 53:154–163. https://doi.org/10.1016/j.baae.2021.03.011

Boyd CS, Davies KW, Collins GH (2017) Impacts of feral horse use on herbaceous riparian vegetation within a sagebrush steppe ecosystem. Rangel Ecol Manag 70:411–417. https://doi.org/10. 1016/j.rama.2017.02.001

Bureau of Land Management [BLM] (2015) 2015 Soda Fire emergency wild horse gather. https:// www.blm.gov/programs/wild-horse-and-burro/herd-management/gathers-and-removals/idaho/ 2015-soda-fire-wild-horse-gather

Bureau of Land Management [BLM] (2018) Report to congress: management options for a sustainable Wild Horse and Burro Program. https://www.blm.gov/sites/blm.gov/files/wildhorse_201 8ReporttoCongress.pdf

Bureau of Land Management [BLM] (2020b) Report to congress: an analysis of achieving a sustainable Wild Horse and Burro Program. https://www.blm.gov/sites/blm.gov/files/WHB-Report-2020-NewCover-051920-508.pdf

Buck CE, Bard E (2007) A calendar chronology for Pleistocene mammoth and horse extinction in North America based on Bayesian radiocarbon calibration. Quat Sci Rev 26:2031–2035. https:/ /doi.org/10.1016/j.quascirev.2007.06.013

Coates PS, O'Neil ST, Munoz DA, Dwight IA, Tull JC (2021) Sage-grouse population dynamics are adversely affected by overabundant feral horses. J Wildlife Manage 85(6):1132–1149

Cothran EG, van Dyk E, van der Merwe FJ (2001) Genetic variation in the feral horses of the Namib Desert, Namibia. J S Afr Vet Assoc 72(1):18–22. https://doi.org/10.4102/jsava.v72i1.603

Crane KK, Smith MA, Reynolds D (1997) Habitat selection patterns of feral horses in southcentral Wyoming. Rangel Ecol Manag 50:374–380

Danvir RE (2018) Multiple-use management of western U.S. rangelands: wild horses, wildlife, and livestock. Hum Wildl Interact 12:5–17. https://doi.org/10.26077/cz0b-6261

Davies KW, Boyd CS (2019) Ecological effects of free-roaming horses in North American rangelands. Biosci 69(7):558–565. https://doi.org/10.1093/biosci/biz060

Debaffe L, McLoughlin PD, Medill SA, Stewart K, Andres D, Shury T, Wagner B, Jenkins E, Gilleard JS, Poiss J (2016) Negative covariance between parasite load and body condition in a population of feral horses. Parasitology 143:983–997. https://doi.org/10.1017/S0031182016000408

Dore MHI (2005) Climate change and changes in global precipitation patterns: what do we know? Environ Int 31:1167–1181. https://doi.org/10.1016/j.envint.2005.03.004

Edouard N, Fleurance G, Dumont B, Baumont R, Duncan P (2009) Does sward height affect feeding patch choice and voluntary intake in horses? Appl Anim Behav Sci 119:219–228. https://doi.org/10.1016/j.applanim.2009.03.017

Eldridge DJ, Jing D, Travers S (2020) Feral horse activity reduces environmental quality in ecosystems globally. Biol Conserv 241:108367. https://doi.org/10.1016/j.biocon.2019.108367

Elizondo V, Fitzgerald T, Rucker RR (2016) You Can't Drag Them Away: an economic analysis of the wild horse and burro program. J Agric Res Econ 41:1–24.

Esmaeili S, Jesmer BR, Albeke SE, Aikens EO, Schoenecker KA, King SRB, Abrahms B, Buuveibaatar B, Beck JL, Boone JB, Cagnacci F, Chamaillé-Jammes S, Chimeddorj B, Cross PC, Dejid N, Enkhbyar J, Fischhoff IR, Ford AT, Hemami KJM, Hennig JD, Petra TYI, Kaczensky, Kauffman, MJ, Linnell JDB, Lkhagvasuren B, McEvoy JF, Melzheimer J, Merkle JA, Mueller T, Muntifering J, Mysterud A, Olson KA, Panzacchi M, Payne JC, Pedrotti L, Rauset GR, Rubenstein DI, Hall S, Scasta JD, Signer J, Songer M, Stabach JA, Stapleton S, Strand O, Sundaresan SR, Usukhjargal D, Uuganbayar G, Fryxell JM, Goheen JR (2021) Body size and digestive system shape resource selection by ungulates: A cross-taxa test of the forage maturation hypothesis. Ecol Lett 24(10):2178–2191. https://doi.org/10.1111/ele.v24.10, https://doi.org/10.1111/ele.13848

Fages A, Hanghøj K, Khan N, Gaunitz C, Seguin-Orlando A, Leonardi M (2019) Tracking five millennia of horse management with extensive ancient genome time series. Cell 177(6):1419-1435.e31. https://doi.org/10.1016/j.cell.2019.03.049

Fuller A, Mitchell D, Maloney SK, Hetem RS (2016) Towards a mechanistic understanding of the responses of large terrestrial mammals to heat and aridity associated with climate change. Climate Chang Respons 3:1–19. https://doi.org/10.1186/s40665-016-0024-1

Ganskopp D, Vavra M (1986) Habitat use by feral horses in the northern sagebrush steppe. J Range Manag 39:207–212. https://doi.org/10.2307/3899050

Garrott RA (2018) Wild horse demography: Implications for sustainable management within economic constraints. Human-Wildlife Interact 12(1):46–57

Garrott RA, Oli MK (2013) A critical crossroad for BLM's Wild Horse Program. Science 341:847–848. https://doi.org/10.1126/science.1240280

Gilchrist P, Sergeant ESG (2011) Risk of an equine influenza virus reservoir establishing in wild horses in New South Wales during the Australian epidemic. Aust Vet J 1:75–78. https://doi.org/10.1111/j.1751-0813.2011.00752.x

Girard TL, Bork EW, Nielsen SE, Alexander MJ (2013) Seasonal variation in habitat selection by free-ranging feral horses within Alberta's forest reserve. Rangel Ecol Manag 66(4):428–437. https://doi.org/10.2111/REM-D-12-00081.1

Gooch AMJ, Petersen SL, Collins GH, Smith TS, McMillan BR, Eggett DL (2017) The impact of feral horses on pronghorn behavior at water sources. J Arid Environ 138:38–43. https://doi.org/10.1016/j.jaridenv.2016.11.012

Griffin PC, Ekernas LS, Schoenecker KA, Lubow BC (2020) Standard operating procedures for wild horse and burro double-observer aerial surveys: U.S. Geological Survey Techniques and Methods, Book 2, Chap. A16

Groves CP (1994) Morphology, habitat and taxonomy. In: Boyd L, Houpt KA (eds) Przewalski's horse. The history and biology of an endangered species. State University of New York Press, Albany, pp 39–60

Hall LK, Larsen RT, Westover MD, Day CC, Knight RN, McMillan BR (2016) Influence of exotic horses on the use of water by communities of native wildlife in a semi-arid environment. J Arid Environ 127:100–105. https://doi.org/10.1016/j.jaridenv.2015.11.008

Hall LK, Larsen RT, Knight RN, McMillan BR (2018) Feral horses influence both spatial and temporal patterns of water use by native ungulates in a semi-arid environment. Ecosphere 9:e02096. https://doi.org/10.1002/ecs2.2096

Hampson BA, Ramsey G, Macintosh AMH, Mills PC, De Laat M, Pollitt CC (2010a) Morphometry and abnormalities of the feet of Kaimanawa feral horses in New Zealand. Aust Vet J 88:124–131. https://doi.org/10.1111/j.1751-0813.2010.00554.x

Hampson BA, de Laat MA, Mills PC, Pollitt CC (2010b) Distances travelled by feral horses in 'outback' Australia. Equine Vet J 42:582–586. https://doi.org/10.1111/j.2042-3306.2010.00203.x

Hampson BA, de Laat M, Mills PC, Walsh DM, Pollitt CC (2013) The feral horse foot. Part B: radiographic, gross visual and histopathological parameters of foot health in 100 Australian feral horses. Aust Vet J 91:23–30. https://doi.org/10.1111/avj.12017

Hanley TA (1982) The nutritional basis for food selection by ungulates. J Range Manag 35:146–151

Harvey AM, Meggiolaro MN, Hall E, Watts ET, Ramp D, Slapeta J (2019) Wild horse populations in south-east Australia have a high prevalence of *Stongylus vulgaris* and may act as a reservoir of infection for domestic horses. Int J Parasitol Parasites Wildl 8:156–163. https://doi.org/10.1016/j.ijppaw.2019.01.008

Hendrickson C (2018) Managing healthy wild horses and burros on healthy rangelands: tools and the tool box. Hum Wildl Interact 12, Article 15. https://doi.org/10.26077/tmnk-7f46

Hennig JD (2021) Feral horse movement, habitat selection, and effects on pronghorn and greater sage-grouse habitat in cold-arid-steppe. Dissertation, University of Wyoming

Hennig JD, Beck JL, Scasta JD (2018) Spatial ecology observations from feral horses equipped with global positioning system transmitters. Hum Wildl Interact 12:75–84. https://doi.org/10.26077/z9cn-4h37

Hennig JD, Schoenecker KA, Cain III JW, Roemer GW, Laake JL (2022) Accounting for residual heterogeneity in double-observer sightability models decreases bias in burro abundance estimates. J Wildl Manag 86: e22239. https://doi.org/10.1002/jwmg.22239

Hurlbert RC Jr (1993) Taxonomic evolution in North American Neogene horses (subfamily Equinae): the rise and fall of an adaptive radiation. Paleobiology 19:216–234. https://doi.org/10.1017/S0094837300015888

Hurwitt MC (2017) Freedom versus forage: balancing wild horses and livestock grazing on the public lands. Ida Law Rev 53:425

Irving BD (2001) The impacts of horse grazing on conifer regeneration in west-central alberta. PhD Dissertation, University of Alberta, Edmonton, Alberta, Canada

Janis C (1976) The evolutionary strategy of the Equidae and the origins of rumen and cecal digestion. Evolution 30:757–774. https://doi.org/10.2307/2407816

Jones A (2000) Effects of cattle grazing on North America arid ecosystems: a quantitative review. West N Am Nat 60:155–164. https://www.jstor.org/stable/41717026

Kaczensky P, Ganbaatar O, Von Wehrden H, Walzer C (2008) Resource selection by sympatric wild equids in the Mongolian Gobi. J Appl Ecol 45:1762–1769. https://doi.org/10.1111/j.1365-2664.2008.01565.x

Kane AJ (2018) A review of contemporary contraceptives and sterilization techniques for feral horses. Human-Wildlife Interact 12(1):111–116

Kelekna P (2009) The horse in human history. Cambridge University Press, Cambridge

King SRB (2002) Home range and habitat use of free-ranging Przewalski horses at Hustai National Park, Mongolia. Appl Anim Behav Sci 78:103–113. https://doi.org/10.1016/S0168-1591(02)00087-4

King SRB, Gurnell J (2005) Habitat use and spatial dynamics of takhi introduced to Hustai National Park, Mongolia. Biol Conserv 124:277–279. https://doi.org/10.1016/j.biocon.2005.01.034

King SRB, Schoenecker KA, Manier D (2019) Potential spread of cheatgrass (*Bromus tectorum*) and other invasive species by feral horses (*Equus ferus caballus*) in western Colorado. Rangel Ecol Manag 72:706–710. https://doi.org/10.1016/j.rama.2019.02.006

Leverkus SER, Fuhlendorf SD, Geertsma M, Allred BW, Gergory M, Bevington AR, Engle DM, Scasta JD (2018) Resource selection of free-ranging horses influenced by fire in northern Canada. Hum Wildl Interact 12:85–101. https://doi.org/10.26077/j5px-af63

Levine MA (1999) Botai and the origins of horse domestication. J Anthropol Archaeol 18:29–78. https://doi.org/10.1006/jaar.1998.0332

Linnartz L, Meissner R (2014) Rewilding horses in Europe. Background and guidelines—a living document. Publication by Rewilding Europe, Nijmegen, Netherlands

Loydi A, Zalba SM, Distel RA (2012) Viable seed banks under grazing and exclosure conditions in montane mesic grasslands of Argentina. Acta Oecol 43:8–15. https://doi.org/10.1016/j.actao.2012.05.002

Lubow BC, Ransom JI (2009) Validating aerial photographic mark-recapture for naturally marked feral horses. J Wildl Manag 73:1420–1429. https://doi.org/10.2193/2008-538

Lubow BC, Ransom JI (2016) Practical bias correction in aerial surveys of large mammals: validation of hybrid double-observer with sightability method against known abundance of feral horse (*Equus caballus*) populations. PlosOne 11:e0154902. https://doi.org/10.1371/journal.pone.0154902

McKnight TL (1958) The feral burro in the United States: distribution and problems. J Wildl Manag 22:163–179. https://doi.org/10.2307/3797325

Mejdell CM, Boe KE (2005) Responses to climatic variables of horses housed outdoors under Nordic winter conditions. Can J Anim Sci 85(3):307–308 https://doi.org/10.4141/A04-066

Menard C, Duncan P, Fleurance G, Georges J, Lila M (2002) Comparative foraging and nutrition of horses and cattle in European wetlands. J Appl Ecol 39:120–133. https://doi.org/10.1046/j.1365-2664.2002.00693.x

Mitchell P (2015) Horse nations. Oxford University Press, Oxford

Messmer T (2017) Call for Papers: Special Topic: Wild Horse and Burro Management. Hum–Wildl Interact 11(2):17. https://doi.org/10.26077/tr8k-xw31

Norris KA (2018) A review of contemporary U.S. wild horse and burro management policies relative to desired management outcomes. Hum Wildl Interact 12:18–30. https://doi.org/10.26077/p9b6-6375

Osterman-Kelm S, Atwill ER, Rubin ES, Jorgensen MC, Boyce WM (2008) Interactions between feral horses and desert bighorn sheep at water. J Mammal 89:459–466. https://doi.org/10.1644/07-MAMM-A-075R1.1

Outram AK, Stear NA, Bendrey R, Olsen S, Kasparov A, Zaibert V, Thorpe N, Evershed RP (2009) The earliest horse harnessing and milking. Science 323:1332–1335. https://doi.org/10.1126/science.1168594

Pihl TH, Nielsen MK, Olsen SN, Leifsson PS, Jacobsen S (2018) Nonstrangulating intestinal infarctions associated with Strongylus vulgaris: clinical presentation and treatment outcomes of 30 horses (2008–2016). Equine Vet J 50:474–480. https://doi.org/10.1111/evj.12779

Pokhrel Y, Felfelani F, Satoh Y, Boulange J, Burek P, Gädeke A et al (2021) Global terrestrial water storage and drought severity under climate change. Nat Clim Change 11:226–233. https://doi.org/10.1038/s41558-020-00972-w

Ransom JI, Kaczensky P (2016) Equus: an ancient genus surviving the modern world. In: Ransom JI, Kaczensky P (eds) Wild equids—ecology, management, and conservation. Johns Hopkins University Press, Baltimore

Roelle JE, Singer FJ, Zeigenfuss LC, Ransom JI, Coates-Markle L, Schoenecker KA (2010) Demography of the Pryor Mountain Wild Horses 1993–2007. USGS Scientific Investigations Report 2010-5125

Salter RE, Hudson RJ (1979) Feeding ecology of feral horses in western Alberta. J Range Manag 32:221–225. https://doi.org/10.2307/3897127

Salter RE, Hudson RJ (1982) Social organization of feral horses in western Canada. Appl Anim Ethol 8:207–223. https://doi.org/10.1016/0304-3762(82)90205-X

Scasta JD (2014) Dietary composition and conflicts of livestock and wildlife on rangeland. University of Wyoming Extension Bulletin B-1260

Scasta JD, Beck JL, Angwin CJ (2016) Meta-analysis of diet composition and potential conflict of wild horses with livestock and wild ungulates on western rangelands of North America. Rangel Ecol Manag 69:310–318. https://doi.org/10.1016/j.rama.2016.01.001

Scasta JD, Hennig JD, Beck JL (2018) Framing contemporary U.S. wild horse and burro management processes in a dynamic ecological, sociological, and political environment. Hum Wildl Interact 12:31–45. https://doi.org/10.26077/2fhw-fz24

Schoenecker KA, King SRB, Nordquist M, Deitich N, Kao Q (2016) Habitat selection and diet of equids. In: Ransom JI, Kaczensky P (eds) Wild equids—ecology, management, and conservation. Johns Hopkins University Press, Baltimore

Schoenecker KA, Doherty P, Hourt J, Romero J (2018) Testing infrared camera aerial surveys and distance sampling to estimate feral horse abundance in a known population. Wildl Soc Bull 42:452–459. https://doi.org/10.1002/wsb.912

Schoenecker KA, King SRB, Ekernas LS, Oyler-McCance SJ (2021a) Using fecal DNA and closed-capture models to estimate feral horse population size. J Wildl Manag 85. https://doi.org/10.1002/jwmg.22056

Schoenecker KA, King SRB, Messmer TA (2021) The Wildlife profession's duty in achieving science-based sustainable management of free-roaming equids. J Wildl Manag 85(6):1057–1061. https://doi.org/10.1002/jwmg.v85.6, https://doi.org/10.1002/jwmg.22091

Schoenecker KA, King SRB, Esmaeili S (2023) Seasonal resource selection and movement ecology of free-ranging horses in the western USA. J Wildl Manag 87. https://doi.org/10.1002/jwmg.22341

Scorolli AL (2018) Feral horse management in Parque Provincial Ernesto Tornquist, Argentina. Hum Wildl Interact 12:102–111. https://doi.org/10.26077/xpbm-6825

Smith MA (1986) Impacts of feral horses grazing on rangelands: an overview. J Equine Vet Sci 6:236–238. https://doi.org/10.1016/S0737-0806(86)80047-8

Stoppelaire GH, Gillespie TW, Brock JC, Tobin GA (2004) Use of remote sensing techniques to determine the effects of grazing on vegetation cover and dune elevation at assateague island national seashore: impact of horses. Environ Manage 34:642–649

Tietjen B, Jeltsch F (2007) Semi-arid grazing systems and climate change: a survey of present modelling potential and future needs. J Appl Ecol 44(2):425–434. https://doi.org/10.1111/jpe.2007.44.issue-2, https://doi.org/10.1111/j.1365-2664.2007.01280.x

van Beest FM, Uzal A, Wal EV, Laforge MP, Contasti AL, Colville D, McLoughlin PD (2014) Increasing density leads to generalization in both coarse-grained habitat selection and fine-grained resource selection in a large mammal. J Anim Ecol 83:147–156. https://doi.org/10.1111/1365-2656.12115

Van Soest PJ (1994) Nutritional ecology of the ruminant, 2nd edn. Cornell University Press, Ithaca

Wallace ZP, Nielson RM, Stahlecker DW, DiDonato GT, Ruehmann MB, Cole J (2021) An abundance estimate of free-roaming horses on the Navajo Nation. Rangeland Ecol Manag 74 100–109. https://doi.org/10.1016/j.rama.2020.10.003

Wildlife Society (2021) Final position statement on invasive and feral species. https://wildlife.org/wp-content/uploads/2014/05/PS_InvasiveFeralSpecies2.pdf

Zeigenfuss LC, Schoenecker KA, Ransom JI, Ignizio DA, Mask T (2014) Influence of nonnative and native ungulate biomass and seasonal precipitation on vegetation production in a Great Basin ecosystem. West N Am Nat 74:286–298 https://doi.org/10.3398/064.074.0304

Chapter 22
Bighorn Sheep and Mountain Goats

Jericho C. Whiting, Vernon C. Bleich, R. Terry Bowyer, Kezia Manlove, and Kevin White

Abstract Bighorn sheep (*Ovis canadensis*), and to a lesser extent mountain goats (*Oreamanos americanus*), historically occupied much of the mountainous rangelands of western North America. Both ungulates inhabit rugged terrain and feed on grasses, forbs, and browse. Bighorn sheep and mountain goats are widely recognized for their consumptive and non-consumptive value. Indigenous peoples valued these species for cultural and subsistence purposes. Populations of these ungulates have declined since the latter part of the nineteenth century—for mountain goats, this decline has occurred particularly in the southern portion of their distribution. Historical declines have been attributed to unregulated harvest, habitat loss, competition with non-native ungulates, and disease contracted from domestic livestock. Regulated hunting has played an important role in the conservation of bighorn sheep, and recent reintroductions of these ungulates have bolstered current populations in rangelands of western North America. Although competition for habitat is minimal for bighorn sheep and mountain goats with domestic livestock (compared with other wild ruminants or feral equids), diseases of domestic sheep and domestic or exotic

J. C. Whiting (✉)
Department of Biology, Brigham Young University-Idaho, 241 Benson Building, Rexburg, ID 83460, USA
e-mail: whitingj@byui.edu

V. C. Bleich
Department of Natural Resources and Environmental Science, University of Nevada Reno, 1660 N. Virginia St., Mail Stop 186, Reno, NV 89557, USA

R. T. Bowyer
Institute of Arctic Biology, University of Alaska Fairbanks, Fairbanks, AK 99775, USA
e-mail: rbowyer@isu.edu

K. Manlove
Department of Wildland Resources and Ecology Center, Utah State University, 5200 Old Main Hill, Logan, UT 84322, USA
e-mail: kezia.manlove@usu.edu

K. White
Department of Natural Sciences, University of Alaska Southeast, 11066 Auke Lake Way, Juneau, AK 99801, USA
e-mail: kevin.white@alaska.gov

© The Author(s) 2023   759
L. B. McNew et al. (eds.), *Rangeland Wildlife Ecology and Conservation*,
https://doi.org/10.1007/978-3-031-34037-6_22

goats have long posed challenges to the conservation of bighorn sheep. In parts of their distributions, mountain goats and bighorn sheep are sympatric, and both species may encounter domestic livestock on grazing allotments on public or private rangelands. If management of bighorn sheep and mountain goats is the goal, spatial and temporal separation is recommended between these species and domestic sheep and goats; doing so will improve the conservation of populations of bighorn sheep and mountain goats and their habitat on rangelands of western North America.

Keywords Grazing · Mountain sheep · *Oreamnos americanus* · *Ovis canadensis* · Rangelands

22.1 Introduction

Bighorn sheep (*Ovis canadensis*) and mountain goats (*Oreamnos americanus*) are herbivores in the family Bovidae (Feldhamer et al. 2020). Other mountain ungulates such as Dall's (*O. dalli*) and Stone's sheep (*O. d. stonei*), and mountain caribou (*Rangifer tarandus*) are not considered in this chapter, because they seldom occur on western rangelands. Ancestors of North American mountain sheep arose in Asia about 2.5 million years ago during the Villafranchian (Geist 1971; Valdez and Krausman 1999) and dispersed to North America via the Bering Land Bridge (Cowan 1940; Péwé and Hopkins 1967). The systematics and taxonomy of bighorn sheep are complex, but three clades currently are recognized: Sierra Nevada bighorn sheep (*O. c. sierrae*), desert bighorn sheep (*O. c. nelsoni, O. c. mexicana*), and Rocky Mountain bighorn sheep (*O. c. canadensis*) (Buchalski et al. 2016). Sierra Nevada and Rocky Mountain bighorn sheep diverged from desert bighorn sheep prior to or during the Illinoian glaciation ~ 315,000–94,000 years ago (Buchalski et al. 2016). By the Wisconsin glaciation (~ 40,000–23,000 years ago), fossils of *Ovis* were common (Guthrie 1968). Ancestors of mountain goats are also believed to have colonized western North America from Asia via the Bering Land Bridge during the Wisconsin glaciation (Rideout and Hoffman 1975). During the last glacial maximum, mountain goats were separated into northern, southern, and coastal refugial subpopulations (Nagorsen and Keddie 2000; Shafer et al. 2011b). Unlike bighorn sheep, subspecies have not been designated for mountain goats.

Bighorn sheep and mountain goats historically occupied suitable habitat across much of western North America; however, populations of these ungulates have declined since the latter part of the nineteenth century (Buechner 1960; Geist 1971). A downward trend in numbers of both species likely began with Euro-American settlement of western North America, and much attention has focused on unregulated market hunting, habitat loss or modification, and diseases contracted from domestic livestock as causes of that decline (Buechner 1960; Smith et al. 1991; Singer et al. 2000); some of these concerns remain. Primary challenges to conserving North American wild sheep on a continent-wide basis are maintaining habitat quality, reducing habitat loss, and managing disease (Krausman 2000; Bleich 2009b; Krausman and

Bleich 2013). In this chapter, we discuss these mountain ungulates in areas where they overlap rangelands of western North America.

22.2 General Life History and Population Dynamics

22.2.1 *Bighorn Sheep*

Bighorn sheep are sexually dimorphic in size with males larger than females (Fig. 22.1; Weckerly 1998; Loison et al. 1999). Weight of adult male bighorn sheep from northern regions averages ~ 102 kg and adult females ~ 72 kg (Geist 1971; Festa-Bianchet et al. 1997; Shackleton et al. 1999; Krausman and Bowyer 2003), whereas desert-dwelling bighorn sheep are smaller (adult males = ~ 70 kg; adult females = ~ 48 kg) in size (Bleich et al. 1997; Krausman et al. 1999). Adult male bighorn sheep have large, curled horns used for ramming, head-to-head clashes, and for display to intimidate rivals, whereas horns of females are much smaller and not as strongly curled as those of males (Fig. 22.1; Geist 1971; Shackleton et al. 1999; Coltman et al. 2003). Bighorn sheep also possess conspicuous rump patches, which are thought to be used as an alarm signal and function primarily to promote group cohesion among conspecifics or as a signal to elicit predator evasion behavior within social groups (Hirth and McCullough 1977; Caro 2005). Additionally, bighorn sheep possess small litters with large-bodied precocial young—i.e., active and able to move independently shortly after birth (Fig. 22.1; Festa-Bianchet 1988b), are long-lived with long generation times, provide high maternal investment in young, and exhibit a low intrinsic rate of population increase (Festa-Bianchet 1988a; Shackleton et al. 1999; Gaillard et al. 2000). This suite of attributes responds strongly in a density-dependent manner, wherein reproduction and survival are negatively associated with population density in relation to the ecological carrying capacity of the environment (K; the number of individuals a particular area can support); as the population approaches K, reproduction and recruitment decline (Swenson 1985; Festa-Bianchet and Jorgenson 1998; Bowyer et al. 2014).

The sexes of bighorn sheep spatially segregate from one another for a portion of the year, thus using different areas in mountainous rangelands (Bleich et al. 1997; Bowyer 2004; Whiting et al. 2010a). Indeed, sexes of desert-dwelling bighorn sheep may segregate into mountain ranges separated by ~ 15 km to balance the needs for crucial resources against risk of predation (Bleich et al. 1997). Young typically are born in the spring while the sexes are segregated (Fig. 22.1; Whiting et al. 2011b, 2012); desert-dwelling bighorn sheep, however, have a protracted birthing period (Bleich et al. 1997; Rubin et al. 2000). Females allocate substantial maternal care to their single young, which they birth and rear in precipitous terrain (Geist 1971; Festa-Bianchet 1988c; Bleich et al. 1997) that contains fewer predators than areas occupied by males during sexual segregation (Bleich et al. 1997). Females also may defend young by attacking predators, especially coyotes (*Canis latrans*; Berger

Fig. 22.1 Two adult male Rocky Mountain bighorn sheep (top), and an adult female Rocky Mountain bighorn sheep and lamb (bottom) during spring, Utah

1978b; Bleich 1999). The sexes of bighorn sheep follow differing strategies for lowering the risk of predation—males increase group size and females move closer to escape terrain (e.g., steep slopes, cliffs, and rock outcroppings) to lower predation risk (Bleich et al. 1997; Bowyer 2004; Schroeder et al. 2010). Tradeoffs between acquiring essential resources and avoiding predation are well-documented for bighorn sheep; these ungulates, especially females, may forego areas of high-quality forage to avoid predators (Festa-Bianchet 1988d; Berger 1991; Bleich et al. 1997). Mountain lions (*Puma concolor*) also are an important predator and can have substantial effects on survival and population growth in small populations of bighorn sheep (Ross et al. 1997; Johnson et al. 2013; Rominger 2018).

Male and female bighorn sheep exhibit important differences in the morphology and physiology of their digestive tracts that lead to males having larger rumens than females, and, as a result, are better adapted to digesting less-nutritious forages. Females, with smaller rumens than males, require high-quality forages necessary to support the high costs of late gestation and lactation; such differences foster sexual segregation (Barboza and Bowyer 2000, 2001). These differences and many other life-history characteristics of bighorn sheep are associated with their population ecology.

22.2.2 Mountain Goats

Mountain goats are sexually dimorphic in size with males larger than females. Adult male mountain goats weigh 90–181 kg and adult females weigh 59–111 kg (Côté and Festa-Bianchet 2003). Mountain goats exhibit specialized morphological and behavioral adaptations that enable them to inhabit steep and rugged environments characterized by severe climatic conditions (Fig. 22.2). For example, soft padded hooves surrounded by a hard keratinous sheath combined with a vertically oriented narrow body and muscular shoulders enable athletic and sure-footed locomotion in rugged, cliffy terrain—habitat that is preferentially used to reduce the risk of predation (Festa-Bianchet and Côté 2008). Like bighorn sheep, the population biology of mountain goats is linked to the seasonal availability of nutritional resources, and this species also exhibits sexual segregation (Festa-Bianchet and Côté 2008). For example, mountain goat parturition occurs during late May and early June and coincides with green-up of highly nutritious forage (Pettorelli et al. 2007; Festa-Bianchet and Côté 2008). During the summer growing season mountain goats accumulate fat and protein reserves needed to survive long winters characterized by severe nutritional deficiency. Thus, summer and winter weather can play an important role in mediating nutritional condition and can exert strong effects on individual growth, reproduction, and survival (Pettorelli et al. 2007; Festa-Bianchet and Côté 2008; White et al. 2011). Consequently, these specialized ungulates exhibit a slow life-history strategy with late age of maturity (age at first reproduction = 4–5 years, body mass asymptote = 4–6 years) and may not reproduce annually to mitigate the effects of reproductive costs on probability of survival (Festa-Bianchet and Côté 2008); such relationships can be associated with density-dependent processes (Houston and Stevens 1988; Bowyer et al. 2014). Consequently, mountain goat populations have low growth rates—i.e., 1–4% (Hamel et al. 2006; Rice and Gay 2010; White et al. 2021a) and are sensitive to weather conditions, especially in northern coastal environments that can be prone to episodic, severe snowfall (White et al. 2011).

Mountain goats are vulnerable to predation by large carnivores, such as wolves (*C. lupus*) and brown (grizzly, *Ursus arctos*) or black bears (*U. americanus*; Fox and Streveler 1986; Festa-Bianchet and Côté 2008), but the specialized adaptations of

Fig. 22.2 An adult female mountain goat and kid during late-winter (top) and an adult male mountain goat on low-elevation winter range (bottom), Alaska

these ungulates for using rugged mountain terrain mitigate predation-risk. Nonetheless, inhabiting rugged terrain can involve nutritional costs leading to trade-offs between safety and acquisition of forage resources (Hamel and Côté 2007). The presence of large carnivores also can elicit indirect effects including increases in endocrine stress responses that can negatively influence reproduction (Dulude-de Broin et al. 2020). Life in extreme environments can also lead to increased sensitivity to stochastic factors, with events such as avalanches as an important source of mortality in some areas of coastal Alaska (White et al. 2011).

22.3 Population Status

22.3.1 Bighorn Sheep

Bighorn sheep are associated with habitats as diverse as the frigid and wind-swept ridges in the alpine regions of the highest mountains in North America to hot, arid areas below sea level in some inland desert basins (Fig. 22.3). Historically, the distribution of bighorn sheep extended eastward from British Columbia (Cowan 1940; Buechner 1960) to the badlands of North Dakota and South Dakota and southward into Mexico (Krausman et al. 1999). The distribution of ~ 48,000 Rocky Mountain bighorn sheep closely follows the Rocky Mountains from northern British Columbia southward to northern New Mexico (Krausman and Bowyer 2003). The ~ 39,000 desert bighorn sheep occupy habitat across much of the Great Basin, Mojave, Sonoran, and Chihuahuan deserts. Sierra Nevada bighorn sheep have a restricted distribution and are endemic to the Sierra Nevada of eastern California (Wehausen and Ramey 2000). Bighorn sheep occupying the peninsular ranges of southern California are considered a distinct population segment that is listed as endangered by the federal government; Sierra Nevada bighorn sheep are recognized as a valid subspecies of bighorn sheep and also are listed as endangered by the federal government (USFWS 2000, 2007).

22.3.2 Mountain Goats

The current distribution of mountain goats (80,000–120,000 individuals) extends eastward from coastal Alaska to the Rocky Mountains and south from Alaska, Yukon and the Northwest Territories to Montana, Idaho, and Washington, and includes a northernmost and geographically isolated population of native mountain goats in the Mackenzie Mountains of Yukon and Northwest Territories (Fig. 22.3; Festa-Bianchet and Côté 2008). In coastal Alaska and British Columbia, mountain goat populations almost exclusively occur on mainland mountain ranges; but apparently native populations historically and currently occur on some islands (Shafer et al.

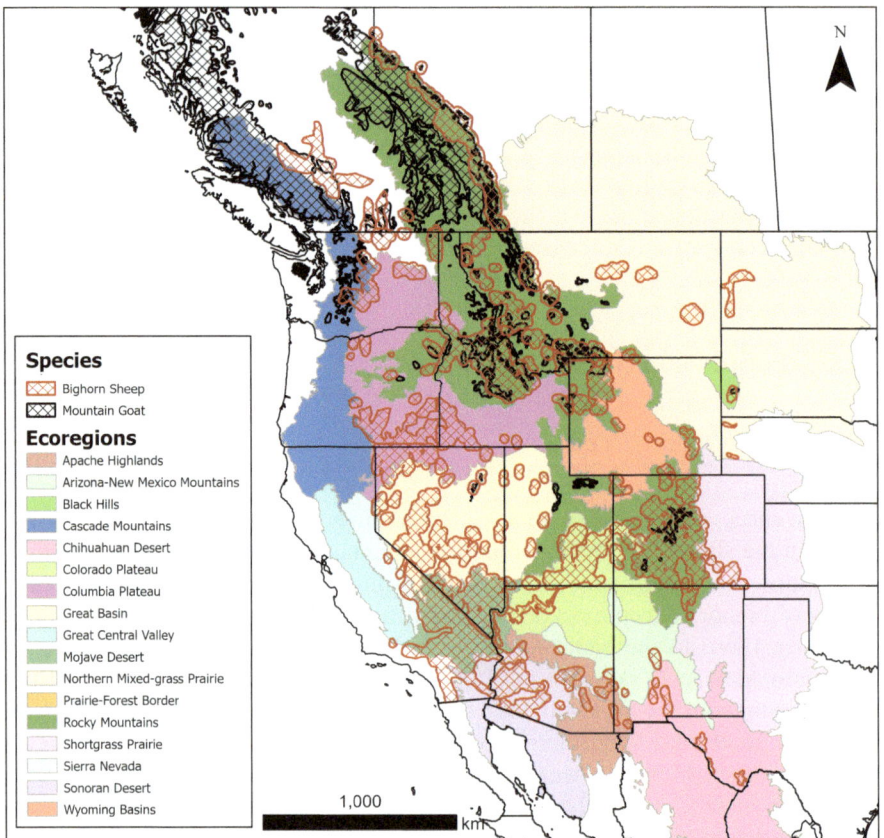

Fig. 22.3 Distributions of bighorn sheep and mountain goats overlain on rangeland ecoregions in the western USA and Canada (Map credit: M. Solomon)

2011a, b). Mountain goats have been successfully introduced into non-native ranges in the western USA (Montana, Oregon, Colorado, Utah, Nevada, South Dakota, Wyoming, and Washington–Olympic Peninsula) as well as into several non-native ranges in Alaska (Kodiak Island, Revillagigedo Island).

22.3.3 Population Monitoring

Two common methods for estimating population abundance of bighorn sheep are aerial surveys (Bleich et al. 1990a; Stockwell et al. 1991; Bates et al. 2021) and resight surveys performed from the ground (McClintock and White 2007; Johnson et al. 2010; Taylor et al. 2020). Helicopter surveys have been used increasingly during the past 20 years to monitor populations of bighorn sheep (Krausman and Hervert

1983; Bleich et al. 1994; McClintock and White 2007). Additionally, photographs of collared bighorn sheep from motion-sensor cameras set at water sources can be used to estimate population abundance (Perry et al. 2010; Taylor et al. 2020, 2022). Mark-recapture methods based on collection of fecal DNA also have been used to estimate population abundance (Schoenecker et al. 2015). Reproduction in bighorn sheep can be estimated by visual observation during the birthing period (Festa-Bianchet et al. 2000; Whiting et al. 2010b, 2011b). Survival often is quantified from animals with radio collars and by using mark-resight or known-fate analyses (Neal et al. 1993; Shannon et al. 2014).

Mountain goats are challenging to monitor because of the rugged and often remote environments they inhabit. Size and composition of mountain goat populations are often estimated using aerial survey techniques (fixed- and rotor-wing aircraft) and have involved use of mark-resight, sightability, and distance-sampling models to derive estimates (Poole 2007; Rice et al. 2009; Schmidt et al. 2019), but uncorrected minimum counts also have been used (McDonough and Selinger 2008). In highly accessible areas, ground-based methods involving direct observation or genetic mark-recapture (i.e., fecal DNA analyses) have been used to derive population estimates (Gonzalez-Voyer et al. 2001; Poole et al. 2011; Belt and Krausman 2012). Survival and reproduction typically are estimated using mark-resight or known-fates analyses involving marked animals (Smith 1986; Festa-Bianchet and Côté 2008; White et al. 2011, 2021b).

22.4 Habitat Associations

22.4.1 Bighorn Sheep

Bighorn sheep are well known for their dependence on steep, rugged terrain of variable elevations and ecoregions in western rangelands, whether in mountains or major river canyons, and adjacent foothills, all of which are generally characterized by sparse vegetation (Krausman and Bowyer 2003). Often bighorn sheep use habitat that is characterized by slopes > 20%, within 1000 m of escape terrain, and in areas of limited vegetational cover (Smith et al. 1991; Bleich et al. 1997; Andrew et al. 1999; Robinson et al. 2020; Lowrey et al. 2021). Bighorn sheep select the most appropriate terrain available in a particular area, and managers view scores derived from habitat models in a relative, rather than in an absolute, context (Andrew et al. 1999). These ungulates rely heavily on their visual acuity and open terrain to detect predators (Geist 1971; Risenhoover and Bailey 1985), and typically occupy areas in which they are well-adapted to detect and evade, or less apt to encounter, predators (Berger 1978a; Bleich et al. 1997). Hence, the distribution of bighorn sheep is restricted largely to mountains, canyons, and river corridors across the western portion of North America (Krausman et al. 1999; Krausman and Bowyer 2003). Migration to and from seasonal ranges is important for this species (Geist 1971; Jesmer et al. 2018; Spitz et al. 2020).

Depleted abundance and distribution compared with pre-European settlement and close association with steep, rugged, and sparsely vegetated areas has resulted in bighorn sheep having a naturally fragmented distribution across mountainous and canyon areas of western North America (Schwartz et al. 1986; Bleich et al. 1990b). As a result, bighorn sheep populations are typically small (e.g., 30 animals) but may number up to several hundred or more individuals occurring in remote and spatially isolated areas (Berger 1990; Epps et al. 2005; Donovan et al. 2020). Metapopulations are the primary foundation for habitat management and conservation of bighorn sheep (Bleich et al. 1990b, 1996; DeCesare and Pletscher 2006). A metapopulation is defined as the total population in a geographic area that is comprised of smaller subpopulations that are interconnected genetically and demographically by periodic movements of individual bighorn sheep (DeCesare and Pletscher 2006; Malaney et al. 2015; Epps et al. 2018). The subpopulations that comprise a metapopulation are expected to exhibit population dynamics independent of each other, and local extinctions are expected to occur; these are offset by colonization events involving individuals that move among isolated habitats, whether occupied or not, within the metapopulation. Thus, the viability of a bighorn sheep metapopulation depends upon the persistence of the subpopulations of which it is comprised (Bleich et al. 1996; DeCesare and Pletscher 2006), and colonization events must occur more frequently than extinction events.

Bighorn sheep diets are dominated by grasses and sedges; however, these ungulates exhibit seasonal variation in diet composition including browse (Fig. 22.4; Bleich et al. 1997; Krausman et al. 1999; Shackleton et al. 1999). In spring and summer, bighorn sheep eat mostly forbs, sedges, and grasses (Wikeem and Pitt 1992; Krausman et al. 1999; Shackleton et al. 1999). During winter, consumption of shrubs and senescent grasses also occurs (Singer and Norland 1994; Shackleton et al. 1999). Desert bighorn sheep also forage on prickly pear (*Opuntia* spp.) and other cactus species (*Mammillaria* spp. and *Ferocactus* spp.). Also, differential use of forage occurs between male and female bighorn sheep, especially when the sexes are segregated, with males consuming more graminoids (Bleich et al. 1997). Bighorn sheep may consume soil during spring and summer to acquire sodium, calcium, magnesium, and other minerals (Holl and Bleich 1987; Krausman et al. 1999). Among trace minerals, selenium may be especially important, given its fundamental role in virtually all physiological processes and because it varies widely in abundance across geographic areas (Flueck et al. 2012; Bleich et al. 2017). Additionally, water sources (artificial and natural) are important features in areas occupied by bighorn sheep (Fig. 22.4; Bleich et al. 2006; Whiting et al. 2009, 2011a), but the development of artificial water sources is a contentious issue for bighorn sheep management on rangelands of the western USA (Rosenstock et al. 1999, 2001; Bleich 2009a). Much of the opposition to provision water sources has its origin in the 1964 Wilderness Act, which opponents of water developments invariably invoke to prevent development of this essential resource because it 'degrades' legislated wilderness (Bleich 2005, 2016). Ironically, grazing and water developments for domestic livestock in wilderness areas are acceptable, and bighorn sheep and many other species of wildlife are dependent on those surface waters. Water development specifically to benefit bighorn sheep, however, has been opposed at virtually every opportunity (Bleich 2009a, 2016), in large part because of the failure of wilderness legislation to have been based more on ecological values than on less tangible benefits (Bleich 2016).

Fig. 22.4 Bighorn sheep foraging on low-elevation shrub and grass winter range (top), and bighorn sheep waiting to access a small, natural water seep (underneath the large rock at the right) in Utah (bottom)

22.4.2   Mountain Goats

Mountain goats exhibit strong selection for steep, rugged habitats proximal to escape terrain (i.e., slopes > 40–50 degrees), provided adequate forage resources are available; mountain goats uncommonly use habitats greater than 400 m from escape terrain (Festa-Bianchet and Côté 2008; Shafer et al. 2012; White and Gregovich 2017; Lowrey et al. 2018). This strategy is well-suited to minimize risk of predation from wolves and brown or black bears (Fox and Streveler 1986; Sarmento and Berger 2020). Mountain goats exhibit ecotypic variation in seasonal migratory behavior and habitat selection (Hebert and Turnbull 1977). For example, in the north Pacific coastal regions, mountain goats generally migrate from alpine summer ranges to low-elevation, forested winter ranges because of the wet, heavy snowpack that occurs at high elevations within this region (Shafer et al. 2012; White and Gregovich 2017). In drier and colder interior regions, however, mountain goat seasonal movements are limited, and animals tend to winter in high-elevation alpine habitats and use wind-blown ridges with exposed vegetation or tree-line habitats (Festa-Bianchet and Côté 2008; Poole et al. 2009; Richard and Côté 2016). In interior regions, mountain goats often are sympatric with bighorn sheep, and can exhibit substantial niche overlap (DeVoe et al. 2015; Lowrey et al. 2018).

Mountain goats consume a wide variety of forage types (Côté and Festa-Bianchet 2003) but exhibit distinct seasonal variation in diet composition (Saunders 1955). Following green-up, mountain goats commonly consume forbs, sedges, and grasses in alpine summer ranges. During winter, however, shrubs, lichen litterfall, and even conifer needles are consumed when other lower-growing forages are buried under snow. In some interior ranges, senesced grasses and sedges also can be used on wind-blown alpine slopes. During spring and summer, mineral licks represent an important resource for mountain goats in interior mountain ranges (Hebert and Cowan 1971; Singer 1978; Ayotte et al. 2008; Poole et al. 2010); use of mineral licks is rarer in more nutritionally productive coastal areas. Use of mineral licks is primarily driven by the need to acquire sodium, although other nutrients (i.e., selenium, calcium, and magnesium) also may be important (Hebert and Cowan 1971; Ayotte et al. 2006). Because mineral licks are uncommon on the landscape, mountain goats may undertake substantial seasonal movements through atypical habitats to access these critical nutritional resources (Rice 2010).

22.5   Interaction with Livestock

22.5.1   Bighorn Sheep

Competition for forage and spatial interactions can occur seasonally between livestock and bighorn sheep (Chap. 4). Bighorn sheep and cattle generally eat grass-dominated diets, and dietary overlap can be high, especially during drought or other times of reduced forage abundance (Coughenour 1991; Bailey 2004; Chaikina and Ruckstuhl 2006; Garrison et al. 2016). Also, spatial competition between livestock and bighorn sheep can occur (Risenhoover et al. 1988). Bite rates of forage can decrease, and vigilance rates can increase for bighorn sheep when cattle are near; also vigilance rates were higher for females than for males with cattle nearby (Brown et al. 2010). Bighorn sheep avoided cattle and decreased use of areas when cattle were in proximity (Bissonette and Steinkamp 1996). Grazing of domestic cattle was negatively correlated with rate of population increase for translocated populations of bighorn sheep (Singer et al. 2000). Also, sharing of ranges by domestic cattle and bighorn sheep ostensibly has led to mountain lions switching from bighorn sheep to livestock predation (Rominger 2018).

22.5.2 Mountain Goats

Interactions between livestock and mountain goats can occur in high-elevation alpine meadows and associated habitats, but most mountain goat habitat is unsuitable for livestock grazing because of its rugged terrain. Most potential for co-occurrence is limited to the southern latitudes of mountain goat range—predominately where mountain goats were introduced. For example, non-native mountain goats in the East Humboldt Mountains, Nevada, may contact domestic livestock on public grazing allotments or on private lands (Wolff et al. 2019).

22.6 Effects of Disease

Risk of pathogen spillover is a major force shaping rangeland dynamics and management of bighorn sheep and mountain goats. Pathogen spillover is a concern between livestock and bighorn sheep, between livestock and mountain goats, between populations of bighorn sheep, between populations of mountain goats, and between populations of bighorn sheep and mountain goats.

Bighorn sheep and mountain goats are vulnerable to a suite of pathogens, including contagious ecythma (Samuel et al. 1975; Tryland et al. 2018), Johne's disease (Williams et al. 1979), bovine viral diarrhea (Wolff et al. 2016), and a variety of helminths and ectoparasites. Epizootic hemorrhagic disease and bluetongue also pose threats to bighorn sheep, though their effects on mountain goats are likely more limited (Ruder et al. 2015). Infectious pneumonia often associated with the bacterial pathogen *Mycoplasma ovipneumoniae* (Besser et al. 2008; Cassirer et al. 2018) can result in extensive, all-age mortality, and place serious constraints on bighorn sheep population growth (Besser et al. 2012), and the same pathogen also may be problematic for mountain goats (Blanchong et al. 2018) and thinhorn sheep (Black et al. 1988). Although *M. ovipneumoniae* is not detected universally in bighorn sheep disease events, and other bacteria can produce sporadic acute pneumonia—e.g., leukotoxin-positive *Pasteurellas* (Shanthalingam et al. 2014)—*M. ovipneumoniae* appears to be a common player in the preponderance of well-documented disease events.

The distribution and demographic structure of bighorn sheep and mountain goat populations has important implications for disease transmission and risk. A disease outbreak in one population may not spread rapidly to nearby populations, despite proximity (Flesch et al. 2020), a somewhat atypical scenario compared with other ungulate species that exhibit more complete mixing patterns. Yet, because of the gregarious nature of bighorn sheep and mountain goats, particularly within female-offspring nursery groups, within-population rates of pathogen transmission can be high. Infected bighorn herds can also pose transmission risks to healthy neighboring herds, emphasizing the fundamental need for separation of bighorn sheep from domestic sheep and from infected bighorn and mountain goat herds as a core component of species conservation.

22.6.1 Bighorn Sheep

M. ovipneumoniae can be carried at high prevalence (Manlove et al. 2019) and genotypic diversities (Kamath et al. 2019) in large flocks of domestic sheep; accordingly, domestic sheep pose serious disease-mediated risks to bighorn sheep. This pathogen is not particularly troublesome in domestic sheep (Besser et al. 2019; Manlove et al. 2019), but it can persist and cause damage to bighorn herds for many years following exposure (Cassirer et al. 2018).

M. ovipneumoniae is primarily transmitted through respiratory droplets. When the bacteria encounter a new host, the pathogen takes up residence in the upper respiratory tract of the host, where it can proliferate and impede motion of the host's cilia. This allows a diverse suite of bacteria that are commensal in the upper respiratory tract to gain access to the lower respiratory tract where they can become pathogenic (Besser et al. 2008). The acute phase of an *M. ovipneumoniae* infection is characterized by symptoms like coughing, which likely facilitates pathogen spread. Animals either resolve their lower respiratory tract infections or succumb to disease. Spillover events vary in their severity—documented die-offs range from 10 to 90% of the infected herd (Cassirer et al. 2018) and have occurred regularly for as long as detailed records exist (Marsh 1938; Buechner 1960).

A small subset of chronic-carrier hosts can continue to harbor *M. ovipneumoniae* in their nostrils even after acute respiratory symptoms decline (Plowright et al. 2017). Chronically infected animals appear to be less apt to transmit the pathogen, and intense contact may be required to generate new infections in previously unexposed conspecifics. Chronically infected females, however, are thought to transmit *M. ovipneumoniae* to susceptible offspring, which then develop acute infections and effectively transmit the pathogen to other susceptible young in their nursery groups. In this way, a small number of chronically infected individuals can affect recruitment for the entire herd. Chronic infection may be facilitated by paranasal sinus tumors that have recently been detected in multiple bighorn herds and have been associated with the presence of *M. ovipneumoniae* and *P. multocida* (Fox et al. 2011, 2015, 2016).

Some habitat manipulations could limit the risk of contact between host animals, but designing appropriate manipulations requires a strong understanding of factors that motivate bighorn sheep movements, and, in particular, forays—i.e., short-term movements of animals that begin and end within an established home range (Singer et al. 2001; Carpenter et al. 2014). Both sexes go on forays, though the distances and frequencies vary by sex. Some herds exhibit higher rates of such movements than others (Singer et al. 2001), and there are many hypotheses regarding the factors that encourage these events (Lassis et al. 2022). Commonly postulated drivers are herd density and sex ratio, habitat structure and viewshed, location of attractive resources (e.g., mineral licks, water, other bighorn sheep, mountain goats, domestic sheep, or domestic goats), rut, and individual age. Which factors are most important in particular contexts remain an open question that if addressed, may help alleviate some of the conflict associated with pathogen transmission to bighorn sheep occupying North American rangelands.

Understanding movements and migrations of bighorn sheep is critically important, as is the proximity of release areas for translocated bighorn sheep to other bighorn sheep, mountain goats, and domestic sheep or goat grazing allotments (Clifford et al. 2009; Shannon et al. 2014). Also, consideration should be given to the presence of hobby farms and trailing operations of domestic sheep and goats in locations adjacent to areas occupied by bighorn sheep (Shannon et al. 2014). If conservation of bighorn populations is the goal, spatial and temporal separation of bighorn and domestic sheep should occur wherever possible (Schommer and Woolever 2008; Wehausen et al. 2011; Besser et al. 2013).

22.6.2 Mountain Goats

Current knowledge of mountain goat disease risk and parasitology is limited when compared with bighorn sheep. Among the most documented diseases reported in mountain goats is contagious ecthyma, a viral disease that causes lesions to eyes, nose and mouth that can be severely debilitating, sometimes leading to death (Samuel et al. 1975; Tryland et al. 2018). While *M. ovipneumoniae* has been documented in mountain goats (Lowrey et al. 2018; Wolff et al. 2019), extreme mortality events commensurate with those observed in bighorn sheep have not been reported. Nonetheless, recent studies of sympatric mountain goat and bighorn sheep populations in Nevada documented extensive *M. ovipneumoniae* related mortality among mountain goat young leading to significant reductions in population recruitment (Blanchong et al. 2018; Wolff et al. 2019). Whether adult mountain goats are similarly vulnerable and the extent to which they are capable of being sources of disease for bighorn sheep populations is unclear. Again, if management of mountain goat populations is the goal, spatial and temporal separation of these ungulates and domestic sheep should occur. Considering the propensity of mountain goats to occupy steep and rugged habitats, reducing livestock interactions with mountain goats may be easier to accommodate than with bighorn sheep (Bailey et al. 2001).

22.7 Ecosystem Threats

22.7.1 Bighorn Sheep

Wild asses (*Equus asinus*), wild horses (*E. caballus*), introduced mountain goats, and introduced aoudads (*Ammotragus lervia*) all present issues for bighorn sheep in one or more ways. Specifically, wild asses are known to compete with bighorn sheep for forage or water and to foul water sources in western North American rangelands (Weaver et al. 1959; Seegmiller and Ohmart 1981; Marshal et al. 2008). Wild horses, although not widely sympatric with bighorn sheep, may influence use of water sources by bighorn sheep through interference competition, by competing directly with bighorn sheep for forage or water, or by altering ecosystem processes through trampling of vegetation (Ostermann-Kelm et al. 2008, 2009).

Mountain goats and aoudads have been introduced outside of their native distributions and are sympatric with bighorn sheep in some locations. Although naturally sympatric with bighorn sheep in some areas, introduced populations of mountain goats are viewed as potential competitors with bighorn sheep for forage or space, and as possible vectors of disease (Reed 1986; Blanchong et al. 2018; Lowrey et al. 2018). Aoudads are native to North Africa and occur in bighorn sheep habitat in parts of western Texas, New Mexico, and northern Mexico. Aoudads use habitat similar to that occupied by bighorn sheep, compete with bighorn sheep for the same resources, and are agonistic or otherwise behaviorally incompatible with bighorn sheep (Seegmiller and Simpson 1979; Brewer and Hernandez 2011). Recently, concerns have arisen about the potential for pathogen transfer from aoudads to bighorn sheep (Wiedmeier 2021).

Bighorn sheep were categorized as "wilderness game" by Leopold (1933), because they may not thrive in contact with human settlement, but some populations continue to do well near urban areas. Investigators have examined effects of recreational activities (Papouchis et al. 2001; Longshore et al. 2013; Wiedmann and Bleich 2014), mineral extraction (Oehler et al. 2005; Jansen et al. 2006, 2007), and road or highway development (Epps et al. 2005; Bleich et al. 2016) on populations of bighorn sheep. Expansion of renewable energy infrastructure is of increasing concern (Kuvlesky Jr et al. 2007; Lovich and Ennen 2011), especially as it relates to negative influences on desert bighorn sheep.

Ecosystem threats to bighorn sheep have been variable and have expanded substantially in recent years; considerable research has been conducted to ascertain the influence of recreational activities. Responses of bighorn sheep to recreational disturbance have ranged from little response (Hicks and Elder 1979; Bates et al. 2021) to temporary displacement (Papouchis et al. 2001; Longshore et al. 2013; Bates et al. 2021), permanent abandonment of previously occupied habitat (Wiedmann and Bleich 2014), and altered foraging regimes (Sproat et al. 2020). Some forms of recreation affect males differently than females. For example, male bighorn sheep respond differently to shed antler hunting than did females (Bates et al. 2021). Although mineral extraction has the potential to modify habitat, negative effects on bighorn sheep have been benign aside from the net loss of habitat associated with mine development; despite this outcome, activities associated with mining can have a positive effect in terms of landscape architecture, forage availability at revegetation sites, or deterring predation (Jansen et al. 2007; Bleich et al. 2009; Anderson et al. 2017). Further, mine reclamation can enhance per capita nutrient availability and increase population size following cessation of extraction activities (MacCallum 1992; MacCallum and Geist 1992).

Development of linear features including canals and highways likely has altered metapopulation processes by affecting movement corridors between sub populations, particularly in areas inhabited by desert bighorn sheep (Schwartz et al. 1986; Epps et al. 2005; Bleich et al. 2016). Such linear features have implications for genetic exchange between bighorn sheep populations, even though they may not be impervious barriers to movement by bighorn sheep (Epps et al. 2018). Continued fragmentation of ecosystems occupied by bighorn sheep, whether the result of infrastructure development for transportation or solar energy, will be problematic (Schwartz et al. 1986; Bleich et al. 1996).

Bighorn sheep occupy habitats ranging in elevation from below sea level to nearly 4500 m; as such they are adapted to a wide variety of environmental conditions. Thus, a changing climate has ecosystem-level implications for population persistence and habitat quality for this species. Vegetation changes resulting from a changing climate will affect distribution and habitat use by bighorn sheep (Epps et al. 2004) and will have evolutionary implications (Bleich 2017) and potential physiological challenges. Nevertheless, responses of bighorn sheep to changes in ecosystem structure or function are influenced greatly by the consistency, predictability, and level of threat associated with each disruption rather than the mere presence of people or other perturbations perceived as benign by these large ungulates (Wiedmann and Bleich 2014). Ultimately, the fate of bighorn sheep is tied to the size and needs of the human population (Bowyer et al. 2019).

22.7.2 Mountain Goats

Landscapes used by mountain goats are subject to a variety of conventional and non-conventional threats. Timber harvest (with its associated roads and infrastructure), mining, and hydroelectric development can have negative effects on mountain goats because of habitat removal or disturbance (Hebert and Turnbull 1977; Foster and Rahs 1985; Joslin 1986). Mining activity at a site in coastal Alaska resulted in a 42% reduction in carrying capacity of winter range habitat for a local population because of apparent displacement effects (White and Gregovich 2017). In other areas, logging of forested winter range resulted in direct removal of important winter habitat, or indirect effects because of disturbance or increased access and subsequent harvest (Hebert and Turnbull 1977).

Mountain goats are obligates of steep terrain and thus sensitive to climate-induced changes in high-elevation environments, particularly heat stress during summer (Sarmento et al. 2019) or severe snow conditions during winter (White et al. 2011; Richard et al. 2014). Climate change may have negative effects on mountain goat populations because of shrinkage of alpine habitats and through indirect effects associated with thermal stress or deleterious change in nutritional characteristics of summer foraging ranges (White et al. 2018). Although changes in climate may negatively influence population dynamics of mountain goats in some regions, further study is needed to assess how dynamics vary across the broad distributional range of the species and whether populations respond more strongly in some areas, as compared with others (White et al. 2018).

22.8   Conservation and Management Actions

22.8.1   Bighorn Sheep

Regulated hunting has played an important role in the conservation and reintroduction of bighorn sheep into rangelands of western North America (Monteith et al. 2013; Hurley et al. 2015). Economic considerations, largely in response to demand for hunting opportunities, have been an important force driving the restoration of bighorn sheep (Lee 2011; Gonzalez-Rebeles Islas et al. 2019). Much of the money garnered through the sale of bighorn sheep hunting tags is used for restoring populations of bighorn sheep to rangelands in western North America (Krausman 2000).

Active restoration of bighorn sheep to their historical distribution has been ongoing for about 100 years. Reintroductions and translocations remain an essential component of bighorn sheep management and conservation (Krausman 2000; Whiting et al. 2012; Sandoval et al. 2019). Recovery of populations of bighorn sheep largely has been a function of successful programs to return these mountain ungulates to their historical ranges, and translocations have contributed to the restoration or maintenance of ecosystem function in alpine or desert regions in much of western North America (Kie et al. 2003; Flesch et al. 2020). Past efforts to restore bighorn sheep to historical habitat have involved extensive efforts by resource-management agencies and conservation organizations, and tremendous financial commitments (Hurley et al. 2015; Donovan et al. 2020). Although translocation has been the primary tool used to reestablish bighorn sheep in rangelands across western North America, use of that method may become more limited in the foreseeable future, in part because of growing recognition that moving animals always includes risk of potentially moving diseases or exposing individuals to disease at the release site.

Management efforts surrounding infectious disease fall into one of two broad categories: actions to limit risk of pathogen spillover, and actions to limit pathogen burden following its introduction. Bighorn sheep often are culled by state wildlife agencies when they are discovered wandering outside of their established ranges to keep them from carrying pathogens back to their herd. At the same time, domestic sheep producers have experienced increasing restrictions on public land grazing allotments near bighorn habitat, leaving federal land-management agencies caught between maintaining healthy bighorn herds and maintaining grazing permits. Formal risk assessment tools exist (O'Brien et al. 2014), but an ongoing evolution in wildlife tracking technology means that the precise methods on which the tools rely are subject to regular revision and updating. Both culling and loss or modification of grazing permits engender frustration within their respective communities, but in the absence of effective treatments, limiting spillover risk through species separation remains the most effective strategy for protecting bighorn sheep (Brewer et al. 2014; Jex et al. 2016).

A suite of new tools is emerging to manage populations struggling to rebound from pathogen introductions. Wildlife management agencies have employed strategies ranging from complete depopulation followed by reintroduction to selective culling of individuals. Although efficacy of these actions has varied, test-and-remove (Garwood et al. 2020) and range expansion (Lula et al. 2020) strategies appear to have promise (Almberg et al. 2021). Under test-and-remove, managers trap and test as many (typically female) individuals as possible within a population, identify chronic carriers, and remove these animals. This option has yielded encouraging results, but it is labor intensive, and of variable efficacy (Paterson et al. 2020). Range expansion involves splitting an infected herd into multiple subunits to reduce densities and sequester the pathogen into unique subunits of the herd. The premise is that sequestration will facilitate local fade-out of the disease. Range expansion has been associated with improved demographic responses in at least one well-studied herd and is currently being tested in several other settings.

22.8.2 Mountain Goats

Mountain goats are widely appreciated as big game for their consumptive and non-consumptive value. Indigenous peoples valued this species for subsistence purposes including the use of wool, horns, and hooves in culturally significant ways (Rofkar 2014). The viewing and hunting of mountain goats generate substantial economic returns and re-investment into species conservation. Native populations of mountain goats exhibit low population growth rates and are sensitive to overharvest, especially if females are removed (Hamel et al. 2006; Rice and Gay 2010; White et al. 2021a); in some instances, introduced populations may be more productive, resilient, and able to sustain higher harvest rates, particularly during initial phases of establishment and expansion (Williams 1999; DeCesare and Smith 2018), but contrary results exist (Côté et al. 2001).

Relative to other ungulates, mountain goats are particularly sensitive to mechanized disturbance associated with commercial and recreational activities (Côté 1996; NWSGC 2020). For example, helicopter overflights or other forms of mechanized disturbance (energy development, blasting, and all-terrain vehicle use) can negatively affect mountain goat foraging behavior, movement patterns, and population dynamics, and mountain goats do not typically habituate to human disturbance (Joslin 1986; Côté et al. 2013; St-Louis et al. 2013). In places where industrial-scale mechanized disturbance occurs, mitigation to lessen or avoid negative effects is important to ensure population sustainability and persistence (NWSGC 2020).

22.9   Research and Management Needs

22.9.1   Bighorn Sheep

Historically, translocations and reintroduction of bighorns sheep to rangelands has been somewhat problematic. These problems have stemmed from issues related to habitat suitability, lack of migration opportunities, genetic issues, lack of understanding of ecotypic or phenotypic adaptation, predation, and disease transmission (Risenhoover et al. 1988; Rominger et al. 2004; Whiting et al. 2011b; Bleich et al. 2018). During recent years, disease concerns have been at the forefront of investigations or concern, and likely will remain so. Although there is general concurrence that fires enhance quality of bighorn sheep habitat through increased visibility or forage quality, responses of bighorn sheep to various fire-management strategies (e.g., suppression, wildfire, and prescribed fire) is a meaningful field in need of further inquiry. The utility of natural or artificial barriers that could provide a hedge against pathogen transfer among populations separated by those barriers is worthy of investigation, particularly from a cost–benefit perspective. For example, "What are the evolutionary consequences of maintaining artificial separation using barriers relative to the costs of pathogen spillover and its potential to affect, or perhaps even to decimate, nearby populations, and over what period of time would such costs accrue?" Related to this issue are questions about which bighorn sheep are most apt to make exploratory movements, or to pioneer unoccupied areas. The traditional thinking has been that young males are most apt to do so, but mature males and females also make such moves. The sex and age of the animals involved in such forays has important implications for demography and formulation of hunting regulations.

22.9.2   Mountain Goats

Mountain goats are among the least-studied large mammals in North America because of the difficulty, expense, and inherent danger of studying a species in remote and rugged landscapes. Although long-term and detailed studies have been conducted in specific areas resulting in substantial advancement of our knowledge of mountain goat ecology (Festa-Bianchet and Côté 2008), key knowledge gaps continue to limit our understanding about how population biology varies across the range of ecological settings inhabited by the species, including neonate survival, density-dependent effects, proportional causes of mortality, predator–prey relationships (including apparent competition), and small population-size effects. For example, recent mountain goat demographic studies have demonstrated reduced resilience and increased risk of extirpation among small populations, as compared with large populations (Hamel et al. 2006; White et al. 2021a). Improved understanding of the relative importance of underlying mechanisms, however, would aid in refining fine-scale conservation strategies. More broadly, detailed understanding of the mechanistic effects of weather and climate, specifically heat stress, represents an important need. Additionally, further study is needed to better understand how industrial or recreational disturbance influences behavior, vital rates, and resultant population productivity (NWSGC 2020).

References

Almberg ES, Manlove KR, Cassirer EF, Ramsey J, Carson K, Gude J, Plowright RK (2021) Modelling management strategies for chronic disease in wildlife: predictions for the control of respiratory disease in bighorn sheep. J Appl Ecol 59:693–703. https://doi.org/10.1111/1365-2664.14084

Anderson DJ, Villepique JT, Bleich VC (2017) Resource selection by desert bighorn relative to limestone mines. Desert Bighorn Council Trans 54:13–30

Andrew NG, Bleich VC, August PV (1999) Habitat selection by mountain sheep in the Sonoran Desert: implications for conservation in the United States and Mexico. Calif Wildl Conserv Bull 12:1–30

Ayotte JB, Parker KL, Arocena JM, Gillingham MP (2006) Chemical composition of lick soils: functions of soil ingestion by four ungulate species. J Mammal 87:878–888. https://doi.org/10.1644/06-MAMM-A-055R1.1

Ayotte JB, Parker KL, Gillingham MP (2008) Use of natural licks by four species of ungulates in northern British Columbia. J Mammal 89:1041–1050. https://doi.org/10.1644/07-MAMM-A-345.1

Bailey D (2004) Management strategies for optimal grazing distribution and use of arid rangelands. J Anim Sci 82:E147–E153. https://doi.org/10.2527/2004.8213_supplE147x

Bailey D, Kress D, Anderson D, Boss D, Miller E (2001) Relationship between terrain use and performance of beef cows grazing foothill rangeland. J Anim Sci 79:1883–1891. https://doi.org/10.2527/2001.7971883x

Barboza PS, Bowyer RT (2000) Sexual segregation in dimorphic deer: a new gastrocentric hypothesis. J Mammal 81:473–489. https://doi.org/10.1644/1545-1542(2000)081%3c0473:SSIDDA%3e2.0.CO;2

Barboza PS, Bowyer RT (2001) Seasonality of sexual segregation in dimorphic deer: extending the gastrocentric model. Alces 37:275–292

Bates SB, Whiting JC, Larsen RT (2021) Comparison of effects of shed antler hunting and helicopter surveys on ungulate movements and space use. J Wildl Manage 85:437–448. https://doi.org/10.1002/jwmg.22008

Belt JJ, Krausman PR (2012) Evaluating population estimates of mountain goats based on citizen science. Wildl Soc B 36:264–276. https://doi.org/10.1002/wsb.139

Berger J (1978a) Group size, foraging, and antipredator ploys: analysis of bighorn sheep decisions. Behav Ecol Sociobiol 4:91–99

Berger J (1978b) Maternal defensive behavior in bighorn sheep. J Mammal 59:620–621

Berger J (1990) Persistence of different-sized populations: an empirical assessment of rapid extinctions in bighorn sheep. Conserv Biol 4:91–98. https://doi.org/10.1111/j.1523-1739.1990.tb00271.x

Berger J (1991) Pregnancy incentives, predation constraints and habitat shifts: experimental and field evidence for wild bighorn sheep. Anim Behav 41:61–71. https://doi.org/10.1016/S0003-3472(05)80503-2

Besser TE, Cassirer EF, Potter KA, VanderSchalie J, Fischer A, Knowles DP, Herndon DR, Rurangirwa FR, Weiser GC, Srikumaran S (2008) Association of *Mycoplasma ovipneumoniae* infection with population-limiting respiratory disease in free-ranging Rocky Mountain bighorn sheep (*Ovis canadensis canadensis*). J Clin Microbiol 46:423–430. https://doi.org/10.1128/jcm.01931-07

Besser TE, Highland MA, Baker K, Cassirer EF, Anderson NJ, Ramsey JM, Mansfield K, Bruning DL, Wolff P, Smith JB, Jenks JA (2012) Causes of pneumonia epizootics among bighorn sheep, western United States, 2008–2010. Emerg Infect Dis 18:406–414. https://doi.org/10.3201/eid1803.111554

Besser TE, Cassirer EF, Highland MA, Wolff P, Justice-Allen A, Mansfield K, Davis MA, Foreyt W (2013) Bighorn sheep pneumonia: sorting out the cause of a polymicrobial disease. Prev Vet Med 108:85–93. https://doi.org/10.1016/j.prevetmed.2012.11.018

Besser TE, Levy J, Ackerman M, Nelson D, Manlove K, Potter KA, Busboom J, Benson M (2019) A pilot study of the effects of *Mycoplasma ovipneumoniae* exposure on domestic lamb growth and performance. PLoS ONE 14:e0207420. https://doi.org/10.1371/journal.pone.0207420

Bissonette JA, Steinkamp MJ (1996) Bighorn sheep response to ephemeral habitat fragmentation by cattle. Great Basin Nat 56:319–325

Black SR, Barker IK, Mehren KG, Crawshaw GJ, Rosendal S, Ruhnke L, Thorsen J, Carman PS (1988) An epizootic of *Mycoplasma ovipneumoniae* infection in captive Dall's sheep (*Ovis dalli dalli*). J Wildl Dis 24:627–635

Blanchong JA, Anderson CA, Clark NJ, Klaver RW, Plummer PJ, Cox M, McAdoo C, Wolff PL (2018) Respiratory disease, behavior, and survival of mountain goat kids. J Wildl Manage 82:1243–1251. https://doi.org/10.1002/jwmg.21470

Bleich VC (1999) Mountain sheep and coyotes: patterns of predator evasion in a mountain ungulate. J Mammal 80:283–289

Bleich VC (2005) In my opinion: politics, promises, and illogical legislation confound wildlife conservation. Wildl Soc B 33:66–73. https://doi.org/10.2193/0091-7648(2005)33[66:IMOPPA]2.0.CO;2

Bleich VC (2009a) Factors to consider when reprovisioning water developments used by mountain sheep. Calif Fish Game 95:153–159

Bleich VC (2009b) Perceived threats to wild sheep: levels of concordance among states, provinces, and territories. Trans Desert Bighorn Council 50:32–39

Bleich VC (2016) Wildlife conservation and wilderness: wishful thinking? Nat Areas J 36:202–206. https://doi.org/10.3375/043.036.0213

Bleich VC (2017) Leucism in bighorn sheep (*Ovis canadensis*), with special reference to the eastern Mojave Desert, California and Nevada, USA. Desert Bighorn Council Trans 54:31–47

Bleich VC, Bowyer RT, Pauli AM, Vernoy RL, Anthes RW (1990a) Responses of mountain sheep to helicopter surveys. Calif Fish Game 76:197–204

Bleich VC, Wehausen JD, Holl SA (1990b) Desert-dwelling mountain sheep: conservation implications of a naturally fragmented distribution. Conserv Biol 4:383–390

Bleich VC, Bowyer RT, Pauli AM, Nicholson MC, Anthes RW (1994) Mountain sheep (*Ovis canadensis*) and helicopter surveys: ramifications for the conservation of large mammals. Biol Conserv 70:1–7

Bleich VC, Wehausen JD, Ramey RR, Rechel JL (1996) Metapopulation theory and mountain sheep: implications for conservation. In: McCullough DR (ed) Metapopulations and wildlife conservation. Island Press, Washington, pp 353–373

Bleich VC, Bowyer RT, Wehausen JD (1997) Sexual segregation in mountain sheep: resources or predation? Wildl Monogr 134:1–50

Bleich VC, Andrew NG, Martin MJ, Mulcahy GP, Pauli AM, Rosenstock SS (2006) Quality of water available to wildlife in desert environments: comparisons among anthropogenic and natural sources. Wildl Soc B 34:627–632. https://doi.org/10.2193/0091-7648(2006)34[627:QOWATW]2.0.CO;2

Bleich VC, Davis JH, Marshal JP, Torres SG, Gonzales BJ (2009) Mining activity and habitat use by mountain sheep (*Ovis canadensis*). Eur J Wildl Res 55:183–191

Bleich VC, Whiting JC, Kie JG, Bowyer RT (2016) Roads, routes and rams: does sexual segregation contribute to anthropogenic risk in a desert-dwelling ungulate? Wildl Res 43:380–388. https://doi.org/10.1071/WR15231

Bleich VC, Oehler MW, Bowyer RT (2017) Mineral content of forage plants of mountain sheep, Mojave Desert, USA. Calif Fish Game 103:55–65

Bleich VC, Sargeant GA, Wiedmann BP (2018) Ecotypic variation in population dynamics of reintroduced bighorn sheep: implications for management. J Wildl Manage 82:8–18. https://doi.org/10.1002/jwmg.21381

Bowyer RT (2004) Sexual segregation in ruminants: definitions, hypotheses, and implications for conservation and management. J Mammal 85:1039–1052. https://doi.org/10.1644/BBL-002.1

Bowyer RT, Bleich VC, Stewart KM, Whiting JC, Monteith KL (2014) Density dependence in ungulates: a review of causes, and concepts with some clarifications. Calif Fish Game 100:550–572

Bowyer RT, Boyce MS, Goheen JR, Rachlow JL (2019) Conservation of the world's mammals: status, protected areas, community efforts, and hunting. J Mammal 100:923–941. https://doi.org/10.1093/jmammal/gyy180

Brewer CE, Hernandez F (2011) Status of desert bighorn sheep in Texas, 2009–2010. Desert Bighorn Council Trans 51:76–79

Brewer C, Bleich VC, Foster J, Hosch-Hebdon T, McWhirter D, Rominger E, Wagner M, Wiedmann B (2014) Bighorn sheep: conservation challenges and management strategies for the 21st century. Western Association of Fish and Wildlife Agencies

Brown NA, Ruckstuhl KE, Donelon S, Corbett C (2010) Changes in vigilance, grazing behaviour and spatial distribution of bighorn sheep due to cattle presence in Sheep River Provincial Park, Alberta. Agric Ecosyst Environ 135:226–231. https://doi.org/10.1016/j.agee.2009.10.001

Buchalski MR, Sacks BN, Gille DA, Penedo MCT, Ernest HB, Morrison SA, Boyce WM (2016) Phylogeographic and population genetic structure of bighorn sheep (*Ovis canadensis*) in North American deserts. J Mammal 97:823–838. https://doi.org/10.1093/jmammal/gyw011

Buechner HK (1960) The bighorn sheep in the United States, its past, present, and future. Wildl Monogr 4:1–174

Caro T (2005) Antipredator defenses in birds and mammals. University of Chicago Press, Chicago

Carpenter TE, Coggins VL, McCarthy C, O'Brien CS, O'Brien JM, Schommer TJ (2014) A spatial risk assessment of bighorn sheep extirpation by grazing domestic sheep on public lands. Prev Vet Med 114:3–10. https://doi.org/10.1016/j.prevetmed.2014.01.008

Cassirer EF, Manlove KR, Almberg ES, Kamath PL, Cox M, Wolff P, Roug A, Shannon J, Robinson R, Harris RB (2018) Pneumonia in bighorn sheep: risk and resilience. J Wildl Manage 82:32–45. https://doi.org/10.1002/jwmg.21309

Chaikina NA, Ruckstuhl KE (2006) The effect of cattle grazing on native ungulates: the good, the bad, and the ugly. Rangelands 28:8–14. https://doi.org/10.2111/1551-501X(2006)28[8:TEOCGO]2.0.CO;2

Clifford DL, Schumaker BA, Stephenson TR, Bleich VC, Cahn ML, Gonzales BJ, Boyce WM, Mazet JAK (2009) Assessing disease risk at the wildlife-livestock interface: a study of Sierra Nevada bighorn sheep. Biol Conserv 142:2559–2568. https://doi.org/10.1016/j.biocon.2009.06.001

Coltman DW, O'Donoghue P, Jorgenson JT, Hogg JT, Strobeck C, Festa-Bianchet M (2003) Undesirable evolutionary consequences of trophy hunting. Nature 426:655–658

Côté SD (1996) Mountain goat responses to helicopter disturbance. Wildl Soc B 24:681–685

Côté SD, Festa-Bianchet M (2003) Mountain goat,*Oreamnos americanus*. In: Feldhamer GA, Thompson BC, Chapman JA (eds) Wild mammals of North America: biology, management and conservation, vol 2. Johns Hopkins University Press, Baltimore, pp 1061–1075

Côté SD, Festa-Bianchet M, Smith KG (2001) Compensatory reproduction in harvested mountain goat populations: a word of caution. Wildl Soc Bull 29:726–730

Côté SD, Hamel S, St-Louis A, Mainguy J (2013) Do mountain goats habituate to helicopter disturbance? J Wildl Manage 77:1244–1244. https://doi.org/10.1002/jwmg.565

Coughenour MB (1991) Spatial components of plant-herbivore interactions in pastoral, ranching, and native ungulate ecosystems. J Range Manage 44:530–542

Cowan IM (1940) Distribution and variation in the native sheep of North America. Amer Midl Nat 24:505–580

DeCesare NJ, Pletscher DH (2006) Movements, connectivity, and resource selection of Rocky Mountain bighorn sheep. J Mammal 87:531–538. https://doi.org/10.1644/05-MAMM-A-259 R1.1

DeCesare NJ, Smith BL (2018) Contrasting native and introduced mountain goat populations in Montana. Proc Biennial Symp Northern Wild Sheep Goat Council 21:80–104

DeVoe JD, Garrott RA, Rotella JJ, Challender S, White PJ, O'Reilly M, Butler CJ (2015) Summer range occupancy modeling of non-native mountain goats in the Greater Yellowstone Area. Ecosphere 6:1–20. https://doi.org/10.1890/ES15-00273.1

Donovan VM, Roberts CP, Wonkka CL, Beck JL, Popp JN, Allen CR, Twidwell D (2020) Range-wide monitoring of population trends for Rocky Mountain bighorn sheep. Biol Conserv 248:108639. https://doi.org/10.1016/j.biocon.2020.108639

Dulude-de Broin F, Hamel S, Mastromonaco GF, Côté SD (2020) Predation risk and mountain goat reproduction: evidence for stress-induced breeding suppression in a wild ungulate. Funct Ecol 34:1003–1014. https://doi.org/10.1111/1365-2435.13514

Epps CW, McCullough DR, Wehausen JD, Bleich VC, Rechel JL (2004) Effects of climate change on population persistence of desert-dwelling mountain sheep in California. Conserv Biol 18:102–113. https://doi.org/10.1111/j.1523-1739.2004.00023.x

Epps CW, Palsboll PJ, Wehausen JD, Roderick GK, Ramey RR, McCullough DR (2005) Highways block gene flow and cause a rapid decline in genetic diversity of desert bighorn sheep. Ecol Letters 8:1029–1038. https://doi.org/10.1111/j.1461-0248.2005.00804.x

Epps CW, Crowhurst RS, Nickerson BS (2018) Assessing changes in functional connectivity in a desert bighorn sheep metapopulation after two generations. Mol Ecol 27:2334–2346. https://doi.org/10.1111/mec.14586

Feldhamer GA, Merritt JF, Krajewski C, Rachlow JL, Stewart KM (2020) Mammalogy: adaptation, diversity, ecology. Johns Hopkins University Press, Baltimore

Festa-Bianchet M (1988a) Age-specific reproduction of bighorn ewes in Alberta, Canada. J Mammal 69:157–160

Festa-Bianchet M (1988b) Birthdate and survival in bighorn lambs (*Ovis canadensis*). J Zool 214:653–661

Festa-Bianchet M (1988c) Nursing behavior of bighorn sheep: correlates of ewe age, parasitism, lamb age, birthdate and sex. Anim Behav 36:1445–1454

Festa-Bianchet M (1988d) Seasonal range selection in bighorn sheep: conflicts between forage quality, forage quantity, and predator avoidance. Oecologia 75:580–586

Festa-Bianchet M, Côté SD (2008) Mountain goats: ecology, behavior, and conservation of an alpine ungulate. Island Press, California

Festa-Bianchet M, Jorgenson JT (1998) Selfish mothers: reproductive expenditure and resource availability in bighorn ewes. Behav Ecol 9:144–150

Festa-Bianchet M, Jorgenson JT, Bérubé CH, Portier C, Wishart WD (1997) Body mass and survival of bighorn sheep. Can J Zool 75:1372–1379

Festa-Bianchet M, Jorgenson JT, Réale D (2000) Early development, adult mass, and reproductive success in bighorn sheep. Behav Ecol 11:633–639. https://doi.org/10.1093/beheco/11.6.633

Flesch EP, Graves TA, Thomson JM, Proffitt KM, White P, Stephenson TR, Garrott RA (2020) Evaluating wildlife translocations using genomics: a bighorn sheep case study. Ecol Evol 10:13687–13704. https://doi.org/10.1002/ece3.6942

Flueck WT, Smith-Flueck J, Mionczynski J, Mincher B (2012) The implications of selenium deficiency for wild herbivore conservation: a review. Eur J Wildl Res 58:761–780

Foster BR, Rahs EY (1985) A study of canyon-dwelling mountain goats in relation to proposed hydroelectric development in northwestern British Columbia, Canada. Biol Conserv 33:209–228

Fox JL, Streveler GP (1986) Wolf predation on mountain goats in southeastern Alaska. J Mammal 67:192–195

Fox K, Wootton S, Quackenbush S, Wolfe L, Levan I, Miller M, Spraker T (2011) Paranasal sinus masses of Rocky Mountain bighorn sheep (*Ovis canadensis canadensis*). Vet Pathol 48:706–712. https://doi.org/10.1177/0300985810383873

Fox KA, Rouse NM, Huyvaert KP, Griffin KA, Killion HJ, Jennings-Gaines J, Edwards WH, Quackenbush SL, Miller MW (2015) Bighorn sheep (*Ovis canadensis*) sinus tumors are associated with coinfections by potentially pathogenic bacteria in the upper respiratory tract. J Wildl Dis 51:19–27. https://doi.org/10.7589/2014-05-130

Fox K, Wootton S, Marolf A, Rouse N, LeVan I, Spraker T, Miller M, Quackenbush S (2016) Experimental transmission of bighorn sheep sinus tumors to bighorn sheep (*Ovis canadensis canadensis*) and domestic sheep. Vet Pathol 53:1164–1171. https://doi.org/10.1177/030098581 6634810

Gaillard J-M, Festa-Bianchet M, Yoccoz N, Loison A, Toigo C (2000) Temporal variation in fitness components and population dynamics of large herbivores. Annu Rev Ecol Syst 31:367–393

Garrison KR, Cain JW III, Rominger EM, Goldstein EJ (2016) Sympatric cattle grazing and desert bighorn sheep foraging. J Wildl Manage 80:197–207. https://doi.org/10.1002/jwmg.1014

Garwood TJ, Lehman CP, Walsh DP, Cassirer EF, Besser TE, Jenks JA (2020) Removal of chronic *Mycoplasma ovipneumoniae* carrier ewes eliminates pneumonia in a bighorn sheep population. Ecol Evol 10:3491–3502. https://doi.org/10.1002/ece3.6146

Geist V (1971) Mountain sheep: a study in behavior and evolution. The University of Chicago Press, Chicago

Gonzalez-Voyer A, Festa-Bianchet M, Smith KG (2001) Efficiency of aerial surveys of mountain goats. Wildl Soc B 29:140–144

Gonzalez-Rebeles Islas C, Mendez M, Valdez R (2019) Evolution of wildlife laws and policy in Mexico. In: Valdez R, Ortega-S J (eds) Wildlife ecology and management in Mexico. Texas A&M University Press, College Station, pp 366–377

Guthrie RD (1968) Paleoecology of the large-mammal community in interior Alaska during the late Pleistocene. Am Midl Nat 79:346–363

Hamel S, Côté S (2007) Habitat use patterns in relation to escape terrain: are alpine ungulate females trading off better foraging sites for safety? Can J Zool 85:933–943. https://doi.org/10.1139/Z07-080

Hamel S, Cote SD, Smith KG, Festa-Bianchet M (2006) Population dynamics and harvest potential of mountain goat herds in Alberta. J Wildl Manage 70:1044–1053. https://doi.org/10.2193/0022-541X(2006)70[1044:PDAHPO]2.0.CO;2

Hebert D, Cowan IM (1971) Natural salt licks as a part of the ecology of the mountain goat. Can J Zool 49:605–610

Hebert D, Turnbull W (1977) A description of southern interior and coastal mountain goat ecotypes in British Columbia. Proc Int Mountain Goat Symp 1:126–146

Hicks LL, Elder JM (1979) Human disturbance of Sierra Nevada bighorn sheep. J Wildl Manage 43:909–915

Hirth DH, McCullough DR (1977) Evolution of alarm signals in ungulates with special reference to white-tailed deer. Am Nat 111:31–42

Holl SA, Bleich VC (1987) Mineral lick use by mountain sheep in the San Gabriel Mountains, California. J Wildl Manage 51:383–385

Houston DB, Stevens V (1988) Resource limitation in mountain goats: a test by experimental cropping. Can J Zool 66:228–238

Hurley K, Brewer C, Thornton GN (2015) The role of hunters in conservation, restoration, and management of North American wild sheep. Int J Env Stud 72:784–796. https://doi.org/10.1080/00207233.2015.1031567

Jansen BD, Krausman PR, Heffelfinger JR, deVos JC (2006) Bighorn sheep selection of landscape features in an active copper mine. Wildl Soc B 34:1121–1126. https://doi.org/10.2193/0091-7648(2006)34[1121:BSSOLF]2.0.CO;2

Jansen BD, Krausman PR, Heffelfinger JR, deVos JC (2007) Influence of mining on behavior of bighorn sheep. Southwest Nat 52:418–423. https://doi.org/10.1894/0038-4909(2007)52[418:IOMOBO]2.0.CO;2

Jesmer BR, Merkle JA, Goheen JR, Aikens EO, Beck JL, Courtemanch AB, Hurley MA, McWhirter DE, Miyasaki HM, Monteith KL, Kauffman MJ (2018) Is ungulate migration culturally transmitted? Evidence of social learning from translocated animals. Science 361:1023–1025. https://doi.org/10.1126/science.aat0985

Jex BA, Ayotte JB, Bleich VC, Brewer CE, Bruning DL, Hegel TM, Larter NC, Schwanke RA, Schwantje HM, Wagner MW (2016) Thinhorn sheep: conservation challenges and management strategies for the 21st century. Western Association of Fish and Wildlife Agencies, Boise, Idaho

Johnson HE, Mills LS, Wehausen JD, Stephenson TR (2010) Combining ground count, telemetry, and mark–resight data to infer population dynamics in an endangered species. J Appl Ecol 47:1083–1093. https://doi.org/10.1111/j.1365-2664.2010.01846.x

Johnson HE, Hebblewhite M, Stephenson TR, German DW, Pierce BM, Bleich VC (2013) Evaluating apparent competition in limiting the recovery of an endangered ungulate. Oecologia 171:295–307

Joslin G (1986) Mountain goat population changes in relation to energy exploration along Montana's Rocky Mountain Front. Proc Biennial Symp Northern Wild Sheep Goat Council 5:253–271

Kamath PL, Manlove K, Cassirer EF, Cross PC, Besser TE (2019) Genetic structure of *Mycoplasma ovipneumoniae* informs pathogen spillover dynamics between domestic and wild Caprinae in the western United States. Sci Rep 9:1–14

Kie JG, Bowyer RT, Stewart KM (2003) Ungulates in western coniferous forests: habitat relationships, population dynamics, and ecosystem processes. In: Zabel CJ, Anthony RG (eds) Mammal community dynamics: management and conservation in the coniferous forests of western North America. Cambridge University Press, New York, pp 296–340

Krausman PR (2000) An introduction to the restoration of bighorn sheep. Restor Ecol 8:3–5

Krausman PR, Bleich VC (2013) Conservation and management of ungulates in North America. Int J Environ Stud 70:372–382. https://doi.org/10.1080/00207233.2013.804748

Krausman PR, Bowyer RT (2003) Mountain sheep (*Ovis canadensis* and *O. dalli*). In: Feldhamer GA, Thompson BC, Chapman JA (eds) Wild mammals of north america: biology, management, and conservation, 2nd edn. John Hopkins University Press, Baltimore, pp 1095–1115

Krausman PR, Hervert JJ (1983) Mountain sheep responses to aerial surveys. Wildl Soc B 11:372–375

Krausman PR, Sandoval AV, Etchberger RC (1999) Natural history of desert bighorn sheep. In: Valdez R, Krausman PR (eds) Mountain sheep of North America. University of Arizona Press, Tucson, pp 139–191

Kuvlesky WP Jr, Brennan LA, Morrison ML, Boydston KK, Ballard BM, Bryant FC (2007) Wind energy development and wildlife conservation: challenges and opportunities. J Wildl Manage 71:2487–2498. https://doi.org/10.2193/2007-248

Lassis R, Festa-Bianchet M, Pelletier F (2022) Breeding migrations by bighorn sheep males are driven by mating opportunities. Ecol Evol 12:e8692. https://doi.org/10.1002/ece3.8692

Lee R (2011) Economic aspects of and the market for desert bighorn sheep. Desert Bighorn Council Trans 51:46–49

Leopold A (1933) Game management. Charles Scribner's Sons, New York

Loison A, Gaillard JM, Pelabon C, Yoccoz NG (1999) What factors shape sexual size dimorphism in ungulates? Evol Ecol Res 1:611–633

Longshore K, Lowrey C, Thompson DB (2013) Detecting short-term responses to weekend recreation activity: desert bighorn sheep avoidance of hiking trails. Wildl Soc B 37:698–706. https://doi.org/10.1002/wsb.349

Lovich JE, Ennen JR (2011) Wildlife conservation and solar energy development in the desert southwest, United States. BioSci 61:982–992. https://doi.org/10.1525/bio.2011.61.12.8

Lowrey B, Garrott RA, McWhirter DE, White PJ, DeCesare NJ, Stewart ST (2018) Niche similarities among introduced and native mountain ungulates. Ecol Appl 28:1131–1142. https://doi.org/10.1002/eap.1719

Lowrey B, DeVoe J, Proffitt K, Garrott R (2021) Behavior-specific habitat models as a tool to inform ungulate restoration. Ecosphere 12:e03687. https://doi.org/10.1002/ecs2.3687

Lula ES, Lowrey B, Proffitt KM, Litt AR, Cunningham JA, Butler CJ, Garrott RA (2020) Is habitat constraining bighorn sheep restoration? A case study. J Wildl Manage 84:588–600. https://doi.org/10.1002/jwmg.21823

MacCallum B (1992) Population dynamics of bighorn sheep using reclaimed habitat in open pit coal mines in west-central Alberta. Proc Biennial Symp Northern Wild Sheep Goat Council 8:374

MacCallum B, Geist V (1992) Mountain restoration: soil and surface wildlife habitat. GeoJ 27:23–46

Malaney JL, Feldman CR, Cox M, Wolff P, Wehausen JD, Matocq MD (2015) Translocated to the fringe: genetic and niche variation in bighorn sheep of the Great Basin and northern Mojave deserts. Divers Distrib 21:1063–1074. https://doi.org/10.1111/ddi.12329

Manlove K, Branan M, Baker K, Bradway D, Cassirer EF, Marshall KL, Miller RS, Sweeney S, Cross PC, Besser TE (2019) Risk factors and productivity losses associated with *Mycoplasma ovipneumoniae* infection in United States domestic sheep operations. Prev Vet Med 168:30–38. https://doi.org/10.1016/j.prevetmed.2019.04.006

Marsh H (1938) Pneumonia in Rocky Mountain bighorn sheep. J Mammal 19:214–219

Marshal JP, Bleich VC, Andrew NG (2008) Evidence for interspecific competition between feral ass *Equus asinus* and mountain sheep *Ovis canadensis* in a desert environment. Wildl Biol 14:228–236. https://doi.org/10.2981/0909-6396(2008)14[228:EFICBF]2.0.CO;2

McClintock BT, White GC (2007) Bighorn sheep abundance following a suspected pneumonia epidemic in Rocky Mountain National Park. J Wildl Manage 71:183–189. https://doi.org/10.2193/2006-336

McDonough TJ, Selinger JS (2008) Mountain goat management on the Kenai Peninsula, Alaska: a new direction. Proc Biennial Symp Northern Wild Sheep Goat Council 16:50–67

Monteith KL, Long RA, Bleich VC, Heffelfinger JR, Krausman PR, Bowyer RT (2013) Effects of harvest, culture, and climate on trends in size of horn-like structures in trophy ungulates. Wildl Monogr 183:1–28. https://doi.org/10.1002/wmon.1007

Nagorsen DW, Keddie G (2000) Late Pleistocene mountain goats (*Oreamnos americanus*) from Vancouver Island: biogeographic implications. J Mammal 81:666–675

Neal AK, White GC, Gill RB, Reed DF, Olterman JH (1993) Evaluation of mark-resight model assumptions for estimating mountain sheep numbers. J Wildl Manage 57:436–450. https://doi.org/10.2307/3809268

NWSGC (2020) Northern Wild Sheep and Goat Council position statement on commercial and recreational disturbance of mountain goats: recommendations for management. Proc Biennial Symp Northern Wild Sheep Goat Council 22:1–15

O'Brien JM, O'Brien CS, McCarthy C, Carpenter TE (2014) Incorporating foray behavior into models estimating contact risk between bighorn sheep and areas occupied by domestic sheep. Wildl Soc B 38:321–331. https://doi.org/10.1002/wsb.387

Oehler MW, Bleich VC, Bowyer RT, Nicholson MC (2005) Mountain sheep and mining: implications for conservation and management. Calif Fish Game 91:149–178

Ostermann-Kelm S, Atwill ER, Rubin ES, Jorgensen MC, Boyce WM (2008) Interactions between feral horses and desert bighorn sheep at water. J Mammal 89:459–466. https://doi.org/10.1644/07-MAMM-A-075R1.1

Ostermann-Kelm SD, Atwill EA, Rubin ES, Hendrickson LE, Boyce WM (2009) Impacts of feral horses on a desert environment. BMC Ecol 9:1–10

Papouchis CM, Singer FJ, Sloan WB (2001) Responses of desert bighorn sheep to increased human recreation. J Wildl Manage 65:573–582

Paterson JT, Butler C, Garrott R, Proffitt K (2020) How sure are you? A web-based application to confront imperfect detection of respiratory pathogens in bighorn sheep. PLoS ONE 15:e0237309. https://doi.org/10.1371/journal.pone.0237309

Perry TW, Newman T, Thibault KM (2010) Evaluation of methods to estimate size of a population of desert bighorn sheep (*Ovis canadensis mexicana*) in New Mexico. Southwest Nat 55:517–524. https://doi.org/10.1894/sgm-07.1

Pettorelli N, Pelletier F, von Hardenberg A, Festa-Bianchet M, Côté SD (2007) Early onset of vegetation growth vs. rapid green-up: impacts on juvenile mountain ungulates. Ecol 88:381–390. https://doi.org/10.1890/06-0875

Péwé TL, Hopkins DM (1967) Mammal remains of Pre-Wisconsin Age in Alaska. In: Hopkins DM (ed) The Bering Land Bridge. Stanford University Press, California, pp 266–270

Plowright RK, Manlove KR, Besser TE, Páez DJ, Andrews KR, Matthews PE, Waits LP, Hudson PJ, Cassirer EF (2017) Age-specific infectious period shapes dynamics of pneumonia in bighorn sheep. Ecol Letters 20:1325–1336. https://doi.org/10.1111/ele.12829

Poole KG (2007) Does survey effort influence sightability of mountain goats *Oreamnos americanus* during aerial surveys? Wildl Biol 113–119. https://doi.org/10.2981/0909-6396(2007)13[113:DSEISO]2.0.CO;2

Poole KG, Stuart-Smith K, Teske IE (2009) Wintering strategies by mountain goats in interior mountains. Can J Zool 87:273–283. https://doi.org/10.1139/Z09-009

Poole KG, Bachmann KD, Teske IE (2010) Mineral lick use by GPS radio-collared mountain goats in southeastern British Columbia. West N Am Nat 70:208–217. https://doi.org/10.3398/064.070.0207

Poole KG, Reynolds DM, Mowat G, Paetkau D (2011) Estimating mountain goat abundance using DNA from fecal pellets. J Wildl Manage 75:1527–1534. https://doi.org/10.1002/jwmg.184

Reed D (1986) Alpine habitat selection in sympatric mountain goats and mountain sheep. Proc Biennial Symp Northern Wild Sheep Goat Council 5:421–422

Rice CG (2010) Mineral lick visitation by mountain goats, *Oreamnos americanus*. Can Field-Nat 124:225–237. https://doi.org/10.22621/cfn.v124i3.1078

Rice CG, Gay D (2010) Effects of mountain goat harvest on historic and contemporary populations. Northwest Nat 91:40–57. https://doi.org/10.1898/NWN08-47.1

Rice CG, Jenkins KJ, Chang WY (2009) A sightability model for mountain goats. J Wildl Manage 73:468–478. https://doi.org/10.2193/2008-196

Richard JH, Côté SD (2016) Space use analyses suggest avoidance of a ski area by mountain goats. J Wildl Manage 80:387–395. https://doi.org/10.1002/jwmg.1028

Richard JH, Wilmshurst J, Côté SD (2014) The effect of snow on space use of an alpine ungulate: recently fallen snow tells more than cumulative snow depth. Can J Zool 92:1067–1074. https://doi.org/10.1139/cjz-2014-0118

Rideout CB, Hoffman RS (1975) *Oreamnos americanus*. Mammalian Species 63:1–6

Risenhoover KL, Bailey JA (1985) Foraging ecology of mountain sheep: implications for habitat management. J Wildl Manage 49:797–804

Risenhoover KL, Bailey JA, Wakelyn LA (1988) Assessing the Rocky Mountain bighorn sheep management problem. Wildl Soc B 16:346–352

Robinson RW, Smith TS, Whiting JC, Larsen RT, Shannon JM (2020) Determining timing of births and habitat selection to identify lambing period habitat for bighorn sheep. Front Ecol Evol 8:97. https://doi.org/10.3389/fevo.2020.00097

Rofkar T (2014) Managing and harvesting mountain goats for traditional purposes by indigenous user groups. Proc Biennial Symp Northern Wild Sheep Goat Council 19:37–41

Rominger EM (2018) The Gordian knot of mountain lion predation and bighorn sheep. J Wildl Manage 82:19–31. https://doi.org/10.1002/jwmg.21396

Rominger EM, Whitlaw HA, Weybright DL, Dunn WC, Ballard WB (2004) The influence of mountain lion predation on bighorn sheep translocations. J Wildl Manage 68:993–999. https://doi.org/10.2193/0022-541X(2004)068[0993:TIOMLP]2.0.CO;2

Rosenstock SS, Ballard WB, deVos Jr JC (1999) Viewpoint: benefits and impacts of wildlife water developments. J Range Manage 52:302–311

Rosenstock SS, Hervert JJ, Bleich VC, Krausman PR (2001) Muddying the water with poor science: a reply to Broyles and Cutler. Wildl Soc B 29:734–738

Ross PI, Jalkotzy MG, Festa-Bianchet M (1997) Cougar predation on bighorn sheep in southwestern Alberta during winter. Can J Zool 75:771–775

Rubin ES, Boyce WM, Bleich VC (2000) Reproductive strategies of desert bighorn sheep. J Mammal 81:769–786

Ruder MG, Lysyk TJ, Stallknecht DE, Foil LD, Johnson DJ, Chase CC, Dargatz DA, Gibbs EPJ (2015) Transmission and epidemiology of bluetongue and epizootic hemorrhagic disease in North America: current perspectives, research gaps, and future directions. Vector Borne Zoonotic Dis 15:348–363. https://doi.org/10.1089/vbz.2014.1703

Samuel W, Chalmers G, Stelfox J, Loewen A, Thomsen J (1975) Contagious ecthyma in bighorn sheep and mountain goat in western Canada. J Wildl Dis 11:26–31

Sandoval AV, Valdez R, Espinosa-T A (2019) Desert bighorn sheep in Mexico. In: Valdez R, Ortega-S J (eds) Wildlife ecology and management in Mexico. Texas A&M University Press, College Station, pp 350–365

Sarmento W, Berger J (2020) Conservation implications of using an imitation carnivore to assess rarely used refuges as critical habitat features in an alpine ungulate. PeerJ 8:e9296

Sarmento W, Biel M, Berger J (2019) Seeking snow and breathing hard–behavioral tactics in high elevation mammals to combat warming temperatures. PLoS ONE 14:e0225456. https://doi.org/10.1371/journal.pone.0225456

Saunders JK (1955) Food habits and range use of the Rocky Mountain goat in the Crazy Mountains, Montana. J Wildl Manage 19:429–437

Schmidt JH, Reynolds JH, Rattenbury KL, Phillips LM, White KS, Schertz D, Morton JM, Kim HS (2019) Integrating distance sampling with minimum counts to improve monitoring. J Wildl Manage 83:1454–1465. https://doi.org/10.1002/jwmg.21691

Schoenecker KA, Watry MK, Ellison LE, Schwartz MK, Luikart G (2015) Estimating bighorn sheep (*Ovis canadensis*) abundance using noninvasive sampling at a mineral lick within a national park wilderness area. West North Am Nat 75:181–191. https://doi.org/10.3398/064.075.0206

Schommer TJ, Woolever MM (2008) A review of disease related conflicts between domestic sheep and goats and bighorn sheep. US Forest Service Rocky Mountain Research Station General Technical Report RMRS-GTR-209. US Department of Agriculture, Forest Service, Rocky Mountain Research Station, Fort Collins

Schroeder CA, Bowyer RT, Bleich VC, Stephenson TR (2010) Sexual segregation in Sierra Nevada bighorn sheep, *Ovis canadensis sierrae*: ramifications for conservation. Arct Antarct Alp Res 42:476–489. https://doi.org/10.1657/1938-4246-42.4.476

Schwartz OA, Bleich VC, Holl SA (1986) Genetics and the conservation of mountain sheep *Ovis canadensis nelsoni*. Biol Conserv 37:179–190

Seegmiller R, Simpson C (1979) The Barbary sheep: some conceptual implications of competition with desert bighorn. Desert Bighorn Council Trans 23:47–49

Seegmiller RF, Ohmart RD (1981) Ecological relationships of feral burros and desert bighorn sheep. Wildl Monogr 78:1–58

Shackleton DM, Shank CC, Wikeem BM (1999) Natural history of Rocky Mountain and California bighorn sheep. In: Valdez R, Krausman PR (eds) Mountain sheep of North America. University of Arizona Press, Tucson, pp 78–138

Shafer AB, Côté SD, Coltman DW (2011a) Hot spots of genetic diversity descended from multiple Pleistocene refugia in an alpine ungulate. Evol 65:125–138. https://doi.org/10.1111/j.1558-5646.2010.01109.x

Shafer AB, White KS, Côté SD, Coltman DW (2011b) Deciphering translocations from relicts in Baranof Island mountain goats: is an endemic genetic lineage at risk? Conserv Genet 12:1261–1268

Shafer AB, Northrup JM, White KS, Boyce MS, Côté SD, Coltman DW (2012) Habitat selection predicts genetic relatedness in an alpine ungulate. Ecol 93:1317–1329. https://doi.org/10.1890/11-0815.1

Shannon JM, Whiting JC, Larsen RT, Olson DD, Flinders JT, Smith TS, Bowyer RT (2014) Population response of reintroduced bighorn sheep after observed commingling with domestic sheep. Eur J Wildl Res 60:737–748. https://doi.org/10.1007/s10344-014-0843-y

Shanthalingam S, Goldy A, Bavananthasivam J, Subramaniam R, Batra SA, Kugadas A, Raghavan B, Dassanayake RP, Jennings-Gaines JE, Killion HJ (2014) PCR assay detects *Mannheimia haemolytica* in culture-negative pneumonic lung tissues of bighorn sheep (*Ovis canadensis*) from outbreaks in the western USA, 2009–2010. J Wildl Dis 50:1–10. https://doi.org/10.7589/2012-09-225

Singer FJ (1978) Behavior of mountain goats in relation to US Highway 2, Glacier National Park, Montana. J Wildl Manage 42:591–597

Singer FJ, Norland JE (1994) Niche relationships within a guild of ungulate species in Yellowstone National Park, Wyoming, following release from artificial controls. Can J Zool 72:1383–1394

Singer FJ, Papouchis CM, Symonds KK (2000) Translocations as a tool for restoring populations of bighorn sheep. Restor Ecol 8:6–13

Singer FJ, Zeigenfuss LC, Spicer L (2001) Role of patch size, disease, and movement in rapid extinction of bighorn sheep. Conserv Biol 15:1347–1354

Smith CA (1986) Rates and causes of mortality in mountain goats in southeast Alaska. J Wildl Manage 50:743–746

Smith TS, Flinders JT, Winn DS (1991) A habitat evaluation procedure for Rocky Mountain bighorn sheep in the Intermountain West. Great Basin Nat 51:205–225

Spitz DB, Hebblewhite M, Stephenson TR (2020) Habitat predicts local prevalence of migratory behaviour in an alpine ungulate. J Anim Ecol 89:1032–1044. https://doi.org/10.1111/1365-2656.13167

Sproat KK, Martinez NR, Smith TS, Sloan WB, Flinders JT, Bates JW, Cresto JG, Bleich VC (2020) Desert bighorn sheep responses to human activity in south-eastern Utah. Wildl Res 47:16–24. https://doi.org/10.1071/WR19029

St-Louis A, Hamel S, Mainguy J, Côté SD (2013) Factors influencing the reaction of mountain goats towards all-terrain vehicles. J Wildl Manage 77:599–605. https://doi.org/10.1002/jwmg.488

Stockwell CA, Bateman GC, Berger J (1991) Conflicts in national parks: a case study of helicopters and bighorn sheep time budgets at the Grand Canyon. Biol Conser 56:317–328

Swenson JE (1985) Compensatory reproduction in an introduced mountain goat population in the Absaroka Mountains, Montana. J Wildl Manage 49:837–843

Taylor JC, Bates SB, Whiting JC, McMillan BR, Larsen RT (2020) Optimising deployment time of remote cameras to estimate abundance of female bighorn sheep. Wildl Res 48:127–133. https://doi.org/10.1071/WR20069

Taylor JC, Bates SB, Whiting JC, McMillan BR, Larsen RT (2022) Using camera traps to estimate ungulate abundance: a comparison of mark–resight methods. Remote Sens Ecol Conserv 8:32–44. https://doi.org/10.1002/rse2.226

Tryland M, Beckmen KB, Burek-Huntington KA, Breines EM, Klein J (2018) Orf virus infection in Alaskan mountain goats, Dall's sheep, muskoxen, caribou and Sitka black-tailed deer. Acta Vet Scand 60:1–11

USFWS (2000) Recovery plan for bighorn sheep in the Pennisular Ranges, California. US Fish and Wildlife Service, Portland

USFWS (2007) Recovery plan for the Sierra Nevada bighorn sheep. US Fish and Wildlife Service, Sacramento

Valdez R, Krausman PR (1999) Description, distribution, and abundance of mountain sheep in North America. In: Valdez R, Krausman PR (eds) Mountain sheep of North America. University of Arizona Press, Arizona, pp 3–22

Weaver RA, Vernoy F, Craig B (1959) Game water development on the desert. Calif Fish Game 45:333–342

Weckerly FW (1998) Sexual-size dimorphism: influence of mass and mating systems in the most dimorphic mammals. J Mammal 79:33–52

Wehausen JD, Ramey RR (2000) Cranial morphometric and evolutionary relationships in the northern range of *Ovis canadensis*. J Mammal 81:145–161

Wehausen JD, Kelley ST, Ramey RR (2011) Domestic sheep, bighorn sheep, and respiratory disease: a review of the experimental evidence. Calif Fish Game 97:7–24

White KS, Gregovich DP (2017) Mountain goat resource selection in relation to mining-related disturbance. Wildl Biol 1:1–12. https://doi.org/10.2981/wlb.00277

White KS, Pendleton GW, Crowley D, Griese HJ, Hundertmark KJ, Mcdonough T, Nichols L, Robus M, Smith CA, Schoen JW (2011) Mountain goat survival in coastal Alaska: effects of age, sex, and climate. J Wildl Manage 75:1731–1744. https://doi.org/10.1002/jwmg.238

White KS, Gregovich DP, Levi T (2018) Projecting the future of an alpine ungulate under climate change scenarios. Glob Chang Biol 24:1136–1149. https://doi.org/10.1111/gcb.13919

White KS, Levi T, Breen J, Britt M, Meröndun J, Martchenko D, Shakeri YN, Porter B, Shafer AB (2021a) Integrating genetic data and demographic modeling to facilitate conservation of small, isolated mountain goat populations. J Wildl Manage 85:271–282. https://doi.org/10.1002/jwmg.21978

White KS, Watts DE, Beckmen KB (2021b) Helicopter-based chemical immobilization of mountain goats in coastal Alaska. Wildl Soc B 45:670–681. https://doi.org/10.1002/wsb.1229

Whiting JC, Bowyer RT, Flinders JT (2009) Annual use of water sources by reintroduced Rocky Mountain bighorn sheep *Ovis canadensis canadensis*: effects of season and drought. Acta Theriol 54:127–136. https://doi.org/10.1007/bf03193168

Whiting JC, Bowyer RT, Flinders JT, Bleich VC, Kie JG (2010a) Sexual segregation and use of water by bighorn sheep: implications for conservation. Anim Conserv 13:541–548. https://doi.org/10.1111/j.1469-1795.2010.00370.x

Whiting JC, Stewart KM, Bowyer RT, Flinders JT (2010b) Reintroduced bighorn sheep: do females adjust maternal care to compensate for late-born young? Eur J Wildl Res 56:349–357. https://doi.org/10.1007/s10344-009-0323-y

Whiting JC, Bleich VC, Bowyer RT, Larsen RT (2011a) Water availability and bighorn sheep: life-history characteristics and persistence of populations. In: Daniels JA (ed) Advances in environmental research, vol 21. Nova Publishers, Inc., pp 131–163

Whiting JC, Bowyer RT, Flinders JT, Eggett DL (2011b) Reintroduced bighorn sheep: fitness consequences of adjusting parturition to local environments. J Mammal 92:213–220. https://doi.org/10.1644/10-mamm-a-145.1

Whiting JC, Olson DD, Shannon JM, Bowyer RT, Klaver RW, Flinders JT (2012) Timing and synchrony of births in bighorn sheep: implications for reintroduction and conservation. Wildl Res 39:565–572. https://doi.org/10.1071/WR12059

Wiedmann BP, Bleich VC (2014) Demographic responses of bighorn sheep to recreational activities: a trial of a trail. Wildl Soc B 38:773–782. https://doi.org/10.1002/wsb.463

Wiedmeier RC (2021) Characterization of aoudad and desert bighorn sheep microbiomes in association to disease risk. Texas Tech University, Lubbock

Wikeem BM, Pitt MD (1992) Diet of California bighorn sheep, *Ovis canadensis californiana*, in British Columbia: assessing optimal foraging habitat. Can Field-Nat 106:327–335

Williams ES, Spraker TR, Schoonveld G (1979) Paratuberculosis (Johne's disease) in bighorn sheep and a Rocky Mountain goat in Colorado. J Wildl Dis 15:221–227

Williams JS (1999) Compensatory reproduction and dispersal in an introduced mountain goat population in central Montana. Wildl Soc B 27:1019–1024

Wolff PL, Schroeder C, McAdoo C, Cox M, Nelson DD, Evermann JF, Ridpath JF (2016) Evidence of bovine viral diarrhea virus infection in three species of sympatric wild ungulates in Nevada: life history strategies may maintain endemic infections in wild populations. Front Microbiol 7:292. https://doi.org/10.3389/fmicb.2016.00292

Wolff PL, Blanchong JA, Nelson DD, Plummer PJ, McAdoo C, Cox M, Besser TE, Muñoz-Gutiérrez J, Anderson CA (2019) Detection of *Mycoplasma ovipneumoniae* in pneumonic mountain goat (*Oreamnos americanus*) kids. J Wildl Dis 55:206–212. https://doi.org/10.7589/2018-02-052

Chapter 23
American Bison (*Bison bison*): A Rangeland Wildlife Continuum

Dustin H. Ranglack, Glenn E. Plumb, and Luke R. Rogers

Abstract American bison (*Bison bison*) are the largest extant land animal in North America and have an important history and contemporary role in modern conservation. Bison historically had the widest continental distribution of all native ungulates but now only function as wildlife under natural selection on < 1.2% of the original range. Bison as rangeland wildlife occur on an array of exclusive and overlapping governance jurisdictions (e.g., Federal, State, Provincial, County, and Tribes and First Nations), private not-for-profit conservation lands enterprises, zoo and education enterprises, and for-profit commodity production. The historical and prevailing relationships within and between these higher order sectors are very complex and often conflicting, yet each sector has invested tremendous effort and public and private resources to increase the total abundance of bison to present levels. Despite long-term public investment in wild bison conservation, the private sector has far outstripped wild bison, resulting in a potentially divergent evolution trajectory towards species domestication. The primary ecosystem function of plains bison on rangelands is contributing to plant community heterogeneity through patchily distributed grazing events that create mosaics of grazing pressure. Additionally, bison exhibit a myriad of other roles in their environment through direct and indirect interactions. Perhaps more than with other rangeland wildlife species, genetics play an outsized role in current bison population management given historical bottlenecks and intentional cross breeding of bison and cattle. However, moving forward the interplay between population size, isolation, and genetic diversity is more important. Along the continuum of bison management there exist a wide variety of rangeland management techniques. However, as a wildlife species, the rangeland management practices associated with bison have generally focused on disturbance ecology with a more recent push to understand the impacts of bison grazing at scale. The question of scale is important given that every bison is behind a barrier, thus restricting their impacts on rangeland ecology and processes. Bison and cattle are considered by many to be

D. H. Ranglack (✉) · L. R. Rogers
Biology Department, University of Nebraska at Kearney, 2401 11th Ave., Kearney, NE 68849, USA
e-mail: dustin.ranglack@usda.gov

D. H. Ranglack · G. E. Plumb · L. R. Rogers
IUCN SSC Bison Specialist Group, 319 South 9th Street, Livingston, MT 59047, USA

© The Author(s) 2023
L. B. McNew et al. (eds.), *Rangeland Wildlife Ecology and Conservation*,
https://doi.org/10.1007/978-3-031-34037-6_23

potential competitors, due to large overlaps in diet and body size, and much research has focused on the ecological equivalence of the two species. While this is still not without controversy, bison and cattle are not incompatible when properly managed. Chronic infection of wild bison populations with diseases that can be transmitted to livestock and humans is an important factor affecting potential recovery of bison outside existing reserve boundaries. Climate change may represent the next major challenge to bison, as it is expected to directly affect bison through decreased forage and water availability and increased thermal stress. These threats, combined with the differences in bison management practices between sectors have led some to classify bison as moderately vulnerable to climate change, recommending the creation of a 'bison coalition' that could seek climate change adaptation solutions through shared stewardship. While much of the continental historical range is no longer available for bison restoration, there are exciting conservation opportunities that are finding voice through the vision of "Shared Stewardship" that embraces innovative collaboration to work together across jurisdictions and sectors to successfully address the scale, complexity, and ecological and cultural significance of wild bison.

Keywords American bison · Climate change · Competition · Conservation · Disease · Ecological functions · Fire · Genetic diversity · Jurisdiction · Shared stewardship

23.1 Introduction

American bison (*Bison bison*) are the largest extant land animal in North America and have an important history and contemporary role in modern conservation. The determined persecution of the American bison during the nineteenth century across the species' once continental abundance and distribution, followed by a narrow escape from extinction and the loss of many native people's lifeways and traditional homelands, is now entrenched in the narratives of the peoples and nations of North America (Aune and Plumb 2019). A diverse suite of enterprises emerged in the early twentieth century to underwrite the species' increase across four orders of magnitude (e.g., from 100 s to 100,000 s) so that the species is once again widely distributed, albeit extremely patchily, across much of the breadth of its historical range (Plumb et al. 2014; Aune et al. 2017). Rogers (2021) estimates that approximately 350,000–400,000 bison now exist within at least 2,500 distinct herds ranging in size from 10s to 1,000s on a wide array of Tribal, national, state, province, county public, private for-profit and nonprofit lands, and accredited zoos in Mexico, Canada, and the United States (Fig. 23.1). This trajectory has now placed the American bison in an unprecedented position across the full scope of North American rangeland wildlife, with less than 2% of the total abundance functioning as rangeland wildlife. Thus, while the American bison may seem to be an iconic and ubiquitous species, worthy of their designation as the United States national mammal, recovery as a wildlife species is not yet assured. There is hope for ecological restoration of the species (Redford et al.

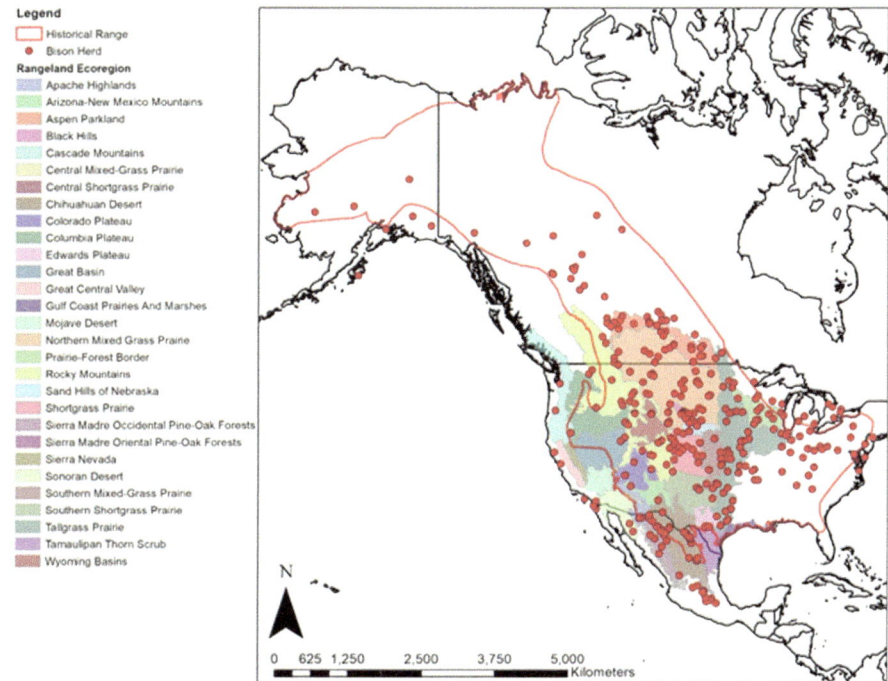

Fig. 23.1 North American private and public bison herds surveyed by Rogers (2021). Historical range is based on Sanderson et al. (2008), Plumb and McMullen (2018) and updated with information from the Mexican Bison Working group (List pers. comm. 2021). Rangeland ecoregions are Environmental Protection Agency level III ecoregions

2016), but time is running out. This chapter focuses on recent advances in understanding of life history and synecology from across an array of herd sizes, rangelands and management approaches, and the challenges and opportunities remaining for full ecological restoration of the American bison as wildlife.

23.2 Species and Population Status

23.2.1 Historic Range

American bison historically had the widest continental distribution of all native ungulates (Roe 1970; Gates and Ellison 2010; Plumb et al. 2014; Aune et al. 2017), and now only functions as wildlife under natural selection on < 1.2% of the original range (Sanderson et al. 2008; Aune et al. 2017). The maximum distribution was from southwest desert grasslands (~30° N × 110° W) to the floodplain meadows and northern forests of interior Alaska and Canada (~65° N × 155° W) and from western basin-range systems (~120° W) eastward across the Rocky Mountain cordillera, and then more continuous across the breadth of the Great Plains and Midwest deciduous forest savannahs, eventually dispersing across the Appalachian Mountains into the eastern coastal plains (~75° W), and from near sea level up to 4,000 m elevation (Gates and Ellison 2010; Bailey 2016; Cannon 2018) (Fig. 23.1). Martin et al (2017) and Plumb and McMullen (2018) reassessed multi-disciplinary evidence and concluded that the Colorado Plateau should be included within the southwestern periphery of the species historic range.

The plains bison (*B. b. bison*) proliferated in North America to an estimated 60 million individuals prior to European contact (Gates et al. 2010; Aune and Plumb 2019). Despite their prolific abundance and vast range in the early 1800s, extreme hunting pressure for profit and intentional subjugation of native peoples resulted in the total bison continental meta-population dwindling to less than 1,000 individuals by the late nineteenth Century (Shaw and Lee 1997). Wood bison (*B. b. athabascae*) were historically less plentiful than their southern plains relatives. Despite an expansive range from mid-Alberta into interior Alaska, initial estimates based on available suitable habitat suggest a pre-European settlement wood bison population of 168,000 individuals (Soper 1941). Further analysis of historically suitable habitat in Alaska identified a significantly larger range than previously estimated by Soper (1941), suggesting that the population size could have been larger (Stephenson et al. 2001). Wood bison were initially able to avoid extirpation due to their northern distribution (Soper 1941), but by 1840, wood bison began to suffer significant declines caused primarily by human overharvest (Gates et al. 2010). By the early twentieth century, only approximately 300 wood bison survived in the wild near the Great Slave Lakes, Northwest Territories, Canada (Fuller 2002).

There is clear evidence that within this continental historical range, the distribution and abundance of the American bison varied widely and included extensive areas with near-continual presence to areas with only intermittent low-density abundance (Isenberg 2000; Stephenson et al. 2001; List et al. 2007; Gates et al. 2010; Plumb et al. 2014; Flores 2016). Flores (1991) illustrates how extensive droughts resulted in multiple intervals spanning centuries when bison were almost totally absent from the southern Great Plains, including between BCE 5000–2500 and AD 500–1300. Given the highly variable abundance and density of bison across its historical range, it is unclear whether low-density peripheral populations were indeed less viable compared to higher density central populations in accordance with MacArthur and Wilson (1967) and Brown and Kodric-Brown (1977), or whether smaller peripheral populations contributed to species viability via local adaptation (Nielsen et al. 2001; Eckert et al. 2008); and whether the edge of historical range was simply an "expansion threshold" wherein peripherally dispersed small populations persisted through local adaptation or failed either through catastrophic mortality events or when genetic drift reduced genetic diversity below that required for adaptation to a heterogeneous environment (Polechová 2018). In summary, there is strong evidence that extensive variability in temporal and spatial patterns of abundance and distribution occurred throughout the historical range (Gates et al. 2010; Plumb et al. 2014; Aune et al. 2017;

Plumb and McMullen 2018), and that these variable patterns should be critically considered anywhere wild bison conservation now occurs, especially at the range periphery (Gates et al. 2010; Plumb and McMullen 2018).

23.2.2 The Continuum

The resurgence of American bison across the historical range in the past 120 years has occurred through a continuum of diverse sectors of legal status and purposes. Bison as rangeland wildlife occur on an array of exclusive and overlapping governance jurisdictions (e.g., Federal, State, Provincial, County, and Tribes and First Nations). Privately-owned bison have also become core assets of private not-for-profit conservation lands enterprises, zoo and education enterprises. Most bison are now living as privately-held for-profit assets on private lands. The historical and prevailing relationships within and between these higher order sectors are very complex and often conflicting, yet each sector has invested tremendous effort and public and private resources to increase the total abundance of bison to present levels. While there is not yet a comprehensive, rigorous, and comparable accounting or long-term monitoring of the total abundance of the American bison across all sectors and jurisdictions, there are several reliable assessments within key sectors. What we do know is that there is now a continuum of wildness from free-ranging herds, under an array of strong natural selection forces, to captive herds exclusively under an array of non-natural selection forces. These two divergent conditions bracket a continuum of diminishing wildness within bison management from wild to the edge of domestication (Gates 2014; Plumb et al. 2014; Aune et al. 2017). The emergent disparity in abundance over the past 50 years between wild bison (10,000s) to not-wild bison (100,000s) is staggering. Indeed, despite an increase in the number of fenced, conservation-focused herds in the past 50 years, the abundance of free-ranging wild bison has remained relatively constant.

Through the Bison Specialist Group (BSG) of the International Union for Conservation of Nature (IUCN), Gates et al. (2010) completed a comprehensive continental species assessment and conservation guidelines, and Aune et al. (2017) completed an updated Red List Assessment for the species that considers the likelihood of species extinction through accounting of all wild bison within historical range. The United States Department of Interior (DOI) maintains records of bison on lands managed by the National Park Service, U.S. Fish and Wildlife Service, and Bureau of Land Management (see Department of the Interior 2014) and recently completed a comprehensive meta-population viability assessment of all bison herds on DOI lands (Hartway et al. 2020). The Committee on the Status of Endangered Wildlife in Canada published a comprehensive assessment and status report on the plains bison in Canada (COSEWIC 2004, 201). The Mexico National Commission of Natural Protected Areas (CONANP) works with partners to maintain information on public and private bison throughout Mexico. Individual state or provincial wildlife management agencies monitor and maintain information on bison managed as wildlife under their jurisdiction. The National Bison Association (United States) primarily represents private producers and maintains updated information on bison ranching in North America (National Bison Association 2021), along with their northern counterpart, the Canadian Bison Association, and a variety of regional/state/provincial associations. The United States Department of Agriculture and Statistics Canada (the national statistical office) track their respective private bison sectors; the Inter-Tribal Buffalo Council maintains records about how many bison are living on member tribal lands, and the American Zoo Association maintains information on bison in member zoos.

23.2.3 Publicly-Owned Wildlife Conservation Status

In the most recent IUCN Red List Assessment, Aune et al. (2017) reported an approximate total of 31,000 bison in 68 conservation herds that are managed in the public interest by federal, state, and provincial governments and non-profit environmental organizations across North America, including 20,000 Plains Bison and 11,000 Wood Bison. A Red List Assessment focuses on a species in the wild under natural selection forces, and accordingly Aune et al. (2017) denoted three categories of conservation bison: (1) functioning as wild, (2) functioning as wild with limitations, and (3) not functioning as wild. Aune et al. (2017) classified the species in the wild as "Near-Threatened" and nearly qualifying as "Vulnerable" because the wild species is entirely conservation dependent, e.g., while the wild species is not currently in decline, the number of wild mature individuals could be greatly reduced if current management regimes are changed or removed. Nearly half (30 of 68) of the conservation bison herds, totaling ~ 2,700 bison combined, were denoted as not functioning as wild due to very small (< 300) population size on small, fenced landscapes (< 10,000 acres) for education, public viewing, and research. Another 18 herds, totaling ~ 9,500 bison combined, were denoted as wild with limitation because they are intensely managed behind fences and culled by artificial selection. Thus, only ~ 18,800 bison in 20 herds were denoted as functioning as wild under a range of natural selection forces (Fig. 23.2), and 4000 of these bison live in 12 herds containing < 400 total individuals, the lowest estimate of a minimum viable population size (MVP; Gross and Wang 2005), though it is likely closer to 1,000 (Hedrick 2009; Gates et al. 2010). As juveniles account for ~ 35% of a wild herd, Aune et al. (2017) estimated the total mature wild bison population in North America to be only ~ 12,000 animals. When we focus further on bison in rangeland habitats, that number is even smaller, with only 6 free-ranging wild herds totaling approximately 4000 adults that are subject to natural selection pressures. These 6 herds all occur outside of the historical bison strongholds of the central and northern Great Plains (Fig. 23.2).

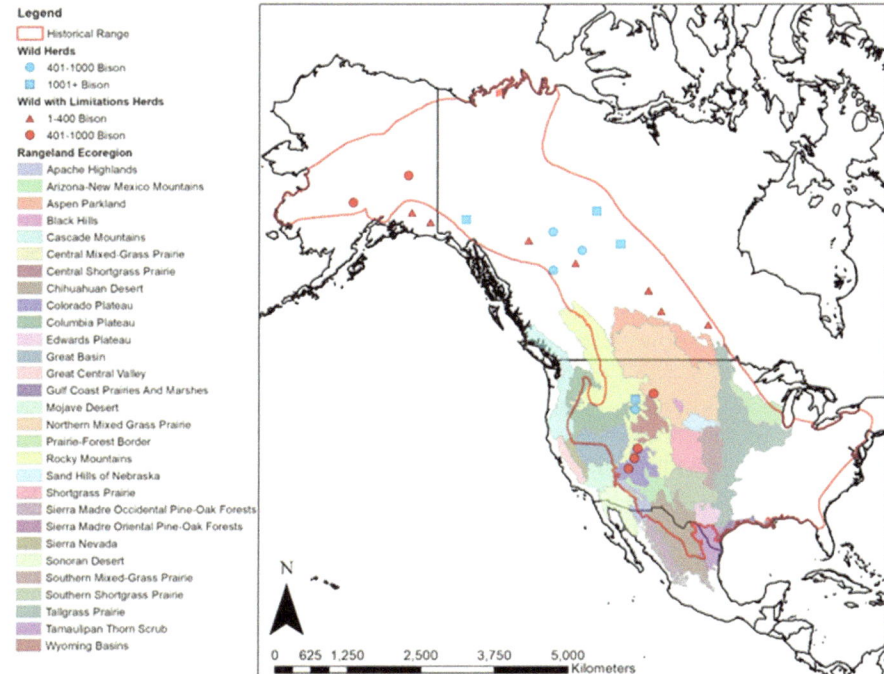

Fig. 23.2 Wild North American bison herds according to Aune et al. (2017). Historical range is based on Sanderson et al. (2008), Plumb and McMullen (2018) and updated with information from the Mexican Bison Working group (List pers. comm. 2021). Wild herds are free-ranging, managed as wildlife under natural selection, on large landscapes (> 10,000 acres), and greater than 400 individuals. Wild with limitations herds are limited in population (< 400), have limited predation from large carnivores, and/or are a subspecies outside of the historical range, but function otherwise similarly to wild herds. Rangeland ecoregions are Environmental Protection Agency level III ecoregions

23.2.4 Tribes and First Nations

The InterTribal Buffalo Council (ITBC) was convened in 1991 to restore the American bison to Indian Country, and now includes 69 federally recognized Tribes from 19 states with ~ 20,000 Plains Bison on Tribal lands in the United States. The ITBC vision is that reestablishing healthy buffalo populations on Tribal lands will reestablish hope for Indian people; and that returning bison to Tribal lands will help heal the land, the animal, and the spirit of the Indian people. In 2014, dignitaries from U.S. Tribes and Canadian First Nations signed the "Northern Tribes Buffalo Treaty" to establish an inter-tribal alliance cooperating to restore bison on Tribal/First Nations Reserves or co-managed lands within the U.S. and Canada. Collectively these treaty Tribes own and manage ~ 6.3 million acres of grassland and prairie habitats in the United States and Canada and have articulated a goal to achieve ecological restoration of the bison, and, in so doing, reaffirm and strengthen ties that formed the basis for traditions thousands of years old, including youth education and cultural restoration among the tribes.

23.2.5 Private Lands Commodity

There are now roughly 300,000 American bison under private ownership under agriculture laws and policies of Canada, United States, and Mexico (Statistics Canada 2016; United States Department of Agriculture 2017; National Bison Association 2021), with an unknown relatively small number in private ownership in Europe and Australia (Rogers 2021). Combined, the private bison producer sector slaughters ~ 70,000 bison annually, compared to the annual North American beef slaughter of ~ 45 million cattle (National Bison Association 2021). The National Bison Association (NBA) has initiated a web-based Conservation Management Program for bison farmers and ranchers to monitor conservation practices to improve their overall stewardship outcomes for bison, the land, and surrounding communities, though it is not required of all members. NBA also has announced a "Million Bison" marketing campaign to triple the number of bison across all sectors, with the private sector playing a major role (National Bison Association 2021).

23.3 Population Monitoring

Population abundance and demography monitoring occurs at the individual herd level across all proprietorship sectors. Most privately-owned bison herds are monitored through direct observation in the field and during annual round-ups, in which vaccinations, disease tests, and pregnancy tests are often administered to assess individual condition and herd health (Rogers 2021). During these round-ups, individuals are often separated for transfer between herds, sale for slaughter, or culling to manage population abundance. Free-ranging herds are monitored through traditional wildlife survey techniques, including ground and aerial surveys. Aerial surveys generally incorporate sightability indices based upon mark-recapture estimates derived from radio-collared animals (see Hess 2002). Distinct age- and sex-specific traits readily allow direct ground sex classification of individual adults and juveniles (Gates et al. 2010). Demographic classification via aerial surveys is possible by visual observation, though use of stabilized digital photography yields higher resolution information that can be assessed following the aerial survey. The free-ranging Henry Mountains bison herd in Utah, for example, does not have an annual round up, and relies purely on ground and aerial surveys (Bates and Hersey 2016; Terletzky and Koons 2016), along with hunter harvest data managed by the Utah Division of Wildlife Resources. Yellowstone National Park has a systemic approach to its annual population surveys, conducting intensive aerial surveys at the beginning of the summer and winter seasons, followed by duplicate aerial surveys, and ground classification surveys to refine estimates of population abundance and demography (Hess 2002).

At the species level, assessments for genetic integrity and extinction risk are completed at semi-regular intervals by government and non-government agencies. Recently the DOI conducted a population viability analysis for all bison herds on federal lands to assess their long-term ability to persist into the future (Hartway et al. 2020). The IUCN BSG conducts Red List Assessments for the American bison based upon data contributed and collated across all conservation herds (see Aune et al. (2017) and plans to publish the next assessment in 2024 (Greg Wilson, personal communication). In addition to Red List Assessments, the IUCN has recently developed the Green Status Assessment, which is a tool that assesses the ecological recovery legacy (e.g., late 1800s–2021) and potential for additional recovery a species has at specific future time intervals (Akçakaya et al. 2018; Grace et al. 2021). The first Green Status Assessment for bison was completed in 2022 (Rogers et al. 2022).

23.4 Life History and Population Dynamics

23.4.1 Description

A compact, large body and a large head set on a strong neck, combined with a pronounced hump and horns curving inwards give the American bison a widely recognized iconic appearance (Fig. 23.3). Weighing up to 1000 kg and with body length of 2.1–3.5 m, and shoulder height 1.5–2 m, it is the largest terrestrial mammal of the Western Hemisphere (Nowak 1991; Shaw and Meagher 2000; Reynolds et al. 2003). Sexual dimorphism occurs among adults with males ~ 20–30% larger, yet females resemble males in color, general body configuration, and presence of permanent horns that are short and sharp from the side of the head that laterally curve upward over the head, with female horns slenderer and showing a greater tendency to curve inward toward the tips (Reynolds et al. 1982, 2003; Nowak 1991; Shaw and Meagher 2000). Allen (1876) first described the species with a narrow muzzle with long pointed nasal bones composed of premaxillae, maxillae, and nasals, with tubular orbits composed of frontals, lacrimals and jugals without preorbital vacuities in the skull. Pelage of the head, neck, shoulders, and front legs is brownish-black, long, and shaggy; while the rest of the body is covered with shorter brown hairs that lighten through sun bleaching (Nowak 1991; Shaw and Meagher 2000). Albino or white/gray pelage is rare but known to occur (Meagher 1973). The bison has a distinctive tufted tail and chin hair that usually resembles a goatee-type beard (Banfield 1974). Young calves are orange-brown to reddish-brown "buff" color that gradually darkens to adult coloration by 4 months (Fig. 23.4; Nowak 1991; Shaw and Meagher 2000). Males present a distinctive shoulder hump suggesting forequarters out of proportion to smaller appearing hindquarters (Reynolds et al. 1982). Bison produce a variety of sounds, including a male "bellow" heard most frequently during the breeding season, and a "snort" and "cough" associated with antagonistic behavior. Cows searching for calves will exhibit a series of snorts, and calves can exhibit bawling (Fuller 1960).

Fig. 23.3 Young adult male bison in the Henry Mountains of southern Utah (photo: Dustin H. Ranglack)

Fig. 23.4 Bison calf in Yellowstone National Park, Wyoming (photo: Dustin H. Ranglack)

23.4.2 Growth

Typically, a single calf is born between 15 and 30 kg and begins grazing and drinking water within a week, while continuing to nurse for approximately 7–12 months (McHugh 1958; Fuller 1960; Halloran 1961; Meagher 1986). Calves double their body mass by 3 months of age, and weigh between 135 and 180 kg by 8–9 months of age (McHugh 1958; Halloran 1961; Meagher 1973, 1986; Gogan et al. 2010). The general age of bison may be determined in the field by body size, and horn size and shape. For both sexes, there is strong growth of body mass and inward-curving horns by 4 years of age, with female bison fully grown by 4 years, and males fully grown by 6 years (Banfield 1974). Distinct sex differences in horn size and shape occur earlier than differences in body size and mass. Male horns grow continuously until full development by 7–8 years of age, with horn tips then frequently becoming worn down and rounded, due to rubbing against trees and aggression with other males, and horn bases larger in diameter than their eyes. Female horns grow longer and more curved inward with age, with 20+ year old females retaining sharp horn tips which are generally smaller in diameter.

23.4.3 Reproduction

Like the European bison (*Bison bonasus*), the American bison exhibits a polygynous tending-bond mating system wherein non-territorial males court individual oestrous females for up to 3–4 days during an annual rut concentrated in August–September, with July seeing initial increased time spent by mature males beginning to search for receptive females (Plumb et al. 2014). The annual rut often starts later at higher latitudes (Fuller 1962). In herds with an even sex ratio, males between 3 and 6 years old are capable of breeding, but generally are prevented by older, larger, and more experienced males (McHugh 1958; Halloran 1961; Lott 1981; King et al. 2019). Females reach sexual maturity between 2 and 4 years old with a first calf often at 3 years of age (McHugh 1958; Halloran 1961; Fuller 1962; Meagher 1973; Gogan et al. 2010). Estrous lasts 19–26 days with females being receptive for 1–2 days, with only 1–2 ovulations during an annual breeding season (Haugen 1974; Rutberg 1986; Kirkpatrick et al. 1991). Females are often fertile up to 16 years of age (Green 1990) and produce young every 1–3 years depending on their age and physical condition (Kirkpatrick et al. 1993). Gestation is 285 days and fetal sex ratios are often male-biased (Fuller 1960; Meagher 1973; Haugen 1974; Rutberg 1986). Birth synchrony is common with 80% of births occurring during April–June (Haugen 1974; Rutberg 1984; Jones et al. 2010), with synchrony especially noticeable in populations where predation on calves occurs (Gates and Larter 1990). Earlier onset of birth synchrony has been observed in landscapes with earlier onset of spring vegetation growth, and it is thought this adaptation yields increased lactation quality and neonate survival based on higher vegetation quality (Gogan et al. 2005). Conversely, females in poor nutritional condition, with debilitating diseases, or of lower rank may calve later and/ or show low synchrony in birthing (Berger 1992; Green and Rothstein 1993a; Berger and Cain 1999; King et al. 2019).

Green and Rothstein (1993b) observed that females born early are more fecund throughout their life compared to calves born towards the end of birth synchrony. Most mature males spend the balance of the year solitary or in small bachelor groups, only joining mixed age-sex herds during the annual rut (Meagher 1973). Competitive mate selection is driven by male competitive dominance through threat displays and short-duration violent pair-wise matches (McHugh 1958; Lott 1981). Females often segregate themselves from the herd or group prior to parturition while lying down, followed by freeing the calf by consumption of the placental membrane and licking amniotic fluid from the calf's fur. Suckling often initiates within 10 min, and newborn calves can stand and continue to nurse within 30 min of birth (McHugh 1958; Meagher 1986; Lott 2002).

23.5 Population Dynamics

American bison are generally long-lived, with females occasionally reaching 25 years age and males rarely exceeding 20 years (Gates et al. 2010; Aune et al. 2017; Hartway et al. 2020). Overall adult annual survival rate generally approaches 90% in fenced-protected herds (Gates et al. 2010). In free-ranging populations below carrying capacity, the annual adult survival rate is variable across the continental distribution, wherein an adult survival rate of 75% was observed for wood bison at the Mackenzie Bison Sanctuary, and 95% for plains bison at the Jackson Hole National Wildlife Refuge and Grand Teton National Park (Larter et al. 2000; US Fish and Wildlife Service and National Park Service 2007). Age-specific annual survival rate slightly favors females over males and declines linearly for both sexes from > 75% up to 3 years age to ~ 50% by 12–15 years age, with a subsequent sharply punctuated decline to < 5% between 16 and 20 years age (Hartway et al. 2020). Individual bison infected with brucellosis and tuberculosis at Wood Buffalo National Park exhibited lower age- and sex-specific survival rate than bison with only one of the two diseases, or not infected at all (Joly and Messier 1999, 2004a, b, 2005; Bradley and Wilmshurst 2005).

Annual population growth rate for free-ranging populations also is variable across the continental distribution, from r = 0.08 with an 8-year generation length for relatively wild conditions with predation pressure; to r = 0.15–0.19 with a 9–10 year generation length for herds without dramatic environmental stochasticity and little predation pressure (Hartway et al. 2020). Under stochastic environmental conditions (e.g., drought, wildland fire, winter severity) or reduced genetic variability, growth rates will be lower, especially for smaller herds (Turner et al. 1994; Green et al. 1997; Wallace et al. 2004; Geremia et al. 2008; Hartway et al. 2020). Large mortality events are known to have occurred historically due to wildland fire (see Haley 1936; Hart and Hart 1997) and occasionally still occur at higher latitudes when bison drown after falling through thin ice in spring and fall (Roe 1970; Gates et al. 1991; Mech et al. 1995). Otherwise, bison are capable of swimming short and long distances. Larter et al. (2003) observed bison swimming across a 1.7 km-wide section of the Liard River, taking 27 min with downstream movement to swim a total of 3.6 km.

23.6 Habitat Associations

As their name suggests, the plains bison is the sub-species that historically occurred across a diverse array of North American rangeland ecoregions (Fig. 23.1). While the plains bison was a dominant keystone species amongst the rangeland ecoregions of the Great Plains (Knapp et al. 1999), the species was also widely distributed across non-rangeland ecoregions throughout much of the North America (Gates et al. 2010; Plumb et al. 2014; Aune et al. 2017). Not-wild plains bison now occur in many diverse rangeland ecoregions, albeit at severely fragmented and significantly reduced spatial scales behind fences subject to agricultural laws and policies (Fig. 23.1).

Free-ranging wild bison are very limited in abundance and distribution throughout their historical range (Rogers 2021), with the IUCN Green Status designating bison as Critically Depleted (Rogers et al. 2022). Out of the 21 free-ranging wild bison herds considered in the most recent IUCN Red List Assessment (Aune et al. 2017), only six were present within western rangelands: the Northern Mixed Grass Prairie, Rocky Mountain, and Colorado Plateau rangeland ecoregions (Fig. 23.2). The other 15 free-ranging wild bison herds outside of western rangelands are present in ecoregions consisting predominantly of aspen parkland, boreal forests, and wetlands in Canada and Alaska (Aune et al. 2017). Concomitant with limited abundance and distribution, the few free-ranging populations are managed to restrict large scale dispersal, range expansion or migratory movement patterns beyond their designated reserve landscapes (Plumb et al. 2014; Aune et al. 2017). Restricting the larger scales of bison habitat associations may thus restrict the fundamental ecological functionality of bison across rangeland types, often yielding continuous smaller-scale grazing patterns inside restricted landscapes rather than large-scale high-intensity and short-duration grazing events (i.e., with potential for prolonged periods of absence) (Gates et al. 2010). Indeed, it has been suggested that American bison are 'terrestrial castaways', stranded on 'island' ranges within a matrix of inaccessible habitat (Ritson 2019). As such, Augustine et al. (2019) suggest increasing spatial scales for rangeland wildlife through novel partnerships for cross-jurisdiction management that could then support large-scale bison movements.

23.6.1 Bison Diet

Diet varies by rangeland ecoregion, climate regime and time of year, with overall average peak dietary quality at the height of summer in June (Bermann et al. 2015;

Craine 2021). Cool, wet climates produce the highest amount of crude protein and digestible plant organic matter, resulting in larger average body mass compared to hot and dry climates (Craine 2021). Plains bison are predominantly grazers and exhibit optimal foraging ecology in response to dynamic temporal and spatial patterns of graminoid forage availability and quality (Plumb and Dodd 1993; Knapp et al. 1999), typified by shifts from cool season (C3) graminoids during spring to warm season graminoids (C4) during peak summer primary production, and back to cool season graminoids depending on late-summer and early-fall precipitation. Consumption of herbaceous forbs, legumes, and woody half-shrubs is exhibited but is generally indicative of non-selective foraging in relation to total availability (Plumb and Dodd 1993; Knapp et al. 1999), and more common in the spring and fall (Begmann et al. 2015).

23.6.2 Ecosystem Influences

The primary ecosystem function of plains bison on rangelands is contributing to plant community heterogeneity through patchily distributed short-duration, high-intensity grazing events that create mosaics of grazing pressure (Jonas and Joern 2007; Gates et al. 2010; Tastad 2014). At larger spatial scales, plains bison can contribute to enhanced total primary productivity by stimulating compensatory vegetative growth characterized by seasonal grazing lawns (Coppock et al. 1983; McNaughton 1984; Ranglack and du Toit 2015b; Merkle et al. 2016; Geremia et al. 2019). Geremia et al. (2019) demonstrated how plains bison on montane grasslands along a strong elevational gradient at Yellowstone National Park not only respond to the onset of spring phenology at lower elevations and continue to "surf" the leading edge of high-quality forage "green wave;" but also create large scale grazing lawns along the elevational gradient that optimizes foraging efficiency and quality. When in high enough abundance, short duration intensive bison grazing stimulates plant material regrowth and delays maturation, allowing bison to continue to consume high-quality plant protein even after they fall behind the leading edge of vegetation green-up (Merkle et al. 2016; Geremia et al. 2019).

Beyond their primary ecosystem function, bison exhibit a myriad of other roles in their environment through direct and indirect interactions. Where it is allowed to occur, the decomposition of bison carcasses at the site of mortality produce biochemical hotspots. Rich in calcium and with elevated pH levels, these biochemical hotspots promote plant growth on the landscape for two to four years after mortality (Knapp et al. 1999; Towne 2000; Melis et al. 2007; Bump 2008; Bump et al. 2009). Additionally, these biochemical hotspots can facilitate heterogeneity and reduce forest expansion, as observed by reduction in aspen expansion on fescue grasslands in Riding Mountain National Park of Canada (Knapp et al. 1999; Bump 2008). Bison can also contribute to dispersal of forb and graminoid seeds through shed hair and feces (Dinerstein 1989; Rosas et al. 2008). Both subspecies of bison can create significant abundance of small-scale disturbance within small and large landscapes through wallows, trampling, horning, and grazing (Coppedge and Shaw 1997; Fox et al. 2012). Wallow pits, when filled with water from rain or flooding events, have been identified as essential breeding sites and aquatic habitat for anurans and invertebrates (Gerlanc and Kaufman 2003). Obligate shortgrass prairie bird species' populations respond positively following bison restoration (Wilkins et al. 2019). Prairie dogs (*Cynomys* spp.) have a mutually beneficial relationship with bison; bison prefer to occupy sites near prairie dog towns, which provide higher crude protein and nitrogen content, and bison facilitate high plant productivity through fecal deposit, improving later foraging for prairie dogs (Coppock et al. 1983; Krueger 1986; Cid et al. 1991). Bison presence may be critical to restoring and sustaining some unique habitats such as the Canadian Sandhills, which supports several endangered and threatened species (Fox et al. 2012). In the Great Plains, grasshopper and other herbivorous insect species richness is directly and positively related to bison grazing pressure (Joern 2005). Bison landscape disturbance also has a negative impact on woody vegetation and positive impact facilitating the growth of grasses, sedges, and other graminoids in a prairie landscape (Coppedge et al. 1998). Bison fur is utilized by many species including red squirrels (*Tamiasciurus hudsonicus*) to insulate nests and burrows (Jung et al. 2010). The breadth and importance of roles bison exhibit within their habitat have led some to classify the plains bison as a keystone species in prairie ecosystems (Knapp et al. 1999; Fuhlendorf et al. 2010).

23.7 Genetics

Perhaps more than with other rangeland wildlife species, genetics play an outsized role in current bison population management. Two major forces have shaped the genetic make-up of all modern-day bison, both being human caused and irreversible. The first of these, and largest in terms of impact, is intentional hybridization of bison and cattle, as early as the late 1800s, that created viable offspring, which were then backcrossed into "pure" bison populations, leaving a legacy of cattle genetics in the resultant bison genome. This has long been of concern in the mitochondrial genome, as female plains bison with cattle mitochondrial introgression may have some indeterminant degree of reduced fitness (Derr et al. 2012). Nuclear cattle genes are also known to have also been perpetuated in the bison genome, and using a panel of 15 microsatellite loci, Halbert and Derr (2008) and Ranglack et al. (2015a) reported that cattle nuclear or mitochondrial introgression had been detected in all U.S. plains bison conservations herds except Yellowstone National Park, Wind Cave National Park, and the Henry Mountain bison herd in Southern Utah. In Canada, plains bison in Elk Island National Park are also considered to be free of cattle introgression (COSEWIC 2013). However, the technological scope of microsatellite testing also limits its inferential power, in that bison have 29 autosomes, and a panel of 15 microsatellite loci leaves entire chromosomes overlooked.

New research, using single nucleotide polymorphism (SNP) tests that examine the entire bison genome, has dramatically restructured our understanding of the consequential scope of historical hybridization, and now confirms that all contemporary plains bison herds have encountered some downstream cattle introgression in the nuclear genome (Stroup et al. 2022); including Wind Cave and Yellowstone National Parks (and therefore also the Henry Mountains herd that was established using individuals from Yellowstone in the early 1940s). Evidence now strongly indicates that 3 male bison that were introduced into the Yellowstone National Park bison herd in 1902 from the Goodnight herd (Texas), as part of an effort to boost the dwindling population in the park, were in fact hybrids that contributed to some unknown scope of breeding and explains the lack of mitochondrial introgression found in the Yellowstone population. As such, evidence now indicates that all modern plains bison herds possess variable levels of artifactual cattle introgression event(s) that occurred over 100 years ago (Stroup et al. 2022). Alternatively, Wang et al. (2018) conclude that introgression in the European bison may represent natural evolutionary legacies, whereas introgression represents incomplete lineage sorting of shared common ancestry with cattle. Still, such introgressed genetic artifacts, from the time of near-extinction of the wild American bison, do not now appear to restrain free-ranging wild bison reproductive processes and phenotypic expression (Dratch and Gogan 2010). Contemporary wild free-ranging bison with relatively higher levels of introgression, such as the Northern Rim population in northern Arizona (Hedrick 2010), exhibit no observable phenotypic traits of cattle (pelage, body conformation, etc.); they look and behave like bison, and they produce viable male and female offspring, indicating essential functionality of chromosomal DNA (Plumb and McMullen 2018).

As we look forward, the interplay between population size, isolation, and genetics is more important than the residual artifacts of a history that cannot be rewritten. A vibrant life history that is rooted in competitive mate selection and natural selection pressures remains the key to avoiding breeding dominance by individual males leading to reduced genetic diversity (Gates et al. 2010).

23.7.1 Genetic Bottlenecks

The second major impact humans have had on bison is the result of the bottleneck following the near extinction of the species by the end of the nineteenth century, and management practices related to their subsequent recovery (Pertoldi et al 2010). Historical records and genetic analyses indicate that all plains bison today may be descended from only 30 to 50 individuals from 6 captive herds (Hedrick 2009), and the estimated 25 wild bison that remained in Yellowstone National Park (Meagher 1973). Given that these captive herds were typically privately owned, a considerable exchange of animals between herds took place, leading to a relative homogenization of bison genetically. While in some regards this genetic exchange may have enhanced genetic diversity within those source herds, there is no doubt that there was an enormous loss of genetic diversity during this time, and that there were distinct historical lineages of bison that are now extinct (Stroup et al. 2022). While bison have recovered numerically, subsequent bottlenecks resulting from founding new herds with small numbers of individuals (Halbert and Derr 2008) have led to significant concern for the genetic health and long-term survival of certain bison populations (Hedrick 2009; Dratch and Gogan 2010; Hartway et al. 2020). This is exacerbated by the fact that most bison herds have been restricted in size and geographic range, with limited genetic exchange taking place between conservation herds (though exchange of individuals within the private sector is common). There is specific concern over the loss of genetic diversity due to genetic drift, which is the random loss of genetic material from generation to generation (Allendorf et al. 2013). This loss of genetic diversity can have very specific short- and long-term impacts on bison population viability, due to increased risk of inbreeding depression, which was documented in the decline of the Texas state bison herd (Halbert et al. 2004, 2005).

A recent population viability analysis conducted on 12 bison herds by DOI and 2 bison herds managed by Parks Canada revealed that all herds are predicted to lose genetic diversity (measured as both heterozygosity and allelic diversity) over the next 200 years due to genetic drift (Hartway et al. 2020), with smaller herds losing diversity faster than larger herds, as would be expected. This is concerning, as decreases in genetic diversity could also decrease the ability of herds to adapt to changing environmental conditions (Weeks et al. 2011; Ralls et al. 2018). Given that all these herds are subject to management removals to control the herd size due to social, political, or biological constraints, the management strategy directing removals were shown to be particularly important for reducing the amount of diversity loss, with strategies that target younger animals or using mean kinship for removal resulting in lower levels of heterozygosity loss (Hartway et al. 2020). Herd size, however, is the most important driver of genetic diversity loss, with larger herds losing diversity at a lower rate (Gross and Wang 2005; Hartway et al. 2020), however, this is often limited by other constraints, biological or otherwise, highlighting the need for a more comprehensive metapopulation management strategy that re-establishes gene flow between populations, through natural movements and/or translocation of individuals.

23.7.2 Genetic Augmentation

Genetic augmentation (Frankham et al. 2017) has been shown to be a successful strategy in reversing the effects of inbreeding depression in a variety of species (Bouzat et al. 2009; Johnson et al. 2010). This strategy of increasing gene flow to isolated populations is valuable (Whiteley et al. 2015; Frankham et al. 2017), but not without risk. There must be a balance both in the number individuals and frequency of translocations to ensure that there is enough genetic material transferred as to increase the genetic diversity in the recipient herd (Frankham et al. 2017), but not so much as to swamp out local adaptation or rare alleles (Edmands 2006; Allendorf et al. 2013).

Hartway et al. (2020) evaluated five different scenarios for bison metapopulation management, varying which herds were used as sources and recipients in each scenario. They suggested that smaller, less frequent translocations (i.e., 2 individuals every 10 years, 3 individuals every 7 years) using either the least-related herds as source populations, or alternating which herds are used as the source herd at each translocation event is adequate for increasing genetic diversity in most herds, while minimizing the loss of diversity at the metapopulation level. Smaller, less genetically diverse herds may benefit from more frequent translocations (Hartway et al. 2020). It is now clear that a metapopulation management strategy to re-connect isolated herds through natural or human-facilitated movement of individuals is required for the long-term conservation of bison. Moving forward, bison conservation genetics need to (1) maintain stable population size and avoid large fluctuations in abundance, (2) encourage competitive mate selection by maintaining adult breeding males approaching a 1:1 sex ratio, and (3) mitigate genetic drift by periodically augmenting isolated herds with additional animals as a part of a larger metapopulation (Dratch and Gogan 2010; Hartway et al. 2020).

23.8   Rangeland Management

Along the continuum of bison management there exists a wide variety of rangeland management techniques, from intensive management of bison movements through the use of adaptive multi-paddock grazing (Hillenbrand et al. 2019) to year-long continuous grazing, with every variation in between. However, as a wildlife species, the rangeland management practices associated with bison have generally focused on disturbance ecology, with an emphasis on bison's relationship with fire (Fuhlendorf et al. 2009), with a more recent push to understand the impacts of bison grazing at scale (Augustine et al. 2019; Geremia et al. 2019). This question of scale is particularly important given that every bison is found behind a barrier—be it biological, social, physical, political, or otherwise—the movements of bison on the modern landscape are greatly restricted, thus restricting their impacts on rangeland ecology and processes.

23.8.1   Bison and Fire

The relationship between bison and fire is well documented for plains ecosystems (Fuhlendorf et al. 2009), with bison exhibiting a strong preference for recently burned areas, attracted by the high green vegetation: senescent vegetation ratios and high-quality forage that emerges due to nutrient release follow fires (Allred et al. 2011). Before European settlement, wildfire would have been common on the Great Plains, but in other parts of the historic bison range, the fire return interval would have been longer and more sporadic, with fire return interval estimates ranging from every 8 years to fire being rare (Anderson 2002). These periodic fires would have prevented shrub and conifer encroachment into open habitat types and maintained piñon–juniper (*Pinus edulis–Juniperus osteosperma*), ponderosa pine (*Pinus ponderosa*), and other woodlands in a more savanna-like state (West 1984), except in steep and rocky areas that were unlikely to provide significant foraging opportunities for bison. Prescribed burning is commonly used to replicate natural fire return intervals in rangelands used by free-ranging bison, as well as conservation herds on conservation lands behind fences (Plumb et al. 2014). Ranglack and du Toit (2015b) showed that forage quality, as indexed by fecal N, was higher, and that bison spent more time feeding relative to moving in previously burned areas, while mechanically treated areas were more like naturally occurring open habitat types. This response was detected even 10 years after a fire, indicating that bison were likely creating grazing lawns in these areas, thus maintaining the forage in a high-quality state (Ranglack and du Toit 2015b). Indeed, bison not only track plant phenology and respond to differences in forage quality at the landscape scale, but also modify and engineer plant phenological responses through their grazing (Geremia et al. 2019).

23.8.2 Bison and Cattle

Bison are largely absent from traditional range management systems, as bison and cattle are considered by many to be potential competitors, due to large overlaps in diet and body size (Van Vuren and Bray 1983; Plumb and Dodd 1993), and much research has focused on the ecological equivalence of the two species (Steuter and Hidinger 1999; Fuhlendorf et al. 2010; Allred et al. 2011, 2013; Kohl et al. 2013), given how cattle have largely replaced bison in rangeland systems. Bison dietary overlap with cattle (Vuren and Bray 1983; Plumb and Dodd 1993), combined with the conspicuous nature of bison, underpins continuing concerns over whether the two species should be allowed to share common rangelands (Ranglack et al. 2015b). In the Henry Mountains of Utah, one of the few places where bison and cattle co-mingle on shared rangeland, lagomorphs were found to have twice the impact on forage reduction than bison (Ranglack et al. 2015b), thus presenting a far greater competitive threat to cattle than bison. Indeed, bison grazing caused no significant impact on plant species composition (Ware et al. 2014) and relatively small reductions in forage availability (Ranglack et al. 2015b). This is likely because bison and cattle tend to spatially segregate on shared rangelands, as bison are more likely to range widely across the landscape, using steep slopes and venturing farther from water sources, while cattle focus grazing efforts in areas near water and on more flat terrain (Van Vuren 2001; Allred et al. 2011; Ranglack 2014). Thus, while there is potential for exclusionary competition, at this time there is little evidence that it occurs on shared rangelands, even in areas along the edges of the species range where resources are most limiting, and you would therefore expect competition to be most severe (Ranglack 2014; Ranglack et al. 2015b). While this is still not without controversy, bison and cattle are not incompatible when properly managed, and there may be opportunities to integrate the ranching and wildlife sectors on western public lands through managed bison populations (Ranglack and du Toit 2015a).

23.8.3   Spatial Scale, Distribution, and Abundance

One of the largest management concerns with bison on modern rangelands is the manipulation and maintenance of appropriate bison distribution and abundances. Understanding the ecological processes, impacts, and interactions (movement, activity and behavior, habitat interactions, population demography, and gene flow/introgression) of bison across time and space is critical to our evaluation of bison as a keystone species and the management of the species. While some of these processes take place over short time scales (foraging decisions; Gogan et al. 2010; Plumb and Dodd 1993) and/or small spatial extents (movements within or between feeding patches; Meagher 1989; Plumb and Dodd 1993), many require large spatial scales (migration, dispersal, range expansion; Gates et al. 2005; Plumb et al. 2009) and/or only occur at decadal time scales or longer (between population gene flow; Dratch and Gogan 2010; Halbert and Derr 2008). Thus, managers need to understand the spatiotemporal scale of the various bison ecological processes to determine what aspects of bison ecology are being conserved at the available spatial scale, and where management needs to allow ecological processes to operate are larger scales than currently available (Augustine et al. 2019).

In wild bison, spatiotemporal variability in overall habitat quality influences many aspects of the behavior of animals in groups, such as group size, composition, and behavior within groups including where, when, and for how long group members forage (Lima and Zollner 1996; WallisDeVries 1996). Optimal foraging theory (OFT) predicts that higher quality foraging patches will lead to larger group sizes (Schoener 1971; Hirth 1977) and more time spent feeding versus vigilance (Lima and Dill 1990; Lima 1995) or moving (Ranglack and du Toit 2015c). Also, ideal free distribution (IFD) theory predicts that the equilibrium distribution of organisms among habitats of different quality, such as results after some patches of rangeland have, or have not, been subjected to natural disturbances or habitat manipulation (e.g., fire or mechanical treatment), will indicate the relative resource qualities of those habitats (Fretwell and Lucas 1969; Fretwell 1972).

Bison meet the main assumptions of IFD theory (Fretwell 1972) in that they are energy maximizers (Van Vuren 2001) and they are long-lived animals that, in most cases, have been present on the landscape of interest for several generations, allowing many foraging patches to be discovered and known. However, learning and memory have also been shown to be important in bison habitat selection (Merkle et al. 2014, 2015a, b; Sigaud et al. 2017). Bison can remember pertinent information about the location and quality of different foraging sites, and thus use that information to choose foraging areas where energy gains could be maximized (Merkle et al. 2014). However, site fidelity still plays an important role, as bison may not always choose the most productive foraging sites, as predicted by OFT and IFD, but show fidelity to previously visited foraging sites (Merkle et al. 2015a). The fusion-fission society of bison herds, however, creates a situation where individuals within a group may have different knowledge of foraging sites, and group familiarity combined with individual knowledge influence decisions on whether to follow the group to a foraging area that may be unknown to the individual but is known by the group, or to leave the group and return to a familiar foraging patch (Merkle et al. 2015b). This memory based foraging strategy allows bison to sample new foraging areas, whether being led there by individuals who have already have knowledge of the site (Merkle et al. 2015b) or through random patch use (Sigaud et al. 2017) that allows for higher energy gains in bison (Merkle et al. 2017). Rangeland managers may therefore use passive techniques for managing distribution through manipulation of habitat quantity and quality across the rangeland, and active techniques such as herding or fencing to manipulate bison distribution both in the present and future.

Abundance is a special consideration for bison, given their unique situation behind barriers—biological, social, physical, political, or otherwise. Rangeland managers must take special care to understand the carrying capacity of the range, both biological and social, and then maintain the appropriate density (Plumb et al. 2009; Steenweg et al. 2016; Cherry et al. 2019), which, given the extremely broad distribution of bison (Figs. 23.1 and 23.2) will vary dramatically and be unique to each bison herd. Unlike many other rangeland wildlife species whose abundance is maintained either largely through sport hunting or natural processes, bison abundances are managed primarily through human activities. In many areas bison populations are maintained through sport hunting (Ranglack and du Toit 2015a), but the most common tool for removing excess bison is using round-ups with excess animals being sold, donated, or culled (Millspaugh et al. 2008; White et al. 2011; Giglio et al. 2018).

23.9   Disease

Following the catastrophic decimation of wild bison by the late nineteenth century, the remaining few wild bison increasingly encountered domestic cattle and their diseases. By 1917, bison at Yellowstone National Park were infected with the non-native disease brucellosis via contagion from domestic cattle that were kept in the park to provide milk and meat for the US Army stationed at Fort Yellowstone (Meagher and Meyer 1994). Brucellosis and tuberculosis were subsequently introduced into the greater Wood Buffalo National Park area in Canada when plains bison, previously infected via cattle, were relocated there during 1925–1928, and through contagion from local cattle herds (Tessaro 1992). Currently, brucellosis, tuberculosis, and anthrax are focal diseases in some populations of wild bison (Aune et al. 2010, 2017; Plumb et al. 2014; National Academies of Sciences Engineering and Medicine 2020). Malignant catarrhal fever can also occur as acute and chronic cases in individual bison, most often associated with mixing with a carrier such as domestic sheep (Schultheiss et al. 1998, 2001).

23.9.1   Brucellosis

Bovine brucellosis is caused by the bacterium *Brucella abortus* that lives as a faculta-tive intracellular parasite (Thorne 2001). Brucellosis causes abortions, still births, and can cause crippling arthritis in infected joints (Williams et al. 1997; Thorne 2001; Geremia et al. 2008; National Academies of Sciences Engineering and Medicine 2020). Only infectious pregnant females have a high probability of shedding live *Brucella* bacteria, that can be transmitted intra-specifically from fecund females to their offspring or through lactation, and inter-specifically through the ingestion of live bacteria from infected birth tissues (Cheville et al. 1998; Rhyan et al. 2009). Sexual transmission from males to females is rare in bison (Robison 1994). Brucellosis can result in lower pregnancy rates and population growth rates (Geremia et al. 2008), but otherwise does not affect adult wild bison survival or limit population increase in the wild (Fuller et al. 2007a, b; National Academies of Sciences Engineering and Medicine 2020).

23.9.2   Tuberculosis and Anthrax

Bovine tuberculosis is caused by the bacterium *Mycobacterium bovis* and is presented as acute debilitating pathology to respiratory, digestive, urinary, nervous, skeletal, and reproductive systems, with transmission by ingestion or inhalation, and from mother to offspring through the placental connection or contaminated milk (Tessaro et al. 1990). Tuberculosis can contribute to reduced population growth rate in wild bison through fetal losses and decreased pregnancy rates due to poorer condition and increased vulnerability of older animals to predation (Tessaro et al. 1990; Joly and Messier 2005). Anthrax is caused by the bacterium *Bacillus anthracis* and is transmitted by inhalation or ingestion of endospores, that are non-reproductive forms of the bacterium that can remain viable but dormant for decades before reactivating when environmental conditions become favorable (Aune et al. 2010). Anthrax is detected primarily in mature male bison (Dragon et al. 1999) and thus this disease appears to have little influence on bison population dynamics unless operating in conjunction with other limiting factors (Aune et al. 2010).

23.9.3   Disease Management

Chronic infection of wild bison populations with diseases that can be transmitted to livestock and humans is an important factor affecting potential recovery of bison outside existing reserve boundaries (Gates et al. 2001; Plumb et al. 2009). While the management of brucellosis and tuberculosis in wild bison is warranted based on risk to livestock and human health (Aune et al. 2010), management authorities face difficult challenges in restoring bison and their ecological processes (e.g., migration, dispersal, grazing influences) while preventing transmission to domestic cattle near reserve boundaries (White et al. 2011). In the Greater Yellowstone Area (GYA) of Montana, Wyoming and Idaho, bison were likely the first chronic wild reservoir of brucellosis, yet wild elk are now the primary wildlife host and all recent cases of brucellosis in GYA cattle are traceable to elk, not bison (Scurlock and Edwards 2010; National Academies of Sciences Engineering and Medicine 2020). Elk now maintain chronic infection at the population level, even when there is no direct contact with feed grounds or with infected bison (Cross et al. 2010a, b, 2013; National Academies of Sciences Engineering and Medicine 2020). Despite chronic brucellosis infection at the population level, lack of transmission from bison to cattle is likely a result of spatial and temporal separation management practices outlined in the Interagency Bison Management Plan (2016) combined with fewer cattle operations on some lands adjacent to areas used by bison during late winter, e.g., during the third trimester of pregnancy when potential shedding of the bacteria into the open environment is highest; thereby effectively managing the transmission risk, as opposed to a lack of transmission risk itself (Rhyan et al. 2009; Treanor et al. 2015; National Academies of Sciences Engineering and Medicine 2020). Detection of brucellosis in domestic bison and cattle automatically invokes mandatory national and state domestic animal health regulations including test-slaughter of infected individuals and potential for depopulation of entire herds (Cheville et al. 1998; National Academies of Sciences Engineering and Medicine 2020). However, these domestic livestock regulations do not automatically apply to free-ranging bison managed under federal or state wildlife authorities. In circumstances like the Greater Yellowstone Area, where wild bison and elk are the last remaining chronic reservoir of brucellosis across the United States, federal and state livestock and wildlife authorities are developing and implementing cooperative long-term adaptive risk management strategies to inform timely and evidence-based decisions for reducing the risk of transmission, including risk management through spatial and temporal separation of bison and cattle, iterative hypothesis testing and periodic scientific assessments (Cheville et al. 1998; Plumb et al. 2007; Gates et al. 2010; Nishi 2010; Geremia et al. 2011; Hobbs et al. 2015; National Academies of Sciences Engineering and Medicine 2020).

23.10 Ecosystem Threats and Conservation Actions

23.10.1 Ecosystem Threats

The most recent IUCN Red List Assessment for the wild American bison classified the species as "Near-Threatened," and determined that as the wild species is completely dependent on active conservation protection, as it would likely revert to "Vulnerable" status if government protection was reduced (Aune et al. 2017). Additionally, the first IUCN Green Status Assessment for bison found bison to be critically depleted (Rogers et al. 2022). In Canada, an assessment of plains bison on rangelands in Canada recommended listing them as 'Threatened' (COSEWIC 2004, 2013). Otherwise, wild plains bison on rangelands have no formal protected status outside of the authorities of the agency jurisdiction, e.g., wild bison can be hunted on some US Forest Service lands but not on adjacent National Park Service lands. The plains bison has been unsuccessfully proposed several times for designation under endangered species authorities in Canada and the United States (Gates et al. 2010). Key threats to the species in the wild include habitat loss and fragmentation, genetic manipulation of commercial bison for market traits, small population effects in most conservation herds, few herds that are exposed to a wide range of natural selection factors, contemporary effects from historical cattle gene introgression, and the threat of depopulation as a management response to infection of some wild populations hosting reportable cattle disease (Gates et al. 2010; Plumb et al. 2014; Aune et al. 2017).

Approximately 93% of all American bison living on rangelands are legally managed as domestic livestock (Gates et al. 2010), with only four states in the U.S., two provinces in Canada, and one state in Mexico (Arizona, Utah, Montana, Wyoming, Alberta, Saskatchewan, and Chihuahua) that manage free-ranging bison as wildlife. Idaho, Missouri, New Mexico, and Texas designate bison as wildlife, but do not have any free ranging populations, and all other states and provinces across the continental historical range designate the species as domestic livestock. Hunting of wild bison on rangelands as a public trust wildlife resource occurs only in Arizona, Utah, Montana, Wyoming, Alberta, and Saskatchewan (Gates et al. 2010). Alarmingly, across all jurisdictions and legal authorities, less than 3% of all American bison on rangelands are managed as wildlife under a meaningful array of natural selection factors (e.g., competitive mate selection, predation, winter kill); while the vast majority of all bison on rangelands are otherwise subjected to anthropogenic selection for preferred population size and demography, body conformation, ease of management handling, and/or genetic manipulation to enhance profitable commercial bison market traits (Gates et al. 2010; Plumb et al. 2014; Aune et al. 2017). Despite long-term public investment in wild bison conservation, the private sector has far out-stripped wild bison, resulting in a potentially divergent evolution trajectory towards species domestication.

Sanderson et al. (2008) found the full continuum of the species now exists on < 2% of its historic range, and most of these sub-populations are managed in isolation with surplus offtake going to start new small, isolated herds or into food markets. There are only three free-ranging, wild, American bison sub-populations on rangelands that are greater than minimum viable population size (e.g., 400–1,000 total animals), that include only ~ 4,200 mature animals in total (Aune et al. 2017). All other herds managed as wildlife on rangelands live behind fences or are less than 400 animals. Despite this greatly unbalanced meta-population structure, there are no formal, long-term, wild bison meta-population viability strategies in place by any major stewardship sectors. Hartway et al. (2020) modeled several meta-population management strategies and found that translocations of selected age-sex class individuals between herds at the decadal scale could buffer these isolated herds against genetic drift at the century scale. This investigation also highlighted the critical importance of creating and maintaining additional large sub-populations under natural selection as core reserves for long-term genetic viability of the species.

Climate change may represent the next major challenge to bison, as it is expected to directly affect bison through decreased forage and water availability and increased thermal stress (Craine et al. 2009, 2013; Craine 2013; Martin and Barboza 2020a). There are also indirect effects of climate, through changes in the distribution and intensity of parasites (Patz et al. 2000; Kutz et al. 2005; Morgan and Wall 2009) and diseases (Janardhan et al. 2010) that have been shown to reduce reproductive success in bison (Fuller et al. 2007a). It is estimated that, with a 4 °C increase in global temperatures, these stressors could reduce bison body size by as much as 50% by the end of the twenty-first century (Craine 2013; Martin et al. 2018; Martin and Barboza 2020b). This could be compounded by the impacts of climate change on agriculture intensification, land use, and woody plant encroachment (Knapp et al. 1999; Allred et al. 2013; Bowler et al. 2020; Klemm et al. 2020). These threats, combined with the differences in bison management practices between sectors have led Martin et al. (2021) to classify bison as moderately vulnerable to climate change, recommending the creation of a 'bison coalition' that could seek climate change adaptation solutions through shared stewardship. That said, changing climate may also open new areas of habitat for bison north of their historical range, leading some to believe that bison may be able to occupy niches currently occupied by moose in boreal regions, though this has yet to be formally evaluated.

23.10.2   Conservation Actions

While much of the continental historical range is no longer available for recovery due to land use conversion as well as concerns about human safety and property damage, lack of local public support, and lack of funds for management as publicly owned wildlife (Boyd 2003; Plumb et al. 2009; Gates et al. 2010; Ranglack and du Toit 2015a), there are exciting conservation opportunities that are finding voice through a new vision of "Shared Stewardship" that embraces innovative collaboration to work together across jurisdictions and sectors to successfully address the scale, complexity, and ecological and cultural significance of wild bison conservation and restoration (Sanderson et al. 2008; Aune et al. 2017; Aune and Plumb 2018; Augustine et al. 2019; Martin et al. 2021; Pejchar et al. 2021). The US Department of Interior recently updated the charter for the federal "Bison Conservation Initiative" with a vision and commitment to leadership and alliances to ensure the conservation and restoration of wild American bison, focusing on wild, healthy bison herds, genetic conservation, and shared stewardship for ecological and cultural restoration. In Canada, a new free-ranging wild bison population was recently created in the foothill rangelands of Banff National Park, and a new expansive meta-population viability assessment is now being conducted for all federal and provincial bison herds in Canada (Wilson, personal communication). In 2014, the Blackfeet Nation, Blood

Tribe, Siksika Nation, Piikani Nation, the Assiniboine and Gros Ventre Tribes of Fort Belknap Indian Reservation, the Assiniboine and Sioux Tribes of Fort Peck Indian Reservation, the Salish and Kootenai Tribes of the Confederated Salish and Kootenai Indian Reservation, and the Tsuu T'ina Nation came together to sign the "Northern Tribes Buffalo Treaty" that formally establishes intertribal alliances for cooperation in the restoration of American bison on Tribal/First Nations Reserves or co-managed lands within the U.S. and Canada. Collectively, these Tribes/First Nations own and manage about 6.3 million acres of grassland and prairie habitats. The Northern Tribes Buffalo Treaty is a formal expression of political unity to achieve ecological restoration of the buffalo tribal lands, and in so doing to reaffirm and strengthen ties that formed the basis for traditions thousands of years old. In Montana, the Blackfeet Nation has recently embarked upon the Iinnii Initiative to restore wild bison to Blackfeet lands in partnership with Glacier and Waterton National Parks in Montana and Alberta. Another example of "Shared Stewardship" is the Sicangu Lakota Oyate nation living on the Rosebud Indian Reservation, South Dakota, who is partnering with the U.S. Department of the Interior and World Wildlife Fund for a herd of 1500 buffalo on tribal lands, which would make it the largest owned by a Native nation.

The Nature Conservancy (TNC) initiated a long-term program of bison stewardship in 1984 at the Samuel H. Ordway, Jr. Memorial Prairie on ~ 8,000 acres of virgin unplowed prairie on the Missouri Coteau in north-central South Dakota. This TNC program has now been expanded to include bison under conservation management on 13 preserves across the Great Plains. The American Prairie Reserve in northeastern Montana is a private lands project of the non-profit American Prairie Foundation, that is envisioned to include over 3 million contiguous acres (12,000 km^2) through a combination of both private and public lands to establish a fully functioning mixed grass prairie ecosystem, complete with several thousand migratory bison. In recognition of the ultimately potentially critical role of private lands for bison conservation, the National Bison Association published Conservation Management Guidelines for Herd Managers in partnership with World Wildlife Fund, with the goal to conserve the wild characteristics of bison on private lands through the conservation of the species' genetic and behavioral traits while at the same time supporting ecosystem function and biodiversity conservation goals on the range the herd inhabits (World Wildlife Fund 2013).

23.11 Research/Management Needs

The biggest challenges facing the species have been entrenched for decades and are largely driven by socioeconomic and political forces. The continuum of bison ranging from free-ranging wildlife to domestic livestock is structured around multiple legal designations of the species, which also ranges from wildlife to domestic livestock, or in some cases both depending on ownership. Only 10 U.S. states, 4 Canadian provinces, and one Mexican state legally classify bison as wildlife, with all other states and provinces within the historic range designating bison as a livestock species only (Aune and Wallen 2010).

All bison, whether wild or wild with limitations are ultimately restricted in their large landscape movements by a barrier—biological, social, physical, political, or otherwise—as "terrestrial castaways" (Ritson 2019). Fragmentation of the entire species into a disjunct metapopulation needs to be better understood and should be at the forefront of all management decisions. In the private sector where live bison are regularly bought and sold, the mixing of animals of different genetic lineages is common, but the public conservation sector has seen little of this. The recent population viability analysis previously discussed highlighted the need for management of DOI herds within a metapopulation framework to maintain genetic viability over the next 200 years (Hartway et al. 2020). Questions of genetic diversity and viability have previously had to be balanced by concerns over introgression of cattle genes into the bison genome, necessitating that the "pure" bison be managed in a separate metapopulation from the introgressed bison. However, given recent advances in genetics, it appears that cattle introgression is more common and widespread than previously thought (Stroup et al. 2022). As such, maintaining genetic diversity and population viability should be the guiding factor in metapopulation management, rather than attempting to eliminate introgression (Dratch and Gogan 2010). With wild bison numbering < 5% of the total bison abundance, it is crucial we monitor the abundance, distribution, and demographics of "wild" and "wild with limitations" bison (see Aune et al. 2017). New quantitative science based upon population viability analyses is needed to identify how many herds of variable size and level of isolation are needed for the long-term conservation of the species as wildlife. These are questions that are rarely asked of other rangeland wildlife, but for bison they are crucial for understanding the status of the species and planning for long-term conservation of bison into an uncertain future, with the ultimate goal of moving bison from "Near Threatened" to "Least Concern" IUCN Red List status (Aune et al. 2017).

23.12   Summary

Innovative approaches to "Shared Stewardship" of bison across sectors will likely be required to change current paradigms and limitations that imply that the only bison that really "matter" are those under explicit conservation management and that bison must always be behind fences and separate from cattle. Changing these paradigms requires a conscious effort to reach beyond what has typically been considered as conservation and recognize the strength provided by a diverse portfolio of management goals and strategies, along with a recognition, both legally and otherwise, that bison are a valuable native wildlife species that should be allowed to exist in the same manner as other native rangeland wildlife species. At an ecoregional scale, Plumb and McMullen (2018) reviewed whether the Colorado Plateau in northern Arizona should be included within the species historical range and highlighted how difficult it can be to challenge entrenched dogma and move forward with updated perspectives while avoiding confusing scientific inferences with societal value judgments. They further characterized sustainable bison conservation as occurring at the intersection of best available multidisciplinary science, compliance with law and policy, and long-term public interest; so that sustainability embraces historical reference conditions and a balance of local social and ecological concerns, all while demonstrably contributing to continental-scale bison conservation.

Bison have one of the strongest links to human culture of any species in North America, and those cultural connections are important for the restoration of both bison and native peoples, but also allows for bison to serve as an icon for all people. Sanderson et al. (2008: 252) presented an overarching continental vision statement "Over the next century, the ecological recovery of the North American bison will occur when multiple large herds move freely across extensive landscapes within all major habitats of their historic range, interacting in ecologically significant ways with the fullest possible set of other native species, and inspiring, sustaining and connecting human cultures." In response to scientific evidence and calls for a more inclusive approach to long-term bison conservation, the DOI Bison Conservation Initiative was recently updated with a bold and expansive approach to what will constitute the Department's bison portfolio moving forward (Department of the Interior 2020). The five main goals of the Bison Conservation Initiative capture key issues for long-term conservation of wild bison on rangelands of North America: (1) conserving bison as wildlife by minimizing artificial selection and allowing natural selection to operate, (2) genetic conservation through metapopulation management, (3) shared stewardship with states, tribes, and other stakeholders, (4) ecological restoration achieved through shared stewardship to establish and maintain large, wide-ranging bison herds and large landscapes where the full ecology of the species can function, and (5) cultural restoration through collaboration with Tribes and First Nations (see Shamon et al. 2022). These goals are universally applicable across the bison continuum, and although ecological recovery of wild free-ranging bison has been relatively static for many decades, there are now new and exciting alliances and opportunities that suggest the era of big conservation is not over; that indeed there is still hope, and after all, hope is a bison (Redford et al. 2016).

References

Akçakaya HR, Bennett EL, Brooks TM et al (2018) Quantifying species recovery and conservation success to develop an IUCN Green List of Species. Conserv Biol 32:1128–1138

Allen JA (1876) American bisons, living and extinct. Cambridge, Massachusetts, USA

Allendorf FW, Luikart G, Aitken SN (2013) Genetics and the conservation of populations, 2nd edn. Wiley-Blackwell, Oxford

Allred BW, Fuhlendorf SD, Hamilton RG (2011) The role of herbivores in Great Plains conservation: comparative ecology of bison and cattle. Ecosphere 2:26

Allred BW, Fuhlendorf SD, Hovick TJ et al (2013) Conservation implications of native and introduced ungulates in a changing climate. Glob Chang Biol 19:1875–1883

Anderson MD (2002) Pinus edulis. Fire Effects Information System

Augustine D, Davidson A, Dickinson K, Van Pelt B (2019) Thinking like a grassland: challenges and opportunities for biodiversity conservation in the Great Plains of North America. Rangel Ecol and Manag 78:281–295

Aune K, Gates CC, Boyd DP (2010) Reportable or notifiable diseases. In: Gates CC, Freese CH, Gogan PJP, Kotzman M (eds) American bison: status survey and conservation guidelines. International Union for Conservation of Nature, Gland, Switzerland, pp 27–38

Aune K, Jorgensen D, Gates C (2017) The IUCN Red List OF Threatened Species™

Aune K, Plumb GE (2019) Theodore Roosevelt and restoration of bison on the Great Plains. History Press, Charleston, SC, USA

Aune KE, Wallen RL (2010) Legal status, policy issues, and listings. In: Gates CC, Freese CH, Gogan PJP, Kotzman M (eds) American bison: status survey and conservation guidelines. International Union for Conservation of Nature, Gland, Switzerland, pp 63–84

Bailey JAA (2016) Historic distribution and abundance of bison in the Rocky Mountains of the United States. Intermountain Journal of Sciences 22:36–53

Banfield AWF (1974) The mammals of Canada. University of Toronto Press, Toronto

Bates B, Hersey K (2016) Lessons learned from bison restoration efforts in Utah on Western Rangelands. Rangelands 38:256–265

Berger J (1992) Facilitation of reproductive synchrony by gestation adjustment in gregarious mammals: a new hypothesis. Ecology 73:323–329

Berger J, Cain SL (1999) Reproductive synchrony in brucellosis-exposed bison in the southern Greater Yellowstone Ecosystem and in noninfected populations. Conserv Biol 13:357–366

Bermann GT, Craine JM, Robeson MS II, Fierer N (2015) Seasonal shirts in diet and gut microbiota of the American bison (*Bison bison*). PLoS ONE 10:e0142409

Bouzat JL, Ae JAJ, Toepfer JE et al (2009) Beyond the beneficial effects of translocations as an effective tool for the genetic restoration of isolated populations. Conserv Genet 10:191–201

Bowler DE, Bjorkman AD, Dornelas M et al (2020) Mapping human pressures on biodiversity across the planet uncovers anthropogenic threat complexes. People Nat 2:380–394

Boyd D (2003) Conservation of North American bison: status and recommendations. University of Calgary

Bradley M, Wilmshurst J (2005) The fall and rise of bison populations in Wood Buffalo National Park: 1971 to 2003. Can J Zool 83:1195–1205

Brown JH, Kodric-Brown A (1977) Turnover rates in insular biogeography: effect of immigration on extinction. Ecology 58:445–449

Bump JK (2008) Large predators, prey carcasses, resource pulses, and heterogeneity in terrestrial ecosystems. Michigan Technological University

Bump JK, Webster CR, Vucetich JA et al (2009) Ungulate carcasses perforate ecological filters and create biogeochemical hotspots in forest herbaceous layers allowing trees a competitive advantage. Ecosystems 12:996–1007

Cannon KP (2018) "They went as high as they choose:" what an isolated skull can tell us about the biogeography of high-altitude bison. Arctic, Antarct Alp Res 39:44–56

Cherry SG, Merkle JA, Sigaud M et al (2019) Managing genetic diversity and extinction risk for a rare Plains bison (*Bison bison bison*) population. Environ Manage 64:553–563

Cheville NF, McCullough DR, Paulson LR (1998) Brucellosis in the Greater Yellowstone Area. National Academy Press, Washington

Cid MS, Detling JK, Whicker AD, Brizuela MA (1991) Vegetational responses of a mixed-grass prairie site following exclusion of prairie dogs and bison. J Range Manag 44:100–105. https://doi.org/10.2307/4002305

Coppedge BR, Engle DM, Toepfer CS, Shaw JH (1998) Effects of seasonal fire, bison grazing and climatic variation on tallgrass prairie vegetation. Plant Ecol 139:235–246

Coppedge BR, Shaw JH (1997) Effects of horning and rubbing behavior by bison (Bison bison) on woody vegetation in a tallgrass prairie landscape. Am Midl Nat 138:189–196

Coppock DL, Ellis JE, Detling JK, Dyer MI (1983) Plant-herbivore interactions in a North American mixed-grass prairie—II. Responses of bison to modification of vegetation by prairie dogs. Oecologia 56:10–15

COSEWIC (2004) COSEWIC assessment and status report on the plains bison *Bison bison bison*. Canada, Ottawa

COSEWIC (2013) COSEWIC assessment and status report on the plains bison *Bison bison bison* and the wood bison *Bison bison athabascae* in Canada. Canada, Ottawa

Craine JM (2013) Long-term climate sensitivity of grazer performance: a cross-site study. PLoS ONE 8:e67065

Craine JM (2021) Seasonal patterns of bison diet across climate gradients in North America. Sci Rep 11:6829

Craine JM, Towne EG, Joern A, Hamilton RG (2009) Consequences of climate variability for the performance of bison in tallgrass prairie. Glob Change Biol 15:772–779

Craine JM, Towne EG, Tolleson D, Nippert JB (2013) Precipitation timing and grazer performance in a tallgrass prairie. Oikos 122:191–198

Cross PC, Cole EK, Dobson AP et al (2010a) Probable causes of increasing brucellosis in free-ranging elk of the Greater Yellowstone Ecosystem. Ecol Appl 20:278–288

Cross PC, Heisey DM, Scurlock BM et al (2010b) Mapping brucellosis increases relative to elk density using hierarchical Bayesian models. PLoS ONE 5:e10322

Cross PC, Maichak EJ, Brennan A et al (2013) An ecological perspective on Brucella abortus in the western United States. OIE Rev Sci Tech 32:79–87

Department of the Interior (2014) DOI bison report: looking forward. Fort Collins, Colorado

Department of the Interior (2020) Bison Conservation Initiative 2020. Fort Collins, Colorado

Derr JN, Hedrick PW, Halbert ND et al (2012) Phenotypic effects of cattle mitochondrial DNA in American bison. Conserv Biol 26:1130–1136

Dinerstein E (1989) The foliage-as-fruit hypothesis and the feeding behavior of South Asian ungulates. Biotropica 21:218

Dragon DC, Elkin BT, Nishi JS, Ellsworth TR (1999) A review of anthrax in Canada and implications for research on the disease in northern bison. J Appl Microbiol 87:208–213

Dratch PA, Gogan PJP (2010) DOI bison conservation genetics workshop report and recommendations. Fort Collins, Colorado

Eckert CG, Samis KE, Lougheed SC (2008) Genetic variation across species' geographical ranges: the central–marginal hypothesis and beyond. Mol Ecol 17:1170–1188

Edmands S (2006) Between a rock and a hard place: evaluating the relative risks of inbreeding and outbreeding for conservation and management. Mol Ecol 16:463–475

Flores D (2016) American serengeti: the last big animals of the great plains. University of Kansas Press, Lawrence

Flores D (1991) Bison ecology and bison diplomacy: the southern plains from 1800 to 1850. J Am Hist 78:465

Fox TA, Hugenholtz CH, Bender D, Gates CC (2012) Can bison play a role in conserving habitat for endangered sandhills species in Canada? Biodivers Conserv 21:1441–1455

Frankham R, Ballou JD, Ralls K et al (2017) Genetic management of fragmented animal and plant populations. Oxford University Press, Oxford

Fretwell SD (1972) Populations in a seasonal environment. Monogr Popul Biol 5:1–217

Fretwell SD, Lucas HL (1969) On territorial behavior and other factors influencing habitat distribution in birds. Acta Biotheor 19:16–36

Fuhlendorf SD, Allred BW, Hamilton RG (2010) Bison as keystone herbivores on the Great Plains: can cattle serve as proxy for evolutionary grazing patterns? Am Bison Soc Work Pap Ser ISSN 2153-3008

Fuhlendorf SD, Engle DM, Kerby J, Hamilton R (2009) Pyric herbivory: rewilding landscapes through the recoupling of fire and grazing. Conserv Biol 23:588–598

Fuller JA, Garrott RA, White PJ et al (2007a) Reproduction and survival of Yellowstone bison. J Wildl Manage 71:2365–2372

Fuller JA, Garrott RA, White PJ (2007b) Emigration and density dependence in Yellowstone bison. J Wildl Manage 71:1924–1933

Fuller WA (2002) Canada and the "buffalo", Bison bison: a tale of two herds. Can Field-Naturalist 116:141–159

Fuller WA (1960) Behaviour and social organization of the wild bison of Wood Buffalo National Park, Canada. Arctic 13:2–19

Fuller WA (1962) The biology and management of the bison of Wood Buffalo National Park. Can Wildl Serv Wildl Manag Bull 16:1–52

Gates C, Ellison K (2010) Numerical and geographic status. In: Gates C, Freese CH, Gogan PJP, Kotzman M (eds) American bison: status survey and conservation guidelines. International Union for Conservation of Nature, Gland, Switzerland, pp 55–62

Gates C, Freese CH, Gogan PJP, Kotzman M (2010) American bison: status survey and conservation guidelines 2010. Gland, Switzerland

Gates CC (2014) What is a wild bison? A case study of plains bison conservation in Canada. In: Melletti M, Burton J (eds) Ecology, evolution and behaviour of wild cattle: implications for conservation. Cambridge University Press, Cambridge, pp 373–384

Gates CC, Larter NC (1990) Growth and dispersal of an erupting large herbivore population in northern Canada: the Mackenzie wood bison (Bison bison athabascae). Arctic 43:231–238

Gates CC, Larter NC, Komers PK (1991) Size and composition of the Mackenzie bison population in 1989. Yellowknife, Northwest Territories, Canada

Gates CC, Stelfox B, Muhly T, Chowns T, Hudson RJ (2005) The ecology of bison movements and distribution in and beyond Yellowstone National Park: a critical review with implications for winter use and transboundary population management. Calgary, Alberta, Canada

Gates CC, Stephenson RO, Reynolds HW (2001) National recovery plan for the wood bison (Bison bison athabascae). Ontario, Canada, Ottawa

Geremia C, White PJ, Garrott RA et al (2008) Demography of central Yellowstone bison: effects of climate, density, and disease. Terr. Ecol. 3:255–279

Geremia C, White PJ, Wallen RL et al (2011) Predicting bison migration out of Yellowstone National Park using Bayesian models. PLoS ONE 6:e16848

Geremia C, Merkle JA, Eacker DR et al (2019) Migrating bison engineer the green wave. Proc Natl Acad Sci USA 116:25707–25713

Gerlanc NM, Kaufman GA (2003) Use of bison wallows by Anurans on Konza Prairie. Am Midl Nat 150:158–168

Giglio RM, Ivy JA, Jones LC, Latch EK (2018) Pedigree-based genetic management improves bison conservation. J Wildl Manage 82:766–774

Gogan PJP, Podruzny KM, Olexa EM et al (2005) Yellowstone bison fetal development and phenology of parturition. J Wildl Manage 69:1716–1730

Gogan PJP, Larter NC, Shaw JH et al (2010) General biology, ecology, and demographics. In: Gates C, Freese CH, Gogan PJP, Kotzman M (eds) American bison: status survey and conservation guidelines. International Union for Conservation of Nature, Gland, Switzerland, pp 39–54

Grace MK, Akcakaya HR, Bennett EL et al (2021) Testing a global standard for quantifying species recovery and assessing conservation impact. Con Bio

Green GI, Mattson DJ, Peek JM (1997) Spring feeding on ungulate carcasses by Grizzly Bears in Yellowstone National Park. J Wildl Manage 61:1040–1055

Green WCH (1990) Reproductive effort and associated costs in bison (Bison bison): do older mothers try harder? Behav Ecol 1:148–160

Green WCH, Rothstein A (1993a) Asynchronous parturition in bison: implications for the hider-follower dichotomy. J Mammal 74:920–925

Green WCH, Rothstein A (1993b) Persistent influences of birth date on dominance, growth and reproductive success in bison. J Zool 230:177–186

Gross JE, Wang G (2005) Effects of population control strategies on retention of genetic diversity in National Park Service bison (Bison bison) herds. Bozeman, Montana, USA

Halbert ND, Derr JN (2008) Patterns of genetic variation in US federal bison herds. Mol Ecol 17:4963–4977

Halbert ND, Raudsepp T, Chowdhary BP, Derr JN (2004) Conservation genetic analysis of the Texas state bison herd. J Mammal 85:924–931

Halbert ND, Ward TJ, Schnabel RD et al (2005) Conservation genomics: disequilibrium mapping of domestic cattle chromosomal segments in North American bison populations. Mol Ecol 14:2343–2362

Haley JE (1936) Charles goodnight: cowman & plainsman. Houghton Mifflin Company, Boston, Massachusetts

Halloran AF (1961) American bison weights and measurements from the Wichita Mountains Refuge. Proc Oklahoma Acad Sci 41:212–218

Hart RH, Hart JA (1997) Rangelands of the Great Plains before European settlement. Rangelands 19:4–11

Hartway C, Hardy A, Jones L et al (2020) Long-term viability of department or interior bison under current management and potential metapopulation strategies. Fort Collins, Colorado, USA

Haugen A (1974) Reproduction in the plains bison. Iowa State J Res 49:1–8

Hedrick PW (2009) Conservation genetics and North American bison (Bison bison). J Hered 100:411–420

Hedrick PW (2010) Cattle ancestry in bison: Explanations for higher mtDNA than autosomal ancestry. Mol Ecol 19:3328–3335

Hess SC (2002) Aerial survey methodology for bison population estimation in Yellowstone National Park. Montana State University

Hillenbrand M, Thompson R, Wang F et al (2019) Impacts of holistic planned grazing with bison compared to continuous grazing with cattle in South Dakota shortgrass prairie. Agr Ecosyst Environ 279:156–168

Hirth D (1977) Social behavior of white-tailed deer in relation to habitat. Wildl Monogr 53:3–55

Hobbs NT, Geremia C, Treanor J et al (2015) State-space modeling to support management of brucellosis in the Yellowstone bison population. Ecol Monogr 85:525–556

Interagency Bison Management Plan (2016) IBMP adaptive management. National Park Service, USDA-Forest Service, USDA-Animal & Plant Health Inspection Service, Montana Department of Livestock and Montana Fish, Wildlife & Parks. http://www.ibmp.info/adaptivemgmt.php. Accessed 2 May 2021

Isenberg AC (2000) The destruction of the bison. Cambridge University Press, Cambridge

Janardhan KS, Hays M, Dyer N et al (2010) Mycoplasma bovis outbreak in a herd of North American bison (Bison bison). J Vet Diagnostic Investig 22:797–801

Joern A (2005) Disturbance by fire frequency and bison grazing modulate grasshopper assemblages in tallgrass prairie. Ecology 86:861–873

Johnson WE, Onorato DP, Roelke ME et al (2010) Genetic restoration of the Florida panther. Science 329(5999):1641–1645

Joly D, Messier F (1999) The limiting effects of bovine brucellosis and tuberculosis on wood bison ecology within Wood Buffalo National Park. F. Rep. (Nov 1999)

Joly DO, Messier F (2004a) Factors affecting apparent prevalence of tuberculosis and brucellosis in wood bison. J Anim Ecol 73:623–631

Joly DO, Messier F (2004b) Testing hypotheses of bison population decline (1970–1999) in Wood Buffalo National Park: synergism between exotic disease and predation. Can J Zool 82:1165–1176

Joly DO, Messier F (2005) The effect of bovine tuberculosis and brucellosis on reproduction and survival of wood bison in Wood Buffalo National Park. J Anim Ecol 74:543–551

Jonas JL, Joern A (2007) Grasshopper (Orthoptera: Acrididae) communities respond to fire, bison grazing and weather in North American tallgrass prairie: a long-term study. Oecologia 153:699–711

Jones JD, Treanor JJ, Wallen RL, White PJ (2010) Timing of parturition events in Yellowstone bison Bison bison: implications for bison conservation and brucellosis transmission risk to cattle. Wildlife Biol 16:333–339

Jung TS, Kukka PM, Milani A (2010) Bison (*Bison bison*) fur used as drey material by red squirrels (*Tamiasciurus hudsonicus*): an indication of ecological restoration. Northwest Nat 91:220–222

King KC, Caven AJ, Leung KG et al (2019) High society: behavioral patterns as a feedback loop to social structure in plains bison (*Bison bison bison*). Mammal Res 64:1–12

Kirkpatrick JF, Kincy V, Bancroft K et al (1991) Oestrous cycle of the North American bison (*Bison bison*) characterized by urinary pregnanediol-3-glucuronide. J Reprod Fertil 93:541–547

Kirkpatrick JF, Gudermuth DF, Flagan RL et al (1993) Remote monitoring of ovulation and pregnancy of Yellowstone Bison. J Wildl Manage 57:407–412

Klemm T, Briske DD, Reeves MC (2020) Potential natural vegetation and NPP responses to future climates in the US Great Plains. Ecosphere 11:e03264

Knapp AK, Blair JM, Briggs JM et al (1999) The keystone role of bison in North American tallgrass prairie. Bioscience 49:39–50

Kohl MT, Krausman PR, Kunkel K, Williams DM (2013) Bison versus cattle: are they ecologically synonymous? Rangel Ecol Manag 66:721–731

Krueger K (1986) Feeding relationships among bison, pronghorn, and prairie dogs: an experimental analysis. Ecology 67:760–770

Kutz SJ, Hoberg EP, Polley L, Jenkins EJ (2005) Global warming is changing the dynamics of Arctic host-parasite systems. Proc R Soc B Biol Sci 272:2571–2576

Larter NC, Sinclair ARE, Ellsworth T et al (2000) Dynamics of reintroduction in an indigenous large ungulate: the wood bison of Northern Canada. Anim Conserv 3:299–309

Larter NC, Nishi JS, Ellsworth T et al (2003) Observations of wood bison swimming across the Liard River, Northwest Territories, Canada. Arctic 56:408–412

Lima SL (1995) Back to the basics of anti-predatory vigilance: the group-size effect. Anim Behav 49:11–20

Lima SL, Dill LM (1990) Behavioral decisions made under the risk of predation: a review and prospectus. Can J Zool 68:619–640

Lima SL, Zollner PA (1996) Towards a behavioral ecology of ecological landscapes. Trends Ecol Evol 11:131–135

List R, Ceballos G, Curtin C et al (2007) Historic distribution and challenges to bison recovery in the northern Chihuahuan Desert. Conserv Biol 21:1487–1494

Lott DF (1981) Sexual behavior and intersexual strategies in American bison. Z Tierpsychol 56:97–114

Lott DF (2002) American bison: a natural history. University of California Press

MacArthur RH, Wilson EO (1967) The theory of island biogeography. Monographs in Population Biology, Princeton

Martin JM, Barboza PS (2020a) Thermal biology and growth of bison (Bison bison) along the Great Plains: examining four theories of endotherm body size. Ecosphere 11:e03176

Martin JM, Barboza PS (2020b) Decadal heat and drought drive body size of North American bison (Bison bison) along the Great Plains. Ecol Evol 10:336–349

Martin JM, Martin RA, Mead JI (2017) Late Pleistocene and Holocene *Bison* of the Colorado Plateau. Southwest Nat 62:14–28

Martin JM, Mead JI, Barboza PS (2018) Bison body size and climate change. Ecol Evol 8:4564–4574

Martin JM, Zarestky J, Briske DD et al (2021) Vulnerability assessment of the multi-sector North American bison (Bison bison) management system to climate change. People Nat 00(pan3):10209

McHugh T (1958) Social behavior of the American buffalo (Bison bison bison). Zoologica 43:1–40

McNaughton SJ (1984) Grazing lawns: animals in herds, plant form, and coevolution. Am Nat 124:863–886

Meagher M (1973) The bison of Yellowstone National Park. National Park Service Monograph Series Number 1

Meagher M (1986) Bison. Mamm Species 16:1–8

Meagher M (1989) Range expansion by bison of Yellowstone National Park. J Mammal 70:670–675

Meagher M, Meyer ME (1994) On the origin of brucellosis in bison of Yellowstone National Park: a review. Conserv Biol 8:645–653

Mech LD, Carbyn LN, Oosenbrug SM, Anions DW (1995) Wolves, bison and the dynamics related to the peace-athabasca delta in Canada's Wood Buffalo National Park. J Wildl Manage 59:189

Melis C, Selva N, Teurlings I et al (2007) Soil and vegetation nutrient response to bison carcasses in Białowieża Primeval Forest, Poland. Ecol Res 22:807–813

Merkle JA, Fortin D, Morales JM (2014) A memory-based foraging tactic reveals an adaptive mechanism for restricted space use. Ecol Lett 17:924–931

Merkle JA, Cherry SG, Fortin D, Kotler BP (2015a) Bison distribution under conflicting foraging strategies: site fidelity vs. energy maximization. Ecology 96:1793–1801

Merkle JA, Sigaud M, Fortin D (2015b) To follow or not? How animals in fusion-fission societies handle conflicting information during group decision-making. Ecol Lett 18:799–806

Merkle JA, Monteith KL, Aikens EO et al (2016) Large herbivores surf waves of green-up during spring. Proc R Soc B 283:20160456

Merkle JA, Potts JR, Fortin D (2017) Energy benefits and emergent space use patterns of an empirically parameterized model of memory-based patch selection. Oikos 126

Millspaugh JJ, Gitzen RA, Licht DS et al (2008) Effects of culling on bison demographics in Wind Cave National Park. South Dakota. Natural Areas J 28(3):240–250

Morgan ER, Wall R (2009) Climate change and parasitic disease: farmer mitigation? Trends Parasitol 25:308–313

National Academies of Sciences Engineering and Medicine (2020) Revisiting brucellosis in the greater Yellowstone Area. National Academy Press, Washington, DC

National Bison Association (2021) Bison by the numbers. https://bisoncentral.com/bison-by-the-numbers/. Accessed 2 May 2021

Nielsen JL, Scott JM, Aycrigg JL (2001) Letter to the editor: endangered species and peripheral populations: cause for conservation. Endangered Species Update 18:194–196

Nishi JS (2010) A review of best practices and principles for bison disease issues: greater Yellowstone and wood buffalo areas. Am Bison Soc Work Pap Ser 3:1–101

Nowak RM (1991) Walker's mammals of the world. Johns Hopkins University Press, Baltimore

Patz JA, Graczyk TK, Geller N, Vittor AY (2000) Effects of environmental change on emerging parasitic diseases. Int J Parasitol 30:1395–1405

Pejchar L, Medrano L, Niemiec RM et al (2021) Challenges and opportunities for cross-jurisdictional bison conservation in North America. Biol Cons 256:109029

Pertoldi C, Tokarska M, Wojcik JM et al (2010) Phylogenetic relationships among the European and American bison and seven cattle breeds reconstructed using the BovineSNP50 Illumina Genotyping BeadChip. Acta Theriol 55:97–108

Plumb G, Babiuk L, Mazet J et al (2007) Vaccination in conservation medicine. OIE Rev Sci Tech 26:229–241

Plumb G, McMullen C (2018) Bison on the Southwest Colorado Plateau: conservation at the edge. Southwest Nat 63:42–48

Plumb GE, Dodd JL (1993) Foraging ecology of bison and cattle on a mixed prairie: implications for natural area management. Ecol Appl 3:631–643

Plumb GE, White PJ, Coughenour MB, Wallen RL (2009) Carrying capacity, migration, and dispersal in Yellowstone bison. Biol Conserv 142:2377–2387

Plumb GE, White PJ, Aune K (2014) American bison bison bison (Linnaeus, 1758). Ecology, evolution and behaviour of wild cattle: implications for conservation. Cambridge University Press, Cambridge, pp 83–114

Polechová J (2018) Is the sky the limit? On the expansion threshold of a species' range. PLoS Biol 16:e2005372

Ralls K, Ballou JD, Dudash MR et al (2018) Call for a paradigm shift in the genetic management of fragmented populations. Conserv Lett 11:e12412

Ranglack DH (2014) American bison ecology and bison-cattle interactions in an isolated montane environment. Utah State University

Ranglack DH, du Toit JT (2015a) Bison with benefits: towards integrating wildlife and ranching sectors on a public rangeland in the western USA. Oryx 50:1–6

Ranglack DH, du Toit JT (2015b) Wild bison as ecological indicators of the effectiveness of management practices to increase forage quality on open rangeland. Ecol Indic 56:145–151

Ranglack DH, Du Toit JT (2015c) Habitat selection by free-ranging bison in a mixed grazing system on public land. Rangel Ecol Manage 68:349–353

Ranglack DH, Dobson LK, Du Toit JT, Derr J (2015a) Genetic analysis of the Henry Mountains Bison Herd. PLoS ONE 10:e0144239

Ranglack DH, Durham S, du Toit JT (2015b) Competition on the range: science vs. perception in a bison-cattle conflict in the western USA. J Appl Ecol 52:467–474

Redford KH, Aune K, Plumb G (2016) Hope is a bison. Conserv Biol 30:689–691

Reynolds HW, Glaholt RD, Hawley AWL (1982) Bison. In: Chapman JA, Feldhammer GA (eds) Wild mammals of North America: biology, management, and economics. Johns Hopkins University Press, Baltimore, pp 972–1007

Reynolds HW, Gates CC, Glaholt RD (2003) Bison (Bison bison). In: Chapman JA, Feldhammer GA (eds) Wild mammals of North America: biology, management, and conservation. Johns Hopkins University Press, Baltimore, pp 1009–1060

Rhyan JC, Aune K, Roffe T et al (2009) Pathogenesis and epidemiology of brucellosis in Yellowstone bison: serologic and culture results from adult females and their progeny. J Wildl Dis 45:729–739

Ritson RJ (2019) The spatial ecology of bison (Bison Bison) in multiple conservation herds across the American West. University of Nebraska at Kearney

Robison CD (1994) Conservation of germplasm from Brucella abortus-infected bison (Bison bison) using natural service, embryo transfer, and in vitro maturation/in vitro fertilization. Texas A&M University

Roe FG (1970) The North American Buffalo. A critical study of the species in its wild state, 2nd edn. University of Toronto Press, Toronto

Rogers LR (2021) Variation in American bison (*Bison bison*) ecological functionality across management sectors: Implications for conducting conservation assessments. University of Nebraska, Kearney

Rogers LR, Ranglack DH, Plumb G (2022) Bison bison (Green Status assessment). The IUCN Red List of Threatened Species 2022: e.T2815A281520221

Rosas CA, Engle DM, Shaw JH, Palmer MW (2008) Seed dispersal by Bison bison in a tallgrass prairie. J Veg Sci 19:769–778

Rutberg AT (1986) Lactation and fetal sex ratios in American bison. Am Nat 127:89–94

Rutberg AT (1984) Birth synchrony in American bison (*Bison bison*): response to predation or season? J Mammal 65:418–423

Sanderson EW, Redford KH, Weber B et al (2008) The ecological future of the North American bison: conceiving long-term, large-scale conservation of wildlife. Conserv Biol 22:252–266

Schoener T (1971) Theory of feeding strategies. Annu Rev Ecol Syst 2:369–404

Schultheiss PC, Collins JK, Austgen LE, DeMartini JC (1998) Malignant catarrhal fever in bison, acute and chronic cases. J Vet Diagnostic Investig 10:255–262

Schultheiss PC, Collins JK, Miller MW et al (2001) Malignant catarrhal fever in bison. Vet Pathol 38:577

Scurlock BM, Edwards WH (2010) Status of brucellosis in free-ranging elk and Bison in Wyoming. J Wildl Dis 46:442–449

Shamon H, Cosby OG, Andersen CL et al (2022) The Potential of bison restoration as an ecological approach to future tribal food sovereignty on the Northern Great Plains. Front Ecol Evol 10:826282. https://doi.org/10.3389/fevo.2022.826282

Shaw JH, Lee M (1997) Relative abundance of Bison, Elk, and Pronghorn on the Southern Plains, 1806–1857. Plains Anthropol 42:163–172

Shaw JH, Meagher M (2000) Bison. In: Desmarais S, Krausman PR (eds) Ecology and management of large mammals in North America. Prentice-Hall, Upper Saddle River, pp 447–466

Sigaud M, Merkle JA, Cherry SG, Fryxell JM et al (2017) Collective decision-making promotes fitness loss in a fusion-fission society. Ecol Lett 20:33–40

Soper JD (1941) History, range, and home life of the Northern Bison. Ecol Monogr 11:347–412

Statistics Canada (2016) Total bison (buffalo) by census division. Thematic maps from the Census of Agriculture. https://www150.statcan.gc.ca/n1/pub/95-634-x/2017001/article/54906/catm-ctra-324-eng.htm. Accessed 2 May 2021

Stephenson RO, Gerlach SC, Guthrie RD, et al (2001) Wood bison in late Holocene Alaska and adjacent Canada: paleontological, archeological and historical records. In: Gerlach SC, Murray MS (eds) People and wildlife in northern North America: essays in honor of R. Dale Guthrie, BAR Intern. British Archeological Reports, Oxford, pp 124–158

Steenweg R, Hebblewhite H, Gummer D et al (2016) Assessing potential habitat and carrying capacity for reintroduction of Plains bison (*Bison bison bison*) in Banff National Park. PLoS ONE 11(2):e0150065

Steuter AA, Hidinger L (1999) Comparative ecology of bison and cattle on mixed-grass prairie. Gt Plains Res 9:329–342

Stroup S, Forgacs D, Harris A, Derr JN, Davis BW (2022) Genomic evaluation of hybridization in historic and modern North American Bison (*Bison bison*). Sci Rep 12:6397

Tastad AC (2014) The relative effects of grazing by bison and cattle on plant community heterogeneity in Northern Mixed Prairie. University of Manitoba

Terletzky PA, Koons DN (2016) Estimating ungulate abundance while accounting for multiple sources of observation error. Wildl Soc Bull 40:525–536

Tessaro SV (1992) Bovine tuberculosis and brucellosis in animals, including man. In: Foster J, Harrison D, MacLaren IS (eds) Buffalo. University of Alberta Press, Edmonton, pp 207–224

Tessaro S V, Forbes LB, Turcotte C (1990) A survey of brucellosis and tuberculosis in bison in and around Wood Buffalo National Park, Canada. Can Vet J = La Rev Vet Can 31:174

Towne EG (2000) Prairie vegetation and soil nutrient responses to ungulate carcasses. Oecologia 122:232–239

Thorne ET (2001) Brucellosis. In: Williams ES, Baker IK (eds) Infectious diseases of wild mammals. Iowa State University Press, Ames, pp 372–395

Treanor JJ, Geremia C, Ballou MA et al (2015) Maintenance of brucellosis in Yellowstone bison: linking seasonal food resources, host-pathogen interaction, and life-history trade-offs. Ecol Evol 5:3783–3799

Turner MG, Wu Y, Wallace LL et al (1994) Simulating winter interactions among ungulates, vegetation, and fire in northern Yellowstone Park. Ecol Appl 4:2931–2942

United States Department of Agriculture (2017) Census of agriculture. Other animals inventory: 2012 and 2017. https://www.nass.usda.gov/Publications/AgCensus/2017/Full_Report/Volume_1,_Chapter_1_US/st99_1_0032_0034.pdf. Accessed 2 May 2021

US Fish and Wildlife Service, National Park Service (2007) Final bison and elk management plan and environmental impact statement. Washington, DC, USA

Van Vuren D, Bray M (1983) Diets of bison and cattle on a seeded range in southern Utah. J Range Manag 36:499–500

Van Vuren DH (2001) Spatial relations of American bison Bison bison and domestic cattle in a montane environment. Anim Biodivers Conserv 24:117–124

Wallace LL, Coughenour MB, Turner MG, Romme WH (2004) Fire patterns and ungulate survival in northern Yellowstone park: the results of two independent models. In: After the fires the ecology of change in Yellowstone National Park

WallisDeVries MF (1996) Effects of resource distribution patterns on ungulate foraging behaviour: a modeling approach. For Ecol Manage 88:167–177

Wang K, Lenstra JA, Liu L, Hu Q, Ma T, Qiu Q, Liu J (2018) Incomplete lineage sorting rather than hybridization explains the inconsistent phylogeny of the wisent. Comm Bio 1:169

Ware IM, Terletzky P, Adler PB (2014) Conflicting management objectives on the Colorado Plateau: understanding the effects of bison and cattle grazing on plant community composition. J Nat Conserv 22:293–301

Weeks AR, Sgro CM, Young AG et al (2011) Assessing the benefits and risks of translocations in changing environments: a genetic perspective. Evol Appl 4:709–725

West NE (1984) Successional patterns and productivity potentials of pinyon- juniper ecosystems. Dev Strateg Rangel Manag NRC/NAS report (Westview Press Distrib Bowker)

White PJ, Wallen RL, Geremia C et al (2011) Management of Yellowstone bison and brucellosis transmission risk—implications for conservation and restoration. Biol Conserv 144:1322

Whiteley AR, Fitzpatrick SW, Funk WC, Tallmon DA (2015) Genetic rescue to the rescue. Trends Ecol Evol 30:42–49

Wilkins K, Pejchar L, Garvoille R (2019) Ecological and social consequences of bison reintroduction in Colorado. Conserv Sci Pract 1:e9

Williams ES, Cain SL, Davis DS (1997) Brucellosis—the disease in bison. In: Thorne ET, Boyce MS, Nicoletti P, Kreeger TJ (eds) Brucellosis, bison, elk and cattle in the Greater Yellowstone area: defining the problem, exploring the solutions. Cheyenne, Wyoming, USA, pp 7–19

World Wildlife Fund (2013) Bison conservation management guidelines for herd managers. https://www.researchgate.net/publication/270892781_Bison_Conservation_Management_Guidelines_for_Herd_Managers/link/54b830570cf28faced6205d1/download. Accessed 2 May 2021

Chapter 24
Large Carnivores

Daniel J. Thompson and Thomas J. Ryder

Abstract Following historical efforts to eradicate them, large carnivores including gray wolves (*Canis lupus*), mountain lions (*Puma concolor*), black bears (*Ursus americanus*), and grizzly bears (*U. arctos*), have demonstrated an ability to recover across rangeland habitats in western North America during the last 50 years. While former distributions of these species were greatly reduced by the early-1960s, all are exhibiting range expansion and population increase across much (e.g., mountain lion and black bear) or portions (e.g., wolf and grizzly bear) of their historical range. This recovery of large carnivores in western landscapes has led to increased conflict with humans and a greater need for science-based management strategies by agencies with statutory responsibility for wildlife conservation. As conflict potential with large carnivores has increased, so have proactive and reactive conflict management programs for those impacted by large carnivores. Imperative to any successful large carnivore conflict mitigation is a focused outreach and education program for those who live, work, and recreate in habitats where wolves, mountain lions, and bears occur. Managers are continually evaluating the challenges and realities of intact large carnivore guilds within rangeland settings. Research and monitoring furthers our understanding and efficacy of management strategies for large carnivores now and into the future, striving to build on knowledge regarding the intricacies of population dynamics among predators and prey, including domestic species and humans.

Keywords Black bear · Gray wolf · Grizzly bear · Human-wildlife conflict · Large carnivore · Livestock depredation · Mountain lion

D. J. Thompson (✉)
Wyoming Game and Fish Department, 260 Buena Vista Dr., Lander, WY, USA
e-mail: daniel.thompson@wyo.gov

T. J. Ryder
22 Meandering Way, Lander, WY, USA

24.1   Life History and Population Dynamics of Large Carnivores

Throughout this chapter, we refer to gray wolves (*Canis lupus*), mountain lions (*Puma concolor*), black bears (*Ursus americanus*), and grizzly bears (*U. arctos*) as "large carnivores." Mountain lions and wolves are obligate carnivores, whereas black and grizzly bears are omnivores. Mountain lions, in fact, require the consumption of animal tissue to obtain taurine, an essential amino acid (Allen et al. 1997). These four species occur at much lower densities than their primary ungulate prey, and overall predator abundance is dictated by prey/food availability and competition between or among other large carnivores (Griffin et al. 2011; Hurley et al. 2011). We list morphological attributes and general life history characteristics of wolves, mountain lions, and black and grizzly bears in Table 24.1.

24.1.1   Gray Wolves

Wolves are the only gregarious species of large carnivore in North America, utilizing a hierarchical social system of pack dynamics with a dominant breeding pair and subordinate wolves that share food acquisition and pup-rearing duties. Wolves are cursorial predators that chase and attempt to single out prey from a group by identifying an individual that exhibits vulnerability to predation. This strategy is sometimes misconstrued as wolves only killing the "sick and weak," but it is more appropriate to state they take the most vulnerable prey items available (Mech 1970; Hebblewhite et al. 2003).

The reproductive strategies of wolves differ substantially from bears and mountain lions in that a dominant, or alpha breeding pair produce pups annually (Table 24.1). Pup production varies based on prey availability and local wolf densities. In areas of high prey availability and little competition, wolves can produce large litters or more than one litter/pack/year and sustain the ability to feed pups to independence (Boertje and Stephenson 1992).

24.1.2   Mountain Lions

Mountain lions are stalk and ambush predators, employing stealth to hunt prey across a breadth of habitats. As solitary hunters, mountain lions are the least observable, but most ubiquitous of the four large carnivore species. Mountain lions can produce kittens at any time of year (Jansen and Jenks 2012), unlike many similar-sized mammalian species. However, most research suggests mountain lions exhibit a birth pulse synchronicity in late-summer/early-fall when 2–4 kittens/litter are born (Jansen and Jenks 2012). Following birth, kittens remain in natal dens for 2–3 months, at

Table 24.1 Comprehensive comparisons of general morphology, reproductive, social structure and dietary characteristics of large carnivore species in North America

Morphology				Reproduction				Social structure	Food habits
Scientific name	Common name(s)	Adult weight range	Color phases	Gestation	Litter size	Sexual maturity			
Puma concolor[1]	Cougar, mountain lion, cucuagurana, panther, painter, puma	Male: 68–82 kg; Female 40–55 kg	Tawny, light brown, auburn, spotted as kittens	~ 90–93 days; birth pulse late summer	2–4 kittens; kittens remain with maternal female 13–16 months prior to dispersal	Approximately 3 years of age		Generally solitary, home ranges delineated primarily through scent marking by resident males and females, not considered territorial	Carnivore; primary prey consists of ungulate species (*Odocoileus* spp., *Cervus elaphus*), but also opportunistic and/or specialized for smaller/medium sized prey species
Canis lupus[2]	Gray wolf, wolf	Male: 40–55 kg; Female 36–45 kg	Gray; black; white phases in northern latitudes	~ 62–75 days; annually born in spring to breeding pair	Variable depending on resource availability and competition; 2–9 pups/pack	2–3 years of age depending on pack density		Territorial pack structure, hierarchical dominance system of alpha male and female and subordinates	Carnivore; primary prey consists of ungulate species (*Cervus elaphus*, *Alces alces*)
Ursus americanus[3]	Black bear	Male: 90–225 kg; Female 68–136 kg	Black, cinnamon, chocolate, brown	~ 180–240 days, includes delayed implantation	2–4 cubs born in the den in late January; cubs disperse from maternal female as yearlings after leaving the den	3–4 years of age		Generally solitary, home ranges delineated, males defend resources and mates from intraspecifics	Omnivore, diet dependent on local and annual availability, opportunistic predator on ungulate neonates and other species depending on availability

(continued)

Table 24.1 (continued)

Morphology				Reproduction					
Scientific name	Common name(s)	Adult weight range	Color phases	Gestation	Litter size	Sexual maturity	Social structure	Food habits
Ursus arctos[4]	Grizzly bear, brown bear	Male: 114–320 kg; Female: 90–180 kg; (Canadian and Alaskan brown bears can reach much higher weights in excess of 450 kg)	"Grizzled", black, chocolate; guard hair tips are light tipped	~ 180–250 days, includes delayed implantation	2–4 cubs born in the den in late January; cubs disperse from maternal female as two-year olds after leaving the den	4–5 years of age	Generally solitary, home ranges delineated, males defend resources and mates from intraspecifcs; pseudo-gregarious on abundant food sources	Omnivore, high dietary plasticity, opportunistic predator on ungulate neonates and other species depending on availability. Feeding behavior is taught from maternal female

Life history metrics citations by species: [1] (Hornocker and Negri 2010; Jenks 2011); [2] (USFWS et al. 2011; USFWS 2020); [3] (Powell et al. 1997; Hristienko and McDonald 2007; Griffin et al. 2011); [4] (Haroldson et al. 2021)

which point they become mobile and begin traveling with their mother to feed on kills. Because most litters are born in late-summer, mountain lions are able to take advantage of peak ungulate birth periods in mid- to late-June when prey abundance is greatest. Mountain lions do not exhibit classic territorial behavior, but do scent-mark home range boundaries. Males also defend female breeding rights within their home ranges.

24.1.3 Black and Grizzly Bears

Black and grizzly bears maintain high levels of dietary plasticity and are considered true omnivores (Hristienko and McDonald 2007; van Manen et al. 2016). Even so, all bears are opportunistic predators and, in fact, certain individuals can become adept neonate predators (Barber-Meyer et al. 2008; Griffin et al. 2011).

Bears are unique among other North American large carnivores in that they spend approximately half the year hibernating in dens. While there remains scientific debate as to whether bears exhibit true hibernation, both black and grizzly bears spend most of the winter in a reduced state of physiological activity between November and March/April of the following year. The length of time spent within dens varies depending on sex and age of the individual and environmental conditions including latitude and elevation.

Most breeding among bears occurs in June, but the resulting fertilized egg undergoes a process termed "blastocyst arrest" where further embryonic development ceases until November. At that time, the egg implants on the female's uterine wall and rapid fetal development begins (Haroldson et al. 2021). Because of this "delayed implantation," of the embryo, all cub births occur at approximately the same time within the natal den, usually in late-January. Bears, especially grizzly bears, are also less fecund than other large mammals in North America (Haroldson et al. 2021; Table 24.1). Where most ungulates begin producing young by 1 1/2–2 years of age, grizzly bears generally do not reach sexual maturity until approximately 5 years of age and once cubs are born, they spend 2 full summers with the maternal female. Thus, grizzly bears produce on average only 2–3 cubs every 3 years after the age of 5 (Haroldson et al. 2021).

24.2 History of Large Carnivores in North America

Human emotions concerning large carnivores have always varied from complete hatred to that of idolatry and worship (Hovardas 2018). Thus, to provide a representative and accurate account of large carnivores in rangeland settings today, it is necessary to delve into the historical record and describe human-caused perturbations that affected these species from the period prior to European settlement of the western U.S. through today. Before Western expansion, Native American peoples long maintained traditional oral accounts of their interactions with large carnivores. Many of their stories described the mystical kinship between human beings and these animals and many behavioral attributes of bears and wolves were incorporated into tribal dances and ceremonials (Neihardt 1932). Often, tribes viewed bears (both black and grizzlies) as familial, but Native Americans also actively hunted them depending on the tribe or use of the harvested animal (i.e., meat, clothing, ceremonial; Young and Goldman 1946). In fact, the ability to successfully take one of these animals often resulted in great credit and adulation to the successful hunter from both tribal members and enemies alike.

As westward European settlement progressed during the middle nineteenth century, attitudes of many pioneers toward large carnivores were vastly different from the country's original Native American inhabitants (Thirgood et al. 2005). Some of the more famous/infamous early accounts of human interactions with carnivores, especially bears, were lavishly recounted in the journals of Lewis and Clark (Leopold 1933; various journals of Lewis and Clark) and countless reports, stories, and representations (both factual and sensationalized) from the era's iconic mountain men and early naturalists. Wolves, mountain lions, and grizzly bears were perceived as direct threats to human safety, with accounts of humans being killed by mountain lions recorded as early as 1747 (Young and Goldman 1946). Following the West-wide American bison (*Bison bison*) slaughter of the 1870s, domestic livestock took their place. The behavioral naivety of domestic cattle and sheep allowed for easy exploitation by native carnivores (Riley et al. 2004). Direct (e.g., human safety) and indirect (i.e., livestock, property, and ungulate population) impacts from large carnivores were met with aggressive removal actions. Similar reductions in populations of other big game species in addition to bison were occurring as the result of uncontrolled market hunting and large-scale habitat destruction.

While the hunting and/or killing of carnivores by individual settlers occurred throughout westward expansion, the enactment of bounties and regimented governmental removal likely resulted in the greatest impact to large carnivore abundance and distribution in North America (Fig. 24.1; Leopold 1933; Caughley 1977; Cougar Management Guidelines Working Group 2005). For example, the Federal government systematically promoted and engaged in wolf eradication efforts across the lower 48 states (Mech 1970; US Fish and Wildlife Service et al. 2011).

Following establishment of Yellowstone National Park in 1872, a slow evolving change in viewpoints and attitudes toward wildlife and wildlands began. At first, only a small number of sportsmen-conservationists, including George Bird Grinnell and others, raised the alarm. They understood the finite nature of wildlife after experiencing first-hand the dramatic reduction of bison and other big game species. Later, hunters including President Theodore Roosevelt, Forest Service Chief Gifford Pinchot, and Congressman John Lacey, along with Grinnell and others, made the first substantive reversals of the environmental destruction that occurred during the late-1800s. Sportsmen's groups and management agencies then began the long process of reestablishing game populations across the West.

However, adoption of regulated management of carnivorous animals lagged well behind that of game species such as deer (*Odocoileus* spp.) and elk (*Cervus elaphus*).

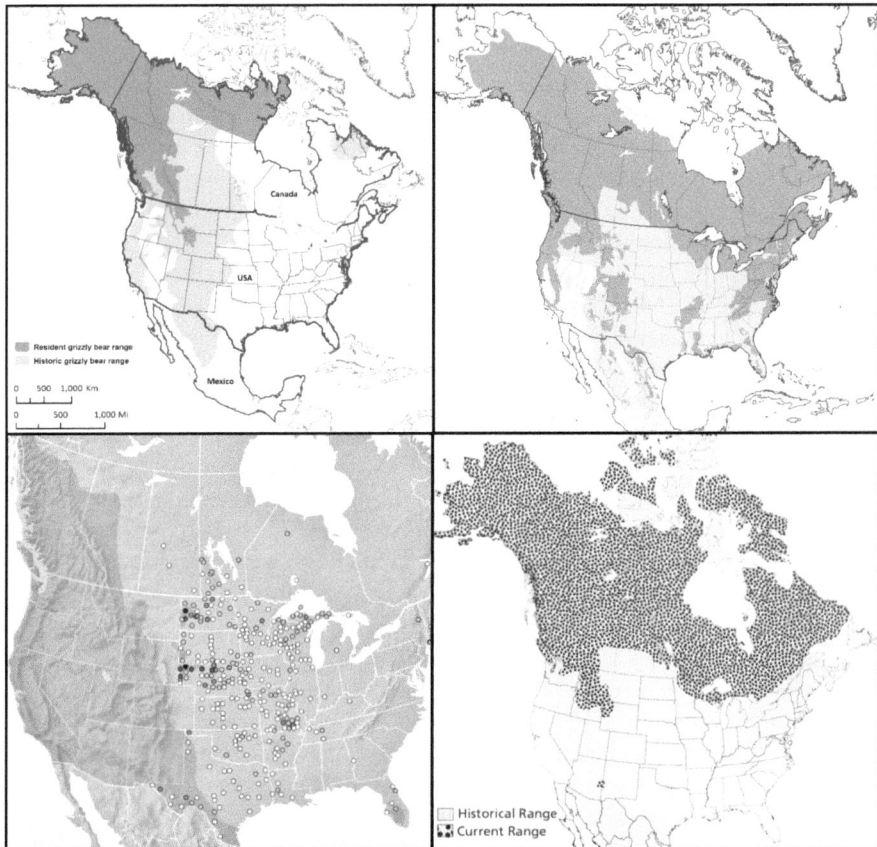

Fig. 24.1 Comparative distributions of large carnivore species (clockwise from top left; grizzly bear [*Ursus arctos*], black bear [*U. americanus*], gray wolf [*Canis lupus*] and cougar [*Puma concolor*]) in North America, demonstrating changes from historical to current range for bears and wolves whereby darker shades represent current distribution. Top left; historical and current grizzly bear range in North America (Haroldson et al. 2021); top right, historical and current black bear range in North America (adapted from International Union for Conservation of Nature Archives); bottom right, historical and current range of gray wolves in North America (adapted from Wyoming Game and Fish Commission 2011), and bottom left; current mountain lion range (darker shaded area) and documented range expansion in North America (adapted from LaRue 2018), shaded dots represent verified mountain lion presence by county outside of current distribution indicative of recolonization of historic mountain lion range (LaRue 2018)

As a result, wolves were reduced to a few remnant individuals and/or packs in the most rugged, inaccessible terrain. In most of their historic ranges, they were considered to be functionally extirpated by the mid-1900s (Mech 1970; Riley et al. 2004). Similarly, grizzly bears were reduced to less than 2% of their original range in the lower 48 states, with extant populations found only in remote/protected areas such as Yellowstone and Glacier National Parks and surrounding wilderness areas (White et al. 2017;

Haroldson et al. 2021). While mountain lion populations were greatly reduced as well, their secretive and elusive behavioral characteristics made them less susceptible to large-scale harvest and allowed them to maintain breeding populations in the more mountainous and rugged areas across the western U.S. and Canada (Riley et al. 2004; Hornocker and Negri 2010). Black bear populations were reduced through direct human harvest or widespread habitat perturbations (i.e., logging, wildfire, intensive livestock grazing, and the resulting erosion) occurring concurrently with wide scale predator reduction activities recounted above (Hristienko and McDonald 2007).

Persecution of large carnivores continued unabated throughout the first three quarters of the twentieth century. However, changing public attitudes during the environmental awakening of the 1960s eventually resulted in the termination of bounty payments and wide-scale poisoning of predatory species by the 1970s in most portions of North America. In addition, changing land use practices and adoption of new livestock husbandry practices (referenced throughout this book) often times reduced the necessity to lethally control large carnivores.

The rebound of large carnivores is one of the greatest conservation success stories of the twentieth century (Bennett 1998; Pyare et al. 2004; US Fish and Wildlife Service 2017a). Populations of wolves, mountain lions, and black and grizzly bears that were on the brink of extinction and/or extirpated across wide swaths of their historical ranges are now stable to increasing and expanding back into formerly occupied habitats throughout North America (Fig. 24.1). For instance, gray wolves in the western U.S. have expanded their range since being reintroduced to the Greater Yellowstone Ecosystem (GYE) and wilderness areas of central Idaho in the mid-1990s in conjunction with natural dispersal into suitable habitats in other portions of the U.S. and Canada (Cullingham et al. 2016; United States Fish and Wildlife Service et al. 2020). Similarly, mountain lions and black bears have naturally re-colonized large areas of their historical ranges in western North America, with mountain lions even expanding eastward into areas bereft of the big cats for the past 150 years (Thompson and Jenks 2010; LaRue and Nielsen 2011). Lastly, grizzly bear populations in the GYE and Northern Continental Divide Ecosystem (NCDE) have experienced a dramatic increase in abundance and distribution, surpassing biologically suitable and socially acceptable habitats in each area (Bjornlie et al. 2014; US Fish and Wildlife Service 2017a, b).

As public attitudes toward large carnivores changed, many aspects of their behavior and biology remained unknown. The implementation of new and innovative techniques to capture, handle, and monitor large carnivores, in conjunction with development of radio telemetry in the 1960s, opened up many opportunities to better understand large carnivore biology (Fuller et al. 2005). The ability to affix radio transmitters on these species provided insight into the movements of large carnivores that was previously unavailable. Additionally, radio monitoring led to a greater understanding of large carnivore life histories and provided important empirical data illustrating how they interact with humans, livestock, and native prey across the largely intact wild rangeland settings of the American West.

Research pioneers Frank and John Craighead were the first to fit a grizzly bear with a radio collar to track its movements in the GYE (Craighead et al. 1974; for an amazing representation of initial marking of a grizzly bear in Yellowstone National Park by the Craighead's search for *"Craighead's grizzly bear"* on Youtube™). At the same time, Maurice Hornocker and his renowned houndsman Wilbur Wiles were the first persons to fit a mountain lion with a radio-tracking collar in the Idaho Primitive Area (Hornocker 1969; Seidensticker et al. 1973) while L. David Mech was gaining valuable insight into wolf ecology through the use of radio telemetry in northern Minnesota (Mech 1970). As wildlife professionals gathered more detailed biological information concerning large carnivores, agencies began developing management plans and strategies for black bears and mountain lions, and later grizzly bears and wolves using these new data.

Today, because grizzly bears are considered a threatened species under the Endangered Species Act of 1973 (ESA) in the lower 48 states, they are somewhat compartmentalized from a management planning perspective (US Fish and Wildlife Service 2013, 2017a). Gray wolves are also unique because, despite being functionally extirpated from the West, the U.S. Fish and Wildlife Service reintroduced wolves into the GYE and central Idaho in 1995–1996 (US Fish and Wildlife Service et al. 2011). After decades of effort, gray wolves are now abundant in many areas across Idaho, Montana, and Wyoming (US Fish and Wildlife Service 2020), with continued expansion of the species into Oregon, Washington, California and Colorado (US Fish and Wildlife Service 2020). Mountain lions and black bears remain solely under the purview of state management planning efforts.

24.3 Understanding Large Carnivores Through Research and Monitoring

With the possible exception of wolves, large carnivores on western rangelands are very difficult to enumerate because of their cryptic nature and relatively low abundance. However, recent technological advances in telemetry, remote sensing, and statistical analyses have resulted in more applied research and monitoring studies and a greater ecological understanding of these species. Additionally, improved data analyses have allowed managers to implement more science-based decisions and strategies to inform conservation and management of these apex predators.

Entire books and numerous book chapters have been devoted toward research rigorously evaluating wolves (Mech 1970; Carbyn et al. 1995), mountain lions (Anderson et al. 2010; Jenks 2011), and bears (Powell et al. 1997; White et al. 2017; Haroldson et al. 2021) and should be reviewed by any serious student of large carnivore ecology. For purposes of this chapter, the following sections highlight the most widely-employed research and monitoring techniques for individual large carnivore species.

24.3.1 Gray Wolves

In Canada and Alaska, where wolf populations are stable to increasing across the majority of their range, rigorous research and management programs are regularly implemented to support continuing hunting and trapping activities. Within the Continental U.S., however, wolf populations must be monitored because either they are currently listed as Threatened/Endangered under ESA, or have been recently delisted, and agencies must provide data to support continued state-controlled management. For example, the Northern Rocky Mountain states of Idaho, Wyoming, and Montana are each required to maintain an annual population or at least 150 wolves and 15 breeding pairs (defined as two adult wolves with at least two pups, US Fish and Wildlife Service et al. 2011) to maintain federal recovered status. These recovery criterion must be achieved even when dealing with conflict issues such as livestock depredation.

To ensure these types of fine-scale recovery criterion are achieved, agencies utilize radio-marking of select individuals within packs. By marking one or more individuals within a pack, annual dynamics of the entire pack can be monitored (Hebblewhite et al. 2003). Through subsequent ground, aerial, and global positioning system (GPS) tracking, den sites can be determined, accurate pup counts can be taken, and rendezvous sites can be identified to assist in monitoring changes in pack size and pup survival throughout summer and fall. Maintaining radio-marked individuals within packs can also assist in determining wolves responsible for depredation or other conflicts (Breck et al. 2011; Bradley et al. 2015) in areas overlapping livestock production. This type of research and monitoring is expensive and time consuming. However, managers must often weigh these costs against recovery/research objectives and/or public and agency scrutiny.

The vagile nature of wolves requires constant research/monitoring adaptability. As wolf populations increase and expand, research efforts are beginning to focus on questions regarding livestock predation, impacts to ungulates, and interactions among mountain lions (Griffin et al. 2011; Hurley et al. 2011) and bears (Barber-Meyer et al. 2008) in multi-prey/multi-carnivore systems. Additionally, GPS technology allows fine-scale assessment of wolf movements and can be used to pinpoint locations of kills and feeding sites [originally developed and widely used for mountain lions (Anderson and Lindzey 2003; Knopff et al. 2010; Wilckens et al. 2015)]. Kill site clusters provide empirical data used to determine kill rates and prey composition (Clapp et al. 2021), and identify den and rendezvous sites. GPS data and cluster analyses are being employed across all carnivore taxa, including the 4 species discussed in this chapter.

24.3.2 Mountain Lions

Because mountain lions are the most reclusive of North American large carnivores, there are techniques that enable agencies to monitor population abundance and/or trend when used in tandem with management and harvest strategies,. The "gold standard" for evaluating abundance of mountain lion populations is through intensive capture, radio-marking, and monitoring techniques (Cougar Management Guidelines Working Group 2005; Jenks 2011). While costly, marking a representative sample of a lion population provides baseline information regarding abundance, survival, fecundity, natality, and recruitment. Further, evaluation of fine-scale movement patterns can provide detailed insight into resource selection and interactions with other species; including use of human-occupied areas (Anderson and Lindzey 2003; Fuller et al. 2005; Knopff et al. 2009; Tomkiewicz et al. 2010).

Realistically, however, not all mountain lion populations can, or need be studied this intensively. The cost-prohibitive nature of capture-collar-monitor techniques have often required managers to develop more creative, less expensive, and non-invasive monitoring techniques; ranging from track and hair surveys (ground and aerial, Van Sickle and Lindzey 1991; Sawaya et al. 2010), remote-camera surveys (Karanth and Nichols 1998; Kelly et al. 2008; Hughson et al. 2010), scent-station surveys, or combined iterations of these methods (Choate et al. 2009; Russell et al. 2012) with differential success depending on the size of the study area and research objectives.

The use of "biopsy darts" has also been used to assess population status in many areas of mountain lion range (Beausoleil et al. 2016; Proffitt et al. 2020), whereby an animal is pursued by trained dogs, brought to bay, and then "marked" by obtaining a sample of tissue from a biopsy dart. Animals are "marked" in this fashion to obtain a genetic subset of the entire population. Then as mountain lions are subsequently harvested or lost through other forms of mortality (i.e., highway strikes, conflict removals, etc.), these mortality "recaptures" are used to derive an estimate of a local population using a modeling technique referred to as "mark-recapture/mark-resight and/or capture/mark/recapture" (White and Burnham 1999; Buckland et al. 2001). The implementation of these types of models has grown exponentially in recent years (Burnham and Anderson 2002; McClintock et al. 2006; R Development Core Team 2019) and will be discussed in greater detail in Sect. 24.3.5 of this chapter.

Another genetics-based, noninvasive monitoring technique is the use of scat-detection dogs (Wasser et al. 2004). These highly-trained canines can differentiate mountain lion scat from other species along systematic transects. The resulting scat "detections" are used to differentiate individual lions and calculate local population density. Scat detection is also very useful to document presence/absence of these large felines in areas of range expansion.

24.3.3 Black Bears

Agencies use capture/collar/monitoring techniques to gather baseline information concerning black bear populations. The method, in addition to the techniques previously described for wolves and mountain lions, is particularly useful to determine black bear range expansion.

Also similar to wolves and mountain lions, the use of genetic monitoring techniques is gaining currency in the development of black bear population estimates. In particular, hair-snare surveys employ a relatively inexpensive, non-invasive sampling technique; whereby a barbed-wire fence is erected around a non-food suspended lure. As a bear visits and investigates the lure, it passes over or under the fencing, shedding a convenient clump of hair, thus providing a genetic sample (Gardner et al. 2010; Gurney et al. 2020). Samples obtained in hair snares are then analyzed using an additional iteration of the mark-recapture technique (Borchers and Efford 2008). Data gathered in this manner have been used to develop clustered sampling techniques (Humm et al. 2017) that provide less logistically restrictive, yet accurate estimates of abundance and population trend, and can be very helpful when designing management strategies for this species. When implemented in a systematic approach across years, managers can derive point population estimates and provide statistically robust evaluation of management strategies to document population increase, stability, or decline depending on objectives.

24.3.4 Grizzly Bears

Techniques discussed throughout this section have been employed to better understand grizzly bear population dynamics and genetic diversity, with hair-snare and other genetic mark-recapture methods being employed throughout western Canada and Alaska (Paetkau et al. 1998; Boulanger et al. 2004). Additionally, advances in GPS technologies have been used for fine-scale analysis of brown bear movements and been used to determine potential barriers to gene flow in western Canadian provinces (Proctor et al. 2002; Boulanger and Stenhouse 2014).

Because grizzly bears have been listed as a threatened species in the lower 48 states since 1973, research and monitoring efforts are generally more intensive in the Continental U.S. (US Fish and Wildlife Service 2017a). As a result, the GYE and NCDE populations have been intensively studied for decades (Apps et al. 2004; Mace 2004; Schwartz et al. 2014; van Manen et al. 2016; White et al. 2017). In the GYE, researchers/managers retain an annual representative sample of marked animals [using both Very High Frequency (VHF) and GPS radio-collars] to gather data on survival, mortality, reproduction, dispersal, movements, habitat use, and overall population dynamics. Select areas within the GYE are very conducive to aerial observations of bears due to favorable topographic and vegetative features. Thus, aerial observation surveys are used in conjunction with telemetry data and ground observations to estimate annual abundance of the population. Each year, 70–90 grizzly bears are tracked through telemetry and methodical records of bear mortality, conflicts with humans, and other myriad parameters that allow for fine-scale evaluation of this increasing grizzly bear population and its potential hazards and threats (Schwartz et al. 2010). Similar techniques have been employed in the NCDE.

The commitment and foresight of attaining long-term, systematically-collected datasets of these intensively-monitored grizzly populations has enabled researchers to continually re-examine these data while taking advantage of the evolving spectrum of advances in statistical and technological analyses. Program MARK™, R™, and derivations of this program (White and Burnham 1999; R Development Core Team 2019), for example, have increased the capability and statistical rigor of analyses for wildlife populations (Sells et al. 2018; Bissonette 2019).

24.3.5 Capture/Mark/Recapture Estimation Techniques

Capture/Mark/Recapture (CMR) methodologies have become the foundation for most population abundance and density estimates for large carnivores and many other wildlife populations (Amstrup et al. 2006; Batemen et al. 2013; Borchers et al. 2015). Techniques developed to noninvasively collect deoxyribonucleic acid (DNA) from wild animals has led to rigorous systematic monitoring programs that provide local density estimates. As previously mentioned, hair snares have been widely used for both black and grizzly bears and efficiency has increased to the point that this method has been implemented across North America. Spatially explicit capture-recapture (SECR; Kristensen and Kovach 2018) sampling has increased the efficacy and applicability of hair snare sampling on a larger scale for ursids and other wildlife populations (Morehouse and Boyce 2016; Humm et al. 2017). Additional creative methods to obtain DNA have been employed for mountain lions (Sawaya et al. 2010; Beausoleil et al. 2016) and less frequently for wolves (Caniglia et al. 2012; Stansbury et al. 2014). The common denominator among all these techniques is the importance of maintaining an adequate marked sample of the population to ensure statistical rigor for local population estimates (Borchers et al. 2015).

Biologists also use individual animal characteristics to estimate localized abundance through the use of camera-traps (Davidson et al. 2014; Mattioli et al. 2018), observations of verified tracks or individually identifiable animals, noninvasive genetic techniques including hair snares and scat detection dogs, and combinations thereof (Davidson et al. 2014; Stone et al. 2017; Mattioli et al. 2018; Murphy et al. 2018).

All aforementioned sampling methods are employed through some type of measured grid and/or transect system to extrapolate abundance/densities on a local scale without expanding results beyond the merits of the study design (Newey et al. 2015). Estimation of population trends allow researchers and managers to evaluate various management strategies/harvest regimes and answer questions salient to the public and/or those individuals impacted (directly or indirectly) by large carnivores.

Recently, integrated population modeling (IPM) techniques have been implemented for black bears, grizzly bears, and mountain lions (Arnold et al. 2018). IPMs were initially developed for ungulates, but the technique has been successfully adapted for large carnivores and other wildlife species. The interpolated model allows input parameters that implement Bayesian statistical probabilistic methodologies to provide insight into particular components of a wildlife population (Arnold et al. 2018) and projected perturbations to calculate potential outcomes for the population(s) in question. The more accurate the input parameters, the more self-correcting and reliable the model outputs become. This technique was used to evaluate mountain lion abundance and predation on elk in the Bitterroot Mountains of Montana (Proffitt et al. 2020). IPM also has merit for bear populations across the West.

The intra- and interspecific interactions among carnivores and their relationships with multiple prey species has fostered a plethora of attempts to design research to better describe and quantify predator–prey dynamics (Moll et al. 2016; Montgomery et al. 2019). Because many Western rangeland ecosystems contain intact mammalian guilds, they, in particular, provide a robust opportunity to examine these interactions. Multiple predator–prey systems exhibit social dynamics and hierarchies between and among carnivores, with mountain lions and black bears serving as subordinate predators to wolves and grizzly bears (Griffin et al. 2011; Elbroch et al. 2020). Because of dwindling mule deer (*O. hemionus*) and moose (*Alces alces*) populations in localized situations, researchers are also attempting to quantify whether predation is acting as an additive or compensatory form of mortality (Griffin et al. 2011; Pierce et al. 2012; Proffitt et al. 2020).

Each land system experiences its own unique dynamic, fluctuating as populations move toward a natural state and homeostasis. However, applied research suggests multiple carnivore systems may impact prey populations by reducing offspring/maternal female ratios (Barber-Meyer et al. 2008; Proffitt et al. 2020) and causing prey to increase vigilance, select for smaller group sizes, and use the landscape in a dispersed distribution (Griffin et al. 2011; Elbroch et al. 2020). All of these evolving selective pressures are directly influenced by livestock grazing and changing land use practices, regardless of area. Agencies and individuals responsible for managing the land, the livestock, and the wildlife must continually work to understand these ever-changing targets.

24.3.6 *Management Strategies*

As stated previously, research and monitoring efforts provide data to understand the daily lives of large carnivores and enable managers to develop management strategies that maintain and/or regulate population densities and abundance; depending on local management objectives and public opinion. This section briefly highlights some of the tools used to manage large carnivore populations in rangeland settings. However, it is important to emphasize that "management" is more encompassing than just developing hunting seasons or population reduction/augmentation programs. In the context of large carnivores, conflict management must be intricately interwoven with harvest-driven management/conservation strategies. Our repetition of holistic management programs is purposeful in that all information discussed within this chapter regarding wolves, mountain lions, and bears are integral components of an overall "successful" management program for these species and/or populations.

Regulated hunting, or harvest, is a primary method of population management for many wildlife species. Currently, all states and provincial agencies that use hunting as a management tool for wolves, mountain lions, and black bears rely on some form of annual monitoring, including harvest data, to evaluate population status. Data gathered from harvested animals provide inferences of the population's sex and age composition, density of mortality per unit area, and other useful information for managers (i.e., hunter success, harvest rates, hunter satisfaction). These data are ultimately used to evaluate efficacy of management strategies.

Wildlife in the U.S. is a publicly-owned resource and managed under the North American Model of Wildlife Conservation, where stakeholder input, science-based management techniques, and professional expertise combine to develop effective hunting strategies that attain desired population densities for game species (Bleich and Thompson 2018). While the North American Model may have been developed with ungulate species in mind, it is wholly applicable to large carnivores as well because they are a vital component of the public trust. The driving factor for how hunting may be used to achieve population reduction/stability/increase is rooted in the species' population density and socio-political parameters that influence localized objectives for population abundance (i.e., public desires, livestock density, proximity to urban/suburban areas, ungulate herds, and habitat quality). For example, Hristienko and McDonald (2007) suggested black bear management objectives should emphasize maintaining viable black bear populations, safeguarding human safety and livelihood, and satisfying the needs of various stakeholders; while considering fiduciary responsibilities and accountability.

By and large, agencies responsible for large carnivore management employ the use of quotas, mortality limits, and season length/timing restrictions to move a population toward specific population objectives. Large carnivore management units and hunt areas are often developed using local population densities, topography, the amount of contiguous habitat present, and other landscape features representing a localized population. Historical population data and previous research/professional expertise are vitally important to devise hunt areas that accurately represent local carnivore and prey populations. The use of management units and hunt areas allow managers to direct harvest to specific areas on a localized level. Life history and movement patterns of carnivores are enveloped into these strategies, whereby animals removed from the population will be replaced by other animals within the population or from surrounding populations [sometimes referred to as source/sink management (Robinson et al. 2008)]. Managers attempt to have a mixture of harvest objectives for hunt areas (i.e., reduce, stabilize, augment) within larger management units. Most states and provinces require mandatory checks of harvested wolves, mountain lions, and bears to ensure sex and age data are collected from animals taken by hunters. Data acquired during mandatory checks are used to evaluate harvest trend, hunter satisfaction, and population composition from each animal.

The importance of utilizing human dimensions methodology in large carnivore management is discussed later in this chapter. However, when evaluating hunting programs, most agencies adopt some form of harvest survey to assess the quality of local wildlife populations from the sportspersons' perspective. The use of harvest surveys assists managers in evaluating the perceptions of successful and non-successful hunters by providing information on their overall effort, species observed while afield, and intangibles otherwise difficult to gather. While this information is not analogous to the rigorous quantitative data acquired from intensive research or monitoring efforts, it is an invaluable introspective into wide-ranging public attitudes from stakeholders that are personally invested in large carnivore hunting and management.

24.4 Large Carnivore Interactions with Humans and Livestock

Wolves and grizzly bears have had an increasing impact on rangeland livestock operations as populations have expanded beyond established recovery zones over the last 25 years. For example, in Wyoming, Montana, and Idaho, grizzly bears now persist in human-dominated agricultural landscapes (Bjornlie et al. 2014) that incur much higher conflict potential in the forms of livestock depredation, property damage, and human safety. Similarly, mountain lion and black bear conflict has increased as their populations have increased in rangeland settings.

Although addressed elsewhere in the book (Chap. 28: Living with Predators), it is important to address specific issues regarding large carnivore conflicts in this chapter. One of the more vital components of large carnivore conservation is the importance of managing and reducing conflict potential from large carnivores to increase human tolerance for, and promote long-term conservation of these species on the landscape. In addition, active large carnivore conflict management programs provide vital techniques and promote methodologies for the public to proactively reduce conflict. These programs can also provide an avenue for producers to reduce the potential and actual predation of their livestock.

There is not a standardized definition of conflict, but it involves *interactions between humans and wildlife that result in property damage, agricultural damage, or public safety issues*, and usually requires verification from a trained professional. Verification of damage is important because it provides an empirical way to track annual trends in conflict and also assists in identifying potential mitigation or resolutions in regards to conflict management.

Approaches to mitigating large carnivore conflicts can be both proactive (see outreach and education later in the chapter) and reactive. When proactive approaches are unsuccessful and conflict occurs, managers must respond rapidly to reduce or eliminate the impact using multiple methods. Professionals that deal with large carnivores realize that controversy is inherent from virtually all viewpoints and most state and provincial agencies have protocols and guidelines in place to deal directly with human-carnivore conflicts. The high level of public scrutiny and opinion regarding large carnivore conflict and agency response to that conflict always requires a high level of professionalism and consistency in approach. Whether dealing with a live-stock producer that has lost calves to wolves or an advocacy group outraged by the lethal removal of a bear that has killed livestock, consistency projects objectivity when assessing conflict management.

24.4.1 Human Safety

The reality of large carnivores and humans traversing the same habitat means there are factual human safety risks that must be considered and addressed. Encounters that result in human injuries or fatalities are extremely rare, but when they do occur, they directly affect those close to the person attacked and heavily impact the human psyche and overall tolerance for carnivores (Herrero 1985; McNay 2002; Quigley and Herrero 2005). Due to the high priority and contentious nature of these events, most agencies have some type of team specially trained to deal with them. State and Provincial "Wildlife Human Attack Response Teams" (WHART) and close deriva-tives consist of highly specialized professionals that respond to wildlife attacks with a multi-faceted systematic approach as to how the investigation is handled. The team operates with a system designed to cover jurisdictional responsibilities, media coverage/control, and logistics in regards to wildlife capture. All human injury/fatality situations are dynamic and very specific to the individual causation of the event, but the consistent priority is immediate response to these situations in the name of human safety.

24.4.2 Livestock Depredation

Livestock depredation is the most relevant form of human/large carnivore conflict to be addressed in this chapter given *Rangeland Wildlife Ecology and Conserva-tion* deals specifically with rangeland management. From a pragmatic perspective, increased depredation of livestock has occurred in many areas of the western U.S. and Canada in recent decades, likely a function of greater abundance and wider distri-bution of carnivore species coupled with an increase in smaller "hobby farms" that acquire more atypical domestic species prone to depredation (i.e., llamas, alpacas, poultry, swine, etc.). While this chapter focuses mainly on depredation of domestic cattle and sheep (US Department of Agriculture et al. 2019a, b), several of the strate-gies to reduce conflict potential with these species are directly applicable to multiple other domestic species.

There are multiple proactive infrastructural updates that can be implemented to reduce large carnivore conflict, depending on livestock type and herd size and the predatory behavior of the specific carnivore. If producers have the ability to night-pen or use electric fencing around livestock, it can be a very effective tool to reduce depredation from canids and ursids (Breck et al. 2011). The use of electric fencing has also been readily employed for poultry, swine, sheep, and smaller cattle/horse operations (Bodenchuck 2011). In addition to reducing livestock depredation, electric fencing has been used to alleviate crop damage from black and grizzly bears. There are temporary approaches such as fladry/turbo-fladry (Davidson-Nelson and Gehring 2010) that can be used at specific times of year (calving/lambing season) to deter wolf depredation, but should be considered a temporary fix as wolves acclimate to the presence of fladry. Free-range grazing and/or larger cattle/sheep operations generally do not have these options; however, there has been success with electric-fence night penning of sheep depending on sheep band-size (Wyoming Game and Fish Department unpublished data). When using any type of electrical deterrence is it imperative to repeatedly evaluate amperage and efficacy of the system, as many large carnivores will continue to test these types of devices (Smith et al. 2018). When operating properly, these methods provide a very efficacious means of reducing conflict, but they are obviously not pertinent to all livestock producers.

For larger livestock herds in open range settings, depredation risk is largely related to abundance of livestock and abundance of large carnivores (Wells et al. 2019). Most open range livestock producers employ the use of riders to stay with or check their sheep and cattle herds, The efficacy of range riding depends on herd size and rider experience (Eklund et al. 2017; Wells et al. 2019). Range riders do not eliminate depredation, especially in rough terrain with high carnivore densities and/or multiple carnivore species. However, adept riders can locate depredated animals and alert management agencies to evaluate potential strategies to reactively mitigate conflict.

Many times, domestic sheep producers employ the use of guard dogs in open range bands/herds for protection from wild canids, felids, and ursids. Dogs can be highly effective and work well when used with other techniques for certain species of predators. However, some precautions may be necessary for use of guard dogs (species of dog, total number of dogs/sheep band) in areas of wolf depredation as wolves may actively seek out and kill livestock guard dogs (Mosley et al. 2020).

Depending on the depredating species involved, additional strategies may be used including hazing/aversive conditioning, capture and relocation, and lethal removal (Bradley et al. 2005; Karlsson and Johansson 2010). While not palatable to some (Slagle et al. 2017), targeting and removing offending individual(s) responsible for livestock depredation is an effective and viable approach to reduce further depredation (Anderson et al. 2002) and is a methodology often used within a suite of options by many management entities. It should be noted that certain species or even individuals of a species may incur localized interest and sometimes federal protections (e.g., grizzly bears in the lower 48 states). Therefore, lethal removal is heavily scrutinized and sometimes disparaged (Slagle et al. 2017).

In terms of cause-specific mortality, mountain lions are generally more associated with sheep depredation, although cattle depredations occur in the southwestern U.S. and Mexico (Bodenchuck 2011; US Department of Agriculture et al. 2019a). Wolves and grizzly bears will depredate cattle and sheep when sympatric depending on vulnerability, density, and/or pack size of carnivore species present. Intensive herding of sheep on a 24-h basis has been effective at reducing predation by wolves, but the technique is obviously labor intensive (Stone et al. 2017).

Another component of conflict mitigation entails reducing the financial impacts of livestock depredation to producers through damage compensation programs. Many states and provinces provide some type of compensation for verified livestock depredation depending on the depredating species involved, but methodologies among agencies vary. Compensation programs do not provide a source of income for producers, but rather attempt to offset some of the costs of maintaining species such as wolves and grizzly bears where they overlap with livestock production (Jacobs and Main 2015). Maintaining wild open landscapes is critical for both livestock producers, large carnivores, and other wildlife. Building or at least maintaining a tolerance for carnivore presence can serve as a mutualistic benefit for carnivorous species and human counterparts alike. Perhaps the most significant point in regards to reducing large carnivore depredation potential is that there is no single technique that eliminates depredation outside of removing every carnivore and every domestic livestock species on the landscape (unrealistic); however, implementing multiple nonlethal/lethal techniques (as described above) can mitigate conflict to the point that both wild and domestic species persist.

24.4.3   Property Damage

Black bears can cause substantial property damage while obtaining anthropogenic foods (Messmer 2009; Lackey et al. 2018). Black and grizzly bears can also damage property while depredating or attempting to depredate domestic species (i.e., poultry, swine, sheep, goats). A significant source of damage from bears can occur in regards to honeybee (*Apis* spp.) apiaries (Messmer 2009). The most reliable method to reduce apiary damage is the use of well-maintained electric fencing (whether permanent or temporary). Since apiary damage can cause tens of thousands of dollars of damage, many agencies work closely with honeybee producer to proactively reduce that potential. Property damage caused by mountain lions and wolves rarely occurs. However, there have been instances of mountain lions breaking into chicken coops or similar outbuildings, and the authors are aware of one instance where a basement screen door was destroyed by a mountain lion in an attempt to depredate domestic housecats (*Felis catus*).

24.5   Large Carnivore Conflict Resolution

The previous section highlighted management strategies to mitigate and reduce conflict between large carnivores and humans, but managers always realize that before the first electric fence is erected or a capture effort initiated on a depredating animal, the fundamental foundation for effective large carnivore management lies within an active outreach and education program. While large carnivores play an important role in ecosystem function on western rangelands and provide intrigue and interest for the general public, they also cause consternation and can result in significant property or livestock damage for people that live and work in areas where large carnivores occur. Thus, information and educational programs are considered to be the foundation of a successful conflict resolution/mitigation program. Similarly, educational programs have evolved to increase efficacy of management actions and proactively resolve human dimension issues (Bennett 1998).

24.5.1 Information and Outreach

There are many options to efficaciously provide public outreach concerning large carnivores. For decades, nearly all public outreach was done via face-to-face inter-actions in classrooms, workshops, and symposia, or while visiting with people in coffeeshops, bars, or other social settings. Some gatherings were large and some small, but all were held with the goal of educating people about large carnivores and their management (Johnson et al. 1993; Bennett 1998). Additionally, most states and provinces had well-developed public meeting formats, as these techniques are still commonly used to gather input on annual hunting seasons and other management regulations for big game species. Prior to the advent of social media and virtual meetings, many citizens considered attendance at in-person public meetings as the best way to learn what was going on and voicing opinions to their local management agency.

As large carnivores expanded in number and distribution, most states and provinces adopted similar meeting formats to inform the public about pertinent management issues. In many cases, the public's understanding and knowledge of large carnivore issues was based more on preconceived notions, hearsay, and hyper-bole than reality. During the aforesaid understanding and knowledge of the ecology of the species was brought into these safety curriculums as well as building in a behavioral component on what one can do when recreating, working, and/or living in "large carnivore country" to increase human safety and reduce the risk of conflict.

As agencies built upon the strengths and weaknesses of previous work, many meetings and workshops turned into "community programs," which greatly increased their efficacy by stressing that homeowners and private landowners were part of the solution rather than being told what to do by governmental agencies. An example of this approach includes implementation of *Bear Wise* and *Bear Aware* programs in the Northern Rocky Mountains (IGBC 2019), especially in areas where grizzly bears occur. These types of programs focus on using improved community aware-ness, public education, and creative problem solving to deal with issues such as grizzly bears in cornfields and chicken coops. Public ownership facilitates active coordination, thereby increasing tolerance for large carnivores and furthering trust and collaboration between the public and governmental agencies (Guynn and Landry 1997). In recent years, many non-governmental agencies have become involved in safety programs, especially with grizzly bears and wolves. The key to success in any outreach program is communication and coordination among all parties involved in issues related to large carnivore conflict—where efforts have failed is where a key landowner or agency component is left out, which can create animosity (Dickman et al. 2013).

Technology has grown exponentially in regards to relaying information, as have outreach and education efforts. Website/webpage development is now a major focus for agencies that deal with large carnivore conflicts, with interactive links accessible to anyone in the world to garner information about safety in landscapes inhabited by carnivores. Social media have proven extremely effective in providing "virtual" safety workshops and community talk sessions to discuss ongoing issues and work being done to rectify those situations. Facebook Live™, Google Meetings™, and Zoom™ formats have evolved and their use and increased exponentially since the global Coronavirus pandemic of 2020–2022. Creative natural resource agencies took advantage of the amplified use of virtual meetings to continue interacting with an interested public, even when the ability to meet in-person was no longer an option. While there is no single format that provides the most effective outreach and education, a combination of multiple techniques can be beneficial. It is critical managers and educators know the stakeholders they are dealing with to determine the best method to reach their intended audience.

24.5.2 Producer Interaction and Communication

When dealing specifically with livestock depredation issues, it is critical that agency personnel and producers develop and maintain active communication channels and cooperative relationships to successfully mitigate problems as they develop. Fostering a trusting, face-to-face relationship based on consistency, honesty, and reliability between livestock producers and agencies when dealing with depredation is by far the most effective methodology for long-term, effective livestock depredation management (Fig. 24.2). Every state and provincial agency have programs to deal with livestock depredation, but many of these programs have subtle nuances that can be confusing to a producer when their operation crosses jurisdictional boundaries. Similarly, because managers must follow state/provincial statutes and commission regulations when verifying livestock losses, miscommunication can often result in anger and frustration for the producer, in particular when compensation is only given for confirmed depredation/damage from large carnivores (Fig. 24.2; Bruscino and Cleveland 2004; Thirgood et al. 2005; Morehouse et al. 2018). Contentious interactions often occur when livestock are killed by a large carnivore, because the toll on an individual's ranch operation can be substantial and livestock producers want the situation remedied immediately. There will also always be instances where disagreement occurs between agency personnel verifying damage and the effected producers. Lethal removal of wolves, mountain lions, or bears is often controversial and can result in litigation and changes in agency policies (Slagle et al. 2017). As stressed before, these volatile situations illustrate why consistency is critical; telling

Fig. 24.2 When verifying cause of death and/or depredation, cooperation and communication with the producer(s) is critical to walk them through what may be observed as the entity responsible for verification. In situations of dead livestock, it is best to skin the animal throughout to look for diagnostic signs of depredation by species. In the case of injured livestock, working with the producer to verify suspected cause of injury will assist to rectify future damage. These data can contribute to potential damage compensation programs if applicable. Photo by Wyoming Game and Fish Department

a producer what they want to hear will only make things more difficult and can have a "snowball effect" in the future.

Therefore, an active damage management program should include everything from proactive conflict reduction measures to outreach and education programs to active on-the-ground conflict resolution (i.e., capture, relocation, removal). Furthermore, an agency that instills the importance of their damage program into its institutional knowledge and trains field personnel to conduct the program with understanding and professionalism is the best way to timely and effective conflict resolution programs (Bradley et al. 2005; Breck et al. 2011).

24.6 Large Carnivore Ecosystem Threats

Section 24.2 addressed the major anthropogenic factors that resulted in population declines of big game and large carnivores across North America during westward European settlement. Direct human persecution, indirect prey reductions, landscape conversion toward tilled agriculture, and urbanization negatively impacted wolves, mountain lions, and black and grizzly bears. Many of these historic impacts have been identified and addressed from a conservation/management perspective and large carnivores and other species of wildlife have responded positively (Jenks 2011; US Fish and Wildlife Service 2011, 2017b).

However, the greatest potential future threat to large carnivores and most wildlife species is the increasing loss and/or fragmentation of rangeland habitats due to subdivisions and other human developments (Lindenmayer and Fischer 2006). An often-overlooked aspect in the recovery of large carnivores has been retention of large tracts of open rangeland within landscape level ranching operations. Through maintenance of these open rangeland settings and adjacent forested habitats, longevity of large carnivores and their principal prey species have been restored. Continued retention of these vital areas, in conjunction with the regulated management and protection of core habitat(s) will be crucial to continued resurgence of all large carnivore species in western North America.

24.7   Large Carnivore Research and Management Needs

Increased research and monitoring lead to more accurate and accountable management plans. These plans, conveyed in a number of ways to the public and cooperating agencies, inform how biologists and livestock producers can assess and deal with large carnivore conflict and conserve species on the landscape through translocation, hunting, or other methods. Because there is great international scrutiny in how carnivores are managed, acquisition of accurate data and strict adherence to plan details is imperative when justifying management actions and harvest regimes (Hovardas 2018). Justification of management strategies and objectives using solid, defensible data and transparency maintains public trust and may belay critics against certain management practices (Martin et al. 2009; US Fish and Wildlife Service 2017b). When being questioned concerning particular management actions (such as lethal removal or no-action taken), reliance on science-based actions and open, frank communication can assist to refute criticism. A regimented approach can also be relayed to the public so there is a transparent understanding of why a particular choice was made.

Management plans and situational guidelines that include the data-driven approaches referenced throughout this chapter provide further foundation for agency reliability and also serve to refute perceived instances of unfounded liability and the possibility of future litigation. While it is impossible to accurately prognosticate the future of large carnivore management and conflict mitigation, it is safe to say that the issues managers and agencies are dealing with today will only increase in intensity and scrutiny. Human occupation of the land continues to increase as does interest in recreational activities on public lands that are in essence prime habitat for large carnivore species. Outreach and educational programs focused on proactive efforts to reduce conflict are the future of successful programs aimed to maintain large carnivores and human livelihoods in areas of overlap. The successful recovery, conservation, and management of wolves, mountain lions, and bears will continue, and therein lies the challenge. It is imperative that management agencies utilize data-driven management strategies and a proactive/reactive conflict management program to reduce conflict and maintain tolerance of large carnivores in rangeland settings.

The intricate interactions among intact carnivore guilds and multiple prey species in natural systems are just beginning to be examined (Griffin et al. 2011; Eaker et al. 2016). The evolving interactions between predators and prey in North America provides an abundance of research opportunities, and using more holistic long-term studies allows insight in ecological phenomena that are of key interest to multiple stakeholders. There is continued interest in the ecosystem role of carnivores and conversely there is keen scrutiny on predation impacts to prey populations, which are research needs throughout all areas where the species are sympatric with multiple prey species. When humans are included in the equation, overall comprehension of these dynamics is even less understood. There is increasing societal pressure to provide instantaneous solutions to problems involving large carnivores and many of the publics interested in these animals expect wildlife managers to be omniscient. Research and monitoring strategies have been designed in recent years to evaluate the interactions between this new norm of intact carnivore guilds (Atwood et al. 2009) and their potential impacts to wild and domestic prey (Barber-Meyer et al. 2008), interspecific carnivore interactions (Bartnick et al. 2013; Elbroch et al. 2015) and inclusion of human population dynamics. Agency personnel often hear the refrain, "they were here first" (referring to wolves, mountain lions, and bears) from individuals or advocacy groups that demand a hands-off approach for native carnivore species. Nevertheless, a more pragmatic approach is to consider humans as integral to the ecosystems we share with these species, thus facilitating the concept of continued cohabitation of western landscapes for all species wild, domestic, and human into the future.

24.8 Conclusion

During the last 4–5 decades, large carnivores have increased in abundance and distribution following more than a century of determined effort to exterminate them. In this chapter, we have presented a brief overview of ways people may gain a better understanding of the ecology and management of wolves, mountain lions, and bears and outlined a number of methods to reduce conflict between these species and people in areas where they overlap. It is impossible to eliminate all conflict between humans and large carnivores because of the continuing intrusion of humans into habitats occupied by large carnivores, and concurrent expansion of these species into rangeland settings dominated by human use (Wilson et al. 2005; Wells et al. 2019). Additionally, outdoor recreationists are increasing their use of public lands, especially in areas that are remote rangeland and wilderness habitats, placing this segment of the public in direct conflict with both existing livestock operations and large carnivore populations. Communication and collaboration among multiple user groups must continue to improve understanding of the varied and sometimes conflicting needs and desires of these groups while recreating or working within rangeland habitats inhabited by large carnivores. Only by combining the dedication and knowledge of wildlife management agencies with the passion of people who live, work, and/

or recreate in rangeland settings can a sympatric land use ethic be realized. The continued persistence or these charismatic megafauna depend on human tolerance, understanding, and acceptance among all resource users—now more than ever.

References

Allen ME, Oftedal OT, Baer JJ (1997) The feeding and nutrition of carnivores. In Kleiman DG, Allen ME, Thompson KV, Lumpkin S (eds) Wild mammals in captivity. University of Chicago Press, Chicago

Amstrup SC, McDonald TL, Manly BFJ (2006) Handbook of capture-recapture analysis. Princeton University Press, Princeton

Anderson CR, Lindzey FG (2003) Estimating cougar predation rates from GPS location clusters. J Wildl Manage 67:307–316. https://doi.org/10.2307/3802772

Anderson CR, Ternent MA, Moody DS (2002) Grizzly bear-cattle interactions on two grazing allotments in northwest Wyoming. Ursus 13:247–256. https://www.jstor.org/stable/3873205

Anderson CR, Lindzey FG, Knopff KH, Jalkotzy MG, Boyce MS (2010) Cougar management in North America. In: Hornocker M, Negri S (eds) Cougar ecology and conservation. University of Chicago Press, Chicago

Apps CD, McLellan BN, Woods JG, Proctor MF (2004) Estimating grizzly bear distribution and abundance relative to habitat and human influence. J Wildl Manage 68:138–152. https://www.jstor.org/stable/3803777

Arnold TW, Clark RG, Koons DN, Schaub M (2018) Integrated population models facilitate ecological understanding and improved management decisions. J Wildl Manage 82:266–274. https://doi.org/10.1002/jwmg.21404

Atwood TC, Gese EM, Kunkel KE (2009) Spatial partitioning of predation risk in a multiple predator-multiple prey system. J Wildl Manage 73:876–884. https://doi.org/10.2193/2008-325

Barber-Meyer SM, Mech LD, White PJ (2008) Elk calf survival and mortality following wolf restoration to Yellowstone National Park. Wildlife Monographs 169. https://doi.org/10.2193/2008-004

Bartnick TD, Van Deelen TR, Quigley HB, Craighead D (2013) Variation in cougar (*Puma concolor*) predation habits during wolf (*Canis lupus*) recovery in the southern Greater Yellowstone Ecosystem. Can J Zool 91:82–93. https://doi.org/10.1139/cjz-2012-0147

Bateman HL, Lindquist TE, Whitehouse R, Gonzalez MM (2013) Mobile application for wildlife capture-mark-recapture data collection and query. Wild Soc Bull 37:838–845. https://doi.org/10.1002/wsb.322

Beausoleil RA, Clark JD, Maletzke BT (2016) A long-term evaluation of biopsy darts and DNA to estimate cougar density: an agency-citizen science collaboration. Wildl Soc Bull 40:583–592. https://doi.org/10.1002/wsb.675

Bennett J (1998) Can state regulatory agencies resolve controversial wildlife management issues involving the broad general public? Trans North Amer Wildl and Nat Res Conf 63:556–562

Bissonette JA (2019) Additional thoughts on rigor in wildlife science: unappreciated impediments. J Wildl Manage 83:1017–1021. https://doi.org/10.1002/jwmg.21677

Bjornlie DD, Thompson DJ, Haroldson MA, Schwartz CC, Gunther KA, Cain SL, Tyers DB, Frey KL, Aber BC (2014) Methods to estimate distribution and range extent of grizzly bears in the Greater Yellowstone Ecosystem. Wildl Soc Bull 38:182–187. https://doi.org/10.1002/wsb.368

Bleich VC, Thompson DJ (2018) State management of big game. In: Ryder TJ (ed) State wildlife management and conservation. Johns Hopkins University Press, Baltimore

Bodenchuck MJ (2011) Population management: depredation. In: Jenks JA (ed) Managing cougars in North America. Jack H. Berryman Institute, Utah State University, Logan

Boertje RD, Stephenson RO (1992) Effects of ungulate availability on wolf reproductive potential in Alaska. Can J Zool 70:2441–2443. https://doi.org/10.1139/z92-328

Borchers DL, Efford MG (2008) Spatially explicit maximum likelihood methods for capture recapture studies. Biometrics 64(2):377–385. https://doi.org/10.1111/j.1541-0420.2007.009 27.x

Borchers DL, Stevenson BC, Kidney D, Thomas L, Marques TA (2015) A unifying model for capture-recapture and distance sampling surveys of wildlife populations. J Amer Stat Assoc 110:195–294. https://doi.org/10.1080/01621459.2014.893884

Boulanger J, Himmer S, Swan C (2004) Monitoring of grizzly bear population trends and demography using DNA mark-recapture methods in the Owikeno Lake area of British Columbia. Can J Zool 82:1267–1277. https://doi.org/10.1139/Z04-100

Boulanger J, Stenhouse GB (2014) The impacts of roads on the demography of grizzly bears in Alberta. PlosOne. https://doi.org/10.1371/journal.pone.0115535

Bradley EH, Pletscher DH, Bangs EE, Kunkel KE, Smith DW, Mack CM, Meier TJ, Fontaine JA, Neimeyer CC, Jimenez MD (2005) Evaluating wolf translocation as a nonlethal method to reduce livestock conflicts in the northwestern United States. Conserv Biol 19:1498–1508. https://doi.org/10.1111/j.1523-1739.2005.00102.x

Bradley EH, Robinson HS, Bangs EE, Kunkel K, Jimenez MD, Gude JA, Grimm T (2015) Effects of wolf removal on livestock depredation recurrence and wolf recovery in Montana, Idaho, and Wyoming. J Wildl Manage 79:1337–1346. https://doi.org/10.1002/jwmg.948

Breck SW, Kluever BM, Panasci M, Oakleaf J, Johnson TB, Ballard W, Howery L, Bergman DL (2011) Domestic calf mortality and producer detection rates in the Mexican wolf recovery area: implications for livestock management and carnivore compensation schemes. Biol Conserv 144:930–936. https://doi.org/10.1016/j.biocon.2010.12.014

Bruscino MT, Cleveland TL (2004) Compensation programs in Wyoming for livestock depredation by large carnivores. Sheep Goat Res J 5. https://digitalcommons.unl.edu/icwdmsheepgoat

Buckland ST, Anderson DR, Burnham KP, Laake JL, Borchers DL, Thomas L (2001) Introduction to distance sampling: estimating abundance of biological populations. Oxford University Press, Oxford, United Kingdom. http://distancesampling.org/whatisds.html

Burnham KP, Anderson DR (2002) Model selection and multi-model inference: a practical information-theoretic approach, 2nd edn. Springer, New York

Caniglia REF, Cubaynes S, Gimenez O, Lebreton JD, Rando E (2012) An improved procedure to estimate wolf abundance using non-invasive genetic sampling and capture-recapture mixture models. Conserv Genetics 13:53–64. https://doi.org/10.1007/s10592-011-0266-1

Carbyn LW, Fritts SH, Seip DR (1995) Ecology and conservation of wolves in a changing world. Canadian Circumpolar Institute, University of Alberta, Edmonton, Alberta

Caughley G (1977) Analysis of vertebrate populations. Wiley, New York

Choate DM, Wolfe ML, Stoner DC (2009) Evaluation of cougar population estimators in Utah. Wildl Soc Bull 34:782–799

Clapp JG, Holbrook JD, Thompson DJ (2021) GPSeqClus: an R package for sequential clustering of animal location data for model building, model application and field site investigations. Methods Ecol Evol. https://doi.org/10.1111/2041-210X.13572

Cougar Management Guidelines Working Group (2005) Cougar management guidelines. Wild Futures, Bainbridge Island, Washington, USA

Craighead JJ, Varney JR, Craighead Jr FC (1974) A population analysis of the Yellowstone grizzly bears. Montana Forest and Conservation Experiment Station and Montana Cooperative Wildlife Research Unit, University of Montana, Missoula, Montana, USA

Cullingham CI, Thiessen CD, Derocher AE, Paquet PC, Miller JM, Hamilton JA, Coltmant DW (2016) Population structure and dispersal of wolves in the Canadian Rocky Mountains. J Mamm 97:839–851. https://doi.org/10.1093/jmammal/gyw015

Davidson GA, Clark DA, Johnson BK, Waits LP, Adams JR (2014) Estimating cougar densities in northeast Oregon using conservation detection dogs. J Wildl Manage 78:1104–1114. https://doi.org/10.1002/jwmg.758

Davidson-Nelson SJ, Gehring TM (2010) Testing fladry as a nonlethal management tool for wolves and coyotes in Michigan. Hum Wildl Interact. https://doi.org/10.26077/mdky-bs63

Dickman A, Marchini S, Manfredo MA (2013) The human dimension in addressing conflict with large carnivores. https://doi.org/10.1002/9781118520178.ch7

Eaker DR, Hebblewhite M, Proffitt KM, Jimenez BS, Mitchell MS, Robinson HS (2016) Annual elk calf survival in a multiple carnivore system. J Wildl Manage 80:1345–1359. https://doi.org/10.1002/JWMG.21133

Eklund A, Poez-Bao JV, Tourani M, Chapron G (2017) Limited evidence on the effectiveness of interventions to reduce livestock predation by large carnivores. Sci Rep 7:2097. https://doi.org/10.1038/s41598-017-02323-w

Elbroch LM, Lendrum PE, Newby J, Quigley H, Thompson DJ (2015) Recolonizing wolves impact the realized niche of resident cougars. Zool Stud. https://doi.org/10.1186/s40555-015-0122-y

Elbroch LM, Ferguson JM, Quigley H, Craighead D, Thompson DJ, Wittmer HU (2020) Reintroduced wolves and hunting limit the abundance of a subordinate apex predator in a multi-use landscape. Proc R Soc B 287:20202202. https://doi.org/10.1098/rspb.2020.2202

Fuller MR, Millspaugh JJ, Church KE, Kenward RE (2005) Wildlife radiotelemetry. In: Braun CE (ed) Techniques for wildlife investigation and management, 6th edn. The Wildlife Society, Bethesda

Gardner B, Royle JA, Wegan MT, Rainbolt RE, Curtis PD (2010) Estimating black bear density using DNA data from hair snares. J Wildl Manage 74:318–325. https://doi.org/10.2193/2009-101

Griffin KA, Hebblewhite M, Robinson HS, Zager P, Barber-Meyer SM, Christianson D, Creel S, Harris NC, Hurley MA, Jackson DH, Johnson BK, Myers WL, Raithel JD, Schlegel M, Smith BL, White C, White PJ (2011) Neonatal mortality of elk driven by climate, predator phenology and predator community composition. J Anim Ecol 80:1246–1257. https://doi.org/10.1111/j.1365-2656.2011.01856.x

Gurney SM, Smith JB, Etter DR, Williams DM (2020) American black bears and hair snares: a behavioral analysis. Ursus 39:1–9. https://doi.org/10.2192/URSUS-D-18-00020.2

Guynn DE, Landry MK (1997) A case study of citizen participation as a success model for innovative solutions for natural resource problems. Wildl Soc Bull 25:392–398

Haroldson MA, Clapham M, Costello CC, Gunther KA, Kendall KC, Miller SD, Pigeon KE, Proctor MF, Rode KD, Servheen C, Stenhouse GB, van Manen FT (2021) Brown bear (*Ursus arctos*; North America). In: Penteriani V, Melletti M (eds) Bears of the world. Cambridge University Press, Cambridge

Hebblewhite M, Paquet PC, Pletscher DH, Lessard RB, Callaghan CJ (2003) Development and application of a ratio estimator to estimate wolf kill rates and variance in a multi-prey system. Wildl Soc Bull 31:933–946

Herrero S (1985) Bear attacks: their causes and avoidance. Nick Lyons Books/Winchester Press, New York

Hornocker MG (1969) Winter territoriality in mountain lions. J Wildl Manage 33:457–464

Hornocker MG, Negri S (2010) Cougar: ecology and conservation. The University of Chicago Press, Chicago

Hovardas T (2018) Large carnivore conservation and management: human dimensions. Routledge Taylor and Francis Group, New York

Hristienko H, McDonald JE Jr (2007) Going into the 21st century: a perspective on trends and controversies in the management of the American black bear. Ursus 18:72–88. https://doi.org/10.2192/1537-6176(2007)18[72:GITSCA]2.0.CO;2

Hughson DL, Darby NW, Dungan JD (2010) Comparison of motion-activated cameras for wildlife investigations. California Fish and Game 96:101–109

Humm JM, McCown JW, Scheick BK, Clark JD (2017) Spatially explicit population estimates for black bears based on cluster sampling. J Wildl Manage 81:1187–1201. https://doi.org/10.1002/jwmg.21294

Hurley MA, Unsworth JW, Zager P, Hebblewhite M, Garton EO, Montgomery DM, Skalski JR, Maycock CL (2011) Demographic response of mule deer to experimental reduction of coyotes and mountain lions in southeastern Idaho. Wildl Monographs 178. https://doi.org/10.1002/wmon.4

Interagency Grizzly Bear Committee (IGBC) (2019) Interagency grizzly bear committee charter. IGBC Executive Committee, Missoula

Jansen BD, Jenks JA (2012) Birth timing for mountain lions (*Puma concolor*); testing the prey availability hypothesis. PlosOne https://doi.org/10.1371/journal.pone.0044625

Jacobs CE, Main MB (2015) A conservation-based approach to compensation for livestock depredation: the Florida panther case study. PlosOne. https://doi.org/10.1371/journalpone.0139203

Jenks JA (ed) (2011) Managing cougars in North America. Jack H. Berryman Institute, Utah State University, Logan

Johnson KN, Johnson RL, Edwards DK, Wheaton CA (1993) Public participation in wildlife management: opinions from public meetings and random surveys. Wildl Soc Bull 21:218–225

Karanth KU, Nichols JD (1998) Estimation of tiger densities in India using photographic captures and recaptures. Ecology 79:2852–2862. https://doi.org/10.1890/0012-9658

Karlsson J, Johansson Ö (2010) Predictability of repeated carnivore attacks on livestock favours reactive use of mitigation measures. J Appl Ecol 47:166–171. https://doi.org/10.1111/J.1365-2664.2009.01747.x

Kelly MJ, Noss AJ, DiBitetti MS, Maffei L, Arispe RL, Paviolo A, De Angelo CD, Di Blanco YE (2008) Estimating puma densities from camera trapping across three study sites: Bolivia, Argentina, and Belize. J Mammal 89:408–418. https://doi.org/10.1644/06-MAMM-A-42R.1

Knopff KH, Knopff AA, Warren MB, Boyce MS (2009) Evaluating global positioning system telemetry techniques for estimating cougar predation parameters. J Wildl Manage 73:586–597. https://doi.org/10.2193/2008-294

Knopff KH, Knopff AA, Kortello A, Boyce MS (2010) Cougar kill rate and prey composition in a multiprey system. J Wildl Manage 74:1435–1447. https://doi.org/10.2193/2009-314

Kristensen TV, Kovach AI (2018) Spatially explicit abundance estimation of a rare habitat specialist: implications for SECR study design. Ecosphere 9. https://doi.org/10.1002/ecs2.2217

Lackey CW, Breck SW, Wakeling BF, White B (2018) Human-black bear conflicts: a review of common management practices. Human Wildl Interact Mono 2:1–68

LaRue MA, Nielson CK (2011) Population viability of recolonizing cougars in Midwestern North America. Ecol Model 321:121–129. https://doi.org/10.1016/j.ecolmodel.2015.09.026

LaRue MA (2018) America's cat is on the comeback. Amer Scien 106. https://doi.org/10.1511/2018.106.6.352

Leopold A (1933) Game management. Charles Scribner's Sons, New York

Lindenmayer JB, Fischer J (2006) Habitat fragmentation and landscape change, an ecological and conservation synthesis. Island Press, Washington, DC

Mace RD (2004) Integrating science and road access management lessons from the Northern Continental Divide Ecosystem. Ursus 15:129–136. https://doi.org/10.2192/1537-6176(2004)015%3c0129:ISARAM%3e2.0.CO;2

Martin J, Runge MC, Nichols JD, Lubow BC, Kendall WL (2009) Structured decision making as a conceptual framework to identify thresholds for conservation and management. Ecol Appl 19:1079–1090. https://doi.org/10.1890/08-0255.1

Mattioli LA, Canu D, Passilongo D, Scandura M, Appolonio M (2018) Estimation of pack density in grey wolf (*Canis lupus*) by applying spatially explicit capture-recapture models to camera trap data supported by genetic monitoring. Frontier Zool 15:38. https://doi.org/10.1186/s12983-018-0281-x

McClintock BT, White GC, Burnham KP (2006) A robust design mark-resight abundance estimator allowing heterogeneity in resighting probabilities. J Agric Biol Environ Stat 11:231–248. https://doi.org/10.1198/108571106X129171

McNay ME (2002) A case history of wolf-human encounters in Alaska and Canada. Alaska Department of Fish and Game, Wildlife Technical Bulletin 13. Juneau, AK, USA

Mech LD (1970) The wolf: the ecology and behavior of an endangered species. The Natural History Press, Garden City

Messmer TA (2009) Human-wildlife conflicts: emerging challenges and opportunities. Human Wildl Interact 3:10–17. http://www.berrymaninstitute.org/journal/index.html

Moll R, Killion A, Montgomer R, Tambling CJ (2016) Spatial patterns of African ungulate aggregation reveal complex but limited risk effects from reintroduced carnivores. Ecol 95:1123–1134. https://doi.org/10.1890/15-0707.1

Montgomery RA, Moll RJ, Say-Saliz E, Valeix M, Prugh LR (2019) A tendency to simplify complex systems. Biol Conserv 233:1–11. https://doi.org/10.1016/j.biocon.2019.02.001

Morehouse AT, Boyce MS (2016) Grizzly bears without borders: spatially explicit capture-recapture in southwestern Alberta. J Wild Manag 80:1152–1166. https://doi.org/10.1002/jwmg.21104

Morehouse AT, Tigner J, Boyce MS (2018) Coexistence with large carnivores supported by a predator-compensation program. Environ Manage 61:719–731. https://doi.org/10.1007/s00267-017-0994-1

Mosley JC, Roeder BL, Frost RA, Wells SL, McNew LB, Clark PE (2020) Mitigating human conflicts with livestock guardian dogs in extensive sheep grazing systems. Rangel Ecol Manage 73:724–732. https://doi.org/10.1016/j.rama.2020.04.009

Murphy SM, Augustine BC, Adams JR, Waits LP, Cox JJ (2018) Integrating multiple genetic detection methods to estimate population density of social and territorial carnivores. Ecosphere 9(10). https://doi.org/10.1002/ecs2.2479

Neihardt JG (1932) Black elk speaks. Pocket Books, New York

Newey S, Davidson P, Nazir S, Fairhurst G, Verdicchio F, Irvine RJ, van der Wal R (2015) Limitations of recreational camera traps for wildlife management and conservation research: a practitioner's perspective. Ambio 44:624–635. https://doi.org/10.1007/s13280-015-0713-1

Paetkau D, Shields GF, Strobeck C (1998) Gene flow between insular, coastal and interior populations of brown bears in Alaska. Mol Ecol 7:1283–1292. https://doi.org/10.1046/j.1365-294x.1998.00440.x

Pierce BM, Bleich VC, Monteith KL, Bowyer RT (2012) Top-down versus bottom-up forcing: evidence from mountain lions and mule deer. J Mammal 93:977–988. https://doi.org/10.1644/12-mamm-a-014.1

Powell RA, Zimmerman JW, Seaman DE (1997) Ecology and behavior of North American black bears: home ranges, habitat and social organization, 1st edn. Chapman and Hall, Chicago

Proctor MF, McLellan BN, Strobeck C (2002) Population fragmentation of grizzly bears in southeastern British Columbia, Canada. Ursus 12:153–161. https://www.jstor.org/stable/3873196

Proffitt KM, Garrott R, Gude JA, HebblewhiteM, Jimenez B, Paterson JT, Rotella J (2020) Integrated carnivore-ungulate management: a case study in West-Central Montana. Wildlife Monographs 206. https://doi.org/10.1002/wmon.1056

Pyare S, Cain S, Moody D, Schwartz C, Berger J (2004) Carnivore re-colonization: reality, possibility and anon-equilibrium century for grizzly bears in the southern Yellowstone Ecosystem. Animal Cons 7:1–7. https://doi.org/10.1017/S1367943003001203

Quigley H, Herrero S (2005) Characterization and prevention of attacks on humans. In Woodroffe R, Thirgood S, Rabinowitz A (eds) People and wildlife: conflict or coexistence? Cambridge University Press, New York

R Development Core Team (2019) R: a language and environment for statistical computing. Vienna, Austria. http://www.R-project.org

Riley SJ, Nesslage GM, Maurer BA (2004) Dynamics of early wolf and cougar eradication efforts in Montana: implications for conservation. Biol Conserv 119:575–579. https://doi.org/10.1016/j.biocon.2004.01.019

Robinson JS, Wielgus RB, Cooley HS, Cooley SW (2008) Sink populations in carnivore management: cougar demography and immigration in a hunted population. Ecol Appl 18:1028–1037. https://doi.org/10.1890/07-0352.1

Russell RE, Royle JA, Desimone R, Schwartz MK, Edwards VL, Pilgrim KP, McKelvey KS (2012) Estimating abundance of mountain lions from unstructured spatial sampling. J Wildl Manage 76:1551–1561. https://doi.org/10.1002/jwmg.412

Sawaya MA, Ruth TK, Creel S, Rotella JJ, Quigley HB, Stetz JB, Kalinowski ST (2010) Evaluation of noninvasive genetic sampling methods for cougars using a radio-collared population in Yellowstone National Park. J Wildl Manage 75:612–622. https://doi.org/10.1002/jwmg.92

Schwartz CC, Fortin JK, Teisberg JE, Haroldson MA, Servheen C, Robbins CT, van Manen FT (2014) Body and diet composition of sympatric black and grizzly bears in the Greater Yellowstone Ecosystem. J Wildl Manage 78:68–78. https://doi.org/10.1002/jwmg.633

Schwartz CC, Haroldson MA, White GC (2010) Hazards affecting grizzly bear survival in the Greater Yellowstone Ecosystem. J Wildl Manage 74:654–667. https://doi.org/10.2193/2009-206

Seidensticker IV JC, Hornocker MG, Wiles WV, Messick JP (1973) Mountain lion social organization in the Idaho Primitive Area. Wildlife Monographs 35

Sells SN, Bassing SB, Barker KJ, Foreshee SC, Keever AC, Goerz JW, Mitchell MS (2018) Increased scientific rigor will improve reliability of research and effectiveness of management. J Wildl Manage 82:485–494. https://doi.org/10.1002/jwmg.21413

Slagle KJ, Bruskotter T, Singh AS, Schmidt RH (2017) Attitudes towards predator control in the United States: 1995 and 2014. J Mammal 98:7–16. https://doi.org/10.1093/jmammal/gyw144

Smith TS, Gookin J, Hopkins BG, Thompson SH (2018) Portable electric fencing for bear deterrence and conservation. Human-Wildl Interac 12:309–321. https://doi.org/10.26077/h9sw-qg28

Stansbury CR, Ausband DE, Zager P, Mack CM, Miller CR, Pennell MW, Waits LP (2014) A long-term population monitoring approach for a wide-ranging carnivore: noninvasive genetic sampling of gray wolf rendezvous sites in Idaho, USA. J Wildl Manage 78:1040–1049. https://doi.org/10.1002/jwmg.736

Stone SA, Breck SW, Timberlake J, Haswell PM, Najem F, Bean BS, Thornhill DJ (2017) Adaptive use of nonlethal strategies for minimizing wolf-sheep conflict in Idaho. J Mammal 98:33–44. https://doi.org/10.1093/jmammal/gyw188

Thirgood S, Woodroffe R, Rabinowitz A (2005) The impact of human wildlife conflict on human lives and livelihoods. In: Woodroffe R, Thirgood S, Rabinowitz A (eds) People and wildlife: conflict or coexistence? Cambridge University Press, New York

Thompson DJ, Jenks JA (2010) Dispersal movements of subadult cougars from the Black Hills: the notions of range expansion and recolonization. Ecosphere 1:1–11. https://doi.org/10.1890/ES10-00028.1

Tomkiewicz SM, Fuller MR, Kie JG, Bates KK (2010) Global positioning system and associated technologies in animal behaviour and ecological research. Phil Trans Roy Soc b. 365:2163–2176. https://doi.org/10.1098/rstb.2010.0090

US Department of Agriculture, Animal and Plant Health Inspection Service, Veterinary Services, National Animal Health Management System (USDA APHIS VS NAHMS) (2019a) Sheep and lamb predator and nonpredator death loss in the United States, 2018. USDA–APHIS–VS–CEAH–NAHMS, Fort Collins

US Department of Agriculture, Animal and Plant Health Inspection Service, Veterinary Services, National Animal Health Management System (USDA APHIS VS NAHMS) (2019b) Death loss in U.S. cattle and calves due to predator and nonpredator causes, 2018. USDA–APHIS–VS–CEAH–NAHMS, Fort Collins

US Fish and Wildlife Service, Montana Fish, Wildlife & Parks, Nez Perce Tribe, National Park Service, Blackfeet Nation, Confederated Salish and Kootenai Tribes, Wind River Tribes, Washington Department of Wildlife, Oregon Department of Wildlife, Utah Department of Natural Resources, and USDA Wildlife Services (2011) Rocky Mountain wolf recovery 2010 interagency annual report. Sime CA, Bangs EE (eds) U.S. Fish and Wildlife Service, Helena

US Fish and Wildlife Service (2013) Grizzly bear recovery plan draft revised supplement: proposed revisions to the demographic recovery criteria for the grizzly bear population in the Greater Yellowstone Area. Missoula, Montana, USA

US Fish and Wildlife Service (2017a) Final conservation strategy for the grizzly bear in the Greater Yellowstone Area. Interagency Conservation Strategy Team, Missoula, Montana, USA

US Fish and Wildlife Service (2017b) Final rule removing the Greater Yellowstone Ecosystem population of grizzly bears from the federal list of endangered and threatened wildlife. 82 Federal Register 30502

US Fish and Wildlife Service (2020) Endangered and threatened wildlife and plants; removing the gray wolf (*Canis lupus*) from the list of endangered and threatened wildlife. 85 FR 69778:69778–69895

van Manen FT, Haroldson MA, Bjornlie DD, Ebinger MR, Thompson DJ, Costello CM, White GC (2016) Density dependence, whitebark pine, and vital rates of grizzly bears. J Wildl Manage 80:300–313. https://doi.org/10.1002/jwmg.1005

Van Sickle WD, Lindzey FG (1991) Evaluation of a cougar population estimator based on probability sampling. J Wildl Manage 55:738–743. https://doi.org/10.2307/3809526

Wasser SK, Davenport B, Ramage ER, Hunt KE, Parker M, Clark C, Stenhouse G (2004) Scat detection dogs in wildlife research and management: application to grizzly and black bears in the Yellowhead Ecosystem, Alberta, Canada. Can J Zool 82:474–492. https://doi.org/10.1139/z04-020

Wells SL, McNew LB, Tyers DB, van Manen FT, Thompson DJ (2019) Grizzly bear depredation on grazing allotments in the Yellowstone Ecosystem. J Wildl Manage 83:556–566. https://doi.org/10.1002/jwmg.21618

Wilson SM, Madel MJ, Mattson DJ, Graham JM, Burchfiled JA, Belsky JM (2005) Natural landscape features, human-related attractants, and conflict hotspots: a spatial analysis of human-grizzly bear conflicts. Ursus 16:117–129. https://doi.org/10.2192/1537-6176(2005)016[0117:NLFHAA]2.0.CO;2

White GC, Burnham KP (1999) Program MARK: survival estimation from populations of marked animals. Bird Study 46:S120-139. https://doi.org/10.1080/00063659909477239

White PJ, Gunther KA, van Manen FT (eds) (2017) Yellowstone grizzly bears: ecology and conservation of an icon of wildness. National Park Service, US Geological Survey, Library of Congress, Washington, DC

Wilckens DT, Smith JB, Tucker SA, Thompson DJ, Jenks JA (2015) Mountain lion (*Puma concolor*) feeding behavior in the Little Missouri Badlands, North Dakota. J Mammal 97:373–385. https://doi.org/10.1093/jmammal/gyv183

Wyoming Game and Fish Commission (2011) Wyoming gray wolf management plan. 14 Sept 2011. Wyoming Game and Fish Commission, Cheyenne. http://gf.state.wy.us/web2011/Departments/Wildlife/pdfs/WOLF_MANAGEMENT_PLAN_FINAL0000348.pdf

Young SP, Goldman EA (1946) The puma, mysterious American cat. The American Wildlife Institute, Washington, DC

Chapter 25
Amphibians and Reptiles

David S. Pilliod and Todd C. Esque

Abstract Amphibians and reptiles are a diverse group of ectothermic vertebrates that occupy a variety of habitats in rangelands of North America, from wetlands to the driest deserts. These two classes of vertebrates are often referred to as herpetofauna and are studied under the field of herpetology. In U.S. rangelands, there are approximately 66 species of frogs and toads, 58 salamanders, 98 lizards, 111 snakes, and 27 turtles and tortoises. Herpetofauna tend to be poorly studied compared with other vertebrates, which creates a challenge for biologists and landowners who are trying to manage rangeland activities for this diverse group of animals and their habitats. Degradation of habitats from human land use and alteration of natural processes, like wildfire, are primary threats to herpetofauna populations. Disease, non-native predators, collection for the pet trade, and persecution are also conservation concerns for some species. Properly managed livestock grazing is generally compatible with herpetofauna conservation, and private and public rangelands provide crucial habitat for many species. Climate change also poses a threat to herpetofauna, but we have an incomplete understanding of the potential effects on species. Dispersal and adaptation could provide some capacity for species to persist on rangelands as climates, disturbance regimes, and habitats change. However, inadequate information and considerable uncertainty will make climate mitigation planning difficult for the foreseeable future. Planning for and mitigating effects of climate change, and interactions with other stressors, is an urgent area for research. Maintaining large, heterogeneous land areas as rangelands will certainly be an important part of the conservation strategy for herpetofauna in North America.

D. S. Pilliod (✉)
U.S. Geological Survey, Forest and Rangeland Ecosystem Science Center, 230 N Collins Rd., Bldg 3, Boise, ID 83702, USA
e-mail: dpilliod@usgs.gov

T. C. Esque
U.S. Geological Survey, Western Ecological Research Center, PO Box 60640, Boulder City, NV 90005, USA
e-mail: tesque@usgs.gov

L. B. McNew et al. (eds.), *Rangeland Wildlife Ecology and Conservation*,
https://doi.org/10.1007/978-3-031-34037-6_25

861

Keywords Amphibia · Frog · Grazing · Herpetofauna · Land management · Lizard · Rangelands · Reptilia · Salamander · Sauropsida · Snake · Toad · Tortoise · Turtle · Wildlife

25.1 General Life History and Population Dynamics

Amphibians and reptiles are diverse classes of vertebrates. Amphibia are organized taxonomically into three orders: Anura (frogs and toads); Caudata or Urodela (salamanders); and Apoda or Gymnophiona (caecilians). Reptilia are organized into four orders: Squamata (lizards and snakes); Testudines (turtles and tortoises); Crocodylia (alligators and their allies); and Rhynchocephalia (tuatara). Amphibians and reptiles were combined historically and studied in the field of herpetology, but their evolutionary history is not so tidy. Modern cladistics even abandons the term reptile in favor of the clade Sauropsida, which includes birds. This chapter, however, follows a traditional taxonomy of amphibians and non-avian reptiles. North America is home to more than 733 amphibian and reptile species (Crother 2017). According to our analyses, there are about 124 amphibian species, composed of frogs and toads (N = 66) and salamanders (N = 58), whose distributions overlap by at least 10% with the rangeland ecoregions described in this book (Table 25.1). The United States (U.S.) is a global hotspot of salamander diversity, but salamanders are much less common in U.S. rangelands. About 89% of U.S. rangeland reptile species are lizards (N = 98) and snakes (N = 111), with the remaining diversity composed of turtles and tortoises (N = 27) (Table 25.2). One crocodilian, the American alligator (*Alligator mississippiensis*), occurs on the periphery of southeastern U.S. rangelands, but its distribution was below the 10% threshold for inclusion in this chapter. Keep in mind that diversity estimates of herpetofauna are dynamic as new species continue to be discovered or described. For example, using molecular and morphological evidence, two new toad species were described in central Nevada in 2019 (Gordon et al. 2020). Taxonomy of amphibians and reptiles is somewhat unresolved and often disputed, which creates challenges for communication in conservation. In this chapter, we use taxonomy from Crother (2017) and the Integrated Taxonomic Information System (ITIS; www.itis.gov, accessed 13 July 2021). Species counts, however, were generated from distribution maps in the USGS Gap Analysis Project (USGS GAP 2018a), which used an older taxonomy (Crother et al. 2003). Crother (2017) is currently the most widely accepted taxonomy for North America and thus we recommend checking this reference and consulting with state herpetologists for the latest taxonomic information about species in your area.

Some life history characteristics are shared by amphibians and reptiles and these are important to consider when characterizing their ecology and understanding their habitat use patterns in rangelands. First, amphibians and reptiles are both ectothermic, meaning they cannot regulate body temperatures through metabolism. Instead, their body temperature tracks the environmental temperatures of their surroundings, although they can influence this process behaviorally. For example, amphibians and

Table 25.1 Count of amphibian species within U.S. rangelands grouped by family

Order	Family	Species count
Anura	Ascaphidae	2
Anura	Bufonidae	17
Anura	Craugastoridae	1
Anura	Eleutherodactylidae	3
Anura	Hylidae	15
Anura	Leptodactylidae	1
Anura	Microhylidae	3
Anura	Ranidae	17
Anura	Rhinophrynidae	1
Anura	Scaphiopodidae	6
Caudata	Ambystomatidae	9
Caudata	Plethodontidae	42
Caudata	Proteidae	1
Caudata	Rhyacotritonidae	2
Caudata	Salamandridae	4
Total		124

Anura are frogs and toads, and Caudata are salamanders and newts. *Data source* USGS GAP (2018a)

reptiles raise their body temperatures by exposure to solar or thermal radiation. This is accomplished by a variety of mechanisms, especially darkening their skin through pigmentation, basking, and pressing their bodies against warm surfaces. This process, known as behavioral thermoregulation, explains why lizards are frequently seen basking in morning sunlight or hugging rocks on a cool day. This also explains why snakes are often encountered (and unfortunately killed) on asphalt roads. Behavioral thermoregulation allows herpetofauna to accelerate temperature increases for activity and maintain optimal body temperatures for more hours of the day, including into the night. Amphibians and reptiles can also lower their body temperatures by evaporative cooling or re-radiating body heat into a cooler surrounding environment, such as water, shade, or burrows (Figs. 25.1 and 25.2). Spadefoot toads (*Spea* and *Scaphiopus* spp.), for example, are some of the most widespread amphibian inhabitants of U.S. rangelands and use a hardened skin spur on their hind feet to dig burrows into sandy soils to escape dangerously hot, dry surface conditions. Thus, amphibians also select specific microsites to maintain preferred body temperatures, but at the expense of water loss and thus strike a fine balance between temperature regulation and dehydration (Bartelt et al. 2010). Scales, shells, and thickened skin protect reptiles from dehydration. These and other anatomical features, traits, and adaptations enable reptiles to use a wider range of terrestrial locations than amphibians to optimize body temperatures to meet physiological needs.

Table 25.2 Count of reptile species within U.S. rangelands grouped by family

Order	Family	Species count
Squamata (L)	Anguidae	7
Squamata (S)	Charinidae	2
Squamata (S)	Colubridae	87
Squamata (L)	Crotaphytidae	7
Squamata (S)	Elapidae	2
Squamata (L)	Eublepharidae	4
Squamata (L)	Helodermatidae	1
Squamata (L)	Iguanidae	24
Squamata (S)	Leptotyphlopidae	3
Squamata (L)	Phrynosomatidae	19
Squamata (L)	Phyllodactylidae	1
Squamata (L)	Scincidae	8
Squamata (L)	Teiidae	20
Squamata (S)	Viperidae	17
Squamata (L)	Xantusiidae	7
Testudines	Chelydridae	2
Testudines	Emydidae	15
Testudines	Kinosternidae	5
Testudines	Testudinidae	2
Testudines	Trionychidae	3
Total		236

Data source USGS GAP (2018a, b). Squamata are lizards (L) and snakes (S), and Testudines are turtles and tortoises. We intentionally did not include any of the five marine turtles

Reproduction and development are life history characteristics where amphibians and reptiles diverge (Pough et al. 1998). Amphibians produce eggs that are not protected by shells and thus must be deposited in water or very moist environments. Like fishes, most frogs and toads fertilize their eggs externally whereby a female deposits her eggs directly into the water and the male releases sperm onto them. Most amphibian embryos develop gills and become free-swimming tadpoles (frogs and toads) or larvae (salamanders). Some terrestrial species of lungless salamanders (family Plethodontidae) skip the larval stage and embryos develop directly into the adult body form, albeit a tiny version. Most tadpoles and larvae go through metamorphosis, which is the developmental transformation from an aquatic, gilled life stage to terrestrial juveniles that have the adult body form and use lungs for respiration. Amphibians are among only a handful of vertebrates that go through metamorphosis (Laudet 2011). The duration of the larval stage and timing of metamorphosis varies considerably by species and is dependent on both genetic and environmental factors. A few salamander species, including tiger salamanders (*Ambystoma* spp.) which are common in U.S. rangelands (Fig. 25.1), can retain their gills and remain aquatic as sexually mature adults. Finally, all amphibians have retained some capacity to respire through their skin, although this inefficient form of respiration usually only occurs for animals overwintering under water and obligatorily among the lungless salamanders, which lack both lungs and gills.

Fig. 25.1 Photographs of some of the amphibian species found in rangelands. Photographs by Alan St. John, Ryan Hagerty, Mindy Meade, Chad Mellison, Charles R. Peterson, and Alan Schmierer

Like birds and a few mammals, reptiles produce eggs that have multiple membranes external to the embryo and a protective outer shell that is either parchment-like and leathery, or hard and calcified similar to a chicken egg. Reptiles use one of three strategies for reproduction: ovipary, vivipary, and ovovivipary. In ovipary, the egg must be fertilized internally by copulation between the male and female prior to eggshell formation. Oviparous embryos partially develop inside the female and eggs are laid in microsites with specific soil substrate and moisture and temperature conditions for development and hatching. Oviparous reptiles hatch fully formed as miniature adults, although they may carry the remainder of the egg yolk as a 'sack-lunch' during their first season. In ovovivipary, embryos may acquire their sustenance from a yolk that remains inside the female during development, or embryos may be connected to the female by a placenta (i.e., true vivipary). Rattlesnakes (*Crotalus* spp.), boas (*Charina* spp.), and gartersnakes (*Thamnophis* spp.) are all ovoviviparous species from U.S. rangelands (Fig. 25.2).

25.2 Species Status

25.2.1 Historical Versus Current Distributions

The diversity of rangeland herpetofauna presently found across North America can be linked to the environments their ancestors experienced and numerous vicariance events (Pyron 2014; Modesto et al. 2015; Wollenberg Valero et al. 2019). Over the eons, the configurations, sizes, and positions of drifting continents shaped the habitats available to herpetofauna with changes in latitude (i.e., tropical versus polar conditions), climates, sea levels, and formations of lava flows, mountain ranges, deserts, and inland seas. The uplift of mountain ranges in western North America (e.g., most recently the Cascade Range and Sierra Nevada around 4–7 mya) produced rain shadows that drastically altered the climates and vegetation of western rangelands. These deserts influenced the evolution, speciation, and adaptations of modern rangeland herpetofauna (Bryson et al. 2012; Bouzid et al. 2021). Glacial cycles and the formation and draining of inland lakes (e.g., Bonneville, Missoula) during the past 15,000–25,000 yrs also influenced diversification of the species we know today (Thompson and Russell 2005; Funk et al. 2008; Kimberly and Fender 2020). Vicariance and introgression of rangeland species is ongoing with modern processes like anthropogenically induced climate change and fragmentation of habitat.

Amphibians and reptiles are generally understudied, even in rangelands where diversity is comparable to, or higher than, other vertebrate groups (Qian 2009). Therefore, information about historical distribution is severely lacking. A logical assumption is that the historical distribution of amphibians in rangelands was probably determined by the availability of surface water and we know surface waters have changed dramatically over contemporary times (Qian 2010). Part of that change is attributed to intensive trapping of North American beaver (*Castor canadensis*) for

Fig. 25.2 Photographs of some of the reptile species found in rangelands. Photographs by Patrick Alexander, Courtney Celley, Michelle Jeffries, Jerry Kirkhart, Gavin O'Leary, Peter Paplanus, and Charles R. Peterson

pelts and draining of wetlands for cropland agriculture and pasture (Gibson and Olden 2014; Grudzinski et al. 2020; Wohl 2021). We suspect that loss of amphibian habitat must have been enormous because beaver activity in the western U.S. today is strongly associated with amphibian occupancy patterns, especially for frogs and toads (Arkle and Pilliod 2015; Hossack et al. 2015; Zero and Murphy 2016). Other novel water sources were American bison (*Bison bison*) wallows, which must have once been numerous and extensive across the Great Plains (Meagher 1986). Remnant wallows were still found during the 1940s in grasslands where wallows had not been destroyed by cultivation. Some of those remnant wallows were about 6 m (20 ft) wide and 2.5 m (8 ft) deep and were used as breeding sites by Great Plains toads (*Anaxyrus cognatus*; Bragg 1940). Western chorus frogs (*Pseudacris triseriata*) and northern cricket frogs (*Acris crepitans*) started using bison wallows for breeding at the Konza Prairie in Kansas after bison were reintroduced in 1987 (Gerlanc and Kaufman 2003). Similarly, western chorus frog choruses can be heard from bison wallows at Theodore Roosevelt National Park, North Dakota where bison were reintroduced in 1956 (Hossack et al. 2005). Hence, evidence suggests that bison wallows were once important breeding sites for prairie amphibians, even though successful meta-morphosis may have only occurred in wetter years that provided sustained surface water, or what is often referred to as hydroperiods that are long enough for successful reproduction (Gerlanc and Kaufman 2003). Bison wallow abundance and distributions in rangelands are certainly much reduced today and we know little about the consequences for prairie amphibians.

Between 1780 and 1980, an estimated 53% of 894,355 km^2 (221 million acres) of wetlands were intentionally or unintentionally drained in the contiguous United States, especially freshwater emergent marshes that are so important to wildlife (Dahl 1990). In Nevada, for example, over half of its original 1971 km^2 (487,000 acres) of wetlands were lost in that 200-yr span. These losses were partially offset by the creation of water impoundments, such as stock ponds and reservoirs, which are common in rangelands. For example, a state-wide inventory in the early twenty-first century found that more than 70% of lentic wetlands in eastern Montana were human-created (Maxell 2009). Both the loss and creation of wetlands has influenced the contemporary distribution of herpetofauna across U.S. rangelands. Species such as the painted turtle (*Chrysemys picta*; Fig. 25.2), Woodhouse's toad (*Anaxyrus wood-housii*), and tiger salamander (*Ambystoma* spp.) may have increased their distribution in some places because of water impoundments. In other cases, stock ponds may be the only habitat remaining in otherwise cropland-dominated landscapes (Knutson et al. 2004). Regardless, anthropogenic changes in the type, size, and depth of wetlands in rangelands has influenced herpetofauna distributions in all like-lihood, and this may have implications for persistence as climates change. The Great Basin, for example, has been getting warmer and drier in the last century resulting in increased isolation of amphibian populations as detectable in the genetic structure of Columbia spotted frogs (*Rana luteiventris*; Fig. 25.1; Pilliod et al. 2015; Robertson et al. 2018). In the Great Plains, connectivity among > 80,000 playas from Nebraska to Texas may have been reduced beyond levels needed to support movements for many amphibian species (Heintzman and McIntyre 2021). Except for a few aquatic species, such as the painted turtle and common water snake (*Nerodia sipedon*), that require surface water to meet their life history needs, reptile distributions and threats are more subtle with regard to water.

Reptiles are much more tolerant of aridity than other vertebrates, enabling them to inhabit a diversity of upland habitats (Fig. 25.3) as long as temperatures are not too cold (Qian 2010). In the last several hundred years, however, reptiles and amphibians have been subjected to large scale land use changes, such as cropland agriculture, livestock production, timber harvest, and urbanization, all of which have influenced species distributions to varying extents (Cordier et al. 2021). Conversion of rangelands to hayfields or irrigated croplands is a major modification to potential habitats from the perspective of local herpetofauna (Fig. 25.3). In rangeland landscapes, the interdigitation of cropland fields, right-of-ways, hedgerows, and fencelines alter herpetofauna communities as these modified areas are frequently only inhabited by the more common and adaptable species (Pulsford et al. 2017). Obviously human features on the landscape that destroy habitat for herpetofauna, such as buildings, parking lots, solar installations, roads, railways, and so forth, also have cumulatively large footprints, and their effects extend into surrounding habitats (Averill-Murray et al. 2021).

Species richness of both amphibians and reptiles tends to be higher in the south than the north (Fig. 25.4). Reptile diversity is greatest below the 37th parallel, which is highlighted by the state boundaries between Colorado and New Mexico as well as Utah and Arizona. This latitude coincidentally defines the average solar insolation on earth: below this line and toward the equator solar insolation is greater than average incoming solar radiation, whereas above the 37th parallel, toward the poles, solar insolation decreases. Below the 37th parallel, there are hotspots of reptile diversity in central and eastern Texas, with over 70 species found there. Moving northward, reptile diversity tapers off in a steady gradient (Kiester 1971). Amphibian diversity also decreases from the equator northward (Wiens 2007), but the Pacific Northwest has an unusually high diversity of salamanders. Similar to reptiles, the Texas–Mexico border area is also a hotspot of amphibian diversity (Fig. 25.4). The warm, dry conditions in the desert regions of the southwestern U.S. are ideal for reptiles, whereas this region has strikingly low amphibian diversity. Utah, for example, may only have one native salamander species, the western tiger salamander (Fig. 25.1; *Ambystoma mavortium*). The species was thought to also occur in Nevada, but molecular evidence suggests the only salamander populations in Nevada may be the eastern tiger salamander (*Ambystoma tigrinum*), introduced as bait by fisherman (Johnson et al. 2011). Introduced populations of eastern tiger salamanders have been discovered in other western states as well.

The elevational range of amphibians and reptiles is broad. For example, the sidewinder rattlesnake (*Crotalus cerastes*) and western threadsnake (*Leptotyphlops humilis*) are found below sea level in Death Valley, California whereas western fence lizards (*Sceloporus occidentalis*: Fig. 25.2) and mountain yellow-legged frogs (*Rana muscosa*) can be found above 3300 m (10,827 ft) in the Sierra Nevada of central California (Stebbins 2003). About the only areas devoid of herpetofauna in rangelands are alkali flats (dry desert lake beds) and alpine zones.

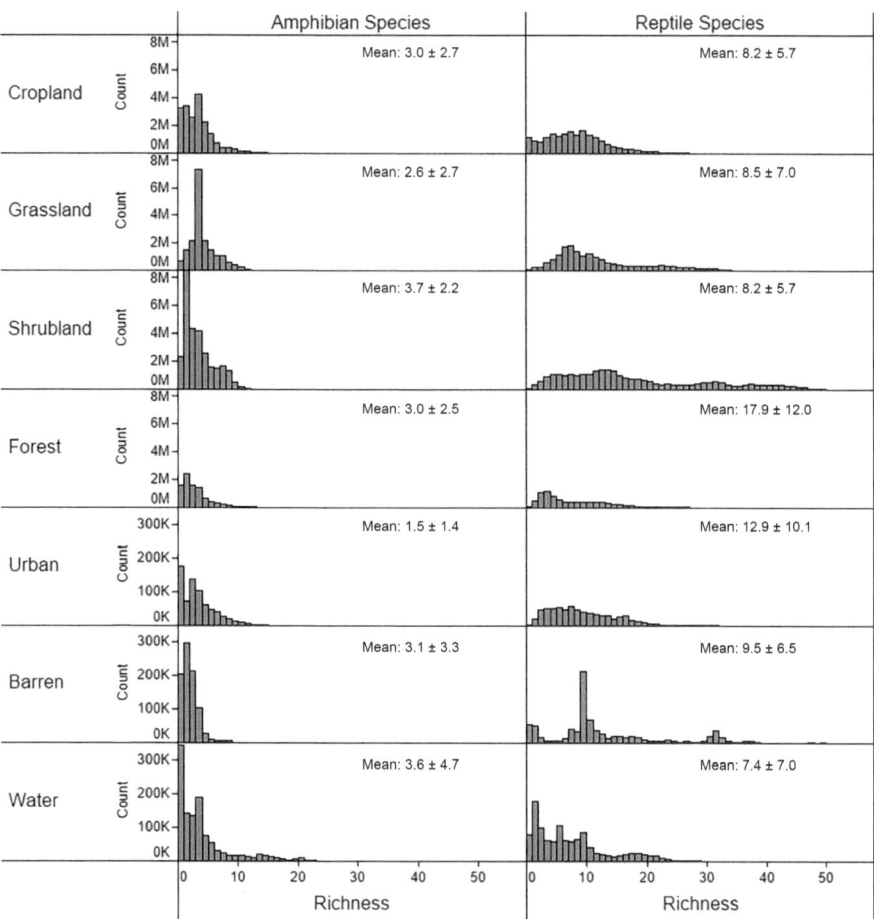

Fig. 25.3 The frequency distribution of predicted richness for amphibians and reptiles across different land cover types in rangelands of the U.S. Cropland and urban are embedded within rangelands and likely represent conversion of former rangelands. Richness data are from GAP predicted species distributions (USGS GAP 2018a). The vegetation cover types come from the North American Atlas Land Cover 2010 Mapping Project (data grain is 250 × 250 m). Metadata about these cover types can be found at: http://www.cec.org/north-american-environmental-atlas/land-cover-2010-modis-250m/. The y-axis is a count of 30-m pixels for each category and group. The pixel count (y-axis) of the top four land cover types (Cropland, Grassland, Shrubland, Forest) have a different range than the three less common land cover types (Barren, Urban, Water)

25.2.2 Population Monitoring

Population monitoring of herpetofauna species is mostly conducted by state and federal agencies, and is usually associated with species listed or petitioned under the Endangered Species Act or those listed as species of greatest conservation need in State Wildlife Action Plans. For amphibians, monitoring focuses almost exclusively

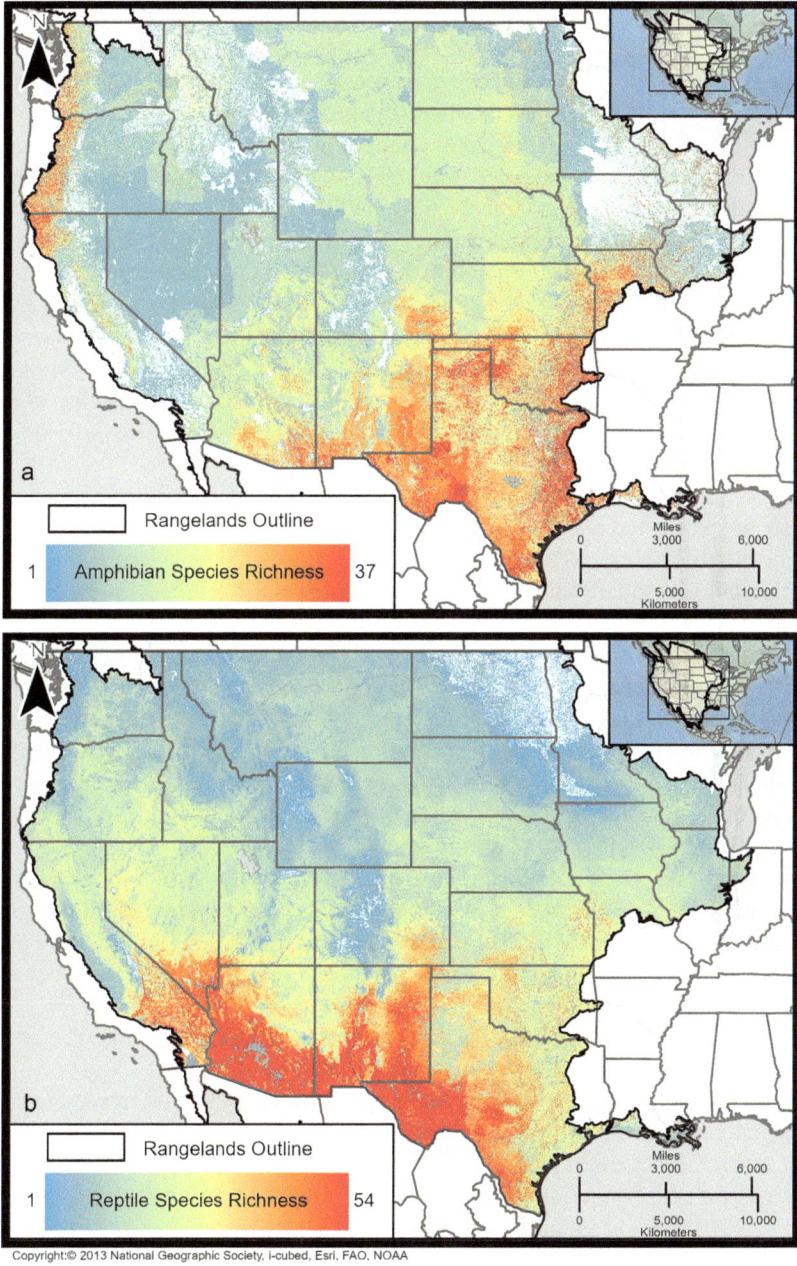

Fig. 25.4 Map of predicted amphibian (upper panel) and reptile (lower panel) species richness in the United States with rangelands delineated based on ecoregions derived from The Nature Conservancy's Geospatial Conservation Atlas (geospatial.tnc.org), modified using ecotype layers downloaded from the EPA Level III ecoregions in the Central Mixed—Grass Prairie region in Nebraska and Texas. The amphibian and reptile richness data came from the Gap Analysis Project (USGS GAP 2018b, c)

on individual breeding sites or groups of sites in a landscape for species that tend to form metapopulations. The gold standard for population monitoring is mark-recapture. Mark-recapture studies involve marking individual animals with a unique identifier so that they can be identified if captured again in subsequent surveys or trapping efforts (Buckland et al. 2000). Common ways of marking animals are with passive integrated transponder (PIT) tags (Fig. 25.5), scale clips, shell notching, and colored paints, inks, and elastomers (Silvy et al. 2012). When conducted over at least three or more years, mark-recapture data can provide valuable estimates of population size and demographic rates. Population demography includes measures of natality or reproduction, recruitment, survival, senescence, and mortality. Demography data can provide more robust measures of trends and responses to environmental stressors or management compared with simple counts of individuals observed (Schmidt 2003). Indirect measures of populations, such as egg mass counts or enumeration of calling frogs and toads (such as with call recorders; Fig. 25.5), can also provide useful information for tracking population trends (Heyer et al. 1994).

Unlike amphibians or turtles that may congregate at water bodies to breed or forage, or snakes that may congregate to breed and overwinter at hibernacula, lizards and tortoises do not congregate and thus must be surveyed intensively over areas spanning hectares to square kilometers. Therefore, for most reptiles, optimal sampling designs include many plots or trapping locations distributed over large areas representing a range or variety of habitats used by a particular species. Search methods and effort must be consistent, or at least accounted for, among surveys and through time. Capture methods vary depending on the target species and include active sampling, such as noosing lizards or visual searching and capturing animals (Fig. 25.5). Sometimes capturing lizards and snakes involves wild chases and long arms to reach animals under rocks or in burrows, while other times it simply involves picking them up, as with tortoises. Passive sampling devices, such as drift fences, pitfall trap arrays, camera traps, or cover boards, are also commonly used and may be necessary for rare, cryptic, or fossorial species. Each method has some sampling bias because of activity patterns and size of target animals and life stages. Many studies are plagued by small sample sizes or high inter-annual variability in capture rates because of strong environmental associations, such as seasonal or annual weather. For community studies, oftentimes one or a few species dominate the capture tally whereas other species are captured too infrequently to even model.

Distance-sampling methods provide population estimates of species over large areas using standardized linear transects traveled by observers (Fig. 25.5; Buckland et al. 2000). This method was adopted to monitor Mojave desert tortoise (*Gopherus agassizii*; Fig. 25.2) population trends across its range since 1999 and continues today (USFWS 2011). Field teams are tested with tortoise models to calculate their ability to detect tortoises at various distances from the transect line and these correction factors are used to reduce error and improve estimates. Radio-tagged tortoises have also been used for this purpose and for validation of population estimates. Distance-sampling estimates of population size and trends, usually averaged over extensive areas, have provided important contributions to population management of the Mojave desert tortoise in Arizona, California, and Utah.

Fig. 25.5 Photographs of common field methods used in herpetological field studies, including: **a** Drift fences with funnel traps; **b** Line transect surveys, including distance sampling; **c** Hand capturing (in this case tubing a Great Basin rattlesnake (*Crotalus oreganus lutosus*) to allow for safe handling; **d** Radio-telemetry; **e** Marking individuals for mark-recapture studies (in this case inserting a passive integrative transponder (PIT) tag); **f** Call recorder for frogs and toads during the breeding season; **g** Environmental DNA sampling for species detection; and **h** PIT tag antenna to record the timing and direction of animal movement. Photographs by Todd Esque, Matthew Laramie, Chad Mellison, Amelia Orton-Palmer, Charles R. Peterson, and David Pilliod

There are other monitoring approaches that are useful for herpetofauna that involve only presence/non-detection data and some of these methods have become quite sophisticated. Occupancy modeling has proven useful for herpetofauna (Bailey et al. 2014), including indirect measures of species occurrence such as environmental DNA (Fig. 25.5; Burian et al. 2021) and open drift-fences with cameras instead of traps (Martin et al. 2017). Occupancy modeling accounts for imperfect detection, which is important for herpetofauna that are often rare, cryptic, fossorial, or otherwise difficult to detect. After accounting for detection probabilities and measured environmental variables, presence and non-detection data from repeated visits are used to create occupancy probabilities for an area or site.

Studies of herpetofauna movements have revealed the complexity of diel, seasonal, and interannual habitat use patterns and the role of migration and dispersal in population dynamics and gene flow (Cayuela et al. 2020). Understanding movement ecology of herpetofauna is crucial for their conservation (Bailey and Muths 2019; Joly 2019). Juveniles are particularly understudied, although, as in other vertebrates, juveniles may represent one of the most important life stages for dispersal, colonization, and gene flow (Petrovan and Schmidt 2019). Movement studies generally involve tagging or marking individual animals and tracking their locations actively using radio-transmitters, passively using trapping, or opportunistically with surveys (Fig. 25.5). Radio-telemetry can also increase the certainty of population and demographic estimations (e.g., Mitchell et al. 2021). Twenty years of monitoring amphibians across the U.S. by the U.S. Geological Survey's Amphibian Research and Monitoring Initiative (ARMI) has provided robust evidence that many amphibian populations are at risk of decline or extinction. An analysis of 83 species revealed that amphibian populations are disappearing from 3.7 to 3.8% of formerly occupied sites annually (Adams et al. 2013; Grant et al. 2016). At this rate, by 2035 many amphibian species will be gone from half of the places where they occurred in 2015 (Grant et al. 2016). These declines are due to a combination of factors driven by habitat loss, invasive predators, disease, and climate change. However, the status and trends of individual amphibian populations depend on many factors and not all species are necessarily at risk (Muths et al. 2018). Monitoring of 14 species of frogs and toads in the southeastern U.S. concluded that seven species were increasing (especially the green treefrog, *Hyla cinerea,* and spring peeper, *Pseudacris crucifer*; Fig. 25.1), while eight species showed a declining trend between 2001 and 2013 (Villena et al. 2016). Comparable regional or national monitoring programs do not exist for reptiles, except for a few species of highest conservation concern, such as the Mojave desert tortoise. Recent analyses showed only one of five of the recovery areas for Mojave desert tortoises had positive population growth after ~ 15 yr of monitoring, and juvenile tortoise numbers were declining (Allison and McLuckie 2018). These results are mostly inconsistent with recovery goals (USFWS 2011).

25.3   Habitat Associations

As ectotherms, climate plays an overarching role in the distribution and habitat associations of herpetofauna. In general, amphibians are limited by environmental temperature and precipitation, whereas reptiles are strongly associated with temperature (Buckley and Jetz 2007; Qian 2010). This explains why we see amphibians and reptiles in specific locations or habitats, including relative to seasons and times of day. As previously described, thermoregulation is crucial for physiological functions (e.g., digestion, metabolism) and performance (locomotion) of herpetofauna. Water balance, or hydroregulation, is also a key process underlying physiological and ecological responses. As might be expected, thermoregulation and hydroregulation are closely linked and thus these physiological and behavioral mechanisms often represent decisions or tradeoffs between optimal body temperature and water loss (Rozen-Rechels et al. 2019). A toad or lizard, for example, may tolerate some dehydration when selecting a warm, dry microsite needed to maintain a higher body temperature necessary for dispersal, digestion, or, in the case of a gravid (pregnant) female, embryonic development.

At the regional or landscape level, availability of freshwater is paramount for amphibians and some reptiles, and species assemblages depend on characteristics of wetland habitats and the spatial distribution and configuration of those wetlands (Mushet et al. 2012). Wetland amphibian habitats in rangelands and croplands are often characterized by the amount and complexity of shoreline, depth of water and availability of shallows, solar insolation, water chemistry, hydrology and hydroperiods, amount of emergent vegetation, and characteristics of riparian and floodplain vegetation (Knutson et al. 2004; Swartz and Miller 2019). Depending upon these habitat characteristics, amphibian communities can also be strongly influenced by predation, especially by salmonids (i.e., trout, char), centrarchids (e.g., bass, bluegill, pumpkinseed, sunfish), gartersnakes (*Thamnophis* spp.), American bullfrogs (*Lithobates catesbeianus*) and various birds (Pilliod et al. 2012; Ford et al. 2013; Rowe et al. 2019). In terrestrial environments, the structure and composition of vegetation have strong influences on herpetofauna habitats, especially related to the thermal environment, food resources, and cover (Fischer et al. 2004). In general, heterogeneous habitats provide more niches and microsites than homogeneous habitats (Fuhlendorf et al. 2017; Londe et al. 2020). Finally, contextual location is important, such as past land uses, elevation, landform, soils, surrounding habitat, and distance to nearest habitat suitable for survival, reproduction, or development (Kay et al. 2017; Sawatzky et al. 2019).

The majority of reptiles are not similarly constrained by water requirements. Although most temperate reptiles drink surface water, they also can temporarily tolerate hyperosmotic states of dehydration, which often occurs seasonally in arid and semi-arid rangelands. Some water can be obtained from food, but reptiles also have several physiological adaptations and behaviors that limit water loss (Dupoué et al. 2017). Furthermore, the diversity of body forms, low energy requirements, and behavioral adaptations to inclement weather and seasons enables reptiles to inhabit nearly all rangeland habitat types including most mesic and aquatic sites, prairies, shrub steppes and shrublands, savannahs, woodlands, and forest (Fig. 25.3). Thus, reptile habitat associations are incredibly varied. Because most species are ground-dwellers, understory vegetation and leaf litter (or inversely, bare ground) are often cited as important variables predicting reptile species occurrence across rangelands (e.g., Lindenmayer et al. 2018). Shrubs and trees are important for some reptile species and these species may disappear if these habitat elements are removed or lost to wildfire (Cossel 2003; James and M'Closkey 2003).

Some habitat selection by herpetofauna is associated with foraging behavior and mate finding. Some species will travel to and forage in locations with higher amounts of food resources, which often varies seasonally. Snakes and lizards can be classified as either active foragers that seek, and sometimes chase, their prey or sit-and-wait predators that opportunistically grab prey that comes close enough. The diet of snakes varies by species and habitat preferences, but can include small mammals, birds, fish, lizards, amphibians, and some invertebrates. A few snakes eat other snakes. Lizards and adult amphibians generally feed on arthropods (insects and spiders), annelids (segmented worms), and gastropods (slugs). Turtles are omnivorous, eating a variety of invertebrates, amphibians, fish, algae, and plants, whereas tortoises are strictly herbivorous.

25.4   Rangeland Management

25.4.1   Livestock Grazing

Excessive livestock grazing can affect amphibians through multiple pathways. First, overgrazing of grasses and forbs during the spring and summer can expose terrestrial amphibians to predators and desiccation in meadows and wetlands by reducing cover and allowing soils to dry (Canals et al. 2011; Pulsford et al. 2019). Second, excessive livestock use in aquatic habitats can increase turbidity and alter water chemistry via deposition of urine and feces (Schmutzer et al. 2008; Smalling et al. 2021). Negative impacts to water quality may affect larval development of amphibians but likely has fewer effects on amphibians compared with other factors such as hydroperiod and predators (Canals et al. 2011; Cole et al. 2016). Larval developmental issues associated with poor water quality, however, may be sublethal and have delayed effects that are only potentially problematic later in an animal's life (Gray and Smith 2005; Chelgren et al. 2006). These time-lagged and carryover effects are particularly difficult to observe or measure but can have consequences at the population level (Babini et al. 2015; Bionda et al. 2018). And, finally, livestock may cause some direct mortality of individuals from trampling, although this is probably not a major source of mortality at the population level.

   Despite these possible impacts from excessive livestock grazing, few studies have documented consistent negative effects of livestock grazing on amphibians and many amphibians breed successfully in stock ponds, even with heavy livestock use. Effects appear to be species-specific and depend upon habitat preferences (Burton et al. 2010). Some of this variability, however, may also be associated with variation in the type of grazers (e.g., cattle, sheep, goats), stocking rates, and timing and duration of grazing. One review of 46 published studies found only 22% demonstrated negative effects on amphibian communities and the remainder had either positive, neutral, or mixed effects (Howell et al. 2019). This meta-analysis indicated that most of the negative consequences of livestock grazing on amphibians occur in closed-canopy habitats whereas well-managed grazing in open habitats is compatible with amphibian conservation objectives. For example, some species, such as tiger salamanders, American toads (*Anaxyrus americana*), and western toads (*A. boreas*), thrive in open, shallow water environments even if used by grazing animals (Pyke and Marty 2005; Burton et al. 2010; Barrile et al. 2021a). A study of livestock-grazed meadows on the western slope of the Sierra Nevada Mountains, California found that Yosemite toads (*Bufo canorus*; Fig. 25.1) occupied pools that tended to be shallower, warmer, and more nitrogen enriched than unoccupied pools, regardless of livestock grazing intensity ranging from heavy to none (Roche et al. 2012). Similarly, Columbia spotted frog populations also do not appear to be impacted by use of breeding ponds by livestock (Adams et al. 2018), even though studies have found that frog survival, recruitment, and reproduction may increase in the first year or two after livestock are fenced out of breeding ponds (Pilliod and Scherer 2015). These short-term benefits to frogs, however, also are known to disappear from ponds as emergent vegetation becomes tall and dense in the absence of any livestock grazing (Pilliod and Scherer 2015).

Livestock may affect reptiles in both negative and positive ways through changes in grazed vegetation, nutrient redistribution, and physical impact of trampling to habitat components (soil, burrows, vegetation). Some research suggests, however, that livestock do not commonly crush reptiles or damage burrows by trampling (Nicholson and Humphreys 1981). In Australia, light to moderate livestock grazing intensities with a wet-season rest supported the most abundant reptile community, but only when compared with heavy, prolonged livestock grazing treatments (Neilly et al. 2018b). Other studies have found that reptile species richness is lower in grazed areas compared with areas where livestock are absent or where livestock have been removed or excluded with fencing (Hellgren et al. 2010; Read and Cunningham 2010). These responses are not universal and depend upon environmental conditions and habitat requirements of species present (Castellano and Valone 2006; Neilly et al. 2021).

Lizards can benefit from habitats opened up by livestock, as they sprint after prey and toward cover from predators, especially when grazing is managed carefully or used to reduce dense stands of invasive annual grasses that dominate formerly open areas and increase fire risk (Barry and Huntsinger 2021). The federally protected blunt-nosed leopard lizard (*Gambelia sila*; Fig. 25.2), for example, increased 500% in areas grazed by cattle in comparison with ungrazed areas dominated by invasive annual grasses in the San Joaquin Desert of southern California (Germano et al. 2012). Furthermore, the benefit of increased solar insolation for thermoregulation of reptiles and their egg temperatures in grazed habitats may confer benefits to reptiles from livestock grazing (Fabricius et al. 2003). The volume of rangeland research on herpetofauna in the last two decades has helped advance livestock grazing strategies that are compatible with reptile conservation. While more research is needed, we have sufficient credible, defensible information to move forward constructively (Barry and Huntsinger 2021).

25.4.2   Other Rangeland Management Actions

Water development, especially in arid and semi-arid environments, has likely influenced the distribution of amphibians. The development of springs, such as installing pipes and pumps, to provide livestock drinking water and other uses may alter the spring such that it no longer provides suitable overwintering habitat for some amphibians. Stock ponds and leaky or overflowing troughs, however, also create surface water in locations that may not have had surface water prior to development. Amphibians may use these artificial sources of water on the landscape to hydrate and occasionally breed (Alvarez et al. 2021). Chorus frogs, tiger salamanders, and other species can be found in water troughs or in their spillage areas in some otherwise dry shrublands and grasslands (Scott 1996). These oases also draw in amphibian predators like gartersnakes (*Thamnophis* spp.). Efforts are underway to help make water developments for livestock more compatible with amphibian and reptile use (Canals et al. 2011).

Vegetation treatments are common throughout rangelands of the western U.S., to improve forage quantity or quality, but also to control or remove non-native plant species, to stabilize soils and reduce erosion, and to rehabilitate recently burned areas, among other intentions (Pilliod et al. 2017). Many of these land treatments have the potential to affect herpetofauna, either positively or negatively (Pilliod et al. 2020). Research on this topic, however, is lacking and thus there are few guidelines to help resource managers design herp-friendly land treatments (but see Pilliod and Wind 2008; Kingsbury and Gibson 2012; Jones et al. 2016).

The thinning and removal of pinyon and juniper trees is a common rangeland management practice in the western U.S., particularly lately in the name of habitat management for the greater sage-grouse (*Centrocercus urophasianus*). What are often called pinyon-juniper or P-J woodlands are a forest type composed of single leaf-pinyon pine (*Pinus monophylla*), Colorado pinyon (*P. edulis*), western juniper (*Juniperus occidentalis*), and Utah juniper (*J. osteosperma*). In the absence of fire and under favorable climatic conditions, these species have expanded their range into grasslands and shrublands, resulting in changes in water availability, soil chemistry, understory vegetation, and animal communities (Miller et al. 2000; Leis et al. 2017). Several lizard species inhabit P-J woodlands and benefit from the woody structure (Morrison and Hall 1999; James and M'Closkey 2003). The lizards use the trees and downed logs for basking, except for the tree lizard (*Urosaurus ornatus*; Fig. 25.2), which is distinctly arboreal and perches at greater heights than the other species (James and M'Closkey 2002). Arboreality may protect some lizard species from typical effects of livestock grazing (Jones 1981; Neilly et al. 2018a). Felling or burning trees might benefit lizards, like the sagebrush lizard (*Sceloporus graciosus*), but removing the dead and downed wood as part of fire management, fuel reduction, or habitat management for shrubland and grassland wildlife species could have negative consequences for tree lizards (Morrison and Hall 1999; James and M'Closkey 2003; Evans et al. 2019). Other lizard species are unlikely to be affected by such activities and ground-dwelling lizard species may benefit from such practices (Radke et al. 2008).

Prescribed fire practices appear to have minimal effects on herpetofauna in range-lands where it is appropriate. Besides concern about causing mortality from combustion or heat stress (Smith et al. 2001), particularly for turtles and tortoises (Larson 2014), most interest in prescribed fire is related to the role of fire in creating or maintaining heterogeneity in vegetation structure and composition that can sustain or enhance herpetofauna diversity (Wilgers and Horne 2006; Larson 2014). In southern Texas, a short-term study concluded that dormant-season fires had little effect on diversity and abundance of herpetofauna, but growing-season fires tended to increase diversity and abundance of grassland species, such as the six-lined racerunner (*Cnemidophorus sexlineatus*; Fig. 25.2; Ruthven et al. 2008). Minimal effects of prescribed fire on herpetofauna also have been reported in other range-lands, including California oak woodlands (Vreeland and Tietje 2002). Prescribed fire, grazing, and herbicide treatments have been used for creating or maintaining habitat heterogeneity for herpetofauna in seasonal wetlands, grasslands, and some woodlands, with mixed success (Jones et al. 2000; Larson 2014; Mester et al. 2015; Wilgers et al. 2006). In general, effects seem to be short-lived as plant communities respond to the disturbance and associated changes in nutrients, light availability, and competition. Even where prescribed fire appears to have negative effects on herpetofauna (e.g., Wilgers et al. 2006; Larson 2014), these effects tend not to persist through time. To optimize diversity, management for habitat mosaics may need to involve rotational burning, sometimes coupled with low-intensity cattle-grazing or herbicide treatments (Mester et al. 2015). This approach may allow species-specific responses in relation to changes in vegetation structure and microhabitat conditions (e.g., temperature, moisture of soil or vegetation) that changes through time (Wilgers and Horne 2006). The winners and losers scenario of wildlife response to local range-land management is a reasonable conservation strategy as long as massive areas are not managed uniformly.

25.5 Impacts of Disease

Several amphibian and reptile diseases may be influenced by human activities and management practices in rangelands (Gray et al. 2017). One of the most notable amphibian diseases is the amphibian chytrid fungus, *Batrachochytrium dendrobatidis* or Bd, which causes chytriodiomycosis and is associated with severe population declines in several North American species (Lips 2016; Scheele et al. 2019). In rangelands, Bd is now thought to have contributed to the near extirpation of two toad species in the mid-1970s: the Wyoming toad (*Anaxyrus hemiophrys baxteri*; Fig. 25.1) and the Yosemite Toad (Kagarise Sherman and Morton 1993; Green and Kagarise Sherman 2001; Fig. 25.1). The Wyoming toad became functionally extinct in the wild by the 1980s (Lewis et al. 1985) and is a case study of the challenges of captive rearing, reintroduction, and species recovery in amphibians (Dreitz 2006). Variants of Bd exist and their pathogenicity are still being studied because not all amphibians in the U.S. are susceptible to Bd, at least under current environmental conditions. Like all wildlife diseases, the contraction of Bd, its prevalence in popula-tions, and its effects on survival and fitness depend on the ecology and evolutionary history of the species with the disease in relation to the environment (Russell et al. 2019). For example, a study of boreal toads (*Anaxyrus boreas*) in western Wyoming revealed that livestock grazing may influence toad-Bd dynamics by creating warmer microclimates from the reduction of vegetation that allow toads to bask and clear themselves of the disease (Barrile et al. 2021a, b). *Batrachochytrium salamandrivo-rans* (Bsal) is a recently discovered disease from Asia that also causes chytrid-iomycosis. It quickly spread across Europe but has yet to arrive in North America (Waddle et al. 2020). The high diversity of North American salamanders puts the U.S. at extreme risk but, like frogs and toads exposed to Bd, some species may have innate protection, such as skin peptide defenses (Pereira and Woodley 2021).

Besides Bd, ranavirus is a major cause of mortality in some populations of amphib-ians (and some reptiles and fishes) around the world (Brunner et al. 2015). Ranavirus is not a single virus, but instead a group of iridoviruses first discovered in the northern leopard frog (*Lithobates pipiens*; Fig. 25.1), a common inhabitant of North Amer-ican rangelands. Besides leopard frogs, it is known to infect the American bullfrog (*Lithobates catesbeianus*) and the commercial sale of leopard frogs and bullfrogs to laboratories and schools across America likely contributed to the spread and conti-nental distribution of the viruses. The most widely known member of this group of viruses is the Ambystoma Tigrinum Virus (ATV), which can cause mortality in three species of tiger salamanders found in U.S. rangelands (Picco et al. 2007; Price et al. 2017). Ranaviruses appear to proliferate under periods of stress for the animals, such as changes in water temperature (Brunner et al. 2015). Ranavirus also may be more prevalent in areas where cattle congregate, possibly due to poor water quality caused by elevated turbidity and ammonia which stresses amphibians, particularly tadpoles and larvae (Hoverman et al. 2012). The creation of permanent ponds as water sources for livestock may also attract American Bullfrogs, which are known vectors of amphibian diseases (Yap et al. 2018; Brunner et al. 2019).

Disease in rangeland reptiles is a growing conservation concern (Fitzgerald et al. 2018; Mendoza-Roldan et al. 2021). Disease agents include microscopic bacteria, viruses, protozoans, and mycoses (fungi), frequently called zoonoses (or zoonotic) when they cause disease in humans and livestock (Mendoza-Roldan et al. 2021). Upper Respiratory Tract Disease Syndrome (URTDS), which causes inflammation and erosion of the nasal cavity and sometimes death, was first described in Mojave desert tortoises in California (Jacobson et al. 1991). The discovery of the disease agents of URTDS, *Mycoplasma agassizii* (Myag) and *M. testudineum* (Myte), was influential in the listing of the Mojave desert tortoises as *Threatened* under the Endan-gered Species Act (Brown et al. 2004; USFWS 2011). *Myag and Myte* are also found in Texas tortoises (*G. berlandieri*), gopher tortoises (*G. polyphemus*), and Sonoran desert tortoises (*G. morafkai*; Weitzman et al. 2017). Snake Fungal Disease is a rapidly emerging mycosis (*Ophidiomyces ophiodiicola*) that has now been found throughout the eastern U.S., and in several rangeland reptile species west of the Mississippi River (Lorch et al. 2016; Allender et al. 2020).

Many macroscopic parasites are also disease agents for reptiles, the most well-known including Arachnida (e.g., ticks, mites) and Diptera (flies and mosquitoes). Ticks are known globally as vectors for diseases hosted by reptiles, other wildlife, livestock, and humans. *Borrelia* spp. are spirochete bacteria carried by ticks and transmitted through the blood of vertebrate hosts. *Borrelia* spp. are causative agents for Lyme disease (Jacobson 2007; Swei et al. 2011) and Tick-Borne Relapsing Fever (TBRF; Forrester et al. 2015; Bechtel et al. 2021). Lyme disease (*B. burgdorfii—Bobu*) is the most common vector-born disease in the United States (CDC 2008). In California, western black-legged ticks (*Ixodes pacificus*) are vectors for *Bobu* among > 55 vertebrate hosts, including nine lizard species (Swei et al. 2011). About 90% of hosts for nymphal and larval ticks are western fence lizards, but the lizards are not very competent hosts because their blood includes borreliacidal components (i.e., when the lizard blood enters the tick during a meal it kills the *Bobu*). Regions with abundant lizards may have a lower proportion of Borrelia-infected tick nymphs and larvae (Ginsberg et al. 2021).

25.6 Ecosystem Threats

25.6.1 General

Some threats to amphibians, such as wetland habitat loss and degradation, non-native predators, and disease, are common to amphibians around the world (Lemckert et al. 2012; Pilliod et al. 2012; Wake and Koo 2018). Much less is known about specific threats to rangeland-associated amphibians, but several warrant consideration even if scientific evidence for their impacts is ambiguous or lacking (Mims et al. 2020). First, changes to hydrology or hydroperiod associated with water pumping, diversions, and dams are concerns. Stable, predictable water levels and flow rates are crucial for the development of amphibian tadpoles and larvae and the survival of post-metamorphic animals during the dry season and drought (Pilliod et al. 2021). Second, intensive human land use puts amphibians at risk because of clearing of vegetation, road construction, culvert installation, wastewater discharge (e.g., from hydrocarbon extraction, concentrated animal feeding operations), and construction of impervious surfaces (i.e., cement, asphalt). Crop production also is an intensive land use, although amphibian responses can be mixed. Some amphibians will venture into fields during pivot and flood irrigation and be attracted to lights when foraging for insects (Hansen et al. 2019) but, in general, homogenization of vegetation and application of chemicals (i.e., fertilizers, herbicides, insecticides) can be detrimental to amphibians or their habitats. Amphibians that forage in moist crop fields may then avoid these same areas after harvest. Fire and its relationship with changes in climate and invasive plant species is also potentially important, but in need of additional study (Mims et al. 2020).

Reptiles face many of the same threats as amphibians in rangelands, especially loss and isolation of suitable habitats, disease, and pollution from animal wastes and agrichemicals (Fitzgerald et al. 2018). A meta-analysis of 56 studies that reported on how habitat modification affected the abundance of 376 reptile species concluded that mining had the most negative impacts, followed by farming, livestock grazing, and tree plantations (Doherty et al. 2020). The mean effect of logging was neutral. Because of their tendency to bask and forage in areas of human use, reptiles may be more prone to direct mortality from human activities than other animals, although this has been difficult to quantify. A study using carcass detection dogs found that 57% of animals killed during typical agricultural mowing were reptiles, especially lizards (Deak et al. 2021). Invasive plants, especially dense annual grasses that cover open areas of bare ground, are known to interfere with lizard and snake movements and foraging ability in desert rangelands (Rieder et al. 2010; Blakemore 2018). The increased frequency of wildfire caused by these grasses also appears to have negative consequences for some reptile species, either through direct mortality (Jolly et al. 2022) or changes in habitat (Woinarski et al. 1999; Cossel 2003). Some species are also the target of exploitation, such as collection and sale in the illicit internet pet trade, whereas others are simply persecuted because of general fear or hatred of snakes, especially rattlesnakes and other pit vipers (Katzner et al. 2020).

The proliferation of transportation and energy infrastructure across rangeland landscapes further increases herpetofauna road mortality, creates barriers to migration and dispersal, and fragments once continuous habitats (Doherty et al. 2021). Road mortality is considered the leading cause of reptile mortality, especially for snakes (Hill et al. 2019). Roads provide attractive surfaces for thermoregulation and movement and collisions with vehicles are rampant, even on rural rangeland roads (Jochimsen et al. 2014; Hubbard et al. 2016). A study in southeastern Ohio found that the amount of pasture within a 100 m buffer of a roadkill was the strongest predictor of road mortality for 14 snake species (Wagner et al 2021). Fencing reptiles out of roadways comes with its own costs for snakes and turtles, including restricting access to seasonal resources and reducing gene flow among populations (Markle et al. 2017). Newly applied genetic tools and analyses provide insight to the influence of geographic factors like roads and railways on reptiles and amphibians. For example, a railway constructed some 120 yr ago bisected a population of Mojave desert tortoises resulting in differences in genetic diversity on either side of the railway after only about eight generations of tortoises (Dutcher et al. 2020). Roads and other human development certainly continue to play a role in shaping the population genetic structure of herpetofauna through reduced movement of individuals and reduced exchange of genes among populations. More research is needed on appropriate methods to avoid herpetofauna mortality and barriers to movement using overpasses or culverts to provide safe passage routes across roads and railways.

25.6.2   Climate Change

Changes in climate across U.S. rangelands may alter environmental conditions to such an extent that many, if not all, aspects of herpetofauna ecology will be affected. Observed changes in climate over the last several decades depend on location, especially latitude and elevation, but also continental position relative to mountain ranges (e.g., rain shadows) and the Pacific coast. Depending upon location, rangelands are experiencing warmer winters, shallower snowpacks, earlier springs, warmer nighttime (i.e., minimum) temperatures, longer and warmer growing seasons, shifts in summer monsoons, and longer, more frequent and severe heat waves and droughts (Polley et al. 2013; McCollum et al. 2017). All of these factors tend to be more variable year to year, and less predictable. These environmental changes will affect herpetofauna reproduction, development, and survival. Changes in wetland hydroperiods, earlier peak flows and more variable intermittency in streams, changes in the insulating capacity of snow in winter, changes in the thermal environment during the active season, and changes in the phenology of plants and prey (insects, small mammals) are most worrisome. Animals will adjust their diel and seasonal activity patterns to a point, but not without consequences. For example, some amphibians are breeding earlier and at smaller body sizes compared with a few decades ago (Li et al. 2013), which may expose some populations to higher mortality and stress (e.g., heightened disease risk) and result in population declines (Miller et al. 2018; Muths et al. 2018). Spiny lizards (*Sceloporus* spp.), which are common in rangelands, may have already experienced widespread population declines associated with climate change. Revisits of 200 sites in Mexico revealed 12% of local populations may have gone extinct since 1975. Using physiological models, the research suggests that thermal niches at these locations may have been altered to the point where lizards can no longer forage adequately to permit viable embryo development (Sinervo et al. 2010). Lizards may be particularly vulnerable to climate change because of their close affiliation with specific soil substrates for thermoregulation and reproduction and relatively limited dispersal abilities, often resulting in small or patchy distributions. Thermal niche modeling suggests that local extinction of lizard populations could reach 39% worldwide and 20% of species may be at risk of extinction by 2080 (Sinervo et al. 2010). Unfortunately, few lizard populations or species are being monitored in rangelands and thus many extinctions may occur quickly and without notice.

Adaptive behavior may enable herpetofauna to cope with climate changes. For example, herpetofauna may find microhabitats that allow them to maintain preferred body temperatures and moisture levels (Long and Prepas 2012). Phenotypic plasticity and genetic adaptation among herpetofauna also may mitigate some of the effects of climate change (Urban et al. 2013) but also may create new challenges. There is concern, for example, that, as temperatures increase, amphibians who rely on terrestrial foraging may need to change their foraging strategies because of the risk of dehydration (Lertzman-Lepofsky et al. 2020). This effect could be worsened in heavily grazed areas where vegetative cover is reduced (Bartelt et al. 2010). A study in California, however, found that vernal pools that were grazed by livestock dried an average of 50 days per year later than ungrazed pools, probably because of increased evapotranspiration from the abundant vegetation in the ungrazed wetlands (Pyke and Marty 2005). This study demonstrates the complex interactions between grazing and climate change and, in this case, climate mitigation strategies for species like the endangered California tiger salamander (*Ambystoma californiense*; Pyke and Marty 2005). Predictions for reptiles are no simpler, because we know little about how these animals are able to adjust their basking and foraging behavior or take advantage of microhabitats. Further, livestock grazing in rangelands may ameliorate or exacerbate the effects of climate change in unforeseen ways, including potential changes in the availability and distribution of thermal refuges (Clayton and Bull 2015; Rutschmann et al. 2016).

25.7   Conservation and Management Actions

Concerns about herpetofauna in the U.S. have stimulated an active community of diverse partners, including federal, state, tribal, NGOs, private landowners, and concerned citizens. These groups and partnerships take many forms. Formal working groups, such as those involved in endangered species conservation, tend to work on single species issues. Examples from U.S. rangelands include the Columbia spotted frog in Nevada and other states (Pilliod, in press). In Nevada, interagency technical teams have met since 1999 and helped the U.S. Fish and Wildlife Service write a conservation plan for this species that balanced species conservation with other rangeland issues. This led to a Conservation Agreement and Strategy for two distinct population segments (Northeast Nevada and Toiyabe subpopulations) that were first implemented in 2003 and then renewed for another 10 yr in 2015 (McAdoo and Mellison 2016). The technical team helps coordinate and implement the conservation plan, recruit assistance from scientists and other stakeholders to evaluate the effectiveness of conservation actions, status, and trends, and change the plan as necessary to meet the stated goals.

Partners in Amphibian and Reptile Conservation (www.parcplace.org, *accessed 14 July 2021*) is another organization that is bringing conservation issues to the forefront and facilitating creative solutions to pressing conservation challenges in rangelands and elsewhere. PARC is an open conservation community with participation, partnerships, and directions determined by current members at state, regional, and national levels. Most importantly, biologists, natural resource specialists, and land managers from public agencies meet with private landowners, concerned citizens, and industry to foster and implement conservation efforts. Simply put, the group forges proactive partnerships to conserve amphibians, reptiles, and the places they live. This inclusive approach to conservation has proven highly successful because it brings diverse perspectives to the table and garners ownership of conservation approaches. An important set of publications produced by PARC is the habitat management guidelines (HMG's). Each volume covers a specific region of the country with range-lands mostly represented in the Northwest and Western Canada (Pilliod and Wind 2008), Midwest (Kingsbury and Gibson 2012), and Southwest (Jones et al. 2016). The HMG's are designed to help managers think about herpetofauna habitat needs from the perspective of specific vegetation types. The guidelines include examples for "maximizing compatibility" whereby landowners and resource managers can contribute to the conservation and stewardship of these animals while managing their land primarily for other uses, such as livestock grazing or farming.

Habitat management guidelines for livestock grazing suggest landowners and managers consider: (1) controlling timing and extent of livestock access to wetlands and streams through fencing, restricted access points, and seasonal use, (2) estab-lishing alternative water sources such as water troughs, (3) carefully developing springs to serve as a source for livestock water without interfering with the spring's ability to provide water to wildlife and hibernacula for amphibians, and (4) managing grazing to maintain a higher stubble height of herbaceous vegetation that could preserve forage quality while maintaining cover from predators and desiccating conditions. More detailed recommendations can be found in the HMGs and other guidelines that are available for specific species or locations (e.g., Ford et al. 2013).

25.8   Research/Management Needs

The research and management needs of herpetofauna in rangelands are considerable because they are some of the least-studied vertebrates and many species lack sufficient information to make informed conclusions about status, trends, and threats, much less decisions about effective management and conservation strategies. Throughout this chapter we have highlighted areas of needed research. We encourage researchers and managers to work together to identify the most pressing and relevant issues to improve conservation actions and outcomes for rangeland amphibians and reptiles. Efficient, timely, co-production of scientific information is urgently needed given the current and forthcoming threats to herpetofauna and rangelands. Public–private engagement and diverse stakeholder partnerships may be the best way to incorpo-rate this information effectively into conservation planning and decision making for herpetofauna and other wildlife across our nation's rangelands.

Acknowledgements We thank Michelle Jeffries for her help developing the tables and figures used. Steve Hromada, Bryce Maxell, and an anonymous reviewer provided helpful comments. Any use of trade, product, or firm names is for descriptive purposes only and does not imply endorsement by the U.S. Government.

References

Adams MJ, Miller DAW, Muths E, Corn PS, Campbell Grant EH, Bailey LL, Fellers GM, Fisher RN, Sadinski WJ, Waddle H, Walls SC (2013) Trends in amphibian occupancy in the United States. PLoS ONE 8:e64347

Adams MJ, Pearl CA, Chambert T, McCreary B, Galvan SK, Rowe J (2018) Effect of cattle exclosures on Columbia spotted frog abundance. Wetlands Ecol Manage 26:627–634

Allender MC, Ravesi MJ, Haynes E, Ospina E, Petersen C, Phillips CA, Lovich R (2020) Ophidiomycosis, an emerging fungal disease of snakes: targeted surveillance on military lands and detection in the western US and Puerto Rico. PLoS ONE 15:e0240415

Allison LJ, McLuckie AM (2018) Population trends in Mojave desert tortoises (*Gopherus agassizii*). Herpetol Conserv Biol 13:433–452

Alvarez JA, Shea MA, Foster SM, Wilcox JT (2021) Use of atypical aquatic breeding habitat by the California tiger salamander. California Fish Wildlife, Special CESA Issue 235–240

Arkle RS, Pilliod DS (2015) Persistence at distributional edges: Columbia spotted frog habitat in the arid Great Basin, USA. Ecol Evol 5:3704–3724

Averill-Murray RC, Esque TC, Allison LJ, Bassett S, Carter SK, Dutcher KE, Hromada SJ, Nussear KE, Shoemaker K (2021) Connectivity of Mojave desert tortoise populations—management implications for maintaining a viable recovery network. U.S. Geological Survey Open-File Report 2021–1033, 23 p. https://doi.org/10.3133/ofr20211033

Babini MS, de Lourdes BC, Salas NE, Martino AL (2015) Health status of tadpoles and metamorphs of *Rhinella arenarum* (Anura, Bufonidae) that inhabit agroecosystems and its implications for land use. Ecotoxicol Environ Saf 118:118–125

Bailey LL, Muths E (2019) Integrating amphibian movement studies across scales better informs conservation decisions. Biol Cons 236:261–268

Bailey LL, MacKenzie DI, Nichols JD (2014) Advances and applications of occupancy models. Methods Ecol Evol 5:1269–1279

Barrile GM, Chalfoun AD, Walters AW (2021a) Livestock grazing, climatic variation, and breeding phenology jointly shape disease dynamics and survival in a wild amphibian. Biol Cons 261:109247

Barrile GM, Chalfoun AD, Walters AW (2021b) Infection status as the basis for habitat choices in a wild amphibian. Am Nat 197:128–137

Barry S, Huntsinger L (2021) Rangeland land-sharing, livestock grazing's role in the conservation of imperiled species. Sustainability 13:4466

Bartelt PE, Klaver RW, Porter WP (2010) Modeling amphibian energetics, habitat suitability, and movements of western toads, *Anaxyrus* (= *Bufo*) *boreas*, across present and future landscapes. Ecol Model 221:2675–2686

Bechtel, MJ, Drake KK, Esque TC, Nieto NC, Foster JT, Teglas MB (2021) Borreliosis transmission from ticks associated with desert tortoise burrows: examples of tick-borne relapsing fever in the Mojave Desert. Vector-Borne Zoonotic Diseases, ahead of print. https://doi.org/10.1089/vbz.2021.0005

Bionda CDL, Babini S, Martino AL, Salas NE, Lajmanovich RC (2018) Impact assessment of agriculture and livestock over age, longevity and growth of populations of common toad *Rhinella arenarum* (anura: Bufonidae), central area of Argentina. Global Ecol Conserv 14:e00398

Blackburn DG, Evans HE, Vitt LJ (1985) The evolution of fetal nutritional adaptations. Fortschritte Der Zoology 30:437–439

Blakemore G (2018) A mechanistic and landscape scale approach quantifying habitat suitability of cheatgrass (*Bromus tectorum*) engineered habitats for Great Basin reptiles. M.S. thesis, University of Nevada Reno, Reno, NV. http://hdl.handle.net/11714/4517

Bouzid NM, Archie JW, Anderson RA, Grummer JA, Leaché AD (2021) Evidence for ephemeral ring species formation during the diversification history of western fence lizards (*Sceloporus occidentalis*). Mol Ecol. https://doi.org/10.1111/mec.15836

Bragg AN (1940) Habits, habitat, and breeding of *Bufo Cognatus* Say. Am Nat 74:322–349

Brown DR, Merritt JL, Jacobson ER, Klein PA, Tully JG (2004) *Mycoplasma testudineum* sp. nov., from a desert tortoise (*Gopherus agassizii*) with upper respiratory tract disease. Int J Syst Evol Microbiol 54:1527–1529

Brunner JL, Storfer A, Gray MJ, Hoverman JT (2015) Ranavirus ecology and evolution: from epidemiology to extinction. In: Gray MJ, Chinchar VG (eds) Ranaviruses: lethal pathogens of ectothermic vertebrates. Springer, Switzerland, pp 71–104

Brunner JL, Olson AD, Rice JG, Meiners SE, Le Sage MJ, Cundiff JA, Goldberg CS, Pessier AP (2019) Ranavirus infection dynamics and shedding in American bullfrogs: consequences for spread and detection in trade. Dis Aquat Org 135:135–150

Bryson RW Jr, Jaeger JR, Lemos-Espinal JA, Lazcano D (2012) A multilocus perspective on the speciation history of a North American aridland toad (*Anaxyrus punctatus*). Mol Phylogenet Evol 64:393–400

Buckland ST, Goudie IBJ, Borchers DL (2000) Wildlife population assessment: past developments and future directions. Biometrics 56:1–12

Buckley LB, Jetz W (2007) Environmental and historical constraints on global patterns of amphibian richness. Proc Royal Soc B: Biol Sci 274(1614):1167–1173

Burian A, Mauvisseau Q, Bulling M, Domisch S, Qian S, Sweet M (2021) Improving the reliability of eDNA data interpretation. Mol Ecol Resour 21:1422–1433

Burton EC, Gray MJ, Schmutzer AC, Miller DL (2010) Differential responses of postmetamorphic amphibians to cattle grazing in wetlands. J Wildl Manag 73:269–277

Canals RM, Ferrer V, Iriarte A, Cárcamo S, San Emeterio L, Villanueva E (2011) Emerging conflicts for the environmental use of water in high-valuable rangelands. Can livestock water ponds be managed as artificial wetlands for amphibians? Ecol Eng 37:1443–1452

Carter SK, Nussear KE, Esque TC, Leinwand IIF, Masters E, Inman RD, Carr NB, Allison LJ (2020) Quantifying development to inform management of Mojave and Sonoran desert tortoise habitat in the American Southwest. Endangered Species Res 42:167–184

Castellano MJ, Valone TJ (2006) Effects of livestock removal and perennial grass recovery on the lizards of a desertified arid grassland. J Arid Environ 66:87–95

Cayuela HA, Teulier L, Martínez-Solano I, Léna J-P, Merilä J, Muths E, Shine R, Quay L, Denoël M, Clobert J, Schmidt BR (2020) Determinants and consequences of dispersal in vertebrates with complex life cycles: a review of pond-breeding amphibians. Q R Biol 95:1–36

CDC [Centers for Disease Control] (2008) Surveillance for Lyme disease—United States, 1992–2006. Morb Mortal Wkly Rep 57:1–9

Chelgren ND, Rosenberg DK, Heppell SS, Gitelman AI (2006) Carryover aquatic effects on survival of metamorphic frogs during pond emigration. Ecol Appl 16:250–261

Clayton J, Bull CM (2015) The impact of sheep grazing on burrows for pygmy bluetongue lizards and on burrow digging spiders. J Zool 297:44–53

Cole EM, Hartman R, North MP (2016) Hydroperiod and cattle use associated with lower recruitment in an r-selected amphibian with a declining population trend in the Klamath Mountains, California. J Herpetol 50:37–43

Cordier JM, Aguilar R, Lescano JN, Leynaud GC, Bonino A, Miloch D, Loyola R, Nori J (2021) A global assessment of amphibian and reptile responses to land-use changes. Biol Cons 253:108863

Cossel Jr JO (2003) Changes in reptile populations in the Snake River Birds of Prey Area, Idaho between 1978–1979 and 1997–1998: the effects of weather, habitat and wildfire. Doctoral dissertation, Idaho State University, Pocatello, Idaho

Crother BI (2017) Scientific and standard English names of amphibians and reptiles of North America north of Mexico, with comments regarding confidence in our understanding, 8th edn. Herpetological Circular 43, Society for the Study of Amphibians and Reptiles, University Heights, Ohio. https://ssarherps.org/wp-content/uploads/2017/10/8th-Ed-2017-Scientific-and-Standard-English-Names.pdf

Crother BI, Boundy J, Campbell JA, DeQuiero K, Frost D, Green DM, Highton R, Iverson JB, McDiarmid RW, Meylan PA, Reeder TA, Seidel ME, Sites JW Jr, Tilley SG, Wake DB (2003) Scientific and standard English names of amphibians and reptiles of North America north of Mexico: update. Herpetol Rev 34:196–203

Dahl TE (1990) Wetlands losses in the United States, 1780's to 1980's. US Department of the Interior, Fish and Wildlife Service, Washington, DC

Deak G, Katona K, Biro Z (2021) Exploring the use of a carcass detection dog to assess mowing mortality in Hungary. J Vertebr Biol 69:20089

Doherty TS, Balouch S, Bell K, Burns TJ, Feldman A, Fist C, Garvey TF, Jessop TS, Meiri S, Driscoll DA (2020) Reptile responses to anthropogenic habitat modification: a global meta-analysis. Glob Ecol Biogeogr 29:1265–1279

Doherty TS, Hays GC, Driscoll DA (2021) Human disturbance causes widespread disruption of animal movement. Nature Ecol Evol 5:513–519

Dreitz VJ (2006) Issues in species recovery: an example based on the Wyoming toad. Bioscience 56:765–771

Dupoué A, Rutschmann A, Le Galliard JF, Miles DB, Clobert J, DeNardo DF, Brusch GA, Meylan S (2017) Water availability and environmental temperature correlate with geographic variation in water balance in common lizards. Oecologia 185:561–571

Dutcher K, Nussear KE, Vandergast A, Esque TC, Mitelberg A, Heaton J (2020) Genes in space: what Mojave desert tortoise genetics can tell us about landscape connectivity. Conserv Genet 21:289–303

Evans MJ, Newport JS, Manning AD (2019) A long-term experiment reveals strategies for the ecological restoration of reptiles in scattered tree landscapes. Biodivers Conserv 28:2825–2843

Fabricius C, Burger M, Hockey PAR (2003) Comparing biodiversity between protected areas and adjacent rangeland in xeric succulent thicket, South Africa: arthropods and reptiles. J Appl Ecol 40:392–403

Fischer J, Lindenmayer D, Cowling A (2004) The challenge of managing multiple species at multiple scales: reptiles in an Australian grazing landscape. J Appl Ecol 41:32–44

Fitzgerald LA, Walkup DK, Chyn K, Buchholtz E, Angeli N, Parker M (2018) The future for reptiles: advances and challenges in the anthropocene. In: Dellasala DA, Goldstein MI (eds) Encyclopedia of the anthropocene. Elsevier Press, Amsterdam, pp 162–174

Ford LD, Van Hoorn PA, Rao DR, Scott NJ, Trenham PC, Bartolome JW (2013) Managing rangelands to benefit California red-legged frogs and California tiger salamanders. Alameda County Resource, Livermore, California

Forrester JD, Kjemtrup AM, Fritz CL, Marsden-Haug N, Nichols JB, Tengelsen LA, Sowadsky R, DeBess E, Cieslak PR, Weiss J, Evert N (2015) Tickborne relapsing fever—United States, 1990–2011. Centers for Disease Control and Prevention (US). MMWR Morb Mortal Weekly Report 64:58–60

Fuhlendorf SD, Fynn RW, McGranahan DA, Twidwell D (2017) Heterogeneity as the basis for rangeland management. In: Briske DD (ed) Rangeland systems: processes, management and challenges. Springer, Switzerland, pp 169–196

Funk WC, Pearl CA, Draheim HM, Adams MJ, Mullins TD, Haig SM (2008) Range-wide phylogeographic analysis of the spotted frog complex (*Rana luteiventris* and *Rana pretiosa*) in northwestern North America. Mol Phylogenet Evol 49:198–210

Gauthier J, Kluge AG, Rowe T (1988) Amniote phylogeny and the importance of fossils. Cladistics 4:105–209

Gerlanc NM, Kaufman GA (2003) Use of bison wallows by anurans on Konza Prairie. Am Midl Nat 150:158–168

Germano DJ, Rathbun GB, Saslaw LR (2012) Effects of grazing and invasive grasses on desert vertebrates in California. J Wildl Manag 76:670–682

Gibson PP, Olden JD (2014) Ecology, management, and conservation implications of North American beaver (*Castor canadensis*) in dryland streams. Aquat Conserv Mar Freshwat Ecosyst 24:391–409

Ginsberg HS, Hickling GJ, Burke RL, Ogden NH, Beati L, LeBrun RA, Arsnoe IM, Gerhold R, Han S, Jackson K, Maestas L (2021) Why Lyme disease is common in the northern US, but rare in the south: the roles of host choice, host-seeking behavior, and tick density. PLoS Biol 19:e3001066

Gordon MR, Simandle ET, Sandmeier FC, Tracy CR (2020) Two new cryptic endemic toads of Bufo discovered in central Nevada, western United States (Amphibia: Bufonidae: *Bufo [Anaxyrus]*). Copeia 108:166–183

Grant EHC, Miller DAW, Schmidt BR, Adams MJ, Amburgey SM, Chambert T, Cruickshank SS, Fisher RN, Green DM, Hossack BR, Johnson PTJ, Joseph MB, Rittenhouse TAG, Ryan ME, Waddle JH, Walls SC, Bailey LL, Fellers GM, Gorman TA, Ray AM, Pilliod DS, Price SJ, Saenz D, Sadinski W, Muths E (2016) Quantitative evidence for the effects of multiple drivers on continental-scale amphibian declines. Sci Rep 6:25625

Gray AL, Duffus J, Haman KH, Harris RN, Allender MC, Thompson TA, Christman MR, Sacerdote-Velat A, Sprague LA, Williams JM, Miller DL (2017) Pathogen surveillance in herpetofaunal populations: guidance on study design, sample collection, biosecurity, and intervention strategies. Herpetological Review 48:334–351

Gray MJ, Smith LM (2005) Influence of land use on postmetamorphic body size of playa lake amphibians. J Wildl Manag 69:515–524

Green DE, Kagarise Sherman C (2001) Diagnostic histological findings in Yosemite toads (*Bufo canorus*) from a die-off in the 1970s. J Herpetol 35:92–103

Grudzinski BP, Cummins H, Vang TK (2020) Beaver canals and their environmental effects. Prog Phys Geogr Earth Environ 44:189–211

Hansen NA, Scheele BC, Driscoll DA, Lindenmayer DB (2019) Amphibians in agricultural landscapes: the habitat value of crop areas, linear plantings and remnant woodland patches. Anim Conserv 22:72–82

Heintzman LJ, McIntyre NE (2021) Assessment of playa wetland network connectivity for amphibians of the south-central Great Plains (USA) using graph-theoretical, least-cost path, and landscape resistance modelling. Landscape Ecol 36:1117–1135

Hellgren EC, Burrow AL, Kazmaier RT, Ruthven DC III (2010) The effects of winter burning and grazing on resources and survival of Texas horned lizards in a thornscrub ecosystem. J Wildl Manag 74:300–309

Heyer R, Donnelly MA, Foster M, Mcdiarmid R (eds) (1994) Measuring and monitoring biological diversity: standard methods for amphibians. Smithsonian Institution, Washington, DC

Hill JE, DeVault TL, Belant JL (2019) Impact of the human footprint on anthropogenic mortality of North American reptiles. Acta Oecologica 101:103486

Hossack BR, Corn PS, Pilliod DS (2005) Lack of significant changes in the herpetofauna of Theodore Roosevelt National Park, North Dakota, since the 1920s. Am Midl Nat 154:423–432

Hossack BR, Gould WR, Patla DA, Muths E, Daley R, Legg K, Corn PS (2015) Trends in Rocky Mountain amphibians and the role of beaver as a keystone species. Biol Cons 187:260–269

Hoverman JT, Gray MJ, Miller DL, Haislip NA (2012) Widespread occurrence of ranavirus in pond-breeding amphibian populations. EcoHealth 9:36–48

Howell HJ, Mothes CC, Clements SL, Catania SV, Rothermel BB, Searcy CA (2019) Amphibian responses to livestock use of wetlands: new empirical data and a global review. Ecol Appl 29:e01976

Hubbard KA, Chalfoun AD, Gerow KG (2016) The relative influence of road characteristics and habitat on adjacent lizard populations in arid shrublands. J Herpetol 50:29–36

Jacobson ER (2007) Chapter 10 Bacterial diseases of reptiles. In: Jacobson ER (ed) Infectious diseases and pathology of reptiles: color atlas and text. CRC Press, New York, NY, pp 461–526

Jacobson ER, Gaskin JM, Brown MB, Harris RK, Gardiner CH, LaPointe JL, Adams JP, Reggiardo C (1991) Chronic upper respiratory tract disease of free-ranging desert tortoises, *Xerobates agassizii*. J Wildl Dis 27:296–316

James SE, M'Closkey RT (2002) Patterns of microhabitat use in a sympatric lizard assemblage. Can J Zool 80:2226–2234

James SE, M'Closkey TR (2003) Lizard microhabitat and fire fuel management. Biol Cons 114:293–297

Jochimsen DM, Peterson CR, Harmon LJ (2014) Influence of ecology and landscape on snake road mortality in a sagebrush-steppe ecosystem. Anim Conserv 17:583–592

Johnson JR, Thomson RC, Micheletti SJ, Shaffer HB (2011) The origin of tiger salamander (*Ambystoma tigrinum*) populations in California, Oregon, and Nevada: introductions or relicts? Conserv Genet 12:355–370

Joly P (2019) Behavior in a changing landscape: using movement ecology to inform the conservation of pond-breeding amphibians. Front Ecol Evol 7:155

Jolly CJ, Dickman CR, Doherty TS, van Eeden LM, Geary WL, Legge SM, Woinarski JC, Nimmo DG (2022) Animal mortality during fire. Global Change Biology

Jones B, Fox SF, Leslie DM, Engle DM, Lochmiller RL (2000) Herpetofaunal responses to brush management with herbicide and fire. Rangel Ecol Manage 53:154–158

Jones KB (1981) Effects of grazing on lizard abundance and diversity in western Arizona. Southwestern Naturalist 26:107–115

Jones LL, Halama KJ, Lovich RE (2016) Habitat management guidelines for amphibians and reptiles of the Southwestern United States. Partners in Amphibian and Reptile Conservation, Technical Publication HMG-5, Birmingham, Alabama

Kagarise Sherman C, Morton ML (1993) Population declines of Yosemite toads in the eastern Sierra Nevada of California. J Herpetol 27:186–198

Katzner TE, Carlisle JD, Poessel SA, Thomason EC, Pauli BP, Pilliod DS, Belthoff JR, Heath JA, Parker KJ, Warner KS, Hayes HM, Aberg MC, Ortiz PA, Amdor SM, Alsup SE, Coates SE, Miller TA, Duran ZK (2020) Illegal killing of non-game wildlife and recreational shooting in conservation areas. Conserv Sci Pract e279

Kay GM, Mortelliti A, Tulloch A, Barton P, Florance D, Cunningham SA, Lindenmayer DB (2017) Effects of past and present livestock grazing on herpetofauna in a landscape-scale experiment. Conserv Biol 31:446–458

Kiester AR (1971) Species density of North American amphibians and reptiles. Syst Zool 20:127–137

Kingsbury BA, Gibson J (2012) Habitat management guidelines for amphibians and reptiles of the Midwestern United States, 2nd edn. Partners in amphibian and reptile conservation, Birmingham, Alabama

Kimberly DA, Fender CL (2020) Amphibians and reptiles of Antelope Island, Great Salt Lake, Utah. In: Baxter BK, Butler JK (eds) Great salt lake biology: a terminal lake in a time of change. Springer, Switzerland, pp 345–367

Knutson MG, Richardson WB, Reineke DM, Gray BR, Parmelee JR, Weick SE (2004) Agricultural ponds support amphibian populations. Ecol Appl 14:669–684

Larson DM (2014) Grassland fire and cattle grazing regulate reptile and amphibian assembly among patches. Environ Manage 54:1434–1444

Laudet V (2011) The origins and evolution of vertebrate metamorphosis. Curr Biol 21:R726–R737

Leis SA, Blocksome CE, Twidwell D, Fuhlendorf SD, Briggs JM, Sanders LD (2017) Juniper invasions in grasslands: research needs and intervention strategies. Rangelands 39:64–72

Lemckert F, Hecnar SJ, Pilliod DS (2012) Loss and modification of habitat. In: Heatwole H, Wilkinson JW (eds) Conservation and decline of amphibians: ecological aspects, effect of humans, and management. Amphibian biology series, vol 10. Surrey Beatty and Sons. NSW, Australia, pp 3291–3342

Lertzman-Lepofsky GF, Kissel AM, Sinervo B, Palen WJ (2020) Water loss and temperature interact to compound amphibian vulnerability to climate change. Glob Change Biol 26:4868–4879

Lewis DL, Baxter GT, Johnson KM, Stone MD (1985) Possible extinction of the Wyoming toad, *Bufo hemiophrys baxteri*. J Herpetol 19:166–168

Li Y, Cohen JM, Rohr JR (2013) Review and synthesis of the effects of climate change on amphibians. Integr Zool 8:145–161

Lindenmayer DB, Blanchard W, Crane M, Michael D, Sato C (2018) Biodiversity benefits of vegetation restoration are undermined by livestock grazing. Restor Ecol 26:1157–1164

Lips KR (2016) Overview of chytrid emergence and impacts on amphibians. Philos Trans Royal Soc B: Biol Sci 371(1709):20150465

Londe DW, Dwayne Elmore R, Davis CA, Fuhlendorf SD, Luttbeg B, Hovick TJ (2020) Structural and compositional heterogeneity influences the thermal environment across multiple scales. Ecosphere 11:e03290

Long ZL, Prepas EE (2012) Scale and landscape perception: the case of refuge use by Boreal Toads (*Anaxyrus boreas boreas*). Can J Zool 90:1015–1022

Lorch JM, Knowles S, Lankton JS, Michell K, Edwards JL, Kapfer JM, Staffen RA, Wild ER, Schmidt KZ, Ballmann AE, Blodgett D, Farrell TM, Glorioso BM, Last LA, Price SJ, Schuler KL, Smith CE, Wellehan JFX Jr, Blehert DS (2016) Snake fungal disease: an emerging threat to wild snakes. Philos Trans R Soc B 371:20150457

Lovreglio R, Meddour-Sahar O, Leone V (2014) Goat grazing as a wildfire prevention tool: a basic review. iForest 7:260–268

Markle CE, Gillingwater SD, Levick R, Chow-Fraser P (2017) The true cost of partial fencing: evaluating strategies to reduce reptile road mortality. Wildl Soc Bull 41:342–350

Martin SA, Rautsaw RM, Robb F, Bolt MR, Parkinson CL, Seigel RA (2017) Set AHDriFT: applying game cameras to drift fences for surveying herpetofauna and small mammals. Wildl Soc Bull 41:804–809

Maxell BA (2009) State-wide assessment of status, predicted distribution, and landscape-level habitat sustainability of amphibians and reptiles in Montana. PhD dissertation, University of Montana, Missoula, MT

McAdoo K, Mellison C (2016) Case study: successful collaboration for Columbia spotted frog conservation in northern and central Nevada. Fact Sheet 16-10. University of Nevada Cooperative Extension, Reno, Nevada

McCollum DW, Tanaka JA, Morgan JA, Mitchell JE, Fox WE, Maczko KA, Hidinger L, Duke CS, Kreuter UP (2017) Climate change effects on rangelands and rangeland management: affirming the need for monitoring. Ecosystem Health Sustain 3:e01264

Meagher M (1986) Bison bison. Mammalian Species 266:1–8

Mendoza-Roldan JA, Mendoza-Roldan MA, Otranto D (2021) Reptile vector-borne diseases of zoonotic concern. Int J Parasitol Parasites Wildlife 15:132–142

Mester B, Szalai M, Mérő T, Puky M, Lengyel S (2015) Spatiotemporally variable management by grazing and burning increases marsh diversity and benefits amphibians: a field experiment. Biol Cons 192:237–246

Miller DA, Grant EHC, Muths E, Amburgey SM, Adams MJ, Joseph MB, Waddle JH, Johnson PT, Ryan ME, Schmidt BR, Calhoun DL (2018) Quantifying climate sensitivity and climate-driven change in North American amphibian communities. Nat Commun 9:1–15

Miller RF, Svejcar TJ, Rose JA (2000) Impacts of western juniper on plant community composition and structure. J Range Manag 53:574–585

Mims MC, Moore CE, Shadle EJ (2020) Threats to aquatic taxa in an arid landscape: Knowledge gaps and areas of understanding for amphibians of the American Southwest. Wiley Interdiscip Rev Water 7:e1449

Mitchell CI, Shoemaker KT, Esque TC, Vandergast AG, Hromada SJ, Dutcher KE, Heaton JS, Nussear KE (2021) Integrating telemetry data at several scales with spatial capture-recapture to improve density estimates for a rare and elusive reptile. Ecography 12:e03689

Modesto SP, Scott DM, MacDougall MJ, Sues H-D, Evans DC, Reisz RR (2015) The oldest parareptile and the early diversification of reptiles. Proc R Soc B 282:20141912

Morrison ML, Hall LS (1999) Habitat characteristics of reptiles in pinyon-juniper woodland. Great Basin Naturalist 59:288–291

Mushet DM, Euliss NH Jr, Stockwell CA (2012) A conceptual model to facilitate amphibian conservation in the northern Great Plains. Great Plains Res 22:45–58

Muths E, Chambert T, Schmidt BR, Miller DAW, Hossack BR, Joly P, Grolet O, Green DM, Pilliod DS, Cheylan M, Fisher R, McCaffery RM, Adams MJ, Palen W, Arntzen JW, Garwood J, Gellers G, Thirion JM, Besnard A, Campbell Grant EH (2018) Heterogeneous responses of amphibian populations to climate change complicates conservation planning. Sci Rep 7:17102

Neilly H, Nordberg EJ, VanDerWal J, Schwarzkopf L (2018a) Arboreality increases reptile community resistance to disturbance from livestock grazing. J Appl Ecol 55:786–799

Neilly H, O'Reagain P, Vanderwal J, Schwarzkopf L (2018b) Profitable and sustainable cattle grazing strategies support reptiles in tropical savanna rangeland. Rangel Ecol Manage 71:205–212

Neilly H, Ward M, Cale P (2021) Converting rangelands to reserves: Small mammal and reptile responses 24 years after domestic livestock grazing removal. Austral Ecol. https://doi.org/10.1111/aec.13047

Nicholson L, Humphreys K (1981) Sheep grazing at the Kramer Study Plot, San Bernadino County, CA. In: Hashagan KA (ed) Proceedings of the 1981 desert tortoise council symposium, Riverside, CA, pp 163–194

Pereira KE, Woodley SK (2021) Skin defenses of North American salamanders against a deadly salamander fungus. Anim Conserv 24:552–567

Petrovan SO, Schmidt BR (2019) Neglected juveniles; a call for integrating all amphibian life stages in assessments of mitigation success (and how to do it). Biol Cons 236:252–260

Picco AM, Brunner JL, Collins JP (2007) Susceptibility of the endangered California tiger salamander, *Ambystoma californiense*, to ranavirus infection. J Wildl Dis 43:286–290

Pilliod DS (In press) Successful outcomes in species conservation: a case study of the Columbia spotted frog (*Rana luteiventris*). In: Walls SC, O'Donnell KM (eds) Strategies for conservation success in herpetology, Society for the Study of Amphibians and Reptiles, University Heights, Ohio, pp 41–52

Pilliod DS, Wind E (2008) Habitat management guidelines for amphibians and reptiles of the north-western United States and western Canada. Partners in Amphibian and Reptile Conservation, Birmingham, Alabama

Pilliod DS, Scherer RD (2015) Managing habitat to slow or reverse population declines of the Columbia spotted frog in the Northern Great Basin. J Wildl Manag 79:579–590

Pilliod DS, Griffiths RA, Kuzmin SL (2012) Ecological impacts of non-native species. In: Heatwole H, Wilkinson JW (eds) Conservation and decline of amphibians: ecological aspects, effect of humans, and management. Amphibian biology series, vol 10. Surrey Beatty and Sons, NSW, Australia, pp 3343–3382

Pilliod DS, Arkle RS, Robertson JM, Murphy MA, Funk WC (2015) Effects of changing climate on aquatic habitat and connectivity for remnant populations of a wide-ranging frog species in an arid landscape. Ecol Evol 5:3979–3994

Pilliod DS, Welty JL, Toevs GR (2017) Seventy-five years of vegetation treatments on public rangelands in the Great Basin of North America. Rangelands 39:1–9

Pilliod DS, Jeffries MI, Arkle RS, Olson DH (2020) Reptiles under the conservation umbrella of the Greater Sage-Grouse. J Wildl Manag 84:478–491

Pilliod DS, Hausner MB, Scherer RD (2021) From satellites to frogs: quantifying ecohydrological change, drought mitigation, and population demography in desert meadows. Sci Total Environ 758:143632

Polley HW, Briske DD, Morgan JA, Wolter K, Bailey DW, Brown JR (2013) Climate change and North American rangelands: trends, projections, and implications. Rangel Ecol Manage 66:493–511

Pough HF, Andrews RM, Cadle JE, Crump ML, Savitzky AH, Wells KD (1998) Herpetology. Prentice Hall, Upper Saddle River, New Jersey

Price SJ, Ariel E, Maclaine A, Rosa GM, Gray MJ, Brunner JL, Garner TW (2017) From fish to frogs and beyond: impact and host range of emergent ranaviruses. Virology 511:272–279

Pulsford SA, Driscoll DA, Barton PS, Lindenmayer DB (2017) Remnant vegetation, plantings and fences are beneficial for reptiles in agricultural landscapes. J Appl Ecol 54:1710–1719

Pulsford SA, Barton PS, Driscoll DA, Lindenmayer DB (2019) Interactive effects of land use, grazing and environment on frogs in an agricultural landscape. Agr Ecosyst Environ 281:25–34

Pyke CR, Marty J (2005) Cattle grazing mediates climate change impacts on ephemeral wetlands. Conserv Biol 19:1619–1625

Pyron RA (2014) Biogeographic analysis reveals ancient continental vicariance and recent oceanic dispersal in amphibians. Syst Biol 63:779–797

Qian H (2009) Global comparisons of beta diversity among mammals, birds, reptiles, and amphibians across spatial scales and taxonomic ranks. J Syst Evol 47:509–514

Qian H (2010) Environment–richness relationships for mammals, birds, reptiles, and amphibians at global and regional scales. Ecol Res 25:629–637

Radke N, Wester D, Perry G, Rideout-Hanzak S (2008) Short-term effects of prescribed fire on lizards in mesquite-ashe juniper vegetation in central Texas. Appl Herpetol 5:281–292

Read JL, Cunningham R (2010) Relative impacts of cattle grazing and feral animals on an Austral arid zone reptile and small mammal assemblage. Austral Ecol 35:314–324

Rieder JP, Newbold TA, Ostoja SM (2010) Structural changes in vegetation coincident with annual grass invasion negatively impacts sprint velocity of small vertebrates. Biol Invasions 12:2429–2439

Robertson JM, Murphy MA, Pearl CA, Adams MJ, Páez-Vacas MI, Haig SM, Pilliod DS, Storfer A, Funk WC (2018) Regional variation in drivers of connectivity for two frog species (*Rana pretiosa* and *R. luteiventris*) from the US Pacific Northwest. Mol Ecol 27:3242–3256

Roche LM, Allen-Diaz B, Eastburn DJ, Tate KW (2012) Cattle grazing and Yosemite toad (*Bufo canorus* Camp) breeding habitat in Sierra Nevada meadows. Rangel Ecol Manage 65:56–65

Rowe JC, Duarte A, Pearl CA, McCreary B, Galvan SK, Peterson JT, Adams MJ (2019) Disentangling effects of invasive species and habitat while accounting for observer error in a long-term amphibian study. Ecosphere 10:e02674

Rozen-Rechels D, Dupoué A, Lourdais O, Chamaillé-Jammes S, Meylan S, Clobert J, Le Galliard JF (2019) When water interacts with temperature: ecological and evolutionary implications of thermo-hydroregulation in terrestrial ectotherms. Ecol Evol 9:10029–10043

Russell RE, Halstead BJ, Mosher BA, Muths E, Adams MJ, Grant EHC, Fisher RN, Kleeman PM, Backlin AR, Pearl CA, Honeycutt RK, Hossack BR (2019) Effect of amphibian chytrid fungus (*Batrachochytrium dendrobatidis*) on apparent survival of frogs and toads in the western USA. Biol Cons 236:296–304

Ruthven DC, Kazmaier RT, Janis MW (2008) Short-term response of herpetofauna to various burning regimes in the south Texas plains. Southwestern Naturalist 53:480–487

Rutschmann A, Miles DB, Le Galliard JF, Richard M, Moulherat S Sinervo, B, Clobert J (2016) Climate and habitat interact to shape the thermal reaction norms of breeding phenology across lizard populations. J Animal Ecol 85:457–466

Sawatzky ME, Martin AE, Fahrig L (2019) Landscape context is more important than wetland buffers for farmland amphibians. Agr Ecosyst Environ 269:97–106

Scheele BC, Pasmans F, Skerratt LF, Berger L, Martel A, Beukema W, Acevedo AA, Burrowes PA, Carvalho T, Catenazzi A, De la Riva I (2019) Amphibian fungal panzootic causes catastrophic and ongoing loss of biodiversity. Science 363(6434):1459–1463

Schmidt B (2003) Declining amphibian populations: the pitfalls of count data in the study of diversity, distribution, dynamics and demography. Herpetol J 14:167–174

Schmutzer AC, Gray MJ, Burton EC, Miller DL (2008) Impacts of cattle on amphibian larvae and the aquatic environment. Freshw Biol 53:2613–2625

Scott NJ (1996) Evolution and management of North American grassland herpetofauna. In: Frey DN (ed) Ecosystem disturbance and wildlife conservation in western grasslands: a symposium proceedings, U.S. Department of Agriculture—Forest Service, Rocky Mountain Forest and Range Research Station, Fort Collins, Colorado, pp 40–53

Silvy NJ, Lopez RR, Peterson MJ (2012) Techniques for marking wildlife. The Wildlife Techniques Manual 1:230–257

Sinervo B, Mendez-De-La-Cruz F, Miles DB, Heulin B, Bastiaans E, Villagrán-Santa Cruz M, Lara-Resendiz R, Martínez-Méndez N, Calderón-Espinosa ML, Meza-Lázaro RN, Gadsden H (2010) Erosion of lizard diversity by climate change and altered thermal niches. Science 328(5980):894–899

Smalling KL, Rowe JC, Pearl CA, Iwanowicz LR, Givens CE, Anderson CW, McCreary B, Adams MJ (2021) Monitoring wetland water quality related to livestock grazing in amphibian habitats. Environ Monit Assess 193:1–17

Smith LJ, Holycross AT, Painter CW, Douglas ME (2001) Montane rattlesnakes and prescribed fire. Southwestern Naturalist 54–61

Stebbins RC (2003) Western reptiles and amphibians, 3rd edn. Houghton Mifflin Company, New York, New York

Swartz TM, Miller JR (2019) Managing farm ponds as breeding sites for amphibians: key trade-offs in agricultural function and habitat conservation. Ecol Appl 29:e01964

Swei A, Ostfeld RS, Lane RS, Briggs CJ (2011) Impact of the experimental removal of lizards on Lyme disease risk. Proc Royal Soc B Biol Sci 278:2970–2978

Thompson MD, Russell AP (2005) Glacial retreat and its influence on migration of mitochondrial genes in the long-toed salamander (*Ambystoma macrodactylum*) in western North America. In: Elewa AMT (ed) Migration of organisms. Springer, Berlin, pp 205–246

Urban MC, Richardson JL, Freidenfelds NA (2013) Plasticity and genetic adaptation mediate amphibian and reptile responses to climate change. Evol Appl 7:88–103

USFWS [U.S. Fish and Wildlife Service] (2011) Revised recovery plan for the Mojave population of the desert tortoise (*Gopherus agassizii*). U.S. Fish and Wildlife Service, California and Nevada Region, Sacramento

USGS GAP (2018a) U.S. geological survey—gap analysis project (GAP) species habitat maps CONUS_2001: U.S. Geological Survey data release, https://doi.org/10.5066/F7V122T2

USGS GAP (2018b) U.S. geological survey—gap analysis project (GAP) amphibian species habitat richness: U.S. Geological Survey, https://doi.org/10.3133/sir20195034. Data available at: https://doi.org/10.5066/P9YW3ZQ2

USGS GAP (2018c) U.S. geological survey—gap analysis project (GAP) reptile species habitat richness: U.S. Geological Survey, https://doi.org/10.3133/sir20195034. Data available at: https://doi.org/10.5066/P9YW3ZQ2

Villena OC, Royle JA, Weir LA, Foreman TM, Gazenski KD, Grant EH (2016) Southeast regional and state trends in anuran occupancy from calling survey data (2001–2013) from the North American Amphibian Monitoring Program. Herpetol Conserv Biol 11:373–385

Vreeland JK, Tietje WD (2002) Numerical response of small vertebrates to prescribed fire in California oak woodland. The role of fire in nongame wildlife management and community restoration: traditional uses and new directions. General technical report NE-GTR-288. US Forest Service, Newton Square, Pennsylvania, pp 100–110

Waddle JH, Grear DA, Mosher BA, Grant EH, Adams MJ, Backlin AR, Barichivich WJ, Brand AB, Bucciarelli GM, Calhoun DL, Chestnut T (2020) *Batrachochytrium salamandrivorans* (Bsal) not detected in an intensive survey of wild North American amphibians. Sci Rep 10(13012):1–7

Wagner RB, Brune CR, Popescu VD (2021) Snakes on a lane: Road type and edge habitat predict hotspots of snake road mortality. J Nat Conserv 61:125978

Wake DB, Koo MS (2018) Amphibians. Curr Biol 28:R1237–R1241

Weitzman CL, Gov R, Sandmeier FC, Snyder SJ, Tracy CR (2017) Co-infection does not predict disease signs in *Gopherus* tortoises. Royal Soc Open Sci 4:171003

Wiens JJ (2007) Global patterns of diversification and species richness in amphibians. Am Nat 170:S86–S106

Wilgers DJ, Horne EA (2006) Effects of different burn regimes on tallgrass prairie herpetofaunal species diversity and community composition in the Flint Hills, Kansas. J Herpetol 40:73–84

Wilgers DJ, Horne EA, Sandercock BK, Volkmann AW (2006) Effects of rangeland management on community dynamics of the herpetofauna of the tallgrass prairie. Herpetologica 4:378–388

Wohl E (2021) Legacy effects of loss of beavers in the continental United States. Environ Res Lett 16:025010

Woinarski JC, Brock C, Fisher A, Milne D, Oliver B (1999) Response of birds and reptiles to fire regimes on pastoral land in the Victoria River District, Northern Territory. Rangeland J 21:24–38

Wollenberg Valero KC, Marshall JC, Bastiaans E, Caccone A, Camargo A, Morando M, Niemiller ML, Pabijan M, Russello MA, Sinervo B, Werneck FP, Sites JW Jr, Wiens JJ, Steinfartz S (2019) Patterns, mechanisms and genetics of speciation in reptiles and amphibians. Genes 10:646

Yap TA, Koo MS, Ambrose RF, Vredenburg VT (2018) Introduced bullfrog facilitates pathogen invasion in the western United States. PLoS ONE 13(4):e0188384

Zero VH, Murphy MA (2016) An amphibian species of concern prefers breeding in active beaver ponds. Ecosphere 7:e01330. https://doi.org/10.1002/ecs2.1330

Chapter 26
Insects in Grassland Ecosystems

Diane M. Debinski

Abstract Insects serve as ecosystem engineers in grasslands. Their impacts are comparable in scale to those of mammals, but because they are so much smaller, their roles and influences are not always as obvious. The roles that insects play in grasslands are as diverse as Class Insecta itself, including herbivory, pollination, seed dispersal, soil profile modification, nutrient cycling, parasitism, and serving as intermediaries between plants and wildlife in food webs. In the context of their effects on grassland wildlife species, insects serve as essential food resources for many species of birds, bats, reptiles, mammals, amphibians, fish, and other insects. Insects also have significant effects on the habitat structure available for wildlife because they can, on the one hand, enhance the productivity of grassland vegetation, but alternatively, they have the power to completely defoliate a grassland. From the perspective of food webs, insects play multiple roles. They can serve as food for wildlife, but they also can serve as parasites, vectors of disease, and decomposers. Ecological changes in grasslands due to events such as fire, grazing, herbicide or insecticide application, and habitat fragmentation or loss can affect both wildlife and insects. For that reason, ecologists are often interested in linking the study of a particular wildlife species to the associated insect community. Insects are simply less visible ecological engineers, continually interacting with wildlife, and modifying the habitat where they coexist with wildlife in grassland ecosystems.

Keywords Insects · Grasslands · Rangelands · Pollinators · Grazing

26.1 General Life History and Population Dynamics

Insects are diverse and abundant within rangelands, including grassland systems. Herein, insects and their ecological interactions in grasslands are addressed. Insects inhabit and occupy the air, soil, vegetation, and aquatic environments. Although they are less conspicuous than other wildlife counterparts, they play a large variety of

D. M. Debinski (✉)
Department of Ecology, Montana State University, Bozeman, MT, USA
e-mail: diane.debinski@montana.edu

© The Author(s) 2023 897
L. B. McNew et al. (eds.), *Rangeland Wildlife Ecology and Conservation*,
https://doi.org/10.1007/978-3-031-34037-6_26

roles in grassland ecosystems. Their development is affected by humidity, rainfall, and temperature (Kremen et al. 1993). They have high reproductive rates, short life spans, and great mobility in the environment (Porrini et al. 2002), and thus can evolve and adapt quickly to environmental changes. Resultant changes in insect distribution and abundance radiate throughout food webs to impact grassland biota.

Insects provide an incredible array of ecosystem services, from dung burial to pest control, pollination, and food sources (Losey and Vaughan 2006). They modify, aerate, and fertilize the soil, which serves as the foundation of grassland habitat, and serve as herbivores, pollinators, and vectors of disease. Insects, in turn, respond to management of grasslands, and are affected by fire, grazing, herbicide or insecticide application, and habitat fragmentation or loss. A model diagram of how insects affect grasslands would be quite complex. In the most simplistic sense (1) insects affect biotic and abiotic components of grasslands, including virtually all associated grassland species, (2) grassland management affects insects at population, community, and ecosystem scales, and (3) there are a plethora of additional interactions, both direct and indirect, that influence each of these primary relationships. This chapter will highlight some of the most important relationships between insects and plants, soil, other insects, and vertebrates of grasslands, and will explain the role of insects as decomposers and biological control agents. The chapter will also address how grassland management affects these relationships, provide some examples of conservation and management issues, and recommend areas for further research.

26.1.1   Insects and Plants

The evolutionary history of insects and plants is intertwined, and grasslands can provide a rich habitat for these interactions. Insects provide pollination services to 80% of all angiosperms (flowering plants), and 35% of the world's food crops depend on animal pollinators (USDA 2022). Many of these co-evolved relationships have become so specialized that in the absence of its pollinator, a plant cannot reproduce. Conversely, in the absence of its food source, many of which are highly specialized, an insect cannot survive. The diversity of flowering plants in a grassland can thus be affected by the corresponding insect community, and vice versa. Floral colors and scents are also a result of this co-evolution (Matthews and Kitching 1984). Grasses are wind pollinated, so they are not dependent on insects for seed production, however, large proportions of grassland communities are forbs, herbaceous flowering plants that are not graminoids (grasses, sedges or rushes). Although the percentage of pollinated forbs in grasslands has not been estimated, it is reasonable that the number is substantial. For context, the estimated proportion of animal-pollinated plant species in temperate zones is 78% (Ollerton et al. 2011). Given that temperate zones include grasslands, shrublands, forest, and other terrestrial communities, it is likely that grassland insects pollinate a large proportion of grassland forb species.

The complex evolutionary history between insects and plants includes a diverse set of interactions, including pollination, herbivory, and parasitism. Insects that consume plants play a variety of functional roles with respect to grasslands, including serving as leaf feeders, leaf miners and gall makers (laying their eggs within the leaf or stem), sapsuckers (feeding on plant juices), and detritivores (feeding on dead plant tissue) (Wilsey 2018). Grasshoppers (Orthoptera) can appear to have a devastating effect on vegetation from an agricultural perspective, but their herbivory also can benefit plant communities because they speed up nutrient cycling by changing the abundance and decomposition rate of plant litter, which increases total plant abundance (Belovsky and Slade 2000). Some insect species consume or parasitize a large variety of plant species, but many insects specialize on utilization of one family of plants or even one species of plant. For example, the grassland obligate regal fritillary butterfly (*Speyeria idalia*) only consumes *Viola* species as a host plant and in some regions only one species of violet is consumed, the blue prairie violet, *Viola pedatifida* (Kelly and Debinski 1998).

The evolutionary pressure of insect herbivory on plants is responsible for the great variety of defensive chemicals plants produce, including nicotine, pyrethrin, and rotenone, which have been exploited for human use as insecticides (Waldbauer 2003). In a complex web of grassland ecosystem interactions, these chemicals influence which insect species can utilize which plant species. In some cases, plant chemicals also make their way into insects (e.g. monarch (*Danaus plexippus*) butterfly larvae eating milkweed *Asclepias* spp. (Petschenka and Agrawal 2015)), thus affecting how other vertebrate wildlife may or may not predate upon them. Relationships at the base of the food chain can impact higher levels, affecting wildlife species that feed on insects. Relatedly, these relationships also influence which parasites or diseases may be transmitted from insects to wildlife.

The value of pollination as an ecosystem service has undergone close examination over the past few decades (Losey and Vaughan 2006; Kremen et al. 2007). Although bees (Hymenoptera) are recognized as the most important pollinators, flies (Diptera) are a close second (Larson et al. 2001), and Lepidoptera (butterflies and moths) serve as pollinators for some grassland obligate plant species (Hendrix and Khyl 2000). The associations between flies and flowers are commonly overlooked, but the role of flies in pollination increases with increasing elevation, and flies are important pollinators especially in montane systems. The honey bee (*Apis mellifera*) is a nonnative insect that is managed to perform pollination services for a broad variety of cultivated fruits, nuts, and vegetables at a continental scale. Honey bees occur in many grasslands and interestingly, native bees can interact synergistically with honey bees to increase the honey bees' pollination efficiency of crops (Brittain et al. 2013), but there may be potentially negative effects of honey bees on native bees. For example, the presence of honey bees in tallgrass prairie could increase wild bee exposure to viruses (Pritchard et al. 2021) or create competition for resources (Cane and Tepedino 2016). So, knowing which pollinators are most important to conserve in a grassland is more complex than at first glance.

Seed dispersal is another example of the delicate symbiosis between plants and insects. Of all the animals that disperse seeds, only birds and mammals are more important than ants (Hymenoptera). In fact, 35% of angiosperms rely completely on ants for seed dispersal (Waldbauer 2003). Ants gather seeds and carry them back to their nests, dropping some along the way, and discarding the rest in a fertile trash pile just outside their nest. Even this short dispersal distance is advantageous to plants because it lessens the competition between a seedling and its parent and sibling plants (Waldbauer 2003).

Finally, rather than just considering specific co-dependent plant and insect species, it is important to understand the larger community of insects, plants, and other taxa interacting in a grassland. Interactions among plants such as grasses and goldenrod (*Solidago* sp.) and spiders (Araneae) and grasshoppers (Orthoptera) have been studied extensively in eastern U.S. grasslands. In this system, the presence of spiders can affect which type of plant grasshoppers feed upon, resulting in differential dominance of goldenrod versus grasses. This occurs because predators cause herbivores to suppress the abundance of a competitively dominant plant species that offers herbivores a refuge from predation risk (Schmitz 2003).

26.1.2   Insects and Soil

Some grassland insects live their lives entirely above ground, whereas other insects spend a portion of their life (usually as eggs or juveniles) in the soil, leaf litter, or aquatic environments and a different portion of their life above ground. Another group of insects spend most of their time living in and on the grassland soil and physically modify the soil profile, improving the habitat for plant growth (Lee and Wood 1971). For example, termites (Blattodea) and ants redistribute soil and nutrients, bringing mineral-rich material from lower soil layers and mixing it with upper layers with high organic matter content, creating a fertile environment rich in carbon, nitrogen, and phosphorus supporting plant growth (Lee and Wood 1971; Waldbauer 2003). Termites and ants build tunnels increasing soil porosity, which facilitates root growth, aeration, and water storage and drainage (Waldbauer 2003). Insects such as springtails (Collembola) and termites shred plant materials into smaller fragments that can be used by bacteria and fungi (Wilsey et al. 2005; Wilsey 2018). Without these types of insect activity, nutrient cycling would be lessened, reducing plant productivity of grasslands. Dung beetles feed on animal excrement as both adults and larvae, thus fertilizing the soil in which they live (Nichols et al. 2008). In addition, without insect activity, lower quality, less fertile soils would become exposed at the surface (Lee and Wood 1971).

In turn, insects are affected by soil conditions. For example, De Bruyn et al. (2001) found that species richness and diversity in flies were affected by soil pH, soil moisture, and the amount of organic matter present in the soil. As such, there was significant feedback between the insect community and grassland soil, and these interactions affected the structure, composition, and nutrient quality of vegetation available for plants and wildlife.

26.1.3 Insects and Wildlife

As noted previously, insects provide a vital connection between plants and wildlife. Herbivorous insects, "play an indispensable and pivotal role as intermediaries in food chains by making the nutrients synthesized by plants available to animals that do not eat plants" (Waldbauer 2003). Insects serve as food for birds, bats, reptiles, amphibians, and fish (Goulson 2019). In aquatic environments, terrestrial invertebrates, most of which are insects, can be a significant source of prey for fish, sometimes providing about 50% of their annual energy (Saunders and Fausch 2007). Similarly, grassland birds and mammals in terrestrial habitats adjacent to aquatic environments harvest rich insect food sources from aquatic and terrestrial environments. As summarized by Malmqvist (2002: 688), "Aquatic insects subsidize terrestrial birds (and other terrestrial predators such as bats, spiders and predacious insects) and terrestrial insects subsidize fish production in the stream habitat."

Grasslands, specifically, produce abundant insects offering a rich food source for wildlife. Kaspari and Joern (1993) conducted a study of three grassland bird species (grasshopper sparrows [*Ammodramus savannarum*], lark sparrows [*Chondestes grammacus*], and western meadowlarks [*Sturnella neglecta*]) in the Nebraska Sandhills and found that grasshoppers and small beetles (Coleoptera) were the primary food source, but these species also consumed other invertebrates, including Homoptera (aphids, scale insects, cicadas, and leafhoppers), Hymenoptera (bees, wasps, ants, and sawflies), and Araneae (spiders). Prey selection in this group of grassland birds was found to be a complex function of prey size, energy, and other nutrients. Most quail, grouse (Tetraoninae), and pheasant (*Phasianus colchicus*) chicks also rely on insects as a source of protein (Losey and Vaughan 2006). For waterfowl, 43% of species are primarily insectivorous (Ehrlich et al. 1988). With respect to raptors, Swainson's Hawks (*Buteo swainsoni*) prey on both insects and mammals, and grasshoppers can comprise a "staple sustenance" of their diet (Cameron 1913). Additionally, insects "bridge the size gap between large predators and unicellular plants or animals too tiny to be profitable eaten by a large animal" (Waldbauer 2003). Without insects, many food chains, including those in grasslands, would collapse.

26.1.4 Insects as Decomposers

Grassland insects aid crucial nutrient cycling by consuming carrion and decomposing organic matter. In this process, dead organic matter is returned to the soil as minerals and to the atmosphere as gases (Waldbauer 2003). In grasslands that support large populations of mammals, decomposition of carrion and feces by insects is significant and if ceased this understated ecosystem service would quickly become apparent. In Colorado, De Jong and Chadwick (1999) reported 53 insect taxa utilizing decaying rabbit (*Oryctolagus cuniculus*) carcasses. Carrion beetles (Silphidae) are especially important in decomposition. For example, Sikes (1994) found that over 50 species of carrion beetles were heavily dependent on the ungulate carcasses present in sagebrush steppe of the northern range of the Greater Yellowstone Ecosystem.

In grasslands that are grazed by cattle (*Bos taurus*), each animal can produce over 9000 kg of solid waste per year (Losey and Vaughan 2006). Beetles in the family Scarabaeidae provide the ecosystem service of decomposing and burying this waste, which reduces the habitat available to parasites such as flies, thus reducing cattle losses due to horn flies (*Haematobia irritans irritans*) and face flies (*Musca autumnalis*) (Losey and Vaughan 2006). In a related manner, the removal of dung beetles (Scarabaeinae) in some grasslands has been experimentally shown to decrease plant productivity (Wilsey 2018). Without ungulates, many species of beetles could not survive, and without beetles, carcasses and feces would decompose much more slowly, changing the rate of nutrient cycling and productivity in grasslands.

26.1.5 *Insects as Biological Control Agents*

Decomposition by insects also serves as an example of how insects act as biological control agents. Insects can serve as biological controls on other insects, plants, and mammals. One insect species can affect the abundance or distribution of another insect species through competition, predation, parasitism, or mutualism. Destructive outbreaks of insects often occur when these relationships have been disrupted by human activity (LaSalle and Gauld 1993). Insects also can keep "pesky" plants at bay, both by herbivory pressures and seed predation. For example, insects have been used to slow the spread of nonnative plant species like leafy spurge (*Euphorbia esula*) in the Northern Great Plains (Butler et al. 2006), prickly pear cactus (*Opuntia*) in Australia, and St. John's wort (*Hypericum perforatum*) in California (Waldbauer 2003). Wildlife population growth can be affected or controlled by insects through insect-borne diseases, parasitism (Mooring and Samuel 1998), and competition for plants if both the insect and the wildlife species are herbivores (Waldbauer 2003). Again, these relationships can be direct or indirect. As an example of both a direct and an indirect relationship caused by one insect species, ticks (Ixodida) may cause significant blood loss to their wildlife host species (direct effect) and influence the foraging ability of elk (*Cervus canadensis*), bison (*Bison bison*), and moose (*Alces alces*) (indirect effect) (Mooring and Samuel 1998).

26.2 Species and Population Status Issues

26.2.1 Historical Versus Current Distributions, Conservation Status

A number of butterfly, bee, and beetle species that historically occurred on U.S. rangelands are now listed as threatened or endangered. Some examples of these species are listed in Table 26.1. However, due to the sheer number of insect species, the historical knowledge of insect species distribution patterns is limited compared to plants or vertebrates. A broad estimate of 800,000 insect species have been named worldwide and, for the majority of these species, the scientific community knows relatively little about their biology, distribution, or abundance (Goulson 2019). There is such a dearth of knowledge that scientists are challenged to assess and quantify even the crudest measures of changes in diversity and distribution over time limiting assessment of conservation status.

Some prominent grassland insect declines across the U.S. have been documented by the U.S. Fish and Wildlife Service (USFWS 2022). We provided examples of endangered or threatened insects under the Endangered Species Act (1973; ESA) and those having state-level designation as "Species of Conservation Concern" (Fig. 26.1 and Table 26.1). The rusty patched bumble bee (*Bombus affinis*) was listed as endangered in 2017 (Lambe 2018). For many of these insects, habitat loss and/or modification are primary threats (USFWS 2022). Conversion of grasslands into row crops has reduced tallgrass and shortgrass prairie upon which many grassland insect species are dependent (see Chaps. 3, 5). The use of fire and/or grazing, can also have detrimental effects, depending upon the insect species (Kral et al. 2017). The endangered American burying beetle (*Nicrophorus americanus*) now occurs in less than 10% of its former range in the grasslands in eastern Oklahoma, central and southern Nebraska, southeastern Kansas, and southcentral South Dakota, but there are multiple possible reasons and the main drivers of decline remain unclear (Sikes and Raithel 2002).

In addition to species formally listed as endangered or threatened under ESA, there are many rare insects that have been proposed for listing but have not yet received status designation. The regal fritillary (*Speyeria idalia*) butterfly has been a "Species of Conservation Concern" in Midwestern grasslands for several decades. It has been lost from much of its historical distribution in Midwestern prairies due to habitat loss and fragmentation (Kelly and Debinski 1998). The monarch butterfly (*Danaus plexippus*) has more recently become a Species of Conservation Concern, both in terms of its eastern and western populations, and this concern has recently advanced to the federal level. The western population dropped by 97% of their average historic abundance between the 1980s and mid-2010s, and during 2018–2019, the population plummeted even farther, to fewer than 30,000 monarchs (Pelton et al. 2019), but then rebounded to 200,000 in 2021–2022 (McKnight 2021). The decline of monarch butterflies in the eastern U.S. has been attributed to herbicide effects and habitat conversion on host plants, poor weather, insecticide exposure and reduced overwintering habitat (Belsky and Joshi 2018). Although the monarch was

Table 26.1 Examples of U.S. threatened and endangered insects associated with rangeland ecoregions

Species	Ecoregion	Habitat	Conservation status	Threats
American burying beetle (*Nicrophorus americanus*)	Shortgrass prairie and sandhills of Nebraska	Primary or virgin forests as well as grasslands	Threatened	Unknown
Dakota skipper (*Hesperia dacotae*)	Northern mixed-grass prairie	Tallgrass prairie and mixed-grass prairie	Threatened	Loss of native prairie, invasion of nonnative plant species Populations may be influenced by grazing, haying, burning, pesticide use, and lack of management
Pawnee montane skipperling (*Hesperia leonardus montana*)	Shortgrass prairie	Open ponderosa pine (*Pinus ponderosa*) woodlands with an understory of blue grama grass (*Bouteloua gracilis*), the larval food plant, and prairie gayfeather (*Liatris spicata*), the primary nectar plant	Threatened	Habitat loss, conifer encroachment, loss of grasses and prairie gayfeather, residential development, mowed pasture, invasion of noxious weeds
Poweshiek skipperling (*Oarisma poweshiek*)	Tallgrass prairie	Prairie fens, grassy lake and stream margins, moist meadows, sedge meadows, and wet-to-dry prairie	Endangered	Habitat loss and degradation of native prairies and prairie fens, dessication of larvae during dry summer months
Rusty patch bumble bee (*Bombus affinis*)	Tallgrass prairie	Grasslands and tallgrass prairies	Endangered	Prairie loss, degradation, or fragmentation by conversion to other uses
Mount Charleston blue butterfly (*Icaricia (Plebejus) shasta charlestonensis*)	Great Basin; endemic to the Spring Mts in southern Nevada	Subalpine, bristlecone, and mixed conifer vegetation communities	Endangered	Loss and degradation of habitat due to changes in fire regimes and succession, recreational development, fuels reduction projects, and increases in nonnative plants

(continued)

Table 26.1   (continued)

Species	Ecoregion	Habitat	Conservation status	Threats
Carson wandering skipper (*Pseudocopaeodes eunus obscurus*)	Great Basin	Lowland grassland habitats on alkaline substrates at elevations of less than 1524 m (5000 ft), with presence of saltgrass (*Distichlis spicata*), the larval hostplant	Endangered	Habitat destruction, degradation, and fragmentation, wetland habitat modification, agricultural practices, gas and geothermal. development, nonnative plant invasion, collecting, livestock trampling/grazing, water exportation, road construction, recreation, pesticide drift
Regal fritillary (*Speyeria idalia*)	Tallgrass prairie	Tallgrass prairies, marshes	Species of conservation concern[a]	Habitat loss and degradation
Monarch butterfly (*Danaus plexippus*)	virtually all ecoregions listed	Grasslands, rangelands, areas adjacent to agriculture, and roadsides	Species of conservation concern[a]	Herbicide effects on host plants, poor weather, and reduced overwintering habitat in Mexico

Data sources The U.S. Fish and Wildlife Service https://ecos.fws.gov/ecp/ and associated U.S. Federal Register links
[a]U.S. state-level designation

Fig. 26.1   Selected rare, threatened, and endangered insects of North American grasslands. **a** The American burying beetle, *Nicrophorus americanus*, is threatened in the shortgrass prairie ecoregion and also in the sandhills of Nebraska. *Photo source* Doug Backlund. **b** The regal fritillary butterfly, *Speyeria idalia*, is a Species of Conservation Concern in tallgrass and shortgrass prairie ecoregions. *Photo source* Raymond Moranz. **c** The rusty patched bumble bee, *Bombus affinis*, is endangered and occurs in the tallgrass prairie ecoregion. *Photo source* https://www.fws.gov/midwest/endang ered/insects/rpbb/rpbbid.html

proposed for listing as a threatened species, a 2020 ruling determined that listing was "warranted but precluded," meaning that the U.S. Fish and Wildlife Service does not have enough resources to complete the listing decision process because of higher-priority reviews (USFWS 2020). However, on June 21, 2022, the International Union for the Conservation of Nature (IUCN) listed migrating monarchs as Endangered on the IUCN Red List (IUCN 2022).

Recent assessments of insect decline at a global level and in regions outside of North America can be used to inform grassland insect conservation within North America (Hallmann et al. 2017; Goulson 2019; Zattara and Aizen 2020). For example, Zattara and Aizen (2020) reported that the number of bee species being collected or observed over time has declined since the 1990s and that these results might, in part, reflect increased impediments to specimen collection and data mobilization, as well as reduced sampling coverage. However, this also could reflect a worldwide decline in bee diversity given that many species are becoming rarer. Similarly, Forister et al. (2021) documented a 1.6% annual reduction in the number of individual butterflies observed over the past four decades in 70 locations within the western U.S. In an alternative approach to species abundance assessments, some scientists have measured trends in total insect biomass (Hallmann et al. 2017). Even these relatively crude assessments reveal declining trends, such as a recent study that reported a 75% decrease in total flying insect biomass over a 27-year period in protected areas of Germany (Hallmann et al. 2017).

Systematists have pointed out that the limited number of taxonomists available to identify many of these species may be another factor influencing these trends in insect abundance, which would make interpretation of declines in species less robust. It is well recognized that a limited number of taxonomists and related jobs is a major challenge (Agnarsson and Kuntner 2007), potentially affecting scientists' ability to assess insect trends. Determining the drivers and the relative significance of these purported temporal trends in insect biomass deserves additional attention in future research.

Changes in insect populations can be reflected in other trophic levels of the ecosystem, and thus have significant relevance to grassland wildlife. In some cases, monitoring insect populations allows the prediction of effects on organisms in higher trophic levels (LaSalle and Gauld 1993). Changes in insect populations also may precede changes in lower trophic levels (Erhardt and Thomas 1991) and provide information about changes in habitat. For example, honey bees have been used as environmental monitors for decades (Devillers 2002). Carabid beetles (Coleoptera, Carabidae) have been used to document long-term (e.g., 100+ yr.) changes in habitats (Turin and Den Boer 1988). Fossil records of arthropod communities even have been used to construct climate histories (Atkinson et al. 1987).

26.2.2 Population and Community Monitoring

A diverse set of field and analytical approaches are used to monitor populations and communities of insects in grasslands. The statistical approaches used for insect population and community analyses are diverse and similar to those used for wildlife species. Numerous books have been devoted to estimating population sizes (Borchers et al. 2004; Mills 2012; Buckland et al. 2015) and analyzing ecological community structure (Magurran 1988; Mittelbach 2018), so these topics will not be covered in this chapter. However, in some cases the methods of data collection for population and community assessment of insects differ from wildlife methods. Below are summarized some of the most frequently used methods for insect population and community monitoring (sweep netting, pan traps, pitfall traps, and individual species netting). Similar to wildlife surveys, video monitoring and quadrat surveys can also be used for insect surveys (Zaller et al. 2015) but are not described here given their less frequent application.

26.2.2.1 Sweep Netting

The most common approach to monitoring insect communities is through sweep netting. Sweep netting is a consistent and reliable survey tool for capturing vegetation dwelling arthropods. This technique is particularly good for medium sized insects but can be challenging for collecting smaller insects. Spafford and Lortie (2013) note the value of sweep netting for Thysanoptera (commonly known as thrips), infrequently collected (i.e., rare) insects, and Arachnida (spiders, ticks, mites, and harvestmen). Other methods may need to be used in combination with sweep netting to assess the entire grassland community such as ground dwelling arthropods.

   Although dimensions may vary, one example of a sweep net has a sturdy canvas net bag attached to a 38 cm (15 inch) diameter ring and a wooden handle 91.5 cm (3 feet) long and 2.5 cm (1 inch) in diameter. Sweep nets are a much more substantial tool than the aerial net used for individual insect surveys described below. For standardized sampling, sweep netting involves a surveyor taking a specific number of steps through the grassland with a canvas sweep net and swinging the net broadly from side to side across ~180° with each step as they walk along a transect of designated length. The number of swings and the distance of the transect are generally standardized so that effort is constant across spatial or temporal replicates. At the end of the transect, the observer grabs the net and closes it with their hand and then carefully turns the netting inside out, placing their catch into a receptacle such as a clear plastic bag. The plastic bag can then be inflated to reduce the chance of crushing the insects. The sample is often taken back to a cooler and then stored in a refrigerator or freezer until the insects can be sorted for identification. With sweep netting, there is often quite a bit of time spent removing the insects from the plant material before the insects can be identified to family, genus, or species with the aid of a dissecting microscope. Alternatively, if the investigator is seeking a particular type of insect group such as bees or ants, these insects can be taken from the sweep net and individually placed in small glass vials filled with alcohol as a preservative.

26.2.2.2 Pan Traps

Pan traps or "bee bowls" are colorful plastic bowls (usually white, yellow, and blue, which colors are visible to bees) filled with a soapy water mixture used to passively sample insects (Baum and Wallen 2011). Pan traps are particularly good at capturing bees and flies that are collecting nectar on flowers, but they also catch a broad variety of insects traveling in the same area. The colors of the bowls mimic the colors of the flowers blooming in the grassland. The bowls can be attached to sticks or posts at various heights within and above the vegetation to select for bees that fly at a particular height, or they can be simply laid on the ground. The height of the pan is set to target the insects moving either through or above the vegetation. However, it is important to note the sampling bias in pan trapping; some of the larger-bodied insects (e.g., bumble bees [*Bombus* spp.], grasshoppers) are less likely to be captured. There is also an issue that pan traps may undersample bee species richness and abundance when floral resources are abundant because bees go to the flowers rather than to the bowls, which can bias estimates of species richness and abundance (Baum and Wallen 2011). Finally, if the traps do not contain a preservative, they will get stinky on warm summer days if not collected within 24 h or less. In such cases, a preservative like propylene glycol be added to the soap water, or traps can be sampled with greater frequency to collect specimens.

26.2.2.3 Pitfall Trapping

Pitfall trapping particularly selects ground-dwelling insects, such as ants, beetles, and a broad range of hopping and walking insects as well as spiders. Note that spiders, although not insects, are often included in the context of insect community surveys. This technique involves digging a small hole in the ground and placing a receptacle, such as a plastic cup, flush with the ground (Zaller et al. 2015). An efficient modification is to insert two cups (e.g., Solo Cups) inside one another and bury them flush to soil surface. When collecting samples, only the inner cup needs to be removed and the dirt hole remains undisturbed. Soapy water, or if a preservative is desired, water mixed with antifreeze, is placed in the cup. Use of a preservative is often preferable in warm environments. Propylene glycol (regular automotive antifreeze) or RV antifreeze are two options; the latter is non-toxic containing alcohol as a preservative. Just water can also be used to avoid captured insect mortality, but traps should be checked relatively frequently. When insects walk by, they fall into the cup and get stuck in the fluid. A consistent diameter of trap should be used among sites so that there is no capture bias and so that comparisons among sites can be valid. Many of the insects that walk in the vicinity of the pitfall trap may not be captured, so if needed additional methods for insect community monitoring should be considered (Zaller et al. 2015).

Similar to pan traps/bee bowls, pitfall traps need to be checked within a day or two or the insects may become unidentifiable. If a preservative is added (e.g., 1:1 propylene glycol:water mix) traps can be left deployed for up to a week. However, sampling periods should be carefully considered because when open pitfall traps and bee bowls continuously collect insects risking oversampling. The liquid and trapped insects are generally collected in small plastic bags and taken to a lab for identification. Refrigerated storage is recommended until samples can be processed.

Generally, if objectives are to sample ground-dwelling arthropods, pitfall traps provide good samples when deployed correctly and efforts are taken to control bias. Leading Coleoptera and Arachnid scientists commonly use pitfall traps to assess community structure around the world. When pitfall traps are combined with sweep nets and/or pan traps, valuable data can be collected to assess insect community structure.

26.2.2.4 Individual Species Netting and Observation

Individual species of butterflies, bees, or dragonflies (suborder Anisoptera) can be surveyed with aerial nets (generally 38.1–45.7 cm [15–18 inch] diameter ring, and 122 cm [48 inch] or longer handle length). Aerial nets are much more lightweight than sweep nets, usually with aluminum handles and lightweight netting material. The netting is somewhat transparent and allows the surveyor to net the insect, handle it carefully within the net, extract it with forceps, and either collect it in an envelope (e.g., butterflies), a cyanide jar (bees), or in a small vial of alcohol (other larger-bodied insects), or release it unharmed. Individual species-focused netting can be used to detect species presence or to conduct mark-recapture surveys to monitor the size of an insect population. For mark-recapture, the insect is carefully handled and, depending upon the type of insect, the wings or body can be marked with a permanent marker or a small sticker. Such markings, if done properly, can have no adverse effects on the insect and, in the case of butterflies, individuals can be released and recaptured multiple times during their lifetime (Auckland et al. 2004).

Finally, there is a growing movement towards the use of visual observation rather than collection for more easily identifiable species such as butterflies and bumble bees. For these easily identifiable species, we are learning a lot about distribution and status trends through well designed observational studies, in many cases using community, amateur, and volunteer scientists to collect data such as the Nebraska Bumble Bee Atlas (Xerces 2022).

26.2.2.5 Analytical Approaches

Statistical analyses for population and community ecology research on insects are similar to those used by wildlife biologists, as noted in Sect. 26.2.2.

However, for insects, there is the additional challenge of accounting for populations capable of large interannual fluctuations, including species such as painted lady butterflies (*Vanessa cardui*) (Vandenbosch 2003), grasshoppers (Kemp 1992), southern pine beetles (*Dendroctonus frontalis*) (Turchin et al. 1999), eastern spruce budworm (*Choristoneura fumiferana*) (Zoladeski and Maycock 1990), and cicadas (Cicadoidea) (Cook et al. 2001). Given these natural fluctuations, changes in numbers—even dramatic at times—are not necessarily an indication that major long-term population-level changes are underway. However, the loss of a subset of the insect community, a major change in geographic distribution patterns, or a downward turn in multiple grassland insect species that exhibit similar sensitivities could be cause for concern. For example, a meta-analysis of species range shifts might detect poleward shifts in geographic distribution patterns associated with climate change (Parmesan et al. 1999). Furthermore, when scientists evaluate insect responses to environmental change, some insect taxa are more sensitive than others. In some insect community analyses, certain species can be classified as either disturbance-tolerant or habitat-restricted (Ries et al. 2001). A large increase in the disturbance-tolerant species or the disappearance of habitat-restricted species would warrant investigation. Habitat-restricted species can be especially valuable indicators when habitat loss or fragmentation is an issue. And, notably, some insects have very short dispersal distances despite that fact that they are winged, making them more vulnerable to habitat loss than might be assumed.

26.3   Habitat Associations

26.3.1   Historical/Evolutionary

As described in the introduction to this chapter, insects have a plethora of specialized ecological roles that they play, whether they live in or on the vegetation, on or within the soil, or within the air or the water. All these associations within a grassland habitat have impacts on grassland wildlife species. And as described in previous chapters, the effects of grazing, fire, and mechanical management have changed the character and ecology of grasslands at large spatial scales.

26.3.2   Contemporary Grasslands

The grasslands that once covered North America have been converted by row crop agriculture, extractive industries, urbanization or impacted by raising domesticated livestock (see Chap. 1). As fragmentation and loss of grasslands becomes a predominant regional driver, keystone wildlife species can be lost, and broad suites of other organisms associated with the ecosystem are affected, including insect communities.

In some regions, grasslands are being restored, fire and grazing regimes are being returned, and models to affect such change include both ecological and sociological approaches (Miller et al. 2012). However, sometimes even if the ecosystem *looks* like a native grassland, it may not yet *act* like a native grassland. For example, restored grasslands (grasslands that are replanted after the native vegetation has been modified or lost due to tilling, development, herbicide use, etc.) can have very different seasonal patterns of abundance of floral resources for pollinators as compared to native grasslands (Delaney et al. 2015). Given that pollinators are dependent upon floral resources for growth and reproduction, such differences in the amount and timing of resource availability could have real consequences on grassland insect abundance and diversity.

Contemporary grasslands frequently contain combinations of native and nonnative species of grasses, remnants of unplowed prairie and restored areas, and forbs and woody plants. These changes in grassland composition can be a result of inter-seeding (seeding within a grassland to enhance forage production or reduce erosion), invasive plants, tillage, grazing, fire, herbicide or fertilizer treatments, and many other forms of management. Resulting differences in the plant community can in turn affect the stature of the vegetation, the ratio of forbs to grasses, and the amount of woody vs. non-woody vegetation. Nonnative plant species can alter the amount of bare ground, the amount of litter that remains at the end of the growing season, and how the vegetation responds to fire. Similarly, the amount of bare ground versus litter cover can affect which insect species inhabit a grassland due to their needs for nesting, overwintering, etc. For example, McGranahan et al. (2012) found that tall fescue (*Lolium arundinaceum*), a grass that is commonly seeded into grasslands in the Midwestern U.S., creates patches of living grass in the early spring within a grassland and limits the ability for fire to spread. Similarly, inter-seeding of grasslands with plant species such as tall fescue can create sweeping effects on plant–herbivore interactions and energy flow through the food web because tall fescue often harbors a fungal endophyte that modifies food web interactions (Rudgers and Clay 2007). Tall fescue, when consumed in large enough quantities, also can be toxic to livestock (Paterson et al. 1995) and may affect wildlife, but less research has been conducted on the latter.

26.4 Rangeland Management

A broad set of rangeland management tools (grazing, fire, and mechanical approaches) have been deployed to manage grasslands in a variety of ecoregions across North America. These tools can have long-lasting legacies, and the history of such management can affect the vegetation composition of the grassland long after a management tool was applied (Moranz et al. 2012). Additionally, the effects of one type of management cannot be expected to result in the same vegetation response for all grasslands. Vegetation composition and the history of previous management can affect how a grassland will respond to management (McGranahan et al. 2012).

For example, as noted in Sect. 26.3.2, the presence of an invasive grass may make it more difficult for a manager to apply fire as a management tool.

26.4.1 Livestock Grazing

As previously noted, grasslands evolved with herbivory and not just by large mammals, but also insect herbivory. Mammalian grazing, whether accomplished by domesticated livestock or wildlife affects the grassland habitat available for insects.

Insect communities can and do have variable responses to grazing and this variation in response also can be influenced by the ecoregion. For example, grassland insect communities in the western deserts of Arizona, which did not evolve with bison herbivory, were found to be sensitive to cattle grazing (Debano 2006). Coleoptera (beetles) had lower species richness, Diptera (flies) were less diverse, and Hymenoptera were less rich and diverse on livestock grazed sites but Hemiptera (true bugs) were more diverse on livestock grazed sites (Debano 2006). In contrast, a grazing study in the shortgrass prairie of central Montana with cattle (Goosey et al. 2019) found that ground-dwelling arthropods that served as bird food (Coleoptera, Lepidoptera, Hymenoptera and Orthoptera) were twice as prevalent in cattle grazed pastures as in ungrazed pastures. Meanwhile, pastures ungrazed by cattle had twice the activity-density (number of beetles that cross the perimeter of the trap opening in a given time (Kromp 1989)) of ground-dwelling arthropods, which was largely driven by increases in detritivores and predators (Goosey et al. 2019). In seeking generalities among livestock grazing studies, plant community, geographic location, as well as the stocking rates, season of use, and grazing regime can all affect insect responses. Similarly, species level results may be different compared to the findings of family level analyses.

In some cases, livestock grazing has been proposed as a tool to control insects, such as grasshopper populations in rangelands. O'Neill et al. (2010) found general support for the hypothesis that grazing could be used to reduce pest grasshopper densities, but there was variation in how specific grasshopper species responded to grazed versus ungrazed treatments of *Agropyron spicatum/Poa sandbergii* pastures in southwestern Montana shortgrass prairie depending upon site, year, and vegetation assemblage.

The intensity and duration of livestock grazing can have effects on adjacent aquatic communities. For example, grassland stream fish communities can be affected when vegetation structure adjacent to streams is modified by grazing so that fewer insects fall into the streams as food for fish. In a study of trout streams in the Wyoming Basin that compared two types of livestock grazing during the summer months (high-density, short-duration grazing versus season-long grazing; see Chap. 4 for grazing-system definitions), the input of terrestrial invertebrates to the riparian areas was two to three times greater in areas managed with high-density, short-duration grazing due to more overhanging vegetation than those managed with season-long grazing management (Saunders and Fausch 2007). Effects of changes in fish communities could impact bird and mammal communities that prey on these fish.

In addition to the intensity of grazing, the combination of grazing intensity and precipitation can create heterogeneity in vegetation structure. Newbold et al. (2014) found that insect responses to grazing in the shortgrass prairie of Colorado were more pronounced in a year when spring and summer rainfall was low, noting that "both exclusion from grazing and precipitation are presumably necessary to create pronounced differences in vegetation structure to which arthropod consumers then respond."

26.4.2 Fire

The effects of fire on both grasslands and insects are diverse, including direct and indirect effects, seasonality, frequency, and fire-grazing interactions. Because grasslands evolved in the context of fire, burning can, in some cases, be effectively used to manage invasive plant species (DiTomaso et al. 2006) and enhance flowering of forb species (Goldas et al. 2021) with potential to indirectly affect insects. Fire is not generally used to directly manage insects, but the effects of fire on insects are often studied along with using fire as a grassland vegetation management tool (Schlicht and Orwig 1998; Kral et al. 2017). A literature review (Kral et al. 2017) of insect responses to fire found that some orders tended to respond negatively (Araneae, Lepidoptera) or positively (Coleoptera, Orthoptera) to fire, but that responses were highly variable among taxa and that characteristics such as life stage, feeding guild and mobility of the insects are key in predicting responses.

The effects of fire in the context of insects can be divided into two categories: (1) the direct effects on the insects (i.e., incineration) and (2) the indirect effects that manifest themselves in the insect populations via the effects on the vegetation that insects use (Fig. 26.2) (Vogel et al. 2010). With respect to direct effects, for any species of insect that predominantly occurs above ground, fire has the potential to directly cause an immediate decrease in the insect population (Vogel et al. 2010). For this reason, there are more management recommendations about the frequency of management for fire as compared to other types of grassland management. Insects may eventually return to near pre-burn levels as vegetation recovers and soil litter stabilizes, but response time can vary. Swengel (2001) noted that many insects decline markedly immediately after fire, with the magnitude of reduction related to the degree of exposure to the flames and mobility of the insect populations. Species that live underground, such as ants, burrowing beetles, or insects overwintering underground may suffer fewer, if any, direct effects of a fire if the fire occurs when they are underground, but this field of inquiry has not been extensive, and the degree of fire effects may vary with fire intensity. Fire had no population or diversity level effects on underground arthropods in one tallgrass prairie study (Pairis et al. 2003), but the lack of impacts could have been due to post-burn recolonization.

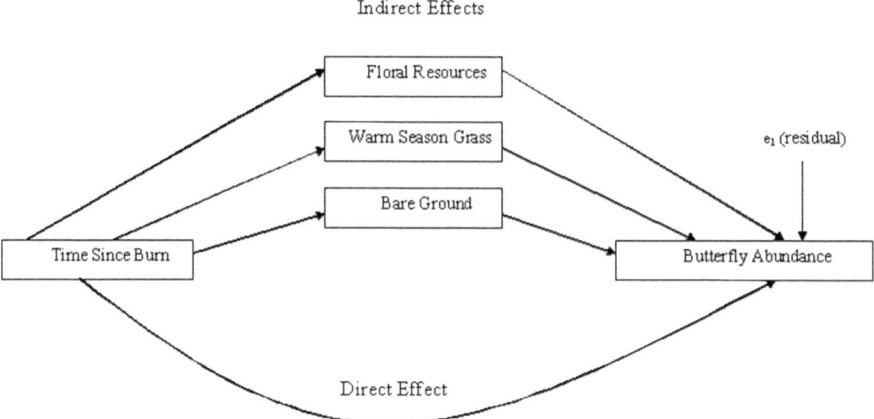

Fig. 26.2 A path diagram depicting direct and indirect relationships between time since burn, vegetation characteristics, and butterfly abundance for butterflies in Iowa grasslands. Independent variables chosen for inclusion in the path models were: floral resources (number of flowering ramets), warm-season grass (percent cover of warm-season grasses), and bare ground (percent cover of bare ground). Residual from the model is designated as e1. Reproduced from (Vogel et al. 2010)

Most fire studies focus largely on the response of insects to changes of vegetation post-fire, i.e., indirect effects. The most dramatic change to vegetation composition is the elimination of the above-ground foliage. Fire changes the composition of the vegetation dramatically by reducing above-ground biomass, removing woody vegetation, releasing nutrients, and increasing the amount of sunlight that reaches the soil (Radho-Toly et al. 2001). Fire also can stimulate the growth of fire-related annual forbs and grasses (Korb et al. 2004). For insect families associated with forbs and grasses, such as Acrididae (grasshoppers, locusts), Gryllidae (crickets), Tettigoniidae (katydids), Aphididae (aphids), Margarodidae (mealybugs), Chrysomelidae (leaf beetles), Carabidae (ground beetles), Cantharidae (soldier beetles), Coccinellidae (ladybird beetles), Asilidae (robber flies), Meloidae (blister beetles), Formicidae (ants) and Lepidoptera (butterflies, moths, cutworms, army worms, skippers), temporary increases in populations, may occur as a result of the increased forb productivity after a fire.

The seasonality of a fire also can have profound effects on the way that environmental change manifests itself in the insect community. Generally, late spring burning of grasslands is thought to reduce insect populations more than early spring burning (Higgins et al. 1987), but species-specific responses may vary. For example, with respect to grasshoppers, Knutson and Campbell (1976) found that early spring burning in Kansas grasslands caused grasshoppers to emerge three weeks earlier than normal, and grasshoppers were higher in number the second year following an early burn. Mid-spring burning produced fewer grasshoppers than early burning, and late spring burning produced fewer grasshoppers than mid- or early spring burning, potentially due to effects on a particular portion of the life-cycle.

The frequency of fire is also critical to the insect response. Frequent or even annual burning of grasslands is part of the culture in some North American grasslands, such as the Flint Hills of Kansas in the tallgrass prairie ecoregion. Welti and Joern (2018) found that fire frequency affected flowering plant and floral visitor community composition in the Flint Hills. In the same system Welti et al. (2019) found that changes in fire frequency affected plant and grasshopper community composition but did not have significant main effects on the plant–grasshopper network structure. The effect of repeated burning on soil arthropods in a Wisconsin tallgrass prairie was studied by Lussenhop (1976), where burning of re-established tallgrass prairie that had been burned biannually for decades was continued for two more burns on one area and discontinued on the other and as a control, a third area was raked to remove the litter. First-year results showed no significant difference in soil microarthropods, but by the fourth year the unburned areas had significantly fewer herbivorous and carnivorous insect species than the burned and control (raked) areas. Lussenhop (1976) concluded that the unburned area was less productive, causing a decrease in soil microarthropods.

Davies et al. (2014) found that in big sagebrush communities (*Artemisia tridentata* Nutt. ssp. vaseyana (Rydb.) Beetle) on the Columbia plateau of Oregon burning altered the arthropod community, which included a doubling of the density of arthropods in the first post-burn year. Some specific groups of arthropods increased, and others decreased with burning. Notably, Hemiptera were 6.6- and 2.1-fold greater during one- and two-years post-burn compared with the control. Changes in the insect community were associated with increases in plant diversity in burned sites in the first post-burn year, but that difference was gone by year two and burned plots actually had lower plant diversity by the third post-burn year (Davies et al. 2014). Changes as a result of fire can have mixed effects on wildlife, in some cases providing additional insect food resources to greater sage-grouse (*Centrocercus urophasianus*), sagebrush lizard (*Sceloporus graciosus*), northern horned lizard (*Phrynosoma platyrhinos platyrhinos*), and sage thrasher (*Oreoscoptes montanus*) (Davies et al. 2014).

Finally, the combination of fire and grazing management also can affect insect abundance and distribution patterns because this combination can create heterogeneity in vegetation structure across the landscape. For example, shorter and sparse vegetation may be found in recently burned patches and taller, denser vegetation in less recently burned patches. Patch-burn grazing, a way of managing a pasture to rotate the patch that is annually burned, has been extensively examined with respect to vegetation and bird responses to landscape heterogeneity (see Chap. 4), and in some cases with respect to insect responses (Debinski et al. 2011). The models of Fuhlendorf et al. (2009) that examine how bird species respond to patch-burn grazing management also can be applied to the insect community. For example, Moranz et al. (2012) found that butterflies in tallgrass prairies such as the regal fritillary, wood nymph (*Cercyonis pegala*), and monarch are more likely to be found in tall-stature grasses, which can be associated with longer intervals between burns, whereas the habitat generalist eastern-tailed blue (*Cupido comyntas*) is more likely to be found in short-statured grasses. In comparing ant species distribution on three combinations of grazed and burned pastures (patch-burn grazed, graze-and-burned, and burn-only),

Moranz et al. (2013) found that "opportunist" and "dominant" ant species in tallgrass prairies were more abundant in burn-only tracts than grasslands that also had grazing treatments. Generalist ant species were more abundant in graze-and-burn tracts than in burn-only tracts. Abundance of *Formica montana*, the dominant ant species, was negatively associated with time since fire, whereas generalist ant abundance was positively correlated with time since fire.

In summary, there are many plant–insect interactions within a grassland ecosystem that can be affected by the seasonality and frequency of fire as a form of grassland management. Responses of particular insect species depend upon their natural history, where they live in the grassland, how specialized they are on particular plants, and how they overwinter. Some insect species evolved to take advantage of recently burned landscapes while others evolved to advance in older, more mature, landscapes. As such, sweeping generalizations about insect species or insect taxonomic group responses to fire are often limited in accuracy.

26.4.3   Herbicide, Pesticide, and Mechanical Treatments

Many grasslands are threatened by the encroachment of woody species and nonnative grasses and forbs, and herbicides are often used to reduce or remove occurrences of these plant species. Chemicals used to control certain plants can also significantly affect the insect community, albeit indirectly. As an example, Taylor et al. (2006) found that weedy plots on a Montana experimental farm contained 12 times more biomass of common insects eaten by nestling birds compared to monoculture plots prior to spraying, but following spraying with bromoxynil and imazamethabenz herbicides, weedy plots contained only 3 times more biomass than monoculture plots. From a wildlife perspective, rangeland modification associated with herbicide application has the potential to have significant effects, particularly for birds, due to the associated changes in the vegetation structure and composition. There is not adequate space in this chapter to thoroughly cover herbicide treatment as a form of grassland management, but in the context of insect changes in the plant community due to herbicides, such chemicals will undoubtedly affect the insect community.

Insecticides are not as frequently used directly on grasslands, but there are some examples that relate to both insect and wildlife responses. In Wyoming, > 1.7 million ha were treated with toxaphene and chlordane for grasshopper control during 1949–50; during the two years afterwards, pesticide poisoning was suspected in 11 of 45 Greater Sage-Grouse mortalities (Rowland 2019). From a landscape context perspective, native grasslands are often adjacent to agricultural lands where insecticides are extensively used. As such, the topic is introduced here, but covered more substantially in Sect. 26.6.2.

Effects of mechanical treatments have not been as extensively examined in terms of insect responses except in the context of being a replacement for fire or grazing. For example, haying is, in some grassland studies, examined as the "third wheel" of management (see Chap. 5). More frequently, the comparison treatment is "ungrazed" or "idled" (not cut, burned, cropped, heavily grazed, cultivated, or otherwise disturbed). In sagebrush systems, mowing and "chaining" (where a large heavy chain is dragged over the ground to clear vegetation) have been used to reduce woody vegetation and increase grasses. There are some studies of chaining on the responses by birds (Castrale 1982), but there are no studies of the effect of chaining on insects in rangelands. Studies of the effects of mowing on bird responses provide some insight into these management effects on insects. For example, Hess and Beck (2014) found that mowing in the Wyoming Basin did not improve Greater Sage-Grouse nesting or early brood-rearing habitat attributes such as cover or nutritional quality of food forbs, or counts of ants, beetles, or grasshoppers. There are ample publications that evaluate responses of insects to grassland mechanical treatments at the fine-grained scale of tens of hectares (Debinski and Babbit 1997; Pairis et al. 2003) but excluding a small number of studies such as Stoner and Joern (2004), studying effects of mechanical treatments on insects at a landscape scale is an area ripe for additional research.

26.5   Insects and Disease

A complete investigation of diseases that affect insects is beyond the scope of this chapter. However, some of the major diseases in which insects serve as vectors of disease to domesticated animals and wildlife are described below.

26.5.1   Insects as Vectors of Disease

Entire books and sections of books have been written about the impacts of insects as vectors of disease on livestock and wildlife (Eldridge et al. 2000; Capinera and Capinera 2010; Botzler and Brown 2014). Suffice it to say that insects can cause a broad set of diseases in livestock and similar relationships apply to many wildlife species. To list a few prominent examples, the following insects are major issues for livestock: black flies (*Simuliidae*), biting midges (*Culicoides*), horn flies (*Haematobia irritans*), stable flies (*Stomoxys calcitrans*), horse flies (Tanabidae) and deer flies (*Chrysops*) as well as mosquitos (Culicidae), face flies, cattle grubs (*Hypoderma*), lice (*Anoplura* and *Mallophaga*), screwworms (*Cochliomyia hominivorax*), ticks (Ixodida), and mites (*Demodex bovis*) (Steelman 1976). These insects can have minor to major impacts on the health and productivity of livestock, affecting their skin, eyes, milk production, weight gain and weight loss, and serving as vectors of disease (Steelman 1976).

Wildlife populations also can be affected by insects as vectors. For example, heavy tick infestation can have significant impacts on moose reproduction and survival of young (Ellingwood et al. 2020). Biting midges (small blood-sucking flies in the family Ceratopogonidae) can transmit viruses, protozoans, and nematodes (Mullen and Murphree 2019). Bluetongue and epizootic hemorrhagic disease virus, which are spread by midges, result in infectious and sometimes fatal diseases of wild ungulates, particularly white-tailed deer (*Odocoileus virginianus*), and also may infect domestic ruminants such as sheep (*Ovis aries*) (Maclachlan et al. 2019). These diseases also may be spread horizontally (i.e., from one individual to another), without the need of a vector (Maclachlan et al. 2019). West Nile Virus, transmitted by mosquitos to birds, has had a particularly negative effect on birds in the crow family (Corvidae) (Kilpatrick et al. 2007). One of the areas of important future research with respect to these issues is how relationships of insects as disease vectors to domestic and wild animals are changing in the context of climate change and in relation to grassland condition.

26.6   Ecosystem Threats

26.6.1   Habitat Conversion

The global intensification of agriculture in recent decades has led to significant grassland conversion to cropland (see Chaps. 3, 5), with associated decreases of grassland bird (−20.8%) and shrubland bird declines (−16.5%) (Stanton et al. 2018). Insect populations also are responsive to changes in micro- and macrohabitats, including fragmentation, ecological disruption, and chemical pollution (Kremen et al. 1993) and all these issues can be associated with land conversion. Chen et al. (1995) documented changes in air temperature, soil temperature, relative humidity, short-wave radiation, and wind speed along the edge of a fragmented area. Fragmentation also may genetically isolate populations, making small populations more prone to extinction (Stacey and Taper 1992). As grasslands are modified, converted, and fragmented, all of these issues must be considered in the context of conservation of insects and wildlife.

26.6.2   Insecticides

The effect of insecticides on grassland insect populations may be significant, but effects are less well understood for insect species that are not pests. Chemicals used to control agricultural pest insects are not always specific in their effects and can affect non-target organisms both in terms of insects, but also with respect to earthworms (*Lumbricus* spp.), birds, and mammals (Sánchez-Bayo 2012). Even "biorational" insecticides (insecticides composed of natural products, including animals, plants, microbes, and minerals, or their derivates), which are considered less toxic than conventional chemical insecticides, can have unintended consequences on non-target insects (Haddi et al. 2020).

Insecticides also can have indirect effects on the wildlife that use insects as food. Swainson's hawks are particularly susceptible to insecticides because they forage for large numbers of insects during insect outbreaks. Their susceptibility was discovered first relative to the use of organochlorides, but they are also vulnerable to other pesticides (Shaffer et al. 2019). Similarly, carbaryl and carbofuran, two insecticides used to control agricultural pests in corn and alfalfa fields, were found to decrease reproductive success of Burrowing Owls (*Athene cunicularia hypugaea*) when sprayed in proximity (50–400 m) to their burrows (Shaffer et al. 2022).

26.6.3 Nonnative Species

Nonnative species, and particularly nonnative plants, may have large, yet undetected effects on grassland insects. Roadways and hiking trails are prime areas for the introduction of nonnative plant species, because they are often transported via humans or horses (Larson 2002; Graves and Shapiro 2003). Some examples of nonnative plant species in western grasslands include cheatgrass (*Bromus tectorum*), Dalmatian toadflax (*Linaria dalmatica*), spotted knapweed (*Centaurea maculosa*), Canada thistle (*Cirsium arvense*), ox-eye daisy (*Chrysanthemum leucanthemum*), houndstongue (*Cynoglossum officinale*), and leafy spurge (*Euphorbia esula*). These species have indirect effects on the insect community by changing the amount and relative proportions of plant biomass and soil nutrients available to insects (Ehrenfeld 2003). This may benefit some insect species by providing additional nectar, food, or host plants (Graves and Shapiro 2003), but others may be negatively affected because their preferred nectar, food, or host plant species are out-competed by nonnatives (Levine et al. 2003).

In addition to nonnative plants, native insects may be affected by nonnative insects that can compete for habitat and resources. As noted previously, *Apis melifera,* the honey bee, is a nonnative species that may be having impacts on native bees in grasslands. In a meta-analysis of invasive species effects on rare and endangered insects worldwide, Wagner and Van Driesche (2010) found that invasive plants, ants, and vertebrate grazers and predators posed major threats to native insect biodiversity. The ecological effects of nonnative insects have only recently been assessed at a broader taxonomic and geographic scale (Kenis et al. 2009; Garnas et al. 2016) and this area of research is ripe for additional work as it relates to grassland conservation.

26.6.4 Climate Change

Climate change models predict warmer temperatures in U.S. rangelands with reduced snowpack and drier conditions in the Northwest, drier conditions in the southern Great Plains and southwest, and wetter conditions in the Great Plains (Briske et al. 2005). A recent model of overall productivity in North American grasslands predicts both earlier spring emergence and delayed autumn senescence of vegetation with climate change, resulting in increased grassland productivity despite some drought in summer seasons (Hufkens et al. 2016). However, there are a variety of interconnected relationships within rangelands that need to be considered. For example, issues of wind and water erosion associated with climate change could have serious implications for both rangeland health and human health (Edwards et al. 2019). Increasing variability of precipitation, woody plant encroachment, heat stress, and threat of drought will influence the future of how rangelands are managed in the context of climate change (Holechek et al. 2020).

Many insect species may be especially responsive to the predicted temperature and precipitation alterations of climate change because of their specialized habitat requirements, potentially making them sensitive indicators for altered ecosystems. Climate change may affect insects that have more specialized host plants more severely than those that can use a broad number of host plants. Within an individual plant species, responses to changes in precipitation, temperature, and carbon dioxide also could affect a grassland's ability to support insect communities. Wenninger and Inouye (2008), in studying sagebrush steppe habitat in the eastern Snake River Plain, found that plants that are less water stressed harbor a greater diversity and abundance of insects. Insect diversity and abundance were positively correlated with both plant diversity and irrigation early in the summer, but by the end of the summer, insect distributions were more strongly influenced by irrigation treatment. In addition to the "filters" that the plant community type and plant physiological condition can have on insect distribution and abundance patterns, responses to climatic variation are also influenced by changes in predators and parasites in the system.

Climate change may be detrimental to some insect species that have narrow niches (i.e., they exist within a narrow set of environmental conditions) and advantageous to others that have broader niches (i.e., more generalist tendencies). Although some insects may respond quickly to climatic changes, their host plants and other members of the ecological community may not necessarily respond at the same rate. So, there is the potential for asynchrony in responses within the ecological community. Asynchrony can be detrimental, for example, if a plant blooms earlier but its pollinator comes out at the same time as in previous years (Maglianesi et al. 2020). The effects could then be manifested at higher levels of the food chain and affect grassland wildlife if plants produce fewer seeds for granivores or less biomass for herbivores. Given that plants and insects develop on a schedule of "degree days" (Sridhar and Reddy 2013), issues of asynchrony could potentially constitute a larger threat in communities where native vegetation has been replaced by introduced species.

Across all insect groups, butterflies have been most extensively studied in the context of climate change and may provide insights on the broader response of the insect community. Many butterfly species are associated with specific grassland types and, as such, butterfly communities have shown changes in distribution and abundance in the context of drought in Rocky Mountain grasslands (Debinski et al. 2013). Butterflies have been shown to exhibit rapid responses to climate change at a global level (Parmesan et al. 1999), and concern has recently been expressed about declines in butterfly species across the warming and drying western U.S. (Forister et al. 2021). Effects of a changing rangeland environment will be manifested in the productivity of the host plants and nectar resources used by many insects, which will impact insect survival and reproductive success. Whereas grasslands in the Rocky Mountains may be more susceptible to summer drought conditions due to reduced snowpack under climate change (Pederson et al. 2011, 2013), tallgrass prairie in Midwestern states may be more susceptible to spring flooding (Wuebbles and Hayhoe 2004). Both stressors will affect insect populations and communities, and this is another area where future research in needed.

26.7 Conservation and Management Actions

The conservation and management of insects in grasslands is still in its infancy compared to other wildlife species described within this volume. Due to high insect diversity and the lack of understanding of their ecosystem roles, only a small number of species have been well-studied in the context of management. Usually these are pest species, rare species, species of conservation concern, or threatened or endangered species, because designating a species in such a category justifies increased research funding for that species. Grasshoppers, due to their potential for creating large defoliation impacts in grasslands and crop fields, have been extensively studied because of apparent damage to crops. However, the benefits of insects are less easily observed. The American burying beetle and monarch butterfly are examples of species of concern that have been well-studied with respect to conservation and management. Bees, butterflies, ants, and, to a lesser degree, beetles, are some of the major groups of insects that have been evaluated for community responses to grassland management due to their importance in conservation education, pollination services, ecosystem engineering, and nutrient cycling respectively. The plethora of other insects are most often surveyed via sweep net and, due to the time-consuming nature of identifying them to species, are often summarized in terms of total biomass or abundance and diversity measure at the family rather than genus or species level, resulting in knowledge gaps in understanding of important relationships that may be occurring at the genus or species level.

Grassland restoration is one of the most effective methods of accomplishing conservation and management for grassland insects because high quality habitat is often a limiting resource for rare grassland insects. However, there are several important caveats to keep in mind when generalizing approaches to grassland restoration: (1) the effects of one type of management cannot be expected to result in the same vegetation response for all grasslands, (2) the effects of grazing, fire, and the combination thereof can vary based upon intensity, duration, timing, location, and year-to-year weather variation, (3) the historical management or use of a piece of land may have long-lasting legacy effects that influence the way that the land can and will respond to management, and (4) some grasslands are much easier to restore or reconstruct than others. Limitations of seed sources or precipitation make restoration and reconstruction of dry western grasslands much more difficult than in Midwestern grasslands. In understanding how insects respond to management, scientists and managers can be informed by experimental studies if sampling sites are well replicated and include experimental and control plots at a landscape scale studied over multiple years. Such experiments can provide guidance in developing an "adaptive management" approach so that prescriptions can be made for grasslands with specific locations, soil types, climate, and land management histories.

Just as grasslands can be managed for conservation, insect species also can be captively reared and reintroduced, and grasslands can be managed specifically for certain insect species. The regal fritillary butterfly was reintroduced to a restored tall-grass prairie in Iowa (Shepherd and Debinski 2005), and some insects are being reared by zoos in conservation efforts. Similarly, grassland habitats, both large and small, in many parts of the U.S. are being managed to improve habitat for monarch butterflies and pollinators. Public awareness has increased in recent decades regarding the value of creating pollinator habitat within grasslands, along roadways, on the edges of row crops and riparian areas, as well as within urban areas. However, conservation of habitat for insects has, to this point, primarily focused on pollinators, endangered species, and to a lesser extent biological control of pest insects. In the future, insect conservation could be expanded to focus on a broader suite of insect taxa, especially those conducive to engaging volunteers, school children, and/or citizen scientists.

26.8   Research and Management Needs

Future research linking insects, wildlife, and rangelands should focus on expanding our knowledge of how insects respond to both naturally occurring and human-induced ecological disturbances to grassland ecosystems including topics such as (1) the linkage between insect life history and responses to fire, and a better understanding of similarities and differences in responses between tallgrass prairie, shortgrass prairie, and Great Basin ecoregions, (2) the potential for differential responses of insects to climate change relative to their niche breadth, (3) the potential for asynchronous responses between insects and plants in the context of climate change, and (4) a broader understanding of the full taxonomic suite of insect population responses to chemicals used in grassland and agricultural management. Understanding the physiological responses, as well as the population, community, and ecosystem-level of responses of insects to fire, grazing, climate change, and chemical management, will allow future rangeland professionals to better predict and understand the complex roles that insects play in rangeland ecosystems and how changes in their distribution and abundance will affect associated wildlife species.

26.9 Summary

Insects have high reproductive rates and short life spans and can evolve and adapt quickly to abiotic and biotic changes in grasslands. Insects may be much smaller than their mammalian and avian counterparts at an individual level, but the total biomass they contribute to an ecosystem warrants notice, and the ecosystem services they provide are significant and critical. Insects are key intermediate components of the rangeland food web. Although the average grassland visitor may be more likely to stop and view the large mammals or birds in a grassland than the ants, beetles, bees, or grasshoppers, it is important to recognize that these "little creatures who run the word" (Wilson 1997) profoundly affect grassland wildlife populations, both directly and indirectly. As such, attention to insect conservation and the effects of management on insect distribution and abundance patterns is an essential component of holistic rangeland management.

Acknowledgements Thanks to H. Goosey, J. Shaffer, and M. Vaughan for comments that improved and broadened the topics covered in this chapter.

References

Agnarsson I, Kuntner M (2007) Taxonomy in a changing world: seeking solutions for a science in crisis. Syst Biol 56:531–539

Atkinson TC, Briffa KR, Coope G (1987) Seasonal temperatures in Britain during the past 22,000 years, reconstructed using beetle remains. Nature 325:587–592

Auckland JN, Debinski DM, Clark WR (2004) Survival, movement, and resource use of the butterfly *Parnassius clodius*. Ecol Entomol 29:139–149

Baum KA, Wallen KE (2011) Potential bias in pan trapping as a function of floral abundance. J Kansas Entomol Soc 84:155–159

Belovsky G, Slade J (2000) Insect herbivory accelerates nutrient cycling and increases plant production. Proc Natl Acad Sci 97:14412–14417

Belsky J, Joshi NK (2018) Assessing role of major drivers in recent decline of monarch butterfly population in North America. Front Environ Sci 6

Borchers DL, Buckland ST, Zucchini W (2004) Estimating animal abundance: closed populations. Springer London Limited, London. https://doi.org/10.1007/978-1-4471-3708-5

Botzler RG, Brown RN (2014) Foundations of wildlife diseases, 1st edn. University of California Press, Berkeley. https://doi.org/10.1525/j.ctt7zw054

Briske DD, Fuhlendorf SD, Smeins FE (2005) State-and-transition models, thresholds, and rangeland health: a synthesis of ecological concepts and perspectives. Rangel Ecol Manage 58:1–10

Brittain C, Williams N, Kremen C et al (2013) Synergistic effects of non-Apis bees and honey bees for pollination services. Proc Royal Soc B Biol Sci 280:20122767

Buckland ST, Rexstad EA, Marques TA et al (2015) Distance sampling. Springer International Publishing AG, Cham

Butler JL, Parker MS, Murphy JT (2006) Efficacy of flea beetle control of Leafy Spurge in Montana and South Dakota. Rangel Ecol Manage 59:453–461

Cameron E (1913) Notes on Swainson's Hawk (*Buteo swainsoni*) in Montana. Auk 30:381–394

Capinera J, Capinera DDJ (2010) Insects and wildlife: arthropods and their relationships with wild vertebrate animals, 1st edn. Wiley, Hoboken

Castrale JS (1982) Effects of two sagebrush control methods on nongame birds. J Wildlife Manage 945–952

Chen J, Franklin JF, Spies TA (1995) Growing-season microclimatic gradients from clearcut edges into old-growth Douglas-fir forests. Ecol Appl 5:74–86

Cook WM, Holt RD, Yao J (2001) Spatial variability in oviposition damage by periodical cicadas in a fragmented landscape. Oecologia 127:51–61

Davies KW, Bates JD, Boyd CS et al (2014) Is fire exclusion in mountain big sagebrush communities prudent? Soil nutrient, plant diversity and arthropod response to burning. Int J Wildland Fire 23:417–424

De Bruyn L, Thys S, Scheirs J et al (2001) Effects of vegetation and soil on species diversity of soil dwelling Diptera in a heathland ecosystem. J Insect Conserv 5:87–97

De Jong GD, Chadwick JW (1999) Decomposition and arthropod succession on exposed rabbit carrion during summer at high altitudes in Colorado, USA. J Med Entomol 36:833–845

Debano SJ (2006) Effects of livestock grazing on aboveground insect communities in semi-arid grasslands of southeastern Arizona. Biodivers Conserv 15:2547

Debinski DM, Babbit AM (1997) Butterfly species in native prairie and restored prairie. Prairie Naturalist 29:219–228

Debinski DM, Moranz RA, Delaney JT et al (2011) A cross-taxonomic comparison of insect responses to grassland management and land-use legacies. Ecosphere 2:art131

Debinski DM, Caruthers JC, Cook D et al (2013) Gradient-based habitat affinities predict species vulnerability to drought. Ecology 94:1036–1045

Delaney JT, Jokela KJ, Debinski DM (2015) Seasonal succession of pollinator floral resources in four types of grasslands. Ecosphere 6:1–14

Devillers J (2002) The ecological importance of honey bees and their relevance to ecotoxicology. Honeybees: estimating the environmental impact of chemicals. Taylor & Francis, London, pp 1–11

DiTomaso JM, Brooks ML, Allen EB et al (2006) Control of invasive weeds with prescribed burning. Weed Technol 20:535–548

Edwards BL, Webb NP, Brown DP et al (2019) Climate change impacts on wind and water erosion on US rangelands. J Soil Water Conserv 74:405–418

Ehrenfeld JG (2003) Effects of exotic plant invasions on soil nutrient cycling processes. Ecosystems 6:503–523

Ehrlich PR, Dobkin DS, Wheye D (1988) The birder's handbook. Simon and Schuster, New York

Eldridge BF, Edman JD, Edman JD (2000) Medical entomology: a textbook on public health and veterinary problems caused by arthropods. Springer, The Netherlands, Dordrecht. https://doi.org/10.1007/978-94-011-6472-6

Ellingwood DD, Pekins PJ, Jones H et al (2020) Evaluating moose *Alces alces* population response to infestation level of winter ticks *Dermacentor albipictus*. Wildlife Biol

Erhardt A, Thomas J (1991) Lepidoptera as indicators of change in the semi-natural grasslands of lowland and upland Europe. Conserv Insects Habitats 112:213–236

Forister M, Halsch C, Nice C et al (2021) Fewer butterflies seen by community scientists across the warming and drying landscapes of the American West. Science 371:1042–1045

Fuhlendorf SD, Engle DM, Kerby J et al (2009) Pyric herbivory: rewilding landscapes through the recoupling of fire and grazing. Conserv Biol 23:588–598

Garnas JR, Auger-Rozenberg M-A, Roques A et al (2016) Complex patterns of global spread in invasive insects: eco-evolutionary and management consequences. Biol Invasions 18:935–952

Goldas CdS, Podgaiski LR, da Silva CVC et al (2021) Burning for grassland pollination: recently burned patches promote plant flowering and insect pollinators. Austral Ecol n/a

Goosey HB, Smith JT, O'Neill KM et al (2019) Ground-dwelling arthropod community response to livestock grazing: implications for avian conservation. Environ Entomol 48:856–866

Goulson D (2019) The insect apocalypse, and why it matters. Curr Biol 29:R967–R971

Graves SD, Shapiro AM (2003) Exotics as host plants of the California butterfly fauna. Biol Cons 110:413–433

Haddi K, Turchen LM, Viteri Jumbo LO et al (2020) Rethinking biorational insecticides for pest management: unintended effects and consequences. Pest Manag Sci 76:2286–2293

Hallmann CA, Sorg M, Jongejans E et al (2017) More than 75 percent decline over 27 years in total flying insect biomass in protected areas. PLoS ONE 12:e0185809

Hendrix SD, Kyhl FJ (2000) Population size and reproduction in *Phlox pilosa*. Conserv Biol 14:304–313

Hess JE, Beck JL (2014) Forb, insect, and soil response to burning and mowing Wyoming big sagebrush in greater sage-grouse breeding habitat. Environ Manage 53:813–822

Higgins KF, Kruse; AD, Piehl. JL (1987) Effects of fire in the Northern Great Plains. South Dakota State University Extension Circulars

Holechek JL, Geli HME, Cibils AF et al (2020) Climate change, rangelands, and sustainability of ranching in the Western United States. Sustainability 12:4942

Hufkens K, Keenan TF, Flanagan LB et al (2016) Productivity of North American grasslands is increased under future climate scenarios despite rising aridity. Nat Clim Chang 6:710–714

IUCN (2022, 07/21/2022) Migratory monarch butterfly now endangered—IUCN red list. https://www.iucn.org/press-release/202207/migratory-monarch-butterfly-now-endangered-iucn-red-list

Kaspari M, Joern A (1993) Prey choice by three insectivorous grassland birds: reevaluating opportunism. Oikos 414–430

Kelly L, Debinski DM (1998) Relationship of host plant density to size and abundance of the regal fritillary *Speyeria idalia* drury (Nymphalidae). J Lepidopterists Soc 52:262–276

Kemp WP (1992) Temporal variation in rangeland grasshopper (Orthoptera: Acrididae) communities in the steppe region of Montana, USA. Can Entomol 124:437–450

Kenis M, Auger-Rozenberg M-A, Roques A et al (2009) Ecological effects of invasive alien insects. Biol Invasions 11:21–45

Kilpatrick AM, LaDeau SL, Marra PP (2007) Ecology of West Nile virus transmission and its impact on birds in the western hemisphere. Auk 124:1121–1136

Knutson H, Campbell JB (1976) Relationships of grasshoppers (Acrididae) to burning, grazing, and range sites of native tallgrass prairie in Kansas. In: Tall timbers conference on ecological animal control habitat management proceedings, Gainesville, FL, 1974. Tall Timbers Research Station, Tallahassee FL pp 107–120, BIBL 1P1/2

Korb JE, Johnson NC, Covington WW (2004) Slash pile burning effects on soil biotic and chemical properties and plant establishment: recommendations for amelioration. Restor Ecol 12:52–62

Kral KC, Limb RF, Harmon JP et al (2017) Arthropods and fire: previous research shaping future conservation. Rangel Ecol Manage 70:589–598

Kremen C, Colwell R, Erwin T et al (1993) Terrestrial arthropod assemblages: their use in conservation planning. Conserv Biol 796–808

Kremen C, Williams NM, Aizen MA et al (2007) Pollination and other ecosystem services produced by mobile organisms: a conceptual framework for the effects of land-use change. Ecol Lett 10:299–314

Kromp B (1989) Carabid beetle communities (Carabidae, coleoptera) in biologically and conventionally farmed agroecosystems. Agr Ecosyst Environ 27:241–251

Lambe CM (2018) What's all the buzz about: analyzing the decision to list the Rusty Patched Bumblebee on the endangered species list. Vill Envtl LJ 29:129

Larson DL (2002) Native weeds and exotic plants: relationships to disturbance in mixed-grass prairie. Plant Ecol 169:317–333

Larson B, Kevan P, Inouye DW (2001) Flies and flowers: taxonomic diversity of anthophiles and pollinators. Can Entomol 133:439–465

LaSalle J, Gauld I (1993) Hymenoptera: their biodiversity, and their impact on the diversity of other organisms. Hymenoptera Biodiversity 1–26

Lee K, Wood T (1971) Physical and chemical effects on soils of some Australian termites, and their pedological significance. Pedobiologia

Levine JM, Vila M, Antonio CMD et al (2003) Mechanisms underlying the impacts of exotic plant invasions. Proc R Soc Lond B 270:775–781

Losey JE, Vaughan M (2006) The economic value of ecological services provided by insects. Bioscience 56:311–323

Lussenhop J (1976) Soil arthropod response to prairie burning. Ecology 57:88–98

Maclachlan NJ, Zientara S, Wilson WC et al (2019) Bluetongue and epizootic hemorrhagic disease viruses: recent developments with these globally re-emerging arboviral infections of ruminants. Curr Opin Virol 34:56–62

Maglianesi MA, Hanson P, Brenes E et al (2020) High levels of phenological asynchrony between specialized pollinators and plants with short flowering phases. Ecology 101:e03162

Magurran AE (1988) Ecological diversity and its measurement. Princeton University Press, Princeton, NJ

Malmqvist B (2002) Aquatic invertebrates in riverine landscapes. Freshw Biol 47:679–694

Matthews EG, Kitching RL (1984). Insect ecology, 2nd ed

McGranahan DA, Engle DM, Fuhlendorf SD et al (2012) An invasive cool-season grass complicates prescribed fire management in a native warm-season grassland. Nat Areas J 32:208–214

McKnight SJ (2021) Resident monarch populations on the rise in California: what does this mean for the western migratory population? https://www.xerces.org/blog/resident-monarch-populations-on-rise-in-california-what-does-this-mean-for-western-migratory. Accessed 1 May 2022

Miller JR, Morton LW, Engle DM et al (2012) Nature reserves as catalysts for landscape change. Front Ecol Environ 10:144–152

Mills LS (2012) Conservation of wildlife populations: demography, genetics, and management, 2, Aufl. edn. Wiley-Blackwell, Hoboken

Mittelbach GG (2018) Community ecology. Sinauer Associates, Sunderland, Massachusetts

Mooring MS, Samuel W (1998) The biological basis of grooming in moose: programmed versus stimulus-driven grooming. Anim Behav 56:1561–1570

Moranz RA, Debinski DM, McGranahan DA et al (2012) Untangling the effects of fire, grazing, and land-use legacies on grassland butterfly communities. Biodivers Conserv 21:2719–2746

Moranz RA, Debinski DM, Winkler L et al (2013) Effects of grassland management practices on ant functional groups in central North America. J Insect Conserv 17:699–713

Mullen GR, Murphree CS (2019a) Biting midges (Ceratopogonidae). In: Medical and veterinary entomology. Elsevier, pp 213–236

Newbold TS, Stapp P, Levensailor KE et al (2014) Community responses of arthropods to a range of traditional and manipulated grazing in shortgrass steppe. Environ Entomol 43:556–568

Nichols E, Spector S, Louzada J et al (2008) Ecological functions and ecosystem services provided by Scarabaeinae dung beetles. Biol Cons 141:1461–1474

O'Neill KM, Olson BE, Wallander R et al (2010) Effects of livestock grazing on grasshopper abundance on a native rangeland in Montana. Environ Entomol 39:775–786

Ollerton J, Winfree R, Tarrant S (2011) How many flowering plants are pollinated by animals? Oikos 120:321–326

Pairis M, Sundermann J, Wang H (2003) Fire, mowing and soil moisture levels have no significant effects on underground arthropod population and diversity. Tillers 4:33–37

Parmesan C, Ryrholm N, Stefanescu C et al (1999) Poleward shifts in geographical ranges of butterfly species associated with regional warming. Nature 399:579–583

Paterson J, Forcherio C, Larson B et al (1995) The effects of fescue toxicosis on beef cattle productivity. J Anim Sci 73:889–898

Pederson GT, Gray ST, Woodhouse CA et al (2011) The unusual nature of recent snowpack declines in the North American cordillera. Science 333:332–335

Pederson GT, Betancourt JL, McCabe GJ (2013) Regional patterns and proximal causes of the recent snowpack decline in the Rocky Mountains, U.S. Geophys Res Lett 40:1811–1816

Pelton EM, Schultz CB, Jepsen SJ et al (2019) Western Monarch population plummets: status, probable causes, and recommended conservation actions. Front Ecol Evol 7

Petschenka G, Agrawal AA (2015) Milkweed butterfly resistance to plant toxins is linked to sequestration, not coping with a toxic diet. Proc Royal Soc B Biol Sci 282:20151865

Porrini C, Ghini S, Girotti S et al (2002) 11 Use of honey bees as bioindicators of environmental pollution in Italy. In: Honey bees: estimating the environmental impact of chemicals, p 186

Pritchard ZA, Hendriksma HP, St Clair AL et al (2021) Do viruses from managed honey bees (Hymenoptera: Apidae) endanger wild bees in native prairies? Environ Entomol

Radho-Toly S, Majer JD, Yates C (2001) Impact of fire on leaf nutrients, arthropod fauna and herbivory of native and exotic eucalypts in Kings Park, Perth, Western Australia. Austral Ecol 26:500–506

Ries L, Debinski DM, Wieland ML (2001) Conservation value of roadside prairie restoration to butterfly communities. Conserv Biol 15:401–411

Rowland MM (2019) The effects of management practices on grassland birds—greater sage-grouse (*Centrocercus urophasianus*). In: Johnson DH, Igl LD, Shaffer JA, DeLong JP (eds) The effects of management practices on grassland birds, chap. B. U.S. Geological Survey Professional Paper 1842. US Geological Survey. https://doi.org/10.3133/pp1842B

Rudgers JA, Clay K (2007) Endophyte symbiosis with tall fescue: how strong are the impacts on communities and ecosystems? Fungal Biol Rev 21:107–124

Sánchez-Bayo F (2012) Insecticides mode of action in relation to their toxicity to non-target organisms. J Environ Anal Toxicol 4:S4-002

Saunders WC, Fausch KD (2007) Improved grazing management increases terrestrial invertebrate inputs that feed trout in Wyoming rangeland streams. Trans Am Fish Soc 136:1216–1230

Schlicht DW, Orwig TT (1998) The status of Iowa's Lepidoptera. J Iowa Acad Sci 105:82–88

Schmitz OJ (2003) Top predator control of plant biodiversity and productivity in an old-field ecosystem. Ecol Lett 6:156–163

Shaffer JA, Igl LD, Johnson DH, Sondreal ML, Goldade CM, Rabie PA, Thiele JP, Euliss BR (2022) The effects of management practices on grassland birds—Burrowing Owl (*Athene cunicularia hypugaea*). US Geological Survey. https://pubs.er.usgs.gov/publication/pp1842P

Shaffer JA, Igl LD, Johnson DH et al (2019) The effects of management practices on grassland birds—Swainson's Hawk (*Buteo swainsoni*). In: The effects of management practices on grassland birds, US Geological Survey, vol Professional Paper 1842. US Geological Survey. https://doi.org/10.3133/pp1842M

Shepherd S, Debinski DM (2005) Reintroduction of Regal Fritillary (*Speyeria idalia*) to a restored prairie. Ecol Restoration 23

Sikes DS (1994) Influences of ungulate carcasses on coleopteran communities in Yellowstone National Park. Montana State University-Bozeman, College of Agriculture, USA

Sikes DS, Raithel CJ (2002) A review of hypotheses of decline of the endangered American burying beetle (Silphidae: *Nicrophorus americanus* Olivier). J Insect Conserv 6:103–113

Spafford RD, Lortie CJ (2013) Sweeping beauty: is grassland arthropod community composition effectively estimated by sweep netting? Ecol Evol 3:3347–3358

Sridhar V, Reddy PVR (2013) Use of degree days and plant phenology: a reliable tool for predicting insect pest activity under climate change conditions. In: Climate-resilient horticulture: adaptation and mitigation strategies. Springer, pp 287–294

Stacey PB, Taper M (1992) Environmental variation and the persistence of small populations. Ecol Appl 2:18–29

Stanton RL, Morrissey CA, Clark RG (2018) Analysis of trends and agricultural drivers of farmland bird declines in North America: a review. Agr Ecosyst Environ 254:244–254

Steelman CD (1976) Effects of external and internal arthropod parasites on domestic livestock production. Annu Rev Entomol 21:155–178

Stoner KJ, Joern A (2004) Landscape vs. local habitat scale influences to insect communities from tallgrass prairie remnants. Ecol Appl 14:1306–1320

Swengel AB (2001) A literature review of insect responses to fire, compared to other conservation managements of open habitat. Biodivers Conserv 10:1141–1169

Taylor RL, Maxwell BD, Boik RJ (2006) Indirect effects of herbicides on bird food resources and beneficial arthropods. Agr Ecosyst Environ 116:157–164

Turchin P, Taylor A, Reeve J (1999) Dynamical role of predators in population cycles of a forest insect: an experimental test. Science 285:1068–1071

Turin H, Den Boer P (1988) Changes in the distribution of carabid beetles in The Netherlands since 1880. II. Isolation of habitats and long-term time trends in the occurence of carabid species with different powers of dispersal (Coleoptera, Carabidae). Biol Cons 44:179–200

USDA (2022) Insects and pollinators. U.S. Department of Agriculture. Available via https://www.nrcs.usda.gov/wps/portal/nrcs/main/national/plantsanimals/pollinate/. Accessed 1 May 2022

USFWS (2020) U.S. fish and wildlife service finds endangered species act listing for monarch butterfly warranted but precluded. Available via https://www.fws.gov/press-release/2020-12/endangered-species-act-listing-monarch-butterfly-warranted-precluded. Accessed 1 May 2022

USFWS (2022) Environmental conservation online system. U.S. Fish and Wildlife Service. https://ecos.fws.gov/ecp/. Accessed 1 May 2022

Vandenbosch R (2003) Fluctuations of *Vanessa cardui* butterfly abundance with El Nino and Pacific Decadal oscillation climatic variables. Glob Change Biol 9:785–790

Vogel JA, Koford RR, Debinski DM (2010) Direct and indirect responses of tallgrass prairie butterflies to prescribed burning. J Insect Conserv 14:663–677

Wagner DL, Van Driesche RG (2010) Threats posed to rare or endangered insects by invasions of nonnative species. Annu Rev Entomol 55:547–568

Waldbauer G (2003) What good are bugs? Insects in the web of life. Harvard University Press

Welti EA, Joern A (2018) Fire and grazing modulate the structure and resistance of plant–floral visitor networks in a tallgrass prairie. Oecologia 186:517–528

Welti EA, Qiu F, Tetreault HM et al (2019) Fire, grazing and climate shape plant–grasshopper interactions in a tallgrass prairie. Funct Ecol 33:735–745

Wenninger EJ, Inouye RS (2008) Insect community response to plant diversity and productivity in a sagebrush–steppe ecosystem. J Arid Environ 72:24–33

Wilsey B (2018) The biology of grasslands. Oxford University Press, Oxford, UK

Wilsey BJ, Chalcraft DR, Bowles CM et al (2005) Relationships among indices suggest that richness is an incomplete surrogate for grassland biodiversity. Ecology 86:1178–1184

Wilson EO (1997) Little creatures who run the word. NOVA PBS Airdate, 12 Aug 1997

Wuebbles DJ, Hayhoe K (2004) Climate change projections for the United States Midwest. Mitig Adapt Strat Global Chang 9:335–363

Xerces (2022) Nebraska bumble bee atlas. https://www.nebraskabumblebeeatlas.org/. Accessed 1 May 2022

Zaller J, Kerschbaumer G, Rizzoli R et al (2015) Monitoring arthropods in protected grasslands: comparing pitfall trapping, quadrat sampling and video monitoring. Web Ecol 15:15–23

Zattara EE, Aizen MA (2020) Worldwide occurrence records reflect a global decline in bee species richness. Available at SSRN 3669390

Zoladeski CA, Maycock PF (1990) Dynamics of the boreal forest in northwestern Ontario. American Midland Naturalist 289–300

Part III
Social–Ecological Considerations

Chapter 27
Wildlife, Rural Communities, and the Rangeland Livelihoods They Share: Opportunities in a Diverse Economies Approach

Julia Hobson Haggerty, Kathleen Epstein, Drew E. Bennett, Bill Milton, Laura Nowlin, and Brian Martin

Abstract Because rangeland ecosystems and the wildlife they support are integral to rural economies, understanding economic trends in rangeland regions is a valuable contribution to wildlife management. This chapter reflects on and synthesizes the experiences of a group of academic and practitioner collaborators working to balance the needs of wildlife and rural ranching communities in a priority conservation region, the central Montana portion of the Northern Great Plains. The chapter summarizes both the challenges facing ranching economies and policy and market strategies available to encourage conservation by private landowners. Its main emphasis, however, is to invite readers into a different kind of conversation about wildlife conservation's role in rangeland economies and livelihoods. The chapter introduces the concept of diverse economies, a way of understanding the economy through social relationships as opposed to merely the exchange of money, with a brief summary of its origins and perspective. It then draws on the theory and practice of diverse economies to

J. H. Haggerty (✉)
Department of Earth Sciences, Montana State University, PO Box 173480, Bozeman, MT 59717, USA
e-mail: julia.haggerty@montana.edu

K. Epstein
Department of Natural Resources and the Environment, Cornell University, 200 Rice Hall, Ithaca, NY 14853, USA

D. E. Bennett
Haub School of Natural Resources, University of Wyoming, 804 E. Fremont St, Laramie, WY 82072, USA

B. Milton
Rancher and ACES Member, Roundup, MT 59072, USA

L. Nowlin
Rancher and ACES Member, Winnett, MT 59087, USA

B. Martin
The Nature Conservancy, 520 E. Babcock St, Bozeman, MT 59715, USA

© The Author(s) 2023   933
L. B. McNew et al. (eds.), *Rangeland Wildlife Ecology and Conservation*,
https://doi.org/10.1007/978-3-031-34037-6_27

map relationships and activities at the intersection of rangeland conservation and community development in central Montana. In emphasizing the diversity of practices that make up "the economy" and the intimate intertwining of the economy with ecologies, diverse economies thinking opens up space to approach the complex ways that the livelihoods of rural residents and rangeland wildlife overlap and the search for adaptive solutions to conservation challenges.

Keywords Collaboration · Grassbanks · Landscape-scale conservation · Northern Great Plains · Ranching

27.1 Introduction

One of the distinguishing features of rangeland ecosystems is their long-standing integration into pastoral economies. In the contemporary rangelands of North America, wildlife shares the landscape with ranching communities, where native rangeland ecosystems support extensive pastoral livestock grazing—largely, but by no means exclusively, by family-scale operations (ranches) (U.S. Dept of Agriculture, Census of Agriculture 2021). Livestock ranching in turn operates as a cornerstone of many rangeland economies and communities. As Chap. 8 suggests, working lands overlap with critical biodiversity habitat. This common geography makes it imperative to identify synergies between ranching and stewarding wildlife, although the subject of whether and how private rangeland management supports or detracts from wildlife conservation goals is an evolving body of research. That said, where rangeland ecosystems are under threat from exurban and residential development or crop conversion pressures, sustaining thriving family ranch operations is currently considered to be an important strategy for preserving intact rangeland wildlife habitat and opportunities for wildlife (Gage et al. 2016; Olimb and Lendrum 2021).

However, sustaining thriving family ranches is no small task. The challenges facing ranching are numerous and diverse, and how ranchers manage these challenges directly influences opportunities to conserve wildlife on private ranchlands (Brunson and Huntsinger 2008; Roche 2021). The uncertainty of weather and markets, high costs of production–especially land–relative to commodity prices, and greater attention to environmental performance by powerful interests, including consumers and regulators, are all concerns voiced by ranchers about the sustainability of ranching (Haggerty et al. 2018a). Ultimately, these challenges point to questions about the viability of ranching livelihoods under changing market, climate, and political conditions.

The goal of this chapter is not just to emphasize the challenges facing ranching to rangeland ecologists, nor is it to summarize policy and market strategies available to encourage conservation by private landowners (although we do both of these things briefly). Rather, by sharing a case study of a region in Montana that exemplifies many of the opportunities and challenges at the nexus of local economies and wildlife (Epstein et al. 2021b), we hope to invite readers into a different kind of conversation about wildlife conservation's role in rangeland economies and livelihoods. This chapter is organized around "diverse economies" thinking (Gibson-Graham and Dombroski 2020), a holistic approach to analyzing the economy as part of a broader set of sustainability challenges and social-ecological dynamics. The goal of the diverse economies school is to broaden conventional discussions of economic activity that focus on things with dollar values to include the diversity of social exchanges and relationships that individuals and groups of people engage in to survive and thrive. It does so by providing a new approach to describing and talking about what makes up an economy. Given the scope of the challenges facing both rangeland ecosystems and rangeland economies, this kind of creativity and experimentation are clearly essential.

Fittingly, we bring a diverse set of perspectives to this chapter. Three of the authors are academics, two are full-time ranchers (who, like almost all ranchers, wear a number of other professional hats as well), and one is a conservation professional. This chapter integrates the perspectives of contributing co-authors through our synthetic discussion and analysis of the regional case study and direct dialogue in the form of quotations and reflections. All of the authors are focused on collaborative solutions to the community development and environmental challenges facing rangeland regions, including in central Montana.

In the next section, we briefly introduce the idea of social-ecological systems, an approach that supports the view that wildlife conservation and ranching economies are linked and interconnected. This lens frames Part 2 of this chapter in which we present our case study region, central Montana, as a place rich in ecological and wildlife value and where range-based livelihoods and economies demonstrate considerable stress and resilience. To that end, part 3 reviews challenges to range-based livelihoods through descriptive statistics and a summary of recent research. Part 4 turns to interventions focused on encouraging conservation activity by private landowners, which we situate relative to the scope of economic and ecological challenges identified in our summary of issues in central Montana. Finally, Part 5 explores developments in central MT from the diverse economies perspectives. The chapter concludes with ideas about how wildlife experts can bring diverse economies thinking into the work of wildlife conservation in rangeland communities.

27.1.1 *Rangelands as Social* and *Ecological Systems*

By addressing questions of wildlife management and conservation through the lens of local ranching economies, this chapter will ask readers to think about rangelands in what are likely new and different ways. For example, we expect that readers of this book will be familiar with ideas related to rangeland ecology because of its importance to wildlife ecology, management, and conservation (Chaps. 1–4). Indeed, questions related to what plant communities are present where, the quality and quantity of water sources, or how nutrients cycle through the soil are expected topics in professional training for wildlife management and conservation. However, rangelands also have social dimensions that influence wildlife and wildlife management (Brunson et al. 2016). Humans and human systems greatly influence biotic and abiotic processes, including wildlife dynamics, on rangelands through land use, management, and policy. Simultaneously, rangeland ecosystems are important to humans as a source of cultural and economic value. Thus, it is important for rangeland scientists and managers to understand rangelands as not just ecological systems, but as social *and* ecological systems (Hruska et al. 2017).

Researchers and managers seeking to adopt a more integrated approach to rangelands and wildlife management are increasingly turning to the concept of social and ecological systems, a way of seeing all the social and ecological dimensions of ecosystems as connected and also interdependent (Colding and Barthel 2019). In fact, much of the science community increasingly views "all humanly used resources as embedded in complex, social-ecological systems (SESs) (Ostrom 2009, p. 419)." Approaching rangelands as an SES allows researchers to ask questions that transcend disciplinary silos by providing a conceptual framework for investigating the various social, political, economic, and ecological aspects of rangelands and rangeland management as part of an integrated system (Hruska et al. 2017; Ojima et al. 2020). For example, how can collaborative processes for rangeland governance improve outcomes for species conservation (Duvall et al. 2017)? How does rancher decision-making influence ecosystem service delivery on rangelands (Roche et al. 2015)? Social-ecological systems analysis has also inspired studies central to the core questions of this chapter, such as how to sustain cattle ranching alongside endangered rangeland species (Charnley et al. 2018).

Interest in understanding SES dynamics across biomes, including rangelands, has grown within the field of environmental management as part of a broader interest in resilience, which addresses how systems or aspects of systems ability of a system (ecological, socio-economical, or social-ecological) or aspects of system to recover from disturbances and return to its pre-disturbed analysis (Folke et al. 2005; Gunderson and Holling 2002; Reid et al. 2014). In the case of this chapter, SES thinking frames our focus on the joint sustainability of rural economies and wildlife populations (Carpenter et al. 2001), or the resilience of different wildlife conservation initiatives to the modern challenges of rural ranching operations. While a comprehensive discussion of rangeland SES dynamics is outside of the scope of this chapter (see Bestelmeyer and Briske (2012), Hruska et al. (2017) for excellent summaries), appreciating the interconnectedness and interdependence of social and ecological dimensions of rangelands is a valuable starting point for our investigation into the economic dimensions of range-based economies and their intersection with wildlife management and conservation.

27.2   Central Montana as Wildlife and Ranching Case Study

One of the last remaining relatively intact temperate native grasslands, the northern mixed-grass prairie ecosystem of the Northern Great Plains are a priority conservation for wildlife scientists, managers, and advocates (Scholtz and Twidwell 2022). This chapter uses a sub-region of the Northern Great Plains in central Montana as a case study perspective on the opportunities and challenges that come with wildlife conservation in rural, ranching communities. Our profile includes six rural Montana counties, all of which surround a major public land complex, the Charles M. Russell (CMR) National Wildlife Refuge (Fig. 27.1). We call this six-million-hectare rangeland expanse the CMR region, acknowledging the refuge's defining presence in regional conservation issues. This portion of the Montana Northern Great Plains features a climate and geomorphology that supports a variety of rangeland vegetation: mixed-grass prairie dominates the glaciated plains in the northern portions of the region and a mix of shrublands and grasslands prevails farther south (Epstein et al. 1996; Rosenberg 1987). As a result, the CMR region is host to a plethora of wildlife including elk (*Cervus canadensis*), American bison (*Bison bison*), golden eagle (*Aquila chrysaetos*), greater sage-grouse (*Centrocercus urophasianus*), pronghorn (*Antilocapra americana*), black-tailed prairie dog (*Cynomys ludovicianus*), and numerous species of songbirds (Lipsey et al. 2015).

The CMR region's diverse wildlife populations and extensive intact habitats (for some species) have motivated interest in conservation initiatives from a variety of agencies and institutions. Public management involves both federal and state actors. The U.S. Fish and Wildlife Service (USFWS) has managed the CMR National Wildlife Refuge, the second largest refuge in the nation, since 1976. The refuge, along with adjacent federally-managed Missouri River Breaks National Monument, features land use explicitly oriented towards wildlife conservation. Other federal lands in the region include those managed by the Bureau of Land Management (BLM) and the U.S. Forest Service (USFS), both of which include wildlife habitat as part of their multiple-use mandates (Wilson 2014). Multiple federal agencies also support private land conservation through cost-share programs and technical assistance such as the U.S. Department of Agriculture's (USDA) Natural Resource Conservation Service (NRCS) and the USFWS Partners for Fish and Wildlife program. The state wildlife agency, the Montana Department of Fish, Wildlife and Parks, manages several Wildlife Management Areas in the region. In addition to state and federal agencies, stakeholder-led working groups and other citizen initiatives are also active in the region. The CMR Community Working Group is a collaborative entity established in 2011 to help facilitate dialogue among the area's many public and private conservation stakeholders.

The area's wildlife values have not gone unnoticed by non-governmental organizations and private actors. A series of ecoregional planning efforts in the 1990s and early 2000s directed attention to the wildlife conservation value and potential of central Montana and the lands surrounding the CMR National Wildlife Refuge

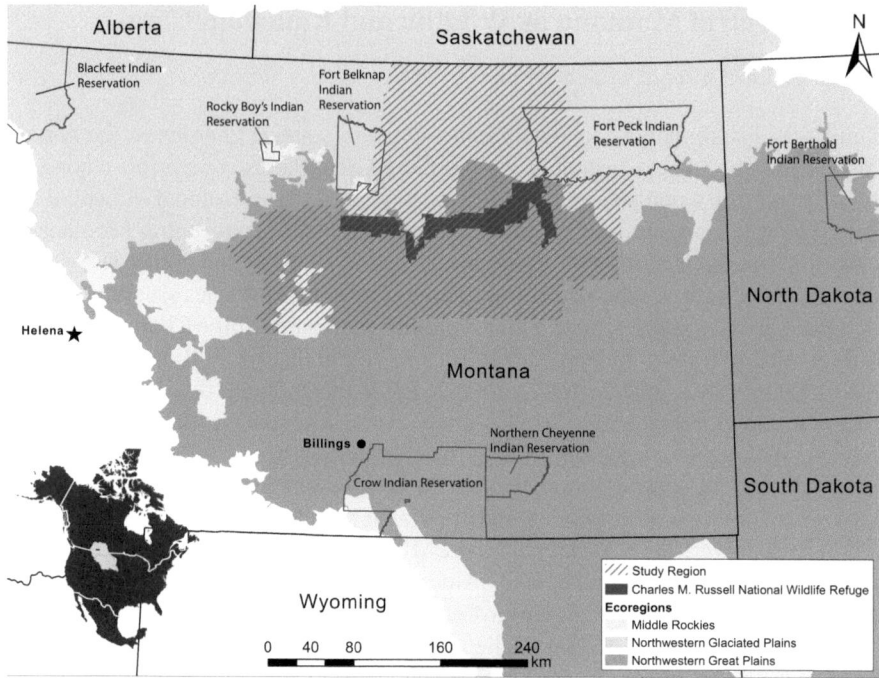

Fig. 27.1 Map of CMR region. *Sources* Montana State Library Clearinghouse, U.S. Geological Survey, U.S. Census Bureau, U.S. Environmental Protection Agency. *Note* Public lands managed as part of the Charles M. Russell (CMR) National Wildlife Refuge include the UL Bend National Wildlife Refuge, which is separately designated but managed by CMR NWR. (Reproduced from Epstein et al. 2021a, b)

(Epstein et al. 2021b). Conservation planners, who referred to the area as "The Big Open" drew inspiration not only from the extent of intact rangeland systems, but also to the limited amount of human influence—declining populations were seen as an opportunity.

A prominent non-profit organization in the area is the American Prairie, formerly American Prairie Reserve (and abbreviated subsequently by its widely-used acronym, APR), that has been purchasing ranch properties with the goal of creating a prairie reserve that extends over 4.1 million acres. Since 2004, the non-profit organization has acquired over 420,000 acres of deeded and leased land (approx. 1700 km^2) strategically acquired to extend the wildlife habitat potential of million-plus acres that make up the nearby CMR refuge Upper Missouri River Breaks National Monument. While several of APR's properties offer leases to local ranching operations for cattle, the primary land use goal on APR's holdings is wildlife conservation, including the restoration of American bison to the northern Great Plains (Bullinger 2017; Davenport 2018). In addition, recreational and amenity-oriented ranch buyers have also begun purchasing properties in the CMR region with increasing frequency (Haggerty et al., in review).

A description of conservation actors in the region would not be complete without including two local groups, the Ranchers' Stewardship Alliance (RSA, established 2003) and the Winnett ACES (Agricultural Community Enhancement and Sustainability, established 2016). Both are grassroots organizations established in the context of economic and social changes in the central Montana region. Their missions focus on sustainable ranching and community development.

In the CMR region, native rangeland, and the wildlife habitat it provides, largely persist because extensive livestock grazing has been the dominant land use for the last century. This land use history and context provides an ideal geography to examine the synergy between ranching economies and wildlife ecologies. The ranching communities we describe in this chapter rely on the CMR refuge and other government-owned multi-use lands for grazing access and serve as a key partner in resource management (e.g., wildland fire management). At the same time, private lands provide a critical extension of the refuge's protected area. Wildlife populations, as we've learned in other chapters in this book, often require a diversity of habitats and resources that can only be realized at the landscape scale. The various landowners in the six counties surrounding the CMR refuge manage acreage that is critical to achieving that scale. Goals and initiatives to secure and improve outcomes for wildlife in the CMR region, thus, inevitably require coordination and cooperation with private landowners, and not just those with an explicit focus on wildlife conservation, but also those that are pursuing range-based livelihoods such as livestock production. In the sections that follow we will explore this important synergy between private landowners and public conservation initiatives, between ranching economies and wildlife ecologies, and the interdependence between rural peoples and wildlife that demands creativity, collaboration, and thinking-at-scale.

27.3 The Role of Ranching in the Regional Economy

As a first step in characterizing the role that range-based livelihoods, especially those related to production agriculture, play in a regional economy, this section profiles key demographic and economic trends in the CMR region. Over the past fifty years, the economic trends of the CMR region characterizes those of most rangeland regions (Goetz et al. 2018): a slow, but steady decline in population, a shrinking role of agricultural income as a share of personal income in the region, volatile returns to farm and ranch proprietors, reliance on non-labor income and public sector employment, and growing recreational and investment interest in private rangeland. Rather than an exhaustive survey of agricultural economics of the region, the following is a profile of key socioeconomic trends that provide a context for thinking about how ranch owners and employees relate to the overall economy, issues for the sustainability of their industry, and by extension, where wildlife management can present an opportunity and challenge.

27.3.1 Population Trends

An estimated 25,798 people occupied the 15.2 million-acre region surrounding the CMR in 2020, about 11,000 or 40% of them in towns, including the area's "larger" towns—Lewistown, MT (~6000), Glasgow (~3000), and Malta, (~2000) and another 2779 spread amongst in the area's small towns, with populations in the hundreds. An estimated 1300 Native Americans occupy the region, including members of the Assiniboine (Nakoda), Gros Ventre (Aaniiih), and Sioux tribes. According to the U.S. Census of Agriculture, 8265 people lived on farms and ranches in the six-county area in 2017, about two-thirds of the area's non-town residents, or one-third of the total population. Considering that less than one percent of the population of Montana is in a farm or ranch household and less than two percent of the national population is, this figure demonstrates the relative significance of farm and ranching lifeways in the CMR region.

Characteristic of many rural rangeland regions in the United States and Canada, the size of the human population in the CMR region has been steadily declining. Over the past half century (1980–2020), the population of the CMR Region declined by 7834 people (23%; Fig. 27.2). By contrast, the state of Montana added nearly 300,000 people in the same period, for a growth rate of just over 33%. As a result, median age in the area is anywhere from 5 to 10 years above that of the state of Montana (39.9; U. S. Census Bureau, n.d.).

27.3.2 Employment and Personal Income

The Census of Agriculture reports 2600 agricultural operations in the 6-county area, of which 44% are livestock operations and 42% are crop-focused, most specializing in grains and cereal crops. As a share of all agricultural land area in the region (9.9 m acres), crop-focused agriculture is 26%, compared to 66% for pasture and rangeland.[1] The number of domestic livestock in the region has fluctuated according to weather and commodity cycles, with producers reporting an estimated 343,000 domestic cattle in 2020 (Fig. 27.2). The important point yielded from looking at Fig. 27.2 is that human and livestock population trends do not show a simple relationship. Cattle trends tend to reflect the combined effects of weather and markets. Human population reflects the shrinking labor force required in agriculture and the limited opportunities in other economic sectors in the region.

---

[1] An important point to remember when interpreting agricultural census data is that all types of rural properties are counted as "farms." From the perspective of intact properties that provide contiguous wildlife habitat for graziers, the number of ranches may differ from (likely be lower than) what the Census of Agriculture reports. Another important piece of information comes from information the Census of Agriculture collects about farm operations—the number of people in each farm household.

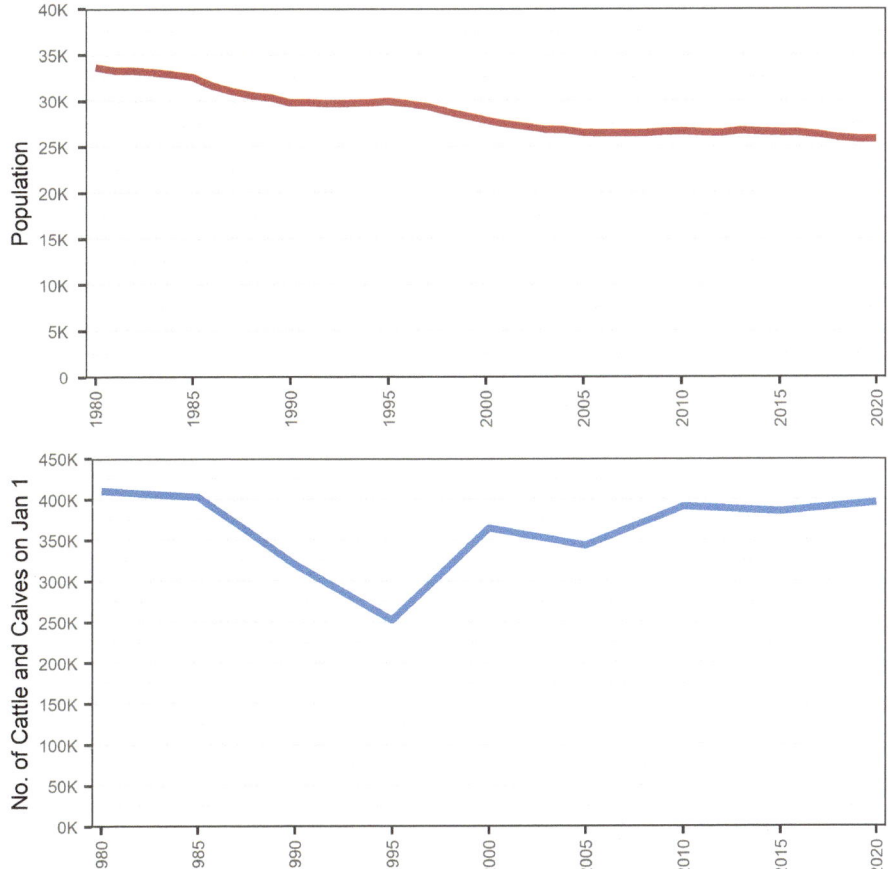

Fig. 27.2 Human Population and Cattle Inventory in CMR Region, 1980–2020. *Sources Population*: U.S Department of Commerce, 2020. U.S. Census Bureau, Decennial Census and American Community Survey, reported by Headwaters Economics' Economic Profile System. *Cattle inventory*: National Agricultural Statistics Service (2021). Cattle inventory is a national sample conducted every month that provides estimates of the number of breeding animals for beef and milk production as well as the number of heifers being held for breeding herd replacement. Data reported are extrapolated from a sampled subset of producers

According to statistics collected by the U.S. Department of Commerce, just under one in five jobs (17%) in the region is in agriculture. Of those, two out of three are farm self-proprietorships (self-employed, non-corporate farm and ranch operators). Income in agriculture is notably variable (Fig. 27.3, top), but agricultural wages have provided about 12% of total labor earnings in the region in recent decades, having declined from close to 40% in the 1970s. Fifteen percent of all jobs in the region in 2019 were in some form of public employment, including federal, military, state, and local governments, with the sector contributing 24% of total labor earnings in the region.

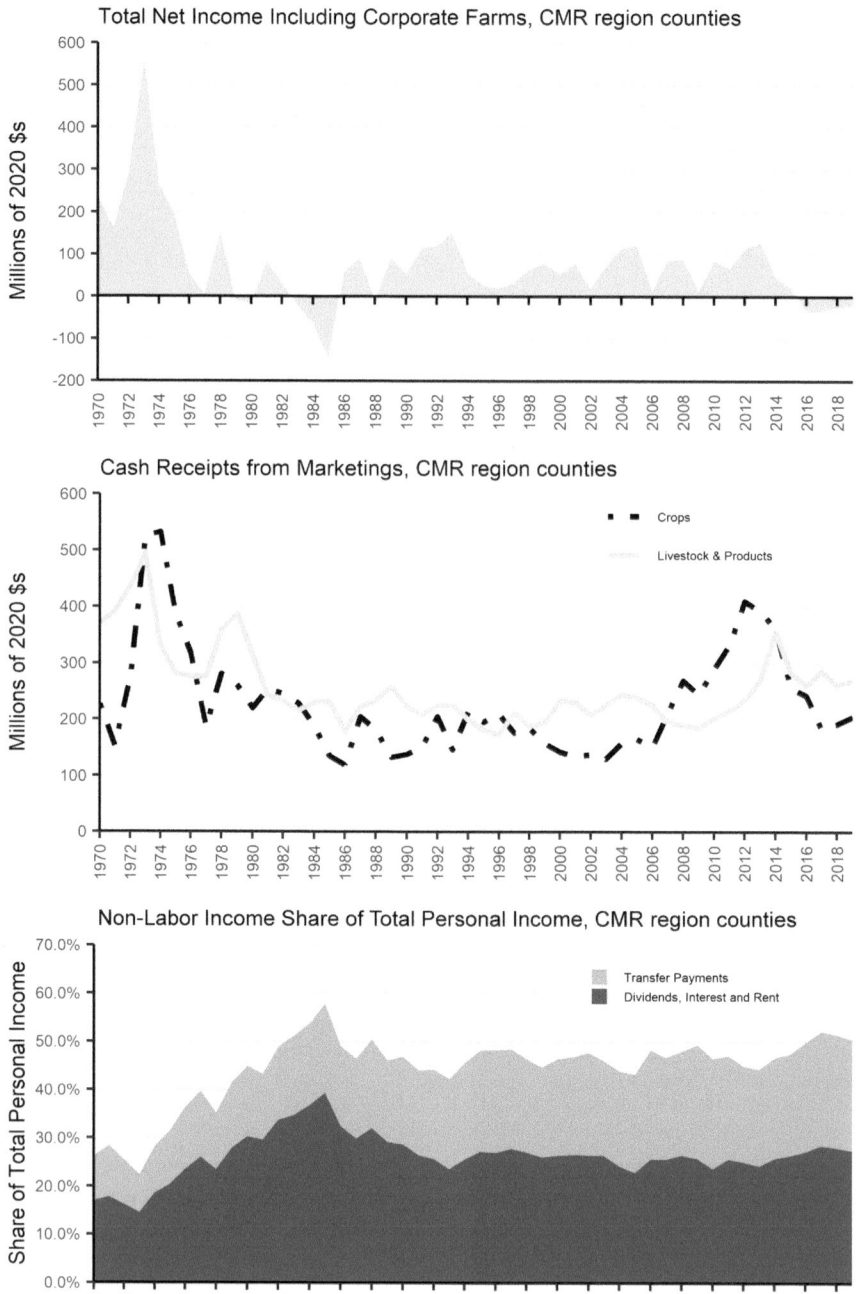

◄**Fig. 27.3** Farm Income and Receipts and Non-Labor Income in CMR Region, 1970–2019. Labor Earnings: Net earnings by place of residence, which is earnings by place of work (the sum of wage and salary disbursements, supplements to wages and salaries, and proprietors' income) less contributions for government social insurance, plus an adjustment to convert earnings by place of work to a place of residence basis. Non-Labor Income: dividends, interest, rent, and transfer payments (includes government retirement and disability insurance benefits, medical payments such as mainly Medicare and Medicaid, income maintenance benefits, unemployment insurance benefits, etc.). Non-labor income is reported by place of residence. *Sources Farm Income and Receipts, 1970–2019*: U.S. Department of Commerce, Bureau of Economic Analysis, 2020. Regional Economic Accounts. As reported by Headwaters Economics Economic Profile System, Accessed 9-06-2021. *Non-Labor Income*: Source. U.S. Department of Commerce, Bureau of Economic Analysis, 2020. Regional Economic Accounts. As reported by Headwaters Economics Economic Profile System, Accessed 9-06-2021

A significant historical change in the region, paralleling rural economies around the nation, is the growth in the importance of non-labor income (Fig. 27.3, bottom). Non-labor income includes private and public sources: it counts rent earned from property, dividends on investments, as well as transfer payments such as public health benefits, social security, unemployment, and veterans benefits. In 1970, non-labor income represented 26% of personal income. Fifteen years later, in the midst of the farm crisis, non-labor income was nearly 60% of personal income. Since then, and in 2019, non-labor income has comprised about half of all personal income in the CMR region, with over half of non-labor income earned as dividends, interest or rent and the remainder dominated by age-related and hardship payments. In other words, in 2019, one in four dollars accruing to residents of the region came from rent, one in five came from a public benefit such as social security, medicare and medicaid, and unemployment payments. This is in contrast to the one in sixteen dollars earned from farming or ranching in 2019.

Figure 27.4 also charts trends in farm earnings and cash receipts from farm products (middle chart). A couple of things are clear from these charts. First, agricultural income is notably volatile, reflecting the dynamics of national and global commodity markets. Second, farm and agricultural income has not grown either as a share of personal income or in absolute volume over the past five decades. This is in contrast to the growth in labor earnings in the state of Montana, which doubled during the same time period.

From the windshield as one passes through the area, ranching looks and feels like the dominant economy in the CMR Region. But measured by standard economic metrics such as income and jobs, ranching is arguably a minority share of economic activity in the area. Farming and ranching depend on, rather than support, local livelihoods. One reading of these economic indicators is that ranching is a highly marginal activity unlikely to support the CMR region. Our research suggests this is somewhat true in the sense that (1) working ranches are continually getting bigger while they support the same or a smaller number of ranching families (see below) and (2) ranches are increasingly owned by people who operate them as recreational investments, do not live on them, and do not depend on them for their livelihood. It is

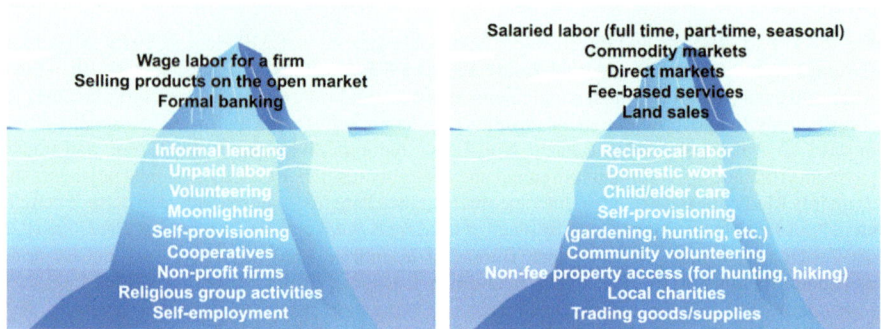

Fig. 27.4 The Diverse Economies Iceberg. A generic version of the diverse economy is shown at the left, with the specific iterations of a diverse economy in a ranching landscape shown on the right. *Source* Diverse Economies Iceberg by Community Economies Collective is licensed under a Creative Commons Attribution-ShareAlike 4.0 International License

a more accurate description of the region to note that ranching is interdependent with other sources of income: rental and interest or government payments, and income earned in other sectors, most likely the public sector. In order to understand the ranching economy as a local economy, then, it is essential to recognize that it is interdependent with other livelihoods, industries and sectors–neither superior nor necessarily subservient to, but rather intimately embedded with them.

27.3.3   Ranch Land Markets

As a final note on the local economy of ranching, it is essential to note the fast-growing interest of recreational buyers and investors in ranch property across North America, a trend that is also at play in the CMR region. Once circumscribed to high amenity areas near ski resorts and national parks and the occasional large property in extensive ranching areas (for example, media mogul Ted Turner has acquired over 1.8 million acres of property across regions in the Southwest, Great Plains, and Rocky Mountains; (Turner Enterprises Inc., 2021), ranch acquisition by the ultra-wealthy has expanded geographically in the past two decades. These ultra-wealthy buyers and other recreational interests are an economic force in the CMR region, meaning that those ranch properties which go on the open market are typically marketed for their diverse recreational, wildlife, and scenic values. For those seeking to own and operate ranches who lack substantial investment capital (i.e., local area producers), high land prices present a major sustainability issue: family-owned operations struggle with estate issues and emerging ranchers cannot afford to pay for land based on returns from ranching (Haggerty et al. 2018a).

Recent research conducted in Montana documents that key features of the amenity ranchland market (Epstein et al. 2021a; Haggerty et al. 2018b) are: increased concentration of farm and ranchland, access to high levels of capital to invest in ranch management on the part of ranch buyers, and wide variation in the key priorities of new buyers with respect to ranch management objectives (Epstein et al. 2021a). Implications for wildlife managers are an important area for future research, but initial work demonstrates that wealthy buyers are part of an increasingly complicated social landscape of wildlife conservation. While their ranches can and often do act as refuges for certain species, the owners themselves do not necessarily have a strong interest in participating in programs for landowner cooperation established with full-time ranchers in mind (Haggerty and Travis 2006). Furthermore, privacy, aesthetics, and recreational opportunities are more likely to play a role in land use decisions—for example, related to public hunting access or lease opportunities that challenge existing norms and relationships between ranch owners and members of the local ranching and resource management community (Epstein et al. 2019, 2021a; Haggerty et al. 2018b).

From the perspective of ranchers who steward a large amount of wildlife habitat in the CMR Region and the wildlife managers who work with them, the demographic, economic, and land market trends discussed here suggest a difficult and uncertain future. Building on our research and experience in central Montana and elsewhere in the northern Great Plains, we have emphasized the difficult times facing new and emerging ranchers and the opportunity and challenge that presents for wildlife conservation (Haggerty et al. 2018a). In terms of working with the owners of large ranch and farm properties, wildlife managers can anticipate a broader range of interests, experience and priorities on the part of landowners. Professional ranch managers are playing a greater role as proxies for landowners in negotiations with neighbors and agencies and emphasizes the complications for the wildlife management profession (Epstein and Haggerty 2022).

In rangeland areas undergoing demographic and land ownership changes, the nature of who is connected to whom through economic relationships, broadly defined, and the nature of those connections continues to evolve, emphasizing the embeddedness and interconnectedness of ranching economies, ranching livelihoods and wildlife habitats. From this perspective, it seems likely that work "in town"—for public agencies and possibly for new kinds of private sector employers—will remain integral to many ranch household economies. Wealthy absentee owners create a demand for certain kinds of services, but whether local businesses will provide those services is an open question. While demographic and land ownership patterns are shifting substantially, the transboundary and interconnected nature of ecological issues persists. Residents and wildlife managers in regions like central Montana will find themselves increasingly attending to the need to identify new kinds of mutually-beneficial economic relationships, raising the question of what role wildlife conservation might play in building a robust and resilient diverse economy.

27.4   Ranching and Wildlife in a Diverse Economies Framework

Although future land ownership trends in the CMR region are hard to predict, the area's rangelands are likely to feature working livestock ranches for many years to come. A premise of this chapter is that (1) among many other factors well-described in this book, rangeland wildlife conservation in the Northern Great Plains also depends on the vitality, resilience, and adaptability of working ranches and (2) such vitality and resilience will depend on the elevation of interdependence and mutualism as key features of the economy—practices that have deep roots in many rural economies, but which are rarely discussed in the context of private land conservation. Economic creativity is all the more important given the increasing heterogeneity in the interests, lifestyles, and capacities of ranch owners.

To that end, we introduce the concept of diverse economies as an alternative approach to thinking about the relationship between livestock ranching and wildlife conservation. First, we define diverse economies. Next, we demonstrate an approach for capturing a diverse economy through two participatory data collection efforts—a community mapping exercise and a "diverse economies inventory." Both efforts were done as service-learning projects to help communities learn about and identify the important relationships and more-than-market based exchanges in their own local community economies. Next, we profile two local initiatives that reflect the core values of diverse economies perspectives by supporting opportunities for local communities to collaborate, build trust, and value interdependence. Lastly, we share a series of perspectives from locals in central MT on the meaning of these collaborative efforts and how they view a more holistic approach to addressing ranching economies.

27.4.1   The Diverse Economies Framework

Diverse economies refers to a subdiscipline of economic geography and, more broadly, a mindset about sustainability. Along with broader critiques of the failure of the current economic system to achieve basic sustainability goals (cf., "Doughnut Economics[2]"; (Raworth 2017), the diverse economies approach seeks to "promote ethical and solidaristic modes of interdependence and help mitigate some of the key challenges of our time (such as environmental destruction and increasing inequality)" (Gibson-Graham and Dombroski 2020, p. 2). A central practice in diverse economies thinking is to broaden what we see and describe as part of the economy. Diverse

---

[2] According to the Doughnut Economics Action Lab, "The Doughnut consists of two concentric rings: a social foundation, to ensure that no one is left falling short on life's essentials, and an ecological ceiling, to ensure that humanity does not collectively overshoot the planetary boundaries that protect Earth's life-supporting systems. Between these two sets of boundaries lies a doughnut-shaped space that is both ecologically safe and socially just: a space in which humanity can thrive." (DEAL 2022).

economies scholars believe that by seeing the economy as broader than just what we get paid to do and sell, we can begin to imagine ways to make the economy better serve sustainability for people and nature—ways that we can't imagine under the very narrow ways we tend to think about what the economy is.

To help academics and practitioners break out of a narrow view of the economy, the diverse economies community has developed a core practice: inventorying economic diversity. This means simply making legible all of the many ways that humans procure and exchange goods and services, broadly defined. The diverse economy is more than just the flow of money, it's also a diverse set of relationships, practices, and activities that link different ranch properties across the landscape. The diverse economies iceberg is a central metaphor used to emphasize that the economy is so much more than things that can be monetized, or counted, and the conventional venues of monetary exchange—selling labor for wages, selling products in formal markets, and providing finance in a mode built to benefit the lender through interest. The iceberg (Fig. 27.4) shows how big the economy is when conceptualized holistically. An example often used to teach the diverse economies idea is to ask whether work done in maintaining and managing a home is an economic act—of course this is labor that ensures a family can survive. But if it is unpaid, it doesn't count as part of the economy. When such "informal" work is counted in a measurement like GDP, the figure instantly multiplies significantly (Coe et al. 2019; Mazzucato 2019).

Indeed it is often heard in rural communities that family operations and the relationships between them are the backbone of a community. While the diverse economies of ranch communities are often well known amongst its residents, it can be hard to see and measure by those of us who are more familiar with thinking about the economy as limited only to those things bought and sold with currency. Below we share two ways that communities can identify and track aspects of their diverse economy. We conducted both exercises with residents and community members in one county in the CMR region. We turn first to the diverse economies inventory and then participatory ownership mapping.

27.4.1.1 Diverse Economies Inventory

On top of, or above the waterline of the diverse economies iceberg are the aspects of the economy related to the capitalist economy. These include wage labor and the production of commodities for market exchange as part of a capitalist firm/ enterprise. In ranching communities, these activities include earning income from a livestock auction, hiring a seasonal employee, or purchasing equipment or feed. However, there are many ways that a ranch-based enterprise might engage with its community. Ranching families might spend the weekend at a neighbor's operation to help brand or move cows. A rancher might lend out her tractor for another to borrow. A landowner might give permission to a local hunter to use their property. While these activities can be essential to the functioning of ranching enterprises and their respective communities, they are largely outside of what we typically think of as the economy.

Conducting a diverse economies inventory is one way to encourage a community dialogue about and awareness of interdependence in the local economy. We conducted one such inventory with seven ranching households in one CMR region county in 2017. Our goal was to understand ways that ranches and other local enterprises exchange goods and services in addition to monetized transactions. The categories of our inventory followed a diverse economies approach. We asked participants about their ranching enterprise, the ways in which their ranching operation and property interacted with the local community; about labor, who they employed or worked for including paid and unpaid labor; and, ranching transactions, who they exchanged goods with and how. These categories included aspects of economy-as-iceberg that were both above and below the water line. We also asked participants about which aspects of the economy they thought created the greatest good for families, which aspects created the most good for their community, and which activities they would do more of if they had more time. The iceberg on the right in Fig. 27.4 shows the types of labor, enterprises, and transactions that were commonly described by participants, with Table 27.1 providing elaboration from the answers to survey questions.

Several things standout from this activity. First, our participants describe an economic life that is more diverse than just wage labor and capitalist production. They report sharing and exchanging labor with neighbors and other community members. As one participant noted: "Most of us can't survive without reciprocal trading. You can't find people to do the work." Taking care of family members and volunteering on local school boards, conservation districts, and with important community institutions was another set of activities recognized as critical to the survival of rural communities. Others emphasized the role of place, and the value of supporting local businesses. As one participated noted. "Shopping locally is very important. Local/traditional owners appreciate that if they want local businesses to survive we have to support them."

Beyond bringing an opportunity to name aspects of the diverse ranching economy, the inventory process was also an opportunity for our participants to reflect on the value of activities pursued for the sake of community. Multiple participants noted

Table 27.1 Perspectives on the diverse economy

Question	Responses
Which activities create the greatest good for your family?	Charity, volunteering, offering/receiving full time salaried positions, self-provisioning, reciprocal labor, community markets, paid labor, credit, direct markets, bull sales
Which activities create the greatest good for your community?	Sales, charity, volunteering, full time salaried positions, reciprocal labor, volunteering, child/elder care, domestic labor, self-provisioning
Which activities would you do more of if you had more time?	Charity, volunteering, direct markets, travel, self-provisioning labor, child care

Source Authors. Data shown here are aggregated from responses to the open-ended questions posed in the diverse economies inventory conducted with CMR region participants. (for a full description of the method, see http://www.communitypartnering.info/local43.html)

Table 27.2 Examples of conservation programs on private lands

Public programs	Examples
Technical assistance	USFWS Partners for Fish and Wildlife University Cooperative Extension
Cost share	USDA's Environmental Quality Incentives Program (EQIP) Conservation districts (varies by district)
Land Rental	USDA's Conservation Reserve Program (CRP) and Grasslands CRP
Conservation Easements	USDA's Agricultural Conservation Easement Program (ACEP) Colorado's Conservation Easement Tax Credit Program
Market-based programs	
Species or habitat banking	Montana Sage Grouse Habitat Conservation Program—Credit Projects Monarch Butterfly Habitat Exchange
Carbon offsetting	Avoided conversion of grasslands projects through protocols established by the American Carbon Registry or Climate Action Reserve
Corporate investments in supply chain sustainability	Ecosystem Services Market Consortium's crediting programs Bayer Carbon Program

the importance of participating in civic life, especially in remote, rural places. "You look around the table here and there's past commissioners, because there's so few of us you feel obligated to take your turn. We feel a stronger sense of connection to our local government and our local school because most of us have been involved."

A final take away from our inventory was affirmation among our participants that private land use plays an important role in both community functioning and wildlife management. For example, all seven of our participants reported providing hunting and fishing access for local recreationists. Because the state game agencies in the U.S. rely on public hunters and hunting to manage wildlife populations, opportunities to access private lands are an essential component of the North American Model of Wildlife Conservation (Organ et al. 2012). The central role that private lands and landowners play in wildlife management have made them a focus of wildlife management initiatives, and in particular, those aiming to increase opportunities for the public to gain access to private lands. While landowners in the region have the opportunity to participate in fee-for-access programs through the state wildlife agency, participants described the opportunity to welcome visitors and recreationists onto their property as a community obligation: "understanding for the generations that grew up here a sign of being a good neighbor or a friend." These sentiments underscore the more-than-financial dimensions of hunting access and demonstrate the value of thinking more holistically about the economy in the context of wildlife management.

27.4.1.2   Participatory Ownership Mapping

Because private land use can strongly influence outcomes for wildlife conservation, ranching operations are a key component of landscape-scale conservation initiatives designed to conserve landscape species (e.g., Chap. 10). Moreover, who owns private lands and how they manage them also matters to the functioning of rural communities beyond just the profitability of individual ranching operations, with further implications for wildlife. Property ownership is an especially compelling, and also tricky, topic from a diverse economies perspective. On the one hand, a diverse economies perspective is likely to focus more on alternatives to formal private property, or ways that people share access and use rights to land and resources, such as commons. On the other hand, and as we show here, making property boundaries visible and the subject of discussion can encourage dialogue and reflection among community members to think broadly about interdependence and connections among properties and neighbors.

In 2018, we worked with a group of local stakeholders from one county in the CMR region to create a dataset describing the ownership and operation of local ranch properties. The goal was to collect information for development of a program to support young ranchers with access to grazing land. We were also interested in understanding the ways that ranch ownership intersected with community dynamics. Together with local partners we agreed on a process that would allow us to count the number of ranch operations in the county and assign attributes to each ranch focused on local employment, land use and social and economic connections in the community. The process involved a careful consolidation and analysis of the publicly-available county cadastral record (see Haggerty et al. 2022). The next step was to work with local experts to characterize the attributes of ranch ownership. Specifically, we collected the group's expert insights about the residence of the ranch owner (local or absentee); their level of activity in the community; the estimated number of families supported by each operation; the dependence of a property on leasing other land; and whether the property was projected to stay in agricultural production in the future. Participants answered based on their own knowledge and group discussion. In early January, 2018, we shared and discussed our draft results with members of the community group. Their feedback was incorporated into our final report.

Taken together, the data generated through this exercise paint a picture of a rangeland region in transition. The vast majority (95%) of private property in the county is held in large units (> 640 acres) and is actively used for farming and ranching. One-quarter of the large farm and ranch private property in the county is owned by absentee owners with 97% of this property remaining in some form of agricultural production. Among 22 absentee owners, seven were known to the local group. About 64% of large farm and ranch owners are considered highly active in the local community. The high level of community engagement among ranchers is a critical asset in the community and one that might be leveraged toward a path to working with other landowners in the future. Locals were uncertain about the future of approximately 1/3rd of the acreage they reviewed.

From a diverse economies perspective, the participatory mapping exercise revealed the importance of looking at property collectively. Building an accurate ownership map is not a small task and discussing private property demands a high level of trust among participants. However, an inventory is critical to taking stock of current ownership demographics and identifying threats and opportunities to working lands. From the perspective of wildlife managers, an inquiry like this can help build a comprehensive picture of current and future local land ownership, one that might be effectively overlaid with spatial data on conservation priorities. In addition, the exercise can generate an appreciation of interdependence, a theme that is relevant both to community development and supporting cooperative wildlife management efforts. One participant in the mapping interviews observed that the exercise gave him a new perspective and appreciation for how interconnected ranch properties are—and by extension, the local economy.

When it comes to conserving working lands and enrolling private landowners in wildlife conservation, there are in fact many public programs underway. The following section provides readers a general survey of landowner-focused conservation initiatives in the United States, as this is a critical set of tools available to wildlife managers, with an additional goal of encouraging consideration of how private land conservation initiatives may or may not help to engender diverse economies.

27.5 Private Land Conservation Initiatives: A Diverse Economies Perspective

A website targeting end users of private land conservation programs (e.g., landowners), the Montana Conservation Menu was established in response to a proliferation of landowner- and producer-focused conservation programs in Montana. While a great many programs exist, they are not always fully enrolled, and it can be a full-time job for a property manager or owner to assemble the programs into something that is a meaningful economic opportunity for their operation. The Montana Conservation Menu website (https://mtconservationmenu.org/), organized by the Soil and Water Conservation Districts of Montana, aims to overcome these hurdles. Landowners can research cost-share programs to enhance pollinator habitat, learn about reimbursements for hosting particular wildlife species, or enroll in specialized conservation markets. The hope is that by aggregating opportunities strategically, the Conservation Menu approach will enhance the value of public programs to landowners and by extension, boost their conservation impact.

The Montana Conservation Menu is a testament to both the increasingly broad landscape of opportunities afforded to rural landowners and the central role that property-specific programs play in the predominant approach for wildlife conservation on private lands. While understanding their scope and structure is important for wildlife managers, so too is an appreciation of their limitations. In this section, we provide a general overview of landowner-focused conservation programs and some of their inherent challenges for achieving wildlife conservation and rural economy goals at scale. In short, while valuable to individual properties, monetizing or subsidizing wildlife stewardship–the thrust of the majority of landowner-focused conservation programs described here–falls short of the demands of building lasting diverse economies, given the many challenges that lie ahead for rangeland communities (human and non-human).

27.5.1 Conservation Programs and Implementation

27.5.1.1 Agencies and Actors

A large network of federal, state, and local natural resource agencies and private non-profit organizations offer a wide range of programs to support landowners with stewardship and conservation on their land. The NRCS within the USDA is the primary federal agency tasked with working with landowners on conservation projects and has offices and agents serving all 50 states and US territories. The USFWS Partners for Fish and Wildlife is another federal agency program, although operating with substantially smaller staff and budget, that works with landowners to implement projects that benefit priority wildlife species in their regions.

At the state level, each state has a wildlife agency (e.g., Montana Fish, Wildlife, and Parks, Wyoming Game and Fish Department) that holds management authority over wildlife (excluding those under federal oversight; i.e. migratory birds). Many wildlife agencies provide technical assistance, cost-sharing, or other programs to assist landowners with management that benefits wildlife species or to provide public access for hunting and fishing on private lands. Some state departments of agriculture also offer programs to support soil and water conservation and address water use and pollution concerns (respectively). Each state also has a university cooperative extension program with a primary objective of translating research into on-the-ground practices. Extension agents are often based in communities around the state and may offer education programs or meet with landowners directly to advise on agricultural or natural resource management issues.

Conservation districts, often called soil and water conservation districts, are one of the primary local entities working on conservation issues on private lands (Roemer et al., in prep). Conservation districts are locally governed, typically at the county level by elected board members, and help facilitate the implementation of conservation programs with local insight and relationships. Some conservation districts are funded through tax levies as well as state and federal funding sources. The budgets and program offerings vary significantly from one conservation district to another even within the same state.

Private entities are also important actors working on private land conservation. Several non-profit conservation organizations are active nationally with local and regional emphases. For instance, Pheasants Forever has a large private lands biologist team that partners with NRCS and conservation district offices to implement

Farm Bill Programs and other efforts on private lands. Other national groups like The Nature Conservancy and Trout Unlimited have priority focus areas where they often partner with ranchers and other landowners to implement projects like conservation easements, riparian restoration, or upgrading irrigation infrastructure to facilitate fish passage. Land trusts are non-profit organizations with a primary mission of conserving land for multiple conservation values. Some land trusts operate throughout a state while others have a more local or regional service area within a state. These groups are governed by a board of directors largely consisting of local community leaders and residents and therefore can be responsive to local needs. Private for-profit consulting groups also have a niche working with private landowners on a range of resource management issues from optimizing grazing plans to developing stream restoration projects to enhance recreational fisheries.

27.5.1.2 Programs

This network of agencies and organizations collectively implement an alphabet soup of conservation programs (Bennett et al. 2018). We characterize these programs along a spectrum from "public" to "market-based". Towards the public end of the spectrum are technical assistance, cost-shares, direct payments or rentals, and conservation easements. In technical assistance programs, agency or organization staff with specific expertise advise landowners on technical aspects of land management, which may take the form of a conservation or management plan. Cost-share programs provide financial support to implement specific conservation practices or to install conservation-oriented infrastructure. For example, an agency may pay for 50% of the cost to install a new "wildlife friendly" fence that allows for migratory ungulates to more easily pass beneath it while the landowner pays the remaining 50%. Other programs, like many of the Farm Bill programs, provide a direct payment to landowners for adopting specific conservation practices. The Conservation Reserve Program, as an example, pays an annual rental payment for each acre of farmland that is taken out of crop production and revegetated with grasses for contract periods of 10 or 15 years.

Conservation easements are another tool that restricts uses of property to conserve values like wildlife habitat or scenic views. Conservation easements are voluntary agreements between a landowner that grants the easement and an agency or organization, often called a land trust, that agrees to accept and hold the easement. The specific terms of each easement are negotiated by the landowner and the organization accepting the easement but typically include restrictions limiting subdivision, development, or preventing conversion of rangelands to crops (i.e., "sodbusting"). After granting an easement, the landowner still continues to own the property and usually can continue existing agricultural activities. Landowners can donate easements, which may qualify for federal and state tax incentives. In some instances where a property is a high conservation priority, landowners have the option to sell a conservation easement and receive a cash payment for all or a portion of the appraised fair market value of the easement (see chapter one in Guillion et al. (2020) for more detail on how easements are appraised). In these instances, agencies or land trusts need to raise grant funds to purchase the easement through programs like the Agricultural Conservation Easement Program administered by NRCS, although a number of states have their own easement funding programs and some private foundations also provide funding towards the purchase of easements.

On the opposite end of the spectrum, a number of market-based opportunities exist. In these types of programs, the landowner typically acts as a "seller" and enters into an agreement with a "buyer" to provide an environmental good or service or a land-use practice that likely results in that environmental good or service. Many of these programs exist within a regulatory framework where buyers need to "offset" an environmental impact by purchasing a "credit" developed by the seller. In the United States, wetland banking is the most established market-based program. When entities like a construction company or a state department of transportation are unable to avoid impacting a wetland, they are required by the US Clean Water Act to mitigate that impact which can be done by purchasing credits established by a seller by creating new wetlands or restoring existing wetlands at a different location. A number of regulation-initiated ecosystem markets exist. These include markets that support flora and fauna protected under the US Endangered Species Act (ESA) as well as pollution credit markets such as California's cap-and-trade program. Under California's Global Warming Solutions Act, enhanced forest stewardship practices can generate credits to be bought and sold.

There are also voluntary markets that operate outside of a regulatory setting. Voluntary carbon markets now allow landowners to develop carbon credits in rangeland systems through "avoided conversion" of grasslands projects. In these projects, landowners agree not to plow or develop rangelands at risk of conversion in order to maintain carbon stocks within the soil. Corporate buyers tend to be the major purchasers of these types of carbon offsets and are motivated by sustainability pledges and climate commitments. Corporate buyers also drive demand in other market-based conservation efforts as these corporations aim to create more sustainable supply chains and mitigate reputational risks (Toombs et al. 2011). While there are a number of voluntary market-based conservation programs, many of these efforts are currently in pilot phases and participation is limited to landowners in specific locations. There is potential for these markets to expand in scope and scale in the future and they may become more influential in how private lands are managed.

27.5.1.3 Landowner Motivations

Research demonstrates that landowners have diverse motivations for participating in conservation programs. Many landowners have strong stewardship values and a connection to their land and their management practices are a reflection of those values (Lien et al. 2017). Conservation programs can assist landowners in exercising their values where it might be otherwise cost prohibitive, such as through cost-share programs. Other landowners may be more economically motivated and conservation may help their bottom line by helping cover the costs of replacing aging infrastructure with more conservation-informed designs. Vegetation treatments like mechanical or chemical removal of woody vegetation may enhance habitat for species like sage-grouse (*Centrocercus* spp.) and mule deer (*Odocoileus hemionus*) while also increasing grass cover that provides forage for livestock (Chaps. 5, 10). Selling a conservation easement is also a strategy used by landowners to extract some of the real estate value of their properties and allow them to reinvest in their agricultural business, pay down debt, or save for retirement. These diverse motivations also reflect the range of landowners in western landscapes where amenity ownership is increasing alongside more traditional livestock operations (Gosnell and Travis 2005).

27.5.2   Challenges and Opportunities in the Existing Conservation Approach

There are a number of challenges associated with the existing portfolio of conservation programs as well as opportunities to build on the current foundation. A major benefit of programs on the public end of the spectrum is that there are programs that most landowners can participate in regardless of what part of the country their property is located. Landowners can contact their local FSA/NRCS office, conservation district, state wildlife agency, or other network actors and begin exploring options currently available for their property. Conversely, many market-based opportunities are geographically limited or limited to specific resources that do not occur on every property (e.g., habitat for an endangered species). The ability to participate in a rangeland avoided conversion carbon market, for example, is limited to properties with soils determined to be at risk of conversion to crops and typically requires enrollment of a substantial amount of acreage to overcome transaction costs and be financially viable (Brammer and Bennett 2022).

Some public conservation programs have been criticized for not being strategic and instead implemented in an ad hoc project-by-project fashion. Some scholars have criticized investments in conservation easements as inefficient and failing to target lands most at risk of development (Merenlender et al. 2009; Rashford et al. 2019). Some efforts now attempt to address this "random acts of conservation" concern such as NRCS's Regional Conservation Partnership Program (RCPP), which focuses funding on specific resource issues in specific geographies. The Sage Grouse Initiative, led by NRCS but with numerous state and private partners, uses a science-based strategy to strategically focus limited budgets on areas and the types of projects that will have the greatest benefit to sage grouse (Naugle et al. 2020). These approaches help bridge the project-by-project nature of working with individual landowners with a landscape focus that is needed for impact at a broader scale.

Public conservation programs typically benefit from relatively secure funding and tend to be more institutionalized. NRCS programs, for example, are funded by the conservation title of the Farm Bill, which drives hundreds of millions of dollars annually towards private lands conservation. State wildlife agencies benefit from the

Pittman-Robertson Act, which returns a portion of taxes levied on guns and ammunition to states for the purpose of managing and restoring wildlife. While states enjoy a great degree of flexibility in how these funds are spent, many wildlife agencies dedicate a portion to conservation programs on private lands. In general, market-based programs do not share this same level of institutionalization and many can be described as "institutionally brittle." Regulatory driven approaches like species banking often require demand for credits created through the listing of a species under the ESA, but court decisions, lawsuits, and political pressures can create dynamic uncertainties over a species' listing status. The USFWS listed the lesser prairie chicken (*Tympanuchus pallidicinctus*) as threatened in 2014, but a federal court vacated the listing decision in 2015 (Chap. 9). These regulatory dynamics have tempered market-based approaches for the species as actors evaluate the situation and weigh costs and benefits of generating and buying credits. Similar regulatory uncertainty plagues market-based approaches for water quality and other ecosystem service markets.

An additional challenge for market-based programs is creating confidence in what is traded within the market. Most market-based programs focus on resources considered public goods that are not directly tradable. Instead, some programs use proxies like land-use practices that are likely to result in the generation of that public good to determine the number of credits created by those actions and available to be traded. Standardized units, like a metric ton of carbon, also need to be established as the "currency" that is traded within a program. Yet standardized units need to be estimated using complex models and protocols to determine, for example, how many credits can be generated from agreeing not to convert an area of grasslands to cropland. These complexities can undermine confidence in market-based programs when the models and protocols do not reliably estimate the ecosystem good, which is common with resources and ecosystems that are highly heterogeneous. In these instances, buyers may question what they get in return, which can result in the collapse of the market, as happened with the Chicago Climate Exchange. This early effort to create a voluntary market for carbon collapsed following, in part, concerns raised about the integrity of credits traded in the market. These technical aspects of creating artificial markets for public goods present significant obstacles to the viability and longevity of market-based approaches.

27.5.2.1 Do Existing Programs Embrace Interdependence and Connectivity?

In addition to their respective challenges, the range of landowner conservation programs above shares one additional limitation relevant for this chapter's focus on the intersection of rangeland economies and wildlife ecologies. Whether subsidized or reimbursed by the public or traded and exchanged within a marketplace, the existing conservation toolkit for supporting wildlife conservation on private ranchlands operates at the level of individual properties and landowners. This emphasis reflects the legal framework of private property in the US—as the property rights of landowners give them the power and responsibility to dictate how land use and management supports stewardship on their respective properties—but also a deeply entrenched assumption about our modern economy and its focus on the individual. While the conservation contributions of single ranch properties can be very meaningful, especially if landowners own and control very large and ecologically significant holdings (Haggerty et al. 2022; Epstein et al. 2021a), wildlife and conservation scientists in this textbook and elsewhere stress the need for conservation practices and habitat restoration at the landscape scale (Brunson and Huntsinger 2008).

The need for large landscape approaches raises important questions about the existing landowner conservation toolkit related to whether individual landowner incentive programs can support the types of continuity of stewardship and connectivity required to mitigate biodiversity loss and support thriving wildlife populations. Because rangeland economies are deeply entangled with wildlife ecologies, answering this question, we argue, requires a re-thinking of our orientation towards the economy, and more specifically, ranchland economies. In the next section, we apply the diverse economies framework to community-based conservation efforts in central Montana.

27.6   Community Collaborations: Diverse Economies on Central Montana Rangelands

The diverse economies material we have presented emphasizes the importance of understanding ranches as interdependent not only with one another, but also with other livelihoods and economic sectors. So where do wildlife fit in this equation? How do local community members view the synergy between ranching economies and wildlife conservation? How well aligned are existing programs targeting wildlife stewardship on private lands with the need to work at scale and to foster diverse and interdependent economies?

27.6.1   Seeing Diverse Economies Practices in Central Montana

> Taking care of little islands in this landscape is great--you look and say, 'oh, yeah, that person over there is really cool. He's doing that, but [it's not enough] … now, the freaking best grass in the world, and if everything's cratering around me, it's useless because I need all these neighbors around me. I need all of these other people, I need a community.—Bill Milton

In this section, we share the ideas, insights, and wisdom of three contributing authors[3] who are community leaders in central Montana dedicated to finding linkages between sustainability of local rangeland economies and wildlife populations.

---

[3] To facilitate the section, the lead authors interviewed the three contributing authors and compiled summary narratives from their direct quotations and supporting material such as public-facing

Together they have decades of personal and professional experience at the nexus of ranching and wildlife. We first offer a portrait of each person and the organizations they help to lead and follow with three key lessons for thinking about rangeland wildlife and local communities based on their experiences. The three community conservation institutions they represent explicitly link conservation—including wildlife, but also water, soil and native plants—with economic opportunities that emphasize the interdependence of ranching operations at local, community, and regional scales: (1) a successful grassbank; (2) a small-town community development organization; and (3) a regional conservation collaborative. The work of these three groups occurs alongside and within the broader network of wildlife conservation stakeholders, actors, and agencies described, yet at the same time, demonstrate key features of the diverse economy approach and thus provide important insights for wildlife managers on the social-ecological-economic dynamics of rural ranching contexts.

27.6.1.1 The Matador Ranch Grassbank: Diverse Economies at the Neighborhood Scale

As one of the world's largest private-land conservation entities, The Nature Conservancy is often associated with conservation easements. TNC is also a leader in the implementation of grassbanks—a term that describes a program which incentivizes ranchers to adopt conservation practices on their property in exchange for grazing access on another property. Qualifying conservation practices may include weed control, removal of fencing, restoring habitat for key species, and granting a conservation easement on their property. Grassbanks benefit ranchers by giving their own lands an opportunity to rest and improve forage quality while their cattle graze elsewhere" (University of Wyoming 2021).

In central Montana, TNC's pioneering grassbank program evolved in 2002 following severe drought, when ranchers were faced with selling off their herds if they couldn't find sufficient forage. After acquiring the 60,000-acre Matador Ranch south of Malta, Montana, TNC established a grassbank in cooperation with local ranchers inspired by successful examples elsewhere. Local ranchers pay discounted fees to graze their cattle on the Matador in exchange for wildlife-friendly practices on their

---

websites. The individuals named here have all agreed to be quoted in this article. Laura Nowlin is the coordinator for the Musselshell Watershed Coalition and serves on the board of the Montana Watershed Coordination Council. She lives on the family ranch with her husband and two children north of Winnett. She is a founding member of the Winnett ACES.

Brian Martin is the Montana Grasslands Conservation Director for the Nature Conservancy—Montana. Educated in Range Science, Brian leads the Conservancy's protection, science, and stewardship efforts in the Northern Great Plains of eastern Montana. Martin leads program efforts focused on working collaboratively with landowners, agencies, and NGOs to conserve natural habitats for the benefit of nature and people. Rancher Bill Milton lives near Roundup, Montana. With his wife, Dana Milton, Bill received the 2019 Montana Leopold Conservation Award. An experienced facilitator, Bill is active in many community and regional groups. He has facilitated the CMR CWG since its inception.

own operations. At a minimum, cooperating ranchers must agree to control noxious weeds and not break any new ground. After that, the lease price drops for additional conservation measures such as protecting prairie dog towns, securing sage-grouse leks, or modifying fences to make them safer for wildlife.

Grassbanks are celebrated as a conservation tool that leverages access to grazing to achieve conservation across multiple properties (White and Conley 2007). This is important. But often overlooked is the idea that a successfully-managed grassbank demands cooperation and integration among a collection of otherwise discrete ranching enterprises—as many as ten or twelve families participate in the Matador Ranch grassbank. While the Matador Ranch has a formal management team to make day-to-day decisions and run the property, participating ranch operators meet on an annual basis to negotiate and prioritize collective practices on the ranch. This kind of required cooperation and coordination emphasizes the necessity of interdependence, first and foremost among livestock producers. In this way, a grassbank represents a diverse economies model operating at a neighborhood/landscape scale—structured to some degree around conventional financial transactions, but also highly dependent on cooperative, non-monetary transactions. The grassbank also connects livestock producers, NGOs, and the broader local economy. As founding member, rancher Dale Veseth has said about TNC's investment: "This has made a huge difference in this community… When you help feed families and cows, they'll remember." (The Nature Conservancy, n.d.).

27.6.1.2 The Winnett ACES: Diverse Economies at the Scale of a Town and Its Hinterland

Winnett, Montana is the main population center in remote Petroleum County, Montana. The Winnett ACES (Agricultural and Community Enhancement and Sustainability) is a local non-profit organization formed in 2016 to strengthen its community so that future generations will live, work, and raise their families there. Initially the group was focused on the challenges of acquiring land for grazing and ranching, particularly for young families just getting their start. The group continues to work to develop a grassbank to this end. They also serve as a hub for local conservation activity by administering grants, many of which align rangeland management and wildlife stewardship objectives.

However, the other projects ACES has pursued are another window into the integral nature of conservation and community development. They focus on enhancing Winnett's public services and core infrastructure. The ACES's first program was Winnett Beef in the School, which serves locally-raised beef to the local K-12 school system. Led by a local producer, area ranchers donated enough beef initially to fill four years' worth of need at the Winnett Schools. Additional volunteers did the brand inspections, hauled the beef to the slaughter facility, and donated freezer space (Sturm 2017). Winnett ACES has also encouraged the building of a community center and is leading an effort to revitalize public and historic buildings to provide housing, office space, and business opportunities. All of these projects acknowledge the importance of the social and physical environment for building community and attracting residents (Western 2021). As founding member Laura Nowlin puts it: "Communities the size of Winnett are really reliant on the agriculture industry and the ranchers. But that is one of the things that ACES has talked a lot about too is that we are just as reliant on the community …"—Laura Nowlin (Beevers 2020). At its core, the Winnett ACES is about interdependent thriving.

Central features of a diverse economies approach to local development, relocalizing food systems, and revitalizing community infrastructure are strategies that support the long-term vitality of rural places, including those where local producers and private landowners are key actors in managing rangelands and conserving wildlife. Thus, the efforts of Winnett ACES sustain interdependent thriving between members of rural communities and between communities and the ecosystems and wildlife they steward. In doing so, the diverse economies approach of the Winnett ACES also illustrates how *social* in social-ecological systems is a vital part of rangeland wildlife conservation and management.

27.6.1.3   The Charles M. Russell National Wildlife Refuge Community Working Group: Diverse Economies at the Regional Scale

The Charles M. Russell National Wildlife Refuge Community Working Group (CMR CWG) was formed to enhance and preserve the ecological, economic, and social well-being of the 6 counties (Fergus, Garfield, McCone, Petroleum, Phillips, and Valley) surrounding the Refuge. Participants in the group include agency representatives, landowners, grazing permittees, county commissioners, conservation districts, interest groups, and engaged citizens. The group has been meeting bi-monthly since July 2010, with the meeting location rotating through the 6 counties.

The CMR CWG collaborated to develop a three-part goal for the region, which focuses on quality of life, production and landscape characteristics (see box). The vision is an excellent example of the degree to which interdependence is recognized as a fundamental characteristic of successful wildlife conservation, including diverse economic interdependence.

> **Title: Charles M. Russell National Wildlife Refuge Community Working Group's 3-part Landscape Vision**
>
> **Describe the quality of life you would like to see be predominant in the region in 5–10 years**
>
> "We want this region to maintain a diversified economy within which a prosperous agriculture industry is sustained and local communities are prosperous with stable populations. We desire an atmosphere where agencies, local government, NGOs, and citizens work together to create positive outcomes for the community and citizens: focusing on common ground, mutual respect, and community-based decision making, where people are committed to the

working group and access to public land is ensured for both the public and producers."

What kind of production will be needed to sustain this quality of life?

"A diversity of unique goods and services to support economic and social values will need to be produced from a working landscape that maintains its scenic value, healthy soils, and ecological integrity. We must also identify and implement best management practices that integrate local ecological knowledge, succession planning in all entities, local working groups to address challenges, incentives to practice conservation, steady tax base to support infrastructure, and responsible, well-educated citizens."

What does the landscape need to look like to obtain your production?

"We desire a landscape that provides habitat for diverse and healthy wildlife populations, where further conversion of native prairie is discouraged, and where the needs of natural resource dependent industries are balanced with conservation. In short, healthy agriculture lands cooperatively managed for the benefit of the resource, wildlife, industry, and community."

The CMR CWG's contribution to a diverse economy may seem less obvious than the grassbank or ACES model. The group has largely functioned as a convening venue for sharing information and building trust and familiarity among diverse conservation stakeholders. However, when regional resource management challenges arise, the substantial economic and community development value of relationships developed in the context of regularly meeting to exchange perspectives becomes clear. For example, when a very large wildfire raced through central Montana in the summer of 2017, the National Wildlife Refuge and local conservation districts coordinated a quick and creative response to the loss of forage on local ranches. Both parties credited the establishment and maintenance of communication channels over the course of years of participating in the CMR CWG as a major factor in their ability to act quickly and effectively, together (Charles M. Russell Community Working Group 2018).

27.6.1.4   Three Lessons from Central Montana

Lesson 1. Everything is connected: Wildlife stewardship must also be community stewardship

First and foremost, our community leaders in the CMR region reiterate the key themes of this chapter, namely that most residents of ranching communities see no separation between rangeland economies and rangeland ecologies. As Bill Milton puts it: "everything is connected, it's all interdependent, you know, you can't separate any of these ideas from each other." Laura Nowlin, rancher, watershed collaborative coordinator, and community development volunteer offers a landowner's perspective on wildlife in this statement:

> Wildlife is part of our identity, both as individual landowners and as a community. We are not disconnected from wildlife, like most (more urban) people are...it's just a part of who we are and what we do...I feel like it's important for managers of any sort to recognize that what we're doing has to encompass everything. And not just wildlife and not just soil and not just economics, because it all has to work together to make everything work.
>
> When you're thinking about wildlife managers, that's their sole job and their sole responsibility. I think the key piece that ranchers play is that we're concerned about all of it because it all affects our business and our land and our communities. So our approach is yes, we want to conserve wildlife, but, for the people who have the capacity to take care of the wildlife they have to have housing, they have to have healthcare, they have to have schools. Fundamental pieces of our community infrastructure have to work, and can't be separated from having the landscape work for wildlife and all these other issues that the rest of the world thinks is so important. What we seek to understand are: which are the programs and approaches to wildlife conservation that benefit us individually as ranchers and benefit the whole community?

Sharing his perspective from three decades working with The Nature Conservancy on the rangelands of the Northern Great Plains, Brian Martin emphasizes the more than monetary economic rewards of conservation practices in terms of enhanced operational resilience:

> Communities benefit from wildlife conservation because it ultimately produces a more diverse experience. Diversity builds resilience for both human and natural communities. There are direct financial benefits that can be derived from wildlife, but the returns are more likely associated with healthier and functional systems that better withstand drought and other climate change driven extremes. In ranch systems, diversity of vegetation likely creates more variable forage for livestock and can help maintain a higher nutritional plane throughout the year.

Recognizing that wildlife and people are interconnected, however, is just the first step. Our local experts also acknowledge a need to think beyond the scale of the individual property. This is important not only for wildlife, but also ranching communities. This leads to lessons 2 and 3.

Lesson 2. Creative solutions spread through relationships, which hinge on trust

The local experts profiled here believe that when place-based institutions like grass-banks, ACES, and the CMR CWG constitute the core of landscape-scale conservation, opportunities to link economic and wildlife stewardship increase. As the conservation menu mentioned in Sect. 27.5 makes clear, there is no shortage of private and public incentive and payment programs available to private rangeland owners. Whether or not these programs reach their maximum potential for wildlife conservation of course depends, among other things, on how wide their uptake is. While there are many factors determining participation in landowner conservation programs, trust is one of the most important (Sketch et al. 2019).

Here, groups like ACES serve as a critical venue for trust and the types of relationships necessary for working beyond individual property lines. Milton observes "land-based, community-based organizations … have the legitimacy and the license to put things in the play… without those groups there's no functional leverage to make serious change happen." In contrast, a lack of trust can stymie even the best and most ambitious of conservation plans (Covey and Merrill 2006). Describing a pathway to grow trust from the "inside" out across the rural landscape, Nowlin offers a metaphor of a tree with many branches growing in many directions, but all reliant on a strong trunk:

> If we all work first from our side of the tree to reach those closest to us and then branch out as we go, then eventually, we should reach everyone. So, ACES will work with our rancher neighbors who are not yet comfortable having conversations and finding common ground with most environmental groups. … We can help ensure that the "trunk" is on solid footing with a high level of trust among the existing partners first and then, when we're all comfortable, we begin to extend to the outliers.

Being locally-based and intertwined with daily lives and projects of local communities are essential when it comes to building trust.

Lesson 3. Wildlife and ranching need scale, scale requires relationships: Enter diverse economies

While rangeland ecologists take the importance of working at scale as practically a given, scaling conservation programs directed to individual landowners up (and out) into a cohesive landscape effect remains a challenge. The emphasis on allowing ideas to diffuse through relationships built on trust is one key platform. A diverse economies strategy of networked economic cooperation helps provide a further solution to this challenge.

Drawing on his experience both as a facilitator of numerous conservation collaboratives and as a founding member of one of the nation's largest beef-marketing cooperatives, Bill Milton articulates this vision of coming together to work at scale through a locally-led vision of economic cooperation:

> Family ranching's role in our economy continues to shrink, yet the resources we steward remain significant and play an outsized role in policy discussions regarding protecting native places and combating climate change. How can ranchers take advantage of this moment in time? I often ask myself, how can family ranchers leverage their resource assets, their wisdom, and their expanding community of partners to better secure a viable livelihood for their fellow family ranchers? I propose the answer to that question is to build a place-based, rancher-owned company to market our collective stewardship of grasslands. Rather than focusing primarily on rewarding benefits to individual ranchers for conservation practices, why not market those practices and benefits at scale within large local regional landscapes?

I think the whole economic model taking care of large landscapes is that local people have to get creative and brave enough to ask to actually create the economic engine to do that. And that's going to involve expanding beyond what we've already done with some really great partners.

27.7 Summary: A Diverse Economies Perspective on Rangeland Wildlife Ecology

For wildlife professionals, understanding ranching communities and range-based livelihoods is important for two reasons. First, because land enrolled in livestock production and other private ranching land uses affects the form and function of wildlife habitat, understanding the constraints and opportunities facing ranchers from the broader economy is key to designing effective conservation and management strategies. At the same time, wildlife professionals will become part of rangeland communities by virtue of living and working in them. A keen understanding of rural places and their economies is a valuable component of the job and are reasons students of wildlife management might be wise to invest in understanding range-based human communities with the same kind of dedication the profession brings to plants and animals.

This chapter invited readers into the world of "diverse economies" thinking (Gibson-Graham and Dombroski 2020), a holistic approach to analyzing the economy as part of a broader set of sustainability challenges and social-ecological dynamics. In rangeland regions, a diverse economies perspective involves appreciating interdependence. This chapter specifically highlighted the interdependence of ranching and agriculture with other formal economic sectors, as well as the many connections among rangeland residents forged through informal economic activities. The chapter offered a survey of public and private programs that attempt to encourage rangeland owners and managers to adopt conservation practices, including wildlife-friendly management choices, and notes that payments, regulations and incentives are not enough to build thriving economies. The local conservation vignettes offered in the chapter's final section encourage readers to think broadly about how economic interdependence—among neighbors, among public and private wildlife management actors, and even between rangeland residents and a broader public—can be leveraged to accomplish landscape-scale objectives of thriving human and wildlife populations.

Acknowledgements This article includes original data collected under protocols approved by Montana State University's Institutional Review Board.

References

Beevers K (2020) Winnett ACES featured on Our Montana TV Program, 3 Sept 2020. Winnett ACES. https://www.winnettaces.org/blog/winnett-aces-featured-on-our-montana-tv-program. Accessed 13 Mar 2021

Bennett DE et al (2018) Using practitioner knowledge to expand the toolbox for private lands conservation. Biol Conserv 227:152–159

Bestelmeyer BT, Briske DD (2012) Grand challenges for resilience-based management of rangelands. Rangeland Ecol Manage 65:654–663. https://doi.org/10.2111/REM-D-12-00072.1

Brammer T, Bennett D (2022, forthcoming) Arriving at a natural solution: bundling credits to access rangeland carbon markets. Rangelands. https://doi.org/10.1016/j.rala.2022.04.001

Brunson MW, Huntsinger L (2008) Ranching as a conservation strategy: can old ranchers save the New West? Rangeland Ecol Manage 61:137–147. https://doi.org/10.2111/07-063.1

Brunson MW et al (2016) Usable socio-economic science for rangelands. Rangelands 38:85–89. https://doi.org/10.1016/J.RALA.2015.08.004

Bullinger J (2017) Montana refuge divides tribes and ranchers. High Country News. https://www.hcn.org/issues/49.9/montana-prairie-refuge-divides-natives-and-ranchers

Carpenter S et al (2001) From metaphor to measurement: resilience of what to what? Ecosystems 4:765–781. https://doi.org/10.1007/s10021-001-0045-9

Charles M. Russell Community Working Group (2018) Charles M. Russell Working Group meeting minutes: Fort Peck. http://www.cmrcwg.org/wp-content/uploads/2018/04/CMR-Notes-2-27-2018.pdf

Charnley S et al (2018) Cattle grazing and fish recovery on US federal lands: can social–ecological systems science help? Front Ecol Environ 16:S11–S22. https://doi.org/10.1002/fee.1751

Coe NM et al (2019) Economic geography: a contemporary introduction. Wiley

Colding J, Barthel S (2019) Exploring the social-ecological systems discourse 20 years later. Ecol Soc 24(1)

Covey SR, Merrill RR (2006) The SPEED of trust: the one thing that changes everything. Simon and Schuster

Davenport J (2018) Making the Buffalo commons new again: rangeland restoration and bison reintroduction in the Montana highline. Great Plains Q 38:199–225. https://doi.org/10.1353/gpq.2018.0024

Doughnut Economics Action Lab (DEAL) (2022) About doughnut economics. https://doughnuteconomics.org/about-doughnut-economics. Accessed 13 Mar 2022

Duvall AL et al (2017) Conserving the greater sage-grouse: a social-ecological systems case study from the California-Nevada Region. Rangeland Ecol Manage 70:129–140. https://doi.org/10.1016/j.rama.2016.08.001

Epstein HE et al (1996) Ecological responses of dominant grasses along two climatic gradients in the Great Plains of the United States. J Veg Sci 7:777–788. https://doi.org/10.2307/3236456

Epstein K et al (2019) Super-rich landowners in social-ecological systems: opportunities in affective political ecology and life course perspectives. Geoforum 105:206–209. https://doi.org/10.1016/j.geoforum.2019.05.007

Epstein K et al (2021a) With, not for, money: ranch management trajectories of the super-rich in greater yellowstone. Ann Assoc Am Geogr 1–17. https://doi.org/10.1080/24694452.2021.1930512

Epstein K et al (2021b) Toward an urgent yet deliberate conservation strategy: sustaining social-ecological systems in rangelands of the Northern Great Plains, Montana. Ecol Soc 26. https://doi.org/10.5751/ES-12141-260110

Epstein K, Haggerty JH (2022) Managing wild emotions: Wildlife managers as intermediaries at the conflictual boundaries of access relations. Geoforum 132(2022):103–112

Folke C et al (2005) Adaptive governance of social-ecological systems. Annu Rev Environ Resour 30:441–473. https://doi.org/10.1146/annurev.energy.30.050504.144511

Gage AM et al (2016) Plowprint: tracking cumulative cropland expansion to target grassland conservation. Great Plains Res 26:107–116. https://doi.org/10.1353/gpr.2016.0019

Gibson-Graham JK, Dombroski K (2020) The handbook of diverse economies. Edward Elgar Publishing

Goetz SJ et al (2018) The economic status of rural America in the President Trump Era and beyond. Appl Econ Perspect Policy 40:97–118. https://doi.org/10.1093/aepp/ppx061

Gosnell H, Travis WR (2005) Ranchland ownership dynamics in the rocky mountain west. Rangeland Ecol Manage 58:191–198. https://doi.org/10.2111/1551-5028

Guillion B et al (2020) Paying for stewardship. Western Landowners Alliance. https://westernlandowners.org/publication/paying-for-stewardship/

Gunderson LH, Holling CS (2002) Panarchy: understanding transformations in human and natural systems, 1st edn. Island Press

Haggerty JH et al (2018a) Ranching sustainability in the Northern Great Plains: an appraisal of local perspectives. Rangelands 40:83–91. https://doi.org/10.1016/j.rala.2018.03.005

Haggerty JH et al (2018b) Land use diversification and intensification on Elk winter range in greater yellowstone: framework and agenda for social-ecological research. Rangeland Ecol Manage 71:171–174. https://doi.org/10.1016/j.rama.2017.11.002

Haggerty JH et al (2022) Rural land concentration & protected areas: Recent trends from Montana and greater Yellowstone. Soc Nat Resour 35(6):692–700

Hruska T et al (2017) Rangelands as social–ecological systems. In: Rangeland systems. Springer, Cham, pp. 263–302

Lien AM et al (2017) The land ethic of ranchers: a core value despite divergent views of government. Rangeland Ecol Manage 70:787–793. https://doi.org/10.1016/j.rama.2017.06.004

Lipsey MK et al (2015) One step ahead of the plow: using cropland conversion risk to guide Sprague's Pipit conservation in the northern Great Plains. Biol Conserv 191:739–749. https://doi.org/10.1016/j.biocon.2015.08.030

Mazzucato M (2019) The value of everything: making and taking in the global economy. Penguin Books

Merenlender AM et al (2009) The importance of incorporating threat for efficient targeting and evaluation of conservation investments. Conserv Lett 2:240–241

Naugle DE et al (2020) Coproducing science to inform working lands: the next frontier in nature conservation. Bioscience 70:90–96. https://doi.org/10.1093/biosci/biz144

Ojima DS et al (2020) A climate change indicator framework for rangelands and pastures of the USA. Clim Change 163:1733–1750. https://doi.org/10.1007/s10584-020-02915-y

Olimb SK, Lendrum PE (2021) Tracking cumulative cropland expansion across the great plains: the Plowprint. Great Plains Res 31:111–114. https://doi.org/10.1353/gpr.2021.0006

Organ JF, Geist V, Mahoney SP, Williams S, Krausman PR, Batcheller GR, Decker TA, Carmichael R, Nanjappa P, Regan R, Medellin RA, Cantu R, McCabe RE, Craven S, Vecellio GM, Decker DJ (2012) The North American model of wildlife conservation. Wildlife Soc Tech Rev 12(04) (The Wildlife Society, Bethesda, Maryland, USA)

Ostrom E (2009) A general framework for analyzing sustainability of social-ecological systems. Science 325:419–422. https://doi.org/10.1126/science.1172133

Rashford BS et al (2019) Assessing economic and biological tradeoffs to target conservation easements in western rangelands. In: Western economics forum, pp 9–23

Raworth K (2017) Donut economics: how to think like a 21st-century economist. Cornerstone Digital, Sevenoaks, UK

Reid RS et al (2014) Dynamics and resilience of rangelands and pastoral peoples around the globe. Annu Rev Environ Resour 39:217–242. https://doi.org/10.1146/annurev-environ-020713-163329

Roche LM (2021) Grand challenges and transformative solutions for rangeland social-ecological systems—emphasizing the human dimensions. Rangelands. https://doi.org/10.1016/j.rala.2021.03.006

Roche LM et al (2015) Sustaining working rangelands: insights from rancher decision making. Rangeland Ecol Manage 68:383–389. https://doi.org/10.1016/j.rama.2015.07.006

Rosenberg NJ (1987) Climate of the great plains region of the United States. Great Plains Quarterly 7:22–32

Scholtz R, Twidwell D (2022) The last continuous grasslands on Earth: Identification and conservation importance. Conserv Sci Pract 4(3). https://doi.org/10.1111/csp2.626

Sketch M et al (2019) Engaging landowners in the conservation conversation through landowner-listening workshops. Soc Nat Resour 1–12. https://doi.org/10.1080/08941920.2019.1657996

Sturm N (2017) Local food systems: case studies exploring opportunities and challenges for local food systems in Eastern Montana. Resources and Communities Research Group. http://resources4communities.org/reports/2017/9/14/local-food-systems-case-studies-exploring-opportunities-and-challenges-for-local-food-systems-in-eastern-montana. Accessed 31 Mar 2022

The Nature Conservancy (n.d.) The matador ranch. Web site. https://www.nature.org/en-us/get-inv olved/how-to-help/places-we-protect/matador-ranch/. Accessed 13 Mar 2022

Toombs T et al (2011) Rangeland ecosystem services, risk management, and the ranch bottom line. Rangelands 33:13–19. https://doi.org/10.2111/1551-501X-33.5.13

Turner Enterprises, Inc. (2021) Turner ranches. https://www.tedturner.com/turner-ranches/

University of Wyoming (2021) Grassbanking. Conservation Toolbox. http://www.uwyo.edu/haub/ ruckelshaus-institute/private-lands-stewardship/conservation-toolbox/grassbanking.html

U.S. Census Bureau (n.d.) CS demographic and housing estimates: 2019 1 year estimates. In: American community survey. https://www.data.census.gov/cedsci/table?q=DP05&g=040000 0US30&tid=ACSDP1Y2019.DP05

U.S. Dept of Agriculture, Census of Agriculture (2021) Summary by farm typology measured by gross cash farm income (GCFI) of family farm producers and non-family farms—United States: 2017. https://www.nass.usda.gov/Publications/AgCensus/2017/Online_Resources/Typ ology/typology_us.pdf

U.S. Department of Agriculture, National Agricultural Statistics Service (2021) NASS—Quick Stats. Accessed 2021-09-06

Western S (2021) Musselshell watershed: centered on the next generation, June 3. Strong Towns. https://www.strongtowns.org/journal/2021/6/3/rbc2021-mussellshell-watershed-centered-on-the-next-generation

White C, Conley C (2007) Grassbank 2.0. Rangelands 29:27–30

Wilson RK (2014) America's public lands: from yellowstone to smokey bear and beyond. Rowman & Littlefield

Chapter 28
Living with Predators: A 20-Year Case Study in the Blackfoot River Watershed of Montana

Seth M. Wilson

Abstract This chapter describes 20 years of efforts to live with large carnivores in the Blackfoot watershed located in western Montana with a focus on the processes and projects that were developed to adapt to the presence of grizzly bears (*Ursus arctos*) and gray wolves (*Canis lupus*) under the capacity of a non-governmental organization called the Blackfoot Challenge. Initial efforts were focused on generating a shared understanding of the problem, engaging community members in the co-generation of data, and designing an inclusive decision-making process that led to the adoption of a suite of tools that represented the values of stakeholders who represented communities of place and interest. Between 2003 and 2018, damages and livestock depredations by grizzlies tended to remain below 10 conflicts per year. Confirmed and probable livestock depredations show a low level of 1.8 annual livestock losses per year to grizzlies. Between 2007 and 2020, the wolf population increased and eventually leveled off, while livestock losses to wolves remained low. Annual confirmed livestock losses to wolves have been 3.3 livestock per year with less than four wolves removed annually due to depredations in the core project area. The central lesson of this effort is that living with large carnivores requires bringing people together to build trust, to generate a shared understanding of the problem using science, and to develop a participatory and equitable approach for changing practices and adopting tools that foster coexistence with carnivores.

Keywords Collaboration · Community-based conservation · Grizzly bears · Partnerships · Wolves

S. M. Wilson (✉)
Blackfoot Challenge, 405 Main Street, Ovando, MT 59854, USA
e-mail: seth@blackfootchallenge.org

© The Author(s) 2023
L. B. McNew et al. (eds.), *Rangeland Wildlife Ecology and Conservation*,
https://doi.org/10.1007/978-3-031-34037-6_28

28.1 Introduction

When large carnivore populations overlap with humans and agricultural activities, interactions can be problematic for people and livestock, and particularly for carnivores. In North America, large carnivores, such as grizzly bears (*Ursus arctos*) and gray wolves (*Canis lupus*), are generally less tolerated outside of designated protected areas (Mattson et al. 1996). When incidents or conflicts occur outside of protected areas, carnivores are often legally trapped, relocated, or removed from populations by wildlife managers. This pattern of conflict is often concentrated when human activities occur on and at the interface of public and private rangelands (Woodruff and Ginsberg 1998; Wilson et al. 2006). Since grizzly bears and gray wolves are generalist species who range widely and use a variety of habitats to satisfy their life history requirements, it is arguable that long-term population persistence is largely governed by human values, behaviors, and land use practices (Mattson et al. 1996; Boitani 2003). With increasing societal demands being placed on natural resources, finding ways to sustain populations of grizzly bears and wolves at landscape scales while incorporating rural livelihoods such as livestock production on rangelands becomes a critical conservation challenge in the American West.

This chapter describes two decades of efforts of living with large carnivores and a suite of projects designed to reduce conflict in the Blackfoot River watershed located in western Montana with a focus on grizzly bears and gray wolves. The purpose of this chapter is to describe how a rural community came together with wildlife agencies and conservation groups to grapple with the complex challenge of adapting to the presence of grizzly bears and wolves over time as both of these large carnivores recolonized this region.

This effort started approximately twenty years ago when Montana Fish, Wildlife and Parks (FWP) and the Blackfoot Challenge (BC)—a grassroots watershed group in the Blackfoot Valley—began meeting to discuss concerns among local residents about increasing grizzly bear activity and conflicts in the watershed and surrounding region. Beginning in the early 2000s, grizzly bears re-expanded their range onto private lands, causing conflict and concern among local residents. By 2007, gray wolves began to establish territories in the watershed after population expansion following reintroduction efforts in the mid-1990s to Yellowstone National Park and central Idaho. There was a clear need to bring people together to determine exactly how to define and address the growing concerns over the renewed carnivore presence.

This chapter emphasizes how a collective decision-making process encouraged diverse local and national stakeholders to engage in a partnership where participatory efforts helped to reduce conflicts for bears, wolves, and people. The capacity of the Blackfoot Challenge was pivotal for forging new connections with state and federal agencies, conservation groups, private landowners, residents, and livestock producers.

The first third of this chapter focuses on the process of community-based conservation that helped generate a shared understanding of problem to facilitate adapting and responding to grizzlies and wolves, how the local community was engaged in co-generation of data, and how an inclusive decision-making process led to the adoption of prevention tools that ultimately serve the common interest. The second section of this chapter discusses specific projects and tools such as electric fencing of calving areas, livestock carcass removal, management of human attractants from bears, and range rider efforts that were developed with community support that have helped reduce conflicts with grizzlies and wolves. This chapter concludes with practical lessons and recommendations useful to practitioners including wildlife managers, livestock producers, landowners, natural resource professionals, conservationists, and rural community members who live with large carnivores.

28.1.1 Conflicts with Grizzly Bears and Wolves

There are a variety of human-grizzly bear conflicts or incidents that typically occur on privately owned rangelands and include bears killing livestock, destroying beehives, foraging for garbage close to homes, or, in rare cases, threatening human safety (Wilson et al. 2006). Often, private rangelands in valley bottoms and foothills adjacent to public lands are problematic zones, especially when available bear attractants coincide with occupied grizzly bear habitat. Researchers have found that ranches in areas close to rivers and streams, with extensive habitat edges and at lower elevations are most susceptible to chronic conflicts with grizzly bears (Wilson et al. 2006; Northrup et al. 2012). Repeated incidents typically lead to more severe conflict, habituation, and eventually to management removal of bears through trapping, relocation or euthanasia by wildlife authorities. Additionally, private rangelands in livestock production adjacent to public lands are typically problematic zones for wolves and livestock since wolves can easily access private agricultural land (Bradley and Pletscher 2005; DeCesare et al. 2018). Repeated incidents with livestock typically lead to wolf removals. In these cases, outcomes are unfortunate for both those losing livestock and for the wolves themselves. One solution to breaking this cycle is to focus efforts on preventative measures that proactively address wolf-livestock and grizzly bear-livestock conflict. This position implicitly recognizes that long-term conservation and management of both grizzly bears wolves in places like Montana will require some level of human acceptance, tolerance and ultimately some changes in husbandry practices that help reduce conflicts with both bears and wolves. The Blackfoot watershed contains landownership patterns common in the West, namely a mix a public and private lands and a tradition of ranching on private and public rangelands and has abundant habitat that supports both grizzly bears and wolves.

28.1.2 Project Area

The project area is located in the Blackfoot watershed of west central Montana (Fig. 28.1). To the north of the 610,000-hectare watershed, United States Forest Service designated wilderness areas, Blackfeet and Flathead Reservations, and Waterton Glacier International Peace Park, comprise slightly more than 4 million hectares of rugged and largely protected landscape that is popularly referred to as the Crown of the Continent (COC) ecosystem (UNESCO 2021). The US Fish and Wildlife Service has designated a large portion of this ecosystem as an official grizzly bear recovery zone since the area has supported a population of grizzlies prior to European settlement. This source population has gradually increased over the past two decades and grizzly bear activity and dispersal events have been on the rise in the watershed since the late 2000s (Kendall et al. 2009). Montana Fish, Wildlife and Parks estimate that the population has grown at approximately 3 percent per year since population trend monitoring began in 2004 (Costello et al. 2016).

The geologic and hydrologic characteristics of the Blackfoot watershed have resulted in extensive community types including wetlands, bogs, fens, spring creeks, riparian swamps, and extensive cottonwood forests. This diverse mosaic of upland foothills, glacial outwash plains, grasslands and extensive creek and river bottoms is quality habitat for a wide array of wildlife including grizzly bears and wolves. The Blackfoot watershed has remained largely undeveloped, is rural in character, and is sparsely populated. Being located at the southern end of the COC, the watershed has been a natural location for grizzly bears to re-colonize former habitat. Additionally, the watershed was recolonized by wolves beginning in 2007.

28.1.3 Communities of Place and Communities of Interest

Stakeholders involved in grizzly bear and wolf conservation and management can be organized into communities of place and communities of interest (Wilson and Clark 2007). These stakeholders include those who live and work in the watershed who hold largely rural values and those communities of interest—often urban-based populations who support nature conservation, wildlife protection, and national goals of species protection as embodied in the Endangered Species Act (Chapter 29). Many of the conservation groups and state and federal agencies involved in bear and wolf recovery can be said to represent regional and national communities of interest who support wildlife conservation. Key groups who have a stake in large carnivores include ranchers, non-ranching residents, state and federal governments, and conservation groups. There are approximately 2,500 households in seven small communities located throughout the watershed. The dominant land use is primarily family-owned cow/calf ranching operations and some small-scale forestry.

Landowners in the Blackfoot have cherished their rural way of life and have worked together for generations to maintain agricultural traditions, open space,

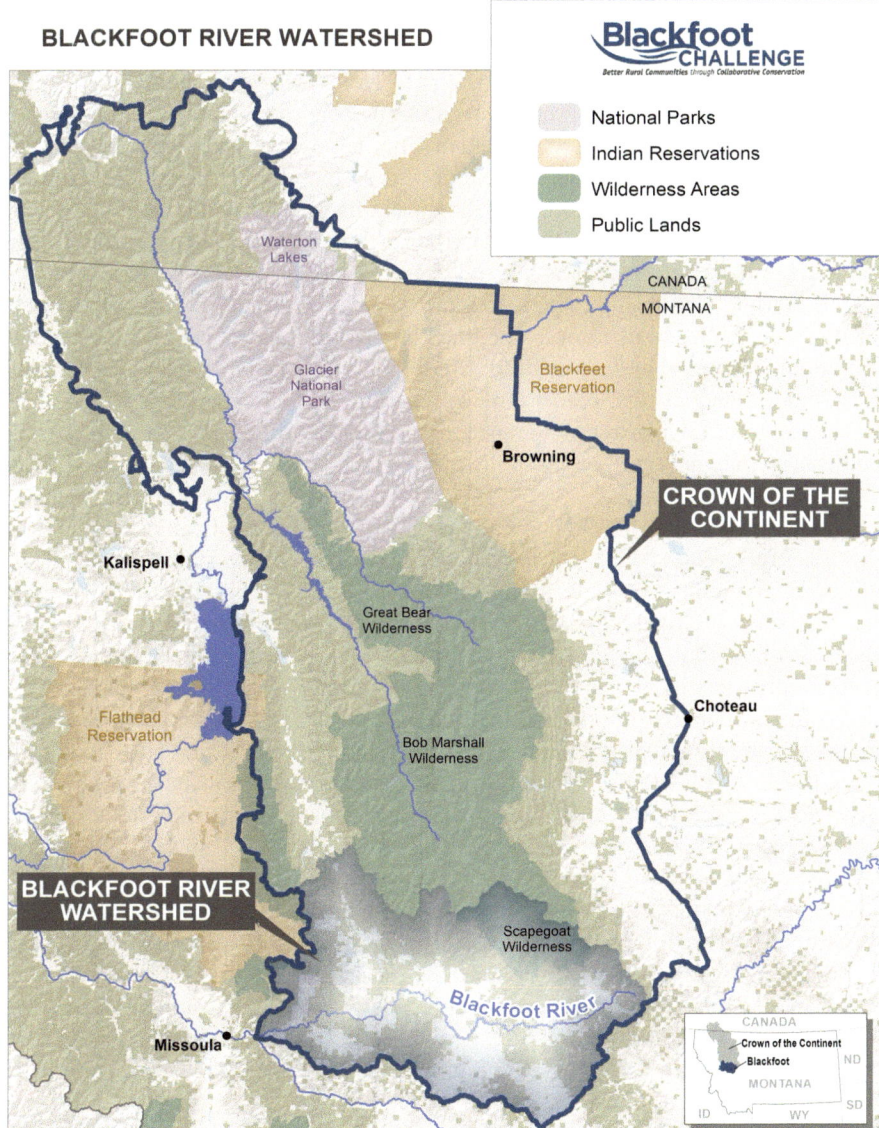

Fig. 28.1 Location of Blackfoot River watershed within western Montana's Crown of the Continent Ecosystem

and rural livelihoods. The ranching community takes pride in an independent life style, yet ranchers also value strong neighborly relationships and sharing seasonal labor demands (e.g., haying) characteristic of North American agrarian communities (Bennett 1967). Private property rights and economic viability are also important values held by the ranching community. Initially with grizzly bears reoccupying the watershed in the early 2000s and followed by wolves, carnivores were initially perceived as unwelcome visitors that threatened livelihoods and human safety.

While ranching is still a dominant land use in the watershed and the cultural norms of ranchers have permeated the general character of the valley, new residents have increasingly moved to the region. In many respects these newcomers are amenity migrants, who have been drawn to the Blackfoot for its beauty, open space, recreation opportunities, and abundant wildlife. These new residents are typically tolerant of grizzly bears and wolves, but in some cases, have limited experience with actually living with them. However, new residents have been willing and, in certain cases, enthusiastic, about participating in the grizzly bear and wolf related projects of the BC.

Currently Montana FWP plays the main role in grizzly bear management in the Blackfoot watershed. FWP is responsible for day-to-day management of grizzlies (e.g., conflict responses, monitoring) in consultation with the USFWS under the Interagency Grizzly Bear Committee (IGBC) guidelines (IGBC 2021). FWP respects traditional ranching livelihoods in Montana, has actively embraced the collaborative nature of the BC partnership concerning grizzly bear management, and actively supports projects that help maintain rural ranching through economic incentives and technical support (Jonkel 2002). Similarly, FWP has current management authority for gray wolves in Montana and coordinates with USDA Wildlife Services for forensic investigations of confirmed and suspected livestock depredations from grizzly bears and wolves. The U.S. Forest Service (USFS), U.S. Bureau of Land Management (BLM), and Montana State Department of Natural Resources and Conservation (DNRC) play minor consultative roles in bear and wolf habitat management and have taken part in the BC efforts on an as needed, project-by-project basis, since the bulk of the BC work on grizzly bear and wolves has focused on private rangelands. Non-government conservation groups (e.g., The Nature Conservancy, Vital Ground, Defenders of Wildlife, Brown Bear Resources, the Great Bear Foundation, and the Living with Wildlife Foundation) have also been active participants in the efforts of the BC partnership.

28.2 The Blackfoot Challenge

The Blackfoot Challenge was incorporated as a non-governmental organization (NGO) in 1993. However, the BC origins date back to the early work of visionary landowners in the 1970s who saw opportunities to conserve and manage land, water, and wildlife in a more holistic way. This approach was based on the premise that collaboration is central to effective conservation (Blackfoot Challenge 2021).

The BC has played a central role as the organization in the 610,000-hectare watershed that has brought people together and facilitated respectful conversations to generate bottom-up solutions embedded in public and private partnerships. The BC mission, "To coordinate efforts to conserve and enhance natural resources and the rural way of life in the Blackfoot watershed for present and future generations," has brought communities of place and interest together to generate a culture of conservation defined by inclusiveness, collaboration, transparency, and most importantly, trust. In turn, these core values provide the overarching framework for a robust, consensus-driven process that committees and work groups use to implement conservation across programs. At the heart of the process is the recognition that conservation rests upon the support of communities of place and communities of interest—where local and broad public values converge (Wilson and Clark 2007). The BC programs reflect this convergence of local and national interests and has allowed robust programs that leverage funding, scientific expertise, technical skills, and local knowledge that generate lasting collective conservation impacts across the watershed. The BC works to build effective partnerships and working relationships based on trust, respect, credibility, and the ability to empathize across a diversity of values. While difficult to measure, these intangibles help build what Robert Putnam terms, social capita and have been benchmarks of the Blackfoot Challenge's success (Putnam 2000). This has allowed the BC and partners to take on complex and difficult conservation challenges like coexisting with grizzly bears and wolves. Additionally, other noteworthy conservation successes include: 52,600 hectares of land permanently protected from development, creation of a 2,300 hectare community-owned and managed forest, restoration of west slope cutthroat (*Oncorhynchus clarkii lewisi*) and bull trout (*Salvelinus confluentus*) habitat, a voluntary drought response plan to maintain in-stream flows in the Blackfoot River, restoration of native trumpeter swans (*Cygnus buccinator*), and improvements to native grasslands and soil health across hundreds of working ranches.

28.2.1 Collaboration Before Conflict

The Blackfoot Challenge has developed a partner-based culture of conservation that helps foster collaboration before conflict. This is achieved in two fundamental ways—non-advocacy and a consensus-driven approach. The BC process led and fostered by local landowners, residents, staff, and an elected Board of Directors has allowed the BC to act as the forum in the watershed for encouraging civic and rational engagement focused on a variety of natural resource issues without taking a particular position. The non-advocacy and non-litigatory stance of the organization has earned the trust and support of local residents who represent diverse values. And like much of the Western landscape, the Blackfoot watershed contains a mix of public and private lands that are cherished and contested by stakeholders who desire resource use, recreation, and non-consumptive ecosystem services.

This framework allows the BC to tackle a range of issues, from simple to complex or contentious. A rule of thumb known as the 80:20 rule encourages stakeholders to work first on the figurative 80% of an issue where agreement may be found and then address the harder 20% of an issue. Or in other cases, set aside those more intractable 20% of issues until simpler problems can be solved. This step-wise process often helps build trust, credibility and goodwill that can allow stakeholders to find mutually beneficial outcomes initially and in turn, help participants address more complex and difficult issues as needed.

The BC is both a process and a project-based organization driven by stakeholder priorities. In a given year, the BC and dozens of key partners are engaged in hundreds of projects from management of a community-owned forest, soil moisture monitoring, reintroduction of trumpeter swans, to removing livestock carcasses from ranches that would otherwise attract grizzlies and wolves into potential conflict. The BC relies on seven committees and respective work groups to address a range of conservation issues. Each committee is chaired by a landowner to ensure that local values are taken into account. Another mechanism that has helped the BC work successfully with local, state, and federal land management agencies has been to invite key people from leadership positions from the various agencies to serve as Board Members and Board Partners. Often these board members are well-placed decision-makers from state and federal agencies whose management jurisdictions fall within the Blackfoot watershed. Committee membership is naturally driven by the specific natural resource issue and interests of stakeholders and has allowed representatives from FWP, the U.S. Forest Service (USFS), Bureau of Land Management (BLM), and FWS to take active roles in issues related to forestry, grazing management, or wildlife.

In many respects the BC serves as a parallel institution of governance within the watershed and is able to harness and engender the collective good will of stakeholders who are willing to take part in the process of collaboration. This capacity has been critical for addressing controversial issues such as grizzly bear population expansion onto private agricultural lands and eventual wolf recolonization of the watershed that began in 2007.

28.2.2 Developing a Shared Understanding of the Problem

Grizzly bears began reoccupying the Blackfoot watershed in the mid-1990s and the first reported and verified conflicts began by 1998 (Jonkel 2002; Fig. 28.2). Conflicts ranged from livestock losses to predation, beehive damage, property damage, sanitation, to human-bear encounters, and bears in close proximity to dwellings. In 2001, a hunter was killed from an encounter with a female grizzly bear with cubs. This event caused widespread concern and anxiety among landowners and residents but also created a point of entry to address the emerging problem of human-grizzly bear conflicts and led to the formation of the Blackfoot Challenge Wildlife Committee in 2001 (Wilson and Clark 2007). BC Wildlife Committee members (approx. 45) included landowners, ranchers, and residents from the Blackfoot watershed and managers from FWP, U.S. Fish and Wildlife Service-Montana (MT) Partners for Fish and Wildlife Program (FWS), U.S. Forest Service, USDA Natural Resource Conservation Services (NRCS), and MT DNRC. Additional NGO members included representatives from Defenders of Wildlife, The Nature Conservancy, and Living with Wildlife Foundation. The committee represented the respective landownerships and management jurisdictions in the watershed both public and private, resulting in a regular dialogue among various stakeholders who represented communities of place and communities of interest (Wilson and Clark 2007).

Stakeholders at that time believed that working under the existing framework of the BC was a pragmatic and thoughtful way to approach the presence of grizzly bears. Instead of trying to build a stand-alone effort, the BC offered a forum for working with ranchers, landowners, conservation groups, and agencies. It was apparent at the time that a partnership-based approach would be needed to respond to increasing conflicts with grizzly bears and that significant decision-making power would need to be in the hands of those landowners and ranchers who confronted daily realities of living with bears (Wilson et al. 2014, 2017). The BC and FWP met to discuss a more formal arrangement where interested parties could directly engage local community members in wildlife management (Wilson et al. 2014). Subsequently the BC agreed

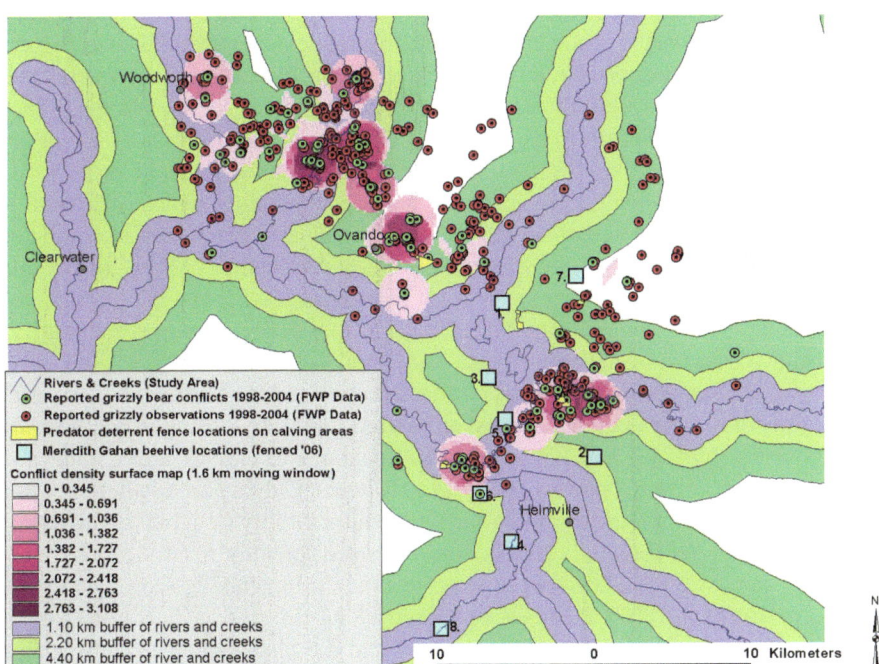

Fig. 28.2 Example of GIS mapping of ranching land uses with calving areas (blue polygons) and boneyards (yellow squares) that are known to be attractive to grizzly bears and increase the risk of conflicts with bears

to form the BC Wildlife Committee with the understanding that the initial focus would respond to grizzly bears.

28.2.3 Livestock Producer Perceptions of Grizzly Bears

A first action of the BC Wildlife Committee was to conduct an in-person survey of livestock producer perceptions of grizzly bears within the Blackfoot watershed. In 2002 the BC Wildlife Committee surveyed thirty-five ranchers, outfitters, and small-scale ranch operators using a pre-tested closed-and-opened ended survey. The responses helped describe how residents' livelihoods could be impacted by grizzly bears (Wilson et al. 2014).

The surveys and discussions in group meetings revealed that local ranchers had varying and complex opinions on living with grizzly bears (Table 28.1). Some respondents took a pragmatic view, with concerns focused on human safety, protecting property and livelihood interests, and the need for more information about bears and bear management—all areas that theoretically could be addressed by understanding the problem as one of risk management and improving information sharing. Other perspectives were more difficult to address. For example, 52% of respondents explained that grizzly bears should be geographically separated from human activities as a way to solve the problem and 71% of respondents felt that there were simply too many bears using private lands. In small group discussions, sentiments such as "environmentalists were the cause of our bear problem" were expressed. These types of problem definitions posed barriers to constructive discussions about how best to respond, since there were no feasible solutions available if the problem was characterized in these ways. This exercise provided a practical pathway forward for framing discussions that enabled the BC Wildlife Committee to co-generate goals that focused on four core issues important to stakeholders: (1) protecting human safety, (2) protecting private property from bear damage, (3) protecting rural livelihoods, and (4) improving information sharing from wildlife managers about grizzly bear behavior and management to stakeholders.

During the survey, data were also collected on the spatial locations of a variety of livestock management practices and attractants that were leading to conflicts with grizzly bears (Jonkel 2002). These included locations of calving areas, boneyards, and beehives relative to habitat used by grizzly bears. Calving areas where newborn calves are born attract grizzly bears due to the vulnerability of calves and afterbirth (Mattson 1990; Wilson et al. 2006). Boneyards are spatially fixed locations where livestock carcasses have been deposited over time during calving season and attract bears (Wilson et al. 2005). Beehives have been a long-time source of conflict for both grizzly and black bears (*Ursus americanus*) and are also spatially fixed (Mattson 1990). Mapping these types of attractants was critical for understanding the scale at which bear conflicts were playing out, where conflict densities were greatest, and helped stakeholders visually register that a collective response across private ranch ownerships and with hundreds of residents would be necessary to address conflicts

Table 28.1 Likert-scaled statements regarding perceptions of grizzly bear activity and appropriate landowner/resident behaviors in the Blackfoot watershed, Montana (adapted from Wilson et al. 2014)

Statement	Agree (%)	Disagree (%)
1. Grizzly bears that use private land are a threat to human safety	71	29
2. I do not feel safe when I am outside on my property because of grizzly bears	45	55
3. There are too many grizzly bears using private lands in this area (Blackfoot)	71	29
4. I am comfortable with the current level of grizzly bear activity in this area (Blackfoot)	32	68
5. Private landowners have a responsibility for protecting grizzly bears	42	58
6. This would be a better place to live if there were no grizzly bears on private lands	52	48
7. People shouldn't have to change their habits to accommodate grizzly bears that use their private land	58	42
8. Private landowners should take precautions to reduce conflicts with grizzly bears	90	10
9. Grizzly bears should remain off limits to hunting	10	90
10. Grizzly bears are a serious threat to my livestock	45	55

at the biological scale of grizzly bear foraging bouts (1.6 km^2 based on Wilson et al. 2005) and bear home ranges.

28.2.4 Co-Generation of Data—Understanding the Scale of Grizzly Bear Conflict

Geographic information systems (GIS) were used to map and analyze ranching land use practices and other ecological features that were known to be associated with an increased risk of conflict (Wilson et al. 2006; Fig. 28.2). Interactive GIS mapping sessions were conducted with 35 active ranchers in the Blackfoot watershed using methods developed by Wilson et al. (2005, 2006). Montana Fish, Wildlife and Parks shared data on reported and verified grizzly bear conflicts and observations (1998–2004) that were then used to analyze and prioritize where in the landscape to focus initial conflict mitigation efforts.

The co-generated GIS maps were shared with the ranching community through the BC Wildlife Committee to address problems over the next decade, namely through continued GIS mapping and monitoring of boneyards and calving areas, electric fencing of high-risk calving areas, and eventual phase out of boneyards that was replaced by a livestock carcass removal program. Conflict reduction efforts were focused on the middle portion of the Blackfoot watershed where there were the greatest densities of bears and past conflicts. Recently the BC Wildlife Committee and FWS have contracted a newly updated GIS spatial hotspot analysis that builds off of our early GIS work (Williams and Hebblewhite 2021) and will help guide another round of conservation investment—beginning with $300,000 in conflict hotspot mitigation with the Natural Resources Conservation Service (NRCS) using electric fencing and a new innovation in drive-over electrified mat system that can replace the need to use a traditional swing gate (Fig. 28.3).

28.2.5   Co-Generation of Data—Estimating Wolf Pack Numbers and Distribution

The first known wolf territory in the Blackfoot watershed was established in 2007 and the first confirmed livestock depredations were recorded in 2008 by Wildlife Services. Rumors circulating in the community regarding the size and distribution of the wolf population were brought to the BC Wildlife Committee meetings and created another point of entry for FWP and the BC Wildlife Committee to address the perception that the "valley was being overrun by wolves", a refrain commonly heard in informal settings. Under the guidance of FWP, the BC Wildlife Committee engaged community members in conducting an annual winter wolf track survey to generate a better collective understanding of wolf abundance, approximate distribution, and general activity within the watershed.

Through the BC Wildlife Committee, permission was granted from dozens of ranchers to conduct a winter wolf track survey across their private lands. Volunteers from FWP, FWS-Partners Program, US Forest Service, BLM, and university collaborators worked together to conduct wolf track surveys during 3 days in late January over a large portion of the watershed identified as wolf habitat. As the effort developed, more than one hundred volunteers took part annually over the next four years. Volunteers recorded and provided sign observations to FWP wolf management specialists who then estimated the annual number of wolf packs, total wolf numbers, and approximate distribution; the annual report was then shared with the community and discussed in subsequent BC Wildlife Committee meetings.

An important benefit from this collaborative effort was that the co-generation of data on wolf activity with community members helped dispel rumors that there were large numbers of wolves in the watershed. Residents could literally see for themselves wolf tracks, sign, and better understand how wolves and respective wolf packs used the landscape. In many cases participants learned firsthand that what they initially believed to be multiple wolf packs was in fact, a single group using a large territory. Simultaneously, the joint effort with the community to conduct winter wolf surveys provided an opportunity for FWP and community members to share knowledge about the behavior and general ecology of wolves. Similar to the collaborative learning process that took place five years earlier with grizzly bears, the information sharing from FWP and community members increased trust and

Driver-Over Electric Mat: Grizzly bear fleeing from drive-over electric mat after receiving shock in successful field test in 2019 (a) and an example of a drive-over electric mat being field tested in a high-use ranch driveway entrance with grizzly bear receiving a shock (b). Drive-over electric mat systems are designed to be used with electric fences to deter bears from calving areas, spring turn-out pastures, and ranch homesites during the season when bears are active. In high-use areas, the drive-over mat can be a convenient way to reduce the need to open and close traditional gates (photo credits: FWS (a) and BC (b)).

Calving Area Electric Fence: An electric fence surrounds a calving area to protect vulnerable newborn calves from grizzly bears and wolves on a ranch in the Blackfoot watershed. The calving season typically overlaps with the emergence of grizzly bears from their dens in the early spring and is a time of high risk for livestock, as the young are small and more vulnerable to predation (photo credit: BC).

Apiary Electric Fence: An electric fence charged by a solar cell surrounds a beehive and protects the hive from grizzly bears and black bears in the Blackfoot watershed. Approximately 50–60 beehives have been secured with permanent electric fences for the commercial beekeepers in the watershed (photo credit: BC).

Range Rider: Range riders increase human presence and livestock herd supervision rates to deter potential wolf depredations. Range riders monitor wolves using radio telemetry, trail cameras, and tracking. Range riders also recover lost, sick, or injured livestock and detect and remove (when possible) naturally occurring livestock carcasses and can detect carcasses for investigation by Wildlife Services for possible compensation (photo credit: Melissa Mylchreest).

Fladry: This type of fencing uses interspersed flagging attached to a line, cord, or electrified poly-wire to create a psychological avoidance response (novel stimuli) in wolves and is effective at deterring wolves when strung around livestock pastures. Electrified fladry, using a line of poly-wire, reinforces a fear response in wolves by adding an electric shock (photo credit: BC).

Fig. 28.3 Examples of tools and technologies used by the BC Wildlife Committee to reduce conflicts with grizzly bears and wolves including drive-over electric mats, calving area electric fences, beehive/apiary electric fences, range rider, and fladry

credibility and the believability of data about wolves among stakeholders—important cornerstones that helped increased participation by livestock producers in future projects (Wilson et al. 2017).

28.2.6 Inclusive Decision-Making—Balancing Communities of Place and Interest

The process of working with stakeholders to orient to the perceived problem of having grizzly bears and wolves back in the Blackfoot watershed was an investment in trust and inclusion, and the process of co-generating data helped bring the scale at which grizzly bears and wolves fulfill life history needs into biological focus. The BC Wildlife Committee provided an inclusive forum that humanized representatives of wildlife agencies, NGOs, and the livestock community through face-to-face meetings among people. Biological information on grizzly bears was readily shared with the community by FWP management specialists and became a critical part of regular meeting updates and helped to reduce anxiety about human safety, bear numbers, densities, and habitat use. Ranchers were also willing to share information about their operations and bear activity they observed—making the overall picture of grizzly bear use in the Blackfoot watershed much clearer and more comprehensive for all stakeholders.

Although grizzly bears were the initial focus of BC Wildlife Committee efforts, with wolves reoccupying the watershed, the BC Wildlife Committee and partners saw opportunities to build off the social capital that had been generated around the response to grizzlies and provided the opportunity to expand and refine on-going projects and develop new projects, anticipating wolf population growth in the watershed (Wilson et al. 2017).

28.3 Participatory Projects to Further Coexistence

The BC Wildlife Committee approached livestock producers, landowners, and residents with the hope that projects would be participatory and pragmatic—in other words, projects would only be useful to stakeholders if each effort was carefully planned and implemented with attention to the needs of the landowner, site conditions, and costs. For nearly all projects the BC Wildlife Committee and partners have provided substantial economic cost-share support and landowners have responded with in-kind labor and in other cases, equipment donations.

28.3.1   Livestock Protection: Electric Fences for Calving Areas and Apiaries; Fladry Fences to Deter Wolves

Cow-calf ranches in the Blackfoot watershed are characterized by winter feeding, centralized and spatially fixed operations, irrigated hay production, and docile breeds of cattle (Dale 1960; Jordan 1993). The calving season typically overlaps with the emergence of grizzly bears from their dens in the early spring. Spring calving/lambing is a time of high risk for livestock, as the young are small and more vulnerable to predation. Bears routinely visit calving areas, and the traditional practice of depositing dead livestock into boneyards (carcass dumps) can lead to chronic livestock-grizzly bear conflicts (Wilson et al. 2006). The first calving area fences in the Blackfoot watershed were built in 2001 as a proven non-lethal method to deter grizzly bears from newborn calves. As of 2021, there are 18 calving area fences constructed on 12 individual ranches and an additional 10 electric fences protecting municipal transfer sites (3) and dwellings (7). The construction of fences was paid for using funds from public and private foundations, FWP, NRCS, and the FWS-Partners Program, and provided ranchers with substantial cost savings on the capital investments. Fences were designed at that time to be both grizzly bear and wolf resistant using a combination of fencing guidelines from FWP, the US Forest Service, and the Province of Alberta where ranchers had long-time experience using electric fences to protect livestock from grizzlies and wolves (Fig. 28.3). Ranchers helped share the total costs through their in-kind donations of labor to prepare sites and remove old fences. At first, some ranchers were concerned that electric fences would require excessive maintenance or would be susceptible to ungulate damage. In some cases, ranchers were unfamiliar with the technical aspects of electric fencing, and the adoption of this new technology challenged norms such as pride in their self-reliance regarding routine work like fixing barbed wire fences (Wilson et al. 2014). Over time, nearly all electric fences have been maintained by ranchers and in only a few cases, did fences fall into disrepair. The BC Wildlife Committee conducts regular inventory of all existing electric fences throughout the watershed and prioritizes maintenance based on sites in high conflict risk areas.

Apiaries, also known as beehives, present food attractants for grizzly bears in the Blackfoot watershed. Early GIS mapping analysis identified beehives that were in high risk areas throughout the watershed and the BC Wildlife Committee has worked closely with two commercial beekeepers over the past two decades to cost-share construction of permanent electric fences (Fig. 28.3). Based on inventories from the two commercial beekeepers who work in the watershed, there are approximately 55–60 apiaries sites with permanent, solar-powered electric fences that have proven extremely effective in preventing damage by grizzly bears and black bears.

Another tool that was useful was fladry. This is a type of fencing that uses interspersed flagging attached to a line, cord, or electrified poly-wire to create a psychological avoidance response (novel stimuli) in wolves and has been shown to be an effective way to deter wolves when strung around livestock pastures (Musiani et al.

2003). Electrified fladry, using a line of poly-wire, reinforces a fear response in wolves by adding an electric shock (Lance et al. 2010; Fig. 28.3).

28.3.2 Managing Agricultural Attractants: Boneyards and Livestock Carcass Removal

Livestock carcasses and boneyards can be an attractant for wolves and grizzly bears and bring them into closer proximity to livestock production areas thereby increasing risk of depredations (Fig. 28.4). Phasing out boneyards and regular carcass removal was designed to remove the cows, calves, ewes, and other livestock that naturally die during the calving and lambing season (mid-February through mid-May), so that carcasses would not be found by foraging grizzly bears and wolves.

In addition, livestock depredations by wolves in the Western United States peak in early spring and fall each year (Musiani et al. 2005). In southwest Alberta where cattle operations, husbandry practices, range use, and terrain are similar to the Blackfoot watershed, researchers found that 85% of all wolf scavenging events occurred on ranchers' boneyards (Morehouse and Boyce 2011). Phasing out boneyards and replacing them with regular carcass removal was designed to remove livestock that naturally die during the calving and lambing season (mid-February through mid-May), so that carcasses would not attract foraging grizzly bears and wolves (Fig. 28.4).

The initial efforts to remove livestock carcasses generated concern from ranchers who did not want to have numbers of livestock deaths on their ranches disclosed to neighbors for fear of being stigmatized as deficient in animal husbandry (Wilson et al. 2014). This concern was addressed by establishing centralized drop-off locations where ranchers could bring carcasses anonymously for pick up. Participation steadily increased in the program in the early 2000s. Today the program covers nearly 4,860 km^2 across four western Montana counties and annually has 110–120 ranches actively participating representing roughly 90% of the total producers in the livestock carcass program area. More than 11,500 carcasses have been removed since the program began in 2003 and approximately 600 carcasses are removed annually. Livestock carcasses are composted at multiple facilities in the region (Fig. 28.4).

The carcass removal program has been enormously successful. In addition to decreasing conflict with carnivores, Montana Department of Transportation, a key partner in the effort, has successfully used the compost by-product on a variety of revegetation projects as well (Fig. 28.4). Composting livestock carcasses has proven to be a highly effective disposal method and has been widely applauded by the ranching community as a more appealing method of disposal than past practices of depositing carcasses on boneyards on their properties or removing carcasses to nearby landfills. The program relies on a mixture of public and private funding and in-kind and cash donations from partners and the ranching community to make the service virtually free to the ranching community.

Livestock Carcass Boneyards: Boneyards are areas where dead livestock are typically deposited (a). Since the calving process is labor intensive and requires constant management, it is typical for boneyards to be located near calving and lambing areas, serving as a labor-saving technique for the disposal of dead animals. However, this increases risk that scavenging grizzly bears (b) and wolves (c) will come into conflict situations on and near ranches (photo credits (a) BC, (b) FWP, (c) BC).

Livestock Carcass Removal and Composting: Livestock carcass removal is designed to remove cows, calves, ewes, and other livestock from ranches that naturally die during the calving and lambing season during mid-February through mid-May so that carcasses are not found by foraging grizzly bears, wolves, or other carnivores (a). Carcasses are brought to composting facilities in the region (b). MT Department of Transportation has partnered with the BC Wildlife Committee to compost carcasses that produce a rich by-product that can be used for revegetation or soil amendment (c) (photo credits: BC).

Managing Household Attractants and Municipal Waste: Household garbage (a) or other attractants can be secured in storage containers (b) and municipal waste transfer stations are electrically fenced to deter grizzly bears from scavenging or posing a risk to human safety (c) (photo credits: BC).

Fig. 28.4 Examples of boneyards, livestock carcass removal and composting, managing household attractants, and an electric fence on a municipal waste transfer station

28.3.3 Managing Household Attractants: Neighbor Networks

In addition to managing agricultural attractants with ranchers, the BC Wildlife Committee worked closely with residents to help increase communication about grizzly bear activity and to provide the means to remove or contain a variety of household attractants that could lead to conflict with grizzlies. As a first step, the BC Wildlife Committee helped to organize residents through a Neighbor Network using the Powell County 911 database to identify occupied households and connect local residents originally through phone trees and later through social media platforms. The network consists of over 120 residents who work together to accomplish the following: (1) minimize the availability of human-related attractants, (2) communicate among neighbors about grizzly bear and wolf activity using phone-trees, e-mail alerts, and social media and (3) provide a centralized reporting location for incidents or observations of bear or wolf behavior that may pose problems. The goal of this effort is to improve communication among neighbors and with Montana FWP in order to prevent conflicts with carnivores from starting in the first place. Nine networks are operational within the project area, each with a coordinator, to help facilitate communication among neighbors and to FWP when there is grizzly bear or wolf activity. A free check-out program administered by the BC Wildlife Committee allows residents to borrow bear resistant trash cans, portable electric fencing, electrified bird feeders, and other non-lethal deterrent tools to prevent conflicts. The BC Wildlife Committee, FWS, Defenders of Wildlife and FWP help provide funding to support the program.

The BC Wildlife Committee also focuses on common sense management of waste and household garbage for all residents of the Blackfoot. The BC Wildlife Committee has worked closely with waste haulers and residents to encourage use of bear resistant garbage cans and to take simple precautions to keep garbage secure from scavengers (Fig. 28.4). Additionally, all rural transfer sites (3) within the watershed have permanent electric fences to deter grizzly bears from scavenging on garbage (Fig. 28.4).

28.3.4 Livestock and Wolf Monitoring Using Range
Riders—2008–2020

Livestock herd supervision, practiced for centuries throughout the world, is a proven tool to help reduce livestock losses to carnivores including wolves (Boitani 2003). Researchers have found that the spatial distributions of predator and prey species vary with human activity levels (Hebblewhite et al. 2005; Muhly et al. 2011). Prey species were more prevalent in areas with high human activity and predator species including wolves avoided high human use areas—hence the justification for increasing herd supervision rates by using range riders (Wells et al. 2019). Ranchers in the Blackfoot watershed were supportive and welcomed the use of range riders as a tool to reduce problems with wolves.

With the arrival of wolves in 2007 and subsequent depredations, several ranchers were concerned, particularly those whose private lands and public grazing allotments fell within the newly established territories. The BC Wildlife Committee responded and worked closely with a prominent ranching family that was concerned and hired a family member and an assistant to pilot test the first range rider effort in the watershed. Using a volunteer agreement with FWP, the range rider was trained in ground-based, VHF telemetry use to detect radio collared wolves and to detect wolf tracks and sign. Livestock were checked daily by the range rider throughout the grazing season (May 1–October 31) on public grazing allotments on horseback, all-terrain vehicles, and a truck. There were no known livestock depredations by wolves on this ranch for that first season.

Piloting the range rider program with a well-respected ranch family and hiring a local community member who was highly competent and well-regarded resulted in a favorable response from the ranching community that range riders could be a workable solution to wolf predation. Additionally, FWP and the BC Wildlife Committee earned credibility from the ranching community by responding to the perceived threat of wolves in a timely manner. As the wolf population increased steadily, and wolf packs became widely distributed throughout the watershed, range riders were perceived as a useful tool for reducing the risk of livestock depredations for dozens of livestock producers (Sime et al. 2011; Hanauska-Brown et al. 2012). At the time of this writing, the BC currently employs (2) full-time and (1) part-time range rider that work closely with the BC Wildlife Committee coordinator. On a given year from 2007–2020, range riders and assistants worked closely with 15–18 ranchers to monitor approximately 4,600 head of livestock, across 78,900 acres in five communities in the Blackfoot watershed (Fig. 28.3). Range riders were in direct contact with another 40–50 livestock producers and ranchers and produced a bi-weekly wolf report that was e-mailed to another 150 interested stakeholders and posted on the Blackfoot Challenge's website.

While range riders helped increase human presence, riders also took proactive actions in cooperation with participating ranchers that included the following: (1) delayed pasture use when wolves were present, (2) detection and recovery of lost livestock, (3) detection and removal of sick / injured livestock, (4) detection and removal (when possible) of naturally occurring livestock carcasses, (5) detection of livestock carcasses from predation for investigation by Wildlife Services for possible compensation by the State of Montana, (6) general herd health surveillance, (7) deployment of fladry when needed, and (8) assisting producers with fall gathering and assessment of cause of death for possible missing livestock (Wilson et al. 2017).

The range rider effort in the Blackfoot watershed has been supported by the livestock community and invested stakeholders. While increased herd supervision rates and human presence may help reduce the frequency of encounter rates between livestock and wolves and subsequent depredations, this metric is difficult to measure without rigorous pre-and-post quasi-experimental design. Nonetheless, a value-added benefit from this effort has been increased and improved communication among stakeholders about wolf activity, wolf pack locations, and the proactive actions that range riders and ranchers collectively take. A researcher who conducted extensive interviews with participating ranchers in the range rider program found similar responses by participants involved in the effort (Parks 2015). Directly engaging ranchers in the effort by the range riders helped producers feel supported by FWP and the BC Wildlife Committee, and having more intensive livestock herd monitoring reduced their anxiety about wolves and potential livestock losses. Additionally, range riders were helpful in detecting livestock killed from natural causes and not from carnivores—an important way to reduce the chances that wolves or other carnivores were blamed for suspected losses (Wilson et al. 2017).

28.4 Conservation Impacts

The willingness of landowners, ranchers, and residents to work with a diversity of stakeholders to reduce conflicts with grizzly bears and wolves was encouraging. According to FWP Region 2 data for the core project area where prevention efforts have been focused over the past two decades, there has been an approximate 71% decrease in property damage and livestock depredations from grizzly bears from 2003 to 2019 with the exception of 2018. Over the past seven years, damages and depredations by grizzlies tended to remain below 10 conflicts per year with the exception of 2018. The Montana Livestock Loss Board's data on confirmed and probable livestock depredations only also suggests a decrease in livestock losses and a low level of 1.8 annual livestock losses per year to grizzlies over the past 23 years (Fig. 28.5). From 1998–2019 there were five confirmed grizzly bear mortalities in this same core area according to these same FWP Region 2 data. Compared to other monitoring units with significant portions of private land in FWP's demographic monitoring area (DMA), grizzly bear mortalities that are caused from repeated conflicts with people in the Blackfoot watershed core project area remain at some of the lowest levels across the Northern Continental Divide Ecosystem (Costello et al. 2016).

The above results are a positive sign that in general, conflicts are relatively low in the core project area in the face of growing and expanding populations of large carnivores in the project area (Costello et al. 2016; Mace et al. 2012; Kendall et al. 2009). The reduction in human-bear conflicts and bear mortality that may have in part, resulted from these efforts had several important outcomes: (1) an increased level of trust and credibility generated among stakeholders as projects produced results, (2) a positive economic impact on livestock producers by minimizing livestock losses to grizzlies, and (3) an impression of overall improvement in community-level acceptance of grizzly bears in the watershed.

For period 2007–2020, livestock losses to wolves have been low while the wolf population increased exponentially and eventually leveled off (Fig. 28.6) (Coltrane et al. 2015, MT Fish, Wildlife and Parks 2021; MT Livestock Loss Board 2021). Wildlife Services provided reports to FWP regarding confirmed livestock losses to wolves and to the MT Livestock Loss Board. Annual confirmed livestock losses

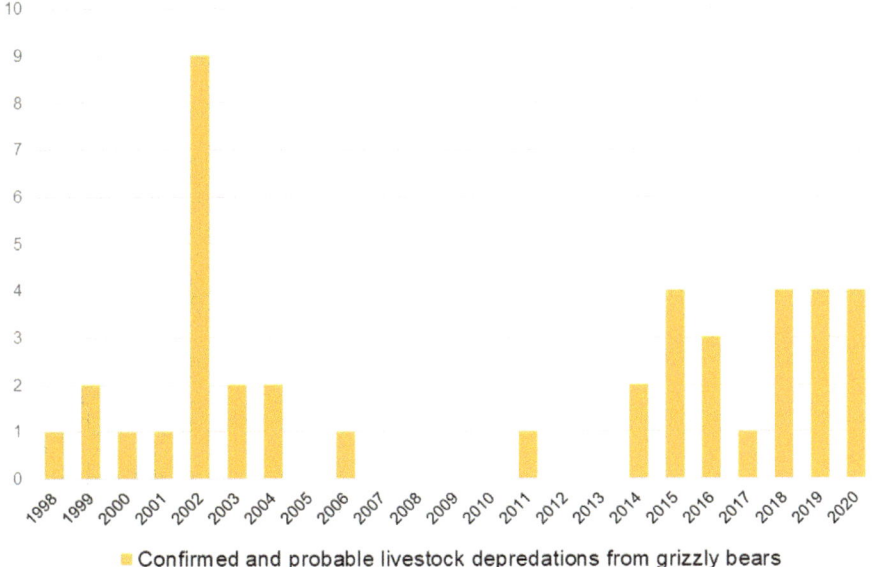

Fig. 28.5 Montana Livestock Loss Board data on confirmed and probable livestock depredations from grizzly bears for Blackfoot Challenge core project area, 1998–2020

(calves and sheep) to wolves have been 3.3 livestock per year. Less than four wolves per year have been removed (3.6 wolves per year) due to these depredations for the same period (Fig. 28.6). The low levels of livestock losses to wolves and the proactive and preventative efforts help balance agricultural needs with those of wildlife. Additionally, the level of livestock losses in the core project area in the Blackfoot watershed is significantly lower that other areas of the state that experience chronic livestock depredations to wolves (DeCesare et al. 2018).

There are ecological and management factors that should be acknowledged when interpreting the above results. These include abundant ungulate populations, small wolf pack sizes likely due to hunting and trapping seasons (2009, 2011–present), seasonally livestock-free areas for several wolf packs, and difficult hunter and trapper access due to private land patterns in the Blackfoot watershed. All of these factors likely contribute to low levels of livestock depredations and may help sustain a population of wolves in the watershed (Wilson et al. 2017).

28.5 Lessons Learned

The major lesson from this case study is that living with large carnivores requires bringing people together to build trust, generate a shared understanding of the problem using science, and to develop a participatory and equitable approach for

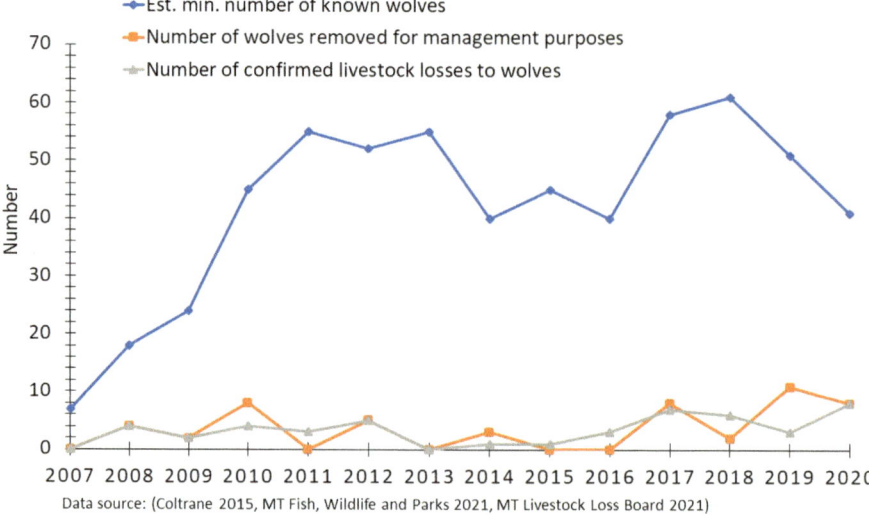

Fig. 28.6 Estimated minimum number of known wolves in the Blackfoot watershed (core project area); wolves removed for management purposes; and confirmed livestock losses to wolves in Blackfoot watershed, 2007–2020

changing practices and adopting tools that foster coexistence with carnivores. The result of this approach was a community-scaled response across public and private ownerships so that conflict reduction efforts were appropriately matched to the scale of bear home ranges and wolf pack territories. Additionally, there are four important pillars to build collaborative and partnership-based efforts. These are: (1) there must be some coordination of resources, (2) efforts should be informed by science, (3) stakeholder values must be incorporated, and (4) a decision-making process must be present in order to rationally discuss the issues, make decisions, and implement actions in a participatory manner with stakeholders.

The existing capacity and support of the BC was critical for coordinating stakeholder values, developing collective goals, and bringing the biological and technical skill sets of key wildlife managers and local knowledge of landowners and ranchers together to implement projects. Second, for both grizzly bears and wolves, existing research and new analysis was used when needed to bring the latest science and management expertise of key partners to address strategies for living with bears and wolves. Third, throughout all this work, keen attention was paid to respecting and incorporating all stakeholder values from those who lived and worked in the watershed to those who from outside the area but who also had keen interest in conservation of the watershed and its wildlife. And fourth, the overall process that defines the BC was critical for managing and integrating these different values using a non-advocacy, non-litigatory, and consensus-driven process through the inclusive forum of the BC Wildlife Committee. This inclusive and creative forum for decision making fostered direct participation of ranchers and landowners in the projects described in this case study.

The efforts described in this chapter are built on trust, credibility, and a reservoir of social capital from the Blackfoot Challenge that helped bring people together. This seemingly simple task for bringing people together was instrumental in generating an inclusive process that allowed stakeholders to work together to successfully respond to and live with grizzly bears and wolves in the American West.

References

Bennett JW (1967) Microcosm-macrocosm relationships in North American agrarian society. Am Anthro 69:441–454

Blackfoot Challenge (2021a) https://blackfootchallenge.org/what-we-do/. Accessed 29 Mar 2021a

Boitani L (2003) Wolf conservation and recovery. In: Mech LD, Boitani L (eds) Wolves: behavior, ecology, and conservation. The University of Chicago Press, Chicago, pp 317–340

Bradley EH, Pletscher DH (2005) Assessing factors related to wolf depredation of cattle in fenced pastures in Montana and Idaho. Wild Soc Bull 33:1256–1265

Coltrane JJ, Gude J, Inman B, Lance N, Laudon K, Messer A, Nelson A, Parks T, Ross M, Smucker T, Steuber J, Vore J (2015) Montana gray wolf conservation and management 2015 annual report. Montana Department of Fish, Wildlife and Parks, Helena

Costello CM, Mace RD, Roberts L (2016) Grizzly bear demographics in the Northern Continental Divide Ecosystem, Montana: research results (2004–2014) and suggested techniques for management of mortality. Montana Department of Fish, Wildlife and Parks, Helena

Dale EE (1960) The range cattle industry: Ranching on the Great Plains from 1865 to 1925 University of Oklahoma Press, Norman

DeCesare NJ, Wilson SM, Bradley EH, Gude JA, Inman RM, Lance NJ, Laudon K, Nelson AA, Ross MS, Smucker TD (2018) Spatial and temporal patterns of wolf-livestock conflict in Montana and the effects of wolf management. J Wild Mng 82:711–722

Hanauska-Brown L, Bradley L, Gude J, Lance N, Laudon K, Messer A, Nelson A, Ross M, Steuber J (2012) Montana gray wolf conservation and management 2011 annual report. Montana Department of Fish, Wildlife and Parks, Helena

Hebblewhite M, White CA, Nietvelt CG, McKenzie JA, Hurd TE, Fryxell JM, Bayley SE, Paquet PC (2005) Human activity mediates a trophic cascade caused by wolves. Ecology 86:2135–2144

Interagency Grizzly Bear Committee (2021) http://igbconline.org/conserving-grizzly-populations-2/. Accessed 29 Mar 2021

Jonkel JJ (2002) Living with black bears, grizzly bears, and lions: project update. Montana Department of Fish, Wildlife and Parks, Region 2, Missoula

Jordan TG (1993) North American cattle-ranching frontiers: Origins, diffusion, and differentiation. University of New Mexico Press, Albuquerque

Kendall KC, Stetz JB, Boulanger J, Macleod AC, Paetkau D, White GC (2009) Demography and genetic structure of a recovering brown bear population. J Wild Mng 73:3–17

Lance NJ, Breck SW, Sime C, Callahan P, Shivik J (2010) Biological, technical, and social aspects of applying electrified fladry for livestock protection from wolves (Canis lupus). Wild Res 37:708–714

Mace RD, Carney DW, Chilton-Radandt T, Courville SA, Haroldson MA, Harris RB, Jonkel J, Mclellan B, Madel M, Manley T, Schwartz C, Servheen C, Stenhouse G, Waller JS, Wenum E (2012) Grizzly bear population vital rates and trend in the Northern Continental Divide Ecosystem, Montana. J Wild Mng 76:119–128

Mattson DJ (1990) Human impacts on bear habitat use. Int Con on Bear Res and Mng 8:33–56

Mattson DJ, Herrero S, Wright RD, Pease CM (1996) Designing and managing protected areas for grizzly bears: How much is enough? In: Wright RG (ed) National parks and protected areas: Their role in environmental protection. Blackwell Science, Cambridge, MA, pp 133–164

Montana Fish, Wildlife and Parks (2021b) Unpublished data. Wolf counts, packs, and sources of mortality for Blackfoot

Montana Livestock Loss Board (2021) Unpublished data. Reported and verified livestock losses for Lewis & Clark, Powell, and Missoula Counties

Morehouse AT, Boyce MS (2011) From venison to beef: seasonal changes in wolf diet composition in a livestock grazing landscape. Fron in Eco and the Env 9:440–445

Muhly TB, Semeniuk C, Massolo A, Hickman L, Musiani M (2011) Human activity helps prey win the predator-prey space race. PLoS ONE 6(3):e17050. https://doi.org/10.1371/journal.pone.001 7050

Musiani M, Mamo C, Boitani L, Callaghan C, Gates CC, Mattei L, Visalberghi E, Breck S, Volpi G (2003) Wolf depredation trends and the use of fladry barriers to protect livestock in western North America. Con Bio 17:1538–1547

Musiani M, Muhly T, Gates C, Callaghan C, Smith C, Tosoni ME, Tosoni E (2005) Seasonality and reoccurrence of depredation and wolf control in western North America. Wild Soc Bull 33:876–887

Northrup JM, Stenhouse G, Boyce M (2012) Agricultural lands as ecological traps for grizzly bears. Anim Con 15:369–377

Parks M (2015) Participant perceptions of range rider programs used to mitigate wolf-livestock conflicts in the Western United States. Utah State University, Logan, Thesis

Putnam RD (2000) Bowling alone: the collapse and revival of American community. Simon and Schuster, New York

Sime CA, Asher V, Bradley L, Lance N, Laudon K, Ross M, Nelson A, Steuber J (2011) Montana gray wolf conservation and management 2010 annual report. Montana Department of Fish, Wildlife and Parks, Helena

United Nations Education Scientific and Cultural Organization—UNESCO (2021) https://en.une sco.org/biosphere/eu-na/crown-continent. Accessed 29 Mar 2021

Wells SL, McNew LB, Tyers DB, Van Manen FT, Thompson DJ (2019) Grizzly bear depredation on grazing allotments in the Yellowstone Ecosystem. J Wild Mng. 83:556–566

Williams S, Hebblewhite M (2021) Spatial hotspot analysis of grizzly bear-human conflicts in the Blackfoot Watershed. Report prepared for the Blackfoot Challenge Contract #2019–37. p 42

Wilson SM, Madel MJ, Mattson DJ, Graham JA, Belsky JM, Burchfield JA (2005) Landscape features, attractants, and conflict hotspots: A spatial analysis of human-grizzly bear conflicts. Ursus 16:117–129

Wilson SM, Clark SG (2007) Resolving human-grizzly bear conflicts: an integrated approach in the common interest. In: Hanna S, Slocombe DS (eds) Fostering integration: concepts and practice in resource and environmental management. Oxford University Press, Ontario, pp 137–163

Wilson SM, Graham JA, Mattson DJ, Madel MJ (2006) Landscape conditions predisposing grizzly bears to conflict on private agricultural lands in the Western USA. Bio Con 130:47–59

Wilson SM, Neudecker GA, Jonkel JJ (2014) Human-grizzly bear coexistence in the Blackfoot River watershed, Montana: Getting ahead of the conflict curve. In: Clark SG, Rutherford MB (eds) Large carnivore conservation: integrating science and policy in the North American West. University of Chicago Press, Chicago, pp 177–214

Wilson SM, Bradley EH, Neudecker GA (2017) Learning to live with wolves through community-based conservation: a case study in the Blackfoot Valley of Montana. Hum-Wild Inter 11:245–257

Woodruff R, Ginsberg JR (1998) Edge effects and the extinction of populations inside protected areas. Science 280:2126–2128

Chapter 29
A Perspective on Rangeland and Wildlife Disciplines: Similarities Over Differences

Eric Thacker, David Dahlgren, David Stoner, and Megan Clayton

Abstract The disciplines of rangeland and wildlife management were born out of necessity to protect dwindling resources during the early twentieth century. The development of the fields followed parallel paths to meet the needs and desires of society. Around the world, rangelands provide habitat for a wide variety of wildlife species. Across North America, wildlife conservation problems have impacted rangelands and thus influenced rangeland management. Wildlife and rangeland professionals often work on the same landscapes to manage related resources. Yet, because of professional traditions and biases, there is potential for misunderstanding of terminology, values, and conflicts. However, these two professions have more in common than differences. For example, early management for both fields revolved around sustainable harvest (i.e., game species and forage) and providing guidance on conserving limited but renewable resources. Although both disciplines were born out of similar needs, they have often been viewed as separate entities. In this chapter, we will attempt to address the differences and parallels between these two disciplines with the objective of finding common ground for future collaboration. We will outline the parallel development of crucial principles of rangeland and wildlife management, and professionals can improve communication and understanding between disciplines.

Keywords Habitat · Land management · Rangeland · Wildlife

E. Thacker (✉) · D. Dahlgren · D. Stoner
Department of Wildland Resources, Utah State University, 5230 Old Main Hill, UMC 5230, Logan, UT 84322, USA
e-mail: eric.thacker@usu.edu

M. Clayton
Texas A&M AgriLife Extension Service, Department of Rangeland, Wildlife, and Fisheries Management, 10345 Hwy 44, Corpus Christi, TX 78406, USA

29.1 Introduction

Rangeland and wildlife professionals manage renewable natural resources valued by society. Early rangeland management was rooted in animal agriculture and centered on stabilizing rangelands to maximize the production of meat (Stoddart and Smith 1943). Wildlife management was developed in response to declining populations of game animals (Trefethen and Corbin 1975). Many species promoted by wildlife biologists were viewed as competitors with livestock on rangelands, naturally leading to some conflicts between professions. Conversely, livestock grazing was often blamed for degrading wildlife habitat for a broad array of taxa (du Toit et al. 2010; Dettenmaier et al. 2017). One of the legacies of this conflict has been the lack of integration between the disciplines. Early rangeland management academic programs were often housed in animal science departments and agricultural colleges. In contrast, early wildlife programs were most often associated with biology and zoology departments. The academic programming of both disciplines influenced their distinct approaches to natural resource management problems and these ecologically interrelated disciplines likely contributed to professional biases, tribalism, and competing "schools of thought". For example, many early grouse publications cited livestock grazing as a potential contributing factor in population declines; however, little empirical data demonstrated a direct linkage between livestock grazing and grouse vital rates until recently (McNew et al. 2015; Dettenmaier et al. 2017; Milligan et al. 2020). Although rangeland management was initially developed to address the negative impacts of overgrazing, in some cases, rangeland managers have overemphasized livestock grazing as a potential tool to fix everything. For example, Twidwell et al. (2013) found that grazing was often the cause of degradation, while also being the cure suggesting rangeland management professionals may have an inherent grazing bias. It is true that grazing management can mitigate some rangeland problems while causing others; thus, livestock grazing can be both the disease and the cure, depending on the problem (Strand et al. 2014).

Unfortunately, these two professions have had limited collaboration, despite the intrinsic dependency of animal-plant relationships for both livestock and wildlife. Although some conflicts have been resolved, there remains a need to increase collaboration and build understanding between rangeland and wildlife disciplines and professionals. Our purpose for this chapter is to (1) provide an in-depth view of the linkages between the rangeland and wildlife professions, (2) demonstrate a much higher degree of commonality and parallelism than most in either camp may have realized, and (3) identify where disparity remains, and (4) offer suggestions for overcoming differences and increasing collaboration that will be needed to meet future challenges facing both disciplines.

29.2   History

The rangeland and wildlife professions emerged in the early twentieth century due to human-induced resource scarcity in the form of rangeland forage and wildlife game species. These plant and animal-focused professions developed independently, driven by seemingly different values and management priorities. Despite the initial distinctions in origination, the developmental trajectory of these disciplines exhibited strong parallels. Early rangeland managers worked to increase forage availability by planting vegetation, adding water sources, and/or fencing to maintain or increase grazing opportunities. This often resulted in direct or indirect impacts that conflicted with wildlife management priorities (Beck and Mitchell 2000). Early European explorers described North American rangelands and wildlife as "inexhaustible, vast, and innumerable" (Trefethen and Corbin 1975). Viewing resources with such descriptors led to over-exploitation by Euro-American colonizers, such that, by the late 1800s, certain game populations were becoming scarce under unregulated market hunting. For example, population estimates of the American bison (*Bison bison*) suggest 40–60 million prior to Euro-American settlement, but bison were nearly extirpated by the 1880s (Roe 1951; Trefethen and Corbin 1975). This extraordinary crash was precipitated by unregulated market hunting and federal policies designed to cripple Native Americans who relied on bison (Irving 2019). Prairie chickens (*Tympanuchus* spp.), another rangeland-dependent species, suffered a similar fate. Early accounts suggested that populations were seemingly unlimited but within a few decades, overharvest and continued loss of rangelands to agriculture resulted in rapid declines in the early 1900s (Grange 1948; Stempel and Rodgers 1961).

During this same period, western rangelands were experiencing significant pressure from increasing numbers of domestic livestock to meet the demand for red meat in the markets of mining boom towns of the West and growing industrial cities in the eastern U.S. Newly constructed railroads linked expanding eastern communities to western suppliers, which allowed livestock operations to move from local subsistence markets to nationwide consumers (Holechek et al. 2011). However, as livestock numbers increased, arid and semi-arid rangelands of the West began to show signs of degradation. By the turn of the twentieth century, livestock numbers were peaking, and drought combined with severe winters resulted in significant livestock die-offs across western rangelands of the US. This led to dramatic swings in the livestock markets, with consequential realizations of the limits to soil fertility and forage availability (Holechek et al. 2011). Despite shocks to agricultural markets, there were no regulatory structures or ecological knowledge for managing livestock grazing on arid rangelands of western North America (Sayre 2017).

In the early twentieth century, many marketable wildlife resources had been depleted, causing increased organizational and political lobbying by recreational hunters to regulate hunting and eliminate wildlife markets. Lobbying efforts led to the passage of wildlife regulations that controlled and limited harvest and set up funding mechanisms to support wildlife management and recovery. The result of the lobbying efforts was that federal and state laws outlawing market hunting and restricting harvest were enacted to conserve wildlife populations as renewable resources (Brown 2010; Trefethen and Corbin 1975). Rangeland management dealt with the overexploitation of western rangelands in principally the same manner, by enacting regulatory mechanisms that limited livestock numbers, especially on public lands. Federal agencies encouraged moderating livestock numbers to manage rangelands properly on private lands. In the West, public land grazing permits were linked to local private lands and rural communities through the Taylor Grazing Act of 1934. The recognition that science-based management was needed resulted in the establishment of The Wildlife Society (TWS) in 1937 and the Society for Range Management (SRM) in 1948. These national societies cemented the professions and formalized the creation of curricula for public universities to produce formal education and training for rangeland and wildlife managers. As a result, rangeland and wildlife disciplines benefited from an increased understanding of science-based management principles. This coupled with federal and state regulations, gave the budding professions the tools needed for the management of rangeland and wildlife resources.

29.3   Parallels

Although developing in relative isolation from each other, similar progressions in ideas, knowledge, and tools led to parallel trajectories of policy and management in both professions. Realizing the widespread degradation of resources but lacking the information or tools to implement management actions, early rangeland and wildlife professionals were focused on stopping the hemorrhaging by enforcing new regulations to curb the over-harvest of resources (Sparling 2014). Over-exploitation of rangeland resources was more pronounced on lands in the public domain and was often motivated by short-term economic incentives by residents of the sparsely populated western rural counties (Rowley 1985). Likewise, the first wildlife legislation and policies in North America regulated harvest, including season dates and bag limits, at the state level (Trefethen and Corbin 1975; Brown 2010). Similar regulatory approaches were used to limit the harvest of forage by creating seasons of use, stocking rates, and grazing allotments on public lands, culminating in the passage of the Taylor Grazing Act (Stoddart and Smith 1943).

Harvest management continues to be one of the most significant focuses for both professions, and there are striking, though not exact, parallels in each discipline's approach to harvest management. As early rangeland and wildlife management continued to develop, harvest regulations assumed that an available surplus (i.e., individual animals or forage) could be removed, and the unharvested stock would provide resource sustainability and produce future surplus available for harvest. In wildlife management, this was first described as "doomed surplus" and later conceptualized within compensatory harvest mortality, theoretically resulting in no net loss to the harvested resource (Errington and Hamerstrom 1935; Errington 1945, 1956). For example, a game bird population might experience limited habitat and food availability during the winter (i.e., winter bottleneck), allowing only a portion of the population to survive to the following breeding season (Errington 1956). Therefore, that portion of the fall population in excess of the limited capacity of the winter bottleneck could be considered a "doomed" surplus that would have been lost with or without hunter harvest and, thus, harvesting the surplus would have no impact on the breeding population. As long as the amount of harvest did not exceed the doomed surplus, mortality due to hunter harvest was considered compensatory and thus sustainable (Reese et al. 2005; Dahlgren et al. 2021). Conversely, additive harvest mortality occurs when loss due to harvested resources are in addition to other natural sources of mortality, resulting in an overall decline of the breeding population. In general, wildlife harvest management continued through the mid-20th Century with the assumption of an available surplus. However, research in the last few decades, specifically for waterfowl and big game, has shown that most harvest impacts fall along a continuum between fully compensatory and fully additive, and harvest management regulations have been adjusted accordingly (Burnham and Anderson 1984; Burnham et al. 1984; Bartmann et al. 1992; Bowyer et al. 2020). Nevertheless, upland game harvest management has largely continued to use an assumption of compensatory harvest mortality with relatively infrequent adjustments to regulations (Reese et al. 2005; Dahlgren et al. 2018, 2021).

Similar harvest principles developed independently within rangeland management. Analogous to doomed surplus, remaining forage at the end of the grazing season was often viewed as wasted. Early range management called for moderate stocking rates that were intended to maintain enough above-ground plant material to support proper root function and plant reproduction, resulting in a general "take half, leave half" approach to grazing management (Shoop and McIlvain 1971; Van Poollen and Lacey 1979; Holechek et al. 1999). Conceptually, leaving half of the available forage is similar to maintaining a wildlife breeding population. Conversely, suppose grazing utilization (i.e., the amount of forage consumed by an herbivore; Chap. 3) exceeds the take-half rule. In that case, it is assumed that grazing has resulted in additive impacts because it may be limiting the ability of plants to maintain adequate root function and could reduce forage biomass and the ability of the plant to reproduce the following growing season (Trlica et al. 1977; Lyons and Hanselka 2001; Sayre 2001). However, if grazing is managed by taking up to some appropriate portion of the above-ground biomass (i.e., analogous to doomed surplus), it is assumed that the plant can compensate for the fraction removed through grazing (Lyons and Hanselka 2001).

For some cases in both disciplines, such as public land grazing and upland game harvest, rigorous evaluations of the assumption of compensatory harvest are currently lacking (Bartolome 1993; Dahlgren et al. 2021). For example, the monitoring of rangelands for livestock grazing on both public and private lands often uses utilization estimates for individual plants of specific "key" species within plant communities and does not evaluate forage availability or what portion of the available forage is harvested. Furthermore, assessment of year-to-year impacts of livestock grazing on forage availability has rarely occurred (Veblen et al. 2014). Uncannily akin to only monitoring forage utilization, the only assessment of harvest for most upland game birds in North America has been post-season hunter surveys (i.e., only monitoring the harvest; Dahlgren et al. 2021).

When considering the similarities between livestock grazing and upland game harvest, the impact of harvest rate on the base resource is often not accounted for. Therefore, there is little understanding of overall harvest impacts, let alone accounting for interactions with environmental variation to inform and adapt future harvest management (Bartolome 1993). Rather, for most of the last century, grazing and upland game management have implemented harvest using a "conservative harvest" approach combined with the assumption of compensatory harvest (Bartolome 1993; Dahlgren et al. 2018). Overall, this approach seems to have successfully provided a more sustainable use of renewable plant and animal resources. However, additional pressures, including anthropogenic development, invasive species, altered fire regimes, competing uses, climate change, and societal tolerance for consumptive uses have created increasing contextual constraints on our rangeland and wildlife resources. It is also likely that confounding factors are often not considered when analyzing and modifying harvest rates. Although the assumption of an available surplus may have been useful as wildlife and rangeland disciplines developed, future management of increasingly dynamic rangeland vegetation communities and associated wildlife will likely require both professions to implement more rigorous scientific evaluations of harvest. Advances in harvest management for other resources, such as waterfowl and fisheries, have used an adaptive harvest management approach, which requires resource assessments based on rigorous scientific methods to help identify more appropriate harvest targets (Hilborn and Sibert 1988; Nichols et al. 2007). Pope and Powell (2021) recognized that new paradigms for wildlife harvest management are needed for sustainable management into the future.

The wildlife profession has expanded from a primary focus on game management to a broader emphasis on wildlife communities and ecosystems (Decker et al. 1992). Contemporary wildlife management has moved toward the conservation of all wildlife species, with a particular focus on maintaining or increasing biodiversity. Conceptually similar, the rangeland profession has moved from managing for maximized livestock production to understanding and sustaining rangelands and their vegetation communities, whether they are grazed by domestic livestock or not (Briske et al. 2017). While each profession has broadened its focus, they both retained harvest management as a central tenant, even though some harvest management approaches have relied on outdated research.

Ultimately, rangeland and wildlife professions would benefit from an interdisciplinary approach to harvest management in the future because they (1) share many ecological underpinnings (e.g., carrying capacity, compensatory harvest assumptions), and (2) manage inextricably interconnected natural resources (e.g., wildlife harvest can be constrained by habitat conditions and land management; wildlife can affect rangeland resources and grazing management through a variety of mechanisms). The wildlife discipline has a comparatively strong background in population ecology, whereas rangeland management has a strong background in vegetation community dynamics and nutrition. The strengths within both professions are needed to assess and guide future harvest management for livestock grazing and wildlife in the context of our rapidly changing environments and contemporary societal values.

29.4 Disparities

Although rangeland and wildlife professions have striking parallels, their independent developments have also led to fundamental differences (Fig. 29.1). Such disparities have given rise to potential conflicts between the professions and are potentially rooted in the genesis of each profession. Because of the profession's agrarian foundation, rangeland managers have been prone to view wildlife issues as stumbling blocks to livestock production or other rangeland management objectives (Stoddart and Smith 1943; du Toit et al. 2017). Conversely, wildlife management originated to stop declines in wildlife populations, and grazing has often been implicated as a detrimental factor to game populations (Trefethen and Corbin 1975; du Toit et al. 2017). We propose that these fundamental differences arise from differing values that helped drive the creation of the disciplines, and the legacy of those differences can still be seen presently.

29.4.1 Terminology

In addition to the fundamental disparities between the rangeland and wildlife professions, terminology can add to the disparity and create barriers to understanding.

"You should be leery of information produced by any biologist who has focused on only one wildlife species for their life's work." *Anonymous Rangeland Professional*

"That publication was from a group of range folks, so you know it includes dogma and bias, especially when they are talking about wildlife habitat." *Anonymous Wildlife Professional*

"Oh, wait, you are a wildlife person! What are you doing here at a Society for Range Management meeting?" *Anonymous Rangeland Professional*

"You are getting a Ph.D. in wildlife. You have gone to the dark side." *Anonymous Rangeland Professional*

"Those are [vegetation monitoring] methods used by range people; we'll stick with the right ones that actually tell us something about cover for wildlife habitat." *Anonymous Wildlife Professional*

Fig. 29.1 Quotes collected to illustrate some of the disparities and tribalism between rangeland and wildlife management

Terminology is a critical but often unexamined foundation for any profession, especially the intended and interpreted underlying meanings. Every profession, whether related to physics, music or natural resources, has a unique vocabulary with specific connotations and implications (Brunson 1992). As members of each profession are educated, they attune to the terminology and its explicit and implicit discipline-based meanings. Brunson (1992) used the word "rotation" as an example. He found that rangeland managers, foresters, and farmers use the term rotation to describe specific management actions, but each has a profession-specific meaning for the term rotation. Rangeland managers rotate livestock grazing among sites/pastures while the managed resource (i.e., forage) remains the same. Farmers rotate by moving the managed resource (i.e., crops) while the sites (i.e., crop fields) stay the same. For foresters, however, the managed resources and sites do not change, but rotation occurs by removing individual plants and restarting growth. Differences in word use and interpretation can lead to confusion, misunderstandings, and inadvertently exercising educational biases (Brunson 1992). Educational biases can become most pronounced when someone communicates within their profession's jargon. Without clear comprehension of the terminology and the specific underlying connotations, communication between professionals from different disciplines can be problematic and lead to misunderstanding, even intense discord, and erroneous conclusions. One of the more challenging issues for the integration of rangeland and wildlife disciplines is using the same terms but with distinctively different meanings. The following are some common, but not exhaustive, examples of shared terms with rangeland and wildlife profession-specific meanings.

29.4.1.1   Habitat

At first consideration, habitat seems like a simple and easily conceptualized term. For the layperson, habitat is simply the characteristics of where something, usually a plant or animal, lives. However, for a wildlife professional, habitat has a much more specialized meaning. Within the wildlife profession, Hall et al. (1997) inextricably linked habitat with one or more wildlife species and includes a time period or season of use for that species. For example, a wildlife biologist could use the term mule deer (*Odocoileus hemionus*) winter "habitat" when describing the environmental conditions used by mule deer during the winter. Although sagebrush (*Artemesia* spp.) is often an important component of mule deer winter habitat, referring to this as "sagebrush habitat" is confusing and meaningless within a professional wildlife context. Conversely, within rangeland science, "sagebrush habitat" is acceptable terminology that refers to the vegetation community present within the focal ecosystem and is not related to a specific wildlife species but to the vegetation type (Daubenmire 1984). So, when a wildlife person hears the term "habitat," they want to know about the wildlife species and season of use, whereas when a rangeland person hears the term "habitat" they want to know what vegetation type is being referred.

On the surface, the differences between disciplines when using the term "habitat" may seem unimportant and inconsequential. However, consider the following hypothetical situation. Several rangeland and wildlife professionals are meeting to discuss management actions to conserve sage-grouse (*Centrocercus* spp.) in their resource area. One of the rangeland professionals says they are considering an area of sagebrush rangeland where the shrub canopy cover has become too high, and they desire to reduce the shrub canopy within the sagebrush habitat to enhance forage for livestock. One of the wildlife biologists then asks about the sage-grouse seasonal habitat(s) included in the sagebrush treatment area. The rangeland manager, unfamiliar with defining habitat as having seasonality for a wildlife species, has difficulty conceptualizing how a vegetation community can be described as seasonal when that vegetation community is always in place and functioning regardless of the time of year. Likely inadvertently and subconsciously based on their own educational bias, the rangeland manager understandably begins to question the validity of the biologist's knowledge about sagebrush ecosystems (Brunson 1992). On the other hand, the wildlife biologist may wonder how the rangeland manager can seemingly ignore the seasonal habitat requirements of sage-grouse using the area, such as the need for higher shrub canopy cover during nesting and wintering periods. In the above scenario, how each profession defines the term "habitat" unintentionally led to misunderstandings and misjudgments between the rangeland manager and biologist. The unfortunate result is that neither person was in error, both had value to add to the management approach, and both misjudged the other based on their own frame of reference and professional bias.

29.4.1.2   Cover

Another term used commonly by both professions is "cover," but with different underlying meanings. In rangeland management, there are multiple definitions related to the term "cover," though generally cover is used to reflect the amount of substrate, most often soil surface, covered by plant or other materials (rocks, litter, etc., Table 29.1). Wildlife management commonly uses the term "cover" to describe wildlife habitat (i.e., vegetation or other structure) as a reflection of hiding or escape cover, meaning covering the animal (Kopp et al.1998; Connelly et al. 2003). Simply put, for the term "cover" wildlife professionals want to know how much structure is available to conceal wildlife, and rangeland professionals want to know how much of the soil is covered, often with the intent of minimizing erosion. The unacknowledged underlying assumptions of these professionally based meanings for the same term "cover" can become problematic when rangeland and wildlife professionals work together.

Greater sage-grouse (*C. urophasianus*) inhabit sagebrush rangelands across western North America. Of necessity, conservation for this species has caused rangeland and wildlife professionals to work closely together and provides an example of differences in discipline-specific meanings of the term "cover." Sagebrush cover is a critical component of sage-grouse habitat because it is used by sage-grouse throughout their life cycle (Connelly et al. 2000a, b; Crawford et al. 2004; Knick and Connelly 2011). Although various methods exist for measuring shrub canopy

Table 29.1 Definitions of different types of "cover" used in range management

Cover	Definition*
Basal cover	Area of plant base. syn. Basal area
Canopy cover	The percentage of ground covered by a vertical projection of the outermost perimeter of the natural spread of foliage of plants. Small openings within the canopy are included. It may exceed 100%. Syn. crown cover
Foliar cover	The percentage of ground covered by the vertical projection of the aerial portion of plants. Small openings in the canopy and intraspecific overlap are excluded. Foliar cover is always less than canopy cover; either may exceed 100%. Syn. Cover
Ground cover	The percentage of material, other than bare ground, covering the land surface. It may include live and standing dead vegetation, litter, cobble, gravel, stones and bedrock. Ground cover plus bare ground would total 100 percent

*All definitions are sourced from the Society for Range Management Glossary (1998)

cover, in reference to sage-grouse habitat, sagebrush canopy cover has largely been assessed using line intercept with the Canfield Method; a method developed within the rangeland discipline (Canfield 1941; Connelly et al. 2000a, b, 2003; Stiver 2006). Sage-grouse biologists have consistently used this method to provide an assessment of available sagebrush cover, in other words to provide hiding cover for sage-grouse (Connelly et al. 2003; Stiver et al. 2006). In this example, the term "cover" seems to work well for both disciplines. After all, sagebrush canopy cover does provide structure for both covering the soil and concealing the grouse. However, considering the discipline-specific intent for the term "cover" may help explain why there is controversy over specific practices in sage-grouse and sagebrush habitat management between rangeland and wildlife professionals. Concerned with the immediate loss of hiding cover for sage-grouse seasonal habitat needs, sage-grouse biologists have implicated sagebrush cover reduction as a rangeland management practice that has a high probability of being detrimental to sage-grouse (Braun et al. 1977; Connelly and Braun 1997; Beck and Mitchell et al. 2000). This view is understandable given historical large-scale sagebrush loss (Vale 1974) and associated sage-grouse population declines (Braun et al. 1977; Connelly and Braun 1997; Braun 1998; Aldridge et al. 2008; Wisdom et al. 2011). For most rangeland ecologists, when sagebrush canopy cover is reduced, the understory herbaceous vegetation remains and often increases, providing the needed ground cover to protect the soil. Rangeland professionals tend to view sagebrush removal as a management tool that has potential benefits for sagebrush communities broadly, with the added advantage of increasing forage for livestock (Vallentine 1971; Crawford et al. 2004).

To further illustrate these underlying meanings for cover and the implications for management approaches, consider two publications meant to provide broad guidance on sage-grouse habitat management; Connelly et al. (2000a, b; authorship with primarily wildlife backgrounds and published in a wildlife journal) and Crawford et al. (2004; authorship with primarily rangeland backgrounds and published in a rangeland science journal). While both publications acknowledge the critical importance of sagebrush cover to sage-grouse, Crawford et al. (2004) emphasize potential benefits when addressing management geared toward reductions in sagebrush cover, and Connelly et al. (2000a, b) focus on maintaining sagebrush cover with strong cautions towards "range management treatments" that reduce sagebrush cover. We suggest that the polarity in the management approach to sagebrush and sage-grouse habitat is related, at least partly, to the discipline-specific connotations attached to the term "cover" and the ultimate differences in what cover is meant to protect, i.e., the animal or soil.

29.4.1.3   Rangeland Condition

The SRM Glossary (Society for Range Management 1998) defines rangeland condition as *"(a) a generic term relating to present status of a unit of rangeland in terms of specific values or potentials. Specific values or potentials must be stated. (b) the present state of vegetation of a rangeland site in relation to the climax (natural potential) plant community for that site"*. Despite a formal definition from the profession, the inconsistent use of rangeland condition increases confusion within and across disciplines. For example, Hervert et al. (2005) evaluated the space use of Sonoran pronghorns (*Antilocapra americana sonoriensis*) in relation to an assessment of "rangeland condition" based on the condition of the vegetation determined by rainfall; they did not specifically outline how they quantified rangeland condition other than relating seasonal rainfall to "rangeland condition". A rangeland manager reading the paper may have assumed that the authors compared the relative space use of pronghorn to some measure of how close used and unused sites were to the expected climax condition of the vegetation communities.

Brunson (1992) suggested that confusion over terminology can lead to mistrust between the professions. Although differences in terminology are not responsible for conflicts, they represent our professional differences. We propose that the differences in terminology can reinforce tribalism and create the illusion of exceptionalism and will stifle the transdisciplinary development of comprehensive solutions to complex ecological problems that impact our collective disciplines and resources. We also do not believe that mandating unified definitions for commonly used terms is realistic or that it would serve to better unify the professionals. Rather, a more suitable solution lies in working toward an understanding of the way different professions use terms and their underlying meanings (Brunson 1992).

29.5   The Big Tent

Ecosystem-level conservation problems now require wildlife and rangeland managers to work closely together to address natural resource management problems. For example, the underlying threats to sage-grouse are related to the degradation of sagebrush communities, which has demanded coordination between rangeland and wildlife managers. Sage-grouse occur in 11 western states within one of North America's largest and most at-risk biomes (Schroeder et al. 2004; Chap. 10). Conservation of sagebrush communities is primarily the responsibility of landowners, private or land management agencies, which generally includes rangeland and wildlife managers, while conserving sage-grouse populations lies with state wildlife managers. The risks to sage-grouse are largely habitat-related and often require management approaches that focus on vegetation communities at varying scales. For instance, sage-grouse conservation has included multiple local working group programs across the distribution of the species, where rangeland and wildlife professionals have worked closely together to develop and implement conservation practices addressing risks, which can include prescribed livestock grazing (Chap. 10). This high degree of collaboration was cited as justification for an unwarranted 2015 ESA listing decision for greater sage-grouse (USFWS 2015), thereby underscoring the significance of combining expertise from plant and animal-focused disciplines.

Recent changes in academic institutional approaches provide an interesting example of the integration of rangeland and wildlife disciplines. There are currently 14 degree programs accredited by the Society for Range Management in North America. Twenty-one percent (3) of the departments are housed within rangeland and wildlife departments, 29% (4) in animal science and rangeland science, 29% (4) in ecosystem or natural resource departments, 14% (2) in forestry and range departments, and 7% (1) in a botany department. Most (79%) of the SRM-accredited programs have wildlife ecologists in their departments. Many of the departments with accredited rangeland programs have appointed faculty with wildlife expertise meant to crossover with rangeland programming in the last 15 years. For example, the Animal and Range (land) Sciences departments at Montana State University and Oregon State University have hired wildlife faculty for teaching and research programs to meet the current demands of rangeland students. We believe that most rangeland programs understand the need to have wildlife management expertise available within their degree programs, even when wildlife programs are not housed within the same department or college. We propose that these changes have created rangeland management students with broader exposure to wildlife-related expertise. For example, John Reese (Kanab, Utah), an alumnus of Utah State University, indicated that the wildlife education he received as part of his Rangeland Ecology and Management degree from USU proved valuable to him as a BLM rangeland specialist.

Similar adaptations have been made within both wildlife and rangeland professional societies. The Rangeland Wildlife Working Group of TWS and the Wildlife Habitat Committee of SRM are two prominent examples. In 2013, the Rangeland

Wildlife Working Group was created to provide a home for professionals who work at the intersection of wildlife and rangeland ecology. The working group has a membership of approximately 100 individuals who are interested in the management and function of rangeland ecosystems that provide value to humans and wildlife. In addition to providing policy statements related to rangeland management, hosting annual Rangeland Wildlife Working Group meetings, and publishing regular newsletters, this group has successfully hosted a symposium or workshop at TWS Annual Conference nearly every year since its inception. A partnership was developed with the SRM's Wildlife Habitat Committee, where dual memberships often exist, and the groups have collaborated by hosting joint symposia at international meetings. Where high-level partnerships between the two organizations have been discussed before, these individuals with complementary interests have organically created a powerful team of professionals advocating for education, proper management, and sound science of rangelands and the wildlife that inhabit them.

29.6   Conclusion

Moving into the future, whether because of budgetary constraints or intentional recognition that our disciplines complement one another, rangeland and wildlife managers will have to continue to work together. We feel there is significant value in the individuality of each discipline and the skills and knowledge that each brings to the table help solve large and complex problems that face multiple-use landscapes in the wake of population growth and climate change. We must recognize the value offered by each specific discipline while embracing the need for cross-pollination of both professions. As this occurs, terminology will always be problematic, but we encourage practitioners of both fields to carefully interpret discipline-specific meanings of terminology. The Society for Rangeland Management has created a glossary (https://rangelandsgateway.org/glossary) of terms available online. Although we have found no such glossary from TWS, there are several articles highlighting the need for greater precision in our professional language (Darracq and Tandy 2019). Careful use of terms from reliable textbooks and publications can help reduce misunderstandings and conflict between disciplines.

Our hope is that wildlife and rangeland professionals will recognize the substantial commonalities shared by both professions, put aside discipline-based tribalism, and seek to first understand and then be understood. The parallel progression of the disciplines shows that we have far more similarities than differences. For example, managing harvest could become a unifying principle that brings rangeland and wildlife professions closer together. As global and political climates continue to change, both disciplines will be faced with reevaluating and justifying harvest to ensure that resources are sustainable into the future. Our problems are too large to tackle alone. Current land use challenges threaten our flora and fauna resources, which were a major part of the motivation for many of us to choose these professions in the first place. Working across the table, or more emphatically, removing the table, is the best way to remain relevant and effective in our rapidly changing times.

References

Aldridge CL, Nielsen SE, Beyer HL et al (2008) Range-wide patterns of greater sage-grouse persistence. Divers Distrib 14:983–994. https://doi.org/10.1111/j.1472-4642.2008.00502.x

Bartmann RM, White GC, Carpenter LH (1992) Compensatory mortality in a Colorado mule deer population. Wildl Monogr 1:3–9

Bartolome JW (1993) Application of herbivore optimization theory to rangelands of the western United States. Ecol Appl 3:27–29

Beck JL, Mitchell DL (2000) Influences of livestock grazing on sage grouse habitat. Wildl Soc Bull:993–1002

Bowyer RT, Stewart KM, Bleich VC et al (2020) Metrics of harvest for ungulate populations: misconceptions, lurking variables, and prudent management. Alces 56:15–38

Braun CE (1998) Sage grouse declines in western North America: what are the problems. In: Proceedings western association of fish and wildlife agencies, Jackson, Wyoming

Braun CE, Britt T, Wallestad RO (1977) Guidelines for maintenance of sage grouse habitats. Wildl Soc Bull 1:99–106

Brown R (2010) A conservation timeline. The Wildl Profe Fall:28–32

Brunson M (1992) Professional bias, public perspectives, and communication pitfalls for natural resource managers. Rangl 14:292–295

Burham KP, Anderson DR (1984) Tests of compensatory vs. additive hypotheses of mortality in mallards. Ecol 65(1):105–112

Burnham KP, White GC, Anderson DR (1984) Estimating the effect of hunting on annual survival rates of adult mallards. J Wildl Manag 1:350–361

Canfield RH (1941) Application of the line interception method in sampling range vegetation. J Forest 39(4):388–394

Connelly JW, Reese KP, Fischer RA et al (2000a) Response of a sage grouse breeding population to fire in southeastern Idaho. Wildl Soc Bull:90–96

Connelly JW, Schroeder M, Sands AR et al (2000b) Guidelines to manage sage grouse populations and their habitats. Wildl Soc Bull 1:967–985

Connelly JW, Reese KP, Schroeder MA (2003) Monitoring of greater sage-grouse habitats and populations. Idaho Forest, Wildlife, and Range Experiment Station Bulletin 80, p 54

Connelly JW, Braun CE (1997) Long-term changes in sage grouse Centrocercus urophasianus populations in western North America. Wildl Biol 3:229–234

Crawford JA, Olson RA, West NE, Mosley JC, Schroeder MA, Whitson TD, Miller RF, Gregg MA, Boyd CS (2004) Ecology and management of sage-grouse and sage-grouse habitat. J Range Manag 57:2–19.https://doi.org/10.2111/1551-5028(2004)057[0002:EAMOSA]2.0.CO;2

Dahlgren, DK, Schroeder MA, Dukes, B (2018) State management of upland and small game. In Ryder TJ (ed) State Wildlife management and conservation. J. Hopkins University Press, Baltimore, MA, USA, pp 96–115

Dahlgren DK, Blomberg EJ, Hagen CA, Elmore RD (2021) Upland game bird harvest management. In: Pope KL, Powell LA (eds) Harvest of fish and wildlife: new paradigms for sustainable management. CRC Press, Boca Raton, FL, USA, pp 307–326

Darracq AK, Tandy J (2019) Misuse of habitat terminology by wildlife educators, scientists, and organizations. J Wildl Manage 83:782–789. https://doi.org/10.1002/jwmg.21660

Daubenmire RF (1984) Ecological site/range/habitat type. Rangel 6:263-264

Decker DJ, Brown Tl, Connelly NA, Enck JW, Pomerantz GA, Purdy KG, Siemer (1992) Toward a comprehensive paradigm of wildlife management: integrating the human and biological dimensions. In: Mangun WR (ed) American fish and wildlife policy: The human dimension. SIU Press. Carbondale, IL, USA, pp 33–54

Dettenmaier SJ, Messmer TA, Hovick TJ, Dahlgren DK (2017) Effects of livestock grazing on rangeland biodiversity: a meta-analysis of grouse populations. Ecol Evol 7:7620–7627. https://doi.org/10.1002/ece3.3287

du Toit JT, Cross PC, Valeix M (2017) Managing the livestock–wildlife interface on rangelands. Rangeland Syst:395–425https://doi.org/10.1007/978-3-319-46709-2

du Toit JT, Kock R, Deutsch J (2010) Wild rangeland: conserving wildlife while maintaining livestock in semi-arid ecosystems. Wiley- Blackwell, London

Errington PL, Hamerstrom FN Jr (1935) Bob-white winter survival on experimentally shot and unshot areas. Iowa State College J Sci 9:625–639

Errington PL (1945) Some contributions of a fifteen-year local study of the northern bobwhite to a knowledge of population phenomena. Ecolo Monog 15:2–34

Errington PL (1956) Factors limiting higher vertebrate populations. J Sci 124:304–307

Grange WB (1948) Wisconsin grouse problems. Wisconsin Conservation Department, Madison

Hall LS, Krausman PR, Morrison ML (1997) The habitat concept and a plea for standard terminology. Wild Soc Bull:173–182

Hervert JJ, Bright JL, Henry RS et al (2005) Home-range and habitat-use patterns of Sonoran pronghorn in Arizona. Wild Soc Bull 33:8–15. https://doi.org/10.2193/0091-7648(2005)33[8:HAHPOS]2.0.CO;2

Hilborn R, Sibert J (1988) Adaptive management of developing fisheries. Mar Policy 12:112–121

Holechek JL, Gomez H, Molinar F, Galt D (1999) Grazing studies: what we've learned. Rangel 21:12–16

Holechek JL, Pieper RD, Herbel CH (2011) Range management: principles and practices. Prentice-Hall, Englewood Cliffs, CA, USA

Irving B (2019) A rebuttal to "reinterpreting the 1882 Bison population collapse." Rangel 41:185–187. https://doi.org/10.1016/j.rala.2019.06.003

Kopp SD, Guthery FS, Forrester ND, Cohen WE (1998) Habitat selection modeling for northern bobwhites on subtropical rangeland. J Wild Manage 1:884–895

Knick S, Connelly JW (eds) (2011) Greater sage-grouse: ecology and conservation of a landscape species and its habitats (Vol. 38). University of California Press

Lyons RK, Hanselka CW (2001) Grazing and browsing: how plants are affected. Texas Farmer Collection. Texas A&M Extension

McNew LB, Winder VL, Pitman JC, Sandercock BK (2015) Alternative rangeland management and the nesting ecology of greater prairie-chickens. Rangel Ecol Manag 68:298–304

Milligan MC, Berkeley LI, McNew LB (2020) Survival of sharp-tailed grouse under variable livestock grazing management. J of Wildl Manag 84:1296–1305

Nichols JD, Runge MC, Johnson FA, Williams BK (2007) Adaptive harvest management of North American waterfowl populations: a brief history and future prospects. J Ornithol 148:343–349. https://doi.org/10.1007/S10336-007-0256-9

Pope KL, Powell LA (2021) Harvest of fish and wildlife: new paradigms for sustainable management. CRC Press, Taylor & Francis Group, Boca Raton, FL, USA

Reese KP, Connelly JW, Garton EO, Commons-Kemner ML (2005) Exploitation and greater sage-grouse Centrocercus urophasianus: a response to Sedinger and Rotella. Wildl Biol 11(4):377–381

Roe FG (1951) The north American buffalo. University of Toronto Press, Toronto, Canada

Rowley WD (1985) U.S. forest service grazing and rangelands: a history. Texas A&M University Press, College Station, TX, USA

Sayre NF (2001) The new ranch handbook: a guide to restoring western rangeland. Quivira Coalition, Santa Fe, New Mexico, USA, p 102

Sayre NF (2017) The politics of scale. University of Chicago Press, Chicago USA

Schroeder MA, Aldridge CL, Apa AD et al (2004) Distribution of sage-grouse in North America. Condor 106:363–376. https://doi.org/10.1093/condor/106.2.363

Shoop MC, McIlvain EH (1971) Why some cattlemen overgraze—and some don't. Rangel Ecol Manag 24:252–257

Sparling D (2014) Natural resource administration wildlife, Fisheries , Forests and Parks. Academic Press, Elsevier, Boston, MA, USA

Stempel ME, Rodgers S (1961) History of prairie chickens in Iowa. In: Proceedings of the Iowa Academy of Science 68:314–322

Stoddart LA Smith AD (1943) Range management. McGraw-Hill Book Company, New York

Strand EK, Launchbaugh KL, Limb RF, Torell LA (2014) Livestock grazing effects on fuel loads for wildland fire in sagebrush dominated ecosystems. J Rangel App 1:35–57

SRM Glossary (1998) https://rangelandsgateway.org/glossary. Accessed 5 July 2022

Stiver SJ (2006) Greater sage-grouse comprehensive conservation strategy

Thacker ET, Messmer TA, Burritt B (2015) Sage-grouse habitat monitoring: Daubenmire versus line-point Intercept. Rangelands 37:7–13. https://doi.org/10.1016/j.rala.2014.12.002

Trefethen JB, Corbin P (1975) American crusade for wildlife. Winchester Press, Winchester, NY USA

Twidwell D, Allred BW, Fuhlendorf SD (2013) National-scale assessment of ecological content in the world's largest land management framework. Ecosphere 4:1–27. https://doi.org/10.1890/ES13-00124.1

Trlica MJ, Buwal M, Menke JW (1977) Effects of rest following defoliations on the recovery of several range species. J Rang Manag 30:21–27

TWS, The Wildlife Society, History and Mission. https://wildlife.org/history-and-mission/. Accessed 15 December 2022

USFWS (2015) Greater sage-grouse 2015. not warranted finding under the Endangered species act. https://www.fws.gov/greaterSageGrouse/PDFs/GrSG_Finding_FINAL.pdf. Accessed 15 December 2022

Vale TR (1974) Sagebrush conversion projects: an element of contemporary environmental change in the western United States. Biol Cons 6:274–284

Vallentine JF (1971) Range development and improvements. Brigham Young University Press, Provo, UT USA

Van Poollen HW, Lacey JR (1979) Herbage response to grazing systems and stocking intensities. J Rang Manag 32:250–253.https://doi.org/10.2307/3897824

Veblen KE, Pyke DA, Aldridge CL, Casazza ML, Assal TJ, Farinha MA (2014) Monitoring of livestock grazing effects on Bureau of Land Management land. Rang Ecol Manag 67(1):68–77

Wisdom MJ, Meinke CW, Knick ST, Schroeder MA (2011) Factors associated with extirpation of sage-grouse. Stud Avian Biol 38:451–472

Chapter 30
The Future of Rangeland Wildlife Conservation—Synopsis

David K. Dahlgren, Lance B. McNew, and Jeffrey L. Beck

Abstract *Rangeland Wildlife Ecology and Conservation* provides a broad array of information on rangeland ecology in association with rangeland-dependent wildlife species. Management of land-use practices from livestock grazing to vegetation manipulation are addressed, as well as ecosystem threats that put the future of rangeland-wildlife at risk. Large-scale pervasive issues, such as climate change and land-use alterations, increase uncertainty for the future of our rangeland resources. Ecosystem services that are essential to sustaining human life may be the most concerning issue as we continue to face further resource degradation. However, such concerns could provide the impetus for general societal support of future conservation actions. Our book addresses emerging topics, such as the interaction of rangelands with riparian habitat, biodiversity, insects, wetland birds, herpetofauna, meso- and large carnivores, and avian predators, subjects that have previously received less attention in relation to rangeland ecosystems. Future conservation of rangeland-wildlife will require more integration from the rangeland and wildlife professions, from academic efforts to individual practitioners. The objective of *Rangeland Wildlife Ecology and Conservation* is to provide a valuable information resource and encourage increased integration for students and early professionals from both disciplines.

Keywords Conservation · Future · Rangeland · Threats · Wildlife

D. K. Dahlgren (✉)
Department of Wildland Resources, S.J. Quinney College of Natural Resources, Utah State University, 5230 Old Main Hill, Logan, UT 84322, USA
e-mail: david.dahlgren@usu.edu

L. B. McNew
Department of Animal and Range Sciences, Montana State University, 103 Animal Biosciences Building, Bozeman, MT 59717-2900, USA

J. L. Beck
Department of Ecosystem Science and Management, University of Wyoming, Agriculture Building 2004, 1000 East University Avenue, Laramie, WY 82071, USA

© The Author(s) 2023   1011
L. B. McNew et al. (eds.), *Rangeland Wildlife Ecology and Conservation*,
https://doi.org/10.1007/978-3-031-34037-6_30

30.1 Introduction

Rangeland Wildlife Ecology and Conservation spans information on the foundations and history of rangeland and wildlife sciences to subject matter on rangeland wildlife taxa and contemporary issues. While thorough published works already exist for such topics as arthropods (Chap. 26), waterfowl (Chap. 13), riparian systems (Chap. 7), raptors (Chap. 14), and herpetofauna (Chap. 25), to our knowledge, these subjects have never been synthesized and presented in the context of rangelands and their management. *Rangeland Wildlife Ecology and Conservation* also provides new insights about taxa that are relatively well understood such as prairie grouse (Chap. 9), sage-grouse (Chap. 10), rangeland ungulates (Chaps. 17–23), and burrowing rodents (Chap. 15). Our extensive authorship consists of the top contemporary professionals across the subject matter expertise, especially in North America.

Both rangeland and wildlife science have undergone parallel changes over time, including a shift from utilitarian resource management to a focus on ecological and ecosystem-based approaches covering a broad context of ecological services and intrinsic values, but still including renewable resource production such as livestock grazing and hunter harvest (Chap. 29). Both rangeland and wildlife professions developed following broad-scale over-exploitation of resources. The rangeland discipline originated from an agrarian need to sustainably manage rangelands for livestock production, whereas the origins of the wildlife profession began with the necessity to regulate sustainable harvest of wild game species.

Even now in the early 21st Century, livestock production remains the dominant and nearly ubiquitous land use within rangelands globally and in North America (Asner et al. 2004; Chap. 4). In some instances, livestock production on rangelands has potential to impact rangeland-dependent wildlife species. This can lead to perceived, and at times real, conflicts between those who see livestock grazing as inherently degrading to rangeland habitat and those who feel that wildlife issues are an impediment to livestock production. These contrasting views will likely remain a significant source of future discord relative to rangeland and wildlife issues. Like most ecological issues, the truth and resolutions are likely found somewhere in the middle.

30.2 Consistent Themes

30.2.1 Management and Conservation

Several management and conservation themes emerged from chapters within this book. Livestock grazing was the most addressed theme. Although livestock grazing has been practically universal on North American rangelands, its application has been highly variable with many operational options (see Chap. 4). Stocking rate has been identified as the most important characteristic of grazing management decisions with potential impacts on rangelands (Briske et al. 2008; 2013). However, there remains debate concerning grazing systems and their implementation across a variety of landscape types, especially areas with low annual precipitation and high variation in topography and vegetation communities (Teague 2014).

Effects of wildfire and prescribed fire was commonly discussed among chapters, with Chapter 6 solely focused on fire effects on rangeland wildlife habitats. For both rangeland and wildlife professionals, understanding first-order (i.e., direct and immediate influences of fire) and second-order (i.e., non-fire factors that influence post-fire ecosystem processes) impacts of fire is critical for future management (Chap. 6). Fire has historically been a major ecosystem driver in rangeland systems. However, the temporal and spatial scale of fire occurrence varies drastically across rangeland types. In some prairie grassland systems, fire can be prescribed in relatively short time scales (i.e., up to annually) while in more arid rangeland systems fire is not generally considered a management tool, but an ecosystem threat. In these drier climates, fire is often intrinsically related to invasion of unwanted plants including non-native annual grasses, which further exacerbates the risk of fire ecologically and as a management tool in these systems (see Chap. 10). Nevertheless, prescribed fire can be a low cost and effective way to increase and maintain heterogeneity in some rangeland ecosystems, and heterogeneity supports increased biodiversity (Chap. 8).

Variability in vegetation or vegetation communities within a system are key characteristics of ecosystem heterogeneity. When management objectives include multi-species and ecosystem services, heterogeneity within ecosystems is crucial. Heterogeneity can include changes in dominant vegetation types across a landscape but could also encompass multiple age structures of a specific vegetation type. Biodiversity within rangeland systems is linked to the degree of heterogeneity. Management actions meant to support heterogeneity and biodiversity should always include specific objectives, even when there is a lack of complete knowledge to consistently predict outcomes. Heterogeneity and biodiversity are contemporary concepts that have been part of the shift in rangeland ecology from a focus on livestock production to a broader ecosystem-based approach to managing rangelands.

Different approaches to wildlife and rangeland management between public and private lands are addressed in several chapters. Both public and private rangelands occur throughout North America, with varying landscape proportions of each depending on location. Most grasslands in central North America are privately owned and managed, while the proportion of public land increases in western shrub steppe and hot deserts. Historically, federal public land grazing permits were tied to deeded private lands in the local area. The idea was that permittees had to own enough private land to support their livestock during the off-season (i.e., winter). Local private land requirements for permittees helped to address the problem of nomadic livestock herds that could remove forage resources in an area leaving local communities and rangeland resources at risk. Most management decisions on private rangelands are at the discretion of the landowner, although available government assistance programs for private rangelands may have specific requirements. Management decisions on federal rangelands include multiple-use and sustained yield mandates and in-depth procedural planning under the National Environmental Policy Act (1970; NEPA).

NEPA usually includes environmental assessments and public input, which generally increases the amount of time needed to make and implement management actions.

30.2.2 Threats

Rangelands of North America are faced with multiple threats that jeopardize their ability to provide wildlife habitat, forage for livestock, and other ecosystem services in the future. Many of these threats are interrelated with compounding impacts, such as wildfire and invasion of non-native annual grasses. The future conservation and management of rangelands by natural resource professionals will largely be oriented toward addressing these threats. As reviewed throughout this book, threats can vary across temporal and spatial scales. Without a clear understanding of the importance of scale, managers may not make optimal decisions even with the best intentions. Specifically, a management action that addresses threats in the context of large intact rangelands might intensify threats in more fragmented landscapes. For example, using vegetation treatments to enhance livestock forage and/or wildlife habitat quality could be a viable and appropriate management alternative in a large intact rangeland, whereas the same actions might be detrimental to the same wildlife species that occupy, but tend to be at more risk in, fragmented rangelands.

Habitat loss and degradation has been and continues to be the most significant threat to rangelands due to multiple factors. Historically, Euro-American settlement of western North America under the Homestead Acts and the resulting conversion of grassland and shrubland into row crop agriculture precipitated the most significant loss of rangelands in any one period. While not quite universal, it is likely that the most arable land, especially in the Great Plains, has been converted to cropland during the past 160 years. In many of these cropland-dominated landscapes, relic rangelands provide the most significant and broadest suite of ecosystem services, including the cleaning and storage of water, sequestration of carbon, habitat for pollinators and other wildlife species, and other critical environmental services. Rangelands provide the bulk of summer forage for livestock production in the shrub steppe and deserts of the western states, and periodic disturbances, such as that provided by livestock grazing, are often critical for the maintenance of functioning, intact rangelands. As such, livestock production provides a market-based incentive for having and maintaining productive intact rangelands. Conversion of rangelands to cropland remains a significant threat, especially as commodity prices increase and more drought-resistant crops are developed.

Energy and exurban development are major threats to rangelands, leading to habitat loss and fragmentation. These anthropogenic developments impact wildlife populations through direct habitat loss and indirect avoidance of developed areas and infrastructure including roads, well pads, and other man-made structures. Many rangeland wildlife species of conservation concern also require large intact contiguous habitat for population persistence. While some opportunities exist to return cropland to rangeland communities, like the U.S.D.A Conservation Reserve

Program (CRP), conservation and management cannot reverse the large-scale conversion of rangelands that occurred during Euro-American settlement. Rather, maintaining remaining rangelands, with emphasis on the largest and most intact areas, is the most significant and highest order of conservation action that can be undertaken at this time. The future of rangeland wildlife and livestock grazing largely rests on society's collective will to keep our remaining rangelands intact and maintain their ecosystem functions.

Fire is an important ecosystem process for rangelands globally. However, the timing of fire within specific rangeland types has often decoupled from the system's historical fire regime. For example, sagebrush (*Artemisia* spp.) systems with high levels of cheatgrass (*Bromus tectorum*) invasion within the Great Basin are burning with much higher frequency, at higher altitudes, and across larger areas compared to the past (Brooks et al. 2015; Smith et al. 2022). Comparatively, many rangelands in prairies of the Great Plains are burning much less frequently, or in the special case of the Flint Hills of Kansas and Oklahoma, they are purposely burned with greater frequency compared to fire periodicity under which these systems evolved (Baldwin et al. 2022). These shifts in fire frequency are severely impacting rangelands across North America, in some areas resulting in an altered state of annual-dominated grasslands or in other areas vegetation communities devoid of non-graminoids and at high risk of tree encroachment (Miller et al. 2017). Rangeland wildlife are effected by these changes in fire frequency, typically through impacts on their habitats, in many cases with negative consequences.

Disease risk to wildlife, livestock, and humans was another common theme throughout the chapters in this book. Most significant was the transference of various diseases between wildlife and livestock, especially for large ungulates. In some interactions the cases are usually infrequent and largely manageable. While in other cases, like domestic and bighorn sheep (*Ovis aries*), disease is a significant issue that has shaped the distribution and persistence of wild sheep populations. Furthermore, disease influences management options such as population augmentation and reintroductions. The interaction of disease among wildlife, livestock, and humans will likely remain a threat to rangeland systems for the foreseeable future.

Climate change has generally compounded the threats described above. Current climate change models suggest continued increases in temperature and higher variability in the amount and timing of precipitation (Melillo et al. 2014). Rangeland systems and their distributions across North America have largely been shaped by both temperature and precipitation regimes over thousands of years (Chap. 3). For example, the Intermountain West has evolved with a pattern of winter-dominated precipitation resulting in high elevation snowpack that provides key water resources to the entire watershed in the drier springs and summers. Rangeland ecosystems, and the services and provisions they provide (e.g., wildlife habitat, livestock production), are highly dependent on snowpack within the region. As snowpack levels become inconsistent, lessen, or precipitation shifts to winter rain, significant impacts to rangelands will occur. Similarly, the Great Plains' grasslands evolved with summer-dominated precipitation, so the region's plant communities and associated wildlife have life histories that are adapted accordingly. Changes in the timing of precipitation in the Great Plains could significantly alter these grassland ecosystems, including wildlife and human food production. Currently, high levels of uncertainty surround our ability to predict the consequences of climate change, making informed projections of conservation and management outcomes extremely challenging. Adaptive management that includes consistent monitoring and science-based research will be needed to address the effects of climate change on rangelands in the future.

30.3   Innovative Topics

Rangeland Wildlife Ecology and Conservation provides coverage of emerging and innovative topics within the context of rangeland systems. Chapters on insects (Chap. 26), amphibians and reptiles (Chap. 25), wetland birds (Chap. 13), and avian predators (Chap. 14) are, to our knowledge, the first syntheses relating these groups to rangeland ecology and management. Additionally, Chap. 7 provides unique perspectives on the management and inter-dependence of riparian areas with adjacent rangelands, whereas Chap. 21 (feral equids), Chap. 24 (large carnivores), and Chap. 28 on living with predators draws attention to contemporary, yet contentious rangeland topics. These innovative chapters speak to the historical shift within the rangeland discipline from a focus on livestock production to broader ecological approaches. Biodiversity (Chap. 8), heterogeneity (Chap. 6), and ecosystem services have become fundamental concepts for both rangeland and wildlife professionals to understand when managing rangelands in the future. Moreover, rangelands are almost always working landscapes that require an understanding of social-economic pressures that constrain land and wildlife management decisions (Chaps. 27, 28). Without a comprehensive understanding, professionals are destined to become overly narrow in their approach to rangeland and wildlife management.

30.4   Current State of Rangeland-Dependent Wildlife

Conservation is a growing concern for many rangeland-associated wildlife species in North America. Rangelands that were once considered "left-over" and of little value during Euro-American settlement and expansion because they were not arable are now viewed through a conservation lens as invaluable landscapes and ecosystems. However, anthropogenic pressures continue to build and are the main source of threats to the future of rangelands and associated wildlife.

Many rangeland-associated wildlife are rangeland obligates, or at least rangeland-dependent, species. For example, many grassland and shrub-steppe passerines rely wholly on rangelands to meet their life-cycle needs (Chap. 12). Pronghorn (*Antilocapra americana*; Chap. 19), prairie dogs (*Cynomys* spp.; Chap. 15), and jackrabbits (*Lepus* spp.), are rangeland-dependent, though not obligated to a specific rangeland type, throughout their entire life-cycle. Prairie grouse (*Tympanuchus* spp.; Chap. 9) and sage-grouse (*Centrocercus* spp.; Chap. 10) are grassland and sagebrush steppe obligates, respectively, with complete dependence on these specific rangeland types to meet all their life-cycle needs. Not only do these grouse rely on rangelands but they are landscape species with populations that require large amounts of intact contiguous habitat space to ensure persistence. Other rangeland-associated wildlife, such as mule deer (*Odocoileus hemionus*; Chap. 17) and elk (*Cervus canadensis*; Chap. 20), similarly require significant space to meet their needs, especially migrating to and using wintering habitat where they exhibit a high degree of rangeland dependence. For many of these rangeland species, future conservation issues will only intensify as threats continue to build over time. Landscape species with low tolerance for habitat fragmentation and other alterations have already, or will shortly, join the first tier of conservation-reliant species in North American rangelands.

30.5   Future Conservation of Rangeland Wildlife

One often overlooked problem, which applies to most ecological conservation concerns, is our state of societal connection, or lack thereof, to wildlands and the ecosystems they support. Humans are inherently connected to and dependent on ecosystem processes through the biosphere (Folke et al. 2011). Human societies increasingly disconnect from ecosystems through use of non-renewable resources and meeting their biophysical needs ex-regionally (Dorninger et al. 2017). Basic processes essential to all life (e.g., clean water, clean air, and food production) are seen as separate or distant operations in relation to society's every-day consciousness. Such disconnect can lead to a lack of understanding and prioritization for the sustainability of our natural resources. This has certainly been the case when we consider the history of rangelands, especially their widespread loss and degradation in North America and globally. Whereas many extant rangelands are society's historical leftovers, the future of rangelands ultimately depends on society's conscious proactivity towards sustainability and conservation.

Ecosystem services can be defined as the services from ecosystems that sustain life. Clean water and air may be the most broadly applicable and important ecosystem services to society. Among others, key ecosystem services include food production, pollination, flood control, and decomposition. For example, pollinators of all kinds (e.g., insects, birds) are crucial to global human food production and are increasingly declining in number and diversity (Chap. 26). Natural ecosystem processes provide flood control when precipitation exceeds normal levels. One of the more significant, but unsung, ecosystem services is the decomposition provided by our natural systems, including carbon storage, the breakdown of pollutants and waste, especially the processing role invertebrates play in decomposition (Chap. 26). Without functioning ecosystems that provide for the disintegration of organic matter, the buildup of waste would quickly become unmanageable on a global scale. For many areas around the world, extant and intact rangelands provide significant ecosystem services as some of the most prevalent undeveloped lands with a full suite of functioning ecological processes. In central North America where large landscapes of historical grasslands have been converted to row crop agriculture, remnant rangelands provide most of the ecosystem services currently available for those regions. Rangeland management, including livestock grazing, is integral to maintaining these undeveloped lands and the services they provide (Chap. 4).

Rangeland wildlife will likely become increasingly more significant to society in the future. Their importance is especially imperative when it comes to prioritizing limited monetary resources towards conservation efforts. However, human societies can either proactively conserve rangelands or they will be forced to retroactively address them due to the loss of essential services that support human life, likely through public policy mandates. Proper and proactive maintenance is almost always less expensive, in most cases orders of magnitude less, than restoration efforts. Rangeland wildlife will benefit from such maintenance, albeit likely with a secondary status compared to ecosystem services and are certainly essential players in those ecosystem services our society requires.

30.5.1 Knowledge Gaps

As demonstrated throughout *Rangeland Wildlife Ecology and Conservation*, many rangeland wildlife species often require large spatial extents to meet their life-history needs. However, we are still lacking key information on spatial and temporal scales of habitat associations and their relative importance for many rangeland-associated species. For example, we are beginning to understand key habitat associations for migrant passerine breeding grounds on which to develop habitat targets for management, yet the relative importance of non-breeding habitats and their management are largely unknown (Chap. 12). A similar lack of key information exists for non-migratory species as well; for example, juvenile survival to recruitment is notoriously difficult to research and understudied in game birds (Chaps. 9, 10, and 11). Identifying limiting factors for wildlife populations could be misguided without an understanding of their full annual life-cycle and habitat requirements.

We need more information concerning the importance of connectivity of intact rangeland habitats for many wildlife species. For many species of conservation concern, there are negative impacts from habitat fragmentation. However, there is also a lack of understanding of the size, spatial arrangement, and connectivity of habitats that would increase the probability of population persistence. Furthermore, we do not understand how habitat quality, or other factors, may interact with the spatial scale of intact habitats required by populations. For example, the size and quality of grassland habitats may constrain or mediate how grassland-obligate species respond to energy development (Lloyd et al. 2022). Knowing species' needs for connectivity, scale of intact habitat, and how these interact with other environmental factors may be critical for future conservation as threats to remaining rangeland habitat increases.

Until recently, wildlife movement and habitat selection has generally been empirically evaluated separately from population demographics and dynamics. The historical lack of integration may be due in part to a deficiency in analytical methods to simultaneously model behavior and vital rates, although *post-hoc* evaluations have been conducted (Kirol et al. 2015; Coates et al. 2017; Sandford et al. 2017). Yet we know that movements, space use, and habitat selection, are linked to survival and reproductive state and vital rates (Dudley et al. 2022; Gelling et al. 2022). Analytical advancements to empirically evaluate the impact of behavior on wildlife vital rates will likely be one of the more significant advances in ecology in the future (Pakanen 2011; Decesare et al. 2013). Understanding of rangeland wildlife, and other species, will increase accordingly and for natural resource managers the effectiveness of conservation actions can be better predicted.

Dietary and nutritional needs for wildlife are closely related to movement and habitat selection. The influence of diet and nutrition on wildlife behavior is a relatively understudied topic but has significant implications, especially for rangeland wildlife. For some prominent species, such as mule deer, research in the last few years has shown that nutritional availability on rangelands drives behavior and resulting body conditions influence survival and reproduction (Tollefson et al. 2010; Merkle et al. 2016). Relatively recent research has linked variation in plant nutrient availability to habitat selection for sage-grouse, with physiological adaptations for local plants (Frye et al. 2013). However, for many rangeland wildlife species there is a paucity of information available concerning the influence of diet and nutrition on behavior and vital rates. Within the rangeland discipline, there has been considerable research concerning nutrition availability related to livestock and their behavior (e.g., Vallentine 2000), but more research in this area is needed for rangeland wildlife.

30.5.2 Integration of Rangeland and Wildlife Ecology

Integration can be defined as bringing people or groups with particular characteristics into equal participation and is increasingly needed for rangeland and wildlife disciplines to direct successful conservation efforts. Much could be done to increase the cross-over of ecological concepts, research questions, and methodologies in both fields. However, the most important integration will require rangeland and wildlife professionals to work collaboratively to address rangeland ecosystem and conservation challenges. Successful integration will come from the willingness of individuals in each profession to build relationships of trust and understanding. While rangeland and wildlife disciplines share much in common, there has been a long-time professional divide with some strongly held biases, with an accompanying assumed superiority, on both sides (Chap. 29). However, in recent years we have been encouraged by the blurring of that line and many examples of both disciplines' scientists and managers working together. One area that could use improvement is when wildlife professionals conduct and publish research that includes or addresses topics from the rangeland discipline. The Society for Range Management produced a glossary of terms commonly used in rangeland management and we encourage its use for consistent terminology (Bedell 1998; rangelandsgateway.org/glossary).

Another area that could use improvement between the professions is more recognition of the validity of prioritized values within the "other" discipline. Although many values are shared between disciplines, the prioritization order for those values can often differ and lead to a sense of disparity. For example, we have found that some rangeland professionals can come across as skeptical of the validity of wildlife conservation issues on rangelands. At times there seems to have been contempt for being "forced" to deal with wildlife issues within the broader field of rangeland management. Similarly, in our experience some wildlife professionals seem to hold the opinion that livestock grazing is ubiquitously detrimental to wildlife and habitat or is of lower importance or consequence compared to wildlife values on rangelands.

We see a need for professionals from both disciplines to show more respect for the values held by one another. For rangeland managers, there is a need to recognize that concerns over wildlife species on rangelands have strong state and federal policies and regulations in place that mandates conservation in addition to great public interest in wildlife. For wildlife managers, there is a need to recognize the legitimate ties livestock production has with both public and private rangelands. For private lands, property taxes must be paid, and for most landowners, monetary resources used to pay taxes must come from the land. In many cases, reductions in ranching profits lead to property sales and land conversion and development that is detrimental to wildlife and their habitats (Plachter and Hampicke 2010). Some natural resource professionals may not realize that most public grazing permits are tied to local private lands and communities. Leases of federal grazing permits include prioritization to specific private entities (e.g., individual permittee, ranch.). The sale of private livestock operations often includes the federal grazing lease, giving prioritization of grazing permits on specified allotments to the buyer. Additionally, producers usually have significant private investment in their publicly permitted allotments, such as water developments, fencing, etc. Livestock grazing on both public and private lands is foundational to the economy of many rural communities (Lewin et al. 2019).

Aldo Leopold wrote that "conservation will ultimately boil down to rewarding the private landowner who conserves the public interest." Perhaps more than any other ecosystem, the goals of livestock producers and conservationists are aligned, because the natural processes that sustain wildlife habitat and functioning rangeland ecosystems are often the same processes that sustain viable livestock production. Recent shifts toward working lands conservation programming that incentivize landowners and producers for conservation-based rangeland management (e.g., federal, state, and NGO working lands conservation programs), have been novel and impactful (NRCS 2020). However, we feel a more direct and explicit integration of the economics of livestock production into adaptive management planning for wildlife would benefit both ends. As most of our remaining rangelands are working lands, two things are needed to conserve rangelands and associated wildlife: (1) economic models that value ecological function, and (2) ecosystem models that incorporate social-economics.

30.6 Summary

Rangeland Wildlife Ecology and Conservation provides a broad array of information on rangeland ecology in association with rangeland-associated wildlife species. Management of land-use practices from livestock grazing to vegetation manipulation are addressed, as well as ecosystem threats that put the future of rangeland-wildlife at risk. Large-scale pervasive issues, such as climate change and land-use alterations, increase uncertainty for the future of our rangeland resources. Ecosystem services that are essential to sustaining human life may be the most concerning issue as we continue to face further resource degradation. However, such concerns could provide the impetus for increased societal interest in future conservation actions. This book addresses emerging and innovative topics, such as the interaction of rangelands with riparian habitat, insects, wetland birds, herpetofauna, and avian predators, subjects that have not been previously well synthesized in relation to rangeland ecosystems. Future conservation of rangeland-wildlife will require more integration from the rangeland and wildlife professions, from academic efforts to individual practitioners. The objective of *Rangeland Wildlife Ecology and Conservation* has been to present a valuable information resource for students and early professionals from both disciplines that also encourages increased integration. We invite readers to integrate rangeland and wildlife science to find creative solutions to the emerging conservation issues presented in this book.

References

Asner GP, Elmore AJ, Olander LP et al (2004) Grazing systems, ecosystem responses, and global change. Annu Rev Environ Resour 29:261–299. https://doi.org/10.1146/annurev.energy.29.062403.102142

Baldwin C, Davidson J, Coleman L (2022) Pyric legacy: prescribed burning in the Flint Hills region, USA. In Weir JR, Scasta JD (eds) Global application of prescribed fire, CRC Press, Boca Raton, FL, USA, p 144

Briske D, Derner J, Brown J et al (2008) Benefits of rotational grazing on rangelands: an evaluation of the experimental evidence. Rangel Ecol Manag 61:3–17

Briske DD, Bestelmeyer BT, Brown JR et al (2013) The Savory method can not green deserts or reverse climate change: a response to Allan Savory TED video. Rangelands 35:72–74

Brooks ML, Matchett JR, Shinneman DJ et al (2015) Fire patterns in the range of greater sage-grouse, 1984–2013—implications for conservation and management. US Geological Survey Open-File Report 2015–1167, Reston, VA

Coates PS, Prochazka BG, Ricca MA et al (2017) Pinyon and juniper encroachment into sagebrush ecosystems impacts distribution and survival of greater sage-grouse. Rangel Ecol Manag 70:25–38

Decesare N, Hebblewhite M, Bradley M et al (2013) Linking habitat selection and predation risk to spatial variation in survival. J Anim Ecol 83:343–252. https://doi.org/10.1111/1365-2656.12144

Dorninger C, Abson DJ, Fischer J et al (2017) Assessing sustainable biophysical human-nature connectedness at regional scales. Environ Res Lett 12.https://doi.org/10.1088/1748-9326/aa68a5

Dudley IA, Coates PS, Prochazka BG, Davis DM, Gardner SC, Delehaty DJ (2022) Maladaptive nest-site selection and reduced nest survival in female sage-grouse following wildfire. Ecosphere 12:e4282. https://doi.org/10.1002/ecs2.4282

Folke C, Jansson A, Rockstrom J et al (2011) Reconnecting to the biosphere. Ambio 40:719–738. https://doi.org/10.1007/s13280-011-0184-y

Frye GG, Connelly JW, Musil DD, Forbey JS (2013) Phytochemistry predicts habitat selection by an avian herbivore at multiple spatial scales. Ecol 94:308–314. https://doi.org/10.1890/12-1313.1

Gelling E, Pratt AC, Beck JL (2022) Linking microhabitat selection, range size, reproductive state, and behavioral state in greater sage-grouse. Wildl Soc Bull 46:e1293. https://doi.org/10.1002/wsb.1293

Kirol CP, Beck JL, Huzurbazar SV et al (2015) Identifying greater sage-grouse source and sink habitats for conservation planning in an energy development landscape. Ecol Appl 25:968–990. https://doi.org/10.6084/m9.figshare.c.3296837.v1

Lewin PA, Wulfhorst JD, Rimbey NR et al (2019) Implications of declining grazing permits on public land: an integrated social and economic impact analysis. Western Economics Forum 17:86–97

Lloyd JD, Aldridge CA, Allison TD, LeBeau CW, McNew LB, Winder VL (2022) Prairie grouse and wind energy: the state of the science and implications for risk management. Wildl Soc Bul 46:e1305. https://doi.org/10.1002/wsb.1305

Melillo JM, Richmond TT, Yohe G (2014) Climate change impacts in the United States: third national climate assessment. U.S. Global Change Research Program

Merkle JA, Monteith KL, Aikens EO et al (2016) Large herbivores surf waves of green-up during spring. Proc of the Royal Soc Bio Sci 283(1833):20160456

Miller RF, Naugle DE, Maestas JD et al (2017) Special issue: targeted woodland removal to recover at-risk grouse and their sagebrush-steppe and prairie ecosystems. Rangel Ecol Manag 70:1–8. https://doi.org/10.1016/j.rama.2016.10.004

Natural Resources Conservation Service [NRCS] (2020) Quantifying outcomes of working lands for wildlife (WLFW) for benefit of landowners and at-risk wildlife. USDA Natural Resources Conservation Service Conservation Effects Assessment Project (CEAP). https://www.nrcs.usda.gov/publications/ceap-wildlife-2020-quantifying-wlfw-benefit.pdf

Pakanen V (2011) Linking demography with dispersal and habitat selection for species conservation. PhD Dissertation, Acta Universitatis Ouluensis A583

Plachter H, Hampicke U (2010) Large-scale livestock grazing: a management tool for nature conservation. Springer-Verlag, Berlin Heidelberg, Germany, p 477

Sandford CP, Kohl MT, Messmer TA et al (2017) Greater sage-grouse resource selection drives reproductive fitness under a conifer removal strategy. Rangel Ecolo Manag 70:59–67

Smith JT, Allred BW, Boyd CS, Davies KW, Jones MO, Kleinhesselink AR, Maestas JD, Naugle DE (2022) Where there's smoke, there's fuel: dynamic vegetation data improve predictions of wildfire hazard in the Great Basin. Rangel Ecol Manage 89:20–32. https://doi.org/10.1016/j.rama.2022.07.005

Teague WR (2014) Deficiencies in the Briske et al. rebuttal of the Savory Method. Rangelands 36:37–38.https://doi.org/10.2111/1551-501X-36.1.37

Tollefson TN, Shipley LA, Myers WL et al (2010) Influence of summer and autumn nutrition on body condition and reproduction in lactating mule deer. J Wildl Manage 74:974–986. https://doi.org/10.2193/2008-529

Vallentine JF (2000) Grazing management. Academic Press, San Diego, CA, USA